Linear Models and the Relevant Distributions and Matrix Algebra

CHAPMAN & HALL/CRC
Texts in Statistical Science Series

Series Editors
Joseph K. Blitzstein, *Harvard University, USA*
Julian J. Faraway, *University of Bath, UK*
Martin Tanner, *Northwestern University, USA*
Jim Zidek, *University of British Columbia, Canada*

Statistical Theory: A Concise Introduction
F. Abramovich and Y. Ritov

Practical Multivariate Analysis, Fifth Edition
A. Afifi, S. May, and V.A. Clark

Practical Statistics for Medical Research
D.G. Altman

Interpreting Data: A First Course in Statistics
A.J.B. Anderson

Introduction to Probability with R
K. Baclawski

Linear Algebra and Matrix Analysis for Statistics
S. Banerjee and A. Roy

Modern Data Science with R
B. S. Baumer, D. T. Kaplan, and N. J. Horton

Mathematical Statistics: Basic Ideas and Selected Topics, Volume I, Second Edition
P. J. Bickel and K. A. Doksum

Mathematical Statistics: Basic Ideas and Selected Topics, Volume II
P. J. Bickel and K. A. Doksum

Analysis of Categorical Data with R
C. R. Bilder and T. M. Loughin

Statistical Methods for SPC and TQM
D. Bissell

Introduction to Probability
J. K. Blitzstein and J. Hwang

Bayesian Methods for Data Analysis, Third Edition
B.P. Carlin and T.A. Louis

Second Edition
R. Caulcutt

The Analysis of Time Series: An Introduction, Sixth Edition
C. Chatfield

Introduction to Multivariate Analysis
C. Chatfield and A.J. Collins

Problem Solving: A Statistician's Guide, Second Edition
C. Chatfield

Statistics for Technology: A Course in Applied Statistics, Third Edition
C. Chatfield

Analysis of Variance, Design, and Regression : Linear Modeling for Unbalanced Data, Second Edition
R. Christensen

Bayesian Ideas and Data Analysis: An Introduction for Scientists and Statisticians
R. Christensen, W. Johnson, A. Branscum, and T.E. Hanson

Modelling Binary Data, Second Edition
D. Collett

Modelling Survival Data in Medical Research, Third Edition
D. Collett

Introduction to Statistical Methods for Clinical Trials
T.D. Cook and D.L. DeMets

Applied Statistics: Principles and Examples
D.R. Cox and E.J. Snell

Multivariate Survival Analysis and Competing Risks
M. Crowder

Statistical Analysis of Reliability Data
M.J. Crowder, A.C. Kimber, T.J. Sweeting, and R.L. Smith

An Introduction to Generalized Linear Models, Third Edition
A.J. Dobson and A.G. Barnett

Nonlinear Time Series: Theory, Methods, and Applications with R Examples
R. Douc, E. Moulines, and D.S. Stoffer

Introduction to Optimization Methods and Their Applications in Statistics
B.S. Everitt

Extending the Linear Model with R: Generalized Linear, Mixed Effects and Nonparametric Regression Models, Second Edition
J.J. Faraway

Linear Models with R, Second Edition
J.J. Faraway

A Course in Large Sample Theory
T.S. Ferguson

Multivariate Statistics: A Practical Approach
B. Flury and H. Riedwyl

Readings in Decision Analysis
S. French

Discrete Data Analysis with R: Visualization and Modeling Techniques for Categorical and Count Data
M. Friendly and D. Meyer

Markov Chain Monte Carlo: Stochastic Simulation for Bayesian Inference, Second Edition
D. Gamerman and H.F. Lopes

Bayesian Data Analysis, Third Edition
A. Gelman, J.B. Carlin, H.S. Stern, D.B. Dunson, A. Vehtari, and D.B. Rubin

Multivariate Analysis of Variance and Repeated Measures: A Practical Approach for Behavioural Scientists
D.J. Hand and C.C. Taylor

Practical Longitudinal Data Analysis
D.J. Hand and M. Crowder

Linear Models and the Relevant Distributions and Matrix Algebra
D.A. Harville

Logistic Regression Models
J.M. Hilbe

Richly Parameterized Linear Models: Additive, Time Series, and Spatial Models Using Random Effects
J.S. Hodges

Statistics for Epidemiology
N.P. Jewell

Stochastic Processes: An Introduction, Third Edition
P.W. Jones and P. Smith

The Theory of Linear Models
B. Jørgensen

Pragmatics of Uncertainty
J.B. Kadane

Principles of Uncertainty
J.B. Kadane

Graphics for Statistics and Data Analysis with R
K.J. Keen

Mathematical Statistics
K. Knight

Introduction to Functional Data Analysis
P. Kokoszka and M. Reimherr

Introduction to Multivariate Analysis: Linear and Nonlinear Modeling
S. Konishi

Nonparametric Methods in Statistics with SAS Applications
O. Korosteleva

Modeling and Analysis of Stochastic Systems, Third Edition
V.G. Kulkarni

Exercises and Solutions in Biostatistical Theory
L.L. Kupper, B.H. Neelon, and S.M. O'Brien

Exercises and Solutions in Statistical Theory
L.L. Kupper, B.H. Neelon, and S.M. O'Brien

Design and Analysis of Experiments with R
J. Lawson

Design and Analysis of Experiments with SAS
J. Lawson

A Course in Categorical Data Analysis
T. Leonard

Statistics for Accountants
S. Letchford

Introduction to the Theory of Statistical Inference
H. Liero and S. Zwanzig

Statistical Theory, Fourth Edition
B.W. Lindgren

Stationary Stochastic Processes: Theory and Applications
G. Lindgren

Statistics for Finance
E. Lindström, H. Madsen, and J. N. Nielsen

The BUGS Book: A Practical Introduction to Bayesian Analysis
D. Lunn, C. Jackson, N. Best, A. Thomas, and D. Spiegelhalter

Introduction to General and Generalized Linear Models
H. Madsen and P. Thyregod

Time Series Analysis
H. Madsen

Pólya Urn Models
H. Mahmoud

Randomization, Bootstrap and Monte Carlo Methods in Biology, Third Edition
B.F.J. Manly

Statistical Regression and Classification: From Linear Models to Machine Learning
N. Matloff

Introduction to Randomized Controlled Clinical Trials, Second Edition
J.N.S. Matthews

Statistical Rethinking: A Bayesian Course with Examples in R and Stan
R. McElreath

Statistical Methods in Agriculture and Experimental Biology, Second Edition
R. Mead, R.N. Curnow, and A.M. Hasted

Statistics in Engineering: A Practical Approach
A.V. Metcalfe

Statistical Inference: An Integrated Approach, Second Edition
H. S. Migon, D. Gamerman, and F. Louzada

Beyond ANOVA: Basics of Applied Statistics
R.G. Miller, Jr.

A Primer on Linear Models
J.F. Monahan

Stochastic Processes: From Applications to Theory
P.D Moral and S. Penev

Applied Stochastic Modelling, Second Edition
B.J.T. Morgan

Elements of Simulation
B.J.T. Morgan

Probability: Methods and Measurement
A. O'Hagan

Introduction to Statistical Limit Theory
A.M. Polansky

Applied Bayesian Forecasting and Time Series Analysis
A. Pole, M. West, and J. Harrison

Statistics in Research and Development, Time Series: Modeling, Computation, and Inference
R. Prado and M. West

Essentials of Probability Theory for Statisticians
M.A. Proschan and P.A. Shaw

Introduction to Statistical Process Control
P. Qiu

Sampling Methodologies with Applications
P.S.R.S. Rao

A First Course in Linear Model Theory
N. Ravishanker and D.K. Dey

Essential Statistics, Fourth Edition
D.A.G. Rees

Stochastic Modeling and Mathematical Statistics: A Text for Statisticians and Quantitative Scientists
F.J. Samaniego

Statistical Methods for Spatial Data Analysis
O. Schabenberger and C.A. Gotway

Bayesian Networks: With Examples in R
M. Scutari and J.-B. Denis

Large Sample Methods in Statistics
P.K. Sen and J. da Motta Singer

Introduction to Statistical Methods for Financial Models
T. A. Severini

Spatio-Temporal Methods in Environmental Epidemiology
G. Shaddick and J.V. Zidek

Decision Analysis: A Bayesian Approach
J.Q. Smith

Analysis of Failure and Survival Data
P. J. Smith

Applied Statistics: Handbook of GENSTAT Analyses
E.J. Snell and H. Simpson

Applied Nonparametric Statistical Methods, Fourth Edition
P. Sprent and N.C. Smeeton

Data Driven Statistical Methods
P. Sprent

Generalized Linear Mixed Models: Modern Concepts, Methods and Applications
W. W. Stroup

Survival Analysis Using S: Analysis of Time-to-Event Data
M. Tableman and J.S. Kim

Applied Categorical and Count Data Analysis
W. Tang, H. He, and X.M. Tu

Elementary Applications of Probability Theory, Second Edition
H.C. Tuckwell

Introduction to Statistical Inference and Its Applications with R
M.W. Trosset

Understanding Advanced Statistical Methods
P.H. Westfall and K.S.S. Henning

Statistical Process Control: Theory and Practice, Third Edition
G.B. Wetherill and D.W. Brown

Generalized Additive Models: An Introduction with R, Second Edition
S. Wood

Epidemiology: Study Design and Data Analysis, Third Edition
M. Woodward

Practical Data Analysis for Designed Experiments
B.S. Yandell

Texts in Statistical Science

Linear Models and the Relevant Distributions and Matrix Algebra

David A. Harville

CRC Press
Taylor & Francis Group
Boca Raton London New York

CRC Press is an imprint of the
Taylor & Francis Group an **informa** business
A CHAPMAN & HALL BOOK

CRC Press
Taylor & Francis Group
6000 Broken Sound Parkway NW, Suite 300
Boca Raton, FL 33487-2742

First issued in paperback 2020

Version Date: 20180131

ISBN-13: 978-0-367-57203-7 (pbk)
ISBN-13: 978-1-138-57833-3 (hbk)

Library of Congress Cataloging-in-Publication Data

Names: Harville, David A., author.
Title: Linear models and the relevant distributions and matrix algebra /
David A. Harville.
Description: Boca Raton : CRC Press, 2018. | Includes bibliographical
references and index.
Identifiers: LCCN 2017046289 | ISBN 9781138578333 (hardback : alk. paper)
Subjects: LCSH: Matrices--Problems, exercises, etc. | Mathematical
statistics--Problems, exercises, etc.
Classification: LCC QA188 .H3798 2018 | DDC 512.9/434--dc23
LC record available at https://lccn.loc.gov/2017046289

Visit the e-resources at: https://www.crcpress.com/9781138578333

Visit the Taylor & Francis Web site at
http://www.taylorandfrancis.com

and the CRC Press Web site at
http://www.crcpress.com

Contents

Preface

Linear statistical models provide the theoretical underpinnings for many of the statistical procedures in common use. In deciding on the suitability of one of those procedures for use in a potential application, it would seem to be important to know the assumptions embodied in the underlying model and the theoretical properties of the procedure as determined on the basis of that model. In fact, the value of such knowledge is not limited to its value in deciding whether or not to use the procedure. When (as is frequently the case) one or more of the assumptions appear to be unrealistic, such knowledge can be very helpful in devising a suitably modified procedure—a situation of this kind is illustrated in Section 7.7f.

Knowledge of matrix algebra has in effect become a prerequisite for reading much of the literature pertaining to linear statistical models. The use of matrix algebra in this literature started to become commonplace in the mid 1900s. Among the early adopters were Scheffé (1959), Graybill (1961), Rao (1965), and Searle (1971). When it comes to clarity and succinctness of exposition, the introduction of matrix algebra represented a great advance. However, those without an adequate knowledge of matrix algebra were left at a considerable disadvantage.

Among the procedures for making statistical inferences are ones that are based on an assumption that the data vector is the realization of a random vector, say $\mathbf{y}$, that follows a linear statistical model. The present volume discusses procedures of that kind and the properties of those procedures. Included in the coverage are various results from matrix algebra needed to effect an efficient presentation of the procedures and their properties. Also included in the coverage are the relevant statistical distributions. Some of the supporting material on matrix algebra and statistical distributions is interspersed with the discussion of the inferential procedures and their properties.

Two classical procedures are the least squares estimator (of an estimable function) and the F test. The least squares estimator is optimal in the sense described in a result known as the Gauss–Markov theorem. The Gauss–Markov theorem has a relatively simple proof. Results on the optimality of the F test are stated and proved herein (in Chapter 7); the proofs of these results are relatively difficult and less "accessible"—reference is sometimes made to Wolfowitz's (1949) proofs of results on the optimality of the F test, which are (at best) extremely terse.

The F test is valid under an assumption that the distribution of the observable random vector $\mathbf{y}$ is multivariate normal. However, that assumption is stronger than necessary. As can be discerned from results like those discussed by Fang, Kotz, and Ng (1990), as has been pointed out by Ravishanker and Dey (2002, sec. 5.5), and is shown herein, the F test and various related procedures depend on $\mathbf{y}$ only through a (possibly vector-valued) function of $\mathbf{y}$ whose distribution is the same for every distribution of $\mathbf{y}$ that is "elliptically symmetric," so that those procedures are valid not only when the distribution of $\mathbf{y}$ is multivariate normal but more generally when the distribution of $\mathbf{y}$ is elliptically symmetric.

The present volume includes considerable discussion of multiple comparisons and simultaneous confidence intervals. At one time, the use of these kinds of procedures was confined to situations where the requisite percentage points were those of a distribution (like the distribution of the Studentized range) that was sufficiently tractable that the percentage points could be computed by numerical means. The percentage points could then be tabulated or could be recomputed on an "as needed" basis. An alternative whose use is not limited by considerations of "numerical tractability" is to determine the percentage points by Monte Carlo methods in the manner described by Edwards and Berry (1987).

The discussion herein of multiple comparisons is not confined to the traditional methods, which serve to control the FWER (familywise error rate). It includes discussion of less conservative methods of the kinds proposed by Benjamini and Hochberg (1995) and by Lehmann and Romano (2005a).

Prerequisites. The reader is assumed to have had at least some exposure to the basic concepts of probability "theory" and to the basic principles of statistical inference. This exposure is assumed to have been of the kind that could have been gained through an introductory course at a level equal to (or exceeding) that of Casella and Berger (2002) or Bickel and Doksum (2001).

The coverage of matrix algebra provided herein is more-or-less self-contained. Nevertheless, some previous exposure of the kind that might have been gained through an introductory course on linear algebra is likely to be helpful. That would be so even if the introductory course were such that the level of abstractness or generality were quite high or (at the other extreme) were such that computations were emphasized at the expense of fundamental concepts, in which case the connections to what is covered herein would be less direct and less obvious.

Potential uses. The book could be used as a reference. Such use has been facilitated by the inclusion of a very extensive and detailed index and by arranging the covered material in a way that allows (to the greatest extent feasible) the various parts of the book to be read more-or-less independently.

Or the book could serve as the text for a graduate-level course on linear statistical models with a secondary purpose of providing instruction in matrix algebra. Knowledge of matrix algebra is critical not only in the study of linear statistical models but also in the study of various other areas of statistics including multivariate analysis. The integration of the instruction in matrix algebra with the coverage of linear statistical models could have a symbiotic effect on the study of both subjects. If desired, topics not covered in the book (either additional topics pertaining to linear statistical models or topics pertaining to some other area such as multivariate analysis) could be included in the course by introducing material from a secondary source.

Alternatively, the book could be used selectively in a graduate-level course on linear statistical models to provide coverage of certain topics that may be covered in less depth (or not covered at all) in another source. It could also be used selectively in a graduate-level course in mathematical statistics to provide in-depth illustrations of various concepts and principles in the context of a relatively important and complex setting.

To facilitate the use of the book as a text, a large number of exercises have been included. A solutions manual is accessible to instructors who have adopted the book at https://www.crcpress.com/9781138578333.

An underlying perspective. A basic problem in statistics (perhaps, *the* basic problem) is that of making inferences about the realizations of some number (assumed for the sake of simplicity to be finite) of unobservable random variables, say $w_1, w_2, \ldots, w_M$, based on the value of an observable random vector $\mathbf{y}$. Let $\mathbf{w} = (w_1, w_2, \ldots, w_M)'$. A *statistical model* might be taken to mean a specification of the joint distribution of $\mathbf{w}$ and $\mathbf{y}$ up to the value of a vector, say $\boldsymbol{\theta}$, of unknown parameters. This definition is sufficiently broad to include the case where $\mathbf{w} = \boldsymbol{\theta}$—when $\mathbf{w} = \boldsymbol{\theta}$, the joint distribution of $\mathbf{w}$ and $\mathbf{y}$ is "degenerate."

In this setting, statistical inference might take the form of a "point" estimate or prediction for the realization of $\mathbf{w}$ or of a set of M-dimensional vectors and might be based on the statistical model (in what might be deemed model-based inference). Depending on the nature of $w_1, w_2, \ldots, w_M$, this activity might be referred to as parametric inference or alternatively as predictive inference.

Let $\tilde{\mathbf{w}}(\mathbf{y})$ represent a point estimator or predictor, and let $A(\mathbf{y})$ represent a set of M-dimensional vectors that varies with the value of $\mathbf{y}$. And consider the use of $\tilde{\mathbf{w}}(\mathbf{y})$ and $A(\mathbf{y})$ in model-based (parametric or predictive) inference. If $\mathrm{E}[\tilde{\mathbf{w}}(\mathbf{y})] = \mathrm{E}(\mathbf{w})$, $\tilde{\mathbf{w}}(\mathbf{y})$ is said to be an unbiased estimator or predictor. And if $\Pr[\mathbf{w} \in A(\mathbf{y})] = 1 - \dot{\gamma}$ for some prespecified constant $\dot{\gamma}$ (and for "every" value of

$\boldsymbol{\theta}$), $A(\mathbf{y})$ is what might be deemed a $100(1-\dot{\gamma})\%$ "confidence" set—depending on the model, such a set might or might not exist.

In the special case where $\boldsymbol{\theta}$ is "degenerate" (i.e., where the joint distribution of $\mathbf{w}$ and $\mathbf{y}$ is known), $\tilde{\mathbf{w}}(\mathbf{y})$ could be taken to be $\mathrm{E}(\mathbf{w}\,|\,\mathbf{y})$ (the so-called posterior mean), in which case $\tilde{\mathbf{w}}(\mathbf{y})$ would be unbiased. And among the choices for the set $A(\mathbf{y})$ in that special case are choices for which $\Pr[\mathbf{w} \in A(\mathbf{y})\,|\,\mathbf{y}] = 1-\dot{\gamma}$ [so-called $100(1-\dot{\gamma})\%$ credible sets].

Other models can be generated from the original model by regarding $\boldsymbol{\theta}$ as a random vector whose distribution is specified up to the value of some parameter vector $\boldsymbol{\phi}$ (of smaller dimension than $\boldsymbol{\theta}$) and by regarding the joint distribution of $\mathbf{w}$ and $\mathbf{y}$ specified by the original model as the conditional distribution of $\mathbf{w}$ and $\mathbf{y}$ given $\boldsymbol{\theta}$. The resultant (hierarchical) models are more parsimonious than the original model, but this (reduction in the number of parameters) comes at the expense of additional assumptions. In the special case where $\boldsymbol{\phi}$ is "degenerate" (i.e., where $\boldsymbol{\theta}$ is regarded as a random vector whose distribution is completely specified and represents what in a Bayesian framework is referred to as the prior distribution), the resultant models are ones in which the joint distribution of $\mathbf{w}$ and $\mathbf{y}$ is completely specified.

As discussed in a 2014 paper (Harville 2014), I regard the division of statistical inference along Bayesian-frequentist lines as unnecessary and undesirable. What in a Bayesian approach is referred to as the prior distribution can simply be regarded as part of a hierarchical model. In combination with the original model, it leads to a new model (in which the joint distribution of $\mathbf{w}$ and $\mathbf{y}$ is completely specified).

In that 2014 paper, it is also maintained that there are many instances (especially in the case of predictive inference) where any particular application of the inferential procedures is one in a sequence of "repeated" applications. In such instances, the "performance" of the procedures in repeated application would seem to be an important consideration. Performance in repeated application can be assessed on the basis of empirical evidence or on the basis of a "model"—for some discussion of performance in repeated application within a rather specific Bayesian framework, refer to Dawid (1982).

As famously stated by George Box, "all models are wrong, but some are useful" (e.g., Box and Draper 1987, p. 424). In fact, a model may be useful for some purposes but not for others. How useful any particular model might be in providing a basis for statistical inference would seem to depend at least in part on the extent to which the relationship between $\mathbf{w}$ and $\mathbf{y}$ implicit in the model is consistent with the "actual relationship"—the more "elaborate" the model, the more opportunities there are for discrepancies. In principle, it would seem that the inferences should be based on a model that reflects all relevant prior information [i.e., the joint distribution of $\mathbf{w}$ and $\mathbf{y}$ should be the conditional (on the prior information) joint distribution]. In practice, it may be difficult to formally account for certain kinds of prior information in a way that seems altogether satisfactory; it may be preferable to account for those kinds of prior information through informal "posterior adjustments."

In devising a model, there is a potential pitfall. It is implicitly assumed that the specification of the joint distribution of $\mathbf{w}$ and $\mathbf{y}$ is not influenced by the observed value of $\mathbf{y}$. Yet, in practice, the model may not be decided upon until after the data become available and/or may undergo modification subsequent to that time. Allowing the observed value of $\mathbf{y}$ to influence the choice of model could introduce subtle biases and distortions into the inferences.

Format. The book is divided into (7) numbered chapters, the chapters into numbered sections, and (in some cases) the sections into lettered subsections. Sections are identified by two numbers (chapter and section within chapter) separated by a decimal point—thus, the fifth section of Chapter 3 is referred to as Section 3.5. Within a section, a subsection is referred to by letter alone. A subsection in a different chapter or in a different section of the same chapter is referred to by referring to the section and by appending a letter to the section number—for example, in Section 6.2, Subsection c of Section 6.1 is referred to as Section 6.1c. An exercise in a different chapter is referred to by the number obtained by inserting the chapter number (and a decimal point) in front of the exercise number.

Some of the subsections are divided into parts. Each such subsection includes two or more parts that begin with a heading and may or may not include an introductory part (with no heading). On the relatively small number of occasions on which reference is made to one or more of the individual parts, the parts that begin with headings are identified as though they had been numbered $1, 2, \ldots$ in order of appearance.

Some of the displayed "equations" are numbered. An equation number consists of two parts (corresponding to section within chapter and equation within section) separated by a decimal point (and is enclosed in parentheses). An equation in a different chapter is referred to by the "number" obtained by starting with the chapter number and appending a decimal point and the equation number—for example, in Chapter 6, result (5.11) of Chapter 3 is referred to as result (3.5.11). For purposes of numbering (and referring to) equations in the exercises, the exercises in each chapter are to be regarded as forming Section E of that chapter.

Notational conventions and issues. The broad coverage of the manuscript (which includes coverage of the statistical distributions and matrix algebra applicable to discussions of linear models) has led to challenges and issues in devising suitable notation. It has sometimes proved necessary to use similar (or even identical) symbols for more than one purpose. In some cases, notational conventions that are typically followed in the treatment of one of the covered topics may conflict with those typically followed in another of the covered topics; such conflicts have added to the difficulties in devising suitable notation.

For example, in discussions of matrix algebra, it is customary (at least among statisticians) to use boldface capital letters to represent matrices, to use boldface lowercase letters to represent vectors, and to use ordinary lowercase letters to represent scalars. And in discussions of statistical distributions and their characteristics, it is customary to distinguish the realization of a random variable or vector from the random variable or vector itself by using a capital letter, say X, to represent the random variable or a boldface capital letter, say $\mathbf{X}$, to represent the random vector and to use the corresponding lowercase letter x or boldface lowercase letter $\mathbf{x}$ to represent its realization. In such a case, the approach taken herein is to use some other device such as an underline to differentiate between the random variable or vector and its realization. Accordingly, x, $\mathbf{x}$, or $\mathbf{X}$ might be used to represent a random variable, vector, or matrix and $\underline{x}$, $\underline{\mathbf{x}}$, or $\underline{\mathbf{X}}$ to represent the realization of x, $\mathbf{x}$, or $\mathbf{X}$. Alternatively, in cases where the intended usage is clear from the context, the same symbol may be used for both.

Credentials. I have brought to the writing of this book an extensive background in the subject matter. On numerous occasions, I have taught graduate-level courses on linear statistical models. Moreover, linear statistical models and their use as a basis for statistical inference has been my primary research interest. My research in that area includes both work that is relatively theoretical in nature and work in which the focus is on applications (including applications in sports and in animal breeding). I am the author of two previous books, both of which pertain to matrix algebra: *Matrix Algebra from a Statistician's Perspective*, which provides coverage of matrix algebra of a kind that would seem to be well-suited for those with interests in statistics and related disciplines, and *Matrix Algebra: Exercises and Solutions*, which provides the solutions to the exercises in *Matrix Algebra from a Statistician's Perspective*.

In the writing of *Matrix Algebra from a Statistician's Perspective*, I adopted the philosophy that (to the greatest extent feasible) the discourse should include the theoretical underpinnings of essentially every result. In the writing of the present volume, I have adopted much the same philosophy. Of course, doing so has a limiting effect on the number of topics and the number of results that can be covered.

Acknowledgments. In the writing of this volume, I have been influenced greatly (either consciously or subconsciously) by insights acquired from others through direct contact or indirectly through exposure to presentations they have given or to documents they have written. Among those from whom I have acquired insights are: Frank Graybill—his 1961 book was an early influence; Justus

Seely (through access to some unpublished class notes from a course he had taught at Oregon State University, as well as through the reading of a number of his published papers); C. R. Henderson (who was my major professor and a source of inspiration and ideas); Oscar Kempthorne (through access to his class notes and through thought-provoking conversations during the time he was a colleague); and Shayle Searle (who was very supportive of my efforts and who was a major contributor to the literature on linear statistical models and the associated matrix algebra). And I am indebted to John Kimmel, who (in his capacity as an executive editor at Chapman and Hall/CRC) has been a source of encouragement, support, and guidance.

David A. Harville
harville@iastate.edu

1

Introduction

This book is about linear statistical models and about the statistical procedures derived on the basis of those models. These statistical procedures include the various procedures that make up a linear regression analysis or an analysis of variance, as well as many other well-known procedures. They have been applied on many occasions and with great success to a wide variety of experimental and observational data.

In agriculture, data on the milk production of dairy cattle are used to make inferences about the "breeding values" of various cows and bulls and ultimately to select breeding stock (e.g., Henderson 1984). These inferences (and the resultant selections) are made on the basis of a linear statistical model. The adoption of this approach to the selection of breeding stock has significantly increased the rate of genetic progress in the affected populations.

In education, student test scores are used in assessing the effectiveness of teachers, schools, and school districts. In the Tennessee value-added assessment system (TVAAS), the assessments are in terms of statistical inferences made on the basis of a linear statistical model (e.g., Sanders and Horn 1994). This approach compares favorably with the more traditional ways of using student test scores to assess effectiveness. Accordingly, its use has been mandated in a number of regions.

In sports such as football and basketball, the outcomes of past and present games can be used to predict the outcomes of future games and to rank or rate the various teams. Very accurate results can be obtained by basing the predictions and the rankings or ratings on a linear statistical model (e.g., Harville 1980, 2003b, 2014). The predictions obtained in this way are nearly as accurate as those implicit in the betting line. And (in the case of college basketball) they are considerably more accurate than predictions based on the RPI (Ratings Percentage Index), which is a statistical instrument used by the NCAA (National Collegiate Athletic Association) to rank teams.

The scope of statistical procedures developed on the basis of linear statistical models can be (and has been) extended. Extensions to various kinds of nonlinear statistical models have been considered by Bates and Watts (1988), Gallant (1987), and Pinheiro and Bates (2000). Extensions to the kinds of statistical models that have come to be known as generalized linear models have been considered by McCullagh and Nelder (1989), Agresti (2013), and McCulloch, Searle, and Neuhaus (2008).

1.1 Linear Statistical Models

A central (perhaps *the* central) problem in statistics is that of using N data points that are to be regarded as the respective values of observable random variables $y_1, y_2, \ldots, y_N$ to make inferences about various future quantities and/or various other quantities that are deemed unobservable. Inference about future quantities is sometimes referred to as predictive inference.

Denote by $\mathbf{y}$ the N-dimensional random (column) vector whose elements are $y_1, y_2, \ldots, y_N$, respectively. In devising or evaluating an inferential procedure, whatever assumptions are made about the distribution of $\mathbf{y}$ play a critical role. This distribution is generally taken to be "conditional" on various kinds of concomitant information. Corresponding to the observed value of y_i may be known values $u_{i1}, u_{i2}, \ldots, u_{iC}$ of C explanatory variables $u_1, u_2, \ldots, u_C$ $(1 \le i \le N)$; these NC values may constitute some or all of the concomitant information. The various assumptions made about the distribution of $\mathbf{y}$ are referred to collectively as a *statistical model* or simply as a *model.*

We shall be concerned herein with what are known as *linear (statistical) models.* These models are relatively tractable and provide the theoretical underpinnings for a broad class of statistical procedures.

What constitutes a linear model? In a linear model, the expected values of $y_1, y_2, \ldots, y_N$ are taken to be linear combinations of some number, say P, of generally unknown parameters $\beta_1, \beta_2, \ldots, \beta_P$. That is, there exist numbers $x_{i1}, x_{i2}, \ldots, x_{iP}$ (assumed known) such that

$$\mathrm{E}(y_i) = \sum_{j=1}^{P} x_{ij}\beta_j \qquad (i = 1, 2, \ldots, N). \tag{1.1}$$

The parameters $\beta_1, \beta_2, \ldots, \beta_P$ may be unrestricted, or (more generally) may be subject to "linear constraints."

For $i = 1, 2, \ldots, N$, the random variable y_i can be reexpressed as

$$y_i = \sum_{j=1}^{P} x_{ij}\beta_j + \Big(y_i - \sum_{j=1}^{P} x_{ij}\beta_j\Big).$$

Accordingly, condition (1.1) is equivalent to the condition

$$y_i = \sum_{j=1}^{P} x_{ij}\beta_j + e_i \qquad (i = 1, 2, \ldots, N), \tag{1.2}$$

where $e_1, e_2, \ldots, e_N$ are random variables, each of which has an expected value of 0. Under condition (1.2), we have that

$$e_i = y_i - \sum_{j=1}^{P} x_{ij}\beta_j = y_i - \mathrm{E}(y_i) \qquad (i = 1, 2, \ldots, N). \tag{1.3}$$

Aside from the two trivial cases $\mathrm{var}(e_i) = 0$ and $x_{i1} = x_{i2} = \cdots = x_{iP} = 0$ and a case where $\beta_1, \beta_2, \ldots, \beta_P$ are subject to restrictions under which $\sum_{j=1}^{P} x_{ij}\beta_j$ is known, e_i is unobservable. The random variables $e_1, e_2, \ldots, e_N$ are sometimes referred to as residual effects or as errors.

In working with linear models, the use of matrix notation is extremely convenient. Note that condition (1.1) can be reexpressed in the form

$$\mathrm{E}(\mathbf{y}) = \mathbf{X}\boldsymbol{\beta}, \tag{1.4}$$

where $\mathbf{X}$ is the $N \times P$ matrix whose ijth element is x_{ij} $(i = 1, 2, \ldots, N;\ j = 1, 2, \ldots, P)$ and $\boldsymbol{\beta}$ is the P-dimensional column vector with elements $\beta_1, \beta_2, \ldots, \beta_P$. And condition (1.2), which is equivalent to condition (1.4), can be restated as

$$\mathbf{y} = \mathbf{X}\boldsymbol{\beta} + \mathbf{e}, \tag{1.5}$$

where $\mathbf{e}$ is a random column vector (the elements of which are $e_1, e_2, \ldots, e_N$) with $\mathrm{E}(\mathbf{e}) = \mathbf{0}$. Further, in matrix notation, result (1.3) becomes

$$\mathbf{e} = \mathbf{y} - \mathbf{X}\boldsymbol{\beta} = \mathbf{y} - \mathrm{E}(\mathbf{y}). \tag{1.6}$$

For a model to qualify as a linear model, we require something more than condition (1.1) or (1.2). Namely, we require that the variance-covariance matrix of $\mathbf{y}$, or equivalently of $\mathbf{e}$, not depend on the elements $\beta_1, \beta_2, \ldots, \beta_P$ of $\boldsymbol{\beta}$—the diagonal elements of the variance-covariance matrix of $\mathbf{y}$ are the variances of the elements $y_1, y_2, \ldots, y_N$ of $\mathbf{y}$, and the off-diagonal elements are the covariances.

This matrix may depend (and typically does depend) on various unknown parameters other than $\beta_1, \beta_2, \ldots, \beta_P$.

For a model to be useful in making inferences about the unobservable quantities of interest, it must be possible to express those quantities in a relevant way. Consider a linear model, in which $\text{E}(y_1), \text{E}(y_2), \ldots, \text{E}(y_N)$ are expressible in the form (1.1) or, equivalently, in which $\mathbf{y}$ is expressible in the form (1.5). This model could be useful in making inferences about a quantity that is expressible as a linear combination, say $\sum_{j=1}^{P} \lambda_j \beta_j$, of the elements $\beta_1, \beta_2, \ldots, \beta_P$ of $\boldsymbol{\beta}$—how useful would depend on $\mathbf{X}$, on the coefficients $\lambda_1, \lambda_2, \ldots, \lambda_P$, and perhaps on various characteristics of the distribution of $\mathbf{e}$. More generally, this model could be useful in making inferences about an unobservable random variable w for which $\text{E}(w) = \sum_{j=1}^{P} \lambda_j \beta_j$ and for which $\text{var}(w)$ and $\text{cov}(w, \mathbf{y})$ do not depend on $\boldsymbol{\beta}$ or, equivalently, an unobservable random variable w of the form $w = \sum_{j=1}^{P} \lambda_j \beta_j + d$, where d is a random variable for which $\text{E}(d) = 0$ and for which $\text{var}(d)$ and $\text{cov}(d, \mathbf{e})$ do not depend on $\boldsymbol{\beta}$—$\text{cov}(w, \mathbf{y})$ and $\text{cov}(d, \mathbf{e})$ are N-dimensional row vectors, the elements of which are the covariances between w and $y_1, y_2, \ldots, y_N$ and the covariances between d and $e_1, e_2, \ldots, e_N$. Strictly speaking, inferences are made about the "realization" of a random variable, not the random variable itself. The model could also be useful in making inferences about a quantity that is expressible in terms of whatever parameters may characterize the distribution of $\mathbf{e}$.

1.2 Regression Models

Suppose that (as in Section 1.1) there are N data points that (for purposes of making inferences about various unobservable quantities) are to be regarded as the respective values of observable random variables $y_1, y_2, \ldots, y_N$. Suppose further that corresponding to the observed value of y_i are known values $u_{i1}, u_{i2}, \ldots, u_{iC}$ of C explanatory variables $u_1, u_2, \ldots, u_C$ $(1 \leq i \leq N)$. For example, in an observational study of how the amount of milk produced by a dairy cow during her first lactation varies with her age and her body weight (recorded at the beginning of her initial pregnancy), y_i might correspond to the amount of milk produced by the ith cow and (taking $C = 2$) u_{i1} and u_{i2} might represent her age and her body weight.

A possible model is that obtained by taking

$$y_i = \alpha_0 + \sum_{j=1}^{C} u_{ij}\alpha_j + e_i \qquad (i = 1, 2, \ldots, N), \tag{2.1}$$

where $\alpha_0, \alpha_1, \ldots, \alpha_C$ are unrestricted parameters (of unknown value) and where $e_1, e_2, \ldots, e_N$ are uncorrelated, unobservable random variables, each with mean 0 and (for a strictly positive parameter σ of unknown value) variance σ^2. Models of the form (2.1) are referred to as *simple* or *multiple* (depending on whether $C = 1$ or $C \geq 2$) *linear regression models*.

As suggested by the name, a linear regression model qualifies as a linear model. Under the linear regression model (2.1),

$$\text{E}(y_i) = \alpha_0 + \sum_{j=1}^{C} u_{ij}\alpha_j \qquad (i = 1, 2, \ldots, N). \tag{2.2}$$

The expected values (2.2) are of the form (1.1), and the expressions (2.1) are of the form (1.2); set $P = C + 1$, $\beta_1 = \alpha_0$, and (for $j = 1, 2, \ldots, C$) $\beta_{j+1} = \alpha_j$, and take $x_{11} = x_{21} = \cdots = x_{N1} = 1$ and (for $i = 1, 2, \ldots, N$ and $j = 1, 2, \ldots, C$) $x_{i,j+1} = u_{ij}$. Moreover, the linear regression model is such that the variance-covariance matrix of $e_1, e_2, \ldots, e_N$ does not depend on the β_j's; it depends only on the parameter σ.

In an application of the multiple linear regression model, we might wish to make inferences about some or all of the individual parameters $\alpha_0, \alpha_1, \ldots, \alpha_C$, and σ. Or we might wish to make inferences about the quantity $\alpha_0 + \sum_{j=1}^{C} u_j\alpha_j$ for various values of the explanatory variables $u_1, u_2, \ldots, u_C$. This quantity could be thought of as representing the "average" value of an infinitely large number of future data points, all of which correspond to the same $u_1, u_2, \ldots, u_C$ values. Also of potential interest are quantities of the form $\alpha_0 + \sum_{j=1}^{C} u_j\alpha_j + d$, where d is an unobservable random variable that is uncorrelated with $e_1, e_2, \ldots, e_N$ and that has mean 0 and variance σ^2. A quantity of this form is a random variable, the value of which can be thought of as representing an individual future data point.

There are potential pitfalls in making predictive inferences on the basis of a statistical model, both in general and in the case of a multiple linear regression model. It is essential that the relevant characteristics of the setting in which the predictive inferences are to be applied be consistent with those of the setting that gives rise to the data. For example, in making predictive inferences about the relationship between a cow's milk production and her age and her body weight, it would be essential that there be consistency with regard to breed and perhaps with regard to various management practices. The use of data collected on a random sample of the population that is the "target" of the predictive inferences can be regarded as an attempt to achieve the desired consistency. In making predictive inferences on the basis of a multiple linear regression model, it is also essential that the model "accurately reflect" the underlying relationships and that it do so over all values of the explanatory variables for which predictive inferences are sought (as well as over all values for which there are data).

For some applications of the multiple linear regression model, the assumption that $e_1, e_2, \ldots, e_N$ (and d) are uncorrelated with each other may be overly simplistic. Consider, for example, an application in which each of the data points represents the amount of milk produced by a cow. If some of the cows are genetically related to others, then we may wish to modify the model accordingly. Any two of the residual effects $e_1, e_2, \ldots, e_N$ (and d) that correspond to cows that are genetically related may be positively correlated (to an extent that depends on the closeness of the relationship). Moreover, the data are likely to come from more than one herd of cows. Cows that belong to the same herd share a common environment, and tend to be more alike than cows that belong to different herds. One way to account for their alikeness is through the introduction of a positive covariance (of unknown value).

In making inferences on the basis of a multiple linear regression model, a possible objective is that of obtaining relevant input to some sort of decision-making process. In particular, when inferences are made about future data points, it may be done with the intent of judging the effects of changes in the values of any of the explanatory variables $u_1, u_2, \ldots, u_C$ that are subject to control. Considerable caution needs to be exercised in making such judgments. There may be variables that are not accounted for in the model but whose values may have "influenced" the values of $y_1, y_2, \ldots, y_N$ and may influence future data points. If the values of any of the excluded variables are related (either positively or negatively) to any of the variables for which changes are contemplated, then the model-based inferences may create a misleading impression of the effects of the changes.

1.3 Classificatory Models

Let us consider further the use (in making statistical inferences) of N data points that are to be regarded as the values of observable random variables $y_1, y_2, \ldots, y_N$. In many applications, the N data points can be partitioned (in a meaningful way) into a number of mutually exclusive and exhaustive subsets or "groups." In fact, the N data points may lend themselves to several such partitionings, each of which is based on a different criterion or "factor." The subsets or groups formed on the basis of any particular factor are sometimes referred to as the "levels" of the factor.

A factor can be converted into an explanatory variable by assigning each of its levels a distinct number. In some cases, the assignment can be done in such a way that the explanatory variable might be suitable for inclusion in a multiple linear regression model. Consider, for example, the case of data on individual animals that have been partitioned into groups on the basis of age or body weight. In a case of this kind, the factor might be referred to as a "quantitative" factor.

There is another kind of situation; one where the data points are partitioned into groups on the basis of a "qualitative" factor and where (regardless of the method of assignment) the numbers assigned to the groups or levels are meaningful only for purposes of identification. For example, in an application where each data point consists of the amount of milk produced by a different one of N dairy cows, the data points might be partitioned into groups, each of which consists of the data points from those cows that are the daughters of a different one of K bulls. The K bulls constitute the levels of a qualitative factor. For purposes of identification, the bulls could be numbered $1, 2, \ldots, K$ in whatever order might be convenient.

In a situation where the N data points have been partitioned into groups on the basis of each of one or more qualitative factors, the data are sometimes referred to as "classificatory data." Among the models that could be applied to classificatory data are what might be called "classificatory models." Suppose (for the sake of simplicity) that there is a single qualitative factor, and that it has K levels numbered $1, 2, \ldots, K$. And (for $k = 1, 2, \ldots, K$) denote by N_k the number of data points associated with level k—clearly, $\sum_{k=1}^{K} N_k = N$.

In this setting, it is convenient to use two subscripts, rather than one, in distinguishing among the random variables $y_1, y_2, \ldots, y_N$ (and among related quantities). The first subscript identifies the level, and the second allows us to distinguish among entities associated with the same level. Accordingly, we write $y_{k1}, y_{k2}, \ldots, y_{kN_k}$ for those of the random variables $y_1, y_2, \ldots, y_N$ associated with the kth level $(k = 1, 2, \ldots, K)$.

As a possible model, we have the classificatory model obtained by taking

$$y_{ks} = \mu + \alpha_k + e_{ks} \qquad (k = 1, 2, \ldots, K; s = 1, 2, \ldots, N_k), \tag{3.1}$$

where $\mu, \alpha_1, \alpha_2, \ldots, \alpha_K$ are unknown parameters and where the e_{ks}'s are uncorrelated, unobservable random variables, each with mean 0 and (for a strictly positive parameter σ of unknown value) variance σ^2. The parameters $\alpha_1, \alpha_2, \ldots, \alpha_K$ are sometimes referred to as effects. And the model itself is sometimes referred to as the one-way-classification model or (to distinguish it from a variation to be discussed subsequently) the one-way-classification fixed-effects model. The parameters $\mu, \alpha_1, \alpha_2, \ldots, \alpha_K$ are generally taken to be unrestricted, though sometimes they are required to satisfy the restriction

$$\sum_{k=1}^{K} \alpha_k = 0 \tag{3.2}$$

or some other restriction (such as $\sum_{k=1}^{K} N_k \alpha_k = 0$, $\mu = 0$, or $\alpha_K = 0$).

Is the one-way-classification model a linear model? The answer is yes, though this may be less obvious than in the case (considered in Section 1.2) of a multiple linear regression model. That the one-way-classification model is a linear model becomes more transparent upon observing that the defining relation (3.1) can be reexpressed in the form

$$y_{ks} = \sum_{j=1}^{K+1} x_{ksj} \beta_j + e_{ks} \qquad (k = 1, 2, \ldots, K; s = 1, 2, \ldots, N_k), \tag{3.3}$$

where $\beta_1 = \mu$ and $\beta_j = \alpha_{j-1}$ $(j = 2, 3, \ldots, K + 1)$ and where (for $k = 1, 2, \ldots, K$; $s = 1, 2, \ldots, N_k$; $j = 1, 2, \ldots, K + 1$)

$$x_{ksj} = \begin{cases} 1, & \text{if } j = 1 \text{ or } j = k + 1, \\ 0, & \text{otherwise.} \end{cases}$$

In that regard, it may be helpful (i.e., provide even more in the way of transparency) to observe that result (3.3) is equivalent to the result

$$y_i = \sum_{j=1}^{K+1} x_{ij}\beta_j + e_i \qquad (i = 1, 2, \ldots, N),$$

where $x_{i1} = 1$ and (for $j = 2, 3, \ldots, K+1$) $x_{ij} = 1$ or $x_{ij} = 0$ depending on whether or not the ith data point is a member of the $(j-1)$th group $(i = 1, 2, \ldots, N)$ and where (as in Section 1.2) $e_1, e_2, \ldots, e_N$ are uncorrelated, unobservable random variables, each with mean 0 and variance σ^2.

In an application of the one-way-classification model, we might wish to make inferences about $\mu + \alpha_1, \mu + \alpha_2, \ldots, \mu + \alpha_K$. For $k = 1, 2, \ldots, K$ (and "all" s)

$$\mathrm{E}(y_{ks}) = \mu + \alpha_k.$$

Accordingly, $\mu + \alpha_k$ can be thought of as representing the average of an infinitely large number of data points, each of which belongs to the kth group.

We might also wish to make inferences about various linear combinations of the quantities $\mu + \alpha_1, \mu + \alpha_2, \ldots, \mu + \alpha_K$, that is, about various quantities of the form

$$\sum_{k=1}^{K} c_k(\mu + \alpha_k). \tag{3.4}$$

When the coefficients $c_1, c_2, \ldots, c_K$ in the linear combination (3.4) are such that $\sum_{k=1}^{K} c_k = 0$, the linear combination is reexpressible as $\sum_{k=1}^{K} c_k\alpha_k$ and is referred to as a contrast. Perhaps the simplest kind of contrast is a difference: $\alpha_{k'} - \alpha_k = \mu + \alpha_{k'} - (\mu + \alpha_k)$ (where $k' \neq k$). Still another possibility is that we may wish to make inferences about the quantity $\mu + \alpha_k + d$, where $1 \leq k \leq K$ and where d is an unobservable random variable that (for $k' = 1, 2, \ldots, K$ and $s = 1, 2, \ldots, N_{k'}$) is uncorrelated with $e_{k's}$ and that has mean 0 and variance σ^2. This quantity can be thought of as representing an individual future data point belonging to the kth group.

As a variation on model (3.1), we have the model

$$y_{ks} = \mu_k + e_{ks} \qquad (k = 1, 2, \ldots, K; s = 1, 2, \ldots, N_k), \tag{3.5}$$

where $\mu_1, \mu_2, \ldots, \mu_K$ are unknown parameters and where the e_{ks}'s are as defined earlier [i.e., in connection with model (3.1)]. Model (3.5), like model (3.1), is a linear model. It is a simple example of what is called a means model or a cell-means model; let us refer to it as the one-way-classification cell-means model. Clearly, $\mu_k = \mathrm{E}(y_{ks})$ $(k = 1, 2, \ldots, K; s = 1, 2, \ldots, N_k)$, so that (for $k = 1, 2, \ldots, K$) μ_k is interpretable as the expected value or the "mean" of an arbitrary one of the random variables $y_{k1}, y_{k2}, \ldots, y_{kN_k}$, whose observed values comprise the kth group or "cell."

In making statistical inferences, it matters not whether the inferences are based on model (3.5) or on model (3.1). Nor does it matter whether the restriction (3.2), or a "similar" restriction, is imposed on the parameters of model (3.1). For purposes of making inferences about the relevant quantities, model (3.5) and the restricted and unrestricted versions of model (3.1) are "interchangeable."

The number of applications for which the one-way-classification model provides a completely satisfactory basis for the statistical inferences is relatively small. Even in those applications where interest centers on a particular factor, the relevant concomitant information is typically not limited to the information associated with that factor. To insure that the inferences obtained in such a circumstance are meaningful, they may need to be based on a model that accounts for the additional information.

Suppose, for example, that each data point consists of the amount of milk produced during the first lactation of a different one of N dairy cows. And suppose that N_k of the cows are the daughters of the kth of K bulls (where $N_1, N_2, \ldots, N_K$ are positive integers that sum to N). Interest might center on differences among the respective "breeding values" of the K bulls, that is, on differences in the "average" amounts of milk produced by infinitely large numbers of future daughters under circumstances that are similar from bull to bull. Any inferences about these differences that are based on a one-way-classification model (in which the factor is that whose levels correspond to the bulls) are likely to be at least somewhat misleading. There are factors of known importance that are not accounted for by this model. These include a factor for the time period during which the lactation was initiated and a factor for the herd to which the cow belongs. The importance of these factors is due to the presence of seasonal differences, environmental and genetic trends, and environmental and genetic differences among herds.

The one-way-classification model may be unsuitable as a basis for making inferences from the milk-production data not only because of the omission of important factors, but also because the assumption that the e_{ks}'s are uncorrelated may not be altogether realistic. Typically, some of the cows will have ancestors in common on the female side of the pedigree, in which case the e_{ks}'s for those cows may be positively correlated.

The negative consequences of not having accounted for a factor that has been omitted from a classificatory model may be exacerbated in a situation in which there is a tendency for the levels of the omitted factor to be "confounded" with those of an included factor. In the case of the milk-production data, there may (in the presence of a positive genetic trend) be a tendency for the better bulls to be associated with the more recent time periods. Moreover, there may be a tendency for an exceptionally large proportion of the daughters of some bulls to be located in "above-average" herds, and for an exceptionally large proportion of the daughters of some other bulls to be located in "below-average" herds.

A failure to account for an important factor may occur for any of several reasons. The factor may have been mistakenly judged to be irrelevant or at least unimportant. Or the requisite information about the factor (i.e., knowledge of which data points correspond to which levels) may be unavailable and may not even have been ascertained (possibly for reasons of cost). Or the factor may be a "hidden" factor.

In some cases, the data from which the inferences are to be made are those from a designed experiment. The incorporation of randomization into the design of the experiment serves to limit the extent of the kind of problematic confounding that may be occasioned by a failure to account for an important factor. This kind of problematic confounding can still occur, but only to the extent that it is introduced by chance during the randomization.

1.4 Hierarchical Models and Random-Effects Models

Let us continue to consider the use (in making statistical inferences) of N data points that are to be regarded as the values of observable random variables $y_1, y_2, \ldots, y_N$. And let us denote by $\mathbf{y}$ the N-dimensional random (column) vector whose elements are $y_1, y_2, \ldots, y_N$, respectively. As before, it is supposed that the inferences are to be based on various assumptions about the distribution of $\mathbf{y}$ that are referred to collectively as a statistical model. It might be assumed that the distribution of $\mathbf{y}$ is known up to the value of a column vector, say $\boldsymbol{\phi}$, of unknown parameters. Or it might simply be assumed that the expected value and the variance-covariance matrix of $\mathbf{y}$ are known up to the value of $\boldsymbol{\phi}$. In either case, it is supposed that the quantities about which the inferences are to be made consist of various functions of $\boldsymbol{\phi}$ or, more generally, various unobservable random variables—a function of $\boldsymbol{\phi}$ can be regarded as a "degenerate" random variable.

The assumptions about the distribution of the observable random variables $y_1, y_2, \ldots, y_N$ (the assumptions that comprise the statistical model) may not in and of themselves provide an adequate basis for the inferences. In general, these assumptions need to be extended to the joint distribution of the observable and unobservable random variables (the unobservable random variables about which the inferences are to be made). It might be assumed that the joint distribution of these random variables is known up to the value of $\boldsymbol{\phi}$, or it might be only the expected values and the variances and covariances of these random variables that are assumed to be known up to the value of $\boldsymbol{\phi}$.

The model consisting of assumptions that define the distribution of $\mathbf{y}$ (or various characteristics of the distribution of $\mathbf{y}$) up to the value of the parameter vector $\boldsymbol{\phi}$ can be subjected to a "hierarchical" approach. This approach gives rise to various alternative models. In the hierarchical approach, $\boldsymbol{\phi}$ is regarded as random, and the assumptions that comprise the original model are reinterpreted as assumptions about the conditional distribution of $\mathbf{y}$ given $\boldsymbol{\phi}$. Further, the distribution of $\boldsymbol{\phi}$ or various characteristics of the distribution of $\boldsymbol{\phi}$ are assumed to be known, or at least to be known up to the value of a column vector, say $\boldsymbol{\theta}$, of unknown parameters. These additional assumptions can be thought of as comprising a model for $\boldsymbol{\phi}$.

As an alternative model for $\mathbf{y}$, we have the model obtained by combining the assumptions comprising the original model for $\mathbf{y}$ with the assumptions about the distribution of $\boldsymbol{\phi}$ (or about its characteristics). This model is referred to as a hierarchical model. In some cases, it can be readily reexpressed in nonhierarchical terms; that is, in terms that do not involve $\boldsymbol{\phi}$. This process is facilitated by the application of some basic results on conditional expectations and on conditional variances and covariances.

For "any" random variable x,

$$\mathrm{E}(x) = \mathrm{E}[\mathrm{E}(x \mid \boldsymbol{\phi})], \tag{4.1}$$

and

$$\mathrm{var}(x) = \mathrm{E}[\mathrm{var}(x \mid \boldsymbol{\phi})] + \mathrm{var}[\mathrm{E}(x \mid \boldsymbol{\phi})]. \tag{4.2}$$

And, for "any" two random variables x and w,

$$\mathrm{cov}(x, w) = \mathrm{E}[\mathrm{cov}(x, w \mid \boldsymbol{\phi})] + \mathrm{cov}[\mathrm{E}(x \mid \boldsymbol{\phi}), \mathrm{E}(w \mid \boldsymbol{\phi})]. \tag{4.3}$$

The unconditional expected values in expressions (4.1), (4.2), and (4.3) and the unconditional variance and covariance in expressions (4.2) and (4.3) are those defined with respect to the (marginal) distribution of $\boldsymbol{\phi}$. In general, the expressions for $\mathrm{E}(x)$, $\mathrm{var}(x)$, and $\mathrm{cov}(x, w)$ given by formulas (4.1), (4.2), and (4.3) depend on $\boldsymbol{\theta}$. Formulas (4.1), (4.2), and (4.3) are obtainable from results presented in Chapter 3.

Formulas (4.1) and (4.2) can be used in particular to obtain expressions for the unconditional expected values and variances of $y_1, y_2, \ldots, y_N$ in terms of their conditional expected values and variances—take $x = y_i$ $(1 \le i \le N)$. Similarly, formula (4.3) can be used to obtain an expression for the unconditional covariance of any two of the random variables $y_1, y_2, \ldots, y_N$—take $x = y_i$ and $w = y_j$ $(1 \le i < j \le N)$. Moreover, if the conditional distribution of $\mathbf{y}$ given $\boldsymbol{\phi}$ has a probability density function, say $f(\mathbf{y} \mid \boldsymbol{\phi})$, then, upon applying formula (4.1) with $x = f(\mathbf{y} \mid \boldsymbol{\phi})$, we obtain an expression for the probability density function of the unconditional distribution of $\mathbf{y}$.

In a typical implementation of the hierarchical approach, the dimension of the vector $\boldsymbol{\theta}$ is significantly smaller than that of the vector $\boldsymbol{\phi}$. The most extreme case is that where the various assumptions about the distribution of $\boldsymbol{\phi}$ do not involve unknown parameters; in that case, $\boldsymbol{\theta}$ can be regarded as "degenerate" (i.e., of dimension 0). The effects of basing the inferences on the hierarchical model, rather than on the original model, can be either positive or negative. If the additional assumptions (i.e., the assumptions about the distribution of $\boldsymbol{\phi}$ or about its characteristics) are at least somewhat reflective of an "underlying reality," the effects are likely to be "beneficial." If the additional assumptions are not sufficiently in conformance with "reality," their inclusion in the model may be "counterproductive."

The hierarchical model itself can be subjected to a hierarchical approach. In this continuation of the hierarchical approach, $\boldsymbol{\theta}$ is regarded as random, and the assumptions that comprise the hierarchical model are reinterpreted as assumptions about the conditional distributions of $\mathbf{y}$ given $\boldsymbol{\phi}$ and $\boldsymbol{\theta}$ and of $\boldsymbol{\phi}$ given $\boldsymbol{\theta}$ or simply about the conditional distribution of $\mathbf{y}$ given $\boldsymbol{\theta}$. And the distribution of $\boldsymbol{\theta}$ or various characteristics of the distribution of $\boldsymbol{\theta}$ are assumed to be known or at least to be known up to the value of a vector of unknown parameters. In general, further continuations of the hierarchical approach are possible. Assuming that each continuation results in a reduction in the number of unknown parameters (as would be the case in a typical implementation), the hierarchical approach eventually (after some number of continuations) results in a model that does not involve any unknown parameters.

In general, a model obtained via the hierarchical approach (like any other model) may not in and of itself provide an adequate basis for the statistical inferences. Instead of applying the approach just to the assumptions (about the observable random variables $y_1, y_2, \ldots, y_N$) that comprise the model, the application of the approach may need to be extended to cover any further assumptions included among those made about the joint distribution of the observable random variables and the unobservable random variables (the unobservable random variables about which the inferences are to be made).

Let us now consider the hierarchical approach in the special case where $\mathbf{y}$ follows a linear model. In this special case, there exist (known) numbers $x_{i1}, x_{i2}, \ldots, x_{iP}$ such that

$$\mathrm{E}(y_i) = \sum_{j=1}^{P} x_{ij}\beta_j \qquad (i = 1, 2, \ldots, N) \tag{4.4}$$

for (unknown) parameters $\beta_1, \beta_2, \ldots, \beta_P$. And the variance-covariance matrix of $y_1, y_2, \ldots, y_N$ is an $N \times N$ matrix, say a matrix $\mathbf{\Sigma}$, with ijth element σ_{ij}, that does not depend on $\beta_1, \beta_2, \ldots, \beta_P$ (though it may depend on unknown parameters other than $\beta_1, \beta_2, \ldots, \beta_P$).

Consider an implementation of the hierarchical approach in which only $\beta_1, \beta_2, \ldots, \beta_P$ are regarded as random. [Think of an implementation of the hierarchical approach in which only some of the unknown parameters are regarded as random as one in which any other unknown parameters are regarded as random variables whose joint distribution is degenerate at (i.e., assigns probability 1 to) a single (unknown) point.] The assumptions about the expected values and variance-covariance matrix of $y_1, y_2, \ldots, y_N$ are now to be interpreted as applying to the conditional expected values and conditional variance-covariance matrix given $\beta_1, \beta_2, \ldots, \beta_P$.

Suppose that the assumptions about the distribution of $\beta_1, \beta_2, \ldots, \beta_P$ are of the same general form as those about the distribution of $y_1, y_2, \ldots, y_N$. More specifically, suppose that the expected values of $\beta_1, \beta_2, \ldots, \beta_P$ are linear combinations of unknown parameters $\tau_1, \tau_2, \ldots, \tau_{P'}$, so that there exist numbers $z_{j1}, z_{j2}, \ldots, z_{jP'}$ (assumed known) such that

$$\mathrm{E}(\beta_j) = \sum_{k=1}^{P'} z_{jk}\tau_k \qquad (j = 1, 2, \ldots, P). \tag{4.5}$$

And suppose that the variance-covariance matrix of $\beta_1, \beta_2, \ldots, \beta_P$ is a $P \times P$ matrix, say a matrix $\mathbf{\Gamma}$ with jsth element γ_{js}, that does not depend on $\tau_1, \tau_2, \ldots, \tau_{P'}$—it may depend on various other unknown parameters, some of which may be among those on which the matrix $\mathbf{\Sigma}$ depends.

Making use of formulas (4.1) and (4.3) along with some basic results on the expected values and the variance and covariances of linear combinations of random variables, we find that, under the hierarchical model,

$$\begin{aligned} \mathrm{E}(y_i) = \mathrm{E}\Big(\sum_{j=1}^{P} x_{ij}\beta_j\Big) = \sum_{j=1}^{P} x_{ij}\,\mathrm{E}(\beta_j) &= \sum_{j=1}^{P} x_{ij}\sum_{k=1}^{P'} z_{jk}\tau_k \\ &= \sum_{k=1}^{P'}\Big(\sum_{j=1}^{P} x_{ij}z_{jk}\Big)\tau_k \end{aligned} \tag{4.6}$$

$(i = 1, 2, \ldots, N)$ and

$$\begin{aligned}\operatorname{cov}(y_i, y_{i'}) &= \mathrm{E}(\sigma_{ii'}) + \operatorname{cov}\Bigl(\sum_{j=1}^{P} x_{ij}\beta_j, \sum_{j'=1}^{P} x_{i'j'}\beta_{j'}\Bigr) \\ &= \sigma_{ii'} + \sum_{j=1}^{P}\sum_{j'=1}^{P} x_{ij}x_{i'j'}\gamma_{jj'} \end{aligned} \tag{4.7}$$

$(i, i' = 1, 2, \ldots, N)$. It follows from results (4.6) and (4.7) that if $\mathbf{\Sigma}$ does not depend on $\tau_1, \tau_2, \ldots, \tau_{P'}$, then the hierarchical model, like the original model, is a linear model.

For any integer j between 1 and P, inclusive, such that $\operatorname{var}(\beta_j) = 0$, we have that $\beta_j = \mathrm{E}(\beta_j)$ (with probability 1). Thus, for any such integer j, the assumption that $\mathrm{E}(\beta_j) = \sum_{k=1}^{P'} z_{jk}\tau_k$ simplifies in effect to an assumption that $\beta_j = \sum_{k=1}^{P'} z_{jk}\tau_k$. In the special case where $\mathrm{E}(\beta_j) = \tau_{k'}$ for some integer k' $(1 \le k' \le P')$, there is a further simplification to $\beta_j = \tau_{k'}$. Thus, the hierarchical approach is sufficiently flexible that some of the parameters $\beta_1, \beta_2, \ldots, \beta_P$ can in effect be retained and included among the parameters $\tau_1, \tau_2, \ldots, \tau_{P'}$.

As indicated earlier (in Section 1.1), it is extremely convenient (in working with linear models) to adopt matrix notation. Let $\boldsymbol{\beta}$ represent the P-dimensional column vector with elements $\beta_1, \beta_2, \ldots, \beta_P$, respectively, and $\boldsymbol{\tau}$ the P'-dimensional column vector with elements $\tau_1, \tau_2, \ldots, \tau_{P'}$, respectively. Then, in matrix notation, equality (4.4) becomes (in the context of the hierarchical approach)

$$\mathrm{E}(\mathbf{y} \mid \boldsymbol{\beta}) = \mathbf{X}\boldsymbol{\beta}, \tag{4.8}$$

where $\mathbf{X}$ is the $N \times P$ matrix with ijth element x_{ij} $(i = 1, 2, \ldots, N;\ j = 1, 2, \ldots, P)$. And equality (4.5) becomes

$$\mathrm{E}(\boldsymbol{\beta}) = \mathbf{Z}\boldsymbol{\tau}, \tag{4.9}$$

where $\mathbf{Z}$ is the $P \times P'$ matrix with jkth element z_{jk} $(j = 1, 2, \ldots, P;\ k = 1, 2, \ldots, P')$. Further, results (4.6) and (4.7) can be recast in matrix notation as

$$\mathrm{E}(\mathbf{y}) = \mathbf{X}\mathbf{Z}\boldsymbol{\tau} \tag{4.10}$$

and

$$\operatorname{var}(\mathbf{y}) = \mathbf{\Sigma} + \mathbf{X}\mathbf{\Gamma}\mathbf{X}' \tag{4.11}$$

(where $\mathbf{X}'$ denotes the transpose of $\mathbf{X}$, i.e., the $P \times N$ matrix with ijth element x_{ji}). Alternatively, by making use of the more general (matrix) versions of formulas (4.1) and (4.2) presented in Chapter 3, results (4.10) and (4.11) can be derived directly from equalities (4.8) and (4.9).

The original (linear) model provides a basis for making inferences about an unobservable quantity that is expressible as a linear combination, say $\sum_{j=1}^{P} \lambda_j\beta_j$, of the parameters $\beta_1, \beta_2, \ldots, \beta_P$. Or, more generally, it provides a basis for making inferences about an unobservable quantity that is expressible as a random variable w for which $\mathrm{E}(w) = \sum_{j=1}^{P} \lambda_j\beta_j$ and for which $\operatorname{var}(w) = \psi$ and $\operatorname{cov}(\mathbf{y}, w) = \boldsymbol{\nu}$ for some number ψ and (column) vector $\boldsymbol{\nu}$ that do not depend on $\beta_1, \beta_2, \ldots, \beta_P$. In the context of the hierarchical approach, $\sum_{j=1}^{P} \lambda_j\beta_j = \mathrm{E}(w \mid \beta_1, \beta_2, \ldots, \beta_P)$, $\psi = \operatorname{var}(w \mid \beta_1, \beta_2, \ldots, \beta_P)$, and $\boldsymbol{\nu} = \operatorname{cov}(\mathbf{y}, w \mid \beta_1, \beta_2, \ldots, \beta_P)$.

Let ν_i represent the ith element of $\boldsymbol{\nu}$ $(i = 1, 2, \ldots, N)$. Then, making use of formulas (4.1), (4.2), and (4.3) and proceeding in much the same way as in the derivation of results (4.6) and (4.7), we find that, under the hierarchical model,

$$\begin{aligned}\mathrm{E}(w) = \mathrm{E}\Bigl(\sum_{j=1}^{P} \lambda_j\beta_j\Bigr) = \sum_{j=1}^{P} \lambda_j \mathrm{E}(\beta_j) &= \sum_{j=1}^{P} \lambda_j \sum_{k=1}^{P'} z_{jk}\tau_k \\ &= \sum_{k=1}^{P'} \Bigl(\sum_{j=1}^{P} \lambda_j z_{jk}\Bigr)\tau_k, \end{aligned} \tag{4.12}$$

$$\text{var}(w) = \text{E}(\psi) + \text{var}\Big(\sum_{j=1}^{P} \lambda_j \beta_j\Big) = \psi + \sum_{j=1}^{P} \sum_{j'=1}^{P} \lambda_j \lambda_{j'} \gamma_{jj'}, \tag{4.13}$$

and (for $i = 1, 2, \ldots, N$)

$$\begin{aligned} \text{cov}(y_i, w) &= \text{E}(\nu_i) + \text{cov}\Big(\sum_{j=1}^{P} x_{ij} \beta_j, \sum_{j'=1}^{P} \lambda_{j'} \beta_{j'}\Big) \\ &= \nu_i + \sum_{j=1}^{P} \sum_{j'=1}^{P} x_{ij} \lambda_{j'} \gamma_{jj'}. \end{aligned} \tag{4.14}$$

Clearly, if ψ and $\boldsymbol{\nu}$ do not depend on $\tau_1, \tau_2, \ldots, \tau_{P'}$, then neither do expressions (4.13) and (4.14).

As in the case of results (4.6) and (4.7), results (4.12), (4.13), and (4.14) can be recast in matrix notation. Denote by $\boldsymbol{\lambda}$ the P-dimensional column vector with elements $\lambda_1, \lambda_2, \ldots, \lambda_P$, respectively—the linear combination $\sum_{j=1}^{P} \lambda_j \beta_j$ is reexpressible as $\sum_{j=1}^{P} \lambda_j \beta_j = \boldsymbol{\lambda}'\boldsymbol{\beta}$. Under the hierarchical model,

$$\text{E}(w) = \boldsymbol{\lambda}'\mathbf{Z}\boldsymbol{\tau}, \tag{4.15}$$

$$\text{var}(w) = \psi + \boldsymbol{\lambda}'\boldsymbol{\Gamma}\boldsymbol{\lambda}, \tag{4.16}$$

and

$$\text{cov}(\mathbf{y}, w) = \boldsymbol{\nu} + \mathbf{X}\boldsymbol{\Gamma}\boldsymbol{\lambda}. \tag{4.17}$$

The hierarchical approach is not the only way of arriving at a model characterized by expected values and variances and covariances of the form (4.6) and (4.7) or, equivalently, of the form (4.10) and (4.11). Under the original model, the distribution of $y_1, y_2, \ldots, y_N$, and w is such that

$$y_i = \sum_{j=1}^{P} x_{ij} \beta_j + e_i \qquad (i = 1, 2, \ldots, N) \tag{4.18}$$

and

$$w = \sum_{j=1}^{P} \lambda_j \beta_j + d, \tag{4.19}$$

where $e_1, e_2, \ldots, e_N$, and d are random variables, each with expected value 0. Or, equivalently, the distribution of $\mathbf{y}$ and w is such that

$$\mathbf{y} = \mathbf{X}\boldsymbol{\beta} + \mathbf{e} \tag{4.20}$$

and

$$w = \boldsymbol{\lambda}'\boldsymbol{\beta} + d, \tag{4.21}$$

where $\mathbf{e}$ is the N-dimensional random (column) vector with elements $e_1, e_2, \ldots, e_N$ and hence with $\text{E}(\mathbf{e}) = \mathbf{0}$. Moreover, under the original model, $\text{var}(\mathbf{e}) = \text{var}(\mathbf{y}) = \boldsymbol{\Sigma}$, $\text{var}(d) = \text{var}(w) = \psi$, and $\text{cov}(\mathbf{e}, d) = \text{cov}(\mathbf{y}, w) = \boldsymbol{\nu}$.

Now, suppose that instead of taking $\beta_1, \beta_2, \ldots, \beta_P$ to be parameters, they are (as in the hierarchical approach) taken to be random variables with expected values of the form (4.5)—in which case, $\boldsymbol{\beta}$ is a random vector with an expected value of the form (4.9)—and with variance-covariance matrix $\boldsymbol{\Gamma}$. Suppose further that each of the random variables $\beta_1, \beta_2, \ldots, \beta_P$ is uncorrelated with each of the random variables $e_1, e_2, \ldots, e_N$, and d [or, equivalently, that $\text{cov}(\boldsymbol{\beta}, \mathbf{e}) = \mathbf{0}$ and $\text{cov}(\boldsymbol{\beta}, d) = \mathbf{0}$] or, perhaps more generally, suppose that each of the quantities $\sum_{j=1}^{P} x_{1j} \beta_j, \sum_{j=1}^{P} x_{2j} \beta_j, \ldots, \sum_{j=1}^{P} x_{Nj} \beta_j$ is uncorrelated with each of the random variables $e_1, e_2, \ldots, e_N$, and d [or, equivalently, that $\text{cov}(\mathbf{X}\boldsymbol{\beta}, \mathbf{e}) = \mathbf{0}$ and $\text{cov}(\mathbf{X}\boldsymbol{\beta}, d) = \mathbf{0}$]. And consider the effect of these suppositions about $\beta_1, \beta_2, \ldots, \beta_P$ on the distribution of the $N+1$ random variables (4.18) and (4.19) (specifically the effect on their expected values and their variances and covariances).

The suppositions about $\beta_1, \beta_2, \ldots, \beta_P$ are equivalent to the supposition that

$$\beta_j = \sum_{k=1}^{P'} z_{jk}\tau_k + \delta_j \qquad (j = 1, 2, \ldots, P), \tag{4.22}$$

where $\delta_1, \delta_2, \ldots, \delta_P$ are random variables with expected values of 0 and variance-covariance matrix $\mathbf{\Gamma}$ and with the property that each of the linear combinations $\sum_{j=1}^{P} x_{1j}\delta_j, \sum_{j=1}^{P} x_{2j}\delta_j, \ldots, \sum_{j=1}^{P} x_{Nj}\delta_j$ is uncorrelated with each of the random variables $e_1, e_2, \ldots, e_N$, and d. Or, in matrix notation,

$$\boldsymbol{\beta} = \mathbf{Z}\boldsymbol{\tau} + \boldsymbol{\delta}, \tag{4.23}$$

where $\boldsymbol{\delta}$ is the P-dimensional random (column) vector with elements $\delta_1, \delta_2, \ldots, \delta_P$ and hence with $\mathrm{E}(\boldsymbol{\delta}) = \mathbf{0}$, $\mathrm{var}(\boldsymbol{\delta}) = \mathbf{\Gamma}$, $\mathrm{cov}(\mathbf{X}\boldsymbol{\delta}, \mathbf{e}) = \mathbf{0}$, and $\mathrm{cov}(\mathbf{X}\boldsymbol{\delta}, d) = \mathbf{0}$.

Upon replacing $\beta_1, \beta_2, \ldots, \beta_P$ in expressions (4.18) and (4.19) with the expressions for $\beta_1, \beta_2, \ldots, \beta_P$ given by result (4.22), we obtain the expressions

$$y_i = \sum_{k=1}^{P'} \Bigl(\sum_{j=1}^{P} x_{ij} z_{jk}\Bigr)\tau_k + f_i \qquad (i = 1, 2, \ldots, N), \tag{4.24}$$

where (for $i = 1, 2, \ldots, N$) $f_i = e_i + \sum_{j=1}^{P} x_{ij}\delta_j$, and the expression

$$w = \sum_{k=1}^{P'} \Bigl(\sum_{j=1}^{P} \lambda_j z_{jk}\Bigr)\tau_k + g, \tag{4.25}$$

where $g = d + \sum_{j=1}^{P} \lambda_j \delta_j$. Results (4.24) and (4.25) can be restated in matrix notation as

$$\mathbf{y} = \mathbf{X}\mathbf{Z}\boldsymbol{\tau} + \mathbf{f}, \tag{4.26}$$

where $\mathbf{f}$ is the N-dimensional random (column) vector with elements $f_1, f_2, \ldots, f_N$ and hence where $\mathbf{f} = \mathbf{e} + \mathbf{X}\boldsymbol{\delta}$, and

$$w = \boldsymbol{\lambda}'\mathbf{Z}\boldsymbol{\tau} + g, \tag{4.27}$$

where $g = d + \boldsymbol{\lambda}'\boldsymbol{\delta}$. Alternatively, expressions (4.26) and (4.27) are obtainable by replacing $\boldsymbol{\beta}$ in expressions (4.20) and (4.21) with expression (4.23).

Clearly, $\mathrm{E}(f_i) = 0$ $(i = 1, 2, \ldots, N)$, or equivalently $\mathrm{E}(\mathbf{f}) = \mathbf{0}$, and $\mathrm{E}(g) = 0$. Further, by making use of some basic results on the variances and covariances of linear combinations of random variables [in essentially the same way as in the derivation of results (4.7), (4.13), and (4.14)], we find that

$$\mathrm{cov}(f_i, f_{i'}) = \sigma_{ii'} + \sum_{j=1}^{P}\sum_{j'=1}^{P} x_{ij}x_{i'j'}\gamma_{jj'} \qquad (i, i' = 1, 2, \ldots, N),$$

$$\mathrm{var}(g) = \psi + \sum_{j=1}^{P}\sum_{j'=1}^{P} \lambda_j\lambda_{j'}\gamma_{jj'},$$

and

$$\mathrm{cov}(f_i, g) = \nu_i + \sum_{j=1}^{P}\sum_{j'=1}^{P} x_{ij}\lambda_{j'}\gamma_{jj'} \qquad (i = 1, 2, \ldots, N),$$

or, equivalently, that

$$\mathrm{var}(\mathbf{f}) = \mathbf{\Sigma} + \mathbf{X}\mathbf{\Gamma}\mathbf{X}',$$

$$\mathrm{var}(g) = \psi + \boldsymbol{\lambda}'\mathbf{\Gamma}\boldsymbol{\lambda},$$

and

$$\mathrm{cov}(\mathbf{f}, g) = \boldsymbol{\nu} + \mathbf{X}\mathbf{\Gamma}\boldsymbol{\lambda}.$$

These results imply that, as in the case of the hierarchical approach, the expected values and the variances and covariances of the random variables $y_1, y_2, \ldots, y_N$, and w are of the form (4.6), (4.12), (4.7), (4.13), and (4.14), or, equivalently, that $\mathrm{E}(\mathbf{y})$, $\mathrm{E}(w)$, $\mathrm{var}(\mathbf{y})$, $\mathrm{var}(w)$, and $\mathrm{cov}(\mathbf{y}, w)$ are of the form (4.10), (4.15), (4.11), (4.16), and (4.17). Let us refer to this alternative way of arriving at a model characterized by expected values and variances and covariances of the form (4.6) and (4.7), or equivalently (4.10) and (4.11), as the random-effects approach.

The assumptions comprising the original model are such that (for $i = 1, 2, \ldots, N$) $\mathrm{E}(y_i) = \sum_{j=1}^{P} x_{ij}\beta_j$ and are such that the variance-covariance matrix of $y_1, y_2, \ldots, y_N$ equals $\boldsymbol{\Sigma}$ (where $\boldsymbol{\Sigma}$ does not vary with $\beta_1, \beta_2, \ldots, \beta_P$). In the hierarchical approach, these assumptions are regarded as applying to the conditional distribution of $y_1, y_2, \ldots, y_N$ given $\beta_1, \beta_2, \ldots, \beta_P$. Thus, in the hierarchical approach, the random vector $\mathbf{e}$ in decomposition (4.20) is such that (with probability 1) $\mathrm{E}(e_i \mid \beta_1, \beta_2, \ldots, \beta_P) = 0$ $(i = 1, 2, \ldots, N)$ and $\mathrm{cov}(e_i, e_{i'} \mid \beta_1, \beta_2, \ldots, \beta_P) = \sigma_{ii'}$ $(i, i' = 1, 2, \ldots, N)$ or, equivalently, $\mathrm{E}(\mathbf{e} \mid \boldsymbol{\beta}) = \mathbf{0}$ and $\mathrm{var}(\mathbf{e} \mid \boldsymbol{\beta}) = \boldsymbol{\Sigma}$.

The random-effects approach results in the same alternative model as the hierarchical approach and does so under less stringent assumptions. In the random-effects approach, it is assumed that the (unconditional) distribution of $\mathbf{e}$ is such that (for $i = 1, 2, \ldots, N$) $\mathrm{E}(e_i) = 0$ or, equivalently, $\mathrm{E}(\mathbf{e}) = \mathbf{0}$. It is also assumed that the joint distribution of $\mathbf{e}$ and of the random vector $\boldsymbol{\delta}$ in decomposition (4.23) is such that (for $i, i' = 1, 2, \ldots, N$) $\mathrm{cov}(\sum_{j=1}^{P} x_{ij}\delta_j, e_{i'}) = 0$ or, equivalently, $\mathrm{cov}(\mathbf{X}\boldsymbol{\delta}, \mathbf{e}) = \mathbf{0}$. By making use of formulas (4.1) and (4.3), these assumptions can be restated as follows: (for $i = 1, 2, \ldots, N$) $\mathrm{E}[\mathrm{E}(e_i \mid \beta_1, \beta_2, \ldots, \beta_P)] = 0$ and (for $i, i' = 1, 2, \ldots, N$) $\mathrm{cov}[\sum_{j=1}^{P} x_{ij}\delta_j, \mathrm{E}(e_{i'} \mid \beta_1, \beta_2, \ldots, \beta_P)] = 0$ or, equivalently, $\mathrm{E}[\mathrm{E}(\mathbf{e} \mid \boldsymbol{\beta})] = \mathbf{0}$ and $\mathrm{cov}[\mathbf{X}\boldsymbol{\delta}, \mathrm{E}(\mathbf{e} \mid \boldsymbol{\beta})] = \mathbf{0}$. Moreover, in the random-effects approach,

$$\begin{aligned}\sigma_{ii'} = \mathrm{cov}(e_i, e_{i'}) = \mathrm{E}[&\mathrm{cov}(e_i, e_{i'} \mid \beta_1, \beta_2, \ldots, \beta_P)] \\ &+ \mathrm{cov}[\mathrm{E}(e_i \mid \beta_1, \beta_2, \ldots, \beta_P), \mathrm{E}(e_{i'} \mid \beta_1, \beta_2, \ldots, \beta_P)]\end{aligned}$$

$(i, i' = 1, 2, \ldots, N)$ or, equivalently,

$$\boldsymbol{\Sigma} = \mathrm{var}(\mathbf{e}) = \mathrm{E}[\mathrm{var}(\mathbf{e} \mid \boldsymbol{\beta})] + \mathrm{var}[\mathrm{E}(\mathbf{e} \mid \boldsymbol{\beta})].$$

Let us now specialize even further by considering an application of the hierarchical approach or random-effects approach in a setting where the N data points have been partitioned into K groups, corresponding to the first through Kth levels of a single qualitative factor. As in our previous discussion of this setting (in Section 1.3), let us write $y_{k1}, y_{k2}, \ldots, y_{kN_k}$ for those of the random variables $y_1, y_2, \ldots, y_N$ associated with the kth level $(k = 1, 2, \ldots, K)$.

A possible model is the one-way-classification cell-means model, in which

$$y_{ks} = \mu_k + e_{ks} \qquad (k = 1, 2, \ldots, K;\, s = 1, 2, \ldots, N_k), \tag{4.28}$$

where $\mu_1, \mu_2, \ldots, \mu_K$ are unknown parameters and where the e_{ks}'s are uncorrelated, unobservable random variables, each with mean 0 and (for a strictly positive parameter σ of unknown value) variance σ^2. As previously indicated (in Section 1.3), this model qualifies as a linear model.

Let us apply the hierarchical approach or random-effects approach to the one-way-classification cell-means model. Suppose that $\mu_1, \mu_2, \ldots, \mu_K$ (but not σ) are regarded as random. Suppose further that $\mu_1, \mu_2, \ldots, \mu_K$ are uncorrelated and that they have a common, unknown mean, say μ, and (for a nonnegative parameter σ_α of unknown value) a common variance σ_α^2. Or, equivalently, suppose that

$$\mu_k = \mu + \alpha_k \qquad (k = 1, 2, \ldots, K), \tag{4.29}$$

where $\alpha_1, \alpha_2, \ldots, \alpha_K$ are uncorrelated random variables having mean 0 and a common variance σ_α^2. (And assume that σ_α and σ are not functionally dependent on μ.)

Under the original model (i.e., the 1-way-classification cell-means model), we have that (for $k = 1, 2, \ldots, K$ and $s = 1, 2, \ldots, N_k$)

$$\mathrm{E}(y_{ks}) = \mu_k \tag{4.30}$$

and that (for $k, k' = 1, 2, \ldots, K$; $s = 1, 2, \ldots, N_k$; and $s' = 1, 2, \ldots, N_{k'}$)

$$\operatorname{cov}(y_{ks}, y_{k's'}) = \begin{cases} \sigma^2, & \text{if } k' = k \text{ and } s' = s, \\ 0, & \text{otherwise.} \end{cases} \tag{4.31}$$

In the hierarchical approach, the expected value (4.30) and the covariance (4.31) are regarded as a conditional (on $\mu_1, \mu_2, \ldots, \mu_K$) expected value and a conditional covariance.

The same model that would be obtained by applying the hierarchical approach (the so-called hierarchical model) is obtainable via the random-effects approach. Accordingly, assume that each of the random variables $\alpha_1, \alpha_2, \ldots, \alpha_K$ [in representation (4.29)] is uncorrelated with each of the N random variables e_{ks} ($k = 1, 2, \ldots, K$; $s = 1, 2, \ldots, N_k$). In this setting, the random-effects approach can be implemented by replacing $\mu_1, \mu_2, \ldots, \mu_K$ in representation (4.28) with the expressions for $\mu_1, \mu_2, \ldots, \mu_K$ comprising representation (4.29). This operation gives

$$y_{ks} = \mu + \alpha_k + e_{ks} \qquad (k = 1, 2, \ldots, K; s = 1, 2, \ldots, N_k) \tag{4.32}$$

or, upon letting $f_{ks} = \alpha_k + e_{ks}$,

$$y_{ks} = \mu + f_{ks} \qquad (k = 1, 2, \ldots, K; s = 1, 2, \ldots, N_k). \tag{4.33}$$

Result (4.32) or (4.33) defines an alternative to the one-way-classification cell-means model. Under the alternative model, we have that (for $k = 1, 2, \ldots, K$ and $s = 1, 2, \ldots, N_k$) $\mathrm{E}(f_{ks}) = 0$ and hence

$$\mathrm{E}(y_{ks}) = \mu$$

and that (for $k, k' = 1, 2, \ldots, K$; $s = 1, 2, \ldots, N_k$; and $s' = 1, 2, \ldots, N_{k'}$)

$$\operatorname{cov}(y_{ks}, y_{k's'}) = \operatorname{cov}(f_{ks}, f_{k's'}) = \begin{cases} \sigma^2 + \sigma_\alpha^2, & \text{if } k' = k \text{ and } s' = s, \\ \sigma_\alpha^2, & \text{if } k' = k \text{ and } s' \neq s, \\ 0, & \text{if } k' \neq k. \end{cases}$$

Thus, under the alternative model, all of the y_{ks}'s have the same expected value, and those of the y_{ks}'s that are associated with the same level may be positively correlated. Under the original model, the expected values of the y_{ks}'s may vary with the level, and none of the y_{ks}'s are correlated.

Representation (4.32) is of the same form as representation (3.1), which is identified with the one-way-classification fixed-effects model. However, the quantities $\alpha_1, \alpha_2, \ldots, \alpha_K$ that appear in representation (4.32) are random variables and are referred to as random effects, whereas the quantities $\alpha_1, \alpha_2, \ldots, \alpha_K$ that appear in representation (3.1) are (unknown) parameters that are referred to as fixed effects. Accordingly, the model obtained from the cell-means model via the random-effects approach (or the corresponding hierarchical approach) is referred to as the one-way-classification random-effects model.

The one-way-classification random-effects model can be obtained not only via an application of the random-effects approach or hierarchical approach to the one-way-classification cell-means model, but also via an application of the random-effects approach or hierarchical approach to the one-way-classification fixed-effects model. In the case of the random-effects approach, simply regard the effects $\alpha_1, \alpha_2, \ldots, \alpha_K$ of the fixed-effects model as random variables rather than (unknown) parameters, assume that $\alpha_1, \alpha_2, \ldots, \alpha_K$ are uncorrelated, each with mean 0 and variance σ_α^2, and assume that each of the α_k's is uncorrelated with each of the e_{ks}'s. Then, by proceeding in much the same way as in the application of the random-effects approach to the one-way-classification cell-means model, we once again arrive at the one-way-classification random-effects model.

We have established that the one-way-classification random-effects model can be obtained by adding (in the context of a random-effects approach or hierarchical approach) to the assumptions that comprise the one-way-classification fixed-effects model or the one-way-classification cell-means model. Under what circumstances are the additional assumptions likely to reflect an underlying reality and hence to be beneficial? The additional assumptions would seem to be warranted in a circumstance where the K levels of the factor can reasonably be envisioned as a random sample from an infinitely large "population" of levels. Or, relatedly, they might be warranted in a circumstance where it is possible to conceive of K infinitely large sets of data points, each of which corresponds to a different one of the K levels, and where the average values of the data points in the K sets can reasonably be regarded as a random sample from an infinitely large population of averages.

Suppose, for example, that each data point consists of the amount of milk produced by a different one of N dairy cows and that N_k of the cows are the daughters of the kth of K bulls ($k = 1, 2, \ldots, K$). Then, as discussed in Section 1.3, interest might center on the differences among the breeding values of the bulls. Among the models that could conceivably serve as a basis for inferences about those differences is the one-way-classification fixed-effects model (in which the factor is that whose levels correspond to the bulls) or the one-way-classification cell-means model. However, under some circumstances, better results are likely to be obtained by basing the inferences on the one-way-classification random-effects model. Those circumstances include ones where the underlying reality is at least reasonably consistent with what might be expected if the K bulls were a random sample from an infinitely large population of bulls.

As a practical matter, the circumstances are likely to be such that the one-way-classification random-effects model is too simplistic to provide a satisfactory basis for the inferences. The circumstances in which the one-way-classification random-effects model is likely to be inadequate include those discussed earlier (in Section 1.3) in which inferences based on a one-way-classification fixed-effects (or cell-means) model are likely to be misleading. They also include other circumstances. Some of the K bulls may have one or more ancestors in common with some of the other bulls. Depending on the extent and closeness of the resultant genetic relationships, it may be important to take those relationships into account. This can be done within the context of the hierarchical or random-effects approach. Instead of taking the random effects to be uncorrelated, they can be taken to be correlated in a way and to an extent that reflects the underlying relationships.

There may exist other information about the K bulls that (like the ancestral information) is "external" to the information provided by the N data points and that is at odds with various of the assumptions of the one-way-classification random-effects model. For example, the information might take the form of results from a statistical analysis of some earlier data. We may wish to base our inferences on a model that accounts for this information. As in the case of the ancestral information, such a model can (at least in principle) be devised within the context of the hierarchical approach or random-effects approach. In our application of the random-effects approach to the one-way-classification cell-means model, we may wish to modify our assumption that the cell means $\mu_1, \mu_2, \ldots, \mu_K$ have a common mean μ as well as our assumption that the random effects $\alpha_1, \alpha_2, \ldots, \alpha_K$ are uncorrelated with a common variance σ_α^2.

In making inferences on the basis of the one-way-classification cell-means model (4.28), the quantities of interest are typically ones that are expressible as a linear combination of the cell means $\mu_1, \mu_2, \ldots, \mu_K$, say a linear combination $\sum_{k=1}^{K} c_k \mu_k$, or, more generally, quantities that are expressible as a random variable w of the form

$$w = \sum_{k=1}^{K} c_k \mu_k + d, \tag{4.34}$$

where d is a random variable for which $\mathrm{E}(d) = 0$ [and for which $\mathrm{var}(d)$ and the covariance of d with each of the e_{ks}'s do not depend on $\mu_1, \mu_2, \ldots, \mu_K$]. Let us consider how w is affected by the application of the random-effects approach (to the one-way-classification cell-means model). In that approach, $\mu_1, \mu_2, \ldots, \mu_K$ are regarded as random variables that are expressible in the form (4.29). And it is supposed that each of the random effects $\alpha_1, \alpha_2, \ldots, \alpha_K$ is uncorrelated with d (as well as with each of the e_{ks}'s).

The random variable w can be reexpressed in terms of the parameter μ and the random effects $\alpha_1, \alpha_2, \ldots, \alpha_K$. Upon replacing $\mu_1, \mu_2, \ldots, \mu_K$ in expression (4.34) with the expressions for $\mu_1, \mu_2, \ldots, \mu_K$ given by representation (4.29), we find that

$$w = \Bigl(\sum_{k=1}^{K} c_k\Bigr)\mu + \sum_{k=1}^{K} c_k\alpha_k + d \tag{4.35}$$

$$= \Bigl(\sum_{k=1}^{K} c_k\Bigr)\mu + g, \tag{4.36}$$

where $g = d + \sum_{k=1}^{K} c_k\alpha_k$.

Expression (4.36) gives w in terms of the parameter μ and a random variable g. Clearly, $\mathrm{E}(g) = 0$. And it follows from a basic formula on the variance of a linear combination of uncorrelated random variables that

$$\mathrm{var}(g) = \mathrm{var}(d) + \sigma_\alpha^2 \sum_{k=1}^{K} c_k^2. \tag{4.37}$$

Recall that (for $k = 1, 2, \ldots, K$ and $s = 1, 2, \ldots, N_k$) $f_{ks} = \alpha_k + e_{ks}$. Under the original (1-way-classification cell-means) model, $\mathrm{cov}(y_{ks}, w) = \mathrm{cov}(e_{ks}, d)$, while under the alternative (1-way-classification random-effects) model, $\mathrm{cov}(y_{ks}, w) = \mathrm{cov}(f_{ks}, g)$. Even if e_{ks} and d are uncorrelated, f_{ks} and g may be (and in many cases are) correlated. In fact,

$$\begin{aligned}\mathrm{cov}(f_{ks}, g) &= \mathrm{cov}(e_{ks}, d) + \mathrm{cov}\Bigl(e_{ks}, \sum_{k'=1}^{K} c_{k'}\alpha_{k'}\Bigr) + \mathrm{cov}(\alpha_k, d) + \mathrm{cov}\Bigl(\alpha_k, \sum_{k'=1}^{K} c_{k'}\alpha_{k'}\Bigr)\\ &= \mathrm{cov}(e_{ks}, d) + 0 + 0 + \mathrm{cov}\Bigl(\alpha_k, \sum_{k'=1}^{K} c_{k'}\alpha_{k'}\Bigr)\\ &= \mathrm{cov}(e_{ks}, d) + \mathrm{cov}\Bigl(\alpha_k, \sum_{k'=1}^{K} c_{k'}\alpha_{k'}\Bigr),\end{aligned} \tag{4.38}$$

as can be readily verified by applying a basic formula for a covariance between sums of random variables. And, in light of the assumption that $\alpha_1, \alpha_2, \ldots, \alpha_K$ are uncorrelated, each with variance σ_α^2, result (4.38) simplifies to

$$\mathrm{cov}(f_{ks}, g) = \mathrm{cov}(e_{ks}, d) + c_k\sigma_\alpha^2. \tag{4.39}$$

As noted earlier in this section, the one-way-classification random-effects model can be obtained by applying a hierarchical approach or random-effects approach to the one-way-classification fixed-effects model (as well as by application to the 1-way-classification cell-means model). A related observation (pertaining to inference about a random variable w) is that the application of the hierarchical or random-effects approach to the one-way-classification fixed-effects model has the same effect on the random variable w defined by $w = \sum_{k=1}^{K} c_k(\mu + \alpha_k) + d$ as the application to the one-way-classification cell-means model has on the random variable $w = \sum_{k=1}^{K} c_k\mu_k + d$.

It is worth noting that results pertaining to inference about a random variable w based on the one-way-classification random-effects model can be readily extended from a random variable w that is expressible in the form (4.35) to one that is expressible in the more general form

$$w = c_0\mu + \sum_{k=1}^{K'} c_k\alpha_k + d,$$

where $K' \geq K$, where $\alpha_{K+1}, \alpha_{K+2}, \ldots, \alpha_{K'}$ (like $\alpha_1, \alpha_2, \ldots, \alpha_K$) are random variables having expected values of 0, and where c_0 and $c_{K+1}, c_{K+2}, \ldots, c_{K'}$ (like $c_1, c_2, \ldots, c_K$) represent arbitrary numbers. Here, $\alpha_{K+1}, \alpha_{K+2}, \ldots, \alpha_{K'}$ are interpretable as random effects that correspond to levels of the factor that are not among those K levels represented in the data. For example, in the case of the milk-production data, the bulls about whose breeding values we might wish to make inferences might include some bulls that do not have any daughters (or at least none who contributed to the data). In the simplest case, the assumption that the α_k's are uncorrelated, each with variance σ_α^2, is not restricted to $\alpha_1, \alpha_2, \ldots, \alpha_K$, but rather is taken to apply to all K' of the random variables $\alpha_1, \alpha_2, \ldots, \alpha_{K'}$. In some settings, it might be appropriate (and fruitful) to allow $\alpha_{K+1}, \alpha_{K+2}, \ldots, \alpha_{K'}$ to be correlated with $\alpha_1, \alpha_2, \ldots, \alpha_K$ (e.g., some or all of the $K'-K$ additional bulls may be related to some of the first K bulls).

1.5 Statistical Inference

Let us continue to consider the use, in making inferences about an unobservable quantity, of N data points that are to be regarded as the respective values of observable random variables $y_1, y_2, \ldots, y_N$. It is supposed that the inferences are to be based on a statistical model, in which the distribution of the N-dimensional random (column) vector $\mathbf{y}$ whose elements are $y_1, y_2, \ldots, y_N$, respectively, is specified up to the value of a (column) vector $\boldsymbol{\phi}$ of unknown parameters or in which various characteristics of the distribution of $\mathbf{y}$ (such as its first and second moments) are specified up to the value of $\boldsymbol{\phi}$. It is supposed further that the quantity of interest is a function, say $\lambda(\boldsymbol{\phi})$, of $\boldsymbol{\phi}$ or, more generally, is an unobservable random variable w whose distribution may depend on $\boldsymbol{\phi}$. (For convenience, we refer to the random variable w as the quantity of interest and speak of inference about w; we do so even though, in any particular application, it is the realization of w that corresponds to the quantity of interest and that is the subject of the inferences.)

Inference about w might consist of point estimation (or prediction), interval (or set) estimation (or prediction), or hypothesis testing. Let $t(\mathbf{y})$ represent a function of $\mathbf{y}$, the realized value of which [i.e., the value of $t(\mathbf{y})$ corresponding to the observed value of $\mathbf{y}$] is to be regarded as an estimate (a so-called point estimate) of the value of w. This function is referred to as a point estimator. Or if w represents a future quantity, $t(\mathbf{y})$ might be referred to as a point predictor.

The difference $t(\mathbf{y}) - w$ between the estimator or predictor and the random variable w is referred to as the error of estimation or prediction. The performance of the estimator or predictor $t(\mathbf{y})$ in repeated applications might be assessed on the basis of various characteristics of the distribution of the estimation or prediction error. In the special case where w is the parametric function $\lambda(\boldsymbol{\phi})$, the distribution of the estimation or prediction error is determined by the distribution of $\mathbf{y}$, and repeated applications refers to repeated draws of $\mathbf{y}$-values. However, in general, the distribution of the estimation or prediction error is that determined by the joint distribution of w and $\mathbf{y}$, and repeated application refers to repeated draws of both w- and $\mathbf{y}$-values.

The function $t(\mathbf{y})$ is said to be an *unbiased estimator* or *predictor* (of w) if $\mathrm{E}[t(\mathbf{y}) - w] = 0$ or, equivalently, if $\mathrm{E}[t(\mathbf{y})] = \mathrm{E}(w)$. An unbiased estimator or predictor is well-calibrated in the sense that, over an infinitely long sequence of repeated applications, the average value of the estimator or predictor would (in theory) be the same as the average value of the quantity being estimated or predicted.

The second moment $\mathrm{E}\{[t(\mathbf{y}) - w]^2\}$ of the distribution of the estimation or prediction error is referred to as the *mean squared error* (MSE) of $t(\mathbf{y})$. And the square root $\sqrt{\mathrm{E}\{[t(\mathbf{y}) - w]^2\}}$ of the MSE of $t(\mathbf{y})$ is referred to as the *root mean squared error* (root MSE) of $t(\mathbf{y})$. The root MSE can be regarded as a (theoretical) measure of the magnitude of the estimation or prediction errors that would be incurred in an infinitely long sequence of repeated applications. Note that if $t(\mathbf{y})$ is an unbiased estimator or predictor, then $\text{E}\{[t(\mathbf{y}) - w]^2\} = \text{var}[t(\mathbf{y}) - w]$, that is, the MSE of $t(\mathbf{y})$ equals the variance of its estimation or prediction error. In the special case where w is a parametric function, the variance $\text{var}[t(\mathbf{y}) - w]$ of the estimation or prediction error equals the variance $\text{var}[t(\mathbf{y})]$ of the estimator or predictor; however, in general, $\text{var}[t(\mathbf{y}) - w]$ is not necessarily equal to $\text{var}[t(\mathbf{y})]$. In the special case where w is a parametric function and where $t(\mathbf{y})$ is an unbiased estimator or predictor, the root MSE of $t(\mathbf{y})$ equals $\sqrt{\text{var}[t(\mathbf{y})]}$ and may be referred to as the *standard error* (SE) of $t(\mathbf{y})$. In the point estimation or prediction of w, it is desirable to include an estimate of the SE or root MSE of the estimator or predictor.

In interval (or set) estimation or prediction, inferences about w are in the form of one or more intervals or, more generally, one or more (1-dimensional) sets. The end points of each interval are functions of $\mathbf{y}$; more generally, the membership of each set varies with $\mathbf{y}$. Associated with each interval or set is a numerical measure that may be helpful in assessing the "chances" of the interval or set including or "covering" w.

Let $S(\mathbf{y})$ represent an arbitrary (1-dimensional) set (the membership of which varies with $\mathbf{y}$). The probability $\Pr[w \in S(\mathbf{y})]$ is referred to as the *probability of coverage*. It is interpretable as the (theoretical) frequency of the event $w \in S(\mathbf{y})$ in an infinitely long sequence of repeated applications (involving repeated draws of both w- and $\mathbf{y}$-values). Clearly, $\Pr[w \in S(\mathbf{y})] = 1 - \Pr[w \notin S(\mathbf{y})]$.

In general, $\Pr[w \notin S(\mathbf{y})]$ may depend on $\boldsymbol{\phi}$ and/or on characteristics of the joint distribution of w and $\mathbf{y}$ not covered by the assumptions that comprise the statistical model. If $S(\mathbf{y})$ is such that $\Pr[w \notin S(\mathbf{y})] = \alpha$ (uniformly for all distributions that conform to the underlying assumptions), then $S(\mathbf{y})$ is said to be a $100(1-\alpha)\%$ confidence set for w [or, if appropriate, a $100(1-\alpha)\%$ confidence interval for w]. More generally, $S(\mathbf{y})$ is said to be a $100(1-\alpha)\%$ confidence set or interval for w if the supremum of $\Pr[w \notin S(\mathbf{y})]$ (over all distributions that conform to the underlying assumptions) equals α. The infimum of the probability of coverage of a $100(1-\alpha)\%$ confidence interval or set equals $1-\alpha$; this number is referred to as the *confidence coefficient* or *confidence level*. When w represents a future quantity, the confidence interval or set might be referred to as a prediction interval or set.

In making statistical inferences about w, it may be instructive to form a $100(1-\alpha)\%$ confidence interval or set for each of several different values of α. Models that provide an adequate basis for point estimation may not provide an adequate basis for obtaining confidence intervals or sets. Rather, we may require a more elaborate model that is rather specific about the form of the distribution of $\mathbf{y}$ (or, in general, the joint distribution of w and $\mathbf{y}$). In the case of a linear model, it is common practice to take the form of this distribution to be multivariate normal.

In making inferences about a parametric function, there are circumstances that would seem to call for the inclusion of hypothesis testing. Suppose that w represents the parametric function $\lambda(\boldsymbol{\phi})$. Suppose further that, for some value $w^{(0)}$ of this function, interest centers on the question of whether the hypothesis $H_0 : w = w^{(0)}$ (called the null hypothesis) is "consistent with the data." Corresponding to a test of the null hypothesis H_0 is a set C of N-dimensional (column) vectors that is referred to as the *critical region* (or the *rejection region*) and the complementary set A (consisting of those N-dimensional vectors not contained in C), which is referred to as the *acceptance region*. The test consists of accepting H_0, if $\mathbf{y} \in A$, and of rejecting H_0, if $\mathbf{y} \in C$.

If (for some number α between 0 and 1) the probability $\Pr(\mathbf{y} \in C)$ of rejecting H_0 equals α whenever $\lambda(\boldsymbol{\phi}) = w^{(0)}$ (i.e., whenever H_0 is true), then the test is said to be a *size-*α test. More generally, the probability $\Pr(\mathbf{y} \in C)$ of rejecting H_0 when H_0 is true may depend on $\boldsymbol{\phi}$ and/or on characteristics of the distribution of $\mathbf{y}$ not covered by the assumptions that comprise the model, in which case the test is said to be a *size-*α test if (under H_0) the supremum of $\Pr(\mathbf{y} \in C)$ equals α. Under either the null hypothesis H_0 or the hypothesis $H_1 : w \neq w^{(0)}$ (called the alternative hypothesis), $\Pr(\mathbf{y} \in C)$ is interpretable as the (theoretical) frequency with which H_0 is rejected in an infinitely long sequence of repeated applications (involving repeated draws from the distribution of $\mathbf{y}$).

Corresponding to any $100(1-\alpha)\%$ confidence interval or set that may exist for the parametric function $w = \lambda(\boldsymbol{\phi})$ is a size-α test of H_0. If $S(\mathbf{y})$ is a $100(1-\alpha)\%$ confidence interval or set for $w = \lambda(\boldsymbol{\phi})$, then a size-$\alpha$ test of H_0 is obtained by taking the acceptance region A to be the set of $\mathbf{y}$-values for which $w^{(0)} \in S(\mathbf{y})$ (e.g., Casella and Berger 2002, sec. 9.2). If the model is to provide an adequate basis for hypothesis testing, it may (as in the case of interval or set estimation) be necessary to include rather specific assumptions about the form of the distribution of $\mathbf{y}$.

There may be a size-α test of H_0 for every α between 0 and 1. Let (for $0 < \alpha < 1$) $C(\alpha)$ represent the critical region of the size-α test. And suppose (as would often be the case) that the critical regions are nested in the sense that, for $0 < \alpha_1 < \alpha_2 < 1$, $C(\alpha_1)$ is a proper subset of $C(\alpha_2)$. Then, the infimum of the set $\{\alpha : \mathbf{y} \in C(\alpha)\}$ (i.e., the infimum of those α-values for which H_0 is rejected by the size-α test) is referred to as the *p-value*. Instead of reporting the results of the size-α test for one or more values of α, it may be preferable to report the p-value—clearly, the p-value conveys more information.

In practice, there will typically be more than one unobservable quantity of interest. Suppose that inferences are to be made about M unobservable quantities, and that these quantities are to be regarded as the respective values of M random variables $w_1, w_2, \ldots, w_M$ (whose joint distribution may depend on $\boldsymbol{\phi}$)—this formulation includes the case where some or all of the unobservable quantities are functions of $\boldsymbol{\phi}$. One approach is to deal with the M unobservable quantities separately, that is, to make inferences about each of these quantities as though that quantity were the only quantity about which inferences were being made. That approach lends itself to misinterpretation of results (more so in the case of interval or set estimation or prediction and hypothesis testing than in the case of point estimation or prediction) and to unwarranted conclusions.

The potential pitfalls can be circumvented by adopting an alternative approach in which the M unobservable quantities of interest are dealt with simultaneously. Let $\mathbf{w}$ represent the M-dimensional unobservable random (column) vector whose elements are $w_1, w_2, \ldots, w_M$, respectively. Inference about the value of the vector $\mathbf{w}$ can be regarded as synonymous with simultaneous inference about $w_1, w_2, \ldots, w_M$. Many of the concepts introduced earlier (in connection with inference about a single unobservable random variable w) can be readily extended to inference about $\mathbf{w}$.

Let $\mathbf{t}(\mathbf{y})$ represent a vector-valued function (in the form of an M-dimensional column vector) of $\mathbf{y}$, the realized value of which is to be regarded as a (point) estimate of the value of the vector $\mathbf{w}$. This function is referred to as a point estimator (or predictor), and the vector $\mathbf{t}(\mathbf{y}) - \mathbf{w}$ is referred to as the error of estimation or prediction. The estimator or predictor $\mathbf{t}(\mathbf{y})$ is said to be *unbiased* if $\mathrm{E}[\mathbf{t}(\mathbf{y}) - \mathbf{w}] = \mathbf{0}$ or, equivalently, if $\mathrm{E}[\mathbf{t}(\mathbf{y})] = \mathrm{E}(\mathbf{w})$.

The elements, say $t_1(\mathbf{y}), t_2(\mathbf{y}), \ldots, t_M(\mathbf{y})$, of $\mathbf{t}(\mathbf{y})$ can be regarded as estimators or predictors of $w_1, w_2, \ldots, w_M$, respectively. Clearly, $\mathbf{t}(\mathbf{y})$ is an unbiased estimator or predictor of the vector $\mathbf{w}$ if and only if, for $i = 1, 2, \ldots, M$, $t_i(\mathbf{y})$ is an unbiased estimator or predictor of w_i. The $M \times M$ matrix $\mathrm{E}\{[\mathbf{t}(\mathbf{y}) - \mathbf{w}][\mathbf{t}(\mathbf{y}) - \mathbf{w}]'\}$ is referred to as the *mean-squared-error* (MSE) *matrix* of $\mathbf{t}(\mathbf{y})$. Its ijth element equals $\mathrm{E}\{[t_i(\mathbf{y}) - w_i][t_j(\mathbf{y}) - w_j]\}$ $(i, j = 1, 2, \ldots, M)$, implying in particular that the diagonal elements of the MSE matrix are the MSEs of $t_1(\mathbf{y}), t_2(\mathbf{y}), \ldots, t_M(\mathbf{y})$, respectively. If $\mathbf{t}(\mathbf{y})$ is an unbiased estimator or predictor of $\mathbf{w}$, then its MSE matrix equals the variance-covariance matrix of the vector $\mathbf{t}(\mathbf{y}) - \mathbf{w}$.

Turning now to the set estimation or prediction of the vector $\mathbf{w}$, let $S(\mathbf{y})$ represent an arbitrary set of M-dimensional (column) vectors (the membership of which varies with $\mathbf{y}$). The terminology introduced earlier (in connection with the interval or set estimation or prediction of an unobservable random variable w) extends in a straightforward way to the set estimation or prediction of $\mathbf{w}$. In particular, $\Pr[\mathbf{w} \in S(\mathbf{y})]$ is referred to as the *probability of coverage*. As in the one-dimensional case, $\Pr[\mathbf{w} \in S(\mathbf{y})] = 1 - \Pr[\mathbf{w} \notin S(\mathbf{y})]$. Further, the set $S(\mathbf{y})$ is said to be a $100(1-\alpha)\%$ confidence set if the supremum of $\Pr[\mathbf{w} \notin S(\mathbf{y})]$ (over all distributions of $\mathbf{w}$ and $\mathbf{y}$ that conform to the underlying assumptions) equals α. And, accordingly, the infimum of the probability of coverage of a $100(1-\alpha)\%$ confidence set equals $1 - \alpha$, a number that is referred to as the *confidence coefficient* or *confidence level*.

Inference about the vector $\mathbf{w}$ can also take the form of M one-dimensional sets, one set for each of its elements. For $j = 1, 2, \ldots, M$, let $S_j(\mathbf{y})$ represent an interval (whose end points depend on $\mathbf{y}$) or, more generally, a one-dimensional set (the membership of which varies with $\mathbf{y}$) of w_j-values. A possible criterion for evaluating these sets (as confidence sets for $w_1, w_2, \ldots, w_M$, respectively) is the probability (for some relatively small integer $k \geq 1$) that $w_j \in S_j(\mathbf{y})$ for at least $M-k+1$ values of j. In the special case where $k = 1$, this probability is referred to as the probability of simultaneous coverage. It is worth noting that the probability of simultaneous coverage can be drastically smaller than any of the M "individual" probabilities of coverage $\Pr[w_1 \in S_1(\mathbf{y})], \Pr[w_2 \in S_2(\mathbf{y})], \ldots, \Pr[w_M \in S_M(\mathbf{y})]$.

Closely related to the sets $S_1(\mathbf{y}), S_2(\mathbf{y}), \ldots, S_M(\mathbf{y})$ is the set $S(\mathbf{y})$ of M-dimensional (column) vectors defined as follows:

$$S(\mathbf{y}) = \{\mathbf{w} : w_j \in S_j(\mathbf{y}) \ (j = 1, 2, \ldots, M)\}. \tag{5.1}$$

When each of the sets $S_1(\mathbf{y}), S_2(\mathbf{y}), \ldots, S_M(\mathbf{y})$ is an interval, the geometrical form of the set $S(\mathbf{y})$ is that of an "M-dimensional rectangle." Clearly, $\Pr[\mathbf{w} \in S(\mathbf{y})]$ equals the probability of simultaneous coverage of the individual sets $S_1(\mathbf{y}), S_2(\mathbf{y}), \ldots, S_M(\mathbf{y})$. Accordingly, the set $S(\mathbf{y})$ is a $100(1-\alpha)\%$ confidence set for $\mathbf{w}$ if (and only if) the probability of simultaneous coverage of $S_1(\mathbf{y}), S_2(\mathbf{y}), \ldots, S_M(\mathbf{y})$ equals $1-\alpha$.

Let us now revisit hypothesis testing. Suppose that every one of the M elements $w_1, w_2, \ldots, w_M$ of $\mathbf{w}$ is a function of $\boldsymbol{\phi}$. Suppose further that, for some values $w_1^{(0)}, w_2^{(0)}, \ldots, w_M^{(0)}$, interest centers on the question of whether the hypothesis $H_0 : w_i = w_i^{(0)}$ $(i = 1, 2, \ldots, M)$ (the so-called null hypothesis) is consistent with the data. The null hypothesis can be restated (in matrix notation) as H_0: $\mathbf{w} = \mathbf{w}^{(0)}$, where $\mathbf{w}^{(0)}$ is the M-dimensional (column) vector with elements $w_1^{(0)}, w_2^{(0)}, \ldots, w_M^{(0)}$.

Various concepts introduced earlier in connection with the testing of the null hypothesis in the case of a single parametric function extend in an altogether straightforward way to the case of an M-dimensional vector of parametric functions. These concepts include those of an acceptance region, of a critical or rejection region, of a size-α test, and of a p-value. Moreover, the relationship between confidence sets and hypothesis tests discussed earlier (in connection with inference about a single parametric function) extends to inference about the M-dimensional vector $\mathbf{w}$ (of parametric functions): if $S(\mathbf{y})$ is a $100(1-\alpha)\%$ confidence set for $\mathbf{w}$, then a size-α test of $H_0 : \mathbf{w} = \mathbf{w}^{(0)}$ is obtained by taking the acceptance region to be the set of $\mathbf{y}$-values for which $\mathbf{w}^{(0)} \in S(\mathbf{y})$.

The null hypothesis $H_0 : \mathbf{w} = \mathbf{w}^{(0)}$ can be regarded as a "composite" of the M "individual" hypotheses $H_0^{(1)} : w_1 = w_1^{(0)}, H_0^{(2)} : w_2 = w_2^{(0)}, \ldots, H_0^{(M)} : w_M = w_M^{(0)}$. Clearly, the composite null hypothesis H_0 is true if and only if all M of the individual null hypotheses $H_0^{(1)}, H_0^{(2)}, \ldots, H_0^{(M)}$ are true and is false if and only if one or more of the individual null hypotheses are false. As a variation on the problem of testing H_0, there is the problem of testing $H_0^{(1)}, H_0^{(2)}, \ldots, H_0^{(M)}$ individually. The latter problem is known as the problem of multiple comparisons—in some applications, $w_1, w_2, \ldots, w_M$ have interpretations relating them to comparisons among various entities. The classical approach to this problem is to restrict attention to tests of $H_0^{(1)}, H_0^{(2)}, \ldots, H_0^{(M)}$ such that the problem of one or more false rejections does not exceed some relatively low level α, for example, $\alpha = 0.05$. For even moderately large values of M, this approach can be quite "conservative." A less conservative approach is obtainable by considering all test procedures for which (for some positive integer $k > 1$) the probability of k or more false rejections does not exceed α (e.g., Lehmann and Romano 2005a). Another such approach is that of Benjamini and Hochberg (1995); it takes the form of controlling the false discovery rate, which by definition is the expected value of the ratio of the number of false rejections to the total number of rejections.

In the testing of a hypothesis about a parametric function or functions and in the point or set estimation or prediction of an unobservable random variable or vector $\mathbf{w}$, the statistical properties of the test or of the estimator or predictor depend on various characteristics of the distribution of $\mathbf{y}$ or, more generally (in the case of the estimator or predictor), on the joint distribution of $\mathbf{w}$ and $\mathbf{y}$. These properties include the probability of acceptance or rejection of a hypothesis (by a hypothesis test), the probability of coverage (by a confidence set), and the unbiasedness and MSE or MSE matrix of a point estimator or predictor. Some or all of the relevant characteristics of the distribution of **y** or of the joint distribution of **w** and **y** are determined by the assumptions (about the distribution of **y**) that comprise the statistical model and by any further assumptions pertaining to the joint distribution of **w** and **y**. By definition, these assumptions pertain to the unconditional distribution of **y** and to the unconditional joint distribution of **w** and **y**.

It can be informative to determine the properties of a hypothesis test or of a point or set estimator or predictor under more than one model (or under more than 1 set of assumptions about the joint distribution of **w** and **y**) and/or to determine the properties of the test or the estimator or predictor conditionally on the values of various random variables (e.g., conditionally on the values of various functions of **y** or even on the value of **y** itself). The appeal of a test or estimation or prediction procedure whose properties have been evaluated unconditionally under a particular model (or particular set of assumptions about the joint distribution of **w** and **y**) can be either enhanced or diminished by evaluating its properties conditionally and/or under an alternative model (or alternative set of assumptions). The relative appeal of alternative procedures may be a matter of emphasis; which procedure has the more favorable properties may depend on whether the properties are evaluated conditionally or unconditionally and under which model. In such a case, it may be instructive to analyze the data in accordance with each of multiple procedures.

1.6 An Overview

This volume provides coverage of linear statistical models and of various statistical procedures that are based on those models. The emphasis is on the underlying theory; however, some discussion of applications and some attempts at illustration are included among the content.

In-depth coverage is provided for a broad class of linear statistical models consisting of what are referred to herein as Gauss–Markov models. Results obtained on the basis of Gauss–Markov models can be extended in a relatively straightforward way to a somewhat broader class of linear statistical models consisting of what are referred to herein as Aitken models. Results on a few selected topics are obtained for what are referred to herein as general linear models, which form a very broad class of linear statistical models and include the Gauss–Markov and Aitken models as special cases.

The models underlying (simple and multiple) linear regression procedures are Gauss–Markov models, and results obtained on the basis of Gauss–Markov models apply more-or-less directly to those procedures. Moreover, many of the procedures that are commonly used to analyze experimental data (and in some cases observational data) are based on classificatory (fixed-effects) models. Like regression models, these models are Gauss–Markov models. However, some of the procedures that are commonly used to analyze classificatory data (such as the analysis of variance) are rather "specialized." For the most part, those kinds of specialized procedures are outside the scope of what is covered herein. They constitute (along with various results that would expand on the coverage of general linear models) potential subjects for a possible future volume.

The organization of the present volume is such that the results that are directly applicable to linear models are presented in Chapters 5 and 7. Chapter 5 provides coverage of (point) estimation and prediction, and Chapter 7 provides coverage of topics related to the construction of confidence intervals and sets and to the testing of hypotheses. Chapters 2, 3, 4, and 6 present results on matrix algebra and the relevant underlying statistical distributions (as well as other supportive material); many of these results are of importance in their own right. Some additional results on matrix algebra and statistical distributions are introduced in Chapters 5 and 7 (as the need for them arises).

2

Matrix Algebra: A Primer

Knowledge of matrix algebra is essential in working with linear models. Chapter 2 provides a limited coverage of matrix algebra, with an emphasis on concepts and results that are highly relevant and that are more-or-less elementary in nature. It forms a core body of knowledge, and, as such, provides a solid foundation for the developments that follow. Derivations or proofs are included for most results. In subsequent chapters, the coverage of matrix algebra is extended (as the need arises) to additional concepts and results.

2.1 The Basics

A *matrix* is a rectangular array of numbers, that is, a collection of numbers, say $a_{11}, a_{12}, \ldots, a_{1N}$, $a_{21}, a_{22}, \ldots, a_{2N}, \ldots, a_{M1}, a_{M2}, \ldots, a_{MN}$, arranged in rows and columns as follows:

$$\begin{pmatrix} a_{11} & a_{12} & \ldots & a_{1N} \\ a_{21} & a_{22} & \ldots & a_{2N} \\ \vdots & \vdots & & \vdots \\ a_{M1} & a_{M2} & \ldots & a_{MN} \end{pmatrix}.$$

The use of the term matrix is restricted herein to real matrices, that is, to rectangular arrays of real numbers. A matrix having M rows and N columns is referred to as an $M \times N$ matrix, and M and N are called the *dimensions* of the matrix. The number located at the intersection of the ith row and the jth column of a matrix is called the ijth *element* or *entry* of the matrix.

Boldface capital letters (e.g., $\mathbf{A}$) are used to represent matrices. The notation $\mathbf{A} = \{a_{ij}\}$ is used in introducing a matrix, the ijth element of which is a_{ij}. Two matrices $\mathbf{A}$ and $\mathbf{B}$ of the same dimensions are said to be *equal* if each element of $\mathbf{A}$ equals the corresponding element of $\mathbf{B}$, in which case we write $\mathbf{A} = \mathbf{B}$ (and are said to be unequal otherwise, i.e., if any element of $\mathbf{A}$ differs from the corresponding element of $\mathbf{B}$, in which case we write $\mathbf{A} \neq \mathbf{B}$).

a. Matrix operations

A matrix can be transformed or can be combined with various other matrices in accordance with operations called scalar multiplication, matrix addition and subtraction, matrix multiplication, and transposition.

Scalar multiplication. The term *scalar* is to be used synonymously with real number. Scalar multiplication is defined for an arbitrary scalar k and an arbitrary $M \times N$ matrix $\mathbf{A} = \{a_{ij}\}$. The *product* of k and $\mathbf{A}$ is written as $k\mathbf{A}$ (or, much less commonly, as $\mathbf{A}k$), and is defined to be the $M \times N$ matrix whose ijth element is ka_{ij}. The matrix $k\mathbf{A}$ is said to be a *scalar multiple* of the matrix $\mathbf{A}$. Clearly, for any scalars c and k and any matrix $\mathbf{A}$,

$$c(k\mathbf{A}) = (ck)\mathbf{A} = (kc)\mathbf{A} = k(c\mathbf{A}). \tag{1.1}$$

It is customary to refer to the product $(-1)\mathbf{A}$ of -1 and $\mathbf{A}$ as the *negative* of $\mathbf{A}$ and to abbreviate $(-1)\mathbf{A}$ to $-\mathbf{A}$.

Matrix addition and subtraction. Matrix addition and subtraction are defined for any two matrices $\mathbf{A} = \{a_{ij}\}$ and $\mathbf{B} = \{b_{ij}\}$ that have the same number of rows, say M, and the same number of columns, say N. The *sum* of the two $M \times N$ matrices $\mathbf{A}$ and $\mathbf{B}$ is denoted by the symbol $\mathbf{A} + \mathbf{B}$ and is defined to be the $M \times N$ matrix whose ijth element is $a_{ij} + b_{ij}$.

Matrix addition is commutative, that is,

$$\mathbf{A} + \mathbf{B} = \mathbf{B} + \mathbf{A}. \tag{1.2}$$

Matrix addition is also associative, that is, taking $\mathbf{C}$ to be a third $M \times N$ matrix,

$$\mathbf{A} + (\mathbf{B} + \mathbf{C}) = (\mathbf{A} + \mathbf{B}) + \mathbf{C}. \tag{1.3}$$

The symbol $\mathbf{A} + \mathbf{B} + \mathbf{C}$ is used to represent the common value of the left and right sides of equality (1.3), and that value is referred to as the *sum* of $\mathbf{A}$, $\mathbf{B}$, and $\mathbf{C}$. This notation and terminology extend in an obvious way to any finite number of $M \times N$ matrices.

Clearly, for any scalar k,

$$k(\mathbf{A} + \mathbf{B}) = k\mathbf{A} + k\mathbf{B}, \tag{1.4}$$

and, for any scalars c and k,

$$(c + k)\mathbf{A} = c\mathbf{A} + k\mathbf{A}. \tag{1.5}$$

Let us write $\mathbf{A} - \mathbf{B}$ for the sum $\mathbf{A} + (-\mathbf{B})$ or, equivalently, for the $M \times N$ matrix whose ijth element is $a_{ij} - b_{ij}$, and refer to this matrix as the *difference* between $\mathbf{A}$ and $\mathbf{B}$.

Matrices having the same number of rows and the same number of columns are said to be *conformal* for addition (and subtraction).

Matrix multiplication. Turning now to matrix multiplication (i.e., the multiplication of one matrix by another), let $\mathbf{A} = \{a_{ij}\}$ represent an $M \times N$ matrix and $\mathbf{B} = \{b_{ij}\}$ a $P \times Q$ matrix. When $N = P$ (i.e., when $\mathbf{A}$ has the same number of columns as $\mathbf{B}$ has rows), the *matrix product* is defined to be the $M \times Q$ matrix whose ijth element is

$$\sum_{k=1}^{N} a_{ik}b_{kj} = a_{i1}b_{1j} + a_{i2}b_{2j} + \cdots + a_{iN}b_{Nj}.$$

The formation of the matrix product $\mathbf{AB}$ is referred to as the *premultiplication* of $\mathbf{B}$ by $\mathbf{A}$ or the *postmultiplication* of $\mathbf{A}$ by $\mathbf{B}$. When $N \neq P$, the matrix product $\mathbf{AB}$ is undefined.

Matrix multiplication is associative. Thus, introducing a third matrix $\mathbf{C}$,

$$\mathbf{A}(\mathbf{BC}) = (\mathbf{AB})\mathbf{C}, \tag{1.6}$$

provided that $N = P$ and that $\mathbf{C}$ has Q rows (so that all relevant matrix products are defined). The symbol $\mathbf{ABC}$ is used to represent the common value of the left and right sides of equality (1.6), and that value is referred to as the *product* of $\mathbf{A}$, $\mathbf{B}$, and $\mathbf{C}$. This notation and terminology extend in an obvious way to any finite number of matrices.

Matrix multiplication is distributive with respect to addition, that is,

$$\mathbf{A}(\mathbf{B} + \mathbf{C}) = \mathbf{AB} + \mathbf{AC}, \tag{1.7}$$

$$(\mathbf{A} + \mathbf{B})\mathbf{C} = \mathbf{AC} + \mathbf{BC}, \tag{1.8}$$

where, in each equality, it is assumed that the dimensions of $\mathbf{A}$, $\mathbf{B}$, and $\mathbf{C}$ are such that all multiplications and additions are defined. Results (1.7) and (1.8) extend in an obvious way to the postmultiplication or premultiplication of a matrix $\mathbf{A}$ or $\mathbf{C}$ by the sum of any finite number of matrices.

In general, matrix multiplication is not commutative. That is, $\mathbf{AB}$ is not necessarily identical to $\mathbf{BA}$. In fact, when $N = P$ but $M \neq Q$ or when $M = Q$ but $N \neq P$, one of the matrix products $\mathbf{AB}$ and $\mathbf{BA}$ is defined, while the other is undefined. When $N = P$ and $M = Q$, $\mathbf{AB}$ and $\mathbf{BA}$ are both defined, but the dimensions $(M \times M)$ of $\mathbf{AB}$ are the same as those of $\mathbf{BA}$ only if $M = N$. Even if $N = P = M = Q$, in which case $\mathbf{A}$ and $\mathbf{B}$ are both $N \times N$ matrices and the two matrix products $\mathbf{AB}$ and $\mathbf{BA}$ are both defined and of the same dimensions, it is not necessarily the case that $\mathbf{AB} = \mathbf{BA}$.

Two $N \times N$ matrices $\mathbf{A}$ and $\mathbf{B}$ are said to *commute* if $\mathbf{AB} = \mathbf{BA}$. More generally, a collection of $N \times N$ matrices $\mathbf{A}_1, \mathbf{A}_2, \ldots, \mathbf{A}_K$ is said to *commute in pairs* if $\mathbf{A}_i\mathbf{A}_j = \mathbf{A}_j\mathbf{A}_i$ for $j > i = 1, 2, \ldots, K$.

For any scalar c, $M \times N$ matrix $\mathbf{A}$, and $N \times P$ matrix $\mathbf{B}$, it is customary to write $c\mathbf{AB}$ for the scalar product $c(\mathbf{AB})$ of c and the matrix product $\mathbf{AB}$. Note that

$$c\mathbf{AB} = (c\mathbf{A})\mathbf{B} = \mathbf{A}(c\mathbf{B}) \tag{1.9}$$

(as is evident from the very definitions of scalar and matrix multiplication). This notation (for a scalar multiple of a product of 2 matrices) and result (1.9) extend in an obvious way to a scalar multiple of a product of any finite number of matrices.

Transposition. Corresponding to any $M \times N$ matrix $\mathbf{A} = \{a_{ij}\}$ is the $N \times M$ matrix obtained by rewriting the columns of $\mathbf{A}$ as rows or the rows of $\mathbf{A}$ as columns. This matrix, the ijth element of which is a_{ji}, is called the *transpose* of $\mathbf{A}$ and is to be denoted herein by the symbol $\mathbf{A}'$.

For any matrix $\mathbf{A}$,

$$(\mathbf{A}')' = \mathbf{A}; \tag{1.10}$$

for any scalar k and any matrix $\mathbf{A}$,

$$(k\mathbf{A})' = k\mathbf{A}'; \tag{1.11}$$

and for any two matrices $\mathbf{A}$ and $\mathbf{B}$ (that are conformal for addition),

$$(\mathbf{A} + \mathbf{B})' = \mathbf{A}' + \mathbf{B}' \tag{1.12}$$

—these 3 results are easily verified. Further, for any two matrices $\mathbf{A}$ and $\mathbf{B}$ (for which the product $\mathbf{AB}$ is defined)

$$(\mathbf{AB})' = \mathbf{B}'\mathbf{A}', \tag{1.13}$$

as can be verified by comparing the ijth element of $\mathbf{B}'\mathbf{A}'$ with that of $(\mathbf{AB})'$. More generally,

$$(\mathbf{A}_1\mathbf{A}_2 \cdots \mathbf{A}_K)' = \mathbf{A}_K' \cdots \mathbf{A}_2'\mathbf{A}_1' \tag{1.14}$$

for any K matrices $\mathbf{A}_1, \mathbf{A}_2, \ldots, \mathbf{A}_K$ of appropriate dimensions.

b. Types of matrices

There are several types of matrices that are worthy of mention.

Square matrices. A matrix having the same number of rows as columns, say N rows and N columns, is referred to as a *square matrix* and is said to be of *order* N. The N elements of a square matrix of order N that lie on an imaginary line (called the diagonal) extending from the upper left corner of the matrix to the lower right corner are called the *diagonal elements*; the other $N(N-1)$ elements of the matrix (those elements that lie above and to the right or below and to the left of the diagonal) are called the *off-diagonal elements*. Thus, the diagonal elements of a square matrix $\mathbf{A} = \{a_{ij}\}$ of order N are a_{ii} $(i = 1, 2, \ldots, N)$, and the off-diagonal elements are a_{ij} $(j \neq i = 1, 2, \ldots, N)$.

Symmetric matrices. A matrix $\mathbf{A}$ is said to be *symmetric* if $\mathbf{A}' = \mathbf{A}$. Thus, a matrix is symmetric if it is square and if (for all i and $j \neq i$) its jith element equals its ijth element.

Diagonal matrices. A *diagonal matrix* is a square matrix whose off-diagonal elements are all equal to 0. Thus, a square matrix $\mathbf{A} = \{a_{ij}\}$ of order N is a diagonal matrix if $a_{ij} = 0$ for $j \neq i = 1, 2, \ldots, N$. The notation $\mathbf{D} = \{d_i\}$ is sometimes used to introduce a diagonal matrix, the ith diagonal element of which is d_i. Also, we may write $\text{diag}(d_1, d_2, \ldots, d_N)$ for such a matrix (where N is the order of the matrix).

Identity matrices. A diagonal matrix $\text{diag}(1, 1, \ldots, 1)$ whose diagonal elements are all equal to 1 is called an *identity matrix*. The symbol $\mathbf{I}_N$ is used to represent an identity matrix of order N. In cases where the order is clear from the context, $\mathbf{I}_N$ may be abbreviated to $\mathbf{I}$.

Triangular matrices. If all of the elements of a square matrix that are located below and to the left of the diagonal are 0, the matrix is said to be *upper triangular*. Similarly, if all of the elements that are located above and to the right of the diagonal are 0, the matrix is said to be *lower triangular*. More formally, a square matrix $\mathbf{A} = \{a_{ij}\}$ of order N is upper triangular if $a_{ij} = 0$ for $j < i = 1, \ldots, N$ and is lower triangular if $a_{ij} = 0$ for $j > i = 1, \ldots, N$. By a *triangular matrix*, we mean a (square) matrix that is upper triangular or lower triangular. An (upper or lower) triangular matrix is called a *unit* (upper or lower) *triangular matrix* if all of its diagonal elements equal 1.

Row and column vectors. A matrix that has only one row, that is, a matrix of the form $(a_1, a_2, \ldots, a_N)$ is called a *row vector*. Similarly, a matrix that has only one column is called a *column vector*. A row or column vector having N elements may be referred to as an N-dimensional row or column vector. Clearly, the transpose of an N-dimensional column vector is an N-dimensional row vector, and vice versa.

Lowercase boldface letters (e.g., $\mathbf{a}$) are used herein to represent column vectors. This notation is helpful in distinguishing column vectors from matrices that may have more than one column. No further notation is introduced for row vectors. Instead, row vectors are represented as the transposes of column vectors. For example, $\mathbf{a}'$ represents the row vector whose transpose is the column vector $\mathbf{a}$. The notation $\mathbf{a} = \{a_i\}$ or $\mathbf{a}' = \{a_i\}$ is used in introducing a column or row vector whose ith element is a_i.

Note that each column of an $M \times N$ matrix $\mathbf{A} = \{a_{ij}\}$ is an M-dimensional column vector, and that each row of $\mathbf{A}$ is an N-dimensional row vector. Specifically, the jth column of $\mathbf{A}$ is the M-dimensional column vector $(a_{1j}, a_{2j}, \ldots, a_{Mj})'$ $(j = 1, \ldots, N)$, and the ith row of $\mathbf{A}$ is the N-dimensional row vector $(a_{i1}, a_{i2}, \ldots, a_{iN})$ $(i = 1, \ldots, M)$.

Null matrices. A matrix all of whose elements are 0 is called a *null matrix*—a matrix having one or more nonzero elements is said to be *nonnull*. A null matrix is denoted by the symbol $\mathbf{0}$—this notation is reserved for use in situations where the dimensions of the null matrix can be ascertained from the context. A null matrix that has one row or one column may be referred to as a *null vector*.

***Matrices of* 1's**. The symbol $\mathbf{1}_N$ is used to represent an N-dimensional column vector all of whose elements equal 1. In a situation where the dimensions of a column vector of 1's is clear from the context or is to be left unspecified, we may simply write $\mathbf{1}$ for such a vector. Note that $\mathbf{1}'_N$ is an N-dimensional row vector, all of whose elements equal 1, and that $\mathbf{1}_M\mathbf{1}'_N$ is an $M \times N$ matrix, all of whose elements equal 1.

c. Submatrices and subvectors

A *submatrix* of a matrix $\mathbf{A}$ is a matrix that can be obtained by striking out rows and/or columns of $\mathbf{A}$. Strictly speaking, a matrix is a submatrix of itself; it is the submatrix obtained by striking out zero rows and zero columns. Submatrices of a row or column vector, that is, of a matrix having one row or one column, are themselves row or column vectors and are customarily referred to as *subvectors*.

A submatrix of a square matrix is called a *principal submatrix* if it can be obtained by striking out the same rows as columns (so that the ith row is struck out whenever the ith column is struck out, and vice versa). The $R \times R$ (principal) submatrix of an $N \times N$ matrix obtained by striking out the last $N - R$ rows and columns is referred to as a *leading principal submatrix* ($R = 1, \ldots, N$). A principal submatrix of a symmetric matrix is symmetric, a principal submatrix of a diagonal matrix is diagonal, and a principal submatrix of an upper or lower triangular matrix is respectively upper or lower triangular, as is easily verified.

2.2 Partitioned Matrices and Vectors

A matrix can be divided or partitioned into submatrices by drawing horizontal or vertical lines between various of its row or columns, in which case the matrix is called a *partitioned matrix* and the submatrices are sometimes referred to as *blocks* (as in blocks of elements). Thus, a partitioned matrix is a matrix, say an $M \times N$ matrix $\mathbf{A}$, that has been expressed in the form

$$\mathbf{A} = \begin{pmatrix} \mathbf{A}_{11} & \mathbf{A}_{12} & \ldots & \mathbf{A}_{1C} \\ \mathbf{A}_{21} & \mathbf{A}_{22} & \ldots & \mathbf{A}_{2C} \\ \vdots & \vdots & & \vdots \\ \mathbf{A}_{R1} & \mathbf{A}_{R2} & \ldots & \mathbf{A}_{RC} \end{pmatrix}. \tag{2.1}$$

Here, R and C are positive integers, and $\mathbf{A}_{ij}$ is an $M_i \times N_j$ matrix ($i = 1, 2, \ldots, R$; $j = 1, 2, \ldots, C$),where $M_1, M_2, \ldots, M_R$ and $N_1, N_2, \ldots, N_C$ are positive integers such that $M_1 + M_2 + \cdots + M_R = M$ and $N_1 + N_2 + \cdots + N_C = N$. Specifically, $\mathbf{A}_{ij}$ is the $M_i \times N_j$ submatrix of $\mathbf{A}$ obtained by striking out all of the rows and columns of $\mathbf{A}$ save the $(M_1 + M_2 + \cdots + M_{i-1} + 1)$th, $(M_1+M_2+\cdots+M_{i-1}+2)$th, $\ldots$, $(M_1+M_2+\cdots+M_i)$th rows and the $(N_1+N_2+\cdots+N_{j-1}+1)$th, $(N_1 + N_2 + \cdots + N_{j-1} + 2)$th, $\ldots$, $(N_1 + N_2 + \cdots + N_j)$th columns. (When $i = 1$ or $j = 1$, interpret the degenerate sum $M_1 + M_2 + \cdots + M_{i-1}$ or $N_1 + N_2 + \cdots + N_{j-1}$ as 0.) Think of a partitioned matrix as an array or "matrix" of matrices.

Note that, by definition, each of the submatrices $\mathbf{A}_{i1}, \mathbf{A}_{i2}, \ldots, \mathbf{A}_{iC}$ in the ith "row" of blocks of the partitioned matrix (2.1) has the same number of rows and that each of the submatrices $\mathbf{A}_{1j}, \mathbf{A}_{2j}, \ldots, \mathbf{A}_{Rj}$ in the jth "column" of blocks has the same number of columns. It is customary to identify each of the blocks in a partitioned matrix by referring to the row of blocks and the column of blocks in which it appears. Accordingly, when a matrix $\mathbf{A}$ is expressed in the partitioned form (2.1), the submatrix $\mathbf{A}_{ij}$ is referred to as the ijth block of $\mathbf{A}$.

Partitioned matrices having one row or one column are customarily referred to as *partitioned* (row or column) *vectors*. Thus, a partitioned column vector is a (column) vector, say an M-dimensional column vector $\mathbf{a}$, that has been expressed in the form

$$\mathbf{a} = \begin{pmatrix} \mathbf{a}_1 \\ \mathbf{a}_2 \\ \vdots \\ \mathbf{a}_R \end{pmatrix}.$$

Here, R is a positive integer, and $\mathbf{a}_i$ is an M_i-dimensional column vector ($i = 1, 2, \ldots, R$), where $M_1, M_2, \ldots, M_R$ are positive integers such that $M_1 + M_2 + \cdots + M_R = M$. Specifically, a_i is the subvector of $\mathbf{a}$ obtained by striking out all of the elements of $\mathbf{a}$ save the $(M_1+M_2+\cdots+M_{i-1}+1)$th, $(M_1 + M_2 + \cdots + M_{i-1} + 2)$th, $\ldots$, $(M_1 + M_2 + \cdots + M_i)$th elements. Similarly, a partitioned row vector is a (row) vector, say the M-dimensional row vector $\mathbf{a}'$, that has been expressed in the form $\mathbf{a}' = (\mathbf{a}_1', \mathbf{a}_2', \ldots, \mathbf{a}_R')$.

a. Matrix operations (as applied to partitioned matrices)

For partitioned matrices, the various matrix operations can be carried out "blockwise" instead of "elementwise." Take $\mathbf{A}$ to be an $M \times N$ matrix that has been expressed in the form (2.1), that is, has been partitioned into R rows and C columns of blocks, the ijth of which is the $M_i \times N_j$ submatrix $\mathbf{A}_{ij}$. Then, clearly, for any scalar k,

$$k\mathbf{A} = \begin{pmatrix} k\mathbf{A}_{11} & k\mathbf{A}_{12} & \dots & k\mathbf{A}_{1C} \\ k\mathbf{A}_{21} & k\mathbf{A}_{22} & \dots & k\mathbf{A}_{2C} \\ \vdots & \vdots & & \vdots \\ k\mathbf{A}_{R1} & k\mathbf{A}_{R2} & \dots & k\mathbf{A}_{RC} \end{pmatrix}. \tag{2.2}$$

Further, it is a simple exercise to show that

$$\mathbf{A}' = \begin{pmatrix} \mathbf{A}'_{11} & \mathbf{A}'_{21} & \dots & \mathbf{A}'_{R1} \\ \mathbf{A}'_{12} & \mathbf{A}'_{22} & \dots & \mathbf{A}'_{R2} \\ \vdots & \vdots & & \vdots \\ \mathbf{A}'_{1C} & \mathbf{A}'_{2C} & \dots & \mathbf{A}'_{RC} \end{pmatrix}, \tag{2.3}$$

that is, $\mathbf{A}'$ is expressible as a partitioned matrix, comprising C rows and R columns of blocks, the ijth of which is the transpose $\mathbf{A}'_{ji}$ of the jith block $\mathbf{A}_{ji}$ of $\mathbf{A}$.

Now, let us consider the sum and the product of the $M \times N$ partitioned matrix $\mathbf{A}$ and a $P \times Q$ partitioned matrix

$$\mathbf{B} = \begin{pmatrix} \mathbf{B}_{11} & \mathbf{B}_{12} & \dots & \mathbf{B}_{1V} \\ \mathbf{B}_{21} & \mathbf{B}_{22} & \dots & \mathbf{B}_{2V} \\ \vdots & \vdots & & \vdots \\ \mathbf{B}_{U1} & \mathbf{B}_{U2} & \dots & \mathbf{B}_{UV} \end{pmatrix},$$

whose ijth block $\mathbf{B}_{ij}$ is of dimensions $P_i \times Q_j$. The matrices $\mathbf{A}$ and $\mathbf{B}$ are conformal for addition provided that $P = M$ and $Q = N$. If $U = R$, $V = C$, $P_i = M_i$ $(i = 1, 2, \dots, R)$, and $Q_j = N_j$ $(j = 1, 2, \dots, C)$, that is, if (besides $\mathbf{A}$ and $\mathbf{B}$ being conformal for addition) the rows and columns of $\mathbf{B}$ are partitioned in the same way as those of $\mathbf{A}$, then

$$\mathbf{A} + \mathbf{B} = \begin{pmatrix} \mathbf{A}_{11} + \mathbf{B}_{11} & \mathbf{A}_{12} + \mathbf{B}_{12} & \dots & \mathbf{A}_{1C} + \mathbf{B}_{1C} \\ \mathbf{A}_{21} + \mathbf{B}_{21} & \mathbf{A}_{22} + \mathbf{B}_{22} & \dots & \mathbf{A}_{2C} + \mathbf{B}_{2C} \\ \vdots & \vdots & & \vdots \\ \mathbf{A}_{R1} + \mathbf{B}_{R1} & \mathbf{A}_{R2} + \mathbf{B}_{R2} & \dots & \mathbf{A}_{RC} + \mathbf{B}_{RC} \end{pmatrix}, \tag{2.4}$$

and the partitioning of $\mathbf{A}$ and $\mathbf{B}$ is said to be *conformal* (for addition). This result and terminology extend in an obvious way to the addition of any finite number of partitioned matrices (and can be readily modified to obtain counterparts for matrix subtraction).

The matrix product $\mathbf{AB}$ is defined provided that $P = N$. If $U = C$ and $P_k = N_k$ $(k = 1, 2, \dots C)$ (in which case all of the products $\mathbf{A}_{ik}\mathbf{B}_{kj}$ $(i = 1, 2, \dots, R;\ j = 1, 2, \dots, V;\ k = 1, 2, \dots, C)$, as well as the product $\mathbf{AB}$, exist), then

$$\mathbf{AB} = \begin{pmatrix} \mathbf{F}_{11} & \mathbf{F}_{12} & \dots & \mathbf{F}_{1V} \\ \mathbf{F}_{21} & \mathbf{F}_{22} & \dots & \mathbf{F}_{2V} \\ \vdots & \vdots & & \vdots \\ \mathbf{F}_{R1} & \mathbf{F}_{R2} & \dots & \mathbf{F}_{RV} \end{pmatrix}, \tag{2.5}$$

where

$$\mathbf{F}_{ij} = \sum_{k=1}^{C} \mathbf{A}_{ik}\mathbf{B}_{kj} = \mathbf{A}_{i1}\mathbf{B}_{1j} + \mathbf{A}_{i2}\mathbf{B}_{2j} + \cdots + \mathbf{A}_{iC}\mathbf{B}_{Cj},$$

and the partitioning of $\mathbf{A}$ and $\mathbf{B}$ is said to be *conformal* (for the premultiplication of $\mathbf{B}$ by $\mathbf{A}$). In the special case where $R = C = U = V = 2$, that is, where

$$\mathbf{A} = \begin{pmatrix} \mathbf{A}_{11} & \mathbf{A}_{12} \\ \mathbf{A}_{21} & \mathbf{A}_{22} \end{pmatrix} \quad \text{and} \quad \mathbf{B} = \begin{pmatrix} \mathbf{B}_{11} & \mathbf{B}_{12} \\ \mathbf{B}_{21} & \mathbf{B}_{22} \end{pmatrix},$$

result (2.5) simplifies to

$$\mathbf{AB} = \begin{pmatrix} \mathbf{A}_{11}\mathbf{B}_{11} + \mathbf{A}_{12}\mathbf{B}_{21} & \mathbf{A}_{11}\mathbf{B}_{12} + \mathbf{A}_{12}\mathbf{B}_{22} \\ \mathbf{A}_{21}\mathbf{B}_{11} + \mathbf{A}_{22}\mathbf{B}_{21} & \mathbf{A}_{21}\mathbf{B}_{12} + \mathbf{A}_{22}\mathbf{B}_{22} \end{pmatrix}. \tag{2.6}$$

b. Block-diagonal and block-triangular matrices

In the special case of a partitioned $M \times N$ matrix $\mathbf{A}$ of the form

$$\mathbf{A} = \begin{pmatrix} \mathbf{A}_{11} & \mathbf{A}_{12} & \dots & \mathbf{A}_{1R} \\ \mathbf{A}_{21} & \mathbf{A}_{22} & \dots & \mathbf{A}_{2R} \\ \vdots & \vdots & \ddots & \vdots \\ \mathbf{A}_{R1} & \mathbf{A}_{R2} & \dots & \mathbf{A}_{RR} \end{pmatrix} \tag{2.7}$$

(for which the number of rows of blocks equals the number of columns of blocks), the ijth block $\mathbf{A}_{ij}$ of $\mathbf{A}$ is called a *diagonal block* if $j = i$ and an *off-diagonal block* if $j \neq i$. If every off-diagonal block of the partitioned matrix (2.7) is a null matrix, that is, if

$$\mathbf{A} = \begin{pmatrix} \mathbf{A}_{11} & \mathbf{0} & \dots & \mathbf{0} \\ \mathbf{0} & \mathbf{A}_{22} & & \mathbf{0} \\ \vdots & & \ddots & \\ \mathbf{0} & \mathbf{0} & & \mathbf{A}_{RR} \end{pmatrix},$$

then $\mathbf{A}$ is said to be *block-diagonal*, and $\text{diag}(\mathbf{A}_{11}, \mathbf{A}_{22}, \dots, \mathbf{A}_{RR})$ is sometimes written for $\mathbf{A}$. If $\mathbf{A}_{ij} = \mathbf{0}$ for $j < i = 1, \dots, R$, that is, if

$$\mathbf{A} = \begin{pmatrix} \mathbf{A}_{11} & \mathbf{A}_{12} & \dots & \mathbf{A}_{1R} \\ \mathbf{0} & \mathbf{A}_{22} & \dots & \mathbf{A}_{2R} \\ \vdots & & \ddots & \vdots \\ \mathbf{0} & \mathbf{0} & & \mathbf{A}_{RR} \end{pmatrix},$$

then $\mathbf{A}$ is called an *upper block-triangular matrix*. Similarly, if $\mathbf{A}_{ij} = \mathbf{0}$ for $j > i = 1, \dots, R$, that is, if

$$\mathbf{A} = \begin{pmatrix} \mathbf{A}_{11} & \mathbf{0} & \dots & \mathbf{0} \\ \mathbf{A}_{21} & \mathbf{A}_{22} & & \mathbf{0} \\ \vdots & \vdots & \ddots & \\ \mathbf{A}_{R1} & \mathbf{A}_{R2} & \dots & \mathbf{A}_{RR} \end{pmatrix},$$

then $\mathbf{A}$ is called a *lower block-triangular matrix*. To indicate that $\mathbf{A}$ is upper or lower block-triangular (without being more specific), $\mathbf{A}$ is referred to simply as *block-triangular*.

c. Matrices partitioned into individual rows or columns

Note that a matrix can be partitioned into its individual rows or its individual columns, and that a (row or column) vector can be partitioned into "subvectors" of one element each. Thus, for an $M \times N$

matrix $\mathbf{A} = (\mathbf{a}_1, \mathbf{a}_2, \ldots, \mathbf{a}_N)$, with columns $\mathbf{a}_1, \mathbf{a}_2, \ldots, \mathbf{a}_N$, and an N-dimensional column vector $\mathbf{x} = (x_1, x_2, \ldots, x_N)'$, with elements $x_1, x_2, \ldots, x_N$, we have, as a special case of result (2.5), that

$$\mathbf{Ax} = x_1\mathbf{a}_1 + x_2\mathbf{a}_2 + \cdots + x_N\mathbf{a}_N. \tag{2.8}$$

Similarly, for an $M \times N$ matrix $\mathbf{A} = \begin{pmatrix} \mathbf{b}'_1 \\ \mathbf{b}'_2 \\ \vdots \\ \mathbf{b}'_M \end{pmatrix}$, with rows $\mathbf{b}'_1, \mathbf{b}'_2, \ldots, \mathbf{b}'_M$, and an M-dimensional row vector $\mathbf{x}' = (x_1, x_2, \ldots, x_M)$, with elements $x_1, x_2, \ldots, x_M$,

$$\mathbf{x}'\mathbf{A} = x_1\mathbf{b}'_1 + x_2\mathbf{b}'_2 + \cdots + x_M\mathbf{b}'_M. \tag{2.9}$$

Note also that an unpartitioned matrix can be regarded as a "partitioned" matrix comprising a single row and a single column of blocks. Thus, letting $\mathbf{A}$ represent an $M \times N$ matrix and taking $\mathbf{X} = (\mathbf{x}_1, \mathbf{x}_2, \ldots, \mathbf{x}_Q)$ to be an $N \times Q$ matrix with columns $\mathbf{x}_1, \mathbf{x}_2, \ldots, \mathbf{x}_Q$ and $\mathbf{Y} = \begin{pmatrix} \mathbf{y}'_1 \\ \mathbf{y}'_2 \\ \vdots \\ \mathbf{y}'_P \end{pmatrix}$ to be a $P \times M$ matrix with rows $\mathbf{y}'_1, \mathbf{y}'_2, \ldots, \mathbf{y}'_P$, result (2.5) implies that

$$\mathbf{AX} = (\mathbf{Ax}_1, \mathbf{Ax}_2, \ldots, \mathbf{Ax}_Q), \tag{2.10}$$

$$\mathbf{YA} = \begin{pmatrix} \mathbf{y}'_1\mathbf{A} \\ \mathbf{y}'_2\mathbf{A} \\ \vdots \\ \mathbf{y}'_P\mathbf{A} \end{pmatrix}, \tag{2.11}$$

and

$$\mathbf{YAX} = \begin{pmatrix} \mathbf{y}'_1\mathbf{Ax}_1 & \mathbf{y}'_1\mathbf{Ax}_2 & \ldots & \mathbf{y}'_1\mathbf{Ax}_Q \\ \mathbf{y}'_2\mathbf{Ax}_1 & \mathbf{y}'_2\mathbf{Ax}_2 & \ldots & \mathbf{y}'_2\mathbf{Ax}_Q \\ \vdots & \vdots & & \vdots \\ \mathbf{y}'_P\mathbf{Ax}_1 & \mathbf{y}'_P\mathbf{Ax}_2 & \ldots & \mathbf{y}'_P\mathbf{Ax}_Q \end{pmatrix}. \tag{2.12}$$

That is, $\mathbf{AX}$ is an $M \times Q$ matrix whose jth column is $\mathbf{Ax}_j$ $(j = 1, 2, \ldots, Q)$; $\mathbf{YA}$ is a $P \times N$ matrix whose ith row is $\mathbf{y}'_i\mathbf{A}$ $(i = 1, 2, \ldots, P)$; and $\mathbf{YAX}$ is a $P \times Q$ matrix whose ijth element is $\mathbf{y}'_i\mathbf{Ax}_j$ $(i = 1, 2, \ldots, P;\ j = 1, 2, \ldots, Q)$.

Representation (2.9) is helpful in establishing the elementary results expressed in the following two lemmas—refer, for instance, to Harville (1997, sec. 2.3) for detailed derivations.

Lemma 2.2.1. For any column vector $\mathbf{y}$ and nonnull column vector $\mathbf{x}$, there exists a matrix $\mathbf{A}$ such that $\mathbf{y} = \mathbf{Ax}$.

Lemma 2.2.2. For any two $M \times N$ matrices $\mathbf{A}$ and $\mathbf{B}$, $\mathbf{A} = \mathbf{B}$ if and only if $\mathbf{Ax} = \mathbf{Bx}$ for every N-dimensional column vector $\mathbf{x}$.

Note that Lemma 2.2.2 implies in particular that $\mathbf{A} = \mathbf{0}$ if and only if $\mathbf{Ax} = \mathbf{0}$ for every $\mathbf{x}$.

2.3 Trace of a (Square) Matrix

The *trace* of a square matrix $\mathbf{A} = \{a_{ij}\}$ of order N is defined to be the sum of the N diagonal elements of $\mathbf{A}$ and is to be denoted by the symbol $\mathrm{tr}(\mathbf{A})$. Thus,

$$\mathrm{tr}(\mathbf{A}) = a_{11} + a_{22} + \cdots + a_{NN}.$$

a. Basic properties

Clearly, for any scalar k and any $N \times N$ matrices $\mathbf{A}$ and $\mathbf{B}$,

$$\mathrm{tr}(k\mathbf{A}) = k\,\mathrm{tr}(\mathbf{A}), \tag{3.1}$$

$$\mathrm{tr}(\mathbf{A} + \mathbf{B}) = \mathrm{tr}(\mathbf{A}) + \mathrm{tr}(\mathbf{B}), \tag{3.2}$$

$$\mathrm{tr}(\mathbf{A}') = \mathrm{tr}(\mathbf{A}). \tag{3.3}$$

Further, for any R scalars $k_1, k_2, \ldots, k_R$ and for any R matrices $\mathbf{A}_1, \mathbf{A}_2, \ldots, \mathbf{A}_R$ of dimensions $N \times N$,

$$\mathrm{tr}\Big(\sum_{i=1}^{R} k_i \mathbf{A}_i\Big) = \sum_{i=1}^{R} k_i\,\mathrm{tr}(\mathbf{A}_i), \tag{3.4}$$

as can be readily verified by, for example, the repeated application of results (3.1) and (3.2). And for a square matrix $\mathbf{A}$ that has been partitioned as

$$\mathbf{A} = \begin{pmatrix} \mathbf{A}_{11} & \mathbf{A}_{12} & \cdots & \mathbf{A}_{1R} \\ \mathbf{A}_{21} & \mathbf{A}_{22} & \cdots & \mathbf{A}_{2R} \\ \vdots & \vdots & \ddots & \vdots \\ \mathbf{A}_{R1} & \mathbf{A}_{R2} & \cdots & \mathbf{A}_{RR} \end{pmatrix}$$

in such a way that the diagonal blocks $\mathbf{A}_{11}, \mathbf{A}_{22}, \ldots, \mathbf{A}_{RR}$ are square,

$$\mathrm{tr}(\mathbf{A}) = \mathrm{tr}(\mathbf{A}_{11}) + \mathrm{tr}(\mathbf{A}_{22}) + \cdots + \mathrm{tr}(\mathbf{A}_{RR}). \tag{3.5}$$

b. Trace of a product

Let $\mathbf{A} = \{a_{ij}\}$ represent an $M \times N$ matrix and $\mathbf{B} = \{b_{ji}\}$ an $N \times M$ matrix. Then,

$$\mathrm{tr}(\mathbf{AB}) = \sum_{i=1}^{M}\sum_{j=1}^{N} a_{ij} b_{ji}, \tag{3.6}$$

as is evident upon observing that the ith diagonal element of $\mathbf{AB}$ is $\sum_{j=1}^{N} a_{ij} b_{ji}$. Thus, since the jith element of $\mathbf{B}$ is the ijth element of $\mathbf{B}'$, the trace of the matrix product $\mathbf{AB}$ can be formed by multiplying the ijth element of $\mathbf{A}$ by the corresponding (ijth) element of $\mathbf{B}'$ and by then summing (over i and j).

A simple (but very important) result on the trace of a product of two matrices is expressed in the following lemma.

Lemma 2.3.1. For any $M \times N$ matrix $\mathbf{A}$ and $N \times M$ matrix $\mathbf{B}$,

$$\mathrm{tr}(\mathbf{AB}) = \mathrm{tr}(\mathbf{BA}). \tag{3.7}$$

Proof. Let a_{ij} represent the ijth element of $\mathbf{A}$ and b_{ji} the jith element of $\mathbf{B}$, and observe that the jth diagonal element of $\mathbf{BA}$ is $\sum_{i=1}^{M} b_{ji}a_{ij}$. Thus, making use of result (3.6), we find that

$$\text{tr}(\mathbf{AB}) = \sum_{i=1}^{M}\sum_{j=1}^{N} a_{ij}b_{ji} = \sum_{j=1}^{N}\sum_{i=1}^{M} a_{ij}b_{ji} = \sum_{j=1}^{N}\sum_{i=1}^{M} b_{ji}a_{ij} = \text{tr}(\mathbf{BA}). \qquad \text{Q.E.D.}$$

Note [in light of results (3.7) and (3.6)] that for any $M \times N$ matrix $\mathbf{A} = \{a_{ij}\}$,

$$\text{tr}(\mathbf{A}'\mathbf{A}) = \text{tr}(\mathbf{AA}') = \sum_{i=1}^{M}\sum_{j=1}^{N} a_{ij}^2 \tag{3.8}$$

$$\geq 0. \tag{3.9}$$

That is, both $\text{tr}(\mathbf{A}'\mathbf{A})$ and $\text{tr}(\mathbf{AA}')$ equal the sum of squares of the MN elements of $\mathbf{A}$, and both are inherently nonnegative.

c. Some equivalent conditions

Note that equality is attained in inequality (3.9) if and only if $\mathbf{A} = \mathbf{0}$ and that as a consequence, we have the following lemma.

Lemma 2.3.2. For any $M \times N$ matrix $\mathbf{A}$, $\mathbf{A} = \mathbf{0}$ if and only if $\text{tr}(\mathbf{A}'\mathbf{A}) = 0$.

As an essentially immediate consequence of Lemma 2.3.2, we have the following corollary.

Corollary 2.3.3. For any $M \times N$ matrix $\mathbf{A}$, $\mathbf{A} = \mathbf{0}$ if and only if $\mathbf{A}'\mathbf{A} = \mathbf{0}$.

The following corollary provides a very useful generalization of Corollary 2.3.3.

Corollary 2.3.4. (1) For any $M \times N$ matrix $\mathbf{A}$ and $N \times S$ matrices $\mathbf{B}$ and $\mathbf{C}$, $\mathbf{AB} = \mathbf{AC}$ if and only if $\mathbf{A}'\mathbf{AB} = \mathbf{A}'\mathbf{AC}$. (2) Similarly, for any $M \times N$ matrix $\mathbf{A}$ and $S \times N$ matrices $\mathbf{B}$ and $\mathbf{C}$, $\mathbf{BA}' = \mathbf{CA}'$ if and only if $\mathbf{BA}'\mathbf{A} = \mathbf{CA}'\mathbf{A}$.

Proof (of Corollary 2.3.4). (1) If $\mathbf{AB} = \mathbf{AC}$, then obviously $\mathbf{A}'\mathbf{AB} = \mathbf{A}'\mathbf{AC}$. Conversely, if $\mathbf{A}'\mathbf{AB} = \mathbf{A}'\mathbf{AC}$, then

$$(\mathbf{AB} - \mathbf{AC})'(\mathbf{AB} - \mathbf{AC}) = (\mathbf{B}' - \mathbf{C}')(\mathbf{A}'\mathbf{AB} - \mathbf{A}'\mathbf{AC}) = \mathbf{0},$$

and it follows from Corollary 2.3.3 that $\mathbf{AB} - \mathbf{AC} = \mathbf{0}$ or, equivalently, that $\mathbf{AB} = \mathbf{AC}$.

(2) To establish Part (2), simply take the transpose of each side of the two equivalent equalities $\mathbf{AB}' = \mathbf{AC}'$ and $\mathbf{A}'\mathbf{AB}' = \mathbf{A}'\mathbf{AC}'$. [The equivalence of these two equalities follows from Part (1).] Q.E.D.

Note that as a special case of Part (1) of Corollary 2.3.4 (the special case where $\mathbf{C} = \mathbf{0}$), we have that $\mathbf{AB} = \mathbf{0}$ if and only if $\mathbf{A}'\mathbf{AB} = \mathbf{0}$, and as a special case of Part (2), we have that $\mathbf{BA}' = \mathbf{0}$ if and only if $\mathbf{BA}'\mathbf{A} = \mathbf{0}$.

2.4 Linear Spaces

A nonempty set, say $\mathcal{V}$, of matrices (all of which have the same dimensions) is called a *linear space* if (1) for every matrix $\mathbf{A}$ in $\mathcal{V}$ and every matrix $\mathbf{B}$ in $\mathcal{V}$, the sum $\mathbf{A} + \mathbf{B}$ is in $\mathcal{V}$, and (2) for every matrix $\mathbf{A}$ in $\mathcal{V}$ and every scalar k, the product $k\mathbf{A}$ is in $\mathcal{V}$. For example, the set consisting of all $M \times N$ matrices is a linear space; and since sums and scalar multiples of symmetric matrices are symmetric, the set of all $N \times N$ symmetric matrices is a linear space. Note that every linear space contains the null matrix $\mathbf{0}$ (of appropriate dimensions), and that the set $\{\mathbf{0}\}$, whose only member is a null matrix, is a linear space. Note also that if a linear space contains a nonnull matrix, then it contains an infinite number of nonnull matrices.

A *linear combination* of matrices $\mathbf{A}_1, \mathbf{A}_2, \ldots, \mathbf{A}_K$ (of the same dimensions) is an expression of the general form

$$x_1\mathbf{A}_1 + x_2\mathbf{A}_2 + \cdots + x_K\mathbf{A}_K,$$

where $x_1, x_2, \ldots, x_K$ are scalars (which are referred to as the coefficients of the linear combination). If $\mathbf{A}_1, \mathbf{A}_2, \ldots, \mathbf{A}_K$ are matrices in a linear space $\mathcal{V}$, then every linear combination of $\mathbf{A}_1, \mathbf{A}_2, \ldots, \mathbf{A}_K$ is also in $\mathcal{V}$.

Corresponding to any finite set of $M \times N$ matrices is the *span* of the set. By definition, the span of a nonempty finite set $\{\mathbf{A}_1, \mathbf{A}_2, \ldots, \mathbf{A}_K\}$ of $M \times N$ matrices is the set consisting of all matrices that are expressible as linear combinations of $\mathbf{A}_1, \mathbf{A}_2, \ldots, \mathbf{A}_K$. (By convention, the span of the empty set of $M \times N$ matrices is the set $\{\mathbf{0}\}$, whose only member is the $M \times N$ null matrix.) The span of a finite set S is denoted herein by the symbol $\text{sp}(S)$; $\text{sp}(\{\mathbf{A}_1, \mathbf{A}_2, \ldots, \mathbf{A}_K\})$, which represents the span of the set comprising the matrices $\mathbf{A}_1, \mathbf{A}_2, \ldots, \mathbf{A}_K$, is typically abbreviated to $\text{sp}(\mathbf{A}_1, \mathbf{A}_2, \ldots, \mathbf{A}_K)$. Clearly, the span of any finite set of $M \times N$ matrices is a linear space.

A finite set S of matrices in a linear space $\mathcal{V}$ is said to *span* $\mathcal{V}$ if $\text{sp}(S) = \mathcal{V}$. Or equivalently [since $\text{sp}(S) \subset \mathcal{V}$], S spans $\mathcal{V}$ if $\mathcal{V} \subset \text{sp}(S)$.

a. Row and column spaces

Corresponding to any $M \times N$ matrix $\mathbf{A}$ are two linear spaces of fundamental importance. The *column space* of $\mathbf{A}$ is the span of the set whose members are the columns of $\mathbf{A}$, that is, the column space of $\mathbf{A}$ is the set consisting of all M-dimensional column vectors that are expressible as linear combinations of the N columns of $\mathbf{A}$. Similarly, the *row space* of $\mathbf{A}$ is the span of the set whose members are the rows of $\mathbf{A}$, that is, the row space of $\mathbf{A}$ is the set consisting of all N-dimensional row vectors that are expressible as linear combinations of the M rows of $\mathbf{A}$.

The column space of a matrix $\mathbf{A}$ is to be denoted by the symbol $\mathcal{C}(\mathbf{A})$ and the row space by the symbol $\mathcal{R}(\mathbf{A})$. The symbol $\mathcal{R}^N$ will be used to denote the set of all N-dimensional column vectors or (depending on the context) the set of all N-dimensional row vectors. Note that $\mathcal{C}(\mathbf{I}_N) = \mathcal{R}^N$ (where $\mathcal{R}^N$ is the set of all N-dimensional column vectors), and $\mathcal{R}(\mathbf{I}_N) = \mathcal{R}^N$ (where $\mathcal{R}^N$ is the set of all N-dimensional row vectors).

In light of result (2.8), it is apparent that an M-dimensional column vector $\mathbf{y}$ is a member of the column space $\mathcal{C}(\mathbf{A})$ of an $M \times N$ matrix $\mathbf{A}$ if and only if there exists an N-dimensional column vector $\mathbf{x}$ for which $\mathbf{y} = \mathbf{A}\mathbf{x}$. And in light of result (2.9), it is apparent that an N-dimensional row vector $\mathbf{y}'$ is a member of $\mathcal{R}(\mathbf{A})$ if and only if there exists an M-dimensional row vector $\mathbf{x}'$ for which $\mathbf{y}' = \mathbf{x}'\mathbf{A}$.

The following lemma relates the column space of a matrix to the row space of its transpose.

Lemma 2.4.1. For any matrix $\mathbf{A}$, $\mathbf{y} \in \mathcal{C}(\mathbf{A})$ if and only if $\mathbf{y}' \in \mathcal{R}(\mathbf{A}')$.

Proof. If $\mathbf{y} \in \mathcal{C}(\mathbf{A})$, then $\mathbf{y} = \mathbf{A}\mathbf{x}$ for some column vector $\mathbf{x}$, implying that $\mathbf{y}' = (\mathbf{A}\mathbf{x})' = \mathbf{x}'\mathbf{A}'$ and hence that $\mathbf{y}' \in \mathcal{R}(\mathbf{A}')$. The converse [that $\mathbf{y}' \in \mathcal{R}(\mathbf{A}') \Rightarrow \mathbf{y} \in \mathcal{C}(\mathbf{A})$] can be established in similar fashion. Q.E.D.

b. Subspaces

A subset $\mathcal{U}$ of a linear space $\mathcal{V}$ (of $M \times N$ matrices) is said to be a *subspace* of $\mathcal{V}$ if $\mathcal{U}$ is itself a linear space. Trivial examples of a subspace of a linear space $\mathcal{V}$ are: (1) the set $\{\mathbf{0}\}$, whose only member is the null matrix, and (2) the entire set $\mathcal{V}$. The column space $\mathcal{C}(\mathbf{A})$ of an $M \times N$ matrix $\mathbf{A}$ is a subspace of $\mathcal{R}^M$ (when $\mathcal{R}^M$ is interpreted as the set of all M-dimensional column vectors), and $\mathcal{R}(\mathbf{A})$ is a subspace of $\mathcal{R}^N$ (when $\mathcal{R}^N$ is interpreted as the set of all N-dimensional row vectors).

We require some additional terminology and notation. Suppose that S and T are subspaces of the linear space of all $M \times N$ matrices or, more generally, that S and T are subsets of a given set. If every member of S is a member of T, then S is said to be *contained* in T (or T is said to contain S), and we write $S \subset T$ (or $T \supset S$). Note that if $S \subset T$ and $T \subset S$, then $S = T$, that is, the two subsets S and T are identical.

Some basic results on row and column spaces are expressed in the following lemmas and corollaries, proofs of which are given by Harville (1997, sec. 4.2).

Lemma 2.4.2. Let $\mathbf{A}$ represent an $M \times N$ matrix. Then, for any subspace $\mathcal{U}$ of $\mathcal{R}^M$, $\mathcal{C}(\mathbf{A}) \subset \mathcal{U}$ if and only if every column of $\mathbf{A}$ belongs to $\mathcal{U}$. Similarly, for any subspace $\mathcal{V}$ of $\mathcal{R}^N$, $\mathcal{R}(\mathbf{A}) \subset \mathcal{V}$ if and only if every row of $\mathbf{A}$ belongs to $\mathcal{V}$.

Lemma 2.4.3. For any $M \times N$ matrix $\mathbf{A}$ and $M \times P$ matrix $\mathbf{B}$, $\mathcal{C}(\mathbf{B}) \subset \mathcal{C}(\mathbf{A})$ if and only if there exists an $N \times P$ matrix $\mathbf{F}$ such that $\mathbf{B} = \mathbf{AF}$. Similarly, for any $M \times N$ matrix $\mathbf{A}$ and $Q \times N$ matrix $\mathbf{C}$, $\mathcal{R}(\mathbf{C}) \subset \mathcal{R}(\mathbf{A})$ if and only if there exists a $Q \times M$ matrix $\mathbf{L}$ such that $\mathbf{C} = \mathbf{LA}$.

Corollary 2.4.4. For any $M \times N$ matrix $\mathbf{A}$ and $N \times P$ matrix $\mathbf{F}$, $\mathcal{C}(\mathbf{AF}) \subset \mathcal{C}(\mathbf{A})$. Similarly, for any $M \times N$ matrix $\mathbf{A}$ and $Q \times M$ matrix $\mathbf{L}$, $\mathcal{R}(\mathbf{LA}) \subset \mathcal{R}(\mathbf{A})$.

Corollary 2.4.5. Let $\mathbf{A}$ represent an $M \times N$ matrix, $\mathbf{E}$ an $N \times K$ matrix, $\mathbf{F}$ an $N \times P$ matrix, $\mathbf{L}$ a $Q \times M$ matrix, and $\mathbf{T}$ an $S \times M$ matrix.
(1) If $\mathcal{C}(\mathbf{E}) \subset \mathcal{C}(\mathbf{F})$, then $\mathcal{C}(\mathbf{AE}) \subset \mathcal{C}(\mathbf{AF})$; and if $\mathcal{C}(\mathbf{E}) = \mathcal{C}(\mathbf{F})$, then $\mathcal{C}(\mathbf{AE}) = \mathcal{C}(\mathbf{AF})$.
(2) If $\mathcal{R}(\mathbf{L}) \subset \mathcal{R}(\mathbf{T})$, then $\mathcal{R}(\mathbf{LA}) \subset \mathcal{R}(\mathbf{TA})$; and if $\mathcal{R}(\mathbf{L}) = \mathcal{R}(\mathbf{T})$, then $\mathcal{R}(\mathbf{LA}) = \mathcal{R}(\mathbf{TA})$.

Lemma 2.4.6. Let $\mathbf{A}$ represent an $M \times N$ matrix and $\mathbf{B}$ an $M \times P$ matrix. Then, (1) $\mathcal{C}(\mathbf{A}) \subset \mathcal{C}(\mathbf{B})$ if and only if $\mathcal{R}(\mathbf{A}') \subset \mathcal{R}(\mathbf{B}')$, and (2) $\mathcal{C}(\mathbf{A}) = \mathcal{C}(\mathbf{B})$ if and only if $\mathcal{R}(\mathbf{A}') = \mathcal{R}(\mathbf{B}')$.

c. Linear dependence and independence

Any finite set of row or column vectors, or more generally any finite set of $M \times N$ matrices, is either linearly dependent or linearly independent. A nonempty finite set $\{\mathbf{A}_1, \mathbf{A}_2, \ldots, \mathbf{A}_K\}$ of $M \times N$ matrices is said to be *linearly dependent* if there exist scalars $x_1, x_2, \ldots, x_K$, not all zero, such that

$$x_1\mathbf{A}_1 + x_2\mathbf{A}_2 + \cdots + x_K\mathbf{A}_K = \mathbf{0}.$$

If no such scalars exist, the set is said to be *linearly independent*. The empty set is considered to be linearly independent. Note that if any subset of a finite set of $M \times N$ matrices is linearly dependent, then the set itself is linearly dependent. Note also that if the set $\{\mathbf{A}_1, \mathbf{A}_2, \ldots, \mathbf{A}_K\}$ is linearly dependent, then some member of the set, say the sth member $\mathbf{A}_s$, can be expressed as a linear combination of the other $K-1$ members $\mathbf{A}_1, \mathbf{A}_2, \ldots, \mathbf{A}_{s-1}, \mathbf{A}_{s+1}, \ldots, \mathbf{A}_K$; that is, $\mathbf{A}_s = \sum_{i \neq s} y_i \mathbf{A}_i$ for some scalars $y_1, y_2, \ldots, y_{s-1}, y_{s+1}, \ldots, y_K$.

While technically linear dependence and independence are properties of sets of matrices, it is customary to speak of "a set of linearly dependent (or independent) matrices" or simply of "linearly dependent (or independent) matrices" instead of "a linearly dependent (or independent) set of matrices." In particular, in the case of row or column vectors, it is customary to speak of "linearly dependent (or independent) vectors."

d. Bases

A *basis* for a linear space $\mathcal{V}$ of $M \times N$ matrices is a linearly independent set of matrices in $\mathcal{V}$ that spans $\mathcal{V}$. The empty set is the (unique) basis for the linear space $\{\mathbf{0}\}$ (whose only member is the null matrix). The set whose members are the M columns $(1, 0, \ldots, 0)', \ldots, (0, \ldots, 0, 1)'$ of the $M \times M$ identity matrix $\mathbf{I}_M$ is a basis for the linear space $\mathcal{R}^M$ of all M-dimensional column vectors. Similarly, the set whose members are the N rows $(1, 0, \ldots, 0), \ldots, (0, \ldots, 0, 1)$ of the $N \times N$ identity matrix $\mathbf{I}_N$ is a basis for the linear space $\mathcal{R}^N$ of all N-dimensional row vectors. More generally, letting $\mathbf{U}_{ij}$ represent the $M \times N$ matrix whose ijth element equals 1 and whose remaining $(MN - 1)$ elements equal 0, the set whose members are the MN matrices $\mathbf{U}_{11}, \mathbf{U}_{21}, \ldots, \mathbf{U}_{M1}, \mathbf{U}_{12}, \mathbf{U}_{22}, \ldots, \mathbf{U}_{M2}, \ldots, \mathbf{U}_{1N}, \mathbf{U}_{2N}, \ldots, \mathbf{U}_{MN}$ is a basis (the so-called natural basis) for the linear space of all $M \times N$ matrices (as can be readily verified).

Now, consider the column space $\mathcal{C}(\mathbf{A})$ and row space $\mathcal{R}(\mathbf{A})$ of an $M \times N$ matrix $\mathbf{A}$. By definition, $\mathcal{C}(\mathbf{A})$ is spanned by the set whose members are the columns of $\mathbf{A}$. If this set is linearly independent, it is a basis for $\mathcal{C}(\mathbf{A})$; otherwise, it is not. Similarly, if the set whose members are the rows of $\mathbf{A}$ is linearly independent, it is a basis for $\mathcal{R}(\mathbf{A})$.

Two fundamentally important properties of linear spaces in general and row and column spaces in particular are described in the following two theorems.

Theorem 2.4.7. Every linear space (of $M \times N$ matrices) has a basis.

Theorem 2.4.8. Any two bases for a linear space (of $M \times N$ natrices) contain the same number of matrices

The number of matrices in a basis for a linear space $\mathcal{V}$ (of $M \times N$ matrices) is referred to as the *dimension* of $\mathcal{V}$ and is denoted by the symbol dim $\mathcal{V}$ or dim($\mathcal{V}$). Note that the term dimension is used not only in reference to the number of matrices in a basis, but also in reference to the number of rows or columns in a matrix—which usage is intended is determinable from the context.

Some basic results related to the dimension of a linear space or subspace are presented in the following two theorems.

Theorem 2.4.9. If a linear space $\mathcal{V}$ (of $M \times N$ matrices) is spanned by a set of R matrices, then $\dim \mathcal{V} \leq R$, and if there is a set of K linearly independent matrices in $\mathcal{V}$, then $\dim \mathcal{V} \geq K$.

Theorem 2.4.10. Let $\mathcal{U}$ and $\mathcal{V}$ represent linear spaces of $M \times N$ matrices. If $\mathcal{U} \subset \mathcal{V}$ (i.e., if $\mathcal{U}$ is a subspace of $\mathcal{V}$), then $\dim \mathcal{U} \leq \dim \mathcal{V}$. Moreover, if $\mathcal{U} \subset \mathcal{V}$ and if in addition $\dim \mathcal{U} = \dim \mathcal{V}$, then $\mathcal{U} = \mathcal{V}$.

Two key results pertaining to bases are as follows.

Theorem 2.4.11. Any set of R linearly independent matrices in an R-dimensional linear space $\mathcal{V}$ (of $M \times N$ matrices) is a basis for $\mathcal{V}$.

Theorem 2.4.12. A matrix $\mathbf{A}$ in a linear space $\mathcal{V}$ (of $M \times N$ matrices) has a unique representation in terms of any particular basis $\{\mathbf{A}_1, \mathbf{A}_2, \ldots, \mathbf{A}_R\}$; that is, the coefficients $x_1, x_2, \ldots, x_R$ in the linear combination

$$\mathbf{A} = x_1\mathbf{A}_1 + x_2\mathbf{A}_2 + \cdots + x_R\mathbf{A}_R$$

are uniquely determined.

For proofs of the results set forth in Theorems 2.4.7 through 2.4.12, refer to Harville (1997, sec. 4.3).

e. Rank (of a matrix)

The *row rank* of a matrix $\mathbf{A}$ is defined to be the dimension of the row space of $\mathbf{A}$, and the *column rank* of $\mathbf{A}$ is defined to be the dimension of the column space of $\mathbf{A}$. A fundamental result on row and column spaces is given by the following theorem.

Theorem 2.4.13. The row rank of any matrix $\mathbf{A}$ equals the column rank of $\mathbf{A}$.

Refer to Harville (1997, sec 4.4) for a proof of Theorem 2.4.13. That proof is based on the following result, which is of some interest in its own right.

Theorem 2.4.14. Let $\mathbf{A}$ represent an $M \times N$ nonnull matrix of row rank R and column rank C. Then, there exist an $M \times C$ matrix $\mathbf{B}$ and a $C \times N$ matrix $\mathbf{L}$ such that $\mathbf{A} = \mathbf{BL}$. Similarly, there exist an $M \times R$ matrix K and an $R \times N$ matrix $\mathbf{T}$ such that $\mathbf{A} = \mathbf{KT}$.

Proof (of Theorem 2.4.14). Take $\mathbf{B}$ to be an $M \times C$ matrix whose columns form a basis for $\mathcal{C}(\mathbf{A})$. Then, $\mathcal{C}(\mathbf{A}) = \mathcal{C}(\mathbf{B})$, and consequently it follows from Lemma 2.4.3 that there exists a $C \times N$ matrix $\mathbf{L}$ such that $\mathbf{A} = \mathbf{BL}$. The existence of an $M \times R$ matrix $\mathbf{K}$ and an $R \times N$ matrix $\mathbf{T}$ such that $\mathbf{A} = \mathbf{KT}$ can be established via a similar argument. Q.E.D.

In light of Theorem 2.4.13, it is not necessary to distinguish between the row and column ranks of a matrix $\mathbf{A}$. Their common value is called the *rank* of $\mathbf{A}$ and is denoted by the symbol rank $\mathbf{A}$ or rank$(\mathbf{A})$.

Various of the results pertaining to the dimensions of linear spaces and subspaces can be specialized to row and column spaces and restated in terms of ranks. Since the column space of an $M \times N$ matrix $\mathbf{A}$ is a subspace of $\mathcal{R}^M$ and the row space of $\mathbf{A}$ is a subspace of $\mathcal{R}^N$, the following lemma is an immediate consequence of Theorem 2.4.10.

Lemma 2.4.15. For any $M \times N$ matrix $\mathbf{A}$, $\text{rank}(\mathbf{A}) \le M$ and $\text{rank}(\mathbf{A}) \le N$.

A further implication of Theorem 2.4.10 is as follows.

Theorem 2.4.16. Let $\mathbf{A}$ represent an $M \times N$ matrix, $\mathbf{B}$ an $M \times P$ matrix, and $\mathbf{C}$ a $Q \times N$ matrix. If $\mathcal{C}(\mathbf{B}) \subset \mathcal{C}(\mathbf{A})$, then $\text{rank}(\mathbf{B}) \le \text{rank}(\mathbf{A})$; if $\mathcal{C}(\mathbf{B}) \subset \mathcal{C}(\mathbf{A})$ and if in addition $\text{rank}(\mathbf{B}) = \text{rank}(\mathbf{A})$, then $\mathcal{C}(\mathbf{B}) = \mathcal{C}(\mathbf{A})$. Similarly, if $\mathcal{R}(\mathbf{C}) \subset \mathcal{R}(\mathbf{A})$, then $\text{rank}(\mathbf{C}) \le \text{rank}(\mathbf{A})$; if $\mathcal{R}(\mathbf{C}) \subset \mathcal{R}(\mathbf{A})$ and if in addition $\text{rank}(\mathbf{C}) = \text{rank}(\mathbf{A})$, then $\mathcal{R}(\mathbf{C}) = \mathcal{R}(\mathbf{A})$.

In light of Corollary 2.4.4, we have the following corollary of Theorem 2.4.16.

Corollary 2.4.17. Let $\mathbf{A}$ represent an $M \times N$ matrix and $\mathbf{F}$ an $N \times P$ matrix. Then, $\text{rank}(\mathbf{AF}) \le \text{rank}(\mathbf{A})$ and $\text{rank}(\mathbf{AF}) \le \text{rank}(\mathbf{F})$. Moreover, if $\text{rank}(\mathbf{AF}) = \text{rank}(\mathbf{A})$, then $\mathcal{C}(\mathbf{AF}) = \mathcal{C}(\mathbf{A})$; similarly, if $\text{rank}(\mathbf{AF}) = \text{rank}(\mathbf{F})$, then $\mathcal{R}(\mathbf{AF}) = \mathcal{R}(\mathbf{F})$.

The rank of an $M \times N$ matrix cannot exceed $\min(M, N)$, as is evident from Lemma 2.4.15. An $M \times N$ matrix $\mathbf{A}$ is said to have *full row rank* if $\text{rank}(\mathbf{A}) = M$, that is, if its rank equals the number of rows, and to have *full column rank* if $\text{rank}(\mathbf{A}) = N$. Clearly, an $M \times N$ matrix can have full row rank only if $M \le N$, that is, only if the number of rows does not exceed the number of columns, and can have full column rank only if $N \le M$.

A matrix is said to be *nonsingular* if it has both full row rank and full column rank. Clearly, any nonsingular matrix is square. By definition, an $N \times N$ matrix $\mathbf{A}$ is nonsingular if and only if $\text{rank}(\mathbf{A}) = N$. An $N \times N$ matrix of rank less than N is said to be *singular*.

Any $M \times N$ matrix can be expressed as the product of a matrix having full column rank and a matrix having full row rank as indicated by the following theorem.

Theorem 2.4.18. Let $\mathbf{A}$ represent an $M \times N$ nonnull matrix of rank R. Then, there exist an $M \times R$ matrix $\mathbf{B}$ and $R \times N$ matrix $\mathbf{T}$ such that $\mathbf{A} = \mathbf{BT}$. Moreover, for any $M \times R$ matrix $\mathbf{B}$ and $R \times N$ matrix $\mathbf{T}$ such that $\mathbf{A} = \mathbf{BT}$, $\text{rank}(\mathbf{B}) = \text{rank}(\mathbf{T}) = R$, that is, $\mathbf{B}$ has full column rank and $\mathbf{T}$ has full row rank.

Proof. The existence of an $M \times R$ matrix $\mathbf{B}$ and an $R \times N$ matrix $\mathbf{T}$ such that $\mathbf{A} = \mathbf{BT}$ follows from Theorem 2.4.14. And, letting $\mathbf{B}$ represent any $M \times R$ matrix and $\mathbf{T}$ any $R \times N$ matrix such that $\mathbf{A} = \mathbf{BT}$, we find that $\text{rank}(\mathbf{B}) \ge R$ and $\text{rank}(\mathbf{T}) \ge R$ (as is evident from Corollary 2.4.17) and that $\text{rank}(\mathbf{B}) \le R$ and $\text{rank}(\mathbf{T}) \le R$ (as is evident from Lemma 2.4.15), and as a consequence, we have that $\text{rank}(\mathbf{B}) = R$ and $\text{rank}(\mathbf{T}) = R$. Q.E.D.

The following theorem, a proof of which is given by Harville (1997, sec. 4.4), characterizes the rank of a matrix in terms of the ranks of its submatrices.

Theorem 2.4.19. Let $\mathbf{A}$ represent an $M \times N$ matrix of rank R. Then, $\mathbf{A}$ contains R linearly independent rows and R linearly independent columns. And for any R linearly independent rows and R linearly independent columns of $\mathbf{A}$, the $R \times R$ submatrix, obtained by striking out the other $M - R$ rows and $N - R$ columns, is nonsingular. Moreover, any set of more than R rows or more than R columns (of $\mathbf{A}$) is linearly dependent, and there exists no submatrix of $\mathbf{A}$ whose rank exceeds R.

As applied to symmetric matrices, Theorem 2.4.19 has the following implication.

Corollary 2.4.20. Any symmetric matrix of rank R contains an $R \times R$ nonsingular principal submatrix.

The rank of a matrix was defined in terms of the dimension of its row and column spaces. Theorem 2.4.19 suggests some equivalent definitions. The rank of a matrix $\mathbf{A}$ is interpretable as the size of the largest linearly independent set that can be formed from the rows of $\mathbf{A}$. Similarly, it is interpretable as the size of the largest linearly independent set that can be formed from the columns of $\mathbf{A}$. The rank of $\mathbf{A}$ is also interpretable as the size (number of rows or columns) of the largest nonsingular (square) submatrix of $\mathbf{A}$.

Clearly, an $M \times N$ matrix has full row rank if and only if all M of its rows are linearly independent, and has full column rank if and only if all N of its columns are linearly independent. An $N \times N$ matrix is nonsingular if and only if all of its rows are linearly independent; similarly, it is nonsingular if and only if all of its columns are linearly independent.

It is a simple exercise to show that, for any matrix $\mathbf{A}$,

$$\text{rank}(\mathbf{A}') = \text{rank}(\mathbf{A}) \tag{4.1}$$

and that, for any matrix $\mathbf{A}$ and nonzero scalar k,

$$\text{rank}(k\mathbf{A}) = \text{rank}(\mathbf{A}). \tag{4.2}$$

As a special case of result (4.2), we have that

$$\text{rank}(-\mathbf{A}) = \text{rank}(\mathbf{A}). \tag{4.3}$$

f. Orthogonal and orthonormal sets

Corresponding to an arbitrary pair of matrices, say $\mathbf{A}$ and $\mathbf{B}$, in a linear space $\mathcal{V}$ of $M \times N$ matrices is a scalar that is denoted by the symbol $\mathbf{A} \bullet \mathbf{B}$ and that is referred to as the *inner product* (or dot product) of $\mathbf{A}$ and $\mathbf{B}$. The inner product of $\mathbf{A}$ and $\mathbf{B}$ can be regarded as the value assigned to $\mathbf{A}$ and $\mathbf{B}$ by a function whose domain consists of all (ordered) pairs of matrices in $\mathcal{V}$. This function is required to have the following four properties (but is otherwise subject to choice):

(1) $\mathbf{A} \bullet \mathbf{B} = \mathbf{B} \bullet \mathbf{A}$;

(2) $\mathbf{A} \bullet \mathbf{A} \geq 0$, with equality holding if and only if $\mathbf{A} = \mathbf{0}$;

(3) $(k\mathbf{A}) \bullet \mathbf{B} = k(\mathbf{A} \bullet \mathbf{B})$;

(4) $(\mathbf{A} + \mathbf{B}) \bullet \mathbf{C} = (\mathbf{A} \bullet \mathbf{C}) + (\mathbf{B} \bullet \mathbf{C})$

(where $\mathbf{A}$, $\mathbf{B}$, and $\mathbf{C}$ represent arbitrary matrices in $\mathcal{V}$ and k represents an arbitrary scalar).

The term inner product is used not only in referring to the values assigned by the function to the various pairs of matrices, but also in referring to the function itself. The *usual inner product* for a linear space $\mathcal{V}$ of $M \times N$ matrices assigns to each pair of matrices $\mathbf{A}$ and $\mathbf{B}$ in $\mathcal{V}$ the value

$$\mathbf{A} \bullet \mathbf{B} = \text{tr}(\mathbf{A}\mathbf{B}') = \sum_{i=1}^{M} \sum_{j=1}^{N} a_{ij} b_{ij}. \tag{4.4}$$

It is a simple exercise to verify that the function defined by expression (4.4) has the four properties required of an inner product. In the special case of a linear space $\mathcal{V}$ of M-dimensional column vectors, the value assigned by the usual inner product to each pair of vectors $\mathbf{x} = \{x_i\}$ and $\mathbf{y} = \{y_i\}$ in $\mathcal{V}$ is expressible as

$$\mathbf{x} \bullet \mathbf{y} = \text{tr}(\mathbf{x}\mathbf{y}') = \text{tr}(\mathbf{y}'\mathbf{x}) = \mathbf{y}'\mathbf{x} = \mathbf{x}'\mathbf{y} = \sum_{i=1}^{M} x_i y_i. \tag{4.5}$$

And in the special case of a linear space $\mathcal{V}$ of N-dimensional row vectors, the value assigned by the usual inner product to each pair of vectors $\mathbf{x}' = \{x_i\}$ and $\mathbf{y}' = \{y_i\}$ in $\mathcal{V}$ is expressible as

$$\mathbf{x}' \bullet \mathbf{y}' = \text{tr}[\mathbf{x}'(\mathbf{y}')'] = \text{tr}(\mathbf{x}'\mathbf{y}) = \mathbf{x}'\mathbf{y} = \sum_{j=1}^{N} x_j y_j. \tag{4.6}$$

The four basic properties of an inner product for a linear space $\mathcal{V}$ (of $M \times N$ matrices) imply various additional properties. We find, in particular, that (for any matrix $\mathbf{A}$ in $\mathcal{V}$)

$$\mathbf{0} \bullet \mathbf{A} = 0, \tag{4.7}$$

as is evident from Property (3) upon observing that

$$\mathbf{0} \bullet \mathbf{A} = (0\mathbf{A}) \bullet \mathbf{A} = 0(\mathbf{A} \bullet \mathbf{A}) = 0.$$

And by making repeated use of Properties (3) and (4), we find that (for any matrices $\mathbf{A}_1, \mathbf{A}_2, \ldots, \mathbf{A}_K$, and $\mathbf{B}$ in $\mathcal{V}$ and any scalars $x_1, x_2, \ldots, x_K$),

$$(x_1\mathbf{A}_1 + x_2\mathbf{A}_2 + \cdots + x_K\mathbf{A}_K) \bullet \mathbf{B} = x_1(\mathbf{A}_1 \bullet \mathbf{B}) + x_2(\mathbf{A}_2 \bullet \mathbf{B}) + \cdots + x_K(\mathbf{A}_K \bullet \mathbf{B}). \tag{4.8}$$

Corresponding to an arbitrary matrix, say $\mathbf{A}$, in the linear space $\mathcal{V}$ of $M \times N$ matrices is the scalar $(\mathbf{A} \bullet \mathbf{A})^{1/2}$. This scalar is called the *norm* of $\mathbf{A}$ and is denoted by the symbol $\|\mathbf{A}\|$. The norm depends on the choice of inner product; when the inner product is taken to be the usual inner product, the norm is referred to as the *usual norm.*

An important and famous inequality, known as the Schwarz inequality or Cauchy–Schwarz inequality, is set forth in the following theorem, a proof of which is given by Harville (1997, sec 6.3).

Theorem 2.4.21 (Cauchy–Schwarz inequality). For any two matrices $\mathbf{A}$ and $\mathbf{B}$ in a linear space $\mathcal{V}$,

$$|\mathbf{A} \bullet \mathbf{B}| \leq \|\mathbf{A}\| \|\mathbf{B}\|, \tag{4.9}$$

with equality holding if and only if $\mathbf{B} = \mathbf{0}$ or $\mathbf{A} = k\mathbf{B}$ for some scalar k.

As a special case of Theorem 2.4.21, we have that for any two M-dimensional column vectors $\mathbf{x}$ and $\mathbf{y}$,

$$|\mathbf{x}'\mathbf{y}| \leq (\mathbf{x}'\mathbf{x})^{1/2}(\mathbf{y}'\mathbf{y})^{1/2}, \tag{4.10}$$

with equality holding if and only if $\mathbf{y} = \mathbf{0}$ or $\mathbf{x} = k\mathbf{y}$ for some scalar k.

Two vectors $\mathbf{x}$ and $\mathbf{y}$ in a linear space $\mathcal{V}$ of M-dimensional column vectors are said to be *orthogonal* to each other if $\mathbf{x} \bullet \mathbf{y} = 0$. More generally, two matrices $\mathbf{A}$ and $\mathbf{B}$ in a linear space $\mathcal{V}$ are said to be *orthogonal* to each other if $\mathbf{A} \bullet \mathbf{B} = 0$. The statement that two matrices $\mathbf{A}$ and $\mathbf{B}$ are orthogonal to each other is sometimes abbreviated to $\mathbf{A} \perp \mathbf{B}$. Whether two matrices are orthogonal to each other depends on the choice of inner product; two matrices that are orthogonal (to each other) with respect to one inner product may not be orthogonal (to each other) with respect to another inner product.

A finite set of matrices in a linear space $\mathcal{V}$ of $M \times N$ matrices is said to be *orthogonal* if every matrix in the set is orthogonal to every other matrix in the set. Thus, the empty set and any set containing only one matrix are orthogonal sets. And a finite set $\{\mathbf{A}_1, \mathbf{A}_2, \ldots, \mathbf{A}_K\}$ of two or more matrices in $\mathcal{V}$ is an orthogonal set if $\mathbf{A}_i \bullet \mathbf{A}_j = 0$ for $j \neq i = 1, 2, \ldots, K$. A finite set of matrices in $\mathcal{V}$ is said to be *orthonormal* if the set is orthogonal and if the norm of every matrix in the set equals 1. In the special case of a set of (row or column) vectors, the expression "set of orthogonal (or orthonormal) vectors," or simply "orthogonal (or orthonormal) vectors," is often used in lieu of the technically more correct expression "orthogonal (or orthonormal) set of vectors."

The following lemma establishes a connection between orthogonality and linear independence.

Lemma 2.4.22. An orthogonal set of nonnull matrices is linearly independent.

Proof. If the orthogonal set is the empty set, then the result is clearly true (since, by convention, the empty set is linearly independent). Suppose then that $\{\mathbf{A}_1, \mathbf{A}_2, \ldots, \mathbf{A}_K\}$ is any nonempty orthogonal set of nonnull matrices. And let $x_1, x_2, \ldots, x_K$ represent arbitrary scalars such that $x_1\mathbf{A}_1 + x_2\mathbf{A}_2 + \cdots + x_K\mathbf{A}_K = \mathbf{0}$. For $i = 1, 2, \ldots, K$, we find [in light of results (4.7) and (4.8)] that

$$\begin{aligned} 0 = \mathbf{0}\bullet\mathbf{A}_i &= (x_1\mathbf{A}_1 + x_2\mathbf{A}_2 + \cdots + x_K\mathbf{A}_K)\bullet\mathbf{A}_i \\ &= x_1(\mathbf{A}_1\bullet\mathbf{A}_i) + x_2(\mathbf{A}_2\bullet\mathbf{A}_i) + \cdots + x_K(\mathbf{A}_K\bullet\mathbf{A}_i) \\ &= x_i(\mathbf{A}_i\bullet\mathbf{A}_i), \end{aligned}$$

implying (since $\mathbf{A}_i$ is nonnull) that $x_i = 0$. We conclude that the set $\{\mathbf{A}_1, \mathbf{A}_2, \ldots, \mathbf{A}_K\}$ is linearly independent. Q.E.D.

Note that Lemma 2.4.22 implies in particular that any orthonormal set of matrices is linearly independent. Note also that the converse of Lemma 2.4.22 is not necessarily true; that is, a linearly independent set is not necessarily orthogonal. For example, the set consisting of the two 2-dimensional row vectors $(1, 0)$ and $(1, 1)$ is linearly independent but is not orthogonal (with respect to the usual inner product).

Suppose now that $\mathbf{A}_1, \mathbf{A}_2, \ldots, \mathbf{A}_K$ are linearly independent matrices in a linear space $\mathcal{V}$ of $M \times N$ matrices. There exists a recursive procedure, known as Gram–Schmidt orthogonalization, that when applied to $\mathbf{A}_1, \mathbf{A}_2, \ldots, \mathbf{A}_K$, generates an orthonormal set of $M \times N$ matrices $\mathbf{B}_1, \mathbf{B}_2, \ldots, \mathbf{B}_K$ (the jth of which is a linear combination of $\mathbf{A}_1, \mathbf{A}_2, \ldots, \mathbf{A}_j$)—refer, for example, to Harville (1997, sec. 6.4) for a discussion of Gram–Schmidt orthogonalization. In combination with Theorems 2.4.7 and 2.4.11 and Lemma 2.4.22, the existence of such a procedure leads to the following conclusion.

Theorem 2.4.23. Every linear space (of $M \times N$ matrices) has an orthonormal basis.

g. Some results on the rank of a matrix partitioned into blocks of rows or columns and on the rank and row or column space of a sum of matrices

A basic result on the rank of a matrix that has been partitioned into two blocks of rows or columns is as follows.

Lemma 2.4.24. For any $M \times N$ matrix $\mathbf{A}$, $M \times P$ matrix $\mathbf{B}$, and $Q \times N$ matrix $\mathbf{C}$,

$$\text{rank}\,(\mathbf{A}, \mathbf{B}) \le \text{rank}(\mathbf{A}) + \text{rank}(\mathbf{B}) \tag{4.11}$$

and

$$\text{rank}\begin{pmatrix}\mathbf{A}\\ \mathbf{C}\end{pmatrix} \le \text{rank}(\mathbf{A}) + \text{rank}(\mathbf{C}). \tag{4.12}$$

Proof. Let $R = \text{rank}(\mathbf{A})$ and $S = \text{rank}(\mathbf{B})$. Then, there exist R M-dimensional column vectors, say $\mathbf{x}_1, \mathbf{x}_2, \ldots, \mathbf{x}_R$, that form a basis for $\mathcal{C}(\mathbf{A})$ and S M-dimensional column vectors, say $\mathbf{y}_1, \mathbf{y}_2, \ldots, \mathbf{y}_S$ that form a basis for $\mathcal{C}(\mathbf{B})$. Clearly, any vector in the column space of the partitioned matrix $(\mathbf{A}, \mathbf{B})$ is expressible in the form $\mathbf{A}\boldsymbol{\ell}_1 + \mathbf{B}\boldsymbol{\ell}_2$ for some N-dimensional column vector $\boldsymbol{\ell}_1$ and some P-dimensional column vector $\boldsymbol{\ell}_2$. Moreover, $\mathbf{A}\boldsymbol{\ell}_1$ is expressible as a linear combination of $\mathbf{x}_1, \mathbf{x}_2, \ldots, \mathbf{x}_R$ and $\mathbf{B}\boldsymbol{\ell}_2$ as a linear combination of $\mathbf{y}_1, \mathbf{y}_2, \ldots, \mathbf{y}_S$, so that $\mathbf{A}\boldsymbol{\ell}_1 + \mathbf{B}\boldsymbol{\ell}_2$ is expressible as a linear combination of $\mathbf{x}_1, \mathbf{x}_2, \ldots, \mathbf{x}_R, \mathbf{y}_1, \mathbf{y}_2, \ldots, \mathbf{y}_S$. Thus, adopting an abbreviated notation in which $\mathcal{C}(\mathbf{A}, \mathbf{B})$ is written for the column space $\mathcal{C}[(\mathbf{A}, \mathbf{B})]$ of the partitioned matrix $(\mathbf{A}, \mathbf{B})$, $\mathcal{C}(\mathbf{A}, \mathbf{B})$ is spanned by the set $\{\mathbf{x}_1, \mathbf{x}_2, \ldots, \mathbf{x}_R, \mathbf{y}_1, \mathbf{y}_2, \ldots, \mathbf{y}_S\}$, implying (in light of Theorem 2.4.9) that

$$\text{rank}\,(\mathbf{A}, \mathbf{B}) = \dim\,\mathcal{C}(\mathbf{A}, \mathbf{B}) \le R + S,$$

which establishes inequality (4.11). Inequality (4.12) can be established via an analogous argument. Q.E.D.

Upon the repeated application of result (4.11), we obtain the more general result that for any matrices $\mathbf{A}_1, \mathbf{A}_2, \ldots, \mathbf{A}_K$ having M rows,

$$\operatorname{rank}(\mathbf{A}_1, \mathbf{A}_2, \ldots, \mathbf{A}_K) \leq \operatorname{rank}(\mathbf{A}_1) + \operatorname{rank}(\mathbf{A}_2) + \cdots + \operatorname{rank}(\mathbf{A}_K). \tag{4.13}$$

And, similarly, upon the repeated application of result (4.12), we find that for any matrices $\mathbf{A}_1, \mathbf{A}_2, \ldots, \mathbf{A}_K$ having N columns,

$$\operatorname{rank}\begin{pmatrix} \mathbf{A}_1 \\ \mathbf{A}_2 \\ \vdots \\ \mathbf{A}_K \end{pmatrix} \leq \operatorname{rank}(\mathbf{A}_1) + \operatorname{rank}(\mathbf{A}_2) + \cdots + \operatorname{rank}(\mathbf{A}_K). \tag{4.14}$$

Every linear space of $M \times N$ matrices contains the $M \times N$ null matrix $\mathbf{0}$. When the intersection $\mathcal{U} \cap \mathcal{V}$ of two linear spaces $\mathcal{U}$ and $\mathcal{V}$ of $M \times N$ matrices contains no matrices other than the $M \times N$ null matrix, $\mathcal{U}$ and $\mathcal{V}$ are said to be *essentially disjoint*. The following theorem gives a necessary and sufficient condition for equality to hold in inequality (4.11) or (4.12) of Lemma 2.4.24.

Theorem 2.4.25. Let $\mathbf{A}$ represent an $M \times N$ matrix, $\mathbf{B}$ an $M \times P$ matrix, and $\mathbf{C}$ a $Q \times N$ matrix. Then,

$$\operatorname{rank}(\mathbf{A}, \mathbf{B}) = \operatorname{rank}(\mathbf{A}) + \operatorname{rank}(\mathbf{B}) \tag{4.15}$$

if and only if $\mathcal{C}(\mathbf{A})$ and $\mathcal{C}(\mathbf{B})$ are essentially disjoint, and, similarly,

$$\operatorname{rank}\begin{pmatrix} \mathbf{A} \\ \mathbf{C} \end{pmatrix} = \operatorname{rank}(\mathbf{A}) + \operatorname{rank}(\mathbf{C}) \tag{4.16}$$

if and only if $\mathcal{R}(\mathbf{A})$ and $\mathcal{R}(\mathbf{C})$ are essentially disjoint.

Proof. Let $R = \operatorname{rank} \mathbf{A}$ and $S = \operatorname{rank} \mathbf{B}$. And take $\mathbf{a}_1, \mathbf{a}_2, \ldots, \mathbf{a}_R$ to be any R linearly independent columns of $\mathbf{A}$ and $\mathbf{b}_1, \mathbf{b}_2, \ldots, \mathbf{b}_S$ any S linearly independent columns of $\mathbf{B}$—their existence follows from Theorem 2.4.19. Clearly,

$$\mathcal{C}(\mathbf{A}, \mathbf{B}) = \operatorname{sp}(\mathbf{a}_1, \mathbf{a}_2, \ldots, \mathbf{a}_R, \mathbf{b}_1, \mathbf{b}_2, \ldots, \mathbf{b}_S).$$

Thus, to establish the first part of Theorem 2.4.25, it suffices to show that the set $\{\mathbf{a}_1, \mathbf{a}_2, \ldots, \mathbf{a}_R, \mathbf{b}_1, \mathbf{b}_2, \ldots, \mathbf{b}_S\}$ is linearly independent if and only if $\mathcal{C}(\mathbf{A})$ and $\mathcal{C}(\mathbf{B})$ are essentially disjoint.

Accordingly, suppose that $\mathcal{C}(\mathbf{A})$ and $\mathcal{C}(\mathbf{B})$ are essentially disjoint. Then, for any scalars $c_1, c_2, \ldots, c_R$ and $k_1, k_2, \ldots, k_S$ such that $\sum_{i=1}^R c_i \mathbf{a}_i + \sum_{j=1}^S k_j \mathbf{b}_j = \mathbf{0}$, we have that $\sum_{i=1}^R c_i \mathbf{a}_i = -\sum_{j=1}^S k_j \mathbf{b}_j$, implying (in light of the essential disjointness) that $\sum_{i=1}^R c_i \mathbf{a}_i = \mathbf{0}$ and $\sum_{j=1}^S k_j \mathbf{b}_j = \mathbf{0}$ and hence that $c_1 = c_2 = \cdots = c_R = 0$ and $k_1 = k_2 = \cdots = k_S = 0$. And we conclude that the set $\{\mathbf{a}_1, \mathbf{a}_2, \ldots, \mathbf{a}_R, \mathbf{b}_1, \mathbf{b}_2, \ldots, \mathbf{b}_S\}$ is linearly independent.

Conversely, if $\mathcal{C}(\mathbf{A})$ and $\mathcal{C}(\mathbf{B})$ were essentially disjoint, there would exist scalars $c_1, c_2, \ldots, c_R$ and $k_1, k_2, \ldots, k_S$, not all of which are 0, such that $\sum_{i=1}^R c_i \mathbf{a}_i = \sum_{j=1}^S k_j \mathbf{b}_j$ or, equivalently, such that $\sum_{i=1}^R c_i \mathbf{a}_i + \sum_{j=1}^S (-k_j) \mathbf{b}_j = \mathbf{0}$, in which case the set $\{\mathbf{a}_1, \mathbf{a}_2, \ldots, \mathbf{a}_R, \mathbf{b}_1, \mathbf{b}_2, \ldots, \mathbf{b}_S\}$ would be linearly dependent.

The second part of Theorem 2.4.25 can be proved in similar fashion. Q.E.D.

Suppose that the $M \times N$ matrix $\mathbf{A}$ and the $M \times P$ matrix $\mathbf{B}$ are such that $\mathbf{A}'\mathbf{B} = \mathbf{0}$. Then, for any $N \times 1$ vector $\mathbf{k}$ and any $P \times 1$ vector $\boldsymbol{\ell}$ such that $\mathbf{A}\mathbf{k} = \mathbf{B}\boldsymbol{\ell}$,

$$\mathbf{A}'\mathbf{A}\mathbf{k} = \mathbf{A}'\mathbf{B}\boldsymbol{\ell} = \mathbf{0},$$

implying (in light of Corollary 2.3.4) that $\mathbf{A}\mathbf{k} = \mathbf{0}$ and leading to the conclusion that $\mathcal{C}(\mathbf{A})$ and $\mathcal{C}(\mathbf{B})$ are essentially disjoint. Similarly, if $\mathbf{A}\mathbf{C}' = \mathbf{0}$, then $\mathcal{R}(\mathbf{A})$ and $\mathcal{R}(\mathbf{C})$ are essentially disjoint. Thus, we have the following corollary of Theorem 2.4.25.

Corollary 2.4.26. Let $\mathbf{A}$ represent an $M \times N$ matrix. Then, for any $M \times P$ matrix $\mathbf{B}$ such that $\mathbf{A}'\mathbf{B} = \mathbf{0}$,

$$\text{rank}\,(\mathbf{A},\ \mathbf{B}) = \text{rank}(\mathbf{A}) + \text{rank}(\mathbf{B}).$$

And for any $Q \times N$ matrix $\mathbf{C}$ such that $\mathbf{AC}' = \mathbf{0}$,

$$\text{rank}\begin{pmatrix}\mathbf{A}\\ \mathbf{C}\end{pmatrix} = \text{rank}(\mathbf{A}) + \text{rank}(\mathbf{C}).$$

For any two $M \times N$ matrices $\mathbf{A}$ and $\mathbf{B}$,

$$\mathbf{A} + \mathbf{B} = (\mathbf{A},\ \mathbf{B})\begin{pmatrix}\mathbf{I}\\ \mathbf{I}\end{pmatrix}.$$

Thus, in light of Corollaries 2.4.4 and 2.4.17 and Lemma 2.4.24, we have the following lemma and corollary.

Lemma 2.4.27. For any two $M \times N$ matrices $\mathbf{A}$ and $\mathbf{B}$,

$$\mathcal{C}(\mathbf{A} + \mathbf{B}) \subset \mathcal{C}(\mathbf{A},\ \mathbf{B}), \qquad \text{rank}(\mathbf{A} + \mathbf{B}) \le \text{rank}\,(\mathbf{A},\ \mathbf{B}), \tag{4.17}$$

$$\mathcal{R}(\mathbf{A} + \mathbf{B}) \subset \mathcal{R}\begin{pmatrix}\mathbf{A}\\ \mathbf{B}\end{pmatrix}, \qquad \text{rank}(\mathbf{A} + \mathbf{B}) \le \text{rank}\begin{pmatrix}\mathbf{A}\\ \mathbf{B}\end{pmatrix}. \tag{4.18}$$

Corollary 2.4.28. For any two $M \times N$ matrices $\mathbf{A}$ and $\mathbf{B}$,

$$\text{rank}(\mathbf{A} + \mathbf{B}) \le \text{rank}(\mathbf{A}) + \text{rank}(\mathbf{B}). \tag{4.19}$$

Upon the repeated application of results (4.17), (4.18), and (4.19), we obtain the more general results that for any K $M \times N$ matrices $\mathbf{A}_1, \mathbf{A}_2, \ldots, \mathbf{A}_K$,

$$\mathcal{C}(\mathbf{A}_1 + \mathbf{A}_2 + \cdots + \mathbf{A}_K) \subset \mathcal{C}(\mathbf{A}_1, \mathbf{A}_2, \ldots, \mathbf{A}_K), \tag{4.20}$$

$$\text{rank}(\mathbf{A}_1 + \mathbf{A}_2 + \cdots + \mathbf{A}_K) \le \text{rank}\,(\mathbf{A}_1, \mathbf{A}_2, \ldots, \mathbf{A}_K), \tag{4.21}$$

$$\mathcal{R}(\mathbf{A}_1 + \mathbf{A}_2 + \cdots + \mathbf{A}_K) \subset \mathcal{R}\begin{pmatrix}\mathbf{A}_1\\ \mathbf{A}_2\\ \vdots\\ \mathbf{A}_K\end{pmatrix}, \tag{4.22}$$

$$\text{rank}(\mathbf{A}_1 + \mathbf{A}_2 + \cdots + \mathbf{A}_K) \le \text{rank}\begin{pmatrix}\mathbf{A}_1\\ \mathbf{A}_2\\ \vdots\\ \mathbf{A}_K\end{pmatrix}, \tag{4.23}$$

and

$$\text{rank}(\mathbf{A}_1 + \mathbf{A}_2 + \cdots + \mathbf{A}_K) \le \text{rank}(\mathbf{A}_1) + \text{rank}(\mathbf{A}_2) + \text{rank}(\mathbf{A}_K). \tag{4.24}$$

2.5 Inverse Matrices

A *right inverse* of an $M \times N$ matrix $\mathbf{A}$ is an $N \times M$ matrix $\mathbf{R}$ such that $\mathbf{AR} = \mathbf{I}_M$. Similarly, a *left inverse* of an $M \times N$ matrix $\mathbf{A}$ is an $N \times M$ matrix $\mathbf{L}$ such that $\mathbf{LA} = \mathbf{I}_N$ (or, equivalently, such that $\mathbf{A}'\mathbf{L}' = \mathbf{I}_N$). A matrix may or may not have a right or left inverse, as indicated by the following lemma.

Lemma 2.5.1. An $M \times N$ matrix $\mathbf{A}$ has a right inverse if and only if $\text{rank}(\mathbf{A}) = M$ (i.e., if and only if $\mathbf{A}$ has full row rank) and has a left inverse if and only if $\text{rank}(\mathbf{A}) = N$ (i.e., if and only if $\mathbf{A}$ has full column rank).

Proof. If $\text{rank}(\mathbf{A}) = M$, then $\mathcal{C}(\mathbf{A}) = \mathcal{C}(\mathbf{I}_M)$ [as is evident from Theorem 2.4.16 upon observing that $\mathcal{C}(\mathbf{A}) \subset \mathcal{R}^M = \mathcal{C}(\mathbf{I}_M)$], implying (in light of Lemma 2.4.3) that there exists a matrix $\mathbf{R}$ such that $\mathbf{AR} = \mathbf{I}_M$ (i.e., that $\mathbf{A}$ has a right inverse). Conversely, if there exists a matrix $\mathbf{R}$ such that $\mathbf{AR} = \mathbf{I}_M$, then

$$\text{rank}(\mathbf{A}) \geq \text{rank}(\mathbf{AR}) = \text{rank}(\mathbf{I}_M) = M,$$

implying [since, according to Lemma 2.4.15, $\text{rank}(\mathbf{A}) \leq M$] that $\text{rank}(\mathbf{A}) = M$. That $\mathbf{A}$ has a left inverse if and only if $\text{rank}(\mathbf{A}) = N$ is evident upon observing that $\mathbf{A}$ has a left inverse if and only if $\mathbf{A}'$ has a right inverse [and recalling that $\text{rank}(\mathbf{A}') = \text{rank}(\mathbf{A})$]. Q.E.D.

As an almost immediate consequence of Lemma 2.5.1, we have the following corollary.

Corollary 2.5.2. A matrix $\mathbf{A}$ has both a right inverse and a left inverse if and only if $\mathbf{A}$ is a (square) nonsingular matrix.

If there exists a matrix $\mathbf{B}$ that is both a right and left inverse of a matrix $\mathbf{A}$ (so that $\mathbf{AB} = \mathbf{I}$ and $\mathbf{BA} = \mathbf{I}$), then $\mathbf{A}$ is said to be *invertible* and $\mathbf{B}$ is referred to as an *inverse* of $\mathbf{A}$. Only a (square) nonsingular matrix can be invertible, as is evident from Corollary 2.5.2.

The following lemma and theorem include some basic results on the existence and uniqueness of inverse matrices.

Lemma 2.5.3. If a square matrix $\mathbf{A}$ has a right or left inverse $\mathbf{B}$, then $\mathbf{A}$ is nonsingular and $\mathbf{B}$ is an inverse of $\mathbf{A}$.

Proof. Suppose that $\mathbf{A}$ has a right inverse $\mathbf{R}$. Then, it follows from Lemma 2.5.1 that $\mathbf{A}$ is nonsingular and further that $\mathbf{A}$ has a left inverse $\mathbf{L}$. Observing that

$$\mathbf{L} = \mathbf{LI} = \mathbf{LAR} = \mathbf{IR} = \mathbf{R}$$

and hence that $\mathbf{RA} = \mathbf{LA} = \mathbf{I}$, we conclude that $\mathbf{R}$ is an inverse of $\mathbf{A}$.

A similar argument can be used to show that if $\mathbf{A}$ has a left inverse $\mathbf{L}$, then $\mathbf{A}$ is nonsingular and $\mathbf{L}$ is an inverse of $\mathbf{A}$. Q.E.D.

Theorem 2.5.4. A matrix is invertible if and only if it is a (square) nonsingular matrix. Further, any nonsingular matrix has a unique inverse $\mathbf{B}$ and has no right or left inverse other than $\mathbf{B}$.

Proof. Suppose that $\mathbf{A}$ is a nonsingular matrix. Then, it follows from Lemma 2.5.1 that $\mathbf{A}$ has a right inverse $\mathbf{B}$ and from Lemma 2.5.3 that $\mathbf{B}$ is an inverse of $\mathbf{A}$. Thus, $\mathbf{A}$ is invertible. Moreover, for any inverse $\mathbf{C}$ of $\mathbf{A}$, we find that

$$\mathbf{C} = \mathbf{CI} = \mathbf{CAB} = \mathbf{IB} = \mathbf{B},$$

implying that $\mathbf{A}$ has a unique inverse and further—in light of Lemma 2.5.3—that $\mathbf{A}$ has no right or left inverse other than $\mathbf{B}$.

That any invertible matrix is nonsingular is (as noted earlier) evident from Corollary 2.5.2. Q.E.D.

The symbol $\mathbf{A}^{-1}$ is used to denote the inverse of a nonsingular matrix $\mathbf{A}$. By definition,

$$\mathbf{AA}^{-1} = \mathbf{A}^{-1}\mathbf{A} = \mathbf{I}.$$

A 1×1 matrix $\mathbf{A} = (a_{11})$ is invertible if and only if its element a_{11} is nonzero, in which case

$$\mathbf{A}^{-1} = (1/a_{11}).$$

For a 2×2 matrix $\mathbf{A} = \begin{pmatrix} a_{11} & a_{12} \\ a_{21} & a_{22} \end{pmatrix}$, we find that

$$\mathbf{AB} = k\mathbf{I},$$

where $\mathbf{B} = \begin{pmatrix} a_{22} & -a_{12} \\ -a_{21} & a_{11} \end{pmatrix}$ and $k = a_{11}a_{22} - a_{12}a_{21}$. If $k = 0$, then $\mathbf{AB} = \mathbf{0}$, implying that the columns of $\mathbf{A}$ are linearly dependent, in which case $\mathbf{A}$ is singular and hence not invertible. If $k \neq 0$, then $\mathbf{A}[(1/k)\mathbf{B}] = \mathbf{I}$, in which case $\mathbf{A}$ is invertible and

$$\mathbf{A}^{-1} = (1/k)\mathbf{B}. \tag{5.1}$$

a. Basic results on inverses and invertibility

For any nonsingular matrix $\mathbf{A}$ and any nonzero scalar k, $k\mathbf{A}$ is nonsingular and

$$(k\mathbf{A})^{-1} = (1/k)\mathbf{A}^{-1}, \tag{5.2}$$

as is easily verified. In the special case $k = -1$, equality (5.2) reduces to

$$(-\mathbf{A})^{-1} = -\mathbf{A}^{-1}. \tag{5.3}$$

It is easy to show that, for any nonsingular matrix $\mathbf{A}$, $\mathbf{A}'$ is nonsingular, and

$$(\mathbf{A}')^{-1} = (\mathbf{A}^{-1})'. \tag{5.4}$$

In the special case of a symmetric matrix $\mathbf{A}$, equality (5.4) reduces to

$$\mathbf{A}^{-1} = (\mathbf{A}^{-1})'. \tag{5.5}$$

Thus, the inverse of any nonsingular symmetric matrix is symmetric.

The inverse $\mathbf{A}^{-1}$ of an $N \times N$ nonsingular matrix $\mathbf{A}$ is invertible, or equivalently (in light of Theorem 2.5.4)

$$\text{rank}(\mathbf{A}^{-1}) = N, \tag{5.6}$$

and

$$(\mathbf{A}^{-1})^{-1} = \mathbf{A}, \tag{5.7}$$

that is, the inverse of $\mathbf{A}^{-1}$ is $\mathbf{A}$ (as is evident from the very definition of $\mathbf{A}^{-1}$).

For any two $N \times N$ nonsingular matrices $\mathbf{A}$ and $\mathbf{B}$,

$$\text{rank}(\mathbf{AB}) = N, \tag{5.8}$$

that is, $\mathbf{AB}$ is nonsingular, and

$$(\mathbf{AB})^{-1} = \mathbf{B}^{-1}\mathbf{A}^{-1}. \tag{5.9}$$

Results (5.8) and (5.9) can be easily verified by observing that $\mathbf{ABB}^{-1}\mathbf{A}^{-1} = \mathbf{I}$ (so that $\mathbf{B}^{-1}\mathbf{A}^{-1}$ is a right inverse of $\mathbf{AB}$) and applying Lemma 2.5.3. (If either or both of two $N \times N$ matrices $\mathbf{A}$ and $\mathbf{B}$ are singular, then their product $\mathbf{AB}$ is singular, as is evident from Corollary 2.4.17.) Repeated application of results (5.8) and (5.9) leads to the conclusion that, for any K nonsingular matrices $\mathbf{A}_1, \mathbf{A}_2, \ldots, \mathbf{A}_K$ of order N,

$$\text{rank}(\mathbf{A}_1\mathbf{A}_2 \cdots \mathbf{A}_K) = N \tag{5.10}$$

and

$$(\mathbf{A}_1\mathbf{A}_2 \cdots \mathbf{A}_K)^{-1} = \mathbf{A}_K^{-1} \cdots \mathbf{A}_2^{-1}\mathbf{A}_1^{-1}. \tag{5.11}$$

b. Some results on the ranks and row and column spaces of matrix products

The following lemma gives some basic results on the effects of premultiplication or postmultiplication by a matrix of full row or column rank.

Lemma 2.5.5. Let $\mathbf{A}$ represent an $M \times N$ matrix and $\mathbf{B}$ an $N \times P$ matrix. If $\mathbf{A}$ has full column rank, then

$$\mathcal{R}(\mathbf{AB}) = \mathcal{R}(\mathbf{B}) \quad \text{and} \quad \text{rank}(\mathbf{AB}) = \text{rank}(\mathbf{B}).$$

Similarly, if $\mathbf{B}$ has full row rank, then

$$\mathcal{C}(\mathbf{AB}) = \mathcal{C}(\mathbf{A}) \quad \text{and} \quad \text{rank}(\mathbf{AB}) = \text{rank}(\mathbf{A}).$$

Proof. It is clear from Corollary 2.4.4 that $\mathcal{R}(\mathbf{AB}) \subset \mathcal{R}(\mathbf{B})$ and $\mathbf{C}(\mathbf{AB}) \subset \mathcal{C}(\mathbf{A})$. If $\mathbf{A}$ has full column rank, then (according to Lemma 2.5.1) it has a left inverse $\mathbf{L}$, implying that

$$\mathcal{R}(\mathbf{B}) = \mathcal{R}(\mathbf{IB}) = \mathcal{R}(\mathbf{LAB}) \subset \mathcal{R}(\mathbf{AB})$$

and hence that $\mathcal{R}(\mathbf{AB}) = \mathcal{R}(\mathbf{B})$ [which implies, in turn, that $\text{rank}(\mathbf{AB}) = \text{rank}(\mathbf{B})$]. Similarly, if $\mathbf{B}$ has full row rank, then it has a right inverse $\mathbf{R}$, implying that $\mathcal{C}(\mathbf{A}) = \mathcal{C}(\mathbf{ABR}) \subset \mathcal{C}(\mathbf{AB})$ and hence that $\mathcal{C}(\mathbf{AB}) = \mathcal{C}(\mathbf{A})$ [and $\text{rank}(\mathbf{AB}) = \text{rank}(\mathbf{A})$]. Q.E.D.

As an immediate consequence of Lemma 2.5.5, we have the following corollary.

Corollary 2.5.6. If $\mathbf{A}$ is an $N \times N$ nonsingular matrix, then for any $N \times P$ matrix $\mathbf{B}$,

$$\mathcal{R}(\mathbf{AB}) = \mathcal{R}(\mathbf{B}) \quad \text{and} \quad \text{rank}(\mathbf{AB}) = \text{rank}(\mathbf{B}).$$

Similarly, if $\mathbf{B}$ is an $N \times N$ nonsingular matrix, then for any $M \times N$ matrix $\mathbf{A}$,

$$\mathcal{C}(\mathbf{AB}) = \mathcal{C}(\mathbf{A}) \quad \text{and} \quad \text{rank}(\mathbf{AB}) = \text{rank}(\mathbf{A}).$$

2.6 Ranks and Inverses of Partitioned Matrices

Expressions can be obtained for the ranks and inverses of partitioned matrices in terms of their constituent blocks. In the special case of block-diagonal and block-triangular matrices, these expressions are relatively simple.

a. Special case: block-diagonal matrices

The following lemma relates the rank of a block-diagonal matrix to the ranks of its diagonal blocks.

Lemma 2.6.1. For any $M \times N$ matrix $\mathbf{A}$ and $P \times Q$ matrix $\mathbf{B}$,

$$\text{rank}\begin{pmatrix} \mathbf{A} & \mathbf{0} \\ \mathbf{0} & \mathbf{B} \end{pmatrix} = \text{rank}(\mathbf{A}) + \text{rank}(\mathbf{B}). \tag{6.1}$$

Proof. Let $R = \text{rank}\,\mathbf{A}$ and $S = \text{rank}\,\mathbf{B}$. And suppose that $R > 0$ and $S > 0$—if R or S equals 0 (in which case $\mathbf{A} = \mathbf{0}$ or $\mathbf{B} = \mathbf{0}$), equality (6.1) is clearly valid. Then, according to Theorem 2.4.18, there exist an $M \times R$ matrix $\mathbf{A}_*$ and an $R \times N$ matrix $\mathbf{E}$ such that $\mathbf{A} = \mathbf{A}_*\mathbf{E}$, and, similarly, there exist a $P \times S$ matrix $\mathbf{B}_*$ and an $S \times Q$ matrix $\mathbf{F}$ such that $\mathbf{B} = \mathbf{B}_*\mathbf{F}$. Moreover, $\text{rank}\,\mathbf{A}_* = \text{rank}\,\mathbf{E} = R$ and $\text{rank}\,\mathbf{B}_* = \text{rank}\,\mathbf{F} = S$.

We have that

$$\begin{pmatrix} \mathbf{A} & \mathbf{0} \\ \mathbf{0} & \mathbf{B} \end{pmatrix} = \begin{pmatrix} \mathbf{A}_* & \mathbf{0} \\ \mathbf{0} & \mathbf{B}_* \end{pmatrix} \begin{pmatrix} \mathbf{E} & \mathbf{0} \\ \mathbf{0} & \mathbf{F} \end{pmatrix}.$$

Further, the columns of $\mathrm{diag}(\mathbf{A}_*, \mathbf{B}_*)$ are linearly independent, as is evident upon observing that, for any R-dimensional column vector $\mathbf{c}$ and S-dimensional column vector $\mathbf{d}$ such that

$$\begin{pmatrix} \mathbf{A}_* & \mathbf{0} \\ \mathbf{0} & \mathbf{B}_* \end{pmatrix} \begin{pmatrix} \mathbf{c} \\ \mathbf{d} \end{pmatrix} = \mathbf{0},$$

$\mathbf{A}_*\mathbf{c} = \mathbf{0}$ and $\mathbf{B}_*\mathbf{d} = \mathbf{0}$, implying (since the columns of $\mathbf{A}_*$ are linearly independent) that $\mathbf{c} = \mathbf{0}$ and likewise that $\mathbf{d} = \mathbf{0}$. Similarly, the rows of $\mathrm{diag}(\mathbf{E}, \mathbf{F})$ are linearly independent. Thus, $\mathrm{diag}(\mathbf{A}_*, \mathbf{B}_*)$ has full column rank and $\mathrm{diag}(\mathbf{E}, \mathbf{F})$ has full row rank. And, recalling Lemma 2.5.5, we conclude that

$$\mathrm{rank}\begin{pmatrix} \mathbf{A} & \mathbf{0} \\ \mathbf{0} & \mathbf{B} \end{pmatrix} = \begin{pmatrix} \mathbf{A}_* & \mathbf{0} \\ \mathbf{0} & \mathbf{B}_* \end{pmatrix} = R + S. \qquad \text{Q.E.D.}$$

Repeated application of result (6.1) gives the following formula for the rank of a block-diagonal matrix with diagonal blocks $\mathbf{A}_1, \mathbf{A}_2, \ldots, \mathbf{A}_K$:

$$\mathrm{rank}[\mathrm{diag}(\mathbf{A}_1, \mathbf{A}_2, \ldots, \mathbf{A}_K)] = \mathrm{rank}(\mathbf{A}_1) + \mathrm{rank}(\mathbf{A}_2) + \cdots + \mathrm{rank}(\mathbf{A}_K). \tag{6.2}$$

Note that result (6.2) implies in particular that the rank of a diagonal matrix $\mathbf{D}$ equals the number of nonzero diagonal elements in $\mathbf{D}$.

Let $\mathbf{T}$ represent an $M \times M$ matrix and $\mathbf{W}$ an $N \times N$ matrix. Then, the $(M + N) \times (M + N)$ block-diagonal matrix $\mathrm{diag}(\mathbf{T}, \mathbf{W})$ is nonsingular if and only if both $\mathbf{T}$ and $\mathbf{W}$ are nonsingular, as is evident from Lemma 2.6.1. Moreover, if both $\mathbf{T}$ and $\mathbf{W}$ are nonsingular, then

$$\begin{pmatrix} \mathbf{T} & \mathbf{0} \\ \mathbf{0} & \mathbf{W} \end{pmatrix}^{-1} = \begin{pmatrix} \mathbf{T}^{-1} & \mathbf{0} \\ \mathbf{0} & \mathbf{W}^{-1} \end{pmatrix}, \tag{6.3}$$

as is easily verified.

More generally, for any square matrices $\mathbf{A}_1, \mathbf{A}_2, \ldots, \mathbf{A}_K$, the block-diagonal matrix $\mathrm{diag}(\mathbf{A}_1, \mathbf{A}_2, \ldots, \mathbf{A}_K)$ is nonsingular if and only if $\mathbf{A}_1, \mathbf{A}_2, \ldots, \mathbf{A}_K$ are all nonsingular [as is evident from result (6.2)], in which case

$$[\mathrm{diag}(\mathbf{A}_1, \mathbf{A}_2, \ldots, \mathbf{A}_K)]^{-1} = \mathrm{diag}\left(\mathbf{A}_1^{-1}, \mathbf{A}_2^{-1}, \ldots, \mathbf{A}_K^{-1}\right). \tag{6.4}$$

As what is essentially a special case of this result, we have that an $N \times N$ diagonal matrix $\mathrm{diag}(d_1, d_2, \ldots, d_N)$ is nonsingular if and only if its diagonal elements $d_1, d_2, \ldots, d_N$ are all nonzero, in which case

$$[\mathrm{diag}(d_1, d_2, \ldots, d_N)]^{-1} = \mathrm{diag}(1/d_1,\ 1/d_2, \ldots,\ 1/d_N). \tag{6.5}$$

b. Special case: block-triangular matrices

Consider a block-triangular matrix of the simple form $\begin{pmatrix} \mathbf{I}_M & \mathbf{0} \\ \mathbf{V} & \mathbf{I}_N \end{pmatrix}$ or $\begin{pmatrix} \mathbf{I}_N & \mathbf{V} \\ \mathbf{0} & \mathbf{I}_M \end{pmatrix}$, where $\mathbf{V}$ is an $N \times M$ matrix. Upon recalling (from Theorem 2.5.4) that an invertible matrix is nonsingular and observing that

$$\begin{pmatrix} \mathbf{I} & \mathbf{0} \\ -\mathbf{V} & \mathbf{I} \end{pmatrix} \begin{pmatrix} \mathbf{I} & \mathbf{0} \\ \mathbf{V} & \mathbf{I} \end{pmatrix} = \begin{pmatrix} \mathbf{I} & \mathbf{0} \\ \mathbf{0} & \mathbf{I} \end{pmatrix} \quad \text{and} \quad \begin{pmatrix} \mathbf{I} & -\mathbf{V} \\ \mathbf{0} & \mathbf{I} \end{pmatrix} \begin{pmatrix} \mathbf{I} & \mathbf{V} \\ \mathbf{0} & \mathbf{I} \end{pmatrix} = \begin{pmatrix} \mathbf{I} & \mathbf{0} \\ \mathbf{0} & \mathbf{I} \end{pmatrix},$$

we obtain the following result.

Lemma 2.6.2. For any $N \times M$ matrix $\mathbf{V}$, the $(M+N) \times (M+N)$ partitioned matrices $\begin{pmatrix} \mathbf{I}_M & \mathbf{0} \\ \mathbf{V} & \mathbf{I}_N \end{pmatrix}$ and $\begin{pmatrix} \mathbf{I}_N & \mathbf{V} \\ \mathbf{0} & \mathbf{I}_M \end{pmatrix}$ are nonsingular, and

$$\begin{pmatrix} \mathbf{I}_M & \mathbf{0} \\ \mathbf{V} & \mathbf{I}_N \end{pmatrix}^{-1} = \begin{pmatrix} \mathbf{I}_M & \mathbf{0} \\ -\mathbf{V} & \mathbf{I}_N \end{pmatrix} \quad \text{and} \quad \begin{pmatrix} \mathbf{I}_N & \mathbf{V} \\ \mathbf{0} & \mathbf{I}_M \end{pmatrix}^{-1} = \begin{pmatrix} \mathbf{I}_N & -\mathbf{V} \\ \mathbf{0} & \mathbf{I}_M \end{pmatrix}.$$

Formula (6.1) for the rank of a block-diagonal matrix can be extended to certain block-triangular matrices, as indicated by the following lemma.

Lemma 2.6.3. Let $\mathbf{T}$ represent an $M \times P$ matrix, $\mathbf{V}$ an $N \times P$ matrix, and $\mathbf{W}$ an $N \times Q$ matrix. If $\mathbf{T}$ has full column rank or $\mathbf{W}$ has full row rank, that is, if $\text{rank}(\mathbf{T}) = P$ or $\text{rank}(\mathbf{W}) = N$, then

$$\text{rank}\begin{pmatrix} \mathbf{T} & \mathbf{0} \\ \mathbf{V} & \mathbf{W} \end{pmatrix} = \text{rank}\begin{pmatrix} \mathbf{W} & \mathbf{V} \\ \mathbf{0} & \mathbf{T} \end{pmatrix} = \text{rank}(\mathbf{T}) + \text{rank}(\mathbf{W}). \tag{6.6}$$

Proof. Suppose that $\text{rank}(\mathbf{T}) = P$. Then, according to Lemma 2.5.1, there exists a matrix $\mathbf{L}$ that is a left inverse of $\mathbf{T}$, in which case

$$\begin{pmatrix} \mathbf{I} & -\mathbf{VL} \\ \mathbf{0} & \mathbf{I} \end{pmatrix}\begin{pmatrix} \mathbf{W} & \mathbf{V} \\ \mathbf{0} & \mathbf{T} \end{pmatrix} = \begin{pmatrix} \mathbf{W} & \mathbf{0} \\ \mathbf{0} & \mathbf{T} \end{pmatrix} \text{ and } \begin{pmatrix} \mathbf{I} & \mathbf{0} \\ -\mathbf{VL} & \mathbf{I} \end{pmatrix}\begin{pmatrix} \mathbf{T} & \mathbf{0} \\ \mathbf{V} & \mathbf{W} \end{pmatrix} = \begin{pmatrix} \mathbf{T} & \mathbf{0} \\ \mathbf{0} & \mathbf{W} \end{pmatrix}.$$

Since (according to Lemma 2.6.2) $\begin{pmatrix} \mathbf{I} & -\mathbf{VL} \\ \mathbf{0} & \mathbf{I} \end{pmatrix}$ and $\begin{pmatrix} \mathbf{I} & \mathbf{0} \\ -\mathbf{VL} & \mathbf{I} \end{pmatrix}$ are nonsingular, we conclude (on the basis of Corollary 2.5.6) that

$$\text{rank}\begin{pmatrix} \mathbf{W} & \mathbf{V} \\ \mathbf{0} & \mathbf{T} \end{pmatrix} = \text{rank}\begin{pmatrix} \mathbf{W} & \mathbf{0} \\ \mathbf{0} & \mathbf{T} \end{pmatrix} \quad \text{and} \quad \text{rank}\begin{pmatrix} \mathbf{T} & \mathbf{0} \\ \mathbf{V} & \mathbf{W} \end{pmatrix} = \text{rank}\begin{pmatrix} \mathbf{T} & \mathbf{0} \\ \mathbf{0} & \mathbf{W} \end{pmatrix}$$

and hence (in light of Lemma 2.6.1) that

$$\text{rank}\begin{pmatrix} \mathbf{T} & \mathbf{0} \\ \mathbf{V} & \mathbf{W} \end{pmatrix} = \text{rank}\begin{pmatrix} \mathbf{W} & \mathbf{V} \\ \mathbf{0} & \mathbf{T} \end{pmatrix} = \text{rank}(\mathbf{T}) + \text{rank}(\mathbf{W}).$$

That result (6.6) holds if $\text{rank}(\mathbf{W}) = N$ can be established via an analogous argument. Q.E.D.

The results of Lemma 2.6.2 can be extended to additional block-triangular matrices, as detailed in the following lemma.

Lemma 2.6.4. Let $\mathbf{T}$ represent an $M \times M$ matrix, $\mathbf{V}$ an $N \times M$ matrix, and $\mathbf{W}$ an $N \times N$ matrix.

(1) The $(M+N) \times (M+N)$ partitioned matrix $\begin{pmatrix} \mathbf{T} & \mathbf{0} \\ \mathbf{V} & \mathbf{W} \end{pmatrix}$ is nonsingular if and only if both $\mathbf{T}$ and $\mathbf{W}$ are nonsingular, in which case

$$\begin{pmatrix} \mathbf{T} & \mathbf{0} \\ \mathbf{V} & \mathbf{W} \end{pmatrix}^{-1} = \begin{pmatrix} \mathbf{T}^{-1} & \mathbf{0} \\ -\mathbf{W}^{-1}\mathbf{V}\mathbf{T}^{-1} & \mathbf{W}^{-1} \end{pmatrix}. \tag{6.7}$$

(2) The $(M+N) \times (M+N)$ partitioned matrix $\begin{pmatrix} \mathbf{W} & \mathbf{V} \\ \mathbf{0} & \mathbf{T} \end{pmatrix}$ is nonsingular if and only if both $\mathbf{T}$ and $\mathbf{W}$ are nonsingular, in which case

$$\begin{pmatrix} \mathbf{W} & \mathbf{V} \\ \mathbf{0} & \mathbf{T} \end{pmatrix}^{-1} = \begin{pmatrix} \mathbf{W}^{-1} & -\mathbf{W}^{-1}\mathbf{V}\mathbf{T}^{-1} \\ \mathbf{0} & \mathbf{T}^{-1} \end{pmatrix}. \tag{6.8}$$

Proof. Suppose that $\begin{pmatrix} \mathbf{T} & \mathbf{0} \\ \mathbf{V} & \mathbf{W} \end{pmatrix}$ is nonsingular (in which case its rows are linearly independent). Then, for any M-dimensional column vector $\mathbf{c}$ such that $\mathbf{c}'\mathbf{T} = \mathbf{0}$, we find that

$$\begin{pmatrix} \mathbf{c} \\ \mathbf{0} \end{pmatrix}' \begin{pmatrix} \mathbf{T} & \mathbf{0} \\ \mathbf{V} & \mathbf{W} \end{pmatrix} = (\mathbf{c}'\mathbf{T},\ \mathbf{0}) = \mathbf{0}$$

and hence that $\mathbf{c} = \mathbf{0}$—if $\mathbf{c}$ were nonnull, the rows of $\begin{pmatrix} \mathbf{T} & \mathbf{0} \\ \mathbf{V} & \mathbf{W} \end{pmatrix}$ would be linearly dependent. Thus, the rows of $\mathbf{T}$ are linearly independent, which implies that $\mathbf{T}$ is nonsingular. That $\mathbf{W}$ is nonsingular can be established in similar fashion.

Conversely, suppose that both $\mathbf{T}$ and $\mathbf{W}$ are nonsingular. Then,

$$\begin{pmatrix} \mathbf{T} & \mathbf{0} \\ \mathbf{V} & \mathbf{W} \end{pmatrix} = \begin{pmatrix} \mathbf{I}_M & \mathbf{0} \\ \mathbf{0} & \mathbf{W} \end{pmatrix} \begin{pmatrix} \mathbf{I}_M & \mathbf{0} \\ \mathbf{W}^{-1}\mathbf{V}\mathbf{T}^{-1} & \mathbf{I}_N \end{pmatrix} \begin{pmatrix} \mathbf{T} & \mathbf{0} \\ \mathbf{0} & \mathbf{I}_N \end{pmatrix},$$

as is easily verified, and (in light of Lemmas 2.6.1 and 2.6.2 or Lemma 2.6.3) $\begin{pmatrix} \mathbf{I} & \mathbf{0} \\ \mathbf{0} & \mathbf{W} \end{pmatrix}$, $\begin{pmatrix} \mathbf{I} & \mathbf{0} \\ \mathbf{W}^{-1}\mathbf{V}\mathbf{T}^{-1} & \mathbf{I} \end{pmatrix}$, and $\begin{pmatrix} \mathbf{T} & \mathbf{0} \\ \mathbf{0} & \mathbf{I} \end{pmatrix}$ are nonsingular. Further, it follows from result (5.10) (and also from Lemma 2.6.3) that $\begin{pmatrix} \mathbf{T} & \mathbf{0} \\ \mathbf{V} & \mathbf{W} \end{pmatrix}$ is nonsingular and from results (5.11) and (6.3) and Lemma 2.6.2 that

$$\begin{aligned} \begin{pmatrix} \mathbf{T} & \mathbf{0} \\ \mathbf{V} & \mathbf{W} \end{pmatrix}^{-1} &= \begin{pmatrix} \mathbf{T} & \mathbf{0} \\ \mathbf{0} & \mathbf{I} \end{pmatrix}^{-1} \begin{pmatrix} \mathbf{I} & \mathbf{0} \\ \mathbf{W}^{-1}\mathbf{V}\mathbf{T}^{-1} & \mathbf{I} \end{pmatrix}^{-1} \begin{pmatrix} \mathbf{I} & \mathbf{0} \\ \mathbf{0} & \mathbf{W} \end{pmatrix}^{-1} \\ &= \begin{pmatrix} \mathbf{T}^{-1} & \mathbf{0} \\ \mathbf{0} & \mathbf{I} \end{pmatrix} \begin{pmatrix} \mathbf{I} & \mathbf{0} \\ -\mathbf{W}^{-1}\mathbf{V}\mathbf{T}^{-1} & \mathbf{I} \end{pmatrix} \begin{pmatrix} \mathbf{I} & \mathbf{0} \\ \mathbf{0} & \mathbf{W}^{-1} \end{pmatrix} \\ &= \begin{pmatrix} \mathbf{T}^{-1} & \mathbf{0} \\ -\mathbf{W}^{-1}\mathbf{V}\mathbf{T}^{-1} & \mathbf{W}^{-1} \end{pmatrix}. \end{aligned}$$

The proof of Part (1) is now complete. Part (2) can be proved via an analogous argument. Q.E.D.

The results of Lemma 2.6.4 can be extended (by repeated application) to block-triangular matrices having more than two rows and columns of blocks. Let $\mathbf{A}_1, \mathbf{A}_2, \ldots, \mathbf{A}_R$ represent square matrices, and take $\mathbf{A}$ to be an (upper or lower) block-triangular matrix whose diagonal blocks are respectively $\mathbf{A}_1, \mathbf{A}_2, \ldots, \mathbf{A}_R$. Then, $\mathbf{A}$ is nonsingular if and only if $\mathbf{A}_1, \mathbf{A}_2, \ldots, \mathbf{A}_R$ are all nonsingular (and, as what is essentially a special case, a triangular matrix is nonsingular if and only if its diagonal elements are all nonzero). Further, $\mathbf{A}^{-1}$ is block-triangular (lower block-triangular if $\mathbf{A}$ is lower block-triangular and upper block-triangular if $\mathbf{A}$ is upper block-triangular). The diagonal blocks of $\mathbf{A}^{-1}$ are $\mathbf{A}_1^{-1}, \mathbf{A}_2^{-1}, \ldots, \mathbf{A}_R^{-1}$, respectively, and the off-diagonal blocks of $\mathbf{A}^{-1}$ are expressible in terms of recursive formulas given by, for example, Harville (1997, sec. 8.5).

c. General case

The following theorem can (when applicable) be used to express the rank of a partitioned matrix in terms of the rank of a matrix of smaller dimensions.

Theorem 2.6.5. Let $\mathbf{T}$ represent an $M \times M$ matrix, $\mathbf{U}$ an $M \times Q$ matrix, $\mathbf{V}$ an $N \times M$ matrix, and $\mathbf{W}$ an $N \times Q$ matrix. If $\text{rank}(\mathbf{T}) = M$, that is, if $\mathbf{T}$ is nonsingular, then

$$\text{rank}\begin{pmatrix} \mathbf{T} & \mathbf{U} \\ \mathbf{V} & \mathbf{W} \end{pmatrix} = \text{rank}\begin{pmatrix} \mathbf{W} & \mathbf{V} \\ \mathbf{U} & \mathbf{T} \end{pmatrix} = M + \text{rank}(\mathbf{W} - \mathbf{V}\mathbf{T}^{-1}\mathbf{U}). \tag{6.9}$$

Proof. Suppose that $\text{rank}(\mathbf{T}) = M$. Then,

$$\begin{pmatrix} \mathbf{I}_M & \mathbf{0} \\ -\mathbf{V}\mathbf{T}^{-1} & \mathbf{I}_N \end{pmatrix} \begin{pmatrix} \mathbf{T} & \mathbf{U} \\ \mathbf{V} & \mathbf{W} \end{pmatrix} = \begin{pmatrix} \mathbf{T} & \mathbf{U} \\ \mathbf{0} & \mathbf{W} - \mathbf{V}\mathbf{T}^{-1}\mathbf{U} \end{pmatrix},$$

as is easily verified. Thus, observing that $\begin{pmatrix} \mathbf{I} & \mathbf{0} \\ -\mathbf{V}\mathbf{T}^{-1} & \mathbf{I} \end{pmatrix}$ is nonsingular (as is evident from Lemma 2.6.2) and making use of Corollary 2.5.6 and Lemma 2.6.3, we find that

$$\begin{aligned} \text{rank}\begin{pmatrix} \mathbf{T} & \mathbf{U} \\ \mathbf{V} & \mathbf{W} \end{pmatrix} &= \text{rank}\begin{pmatrix} \mathbf{T} & \mathbf{U} \\ \mathbf{0} & \mathbf{W} - \mathbf{V}\mathbf{T}^{-1}\mathbf{U} \end{pmatrix} \\ &= \text{rank}(\mathbf{T}) + \text{rank}(\mathbf{W} - \mathbf{V}\mathbf{T}^{-1}\mathbf{U}) \\ &= M + \text{rank}(\mathbf{W} - \mathbf{V}\mathbf{T}^{-1}\mathbf{U}). \end{aligned}$$

And it can be shown in similar fashion that

$$\text{rank}\begin{pmatrix} \mathbf{W} & \mathbf{V} \\ \mathbf{U} & \mathbf{T} \end{pmatrix} = M + \text{rank}(\mathbf{W} - \mathbf{V}\mathbf{T}^{-1}\mathbf{U}).$$ Q.E.D.

Results (6.7) and (6.8) on the inverse of a block-triangular matrix can be extended to a more general class of partitioned matrices, as described in the following theorem.

Theorem 2.6.6. Let $\mathbf{T}$ represent an $M \times M$ matrix, $\mathbf{U}$ an $M \times N$ matrix, $\mathbf{V}$ an $N \times M$ matrix, and $\mathbf{W}$ an $N \times N$ matrix. Suppose that $\mathbf{T}$ is nonsingular, and define

$$\mathbf{Q} = \mathbf{W} - \mathbf{V}\mathbf{T}^{-1}\mathbf{U}.$$

Then, the partitioned matrix $\begin{pmatrix} \mathbf{T} & \mathbf{U} \\ \mathbf{V} & \mathbf{W} \end{pmatrix}$ is nonsingular if and only if $\mathbf{Q}$ is nonsingular, in which case

$$\begin{pmatrix} \mathbf{T} & \mathbf{U} \\ \mathbf{V} & \mathbf{W} \end{pmatrix}^{-1} = \begin{pmatrix} \mathbf{T}^{-1} + \mathbf{T}^{-1}\mathbf{U}\mathbf{Q}^{-1}\mathbf{V}\mathbf{T}^{-1} & -\mathbf{T}^{-1}\mathbf{U}\mathbf{Q}^{-1} \\ -\mathbf{Q}^{-1}\mathbf{V}\mathbf{T}^{-1} & \mathbf{Q}^{-1} \end{pmatrix}. \tag{6.10}$$

Similarly, the partitioned matrix $\begin{pmatrix} \mathbf{W} & \mathbf{V} \\ \mathbf{U} & \mathbf{T} \end{pmatrix}$ is nonsingular if and only if $\mathbf{Q}$ is nonsingular, in which case

$$\begin{pmatrix} \mathbf{W} & \mathbf{V} \\ \mathbf{U} & \mathbf{T} \end{pmatrix}^{-1} = \begin{pmatrix} \mathbf{Q}^{-1} & -\mathbf{Q}^{-1}\mathbf{V}\mathbf{T}^{-1} \\ -\mathbf{T}^{-1}\mathbf{U}\mathbf{Q}^{-1} & \mathbf{T}^{-1} + \mathbf{T}^{-1}\mathbf{U}\mathbf{Q}^{-1}\mathbf{V}\mathbf{T}^{-1} \end{pmatrix}. \tag{6.11}$$

Proof. That $\begin{pmatrix} \mathbf{T} & \mathbf{U} \\ \mathbf{V} & \mathbf{W} \end{pmatrix}$ is nonsingular if and only if $\mathbf{Q}$ is nonsingular and that $\begin{pmatrix} \mathbf{W} & \mathbf{V} \\ \mathbf{U} & \mathbf{T} \end{pmatrix}$ is nonsingular if and only if $\mathbf{Q}$ is nonsingular are immediate consequences of Theorem 2.6.5.

Suppose now that $\mathbf{Q}$ is nonsingular, and observe that

$$\begin{pmatrix} \mathbf{T} & \mathbf{0} \\ \mathbf{V} & \mathbf{Q} \end{pmatrix} = \begin{pmatrix} \mathbf{T} & \mathbf{U} \\ \mathbf{V} & \mathbf{W} \end{pmatrix} \begin{pmatrix} \mathbf{I} & -\mathbf{T}^{-1}\mathbf{U} \\ \mathbf{0} & \mathbf{I} \end{pmatrix}. \tag{6.12}$$

Then, in light of Corollary 2.5.6 and Lemma 2.6.2, $\begin{pmatrix} \mathbf{T} & \mathbf{0} \\ \mathbf{V} & \mathbf{Q} \end{pmatrix}$, like $\begin{pmatrix} \mathbf{T} & \mathbf{U} \\ \mathbf{V} & \mathbf{W} \end{pmatrix}$, is nonsingular. Premultiplying both sides of equality (6.12) by $\begin{pmatrix} \mathbf{T} & \mathbf{U} \\ \mathbf{V} & \mathbf{W} \end{pmatrix}^{-1}$ and postmultiplying both sides by $\begin{pmatrix} \mathbf{T} & \mathbf{0} \\ \mathbf{V} & \mathbf{Q} \end{pmatrix}^{-1}$

and making use of Lemma 2.6.4, we find that

$$\begin{aligned}\begin{pmatrix}\mathbf{T} & \mathbf{U}\\ \mathbf{V} & \mathbf{W}\end{pmatrix}^{-1} &= \begin{pmatrix}\mathbf{I} & -\mathbf{T}^{-1}\mathbf{U}\\ \mathbf{0} & \mathbf{I}\end{pmatrix}\begin{pmatrix}\mathbf{T} & \mathbf{0}\\ \mathbf{V} & \mathbf{Q}\end{pmatrix}^{-1}\\ &= \begin{pmatrix}\mathbf{I} & -\mathbf{T}^{-1}\mathbf{U}\\ \mathbf{0} & \mathbf{I}\end{pmatrix}\begin{pmatrix}\mathbf{T}^{-1} & \mathbf{0}\\ -\mathbf{Q}^{-1}\mathbf{V}\mathbf{T}^{-1} & \mathbf{Q}^{-1}\end{pmatrix}\\ &= \begin{pmatrix}\mathbf{T}^{-1}+\mathbf{T}^{-1}\mathbf{U}\mathbf{Q}^{-1}\mathbf{V}\mathbf{T}^{-1} & -\mathbf{T}^{-1}\mathbf{U}\mathbf{Q}^{-1}\\ -\mathbf{Q}^{-1}\mathbf{V}\mathbf{T}^{-1} & \mathbf{Q}^{-1}\end{pmatrix},\end{aligned}$$

which establishes formula (6.10). Formula (6.11) can be derived in similar fashion. Q.E.D.

In proving Theorem 2.6.6, the approach taken to the verification of equality (6.10) was to relate the inverse of $\begin{pmatrix}\mathbf{T} & \mathbf{U}\\ \mathbf{V} & \mathbf{W}\end{pmatrix}$ to that of $\begin{pmatrix}\mathbf{T} & \mathbf{0}\\ \mathbf{V} & \mathbf{Q}\end{pmatrix}$ and to then apply formula (6.7) for the inverse of a lower block-triangular matrix. Alternatively, equality (6.10) could be verified by premultiplying or postmultiplying the right side of the equality by $\begin{pmatrix}\mathbf{T} & \mathbf{U}\\ \mathbf{V} & \mathbf{W}\end{pmatrix}$ and by confirming that the resultant product equals $\mathbf{I}_{M+N}$. And equality (6.11) could be verified in much the same way.

When $\mathbf{T}$ is nonsingular, the matrix $\mathbf{Q} = \mathbf{W} - \mathbf{V}\mathbf{T}^{-1}\mathbf{U}$, which appears in expressions (6.10) and (6.11) for the inverse of a partitioned matrix, is called the *Schur complement* of $\mathbf{T}$ in $\begin{pmatrix}\mathbf{T} & \mathbf{U}\\ \mathbf{V} & \mathbf{W}\end{pmatrix}$ or the Schur complement of $\mathbf{T}$ in $\begin{pmatrix}\mathbf{W} & \mathbf{V}\\ \mathbf{U} & \mathbf{T}\end{pmatrix}$. Moreover, when the context is clear, it is sometimes referred to simply as the Schur complement of $\mathbf{T}$ or even more simply as the Schur complement. Note that if $\begin{pmatrix}\mathbf{T} & \mathbf{U}\\ \mathbf{V} & \mathbf{W}\end{pmatrix}$ or $\begin{pmatrix}\mathbf{W} & \mathbf{V}\\ \mathbf{U} & \mathbf{T}\end{pmatrix}$ is symmetric (in which case $\mathbf{T}' = \mathbf{T}$, $\mathbf{W}' = \mathbf{W}$, and $\mathbf{V} = \mathbf{U}'$), then the Schur complement of $\mathbf{T}$ is also symmetric.

2.7 Orthogonal Matrices

A (square) matrix $\mathbf{A}$ is said to be *orthogonal* if

$$\mathbf{A}\mathbf{A}' = \mathbf{A}'\mathbf{A} = \mathbf{I},$$

or, equivalently, if $\mathbf{A}$ is nonsingular and $\mathbf{A}^{-1} = \mathbf{A}'$. To show that a (square) matrix $\mathbf{A}$ is orthogonal, it suffices (in light of Lemma 2.5.3) to demonstrate that $\mathbf{A}'\mathbf{A} = \mathbf{I}$ or, alternatively, that $\mathbf{A}\mathbf{A}' = \mathbf{I}$.

For any $N \times N$ matrix $\mathbf{A}$, $\mathbf{A}'\mathbf{A} = \mathbf{I}$ if and only if the columns $\mathbf{a}_1, \mathbf{a}_2, \ldots, \mathbf{a}_N$ of $\mathbf{A}$ are such that

$$\mathbf{a}_i'\mathbf{a}_j = \begin{cases} 1, & \text{for } j = i = 1, 2, \ldots, N, \\ 0, & \text{for } j \neq i = 1, 2, \ldots, N \end{cases} \tag{7.1}$$

(as is evident upon observing that $\mathbf{a}_i'\mathbf{a}_j$ equals the ijth element of $\mathbf{A}'\mathbf{A}$). Thus, a square matrix is orthogonal if and only if its columns form an orthonormal (with respect to the usual inner product) set of vectors. Similarly, a square matrix is orthogonal if and only if its rows form an orthonormal set of vectors.

Note that if $\mathbf{A}$ is an orthogonal matrix, then its transpose $\mathbf{A}'$ is also orthogonal. Note also that in using the term orthogonal in connection with one or more matrices, say $\mathbf{A}_1, \mathbf{A}_2, \ldots, \mathbf{A}_K$, care must be exercised to avoid confusion. Under a strict interpretation, saying that $\mathbf{A}_1, \mathbf{A}_2, \ldots, \mathbf{A}_K$ are orthogonal matrices has an entirely different meaning than saying that the set $\{\mathbf{A}_1, \mathbf{A}_2, \ldots, \mathbf{A}_K\}$, whose members are $\mathbf{A}_1, \mathbf{A}_2, \ldots, \mathbf{A}_K$, is an orthogonal set.

If $\mathbf{P}$ and $\mathbf{Q}$ are both $N \times N$ orthogonal matrices, then [in light of result (1.13)]

$$(\mathbf{PQ})'\mathbf{PQ} = \mathbf{Q}'\mathbf{P}'\mathbf{PQ} = \mathbf{Q}'\mathbf{IQ} = \mathbf{Q}'\mathbf{Q} = \mathbf{I}.$$

Thus, the product of two $N \times N$ orthogonal matrices is another $(N \times N)$ orthogonal matrix. The repeated application of this result leads to the following lemma.

Lemma 2.7.1. If each of the matrices $\mathbf{Q}_1, \mathbf{Q}_2, \ldots, \mathbf{Q}_K$ is an $N \times N$ orthogonal matrix, then the product $\mathbf{Q}_1\mathbf{Q}_2 \cdots \mathbf{Q}_K$ is an $(N \times N)$ orthogonal matrix.

2.8 Idempotent Matrices

A (square) matrix $\mathbf{A}$ is said to be *idempotent* if $\mathbf{A}^2 = \mathbf{A}$—for any (square) matrix $\mathbf{A}$, $\mathbf{A}^2$ represents the product $\mathbf{AA}$ (and, for $k = 3, 4, \ldots$, $\mathbf{A}^k$ represents the product defined recursively by $\mathbf{A}^k = \mathbf{AA}^{k-1}$). Examples of $N \times N$ idempotent matrices are the identity matrix $\mathbf{I}_N$, the $N \times N$ null matrix $\mathbf{0}$, and the matrix $(1/N)\mathbf{1}_N\mathbf{1}_N'$, each element of which equals $1/N$.

If a square matrix $\mathbf{A}$ is idempotent, then

$$(\mathbf{A}')^2 = (\mathbf{AA})' = \mathbf{A}'$$

and

$$(\mathbf{I} - \mathbf{A})^2 = \mathbf{I} - 2\mathbf{A} + \mathbf{A}^2 = \mathbf{I} - 2\mathbf{A} + \mathbf{A} = \mathbf{I} - \mathbf{A}.$$

Thus, upon observing that $\mathbf{A} = (\mathbf{A}')'$ and $\mathbf{A} = \mathbf{I} - (\mathbf{I} - \mathbf{A})$, we have the following lemma.

Lemma 2.8.1. Let $\mathbf{A}$ represent a square matrix. Then, (1) $\mathbf{A}'$ is idempotent if and only if $\mathbf{A}$ is idempotent, and (2) $\mathbf{I} - \mathbf{A}$ is idempotent if and only if $\mathbf{A}$ is idempotent.

Now, suppose that $\mathbf{A}$ is a square matrix such that

$$\mathbf{A}^2 = k\mathbf{A}$$

for some nonzero scalar k. Or, equivalently, suppose that

$$[(1/k)\mathbf{A}]^2 = (1/k)\mathbf{A},$$

that is, suppose that $(1/k)\mathbf{A}$ is an idempotent matrix (so that, depending on whether $k = 1$ or $k \neq 1$, $\mathbf{A}$ is either an idempotent matrix or a scalar multiple of an idempotent matrix). Then, in light of the following theorem,

$$\text{rank}(\mathbf{A}) = (1/k)\,\text{tr}(\mathbf{A}), \tag{8.1}$$

and, consequently, the rank of $\mathbf{A}$ is determinable from the trace of $\mathbf{A}$.

Theorem 2.8.2. For any square matrix $\mathbf{A}$ such that $\mathbf{A}^2 = k\mathbf{A}$ for some scalar k,

$$\text{tr}(\mathbf{A}) = k\,\text{rank}(\mathbf{A}). \tag{8.2}$$

Proof. Let us restrict attention to the case where $\mathbf{A}$ is nonnull. [The case where $\mathbf{A} = \mathbf{0}$ is trivial—if $\mathbf{A} = \mathbf{0}$, then $\text{tr}(\mathbf{A}) = 0 = k\,\text{rank}(\mathbf{A})$.]

Let N denote the order of $\mathbf{A}$, and let $R = \text{rank}(\mathbf{A})$. Then, according to Theorem 2.4.18, there exist an $N \times R$ matrix $\mathbf{B}$ and an $R \times N$ matrix $\mathbf{T}$ such that $\mathbf{A} = \mathbf{BT}$. Moreover, $\text{rank}(\mathbf{B}) = \text{rank}(\mathbf{T}) = R$, implying (in light of Lemma 2.5.1) the existence of a matrix $\mathbf{L}$ such that $\mathbf{LB} = \mathbf{I}_R$ and a matrix $\mathbf{H}$ such that $\mathbf{TH} = \mathbf{I}_R$—$\mathbf{L}$ is a left inverse of $\mathbf{B}$ and $\mathbf{H}$ a right inverse of $\mathbf{T}$.

We have that

$$\mathbf{BTBT} = \mathbf{A}^2 = k\mathbf{A} = k\mathbf{BT} = \mathbf{B}(k\mathbf{I}_R)\mathbf{T}.$$

Thus,

$$\mathbf{TB} = \mathbf{I}_R\mathbf{TBI}_R = \mathbf{LBTBTH} = \mathbf{L}(\mathbf{BTBT})\mathbf{H} = \mathbf{LB}(k\mathbf{I}_R)\mathbf{TH} = \mathbf{I}_R(k\mathbf{I}_R)\mathbf{I}_R = k\mathbf{I}_R.$$

And making use of Lemma 2.3.1, we find that

$$\mathrm{tr}(\mathbf{A}) = \mathrm{tr}(\mathbf{BT}) = \mathrm{tr}(\mathbf{TB}) = \mathrm{tr}(k\mathbf{I}_R) = k\,\mathrm{tr}(\mathbf{I}_R) = kR. \qquad \text{Q.E.D.}$$

In the special case where $k = 1$, Theorem 2.8.2 can be restated as follows.

Corollary 2.8.3. For any idempotent matrix $\mathbf{A}$,

$$\mathrm{rank}(\mathbf{A}) = \mathrm{tr}(\mathbf{A}). \tag{8.3}$$

By making use of Lemma 2.8.1 and Corollary 2.8.3, we find that for any $N \times N$ idempotent matrix $\mathbf{A}$,

$$\mathrm{rank}(\mathbf{I} - \mathbf{A}) = \mathrm{tr}(\mathbf{I} - \mathbf{A}) = \mathrm{tr}(\mathbf{I}_N) - \mathrm{tr}(\mathbf{A}) = N - \mathrm{rank}(\mathbf{A}),$$

thereby establishing the following, additional result.

Lemma 2.8.4. For any $N \times N$ idempotent matrix $\mathbf{A}$,

$$\mathrm{rank}(\mathbf{I} - \mathbf{A}) = \mathrm{tr}(\mathbf{I} - \mathbf{A}) = N - \mathrm{rank}(\mathbf{A}). \tag{8.4}$$

2.9 Linear Systems

Consider a set of M equations of the general form

$$\begin{aligned} a_{11}x_1 + a_{12}x_2 + \cdots + a_{1N}x_N &= b_1 \\ a_{21}x_1 + a_{22}x_2 + \cdots + a_{2N}x_N &= b_2 \\ &\vdots \\ a_{M1}x_1 + a_{M2}x_2 + \cdots + a_{MN}x_N &= b_M, \end{aligned}$$

where $a_{11}, a_{12}, \ldots, a_{1N}, a_{21}, a_{22}, \ldots, a_{2N}, \ldots, a_{M1}, a_{M2}, \ldots, a_{MN}$ and $b_1, b_2, \ldots, b_M$ represent "fixed" scalars and $x_1, x_2, \ldots, x_N$ are scalar-valued unknowns or variables. The left side of each of these equations is a linear combination of the unknowns $x_1, x_2, \ldots, x_N$. Collectively, these equations are called a *system of linear equations* (in unknowns $x_1, x_2, \ldots, x_N$) or simply a *linear system* (in $x_1, x_2, \ldots, x_N$).

The linear system can be rewritten in matrix form as

$$\mathbf{Ax} = \mathbf{b}, \tag{9.1}$$

where $\mathbf{A}$ is the $M \times N$ matrix whose ijth element is a_{ij} $(i = 1, 2, \ldots, M;\ j = 1, 2, \ldots, N)$, $\mathbf{b} = (b_1, b_2, \ldots, b_M)'$, and $\mathbf{x} = (x_1, x_2, \ldots, x_N)'$. The matrix $\mathbf{A}$ is called the *coefficient matrix* of the linear system, and $\mathbf{b}$ is called the *right side* (or right-hand side). Any value of the vector $\mathbf{x}$ of unknowns that satisfies $\mathbf{Ax} = \mathbf{b}$ is called a *solution* to the linear system, and the process of finding a solution (when one exists) is called solving the linear system.

There may be occasion to solve the linear system for more than one right side, that is, to solve each of P linear systems

$$\mathbf{Ax}_k = \mathbf{b}_k \quad (k = 1, 2, \ldots, P) \tag{9.2}$$

(in vectors $\mathbf{x}_1, \mathbf{x}_2, \ldots, \mathbf{x}_P$, respectively, of unknowns) that have the same coefficient matrix $\mathbf{A}$ but right sides $\mathbf{b}_1, \mathbf{b}_2, \ldots, \mathbf{b}_P$ that may differ. By forming an $N \times P$ matrix $\mathbf{X}$ whose first, second, ..., Pth columns are $\mathbf{x}_1, \mathbf{x}_2, \ldots, \mathbf{x}_P$, respectively, and an $M \times P$ matrix $\mathbf{B}$ whose first, second, ..., Pth columns are $\mathbf{b}_1, \mathbf{b}_2, \ldots, \mathbf{b}_P$, respectively, the P linear systems (9.2) can be rewritten collectively as

$$\mathbf{AX} = \mathbf{B}. \tag{9.3}$$

As in the special case (9.1) where $P = 1$, $\mathbf{AX} = \mathbf{B}$ is called a *linear system* (in $\mathbf{X}$), $\mathbf{A}$ and $\mathbf{B}$ are called the *coefficient matrix* and the *right side*, respectively, and any value of $\mathbf{X}$ that satisfies $\mathbf{AX} = \mathbf{B}$ is called a *solution*.

In the special case $\mathbf{AX} = \mathbf{0}$, where the right side $\mathbf{B}$ of linear system (9.3) is a null matrix, the linear system is said to be *homogeneous*. If $\mathbf{B}$ is nonnull, linear system (9.3) is said to be *nonhomogeneous*.

a. Consistency

A linear system is said to be *consistent* if it has one or more solutions. If no solution exists, the linear system is said to be *inconsistent*.

Every homogeneous linear system is consistent—one solution to a homogeneous linear system is the null matrix (of appropriate dimensions). A nonhomogeneous linear system may be either consistent or inconsistent.

Let us determine the characteristics that distinguish the coefficient matrix and right side of a consistent linear system from those of an inconsistent linear system. In doing so, let us adopt an abbreviated notation for the column space of a partitioned matrix of the form $(\mathbf{A}, \mathbf{B})$ by writing $\mathcal{C}(\mathbf{A}, \mathbf{B})$ for $\mathcal{C}[(\mathbf{A}, \mathbf{B})]$. And as a preliminary step, let us establish the following lemma.

Lemma 2.9.1. Let $\mathbf{A}$ represent an $M \times N$ matrix and $\mathbf{B}$ an $M \times P$ matrix. Then,

$$\mathcal{C}(\mathbf{A}, \mathbf{B}) = \mathcal{C}(\mathbf{A}) \quad \Leftrightarrow \quad \mathcal{C}(\mathbf{B}) \subset \mathcal{C}(\mathbf{A}); \tag{9.4}$$

$$\text{rank}(\mathbf{A}, \mathbf{B}) = \text{rank}(\mathbf{A}) \quad \Leftrightarrow \quad \mathcal{C}(\mathbf{B}) \subset \mathcal{C}(\mathbf{A}). \tag{9.5}$$

Proof. We have that $\mathcal{C}(\mathbf{A}) \subset \mathcal{C}(\mathbf{A}, \mathbf{B})$ and $\mathcal{C}(\mathbf{B}) \subset \mathcal{C}(\mathbf{A}, \mathbf{B})$, as is evident from Corollary 2.4.4 upon observing that $\mathbf{A} = (\mathbf{A}, \mathbf{B}) \begin{pmatrix} \mathbf{I} \\ \mathbf{0} \end{pmatrix}$ and $\mathbf{B} = (\mathbf{A}, \mathbf{B}) \begin{pmatrix} \mathbf{0} \\ \mathbf{I} \end{pmatrix}$.

Now, suppose that $\mathcal{C}(\mathbf{B}) \subset \mathcal{C}(\mathbf{A})$. Then, according to Lemma 2.4.3, there exists a matrix $\mathbf{F}$ such that $\mathbf{B} = \mathbf{AF}$ and hence such that $(\mathbf{A}, \mathbf{B}) = \mathbf{A}(\mathbf{I}, \mathbf{F})$. Thus, $\mathcal{C}(\mathbf{A}, \mathbf{B}) \subset \mathcal{C}(\mathbf{A})$, implying [since $\mathcal{C}(\mathbf{A}) \subset \mathcal{C}(\mathbf{A}, \mathbf{B})$] that $\mathcal{C}(\mathbf{A}, \mathbf{B}) = \mathcal{C}(\mathbf{A})$.

Conversely, suppose that $\mathcal{C}(\mathbf{A}, \mathbf{B}) = \mathcal{C}(\mathbf{A})$. Then, since $\mathcal{C}(\mathbf{B}) \subset \mathcal{C}(\mathbf{A}, \mathbf{B})$, $\mathcal{C}(\mathbf{B}) \subset \mathcal{C}(\mathbf{A})$, and the proof of result (9.4) is complete.

To prove result (9.5), it suffices [having established result (9.4)] to show that

$$\text{rank}(\mathbf{A}, \mathbf{B}) = \text{rank}(\mathbf{A}) \quad \Leftrightarrow \quad \mathcal{C}(\mathbf{A}, \mathbf{B}) = \mathcal{C}(\mathbf{A}).$$

That $\mathcal{C}(\mathbf{A}, \mathbf{B}) = \mathcal{C}(\mathbf{A}) \Rightarrow \text{rank}(\mathbf{A}, \mathbf{B}) = \text{rank}(\mathbf{A})$ is clear. And since $\mathcal{C}(\mathbf{A}) \subset \mathcal{C}(\mathbf{A}, \mathbf{B})$, it follows from Theorem 2.4.16 that $\text{rank}(\mathbf{A}, \mathbf{B}) = \text{rank}(\mathbf{A}) \Rightarrow \mathcal{C}(\mathbf{A}, \mathbf{B}) = \mathcal{C}(\mathbf{A})$. Q.E.D.

We are now in a position to establish the following result on the consistency of a linear system.

Theorem 2.9.2. *Each* of the following conditions is necessary and sufficient for a linear system $\mathbf{AX} = \mathbf{B}$ (in $\mathbf{X}$) to be consistent:

(1) $\mathcal{C}(\mathbf{B}) \subset \mathcal{C}(\mathbf{A})$;

(2) every column of $\mathbf{B}$ belongs to $\mathcal{C}(\mathbf{A})$;

(3) $\mathcal{C}(\mathbf{A}, \mathbf{B}) = \mathcal{C}(\mathbf{A})$;

(4) $\text{rank}(\mathbf{A}, \mathbf{B}) = \text{rank}(\mathbf{A})$.

Proof. That Condition (1) is necessary and sufficient for the consistency of $\mathbf{AX} = \mathbf{B}$ is an immediate consequence of Lemma 2.4.3. Further, it follows from Lemma 2.4.2 that Condition (2) is equivalent to Condition (1), and from Lemma 2.9.1 that each of Conditions (3) and (4) is equivalent to Condition (1). Thus, like Condition (1), each of Conditions (2) through (4) is necessary and sufficient for the consistency of $\mathbf{AX} = \mathbf{B}$. Q.E.D.

A sufficient (but in general not a necessary) condition for the consistency of a linear system is given by the following theorem.

Theorem 2.9.3. If the coefficient matrix $\mathbf{A}$ of a linear system $\mathbf{AX} = \mathbf{B}$ (in $\mathbf{X}$) has full row rank, then $\mathbf{AX} = \mathbf{B}$ is consistent.

Proof. Suppose that $\mathbf{A}$ has full row rank. Then, it follows from Lemma 2.5.1 that there exists a matrix $\mathbf{R}$ (a right inverse of $\mathbf{A}$) such that $\mathbf{AR} = \mathbf{I}$ and hence such that $\mathbf{ARB} = \mathbf{B}$. Thus, setting $\mathbf{X} = \mathbf{RB}$ gives a solution to the linear system $\mathbf{AX} = \mathbf{B}$, and we conclude that $\mathbf{AX} = \mathbf{B}$ is consistent. Q.E.D.

b. Solution set

The collection of all solutions to a linear system $\mathbf{AX} = \mathbf{B}$ (in $\mathbf{X}$) is called the *solution set* of the linear system. Clearly, a linear system is consistent if and only if its solution set is nonempty.

Is the solution set of a linear system $\mathbf{AX} = \mathbf{B}$ (in $\mathbf{X}$) a linear space? The answer depends on whether the linear system is homogeneous or nonhomogeneous, that is, on whether the right side $\mathbf{B}$ is null or nonnull.

Consider first the solution set of a homogeneous linear system $\mathbf{AX} = \mathbf{0}$. A homogeneous linear system is consistent, and hence its solution set is nonempty—its solution set includes the null matrix $\mathbf{0}$. Furthermore, if $\mathbf{X}_1$ and $\mathbf{X}_2$ are solutions to $\mathbf{AX} = \mathbf{0}$ and k is a scalar, then $\mathbf{A}(\mathbf{X}_1 + \mathbf{X}_2) = \mathbf{AX}_1 + \mathbf{AX}_2 = \mathbf{0}$ and $\mathbf{A}(k\mathbf{X}_1) = k(\mathbf{AX}_1) = \mathbf{0}$, so that $\mathbf{X}_1 + \mathbf{X}_2$ and $k\mathbf{X}_1$ are solutions to $\mathbf{AX} = \mathbf{0}$. Thus, the solution set of a homogeneous linear system is a linear space. Accordingly, the solution set of a homogeneous linear system $\mathbf{AX} = \mathbf{0}$ may be called the *solution space* of $\mathbf{AX} = \mathbf{0}$.

The solution space of a homogeneous linear system $\mathbf{Ax} = \mathbf{0}$ (in a column vector $\mathbf{x}$) is called the *null space* of the matrix $\mathbf{A}$ and is denoted by the symbol $\mathcal{N}(\mathbf{A})$. Thus, for any $M \times N$ matrix $\mathbf{A}$,

$$\mathcal{N}(\mathbf{A}) = \{\mathbf{x} \in \mathcal{R}^N : \mathbf{Ax} = \mathbf{0}\}.$$

The solution set of a nonhomogeneous linear system is not a linear space (as can be easily seen by, e.g., observing that the solution set does not contain the null matrix).

2.10 Generalized Inverses

There is an intimate relationship between the inverse $\mathbf{A}^{-1}$ of a nonsingular matrix $\mathbf{A}$ and the solution of linear systems whose coefficient matrix is $\mathbf{A}$. This relationship is described in the following theorem.

Theorem 2.10.1. Let $\mathbf{A}$ represent any $N \times N$ nonsingular matrix, $\mathbf{G}$ any $N \times N$ matrix, and P any positive integer. Then, $\mathbf{GB}$ is a solution to a linear system $\mathbf{AX} = \mathbf{B}$ (in $\mathbf{X}$) for every $N \times P$ matrix $\mathbf{B}$ if and only if $\mathbf{G} = \mathbf{A}^{-1}$.

Is there a matrix that relates to the solution of linear systems whose coefficient matrix is an $M \times N$ matrix $\mathbf{A}$ of arbitrary rank and that does so in the same way that $\mathbf{A}^{-1}$ relates to their solution in the special case where $\mathbf{A}$ is nonsingular? The following theorem serves to characterize any such matrix.

Theorem 2.10.2. Let $\mathbf{A}$ represent any $M \times N$ matrix, $\mathbf{G}$ any $N \times M$ matrix, and P any positive integer. Then, $\mathbf{GB}$ is a solution to a linear system $\mathbf{AX} = \mathbf{B}$ (in $\mathbf{X}$) for every $M \times P$ matrix $\mathbf{B}$ for which the linear system is consistent if and only if $\mathbf{AGA} = \mathbf{A}$.

Note that if the matrix $\mathbf{A}$ in Theorem 2.10.2 is nonsingular, then the linear system $\mathbf{AX} = \mathbf{B}$ is consistent for every $M \times P$ matrix $\mathbf{B}$ (as is evident from Theorem 2.9.3) and $\mathbf{AGA} = \mathbf{A} \Leftrightarrow \mathbf{G} = \mathbf{A}^{-1}$ (as is evident upon observing that if $\mathbf{A}$ is nonsingular, then $\mathbf{AA}^{-1}\mathbf{A} = \mathbf{AI} = \mathbf{A}$ and $\mathbf{AGA} = \mathbf{A} \Rightarrow$

$\mathbf{A}^{-1}\mathbf{AGAA}^{-1} = \mathbf{A}^{-1}\mathbf{AA}^{-1} \Rightarrow \mathbf{G} = \mathbf{A}^{-1}$). Thus, Theorem 2.10.2 can be regarded as a generalization of Theorem 2.10.1.

Proof (of Theorem 2.10.2). Suppose that $\mathbf{AGA} = \mathbf{A}$. And let $\mathbf{B}$ represent any $M \times P$ matrix for which $\mathbf{AX} = \mathbf{B}$ is consistent, and take $\mathbf{X}_*$ to be any solution to $\mathbf{AX} = \mathbf{B}$. Then,

$$\mathbf{A}(\mathbf{GB}) = (\mathbf{AG})\mathbf{B} = \mathbf{AGAX}_* = \mathbf{AX}_* = \mathbf{B},$$

so that $\mathbf{GB}$ is a solution to $\mathbf{AX} = \mathbf{B}$.

Conversely, suppose that $\mathbf{GB}$ is a solution to $\mathbf{AX} = \mathbf{B}$ (i.e., that $\mathbf{AGB} = \mathbf{B}$) for every $M \times P$ matrix $\mathbf{B}$ for which $\mathbf{AX} = \mathbf{B}$ is consistent. Letting $\mathbf{a}_i$ represent the ith column of $\mathbf{A}$, observe that $\mathbf{AX} = \mathbf{B}$ is consistent in particular for $\mathbf{B} = (\mathbf{a}_i, \mathbf{0}, \ldots, \mathbf{0})$—for this $\mathbf{B}$, one solution to $\mathbf{AX} = \mathbf{B}$ is the matrix $(\mathbf{u}_i, \mathbf{0}, \ldots, \mathbf{0})$, where $\mathbf{u}_i$ is the ith column of $\mathbf{I}_N$ $(i = 1, 2, \ldots, N)$. It follows that

$$\mathbf{AG}(\mathbf{a}_i, \mathbf{0}, \ldots, \mathbf{0}) = (\mathbf{a}_i, \mathbf{0}, \ldots, \mathbf{0})$$

and hence that $\mathbf{AGa}_i = \mathbf{a}_i$ $(i = 1, 2, \ldots, N)$. Thus,

$$\mathbf{AGA} = \mathbf{AG}(\mathbf{a}_1, \ldots, \mathbf{a}_N) = (\mathbf{AGa}_1, \ldots, \mathbf{AGa}_N) = (\mathbf{a}_1, \ldots, \mathbf{a}_N) = \mathbf{A}. \qquad \text{Q.E.D.}$$

An $N \times M$ matrix $\mathbf{G}$ is said to be a *generalized inverse* of an $M \times N$ matrix $\mathbf{A}$ if it satisfies the condition

$$\mathbf{AGA} = \mathbf{A}.$$

For example, each of the two 3×2 matrices

$$\begin{pmatrix} 1 & 0 \\ 0 & 0 \\ 0 & 0 \end{pmatrix} \quad \text{and} \quad \begin{pmatrix} -42 & -1 \\ 5 & 3 \\ 2 & 2 \end{pmatrix}$$

is a generalized inverse of the 2×3 matrix

$$\begin{pmatrix} 1 & 3 & 2 \\ 2 & 6 & 4 \end{pmatrix}.$$

The following lemma and corollary pertain to generalized inverses of matrices of full row or column rank and to generalized inverses of nonsingular matrices.

Lemma 2.10.3. Let $\mathbf{A}$ represent a matrix of full column rank and $\mathbf{B}$ a matrix of full row rank. Then, (1) a matrix $\mathbf{G}$ is a generalized inverse of $\mathbf{A}$ if and only if $\mathbf{G}$ is a left inverse of $\mathbf{A}$. And (2) a matrix $\mathbf{G}$ is a generalized inverse of $\mathbf{B}$ if and only if $\mathbf{G}$ is a right inverse of $\mathbf{B}$.

Proof. Let $\mathbf{L}$ represent a left inverse of $\mathbf{A}$—that $\mathbf{A}$ has a left inverse is a consequence of Lemma 2.5.1. Then, $\mathbf{ALA} = \mathbf{AI} = \mathbf{A}$, so that $\mathbf{L}$ is a generalized inverse of $\mathbf{A}$. And the proof of Part (1) of the lemma is complete upon observing that if $\mathbf{G}$ is a generalized inverse of $\mathbf{A}$, then

$$\mathbf{GA} = \mathbf{IGA} = \mathbf{LAGA} = \mathbf{LA} = \mathbf{I},$$

so that $\mathbf{G}$ is a left inverse of $\mathbf{A}$. The validity of Part (2) can be established via an analogous argument. Q.E.D.

Corollary 2.10.4. The "ordinary" inverse $\mathbf{A}^{-1}$ of a nonsingular matrix $\mathbf{A}$ is a generalized inverse of $\mathbf{A}$. And a nonsingular matrix $\mathbf{A}$ has no generalized inverse other than $\mathbf{A}^{-1}$.

Corollary 2.10.4 follows from either part of Lemma 2.10.3. That it follows in particular from Part (1) of Lemma 2.10.3 is evident upon observing that (by definition) $\mathbf{A}^{-1}$ is a left inverse of a nonsingular matrix $\mathbf{A}$ and upon recalling (from Theorem 2.5.4) that a nonsingular matrix $\mathbf{A}$ has no left inverse other than $\mathbf{A}^{-1}$.

Let us now consider the existence of generalized inverses. Together, Lemmas 2.10.3 and 2.5.1 imply that matrices of full row rank or full column rank have generalized inverses. Does every matrix have at least one generalized inverse? The answer to that question is yes, as can be shown by making use of the following theorem, which is of some interest in its own right.

Theorem 2.10.5. Let $\mathbf{B}$ represent an $M \times K$ matrix of full column rank and $\mathbf{T}$ a $K \times N$ matrix of full row rank. Then, $\mathbf{B}$ has a left inverse, say $\mathbf{L}$, and $\mathbf{T}$ has a right inverse, say $\mathbf{R}$; and $\mathbf{RL}$ is a generalized inverse of $\mathbf{BT}$.

Proof. That $\mathbf{B}$ has a left inverse $\mathbf{L}$ and $\mathbf{T}$ a right inverse $\mathbf{R}$ is an immediate consequence of Lemma 2.5.1. Moreover,

$$\mathbf{BT(RL)BT} = \mathbf{B(TR)(LB)T} = \mathbf{BIIT} = \mathbf{BT}.$$

Thus, $\mathbf{RL}$ is a generalized inverse of $\mathbf{BT}$. Q.E.D.

Now, consider an arbitrary $M \times N$ matrix $\mathbf{A}$. If $\mathbf{A} = \mathbf{0}$, then clearly any $N \times M$ matrix is a generalized inverse of $\mathbf{A}$. If $\mathbf{A} \neq \mathbf{0}$, then (according to Theorem 2.4.18) there exist a matrix $\mathbf{B}$ of full column rank and a matrix $\mathbf{T}$ of full row rank such that $\mathbf{A} = \mathbf{BT}$, and hence (in light of Theorem 2.10.5) $\mathbf{A}$ has a generalized inverse. Thus, we arrive at the following conclusion.

Corollary 2.10.6. Every matrix has at least one generalized inverse.

The symbol $\mathbf{A}^-$ is used to denote an arbitrary generalized inverse of an $M \times N$ matrix $\mathbf{A}$. By definition,

$$\mathbf{AA^-A} = \mathbf{A}.$$

a. General form and nonuniqueness of generalized inverses

A general expression can be obtained for a generalized inverse of an $M \times N$ matrix $\mathbf{A}$ in terms of any particular generalized inverse of $\mathbf{A}$, as described in the following theorem.

Theorem 2.10.7. Let $\mathbf{A}$ represent an $M \times N$ matrix, and $\mathbf{G}$ any particular generalized inverse of $\mathbf{A}$. Then, an $N \times M$ matrix $\mathbf{G}^*$ is a generalized inverse of $\mathbf{A}$ if and only if

$$\mathbf{G}^* = \mathbf{G} + \mathbf{Z} - \mathbf{GAZAG} \tag{10.1}$$

for some $N \times M$ matrix $\mathbf{Z}$. Also, $\mathbf{G}^*$ is a generalized inverse of $\mathbf{A}$ if and only if

$$\mathbf{G}^* = \mathbf{G} + (\mathbf{I} - \mathbf{GA})\mathbf{T} + \mathbf{S}(\mathbf{I} - \mathbf{AG}) \tag{10.2}$$

for some $N \times M$ matrices $\mathbf{T}$ and $\mathbf{S}$.

Proof. It is a simple exercise to verify that any matrix $\mathbf{G}^*$ that is expressible in the form (10.1) or (10.2) is a generalized inverse of $\mathbf{A}$. Conversely, if $\mathbf{G}^*$ is a generalized inverse of $\mathbf{A}$, then

$$\begin{aligned}\mathbf{G}^* &= \mathbf{G} + (\mathbf{G}^* - \mathbf{G}) - \mathbf{GA}(\mathbf{G}^* - \mathbf{G})\mathbf{AG} \\ &= \mathbf{G} + \mathbf{Z} - \mathbf{GAZAG},\end{aligned}$$

where $\mathbf{Z} = \mathbf{G}^* - \mathbf{G}$, and

$$\begin{aligned}\mathbf{G}^* &= \mathbf{G} + (\mathbf{I} - \mathbf{GA})\mathbf{G}^*\mathbf{AG} + (\mathbf{G}^* - \mathbf{G})(\mathbf{I} - \mathbf{AG}) \\ &= \mathbf{G} + (\mathbf{I} - \mathbf{GA})\mathbf{T} + \mathbf{S}(\mathbf{I} - \mathbf{AG}),\end{aligned}$$

where $\mathbf{T} = \mathbf{G}^*\mathbf{AG}$ and $\mathbf{S} = \mathbf{G}^* - \mathbf{G}$. Q.E.D.

All generalized inverses of the $M \times N$ matrix $\mathbf{A}$ can be generated from expression (10.1) by letting $\mathbf{Z}$ range over all $N \times M$ matrices. Alternatively, all generalized inverses of $\mathbf{A}$ can be generated from expression (10.2) by letting both $\mathbf{T}$ and $\mathbf{S}$ range over all $N \times M$ matrices. Note that distinct choices for $\mathbf{Z}$ or for $\mathbf{T}$ and/or $\mathbf{S}$ may or may not result in distinct generalized inverses.

How many generalized inverses does an $M \times N$ matrix $\mathbf{A}$ possess? If $\mathbf{A}$ is nonsingular, it has a unique generalized inverse, namely $\mathbf{A}^{-1}$. Now, suppose that $\mathbf{A}$ is not nonsingular (i.e., that $\mathbf{A}$ is either not square or is square but singular). Then, $\text{rank}(\mathbf{A}) < M$ (in which case $\mathbf{A}$ does not have a right inverse) and/or $\text{rank}(\mathbf{A}) < N$ (in which case $\mathbf{A}$ does not have a left inverse). Thus, for any generalized inverse $\mathbf{G}$ of $\mathbf{A}$, $\mathbf{I} - \mathbf{AG} \neq \mathbf{0}$ and/or $\mathbf{I} - \mathbf{GA} \neq \mathbf{0}$. And, based on the second part of Theorem 2.10.7, we conclude that $\mathbf{A}$ has an infinite number of generalized inverses.

b. Some basic results on generalized inverses

Let us consider the extent to which various basic results on "ordinary" inverses extend to generalized inverses. It is easy to verify the following lemma, which is a generalization of result (5.2).

Lemma 2.10.8. For any matrix $\mathbf{A}$ and any nonzero scalar k, $(1/k)\mathbf{A}^-$ is a generalized inverse of $k\mathbf{A}$.

Upon setting $k = -1$ in Lemma 2.10.8, we obtain the following corollary, which is a generalization of result (5.3).

Corollary 2.10.9. For any matrix $\mathbf{A}$, $-\mathbf{A}^-$ is a generalized inverse of $-\mathbf{A}$.

For any matrix $\mathbf{A}$, we find that

$$\mathbf{A}'(\mathbf{A}^-)'\mathbf{A}' = (\mathbf{A}\mathbf{A}^-\mathbf{A})' = \mathbf{A}'.$$

Thus, we have the following lemma, which is a generalization of result (5.4).

Lemma 2.10.10. For any matrix $\mathbf{A}$, $(\mathbf{A}^-)'$ is a generalized inverse of $\mathbf{A}'$.

While [according to result (5.5)] the inverse of a nonsingular symmetric matrix is symmetric, a generalized inverse of a singular symmetric matrix (of order greater than 1) need not be symmetric. For example, the matrix $\begin{pmatrix} 1 & 1 \\ 0 & 0 \end{pmatrix}$ is a generalized inverse of the matrix $\begin{pmatrix} 1 & 0 \\ 0 & 0 \end{pmatrix}$. However, Lemma 2.10.10 implies that a generalized inverse of a singular symmetric matrix has the following, weaker property.

Corollary 2.10.11. For any symmetric matrix $\mathbf{A}$, $(\mathbf{A}^-)'$ is a generalized inverse of $\mathbf{A}$.

For any $M \times N$ matrix $\mathbf{A}$, the $M \times M$ matrix $\mathbf{A}\mathbf{A}^-$ and $N \times N$ matrix $\mathbf{A}^-\mathbf{A}$ are as described in the following two lemmas—in the special case where $\mathbf{A}$ is nonsingular, $\mathbf{A}\mathbf{A}^{-1} = \mathbf{A}^{-1}\mathbf{A} = \mathbf{I}$.

Lemma 2.10.12. Let $\mathbf{A}$ represent an $M \times N$ matrix. Then, the $M \times M$ matrix $\mathbf{A}\mathbf{A}^-$ and the $N \times N$ matrix $\mathbf{A}^-\mathbf{A}$ are both idempotent.

Proof. Clearly, $\mathbf{A}\mathbf{A}^-\mathbf{A}\mathbf{A}^- = \mathbf{A}\mathbf{A}^-$ and $\mathbf{A}^-\mathbf{A}\mathbf{A}^-\mathbf{A} = \mathbf{A}^-\mathbf{A}$. Q.E.D.

Lemma 2.10.13. For any matrix $\mathbf{A}$, $\mathcal{C}(\mathbf{A}\mathbf{A}^-) = \mathcal{C}(\mathbf{A})$, $\mathcal{R}(\mathbf{A}^-\mathbf{A}) = \mathcal{R}(\mathbf{A})$, and

$$\text{rank}(\mathbf{A}\mathbf{A}^-) = \text{rank}(\mathbf{A}^-\mathbf{A}) = \text{rank}(\mathbf{A}). \tag{10.3}$$

Proof. It follows from Corollary 2.4.4 that $\mathcal{C}(\mathbf{A}\mathbf{A}^-) \subset \mathcal{C}(\mathbf{A})$ and also, since $\mathbf{A} = \mathbf{A}\mathbf{A}^-\mathbf{A} = (\mathbf{A}\mathbf{A}^-)\mathbf{A}$, that $\mathcal{C}(\mathbf{A}) \subset \mathcal{C}(\mathbf{A}\mathbf{A}^-)$. Thus, $\mathcal{C}(\mathbf{A}\mathbf{A}^-) = \mathcal{C}(\mathbf{A})$. That $\mathcal{R}(\mathbf{A}^-\mathbf{A}) = \mathcal{R}(\mathbf{A})$ follows from an analogous argument. And since $\mathbf{A}\mathbf{A}^-$ has the same column space as $\mathbf{A}$ and $\mathbf{A}^-\mathbf{A}$ the same row space as $\mathbf{A}$, $\mathbf{A}\mathbf{A}^-$ and $\mathbf{A}^-\mathbf{A}$ have the same rank as $\mathbf{A}$. Q.E.D.

It follows from Corollary 2.4.17 and Lemma 2.10.13 that, for any matrix $\mathbf{A}$,

$$\text{rank}(\mathbf{A}^-) \geq \text{rank}(\mathbf{A}\mathbf{A}^-) = \text{rank}(\mathbf{A}).$$

Thus, we have the following lemma, which can be regarded as an extension of result (5.6).

Lemma 2.10.14. For any matrix $\mathbf{A}$, $\text{rank}(\mathbf{A}^-) \geq \text{rank}(\mathbf{A})$.

The following lemma extends to generalized inverses some results on the inverses of products of matrices—refer to results (5.9) and (5.11).

Lemma 2.10.15. Let $\mathbf{B}$ represent an $M \times N$ matrix and $\mathbf{G}$ an $N \times M$ matrix. Then, for any $M \times M$ nonsingular matrix $\mathbf{A}$ and $N \times N$ nonsingular matrix $\mathbf{C}$, (1) $\mathbf{G}$ is a generalized inverse of $\mathbf{AB}$ if and only if $\mathbf{G} = \mathbf{H}\mathbf{A}^{-1}$ for some generalized inverse $\mathbf{H}$ of $\mathbf{B}$, (2) $\mathbf{G}$ is a generalized inverse of $\mathbf{BC}$ if and only if $\mathbf{G} = \mathbf{C}^{-1}\mathbf{H}$ for some generalized inverse $\mathbf{H}$ of $\mathbf{B}$, and (3) $\mathbf{G}$ is a generalized inverse of $\mathbf{ABC}$ if and only if $\mathbf{G} = \mathbf{C}^{-1}\mathbf{H}\mathbf{A}^{-1}$ for some generalized inverse $\mathbf{H}$ of $\mathbf{B}$.

Proof. Parts (1) and (2) are special cases of Part (3) (those where $\mathbf{C} = \mathbf{I}$ and $\mathbf{A} = \mathbf{I}$, respectively). Thus, it suffices to prove Part (3).

By definition, $\mathbf{G}$ is a generalized inverse of $\mathbf{ABC}$ if and only if

$$\mathbf{ABCGABC} = \mathbf{ABC}. \tag{10.4}$$

Upon premultiplying both sides of equality (10.4) by $\mathbf{A}^{-1}$ and postmultiplying both sides by $\mathbf{C}^{-1}$, we obtain the equivalent equality

$$\mathbf{BCGAB} = \mathbf{B}.$$

Thus, $\mathbf{G}$ is a generalized inverse of $\mathbf{ABC}$ if and only if $\mathbf{CGA} = \mathbf{H}$ for some generalized inverse $\mathbf{H}$ of $\mathbf{B}$ or, equivalently, if and only if $\mathbf{G} = \mathbf{C}^{-1}\mathbf{H}\mathbf{A}^{-1}$ for some generalized inverse $\mathbf{H}$ of $\mathbf{B}$. Q.E.D.

2.11 Linear Systems Revisited

Having introduced (in Section 2.10) the concept of a generalized inverse, we are now in a position to add to the results obtained earlier (in Section 2.9) on the consistency and the solution set of a linear system.

a. More on the consistency of a linear system

Each of the four conditions of Theorem 2.9.2 is necessary and sufficient for a linear system $\mathbf{AX} = \mathbf{B}$ (in $\mathbf{X}$) to be consistent. The following theorem describes another such condition.

Theorem 2.11.1. A linear system $\mathbf{AX} = \mathbf{B}$ (in $\mathbf{X}$) is consistent if and only if $\mathbf{AA}^{-}\mathbf{B} = \mathbf{B}$ or, equivalently, if and only if $(\mathbf{I} - \mathbf{AA}^{-})\mathbf{B} = \mathbf{0}$.

With the establishment of the following lemma, Theorem 2.11.1 becomes an immediate consequence of Theorem 2.9.2.

Lemma 2.11.2. Let $\mathbf{A}$ represent an $M \times N$ matrix. Then, for any $M \times P$ matrix $\mathbf{B}$, $\mathcal{C}(\mathbf{B}) \subset \mathcal{C}(\mathbf{A})$ if and only if $\mathbf{B} = \mathbf{AA}^{-}\mathbf{B}$ or, equivalently, if and only if $(\mathbf{I} - \mathbf{AA}^{-})\mathbf{B} = \mathbf{0}$. And, for any $Q \times N$ matrix $\mathbf{C}$, $\mathcal{R}(\mathbf{C}) \subset \mathcal{R}(\mathbf{A})$ if and only if $\mathbf{C} = \mathbf{CA}^{-}\mathbf{A}$ or, equivalently, if and only if $\mathbf{C}(\mathbf{I} - \mathbf{A}^{-}\mathbf{A}) = \mathbf{0}$.

Proof (of Lemma 2.11.2). If $\mathbf{B} = \mathbf{AA}^{-}\mathbf{B}$, then it follows immediately from Corollary 2.4.4 that $\mathcal{C}(\mathbf{B}) \subset \mathcal{C}(\mathbf{A})$. Conversely, if $\mathcal{C}(\mathbf{B}) \subset \mathcal{C}(\mathbf{A})$, then, according to Lemma 2.4.3, there exists a matrix $\mathbf{F}$ such that $\mathbf{B} = \mathbf{AF}$, implying that

$$\mathbf{B} = \mathbf{AA}^{-}\mathbf{AF} = \mathbf{AA}^{-}\mathbf{B}.$$

Thus, $\mathcal{C}(\mathbf{B}) \subset \mathcal{C}(\mathbf{A})$ if and only if $\mathbf{B} = \mathbf{AA}^{-}\mathbf{B}$. That $\mathcal{R}(\mathbf{C}) \subset \mathcal{R}(\mathbf{A})$ if and only if $\mathbf{C} = \mathbf{CA}^{-}\mathbf{A}$ follows from an analogous argument. Q.E.D.

According to Theorem 2.11.1, either of the two matrices $\mathbf{AA}^{-}$ or $\mathbf{I} - \mathbf{AA}^{-}$ can be used to determine whether a linear system having $\mathbf{A}$ as a coefficient matrix is consistent or inconsistent. If the right side of the linear system is unaffected by premultiplication by $\mathbf{AA}^{-}$, then the linear system is consistent; otherwise, it is inconsistent. Similarly, if the premultiplication of the right side by $\mathbf{I} - \mathbf{AA}^{-}$ produces a null matrix, then the linear system is consistent; otherwise, it is inconsistent.

Consider, for example, the linear system $\mathbf{Ax} = \mathbf{b}$ (in $\mathbf{x}$), where

$$\mathbf{A} = \begin{pmatrix} -6 & 2 & -2 & -3 \\ 3 & -1 & 5 & 2 \\ -3 & 1 & 3 & -1 \end{pmatrix}.$$

One generalized inverse of $\mathbf{A}$ is

$$\mathbf{G} = \begin{pmatrix} 0 & 0 & 0 \\ -1 & 0 & 3 \\ 0 & 0 & 0 \\ -1 & 0 & 2 \end{pmatrix},$$

as can be easily verified. Clearly,

$$\mathbf{AG} = \begin{pmatrix} 1 & 0 & 0 \\ -1 & 0 & 1 \\ 0 & 0 & 1 \end{pmatrix}.$$

If $\mathbf{b} = (3, 2, 5)'$, then

$$\mathbf{AGb} = (3, 2, 5)' = \mathbf{b},$$

in which case the linear system $\mathbf{Ax} = \mathbf{b}$ is consistent. However, if $\mathbf{b} = (1, 2, 1)'$, then

$$\mathbf{AGb} = (1, 0, 1)' \neq \mathbf{b},$$

in which case $\mathbf{Ax} = \mathbf{b}$ is inconsistent.

b. General form of a solution to a linear system

The following theorem gives an expression for the general form of a solution to a homogeneous linear system in terms of any particular generalized inverse of the coefficient matrix.

Theorem 2.11.3. A matrix $\mathbf{X}^*$ is a solution to a homogeneous linear system $\mathbf{AX} = \mathbf{0}$ (in $\mathbf{X}$) if and only if

$$\mathbf{X}^* = (\mathbf{I} - \mathbf{A}^-\mathbf{A})\mathbf{Y}$$

for some matrix $\mathbf{Y}$.

Proof. If $\mathbf{X}^* = (\mathbf{I} - \mathbf{A}^-\mathbf{A})\mathbf{Y}$ for some matrix $\mathbf{Y}$, then

$$\mathbf{AX}^* = (\mathbf{A} - \mathbf{AA}^-\mathbf{A})\mathbf{Y} = (\mathbf{A} - \mathbf{A})\mathbf{Y} = \mathbf{0},$$

so that $\mathbf{X}^*$ is a solution to $\mathbf{AX} = \mathbf{0}$. Conversely, if $\mathbf{X}^*$ is a solution to $\mathbf{AX} = \mathbf{0}$, then

$$\mathbf{X}^* = \mathbf{X}^* - \mathbf{A}^-\mathbf{0} = \mathbf{X}^* - \mathbf{A}^-\mathbf{AX}^* = (\mathbf{I} - \mathbf{A}^-\mathbf{A})\mathbf{X}^*,$$

so that $\mathbf{X}^* = (\mathbf{I} - \mathbf{A}^-\mathbf{A})\mathbf{Y}$ for $\mathbf{Y} = \mathbf{X}^*$. Q.E.D.

According to Theorem 2.11.3, all solutions to the homogeneous linear system $\mathbf{AX} = \mathbf{0}$ can be generated by setting

$$\mathbf{X} = (\mathbf{I} - \mathbf{A}^-\mathbf{A})\mathbf{Y}$$

and allowing $\mathbf{Y}$ to range over all matrices (of the appropriate dimensions).

As a special case of Theorem 2.11.3, we have that a column vector $\mathbf{x}^*$ is a solution to a homogeneous linear system $\mathbf{Ax} = \mathbf{0}$ (in a column vector $\mathbf{x}$) if and only if

$$\mathbf{x}^* = (\mathbf{I} - \mathbf{A}^-\mathbf{A})\mathbf{y}$$

for some column vector $\mathbf{y}$. Thus, we have the following corollary of Theorem 2.11.3.

Corollary 2.11.4. For any matrix $\mathbf{A}$,

$$\mathcal{N}(\mathbf{A}) = \mathcal{C}(\mathbf{I} - \mathbf{A}^-\mathbf{A}).$$

How "large" is the solution space of a homogeneous linear system $\mathbf{Ax} = \mathbf{0}$ (in an N-dimensional column vector $\mathbf{x}$)? The answer is given by the following lemma.

Lemma 2.11.5. Let $\mathbf{A}$ represent an $M \times N$ matrix. Then,

$$\dim[\mathcal{N}(\mathbf{A})] = N - \text{rank}(\mathbf{A}).$$

That is, the dimension of the solution space of the homogeneous linear system $\mathbf{Ax} = \mathbf{0}$ (in an N-dimensional column vector $\mathbf{x}$) equals $N - \text{rank}(\mathbf{A})$.

Proof. Recalling (from Lemma 2.10.12) that $\mathbf{A}^-\mathbf{A}$ is idempotent and making use of Corollary 2.11.4 and Lemmas 2.8.4 and 2.10.13, we find that

$$\dim[\mathcal{N}(\mathbf{A})] = \dim[\mathcal{C}(\mathbf{I}_N - \mathbf{A}^-\mathbf{A})] = \text{rank}(\mathbf{I}_N - \mathbf{A}^-\mathbf{A}) = N - \text{rank}(\mathbf{A}^-\mathbf{A}) = N - \text{rank}(\mathbf{A}).$$

Q.E.D.

The solution space of a homogeneous linear system $\mathbf{Ax} = \mathbf{0}$ (in an N-dimensional column vector $\mathbf{x}$) is a subspace of the linear space $\mathcal{R}^N$ of all N-dimensional column vectors. According to Lemma 2.11.5, the dimension of this subspace equals $N - \text{rank}(\mathbf{A})$. Thus, if $\text{rank}(\mathbf{A}) = N$, that is, if $\mathbf{A}$ is of full column rank, then the homogeneous linear system $\mathbf{Ax} = \mathbf{0}$ has a unique solution, namely, the null vector $\mathbf{0}$. And if $\text{rank}(\mathbf{A}) < N$, then $\mathbf{Ax} = \mathbf{0}$ has an infinite number of solutions.

The following theorem relates the solutions of an arbitrary (consistent) linear system $\mathbf{AX} = \mathbf{B}$ (in $\mathbf{X}$) to those of the homogeneous linear system $\mathbf{AZ} = \mathbf{0}$ (in $\mathbf{Z}$). (To avoid confusion, the matrix of unknowns of the homogeneous linear system is being denoted by a different symbol than that of the linear system $\mathbf{AX} = \mathbf{B}$.)

Theorem 2.11.6. Let $\mathbf{X}_0$ represent any particular solution to a consistent linear system $\mathbf{AX} = \mathbf{B}$ (in $\mathbf{X}$). Then, a matrix $\mathbf{X}^*$ is a solution to $\mathbf{AX} = \mathbf{B}$ if and only if

$$\mathbf{X}^* = \mathbf{X}_0 + \mathbf{Z}^*$$

for some solution $\mathbf{Z}^*$ to the homogeneous linear system $\mathbf{AZ} = \mathbf{0}$ (in $\mathbf{Z}$).

Proof. If $\mathbf{X}^* = \mathbf{X}_0 + \mathbf{Z}^*$ for some solution $\mathbf{Z}^*$ to $\mathbf{AZ} = \mathbf{0}$, then

$$\mathbf{AX}^* = \mathbf{AX}_0 + \mathbf{AZ}^* = \mathbf{B} + \mathbf{0} = \mathbf{B},$$

so that $\mathbf{X}^*$ is a solution to $\mathbf{AX} = \mathbf{B}$. Conversely, if $\mathbf{X}^*$ is a solution to $\mathbf{AX} = \mathbf{B}$, then, defining $\mathbf{Z}^* = \mathbf{X}^* - \mathbf{X}_0$, we find that $\mathbf{X}^* = \mathbf{X}_0 + \mathbf{Z}^*$ and that

$$\mathbf{AZ}^* = \mathbf{AX}^* - \mathbf{AX}_0 = \mathbf{B} - \mathbf{B} = \mathbf{0}$$

(so that $\mathbf{Z}^*$ is a solution to $\mathbf{AZ} = \mathbf{0}$). Q.E.D.

The upshot of Theorem 2.11.6 is that all of the matrices in the solution set of a consistent linear system $\mathbf{AX} = \mathbf{B}$ (in $\mathbf{X}$) can be generated from any particular solution $\mathbf{X}_0$ by setting

$$\mathbf{X} = \mathbf{X}_0 + \mathbf{Z}$$

and allowing $\mathbf{Z}$ to range over all of the matrices in the solution space of the homogeneous linear system $\mathbf{AZ} = \mathbf{0}$ (in $\mathbf{Z}$).

It follows from Theorem 2.10.2 that one solution to a consistent linear system $\mathbf{AX} = \mathbf{B}$ (in $\mathbf{X}$) is the matrix $\mathbf{A}^-\mathbf{B}$. Thus, in light of Theorem 2.11.6, we have the following extension of Theorem 2.11.3.

Theorem 2.11.7. A matrix $\mathbf{X}^*$ is a solution to a consistent linear system $\mathbf{AX} = \mathbf{B}$ (in $\mathbf{X}$) if and only if

$$\mathbf{X}^* = \mathbf{A}^-\mathbf{B} + (\mathbf{I} - \mathbf{A}^-\mathbf{A})\mathbf{Y} \tag{11.1}$$

for some matrix $\mathbf{Y}$.

As a special case of Theorem 2.11.7, we have that a column vector $\mathbf{x}^*$ is a solution to a consistent linear system $\mathbf{Ax} = \mathbf{b}$ (in a column vector $\mathbf{x}$) if and only if

$$\mathbf{x}^* = \mathbf{A}^-\mathbf{b} + (\mathbf{I} - \mathbf{A}^-\mathbf{A})\mathbf{y} \tag{11.2}$$

for some column vector $\mathbf{y}$.

Consider, for example, expression (11.2) as applied to the linear system

$$\begin{pmatrix} -6 & 2 & -2 & -3 \\ 3 & -1 & 5 & 2 \\ -3 & 1 & 3 & -1 \end{pmatrix} \mathbf{x} = \begin{pmatrix} 3 \\ 2 \\ 5 \end{pmatrix}, \tag{11.3}$$

the consistency of which was established earlier (in Subsection a). Taking $\mathbf{A}$ to be the coefficient matrix and $\mathbf{b}$ the right side of linear system (11.3), choosing

$$\mathbf{A}^- = \begin{pmatrix} 0 & 0 & 0 \\ -1 & 0 & 3 \\ 0 & 0 & 0 \\ -1 & 0 & 2 \end{pmatrix},$$

and denoting the elements of $\mathbf{y}$ by y_1, y_2, y_3, and y_4, respectively, we find that

$$\mathbf{A}^-\mathbf{b} + (\mathbf{I} - \mathbf{A}^-\mathbf{A})\mathbf{y} = \begin{pmatrix} y_1 \\ 12 + 3y_1 - 11y_3 \\ y_3 \\ 7 - 8y_3 \end{pmatrix}. \tag{11.4}$$

Thus, the members of the solution set of linear system (11.3) consist of all vectors of the general form (11.4).

A possibly nonhomogeneous linear system $\mathbf{AX} = \mathbf{B}$ (in an $N \times P$ matrix $\mathbf{X}$) has 0 solutions, 1 solution, or an infinite number of solutions. If $\mathbf{AX} = \mathbf{B}$ is inconsistent, then, by definition, it has 0 solutions. If $\mathbf{AX} = \mathbf{B}$ is consistent and $\mathbf{A}$ is of full column rank (i.e., of rank N), then $\mathbf{I} - \mathbf{A}^-\mathbf{A} = \mathbf{0}$ (as is evident from Lemma 2.10.3), and it follows from Theorem 2.11.7 that $\mathbf{AX} = \mathbf{B}$ has 1 solution. If $\mathbf{AX} = \mathbf{B}$ is consistent and $\text{rank}(\mathbf{A}) < Ns$, then $\mathbf{I} - \mathbf{A}^-\mathbf{A} \neq \mathbf{0}$ (since otherwise we would arrive at a contradiction of Lemma 2.5.1), and it follows from Theorem 2.11.7 that $\mathbf{AX} = \mathbf{B}$ has an infinite number of solutions.

In the special case of a linear system with a nonsingular coefficient matrix, we obtain the following, additional result (that can be regarded as a consequence of Theorems 2.9.3 and 2.11.7 or that can be verified directly).

Theorem 2.11.8. If the coefficient matrix $\mathbf{A}$ of a linear system $\mathbf{AX} = \mathbf{B}$ (in $\mathbf{X}$) is nonsingular, then $\mathbf{AX} = \mathbf{B}$ has a unique solution and that solution equals $\mathbf{A}^{-1}\mathbf{B}$.

2.12 Projection Matrices

Let $\mathbf{X}$ represent an arbitrary matrix. It follows from the very definition of a generalized inverse that

$$\mathbf{X}'\mathbf{X}(\mathbf{X}'\mathbf{X})^-\mathbf{X}'\mathbf{X} = \mathbf{X}'\mathbf{X}.$$

And, upon applying Part (1) of Corollary 2.3.4 [with $\mathbf{A} = \mathbf{X}$, $\mathbf{B} = (\mathbf{X}'\mathbf{X})^-\mathbf{X}'\mathbf{X}$, and $\mathbf{C} = \mathbf{I}$], we find that

$$\mathbf{X}(\mathbf{X}'\mathbf{X})^-\mathbf{X}'\mathbf{X} = \mathbf{X}. \tag{12.1}$$

Similarly, applying Part (2) of Corollary 2.3.4 [with $\mathbf{A} = \mathbf{X}$, $\mathbf{B} = \mathbf{X}'\mathbf{X}(\mathbf{X}'\mathbf{X})^-$, and $\mathbf{C} = \mathbf{I}$], we find that

$$\mathbf{X}'\mathbf{X}(\mathbf{X}'\mathbf{X})^-\mathbf{X}' = \mathbf{X}'. \tag{12.2}$$

In light of results (12.1) and (12.2), Corollary 2.4.4 implies that $\mathcal{R}(\mathbf{X}) \subset \mathcal{R}(\mathbf{X}'\mathbf{X})$ and that $\mathcal{C}(\mathbf{X}') \subset \mathcal{C}(\mathbf{X}'\mathbf{X})$. Since Corollary 2.4.4 also implies that $\mathcal{R}(\mathbf{X}'\mathbf{X}) \subset \mathcal{R}(\mathbf{X})$ and $\mathcal{C}(\mathbf{X}'\mathbf{X}) \subset \mathcal{C}(\mathbf{X}')$, we conclude that

$$\mathcal{R}(\mathbf{X}'\mathbf{X}) = \mathcal{R}(\mathbf{X}) \quad \text{and} \quad \mathcal{C}(\mathbf{X}'\mathbf{X}) = \mathcal{C}(\mathbf{X}').$$

Moreover, matrices having the same row space have the same rank, so that

$$\text{rank}(\mathbf{X}'\mathbf{X}) = \text{rank}(\mathbf{X}).$$

Now, let $\mathbf{P_X}$ represent the (square) matrix $\mathbf{X}(\mathbf{X}'\mathbf{X})^-\mathbf{X}'$. Then, results (12.1) and (12.2) can be restated succinctly as

$$\mathbf{P_X X} = \mathbf{X} \tag{12.3}$$

and

$$\mathbf{X}'\mathbf{P_X} = \mathbf{X}'. \tag{12.4}$$

For any generalized inverses $\mathbf{G}_1$ and $\mathbf{G}_2$ of $\mathbf{X}'\mathbf{X}$, we have [in light of result (12.1)] that

$$\mathbf{X}\mathbf{G}_1\mathbf{X}'\mathbf{X} = \mathbf{X} = \mathbf{X}\mathbf{G}_2\mathbf{X}'\mathbf{X}.$$

And, upon applying Part (2) of Corollary 2.3.4 (with $\mathbf{A} = \mathbf{X}$, $\mathbf{B} = \mathbf{X}\mathbf{G}_1$, and $\mathbf{C} = \mathbf{X}\mathbf{G}_2$), we find that

$$\mathbf{X}\mathbf{G}_1\mathbf{X}' = \mathbf{X}\mathbf{G}_2\mathbf{X}'.$$

Thus, $\mathbf{P_X}$ is invariant to the choice of the generalized inverse $(\mathbf{X}'\mathbf{X})^-$.

There is a stronger version of this invariance property. Consider the linear system $\mathbf{X}'\mathbf{X}\mathbf{B} = \mathbf{X}'$ (in $\mathbf{B}$). A solution to this linear system can be obtained by taking $\mathbf{B} = (\mathbf{X}'\mathbf{X})^-\mathbf{X}'$ [as is evident from result (12.2)], and, for $\mathbf{B} = (\mathbf{X}'\mathbf{X})^-\mathbf{X}'$, $\mathbf{P_X} = \mathbf{X}\mathbf{B}$. Moreover, for any two solutions $\mathbf{B}_1$ and $\mathbf{B}_2$ to $\mathbf{X}'\mathbf{X}\mathbf{B} = \mathbf{X}'$, we have that $\mathbf{X}'\mathbf{X}\mathbf{B}_1 = \mathbf{X}' = \mathbf{X}'\mathbf{X}\mathbf{B}_2$, and, upon applying Part (1) of Corollary 2.3.4 (with $\mathbf{A} = \mathbf{X}$, $\mathbf{B} = \mathbf{B}_1$, and $\mathbf{C} = \mathbf{B}_2$), we find that $\mathbf{X}\mathbf{B}_1 = \mathbf{X}\mathbf{B}_2$. Thus, $\mathbf{P_X} = \mathbf{X}\mathbf{B}$ for every solution to $\mathbf{X}'\mathbf{X}\mathbf{B} = \mathbf{X}'$.

Note that $\mathbf{P}'_\mathbf{X} = \mathbf{X}[(\mathbf{X}'\mathbf{X})^-]'\mathbf{X}'$. According to Corollary 2.10.11, $[(\mathbf{X}'\mathbf{X})^-]'$, like $(\mathbf{X}'\mathbf{X})^-$ itself, is a generalized inverse of $\mathbf{X}'\mathbf{X}$, and, since $\mathbf{P_X}$ is invariant to the choice of the generalized inverse $(\mathbf{X}'\mathbf{X})^-$, it follows that $\mathbf{P_X} = \mathbf{X}[(\mathbf{X}'\mathbf{X})^-]'\mathbf{X}'$. Thus, $\mathbf{P}'_\mathbf{X} = \mathbf{P_X}$; that is, $\mathbf{P_X}$ is symmetric. And $\mathbf{P_X}$ is idempotent, as is evident upon observing [in light of result (12.3)] that

$$\mathbf{P}^2_\mathbf{X} = \mathbf{P_X}\mathbf{X}(\mathbf{X}'\mathbf{X})^-\mathbf{X}' = \mathbf{X}(\mathbf{X}'\mathbf{X})^-\mathbf{X}' = \mathbf{P_X}.$$

Moreover, $\text{rank}(\mathbf{P_X}) = \text{rank}(\mathbf{X})$, as is evident upon observing [in light of Corollary 2.4.17 and result (12.3)] that

$$\text{rank}\,\mathbf{P_X} \le \text{rank}(\mathbf{X}) = \text{rank}(\mathbf{P_X X}) \le \text{rank}(\mathbf{P_X}).$$

Summarizing, we have the following lemma and theorem.

Lemma 2.12.1. For any matrix $\mathbf{X}$,

$$\mathcal{R}(\mathbf{X}'\mathbf{X}) = \mathcal{R}(\mathbf{X}), \quad \mathcal{C}(\mathbf{X}'\mathbf{X}) = \mathcal{C}(\mathbf{X}'), \quad \text{and} \quad \text{rank}(\mathbf{X}'\mathbf{X}) = \text{rank}(\mathbf{X}).$$

Theorem 2.12.2. Take $\mathbf{X}$ to be an arbitrary matrix, and let $\mathbf{P_X} = \mathbf{X}(\mathbf{X}'\mathbf{X})^-\mathbf{X}'$. Then,

(1) $\mathbf{P_X X} = \mathbf{X}$, that is, $\mathbf{X}(\mathbf{X}'\mathbf{X})^-\mathbf{X}'\mathbf{X} = \mathbf{X}$;
(2) $\mathbf{X}'\mathbf{P_X} = \mathbf{X}'$, that is, $\mathbf{X}'\mathbf{X}(\mathbf{X}'\mathbf{X})^-\mathbf{X}' = \mathbf{X}'$;
(3) $\mathbf{P_X}$ is invariant to the choice of the generalized inverse $(\mathbf{X}'\mathbf{X})^-$;
(3′) $\mathbf{P_X} = \mathbf{X}\mathbf{B}^*$ for any solution $\mathbf{B}^*$ to the (consistent) linear system $\mathbf{X}'\mathbf{X}\mathbf{B} = \mathbf{X}'$ (in $\mathbf{B}$);
(4) $\mathbf{P}'_\mathbf{X} = \mathbf{P_X}$, that is, $\mathbf{P_X}$ is symmetric;
(5) $\mathbf{P}^2_\mathbf{X} = \mathbf{P_X}$, that is, $\mathbf{P_X}$ is idempotent;
(6) $\text{rank}(\mathbf{P_X}) = \text{rank}(\mathbf{X})$.

Subsequently, we continue to use (for any matrix $\mathbf{X}$) the symbol $\mathbf{P_X}$ to represent the matrix $\mathbf{X}(\mathbf{X}'\mathbf{X})^-\mathbf{X}'$. For reasons that will eventually become apparent, a matrix of the general form $\mathbf{P_X}$ is referred to as a *projection matrix*.

Upon applying Corollary 2.4.5, we obtain the following generalization of Lemma 2.12.1.

Lemma 2.12.3. For any $M \times N$ matrix $\mathbf{X}$ and for any $P \times N$ matrix $\mathbf{S}$ and $N \times Q$ matrix $\mathbf{T}$,

$$\mathcal{C}(\mathbf{S}\mathbf{X}'\mathbf{X}) = \mathcal{C}(\mathbf{S}\mathbf{X}') \quad \text{and} \quad \text{rank}(\mathbf{S}\mathbf{X}'\mathbf{X}) = \text{rank}(\mathbf{S}\mathbf{X}')$$

and

$$\mathcal{R}(\mathbf{X}'\mathbf{X}\mathbf{T}) = \mathcal{R}(\mathbf{X}\mathbf{T}) \quad \text{and} \quad \text{rank}(\mathbf{X}'\mathbf{X}\mathbf{T}) = \text{rank}(\mathbf{X}\mathbf{T}).$$

2.13 Quadratic Forms

Let $\mathbf{A} = \{a_{ij}\}$ represent an arbitrary $N \times N$ matrix, and consider the function that assigns to each N-dimensional column vector $\mathbf{x} = (x_1, x_2, \ldots, x_N)'$ in $\mathcal{R}^N$ the value

$$\mathbf{x}'\mathbf{A}\mathbf{x} = \sum_{i,j} a_{ij}x_ix_j = \sum_i a_{ii}x_i^2 + \sum_{i,\, j \neq i} a_{ij}x_ix_j.$$

A function of $\mathbf{x}$ that is expressible in the form $\mathbf{x}'\mathbf{A}\mathbf{x}$ is called a *quadratic form* (in $\mathbf{x}$). It is customary to refer to $\mathbf{A}$ as the *matrix of the quadratic form* $\mathbf{x}'\mathbf{A}\mathbf{x}$.

Let $\mathbf{B} = \{b_{ij}\}$ represent a second $N \times N$ matrix. Under what circumstances are the two quadratic forms $\mathbf{x}'\mathbf{A}\mathbf{x}$ and $\mathbf{x}'\mathbf{B}\mathbf{x}$ identically equal (i.e., equal for every value of $\mathbf{x}$)? Clearly, a sufficient condition for them to be identically equal is that $\mathbf{A} = \mathbf{B}$. However, except in the special case where $N = 1$, $\mathbf{A} = \mathbf{B}$ is not a necessary condition.

For purposes of establishing a necessary condition, suppose that $\mathbf{x}'\mathbf{A}\mathbf{x}$ and $\mathbf{x}'\mathbf{B}\mathbf{x}$ are identically equal. Then, setting $\mathbf{x}$ equal to the ith column of $\mathbf{I}_N$, we find that

$$a_{ii} = \mathbf{x}'\mathbf{A}\mathbf{x} = \mathbf{x}'\mathbf{B}\mathbf{x} = b_{ii} \qquad (i = 1, 2, \ldots, N). \tag{13.1}$$

That is, the diagonal elements of $\mathbf{A}$ are the same as those of $\mathbf{B}$. Consider now the off-diagonal elements of $\mathbf{A}$ and $\mathbf{B}$. Setting $\mathbf{x}$ equal to the N-dimensional column vector whose ith and jth elements equal 1 and whose remaining elements equal 0, we find that

$$a_{ii} + a_{ij} + a_{ji} + a_{jj} = \mathbf{x}'\mathbf{A}\mathbf{x} = \mathbf{x}'\mathbf{B}\mathbf{x} = b_{ii} + b_{ij} + b_{ji} + b_{jj} \quad (j \neq i = 1, 2, \ldots, N). \tag{13.2}$$

Together, results (13.1) and (13.2) imply that

$$a_{ii} = b_{ii} \ \text{ and } \ a_{ij} + a_{ji} = b_{ij} + b_{ji} \qquad (j \neq i = 1, 2, \ldots, N)$$

or, equivalently, that

$$\mathbf{A} + \mathbf{A}' = \mathbf{B} + \mathbf{B}'. \tag{13.3}$$

Thus, condition (13.3) is a necessary condition for $\mathbf{x}'\mathbf{A}\mathbf{x}$ and $\mathbf{x}'\mathbf{B}\mathbf{x}$ to be identically equal. It is also a sufficient condition. To see this, observe that (since a 1×1 matrix is symmetric) condition (13.3) implies that

$$\begin{aligned}
\mathbf{x}'\mathbf{A}\mathbf{x} &= (1/2)[\mathbf{x}'\mathbf{A}\mathbf{x} + (\mathbf{x}'\mathbf{A}\mathbf{x})'] \\
&= (1/2)(\mathbf{x}'\mathbf{A}\mathbf{x} + \mathbf{x}'\mathbf{A}'\mathbf{x}) \\
&= (1/2)\mathbf{x}'(\mathbf{A} + \mathbf{A}')\mathbf{x} \\
&= (1/2)\mathbf{x}'(\mathbf{B} + \mathbf{B}')\mathbf{x} = (1/2)[\mathbf{x}'\mathbf{B}\mathbf{x} + (\mathbf{x}'\mathbf{B}\mathbf{x})'] = \mathbf{x}'\mathbf{B}\mathbf{x}.
\end{aligned}$$

In summary, we have the following lemma.

Lemma 2.13.1. Let $\mathbf{A} = \{a_{ij}\}$ and $\mathbf{B} = \{b_{ij}\}$ represent arbitrary $N \times N$ matrices. The two quadratic forms $\mathbf{x}'\mathbf{A}\mathbf{x}$ and $\mathbf{x}'\mathbf{B}\mathbf{x}$ (in $\mathbf{x}$) are identically equal if and only if, for $j \neq i = 1, 2, \ldots, N$, $a_{ii} = b_{ii}$ and $a_{ij} + a_{ji} = b_{ij} + b_{ji}$ or, equivalently, if and only if $\mathbf{A} + \mathbf{A}' = \mathbf{B} + \mathbf{B}'$.

Note that Lemma 2.13.1 implies in particular that the quadratic form $\mathbf{x}'\mathbf{A}'\mathbf{x}$ (in $\mathbf{x}$) is identically equal to the quadratic form $\mathbf{x}'\mathbf{A}\mathbf{x}$ (in $\mathbf{x}$). When $\mathbf{B}$ is symmetric, the condition $\mathbf{A} + \mathbf{A}' = \mathbf{B} + \mathbf{B}'$ is equivalent to the condition $\mathbf{B} = (1/2)(\mathbf{A} + \mathbf{A}')$, and when both $\mathbf{A}$ and $\mathbf{B}$ are symmetric, the condition $\mathbf{A} + \mathbf{A}' = \mathbf{B} + \mathbf{B}'$ is equivalent to the condition $\mathbf{A} = \mathbf{B}$. Thus, we have the following two corollaries of Lemma 2.13.1.

Corollary 2.13.2. Corresponding to any quadratic form $\mathbf{x}'\mathbf{A}\mathbf{x}$ (in $\mathbf{x}$), there is a unique symmetric matrix $\mathbf{B}$ such that $\mathbf{x}'\mathbf{B}\mathbf{x} = \mathbf{x}'\mathbf{A}\mathbf{x}$ for all $\mathbf{x}$, namely, the matrix $\mathbf{B} = (1/2)(\mathbf{A} + \mathbf{A}')$.

Corollary 2.13.3. For any pair of $N \times N$ symmetric matrices $\mathbf{A}$ and $\mathbf{B}$, the two quadratic forms $\mathbf{x}'\mathbf{A}\mathbf{x}$ and $\mathbf{x}'\mathbf{B}\mathbf{x}$ (in $\mathbf{x}$) are identically equal (i.e., $\mathbf{x}'\mathbf{A}\mathbf{x} = \mathbf{x}'\mathbf{B}\mathbf{x}$ for all $\mathbf{x}$) if and only if $\mathbf{A} = \mathbf{B}$.

As a special case of Corollary 2.13.3 (that where $\mathbf{B} = \mathbf{0}$), we have the following additional corollary.

Corollary 2.13.4. Let $\mathbf{A}$ represent an $N \times N$ symmetric matrix. If $\mathbf{x}'\mathbf{A}\mathbf{x} = 0$ for every $(N \times 1)$ vector $\mathbf{x}$, then $\mathbf{A} = \mathbf{0}$.

a. Nonnegative definiteness and positive definiteness or semidefiniteness

A quadratic form $\mathbf{x}'\mathbf{A}\mathbf{x}$ [in an N-dimensional column vector $\mathbf{x} = (x_1, x_2, \ldots, x_N)'$] is said to be *nonnegative definite* if $\mathbf{x}'\mathbf{A}\mathbf{x} \geq 0$ for every $\mathbf{x}$ in $\mathbb{R}^N$. Note that $\mathbf{x}'\mathbf{A}\mathbf{x} = 0$ for at least one value of $\mathbf{x}$, namely, $\mathbf{x} = \mathbf{0}$. If $\mathbf{x}'\mathbf{A}\mathbf{x}$ is nonnegative definite and if, in addition, the null vector $\mathbf{0}$ is the only value of $\mathbf{x}$ for which $\mathbf{x}'\mathbf{A}\mathbf{x} = 0$, then $\mathbf{x}'\mathbf{A}\mathbf{x}$ is said to be *positive definite*. That is, $\mathbf{x}'\mathbf{A}\mathbf{x}$ is positive definite if $\mathbf{x}'\mathbf{A}\mathbf{x} > 0$ for every $\mathbf{x}$ except $\mathbf{x} = \mathbf{0}$. A quadratic form that is nonnegative definite, but not positive definite, is called *positive semidefinite*. Thus, $\mathbf{x}'\mathbf{A}\mathbf{x}$ is positive semidefinite if $\mathbf{x}'\mathbf{A}\mathbf{x} \geq 0$ for every $\mathbf{x} \in \mathbb{R}^N$ and $\mathbf{x}'\mathbf{A}\mathbf{x} = 0$ for some nonnull $\mathbf{x}$.

Consider, for example, the two quadratic forms $\mathbf{x}'\mathbf{I}_N\mathbf{x} = x_1^2 + x_2^2 + \cdots + x_N^2$ and $\mathbf{x}'\mathbf{1}_N\mathbf{1}_N'\mathbf{x} = (\mathbf{1}_N'\mathbf{x})'\mathbf{1}_N'\mathbf{x} = (x_1 + x_2 + \cdots + x_N)^2$. Clearly, $\mathbf{x}'\mathbf{I}_N\mathbf{x}$ and $\mathbf{x}'\mathbf{1}_N\mathbf{1}_N'\mathbf{x}$ are both nonnegative definite. Moreover, $\mathbf{x}'\mathbf{I}_N\mathbf{x} > 0$ for all nonnull $\mathbf{x}$, while (assuming that $N \geq 2$) $\mathbf{x}'\mathbf{1}_N\mathbf{1}_N'\mathbf{x} = 0$ for the nonnull vector $\mathbf{x} = (1-N, 1, 1, \ldots, 1)'$. Accordingly, $\mathbf{x}'\mathbf{I}_N\mathbf{x}$ is positive definite, and $\mathbf{x}'\mathbf{1}_N\mathbf{1}_N'\mathbf{x}$ is positive semidefinite.

The terms nonnegative definite, positive definite, and positive semidefinite are applied to matrices as well as to quadratic forms. An $N \times N$ matrix $\mathbf{A}$ is said to be *nonnegative definite*, *positive definite*, or *positive semidefinite* if the quadratic form $\mathbf{x}'\mathbf{A}\mathbf{x}$ (in $\mathbf{x}$) is nonnegative definite, positive definite, or positive semidefinite, respectively.

It is instructive to consider the following lemma, which characterizes the concepts of nonnegative definiteness, positive definiteness, and positive semidefiniteness as applied to diagonal matrices and which is easy to verify.

Lemma 2.13.5. Let $\mathbf{D} = \{d_i\}$ represent an $N \times N$ diagonal matrix. Then, (1) $\mathbf{D}$ is nonnegative definite if and only if $d_1, d_2, \ldots, d_N$ are nonnegative; (2) $\mathbf{D}$ is positive definite if and only if $d_1, d_2, \ldots, d_N$ are (strictly) positive; and (3) $\mathbf{D}$ is positive semidefinite if and only if $d_i \geq 0$ for $i = 1, 2, \ldots, N$ with equality holding for one or more values of i.

The following two lemmas give some basic results on scalar multiples and sums of nonnegative definite matrices.

Lemma 2.13.6. Let k (> 0) represent a (strictly) positive scalar, and $\mathbf{A}$ an $N \times N$ matrix. If $\mathbf{A}$ is positive definite, then $k\mathbf{A}$ is also positive definite. Similarly, if $\mathbf{A}$ is positive semidefinite, then $k\mathbf{A}$ is also positive semidefinite.

Proof. Consider the two quadratic forms $\mathbf{x}'\mathbf{A}\mathbf{x}$ and $\mathbf{x}'(k\mathbf{A})\mathbf{x}$ (in $\mathbf{x}$). Clearly, $\mathbf{x}'(k\mathbf{A})\mathbf{x} = k\mathbf{x}'\mathbf{A}\mathbf{x}$.

Thus, if $\mathbf{x}'\mathbf{A}\mathbf{x}$ is positive definite, then $\mathbf{x}'(k\mathbf{A})\mathbf{x}$ is positive definite; similarly, if $\mathbf{x}'\mathbf{A}\mathbf{x}$ is positive semidefinite, then $\mathbf{x}'(k\mathbf{A})\mathbf{x}$ is positive semidefinite. Or, equivalently, if $\mathbf{A}$ is positive definite, then $k\mathbf{A}$ is positive definite; and if $\mathbf{A}$ is positive semidefinite, then $k\mathbf{A}$ is positive semidefinite. Q.E.D.

Lemma 2.13.7. Let $\mathbf{A}$ and $\mathbf{B}$ represent $N \times N$ matrices. If $\mathbf{A}$ and $\mathbf{B}$ are both nonnegative definite, then $\mathbf{A} + \mathbf{B}$ is nonnegative definite. Moreover, if either $\mathbf{A}$ or $\mathbf{B}$ is positive definite and the other is nonnegative definite (i.e., either positive definite or positive semidefinite), then $\mathbf{A} + \mathbf{B}$ is positive definite.

Proof. Suppose that one of the matrices, say $\mathbf{A}$, is positive definite and that the other ($\mathbf{B}$) is nonnegative definite. Then, for every nonnull vector $\mathbf{x}$ in $\mathcal{R}^N$, $\mathbf{x}'\mathbf{A}\mathbf{x} > 0$ and $\mathbf{x}'\mathbf{B}\mathbf{x} \geq 0$, and hence

$$\mathbf{x}'(\mathbf{A} + \mathbf{B})\mathbf{x} = \mathbf{x}'\mathbf{A}\mathbf{x} + \mathbf{x}'\mathbf{B}\mathbf{x} > 0.$$

Thus, $\mathbf{A} + \mathbf{B}$ is positive definite. A similar argument shows that if $\mathbf{A}$ and $\mathbf{B}$ are both nonnegative definite, then $\mathbf{A} + \mathbf{B}$ is nonnegative definite. Q.E.D.

The repeated application of Lemma 2.13.7 leads to the following generalization.

Lemma 2.13.8. Let $\mathbf{A}_1, \mathbf{A}_2, \ldots, \mathbf{A}_K$ represent $N \times N$ matrices. If $\mathbf{A}_1$, $\mathbf{A}_2$, $\ldots$, $\mathbf{A}_K$ are all nonnegative definite, then their sum $\mathbf{A}_1 + \mathbf{A}_2 + \cdots + \mathbf{A}_K$ is also nonnegative definite. Moreover, if one or more of the matrices $\mathbf{A}_1, \mathbf{A}_2, \ldots, \mathbf{A}_K$ are positive definite and the others are nonnegative definite, then $\mathbf{A}_1 + \mathbf{A}_2 + \cdots + \mathbf{A}_K$ is positive definite.

A basic property of positive definite matrices is described in the following lemma.

Lemma 2.13.9. Any positive definite matrix is nonsingular.

Proof. Let $\mathbf{A}$ represent an $N \times N$ positive definite matrix. For purposes of establishing a contradiction, suppose that $\mathbf{A}$ is singular or, equivalently, that $\text{rank}(\mathbf{A}) < N$. Then, the columns of $\mathbf{A}$ are linearly dependent, and hence there exists a nonnull vector $\mathbf{x}_*$ such that $\mathbf{A}\mathbf{x}_* = \mathbf{0}$. We find that

$$\mathbf{x}_*'\mathbf{A}\mathbf{x}_* = \mathbf{x}_*'(\mathbf{A}\mathbf{x}_*) = \mathbf{x}_*'\mathbf{0} = 0,$$

which (since $\mathbf{A}$ is positive definite) establishes the desired contradiction. Q.E.D.

Additional basic properties of nonnegative definite matrices are established in the following theorem and corollaries.

Theorem 2.13.10. Let $\mathbf{A}$ represent an $N \times N$ matrix, and $\mathbf{P}$ an $N \times M$ matrix.

(1) If $\mathbf{A}$ is nonnegative definite, then $\mathbf{P}'\mathbf{A}\mathbf{P}$ is nonnegative definite.

(2) If $\mathbf{A}$ is nonnegative definite and $\text{rank}(\mathbf{P}) < M$, then $\mathbf{P}'\mathbf{A}\mathbf{P}$ is positive semidefinite.

(3) If $\mathbf{A}$ is positive definite and $\text{rank}(\mathbf{P}) = M$, then $\mathbf{P}'\mathbf{A}\mathbf{P}$ is positive definite.

Proof. Suppose that $\mathbf{A}$ is nonnegative definite (either positive definite or positive semidefinite). Then, $\mathbf{y}'\mathbf{A}\mathbf{y} \geq 0$ for every $\mathbf{y}$ in $\mathcal{R}^N$ and in particular for every $\mathbf{y}$ that is expressible in the form $\mathbf{y} = \mathbf{P}\mathbf{x}$. Thus, for every M-dimensional (column) vector $\mathbf{x}$,

$$\mathbf{x}'(\mathbf{P}'\mathbf{A}\mathbf{P})\mathbf{x} = (\mathbf{P}\mathbf{x})'\mathbf{A}\mathbf{P}\mathbf{x} \geq 0, \tag{13.4}$$

which establishes that $\mathbf{P}'\mathbf{A}\mathbf{P}$ is nonnegative definite, thereby completing the proof of Part (1).

If $\text{rank}(\mathbf{P}) < M$, then

$$\text{rank}(\mathbf{P}'\mathbf{A}\mathbf{P}) \leq \text{rank}(\mathbf{P}) < M,$$

which (in light of Lemma 2.13.9) establishes that $\mathbf{P}'\mathbf{A}\mathbf{P}$ is not positive definite and hence (since $\mathbf{P}'\mathbf{A}\mathbf{P}$ is nonnegative definite) that $\mathbf{P}'\mathbf{A}\mathbf{P}$ is positive semidefinite, thereby completing the proof of Part (2).

If $\mathbf{A}$ is positive definite, then equality is attained in inequality (13.4) only when $\mathbf{P}\mathbf{x} = \mathbf{0}$. Moreover, if $\text{rank}(\mathbf{P}) = M$ (so that the columns of $\mathbf{P}$ are linearly independent), then $\mathbf{P}\mathbf{x} = \mathbf{0}$ implies $\mathbf{x} = \mathbf{0}$. Thus, if $\mathbf{A}$ is positive definite and $\text{rank}(\mathbf{P}) = M$, then equality is attained in inequality (13.4) only when $\mathbf{x} = \mathbf{0}$, implying (since $\mathbf{P}'\mathbf{A}\mathbf{P}$ is nonnegative definite) that $\mathbf{P}'\mathbf{A}\mathbf{P}$ is positive definite. Q.E.D.

Corollary 2.13.11. Let $\mathbf{A}$ represent an $N \times N$ matrix and $\mathbf{P}$ an $N \times N$ nonsingular matrix.

(1) If $\mathbf{A}$ is positive definite, then $\mathbf{P}'\mathbf{AP}$ is positive definite.
(2) If $\mathbf{A}$ is positive semidefinite, then $\mathbf{P}'\mathbf{AP}$ is positive semidefinite.

Proof. (1) That $\mathbf{P}'\mathbf{AP}$ is positive definite if $\mathbf{A}$ is positive definite is a direct consequence of Part (3) of Theorem 2.13.10.

(2) Suppose that $\mathbf{A}$ is positive semidefinite. Then, according to Part (1) of Theorem 2.13.10, $\mathbf{P}'\mathbf{AP}$ is nonnegative definite. Moreover, there exists a nonnull vector $\mathbf{y}$ such that $\mathbf{y}'\mathbf{Ay} = 0$. Now, let $\mathbf{x} = \mathbf{P}^{-1}\mathbf{y}$. Accordingly, $\mathbf{y} = \mathbf{Px}$, and we find that $\mathbf{x} \neq \mathbf{0}$ (since, otherwise, we would have that $\mathbf{y} = \mathbf{0}$) and that

$$\mathbf{x}'(\mathbf{P}'\mathbf{AP})\mathbf{x} = (\mathbf{Px})'\mathbf{APx} = \mathbf{y}'\mathbf{Ay} = 0.$$

We conclude that $\mathbf{P}'\mathbf{AP}$ is positive semidefinite. Q.E.D.

Corollary 2.13.12. A positive definite matrix is invertible, and its inverse is positive definite.

Proof. Let $\mathbf{A}$ represent a positive definite matrix. Then, according to Lemma 2.13.9, $\mathbf{A}$ is nonsingular and hence (according to Theorem 2.5.4) invertible. Further, since $(\mathbf{A}^{-1})' = (\mathbf{A}^{-1})'\mathbf{AA}^{-1}$, it follows from Part (1) of Corollary 2.13.11 [together with result (5.6)] that $(\mathbf{A}^{-1})'$ is positive definite. And, upon observing that the two quadratic forms $\mathbf{x}'\mathbf{A}^{-1}\mathbf{x}$ and $\mathbf{x}'(\mathbf{A}^{-1})'\mathbf{x}$ (in $\mathbf{x}$) are identically equal (as is evident from Lemma 2.13.1), we conclude that $\mathbf{A}^{-1}$ is positive definite. Q.E.D.

Corollary 2.13.13. Any principal submatrix of a positive definite matrix is positive definite; any principal submatrix of a positive semidefinite matrix is nonnegative definite.

Proof. Let $\mathbf{A}$ represent an $N \times N$ matrix, and consider the principal submatrix of $\mathbf{A}$ obtained by striking out all of its rows and columns except its $i_1, i_2, \ldots, i_M$th rows and columns (where $i_1 < i_2 < \cdots < i_M$). This submatrix is expressible as $\mathbf{P}'\mathbf{AP}$, where $\mathbf{P}$ is the $N \times M$ matrix whose columns are the $i_1, i_2, \ldots, i_M$th columns of $\mathbf{I}_N$. Since $\text{rank}(\mathbf{P}) = M$, it follows from Part (3) of Theorem 2.13.10 that $\mathbf{P}'\mathbf{AP}$ is positive definite if $\mathbf{A}$ is positive definite. Further, it follows from Part (1) of Theorem 2.13.10 that $\mathbf{P}'\mathbf{AP}$ is nonnegative definite if $\mathbf{A}$ is nonnegative definite (and, in particular, if $\mathbf{A}$ is positive semidefinite). Q.E.D.

Corollary 2.13.14. The diagonal elements of a positive definite matrix are positive; the diagonal elements of a positive semidefinite matrix are nonnegative.

Proof. The corollary follows immediately from Corollary 2.13.13 upon observing (1) that the ith diagonal element of a (square) matrix $\mathbf{A}$ is the element of a 1×1 principal submatrix (that obtained by striking out all of the rows and columns of $\mathbf{A}$ except the ith row and column) and (2) that the element of a 1×1 positive definite matrix is positive and the element of a 1×1 nonnegative definite matrix is nonnegative. Q.E.D.

Corollary 2.13.15. Let $\mathbf{P}$ represent an arbitrary $N \times M$ matrix. The $M \times M$ matrix $\mathbf{P}'\mathbf{P}$ is nonnegative definite. If $\text{rank}(\mathbf{P}) = M$, $\mathbf{P}'\mathbf{P}$ is positive definite; otherwise (if $\text{rank}(\mathbf{P}) < M$), $\mathbf{P}'\mathbf{P}$ is positive semidefinite.

Proof. The corollary follows from Theorem 2.13.10 upon observing that $\mathbf{P}'\mathbf{P} = \mathbf{P}'\mathbf{IP}$ and that (as demonstrated earlier) $\mathbf{I}$ is positive definite. Q.E.D.

Corollary 2.13.16. Let $\mathbf{P}$ represent an $N \times N$ nonsingular matrix and $\mathbf{D} = \{d_i\}$ an $N \times N$ diagonal matrix. Then, (1) $\mathbf{P}'\mathbf{DP}$ is nonnegative definite if and only if $d_1, d_2, \ldots, d_N$ are nonnegative; (2) $\mathbf{P}'\mathbf{DP}$ is positive definite if and only if $d_1, d_2, \ldots, d_N$ are (strictly) positive; and (3) $\mathbf{P}'\mathbf{DP}$ is positive semidefinite if and only if $d_i \geq 0$ for $i = 1, 2, \ldots, N$ with equality holding for one or more values of i.

Proof. Let $\mathbf{A} = \mathbf{P}'\mathbf{DP}$. Then, clearly, $\mathbf{D} = (\mathbf{P}^{-1})'\mathbf{AP}^{-1}$. Accordingly, it follows from Part (1) of Theorem 2.13.10 that $\mathbf{A}$ is nonnegative definite if and only if $\mathbf{D}$ is nonnegative definite; and, it follows, respectively, from Parts (1) and (2) of Corollary 2.13.11 that $\mathbf{A}$ is positive definite if and only if $\mathbf{D}$ is positive definite and that $\mathbf{A}$ is positive semidefinite if and only if $\mathbf{D}$ is positive semidefinite. In light of Lemma 2.13.5, the proof is complete. Q.E.D.

A diagonal element, say the ith diagonal element, of a nonnegative definite matrix can equal 0 only if the other elements in the ith row and the ith column of that matrix satisfy the conditions set forth in the following lemma and corollary.

Lemma 2.13.17. Let $\mathbf{A} = \{a_{ij}\}$ represent an $N \times N$ nonnegative definite matrix. If $a_{ii} = 0$, then, for $j = 1, 2, \ldots, N$, $a_{ij} = -a_{ji}$; that is, if the ith diagonal element of $\mathbf{A}$ equals 0, then $(a_{i1}, a_{i2}, \ldots, a_{iN})$, which is the ith row of $\mathbf{A}$, equals $-(a_{1i}, a_{2i}, \ldots, a_{Ni})$, which is -1 times the transpose of the ith column of $\mathbf{A}$ $(i = 1, 2, \ldots, N)$.

Proof. Suppose that $a_{ii} = 0$, and take $\mathbf{x} = \{x_k\}$ to be an N-dimensional column vector such that $x_i < -a_{jj}$, $x_j = a_{ij} + a_{ji}$, and $x_k = 0$ for k other than $k = i$ and $k = j$ (where $j \neq i$). Then,

$$\begin{aligned} \mathbf{x}'\mathbf{A}\mathbf{x} &= a_{ii}x_i^2 + (a_{ij} + a_{ji})x_i x_j + a_{jj}x_j^2 \\ &= (a_{ij} + a_{ji})^2(x_i + a_{jj}) \\ &\leq 0, \end{aligned} \tag{13.5}$$

with equality holding only if $a_{ij} + a_{ji} = 0$ or, equivalently, only if $a_{ij} = -a_{ji}$. Moreover, since $\mathbf{A}$ is nonnegative definite, $\mathbf{x}'\mathbf{A}\mathbf{x} \geq 0$, which—together with inequality (13.5)—implies that $\mathbf{x}'\mathbf{A}\mathbf{x} = 0$. We conclude that $a_{ij} = -a_{ji}$. Q.E.D.

If the nonnegative definite matrix $\mathbf{A}$ in Lemma 2.13.17 is symmetric, then $a_{ij} = -a_{ji} \Leftrightarrow 2a_{ij} = 0 \Leftrightarrow a_{ij} = 0$. Thus, we have the following corollary.

Corollary 2.13.18. Let $\mathbf{A} = \{a_{ij}\}$ represent an $N \times N$ symmetric nonnegative definite matrix. If $a_{ii} = 0$, then, for $j = 1, 2, \ldots, N$, $a_{ji} = a_{ij} = 0$; that is, if the ith diagonal element of $\mathbf{A}$ equals zero, then the ith column $(a_{1i}, a_{2i}, \ldots, a_{Ni})'$ of $\mathbf{A}$ and the ith row $(a_{i1}, a_{i2} \ldots, a_{iN})$ of $\mathbf{A}$ are null. Moreover, if all N diagonal elements $a_{11}, a_{22}, \ldots, a_{NN}$ of $\mathbf{A}$ equal 0, then $\mathbf{A} = \mathbf{0}$.

b. General form of symmetric nonnegative definite matrices

According to Corollary 2.13.15, every matrix $\mathbf{A}$ that is expressible in the form $\mathbf{A} = \mathbf{P}'\mathbf{P}$ is a (symmetric) nonnegative definite matrix. Is the converse true? That is, is every symmetric nonnegative definite matrix $\mathbf{A}$ expressible in the form $\mathbf{A} = \mathbf{P}'\mathbf{P}$? The answer is yes. In fact, corresponding to any symmetric nonnegative definite matrix $\mathbf{A}$ is an upper triangular matrix $\mathbf{P}$ such that $\mathbf{A} = \mathbf{P}'\mathbf{P}$, as we now proceed to show, beginning with the establishment of the following lemma.

Lemma 2.13.19. Let $\mathbf{A} = \{a_{ij}\}$ represent an $N \times N$ symmetric matrix (where $N \geq 2$). Define $\mathbf{B} = \{b_{ij}\} = \mathbf{U}'\mathbf{A}\mathbf{U}$, where $\mathbf{U}$ is an $N \times N$ unit upper triangular matrix of the form $\mathbf{U} = \begin{pmatrix} 1 & \mathbf{u}' \\ \mathbf{0} & \mathbf{I}_{N-1} \end{pmatrix}$. Partition $\mathbf{B}$ and $\mathbf{A}$ as $\mathbf{B} = \begin{pmatrix} b_{11} & \mathbf{b}' \\ \mathbf{b} & \mathbf{B}_{22} \end{pmatrix}$ and $\mathbf{A} = \begin{pmatrix} a_{11} & \mathbf{a}' \\ \mathbf{a} & \mathbf{A}_{22} \end{pmatrix}$. Suppose that $a_{11} \neq 0$. Then, the $(N-1)$-dimensional vector $\mathbf{u}$ can be chosen so that $\mathbf{b} = \mathbf{0}$; this can be done by taking $\mathbf{u} = -(1/a_{11})\mathbf{a}$.

Proof. We find that

$$\mathbf{b} = (\mathbf{u}, \mathbf{I})\mathbf{A}\begin{pmatrix} 1 \\ \mathbf{0} \end{pmatrix} = a_{11}\mathbf{u} + \mathbf{a},$$

which, for $\mathbf{u} = -(1/a_{11})\mathbf{a}$, gives $\mathbf{b} = \mathbf{a} - \mathbf{a} = \mathbf{0}$. Q.E.D.

We are now in a position to establish the following theorem.

Theorem 2.13.20. Corresponding to any $N \times N$ symmetric nonnegative definite matrix $\mathbf{A}$, there exists a unit upper triangular matrix $\mathbf{Q}$ such that $\mathbf{Q}'\mathbf{A}\mathbf{Q}$ is a diagonal matrix.

Proof. The proof is by mathematical induction. The theorem is clearly valid for any 1×1 (symmetric) nonnegative definite matrix. Suppose now that it is valid for any $(N-1) \times (N-1)$ symmetric nonnegative definite matrix, and consider an arbitrary $N \times N$ symmetric nonnegative definite matrix $\mathbf{A} = \{a_{ij}\}$. For purposes of establishing the existence of a unit upper triangular matrix $\mathbf{Q}$ such that $\mathbf{Q}'\mathbf{A}\mathbf{Q}$ is diagonal, it is convenient to partition $\mathbf{A}$ as $\mathbf{A} = \begin{pmatrix} a_{11} & \mathbf{a}' \\ \mathbf{a} & \mathbf{A}_{22} \end{pmatrix}$ and to consider the case where $a_{11} > 0$ separately from that where $a_{11} = 0$—since $\mathbf{A}$ is nonnegative definite, $a_{11} \geq 0$.

Case (1): $a_{11} > 0$. According to Lemma 2.13.19, there exists a unit upper triangular matrix $\mathbf{U}$ such that $\mathbf{U}'\mathbf{A}\mathbf{U} = \text{diag}(b_{11}, \mathbf{B}_{22})$ for some scalar b_{11} and some $(N-1) \times (N-1)$ matrix $\mathbf{B}_{22}$. Moreover, $\mathbf{B}_{22}$ is symmetric and nonnegative definite (as is evident upon observing that $\mathbf{U}'\mathbf{A}\mathbf{U}$ is symmetric and nonnegative definite). Thus, by supposition, there exists a unit upper triangular matrix $\mathbf{Q}_*$ such that $\mathbf{Q}_*'\mathbf{B}_{22}\mathbf{Q}_*$ is a diagonal matrix. Take $\mathbf{Q} = \mathbf{U}\,\text{diag}(1, \mathbf{Q}_*)$. Then, $\mathbf{Q}$ is a unit upper triangular matrix, as is evident upon observing that it is the product of two unit upper triangular matrices—a product of 2 unit upper triangular matrices is itself a unit upper triangular matrix, as can be readily verified by, e.g., making use of result (2.5). And

$$\mathbf{Q}'\mathbf{A}\mathbf{Q} = \text{diag}(1, \mathbf{Q}_*')\,\text{diag}(b_{11}, \mathbf{B}_{22})\,\text{diag}(1, \mathbf{Q}_*) = \text{diag}(b_{11}, \mathbf{Q}_*'\mathbf{B}_{22}\mathbf{Q}_*),$$

which (like $\mathbf{Q}_*'\mathbf{B}_{22}\mathbf{Q}_*$) is a diagonal matrix.

Case (2): $a_{11} = 0$. The submatrix $\mathbf{A}_{22}$ is an $(N-1) \times (N-1)$ symmetric nonnegative definite matrix. Thus, by supposition, there exists a unit upper triangular matrix $\mathbf{Q}_*$ such that $\mathbf{Q}_*'\mathbf{A}_{22}\mathbf{Q}_*$ is a diagonal matrix. Take $\mathbf{Q} = \text{diag}(1, \mathbf{Q}_*)$. Then, $\mathbf{Q}$ is a unit upper triangular matrix. And, upon observing (in light of Corollary 2.13.18) that $a_{11} = 0 \Rightarrow \mathbf{a} = \mathbf{0}$, we find that $\mathbf{Q}'\mathbf{A}\mathbf{Q} = \text{diag}(a_{11}, \mathbf{Q}_*'\mathbf{A}_{22}\mathbf{Q}_*)$, which (like $\mathbf{Q}_*'\mathbf{A}_{22}\mathbf{Q}_*$) is a diagonal matrix. Q.E.D.

Observe (in light of the results of Section 2.6 b) that a unit upper triangular matrix is nonsingular and that its inverse is a unit upper triangular matrix. And note (in connection with Theorem 2.13.20) that if $\mathbf{Q}$ is a unit upper triangular matrix such that $\mathbf{Q}'\mathbf{A}\mathbf{Q} = \mathbf{D}$ for some diagonal matrix $\mathbf{D}$, then

$$\mathbf{A} = (\mathbf{Q}^{-1})'\mathbf{Q}'\mathbf{A}\mathbf{Q}\mathbf{Q}^{-1} = (\mathbf{Q}^{-1})'\mathbf{D}\mathbf{Q}^{-1}.$$

Thus, we have the following corollary.

Corollary 2.13.21. Corresponding to any $N \times N$ symmetric nonnegative definite matrix $\mathbf{A}$, there exist a unit upper triangular matrix $\mathbf{U}$ and a diagonal matrix $\mathbf{D}$ such that $\mathbf{A} = \mathbf{U}'\mathbf{D}\mathbf{U}$.

Suppose that $\mathbf{A}$ is an $N \times N$ symmetric nonnegative definite matrix, and take $\mathbf{U}$ to be a unit upper triangular matrix and $\mathbf{D}$ a diagonal matrix such that $\mathbf{A} = \mathbf{U}'\mathbf{D}\mathbf{U}$ (the existence of such matrices is established in Corollary 2.13.21). Further, let $R = \text{rank}\,\mathbf{A}$, and denote by $d_1, d_2, \ldots, d_N$ the diagonal elements of $\mathbf{D}$. Then, $d_i \geq 0$ $(i = 1, 2, \ldots, N)$, as is evident from Corollary 2.13.16. And $\mathbf{A}$ is expressible as $\mathbf{A} = \mathbf{T}'\mathbf{T}$, where $\mathbf{T} = \text{diag}(\sqrt{d_1}, \sqrt{d_2}, \ldots, \sqrt{d_N})\,\mathbf{U}$. Moreover, $\text{rank}\,\mathbf{D} = R$, so that R of the diagonal elements of $\mathbf{D}$ are strictly positive and the others are equal to 0. Thus, as an additional corollary of Theorem 2.13.20, we have the following result.

Corollary 2.13.22. Let $\mathbf{A}$ represent an $N \times N$ symmetric nonnegative definite matrix. And let $R = \text{rank}\,\mathbf{A}$. Then, there exists an upper triangular matrix $\mathbf{T}$ with R (strictly) positive diagonal elements and with $N-R$ null rows such that

$$\mathbf{A} = \mathbf{T}'\mathbf{T}. \tag{13.6}$$

Let (for $i, j = 1, 2, \ldots, N$) t_{ij} represent the ijth element of the upper triangular matrix $\mathbf{T}$ in decomposition (13.6) (and observe that $t_{ij} = 0$ for $i > j$). Equality (13.6) implies that

$$t_{11} = \sqrt{a_{11}}, \tag{13.7}$$

$$t_{1j} = \begin{cases} a_{1j}/t_{11}, & \text{if } t_{11} > 0, \\ 0, & \text{if } t_{11} = 0, \end{cases} \tag{13.8}$$

$$t_{ij} = \begin{cases} \left(a_{ij} - \sum_{k=1}^{i-1} t_{ki}t_{kj}\right)/t_{ii}, & \text{if } t_{ii} > 0, \\ 0, & \text{if } t_{ii} = 0 \end{cases} \tag{13.9}$$

$(i = 2, 3, \ldots, j-1)$,

$$t_{jj} = \left(a_{jj} - \sum_{k=1}^{j-1} t_{kj}^2\right)^{1/2} \tag{13.10}$$

$(j = 2, 3, \ldots, N)$.

It follows from equalities (13.7), (13.8), (13.9), and (13.10) that decomposition (13.6) is unique. This decomposition is known as the *Cholesky decomposition*. It can be computed by a method that is sometimes referred to as the square root method. In the square root method, formulas (13.7), (13.8), (13.9), and (13.10) are used to construct the matrix $\mathbf{T}$ in N steps, one row or one column per step. Refer, for instance, to Harville (1997, sec. 14.5) for more details and for an illustrative example.

As a variation on decomposition (13.6), we have the decomposition

$$\mathbf{A} = \mathbf{T}_*'\mathbf{T}_*, \tag{13.11}$$

where $\mathbf{T}_*$ is the $R \times N$ matrix whose rows are the nonnull rows of the upper triangular matrix $\mathbf{T}$. Among the implications of result (13.11) is the following result, which can be regarded as an additional corollary of Theorem 2.13.20.

Corollary 2.13.23. Corresponding to any $N \times N$ (nonnull) symmetric nonnegative definite matrix $\mathbf{A}$ of rank R, there exists an $R \times N$ matrix $\mathbf{P}$ such that $\mathbf{A} = \mathbf{P}'\mathbf{P}$ (and any such $R \times N$ matrix $\mathbf{P}$ is of full row rank R).

The following corollary can be regarded as a generalization of Corollary 2.13.23.

Corollary 2.13.24. Let $\mathbf{A}$ represent an $N \times N$ matrix of rank R, and take M to be any positive integer greater than or equal to R. If $\mathbf{A}$ is symmetric and nonnegative definite, then there exists an $M \times N$ matrix $\mathbf{P}$ such that $\mathbf{A} = \mathbf{P}'\mathbf{P}$ (and any such $M \times N$ matrix $\mathbf{P}$ is of rank R).

Proof. Suppose that $\mathbf{A}$ is symmetric and nonnegative definite. And assume that $R > 0$—when $R = 0$, $\mathbf{A} = \mathbf{0}'\mathbf{0}$. According to Corollary 2.13.23, there exists an $R \times N$ matrix $\mathbf{P}_1$ such that $\mathbf{A} = \mathbf{P}_1'\mathbf{P}_1$. Take $\mathbf{P}$ to be the $M \times N$ matrix of the form $\mathbf{P} = \begin{pmatrix} \mathbf{P}_1 \\ \mathbf{0} \end{pmatrix}$. Then, clearly, $\mathbf{A} = \mathbf{P}'\mathbf{P}$. Moreover, it is clear from Lemma 2.12.1 that $\text{rank}(\mathbf{P}) = R$ for any $M \times N$ matrix $\mathbf{P}$ such that $\mathbf{A} = \mathbf{P}'\mathbf{P}$. Q.E.D.

In light of Corollary 2.13.15, Corollary 2.13.24 has the following implication.

Corollary 2.13.25. An $N \times N$ matrix $\mathbf{A}$ is a symmetric nonnegative definite matrix if and only if there exists a matrix $\mathbf{P}$ (having N columns) such that $\mathbf{A} = \mathbf{P}'\mathbf{P}$.

Further results on symmetric nonnegative definite matrices are given by the following three corollaries.

Corollary 2.13.26. Let $\mathbf{A}$ represent an $N \times N$ nonnegative definite matrix, and let $R = \text{rank}(\mathbf{A} + \mathbf{A}')$. Then, assuming that $R > 0$, there exists an $R \times N$ matrix $\mathbf{P}$ (of full row rank R) such that the quadratic form $\mathbf{x}'\mathbf{A}\mathbf{x}$ (in an N-dimensional vector $\mathbf{x}$) is expressible as the sum $\sum_{i=1}^{R} y_i^2$ of the squares of the elements $y_1, y_2, \ldots, y_R$ of the R-dimensional vector $\mathbf{y} = \mathbf{P}\mathbf{x}$.

Proof. According to Corollary 2.13.2, there is a unique symmetric matrix $\mathbf{B}$ such that $\mathbf{x}'\mathbf{A}\mathbf{x} = \mathbf{x}'\mathbf{B}\mathbf{x}$ for all $\mathbf{x}$, namely, the matrix $\mathbf{B} = (1/2)(\mathbf{A} + \mathbf{A}')$. Moreover, $\mathbf{B}$ is nonnegative definite, and (assuming that $R > 0$) it follows from Corollary 2.13.23 that there exists an $R \times N$ matrix $\mathbf{P}$ (of full row rank R) such that $\mathbf{B} = \mathbf{P}'\mathbf{P}$. Thus, letting $y_1, y_2, \ldots, y_R$ represent the elements of the R-dimensional vector $\mathbf{y} = \mathbf{P}\mathbf{x}$, we find that

$$\mathbf{x}'\mathbf{A}\mathbf{x} = \mathbf{x}'\mathbf{P}'\mathbf{P}\mathbf{x} = (\mathbf{P}\mathbf{x})'\mathbf{P}\mathbf{x} = \sum_{i=1}^{R} y_i^2. \qquad \text{Q.E.D.}$$

Corollary 2.13.27. For any $N \times M$ matrix $\mathbf{X}$ and any $N \times N$ symmetric nonnegative definite matrix $\mathbf{A}$, $\mathbf{A}\mathbf{X} = \mathbf{0}$ if and only if $\mathbf{X}'\mathbf{A}\mathbf{X} = \mathbf{0}$.

Proof. According to Corollary 2.13.25, there exists a matrix $\mathbf{P}$ such that $\mathbf{A} = \mathbf{P}'\mathbf{P}$ and hence such that $\mathbf{X}'\mathbf{A}\mathbf{X} = (\mathbf{P}\mathbf{X})'\mathbf{P}\mathbf{X}$. Thus, if $\mathbf{X}'\mathbf{A}\mathbf{X} = \mathbf{0}$, then (in light of Corollary 2.3.3) $\mathbf{P}\mathbf{X} = \mathbf{0}$, implying that $\mathbf{A}\mathbf{X} = \mathbf{P}'\mathbf{P}\mathbf{X} = \mathbf{P}'\mathbf{0} = \mathbf{0}$. That $\mathbf{X}'\mathbf{A}\mathbf{X} = \mathbf{0}$ if $\mathbf{A}\mathbf{X} = \mathbf{0}$ is obvious. Q.E.D.

Corollary 2.13.28. A symmetric nonnegative definite matrix is positive definite if and only if it is nonsingular (or, equivalently, is positive semidefinite if and only if it is singular).

Proof. Let $\mathbf{A}$ represent an $N \times N$ symmetric nonnegative definite matrix. If $\mathbf{A}$ is positive definite, then we have, as an immediate consequence of Lemma 2.13.9, that $\mathbf{A}$ is nonsingular.

Suppose now that the symmetric nonnegative definite matrix $\mathbf{A}$ is nonsingular, and consider the quadratic form $\mathbf{x}'\mathbf{A}\mathbf{x}$ (in $\mathbf{x}$). If $\mathbf{x}'\mathbf{A}\mathbf{x} = 0$, then, according to Corollary 2.13.27, $\mathbf{A}\mathbf{x} = \mathbf{0}$, and consequently $\mathbf{x} = \mathbf{A}^{-1}\mathbf{A}\mathbf{x} = \mathbf{A}^{-1}\mathbf{0} = \mathbf{0}$. Thus, the quadratic form $\mathbf{x}'\mathbf{A}\mathbf{x}$ is positive definite and hence the matrix $\mathbf{A}$ is positive definite. Q.E.D.

When specialized to positive definite matrices, Corollary 2.13.23, in combination with Lemma 2.13.9 and Corollary 2.13.15, yields the following result.

Corollary 2.13.29. An $N \times N$ matrix $\mathbf{A}$ is a symmetric positive definite matrix if and only if there exists a nonsingular matrix $\mathbf{P}$ such that $\mathbf{A} = \mathbf{P}'\mathbf{P}$.

In the special case of a positive definite matrix, Corollary 2.13.26 can (in light of Corollary 2.13.2 and Lemma 2.13.9) be restated as follows—Corollary 2.13.2 implies that if $\mathbf{A}$ is positive definite, then so is $(1/2)(\mathbf{A} + \mathbf{A}')$.

Corollary 2.13.30. Let $\mathbf{A}$ represent an $N \times N$ positive definite matrix. Then, there exists a nonsingular matrix $\mathbf{P}$ such that the quadratic form $\mathbf{x}'\mathbf{A}\mathbf{x}$ (in an N-dimensional vector $\mathbf{x}$) is expressible as the sum $\sum_{i=1}^{N} y_i^2$ of the squares of the elements $y_1, y_2, \ldots, y_N$ of the transformed vector $\mathbf{y} = \mathbf{P}\mathbf{x}$.

c. Positive definiteness or semidefiniteness of partitioned matrices

Lemma 2.13.5 characterizes the concepts of nonnegative definiteness, positive definiteness, and positive semidefiniteness as applied to diagonal matrices. The following lemma extends the results of Lemma 2.13.5 to block-diagonal matrices.

Lemma 2.13.31. Let $\mathbf{A}_i$ represent an $N_i \times N_i$ matrix ($i = 1, 2, \ldots, K$), let $N = N_1 + N_2 + \cdots + N_K$, and define $\mathbf{A}$ to be the $N \times N$ block-diagonal matrix $\mathbf{A} = \mathrm{diag}(\mathbf{A}_1, \mathbf{A}_2, \ldots, \mathbf{A}_K)$. Then, (1) $\mathbf{A}$ is nonnegative definite if and only if $\mathbf{A}_1, \mathbf{A}_2, \ldots, \mathbf{A}_K$ are nonnegative definite; (2) $\mathbf{A}$ is positive definite if and only if $\mathbf{A}_1, \mathbf{A}_2, \ldots, \mathbf{A}_K$ are positive definite; and (3) $\mathbf{A}$ is positive semidefinite if and only if the diagonal blocks $\mathbf{A}_1, \mathbf{A}_2, \ldots, \mathbf{A}_K$ are nonnegative definite with at least one of the diagonal blocks being positive semidefinite.

Proof. Consider the quadratic form $\mathbf{x}'\mathbf{A}\mathbf{x}$ (in an N-dimensional column vector $\mathbf{x}$) whose matrix is $\mathbf{A}$. Partition $\mathbf{x}'$ as $\mathbf{x}' = (\mathbf{x}_1', \mathbf{x}_2', \ldots, \mathbf{x}_K')$, where (for $i = 1, 2, \ldots, K$) $\mathbf{x}_i$ is an N_i-dimensional (column) vector. Then, clearly,

$$\mathbf{x}'\mathbf{A}\mathbf{x} = \mathbf{x}_1'\mathbf{A}_1\mathbf{x}_1 + \mathbf{x}_2'\mathbf{A}_2\mathbf{x}_2 + \cdots + \mathbf{x}_K'\mathbf{A}_K\mathbf{x}_K.$$

(1) If $\mathbf{A}_1, \mathbf{A}_2, \ldots, \mathbf{A}_K$ are nonnegative definite, then, by definition, $\mathbf{x}_i'\mathbf{A}_i\mathbf{x}_i \geq 0$ for every $\mathbf{x}_i$ ($i = 1, 2, \ldots, K$), implying that $\mathbf{x}'\mathbf{A}\mathbf{x} \geq 0$ for every $\mathbf{x}$ and hence that $\mathbf{A}$ is nonnegative definite. Conversely, if $\mathbf{A}$ is nonnegative definite, then it follows from Corollary 2.13.13 that $\mathbf{A}_1, \mathbf{A}_2, \ldots, \mathbf{A}_K$ are nonnegative definite.

(2) If $\mathbf{A}_1, \mathbf{A}_2, \ldots, \mathbf{A}_K$ are positive definite, then, by definition, $\mathbf{x}_i'\mathbf{A}_i\mathbf{x}_i > 0$ for every nonnull $\mathbf{x}_i$ ($i = 1, 2, \ldots, K$), implying that $\mathbf{x}'\mathbf{A}\mathbf{x} > 0$ for every nonnull $\mathbf{x}$ and hence that $\mathbf{A}$ is positive definite. Conversely, if $\mathbf{A}$ is positive definite, then it follows from Corollary 2.13.13 that $\mathbf{A}_1, \mathbf{A}_2, \ldots, \mathbf{A}_K$ are positive definite.

(3) Suppose that $\mathbf{A}_1, \mathbf{A}_2, \ldots, \mathbf{A}_K$ are nonnegative definite and that for some i, say $i = i^*$, $\mathbf{A}_i$ is positive semidefinite. Then, by definition, $\mathbf{x}_i'\mathbf{A}_i\mathbf{x}_i \geq 0$ for every $\mathbf{x}_i$ ($i = 1, 2, \ldots, K$), and there exists some nonnull value of $\mathbf{x}_{i^*}$, say $\mathbf{x}_{i^*} = \tilde{\mathbf{x}}_{i^*}$, for which $\mathbf{x}_{i^*}'\mathbf{A}_{i^*}\mathbf{x}_{i^*} = 0$. It follows that $\mathbf{x}'\mathbf{A}\mathbf{x} \geq 0$ for every $\mathbf{x}$, with equality holding for $\mathbf{x}' = (\mathbf{0}, \ldots, \mathbf{0}, \tilde{\mathbf{x}}_{i^*}', \mathbf{0}, \ldots, \mathbf{0})$. Thus, $\mathbf{A}$ is positive semidefinite.

Conversely, suppose that $\mathbf{A}$ is positive semidefinite. Then, it follows from Part (1) that $\mathbf{A}_1, \mathbf{A}_2, \ldots, \mathbf{A}_K$ are nonnegative definite. Moreover, it follows from Part (2) that not all of the matrices $\mathbf{A}_1, \mathbf{A}_2, \ldots, \mathbf{A}_K$ are positive definite and hence (since they are nonnegative definite) that at least one of them is positive semidefinite. Q.E.D.

The following theorem relates the positive definiteness or semidefiniteness of a symmetric matrix to the positive definiteness or semidefiniteness of the Schur complement of a positive definite principal submatrix.

Theorem 2.13.32. Let $\mathbf{T}$ represent an $M \times M$ symmetric matrix, $\mathbf{W}$ an $N \times N$ symmetric matrix, and $\mathbf{U}$ an $M \times N$ matrix. Suppose that $\mathbf{T}$ is positive definite, and define

$$\mathbf{Q} = \mathbf{W} - \mathbf{U}'\mathbf{T}^{-1}\mathbf{U}.$$

Then, the partitioned (symmetric) matrix $\begin{pmatrix} \mathbf{T} & \mathbf{U} \\ \mathbf{U}' & \mathbf{W} \end{pmatrix}$ is positive definite if and only if $\mathbf{Q}$ is positive definite and is positive semidefinite if and only if $\mathbf{Q}$ is positive semidefinite. Similarly, the partitioned (symmetric) matrix $\begin{pmatrix} \mathbf{W} & \mathbf{U}' \\ \mathbf{U} & \mathbf{T} \end{pmatrix}$ is positive definite if and only if $\mathbf{Q}$ is positive definite and is positive semidefinite if and only if $\mathbf{Q}$ is positive semidefinite.

Proof. Let $\mathbf{A} = \begin{pmatrix} \mathbf{T} & \mathbf{U} \\ \mathbf{U}' & \mathbf{W} \end{pmatrix}$, and define $\mathbf{P} = \begin{pmatrix} \mathbf{I}_M & -\mathbf{T}^{-1}\mathbf{U} \\ \mathbf{0} & \mathbf{I}_N \end{pmatrix}$. According to Lemma 2.6.2, $\mathbf{P}$ is nonsingular. Moreover,

$$\mathbf{P}'\mathbf{A}\mathbf{P} = \mathrm{diag}(\mathbf{T}, \mathbf{Q}),$$

as is easily verified. And, upon observing that

$$(\mathbf{P}^{-1})' \,\mathrm{diag}(\mathbf{T}, \mathbf{Q})\mathbf{P}^{-1} = \mathbf{A},$$

it follows from Corollary 2.13.11 that $\mathbf{A}$ is positive definite if and only if diag($\mathbf{T}$, $\mathbf{Q}$) is positive definite and is positive semidefinite if and only if diag($\mathbf{T}, \mathbf{Q}$) is positive semidefinite. In light of Lemma 2.13.31, we conclude that $\mathbf{A}$ is positive definite if and only if $\mathbf{Q}$ is positive definite and is positive semidefinite if and only if $\mathbf{Q}$ is positive semidefinite, thereby completing the proof of the first part of Theorem 2.13.32. The second part of Theorem 2.13.32 can be proved in similar fashion. Q.E.D.

As a corollary of Theorem 2.13.32, we have the following result on the positive definiteness of a symmetric matrix.

Corollary 2.13.33. Suppose that a symmetric matrix $\mathbf{A}$ is partitioned as

$$\mathbf{A} = \begin{pmatrix} \mathbf{T} & \mathbf{U} \\ \mathbf{U}' & \mathbf{W} \end{pmatrix}$$

(where $\mathbf{T}$ and $\mathbf{W}$ are square). Then, $\mathbf{A}$ is positive definite if and only if $\mathbf{T}$ and the Schur complement $\mathbf{W} - \mathbf{U}'\mathbf{T}^{-1}\mathbf{U}$ of $\mathbf{T}$ are both positive definite. Similarly, $\mathbf{A}$ is positive definite if and only if $\mathbf{W}$ and the Schur complement $\mathbf{T} - \mathbf{U}\mathbf{W}^{-1}\mathbf{U}'$ of $\mathbf{W}$ are both positive definite.

Proof. If $\mathbf{T}$ is positive definite (in which case $\mathbf{T}$ is nonsingular) and $\mathbf{W} - \mathbf{U}'\mathbf{T}^{-1}\mathbf{U}$ is positive definite, then it follows from the first part of Theorem 2.13.32 that $\mathbf{A}$ is positive definite. Similarly, if $\mathbf{W}$ is positive definite (in which case $\mathbf{W}$ is nonsingular) and $\mathbf{T} - \mathbf{U}\mathbf{W}^{-1}\mathbf{U}'$ is positive definite, then it follows from the second part of Theorem 2.13.32 that $\mathbf{A}$ is positive definite.

Conversely, suppose that $\mathbf{A}$ is positive definite. Then, it follows from Corollary 2.13.13 that $\mathbf{T}$ is positive definite and also that $\mathbf{W}$ is positive definite. And, based on the first part of Theorem 2.13.32, we conclude that $\mathbf{W} - \mathbf{U}'\mathbf{T}^{-1}\mathbf{U}$ is positive definite; similarly, based on the second part of that theorem, we conclude that $\mathbf{T} - \mathbf{U}\mathbf{W}^{-1}\mathbf{U}'$ is positive definite. Q.E.D.

2.14 Determinants

Associated with any square matrix is a scalar that is known as the determinant of the matrix. As a preliminary to defining the determinant, it is convenient to introduce a convention for classifying various pairs of matrix elements as either positive or negative.

Let $\mathbf{A} = \{a_{ij}\}$ represent an arbitrary $N \times N$ matrix. Consider any pair of elements of $\mathbf{A}$ that do not lie either in the same row or the same column, say a_{ij} and $a_{i'j'}$ (where $i' \neq i$ and $j' \neq j$). The pair is said to be a *negative pair* if one of the elements is located above and to the right of the other, or, equivalently, if either $i' > i$ and $j' < j$ or $i' < i$ and $j' > j$. Otherwise (if one of the elements is located above and to the left of the other, or, equivalently, if either $i' > i$ and $j' > j$ or $i' < i$ and $j' < j$), the pair is said to be a *positive pair*. Thus, the pair a_{ij} and $a_{i'j'}$ is classified as positive or negative in accordance with the following two-way table:

	$i' > i$	$i' < i$
$j' > j$	$+$	$-$
$j' < j$	$-$	$+$

For example (supposing that $N \geq 4$), the pair a_{34} and a_{22} is positive, while the pair a_{34} and a_{41} is negative. Note that whether the pair a_{ij} and $a_{i'j'}$ is positive or negative is completely determined by the relative locations of a_{ij} and $a_{i'j'}$ and has nothing to do with whether a_{ij} and $a_{i'j'}$ are positive or negative numbers.

Now, consider N elements of $\mathbf{A}$, no two of which lie either in the same row or the same column, say the $i_1j_1, i_2j_2, \ldots, i_Nj_N$th elements (where both $i_1, i_2, \ldots, i_N$ and $j_1, j_2, \ldots, j_N$ are permutations of the first N positive integers). A total of $\binom{N}{2}$ pairs can be formed from these N elements. The symbol $\sigma_N(i_1, j_1; i_2, j_2; \ldots; i_N, j_N)$ is to be used to represent the number of these $\binom{N}{2}$ pairs that are negative pairs.

Observe that $\sigma_N(i_1, j_1; i_2, j_2; \ldots; i_N, j_N)$ has the following two properties:

(1) the value of $\sigma_N(i_1, j_1; i_2, j_2; \ldots; i_N, j_N)$ is not affected by permuting its N pairs of arguments; in particular, it is not affected if the N pairs are permuted so that they are ordered by row number or by column number [e.g., $\sigma_3(2,3;\ 1,2;\ 3,1) = \sigma_3(1,2;\ 2,3;\ 3,1) = \sigma_3(3,1;\ 1,2;\ 2,3)$];

(2) $$\sigma_N(i_1, j_1; i_2, j_2; \ldots; i_N, j_N) = \sigma_N(j_1, i_1; j_2, i_2; \ldots; j_N, i_N)$$
[e.g., $\sigma_3(2,3;\ 1,2;\ 3,1) = \sigma_3(3,2;\ 2,1;\ 1,3)$].

For any sequence of N distinct integers $i_1, i_2, \ldots, i_N$, define

$$\phi_N(i_1, i_2, \ldots, i_N) = p_1 + p_2 + \cdots + p_{N-1},$$

where (for $k = 1, 2, \ldots, N-1$) p_k represents the number of integers in the subsequence $i_{k+1}, i_{k+2}, \ldots, i_N$ that are smaller than i_k. For example,

$$\phi_5(3,7,2,1,4) = 2 + 3 + 1 + 0 = 6.$$

Then, clearly, for any permutation $i_1, i_2, \ldots, i_N$ of the first N positive integers,

$$\sigma_N(1, i_1; 2, i_2; \ldots; N, i_N) = \sigma_N(i_1, 1; i_2, 2; \ldots; i_N, N) = \phi_N(i_1, i_2, \ldots, i_N). \tag{14.1}$$

The *determinant* of an $N \times N$ matrix $\mathbf{A} = \{a_{ij}\}$, to be denoted by $|\mathbf{A}|$ or (to avoid confusion with the absolute value of a scalar) by $\det \mathbf{A}$ or $\det(\mathbf{A})$, is defined by

$$|\mathbf{A}| = \sum (-1)^{\sigma_N(1,j_1;2,j_2;\ldots;N,j_N)} a_{1j_1} a_{2j_2} \cdots a_{Nj_N} \tag{14.2}$$

or, equivalently, by

$$|\mathbf{A}| = \sum (-1)^{\phi_N(j_1,j_2,\ldots,j_N)} a_{1j_1} a_{2j_2} \cdots a_{Nj_N}, \tag{14.2$'$}$$

where $j_1, j_2, \ldots, j_N$ is a permutation of the first N positive integers and the summation is over all such permutations.

Thus, the determinant of an $N \times N$ matrix $\mathbf{A}$ can (at least in principle) be obtained via the following process:

(1) Form all possible products, each of N factors, that can be obtained by picking one and only one element from each row and column of $\mathbf{A}$.

(2) In each product, count the number of negative pairs among the $\binom{N}{2}$ pairs of elements that can be generated from the N elements that contribute to this particular product. If the number of negative pairs is an even number, attach a plus sign to the product; if it is an odd number, attach a minus sign.

(3) Sum the signed products.

In particular, the determinant of a 1×1 matrix $\mathbf{A} = (a_{11})$ is

$$|\mathbf{A}| = a_{11}; \tag{14.3}$$

the determinant of a 2×2 matrix $\mathbf{A} = \{a_{ij}\}$ is

$$|\mathbf{A}| = (-1)^0 a_{11}a_{22} + (-1)^1 a_{12}a_{21} = a_{11}a_{22} - a_{12}a_{21}; \tag{14.4}$$

and the determinant of a 3×3 matrix $\mathbf{A} = \{a_{ij}\}$ is

$$\begin{aligned}|\mathbf{A}| &= (-1)^0 a_{11}a_{22}a_{33} + (-1)^1 a_{11}a_{23}a_{32} + (-1)^1 a_{12}a_{21}a_{33} \\ &\quad + (-1)^2 a_{12}a_{23}a_{31} + (-1)^2 a_{13}a_{21}a_{32} + (-1)^3 a_{13}a_{22}a_{31} \\ &= a_{11}a_{22}a_{33} + a_{12}a_{23}a_{31} + a_{13}a_{21}a_{32} \\ &\quad - a_{11}a_{23}a_{32} - a_{12}a_{21}a_{33} - a_{13}a_{22}a_{31}. \end{aligned} \tag{14.5}$$

An alternative definition of the determinant of an $N \times N$ matrix $\mathbf{A}$ is

$$|\mathbf{A}| = \sum (-1)^{\sigma_N(i_1,1;i_2,2;\ldots;i_N,N)} a_{i_1 1} a_{i_2 2} \cdots a_{i_N N} \tag{14.6}$$

$$= \sum (-1)^{\phi_N(i_1,i_2,\ldots,i_N)} a_{i_1 1} a_{i_2 2} \cdots a_{i_N N}, \tag{14.6$'$}$$

where $i_1, i_2, \ldots, i_N$ is a permutation of the first N positive integers and the summation is over all such permutations.

Definition (14.6) is equivalent to definition (14.2). To see this, observe that the product $a_{1j_1}a_{2j_2}\cdots a_{Nj_N}$, which appears in definition (14.2), can be reexpressed by permuting the N factors $a_{1j_1}, a_{2j_2}, \ldots, a_{Nj_N}$ so that they are ordered by column number, giving

$$a_{1j_1}a_{2j_2}\cdots a_{Nj_N} = a_{i_1 1}a_{i_2 2}\cdots a_{i_N N},$$

where $i_1, i_2, \ldots, i_N$ is a permutation of the first N positive integers that is defined uniquely by

$$j_{i_1} = 1,\ j_{i_2} = 2,\ \ldots,\ j_{i_N} = N.$$

Further,

$$\begin{aligned}\sigma_N(1, j_1; 2, j_2; \ldots; N, j_N) &= \sigma_N(i_1, j_{i_1}; i_2, j_{i_2}, \ldots; i_N, j_{i_N}) \\ &= \sigma_N(i_1, 1; i_2, 2; \ldots; i_N, N),\end{aligned}$$

so that

$$(-1)^{\sigma_N(1,j_1;2,j_2;\ldots;N,j_N)} a_{1j_1}a_{2j_2}\cdots a_{Nj_N} = (-1)^{\sigma_N(i_1,1;i_2,2;\ldots;i_N,N)} a_{i_1 1}a_{i_2 2}\cdots a_{i_N N}.$$

Thus, we can establish a one-to-one correspondence between the terms of the sum (14.6) and the terms of the sum (14.2) such that the corresponding terms are equal. We conclude that the two sums are themselves equal.

In considering the determinant of a partitioned matrix, say $\begin{pmatrix} \mathbf{A}_{11} & \mathbf{A}_{12} \\ \mathbf{A}_{21} & \mathbf{A}_{22} \end{pmatrix}$, it is customary to abbreviate $\left|\begin{pmatrix} \mathbf{A}_{11} & \mathbf{A}_{12} \\ \mathbf{A}_{21} & \mathbf{A}_{22} \end{pmatrix}\right|$ to $\begin{vmatrix} \mathbf{A}_{11} & \mathbf{A}_{12} \\ \mathbf{A}_{21} & \mathbf{A}_{22} \end{vmatrix}$.

a. Determinants of triangular and diagonal matrices

In the special case of a triangular matrix, the formula for the determinant simplifies greatly, as described in the following lemma.

Lemma 2.14.1. If an $N \times N$ matrix $\mathbf{A} = \{a_{ij}\}$ is (upper or lower) triangular, then

$$|\mathbf{A}| = a_{11}a_{22} \cdots a_{NN}, \tag{14.7}$$

that is, the determinant of a triangular matrix equals the product of its diagonal elements.

Proof. Consider the case where $\mathbf{A}$ is lower triangular, that is, of the form

$$\mathbf{A} = \begin{pmatrix} a_{11} & 0 & \dots & 0 \\ a_{21} & a_{22} & & 0 \\ \vdots & \vdots & \ddots & \\ a_{N1} & a_{N2} & \dots & a_{NN} \end{pmatrix}.$$

That $|\mathbf{A}| = a_{11}a_{22} \cdots a_{NN}$ is clear upon observing that the only term in the sum (14.2) or (14.2′) that can be nonzero is that corresponding to the permutation $j_1 = 1, j_2 = 2, \dots, j_N = N$ [and upon observing that $\phi_N(1, 2, \dots, N) = 0$]. (To verify formally that this is the only term that can be nonzero, let $j_1, j_2, \dots, j_N$ represent an arbitrary permutation of the first N positive integers, and suppose that $a_{1j_1}a_{2j_2} \cdots a_{Nj_N} \neq 0$ or, equivalently, that $a_{ij_i} \neq 0$ for $i = 1, 2, \dots, N$. Then, it is clear that $j_1 = 1$ and that if $j_1 = 1,\ j_2 = 2,\ \dots,\ j_{i-1} = i - 1$, then $j_i = i$. We conclude, on the basis of mathematical induction, that $j_1 = 1,\ j_2 = 2,\ \dots,\ j_N = N$.)

The validity of formula (14.7) as applied to an upper triangular matrix follows from a similar argument. Q.E.D.

Note that Lemma 2.14.1 implies in particular that the determinant of a unit (upper or lower) triangular matrix equals 1. And, as a further implication of Lemma 2.14.1, we have the following corollary.

Corollary 2.14.2. The determinant of a diagonal matrix equals the product of its diagonal elements.

As obvious special cases of Corollary 2.14.2, we have that

$$|\mathbf{0}| = 0, \tag{14.8}$$

$$|\mathbf{I}| = 1. \tag{14.9}$$

b. Some basic results on determinants

The following lemma relates the determinant of a matrix to the determinant of its transpose.

Lemma 2.14.3. For any $N \times N$ matrix $\mathbf{A}$,

$$|\mathbf{A}'| = |\mathbf{A}|. \tag{14.10}$$

Proof. Let a_{ij} and b_{ij} represent the ijth elements of $\mathbf{A}$ and $\mathbf{A}'$, respectively ($i, j = 1, 2, \dots, N$). Then, in light of the equivalence of definitions (14.2′) and (14.6′),

$$\begin{aligned} |\mathbf{A}'| &= \sum (-1)^{\phi_N(j_1, j_2, \dots, j_N)} b_{1j_1} b_{2j_2} \cdots b_{Nj_N} \\ &= \sum (-1)^{\phi_N(j_1, j_2, \dots, j_N)} a_{j_1 1} a_{j_2 2} \cdots a_{j_N N} \\ &= |\mathbf{A}|, \end{aligned}$$

where $j_1, j_2, \ldots, j_N$ is a permutation of the first N positive integers and the summations are over all such permutations. Q.E.D.

As an immediate consequence of the definition of a determinant, we have the following lemma.

Lemma 2.14.4. If an $N \times N$ matrix $\mathbf{B}$ is formed from an $N \times N$ matrix $\mathbf{A}$ by multiplying all of the elements of one row or one column of $\mathbf{A}$ by the same scalar k (and leaving the elements of the other $N-1$ rows or columns unchanged), then

$$|\mathbf{B}| = k|\mathbf{A}|.$$

As a corollary of Lemma 2.14.4, we obtain the following result on the determinant of a matrix having a null row or a null column.

Corollary 2.14.5. If one or more rows (or columns) of an $N \times N$ matrix $\mathbf{A}$ are null, then $|\mathbf{A}| = 0$.

Proof. Suppose that the ith row of $\mathbf{A}$ is null, and let $\mathbf{B}$ represent an $N \times N$ matrix formed from $\mathbf{A}$ by multiplying every element of the ith row of $\mathbf{A}$ by 0. Then, $\mathbf{A} = \mathbf{B}$, and we find that

$$|\mathbf{A}| = |\mathbf{B}| = 0|\mathbf{A}| = 0. \qquad \text{Q.E.D.}$$

The following corollary (of Lemma 2.14.4) relates the determinant of a scalar multiple of a matrix $\mathbf{A}$ to that of $\mathbf{A}$ itself.

Corollary 2.14.6. For any $N \times N$ matrix $\mathbf{A}$ and any scalar k,

$$|k\mathbf{A}| = k^N|\mathbf{A}|. \tag{14.11}$$

Proof. This result follows from Lemma 2.14.4 upon observing that $k\mathbf{A}$ can be formed from $\mathbf{A}$ by successively multiplying the N rows of $\mathbf{A}$ by k. Q.E.D.

As a special case of Corollary 2.14.6, we have the following, additional corollary.

Corollary 2.14.7. For any $N \times N$ matrix $\mathbf{A}$,

$$|-\mathbf{A}| = (-1)^N|\mathbf{A}|. \tag{14.12}$$

The following two theorems describe how the determinant of a matrix is affected by permuting its rows or columns in certain ways.

Theorem 2.14.8. If an $N \times N$ matrix $\mathbf{B} = \{b_{ij}\}$ is formed from an $N \times N$ matrix $\mathbf{A} = \{a_{ij}\}$ by interchanging two rows or two columns of $\mathbf{A}$, then

$$|\mathbf{B}| = -|\mathbf{A}|.$$

Proof. Consider first the case where $\mathbf{B}$ is formed from $\mathbf{A}$ by interchanging two adjacent rows, say the ith and $(i+1)$th rows. Then,

$$\begin{aligned}
|\mathbf{B}| &= \sum (-1)^{\phi_N(j_1,j_2,\ldots,j_N)} b_{1j_1} b_{2j_2} \cdots b_{i-1,j_{i-1}} b_{ij_i} b_{i+1,j_{i+1}} b_{i+2,j_{i+2}} \cdots b_{Nj_N} \\
&= \sum (-1)^{\phi_N(j_1,j_2,\ldots,j_N)} a_{1j_1} a_{2j_2} \cdots a_{i-1,j_{i-1}} a_{i+1,j_i} a_{i,j_{i+1}} a_{i+2,j_{i+2}} \cdots a_{Nj_N} \\
&= -\sum (-1)^{\phi_N(j_1,j_2,\ldots,j_{i-1},j_{i+1},j_i,j_{i+2},\ldots,j_N)} \\
&\qquad \times a_{1j_1} a_{2j_2} \cdots a_{i-1,j_{i-1}} a_{i,j_{i+1}} a_{i+1,j_i} a_{i+2,j_{i+2}} \cdots a_{Nj_N} \\
&\qquad \left[\begin{aligned} &\text{since } \phi_N(j_1, j_2, \ldots, j_{i-1}, j_{i+1}, j_i, j_{i+2}, \ldots, j_N) \\
&\qquad = \begin{cases} \phi_N(j_1, j_2, \ldots, j_N) + 1, & \text{if } j_{i+1} > j_i \\ \phi_N(j_1, j_2, \ldots, j_N) - 1, & \text{if } j_{i+1} < j_i \end{cases} \end{aligned} \right] \\
&= -|\mathbf{A}|,
\end{aligned}$$

where $j_1, j_2, \ldots, j_N$ (and hence $j_1, j_2, \ldots, j_{i-1}, j_{i+1}, j_i, j_{i+2}, \ldots, j_N$) is a permutation of the first N positive integers and the summation is over all such permutations.

Consider now the case where $\mathbf{B}$ is formed from $\mathbf{A}$ by interchanging two not-necessarily-adjacent rows, say the ith and kth rows where $k > i$. Suppose that we successively interchange the kth row of $\mathbf{A}$ with the $k - i$ rows immediately preceding it, putting the N rows of $\mathbf{A}$ in the order $1, 2, \ldots, i-1, k, i, i+1, \ldots, k-1, k+1, \ldots, N$. Suppose that we then further reorder the rows of $\mathbf{A}$ by successively interchanging what was originally the ith row with the $k-i-1$ rows immediately succeeding it, putting the N rows in the order $1, 2, \ldots, i-1, k, i+1, \ldots, k-1, i, k+1, \ldots, N$. Thus, by executing $2(k-i)-1$ successive interchanges of adjacent rows, we have in effect interchanged the ith and kth rows of $\mathbf{A}$. Since each interchange of adjacent rows changes the sign of the determinant, we conclude that

$$|\mathbf{B}| = (-1)^{2(k-i)-1}|\mathbf{A}| = -|\mathbf{A}|.$$

By employing an analogous argument, we find that the interchange of any two columns of $\mathbf{A}$ likewise changes the sign of the determinant. Q.E.D.

Theorem 2.14.9. If $\mathbf{B}$ is an $N \times P$ matrix (where $P < N$) and $\mathbf{C}$ an $N \times Q$ matrix (where $Q = N - P$), then

$$|\mathbf{B}, \mathbf{C}| = (-1)^{PQ}|\mathbf{C}, \mathbf{B}|. \tag{14.13}$$

Similarly, if $\mathbf{B}$ is a $P \times N$ matrix and $\mathbf{C}$ a $Q \times N$ matrix, then

$$\begin{vmatrix} \mathbf{B} \\ \mathbf{C} \end{vmatrix} = (-1)^{PQ} \begin{vmatrix} \mathbf{C} \\ \mathbf{B} \end{vmatrix}. \tag{14.14}$$

Proof. Consider the case where $\mathbf{B}$ is $N \times P$ and $\mathbf{C}$ is $N \times Q$. Let $\mathbf{b}_1, \mathbf{b}_2, \ldots, \mathbf{b}_P$ represent the columns of $\mathbf{B}$ and $\mathbf{c}_1, \mathbf{c}_2, \ldots, \mathbf{c}_Q$ the columns of $\mathbf{C}$. Then, $(\mathbf{C}, \mathbf{B}) = (\mathbf{c}_1, \mathbf{c}_2, \ldots, \mathbf{c}_Q, \mathbf{b}_1, \mathbf{b}_2, \ldots, \mathbf{b}_P)$.

Suppose that in the matrix $(\mathbf{C}, \mathbf{B})$, we successively interchange the column $\mathbf{b}_1$ with the columns $\mathbf{c}_Q, \ldots, \mathbf{c}_2, \mathbf{c}_1$, producing the matrix $(\mathbf{b}_1, \mathbf{c}_1, \mathbf{c}_2, \ldots, \mathbf{c}_Q, \mathbf{b}_2, \ldots, \mathbf{b}_P)$. And suppose that in the latter matrix, we successively interchange the column $\mathbf{b}_2$ with the columns $\mathbf{c}_Q, \ldots, \mathbf{c}_2, \mathbf{c}_1$, producing the matrix $(\mathbf{b}_1, \mathbf{b}_2, \mathbf{c}_1, \mathbf{c}_2, \ldots, \mathbf{c}_Q, \mathbf{b}_3, \ldots, \mathbf{b}_P)$. Continuing in this fashion, we produce (after P steps) the matrix $(\mathbf{b}_1, \mathbf{b}_2, \ldots, \mathbf{b}_P, \mathbf{c}_1, \mathbf{c}_2, \ldots, \mathbf{c}_Q) = (\mathbf{B}, \mathbf{C})$.

It is now clear that we can obtain the matrix $(\mathbf{B}, \mathbf{C})$ from the matrix $(\mathbf{C}, \mathbf{B})$ via a total of PQ successive interchanges of columns. Thus, it follows from Theorem 2.14.8 that

$$|\mathbf{B}, \mathbf{C}| = (-1)^{PQ}|\mathbf{C}, \mathbf{B}|.$$

Result (14.14) can be derived via an analogous approach. Q.E.D.

A (square) matrix that has one or more null rows or columns has (according to Corollary 2.14.5) a zero determinant. Other matrices whose determinants are zero are described in the following two lemmas.

Lemma 2.14.10. If two rows or two columns of an $N \times N$ matrix $\mathbf{A}$ are identical, then $|\mathbf{A}| = 0$.

Proof. Suppose that two rows of $\mathbf{A}$ are identical, say the ith and kth rows, and let $\mathbf{B}$ represent a matrix formed from $\mathbf{A}$ by interchanging its ith and kth rows. Obviously, $\mathbf{B} = \mathbf{A}$ and hence $|\mathbf{B}| = |\mathbf{A}|$. Moreover, according to Theorem 2.14.8, $|\mathbf{B}| = -|\mathbf{A}|$. Thus, $|\mathbf{A}| = |\mathbf{B}| = -|\mathbf{A}|$, implying that $|\mathbf{A}| = 0$.

That the determinant of a (square) matrix having two identical columns equals zero can be proved via an analogous argument. Q.E.D.

Lemma 2.14.11. If a row or column of an $N \times N$ matrix $\mathbf{A}$ is a scalar multiple of another row or column, then $|\mathbf{A}| = 0$.

Proof. Let $\mathbf{a}'_1, \mathbf{a}'_2, \ldots, \mathbf{a}'_N$ represent the rows of $\mathbf{A}$. Suppose that one row is a scalar multiple of another, that is, suppose that $\mathbf{a}'_s = k\mathbf{a}'_i$ for some s and i (with $s \neq i$) and some scalar k. Let $\mathbf{B}$

represent a matrix formed from $\mathbf{A}$ by multiplying the ith row of $\mathbf{A}$ by the scalar k. Then, according to Lemmas 2.14.4 and 2.14.10,

$$k|\mathbf{A}| = |\mathbf{B}| = 0. \tag{14.15}$$

If $k \neq 0$, then it follows from equality (14.15) that $|\mathbf{A}| = 0$. If $k = 0$, then $\mathbf{a}'_s = \mathbf{0}$, and it follows from Corollary 2.14.5 that $|\mathbf{A}| = 0$. Thus, in either case, $|\mathbf{A}| = 0$.

An analogous argument shows that if one column of a (square) matrix is a scalar multiple of another, then again the determinant of the matrix equals zero. Q.E.D.

The transposition of a (square) matrix does not (according to Lemma 2.14.3) affect its determinant. Other operations that do not affect the determinant of a matrix are described in the following two theorems.

Theorem 2.14.12. Let $\mathbf{B}$ represent a matrix formed from an $N \times N$ matrix $\mathbf{A}$ by adding, to any one row or column of $\mathbf{A}$, scalar multiples of one or more other rows or columns. Then, $|\mathbf{B}| = |\mathbf{A}|$.

Proof. Let $\mathbf{a}'_i = (a_{i1}, a_{i2}, \ldots, a_{iN})$ and $\mathbf{b}'_i = (b_{i1}, b_{i2}, \ldots, b_{iN})$ represent the ith rows of $\mathbf{A}$ and $\mathbf{B}$, respectively ($i = 1, 2, \ldots, N$). And suppose that for some integer s ($1 \leq s \leq N$) and some scalars $k_1, k_2, \ldots, k_{s-1}, k_{s+1}, \ldots, k_N$,

$$\mathbf{b}'_s = \mathbf{a}'_s + \sum_{i \neq s} k_i \mathbf{a}'_i \quad \text{and} \quad \mathbf{b}'_i = \mathbf{a}'_i \quad (i \neq s).$$

Then,

$$\begin{aligned}
|\mathbf{B}| &= \sum (-1)^{\phi_N(j_1, j_2, \ldots, j_N)} b_{1j_1} b_{2j_2} \cdots b_{Nj_N} \\
&= \sum (-1)^{\phi_N(j_1, j_2, \ldots, j_N)} a_{1j_1} a_{2j_2} \cdots a_{s-1, j_{s-1}} \Big(a_{sj_s} + \sum_{i \neq s} k_i a_{ij_s}\Big) a_{s+1, j_{s+1}} \cdots a_{Nj_N} \\
&= |\mathbf{A}| + \sum_{i \neq s} \sum (-1)^{\phi_N(j_1, j_2, \ldots, j_N)} a_{1j_1} a_{2j_2} \cdots a_{s-1, j_{s-1}} (k_i a_{ij_s}) a_{s+1, j_{s+1}} \cdots a_{Nj_N} \\
&= |\mathbf{A}| + \sum_{i \neq s} |\mathbf{B}_i|,
\end{aligned}$$

where $\mathbf{B}_i$ is a matrix formed from $\mathbf{A}$ by replacing the sth row of $\mathbf{A}$ with $k_i \mathbf{a}'_i$ and where $j_1, j_2, \ldots, j_N$ is a permutation of the first N positive integers and the (unlabeled) summations are over all such permutations. Since (according to Lemma 2.14.11) $|\mathbf{B}_i| = 0$ ($i \neq s$), we conclude that $|\mathbf{B}| = |\mathbf{A}|$.

An analogous argument shows that $|\mathbf{B}| = |\mathbf{A}|$ when $\mathbf{B}$ is formed from $\mathbf{A}$ by adding, to a column of $\mathbf{A}$, scalar multiples of other columns. Q.E.D.

Theorem 2.14.13. For any $N \times N$ matrix $\mathbf{A}$ and any $N \times N$ unit (upper or lower) triangular matrix $\mathbf{T}$,

$$|\mathbf{AT}| = |\mathbf{TA}| = |\mathbf{A}|. \tag{14.16}$$

Proof. Consider the case where $\mathbf{A}$ is postmultiplied by $\mathbf{T}$ and $\mathbf{T}$ is unit lower triangular. Define $\mathbf{T}_i$ to be a matrix formed from $\mathbf{I}_N$ by replacing the ith column of $\mathbf{I}_N$ with the ith column of $\mathbf{T}$ ($i = 1, 2, \ldots, N$). Then, $\mathbf{T} = \mathbf{T}_1 \mathbf{T}_2 \cdots \mathbf{T}_N$ (as is easily verified), and consequently

$$\mathbf{AT} = \mathbf{AT}_1 \mathbf{T}_2 \cdots \mathbf{T}_N.$$

Now, define $\mathbf{B}_0 = \mathbf{A}$, and $\mathbf{B}_i = \mathbf{AT}_1 \mathbf{T}_2 \cdots \mathbf{T}_i$ ($i = 1, 2, \ldots, N - 1$). Clearly, to show that $|\mathbf{AT}| = |\mathbf{A}|$, it suffices to show that, for $i = 1, 2, \ldots, N$, the postmultiplication of $\mathbf{B}_{i-1}$ by $\mathbf{T}_i$ does not alter the determinant of $\mathbf{B}_{i-1}$. Observe that the columns of $\mathbf{B}_{i-1}\mathbf{T}_i$ are the same as those of $\mathbf{B}_{i-1}$, except for the ith column of $\mathbf{B}_{i-1}\mathbf{T}_i$, which consists of the ith column of $\mathbf{B}_{i-1}$ plus scalar multiples of the $(i + 1), \ldots, N$th columns of $\mathbf{B}_{i-1}$. Thus, it follows from Theorem 2.14.12 that $|\mathbf{B}_{i-1}\mathbf{T}_i| = |\mathbf{B}_{i-1}|$. We conclude that $|\mathbf{AT}| = |\mathbf{A}|$.

The validity of the parts of result (14.16) that pertain to the postmultiplication of $\mathbf{A}$ by a unit upper triangular matrix and the premultiplication of $\mathbf{A}$ by a unit upper or lower triangular matrix can be established via similar arguments. Q.E.D.

c. Determinants of block-triangular matrices

Formula (14.7) for the determinant of a triangular matrix can be extended to a block-triangular matrix based on the following theorem.

Theorem 2.14.14. Let $\mathbf{T}$ represent an $M \times M$ matrix, $\mathbf{V}$ an $N \times M$ matrix, and $\mathbf{W}$ an $N \times N$ matrix. Then,

$$\begin{vmatrix} \mathbf{T} & \mathbf{0} \\ \mathbf{V} & \mathbf{W} \end{vmatrix} = \begin{vmatrix} \mathbf{W} & \mathbf{V} \\ \mathbf{0} & \mathbf{T} \end{vmatrix} = |\mathbf{T}||\mathbf{W}|. \tag{14.17}$$

Proof. Let

$$\mathbf{A} = \begin{pmatrix} \mathbf{T} & \mathbf{0} \\ \mathbf{V} & \mathbf{W} \end{pmatrix},$$

and let a_{ij} represent the ijth element of $\mathbf{A}$ $(i, j = 1, 2, \ldots, M + N)$. Further, denote by t_{ij} the ijth element of $\mathbf{T}$ $(i, j = 1, 2, \ldots, M)$ and by w_{ij} the ijth element of $\mathbf{W}$ $(i, j = 1, 2, \ldots, N)$.

By definition,

$$|\mathbf{A}| = \sum (-1)^{\phi_{M+N}(j_1, \ldots, j_M, j_{M+1}, \ldots, j_{M+N})} \times a_{1j_1} \cdots a_{Mj_M} a_{M+1, j_{M+1}} \cdots a_{M+N, j_{M+N}}, \tag{14.18}$$

where $j_1, \ldots, j_M, j_{M+1}, \ldots, j_{M+N}$ is a permutation of the first $M + N$ positive integers and the summation is over all such permutations. Clearly, the only terms of the sum (14.18) that can be nonzero are those for which $j_1, \ldots, j_M$ constitutes a permutation of the first M positive integers and thus for which $j_{M+1}, \ldots, j_{M+N}$ constitutes a permutation of the integers $M + 1, \ldots, M + N$. For any such permutation, we have that

$$\begin{aligned} a_{1j_1} \cdots a_{Mj_M} a_{M+1, j_{M+1}} \cdots a_{M+N, j_{M+N}} &= t_{1j_1} \cdots t_{Mj_M} w_{1, j_{M+1}-M} \cdots w_{N, j_{M+N}-M} \\ &= t_{1j_1} \cdots t_{Mj_M} w_{1k_1} \cdots w_{Nk_N}, \end{aligned}$$

where $k_1 = j_{M+1} - M, \ldots, k_N = j_{M+N} - M$, and we also have that

$$\begin{aligned} \phi_{M+N}(j_1, \ldots, j_M, j_{M+1}, \ldots, j_{M+N}) &= \phi_M(j_1, \ldots, j_M) + \phi_N(j_{M+1}, \ldots, j_{M+N}) \\ &= \phi_M(j_1, \ldots, j_M) + \phi_N(j_{M+1} - M, \ldots, j_{M+N} - M) \\ &= \phi_M(j_1, \ldots, j_M) + \phi_N(k_1, \ldots, k_N). \end{aligned}$$

Thus,

$$\begin{aligned} |\mathbf{A}| &= \sum \sum (-1)^{\phi_M(j_1, \ldots, j_M) + \phi_N(k_1, \ldots, k_N)} t_{1j_1} \cdots t_{Mj_M} w_{1k_1} \cdots w_{Nk_N} \\ &= \sum (-1)^{\phi_M(j_1, \ldots, j_M)} t_{1j_1} \cdots t_{Mj_M} \sum (-1)^{\phi_N(k_1, \ldots, k_N)} w_{1k_1} \cdots w_{Nk_N} \\ &= |\mathbf{T}||\mathbf{W}|, \end{aligned}$$

where $j_1, \ldots, j_M$ is a permutation of the first M positive integers and $k_1, \ldots, k_N$ a permutation of the first N positive integers and where the respective summations are over all such permutations.

That $\begin{vmatrix} \mathbf{W} & \mathbf{V} \\ \mathbf{0} & \mathbf{T} \end{vmatrix} = |\mathbf{T}||\mathbf{W}|$ can be established via a similar argument. Q.E.D.

The repeated application of Theorem 2.14.14 leads to the following formulas for the determinant of an arbitrary (square) upper or lower block-triangular matrix (with square diagonal blocks):

$$\begin{vmatrix} \mathbf{A}_{11} & \mathbf{A}_{12} & \ldots & \mathbf{A}_{1R} \\ \mathbf{0} & \mathbf{A}_{22} & \ldots & \mathbf{A}_{2R} \\ \vdots & & \ddots & \vdots \\ \mathbf{0} & \mathbf{0} & & \mathbf{A}_{RR} \end{vmatrix} = |\mathbf{A}_{11}||\mathbf{A}_{22}| \cdots |\mathbf{A}_{RR}|; \tag{14.19}$$

$$\begin{vmatrix} \mathbf{B}_{11} & \mathbf{0} & \ldots & \mathbf{0} \\ \mathbf{B}_{21} & \mathbf{B}_{22} & & \mathbf{0} \\ \vdots & \vdots & \ddots & \\ \mathbf{B}_{R1} & \mathbf{B}_{R2} & \ldots & \mathbf{B}_{RR} \end{vmatrix} = |\mathbf{B}_{11}||\mathbf{B}_{22}| \cdots |\mathbf{B}_{RR}|. \tag{14.20}$$

In the special case of a block-diagonal matrix, formula (14.19) becomes

$$\text{diag}(\mathbf{A}_{11}, \mathbf{A}_{22}, \ldots, \mathbf{A}_{RR}) = |\mathbf{A}_{11}||\mathbf{A}_{22}| \cdots |\mathbf{A}_{RR}|. \tag{14.21}$$

Formulas (14.19), (14.20), and (14.21) generalize the results of Lemma 2.14.1 and Corollary 2.14.2 on the determinants of triangular and diagonal matrices.

As an immediate consequence of Theorem 2.14.9, we have the following corollary of Theorem 2.14.14.

Corollary 2.14.15. Let $\mathbf{T}$ represent an $M \times M$ matrix, $\mathbf{V}$ an $N \times M$ matrix, and $\mathbf{W}$ an $N \times N$ matrix. Then,

$$\begin{vmatrix} \mathbf{0} & \mathbf{T} \\ \mathbf{W} & \mathbf{V} \end{vmatrix} = \begin{vmatrix} \mathbf{V} & \mathbf{W} \\ \mathbf{T} & \mathbf{0} \end{vmatrix} = (-1)^{MN}|\mathbf{T}||\mathbf{W}|. \tag{14.22}$$

The following corollary gives a simplified version of formula (14.22) for the special case where $M = N$ and $\mathbf{T} = -\mathbf{I}_N$.

Corollary 2.14.16. For $N \times N$ matrices $\mathbf{W}$ and $\mathbf{V}$,

$$\begin{vmatrix} \mathbf{0} & -\mathbf{I}_N \\ \mathbf{W} & \mathbf{V} \end{vmatrix} = \begin{vmatrix} \mathbf{V} & \mathbf{W} \\ -\mathbf{I}_N & \mathbf{0} \end{vmatrix} = |\mathbf{W}|. \tag{14.23}$$

Proof (of Corollary 2.14.16). Corollary 2.14.16 can be derived from the special case of Corollary 2.14.15 where $M = N$ and $\mathbf{T} = -\mathbf{I}_N$ by observing that

$$(-1)^{NN}|-\mathbf{I}_N||\mathbf{W}| = (-1)^{NN}(-1)^N|\mathbf{W}| = (-1)^{N(N+1)}|\mathbf{W}|$$

and that either N or $N+1$ is an even number and consequently $N(N+1)$ is an even number. Q.E.D.

d. Determinants of matrix products and inverses

By using Theorems 2.14.14 and 2.14.13 and Corollary 2.14.16, we find that for $N \times N$ matrices $\mathbf{A}$ and $\mathbf{B}$,

$$|\mathbf{A}||\mathbf{B}| = \begin{vmatrix} \mathbf{A} & \mathbf{0} \\ -\mathbf{I} & \mathbf{B} \end{vmatrix} = \left| \begin{pmatrix} \mathbf{A} & \mathbf{0} \\ -\mathbf{I} & \mathbf{B} \end{pmatrix} \begin{pmatrix} \mathbf{I} & \mathbf{B} \\ \mathbf{0} & \mathbf{I} \end{pmatrix} \right| = \begin{vmatrix} \mathbf{A} & \mathbf{AB} \\ -\mathbf{I} & \mathbf{0} \end{vmatrix} = |\mathbf{AB}|,$$

thereby establishing the following, very important result.

Theorem 2.14.17. For $N \times N$ matrices $\mathbf{A}$ and $\mathbf{B}$,

$$|\mathbf{AB}| = |\mathbf{A}||\mathbf{B}|. \tag{14.24}$$

The repeated application of Theorem 2.14.17 leads to the following formula for the determinant of the product of an arbitrary number of $N \times N$ matrices $\mathbf{A}_1, \mathbf{A}_2, \ldots, \mathbf{A}_K$:

$$|\mathbf{A}_1\mathbf{A}_2 \cdots \mathbf{A}_K| = |\mathbf{A}_1||\mathbf{A}_2| \cdots |\mathbf{A}_K|. \tag{14.25}$$

As a special case of this formula, we obtain the following formula for the determinant of the kth power of an $N \times N$ matrix $\mathbf{A}$:

$$|\mathbf{A}^k| = |\mathbf{A}|^k \tag{14.26}$$

$(k = 2, 3, \ldots)$.

In light of Lemma 2.14.3, we have the following corollary of Theorem 2.14.17.

Corollary 2.14.18. For any $N \times N$ matrix $\mathbf{A}$,

$$|\mathbf{A}'\mathbf{A}| = |\mathbf{A}|^2. \tag{14.27}$$

Corollary 2.14.18 gives rise to the following result on the determinant of an orthogonal matrix.

Corollary 2.14.19. For any orthogonal matrix $\mathbf{P}$,

$$|\mathbf{P}| = \pm 1.$$

Proof (of Corollary 2.14.19). Using Corollary 2.14.18 [and result (14.9)], we find that

$$|\mathbf{P}|^2 = |\mathbf{P}'\mathbf{P}| = |\mathbf{I}| = 1. \qquad \text{Q.E.D.}$$

Having established Theorem 2.14.17, we are now in a position to prove the following result on the nonsingularity of a matrix and on the determinant of an inverse matrix.

Theorem 2.14.20. Let $\mathbf{A}$ represent an $N \times N$ matrix. Then, $\mathbf{A}$ is nonsingular (or, equivalently, $\mathbf{A}$ is invertible) if and only if $|\mathbf{A}| \neq 0$, in which case

$$|\mathbf{A}^{-1}| = 1/|\mathbf{A}|. \tag{14.28}$$

Proof. It suffices to show that if $\mathbf{A}$ is nonsingular, then $|\mathbf{A}| \neq 0$ and $|\mathbf{A}^{-1}| = 1/|\mathbf{A}|$ and that if $\mathbf{A}$ is singular, then $|\mathbf{A}| = 0$.

Suppose that $\mathbf{A}$ is nonsingular. Then, according to Theorem 2.14.17 [and result (14.9)],

$$|\mathbf{A}^{-1}||\mathbf{A}| = |\mathbf{A}^{-1}\mathbf{A}| = |\mathbf{I}| = 1,$$

implying that $|\mathbf{A}| \neq 0$ and further that $|\mathbf{A}^{-1}| = 1/|\mathbf{A}|$.

Alternatively, suppose that $\mathbf{A}$ is singular. Then, some column of $\mathbf{A}$, say the sth column $\mathbf{a}_s$, can be expressed as a linear combination of the other $N-1$ columns $\mathbf{a}_1, \mathbf{a}_2, \ldots, \mathbf{a}_{s-1}, \mathbf{a}_{s+1}, \ldots, \mathbf{a}_N$; that is, $\mathbf{a}_s = \sum_{i \neq s} k_i \mathbf{a}_i$ for some scalars $k_1, k_2, \ldots, k_{s-1}, k_{s+1}, \ldots, k_N$. Now, let $\mathbf{B}$ represent a matrix formed from $\mathbf{A}$ by adding the vector $-\sum_{i \neq s} k_i \mathbf{a}_i$ to the sth column of $\mathbf{A}$. Clearly, the sth column of $\mathbf{B}$ is null, and it follows from Corollary 2.14.5 that $|\mathbf{B}| = 0$. And it follows from Theorem 2.14.12 that $|\mathbf{A}| = |\mathbf{B}|$. Thus, $|\mathbf{A}| = 0$. Q.E.D.

Let $\mathbf{A}$ represent an $N \times N$ symmetric positive definite matrix. Then, according to Corollary 2.13.29, there exists a nonsingular matrix $\mathbf{P}$ such that $\mathbf{A} = \mathbf{P}'\mathbf{P}$. Thus, making use of Corollary 2.14.18 and observing (in light of Theorem 2.14.20) that $|\mathbf{P}| \neq 0$, we find that

$$|\mathbf{A}| = |\mathbf{P}'\mathbf{P}| = |\mathbf{P}|^2 > 0.$$

Moreover, the determinant of a symmetric positive semidefinite matrix equals 0, as is evident from Theorem 2.14.20 upon recalling (from Corollary 2.13.28) that a symmetric positive semidefinite matrix is singular. Accordingly, we have the following lemma.

Lemma 2.14.21. The determinant of a symmetric positive definite matrix is positive; the determinant of a symmetric positive semidefinite matrix equals 0.

e. Determinants of partitioned matrices

The following theorem gives formulas for the determinant of a partitioned matrix that are analogous to formulas (6.10) and (6.11) for the inverse of a partitioned matrix.

Theorem 2.14.22. Let $\mathbf{T}$ represent an $M \times M$ matrix, $\mathbf{U}$ an $M \times N$ matrix, $\mathbf{V}$ an $N \times M$ matrix, and $\mathbf{W}$ an $N \times N$ matrix. If $\mathbf{T}$ is nonsingular, then

$$\begin{vmatrix} \mathbf{T} & \mathbf{U} \\ \mathbf{V} & \mathbf{W} \end{vmatrix} = \begin{vmatrix} \mathbf{W} & \mathbf{V} \\ \mathbf{U} & \mathbf{T} \end{vmatrix} = |\mathbf{T}||\mathbf{W} - \mathbf{V}\mathbf{T}^{-1}\mathbf{U}|. \tag{14.29}$$

Proof. Suppose that $\mathbf{T}$ is nonsingular. Then,

$$\begin{pmatrix} \mathbf{T} & \mathbf{U} \\ \mathbf{V} & \mathbf{W} \end{pmatrix} = \begin{pmatrix} \mathbf{I} & \mathbf{0} \\ \mathbf{V}\mathbf{T}^{-1} & \mathbf{W}-\mathbf{V}\mathbf{T}^{-1}\mathbf{U} \end{pmatrix} \begin{pmatrix} \mathbf{T} & \mathbf{U} \\ \mathbf{0} & \mathbf{I} \end{pmatrix}.$$

Applying Theorems 2.14.17 and 2.14.14 and result (14.9), we find that

$$\begin{vmatrix} \mathbf{T} & \mathbf{U} \\ \mathbf{V} & \mathbf{W} \end{vmatrix} = |\mathbf{T}||\mathbf{W} - \mathbf{V}\mathbf{T}^{-1}\mathbf{U}|.$$

That $\begin{vmatrix} \mathbf{W} & \mathbf{V} \\ \mathbf{U} & \mathbf{T} \end{vmatrix} = |\mathbf{T}||\mathbf{W} - \mathbf{V}\mathbf{T}^{-1}\mathbf{U}|$ can be proved in similar fashion. Q.E.D.

f. A necessary and sufficient condition for the positive definiteness of a symmetric matrix

Whether or not a symmetric matrix is positive definite can be ascertained from the determinants of its leading principal submatrices. The following theorem provides the basis for doing so.

Theorem 2.14.23. Let $\mathbf{A} = \{a_{ij}\}$ represent an $N \times N$ symmetric matrix, and, for $k = 1, 2, \ldots, N$, let $\mathbf{A}_k$ represent the leading principal submatrix of $\mathbf{A}$ of order k (i.e., the principal submatrix obtained by striking out the last $N - k$ rows and columns). Then, $\mathbf{A}$ is positive definite if and only if, for $k = 1, 2, \ldots, N$, $\det(\mathbf{A}_k) > 0$, that is, if and only if the determinants of all N of the leading principal submatrices $\mathbf{A}_1, \mathbf{A}_2, \ldots, \mathbf{A}_N$ of $\mathbf{A}$ are positive.

In proving Theorem 2.14.23, it is convenient to make use of the following result, which is of some interest in its own right.

Lemma 2.14.24. Let $\mathbf{A}$ represent an $N \times N$ symmetric matrix (where $N \geq 2$), and partition $\mathbf{A}$ as

$$\mathbf{A} = \begin{pmatrix} \mathbf{A}_* & \mathbf{a} \\ \mathbf{a}' & c \end{pmatrix},$$

where the dimensions of $\mathbf{A}_*$ are $(N-1) \times (N-1)$. Then, $\mathbf{A}$ is positive definite if and only if $\mathbf{A}_*$ is positive definite and $|\mathbf{A}| > 0$.

Proof (of Lemma 2.14.24). If $\mathbf{A}$ is positive definite, then it is clear from Corollary 2.13.13 that $\mathbf{A}_*$ is positive definite and from Lemma 2.14.21 that $|\mathbf{A}| > 0$.

Conversely, suppose that $\mathbf{A}_*$ is positive definite (and hence nonsingular) and that $|\mathbf{A}| > 0$. Then, according to Theorem 2.14.22,

$$|\mathbf{A}| = |\mathbf{A}_*|\,(c - \mathbf{a}'\mathbf{A}_*^{-1}\mathbf{a}).$$

Since (according to Lemma 2.14.21) $|\mathbf{A}_*| > 0$ (and since $|\mathbf{A}| > 0$), we conclude that the Schur complement $c - \mathbf{a}'\mathbf{A}_*^{-1}\mathbf{a}$ of $\mathbf{A}_*$ (like $\mathbf{A}_*$ itself) is positive definite and hence (in light of Corollary 2.13.33) that $\mathbf{A}$ is positive definite. Q.E.D.

Proof (of Theorem 2.14.23). That the determinants of $\mathbf{A}_1, \mathbf{A}_2, \ldots, \mathbf{A}_N$ are positive if $\mathbf{A}$ is positive definite is an immediate consequence of Corollary 2.13.13 and Lemma 2.14.21.

For purposes of proving the converse, suppose that the determinants of $\mathbf{A}_1, \mathbf{A}_2, \ldots, \mathbf{A}_N$ are positive. The proof consists of establishing, via a mathematical induction argument, that $\mathbf{A}_1, \mathbf{A}_2, \ldots, \mathbf{A}_N$ are positive definite, which (since $\mathbf{A} = \mathbf{A}_N$) implies in particular that $\mathbf{A}$ is positive definite.

Clearly, $\mathbf{A}_1$ is positive definite. Suppose now that $\mathbf{A}_{k-1}$ is positive definite (where $2 \leq k \leq N$), and partition $\mathbf{A}_k$ as

$$\mathbf{A}_k = \begin{pmatrix} \mathbf{A}_{k-1} & \mathbf{a}_k \\ \mathbf{a}_k' & a_{kk} \end{pmatrix},$$

where $\mathbf{a}_k = (a_{1k}, a_{2k}, \ldots, a_{k-1,k})'$. Since $\mathbf{A}_{k-1}$ is (by supposition) positive definite (and since $|\mathbf{A}_k| > 0$), it follows from Lemma 2.14.24 that $\mathbf{A}_k$ is positive definite.

We conclude on the basis of the induction argument that $\mathbf{A}_1, \mathbf{A}_2, \ldots, \mathbf{A}_N$ are positive definite and that $\mathbf{A}$ in particular is positive definite. Q.E.D.

Exercises

Exercise 1. Let $\mathbf{A}$ represent an $M \times N$ matrix and $\mathbf{B}$ an $N \times M$ matrix. Can the value of $\mathbf{A} + \mathbf{B}'$ be determined from the value of $\mathbf{A}' + \mathbf{B}$ (in the absence of any other information about $\mathbf{A}$ and $\mathbf{B}$)? Describe your reasoning.

Exercise 2. Show that for any $M \times N$ matrix $\mathbf{A} = \{a_{ij}\}$ and $N \times P$ matrix $\mathbf{B} = \{b_{ij}\}$, $(\mathbf{AB})' = \mathbf{B}'\mathbf{A}'$ [thereby verifying result (1.13)].

Exercise 3. Let $\mathbf{A} = \{a_{ij}\}$ and $\mathbf{B} = \{b_{ij}\}$ represent $N \times N$ symmetric matrices.

(a) Show that in the special case where $N = 2$, $\mathbf{AB}$ is symmetric if and only if $b_{12}(a_{11} - a_{22}) = a_{12}(b_{11} - b_{22})$.

(b) Give a numerical example where $\mathbf{AB}$ is nonsymmetric.

(c) Show that $\mathbf{A}$ and $\mathbf{B}$ commute if and only if $\mathbf{AB}$ is symmetric.

Exercise 4. Let $\mathbf{A}$ represent an $M \times N$ partitioned matrix comprising R rows and U columns of blocks, the ijth of which is an $M_i \times N_j$ matrix $\mathbf{A}_{ij}$ that (for some scalar c_{ij}) is expressible as $\mathbf{A}_{ij} = c_{ij}\mathbf{1}_{M_i}\mathbf{1}'_{N_j}$ (a scalar multiple of an $M_i \times N_j$ matrix of 1's). Similarly, let $\mathbf{B}$ represent an $N \times Q$ partitioned matrix comprising U rows and V columns of blocks, the ijth of which is an $N_i \times Q_j$ matrix $\mathbf{B}_{ij}$ that (for some scalar d_{ij}) is expressible as $\mathbf{B}_{ij} = d_{ij}\mathbf{1}_{N_i}\mathbf{1}'_{Q_j}$. Obtain (in as simple form as possible) the conditions that must be satisfied by the scalars c_{ij} $(i = 1, 2, \ldots, R;\ j = 1, 2, \ldots, U)$ and d_{ij} $(i = 1, 2, \ldots, U;\ j = 1, 2, \ldots, V)$ in order for $\mathbf{AB}$ to equal a null matrix.

Exercise 5. Show that for any $M \times N$ matrix $\mathbf{A}$ and $N \times M$ matrix $\mathbf{B}$,

$$\operatorname{tr}(\mathbf{AB}) = \operatorname{tr}(\mathbf{A}'\mathbf{B}').$$

Exercise 6. Show that for any $M \times N$ matrix $\mathbf{A}$, $N \times P$ matrix $\mathbf{B}$, and $P \times M$ matrix $\mathbf{C}$,

$$\operatorname{tr}(\mathbf{ABC}) = \operatorname{tr}(\mathbf{CAB}) = \operatorname{tr}(\mathbf{BCA})$$

(i.e., the cyclic permutation of the 3 matrices in the product $\mathbf{ABC}$ does not affect the trace of the product).

Exercise 7. Let $\mathbf{A}$, $\mathbf{B}$, and $\mathbf{C}$ represent square matrices of order N.

(a) Using the result of Exercise 5 (or otherwise), show that if $\mathbf{A}$, $\mathbf{B}$, and $\mathbf{C}$ are symmetric, then $\operatorname{tr}(\mathbf{ABC}) = \operatorname{tr}(\mathbf{BAC})$.

(b) Show that [aside from special cases like that considered in Part (a)] $\operatorname{tr}(\mathbf{BAC})$ is not necessarily equal to $\operatorname{tr}(\mathbf{ABC})$.

Exercise 8. Which of the following sets are linear spaces: (1) the set of all $N \times N$ diagonal matrices, (2) the set of all $N \times N$ upper triangular matrices, and (3) the set of all $N \times N$ nonsymmetric matrices?

Exercise 9. Define

$$\mathbf{A} = \begin{pmatrix} 1 & 2 & -1 & 0 \\ 2 & 1 & 1 & 1 \\ 1 & -1 & 2 & 1 \end{pmatrix},$$

and (for $i = 1, 2, 3$) let $\mathbf{a}'_i$ represent the ith row of $\mathbf{A}$.

(a) Show that the set $\{\mathbf{a}'_1, \mathbf{a}'_2\}$ is a basis for $\mathcal{R}(\mathbf{A})$.

(b) Find rank($\mathbf{A}$).

(c) Making use of the answer to Part (b) (or otherwise), find a basis for $\mathcal{C}(\mathbf{A})$.

Exercise 10. Let $\mathbf{A}_1, \mathbf{A}_2, \ldots, \mathbf{A}_K$ represent matrices in a linear space $\mathcal{V}$, and let $\mathcal{U}$ represent a subspace of $\mathcal{V}$. Show that $\text{sp}(\mathbf{A}_1, \mathbf{A}_2, \ldots, \mathbf{A}_K) \subset \mathcal{U}$ if and only if $\mathbf{A}_1, \mathbf{A}_2, \ldots, \mathbf{A}_K$ are contained in $\mathcal{U}$ (thereby establishing what is essentially a generalization of Lemma 2.4.2).

Exercise 11. Let $\mathcal{V}$ represent a K-dimensional linear space of $M \times N$ matrices (where $K \geq 1$). Further, let $\{\mathbf{A}_1, \mathbf{A}_2, \ldots, \mathbf{A}_K\}$ represent a basis for $\mathcal{V}$, and, for arbitrary scalars $x_1, x_2, \ldots, x_K$ and $y_1, y_2, \ldots, y_K$, define $\mathbf{A} = \sum_{i=1}^{K} x_i \mathbf{A}_i$ and $\mathbf{B} = \sum_{j=1}^{K} y_j \mathbf{A}_j$. Show that

$$\mathbf{A} \bullet \mathbf{B} = \sum_{i=1}^{K} x_i y_i$$

for all choices of $x_1, x_2, \ldots, x_K$ and $y_1, y_2, \ldots, y_K$ if and only if the basis $\{\mathbf{A}_1, \mathbf{A}_2, \ldots, \mathbf{A}_K\}$ is orthonormal.

Exercise 12. An $N \times N$ matrix $\mathbf{A}$ is said to be *involutory* if $\mathbf{A}^2 = \mathbf{I}$, that is, if $\mathbf{A}$ is invertible and is its own inverse.

(a) Show that an $N \times N$ matrix $\mathbf{A}$ is involutory if and only if $(\mathbf{I} - \mathbf{A})(\mathbf{I} + \mathbf{A}) = \mathbf{0}$.

(b) Show that a 2×2 matrix $\mathbf{A} = \begin{pmatrix} a & b \\ c & d \end{pmatrix}$ is involutory if and only if (1) $a^2 + bc = 1$ and $d = -a$ or (2) $b = c = 0$ and $d = a = \pm 1$.

Exercise 13. Let $\mathbf{A} = \{a_{ij}\}$ represent an $M \times N$ matrix of full row rank.

(a) Show that in the special case $M = 1$ (i.e., in the special case where $\mathbf{A}$ is an N-dimensional row vector), there exists an N-dimensional column vector $\mathbf{b}$, $N-1$ elements of which are 0, that is a right inverse of $\mathbf{A}$.

(b) Generalize from Part (a) (to an arbitrary value of M) by showing that there exists an $N \times M$ matrix $\mathbf{B}$, $N - M$ rows of which are null vectors, that is a right inverse of $\mathbf{A}$.

Exercise 14. Provide an alternative verification of equality (6.10) by premultiplying or postmultiplying the right side of the equality by $\begin{pmatrix} \mathbf{T} & \mathbf{U} \\ \mathbf{V} & \mathbf{W} \end{pmatrix}$ and by confirming that the resultant product equals $\mathbf{I}_{M+N}$.

Exercise 15. Let $\mathbf{A} = \begin{pmatrix} 2 & 0 & 4 \\ 3 & 5 & 6 \\ 4 & 2 & 12 \end{pmatrix}$. Use the results of Section 2.6 to show that $\mathbf{A}$ is nonsingular and to obtain $\mathbf{A}^{-1}$. (*Hint.* Partition $\mathbf{A}$ as $\mathbf{A} = \begin{pmatrix} \mathbf{T} & \mathbf{U} \\ \mathbf{V} & \mathbf{W} \end{pmatrix}$, where $\mathbf{T}$ is a square matrix of order 2.)

Exercise 16. Let $\mathbf{T} = \{t_{ij}\}$ represent an $N \times N$ triangular matrix. Show that if $\mathbf{T}$ is orthogonal, then $\mathbf{T}$ is diagonal. If $\mathbf{T}$ is orthogonal, what can be inferred about the values of the diagonal elements $t_{11}, t_{22}, \ldots, t_{NN}$ of $\mathbf{T}$?

Exercise 17. Let $\mathbf{A}$ represent an $N \times N$ matrix. Show that for any $N \times N$ nonsingular matrix $\mathbf{B}$, $\mathbf{B}^{-1}\mathbf{A}\mathbf{B}$ is idempotent if and only if $\mathbf{A}$ is idempotent.

Exercise 18. Let $\mathbf{x} = \{x_i\}$ and $\mathbf{y} = \{y_i\}$ represent nonnull N-dimensional column vectors. Show that $\mathbf{x}\mathbf{y}'$ is a scalar multiple of an idempotent matrix (i.e., that $\mathbf{x}\mathbf{y}' = c\mathbf{A}$ for some scalar c and some idempotent matrix $\mathbf{A}$) if and only if $\sum_{i=1}^{N} x_i y_i \neq 0$ (i.e., if and only if $\mathbf{x}$ and $\mathbf{y}$ are not orthogonal with respect to the usual inner product).

Exercise 19. Let $\mathbf{A}$ represent a $4 \times N$ matrix of rank 2, and take $\mathbf{b} = \{b_i\}$ to be a 4-dimensional column vector. Suppose that $b_1 = 1$ and $b_2 = 0$ and that two of the N columns of $\mathbf{A}$ are the vectors $\mathbf{a}_1 = (5, 4, 3, 1)'$ and $\mathbf{a}_2 = (1, 2, 0, -1)'$. Determine for which values of b_3 and b_4 the linear system $\mathbf{Ax} = \mathbf{b}$ (in $\mathbf{x}$) is consistent.

Exercise 20. Let $\mathbf{A}$ represent an $M \times N$ matrix. Show that for any generalized inverses $\mathbf{G}_1$ and $\mathbf{G}_2$ of $\mathbf{A}$ and for any scalars w_1 and w_2 such that $w_1 + w_2 = 1$, the linear combination $w_1\mathbf{G}_1 + w_2\mathbf{G}_2$ is a generalized inverse of $\mathbf{A}$.

Exercise 21. Let $\mathbf{A}$ represent an $N \times N$ matrix.

(a) Using the result of Exercise 20 in combination with Corollary 2.10.11 (or otherwise), show that if $\mathbf{A}$ is symmetric, then $\mathbf{A}$ has a symmetric generalized inverse.

(b) Show that if $\mathbf{A}$ is singular (i.e., of rank less than N) and if $N > 1$, then (even if $\mathbf{A}$ is symmetric) $\mathbf{A}$ has a nonsymmetric generalized inverse. (*Hint.* Make use of the second part of Theorem 2.10.7.)

Exercise 22. Let $\mathbf{A}$ represent an $M \times N$ matrix of rank $N-1$. And let $\mathbf{x}$ represent any nonnull vector in $\mathcal{N}(\mathbf{A})$, that is, any N-dimensional nonnull column vector such that $\mathbf{Ax} = \mathbf{0}$. Show that a matrix $\mathbf{Z}^*$ is a solution to the homogeneous linear system $\mathbf{AZ} = \mathbf{0}$ (in an $N \times P$ matrix $\mathbf{Z}$) if and only if $\mathbf{Z}^* = \mathbf{xk}'$ for some P-dimensional row vector $\mathbf{k}'$.

Exercise 23. Suppose that $\mathbf{AX} = \mathbf{B}$ is a consistent linear system (in an $N \times P$ matrix $\mathbf{X}$).

(a) Show that if $\text{rank}(\mathbf{A}) = N$ or $\text{rank}(\mathbf{B}) = P$, then, corresponding to any solution $\mathbf{X}^*$ to $\mathbf{AX} = \mathbf{B}$, there is a generalized inverse $\mathbf{G}$ of $\mathbf{A}$ such that $\mathbf{X}^* = \mathbf{GB}$.

(b) Show that if $\text{rank}(\mathbf{A}) < N$ and $\text{rank}(\mathbf{B}) < P$, then there exists a solution $\mathbf{X}^*$ to $\mathbf{AX} = \mathbf{B}$ such that there is no generalized inverse $\mathbf{G}$ of $\mathbf{A}$ for which $\mathbf{X}^* = \mathbf{GB}$.

Exercise 24. Show that a matrix $\mathbf{A}$ is symmetric and idempotent if and only if there exists a matrix $\mathbf{X}$ such that $\mathbf{A} = \mathbf{P_X}$.

Exercise 25. Show that corresponding to any quadratic form $\mathbf{x}'\mathbf{Ax}$ (in an N-dimensional vector $\mathbf{x}$), there exists a unique lower triangular matrix $\mathbf{B}$ such that $\mathbf{x}'\mathbf{Ax}$ and $\mathbf{x}'\mathbf{Bx}$ are identically equal, and express the elements of $\mathbf{B}$ in terms of the elements of $\mathbf{A}$.

Exercise 26. Show, via an example, that the sum of two positive semidefinite matrices can be positive definite.

Exercise 27. Let $\mathbf{A}$ represent an $N \times N$ symmetric nonnegative definite matrix (where $N \geq 2$). Define $\mathbf{A}_0 = \mathbf{A}$, and, for $k = 1, 2, \ldots, N-1$, take $\mathbf{Q}_k$ to be an $(N - k + 1) \times (N - k + 1)$ unit upper triangular matrix, $\mathbf{A}_k$ an $(N - k) \times (N - k)$ matrix, and d_k a scalar that satisfy the recursive relationship

$$\mathbf{Q}_k'\mathbf{A}_{k-1}\mathbf{Q}_k = \text{diag}(d_k,\ \mathbf{A}_k) \tag{E.1}$$

—$\mathbf{Q}_k$, $\mathbf{A}_k$, and d_k can be constructed by making use of Lemma 2.13.19 and by proceeding as in the proof of Theorem 2.13.20.

(a) Indicate how $\mathbf{Q}_1, \mathbf{Q}_2, \ldots, \mathbf{Q}_{N-1}, \mathbf{A}_1, \mathbf{A}_2, \ldots, \mathbf{A}_{N-1}$, and $d_1, d_2, \ldots, d_{N-1}$ could be used to form an $N \times N$ unit upper triangular matrix $\mathbf{Q}$ and a diagonal matrix $\mathbf{D}$ such that $\mathbf{Q}'\mathbf{AQ} = \mathbf{D}$.

(b) Taking $\mathbf{A} = \begin{pmatrix} 2 & 0 & 0 & 0 \\ 0 & 4 & -2 & -4 \\ 0 & -2 & 1 & 2 \\ 0 & -4 & 2 & 7 \end{pmatrix}$ (which is a symmetric nonnegative definite matrix), determine unit upper triangular matrices $\mathbf{Q}_1$, $\mathbf{Q}_2$, and $\mathbf{Q}_3$, matrices $\mathbf{A}_1$, $\mathbf{A}_2$, and $\mathbf{A}_3$, and scalars d_1, d_2, and d_3 that satisfy the recursive relationship (E.1), and illustrate the procedure devised in response to Part (a) by using it to find a 4×4 unit upper triangular matrix $\mathbf{Q}$ and a diagonal matrix $\mathbf{D}$ such that $\mathbf{Q}'\mathbf{A}\mathbf{Q} = \mathbf{D}$.

Exercise 28. Let $\mathbf{A} = \{a_{ij}\}$ represent an $N \times N$ symmetric positive definite matrix, and let $\mathbf{B} = \{b_{ij}\} = \mathbf{A}^{-1}$. Show that, for $i = 1, 2, \ldots, N$,

$$b_{ii} \geq 1/a_{ii},$$

with equality holding if and only if $a_{ij} = 0$ for all $j \neq i$.

Exercise 29. Let

$$\mathbf{A} = \begin{pmatrix} a_{11} & a_{12} & a_{13} & \boxed{a_{14}} \\ a_{21} & a_{22} & \boxed{a_{23}} & a_{24} \\ \boxed{a_{31}} & a_{32} & a_{33} & a_{34} \\ a_{41} & \boxed{a_{42}} & a_{43} & a_{44} \end{pmatrix}.$$

(a) Write out all of the pairs that can be formed from the four "boxed" elements of $\mathbf{A}$.

(b) Indicate which of the pairs from Part (a) are positive and which are negative.

(c) Use formula (14.1) to compute the number of pairs from Part (a) that are negative, and check that the result of this computation is consistent with your answer to Part (b).

Exercise 30. Obtain (in as simple form as possible) an expression for the determinant of each of the following two matrices: (1) an $N \times N$ matrix $\mathbf{A} = \{a_{ij}\}$ of the general form

$$\mathbf{A} = \begin{pmatrix} 0 & \ldots & 0 & 0 & a_{1N} \\ 0 & \ldots & 0 & a_{2,N-1} & a_{2N} \\ 0 & & a_{3,N-2} & a_{3,N-1} & a_{3N} \\ & & \vdots & \vdots & \vdots \\ a_{N1} & \ldots & a_{N,N-2} & a_{N,N-1} & a_{NN} \end{pmatrix}$$

(where $a_{ij} = 0$ for $j = 1, 2, \ldots, N-i$; $i = 1, 2, \ldots, N-1$); (2) an $N \times N$ matrix $\mathbf{B} = \{b_{ij}\}$ of the general form

$$\mathbf{B} = \begin{pmatrix} 0 & 1 & 0 & \ldots & 0 \\ 0 & 0 & 1 & & 0 \\ \vdots & \vdots & & \ddots & \\ 0 & 0 & 0 & & 1 \\ -k_0 & -k_1 & -k_2 & \ldots & -k_{N-1} \end{pmatrix}$$

—a matrix of this general form is called a *companion matrix*.

Exercise 31. Verify the part of result (14.16) that pertains to the postmultiplication of a matrix by a unit upper triangular matrix by showing that for any $N \times N$ matrix $\mathbf{A}$ and any $N \times N$ unit upper triangular matrix $\mathbf{T}$, $|\mathbf{AT}| = |\mathbf{A}|$.

Exercise 32. Show that for any $N \times N$ matrix $\mathbf{A}$ and any $N \times N$ nonsingular matrix $\mathbf{C}$,

$$|\mathbf{C}^{-1}\mathbf{A}\mathbf{C}| = |\mathbf{A}|.$$

Exercise 33. Let $\mathbf{A} = \begin{pmatrix} a & b \\ c & d \end{pmatrix}$, where a, b, c, and d are scalars.

(a) Show that in the special case where $\mathbf{A}$ is symmetric (i.e., where $c = b$), $\mathbf{A}$ is nonnegative definite if and only if $a \geq 0$, $d \geq 0$, and $|b| \leq \sqrt{ad}$ and is positive definite if and only if $a > 0$, $d > 0$, and $|b| < \sqrt{ad}$.

(b) Extend the result of Part (a) by showing that in the general case where $\mathbf{A}$ is not necessarily symmetric (i.e., where possibly $c \neq b$), $\mathbf{A}$ is nonnegative definite if and only if $a \geq 0$, $d \geq 0$, and $|b+c|/2 \leq \sqrt{ad}$ and is positive definite if and only if $a > 0$, $d > 0$, and $|b+c|/2 < \sqrt{ad}$. [*Hint*. Take advantage of the result of Part (a).]

Exercise 34. Let $\mathbf{A} = \{a_{ij}\}$ represent an $N \times N$ symmetric matrix. And suppose that $\mathbf{A}$ is nonnegative definite (in which case its diagonal elements are nonnegative). By, for example, making use of the result of Part (a) of Exercise 33, show that, for $j \neq i = 1, 2, \ldots, N$,

$$|a_{ij}| \leq \sqrt{a_{ii} a_{jj}} \leq \max(a_{ii}, a_{jj}),$$

with $|a_{ij}| < \sqrt{a_{ii} a_{jj}}$ if $\mathbf{A}$ is positive definite.

Exercise 35. Let $\mathbf{A} = \{a_{ij}\}$ represent an $N \times N$ symmetric positive definite matrix. Show that

$$\det \mathbf{A} \leq \prod_{i=1}^{N} a_{ii},$$

with equality holding if and only if $\mathbf{A}$ is diagonal.

Bibliographic and Supplementary Notes

Much of what is presented in Chapter 2 is taken from Chapters 1–14 in Harville's (1997) book, *Matrix Algebra from a Statistician's Perspective*, which provides more extensive coverage of the same topics.

§4. What is referred to herein as a linear space is a special case of what is known as a finite-dimensional vector space. A classical reference on that topic is Halmos's (1958) book, *Finite-Dimensional Vector Spaces*. The term linear space is used in lieu of the somewhat more common term vector space. At least in the present setting, this usage is advantageous (especially for less mathematically sophisticated readers). It avoids a dual use of the term vector, in which that term is used at times to refer to a member of a linear space of $M \times N$ matrices and at times to specify a matrix having a single row or column.

§9 and §11. No attempt is made to discuss the computational aspects of solving a linear system. A classical reference on that topic (and on related topics) is Golub and Van Loan's (2013) book, *Matrix Computations*. Another highly regarded source of information on computational issues is Trefethen and Bau's (1997) book, *Numerical Linear Algebra*. A source that emphasizes those computational issues that are highly relevant to statistical applications is Gentle's (1998) book, *Numerical Linear Algebra for Applications in Statistics*.

§13. The usage herein of the terms nonnegative definite, positive definite, and positive semidefinite differs somewhat from that employed in various other presentations. In particular, these terms are applied to both symmetric and nonsymmetric matrices, whereas in many other presentations their application to matrices is confined to symmetric matrices. Moreover, the term positive semidefinite is used in a way that, while not uncommon, is at odds with its use in some other presentations. In some presentations, the term positive semidefinite is used in the same way that nonnegative definite is used herein.

3

Random Vectors and Matrices

In working with linear models, knowledge of basic results on the distribution of random variables is essential. Of particular relevance are various results on expected values and on variances and covariances. Also of relevance are results that pertain to conditional distributions and to the multivariate normal distribution.

In working with a large number (or even a modest number) of random variables, the use of matrix notation can be extremely helpful. In particular, formulas for the expected values and the variances and covariances of linear combinations of random variables can be expressed very concisely in matrix notation. The use of matrix notation is facilitated by the arrangement of random variables in the form of a vector or a matrix. A *random* (row or column) *vector* is a (row or column) vector whose elements are (jointly distributed) random variables. More generally, a *random matrix* is a matrix whose elements are (jointly distributed) random variables.

3.1 Expected Values

The expected value of a random variable x is denoted by the symbol $\mathrm{E}(x)$. The expected value of a nonnegative random variable is well-defined, but not necessarily finite. The expected value $\mathrm{E}(x)$ of a nonnegative random variable x is said to *exist* (or to be integrable) if $\mathrm{E}(x) < \infty$. More generally, the expected value $\mathrm{E}(x)$ of an arbitrary random variable x is said to *exist* (or to be integrable) if $\mathrm{E}(|x|) < \infty$, in which case $\mathrm{E}(x)$ is well-defined and finite. Unless otherwise indicated, results involving the expected values of random variables are to be regarded as including an implicit assumption and/or (depending on the context) claim that the expected values exist.

The expected value $\mathrm{E}(x)$ of a random variable x and the existence or nonexistence of $\mathrm{E}(x)$ are characteristics of the distribution of x. Accordingly, if two random variables x and y have the same distribution, then either $\mathrm{E}(x)$ and $\mathrm{E}(y)$ both exist and are equal or neither $\mathrm{E}(x)$ nor $\mathrm{E}(y)$ exists. In that regard, it is worth noting that if two random variables x and y (defined on the same probability space) are equal with probability 1 (i.e., are equal except possibly on a set of probability 0), then they have the same distribution.

A random variable x (or its distribution) is said to be *discrete* if there exists a finite or countably infinite set of distinct values $x_1, x_2, x_3, \ldots$ of x such that $\sum_i \Pr(x = x_i) = 1$, in which case

$$\mathrm{E}[g(x)] = \sum_i g(x_i) \Pr(x = x_i)$$

for "any" function $g(x)$ of x. More generally, a random vector $\mathbf{x}$ (or its distribution) is said to be *discrete* if there exists a finite or countably infinite set of distinct values $\mathbf{x}_1, \mathbf{x}_2, \mathbf{x}_3, \ldots$ of $\mathbf{x}$ such that $\sum_i \Pr(\mathbf{x} = \mathbf{x}_i) = 1$, in which case

$$\mathrm{E}[g(\mathbf{x})] = \sum_i g(\mathbf{x}_i) \Pr(\mathbf{x} = \mathbf{x}_i) \tag{1.1}$$

for "any" function $g(\mathbf{x})$ of $\mathbf{x}$.

A random variable x (or its distribution) is said to be *absolutely continuous* if there exists a nonnegative function $f(x)$ of x, called a *probability density function*, such that, for an "arbitrary" set A of real numbers, $\Pr(x \in A) = \int_A f(s)\,ds$, in which case

$$E[g(x)] = \int_{-\infty}^{\infty} g(s)f(s)\,ds$$

for "any" function $g(x)$ of x. More generally, an N-dimensional random vector $\mathbf{x}$ (or its distribution) is said to be *absolutely continuous* if there exists a nonnegative function $f(\mathbf{x})$ of $\mathbf{x}$, called a *probability density function* (pdf), such that, for an "arbitrary" subset A of $\mathbb{R}^N$, $\Pr(\mathbf{x} \in A) = \int_A f(\mathbf{s})\,d\mathbf{s}$, in which case

$$E[g(\mathbf{x})] = \int_{\mathbb{R}^N} g(\mathbf{s})f(\mathbf{s})\,d\mathbf{s} \tag{1.2}$$

for "any" function $g(\mathbf{x})$ of $\mathbf{x}$.

If $\mathbf{x}$ is a random vector and $g(\mathbf{x})$ "any" function of $\mathbf{x}$ that is nonnegative [in the sense that $g(\mathbf{x}) \geq 0$ for every value of $\mathbf{x}$] or is nonnegative with probability 1 [in the sense that for some set A of $\mathbf{x}$-values for which $\Pr(\mathbf{x} \in A) = 1$, $g(\mathbf{x}) \geq 0$ for every value of $\mathbf{x}$ in A], then

$$E[g(\mathbf{x})] = 0 \quad \Leftrightarrow \quad g(\mathbf{x}) = 0 \text{ with probability 1.} \tag{1.3}$$

By definition, two random vectors, say $\mathbf{x}$ and $\mathbf{y}$, are *statistically independent* if for "every" set A (of $\mathbf{x}$-values) and "every" set B (of $\mathbf{y}$-values),

$$\Pr(\mathbf{x} \in A,\ \mathbf{y} \in B) = \Pr(\mathbf{x} \in A)\Pr(\mathbf{y} \in B).$$

If $\mathbf{x}$ and $\mathbf{y}$ are statistically independent, then for "any" function $f(\mathbf{x})$ of $\mathbf{x}$ and "any" function $g(\mathbf{y})$ of $\mathbf{y}$ (for which $E[f(\mathbf{x})]$ and $E[g(\mathbf{y})]$ exist),

$$E[f(\mathbf{x})g(\mathbf{y})] = E[f(\mathbf{x})]\,E[g(\mathbf{y})] \tag{1.4}$$

(e.g., Casella and Berger 2002, sec. 4.2; Parzen 1960, p. 361).

The *expected value* of an N-dimensional random row or column vector is the N-dimensional (respectively row or column) vector whose ith element is the expected value of the ith element of the random vector ($i = 1, 2, \ldots, N$). More generally, the *expected value* of an $M \times N$ random matrix is the $M \times N$ matrix whose ijth element is the expected value of the ijth element of the random matrix ($i = 1, 2, \ldots, M$; $j = 1, 2, \ldots, N$). The expected value of a random matrix $\mathbf{X}$ is denoted by the symbol $E(\mathbf{X})$ (and is said to exist if the expected value of every element of $\mathbf{X}$ exists). Thus, for an $M \times N$ random matrix $\mathbf{X}$ with ijth element x_{ij} ($i = 1, 2, \ldots, M$; $j = 1, 2, \ldots, N$),

$$E(\mathbf{X}) = \begin{pmatrix} E(x_{11}) & E(x_{12}) & \ldots & E(x_{1N}) \\ E(x_{21}) & E(x_{22}) & \ldots & E(x_{2N}) \\ \vdots & \vdots & & \vdots \\ E(x_{M1}) & E(x_{M2}) & \ldots & E(x_{MN}) \end{pmatrix}.$$

The expected value of a random variable x is referred to as the *mean* of x (or of the distribution of x). And, similarly, the expected value of a random vector or matrix $\mathbf{X}$ is referred to as the *mean* (or, if applicable, mean vector) of $\mathbf{X}$ (or of the distribution of $\mathbf{X}$).

It follows from elementary properties of the expected values of random variables that for a finite number of random variables $x_1, x_2, \ldots, x_N$ and for nonrandom scalars $c, a_1, a_2, \ldots, a_N$,

$$E\Big(c + \sum_{j=1}^{N} a_j x_j\Big) = c + \sum_{j=1}^{N} a_j E(x_j). \tag{1.5}$$

Letting $\mathbf{x} = (x_1, x_2, \ldots, x_N)'$ and $\mathbf{a} = (a_1, a_2, \ldots, a_N)'$, this equality can be reexpressed in matrix notation as

$$E(c + \mathbf{a}'\mathbf{x}) = c + \mathbf{a}'E(\mathbf{x}). \tag{1.6}$$

As a generalization of equality (1.5) or (1.6), we have that

$$\mathrm{E}(\mathbf{c} + \mathbf{A}\mathbf{x}) = \mathbf{c} + \mathbf{A}\mathrm{E}(\mathbf{x}), \tag{1.7}$$

where $\mathbf{c}$ is an M-dimensional nonrandom column vector, $\mathbf{A}$ an $M \times N$ nonrandom matrix, and $\mathbf{x}$ an N-dimensional random column vector. Equality (1.5) can also be generalized as follows:

$$\mathrm{E}\Big(\mathbf{c} + \sum_{j=1}^{N} a_j \mathbf{x}_j\Big) = \mathbf{c} + \sum_{j=1}^{N} a_j \mathrm{E}(\mathbf{x}_j), \tag{1.8}$$

where $\mathbf{x}_1, \mathbf{x}_2, \ldots, \mathbf{x}_N$ are M-dimensional random column vectors, $\mathbf{c}$ is an M-dimensional nonrandom column vector, and $a_1, a_2, \ldots, a_N$ are nonrandom scalars.

Equality (1.7) can be readily verified by using equality (1.5) [or equality (1.6)] to show that each element of the left side of equality (1.7) equals the corresponding element of the right side. A similar approach can be used to verify equality (1.8). Or equality (1.8) can be derived by observing that $\sum_{j=1}^{N} a_j \mathbf{x}_j = \mathbf{A}\mathbf{x}$, where $\mathbf{A} = (a_1\mathbf{I}, a_2\mathbf{I}, \ldots, a_N\mathbf{I})$ and $\mathbf{x}' = (\mathbf{x}_1', \mathbf{x}_2', \ldots, \mathbf{x}_N')$, and by applying equality (1.7).

Further generalizations are possible. We have that

$$\mathrm{E}(\mathbf{C} + \mathbf{A}\mathbf{X}\mathbf{K}) = \mathbf{C} + \mathbf{A}\mathrm{E}(\mathbf{X})\mathbf{K}, \tag{1.9}$$

where $\mathbf{C}$ is an $M \times Q$ nonrandom matrix, $\mathbf{A}$ an $M \times N$ nonrandom matrix, $\mathbf{K}$ a $P \times Q$ nonrandom matrix, and $\mathbf{X}$ an $N \times P$ random matrix, and that

$$\mathrm{E}\Big(\mathbf{C} + \sum_{j=1}^{N} a_j \mathbf{X}_j\Big) = \mathbf{C} + \sum_{j=1}^{N} a_j \mathrm{E}(\mathbf{X}_j), \tag{1.10}$$

where $\mathbf{X}_1, \mathbf{X}_2, \ldots, \mathbf{X}_N$ are $M \times P$ random matrices, $\mathbf{C}$ is an $M \times P$ nonrandom matrix, and $a_1, a_2, \ldots, a_N$ are nonrandom scalars. Equalities (1.9) and (1.10) can be verified by for instance using equality (1.7) and/or equality (1.8) to show that each column of the left side of equality (1.9) or (1.10) equals the corresponding column of the right side. Or, upon observing that $\sum_{j=1}^{N} a_j \mathbf{X}_j = \mathbf{A}\mathbf{X}\mathbf{K}$, where $\mathbf{A} = (a_1\mathbf{I}, a_2\mathbf{I}, \ldots, a_N\mathbf{I})$, $\mathbf{X}' = (\mathbf{X}_1', \mathbf{X}_2', \ldots, \mathbf{X}_N')$, and $\mathbf{K} = \mathbf{I}$, equality (1.10) can be derived from equality (1.9).

3.2 Variances, Covariances, and Correlations

a. The basics: (univariate) dispersion and pairwise (statistical) dependence

Variance (and standard deviation) of a random variable. The *variance* of a random variable x (whose expected value exists) is (by definition) the expected value $\mathrm{E}\{[x - \mathrm{E}(x)]^2\}$ of the square of the difference between x and its expected value. The variance of x is denoted by the symbol $\mathrm{var}\, x$ or $\mathrm{var}(x)$. The positive square root $\sqrt{\mathrm{var}(x)}$ of the variance of x is referred to as the *standard deviation* of x.

If a random variable x is such that $\mathrm{E}(x^2)$ exists [i.e., such that $\mathrm{E}(x^2) < \infty$], then $\mathrm{E}(x)$ exists and $\mathrm{var}(x)$ also exists (i.e., is finite). That the existence of $\mathrm{E}(x^2)$ implies the existence of $\mathrm{E}(x)$ can be readily verified by making use of the inequality $|x| < 1 + x^2$. That it also implies the existence (finiteness) of $\mathrm{var}(x)$ becomes clear upon observing that

$$[x - \mathrm{E}(x)]^2 = x^2 - 2x\,\mathrm{E}(x) + [\mathrm{E}(x)]^2. \tag{2.1}$$

The existence of $\mathrm{E}(x^2)$ is a necessary as well as a sufficient condition for the existence of $\mathrm{E}(x)$ and $\mathrm{var}(x)$, as is evident upon reexpressing equality (2.1) as

$$x^2 = [x - \mathrm{E}(x)]^2 + 2x\,\mathrm{E}(x) - [\mathrm{E}(x)]^2.$$

In summary, we have that

$$\mathrm{E}(x^2) \text{ exists} \;\Leftrightarrow\; \mathrm{E}(x) \text{ exists and } \mathrm{var}(x) \text{ also exists.} \tag{2.2}$$

Further,

$$\mathrm{var}(x) = \mathrm{E}(x^2) - [\mathrm{E}(x)]^2, \tag{2.3}$$

as can be readily verified by using formula (1.5) to evaluate expression (2.1). Also, it is worth noting that

$$\mathrm{var}(x) = 0 \;\Leftrightarrow\; x = \mathrm{E}(x) \text{ with probability 1.} \tag{2.4}$$

Covariance of two random variables. The *covariance* of two random variables x and y (whose expected values exist) is (by definition) $\mathrm{E}\{[x - \mathrm{E}(x)][y - \mathrm{E}(y)]\}$. The covariance of x and y is denoted by the symbol $\mathrm{cov}(x, y)$. We have that

$$\mathrm{cov}(y, x) = \mathrm{cov}(x, y) \tag{2.5}$$

and

$$\mathrm{var}(x) = \mathrm{cov}(x, x), \tag{2.6}$$

as is evident from the very definitions of a variance and a covariance (and from an elementary property of the expected-value operator).

If two random variables x and y whose expected values exist are such that the expected value of xy also exists, then the covariance of x and y exists and

$$\mathrm{cov}(x, y) = \mathrm{E}(xy) - \mathrm{E}(x)\mathrm{E}(y), \tag{2.7}$$

as becomes clear upon observing that

$$[x - \mathrm{E}(x)][y - \mathrm{E}(y)] = xy - x\,\mathrm{E}(y) - y\,\mathrm{E}(x) + \mathrm{E}(x)\mathrm{E}(y) \tag{2.8}$$

and applying formula (1.5). The existence of the expected value of xy is necessary as well as sufficient for the existence of the covariance of x and y, as is evident upon reexpressing equality (2.8) as

$$xy = [x - \mathrm{E}(x)][y - \mathrm{E}(y)] + x\,\mathrm{E}(y) + y\,\mathrm{E}(x) - \mathrm{E}(x)\mathrm{E}(y).$$

Note that in the special case where $y = x$, formula (2.7) reduces to formula (2.3).

Some fundamental results bearing on the covariance of two random variables x and y (whose expected values exist) and on the relationship of the covariance to the variances of x and y are as follows. The covariance of x and y exists if the variances of x and y both exist, in which case

$$|\mathrm{cov}(x, y)| \le \sqrt{\mathrm{var}(x)}\sqrt{\mathrm{var}(y)} \tag{2.9}$$

or, equivalently,

$$[\mathrm{cov}(x, y)]^2 \le \mathrm{var}(x)\,\mathrm{var}(y) \tag{2.10}$$

or (also equivalently)

$$-\sqrt{\mathrm{var}(x)}\sqrt{\mathrm{var}(y)} \le \mathrm{cov}(x, y) \le \sqrt{\mathrm{var}(x)}\sqrt{\mathrm{var}(y)}. \tag{2.11}$$

Further,

$$\mathrm{var}(x) = 0 \text{ or } \mathrm{var}(y) = 0 \;\Rightarrow\; \mathrm{cov}(x, y) = 0, \tag{2.12}$$

so that when $\mathrm{var}(x) = 0$ or $\mathrm{var}(y) = 0$ or, equivalently, when $x = \mathrm{E}(x)$ with probability 1 or $y = \mathrm{E}(y)$ with probability 1, inequality (2.9) holds as an equality, both sides of which equal 0. And when

$\text{var}(x) > 0$ and $\text{var}(y) > 0$, inequality (2.9) holds as the equality $\text{cov}(x, y) = \sqrt{\text{var}(x)}\sqrt{\text{var}(y)}$ if and only if

$$\frac{y - \text{E}(y)}{\sqrt{\text{var}(y)}} = \frac{x - \text{E}(x)}{\sqrt{\text{var}(x)}} \text{ with probability 1,}$$

and holds as the equality $\text{cov}(x, y) = -\sqrt{\text{var}(x)}\sqrt{\text{var}(y)}$ if and only if

$$\frac{y - \text{E}(y)}{\sqrt{\text{var}(y)}} = -\frac{x - \text{E}(x)}{\sqrt{\text{var}(x)}} \text{ with probability 1.}$$

These results (on the covariance of the random variables x and y) can be inferred from the following results on the expected value of the product of two random variables, say w and z—they are obtained from the results on $\text{E}(wz)$ by setting $w = x - \text{E}(x)$ and $z = y - \text{E}(y)$. The expected value $\text{E}(wz)$ of wz exists if the expected values $\text{E}(w^2)$ and $\text{E}(z^2)$ of w^2 and z^2 both exist, in which case

$$|\text{E}(wz)| \leq \sqrt{\text{E}(w^2)}\sqrt{\text{E}(z^2)} \tag{2.13}$$

or, equivalently,

$$[\text{E}(wz)]^2 \leq \text{E}(w^2)\,\text{E}(z^2) \tag{2.14}$$

or (also equivalently)

$$-\sqrt{\text{E}(w^2)}\sqrt{\text{E}(z^2)} \leq \text{E}(wz) \leq \sqrt{\text{E}(w^2)}\sqrt{\text{E}(z^2)}. \tag{2.15}$$

Further,

$$\text{E}(w^2) = 0 \text{ or } \text{E}(z^2) = 0 \;\Rightarrow\; \text{E}(wz) = 0, \tag{2.16}$$

so that when $\text{E}(w^2) = 0$ or $\text{E}(z^2) = 0$ or, equivalently, when $w = 0$ with probability 1 or $z = 0$ with probability 1, inequality (2.13) holds as an equality, both sides of which equal 0. And when $\text{E}(w^2) > 0$ and $\text{E}(z^2) > 0$, inequality (2.13) holds as the equality $\text{E}(wz) = \sqrt{\text{E}(w^2)}\sqrt{\text{E}(z^2)}$ if and only if $z/\sqrt{\text{E}(z^2)} = w/\sqrt{\text{E}(w^2)}$ with probability 1, and holds as the equality $\text{E}(wz) = -\sqrt{\text{E}(w^2)}\sqrt{\text{E}(z^2)}$ if and only if $z/\sqrt{\text{E}(z^2)} = -w/\sqrt{\text{E}(w^2)}$ with probability 1. A verification of these results on $\text{E}(wz)$ is provided subsequently (in the final part of the present subsection).

Note that inequality (2.9) implies that

$$|\text{cov}(x, y)| \leq \max(\text{var}\, x,\ \text{var}\, y). \tag{2.17}$$

Correlation of two random variables. The *correlation* of two random variables x and y (whose expected values exist and whose variances also exist and are strictly positive) is (by definition)

$$\frac{\text{cov}(x, y)}{\sqrt{\text{var}(x)}\sqrt{\text{var}(y)}},$$

and is denoted by the symbol $\text{corr}(x, y)$. From result (2.5), it is clear that

$$\text{corr}(y, x) = \text{corr}(x, y). \tag{2.18}$$

And, as a consequence of result (2.9), we have that

$$|\text{corr}(x, y)| \leq 1, \tag{2.19}$$

which is equivalent to

$$[\text{corr}(x, y)]^2 \leq 1 \tag{2.20}$$

and also to

$$-1 \leq \text{corr}(x, y) \leq 1. \tag{2.21}$$

Further, inequality (2.19) holds as the equality $\text{corr}(x, y) = 1$ if and only if

$$\frac{y - \text{E}(y)}{\sqrt{\text{var}(y)}} = \frac{x - \text{E}(x)}{\sqrt{\text{var}(x)}} \quad \text{with probability 1,}$$

and holds as the equality $\text{corr}(x, y) = -1$ if and only if

$$\frac{y - \text{E}(y)}{\sqrt{\text{var}(y)}} = -\frac{x - \text{E}(x)}{\sqrt{\text{var}(x)}} \quad \text{with probability 1.}$$

Verification of results on the expected value of the product of two random variables. Let us verify the results (given earlier in the present subsection) on the expected value $\text{E}(wz)$ of the product of two random variables w and z. Suppose that $\text{E}(w^2)$ and $\text{E}(z^2)$ both exist. Then, what we wish to establish are the existence of $\text{E}(wz)$ and the validity of inequality (2.13) and of the conditions under which equality is attained in inequality (2.13).

Let us begin by observing that, for arbitrary scalars a and b,

$$\tfrac{1}{2}(a^2 + b^2) - ab = \tfrac{1}{2}(a - b)^2 \geq 0$$

and

$$-\tfrac{1}{2}(a^2 + b^2) - ab = -\tfrac{1}{2}(a + b)^2 \leq 0,$$

implying in particular that

$$-\tfrac{1}{2}(a^2 + b^2) \leq ab \leq \tfrac{1}{2}(a^2 + b^2) \tag{2.22}$$

or, equivalently, that

$$|ab| \leq \tfrac{1}{2}(a^2 + b^2). \tag{2.23}$$

Upon setting $a = w$ and $b = z$ in inequality (2.23), we obtain the inequality

$$|wz| \leq \tfrac{1}{2}(w^2 + z^2). \tag{2.24}$$

The expected value of the right side of inequality (2.24) exists, implying the existence of the expected value of the left side of inequality (2.24) and hence the existence of $\text{E}(wz)$.

Now, consider inequality (2.13). When $\text{E}(w^2) = 0$ or $\text{E}(z^2) = 0$ or, equivalently, when $w = 0$ with probability 1 or $z = 0$ with probability 1, $wz = 0$ with probability 1 and hence inequality (2.13) holds as an equality, both sides of which equal 0.

Alternatively, suppose that $\text{E}(w^2) > 0$ and $\text{E}(z^2) > 0$. And take $a = w/\sqrt{\text{E}(w^2)}$ and $b = z/\sqrt{\text{E}(z^2)}$. In light of result (2.22), we have that

$$-\text{E}\big[\tfrac{1}{2}(a^2 + b^2)\big] \leq \text{E}(ab) \leq \text{E}\big[\tfrac{1}{2}(a^2 + b^2)\big]. \tag{2.25}$$

Moreover,

$$\text{E}\big[\tfrac{1}{2}(a^2 + b^2)\big] = 1 \quad \text{and} \quad \text{E}(ab) = \frac{\text{E}(wz)}{\sqrt{\text{E}(w^2)}\sqrt{\text{E}(z^2)}}. \tag{2.26}$$

Together, results (2.25) and (2.26) imply that

$$-1 \leq \frac{\text{E}(wz)}{\sqrt{\text{E}(w^2)}\sqrt{\text{E}(z^2)}} \leq 1,$$

which is equivalent to result (2.15) and hence to inequality (2.13). Further, inequality (2.13) holds as the equality $\text{E}(wz) = \sqrt{\text{E}(w^2)}\sqrt{\text{E}(z^2)}$ if and only if $\text{E}\big[\tfrac{1}{2}(a^2 + b^2) - ab\big] = 0$ [as is evident from result (2.26)], or equivalently if and only if $\text{E}\big[\tfrac{1}{2}(a - b)^2\big] = 0$, and hence if and only if $b = a$ with probability 1. And, similarly, inequality (2.13) holds as the equality $\text{E}(wz) = -\sqrt{\text{E}(w^2)}\sqrt{\text{E}(z^2)}$ if and only if $\text{E}[-\tfrac{1}{2}(a^2 + b^2) - ab] = 0$, or equivalently if and only if $\text{E}[-\tfrac{1}{2}(a + b)^2] = 0$, and hence if and only if $b = -a$ with probability 1.

b. Variance-covariance matrices and covariances of random vectors

As multivariate extensions of the variance of a random variable and the covariance of two random variables, we have the variance-covariance matrix of a random vector and the covariance of two random vectors. The *variance-covariance matrix* of an N-dimensional random (row or column) vector with first through Nth elements $x_1, x_2, \ldots, x_N$ (whose expected values exist) is (by definition) the $N \times N$ matrix whose ijth element is $\mathrm{cov}(x_i, x_j)$ $(i, j = 1, 2, \ldots, N)$. Note that the diagonal elements of this matrix equal the variances of $x_1, x_2, \ldots, x_N$. The *covariance* (or covariance matrix) of an N-dimensional random (row or column) vector with first through Nth elements $x_1, x_2, \ldots, x_N$ (whose expected values exist) and a T-dimensional random (row or column) vector with first through Tth elements $y_1, y_2, \ldots, y_T$ (whose expected values exist) is (by definition) the $N \times T$ matrix whose ijth element is $\mathrm{cov}(x_i, y_j)$ $(i = 1, 2, \ldots, N;\ j = 1, 2, \ldots, T)$. The variance-covariance matrix of a random vector is sometimes referred to simply as the variance matrix or the covariance matrix of the vector or, even more simply, as the variance or the covariance of the vector.

Denote by $\mathrm{var}\,\mathbf{x}$ or $\mathrm{var}(\mathbf{x})$ or by $\mathrm{var}(\mathbf{x}')$ the variance-covariance matrix of a random column vector $\mathbf{x}$ or its transpose $\mathbf{x}'$. Similarly, denote by $\mathrm{cov}(\mathbf{x}, \mathbf{y})$, $\mathrm{cov}(\mathbf{x}', \mathbf{y})$, $\mathrm{cov}(\mathbf{x}, \mathbf{y}')$, or $\mathrm{cov}(\mathbf{x}', \mathbf{y}')$ the covariance of $\mathbf{x}$ or $\mathbf{x}'$ and a random column vector $\mathbf{y}$ or its transpose $\mathbf{y}'$. Thus, for an N-dimensional column vector $\mathbf{x} = (x_1, x_2, \ldots, x_N)'$,

$$\mathrm{var}(\mathbf{x}) = \mathrm{var}(\mathbf{x}') = \begin{pmatrix} \mathrm{var}(x_1) & \mathrm{cov}(x_1, x_2) & \ldots & \mathrm{cov}(x_1, x_N) \\ \mathrm{cov}(x_2, x_1) & \mathrm{var}(x_2) & & \mathrm{cov}(x_2, x_N) \\ \vdots & & \ddots & \\ \mathrm{cov}(x_N, x_1) & \mathrm{cov}(x_N, x_2) & & \mathrm{var}(x_N) \end{pmatrix};$$

and for an N-dimensional column vector $\mathbf{x} = (x_1, x_2, \ldots, x_N)'$ and a T-dimensional column vector $\mathbf{y} = (y_1, y_2, \ldots, y_T)'$,

$$\begin{aligned} \mathrm{cov}(\mathbf{x}, \mathbf{y}) &= \mathrm{cov}(\mathbf{x}', \mathbf{y}) = \mathrm{cov}(\mathbf{x}, \mathbf{y}') = \mathrm{cov}(\mathbf{x}', \mathbf{y}') \\ &= \begin{pmatrix} \mathrm{cov}(x_1, y_1) & \mathrm{cov}(x_1, y_2) & \ldots & \mathrm{cov}(x_1, y_T) \\ \mathrm{cov}(x_2, y_1) & \mathrm{cov}(x_2, y_2) & \ldots & \mathrm{cov}(x_2, y_T) \\ \vdots & \vdots & & \vdots \\ \mathrm{cov}(x_N, y_1) & \mathrm{cov}(x_N, y_2) & \ldots & \mathrm{cov}(x_N, y_T) \end{pmatrix}. \end{aligned}$$

For an N-dimensional random column vector $\mathbf{x}$ and a T-dimensional random column vector $\mathbf{y}$,

$$\mathrm{cov}(\mathbf{x}, \mathbf{y}) = \mathrm{E}\{[\mathbf{x} - \mathrm{E}(\mathbf{x})][\mathbf{y} - \mathrm{E}(\mathbf{y})]'\}, \tag{2.27}$$

$$\mathrm{cov}(\mathbf{x}, \mathbf{y}) = \mathrm{E}(\mathbf{x}\mathbf{y}') - \mathrm{E}(\mathbf{x})[\mathrm{E}(\mathbf{y})]', \tag{2.28}$$

and

$$\mathrm{cov}(\mathbf{y}, \mathbf{x}) = [\mathrm{cov}(\mathbf{x}, \mathbf{y})]'. \tag{2.29}$$

Equality (2.27) can be regarded as a multivariate extension of the formula $\mathrm{cov}(x, y) = \mathrm{E}\{[x - \mathrm{E}(x)][y - \mathrm{E}(y)]\}$ for the covariance of two random variables x and y. And equality (2.28) can be regarded as a multivariate extension of equality (2.7), and equality (2.29) as a multivariate extension of equality (2.5). Each of equalities (2.27), (2.28), and (2.29) can be readily verified by comparing each element of the left side with the corresponding element of the right side.

Clearly, for an N-dimensional random column vector $\mathbf{x}$,

$$\mathrm{var}(\mathbf{x}) = \mathrm{cov}(\mathbf{x}, \mathbf{x}). \tag{2.30}$$

Thus, as special cases of equalities (2.27), (2.28), and (2.29), we have that

$$\mathrm{var}(\mathbf{x}) = \mathrm{E}\{[\mathbf{x} - \mathrm{E}(\mathbf{x})][\mathbf{x} - \mathrm{E}(\mathbf{x})]'\}, \tag{2.31}$$

$$\mathrm{var}(\mathbf{x}) = \mathrm{E}(\mathbf{x}\mathbf{x}') - \mathrm{E}(\mathbf{x})[\mathrm{E}(\mathbf{x})]', \tag{2.32}$$

and

$$\text{var}(\mathbf{x}) = [\text{var}(\mathbf{x})]'. \tag{2.33}$$

Equality (2.31) can be regarded as a multivariate extension of the formula $\text{var}(x) = \text{E}\{[x - \text{E}(x)]^2\}$ for the variance of a random variable x, and equality (2.32) can be regarded as a multivariate extension of equality (2.3). Equality (2.33) indicates that a variance-covariance matrix is symmetric.

For an N-dimensional random column vector $\mathbf{x} = (x_1, x_2, \ldots, x_N)'$,

$$\Pr[\mathbf{x} \neq \text{E}(\mathbf{x})] \leq \sum_{i=1}^{N} \Pr[x_i \neq \text{E}(x_i)], \tag{2.34}$$

as is evident upon observing that $\{\mathbf{x} : \mathbf{x} \neq \text{E}(\mathbf{x})\} = \cup_{i=1}^{N} \{\mathbf{x} : x_i \neq \text{E}(x_i)\}$. Moreover, according to result (2.4), $\text{var}(x_i) = 0 \Rightarrow \Pr[x_i \neq \text{E}(x_i)] = 0$ $(i = 1, 2, \ldots, N)$, implying [in combination with inequality (2.34)] that if $\text{var}(x_i) = 0$ for $i = 1, 2, \ldots, N$, then $\Pr[\mathbf{x} \neq \text{E}(\mathbf{x})] = 0$ or equivalently $\Pr[\mathbf{x} = \text{E}(\mathbf{x})] = 1$. Thus, as a generalization of result (2.4), we have [since (for $i = 1, 2, \ldots, N$) $\Pr[\mathbf{x} = \text{E}(\mathbf{x})] = 1 \Rightarrow \Pr[x_i = \text{E}(x_i)] = 1 \Rightarrow \text{var}(x_i) = 0$] that [for an N-dimensional random column vector $\mathbf{x} = (x_1, x_2, \ldots, x_N)'$]

$$\text{var}(x_i) = 0 \text{ for } i = 1, 2, \ldots, N \quad \Leftrightarrow \quad \mathbf{x} = \text{E}(\mathbf{x}) \text{ with probability 1}. \tag{2.35}$$

Alternatively, result (2.35) can be established by observing that

$$\begin{aligned} \text{var}(x_i) = 0 \text{ for } i = 1, 2, \ldots, N \quad &\Rightarrow \quad \textstyle\sum_{i=1}^{N} \text{var}(x_i) = 0 \\ &\Rightarrow \quad \text{E}\{[\mathbf{x} - \text{E}(\mathbf{x})]'[\mathbf{x} - \text{E}(\mathbf{x})]\} = 0 \\ &\Rightarrow \quad \Pr\{[\mathbf{x} - \text{E}(\mathbf{x})]'[\mathbf{x} - \text{E}(\mathbf{x})]\} = 0\} = 1 \end{aligned}$$

and that $[\mathbf{x} - \text{E}(\mathbf{x})]'[\mathbf{x} - \text{E}(\mathbf{x})] = 0 \Leftrightarrow \mathbf{x} - \text{E}(\mathbf{x}) = \mathbf{0}$.

In connection with result (2.35) (and otherwise), it is worth noting that, for an N-dimensional random column vector $\mathbf{x} = (x_1, x_2, \ldots, x_N)'$,

$$\text{var}(x_i) = 0 \text{ for } i = 1, 2, \ldots, N \quad \Leftrightarrow \quad \text{var}(\mathbf{x}) = \mathbf{0}. \tag{2.36}$$

—result (2.36) is a consequence of result (2.12).

For random column vectors $\mathbf{x}$ and $\mathbf{y}$,

$$\text{var}\begin{pmatrix} \mathbf{x} \\ \mathbf{y} \end{pmatrix} = \begin{pmatrix} \text{var}(\mathbf{x}) & \text{cov}(\mathbf{x}, \mathbf{y}) \\ \text{cov}(\mathbf{y}, \mathbf{x}) & \text{var}(\mathbf{y}) \end{pmatrix}, \tag{2.37}$$

as is evident from the very definition of a variance-covariance matrix (and from the definition of the covariance of two random vectors). More generally, for a random column vector $\mathbf{x}$ that has been partitioned into subvectors $\mathbf{x}_1, \mathbf{x}_2, \ldots, \mathbf{x}_R$ [so that $\mathbf{x}' = (\mathbf{x}_1', \mathbf{x}_2', \ldots, \mathbf{x}_R')$],

$$\text{var}(\mathbf{x}) = \begin{pmatrix} \text{var}(\mathbf{x}_1) & \text{cov}(\mathbf{x}_1, \mathbf{x}_2) & \ldots & \text{cov}(\mathbf{x}_1, \mathbf{x}_R) \\ \text{cov}(\mathbf{x}_2, \mathbf{x}_1) & \text{var}(\mathbf{x}_2) & & \text{cov}(\mathbf{x}_2, \mathbf{x}_R) \\ \vdots & & \ddots & \\ \text{cov}(\mathbf{x}_R, \mathbf{x}_1) & \text{cov}(\mathbf{x}_R, \mathbf{x}_2) & & \text{var}(\mathbf{x}_R) \end{pmatrix}. \tag{2.38}$$

Corresponding to the variance-covariance matrix of an N-dimensional random column vector $\mathbf{x}$ (or row vector $\mathbf{x}'$) with first through Nth elements $x_1, x_2, \ldots, x_N$ (whose expected values exist and whose variances exist and are strictly positive) is the $N \times N$ matrix whose ijth element is $\text{corr}(x_i, x_j)$. This matrix is referred to as the *correlation matrix* of $\mathbf{x}$ (or $\mathbf{x}'$). It equals

$$\mathbf{S}^{-1} \text{var}(\mathbf{x})\, \mathbf{S}^{-1},$$

where $\mathbf{S} = \text{diag}(\sqrt{\text{var}\, x_1}, \sqrt{\text{var}\, x_2}, \ldots, \sqrt{\text{var}\, x_N})$. The correlation matrix is symmetric, and each of its diagonal elements equals 1.

c. Uncorrelated random variables or vectors

Two random variables x and y are said to be *uncorrelated* (or one of the random variables x and y is said to be uncorrelated with the other) if $\text{cov}(x, y) = 0$. Two or more random variables $x_1, x_2, \ldots, x_P$ are said to be *pairwise uncorrelated* or simply *uncorrelated* if every two of them are uncorrelated, that is, if $\text{cov}(x_i, x_j) = 0$ for $j > i = 1, 2, \ldots, P$. Accordingly, $x_1, x_2, \ldots, x_P$ are uncorrelated if the variance-covariance matrix of the P-dimensional random vector whose elements are $x_1, x_2, \ldots, x_P$ is a diagonal matrix.

Two random vectors $\mathbf{x}$ (or $\mathbf{x}'$) and $\mathbf{y}$ (or $\mathbf{y}'$) are said to be *uncorrelated* if $\text{cov}(\mathbf{x}, \mathbf{y}) = \mathbf{0}$, that is, if every element of $\mathbf{x}$ is uncorrelated with every element of $\mathbf{y}$. Two or more random vectors $\mathbf{x}_1, \mathbf{x}_2, \ldots, \mathbf{x}_P$ are said to be *pairwise uncorrelated* or simply *uncorrelated* if every two of them are uncorrelated, that is, if $\text{cov}(\mathbf{x}_i, \mathbf{x}_j) = \mathbf{0}$ for $j > i = 1, 2, \ldots, P$. Accordingly, $\mathbf{x}_1, \mathbf{x}_2, \ldots, \mathbf{x}_P$ are uncorrelated if the variance-covariance matrix of the random column vector $\mathbf{x}$ (or the random row vector $\mathbf{x}'$) defined by $\mathbf{x}' = (\mathbf{x}_1', \mathbf{x}_2', \ldots, \mathbf{x}_P')$ is of the block-diagonal form $\text{diag}(\text{var}\,\mathbf{x}_1, \text{var}\,\mathbf{x}_2, \ldots, \text{var}\,\mathbf{x}_P)$.

For statistically independent random variables x and y (whose expected values exist), we find [upon recalling result (1.4)] that

$$\text{cov}(x, y) = \text{E}\{[x - \text{E}(x)][y - \text{E}(y)]\} = \text{E}[x - \text{E}(x)]\,\text{E}[y - \text{E}(y)] = 0.$$

This result can be stated in the form of the following lemma.

Lemma 3.2.1. If two random variables (whose expected values exist) are statistically independent, they are uncorrelated.

In general, the converse of Lemma 3.2.1 is not true. That is, uncorrelated random variables are not necessarily statistically independent.

The repeated application of Lemma 3.2.1 gives rise to the following extension.

Lemma 3.2.2. Let $\mathbf{x}$ represent an N-dimensional random column vector with elements $x_1, x_2, \ldots, x_N$ (whose expected values exist) and $\mathbf{y}$ a T-dimensional random column vector with elements $y_1, y_2, \ldots, y_T$ (whose expected values exist). And suppose that for $i = 1, 2, \ldots, N$ and $j = 1, 2, \ldots, T$, x_i and y_j are statistically independent. Then, $\text{cov}(\mathbf{x}, \mathbf{y}) = \mathbf{0}$, that is, $\mathbf{x}$ and $\mathbf{y}$ are uncorrelated.

d. Variances and covariances of linear combinations of random variables or vectors

Consider now the covariance of $c + \sum_{i=1}^{N} a_i x_i$ and $k + \sum_{j=1}^{T} b_j y_j$, where $x_1, x_2, \ldots, x_N, y_1, y_2, \ldots, y_T$ are random variables (whose expected values exist) and where $c, a_1, a_2, \ldots, a_N, k, b_1, b_2, \ldots, b_T$ are nonrandom scalars. This covariance is expressible as

$$\text{cov}\Big(c + \sum_{i=1}^{N} a_i x_i,\; k + \sum_{j=1}^{T} b_j y_j\Big) = \sum_{i=1}^{N}\sum_{j=1}^{T} a_i b_j \,\text{cov}(x_i, y_j), \tag{2.39}$$

as can be readily verified by making use of result (1.5). As a special case of equality (2.39), we have that

$$\text{var}\Big(c + \sum_{i=1}^{N} a_i x_i\Big) = \sum_{i=1}^{N}\sum_{j=1}^{N} a_i a_j \,\text{cov}(x_i, x_j) \tag{2.40}$$

$$= \sum_{i=1}^{N} a_i^2 \,\text{var}(x_i) + 2 \sum_{i=1}^{N-1}\sum_{j=i+1}^{N} a_i a_j \,\text{cov}(x_i, x_j). \tag{2.41}$$

As in the case of equality (1.5), equalities (2.39) and (2.40) are reexpressible in matrix notation. Upon letting $\mathbf{x} = (x_1, x_2, \ldots, x_N)'$, $\mathbf{a} = (a_1, a_2, \ldots, a_N)'$, $\mathbf{y} = (y_1, y_2, \ldots, y_T)'$, and

$\mathbf{b} = (b_1, b_2, \ldots, b_T)'$, equality (2.39) is reexpressible as

$$\operatorname{cov}(c + \mathbf{a}'\mathbf{x},\ k + \mathbf{b}'\mathbf{y}) = \mathbf{a}'\operatorname{cov}(\mathbf{x}, \mathbf{y})\,\mathbf{b}, \tag{2.42}$$

and equality (2.40) as

$$\operatorname{var}(c + \mathbf{a}'\mathbf{x}) = \mathbf{a}'\operatorname{var}(\mathbf{x})\,\mathbf{a}. \tag{2.43}$$

Note that in the special case where $\mathbf{y} = \mathbf{x}$ (and $T = N$), equality (2.42) simplifies to

$$\operatorname{cov}(c + \mathbf{a}'\mathbf{x},\ k + \mathbf{b}'\mathbf{x}) = \mathbf{a}'\operatorname{var}(\mathbf{x})\,\mathbf{b}. \tag{2.44}$$

Results (2.42), (2.43), and (2.44) can be generalized. Let $\mathbf{c}$ represent an M-dimensional nonrandom column vector, $\mathbf{A}$ an $M \times N$ nonrandom matrix, and $\mathbf{x}$ an N-dimensional random column vector (whose expected value exists). Similarly, let $\mathbf{k}$ represent an S-dimensional nonrandom column vector, $\mathbf{B}$ an $S \times T$ nonrandom matrix, and $\mathbf{y}$ a T-dimensional random column vector (whose expected value exists). Then,

$$\operatorname{cov}(\mathbf{c} + \mathbf{A}\mathbf{x},\ \mathbf{k} + \mathbf{B}\mathbf{y}) = \mathbf{A}\operatorname{cov}(\mathbf{x}, \mathbf{y})\mathbf{B}', \tag{2.45}$$

which is a generalization of result (2.42) and which in the special case where $\mathbf{y} = \mathbf{x}$ (and $T = N$) yields the following generalization of result (2.44):

$$\operatorname{cov}(\mathbf{c} + \mathbf{A}\mathbf{x},\ \mathbf{k} + \mathbf{B}\mathbf{x}) = \mathbf{A}\operatorname{var}(\mathbf{x})\mathbf{B}'. \tag{2.46}$$

When $\mathbf{k} = \mathbf{c}$ and $\mathbf{B} = \mathbf{A}$, result (2.46) simplifies to the following generalization of result (2.43):

$$\operatorname{var}(\mathbf{c} + \mathbf{A}\mathbf{x}) = \mathbf{A}\operatorname{var}(\mathbf{x})\mathbf{A}'. \tag{2.47}$$

Equality (2.45) can be readily verified by comparing each element of the left side with the corresponding element of the right side and by applying result (2.42).

Another sort of generalization is possible. Let $\mathbf{x}_1, \mathbf{x}_2, \ldots, \mathbf{x}_N$ represent M-dimensional random column vectors (whose expected values exist), $\mathbf{c}$ an M-dimensional nonrandom column vector, and $a_1, a_2, \ldots, a_N$ nonrandom scalars. Similarly, let $\mathbf{y}_1, \mathbf{y}_2, \ldots, \mathbf{y}_T$ represent S-dimensional random column vectors (whose expected values exist), $\mathbf{k}$ an S-dimensional nonrandom column vector, and $b_1, b_2, \ldots, b_T$ nonrandom scalars. Then,

$$\operatorname{cov}\Big(\mathbf{c} + \sum_{i=1}^{N} a_i\mathbf{x}_i,\ \mathbf{k} + \sum_{j=1}^{T} b_j\mathbf{y}_j\Big) = \sum_{i=1}^{N}\sum_{j=1}^{T} a_i b_j \operatorname{cov}(\mathbf{x}_i, \mathbf{y}_j), \tag{2.48}$$

which is a generalization of result (2.39). As a special case of equality (2.48) [that obtained by setting $T = N$ and (for $j = 1, 2, \ldots, T$) $\mathbf{y}_j = \mathbf{x}_j$], we have that

$$\operatorname{cov}\Big(\mathbf{c} + \sum_{i=1}^{N} a_i\mathbf{x}_i,\ \mathbf{k} + \sum_{j=1}^{N} b_j\mathbf{x}_j\Big) = \sum_{i=1}^{N}\sum_{j=1}^{N} a_i b_j \operatorname{cov}(\mathbf{x}_i, \mathbf{x}_j). \tag{2.49}$$

And as a further special case [that obtained by setting $\mathbf{k} = \mathbf{c}$ and (for $j = 1, 2, \ldots, N$) $b_j = a_j$], we have that

$$\operatorname{var}\Big(\mathbf{c} + \sum_{i=1}^{N} a_i\mathbf{x}_i\Big) = \sum_{i=1}^{N} a_i^2 \operatorname{var}(\mathbf{x}_i) + \sum_{i=1}^{N}\sum_{\substack{j=1 \\ (j \neq i)}}^{N} a_i a_j \operatorname{cov}(\mathbf{x}_i, \mathbf{x}_j). \tag{2.50}$$

Equality (2.48) can be verified by comparing each element of the left side with the corresponding element of the right side and by applying result (2.39). Alternatively, equality (2.48) can be derived by observing that $\sum_{i=1}^{N} a_i\mathbf{x}_i = \mathbf{A}\mathbf{x}$, where $\mathbf{A} = (a_1\mathbf{I}, a_2\mathbf{I}, \ldots, a_N\mathbf{I})$ and $\mathbf{x}' = (\mathbf{x}_1', \mathbf{x}_2', \ldots, \mathbf{x}_N')$, and that $\sum_{j=1}^{T} b_j\mathbf{y}_j = \mathbf{B}\mathbf{y}$, where $\mathbf{B} = (b_1\mathbf{I}, b_2\mathbf{I}, \ldots, b_T\mathbf{I})$ and $\mathbf{y}' = (\mathbf{y}_1', \mathbf{y}_2', \ldots, \mathbf{y}_T')$, and by applying equality (2.45).

e. Nonnegative definiteness of a variance-covariance matrix

Let $\mathbf{x}$ represent an N-dimensional random column vector (with elements whose expected values and variances exist). Because a variance is inherently nonnegative, it follows immediately from result (2.43) that $\mathbf{a}' \operatorname{var}(\mathbf{x}) \mathbf{a} \ge 0$ for every N-dimensional nonrandom column vector $\mathbf{a}$. Thus, recalling result (2.33), we have the following theorem.

Theorem 3.2.3. The variance-covariance matrix of a random vector is nonnegative definite (and symmetric).

Positive definite vs. positive semidefinite variance-covariance matrices. Let $\mathbf{V}$ represent the variance-covariance matrix of an N-dimensional random column vector $\mathbf{x}$. Theorem 3.2.3 implies that $\mathbf{V}$ is either positive definite or positive semidefinite. Moreover, for an arbitrary N-dimensional nonrandom column vector $\mathbf{a}$, we have [from result (2.43)] that $\operatorname{var}(\mathbf{a}'\mathbf{x}) = \mathbf{a}'\mathbf{V}\mathbf{a}$. Accordingly, if $\mathbf{V}$ is positive semidefinite, there exist nonnull values of the vector $\mathbf{a}$ for which $\operatorname{var}(\mathbf{a}'\mathbf{x}) = 0$ or, equivalently, for which $\mathbf{a}'\mathbf{x} = E(\mathbf{a}'\mathbf{x})$ with probability 1. Alternatively, if $\mathbf{V}$ is positive definite, no such values exist; that is, if $\mathbf{V}$ is positive definite, then $\operatorname{var}(\mathbf{a}'\mathbf{x}) = 0 \Leftrightarrow \mathbf{a} = \mathbf{0}$.

In fact, in light of Corollary 2.13.27 (and the symmetry of $\mathbf{V}$),

$$\operatorname{var}(\mathbf{a}'\mathbf{x}) = 0 \quad \Leftrightarrow \quad \mathbf{V}\mathbf{a} = \mathbf{0} \quad \Leftrightarrow \quad \mathbf{a} \in \mathcal{N}(\mathbf{V}). \tag{2.51}$$

The null space $\mathcal{N}(\mathbf{V})$ of $\mathbf{V}$ is (as discussed in Section 2.9b) a linear space and (according to Lemma 2.11.5) is of dimension $N - \operatorname{rank}(\mathbf{V})$. When $\mathbf{V}$ is positive definite, $\mathbf{V}$ is nonsingular, so that $\dim[\mathcal{N}(\mathbf{V})] = 0$. When $\mathbf{V}$ is positive semidefinite, $\mathbf{V}$ is singular, so that $\dim[\mathcal{N}(\mathbf{V})] \ge 1$.

In light of Lemma 2.14.21, we have that $|\mathbf{V}| \ge 0$ and, more specifically, that

$$|\mathbf{V}| > 0 \quad \Leftrightarrow \quad \mathbf{V} \text{ is positive definite} \tag{2.52}$$

and

$$|\mathbf{V}| = 0 \quad \Leftrightarrow \quad \mathbf{V} \text{ is positive semidefinite} \tag{2.53}$$

An inequality revisited. Consider the implications of results (2.52) and (2.53) in the special case where $N = 2$. Accordingly, suppose that $\mathbf{V}$ is the variance-covariance matrix of a vector of two random variables, say x and y. Then, in light of result (2.14.4),

$$|\mathbf{V}| = \operatorname{var}(x) \operatorname{var}(y) - [\operatorname{cov}(x, y)]^2, \tag{2.54}$$

Thus, it follows from results (2.52) and (2.53 that

$$|\operatorname{cov}(x, y)| \le \sqrt{\operatorname{var}(x)} \sqrt{\operatorname{var}(y)}, \tag{2.55}$$

with equality holding if and only if $|\mathbf{V}| = 0$ and hence if and only if $\mathbf{V}$ is positive semidefinite (or, equivalently, if and only if $\mathbf{V}$ is singular or, also equivalently, if and only if $\dim[\mathcal{N}(\mathbf{V})] \ge 1$).

Note that inequality (2.55) is identical to inequality (2.9), the validity of which was established (in the final part of Subsection a) via a different approach.

3.3 Standardized Version of a Random Variable

Let x represent a random variable, and for an arbitrary nonrandom scalar c and an arbitrary nonzero nonrandom scalar a, define

$$z = \frac{x - c}{a}.$$

Then, x is expressible as

$$x = c + az,$$

and the distribution of x is determinable from c and a and from the distribution of the transformed random variable z

Clearly,

$$\mathrm{E}(z) = \frac{\mathrm{E}(x) - c}{a}.$$

And

$$\mathrm{var}(z) = \frac{\mathrm{var}(x)}{a^2},$$

or more generally, taking y to be a random variable and w to be the transformed random variable defined by $w = (y - k)/b$ (where k is a nonrandom scalar and b a nonzero nonrandom scalar),

$$\mathrm{cov}(z, w) = \frac{\mathrm{cov}(x, y)}{ab}.$$

In the special case where $c = \mathrm{E}(x)$ and $a = \sqrt{\mathrm{var}\, x}$, the random variable z is referred to as the *standardized version* of the random variable x (with the use of this term being restricted to situations where the expected value of x exists and where the variance of x exists and is strictly positive). When $c = \mathrm{E}(x)$ and $a = \sqrt{\mathrm{var}\, x}$,

$$\mathrm{E}(z) = 0 \quad \text{and} \quad \mathrm{var}(z) = 1.$$

Further, if z is the standardized version of x and w the standardized version of a random variable y, then

$$\mathrm{cov}(z, w) = \mathrm{corr}(x, y).$$

a. Transformation of a random vector to a vector of standardized versions

Let $\mathbf{x}$ represent an N-dimensional random column vector with first through Nth elements $x_1, x_2, \ldots, x_N$ (whose expected values exist and whose variances exist and are strictly positive). And take $\mathbf{z}$ to be the N-dimensional random column vector whose ith element is the standardized version $z_i = [x_i - \mathrm{E}(x_i)]/\sqrt{\mathrm{var}\, x_i}$ of x_i $(i = 1, 2, \ldots, N)$; or, equivalently, take

$$\mathbf{z} = \mathbf{S}^{-1}[\mathbf{x} - \mathrm{E}(\mathbf{x})],$$

where $\mathbf{S} = \mathrm{diag}(\sqrt{\mathrm{var}\, x_1}, \sqrt{\mathrm{var}\, x_2}, \ldots, \sqrt{\mathrm{var}\, x_N})$. Then, $\mathbf{x}$ is expressible as

$$\mathbf{x} = \mathrm{E}(\mathbf{x}) + \mathbf{S}^{-1}\mathbf{z},$$

and the distribution of $\mathbf{x}$ is determinable from its mean vector $\mathrm{E}(\mathbf{x})$ and the vector $(\sqrt{\mathrm{var}\, x_1}, \sqrt{\mathrm{var}\, x_2}, \ldots, \sqrt{\mathrm{var}\, x_N})$ of standard deviations and from the distribution of the transformed random vector $\mathbf{z}$.

Clearly,

$$\mathrm{E}(\mathbf{z}) = \mathbf{0}.$$

Further, $\mathrm{var}(\mathbf{z})$ equals the correlation matrix of $\mathbf{x}$, or more generally, taking $\mathbf{y}$ to be a T-dimensional random column vector with first through Tth elements $y_1, y_2, \ldots, y_T$ (whose expected values exist and whose variances exist and are strictly positive) and taking $\mathbf{w}$ to be the T-dimensional random column vector whose jth element is the standardized version of y_j, $\mathrm{cov}(\mathbf{z}, \mathbf{w})$ equals the $N \times T$ matrix whose ijth element is $\mathrm{corr}(x_i, y_j)$.

b. Transformation of a random vector to a vector of uncorrelated random variables having mean 0 and variance 1

Let $\mathbf{x}$ represent an N-dimensional random column vector (with elements whose expected values and variances exist). Further, let $\mathbf{V} = \mathrm{var}(\mathbf{x})$ and $K = \mathrm{rank}(\mathbf{V})$. And suppose that $\mathbf{V}$ is nonnull (so that $K \geq 1$).

Now, take $\mathbf{T}$ to be any $K \times N$ nonrandom matrix such that $\mathbf{V} = \mathbf{T}'\mathbf{T}$, and observe that $\mathrm{rank}(\mathbf{T}) = K$—that such a matrix exists and is of rank K follows from Corollary 2.13.23. And define

$$\mathbf{z} = \mathbf{R}'[\mathbf{x} - \mathrm{E}(\mathbf{x})],$$

where $\mathbf{R}$ is any right inverse of $\mathbf{T}$—the existence of a right inverse follows from Lemma 2.5.1. Then, clearly,

$$\mathrm{E}(\mathbf{z}) = \mathbf{0},$$

and

$$\mathrm{var}(\mathbf{z}) = \mathbf{R}'\mathbf{V}\mathbf{R} = (\mathbf{T}\mathbf{R})'\mathbf{T}\mathbf{R} = \mathbf{I}'\mathbf{I} = \mathbf{I}.$$

Thus, the elements of $\mathbf{z}$ are uncorrelated, and each has a mean of 0 and a variance of 1.

Is the vector $\mathbf{x}$ expressible in terms of the vector $\mathbf{z}$? Consider the vector $\mathrm{E}(\mathbf{x}) + \mathbf{T}'\mathbf{z}$. We find that

$$\mathrm{E}(\mathbf{x}) + \mathbf{T}'\mathbf{z} = \mathrm{E}(\mathbf{x}) + \mathbf{T}'\mathbf{R}'[\mathbf{x} - \mathrm{E}(\mathbf{x})]$$

and, accordingly, that

$$\mathbf{x} - [\mathrm{E}(\mathbf{x}) + \mathbf{T}'\mathbf{z}] = (\mathbf{I} - \mathbf{T}'\mathbf{R}')[\mathbf{x} - \mathrm{E}(\mathbf{x})]. \tag{3.1}$$

In the special case where $K = N$, $\mathbf{R} = \mathbf{T}^{-1}$ (as is evident from Lemma 2.5.3), so that $\mathbf{T}'\mathbf{R}' = (\mathbf{R}\mathbf{T})' = \mathbf{I}$. Thus, in that special case, $\mathbf{x} = \mathrm{E}(\mathbf{x}) + \mathbf{T}'\mathbf{z}$.

More generally (when K is possibly less than N), $\mathbf{x} = \mathrm{E}(\mathbf{x}) + \mathbf{T}'\mathbf{z}$ for those values of $\mathbf{x}$ for which $\mathbf{x} - \mathrm{E}(\mathbf{x}) \in \mathcal{C}(\mathbf{V})$ (but not for the other values of $\mathbf{x}$). To see this, observe [in light of equality (3.1)] that the condition $\mathbf{x} = \mathrm{E}(\mathbf{x}) + \mathbf{T}'\mathbf{z}$ is equivalent to the condition $(\mathbf{I} - \mathbf{T}'\mathbf{R}')[\mathbf{x} - \mathrm{E}(\mathbf{x})] = \mathbf{0}$. And let $\mathbf{d}$ represent the value of $\mathbf{x} - \mathrm{E}(\mathbf{x})$ corresponding to an arbitrary value of $\mathbf{x}$. If $\mathbf{d} \in \mathcal{C}(\mathbf{V})$, then $\mathbf{d} = \mathbf{V}\mathbf{h}$ for some column vector $\mathbf{h}$, and we find that

$$(\mathbf{I} - \mathbf{T}'\mathbf{R}')\mathbf{d} = (\mathbf{I} - \mathbf{T}'\mathbf{R}')\mathbf{T}'\mathbf{T}\mathbf{h} = \mathbf{T}'[\mathbf{I} - (\mathbf{T}\mathbf{R})']\mathbf{T}\mathbf{h} = \mathbf{T}'(\mathbf{I} - \mathbf{I}')\mathbf{T}\mathbf{h} = \mathbf{0}.$$

Conversely, if $(\mathbf{I} - \mathbf{T}'\mathbf{R}')\mathbf{d} = \mathbf{0}$, then (making use of Lemma 2.12.1), we find that

$$\mathbf{d} = \mathbf{T}'\mathbf{R}'\mathbf{d} \in \mathcal{C}(\mathbf{T}') = \mathcal{C}(\mathbf{V}).$$

Thus, $(\mathbf{I} - \mathbf{T}'\mathbf{R}')\mathbf{d} = \mathbf{0}$ if and only if $\mathbf{d} \in \mathcal{C}(\mathbf{V})$.

While (in general) $\mathbf{x}$ is not necessarily equal to $\mathrm{E}(\mathbf{x}) + \mathbf{T}'\mathbf{z}$ for every value of $\mathbf{x}$, it is the case that

$$\mathbf{x} = \mathrm{E}(\mathbf{x}) + \mathbf{T}'\mathbf{z} \ \text{ with probability 1},$$

as is evident upon observing [in light of result (3.1)] that

$$\mathrm{E}\{\mathbf{x} - [\mathrm{E}(\mathbf{x}) + \mathbf{T}'\mathbf{z}]\} = (\mathbf{I} - \mathbf{T}'\mathbf{R}')[\mathrm{E}(\mathbf{x}) - \mathrm{E}(\mathbf{x})] = \mathbf{0}$$

and

$$\begin{aligned}\mathrm{var}\{\mathbf{x} - [\mathrm{E}(\mathbf{x}) + \mathbf{T}'\mathbf{z}]\} &= (\mathbf{I} - \mathbf{T}'\mathbf{R}')\mathbf{V}(\mathbf{I} - \mathbf{T}'\mathbf{R}')' \\ &= (\mathbf{I} - \mathbf{T}'\mathbf{R}')\mathbf{T}'(\mathbf{I} - \mathbf{T}\mathbf{R})\mathbf{T} \\ &= (\mathbf{I} - \mathbf{T}'\mathbf{R}')\mathbf{T}'(\mathbf{I} - \mathbf{I})\mathbf{T} \\ &= \mathbf{0}\end{aligned}$$

and hence that $\mathbf{x} - [\mathrm{E}(\mathbf{x}) + \mathbf{T}'\mathbf{z}] = \mathbf{0}$ with probability 1.

Note that in the "degenerate" special case where $\mathbf{V} = \mathbf{0}$, $\mathbf{x} = \mathrm{E}(\mathbf{x})$ with probability 1.

3.4 Conditional Expected Values and Conditional Variances and Covariances of Random Variables or Vectors

For a random variable y [whose expected value $\mathrm{E}(y)$ exists] and for a random variable, random vector, or more generally random matrix $\mathbf{X}$, let us write $\mathrm{E}(y \mid \mathbf{X})$ for the *expected value of y conditional on* $\mathbf{X}$—refer, e.g., to Bickel and Doksum (2001, app. B.1) or (at a more advanced level) to Feller (1971, chap. V) or Billingsley (1995, sec. 34) for the definition of a conditional expected value. The conditional expected value $\mathrm{E}(y \mid \mathbf{X})$ can be regarded as a function of the random matrix $\mathbf{X}$ and as such has the following basic property:

$$\mathrm{E}(y) = \mathrm{E}[\mathrm{E}(y \mid \mathbf{X})] \tag{4.1}$$

(e.g., Casella and Berger 2002, thm. 4.4.3; Bickel and Doksum 2001, eq. B.1.20; Billingsley 1995, eq. 34.6).

The *variance of y conditional on* $\mathbf{X}$ is the quantity $\mathrm{var}(y \mid \mathbf{X})$ defined as follows:

$$\mathrm{var}(y \mid \mathbf{X}) = \mathrm{E}\{[y - \mathrm{E}(y \mid \mathbf{X})]^2 \mid \mathbf{X}\}.$$

Further, the *covariance of y and a random variable w conditional on* $\mathbf{X}$ is the quantity $\mathrm{cov}(y, w \mid \mathbf{X})$ defined as

$$\mathrm{cov}(y, w \mid \mathbf{X}) = \mathrm{E}\{[y - \mathrm{E}(y \mid \mathbf{X})][w - \mathrm{E}(y \mid \mathbf{X})] \mid \mathbf{X}\}.$$

The following identity relates the (unconditional) variance of the random variable y to its conditional mean and variance:

$$\mathrm{var}(y) = \mathrm{E}[\mathrm{var}(y \mid \mathbf{X})] + \mathrm{var}[\mathrm{E}(y \mid \mathbf{X})]. \tag{4.2}$$

Similarly,

$$\mathrm{cov}(y, w) = \mathrm{E}[\mathrm{cov}(y, w \mid \mathbf{X})] + \mathrm{cov}[\mathrm{E}(y \mid \mathbf{X}),\ \mathrm{E}(w \mid \mathbf{X})]. \tag{4.3}$$

Let us verify equality (4.3)—equality (4.2) can be regarded as a special case of equality (4.3) (that where $w = y$). Starting with the very definition of a covariance between two random variables, we find that

$$\begin{aligned}
\mathrm{cov}(y, w) &= \mathrm{E}\{[y - \mathrm{E}(y)][w - \mathrm{E}(w)]\} \\
&= \mathrm{E}\{[y - \mathrm{E}(y \mid \mathbf{X}) + \mathrm{E}(y \mid \mathbf{X}) - \mathrm{E}(y)][w - \mathrm{E}(w \mid \mathbf{X}) + \mathrm{E}(w \mid \mathbf{X}) - \mathrm{E}(w)]\} \\
&= \mathrm{E}\{[y - \mathrm{E}(y \mid \mathbf{X})][w - \mathrm{E}(w \mid \mathbf{X})]\} \\
&\qquad + \mathrm{E}\{[\mathrm{E}(y \mid \mathbf{X}) - \mathrm{E}(y)][\mathrm{E}(w \mid \mathbf{X}) - \mathrm{E}(w)]\} \\
&\qquad + \mathrm{E}\{[\mathrm{E}(y \mid \mathbf{X}) - \mathrm{E}(y)][w - \mathrm{E}(w \mid \mathbf{X})]\} \\
&\qquad + \mathrm{E}\{[y - \mathrm{E}(y \mid \mathbf{X})][\mathrm{E}(w \mid \mathbf{X}) - \mathrm{E}(w)]\}.
\end{aligned} \tag{4.4}$$

The first term of the sum (4.4) is expressible as

$$\mathrm{E}\big(\mathrm{E}\{[y - E(y \mid \mathbf{X})][w - \mathrm{E}(w \mid \mathbf{X})] \mid \mathbf{X}\}\big) = \mathrm{E}[\mathrm{cov}(y, w \mid \mathbf{X})],$$

and, since $\mathrm{E}(y) = \mathrm{E}[\mathrm{E}(y \mid \mathbf{X})]$ and $\mathrm{E}(w) = \mathrm{E}[\mathrm{E}(w \mid \mathbf{X})]$, the second term equals $\mathrm{cov}[\mathrm{E}(y \mid \mathbf{X}),\ \mathrm{E}(w \mid \mathbf{X})]$. It remains to show that the third and fourth terms of the sum (4.4) equal 0. Using basic properties of conditional expected values (e.g., Bickel and Doksum 2001, app. B.1.3; Billingsley 1995, sec. 34), we find that

$$\begin{aligned}
\mathrm{E}\{[\mathrm{E}(y \mid \mathbf{X}) - \mathrm{E}(y)][w - \mathrm{E}(w \mid \mathbf{X})]\} &= \mathrm{E}\big(\mathrm{E}\{[\mathrm{E}(y \mid \mathbf{X}) - \mathrm{E}(y)][w - \mathrm{E}(w \mid \mathbf{X})] \mid \mathbf{X}\}\big) \\
&= \mathrm{E}\{[\mathrm{E}(y \mid \mathbf{X}) - \mathrm{E}(y)]\,\mathrm{E}[w - \mathrm{E}(w \mid \mathbf{X}) \mid \mathbf{X}]\} \\
&= \mathrm{E}\{[\mathrm{E}(y \mid \mathbf{X}) - \mathrm{E}(y)][\mathrm{E}(w \mid \mathbf{X}) - \mathrm{E}(w \mid \mathbf{X})]\} \\
&= 0.
\end{aligned}$$

Thus, the third term of the sum (4.4) equals 0. That the fourth term equals 0 can be demonstrated in similar fashion.

The definition of the conditional expected value of a random variable can be readily extended to a random row or column vector or more generally to a random matrix. The *expected value of an* $M \times N$ *random matrix* $\mathbf{Y} = \{y_{ij}\}$ *conditional on a random matrix* $\mathbf{X}$ is defined to be the $M \times N$ matrix whose ijth element is the conditional expected value $\mathrm{E}(y_{ij} \mid \mathbf{X})$ of the ijth element of $\mathbf{Y}$. It is to be denoted by the symbol $\mathrm{E}(\mathbf{Y} \mid \mathbf{X})$. As a straightforward extension of the property (4.1), we have that

$$\mathrm{E}(\mathbf{Y}) = \mathrm{E}[\mathrm{E}(\mathbf{Y} \mid \mathbf{X})]. \tag{4.5}$$

The definition of the conditional variance of a random variable and the definition of a conditional covariance of two random variables can also be readily extended. The *variance-covariance matrix of an* M*-dimensional random column vector* $\mathbf{y} = \{y_i\}$ *(or its transpose* $\mathbf{y}'$*) conditional on a random matrix* $\mathbf{X}$ is defined to be the $M \times M$ matrix whose ijth element is the conditional covariance $\mathrm{cov}(y_i, y_j \mid \mathbf{X})$ of the ith and jth elements of $\mathbf{y}$ or $\mathbf{y}'$. It is to be denoted by the symbol $\mathrm{var}(\mathbf{y} \mid \mathbf{X})$ or $\mathrm{var}(\mathbf{y}' \mid \mathbf{X})$. Note that the diagonal elements of this matrix are the N conditional variances $\mathrm{var}(y_1 \mid \mathbf{X})$, $\mathrm{var}(y_2 \mid \mathbf{X})$, $\ldots$, $\mathrm{var}(y_N \mid \mathbf{X})$. Further, the *covariance of an* M*-dimensional random column vector* $\mathbf{y} = \{y_i\}$ *(or its transpose* $\mathbf{y}'$*) and an* N*-dimensional random column vector* $\mathbf{w} = \{w_j\}$ *(or its transpose* $\mathbf{w}'$*) conditional on a random matrix* $\mathbf{X}$ is defined to be the $M \times N$ matrix whose ijth element is the conditional covariance $\mathrm{cov}(y_i, w_j \mid \mathbf{X})$ of the ith element of $\mathbf{y}$ or $\mathbf{y}'$ and the jth element of $\mathbf{w}$ or $\mathbf{w}'$. It is to be denoted by the symbol $\mathrm{cov}(\mathbf{y}, \mathbf{w} \mid \mathbf{X})$, $\mathrm{cov}(\mathbf{y}', \mathbf{w} \mid \mathbf{X})$, $\mathrm{cov}(\mathbf{y}, \mathbf{w}' \mid \mathbf{X})$, or $\mathrm{cov}(\mathbf{y}', \mathbf{w}' \mid \mathbf{X})$.

As generalizations of equalities (4.2) and (4.3), we have that

$$\mathrm{var}(\mathbf{y}) = \mathrm{E}[\mathrm{var}(\mathbf{y} \mid \mathbf{X})] + \mathrm{var}[\mathrm{E}(\mathbf{y} \mid \mathbf{X})] \tag{4.6}$$

and

$$\mathrm{cov}(\mathbf{y}, \mathbf{w}) = \mathrm{E}[\mathrm{cov}(\mathbf{y}, \mathbf{w} \mid \mathbf{X})] + \mathrm{cov}[\mathrm{E}(\mathbf{y} \mid \mathbf{X}),\ \mathrm{E}(\mathbf{w} \mid \mathbf{X})]. \tag{4.7}$$

The validity of equalities (4.6) and (4.7) is evident upon observing that equality (4.6) can be regarded as a special case of equality (4.7) and that equality (4.3) implies that each element of the left side of equality (4.7) equals the corresponding element of the right side.

3.5 Multivariate Normal Distribution

The multivariate normal distribution provides the theoretical underpinnings for many of the procedures devised for making inferences on the basis of a linear statistical model.

a. Standard (univariate) normal distribution

Consider the function $f(\cdot)$ defined by

$$f(z) = \frac{1}{\sqrt{2\pi}} e^{-z^2/2} \qquad (-\infty < z < \infty).$$

Clearly,

$$f(z) = f(-z) \tag{5.1}$$

for all z, that is, $f(\cdot)$ is symmetric about 0. Accordingly, $\int_{-\infty}^{0} f(z)\,dz = \int_{0}^{\infty} f(z)\,dz$ (as can be formally verified by making the change of variable $y = -z$), so that

$$\int_{-\infty}^{\infty} f(z)\,dz = \int_{0}^{\infty} f(z)\,dz + \int_{-\infty}^{0} f(z)\,dz = 2\int_{0}^{\infty} f(z)\,dz. \tag{5.2}$$

Moreover,

$$\int_0^\infty e^{-z^2/2}\,dz = \sqrt{\frac{\pi}{2}}, \tag{5.3}$$

as is well-known and as can be verified by observing that

$$\begin{aligned}\left(\int_0^\infty e^{-z^2/2}\,dz\right)^2 &= \left(\int_0^\infty e^{-z^2/2}\,dz\right)\left(\int_0^\infty e^{-y^2/2}\,dy\right)\\ &= \int_0^\infty\int_0^\infty e^{-(z^2+y^2)/2}\,dz\,dy \qquad (5.4)\end{aligned}$$

and by evaluating the double integral (5.4) by converting to polar coordinates—refer, e.g., to Casella and Berger (2002, sec. 3.3) for the details.

Together, results (5.2) and (5.3) imply that

$$\int_{-\infty}^\infty f(z)\,dz = 1.$$

And, upon observing that $f(z) \geq 0$ for all z, we conclude that the function $f(\cdot)$ can serve as a probability density function. The probability distribution determined by this probability density function is referred to as the *standard normal* (or standard Gaussian) *distribution.*

b. Gamma function

To obtain convenient expressions for the moments of the standard normal distribution, it is helpful to recall (e.g., from Parzen 1960, pp. 161–163, or Casella and Berger 2002, sec. 3.3) the definition and some basic properties of the gamma function. The *gamma function* is the function $\Gamma(\cdot)$ defined by

$$\Gamma(x) = \int_0^\infty w^{x-1}e^{-w}\,dw \qquad (x > 0). \tag{5.5}$$

By integrating by parts, it can be shown that

$$\Gamma(x+1) = x\Gamma(x). \tag{5.6}$$

It is a simple exercise to show that

$$\Gamma(1) = 1. \tag{5.7}$$

And, by repeated application of the recursive formula (5.6), result (5.7) can be generalized to

$$\Gamma(n+1) = n! = n(n-1)(n-2)\cdots 1 \qquad (n = 0, 1, 2, \ldots). \tag{5.8}$$

(By definition, $0! = 1$.)

By making the change of variable $z = \sqrt{2w}$ in integral (5.5), we find that, for $r > -1$,

$$\Gamma\left(\frac{r+1}{2}\right) = \left(\frac{1}{2}\right)^{(r-1)/2}\int_0^\infty z^r e^{-z^2/2}\,dz, \tag{5.9}$$

thereby obtaining an alternative representation for the gamma function. And, upon applying result (5.9) in the special case where $r = 0$ and upon recalling result (5.3), we obtain the formula

$$\Gamma\left(\tfrac{1}{2}\right) = \sqrt{\pi}. \tag{5.10}$$

This result is extended to $\Gamma(n + \frac{1}{2})$ by the formula

$$\Gamma\left(n+\tfrac{1}{2}\right) = \frac{(2n)!\sqrt{\pi}}{4^n\,n!} = \frac{1\cdot 3\cdot 5\cdot 7\cdots(2n-1)}{2^n}\sqrt{\pi} \qquad (n = 0, 1, 2, \ldots), \tag{5.11}$$

the validity of which can be established by making use of result (5.6) and employing mathematical induction.

c. Moments of the standard normal distribution

Denote by $f(\cdot)$ the probability density function of the standard normal distribution, and let z represent a random variable whose distribution is standard normal. For $r = 1, 2, 3, \dots$, the rth absolute moment of the standard normal distribution is

$$E(|z|^r) = \int_{-\infty}^{\infty} |z|^r f(z)\, dz = \int_0^{\infty} z^r f(z)\, dz + (-1)^r \int_{-\infty}^{0} z^r f(z)\, dz. \tag{5.12}$$

[In result (5.12), the symbol z is used to represent a variable of integration as well as a random variable. In circumstances where this kind of dual usage might result in confusion or ambiguity, either altogether different symbols are to be used for a random variable (or random vector or random matrix) and for a related quantity (such as a variable of integration or a value of the random variable), or the related quantity is to be distinguished from the random quantity simply by underlining whatever symbol is used for the random quantity.]

We have that, for $r = 1, 2, 3, \dots$,

$$\int_{-\infty}^{0} z^r f(z)\, dz = (-1)^r \int_0^{\infty} z^r f(z)\, dz \tag{5.13}$$

[as can be readily verified by making the change of variable $y = -z$ and recalling result (5.1)] and that, for $r > -1$,

$$\int_0^{\infty} z^r f(z)\, dz = \frac{1}{\sqrt{2\pi}} \int_0^{\infty} z^r e^{-z^2/2}\, dz = \frac{2^{(r/2)-1}}{\sqrt{\pi}} \Gamma\left(\frac{r+1}{2}\right) \tag{5.14}$$

[as is evident from result (5.9)].

Now, starting with expression (5.12) and making use of results (5.13), (5.14), (5.8), and (5.11), we find that, for $r = 1, 2, 3, \dots$,

$$\begin{aligned} E(|z|^r) &= 2 \int_0^{\infty} z^r f(z)\, dz \\ &= \sqrt{\frac{2^r}{\pi}}\, \Gamma\left(\frac{r+1}{2}\right) && (5.15) \\ &= \begin{cases} \sqrt{\dfrac{2^r}{\pi}}\, [(r-1)/2]! & \text{if } r \text{ is odd,} \\ (r-1)(r-3)(r-5)\cdots 7\cdot 5\cdot 3\cdot 1 & \text{if } r \text{ is even.} \end{cases} && (5.16) \end{aligned}$$

Accordingly, the rth moment of the standard normal distribution exists for $r = 1, 2, 3, \dots$. For $r = 1, 3, 5, \dots$, we find [in light of result (5.13)] that

$$E(z^r) = \int_0^{\infty} z^r f(z)\, dz + (-1)^r \int_0^{\infty} z^r f(z)\, dz = 0. \tag{5.17}$$

And, for $r = 2, 4, 6, \dots$, we have that

$$\begin{aligned} E(z^r) &= E(|z|^r) \\ &= (r-1)(r-3)(r-5)\cdots 7\cdot 5\cdot 3\cdot 1. \end{aligned} \tag{5.18}$$

Thus, the odd moments of the standard normal distribution equal 0, while the even moments are given by formula (5.18).

In particular, we have that

$$E(z) = 0 \quad \text{and} \quad \operatorname{var}(z) = E(z^2) = 1. \tag{5.19}$$

That is, the standard normal distribution has a mean of 0 and a variance of 1. Further, the third and fourth moments of the standard normal distribution are

$$E(z^3) = 0 \quad \text{and} \quad E(z^4) = 3. \tag{5.20}$$

d. Normal distribution (univariate)

Define

$$x = \mu + \sigma z, \tag{5.21}$$

where μ and σ are arbitrary nonrandom scalars and z is a random variable whose distribution is standard normal. Applying formulas (1.5) and (2.41) and making use of result (5.19), we find that

$$E(x) = \mu \quad \text{and} \quad \text{var}(x) = \sigma^2. \tag{5.22}$$

The standard deviation of x is $|\sigma| = \sqrt{\sigma^2}$.

Denote by $h(\cdot)$ the probability density function of the standard normal distribution. If $\sigma^2 > 0$, then the distribution of x is the absolutely continuous distribution with probability density function $f(\cdot)$ defined by

$$f(x) = \left|\frac{1}{\sigma}\right| h\Big(\frac{x-\mu}{\sigma}\Big) = \frac{1}{\sqrt{2\pi\sigma^2}} e^{-(x-\mu)^2/(2\sigma^2)}. \tag{5.23}$$

If $\sigma^2 = 0$, then the distribution of x is not continuous. Rather,

$$\Pr(x = \mu) = 1, \tag{5.24}$$

so that the distribution of x is completely concentrated at a single value, namely μ. Note that the distribution of x depends on σ only through the value of σ^2.

Let us refer to an absolutely continuous distribution having a probability density function of the form (5.23) and also to a "degenerate" distribution of the form (5.24) as a *normal* (or Gaussian) *distribution*. Accordingly, there is a family of normal distributions, the members of which are indexed by the mean and the variance of the distribution. The symbol $N(\mu, \sigma^2)$ is used to denote a normal distribution with mean μ and variance σ^2. Note that the $N(0, 1)$ distribution is identical to the standard normal distribution.

The rth central moment of the random variable x defined by equality (5.21) is expressible as

$$E[(x-\mu)^r] = E[(\sigma z)^r] = \sigma^r E(z^r).$$

Accordingly, it follows from results (5.17) and (5.18) that, for $r = 1, 3, 5, \ldots$,

$$E[(x-\mu)^r] = 0 \tag{5.25}$$

and that, for $r = 2, 4, 6, \ldots$,

$$E[(x-\mu)^r] = \sigma^r (r-1)(r-3)(r-5)\cdots 7 \cdot 5 \cdot 3 \cdot 1. \tag{5.26}$$

We find in particular [upon applying result (5.25) in the special case where $r = 3$ and result (5.26) in the special case where $r = 4$] that the third and fourth central moments of the $N(\mu, \sigma^2)$ distribution are

$$E[(x-\mu)^3] = 0 \quad \text{and} \quad E[(x-\mu)^4] = 3\sigma^4 \tag{5.27}$$

—in the special case where $r = 2$, result (5.26) simplifies to $\text{var}(x) = \sigma^2$.

The form of the probability density function of a (nondegenerate) normal distribution is illustrated in Figure 3.1.

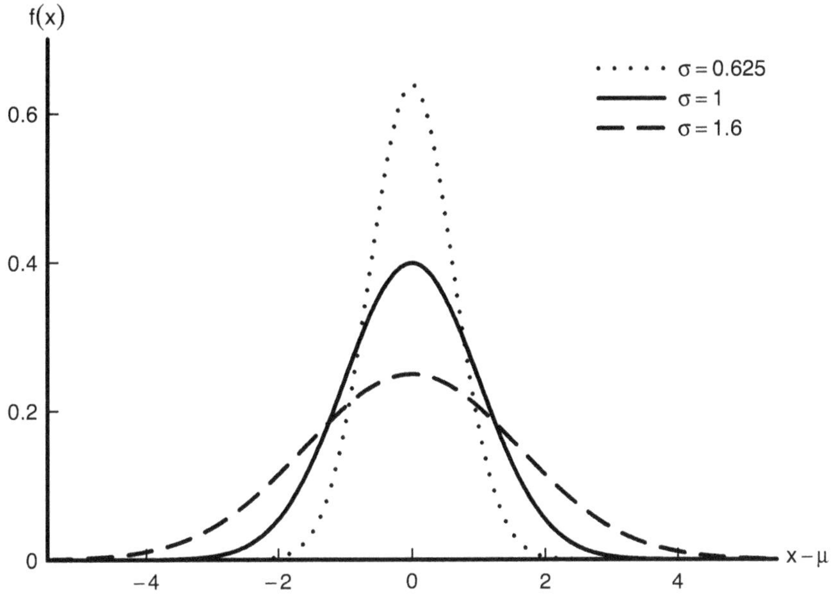

FIGURE 3.1. The probability density function $f(\cdot)$ of a (nondegenerate) $N(\mu, \sigma^2)$ distribution: plot of $f(x)$ against $x - \mu$ for each of 3 values of σ.

e. Multivariate extension

Let us now extend the approach taken in Subsection d [in defining the (univariate) normal distribution] to the multivariate case.

Let us begin by considering the distribution of an M-dimensional random column vector, say $\mathbf{z}$, whose elements are statistically independent and individually have standard normal distributions. This distribution is referred to as the *M-variate* (or multivariate) *standard normal* (or standard Gaussian) *distribution*. It has the probability density function $f(\cdot)$ defined by

$$f(\mathbf{z}) = \frac{1}{(2\pi)^{M/2}} \exp\left(-\tfrac{1}{2}\,\mathbf{z}'\mathbf{z}\right) \qquad (\mathbf{z} \in R^M). \tag{5.28}$$

Its mean vector and variance-covariance matrix are:

$$\mathrm{E}(\mathbf{z}) = \mathbf{0} \quad \text{and} \quad \mathrm{var}(\mathbf{z}) = \mathbf{I}. \tag{5.29}$$

Now, let M and N represent arbitrary positive integers. And define

$$\mathbf{x} = \boldsymbol{\mu} + \boldsymbol{\Gamma}'\mathbf{z}, \tag{5.30}$$

where $\boldsymbol{\mu}$ is an arbitrary M-dimensional nonrandom column vector, $\boldsymbol{\Gamma}$ is an arbitrary $N \times M$ nonrandom matrix, and $\mathbf{z}$ is an N-dimensional random column vector whose distribution is N-variate standard normal. Further, let

$$\boldsymbol{\Sigma} = \boldsymbol{\Gamma}'\boldsymbol{\Gamma}.$$

Then, upon applying formulas (1.7) and (2.47), we find [in light of result (5.29)] that

$$\mathrm{E}(\mathbf{x}) = \boldsymbol{\mu} \quad \text{and} \quad \mathrm{var}(\mathbf{x}) = \boldsymbol{\Sigma}. \tag{5.31}$$

Probability density function: existence, derivation, and geometrical form. Let us consider the distribution of the random vector $\mathbf{x}$ [defined by equality (5.30)]. If $\mathrm{rank}(\boldsymbol{\Gamma}) < M$ or, equivalently, if

$\text{rank}(\boldsymbol{\Sigma}) < M$ (in which case $\boldsymbol{\Sigma}$ is positive semidefinite), then the distribution of $\mathbf{x}$ has no probability density function. Suppose now that $\text{rank}(\boldsymbol{\Gamma}) = M$ or, equivalently, that $\text{rank}(\boldsymbol{\Sigma}) = M$ (in which case $\boldsymbol{\Sigma}$ is positive definite and $N \geq M$). Then, the distribution of $\mathbf{x}$ has a probability density function, which we now proceed to derive.

Take $\boldsymbol{\Lambda}$ to be an $N \times (N - M)$ matrix whose columns form an orthonormal (with respect to the usual inner product) basis for $\mathcal{N}(\boldsymbol{\Gamma}')$—according to Lemma 2.11.5, $\dim[\mathcal{N}(\boldsymbol{\Gamma}')] = N - M$. Then, observing that $\boldsymbol{\Lambda}'\boldsymbol{\Lambda} = \mathbf{I}$ and $\boldsymbol{\Gamma}'\boldsymbol{\Lambda} = \mathbf{0}$ and making use of Lemmas 2.12.1 and 2.6.1, we find that

$$\begin{aligned}\text{rank}\,(\boldsymbol{\Gamma},\ \boldsymbol{\Lambda}) &= \text{rank}[(\boldsymbol{\Gamma},\ \boldsymbol{\Lambda})'(\boldsymbol{\Gamma},\ \boldsymbol{\Lambda})]\\ &= \text{rank}\begin{pmatrix}\boldsymbol{\Sigma} & \mathbf{0}\\ \mathbf{0} & \mathbf{I}_{N-M}\end{pmatrix} = \text{rank}(\boldsymbol{\Sigma}) + N - M = N.\end{aligned}$$

Thus, the $N \times N$ matrix $(\boldsymbol{\Gamma},\ \boldsymbol{\Lambda})$ is nonsingular.

Define $\mathbf{w} = \boldsymbol{\Lambda}'\mathbf{z}$, and denote by $g(\cdot)$ the probability density function of the distribution of $\mathbf{z}$. Because

$$\mathbf{z} = \begin{pmatrix}\boldsymbol{\Gamma}'\\ \boldsymbol{\Lambda}'\end{pmatrix}^{-1}\begin{pmatrix}\mathbf{x}-\boldsymbol{\mu}\\ \mathbf{w}\end{pmatrix},$$

the joint distribution of $\mathbf{x}$ and $\mathbf{w}$ has the probability density function $h(\cdot\,,\cdot)$ given by

$$h(\mathbf{x},\mathbf{w}) = \left|\det\begin{pmatrix}\boldsymbol{\Gamma}'\\ \boldsymbol{\Lambda}'\end{pmatrix}^{-1}\right|\, g\!\left[\begin{pmatrix}\boldsymbol{\Gamma}'\\ \boldsymbol{\Lambda}'\end{pmatrix}^{-1}\begin{pmatrix}\mathbf{x}-\boldsymbol{\mu}\\ \mathbf{w}\end{pmatrix}\right].$$

And, upon observing that

$$\left|\det\begin{pmatrix}\boldsymbol{\Gamma}'\\ \boldsymbol{\Lambda}'\end{pmatrix}\right| = \left[\det\begin{pmatrix}\boldsymbol{\Gamma}'\\ \boldsymbol{\Lambda}'\end{pmatrix}\det(\boldsymbol{\Gamma},\ \boldsymbol{\Lambda})\right]^{1/2} = \left[\det\begin{pmatrix}\boldsymbol{\Sigma} & \mathbf{0}\\ \mathbf{0} & \mathbf{I}\end{pmatrix}\right]^{1/2} = [\det(\boldsymbol{\Sigma})]^{1/2},$$

we find that

$$\begin{aligned}h(\mathbf{x},\mathbf{w}) &= \frac{1}{(2\pi)^{N/2}|\boldsymbol{\Sigma}|^{1/2}}\exp\!\left[-\frac{1}{2}\begin{pmatrix}\mathbf{x}-\boldsymbol{\mu}\\ \mathbf{w}\end{pmatrix}'\begin{pmatrix}\boldsymbol{\Sigma}^{-1} & \mathbf{0}\\ \mathbf{0} & \mathbf{I}\end{pmatrix}\begin{pmatrix}\mathbf{x}-\boldsymbol{\mu}\\ \mathbf{w}\end{pmatrix}\right]\\ &= \frac{1}{(2\pi)^{M/2}|\boldsymbol{\Sigma}|^{1/2}}\exp\!\left[-\tfrac{1}{2}(\mathbf{x}-\boldsymbol{\mu})'\boldsymbol{\Sigma}^{-1}(\mathbf{x}-\boldsymbol{\mu})\right]\\ &\qquad\qquad\times\frac{1}{(2\pi)^{(N-M)/2}}\exp[-\tfrac{1}{2}\mathbf{w}'\mathbf{w}].\end{aligned}$$

Thus, the distribution of $\mathbf{x}$ has the probability density function $f(\cdot)$ given by

$$\begin{aligned}f(\mathbf{x}) &= \int_{-\infty}^{\infty}\cdots\int_{-\infty}^{\infty} h(\mathbf{x},\mathbf{w})\,d\mathbf{w}\\ &= \frac{1}{(2\pi)^{M/2}|\boldsymbol{\Sigma}|^{1/2}}\exp\!\left[-\tfrac{1}{2}(\mathbf{x}-\boldsymbol{\mu})'\boldsymbol{\Sigma}^{-1}(\mathbf{x}-\boldsymbol{\mu})\right].\end{aligned}\qquad(5.32)$$

Each of the contour lines or surfaces of $f(\cdot)$ consists of the points in a set of the form

$$\{\mathbf{x} \,:\, (\mathbf{x}-\boldsymbol{\mu})'\boldsymbol{\Sigma}^{-1}(\mathbf{x}-\boldsymbol{\mu}) = c\},$$

where c is a nonnegative scalar—$f(\cdot)$ has the same value for every point in the set. When $M = 2$, each of these lines or surfaces is an ellipse. More generally (when $M \geq 2$), each is an M-dimensional ellipsoid. In the special case where $\boldsymbol{\Sigma}$ (and hence $\boldsymbol{\Sigma}^{-1}$) is a scalar multiple of $\mathbf{I}_M$, each of the contour lines or surfaces is (when $M = 2$) a circle or (when $M \geq 2$) an M-dimensional sphere.

Uniqueness property. The matrix $\mathbf{\Sigma} = \mathbf{\Gamma}'\mathbf{\Gamma}$ has the same value for various choices of the $N \times M$ matrix $\mathbf{\Gamma}$ that differ with regard to their respective entries and/or with regard to the value of N. However, the distribution of the random vector $\mathbf{x} = \boldsymbol{\mu} + \mathbf{\Gamma}'\mathbf{z}$ is the same for all such choices—it depends on $\mathbf{\Gamma}$ only through the value of $\mathbf{\Sigma} = \mathbf{\Gamma}'\mathbf{\Gamma}$. That this is the case when $\mathbf{\Sigma}$ is positive definite is evident from result (5.32). That it is the case in general (i.e., even if $\mathbf{\Sigma}$ is positive semidefinite) is established in Subsection f.

Definition and notation. Let us refer to the distribution of the random vector $\mathbf{x} = \boldsymbol{\mu} + \mathbf{\Gamma}'\mathbf{z}$ as an *M-variate* (or multivariate) *normal* (or Gaussian) *distribution.* Accordingly, there is a family of M-variate normal distributions, the members of which are indexed by the mean vector and the variance-covariance matrix of the distribution. For every M-dimensional column vector $\boldsymbol{\mu}$ and every $M \times M$ symmetric nonnegative definite matrix $\mathbf{\Sigma}$, there is an M-variate normal distribution having $\boldsymbol{\mu}$ as its mean vector and $\mathbf{\Sigma}$ as its variance-covariance matrix (as is evident upon recalling, from Corollary 2.13.25, that every symmetric nonnegative definite matrix is expressible in the form $\mathbf{\Gamma}'\mathbf{\Gamma}$). The symbol $N(\boldsymbol{\mu}, \mathbf{\Sigma})$ is used to denote an MVN (multivariate normal) distribution with mean vector $\boldsymbol{\mu}$ and variance-covariance matrix $\mathbf{\Sigma}$. Note that the $N(\mathbf{0}, \mathbf{I}_M)$ distribution is identical to the M-variate standard normal distribution.

f. Verification of uniqueness property: general case

Let M represent an arbitrary positive integer. And take $\boldsymbol{\mu}$ to be an arbitrary M-dimensional (nonrandom) column vector, and $\mathbf{\Sigma}$ to be an arbitrary $M \times M$ (nonrandom) symmetric nonnegative definite matrix. Further, denote by $\mathbf{\Gamma}$ and $\mathbf{\Lambda}$ (nonrandom) matrices such that

$$\mathbf{\Sigma} = \mathbf{\Gamma}'\mathbf{\Gamma} = \mathbf{\Lambda}'\mathbf{\Lambda},$$

let N represent the number of rows in $\mathbf{\Gamma}$ and S the number of rows in $\mathbf{\Lambda}$, and let $R = \text{rank}(\mathbf{\Sigma})$. Finally, define

$$\mathbf{x} = \boldsymbol{\mu} + \mathbf{\Gamma}'\mathbf{z} \quad \text{and} \quad \mathbf{w} = \boldsymbol{\mu} + \mathbf{\Lambda}'\mathbf{y},$$

where $\mathbf{z}$ is an N-dimensional random column vector whose distribution is (N variate) standard normal and $\mathbf{y}$ is an S-dimensional random column vector whose distribution is (S variate) standard normal.

Let us verify that $\mathbf{w} \sim \mathbf{x}$, thereby validating the assertion made in the next-to-last paragraph of Subsection e—depending on the context, the symbol $\sim$ means "is (or be) distributed as" or "has the same distribution as." In doing so, let us partition $\mathbf{\Sigma}$ as

$$\mathbf{\Sigma} = \begin{pmatrix} \mathbf{\Sigma}_{11} & \mathbf{\Sigma}_{12} \\ \mathbf{\Sigma}_{21} & \mathbf{\Sigma}_{22} \end{pmatrix},$$

where the dimensions of $\mathbf{\Sigma}_{11}$ are $R \times R$. And let us assume that $\mathbf{\Sigma}_{11}$ is nonsingular. This assumption is convenient and can be made without loss of generality—the matrix $\mathbf{\Sigma}$ contains an $R \times R$ nonsingular principal submatrix, which can be relocated (should it be necessary to satisfy the assumption) by reordering the matrix's rows and columns.

Partition $\mathbf{\Gamma}$ and $\mathbf{\Lambda}$ as

$$\mathbf{\Gamma} = (\mathbf{\Gamma}_1, \mathbf{\Gamma}_2) \quad \text{and} \quad \mathbf{\Lambda} = (\mathbf{\Lambda}_1, \mathbf{\Lambda}_2),$$

where the dimensions of $\mathbf{\Gamma}_1$ are $N \times R$ and those of $\mathbf{\Lambda}_1$ are $S \times R$. Then,

$$\mathbf{\Gamma}_1'\mathbf{\Gamma}_1 = \mathbf{\Sigma}_{11} \quad \text{and} \quad \mathbf{\Gamma}_2'\mathbf{\Gamma}_1 = \mathbf{\Sigma}_{21}.$$

Similarly,

$$\mathbf{\Lambda}_1'\mathbf{\Lambda}_1 = \mathbf{\Sigma}_{11} \quad \text{and} \quad \mathbf{\Lambda}_2'\mathbf{\Lambda}_1 = \mathbf{\Sigma}_{21}.$$

Moreover, in light of Lemma 2.12.1, we have that

$$\text{rank}(\mathbf{\Gamma}_1) = R = \text{rank}(\mathbf{\Gamma})$$

and, similarly, that

$$\text{rank}(\mathbf{\Lambda}_1) = R = \text{rank}(\mathbf{\Lambda}).$$

Thus, the columns of $\mathbf{\Gamma}_1$ form a basis for $\mathcal{C}(\mathbf{\Gamma})$, and the columns of $\mathbf{\Lambda}_1$ form a basis for $\mathcal{C}(\mathbf{\Lambda})$. Accordingly, there exist matrices $\mathbf{A}$ and $\mathbf{B}$ such that

$$\mathbf{\Gamma}_2 = \mathbf{\Gamma}_1\mathbf{A} \quad \text{and} \quad \mathbf{\Lambda}_2 = \mathbf{\Lambda}_1\mathbf{B}.$$

And, taking $\mathbf{A}$ and $\mathbf{B}$ to be any such matrices, we find that

$$\mathbf{A}'\mathbf{\Gamma}_1'\mathbf{\Gamma}_1 = \mathbf{\Gamma}_2'\mathbf{\Gamma}_1 = \mathbf{\Sigma}_{21} = \mathbf{\Lambda}_2'\mathbf{\Lambda}_1 = \mathbf{B}'\mathbf{\Lambda}_1'\mathbf{\Lambda}_1 = \mathbf{B}'\mathbf{\Sigma}_{11} = \mathbf{B}'\mathbf{\Gamma}_1'\mathbf{\Gamma}_1,$$

implying (in light of Corollary 2.3.4) that

$$\mathbf{A}'\mathbf{\Gamma}_1' = \mathbf{B}'\mathbf{\Gamma}_1'.$$

Now, observe that the two random vectors $\mathbf{\Gamma}_1'\mathbf{z}$ and $\mathbf{\Lambda}_1'\mathbf{y}$ have the same probability distribution; according to result (5.32), each of them has the distribution with probability density function $f(\cdot)$ given by

$$f(\mathbf{u}) = \frac{1}{(2\pi)^{R/2}|\mathbf{\Sigma}_{11}|^{1/2}} \exp(-\tfrac{1}{2}\mathbf{u}'\mathbf{\Sigma}_{11}^{-1}\mathbf{u}).$$

We conclude that

$$\mathbf{w} = \boldsymbol{\mu} + \mathbf{\Lambda}'\mathbf{y} = \boldsymbol{\mu} + \begin{pmatrix}\mathbf{I}\\ \mathbf{B}'\end{pmatrix}\mathbf{\Lambda}_1'\mathbf{y} \sim \boldsymbol{\mu} + \begin{pmatrix}\mathbf{I}\\ \mathbf{B}'\end{pmatrix}\mathbf{\Gamma}_1'\mathbf{z} = \boldsymbol{\mu} + \begin{pmatrix}\mathbf{I}\\ \mathbf{A}'\end{pmatrix}\mathbf{\Gamma}_1'\mathbf{z} = \boldsymbol{\mu} + \mathbf{\Gamma}'\mathbf{z} = \mathbf{x}.$$

That is, $\mathbf{w}$ has the same probability distribution as $\mathbf{x}$.

g. Probability density function of a bivariate (2-variate) normal distribution

Let us consider in some detail the bivariate normal distribution. Take x and y to be random variables whose joint distribution is $N(\boldsymbol{\mu}, \mathbf{\Sigma})$, and let $\mu_1 = \text{E}(x)$, $\mu_2 = \text{E}(y)$, $\sigma_1^2 = \text{var}\, x$, $\sigma_2^2 = \text{var}\, y$, and $\rho = \text{corr}(x, y)$ (where $\sigma_1 \geq 0$ and $\sigma_2 \geq 0$), so that

$$\boldsymbol{\mu} = \begin{pmatrix}\mu_1\\ \mu_2\end{pmatrix} \quad \text{and} \quad \mathbf{\Sigma} = \begin{pmatrix}\sigma_1^2 & \rho\sigma_1\sigma_2\\ \rho\sigma_2\sigma_1 & \sigma_2^2\end{pmatrix}.$$

And observe [in light of result (2.14.4)] that

$$|\mathbf{\Sigma}| = \sigma_1^2\sigma_2^2(1-\rho^2). \tag{5.33}$$

The joint distribution of x and y has a probability density function if $\text{rank}(\mathbf{\Sigma}) = 2$, or equivalently if $|\mathbf{\Sigma}| > 0$, or (also equivalently) if

$$\sigma_1 > 0, \;\; \sigma_2 > 0, \;\; \text{and} \;\; -1 < \rho < 1. \tag{5.34}$$

Now, suppose that condition (5.34) is satisfied. Then, in light of result (2.5.1),

$$\mathbf{\Sigma}^{-1} = (1-\rho^2)^{-1}\begin{pmatrix}1/\sigma_1^2 & -\rho/(\sigma_1\sigma_2)\\ -\rho/(\sigma_1\sigma_2) & 1/\sigma_2^2\end{pmatrix}. \tag{5.35}$$

Upon substituting expressions (5.33) and (5.35) into expression (5.32), we obtain the following expression for the probability density function $f(\cdot\,,\cdot)$ of the joint distribution of x and y:

$$f(x, y) = \frac{1}{2\pi\sigma_1\sigma_2\sqrt{1-\rho^2}} \exp\biggl\{-\frac{1}{2(1-\rho^2)}\biggl[\Bigl(\frac{x-\mu_1}{\sigma_1}\Bigr)^2 - 2\rho\Bigl(\frac{x-\mu_1}{\sigma_1}\Bigr)\Bigl(\frac{y-\mu_2}{\sigma_2}\Bigr) + \Bigl(\frac{y-\mu_2}{\sigma_2}\Bigr)^2\biggr]\biggr\} \tag{5.36}$$

$(-\infty < x < \infty;\; -\infty < y < \infty)$.

The form of the probability density function (5.36) and the effect on the probability density function of changes in ρ and σ_2/σ_1 are illustrated in Figure 3.2.

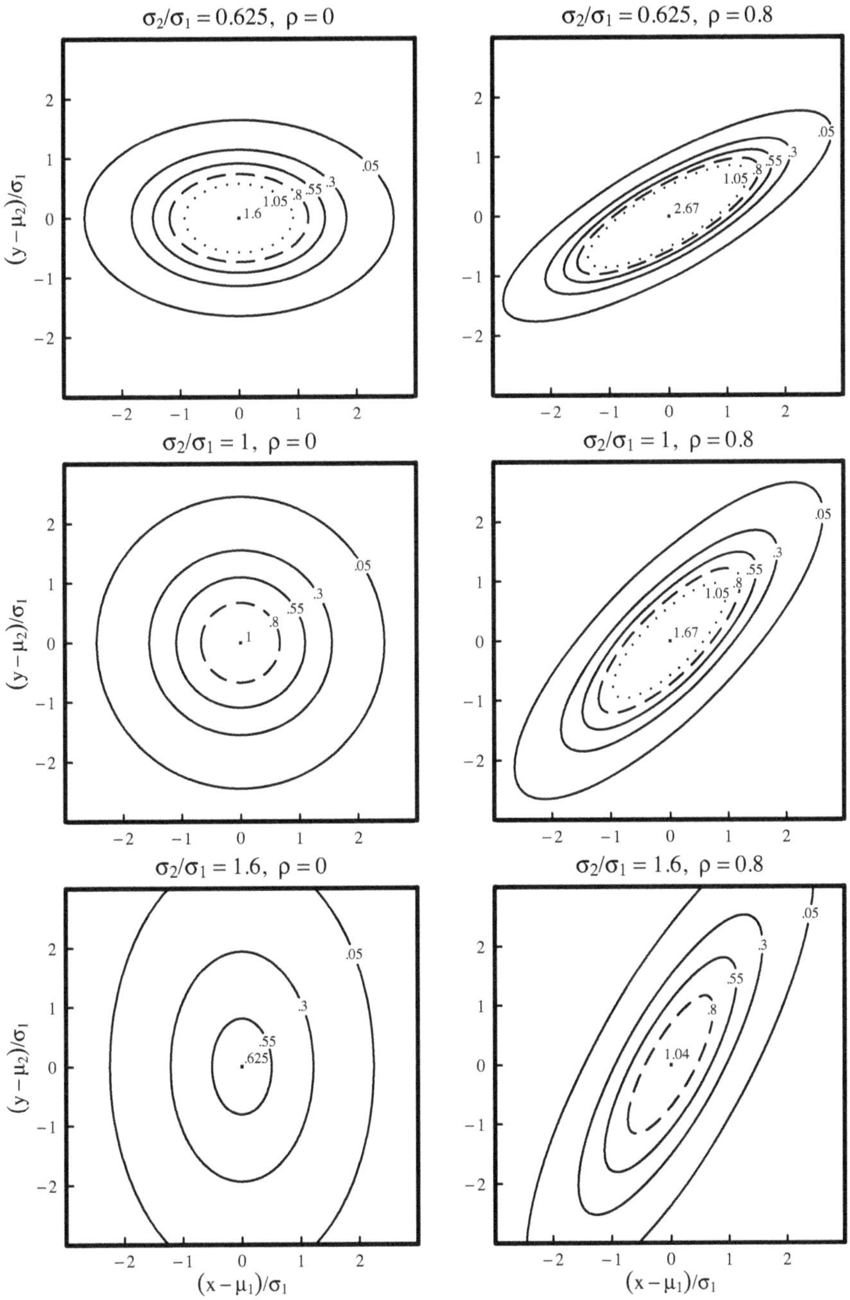

FIGURE 3.2. Contour maps of the probability density function $f(\cdot, \cdot)$ of the distribution of random variables x and y that are jointly normal with $\text{E}(x) = \mu_1$, $\text{E}(y) = \mu_2$, $\text{var}\, x = \sigma_1^2$ $(\sigma_1 > 0)$, $\text{var}\, y = \sigma_2^2$ $(\sigma_2 > 0)$, and $\text{corr}(x, y) = \rho$. The 6 maps are arranged in 3 rows, corresponding to values of σ_2/σ_1 of 0.625, 1, and 1.6, respectively, and in 2 columns, corresponding to values of ρ of 0 and 0.8. The coordinates of the points of each contour line are the values of $(x - \mu_1)/\sigma_1$ and $(y - \mu_2)/\sigma_1$ at which $f(x, y) = k/(2\pi\sigma_1^2)$, where $k = 0.05, 0.3, 0.55, 0.8$, or 1.05. Contour maps corresponding to $\rho = -0.8$ could be obtained by forming the mirror images of those corresponding to $\rho = 0.8$.

h. Linear transformation of a normally distributed random vector

Let $\mathbf{x}$ represent an N-dimensional random column vector whose distribution is $N(\boldsymbol{\mu}, \boldsymbol{\Sigma})$, and consider the distribution of the M-dimensional random column vector $\mathbf{y}$ defined by

$$\mathbf{y} = \mathbf{c} + \mathbf{A}\mathbf{x},$$

where $\mathbf{c}$ is an M-dimensional nonrandom column vector and $\mathbf{A}$ an $M \times N$ nonrandom matrix. By definition, the distribution of $\mathbf{x}$ is that of a random vector $\boldsymbol{\mu} + \boldsymbol{\Gamma}'\mathbf{z}$, where $\boldsymbol{\Gamma}$ is a nonrandom matrix such that $\boldsymbol{\Sigma} = \boldsymbol{\Gamma}'\boldsymbol{\Gamma}$ and where $\mathbf{z} \sim N(\mathbf{0}, \mathbf{I})$. Thus, the distribution of $\mathbf{y}$ is identical to that of the vector

$$\mathbf{c} + \mathbf{A}(\boldsymbol{\mu} + \boldsymbol{\Gamma}'\mathbf{z}) = \mathbf{c} + \mathbf{A}\boldsymbol{\mu} + (\boldsymbol{\Gamma}\mathbf{A}')'\mathbf{z}.$$

Since $(\boldsymbol{\Gamma}\mathbf{A}')'\boldsymbol{\Gamma}\mathbf{A}' = \mathbf{A}\boldsymbol{\Sigma}\mathbf{A}'$, it follows that

$$\mathbf{y} \sim N(\mathbf{c} + \mathbf{A}\boldsymbol{\mu},\ \mathbf{A}\boldsymbol{\Sigma}\mathbf{A}').$$

In summary, we have the following theorem.

Theorem 3.5.1. Let $\mathbf{x} \sim N(\boldsymbol{\mu}, \boldsymbol{\Sigma})$, and let $\mathbf{y} = \mathbf{c} + \mathbf{A}\mathbf{x}$, where $\mathbf{c}$ is a nonrandom vector and $\mathbf{A}$ a nonrandom matrix. Then,

$$\mathbf{y} \sim N(\mathbf{c} + \mathbf{A}\boldsymbol{\mu},\ \mathbf{A}\boldsymbol{\Sigma}\mathbf{A}').$$

i. Symmetry of the MVN distribution

The distribution of an M-dimensional random column vector $\mathbf{x}$ is said to be *symmetric* about an M-dimensional nonrandom column vector $\boldsymbol{\mu}$ if the distribution of $-(\mathbf{x} - \boldsymbol{\mu})$ is the same as that of $\mathbf{x} - \boldsymbol{\mu}$. If $\mathbf{x} \sim N(\boldsymbol{\mu}, \boldsymbol{\Sigma})$, then it follows from Theorem 3.5.1 that the distribution of $\mathbf{x} - \boldsymbol{\mu}$ $(= -\boldsymbol{\mu} + \mathbf{I}\mathbf{x})$ and the distribution of $-(\mathbf{x} - \boldsymbol{\mu})$ $[= \boldsymbol{\mu} + (-\mathbf{I})\mathbf{x}]$ are both $N(\mathbf{0}, \boldsymbol{\Sigma})$. Thus, as a consequence of Theorem 3.5.1, we have the following result.

Theorem 3.5.2. The $N(\boldsymbol{\mu}, \boldsymbol{\Sigma})$ distribution is symmetric about $\boldsymbol{\mu}$.

j. Marginal distributions

Let $\mathbf{x}$ represent an M-dimensional random column vector whose distribution is $N(\boldsymbol{\mu}, \boldsymbol{\Sigma})$, and consider the distribution of a subvector of $\mathbf{x}$, say the M_*-dimensional subvector $\mathbf{x}_*$ obtained by striking out all of the elements of $\mathbf{x}$ except the $j_1, j_2, \ldots, j_{M_*}$th elements. Clearly,

$$\mathbf{x}_* = \mathbf{A}\mathbf{x},$$

where $\mathbf{A}$ is the $M_* \times M$ submatrix of I_M obtained by striking out all of the rows of $\mathbf{I}_M$ except the $j_1, j_2, \ldots, j_{M_*}$th rows—if the elements of $\mathbf{x}_*$ are the first M_* elements of $\mathbf{x}$, then $\mathbf{A} = (\mathbf{I}, \mathbf{0})$. Accordingly, it follows from Theorem 3.5.1 that

$$\mathbf{x}_* \sim N(\mathbf{A}\boldsymbol{\mu}, \mathbf{A}\boldsymbol{\Sigma}\mathbf{A}').$$

Thus, as an additional consequence of Theorem 3.5.1, we have the following result.

Theorem 3.5.3. Let $\mathbf{x} \sim N(\boldsymbol{\mu}, \boldsymbol{\Sigma})$. Further, let $\mathbf{x}_*$ represent a subvector of $\mathbf{x}$, and let $\boldsymbol{\mu}_*$ represent the corresponding subvector of $\boldsymbol{\mu}$ and $\boldsymbol{\Sigma}_*$ the corresponding principal submatrix of $\boldsymbol{\Sigma}$. Then,

$$\mathbf{x}_* \sim N(\boldsymbol{\mu}_*, \boldsymbol{\Sigma}_*).$$

k. Statistical independence

Let $\mathbf{x}_1, \mathbf{x}_2, \ldots, \mathbf{x}_P$ represent random column vectors having expected values $\boldsymbol{\mu}_i = \mathrm{E}(\mathbf{x}_i)$ $(i = 1, 2, \ldots, P)$ and covariances $\boldsymbol{\Sigma}_{ij} = \mathrm{cov}(\mathbf{x}_i, \mathbf{x}_j)$ $(i, j = 1, 2, \ldots, P)$. And for $i = 1, 2, \ldots, P$, denote by M_i the number of elements in $\mathbf{x}_i$.

Let x_{is} represent the sth element of $\mathbf{x}_i$ $(i = 1, 2, \ldots, P;\ s = 1, 2, \ldots, M_i)$. If $\mathbf{x}_i$ and $\mathbf{x}_j$ are statistically independent, in which case x_{is} and x_{jt} are statistically independent for every s and every t, then (according to Lemma 3.2.2) $\boldsymbol{\Sigma}_{ij} = \mathbf{0}$ $(j \neq i = 1, 2, \ldots, P)$. In general, the converse is not true (as is well-known and as could be surmised from the discussion of Section 3.2c). That is, $\mathbf{x}_i$ and $\mathbf{x}_j$ being uncorrelated does not necessarily imply their statistical independence. However, when their joint distribution is MVN, $\mathbf{x}_i$ and $\mathbf{x}_j$ being uncorrelated does imply their statistical independence. More generally, when the joint distribution of $\mathbf{x}_1, \mathbf{x}_2, \ldots, \mathbf{x}_P$ is MVN, $\boldsymbol{\Sigma}_{ij} = \mathbf{0}$ (i.e., $\mathbf{x}_i$ and $\mathbf{x}_j$ being uncorrelated) for $j \neq i = 1, 2, \ldots, P$ implies that $\mathbf{x}_1, \mathbf{x}_2, \ldots, \mathbf{x}_P$ are mutually (jointly) independent.

To see this, let

$$\mathbf{x} = \begin{pmatrix} \mathbf{x}_1 \\ \mathbf{x}_2 \\ \vdots \\ \mathbf{x}_P \end{pmatrix},$$

and observe that $\mathrm{E}(\mathbf{x}) = \boldsymbol{\mu}$ and $\mathrm{var}(\mathbf{x}) = \boldsymbol{\Sigma}$, where

$$\boldsymbol{\mu} = \begin{pmatrix} \boldsymbol{\mu}_1 \\ \boldsymbol{\mu}_2 \\ \vdots \\ \boldsymbol{\mu}_P \end{pmatrix} \quad \text{and} \quad \boldsymbol{\Sigma} = \begin{pmatrix} \boldsymbol{\Sigma}_{11} & \boldsymbol{\Sigma}_{12} & \ldots & \boldsymbol{\Sigma}_{1P} \\ \boldsymbol{\Sigma}_{21} & \boldsymbol{\Sigma}_{22} & \ldots & \boldsymbol{\Sigma}_{2P} \\ \vdots & \vdots & \ddots & \vdots \\ \boldsymbol{\Sigma}_{P1} & \boldsymbol{\Sigma}_{P2} & \ldots & \boldsymbol{\Sigma}_{PP} \end{pmatrix}.$$

And suppose that the distribution of $\mathbf{x}$ is MVN and that

$$\boldsymbol{\Sigma}_{ij} = \mathbf{0} \qquad (j \neq i = 1, 2, \ldots, P).$$

Further, define

$$\boldsymbol{\Gamma} = \mathrm{diag}(\boldsymbol{\Gamma}_1,\ \boldsymbol{\Gamma}_2, \ldots,\ \boldsymbol{\Gamma}_P),$$

where (for $i = 1, 2, \ldots, P$) $\boldsymbol{\Gamma}_i$ is any matrix such that $\boldsymbol{\Sigma}_{ii} = \boldsymbol{\Gamma}_i'\boldsymbol{\Gamma}_i$. Then,

$$\boldsymbol{\Sigma} = \mathrm{diag}(\boldsymbol{\Sigma}_{11}, \boldsymbol{\Sigma}_{22}, \ldots, \boldsymbol{\Sigma}_{PP}) = \mathrm{diag}(\boldsymbol{\Gamma}_1'\boldsymbol{\Gamma}_1,\ \boldsymbol{\Gamma}_2'\boldsymbol{\Gamma}_2, \ldots,\ \boldsymbol{\Gamma}_P'\boldsymbol{\Gamma}_P) = \boldsymbol{\Gamma}'\boldsymbol{\Gamma}.$$

Now, denote by N_i the number of rows in $\boldsymbol{\Gamma}_i$ $(i = 1, 2, \ldots, P)$, and take

$$\mathbf{z} = \begin{pmatrix} \mathbf{z}_1 \\ \mathbf{z}_2 \\ \vdots \\ \mathbf{z}_P \end{pmatrix},$$

where (for $i = 1, 2, \ldots, P$) $\mathbf{z}_i$ is an N_i-dimensional random column vector whose distribution is $N(\mathbf{0}, \mathbf{I})$ and where $\mathbf{z}_1, \mathbf{z}_2, \ldots, \mathbf{z}_P$ are statistically independent. Clearly, $\mathbf{z} \sim N(\mathbf{0}, \mathbf{I})$, so that (by definition) the distribution of $\mathbf{x}$ is identical to that of the random vector

$$\boldsymbol{\mu} + \boldsymbol{\Gamma}'\mathbf{z} = \begin{pmatrix} \boldsymbol{\mu}_1 + \boldsymbol{\Gamma}_1'\mathbf{z}_1 \\ \boldsymbol{\mu}_2 + \boldsymbol{\Gamma}_2'\mathbf{z}_2 \\ \vdots \\ \boldsymbol{\mu}_P + \boldsymbol{\Gamma}_P'\mathbf{z}_P \end{pmatrix},$$

or, equivalently, the joint distribution of $\mathbf{x}_1, \mathbf{x}_2, \ldots, \mathbf{x}_P$ is identical to that of the random vectors $\boldsymbol{\mu}_1 + \boldsymbol{\Gamma}_1'\mathbf{z}_1,\ \boldsymbol{\mu}_2 + \boldsymbol{\Gamma}_2'\mathbf{z}_2,\ \ldots,\ \boldsymbol{\mu}_P + \boldsymbol{\Gamma}_P'\mathbf{z}_P$. Since $\mathbf{z}_1, \mathbf{z}_2, \ldots, \mathbf{z}_P$ are distributed independently, so are the vector-valued functions $\boldsymbol{\mu}_1 + \boldsymbol{\Gamma}_1'\mathbf{z}_1,\ \boldsymbol{\mu}_2 + \boldsymbol{\Gamma}_2'\mathbf{z}_2,\ \ldots,\ \boldsymbol{\mu}_P + \boldsymbol{\Gamma}_P'\mathbf{z}_P$ and hence so are $\mathbf{x}_1, \mathbf{x}_2, \ldots, \mathbf{x}_P$—vector-valued functions of statistically independent random vectors are statistically independent, as is evident, for example, from the discussion of Casella and Berger (2002, sec. 4.6) or Bickel and Doksum (2001, app. A).

In summary, we have the following theorem.

Theorem 3.5.4. Let $\mathbf{x}_1, \mathbf{x}_2, \ldots, \mathbf{x}_P$ represent random column vectors whose joint distribution is MVN. Then, $\mathbf{x}_1, \mathbf{x}_2, \ldots, \mathbf{x}_P$ are distributed independently if (and only if)

$$\text{cov}(\mathbf{x}_i, \mathbf{x}_j) = \mathbf{0} \quad (j > i = 1, 2, \ldots, P).$$

Note that the coverage of Theorem 3.5.4 includes the case where each of the random vectors $\mathbf{x}_1, \mathbf{x}_2, \ldots, \mathbf{x}_P$ is of dimension 1 and hence is in effect a random variable. In the special case where $P = 2$, Theorem 3.5.4 can be restated in the form of the following corollary.

Corollary 3.5.5. Let $\mathbf{x}$ and $\mathbf{y}$ represent random column vectors whose joint distribution is MVN. Then, $\mathbf{x}$ and $\mathbf{y}$ are statistically independent if (and only if) $\text{cov}(\mathbf{x}, \mathbf{y}) = \mathbf{0}$.

As an additional corollary of Theorem 3.5.4, we have the following result.

Corollary 3.5.6. Let $\mathbf{x}$ represent an N-dimensional random column vector whose distribution is MVN; and, for $i = 1, 2, \ldots, P$, let $\mathbf{y}_i = \mathbf{c}_i + \mathbf{A}_i\mathbf{x}$, where $\mathbf{c}_i$ is an M_i-dimensional nonrandom column vector and $\mathbf{A}_i$ is an $M_i \times N$ nonrandom matrix. Then, $\mathbf{y}_1, \mathbf{y}_2, \ldots, \mathbf{y}_P$ are distributed independently if (and only if)

$$\text{cov}(\mathbf{y}_i, \mathbf{y}_j) = \mathbf{0} \quad (j > i = 1, 2, \ldots, P).$$

Proof. Let

$$\mathbf{y} = \begin{pmatrix} \mathbf{y}_1 \\ \mathbf{y}_2 \\ \vdots \\ \mathbf{y}_P \end{pmatrix}, \quad \mathbf{c} = \begin{pmatrix} \mathbf{c}_1 \\ \mathbf{c}_2 \\ \vdots \\ \mathbf{c}_P \end{pmatrix}, \quad \text{and} \quad \mathbf{A} = \begin{pmatrix} \mathbf{A}_1 \\ \mathbf{A}_2 \\ \vdots \\ \mathbf{A}_P \end{pmatrix}.$$

Then,

$$\mathbf{y} = \mathbf{c} + \mathbf{A}\mathbf{x},$$

implying (in light of Theorem 3.5.1) that the joint distribution of $\mathbf{y}_1, \mathbf{y}_2, \ldots, \mathbf{y}_P$ is MVN. Accordingly, it follows from Theorem 3.5.4 that $\mathbf{y}_1, \mathbf{y}_2, \ldots, \mathbf{y}_P$ are distributed independently if (and only if)

$$\text{cov}(\mathbf{y}_i, \mathbf{y}_j) = \mathbf{0} \quad (j > i = 1, 2, \ldots, P). \qquad \text{Q.E.D.}$$

If each of two or more independently distributed random vectors has an MVN distribution, then, as indicated by the following theorem, their joint distribution is MVN.

Theorem 3.5.7. For $i = 1, 2, \ldots, P$, let $\mathbf{x}_i$ represent an M_i-dimensional random column vector whose distribution is $N(\boldsymbol{\mu}_i, \boldsymbol{\Sigma}_i)$. If $\mathbf{x}_1, \mathbf{x}_2, \ldots, \mathbf{x}_P$ are mutually independent, then the distribution of the random vector $\mathbf{x}$ defined by $\mathbf{x}' = (\mathbf{x}_1', \mathbf{x}_2', \ldots, \mathbf{x}_P')$ is $N(\boldsymbol{\mu}, \boldsymbol{\Sigma})$, where $\boldsymbol{\mu}' = (\boldsymbol{\mu}_1', \boldsymbol{\mu}_2', \ldots, \boldsymbol{\mu}_P')$ and $\boldsymbol{\Sigma} = \text{diag}(\boldsymbol{\Sigma}_1, \boldsymbol{\Sigma}_2, \ldots, \boldsymbol{\Sigma}_P)$.

Proof. For $i = 1, 2, \ldots, P$, take $\boldsymbol{\Gamma}_i$ to be a matrix (having M_i columns) such that $\boldsymbol{\Sigma}_i = \boldsymbol{\Gamma}_i'\boldsymbol{\Gamma}_i$—the existence of such a matrix follows from Corollary 2.13.25—and denote by N_i the number of rows in $\boldsymbol{\Gamma}_i$. And let $N = \sum_{i=1}^{P} N_i$. Further, let $\boldsymbol{\Gamma} = \text{diag}(\boldsymbol{\Gamma}_1, \boldsymbol{\Gamma}_2, \ldots, \boldsymbol{\Gamma}_P)$, and define $\mathbf{z}$ by $\mathbf{z}' = (\mathbf{z}_1', \mathbf{z}_2', \ldots, \mathbf{z}_P')$, where (for $i = 1, 2, \ldots, P$) $\mathbf{z}_i$ is an N_i-dimensional random column vector whose distribution is $N(\mathbf{0}, \mathbf{I})$ and where $\mathbf{z}_1, \mathbf{z}_2, \ldots, \mathbf{z}_P$ are statistically independent.

For $i = 1, 2, \ldots, P$, $\boldsymbol{\mu}_i + \boldsymbol{\Gamma}_i'\mathbf{z}_i \sim N(\boldsymbol{\mu}_i, \boldsymbol{\Sigma}_i)$ (so that $\boldsymbol{\mu}_i + \boldsymbol{\Gamma}_i'\mathbf{z}_i$ has the same distribution as $\mathbf{x}_i$). Further, $\boldsymbol{\mu}_1 + \boldsymbol{\Gamma}_1'\mathbf{z}_1,\ \boldsymbol{\mu}_2 + \boldsymbol{\Gamma}_2'\mathbf{z}_2,\ \ldots,\ \boldsymbol{\mu}_P + \boldsymbol{\Gamma}_P'\mathbf{z}_P$ are mutually independent. And since

$$\boldsymbol{\mu} + \boldsymbol{\Gamma}'\mathbf{z} = \begin{pmatrix} \mu_1 + \boldsymbol{\Gamma}_1'\mathbf{z}_1 \\ \mu_2 + \boldsymbol{\Gamma}_2'\mathbf{z}_2 \\ \vdots \\ \mu_P + \boldsymbol{\Gamma}_P'\mathbf{z}_P \end{pmatrix},$$

it follows that $\boldsymbol{\mu} + \boldsymbol{\Gamma}'\mathbf{z}$ has the same distribution as $\mathbf{x}$. Clearly, $\mathbf{z} \sim N(\mathbf{0}, \mathbf{I}_N)$ and $\boldsymbol{\Gamma}'\boldsymbol{\Gamma} = \boldsymbol{\Sigma}$. Thus, $\boldsymbol{\mu} + \boldsymbol{\Gamma}'\mathbf{z} \sim N(\boldsymbol{\mu}, \boldsymbol{\Sigma})$ and hence $\mathbf{x} \sim N(\boldsymbol{\mu}, \boldsymbol{\Sigma})$. Q.E.D.

l. Conditional distributions: a special case

Suppose that $\mathbf{x}$ is an M_1-dimensional random column vector and $\mathbf{y}$ an M_2-dimensional random column vector whose joint distribution is MVN. Let $\boldsymbol{\mu}_1 = E(\mathbf{x})$, $\boldsymbol{\mu}_2 = E(\mathbf{y})$, $\boldsymbol{\Sigma}_{11} = \text{var}(\mathbf{x})$, $\boldsymbol{\Sigma}_{22} = \text{var}(\mathbf{y})$, and $\boldsymbol{\Sigma}_{21} = \text{cov}(\mathbf{y}, \mathbf{x})$. And define

$$\boldsymbol{\Sigma} = \begin{pmatrix} \boldsymbol{\Sigma}_{11} & \boldsymbol{\Sigma}_{21}' \\ \boldsymbol{\Sigma}_{21} & \boldsymbol{\Sigma}_{22} \end{pmatrix}.$$

Let us derive the conditional distribution of $\mathbf{y}$ given $\mathbf{x}$. Assume that $\boldsymbol{\Sigma}$ is positive definite (and hence nonsingular)—consideration of the more general (and more difficult) case where $\boldsymbol{\Sigma}$ may be positive semidefinite is deferred until Subsection m. Denoting by $h(\cdot\,,\cdot)$ the probability density function of the joint distribution of $\mathbf{x}$ and $\mathbf{y}$ and by $h_1(\cdot)$ the probability density function of the marginal distribution of $\mathbf{x}$, letting

$$\boldsymbol{\eta}(\mathbf{x}) = \boldsymbol{\mu}_2 + \boldsymbol{\Sigma}_{21}\boldsymbol{\Sigma}_{11}^{-1}(\mathbf{x} - \boldsymbol{\mu}_1) \quad \text{and} \quad \mathbf{V} = \boldsymbol{\Sigma}_{22} - \boldsymbol{\Sigma}_{21}\boldsymbol{\Sigma}_{11}^{-1}\boldsymbol{\Sigma}_{21}',$$

and making use of Theorems 2.14.22 and 2.6.6, we find that the conditional distribution of $\mathbf{y}$ given $\mathbf{x}$ is the distribution with probability density function $f(\cdot\,|\,\cdot)$ given by

$$f(\mathbf{y}\,|\,\mathbf{x}) = \frac{h(\mathbf{x}, \mathbf{y})}{h_1(\mathbf{x})} = \frac{1}{(2\pi)^{M_2/2}c^{1/2}} \exp[-\tfrac{1}{2}q(\mathbf{x}, \mathbf{y})],$$

where

$$c = |\boldsymbol{\Sigma}|/|\boldsymbol{\Sigma}_{11}| = |\boldsymbol{\Sigma}_{11}||\mathbf{V}|/|\boldsymbol{\Sigma}_{11}| = |\mathbf{V}|$$

and

$$\begin{aligned} q(\mathbf{x}, \mathbf{y}) &= \begin{pmatrix} \mathbf{x} - \boldsymbol{\mu}_1 \\ \mathbf{y} - \boldsymbol{\mu}_2 \end{pmatrix}' \begin{pmatrix} \boldsymbol{\Sigma}_{11} & \boldsymbol{\Sigma}_{21}' \\ \boldsymbol{\Sigma}_{21} & \boldsymbol{\Sigma}_{22} \end{pmatrix}^{-1} \begin{pmatrix} \mathbf{x} - \boldsymbol{\mu}_1 \\ \mathbf{y} - \boldsymbol{\mu}_2 \end{pmatrix} - (\mathbf{x} - \boldsymbol{\mu}_1)'\boldsymbol{\Sigma}_{11}^{-1}(\mathbf{x} - \boldsymbol{\mu}_1) \\ &= \begin{pmatrix} \mathbf{x} - \boldsymbol{\mu}_1 \\ \mathbf{y} - \boldsymbol{\mu}_2 \end{pmatrix}' \begin{pmatrix} \boldsymbol{\Sigma}_{11}^{-1} + \boldsymbol{\Sigma}_{11}^{-1}\boldsymbol{\Sigma}_{21}'\mathbf{V}^{-1}\boldsymbol{\Sigma}_{21}\boldsymbol{\Sigma}_{11}^{-1} & -\boldsymbol{\Sigma}_{11}^{-1}\boldsymbol{\Sigma}_{21}'\mathbf{V}^{-1} \\ -\mathbf{V}^{-1}\boldsymbol{\Sigma}_{21}\boldsymbol{\Sigma}_{11}^{-1} & \mathbf{V}^{-1} \end{pmatrix} \begin{pmatrix} \mathbf{x} - \boldsymbol{\mu}_1 \\ \mathbf{y} - \boldsymbol{\mu}_2 \end{pmatrix} \\ &\qquad - (\mathbf{x} - \boldsymbol{\mu}_1)'\boldsymbol{\Sigma}_{11}^{-1}(\mathbf{x} - \boldsymbol{\mu}_1) \\ &= [\mathbf{y} - \boldsymbol{\eta}(\mathbf{x})]'\,\mathbf{V}^{-1}\,[\mathbf{y} - \boldsymbol{\eta}(\mathbf{x})]. \end{aligned}$$

The probability density function of the conditional distribution of $\mathbf{y}$ given $\mathbf{x}$ is seen to be that of the MVN distribution with mean vector $\boldsymbol{\eta}(\mathbf{x})$ and variance-covariance matrix $\mathbf{V}$. Thus, we have the following theorem.

Theorem 3.5.8. Let $\mathbf{x}$ and $\mathbf{y}$ represent random column vectors whose joint distribution is MVN, and let $\boldsymbol{\mu}_1 = E(\mathbf{x})$, $\boldsymbol{\mu}_2 = E(\mathbf{y})$, $\boldsymbol{\Sigma}_{11} = \text{var}(\mathbf{x})$, $\boldsymbol{\Sigma}_{22} = \text{var}(\mathbf{y})$, and $\boldsymbol{\Sigma}_{21} = \text{cov}(\mathbf{y}, \mathbf{x})$. Then, under the supposition that $\text{var}\begin{pmatrix} \mathbf{x} \\ \mathbf{y} \end{pmatrix} = \begin{pmatrix} \boldsymbol{\Sigma}_{11} & \boldsymbol{\Sigma}_{21}' \\ \boldsymbol{\Sigma}_{21} & \boldsymbol{\Sigma}_{22} \end{pmatrix}$ is positive definite, the conditional distribution of $\mathbf{y}$ given $\mathbf{x}$ is $N[\boldsymbol{\eta}(\mathbf{x}), \mathbf{V}]$, where

$$\boldsymbol{\eta}(\mathbf{x}) = \boldsymbol{\mu}_2 + \boldsymbol{\Sigma}_{21}\boldsymbol{\Sigma}_{11}^{-1}(\mathbf{x} - \boldsymbol{\mu}_1) \quad \text{and} \quad \mathbf{V} = \boldsymbol{\Sigma}_{22} - \boldsymbol{\Sigma}_{21}\boldsymbol{\Sigma}_{11}^{-1}\boldsymbol{\Sigma}_{21}'.$$

The results of the following theorem complement those of Theorem 3.5.8.

Theorem 3.5.9. Let $\mathbf{x}$ and $\mathbf{y}$ represent random column vectors, and let $\boldsymbol{\mu}_1 = \mathrm{E}(\mathbf{x})$, $\boldsymbol{\mu}_2 = \mathrm{E}(\mathbf{y})$, $\boldsymbol{\Sigma}_{11} = \mathrm{var}(\mathbf{x})$, $\boldsymbol{\Sigma}_{22} = \mathrm{var}(\mathbf{y})$, and $\boldsymbol{\Sigma}_{21} = \mathrm{cov}(\mathbf{y}, \mathbf{x})$. Suppose that $\boldsymbol{\Sigma}_{11}$ is nonsingular. And let

$$\boldsymbol{\eta}(\mathbf{x}) = \boldsymbol{\mu}_2 + \boldsymbol{\Sigma}_{21}\boldsymbol{\Sigma}_{11}^{-1}(\mathbf{x} - \boldsymbol{\mu}_1) \quad \text{and} \quad \mathbf{V} = \boldsymbol{\Sigma}_{22} - \boldsymbol{\Sigma}_{21}\boldsymbol{\Sigma}_{11}^{-1}\boldsymbol{\Sigma}_{21}',$$

and define

$$\mathbf{e} = \mathbf{y} - \boldsymbol{\eta}(\mathbf{x}).$$

Then,

(1) $\mathrm{E}[\boldsymbol{\eta}(\mathbf{x})] = \boldsymbol{\mu}_2$, and $\mathrm{var}[\boldsymbol{\eta}(\mathbf{x})] = \mathrm{cov}[\boldsymbol{\eta}(\mathbf{x}), \mathbf{y}] = \boldsymbol{\Sigma}_{21}\boldsymbol{\Sigma}_{11}^{-1}\boldsymbol{\Sigma}_{21}'$;

(2) $\mathrm{E}(\mathbf{e}) = \mathbf{0}$, $\mathrm{var}(\mathbf{e}) = \mathbf{V}$, and $\mathrm{cov}(\mathbf{e}, \mathbf{x}) = \mathbf{0}$; and

(3) under the assumption that the joint distribution of $\mathbf{x}$ and $\mathbf{y}$ is MVN, the distribution of $\mathbf{e}$ is $N(\mathbf{0}, \mathbf{V})$ and $\mathbf{e}$ and $\mathbf{x}$ are statistically independent.

Proof. (1) and (2). Making use of results (1.7) and (1.8), we find that

$$\mathrm{E}[\boldsymbol{\eta}(\mathbf{x})] = \boldsymbol{\mu}_2 + \boldsymbol{\Sigma}_{21}\boldsymbol{\Sigma}_{11}^{-1}\mathrm{E}(\mathbf{x} - \boldsymbol{\mu}_1) = \boldsymbol{\mu}_2$$

and that

$$\mathrm{E}(\mathbf{e}) = \mathrm{E}(\mathbf{y}) - \mathrm{E}[\boldsymbol{\eta}(\mathbf{x})] = \mathbf{0}.$$

Further, in light of result (2.47), we have that

$$\mathrm{var}[\boldsymbol{\eta}(\mathbf{x})] = \boldsymbol{\Sigma}_{21}\boldsymbol{\Sigma}_{11}^{-1}\boldsymbol{\Sigma}_{11}(\boldsymbol{\Sigma}_{21}\boldsymbol{\Sigma}_{11}^{-1})' = \boldsymbol{\Sigma}_{21}(\boldsymbol{\Sigma}_{21}\boldsymbol{\Sigma}_{11}^{-1})' = (\boldsymbol{\Sigma}_{21}\boldsymbol{\Sigma}_{11}^{-1}\boldsymbol{\Sigma}_{21}')'$$

and hence (because a variance-covariance matrix is inherently symmetric) that

$$\mathrm{var}[\boldsymbol{\eta}(\mathbf{x})] = \boldsymbol{\Sigma}_{21}\boldsymbol{\Sigma}_{11}^{-1}\boldsymbol{\Sigma}_{21}'.$$

Similarly, in light of results (2.45) and (2.46), we have that

$$\mathrm{cov}[\boldsymbol{\eta}(\mathbf{x}), \mathbf{y}] = \mathrm{cov}[\boldsymbol{\eta}(\mathbf{x}), \mathbf{I}\mathbf{y}] = \boldsymbol{\Sigma}_{21}\boldsymbol{\Sigma}_{11}^{-1}\boldsymbol{\Sigma}_{21}'\mathbf{I}' = \boldsymbol{\Sigma}_{21}\boldsymbol{\Sigma}_{11}^{-1}\boldsymbol{\Sigma}_{21}'$$

and

$$\mathrm{cov}[\boldsymbol{\eta}(\mathbf{x}), \mathbf{x}] = \mathrm{cov}[\boldsymbol{\eta}(\mathbf{x}), \mathbf{I}\mathbf{x}] = \boldsymbol{\Sigma}_{21}\boldsymbol{\Sigma}_{11}^{-1}\boldsymbol{\Sigma}_{11}\mathbf{I}' = \boldsymbol{\Sigma}_{21}.$$

And, upon recalling results (2.48) and (2.50), it follows that

$$\mathrm{cov}(\mathbf{e}, \mathbf{x}) = \mathrm{cov}(\mathbf{y}, \mathbf{x}) - \mathrm{cov}[\boldsymbol{\eta}(\mathbf{x}), \mathbf{x}] = \mathbf{0}$$

and

$$\begin{aligned}\mathrm{var}(\mathbf{e}) &= \mathrm{var}(\mathbf{y}) + \mathrm{var}[\boldsymbol{\eta}(\mathbf{x})] - \{\mathrm{cov}[\boldsymbol{\eta}(\mathbf{x}), \mathbf{y}]\}' - \mathrm{cov}[\boldsymbol{\eta}(\mathbf{x}), \mathbf{y}] \\ &= \boldsymbol{\Sigma}_{22} + (\boldsymbol{\Sigma}_{21}\boldsymbol{\Sigma}_{11}^{-1}\boldsymbol{\Sigma}_{21}')' - (\boldsymbol{\Sigma}_{21}\boldsymbol{\Sigma}_{11}^{-1}\boldsymbol{\Sigma}_{21}')' - \boldsymbol{\Sigma}_{21}\boldsymbol{\Sigma}_{11}^{-1}\boldsymbol{\Sigma}_{21}' \\ &= \mathbf{V}.\end{aligned}$$

(3) Suppose that the joint distribution of $\mathbf{x}$ and $\mathbf{y}$ is MVN. Then, upon observing that

$$\begin{pmatrix}\mathbf{e}\\ \mathbf{x}\end{pmatrix} = \begin{pmatrix}-\boldsymbol{\mu}_2 + \boldsymbol{\Sigma}_{21}\boldsymbol{\Sigma}_{11}^{-1}\boldsymbol{\mu}_1\\ \mathbf{0}\end{pmatrix} + \begin{pmatrix}-\boldsymbol{\Sigma}_{21}\boldsymbol{\Sigma}_{11}^{-1} & \mathbf{I}\\ \mathbf{I} & \mathbf{0}\end{pmatrix}\begin{pmatrix}\mathbf{x}\\ \mathbf{y}\end{pmatrix},$$

it follows from Theorem 3.5.1 that the joint distribution of $\mathbf{e}$ and $\mathbf{x}$ is MVN. Since [according to Part (2)] $\mathrm{cov}(\mathbf{e}, \mathbf{x}) = \mathbf{0}$, we conclude (on the basis of Corollary 3.5.5) that $\mathbf{e}$ and $\mathbf{x}$ are statistically independent. To establish that the distribution of $\mathbf{e}$ is $N(\mathbf{0}, \mathbf{V})$, it suffices [since it has already been established in Part (2) that $\mathrm{E}(\mathbf{e}) = \mathbf{0}$ and $\mathrm{var}(\mathbf{e}) = \mathbf{V}$] to observe (e.g., on the basis of Theorem 3.5.3) that the distribution of $\mathbf{e}$ is MVN. Q.E.D.

m. Conditional distributions: general case

Let $\mathbf{x}$ represent an M_1-dimensional random column vector and $\mathbf{y}$ an M_2-dimensional random column vector, and let $\boldsymbol{\mu}_1 = \mathrm{E}(\mathbf{x})$, $\boldsymbol{\mu}_2 = \mathrm{E}(\mathbf{y})$, $\boldsymbol{\Sigma}_{11} = \mathrm{var}(\mathbf{x})$, $\boldsymbol{\Sigma}_{22} = \mathrm{var}(\mathbf{y})$, and $\boldsymbol{\Sigma}_{21} = \mathrm{cov}(\mathbf{y}, \mathbf{x})$. And define

$$\boldsymbol{\Sigma} = \begin{pmatrix} \boldsymbol{\Sigma}_{11} & \boldsymbol{\Sigma}_{21}' \\ \boldsymbol{\Sigma}_{21} & \boldsymbol{\Sigma}_{22} \end{pmatrix}.$$

What is the distribution of $\mathbf{y}$ conditional on $\mathbf{x}$ when the joint distribution of $\mathbf{y}$ and $\mathbf{x}$ is MVN? This distribution was derived in Subsection l under the supposition that $\boldsymbol{\Sigma}$ is positive definite. Under that supposition, the conditional distribution of $\mathbf{y}$ given $\mathbf{x}$ is MVN with mean vector $\boldsymbol{\mu}_2 + \boldsymbol{\Sigma}_{21}\boldsymbol{\Sigma}_{11}^{-1}(\mathbf{x} - \boldsymbol{\mu}_1)$ and variance-covariance matrix $\boldsymbol{\Sigma}_{22} - \boldsymbol{\Sigma}_{21}\boldsymbol{\Sigma}_{11}^{-1}\boldsymbol{\Sigma}_{21}'$. What is the conditional distribution of $\mathbf{y}$ given $\mathbf{x}$ in the general case where $\boldsymbol{\Sigma}$ may be positive semidefinite? In what follows, it is established that in the general case, the conditional distribution of $\mathbf{y}$ given $\mathbf{x}$ is MVN with mean vector $\boldsymbol{\mu}_2 + \boldsymbol{\Sigma}_{21}\boldsymbol{\Sigma}_{11}^{-}(\mathbf{x} - \boldsymbol{\mu}_1)$ and variance-covariance matrix $\boldsymbol{\Sigma}_{22} - \boldsymbol{\Sigma}_{21}\boldsymbol{\Sigma}_{11}^{-}\boldsymbol{\Sigma}_{21}'$. Thus, the generalization takes a simple form; it suffices to replace the ordinary inverse of $\boldsymbol{\Sigma}_{11}$ with a generalized inverse.

As a first step in establishing this generalization, let us extend the results of Theorem 3.5.9 (to the general case where $\boldsymbol{\Sigma}$ may be positive semidefinite). Let us take

$$\boldsymbol{\eta}(\mathbf{x}) = \boldsymbol{\mu}_2 + \boldsymbol{\Sigma}_{21}\boldsymbol{\Sigma}_{11}^{-}(\mathbf{x} - \boldsymbol{\mu}_1) \quad \text{and} \quad \mathbf{V} = \boldsymbol{\Sigma}_{22} - \boldsymbol{\Sigma}_{21}\boldsymbol{\Sigma}_{11}^{-}\boldsymbol{\Sigma}_{21}',$$

and define

$$\mathbf{e} = \mathbf{y} - \boldsymbol{\eta}(\mathbf{x}).$$

Observe (in light of Theorem 2.13.25) that $\boldsymbol{\Sigma} = \boldsymbol{\Gamma}'\boldsymbol{\Gamma}$ for some matrix $\boldsymbol{\Gamma}$. Accordingly,

$$\boldsymbol{\Sigma}_{11} = \boldsymbol{\Gamma}_1'\boldsymbol{\Gamma}_1 \quad \text{and} \quad \boldsymbol{\Sigma}_{21} = \boldsymbol{\Gamma}_2'\boldsymbol{\Gamma}_1 \tag{5.37}$$

for a suitable partitioning $\boldsymbol{\Gamma} = (\boldsymbol{\Gamma}_1, \boldsymbol{\Gamma}_2)$. And making use of Theorem 2.12.2 [and equating $(\boldsymbol{\Gamma}_1'\boldsymbol{\Gamma}_1)^-$ and $\boldsymbol{\Sigma}_{11}^-$], we find that

$$\boldsymbol{\Sigma}_{21}\boldsymbol{\Sigma}_{11}^{-}\boldsymbol{\Sigma}_{11} = \boldsymbol{\Gamma}_2'\boldsymbol{\Gamma}_1(\boldsymbol{\Gamma}_1'\boldsymbol{\Gamma}_1)^-\boldsymbol{\Gamma}_1'\boldsymbol{\Gamma}_1 = \boldsymbol{\Gamma}_2'\boldsymbol{\Gamma}_1 = \boldsymbol{\Sigma}_{21}. \tag{5.38}$$

By taking advantage of equality (5.38) and by proceeding in the same fashion as in proving Parts (1) and (2) of Theorem 3.5.9, we obtain the following results:

$$\mathrm{E}[\boldsymbol{\eta}(\mathbf{x})] = \boldsymbol{\mu}_2 \quad \text{and} \quad \mathrm{var}[\boldsymbol{\eta}(\mathbf{x})] = \mathrm{cov}[\boldsymbol{\eta}(\mathbf{x}), \mathbf{y}] = \boldsymbol{\Sigma}_{21}\boldsymbol{\Sigma}_{11}^{-}\boldsymbol{\Sigma}_{21}';$$
$$\mathrm{E}(\mathbf{e}) = \mathbf{0}, \quad \mathrm{var}(\mathbf{e}) = \mathbf{V}, \quad \text{and} \quad \mathrm{cov}(\mathbf{e}, \mathbf{x}) = \mathbf{0}. \tag{5.39}$$

Before proceeding, let us consider the extent to which $\boldsymbol{\eta}(\mathbf{x})$ and $\boldsymbol{\Sigma}_{21}\boldsymbol{\Sigma}_{11}^{-}\boldsymbol{\Sigma}_{21}'$ (and hence $\mathbf{V}$) are invariant to the choice of the generalized inverse $\boldsymbol{\Sigma}_{11}^{-}$. Recalling result (5.37) [and equating $(\boldsymbol{\Gamma}_1'\boldsymbol{\Gamma}_1)^-$ and $\boldsymbol{\Sigma}_{11}^-$], we find that

$$\boldsymbol{\Sigma}_{21}\boldsymbol{\Sigma}_{11}^{-}\boldsymbol{\Sigma}_{21}' = \boldsymbol{\Gamma}_2'\boldsymbol{\Gamma}_1(\boldsymbol{\Gamma}_1'\boldsymbol{\Gamma}_1)^-(\boldsymbol{\Gamma}_2'\boldsymbol{\Gamma}_1)' = \boldsymbol{\Gamma}_2'[\boldsymbol{\Gamma}_1(\boldsymbol{\Gamma}_1'\boldsymbol{\Gamma}_1)^-\boldsymbol{\Gamma}_1']\boldsymbol{\Gamma}_2.$$

And, based on Theorem 2.12.2, we conclude that $\boldsymbol{\Sigma}_{21}\boldsymbol{\Sigma}_{11}^{-}\boldsymbol{\Sigma}_{21}'$ and $\mathbf{V}$ are invariant to the choice of $\boldsymbol{\Sigma}_{11}^{-}$.

With regard to $\boldsymbol{\eta}(\mathbf{x})$, the situation is a bit more complicated. If $\mathbf{x}$ is such that $\mathbf{x} - \boldsymbol{\mu}_1 \in \mathcal{C}(\boldsymbol{\Sigma}_{11})$, then there exists a column vector $\mathbf{t}$ such that

$$\mathbf{x} - \boldsymbol{\mu}_1 = \boldsymbol{\Sigma}_{11}\mathbf{t},$$

in which case it follows from result (5.38) that

$$\boldsymbol{\Sigma}_{21}\boldsymbol{\Sigma}_{11}^{-}(\mathbf{x} - \boldsymbol{\mu}_1) = \boldsymbol{\Sigma}_{21}\boldsymbol{\Sigma}_{11}^{-}\boldsymbol{\Sigma}_{11}\mathbf{t} = \boldsymbol{\Sigma}_{21}\mathbf{t}$$

and hence that $\boldsymbol{\Sigma}_{21}\boldsymbol{\Sigma}_{11}^{-}(\mathbf{x}-\boldsymbol{\mu}_1)$ is invariant to the choice of $\boldsymbol{\Sigma}_{11}^{-}$. Thus, $\boldsymbol{\eta}(\mathbf{x})$ is invariant to the choice of $\boldsymbol{\Sigma}_{11}^{-}$ for every $\mathbf{x}$ such that $\mathbf{x}-\boldsymbol{\mu}_1 \in \mathcal{C}(\boldsymbol{\Sigma}_{11})$. Moreover, $\mathbf{x}-\boldsymbol{\mu}_1 \in \mathcal{C}(\boldsymbol{\Sigma}_{11})$ is an event of probability one. To see this, observe (in light of Lemmas 2.4.2 and 2.11.2) that

$$\mathbf{x}-\boldsymbol{\mu}_1 \in \mathcal{C}(\boldsymbol{\Sigma}_{11}) \;\Leftrightarrow\; \mathcal{C}(\mathbf{x}-\boldsymbol{\mu}_1) \subset \mathcal{C}(\boldsymbol{\Sigma}_{11}) \;\Leftrightarrow\; (\mathbf{I}-\boldsymbol{\Sigma}_{11}\boldsymbol{\Sigma}_{11}^{-})(\mathbf{x}-\boldsymbol{\mu}_1) = \mathbf{0},$$

and observe also that

$$\mathrm{E}[(\mathbf{I}-\boldsymbol{\Sigma}_{11}\boldsymbol{\Sigma}_{11}^{-})(\mathbf{x}-\boldsymbol{\mu}_1)] = \mathbf{0}$$

and

$$\mathrm{var}[(\mathbf{I}-\boldsymbol{\Sigma}_{11}\boldsymbol{\Sigma}_{11}^{-})(\mathbf{x}-\boldsymbol{\mu}_1)] = (\mathbf{I}-\boldsymbol{\Sigma}_{11}\boldsymbol{\Sigma}_{11}^{-})\boldsymbol{\Sigma}_{11}(\mathbf{I}-\boldsymbol{\Sigma}_{11}\boldsymbol{\Sigma}_{11}^{-})' = \mathbf{0}$$

and hence that $\Pr[(\mathbf{I}-\boldsymbol{\Sigma}_{11}\boldsymbol{\Sigma}_{11}^{-})(\mathbf{x}-\boldsymbol{\mu}_1) = \mathbf{0}] = 1$.

Now, returning to the primary development, suppose that the joint distribution of $\mathbf{x}$ and $\mathbf{y}$ is MVN. Then, by making use of result (5.39) and by proceeding in the same fashion as in proving Part (3) of Theorem 3.5.9, we find that $\mathbf{e} \sim N(\mathbf{0}, \mathbf{V})$ and that $\mathbf{e}$ and $\mathbf{x}$ are statistically independent.

At this point, we are in a position to derive the conditional distribution of $\mathbf{y}$ given $\mathbf{x}$. Since $\mathbf{e}$ and $\mathbf{x}$ are distributed independently, the conditional distribution of $\mathbf{e}$ given $\mathbf{x}$ is the same as the marginal distribution of $\mathbf{e}$. Thus, the conditional distribution of $\mathbf{e}$ given $\mathbf{x}$ is $N(\mathbf{0}, \mathbf{V})$. And upon observing that $\mathbf{y} = \boldsymbol{\eta}(\mathbf{x}) + \mathbf{e}$, it follows that the conditional distribution of $\mathbf{y}$ given $\mathbf{x}$ is $N[\boldsymbol{\eta}(\mathbf{x}), \mathbf{V}]$.

In summary, we have the following two theorems, which are generalizations of Theorems 3.5.8 and 3.5.9.

Theorem 3.5.10. Let $\mathbf{x}$ and $\mathbf{y}$ represent random column vectors whose joint distribution is MVN, and let $\boldsymbol{\mu}_1 = \mathrm{E}(\mathbf{x})$, $\boldsymbol{\mu}_2 = \mathrm{E}(\mathbf{y})$, $\boldsymbol{\Sigma}_{11} = \mathrm{var}(\mathbf{x})$, $\boldsymbol{\Sigma}_{22} = \mathrm{var}(\mathbf{y})$, and $\boldsymbol{\Sigma}_{21} = \mathrm{cov}(\mathbf{y}, \mathbf{x})$. Then, the conditional distribution of $\mathbf{y}$ given $\mathbf{x}$ is $N[\boldsymbol{\eta}(\mathbf{x}), \mathbf{V}]$, where

$$\boldsymbol{\eta}(\mathbf{x}) = \boldsymbol{\mu}_2 + \boldsymbol{\Sigma}_{21}\boldsymbol{\Sigma}_{11}^{-}(\mathbf{x}-\boldsymbol{\mu}_1) \quad \text{and} \quad \mathbf{V} = \boldsymbol{\Sigma}_{22} - \boldsymbol{\Sigma}_{21}\boldsymbol{\Sigma}_{11}^{-}\boldsymbol{\Sigma}_{21}'.$$

Theorem 3.5.11. Let $\mathbf{x}$ and $\mathbf{y}$ represent random column vectors, and let $\boldsymbol{\mu}_1 = \mathrm{E}(\mathbf{x})$, $\boldsymbol{\mu}_2 = \mathrm{E}(\mathbf{y})$, $\boldsymbol{\Sigma}_{11} = \mathrm{var}(\mathbf{x})$, $\boldsymbol{\Sigma}_{22} = \mathrm{var}(\mathbf{y})$, and $\boldsymbol{\Sigma}_{21} = \mathrm{cov}(\mathbf{y}, \mathbf{x})$. Further, let

$$\boldsymbol{\eta}(\mathbf{x}) = \boldsymbol{\mu}_2 + \boldsymbol{\Sigma}_{21}\boldsymbol{\Sigma}_{11}^{-}(\mathbf{x}-\boldsymbol{\mu}_1) \quad \text{and} \quad \mathbf{V} = \boldsymbol{\Sigma}_{22} - \boldsymbol{\Sigma}_{21}\boldsymbol{\Sigma}_{11}^{-}\boldsymbol{\Sigma}_{21}',$$

and define

$$\mathbf{e} = \mathbf{y} - \boldsymbol{\eta}(\mathbf{x}).$$

Then,

(0) $\boldsymbol{\Sigma}_{21}\boldsymbol{\Sigma}_{11}^{-}\boldsymbol{\Sigma}_{21}'$ and $\mathbf{V}$ are invariant to the choice of $\boldsymbol{\Sigma}_{11}^{-}$, and for $\mathbf{x}$ such that $\mathbf{x}-\boldsymbol{\mu}_1 \in \mathcal{C}(\boldsymbol{\Sigma}_{11})$, $\boldsymbol{\eta}(\mathbf{x})$ is invariant to the choice of $\boldsymbol{\Sigma}_{11}^{-}$; moreover, $\mathbf{x}-\boldsymbol{\mu}_1 \in \mathcal{C}(\boldsymbol{\Sigma}_{11})$ is an event of probability one;

(1) $\mathrm{E}[\boldsymbol{\eta}(\mathbf{x})] = \boldsymbol{\mu}_2$, and $\mathrm{var}[\boldsymbol{\eta}(\mathbf{x})] = \mathrm{cov}[\boldsymbol{\eta}(\mathbf{x}), \mathbf{y}] = \boldsymbol{\Sigma}_{21}\boldsymbol{\Sigma}_{11}^{-}\boldsymbol{\Sigma}_{21}'$;

(2) $\mathrm{E}(\mathbf{e}) = \mathbf{0}$, $\mathrm{var}(\mathbf{e}) = \mathbf{V}$, and $\mathrm{cov}(\mathbf{e}, \mathbf{x}) = \mathbf{0}$; and

(3) under the assumption that the joint distribution of $\mathbf{x}$ and $\mathbf{y}$ is MVN, the distribution of $\mathbf{e}$ is $N(\mathbf{0}, \mathbf{V})$ and $\mathbf{e}$ and $\mathbf{x}$ are statistically independent.

n. Third- and fourth-order central moments of the MVN distribution

Result (5.27) gives the third and fourth central moments of the univariate normal distribution. Let us extend that result by obtaining the third- and fourth-order central moments of the MVN distribution.

Take $\mathbf{x}$ to be an M-dimensional random column vector whose distribution is $N(\boldsymbol{\mu}, \boldsymbol{\Sigma})$, and denote by x_i and μ_i the ith elements of $\mathbf{x}$ and $\boldsymbol{\mu}$, respectively, and by σ_{ij} the ijth element of $\boldsymbol{\Sigma}$. Further, let $\boldsymbol{\Gamma} = \{\gamma_{ij}\}$ represent any matrix such that $\boldsymbol{\Sigma} = \boldsymbol{\Gamma}'\boldsymbol{\Gamma}$, denote by N the number of rows in $\boldsymbol{\Gamma}$, and take $\mathbf{z} = \{z_i\}$ to be an N-dimensional random column vector whose distribution is $N(\mathbf{0}, \mathbf{I})$, so that $\mathbf{x} \sim \boldsymbol{\mu} + \boldsymbol{\Gamma}'\mathbf{z}$.

As a preliminary step, observe that, for arbitrary nonnegative integers $k_1, k_2, \ldots, k_N$,

$$\mathrm{E}\big(z_1^{k_1} z_2^{k_2} \cdots z_N^{k_N}\big) = \mathrm{E}\big(z_1^{k_1}\big)\,\mathrm{E}\big(z_2^{k_2}\big)\cdots \mathrm{E}\big(z_N^{k_N}\big). \tag{5.40}$$

Observe also that $-\mathbf{z} \sim \mathbf{z}$ (i.e., $\mathbf{z}$ is distributed symmetrically about $\mathbf{0}$), so that

$$\begin{aligned}(-1)^{\sum_i k_i} z_1^{k_1} z_2^{k_2} \cdots z_N^{k_N} &= (-z_1)^{k_1}(-z_2)^{k_2}\cdots(-z_N)^{k_N} \\ &\sim z_1^{k_1} z_2^{k_2} \cdots z_N^{k_N}. \end{aligned} \tag{5.41}$$

If $\sum_i k_i$ is an odd number or, equivalently, if $(-1)^{\sum_i k_i} = -1$, then it follows from result (5.41) that $-\mathrm{E}\big(z_1^{k_1} z_2^{k_2} \cdots z_N^{k_N}\big) = \mathrm{E}\big(z_1^{k_1} z_2^{k_2} \cdots z_N^{k_N}\big)$ and hence that $\mathrm{E}\big(z_1^{k_1} z_2^{k_2} \cdots z_N^{k_N}\big) = 0$. Thus,

$$\mathrm{E}\big(z_1^{k_1} z_2^{k_2} \cdots z_N^{k_N}\big) = 0 \text{ for } k_1, k_2, \ldots, k_N \text{ such that } \textstyle\sum_i k_i = 1, 3, 5, 7, \ldots. \tag{5.42}$$

Now, consider the third-order central moments of the MVN distribution. Making use of result (5.42), we find that, for $i, j, s = 1, 2, \ldots, M$,

$$\begin{aligned}\mathrm{E}[(x_i-\mu_i)(x_j-\mu_j)(x_s-\mu_s)] &= \mathrm{E}\Big[\Big(\sum_{i'} \gamma_{i'i} z_{i'}\Big)\Big(\sum_{j'} \gamma_{j'j} z_{j'}\Big)\Big(\sum_{s'} \gamma_{s's} z_{s'}\Big)\Big] \\ &= \sum_{i',\, j',\, s'} \gamma_{i'i}\gamma_{j'j}\gamma_{s's}\,\mathrm{E}(z_{i'} z_{j'} z_{s'}) \\ &= 0. \end{aligned} \tag{5.43}$$

That is, all of the third-order central moments of the MVN distribution equal zero. In fact, by proceeding in similar fashion, we can establish the more general result that, for every odd positive integer r (i.e., for $r = 1, 3, 5, 7, \ldots$), all of the rth-order central moments of the MVN distribution equal zero.

Turning to the fourth-order central moments of the MVN distribution, we find [in light of results (5.40), (5.19), and (5.20)] that, for $i, j, s, t = 1, 2, \ldots, M$,

$$\begin{aligned}&\mathrm{E}[(x_i-\mu_i)(x_j-\mu_j)(x_s-\mu_s)(x_t-\mu_t)] \\ &\quad= \mathrm{E}\Big[\Big(\sum_{i'} \gamma_{i'i} z_{i'}\Big)\Big(\sum_{j'} \gamma_{j'j} z_{j'}\Big)\Big(\sum_{s'} \gamma_{s's} z_{s'}\Big)\Big(\sum_{t'} \gamma_{t't} z_{t'}\Big)\Big] \\ &\quad= \sum_{i',\, j',\, s',\, t'} \gamma_{i'i}\gamma_{j'j}\gamma_{s's}\gamma_{t't}\,\mathrm{E}(z_{i'} z_{j'} z_{s'} z_{t'}) \\ &\quad= \sum_{i'} \gamma_{i'i}\gamma_{i'j}\gamma_{i's}\gamma_{i't}\,(3) + \sum_{i'} \gamma_{i'i}\Big(\gamma_{i'j}\sum_{s'\neq i'} \gamma_{s's}\gamma_{s't} \\ &\qquad\qquad + \gamma_{i's}\sum_{j'\neq i'} \gamma_{j'j}\gamma_{j't} + \gamma_{i't}\sum_{j'\neq i'} \gamma_{j'j}\gamma_{j's}\Big)\,(1) \\ &\quad= \sum_{i'} \gamma_{i'i}\Big(\gamma_{i'j}\sum_{s'} \gamma_{s's}\gamma_{s't} + \gamma_{i's}\sum_{j'} \gamma_{j'j}\gamma_{j't} + \gamma_{i't}\sum_{j'} \gamma_{j'j}\gamma_{j's}\Big) \\ &\quad= \sigma_{ij}\sigma_{st} + \sigma_{is}\sigma_{jt} + \sigma_{it}\sigma_{js}. \end{aligned} \tag{5.44}$$

In the special case $t = s = j = i$, formula (5.44) reduces to

$$\mathrm{E}[(x_i-\mu_i)^4] = 3\sigma_{ii},$$

in agreement with the expression given earlier [in result (5.27)] for the fourth central moment of the univariate normal distribution.

o. Moment generating function

Consider an M-variate normal distribution with mean vector $\boldsymbol{\mu}$ and variance-covariance matrix $\boldsymbol{\Sigma}$. When $\boldsymbol{\Sigma}$ is positive definite, this distribution can be characterized in terms of its probability density function—when $\boldsymbol{\Sigma}$ is positive semidefinite, there is no probability density function. Alternatively (and regardless of whether $\boldsymbol{\Sigma}$ is positive definite or positive semidefinite), the $N(\boldsymbol{\mu}, \boldsymbol{\Sigma})$ distribution can be characterized in terms of its moment generating function.

Denote by $m(\cdot)$ the moment generating function of the $N(\boldsymbol{\mu}, \boldsymbol{\Sigma})$ distribution. And let $\mathbf{x} = \boldsymbol{\mu} + \boldsymbol{\Gamma}'\mathbf{z}$, where (for some integer N) $\boldsymbol{\Gamma}$ is an $N \times M$ matrix such that $\boldsymbol{\Sigma} = \boldsymbol{\Gamma}'\boldsymbol{\Gamma}$ and where $\mathbf{z}$ is an N-dimensional random column vector whose distribution is $N(\mathbf{0}, \mathbf{I})$. Then, for an arbitrary M-dimensional (nonrandom) column vector $\mathbf{t}$, we have that

$$\begin{aligned} m(\mathbf{t}) &= \mathrm{E}[\exp(\mathbf{t}'\mathbf{x})] = \mathrm{E}\{\exp[\mathbf{t}'(\boldsymbol{\mu} + \boldsymbol{\Gamma}'\mathbf{z})]\} \\ &= \int_{\mathbb{R}^N} (2\pi)^{-N/2} \exp(\mathbf{t}'\boldsymbol{\mu} + \mathbf{t}'\boldsymbol{\Gamma}'\mathbf{z} - \tfrac{1}{2}\mathbf{z}'\mathbf{z})\, d\mathbf{z}. \end{aligned} \tag{5.45}$$

The evaluation of expression (5.45) is facilitated by the identity

$$\mathbf{t}'\boldsymbol{\Gamma}'\mathbf{z} - \tfrac{1}{2}\mathbf{z}'\mathbf{z} = -\tfrac{1}{2}(\mathbf{z} - \boldsymbol{\Gamma}\mathbf{t})'(\mathbf{z} - \boldsymbol{\Gamma}\mathbf{t}) + \tfrac{1}{2}\mathbf{t}'\boldsymbol{\Sigma}\mathbf{t}, \tag{5.46}$$

obtained by "completing the square" and observing that (since $\mathbf{z}'\boldsymbol{\Gamma}\mathbf{t}$ is of dimensions 1×1) $\mathbf{z}'\boldsymbol{\Gamma}\mathbf{t} = (\mathbf{z}'\boldsymbol{\Gamma}\mathbf{t})' = \mathbf{t}'\boldsymbol{\Gamma}'\mathbf{z}$. Upon substituting expression (5.46) into expression (5.45), we find that

$$m(\mathbf{t}) = \exp(\mathbf{t}'\boldsymbol{\mu} + \tfrac{1}{2}\mathbf{t}'\boldsymbol{\Sigma}\mathbf{t}) \int_{\mathbb{R}^N} (2\pi)^{-N/2} \exp[-\tfrac{1}{2}(\mathbf{z} - \boldsymbol{\Gamma}\mathbf{t})'(\mathbf{z} - \boldsymbol{\Gamma}\mathbf{t})]\, d\mathbf{z}.$$

Moreover,

$$\int_{\mathbb{R}^N} (2\pi)^{-N/2} \exp[-\tfrac{1}{2}(\mathbf{z} - \boldsymbol{\Gamma}\mathbf{t})'(\mathbf{z} - \boldsymbol{\Gamma}\mathbf{t})]\, d\mathbf{z} = 1,$$

as is evident upon making a change of variables from $\mathbf{z}$ to $\mathbf{y} = \mathbf{z} - \boldsymbol{\Gamma}\mathbf{t}$ or upon observing that the integrand is a probability density function [that of the $N(\boldsymbol{\Gamma}\mathbf{t}, \mathbf{I})$ distribution].

Thus, the moment generating function $m(\cdot)$ of the M-variate normal distribution with mean vector $\boldsymbol{\mu}$ and variance-covariance matrix $\boldsymbol{\Sigma}$ is given by

$$m(\mathbf{t}) = \exp(\mathbf{t}'\boldsymbol{\mu} + \tfrac{1}{2}\mathbf{t}'\boldsymbol{\Sigma}\mathbf{t}) \qquad (\mathbf{t} \in \mathbb{R}^M). \tag{5.47}$$

Corresponding to the moment generating function $m(\cdot)$ is the cumulant generating function, say $c(\cdot)$, of this distribution, which is given by

$$c(\mathbf{t}) = \log m(\mathbf{t}) = \mathbf{t}'\boldsymbol{\mu} + \tfrac{1}{2}\mathbf{t}'\boldsymbol{\Sigma}\mathbf{t} \qquad (\mathbf{t} \in \mathbb{R}^M). \tag{5.48}$$

p. A univariate characterization of multivariate normality

Let $\mathbf{x} = \{x_i\}$ represent an M-dimensional random column vector. If the distribution of $\mathbf{x}$ is MVN, then it follows from Theorem 3.5.1 that for every M-dimensional nonrandom column vector $\mathbf{a} = \{a_i\}$, the linear combination $\mathbf{a}'\mathbf{x} = \sum_{i=1}^{M} a_i x_i$ has a (univariate) normal distribution.

Is the converse true? That is, if every linear combination of the elements $x_1, x_2, \ldots, x_M$ of $\mathbf{x}$ has a (univariate) normal distribution, is it necessarily the case that the distribution of $\mathbf{x}$ is MVN? In what follows, the answer is shown to be yes.

Let $\mathbf{a}$ represent an arbitrary M-dimensional nonrandom column vector, and suppose that for every $\mathbf{a}$, $\mathbf{a}'\mathbf{x}$ has a (univariate) normal distribution (implying, in particular, that $x_1, x_2, \ldots, x_M$ have normal distributions and hence that the expected values and variances of $x_1, x_2, \ldots, x_M$ exist). Further, let

$\boldsymbol{\mu} = \text{E}(\mathbf{x})$ and $\boldsymbol{\Sigma} = \text{var}(\mathbf{x})$, and observe that (for every $\mathbf{a}$) $\text{E}(\mathbf{a}'\mathbf{x}) = \mathbf{a}'\boldsymbol{\mu}$ and $\text{var}(\mathbf{a}'\mathbf{x}) = \mathbf{a}'\boldsymbol{\Sigma}\mathbf{a}$. Then, recalling the results of Subsection o and denoting by $m^*(\cdot;\, \mathbf{a})$ the moment generating function of the $N(\mathbf{a}'\boldsymbol{\mu},\, \mathbf{a}'\boldsymbol{\Sigma}\mathbf{a})$ distribution, we find that (for every $\mathbf{a}$)

$$\text{E}[\exp(\mathbf{a}'\mathbf{x})] = \text{E}[\exp(1\,\mathbf{a}'\mathbf{x})] = m^*(1;\, \mathbf{a}) = \exp(\mathbf{a}'\boldsymbol{\mu} + \tfrac{1}{2}\mathbf{a}'\boldsymbol{\Sigma}\mathbf{a}). \tag{5.49}$$

And we conclude that the distribution of $\mathbf{x}$ has a moment generating function, say $m(\cdot)$, and that

$$m(\mathbf{a}) = \exp(\mathbf{a}'\boldsymbol{\mu} + \tfrac{1}{2}\mathbf{a}'\boldsymbol{\Sigma}\mathbf{a}) \qquad (\mathbf{a} \in \mathbb{R}^M). \tag{5.50}$$

A comparison of expression (5.50) with expression (5.47) reveals that the moment generating function of the distribution of $\mathbf{x}$ is the same as that of the $N(\boldsymbol{\mu}, \boldsymbol{\Sigma})$ distribution. If two distributions have the same moment generating function, they are identical (e.g., Casella and Berger 2002, p. 65; Bickel and Doksum 2001, pp. 460 and 505). Consequently, the distribution of $\mathbf{x}$ is MVN.

In summary, we have the following characterization of multivariate normality.

Theorem 3.5.12. The distribution of the M-dimensional random column vector $\mathbf{x} = \{x_i\}$ is MVN if and only if, for every M-dimensional nonrandom column vector $\mathbf{a} = \{a_i\}$, the distribution of the linear combination $\mathbf{a}'\mathbf{x} = \sum_{i=1}^{M} a_i x_i$ is (univariate) normal.

Exercises

Exercise 1. Provide detailed verifications for (1) equality (1.7), (2) equality (1.8), (3) equality (1.10), and (4) equality (1.9).

Exercise 2.

(a) Let w and z represent random variables [such that $\text{E}(w^2) < \infty$ and $\text{E}(z^2) < \infty$]. Show that

$$|\text{E}(wz)| \le \text{E}(|wz|) \le \sqrt{\text{E}(w^2)}\sqrt{\text{E}(z^2)}; \tag{E.1}$$

and determine the conditions under which the first inequality holds as an equality, the conditions under which the second inequality holds as an equality, and the conditions under which both inequalities hold as equalities.

(b) Let x and y represent random variables [such that $\text{E}(x^2) < \infty$ and $\text{E}(y^2) < \infty$]. Using Part (a) (or otherwise), show that

$$|\text{cov}(x, y)| \le \text{E}[|x - \text{E}(x)||y - \text{E}(y)|] \le \sqrt{\text{var}(x)}\sqrt{\text{var}(y)}; \tag{E.2}$$

and determine the conditions under which the first inequality holds as an equality, the conditions under which the second inequality holds as an equality, and the conditions under which both inequalities hold as equalities.

Exercise 3. Let $\mathbf{x}$ represent an N-dimensional random column vector and $\mathbf{y}$ a T-dimensional random column vector. And define $\mathbf{x}_*$ to be an R-dimensional subvector of $\mathbf{x}$ and $\mathbf{y}_*$ an S-dimensional subvector of $\mathbf{y}$ (where $1 \le R \le N$ and $1 \le S \le T$). Relate $\text{E}(\mathbf{x}_*)$ to $\text{E}(\mathbf{x})$, $\text{var}(\mathbf{x}_*)$ to $\text{var}(\mathbf{x})$, and $\text{cov}(\mathbf{x}_*, \mathbf{y}_*)$ and $\text{cov}(\mathbf{y}_*, \mathbf{x}_*)$ to $\text{cov}(\mathbf{x}, \mathbf{y})$.

Exercise 4. Let x represent a random variable that is distributed symmetrically about 0 (so that $-x \sim x$); and suppose that the distribution of x is "nondegenerate" in the sense that there exists a nonnegative constant c such that $0 < \Pr(x > c) < \frac{1}{2}$ [and assume that $\text{E}(x^2) < \infty$]. Further, define $y = |x|$.

(a) Show that $\text{cov}(x, y) = 0$.

(b) Are x and y statistically independent? Why or why not?

Exercise 5. Provide detailed verifications for (1) equality (2.39), (2) equality (2.45), and (3) equality (2.48).

Exercise 6.

(a) Let x and y represent random variables. Show that $\text{cov}(x, y)$ can be determined from knowledge of $\text{var}(x)$, $\text{var}(y)$, and $\text{var}(x + y)$, and give a formula for doing so.

(b) Let $\mathbf{x} = (x_1, x_2)'$ and $\mathbf{y} = (y_1, y_2)'$ represent 2-dimensional random column vectors. Can $\text{cov}(\mathbf{x}, \mathbf{y})$ be determined from knowledge of $\text{var}(\mathbf{x})$, $\text{var}(\mathbf{y})$, and $\text{var}(\mathbf{x} + \mathbf{y})$? Why or why not?

Exercise 7. Let $\mathbf{x}$ represent an M-dimensional random column vector with mean vector $\boldsymbol{\mu}$ and variance-covariance matrix $\boldsymbol{\Sigma}$. Show that there exist $M(M+1)/2$ linear combinations of the M elements of $\mathbf{x}$ such that $\boldsymbol{\mu}$ can be determined from knowledge of the expected values of M of these linear combinations and $\boldsymbol{\Sigma}$ can be determined from knowledge of the $[M(M+1)/2]$ variances of these linear combinations.

Exercise 8. Let x and y represent random variables (whose expected values and variances exist), let $\mathbf{V}$ represent the variance-covariance matrix of the random vector (x, y), and suppose that $\text{var}(x) > 0$ and $\text{var}(y) > 0$.

(a) Show that if $|\mathbf{V}| = 0$, then, for scalars a and b,

$$(a, b)' \in \mathcal{N}(\mathbf{V}) \;\Leftrightarrow\; b\sqrt{\text{var}\, y} = ca\sqrt{\text{var}\, x},$$

where $c = \begin{cases} +1, & \text{when } \text{cov}(x, y) < 0, \\ -1, & \text{when } \text{cov}(x, y) > 0. \end{cases}$

(b) Use the result of Part (a) and the results of Section 3.2e to devise an alternative proof of the result (established in Section 3.2a) that $\text{cov}(x, y) = \sqrt{\text{var}\, x}\sqrt{\text{var}\, y}$ if and only if $[y - \text{E}(y)]/\sqrt{\text{var}\, y} = [x - \text{E}(x)]/\sqrt{\text{var}\, x}$ with probability 1 and that $\text{cov}(x, y) = -\sqrt{\text{var}\, x}\sqrt{\text{var}\, y}$ if and only if $[y - \text{E}(y)]/\sqrt{\text{var}\, y} = -[x - \text{E}(x)]/\sqrt{\text{var}\, x}$ with probability 1.

Exercise 9. Let $\mathbf{x}$ represent an N-dimensional random column vector (with elements whose expected values and variances exist). Show that (regardless of the rank of $\text{var}\,\mathbf{x}$) there exist a nonrandom column vector $\mathbf{c}$ and an $N \times N$ nonsingular nonrandom matrix $\mathbf{A}$ such that the random vector $\mathbf{w}$, defined implicitly by $\mathbf{x} = \mathbf{c} + \mathbf{A}'\mathbf{w}$, has mean $\mathbf{0}$ and a variance-covariance matrix of the form $\text{diag}(\mathbf{I}, \mathbf{0})$ [where $\text{diag}(\mathbf{I}, \mathbf{0})$ is to be regarded as including $\mathbf{0}$ and $\mathbf{I}$ as special cases].

Exercise 10. Establish the validity of result (5.11).

Exercise 11. Let $w = |z|$, where z is a standard normal random variable.

(a) Find a probability density function for the distribution of w.

(b) Use the expression obtained in Part (a) (for the probability density function of the distribution of w) to derive formula (5.15) for $\text{E}(w^r)$—in Section 3.5c, this formula is derived from the probability density function of the distribution of z.

(c) Find $\text{E}(w)$ and $\text{var}(w)$.

Exercise 12. Let x represent a random variable having mean μ and variance σ^2. Then, $\text{E}(x^2) = \mu^2 + \sigma^2$ [as is evident from result (2.3)]. Thus, the second moment of x depends on the distribution of x only through μ and σ^2. If the distribution of x is normal, then the third and higher moments of x

also depend only on μ and σ^2. Taking the distribution of x to be normal, obtain explicit expressions for $E(x^3)$, $E(x^4)$, and, more generally, $E(x^r)$ (where r is an arbitrary positive integer).

Exercise 13. Let $\mathbf{x}$ represent an N-dimensional random column vector whose distribution is $N(\boldsymbol{\mu}, \boldsymbol{\Sigma})$. Further, let $R = \text{rank}(\boldsymbol{\Sigma})$, and assume that $\boldsymbol{\Sigma}$ is nonnull (so that $R \geq 1$). Show that there exist an R-dimensional nonrandom column vector $\mathbf{c}$ and an $R \times N$ nonrandom matrix $\mathbf{A}$ such that $\mathbf{c} + \mathbf{A}\mathbf{x} \sim N(\mathbf{0}, \mathbf{I})$ (i.e., such that the distribution of $\mathbf{c} + \mathbf{A}\mathbf{x}$ is R-variate standard normal).

Exercise 14. Let x and y represent random variables, and suppose that $x + y$ and $x - y$ are independently and normally distributed and have the same mean, say γ, and the same variance, say τ^2. Show that x and y are statistically independent, and determine the distribution of x and the distribution of y.

Exercise 15. Suppose that two or more random column vectors $\mathbf{x}_1, \mathbf{x}_2, \ldots, \mathbf{x}_P$ are pairwise independent (i.e., $\mathbf{x}_i$ and $\mathbf{x}_j$ are statistically independent for $j > i = 1, 2, \ldots, P$) and that the joint distribution of $\mathbf{x}_1, \mathbf{x}_2, \ldots, \mathbf{x}_P$ is MVN. Is it necessarily the case that $\mathbf{x}_1, \mathbf{x}_2, \ldots, \mathbf{x}_P$ are mutually independent? Why or why not?

Exercise 16. Let x represent a random variable whose distribution is $N(0, 1)$, and define $y = ux$, where u is a discrete random variable that is distributed independently of x with $\Pr(u = 1) = \Pr(u = -1) = \frac{1}{2}$.

(a) Show that $y \sim N(0, 1)$.

(b) Show that $\text{cov}(x, y) = 0$.

(c) Show that x and y are statistically dependent.

(d) Is the joint distribution of x and y bivariate normal? Why or why not?

Exercise 17. Let $\mathbf{x}_1, \mathbf{x}_2, \ldots, \mathbf{x}_K$ represent N-dimensional random column vectors, and suppose that $\mathbf{x}_1, \mathbf{x}_2, \ldots, \mathbf{x}_K$ are mutually independent and that (for $i = 1, 2, \ldots, K$) $\mathbf{x}_i \sim N(\boldsymbol{\mu}_i, \boldsymbol{\Sigma}_i)$. Derive (for arbitrary scalars $a_1, a_2, \ldots, a_K$) the distribution of the linear combination $\sum_{i=1}^{K} a_i \mathbf{x}_i$.

Exercise 18. Let x and y represent random variables whose joint distribution is bivariate normal. Further, let $\mu_1 = E(x)$, $\mu_2 = E(y)$, $\sigma_1^2 = \text{var}\, x$, $\sigma_2^2 = \text{var}\, y$, and $\rho = \text{corr}(x, y)$ (where $\sigma_1 \geq 0$ and $\sigma_2 \geq 0$). Assuming that $\sigma_1 > 0$, $\sigma_2 > 0$, and $-1 < \rho < 1$, show that the conditional distribution of y given x is $N[\mu_2 + \rho\sigma_2(x - \mu_1)/\sigma_1,\ \sigma_2^2(1 - \rho^2)]$.

Exercise 19. Let x and y represent random variables whose joint distribution is bivariate normal. Further, let $\sigma_1^2 = \text{var}\, x$, $\sigma_2^2 = \text{var}\, y$, and $\sigma_{12} = \text{cov}(x, y)$ (where $\sigma_1 \geq 0$ and $\sigma_2 \geq 0$). Describe (in as simple terms as possible) the marginal distributions of x and y and the conditional distributions of y given x and of x given y. Do so for each of the following two "degenerate" cases: (1) $\sigma_1^2 = 0$; and (2) $\sigma_1^2 > 0$, $\sigma_2^2 > 0$, and $|\sigma_{12}| = \sigma_1\sigma_2$.

Exercise 20. Let $\mathbf{x}$ represent an N-dimensional random column vector, and take $\mathbf{y}$ to be the M-dimensional random column vector defined by $\mathbf{y} = \mathbf{c} + \mathbf{A}\mathbf{x}$, where $\mathbf{c}$ is an M-dimensional nonrandom column vector and $\mathbf{A}$ an $M \times N$ nonrandom matrix.

(a) Express the moment generating function of the distribution of $\mathbf{y}$ in terms of the moment generating function of the distribution of $\mathbf{x}$.

(b) Use the result of Part (a) to show that if the distribution of $\mathbf{x}$ is $N(\boldsymbol{\mu}, \boldsymbol{\Sigma})$, then the moment generating function of the distribution of $\mathbf{y}$ is the same as that of the $N(\mathbf{c} + \mathbf{A}\boldsymbol{\mu},\ \mathbf{A}\boldsymbol{\Sigma}\mathbf{A}')$ distribution, thereby (since distributions having the same moment generating function are identical) providing an alternative way of arriving at Theorem 3.5.1.

Bibliographic and Supplementary Notes

§1. The term probability density function is used in connection with the probability distribution of a random variable or, more generally, that of a random vector, say a random vector of dimension N—refer to Section 3.1. Such use is restricted herein to probability density functions that are probability density functions with respect to N-dimensional Lebesgue measure. Thus, a statement that the probability distribution of an N-dimensional random vector does not have a probability density function means that it does not have a probability density function with respect to N-dimensional Lebesgue measure.

§2. Inequality (2.13) can be regarded as a special case of an inequality, known as the Cauchy–Schwarz inequality (or simply as the Schwarz inequality), that, in a more general form, is applicable to the members of any inner-product space—refer, e.g., to Halmos (1958, sec. 64) for a statement, proof, and discussion of the general version of this inequality. Accordingly, inequality (2.13) [or the equivalent inequality (2.14)] is often referred to as the Cauchy–Schwarz inequality. Earlier (in Section 2.4), the names of Cauchy and Schwarz were applied to inequality (2.4.9). Like inequality (2.13), that inequality can be regarded as a special case of the general version of the Cauchy–Schwarz inequality.

§5p. Moment generating functions are closely related to characteristic functions. For the most part, moment generating functions are adequate for our purposes, and their use results in a presentation suitable for those with little or no knowledge of complex analysis—characteristic functions involve complex numbers, while moment generating functions do not. Distributions having the same moment generating function are identical (e.g., Casella and Berger 2002, p. 65; Bickel and Doksum 2001, pp. 460 and 505). And if two or more random variables or vectors are such that the moment generating function of their joint distribution equals the product of the moment generating functions of their marginal distributions, then those random variables or vectors are statistically independent (e.g., Parzen 1960, p. 364). These two results constitute powerful "tools" for establishing the identity of a distribution and for establishing the statistical independence of two or more random variables or vectors. In fact, the first of the two results is used in Section 3.5p in arriving at Theorem 3.5.12. Unfortunately, there is a downside to their use. Their proofs are relatively difficult and may be unfamiliar to (and possibly "inaccessible" to) many potential readers. Consequently, an attempt is made herein to avoid the use of the aforementioned results (on moment generating functions) in proving other results. Preference is given to the use of results that are relatively elementary and easily proven.

4

The General Linear Model

The first two sections of Chapter 1 provide an introduction to linear statistical models in general and to linear regression models in particular. Let us now expand on that introduction, doing so in a way that facilitates the presentation (in subsequent chapters) of the results on statistical theory and methodology that constitute the primary subject matter of the book.

The setting is one in which some number, say N, of data points are (for purposes of making statistical inferences about various quantities of interest) to be regarded as the respective values of observable random variables $y_1, y_2, \ldots, y_N$. Define $\mathbf{y} = (y_1, y_2, \ldots, y_N)'$. It is supposed that (for $i = 1, 2, \ldots, N$) the ith datum (the observed value of y_i) is accompanied by the corresponding value $\mathbf{u}_i$ of a column vector $\mathbf{u} = (u_1, u_2, \ldots, u_C)'$ of C "explanatory" variables $u_1, u_2, \ldots, u_C$. The observable random vector $\mathbf{y}$ is to be modeled by specifying a "family," say Δ, of functions of $\mathbf{u}$, and by assuming that for some member of Δ (of unknown identity), say $\delta(\cdot)$, the random deviations $y_i - \delta(\mathbf{u}_i)$ $(i = 1, 2, \ldots, N)$ have ("conditionally" on $\mathbf{u}_1, \mathbf{u}_2, \ldots, \mathbf{u}_N$) a joint distribution with certain specified characteristics. In particular, it might be assumed that these random deviations have a common mean of 0 and a common variance σ^2 (of unknown value), that they are uncorrelated, and possibly that they are jointly normal.

The emphasis herein is on models in which Δ consists of some or all of those functions (of $\mathbf{u}$) that are expressible as linear combinations of P specified functions $\delta_1(\cdot), \delta_2(\cdot), \ldots, \delta_P(\cdot)$. When Δ is of that form, the assumption that, for some function $\delta(\cdot)$, the joint distribution of $y_i - \delta(\mathbf{u}_i)$ $(i = 1, 2, \ldots, N)$ has certain specified characteristics can be replaced by the assumption that, for some linear combination in Δ (of unknown identity), say one with coefficients $\beta_1, \beta_2, \ldots, \beta_P$, the joint distribution of $y_i - \sum_{j=1}^{C} \beta_j \delta_j(\mathbf{u}_i)$ $(i = 1, 2, \ldots, N)$ has the specified characteristics. Corresponding to this linear combination is the P-dimensional parameter vector $\boldsymbol{\beta} = (\beta_1, \beta_2, \ldots, \beta_P)'$ of coefficients—in general, this vector is of unknown value.

As what can be regarded as a very special case, we have the kind of situation where (with probability 1) $y_i = \delta(\mathbf{u}_i)$ $(i = 1, 2, \ldots, N)$ for some function $\delta(\cdot)$ whose identity is known. In that special case, Δ has only one member, and (with probability 1) all N of the "random" deviations $y_i - \delta(\mathbf{u}_i)$ $(i = 1, 2, \ldots, N)$ equal 0. An example of that kind of situation is provided by the ideal-gas law of physics:

$$p = rnt/v.$$

Here, p is the pressure within a container of gas, v is the volume of the container, t is the absolute temperature of the gas, n is the number of moles of gas present in the container, and r is the universal gas constant. It has been found that, under laboratory conditions and for any of a wide variety of gases, the pressure readings obtained for any of a broad range of values of n, t, and v conform almost perfectly to the ideal-gas law. Note that (by taking logarithms) the ideal-gas law can be reexpressed in the form of the linear equation

$$\log p = -\log v + \log r + \log n + \log t.$$

In general, Δ consists of a possibly infinite number of functions. And the relationship between $y_1, y_2, \ldots, y_N$ and the corresponding values $\mathbf{u}_1, \mathbf{u}_2, \ldots, \mathbf{u}_N$ of the vector $\mathbf{u}$ of explanatory variables is typically imperfect.

At times (when the intended usage would seem to be clear from the context), resort is made herein to the convenient practice of using the same symbol for a realization of a random variable (or of a random vector or random matrix) as for the random quantity itself. At other times, the realization might be distinguished from the random quantity by means of an underline or by the use of an altogether different symbol. Thus, depending on the context, the N data points might be represented by either $y_1, y_2, \dots, y_N$ or $\underline{y}_1, \underline{y}_2, \dots, \underline{y}_N$, and the N-dimensional column vector comprising these points might be represented by either $\mathbf{y}$ or $\underline{\mathbf{y}}$. Similarly, depending on the context, an arbitrary member of Δ might be denoted either by $\delta(\cdot)$ (the same symbol used to denote the member having the specified characteristics) or by $\underline{\delta}(\cdot)$. And either $\beta_1, \beta_2, \dots, \beta_P$ or $b_1, b_2, \dots, b_P$ might be used to represent the coefficients of an arbitrary linear combination of the functions $\delta_1(\cdot), \delta_2(\cdot), \dots, \delta_P(\cdot)$, and either $\boldsymbol{\beta}$ or $\mathbf{b}$ might be used to represent the P-dimensional column vector comprising these coefficients.

The family Δ of functions of $\mathbf{u}$, in combination with whatever assumptions are made about the joint distribution of the N random deviations $y_i - \delta(\mathbf{u}_i)$ $(i = 1, 2, \dots, N)$, determines the statistical model. Accordingly, it can play a critical role in establishing a basis for the use of the data in making statistical inferences. Moreover, corresponding to an arbitrary member $\underline{\delta}(\cdot)$ of Δ is the approximation to the data vector $\underline{\mathbf{y}} = (\underline{y}_1, \underline{y}_2, \dots, \underline{y}_N)'$ provided by the vector $[\underline{\delta}(\mathbf{u}_1), \underline{\delta}(\mathbf{u}_2), \dots, \underline{\delta}(\mathbf{u}_N)]'$. It may be of direct or indirect interest to determine the member of Δ for which this approximation is the best [best in the sense that the norm of the N-dimensional vector with elements $\underline{y}_1 - \underline{\delta}(\mathbf{u}_1)$, $\underline{y}_2 - \underline{\delta}(\mathbf{u}_2)$, ..., $\underline{y}_N - \underline{\delta}(\mathbf{u}_N)$ is minimized for $\underline{\delta}(\cdot) \in \Delta$]. Note that the solution to this optimization problem may be well-defined even in the absence of any assumptions of a statistical nature.

The data could be either univariate or multivariate. If some of the N data points were "altogether different" in character than some of the others, the data would be regarded as multivariate. Such would be the case if, for example, part of the data consisted of height measurements and part of weight measurements. In some cases, the distinction (between univariate data and multivariate data) might be less than clear-cut. For example, if the data consisted entirely of measurements of the level of a pollutant but some measurements were obtained by different means, at a different time, or under different conditions than others, the data might be regarded as univariate or, alternatively, as multivariate.

4.1 Some Basic Types of Linear Models

Let us continue to take the setting to be one in which there are N data points that are to be regarded as the respective values of observable random variables $y_1, y_2, \dots, y_N$. And let us continue to define $\mathbf{y} = (y_1, y_2, \dots, y_N)'$ and to suppose that (for $i = 1, 2, \dots, N$) the ith datum is accompanied by the corresponding value $\mathbf{u}_i$ of a C-dimensional column vector $\mathbf{u} = (u_1, u_2, \dots, u_C)'$ of explanatory variables. As previously indicated, the observable random vector $\mathbf{y}$ is to be modeled in terms of a specified family Δ of functions of $\mathbf{u}$—it is implicitly assumed that the domain of each of these functions includes $\mathbf{u}_1, \mathbf{u}_2, \dots, \mathbf{u}_N$.

The models to be considered start with the assumption that for some (unidentified) function $\delta(\cdot)$ in Δ,

$$E(y_i) = \delta(\mathbf{u}_i) \quad (i = 1, 2, \dots, N) \tag{1.1}$$

or, equivalently,

$$E[y_i - \delta(\mathbf{u}_i)] = 0 \quad (i = 1, 2, \dots, N). \tag{1.2}$$

Letting $\boldsymbol{\delta} = [\delta(\mathbf{u}_1), \delta(\mathbf{u}_2), \dots, \delta(\mathbf{u}_N)]'$, condition (1.1) is reexpressible as

$$E(\mathbf{y}) = \boldsymbol{\delta}, \tag{1.3}$$

and condition (1.2) as

$$E(\mathbf{y} - \boldsymbol{\delta}) = \mathbf{0}. \tag{1.4}$$

Here, the distributions of $\mathbf{y}$ and $\mathbf{y} - \boldsymbol{\delta}$ [and hence the expected values in conditions (1.1), (1.2), (1.3), and (1.4)] are regarded as "conditional" on $\mathbf{u}_1, \mathbf{u}_2, \ldots, \mathbf{u}_N$. Further, it is assumed that the distribution of $\mathbf{y} - \boldsymbol{\delta}$ does not depend on $\delta(\cdot)$ or, less stringently, that $\text{var}(\mathbf{y} - \boldsymbol{\delta})$ ($= \text{var}\,\mathbf{y}$) does not depend on $\delta(\cdot)$.

As is evident from the simple identity

$$\mathbf{y} = \boldsymbol{\delta} + (\mathbf{y} - \boldsymbol{\delta}),$$

condition (1.3) or (1.4) is equivalent to the condition

$$\mathbf{y} = \boldsymbol{\delta} + \mathbf{e}, \tag{1.5}$$

where $\mathbf{e}$ is an N-dimensional random column vector with $\text{E}(\mathbf{e}) = \mathbf{0}$. Under condition (1.5),

$$\mathbf{e} = \mathbf{y} - \boldsymbol{\delta} = \mathbf{y} - \text{E}(\mathbf{y}),$$

and—aside from a trivial case where $\boldsymbol{\delta}$ has the same value for all $\delta(\cdot) \in \Delta$ (and that value is known) and/or where $\mathbf{e} = \mathbf{0}$ (with probability 1)—$\mathbf{e}$ is unobservable. Moreover, under condition (1.5), the assumption that the distribution of $\mathbf{y} - \boldsymbol{\delta}$ does not depend on $\delta(\cdot)$ is equivalent to an assumption that the distribution of $\mathbf{e}$ does not depend on $\delta(\cdot)$, and, similarly, the less stringent assumption that $\text{var}(\mathbf{y} - \boldsymbol{\delta})$ or $\text{var}(\mathbf{y})$ does not depend on $\delta(\cdot)$ is equivalent to an assumption that $\text{var}(\mathbf{e})$ does not depend on $\delta(\cdot)$.

Suppose now that Δ consists of some or all of those functions (of $\mathbf{u}$) that are expressible as linear combinations of the P specified functions $\delta_1(\cdot), \delta_2(\cdot), \ldots, \delta_P(\cdot)$. Then, by definition, a function $\underline{\delta}(\cdot)$ (of $\mathbf{u}$) is a member of Δ only if (for "all" $\mathbf{u}$) $\underline{\delta}(\mathbf{u})$ is expressible in the form

$$\underline{\delta}(\mathbf{u}) = b_1\delta_1(\mathbf{u}) + b_2\delta_2(\mathbf{u}) + \cdots + b_P\delta_P(\mathbf{u}), \tag{1.6}$$

where $b_1, b_2, \ldots, b_P$ are scalars (that do not vary with $\mathbf{u}$).

Let $x_{ij} = \delta_j(\mathbf{u}_i)$ $(i = 1, 2, \ldots, N)$, and define $\mathbf{x}_j = (x_{1j}, x_{2j}, \ldots, x_{Nj})'$ $(j = 1, 2, \ldots, P)$. Further, take $\mathbf{X} = (\mathbf{x}_1, \mathbf{x}_2, \ldots, \mathbf{x}_P)$, or equivalently take $\mathbf{X}$ to be the matrix with ijth element x_{ij}. And observe that for a function $\underline{\delta}(\cdot)$ (of $\mathbf{u}$) for which $\underline{\delta}(\mathbf{u})$ is expressible in the form (1.6), we have that

$$\underline{\delta}(\mathbf{u}_i) = \sum_{j=1}^{P} x_{ij} b_j \qquad (i = 1, 2, \ldots, N) \tag{1.7}$$

and that the N-dimensional column vector $\underline{\boldsymbol{\delta}}$ with elements $\underline{\delta}(\mathbf{u}_1), \underline{\delta}(\mathbf{u}_2), \ldots, \underline{\delta}(\mathbf{u}_N)$ is expressible as

$$\underline{\boldsymbol{\delta}} = b_1\mathbf{x}_1 + b_2\mathbf{x}_2 + \cdots + b_P\mathbf{x}_P = \mathbf{X}\mathbf{b}, \tag{1.8}$$

where $\mathbf{b} = (b_1, b_2, \ldots, b_P)'$.

In light of result (1.7), the assumption that condition (1.1) or (1.2) is satisfied by some (unidentified) function $\delta(\cdot)$ in Δ can be replaced by the assumption that for parameters $\beta_1, \beta_2, \ldots, \beta_P$ (which are the coefficients of a linear combination in Δ of unknown identity),

$$\text{E}(y_i) = \sum_{j=1}^{P} x_{ij}\beta_j \qquad (i = 1, 2, \ldots, N) \tag{1.9}$$

or, equivalently,

$$\text{E}(y_i - \sum_{j=1}^{P} x_{ij}\beta_j) = 0 \qquad (i = 1, 2, \ldots, N). \tag{1.10}$$

Upon letting $\boldsymbol{\beta} = (\beta_1, \beta_2, \ldots, \beta_P)'$ (which is a parameter vector of unknown value), condition (1.9) is reexpressible as

$$\text{E}(\mathbf{y}) = \mathbf{X}\boldsymbol{\beta}, \tag{1.11}$$

and condition (1.10) as

$$\mathrm{E}(\mathbf{y} - \mathbf{X}\boldsymbol{\beta}) = \mathbf{0}. \tag{1.12}$$

Moreover, the assumption that $\mathrm{var}(\mathbf{y} - \boldsymbol{\delta})$ or $\mathrm{var}(\mathbf{y})$ does not depend on $\boldsymbol{\delta}$ can be replaced by the assumption that $\mathrm{var}(\mathbf{y} - \mathbf{X}\boldsymbol{\beta})$ ($= \mathrm{var}\, \mathbf{y}$) does not depend on $\boldsymbol{\beta}$. [If $\mathrm{rank}(\mathbf{X}) < P$, then, strictly speaking, the assumption that $\mathrm{var}(\mathbf{y} - \mathbf{X}\boldsymbol{\beta})$ does not depend on $\boldsymbol{\beta}$ may be a "bit" stronger than the assumption that $\mathrm{var}(\mathbf{y} - \boldsymbol{\delta})$ does not depend on $\boldsymbol{\delta}$.]

As noted earlier, condition (1.3) or (1.4) is equivalent to condition (1.5). Similarly, condition (1.11) or (1.12) is equivalent to the condition

$$\mathbf{y} = \mathbf{X}\boldsymbol{\beta} + \mathbf{e}, \tag{1.13}$$

where [as in condition (1.5)] $\mathbf{e}$ is an N-dimensional random column vector with $\mathrm{E}(\mathbf{e}) = \mathbf{0}$. In nonmatrix notation, condition (1.13) is

$$y_i = \sum_{j=1}^{P} x_{ij}\beta_j + e_i \qquad (i = 1, 2, \ldots, N),$$

where (for $i = 1, 2, \ldots, N$) e_i is the ith element of $\mathbf{e}$ (which is a random variable with an expected value of 0). Under condition (1.13),

$$\mathbf{e} = \mathbf{y} - \mathbf{X}\boldsymbol{\beta} = \mathbf{y} - \mathrm{E}(\mathbf{y}).$$

Moreover, under condition (1.13), the assumption that $\mathrm{var}(\mathbf{y} - \mathbf{X}\boldsymbol{\beta})$ or $\mathrm{var}(\mathbf{y})$ does not depend on $\boldsymbol{\beta}$ is equivalent to an assumption that $\mathrm{var}(\mathbf{e})$ does not depend on $\boldsymbol{\beta}$.

In what follows, three progressively more general models (for $\mathbf{y}$) are defined. Each of these models starts with the assumption that $\mathbf{y}$ satisfies condition (1.13), and each is such that $\mathrm{var}(\mathbf{e})$ does not depend on $\boldsymbol{\beta}$. And, in each of them, $\boldsymbol{\beta}$ is assumed to be unrestricted; that is, the parameter space for $\boldsymbol{\beta}$ is taken to be the linear space $\mathbb{R}^P$ comprising all P-dimensional column vectors. A model with these characteristics is referred to as a *linear model*. (Depending on the nature of the restrictions, the model may be referred to as a linear model even if $\boldsymbol{\beta}$ is restricted to a subset of $\mathbb{R}^P$.)

In each of the three progressively more general models, the N-dimensional observable random column vector $\mathbf{y} = (y_1, y_2, \ldots, y_N)'$ is assumed to be such that

$$\mathbf{y} = \mathbf{X}\boldsymbol{\beta} + \mathbf{e}, \tag{1.14}$$

where $\mathbf{X}$ is the $N \times P$ nonrandom matrix with ijth element $x_{ij} = \delta_j(\mathbf{u}_i)$, where $\boldsymbol{\beta} = (\beta_1, \beta_2, \ldots, \beta_P)'$ is a P-dimensional column vector of unknown parameters, and where $\mathbf{e} = (e_1, e_2, \ldots, e_N)'$ is an N-dimensional random column vector with $\mathrm{E}(\mathbf{e}) = \mathbf{0}$. Or, equivalently, $\mathbf{y}$ is assumed to be such that

$$y_i = \sum_{j=1}^{P} x_{ij}\beta_j + e_i \qquad (i = 1, 2, \ldots, N). \tag{1.15}$$

The elements $e_1, e_2, \ldots, e_N$ of the vector $\mathbf{e}$ constitute what are sometimes (including herein) referred to as residual effects and sometimes referred to as errors.

When assumption (1.14) or (1.15) is coupled with an assumption that

$$\mathrm{var}(e_i) = \sigma^2 \text{ and } \mathrm{cov}(e_i, e_s) = 0 \qquad (s > i = 1, 2, \ldots, N), \tag{1.16}$$

or equivalently that

$$\mathrm{var}(\mathbf{e}) = \sigma^2\mathbf{I}, \tag{1.17}$$

we arrive at a model that is to be called the *Gauss–Markov model* or (for short) the G–M model. Here, σ is a (strictly) positive parameter that is functionally unrelated to $\boldsymbol{\beta}$; accordingly, the parameter space of the G–M model is

$$\{\boldsymbol{\beta}, \sigma \; : \; \boldsymbol{\beta} \in \mathcal{R}^P, \; \sigma > 0\}. \tag{1.18}$$

Assumption (1.16) or (1.17) indicates that $e_1, e_2, \ldots, e_N$ have a common standard deviation σ and variance σ^2 and that $e_1, e_2, \ldots, e_N$ are uncorrelated with each other.

A generalization of the G–M model can be obtained by replacing assumption (1.16) or (1.17) with the assumption that

$$\text{var}(\mathbf{e}) = \sigma^2 \mathbf{H}, \tag{1.19}$$

where $\mathbf{H} = \{h_{ij}\}$ is a symmetric nonnegative definite matrix (whose value is known). This generalization is to be called the *Aitken generalization* or the *Aitken model.* It is able to allow for the possibility of nonzero correlations and nonhomogeneous variances. However, it includes an implicit assumption that the correlations are known and that the variances are known up to a constant of proportionality, which limits its usefulness. In the special case where $\mathbf{H} = \mathbf{I}$, the Aitken model reduces to the G–M model.

A generalization of the G–M model that is considerably more flexible than the Aitken generalization is obtained by replacing assumption (1.16) or (1.17) with the assumption that

$$\text{var}(\mathbf{e}) = \mathbf{V}(\boldsymbol{\theta}), \tag{1.20}$$

where $\mathbf{V}(\boldsymbol{\theta})$ is an $N \times N$ symmetric nonnegative definite matrix with ijth element $v_{ij}(\boldsymbol{\theta})$ that (for $i, j = 1, 2, \ldots, N$) is functionally dependent on a T-dimensional (column) vector $\boldsymbol{\theta} = (\theta_1, \theta_2, \ldots, \theta_T)'$ of unknown parameters. Here, $\boldsymbol{\theta}$ belongs to a specified subset, say Θ, of $\mathcal{R}^T$. In the special case where $T = 1$, $\Theta = \{\boldsymbol{\theta} \; : \; \theta_1 > 0\}$, and $\mathbf{V}(\boldsymbol{\theta}) = \theta_1^2 \mathbf{H}$, assumption (1.20) reduces to what is essentially assumption (1.19) (with θ_1 in place of σ)—when $\mathbf{H} = \mathbf{I}$, there is a further reduction to what is essentially assumption (1.17). Accordingly, when the assumption that the observable random column vector $\mathbf{y}$ is such that $\mathbf{y} = \mathbf{X}\boldsymbol{\beta} + \mathbf{e}$ is coupled with assumption (1.20), we obtain what is essentially a further generalization of the Aitken generalization of the G–M model. This generalization, whose parameter space is implicitly taken to be

$$\{\boldsymbol{\beta}, \boldsymbol{\theta} \; : \; \boldsymbol{\beta} \in \mathcal{R}^P, \; \boldsymbol{\theta} \in \Theta\}, \tag{1.21}$$

is to be called the *general linear model.*

In effect, the G–M, Aitken, and general linear models serve to specify the form of the first-order moments and second-order (central) moments of the joint distribution of the observable random variables $y_1, y_2, \ldots, y_N$. In the case of the G–M model,

$$\text{E}(\mathbf{y}) = \mathbf{X}\boldsymbol{\beta} \quad \text{and} \quad \text{var}(\mathbf{y}) = \text{var}(\mathbf{e}) = \sigma^2 \mathbf{I}; \tag{1.22}$$

more generally (in the case of the Aitken model),

$$\text{E}(\mathbf{y}) = \mathbf{X}\boldsymbol{\beta} \quad \text{and} \quad \text{var}(\mathbf{y}) = \text{var}(\mathbf{e}) = \sigma^2 \mathbf{H}; \tag{1.23}$$

and still more generally (in the case of the general linear model),

$$\text{E}(\mathbf{y}) = \mathbf{X}\boldsymbol{\beta} \quad \text{and} \quad \text{var}(\mathbf{y}) = \text{var}(\mathbf{e}) = \mathbf{V}(\boldsymbol{\theta}). \tag{1.24}$$

Unlike the G–M model, the Aitken and general linear models are able to allow for the possibility that $y_1, y_2, \ldots, y_N$ are correlated and/or have nonhomogeneous variances, however (as in the special case of the G–M model) the variances and covariances of $y_1, y_2, \ldots, y_N$ are functionally unrelated to the expected values of $y_1, y_2, \ldots, y_N$.

Instead of parameterizing the G–M, Aitken, and general linear models in terms of the P-dimensional vector $\boldsymbol{\beta} = (\beta_1, \beta_2, \ldots, \beta_P)'$ of coefficients of the P specified functions $\delta_1(\cdot), \delta_2(\cdot), \ldots, \delta_P(\cdot)$, they can be parameterized in terms of an N-dimensional (column) vector

$\boldsymbol{\mu} = (\mu_1, \mu_2, \ldots, \mu_N)'$ that "corresponds" to the vector $\mathbf{X}\boldsymbol{\beta}$. In the alternative parameterization, the model equation (1.14) becomes

$$\mathbf{y} = \boldsymbol{\mu} + \mathbf{e}. \tag{1.25}$$

And the parameter space for the G–M model and its Aitken generalization becomes

$$\{\boldsymbol{\mu}, \sigma \ : \ \boldsymbol{\mu} \in \mathcal{C}(\mathbf{X}), \ \sigma > 0\}, \tag{1.26}$$

and that for the general linear model becomes

$$\{\boldsymbol{\mu}, \boldsymbol{\theta} \ : \ \boldsymbol{\mu} \in \mathcal{C}(\mathbf{X}), \ \boldsymbol{\theta} \in \Theta\}. \tag{1.27}$$

Further, specifying (in the case of the G–M model or its Aitken generalization) that $\mathrm{E}(\mathbf{y})$ and $\mathrm{var}(\mathbf{y})$ are expressible in the form (1.22) or (1.23) [subject to the restriction that $\boldsymbol{\beta}$ and σ are confined to the space (1.18)] is equivalent to specifying that they are expressible in the form

$$\mathrm{E}(\mathbf{y}) = \boldsymbol{\mu} \quad \text{and} \quad \mathrm{var}(\mathbf{y}) = \sigma^2\mathbf{I} \tag{1.28}$$

or

$$\mathrm{E}(\mathbf{y}) = \boldsymbol{\mu} \quad \text{and} \quad \mathrm{var}(\mathbf{y}) = \sigma^2\mathbf{H}, \tag{1.29}$$

respectively [subject to the restriction that $\boldsymbol{\mu}$ and σ are confined to the space (1.26)]. Similarly, specifying (in the case of the general linear model) that $\mathrm{E}(\mathbf{y})$ and $\mathrm{var}(\mathbf{y})$ are expressible in the form (1.24) [subject to the restriction that $\boldsymbol{\beta}$ and $\boldsymbol{\theta}$ are confined to the space (1.21)] is equivalent to specifying that they are expressible in the form

$$\mathrm{E}(\mathbf{y}) = \boldsymbol{\mu} \quad \text{and} \quad \mathrm{var}(\mathbf{y}) = \mathbf{V}(\boldsymbol{\theta}) \tag{1.30}$$

[subject to the restriction that $\boldsymbol{\mu}$ and $\boldsymbol{\theta}$ are confined to the space (1.27)].

The equation (1.14) is sometimes referred to as the *model equation*. And the $(N \times P)$ matrix $\mathbf{X}$, which appears in that equation, is referred to as the *model matrix*. What distinguishes one G–M, Aitken, or general linear model from another is the choice of the model matrix $\mathbf{X}$ and (in the case of the Aitken or general linear model) the choice of the matrix $\mathbf{H}$ or the choices made in the specification (up to the value of a vector of parameters) of the variance-covariance matrix $\mathbf{V}(\boldsymbol{\theta})$. Different choices for $\mathbf{X}$ are associated with different choices for the functions $\delta_1(\cdot), \delta_2(\cdot), \ldots, \delta_P(\cdot)$. The number (N) of rows of $\mathbf{X}$ is fixed; it must equal the number of observations. However, the number (P) of columns of $\mathbf{X}$ (and accordingly the dimension of the parameter vector $\boldsymbol{\beta}$) may vary from one choice of $\mathbf{X}$ to another.

The only assumptions inherent in the G–M, Aitken, or general linear model about the distribution of the vector $\mathbf{y}$ are those reflected in result (1.22), (1.23), or (1.24), which pertain to the first- and second-order moments. Stronger versions of these models can be obtained by making additional assumptions. One highly convenient and frequently adopted additional assumption is that of taking the distribution of the vector $\mathbf{e}$ to be MVN. In combination with assumption (1.17), (1.19), or (1.20), this assumption implies that $\mathbf{e} \sim N(\mathbf{0}, \sigma^2\mathbf{I})$, $\mathbf{e} \sim N(\mathbf{0}, \sigma^2\mathbf{H})$, or $\mathbf{e} \sim N[\mathbf{0}, \mathbf{V}(\boldsymbol{\theta})]$. Further, taking the distribution of $\mathbf{e}$ to be MVN implies (in light of Theorem 3.5.1) that the distribution of $\mathbf{y}$ is also MVN; specifically, it implies that $\mathbf{y} \sim N(\mathbf{X}\boldsymbol{\beta}, \sigma^2\mathbf{I})$, $\mathbf{y} \sim N(\mathbf{X}\boldsymbol{\beta}, \sigma^2\mathbf{H})$, or $\mathbf{y} \sim N[\mathbf{X}\boldsymbol{\beta}, \mathbf{V}(\boldsymbol{\theta})]$.

The credibility of the normality assumption varies with the nature of the application. Its credibility would seem to be greatest under circumstances in which each of the quantities $e_1, e_2, \ldots, e_N$ can reasonably be regarded as the sum of a large number of deviations of a similar magnitude. Under such circumstances, the central limit theorem may be "operative." The credibility of the basic assumptions inherent in the G–M, Aitken, and general linear models is affected by the choice of the functions $\delta_1(\cdot), \delta_2(\cdot), \ldots, \delta_P(\cdot)$. The credibility of these assumptions as well as that of the normality assumption can sometimes be enhanced via a transformation of the data.

The coverage in subsequent chapters (and in subsequent sections of the present chapter) includes numerous results on the G–M, Aitken, and general linear models and much discussion pertaining to those results and to various aspects of the models themselves. Except where otherwise indicated, the notation employed in that coverage is implicitly taken to be that employed in the present section (in the introduction of the models). And in connection with the G–M, Aitken, and general linear models, $\delta(\cdot)$ is subsequently taken to be the function (of $\mathbf{u}$) defined by $\delta(\mathbf{u}) = \sum_{j=1}^{P} \beta_j \delta_j(\mathbf{u})$.

In making statistical inferences on the basis of a statistical model, it is necessary to express the quantities of interest in terms related to the model. A G–M, Aitken, or general linear model is often used as a basis for making inferences about quantities that are expressible as linear combinations of $\beta_1, \beta_2, \ldots, \beta_P$, that is, ones that are expressible in the form $\boldsymbol{\lambda}'\boldsymbol{\beta}$, where $\boldsymbol{\lambda} = (\lambda_1, \lambda_2, \ldots, \lambda_P)'$ is a P-dimensional column vector of scalars, or equivalently ones that are expressible in the form $\sum_{j=1}^{P} \lambda_j \beta_j$. More generally, a G–M, Aitken, or general linear model is often used as a basis for making inferences about quantities that are expressible as random variables, each of which has an expected value that is a linear combination of $\beta_1, \beta_2, \ldots, \beta_P$; these quantities may or may not be correlated with $e_1, e_2, \ldots, e_N$ and/or with each other. And a general linear model is sometimes used as a basis for making inferences about quantities that are expressible as functions of $\boldsymbol{\theta}$ (or in the special case of a G–M or Aitken model, for making inferences about σ^2 or a function of σ^2).

4.2 Some Specific Types of Gauss–Markov Models (with Examples)

The different choices for a G–M model are associated with different choices for the functions $\delta_1(\mathbf{u}), \delta_2(\mathbf{u}), \ldots, \delta_P(\mathbf{u})$. These choices determine the form of the function $\delta(\mathbf{u})$; they determine this function up to the values of $\beta_1, \beta_2, \ldots, \beta_P$. In many cases, $\delta(\mathbf{u})$ is taken to be a polynomial in the elements of $\mathbf{u}$ (typically a polynomial of low degree). The appropriateness of taking $\delta(\mathbf{u})$ to be such a polynomial can sometimes be enhanced by introducing a transformation of the vector $\mathbf{u}$ and by redefining $\mathbf{u}$ accordingly [so that $\delta(\mathbf{u})$ is a polynomial in the elements of the transformed vector].

a. Polynomials (in 1 variable)

Suppose (in connection with the G–M model) that $C = 1$, in which case $\mathbf{u} = (u_1)$. And let us write u for the variable u_1 and also for the ($C = 1$)-dimensional vector $\mathbf{u}$. Further, suppose that for $j = 1, 2, \ldots, P$,

$$\delta_j(u) = (u - a)^{j-1},$$

where a is a specified value of u—by convention, $(u - a)^0 = 1$ for all u (including $u = a$). Then,

$$\delta(u) = \beta_1 + \beta_2(u - a) + \beta_3(u - a)^2 + \cdots + \beta_P(u - a)^{P-1}, \tag{2.1}$$

which is a polynomial in $u - a$ (and also in u) of degree $P - 1$.

Now, for $k = 1, 2, \ldots,$ write $\delta^{(k)}(u)$ for the kth-order derivative of $\delta(\cdot)$ at an arbitrary point u [in the interior of the domain of $\delta(\cdot)$]. Then, for $k = 2, 3, \ldots, P$,

$$\delta^{(k-1)}(a) = (k - 1)!\beta_k \tag{2.2}$$

[assuming that a is an interior point of the domain of $\delta(\cdot)$] and, more generally,

$$\delta^{(k-1)}(u) = (k - 1)!\beta_k + \sum_{j=k+1}^{P} (j - 1)(j - 2)\cdots(j - k + 1)(u - a)^{j-k}\beta_j, \tag{2.3}$$

as is easily verified. Thus, the derivatives of $\delta(\cdot)$ at the point a are scalar multiples of the parameters $\beta_2, \beta_3, \ldots, \beta_P$—the parameter β_1 is the value $\delta(a)$ of $\delta(u)$ at $u = a$. And the $(k-1)$th derivative of $\delta(\cdot)$ at an arbitrary interior point u is a linear combination of $\beta_k, \beta_{k+1}, \ldots, \beta_P$.

TABLE 4.1. Lethal doses of ouabain in cats for each of four rates of injection.

Rate	Lethal doses for individual cats
1	5, 9, 11, 13, 14, 16, 17, 20, 22, 28, 31, 31
2	3, 6, 22, 27, 27, 28, 28, 37, 40, 42, 50
4	34, 34, 38, 40, 46, 58, 60, 60, 65
8	51, 56, 62, 63, 70, 73, 76, 89, 92

Taking $\delta(u)$ to be a relatively low-degree polynomial is most likely to be satisfactory in a situation where $\mathbf{u}_1, \mathbf{u}_2, \ldots, \mathbf{u}_N$ (and other "relevant" values of u) are confined to a range that is not overly large. Taylor's theorem (e.g., Bartle 1976, sec. 28; Bartle and Sherbert 2011, sec. 6.4) indicates that, over a limited range, low-degree polynomials provide "reasonable" approximations to a broad class of functions.

In the special case $P = 2$, expression (2.1) simplifies to

$$\delta(u) = \beta_1 + \beta_2(u - a), \tag{2.4}$$

and is reexpressible as

$$\delta(u) = (\beta_1 - \beta_2 a) + \beta_2 u. \tag{2.5}$$

In this special case, $\delta(u)$ is a polynomial of degree 1, and the model is sometimes referred to as first-order. When $a = 0$, expressions (2.4) and (2.5) reduce to

$$\delta(u) = \beta_1 + \beta_2 u. \tag{2.6}$$

There are cases where $\delta(u)$ is not of the form (2.1) but where $\delta(u)$ can be reexpressed in that form by introducing a transformation of u and by redefining u accordingly. Suppose, for example, that u is strictly positive and that

$$\delta(u) = \beta_1 + \beta_2 \log u \tag{2.7}$$

(where β_1 and β_2 are parameters). Clearly, expression (2.7) is not a polynomial in u. However, upon introducing a transformation from u to the variable u^* defined by $u^* = \log u$, $\delta(u)$ can be reexpressed as a function of u^*; specifically, it can be reexpressed in terms of the function $\delta^*(\cdot)$ defined by

$$\delta^*(u^*) = \beta_1 + \beta_2 u^*.$$

Thus, when u is redefined to be u^*, we can take $\delta(u)$ to be of the form

$$\delta(u) = \beta_1 + \beta_2 u.$$

b. Example: ouabain data

Snedecor and Cochran (1989) published data on the dose of ouabain that proves to be lethal when injected intravenously into a cat. To learn how the lethal dose is affected by the rate of injection, each of 41 cats was injected at one of four rates. The data are reproduced in Table 4.1 (with dose and rate being recorded in the same units and same way as by Snedecor and Cochran).

We could consider applying to these data a G–M model in which $N = 41$, in which $y_1, y_2, \ldots, y_N$ are the observable random variables whose values are the lethal doses (or perhaps the logarithms of the lethal doses), in which $C = 1$, and (adopting the same notation for the special case $C = 1$ as in Subsection a) in which u is the rate of injection (or perhaps the logarithm of the rate). And we could consider taking $\delta(u)$ to be the $(P-1)$-degree polynomial (2.1). Different choices for P correspond to different versions of the model, more than one of which may be worthy of consideration—the design of the study that gave rise to these data suggests that a choice for P of no more than 4 would have been regarded as adequate. The choice of the value a in expression (2.1) is more or less a matter of convenience.

Now, let us consider (for purposes of illustration) the matrix representation $\mathbf{y} = \mathbf{X}\boldsymbol{\beta} + \mathbf{e}$ of the observable random column vector $\mathbf{y}$ in an application of the G–M model in which the observed value of $\mathbf{y}$ comprises the lethal doses, in which $C = 1$, in which u is the rate of injection, and in which $\delta(u)$ is the special case of the $(P-1)$-degree polynomial (2.1) (in $u - a$) obtained by setting $P = 5$ and $a = 0$. There are $N = 41$ data points. Suppose that the data points are numbered $1, 2, \ldots, 41$ by proceeding row by row in Table 4.1 from the top to the bottom and by proceeding from left to right within each row. Then, $\mathbf{y}' = (\mathbf{y}_1', \mathbf{y}_2', \mathbf{y}_3', \mathbf{y}_4')$, where

$$\begin{aligned}
\mathbf{y}_1' &= (y_1, y_2, y_3, y_4, y_5, y_6, y_7, y_8, y_9, y_{10}, y_{11}, y_{12}) \\
&= (5, 9, 11, 13, 14, 16, 17, 20, 22, 28, 31, 31), \\
\mathbf{y}_2' &= (y_{13}, y_{14}, y_{15}, y_{16}, y_{17}, y_{18}, y_{19}, y_{20}, y_{21}, y_{22}, y_{23}) \\
&= (3, 6, 22, 27, 27, 28, 28, 37, 40, 42, 50), \\
\mathbf{y}_3' &= (y_{24}, y_{25}, y_{26}, y_{27}, y_{28}, y_{29}, y_{30}, y_{31}, y_{32}) \\
&= (34, 34, 38, 40, 46, 58, 60, 60, 65), \quad \text{and} \\
\mathbf{y}_4' &= (y_{33}, y_{34}, y_{35}, y_{36}, y_{37}, y_{38}, y_{39}, y_{40}, y_{41}) \\
&= (51, 56, 62, 63, 70, 73, 76, 89, 92).
\end{aligned}$$

Further, $\boldsymbol{\beta}' = (\beta_1, \beta_2, \beta_3, \beta_4, \beta_5)$. Because the data are arranged in groups, with each group having a common value of u (corresponding to a common rate of injection), the model matrix $\mathbf{X}$ has a succinct representation. Specifically, this matrix, whose ijth element x_{ij} corresponds to the ith datum and to the jth element β_j of $\boldsymbol{\beta}$, is given by

$$\mathbf{X} = \begin{pmatrix} \mathbf{1}_{12} & \mathbf{1}_{12} & \mathbf{1}_{12} & \mathbf{1}_{12} & \mathbf{1}_{12} \\ \mathbf{1}_{11} & 2\mathbf{1}_{11} & 4\mathbf{1}_{11} & 8\mathbf{1}_{11} & 16\mathbf{1}_{11} \\ \mathbf{1}_9 & 4\mathbf{1}_9 & 16\mathbf{1}_9 & 64\mathbf{1}_9 & 256\mathbf{1}_9 \\ \mathbf{1}_9 & 8\mathbf{1}_9 & 64\mathbf{1}_9 & 512\mathbf{1}_9 & 4{,}096\mathbf{1}_9 \end{pmatrix}.$$

c. Polynomials (in general)

Let us now extend (in connection with the G–M model) the development in Subsection a (which pertains to the special case where $C = 1$) to the general case where C is possibly greater than 1. Taking $a_1, a_2, \ldots, a_C$ to be specified values of $u_1, u_2, \ldots, u_C$, respectively, suppose that, for $j = 1, 2, \ldots, P$,

$$\delta_j(\mathbf{u}) = (u_1 - a_1)^{k_{j1}} (u_2 - a_2)^{k_{j2}} \cdots (u_C - a_C)^{k_{jC}}, \tag{2.8}$$

where $k_{j1}, k_{j2}, \ldots, k_{jC}$ are nonnegative integers. To avoid trivialities (in the form of duplicative expressions), suppose that, for $s > j = 1, 2, \ldots, P$, $k_{st} \neq k_{jt}$ for one or more values of t. The setting is such that the function $\delta(\mathbf{u})$ is a polynomial (in $u_1 - a_1, u_2 - a_2, \ldots, u_C - a_C$), the coefficients of which are $\beta_1, \beta_2, \ldots, \beta_P$. The jth term of this polynomial is of degree $\sum_{t=1}^{C} k_{jt}$, and the polynomial itself is of degree $\max_j \sum_{t=1}^{C} k_{jt}$. Often, the first term is of degree 0, that is, the first term equals β_1.

A simple (but important) special case is that where $P = C + 1$ and where $k_{11} = k_{12} = \cdots = k_{1C} = 0$ and, for $j = 2, 3, \ldots, C+1$,

$$k_{jt} = \begin{cases} 1 & \text{for } t = j - 1, \\ 0 & \text{for } t \neq j - 1. \end{cases} \tag{2.9}$$

In that special case,

$$\delta(\mathbf{u}) = \beta_1 + \beta_2(u_1 - a_1) + \beta_3(u_2 - a_2) + \cdots + \beta_{C+1}(u_C - a_C), \tag{2.10}$$

which is a "polynomial" of degree 1, and the model is sometimes referred to as first-order. When $a_1 = a_2 = \cdots = a_C = 0$, expression (2.10) reduces to

$$\delta(\mathbf{u}) = \beta_1 + \beta_2 u_1 + \beta_3 u_2 + \cdots + \beta_{C+1} u_C. \tag{2.11}$$

Another special case worthy of mention is that where $P = C + 1 + [C(C + 1)/2]$, where (as in the previous special case) $k_{11} = k_{12} = \cdots = k_{1C} = 0$ and, for $j = 2, 3, \ldots, C + 1$, $k_{j1}, k_{j2}, \ldots, k_{jC}$ are given by expression (2.9), and where, for $j = C+2, C+3, \ldots, C+1+[C(C+1)/2]$, $k_{j1}, k_{j2}, \ldots, k_{jC}$ are such that $\sum_{t=1}^{C} k_{jt} = 2$. In that special case, $\delta(\mathbf{u})$ is a polynomial of degree 2. That polynomial is obtainable from the degree-1 polynomial (2.10) by adding $C(C+1)/2$ terms of degree 2. Each of these degree-2 terms is expressible in the form

$$\beta_j(u_t - a_t)(u_{t'} - a_{t'}), \tag{2.12}$$

where j is an integer between $C + 2$ and $C + 1 + [C(C + 1)/2]$, inclusive, and where t is an integer between 1 and C, inclusive, and t' an integer between t and C, inclusive—j, t, and t' are such that either $k_{jt} = 2$ and $t' = t$ or $k_{jt} = k_{jt'} = 1$ and $t' > t$. It is convenient to express the degree-2 terms in the form (2.12) and to adopt a modified notation for the coefficients $\beta_{C+2}, \beta_{C+3}, \ldots, \beta_{C+1+[C(C+1)/2]}$ in which the coefficient in term (2.12) is identified by the values of t and t', that is, in which $\beta_{tt'}$ is written in place of β_j. Accordingly,

$$\delta(\mathbf{u}) = \beta_1 + \sum_{j=1}^{C} \beta_{j+1}(u_j - a_j) + \sum_{t=1}^{C}\sum_{t'=t}^{C} \beta_{tt'}(u_t - a_t)(u_{t'} - a_{t'}). \tag{2.13}$$

When $\delta(\mathbf{u})$ is of the form (2.13), the model is sometimes referred to as second-order. When $a_1 = a_2 = \cdots = a_C = 0$, expression (2.13) reduces to

$$\delta(\mathbf{u}) = \beta_1 + \sum_{j=1}^{C} \beta_{j+1} u_j + \sum_{t=1}^{C}\sum_{t'=t}^{C} \beta_{tt'} u_t u_{t'}. \tag{2.14}$$

d. Example: cement data

Hald (1952, p. 647) published data from an experimental investigation of how the heat that evolves during the hardening of cement varies with the respective amounts (in the clinkers from which the cement is produced) of the following four compounds: tricalcium aluminate, tricalcium silicate, tetracalcium aluminoferrite, and β-dicalcium silicate. Heat was measured as of 180 days and recorded in units of calories per gram of cement, and the amount of each of the four compounds was recorded as a percentage of the weight of the clinkers. This process was carried out for each of 13 batches of cement. These data, the original source of which was Table I of Woods, Steinour, and Starke (1932), are reproduced in Table 4.2. They could possibly be regarded as suitable for the application of a G–M model in which $N = 13$, in which the value of $\mathbf{y}$ comprises the 13 heat measurements, in which $C = 4$, and in which u_1, u_2, u_3, and u_4 are the respective amounts of the first through fourth compounds. Among the various versions of the model that might be applied to the cement data is the first-order model, in which $\delta(\mathbf{u})$ is expressible in the form (2.10) or (2.11).

e. Example: lettuce data

Hader, Harward, Mason, and Moore (1957) reported the results of an experimental study of how the yield of lettuce plants is affected by the amounts of three trace minerals: copper (Cu), molybdenum

TABLE 4.2. The heat evolved during hardening and the respective amounts of four compounds for each of 13 batches of cement (Hald 1952, p. 647).

Heat evolved during hardening	Amount of tricalcium aluminate	Amount of tricalcium silicate	Amount of tetracalcium aluminoferrite	Amount of β-dicalcium silicate
78.5	7	26	6	60
74.3	1	29	15	52
104.3	11	56	8	20
87.6	11	31	8	47
95.9	7	52	6	33
109.2	11	55	9	22
102.7	3	71	17	6
72.5	1	31	22	44
93.1	2	54	18	22
115.9	21	47	4	26
83.8	1	40	23	34
113.3	11	66	9	12
109.4	10	68	8	12

(Mo), and iron (Fe). The plants were grown in a medium in containers of 3 plants each. Each container was assigned one of 5 levels of each of the 3 trace minerals; for Fe, the lowest and highest levels were 0.0025 ppm and 25 ppm, and for both Cu and Mo, the lowest and highest levels were 0.0002 ppm and 2 ppm. The 5 levels of each mineral were reexpressed on a transformed scale: the ppm were replaced by a linear function of the logarithm of the ppm chosen so that the transformed values of the highest and lowest levels were $\pm\sqrt[4]{8}$. Yield was recorded as grams of dry weight. The results of the experimental study are reproduced in Table 4.3.

A G–M model could be applied to these data. We could take $N = 20$, take the values of $y_1, y_2, \ldots, y_N$ to be the yields, take $C = 3$, and take u_1, u_2, and u_3 to be the transformed amounts of Cu, Mo, and Fe, respectively. And we could consider taking $\delta(\mathbf{u})$ to be a polynomial (in u_1, u_2, and u_3). The levels of Cu, Mo, and Fe represented in the study covered what was considered to be a rather wide range. Accordingly, the first-order model, in which $\delta(\mathbf{u})$ is taken to be the degree-1 polynomial (2.10) or (2.11), is not suitable. The second-order model, in which $\delta(\mathbf{u})$ is taken to be the degree-2 polynomial (2.13) or (2.14), would seem to be a much better choice.

4.3 Regression

Suppose there are N data points $y_1, y_2, \ldots, y_N$ and that (for $i = 1, 2, \ldots, N$) y_i is accompanied by the corresponding value $\mathbf{u}_i$ of a vector $\mathbf{u} = (u_1, u_2, \ldots, u_C)'$ of C explanatory variables. Further, regard $y_1, y_2, \ldots, y_N$ as N values of a variable y. In some applications, both $\mathbf{u}$ and y can reasonably be regarded as random, that is, the $(C+1)$-dimensional column vector $\begin{pmatrix} y \\ \mathbf{u} \end{pmatrix}$ can reasonably be regarded as a random vector; and the $(C+1)$-dimensional vectors $\begin{pmatrix} y_1 \\ \mathbf{u}_1 \end{pmatrix}, \begin{pmatrix} y_2 \\ \mathbf{u}_2 \end{pmatrix}, \ldots, \begin{pmatrix} y_N \\ \mathbf{u}_N \end{pmatrix}$ can reasonably be regarded as a random sample of size N from the distribution of $\begin{pmatrix} y \\ \mathbf{u} \end{pmatrix}$. Assume that $\begin{pmatrix} y \\ \mathbf{u} \end{pmatrix}$ and $\begin{pmatrix} y_1 \\ \mathbf{u}_1 \end{pmatrix}, \begin{pmatrix} y_2 \\ \mathbf{u}_2 \end{pmatrix}, \ldots, \begin{pmatrix} y_N \\ \mathbf{u}_N \end{pmatrix}$ can be so regarded. Then, in effect, $\begin{pmatrix} y_1 \\ \mathbf{u}_1 \end{pmatrix}, \begin{pmatrix} y_2 \\ \mathbf{u}_2 \end{pmatrix}, \ldots, \begin{pmatrix} y_N \\ \mathbf{u}_N \end{pmatrix}$

TABLE 4.3. Yields of lettuce from pots containing various amounts of Cu, Mo, and Fe (Hader, Harward, Mason, and Moore 1957, p. 63; Moore, Harward, Mason, Hader, Lott, and Jackson 1957, p. 67).

	Transformed amount		
Yield	Cu	Mo	Fe
21.42	-1	-1	-0.4965
15.92	1	-1	-0.4965
22.81	-1	1	-0.4965
14.90	-1	-1	1
14.95	1	1	-0.4965
7.83	1	-1	1
19.90	-1	1	1
4.68	1	1	1
0.20	$\sqrt[4]{8}$	0	0
17.65	$-\sqrt[4]{8}$	0	0
18.16	0	$\sqrt[4]{8}$	0
25.39	0	$-\sqrt[4]{8}$	0
11.99	0	0	$\sqrt[4]{8}$
7.37	0	0	$-\sqrt[4]{8}$
22.22	0	0	0
19.49	0	0	0
22.76	0	0	0
24.27	0	0	0
27.88	0	0	0
27.53	0	0	0

are the realizations of N $(C+1)$-dimensional statistically independent random vectors, each of which is distributed identically to $\begin{pmatrix} y \\ \mathbf{u} \end{pmatrix}$. In what follows, let us (for the sake of simplicity and as a matter of convenience) use the same notation for these random vectors as for their realizations.

Let us consider the distribution of $y_1, y_2, \ldots, y_N$ conditional on $\mathbf{u}_1, \mathbf{u}_2, \ldots, \mathbf{u}_N$, doing so with the ultimate objective of establishing a connection to the G–M model. Take $\delta(\cdot)$ to be the function (of $\mathbf{u}$) defined by

$$\delta(\mathbf{u}) = \mathrm{E}(y \mid \mathbf{u}).$$

And observe that (with probability 1)

$$\mathrm{E}[y - \delta(\mathbf{u}) \mid \mathbf{u}] = 0. \tag{3.1}$$

Further, take $v(\cdot)$ to be the function (of $\mathbf{u}$) defined by

$$v(\mathbf{u}) = \mathrm{var}[y - \delta(\mathbf{u}) \mid \mathbf{u}]$$

or, equivalently, by $v(\mathbf{u}) = \mathrm{var}(y \mid \mathbf{u})$. The nature of the functions $\delta(\cdot)$ and $v(\cdot)$ depends on the nature of the distribution of the vector $\begin{pmatrix} y \\ \mathbf{u} \end{pmatrix}$.

Corresponding to the function $\delta(\cdot)$ is the decomposition

$$y_i = \delta(\mathbf{u}_i) + e_i,$$

where $e_i = y_i - \delta(\mathbf{u}_i)$ $(i = 1, 2, \ldots, N)$. That $\begin{pmatrix} y_1 \\ \mathbf{u}_1 \end{pmatrix}, \begin{pmatrix} y_2 \\ \mathbf{u}_2 \end{pmatrix}, \ldots, \begin{pmatrix} y_N \\ \mathbf{u}_N \end{pmatrix}$ are distributed independently implies that, conditionally on $\mathbf{u}_1, \mathbf{u}_2, \ldots, \mathbf{u}_N$ (as well as unconditionally), $e_1, e_2, \ldots, e_N$ are distributed independently. A further implication is that (for $i = 1, 2, \ldots, N$) a conditional distribution of e_i given $\mathbf{u}_i$ is a conditional distribution of e_i given $\mathbf{u}_1, \mathbf{u}_2, \ldots, \mathbf{u}_N$. Thus, conditionally on $\mathbf{u}_1, \mathbf{u}_2, \ldots, \mathbf{u}_N$, we have [in light of result (3.1) and the very definition of the function $v(\cdot)$] that

$$\mathrm{E}(e_i) = 0 \quad \text{and} \quad \mathrm{var}(e_i) = v(\mathbf{u}_i) \qquad (i = 1, 2, \ldots, N)$$

and we also have that

$$\mathrm{cov}(e_i, e_s) = 0 \qquad (s > i = 1, 2, \ldots, N).$$

Upon defining $\mathbf{e} = (e_1, e_2, \ldots, e_N)'$, these results can be restated in matrix notation. Conditionally on $\mathbf{u}_1, \mathbf{u}_2, \ldots, \mathbf{u}_N$, we have that

$$\mathrm{E}(\mathbf{e}) = \mathbf{0} \quad \text{and} \quad \mathrm{var}(\mathbf{e}) = \mathrm{diag}[v(\mathbf{u}_1), v(\mathbf{u}_2), \ldots, v(\mathbf{u}_N)].$$

Let us now specialize to the case where the $(C+1)$-dimensional vector $\begin{pmatrix} y \\ \mathbf{u} \end{pmatrix}$ has an MVN distribution (with a nonsingular variance-covariance matrix). In that special case, we have (in light of the results of Section 3.51) that the distribution of y conditional on $\mathbf{u}$ is normal with mean

$$\mathrm{E}(y \mid \mathbf{u}) = \mathrm{E}(y) + \mathrm{cov}(y, \mathbf{u})(\mathrm{var}\,\mathbf{u})^{-1}[\mathbf{u} - \mathrm{E}(\mathbf{u})] \tag{3.2}$$

and variance

$$\mathrm{var}(y \mid \mathbf{u}) = \mathrm{var}(y) - \mathrm{cov}(y, \mathbf{u})(\mathrm{var}\,\mathbf{u})^{-1}\mathrm{cov}(\mathbf{u}, y). \tag{3.3}$$

Further, based on result (3.2), we have that

$$\delta(\mathbf{u}) = \beta_1 + \beta_2(u_1 - a_1) + \beta_3(u_2 - a_2) + \cdots + \beta_{C+1}(u_C - a_C), \tag{3.4}$$

where $a_1, a_2, \ldots, a_C$ are arbitrarily specified values of $u_1, u_2, \ldots, u_C$, respectively, where

$$(\beta_2, \beta_3, \ldots, \beta_{C+1}) = \mathrm{cov}(y, \mathbf{u})(\mathrm{var}\,\mathbf{u})^{-1},$$

and where

$$\beta_1 = \mathrm{E}(y) + \sum_{j=2}^{C+1} \beta_j [a_{j-1} - \mathrm{E}(u_{j-1})].$$

[If (for $j = 1, 2, \ldots, C$) $a_j = \mathrm{E}(u_j)$, then $\beta_1 = \mathrm{E}(y)$; if $a_1 = a_2 = \cdots = a_C = 0$, then $\beta_1 = \mathrm{E}(y) - \sum_{j=2}^{C+1} \beta_j\, \mathrm{E}(u_{j-1})$.] And, based on result (3.3), we have that

$$v(\mathbf{u}) = \sigma^2, \tag{3.5}$$

where

$$\sigma^2 = \mathrm{var}(y) - \mathrm{cov}(y, \mathbf{u})(\mathrm{var}\,\mathbf{u})^{-1}\mathrm{cov}(\mathbf{u}, y).$$

We conclude that when the distribution of the vector $\begin{pmatrix} y \\ \mathbf{u} \end{pmatrix}$ is MVN, the joint distribution of $y_1, y_2, \ldots, y_N$ obtained by regarding $\begin{pmatrix} y_1 \\ \mathbf{u}_1 \end{pmatrix}, \begin{pmatrix} y_2 \\ \mathbf{u}_2 \end{pmatrix}, \ldots, \begin{pmatrix} y_N \\ \mathbf{u}_N \end{pmatrix}$ as a random sample from the distribution of $\begin{pmatrix} y \\ \mathbf{u} \end{pmatrix}$ and by conditioning on $\mathbf{u}_1, \mathbf{u}_2, \ldots, \mathbf{u}_N$ is identical to that obtained by adopting a first-order G–M model and by taking the joint distribution of the residual effects $e_1, e_2, \ldots, e_N$ of the G–M model to be MVN. Moreover, when the distribution of $\begin{pmatrix} y \\ \mathbf{u} \end{pmatrix}$ is MVN, the parameters $\beta_1, \beta_2, \ldots, \beta_{C+1}$ and σ of that first-order G–M model are expressible in terms of the mean vector and the variance-covariance matrix of $\begin{pmatrix} y \\ \mathbf{u} \end{pmatrix}$.

There are other distributions of $\begin{pmatrix} y \\ \mathbf{u} \end{pmatrix}$ (besides the MVN) for which the joint distribution of $y_1, y_2, \ldots, y_N$ obtained by regarding $\begin{pmatrix} y_1 \\ \mathbf{u}_1 \end{pmatrix}, \begin{pmatrix} y_2 \\ \mathbf{u}_2 \end{pmatrix}, \ldots, \begin{pmatrix} y_N \\ \mathbf{u}_N \end{pmatrix}$ as a random sample from the distribution of $\begin{pmatrix} y \\ \mathbf{u} \end{pmatrix}$ and by conditioning on $\mathbf{u}_1, \mathbf{u}_2, \ldots, \mathbf{u}_N$ is consistent with the adoption of a first-order G–M model. Whether or not the joint distribution obtained in that way is consistent with the adoption of a first-order G–M model depends on the nature of the distribution of $\begin{pmatrix} y \\ \mathbf{u} \end{pmatrix}$ solely through the nature of the conditional distribution of y given $\mathbf{u}$; in fact, it depends solely on the nature of the mean and variance of the conditional distribution of y given $\mathbf{u}$. The nature of the marginal distribution of $\mathbf{u}$ is without relevance (to that particular issue).

It is instructive to consider the implications of the expression for $\mathrm{E}(y \mid \mathbf{u})$ given by equation (3.2). Suppose that $C = 1$. Then, writing u for u_1, equation (3.2) can be reexpressed in the form

$$\frac{\mathrm{E}(y \mid u) - \mathrm{E}(y)}{\sqrt{\mathrm{var}\, y}} = \mathrm{corr}(y, u) \frac{u - \mathrm{E}(u)}{\sqrt{\mathrm{var}\, u}}. \tag{3.6}$$

Excluding the limiting case where $|\mathrm{corr}(y, u)| = 1$, there is an implication that, for any particular value of u, $\mathrm{E}(y \mid u)$ is less extreme than u in the sense that (in units of standard deviations) it is closer to $\mathrm{E}(y)$ than u is to $\mathrm{E}(u)$.

Roughly speaking, observations on y corresponding to any particular value of u are on average less extreme than the value of u. This phenomenon was recognized early on by Sir Francis Galton, who determined (from data on a human population) that the heights of the offspring of very tall (or very short) parents, while typically above (or below) average, tend to be less extreme than the heights of the parents. It is a phenomenon that has come to be known as "regression to the mean" or simply as regression. This term evolved from the term "regression (or reversion) towards mediocrity" introduced by Galton.

Some authors reserve the use of the term regression for situations (like that under consideration in the present section) where the explanatory variables can reasonably be regarded as realizations of random variables (e.g., Graybill 1976; Rao 1973). This would seem to be more or less in keeping with the original meaning of the term. However, over time, the term regression has come to be used much more broadly. In particular, it has become common to use the term linear regression almost synonymously with what is being referred to herein as the G–M model, with the possible proviso that the explanatory variables be continuous. This broader usage is inclusive enough to cover a study (like that which produced the lettuce data) where the values of the explanatory variables have been determined systematically as part of a designed experiment.

4.4 Heteroscedastic and Correlated Residual Effects

In the G–M model, the residual effects $e_1, e_2, \ldots, e_N$ are regarded as random variables that, in addition to having expected values of 0, are assumed to be homoscedastic (i.e., to have the same variance) and to be uncorrelated. There are applications for which these assumptions are unrealistic. The Aitken generalization of the G–M model and (to a considerably greater extent) the general linear model are more flexible. The residual effects in the Aitken generalization and in the general linear model can be heteroscedastic or correlated (or both).

There are certain types of heteroscedasticity and certain correlation patterns that are relatively common and that can be readily accommodated within the framework of the Aitken generalization or the general linear model. In the present section, an attempt is made to identify and discuss some of the more basic types of heteroscedasticity and some of the more basic correlation patterns.

In the Aitken generalization or the general linear model, the variance-covariance matrix of the vector $\mathbf{e}$ of residual effects is of the form $\sigma^2\mathbf{H}$ or $\mathbf{V}(\boldsymbol{\theta})$, respectively. By definition, the elements h_{ij} $(i, j = 1, 2, \ldots, N)$ of $\mathbf{H}$ are known constants, and the elements $v_{ij}(\boldsymbol{\theta})$ $(i, j = 1, 2, \ldots, N)$ of $\mathbf{V}(\boldsymbol{\theta})$ are known functions of the parameter vector $\boldsymbol{\theta}$. These constants or functions may depend on the N values $\mathbf{u}_1, \mathbf{u}_2, \ldots, \mathbf{u}_N$ of the vector $\mathbf{u} = (u_1, u_2, \ldots, u_C)'$ of explanatory variables, though any such dependence is suppressed in the notation.

a. Heteroscedastic residual effects

There are situations where the residual effects $e_1, e_2, \ldots, e_N$ in the model equation (1.14) or equations (1.15) can reasonably be regarded as uncorrelated, but cannot reasonably be regarded as homoscedastic (at least not completely so). Three situations of this sort are as follows.

Group averages. Suppose that $y_1, y_2, \ldots, y_N$ follow a G–M model. Suppose also that, among the N rows of the model matrix $\mathbf{X}$, there are only K distinct rows, say the $i_1, i_2, \ldots, i_K$th rows, $(x_{i_k1}, x_{i_k2}, \ldots, x_{i_kP})$ $(k = 1, 2, \ldots, K)$. And, for $k = 1, 2, \ldots, K$, let I_k represent the subset of the integers $1, 2, \ldots, N$ such that $i \in I_k$ if the ith row of $\mathbf{X}$ equals $(x_{i_k1}, x_{i_k2}, \ldots, x_{i_kP})$, and denote by N_k the size of this subset. Further, define $\bar{y}_k = (1/N_k)\sum_{i \in I_k} y_i$.

Now, suppose that the individual observations $y_1, y_2, \ldots, y_N$ are discarded, but that the averages $\bar{y}_1, \bar{y}_2, \ldots, \bar{y}_K$ are retained—if the residual effects are jointly normal, then it could be argued, on the grounds of sufficiency, that there is no need to retain anything other than $\bar{y}_1, \bar{y}_2, \ldots, \bar{y}_K$ and the sum of squares $\sum_{i=1}^N y_i^2$. And suppose that $\bar{y}_1, \bar{y}_2, \ldots, \bar{y}_K$ are then regarded as the data. It is a simple exercise to show that, for $k = 1, 2, \ldots, K$, $\mathrm{E}(\bar{y}_k) = \sum_{j=1}^P x_{i_kj}\beta_j$ and $\mathrm{var}(\bar{y}_k) = \sigma^2/N_k$ and that, for $k' \neq k = 1, 2, \ldots, K$, $\mathrm{cov}(\bar{y}_k, \bar{y}_{k'}) = 0$. Thus, $\bar{y}_1, \bar{y}_2, \ldots, \bar{y}_K$ follow an Aitken generalization of a G–M model in which, taking $\mathbf{y}$ to be the K-dimensional vector $(\bar{y}_1, \bar{y}_2, \ldots, \bar{y}_K)'$, the model matrix is the $K \times P$ matrix whose kth row is the i_kth row of the original model matrix, and the matrix $\mathbf{H}$ is the diagonal matrix $\mathrm{diag}(1/N_1, 1/N_2, \ldots, 1/N_K)$—the parameters $\beta_1, \beta_2, \ldots, \beta_P$ and σ of this model are "identical" to those of the original (G–M) model (i.e., the model for the individual observations $y_1, y_2, \ldots, y_N$).

As a simple example, consider the ouabain data of Section 4.2b. Suppose that the lethal doses for the 41 cats follow a G–M model in which the rate of injection is the sole explanatory variable and in which $\delta(\cdot)$ is a polynomial. Then, the model matrix has 4 distinct rows, corresponding to the 4 rates of injection: 1, 2, 4, and 8. The numbers of cats injected at those 4 rates were 12, 11, 9, and 9, respectively. The average lethal doses for the 4 rates would follow an Aitken model, with a model matrix that has 4 rows (which are the distinct rows of the original model matrix and whose lengths and entries depend on the choice of polynomial) and with $\mathbf{H} = \mathrm{diag}(1/12, 1/11, 1/9, 1/9)$.

Within-group homoscedasticity. There are situations where the residual effects $e_1, e_2, \ldots, e_N$ cannot reasonably be regarded as homoscedastic, but where they can be partitioned into some (hopefully modest) number of mutually exclusive and exhaustive subsets or "groups," each of which consists of residual effects that are thought to be homoscedastic. Suppose there are K such groups and that (for purposes of identification) they are numbered $1, 2, \ldots, K$. And (for $k = 1, 2, \ldots, K$) denote by I_k the subset of the integers $1, 2, \ldots, N$ defined by $i \in I_k$ if the ith residual effect e_i is a member of the kth group.

Let us assume the existence of a function, say $\phi(\mathbf{u})$, of $\mathbf{u}$ (the vector of explanatory variables) whose value $\phi(\mathbf{u}_i)$ for $\mathbf{u} = \mathbf{u}_i$ is as follows: $\phi(\mathbf{u}_i) = k$ if $i \in I_k$. This assumption entails essentially no loss of generality. If necessary, it can be satisfied by introducing an additional explanatory variable. In particular, it can be satisfied by including an explanatory variable whose ith value (the value of the explanatory variable corresponding to y_i) equals k for every $i \in I_k$ $(k = 1, 2, \ldots, K)$. It is worth noting that there is nothing in the formulation of the G–M model (or its Aitken generalization) or in the general linear model requiring that $\delta(\mathbf{u})$ (whose values for $\mathbf{u} = \mathbf{u}_1, \mathbf{u}_2, \ldots, \mathbf{u}_N$ are the expected values of $y_1, y_2, \ldots, y_N$) depend nontrivially on every component of $\mathbf{u}$.

Consider, for example, the case of the ouabain data. For those data, we could conceivably define 4 groups of residual effects, corresponding to the 4 rates of injection, and assume that the residual effects are homoscedastic within a group but (contrary to what is inherent in the G–M model) not necessarily homoscedastic across groups. Then, assuming (as before) that the rate of injection is the sole explanatory variable u (and writing u for $\mathbf{u}$), we could choose the function $\phi(u)$ so that $\phi(1) = 1$, $\phi(2) = 2$, $\phi(4) = 3$, and $\phi(8) = 4$.

The situation is one in which the residual effects in the kth group have a common variance, say

σ_k^2 $(k = 1, 2, \ldots, K)$. One way of proceeding is to regard the standard deviations $\sigma_1, \sigma_2, \ldots, \sigma_K$ as "unrelated," strictly positive parameters (whose values are unknown). This approach is simple and highly "flexible," though in general not very "parsimonious." It can be accommodated within the framework of the general linear model. One way of doing so is to take the parameter vector $\boldsymbol{\theta}$ to be the K-dimensional column vector with kth element $\theta_k = \sigma_k$ $(k = 1, 2, \ldots, K)$, in which case $\Theta = \{\boldsymbol{\theta} : \theta_k > 0 \ (k = 1, 2, \ldots, K)\}$. Then (assuming that the N residual effects are uncorrelated), $\mathbf{V}(\boldsymbol{\theta})$ is the diagonal matrix whose ith diagonal element $v_{ii}(\boldsymbol{\theta})$ is $\theta_k^2 = \sigma_k^2$ for all $i \in I_k$ $(k = 1, 2, \ldots, K)$ or, equivalently, is $\theta_{\phi(\mathbf{u}_i)}^2 = \sigma_{\phi(\mathbf{u}_i)}^2$ (for $i = 1, 2, \ldots, N$).

Dependence of variability on the explanatory variables. In some situations in which the residual effects $e_1, e_2, \ldots, e_N$ are heteroscedastic, the variances of the residual effects may be related to the values of the explanatory variables (related in a more substantial way than in the case of within-group homoscedasticity). Suppose that for some nonnegative function, say $v(\mathbf{u})$, of $\mathbf{u}$ (and for $i = 1, 2, \ldots, N$) $\text{var}(e_i) = v(\mathbf{u}_i)$ [or, equivalently, that the standard deviation of e_i equals $\sqrt{v(\mathbf{u}_i)}$]. Then, assuming that $e_1, e_2, \ldots, e_N$ are uncorrelated,

$$\text{var}(\mathbf{e}) = \text{diag}[v(\mathbf{u}_1),\ v(\mathbf{u}_2), \ldots,\ v(\mathbf{u}_N)]. \tag{4.1}$$

In general, the function $v(\cdot)$ is known only up to the (unknown) values of one or more parameters—the dependence on the parameters is suppressed in the notation. It is implicitly assumed that these parameters are unrelated to the parameters $\beta_1, \beta_2, \ldots, \beta_P$, whose values determine the expected values of $y_1, y_2, \ldots, y_N$. Thus, when they are regarded as the elements of the vector $\boldsymbol{\theta}$, $\text{var}(\mathbf{e})$ is of the form $\mathbf{V}(\boldsymbol{\theta})$ of $\text{var}(\mathbf{e})$ in the general linear model.

In what is a relatively simple special case, $v(\mathbf{u})$ is of the form

$$v(\mathbf{u}) = \sigma^2 h(\mathbf{u}), \tag{4.2}$$

where σ is a strictly positive parameter (of unknown value) and $h(\mathbf{u})$ is a known (nonnegatively valued) function of $\mathbf{u}$. In that special case, formula (4.1) is expressible as

$$\text{var}(\mathbf{e}) = \sigma^2\, \text{diag}[h(\mathbf{u}_1),\ h(\mathbf{u}_2), \ldots,\ h(\mathbf{u}_N)]. \tag{4.3}$$

This expression is of the form $\sigma^2\mathbf{H}$ of $\text{var}(\mathbf{e})$ in the Aitken generalization of the G–M model.

Let us consider some of the more common choices for the function $v(\mathbf{u})$. For the sake of simplicity, let us do so for the special case where $v(\mathbf{u})$ depends on $\mathbf{u}$ only through a single one of its C components. Further, for convenience, let us write u for this component (dropping the subscript) and write $v(u)$ for $v(\mathbf{u})$ [thereby regarding $v(\mathbf{u})$ as a function solely of u]. In the case of the ouabain data, we could take u to be the rate of injection, or, alternatively, we could regard some strictly monotonic function of the rate of injection as an explanatory variable and take it to be u.

One very simple choice for $v(u)$ is

$$v(u) = \sigma^2 |u| \tag{4.4}$$

(where σ is a strictly positive parameter of unknown value). More generally, we could take $v(u)$ to be of the form

$$v(u) = \sigma^2 |u|^{2\alpha}, \tag{4.5}$$

where α is a strictly positive scalar or possibly (if the domain of u does not include the value 0) an unrestricted scalar. And, still more generally, we could take $v(u)$ to be of the form

$$v(u) = \sigma^2 (\gamma + |u|^{\alpha})^2, \tag{4.6}$$

where γ is a nonnegative scalar. While expressions (4.4), (4.5), and (4.6) depend on u only through its absolute value and consequently are well-defined for both positive and negative values of u, choices for $v(u)$ of the form (4.4), (4.5), or (4.6) would seem to be best-suited for use in situations where u is either strictly positive or strictly negative.

Note that taking $v(u)$ to be of the form (4.4), (4.5), or (4.6) is equivalent to taking $\sqrt{v(u)}$ to be $\sigma\sqrt{|u|}$, $\sigma|u|^{\alpha}$, or $\sigma(\gamma + |u|^{\alpha})$, respectively. Note also that expression (4.4) is of the form (4.2);

and recall that when $v(u)$ is of the form (4.2), $\text{var}(\mathbf{e})$ is of the form $\sigma^2\mathbf{H}$ associated with the Aitken generalization of the G–M model. More generally, if α is known, then expression (4.5) is of the form (4.2); and if both γ and α are known, expression (4.6) is of the form (4.2). However, if α or if γ and/or α are (like σ) regarded as unknown parameters, then expression (4.5) or (4.6), respectively, is not of the form (4.2), and, while $\text{var}(\mathbf{e})$ is of the form associated with the general linear model, it is not of the form associated with the Aitken generalization of the G–M model.

As an alternative to taking $v(u)$ to be of the form (4.4), (4.5), or (4.6), we could take it to be of the form

$$v(u) = \sigma^2 e^{2\alpha u}, \tag{4.7}$$

where α is an unrestricted scalar (and σ is a strictly positive parameter of unknown value). Note that taking $v(u)$ to be of the form (4.7) is equivalent to taking $\sqrt{v(u)}$ to be of the form $\sqrt{v(u)} = \sigma e^{\alpha u}$, and is also equivalent to taking $\log \sqrt{v(u)}$ [which equals $\frac{1}{2} \log v(u)$] to be of the form

$$\log \sqrt{v(u)} = \log \sigma + \alpha u.$$

Note also that if the scalar α in expression (4.7) is known, then that expression is of the form (4.2), in which case $\text{var}(\mathbf{e})$ is of the form associated with the Aitken generalization of the G–M model. Alternatively, if α is an unknown parameter, then expression (4.7) is not of the form (4.2) and $\text{var}(\mathbf{e})$ is not of the form associated with the Aitken generalization of the G–M model [though $\text{var}(\mathbf{e})$ is of the form associated with the general linear model].

In some applications, there may not be any choice for $v(u)$ of a relatively simple form [like (4.6) or (4.7)] for which it is realistic to assume that $\text{var}(e_i) = v(\mathbf{u}_i)$ for $i = 1, 2, \ldots, N$. However, in some such applications, it may be possible to partition $e_1, e_2, \ldots, e_N$ into mutually exclusive and exhaustive subsets (perhaps on the basis of the values $\mathbf{u}_1, \mathbf{u}_2, \ldots, \mathbf{u}_N$ of the vector $\mathbf{u}$ of explanatory variables) in such a way that, specific to each subset, there is a choice for $v(\mathbf{u})$ (of a relatively simple form) for which it may be realistic to assume that $\text{var}(e_i) = v(\mathbf{u}_i)$ for those i corresponding to the members of that subset. While one subset may require a different choice for $v(\mathbf{u})$ than another, the various choices (corresponding to the various subsets) may all be of the same general form; for example, they could all be of the form (4.7) (but with possibly different values of σ and/or α).

b. Intraclass correlation: compound symmetry

There are situations where not all of the residual effects $e_1, e_2, \ldots, e_N$ can reasonably be regarded as uncorrelated. In some such situations, the residual effects can be partitioned into some number of mutually exclusive and exhaustive subsets in such a way that the residual effects in any particular subset are thought to be correlated (to an equal extent) while those in different subsets are thought to be uncorrelated. It is customary to refer to these subsets as classes. Residual effects in the same class are assumed to be homoscedastic (i.e., to have the same variance); in the most general case, the variances of the residual effects may (as in Part 2 of Subsection a) differ from class to class.

For random variables x and w (whose variances exist and are strictly positive),

$$\text{var}\left(\frac{x}{\sqrt{\text{var}\, x}} - \frac{w}{\sqrt{\text{var}\, w}}\right) = 2\,[1 - \text{corr}(x, w)].$$

Thus, the variance of the difference between the standardized versions of x and w is a decreasing function of $\text{corr}(x, w)$. It follows that, in a certain sense, positively correlated random variables tend to be more alike, and negatively correlated random variables less alike, than uncorrelated random variables. In the case of the residual effects, there is an implication that, depending on whether the "intraclass correlation" is positive or negative, residual effects in the same class tend to be either more alike or less alike than residual effects in different classes.

Suppose that the N residual effects have been partitioned into K mutually exclusive and exhaustive subsets or classes numbered $1, 2, \ldots, K$. And for $k = 1, 2, \ldots, K$, denote by N_k the number of residual effects in the kth subset or class (so that $\sum_{k=1}^{K} N_k = N$). Further, let us suppose that the numbering of the N data points and residual effects is such that the residual effects numbered $N_1+N_2+\cdots+N_{k-1}+1$ through $N_1+N_2+\cdots+N_k$ are those in the kth subset or class—interpret $N_1+N_2+\cdots+N_0$ as 0. And for the sake of convenience and simplicity, let us use two subscripts instead of one to identify the residual effects. Accordingly, let us write $e_{k1}, e_{k2}, \ldots, e_{kN_k}$ (instead of $e_{N_1+N_2+\cdots+N_{k-1}+1}, e_{N_1+N_2+\cdots+N_{k-1}+2}, \ldots, e_{N_1+N_2+\cdots+N_k}$) for the residual effects in the kth subset or class. Also, define (for $k = 1, 2, \ldots, K$) $\mathbf{e}_k = (e_{k1}, e_{k2}, \ldots, e_{kN_k})'$, and observe that

$$\mathbf{e}' = (\mathbf{e}_1', \mathbf{e}_2', \ldots, \mathbf{e}_K').$$

It is supposed that $\operatorname{cov}(\mathbf{e}_k, \mathbf{e}_j) = \mathbf{0}$ for $j \neq k = 1, 2, \ldots, K$. It is further supposed that, for some scalar ρ_k,

$$\operatorname{corr}(e_{ks}, e_{kt}) = \rho_k \qquad (t \neq s = 1, 2, \ldots, N_k) \tag{4.8}$$

and that, for some strictly positive scalar σ_k,

$$\operatorname{var}(e_{ks}) = \sigma_k^2 \qquad (s = 1, 2, \ldots, N_k), \tag{4.9}$$

so that the correlation matrix of $\mathbf{e}_k$ is the $N_k \times N_k$ matrix

$$\mathbf{R}_k = \begin{pmatrix} 1 & \rho_k & \cdots & \rho_k \\ \rho_k & 1 & & \rho_k \\ \vdots & & \ddots & \\ \rho_k & \rho_k & & 1 \end{pmatrix} \tag{4.10}$$

$$= (1-\rho_k)\mathbf{I}_{N_k} + \rho_k \mathbf{1}_{N_k}\mathbf{1}_{N_k}' \tag{4.11}$$

and the variance-covariance matrix of $\mathbf{e}_k$ is

$$\operatorname{var}(\mathbf{e}_k) = \sigma_k^2 \mathbf{R}_k \tag{4.12}$$

($k = 1, 2, \ldots, K$). It follows that

$$\operatorname{var}(\mathbf{e}) = \operatorname{diag}(\sigma_1^2\mathbf{R}_1, \sigma_2^2\mathbf{R}_2, \ldots, \sigma_K^2\mathbf{R}_K). \tag{4.13}$$

Condition (4.8) stipulates that the correlation of every two residual effects in the kth class equals ρ_k, so that, by definition, ρ_k is the "intraclass correlation." Because ρ_k is a correlation, it is necessarily the case that $-1 \leq \rho_k \leq 1$. However, not every value of ρ_k between ± 1 is a "permissible" value. The permissible values of ρ_k are those for which the matrix (4.11) is nonnegative definite. The determination of those values is greatly facilitated by the introduction of the following lemma.

A matrix lemma.

Lemma 4.4.1. Let $\mathbf{R}$ represent an $M \times M$ matrix of the form

$$\mathbf{R} = a\mathbf{I}_M + b\mathbf{1}_M\mathbf{1}_M',$$

where a and b are scalars and where $M \geq 2$. Then, $\mathbf{R}$ is nonnegative definite if and only if $a \geq 0$ and $a + Mb \geq 0$, and is positive definite if and only if $a > 0$ and $a + Mb > 0$.

Proof. Let $\mathbf{x} = (x_1, x_2, \ldots, x_M)'$ represent an arbitrary M-dimensional column vector, and define $\bar{x} = (1/M)\sum_{i=1}^{M} x_i$. Observe that

$$\begin{aligned} \mathbf{x}'\mathbf{R}\mathbf{x} &= a\sum_{i=1}^{M} x_i^2 + b(M\bar{x})^2 \\ &= a\Big(\sum_{i=1}^{M} x_i^2 - M\bar{x}^2\Big) + (a+Mb)M\bar{x}^2 \\ &= a\sum_{i=1}^{M}(x_i-\bar{x})^2 + (a+Mb)M\bar{x}^2. \end{aligned} \tag{4.14}$$

Observe also that $\sum_{i=1}^{M}(x_i-\bar{x})^2 = 0$ if and only if $x_1 = x_2 = \cdots = x_M$, and that $\sum_{i=1}^{M}(x_i-\bar{x})^2 = 0$ and $M\bar{x}^2 = 0$ if and only if $x_1 = x_2 = \cdots = x_M = 0$ or, equivalently, if and only if $\mathbf{x} = \mathbf{0}$.

If $a \geq 0$ and $a + Mb \geq 0$, then it is clear from result (4.14) that $\mathbf{x}'\mathbf{R}\mathbf{x} \geq 0$ for every $\mathbf{x}$ and hence that $\mathbf{R}$ is nonnegative definite. Further, if $a + Mb < 0$, then $\mathbf{x}'\mathbf{R}\mathbf{x} < 0$ for any $\mathbf{x}$ of the form $\mathbf{x} = c\mathbf{1}_M$, where c is a nonzero scalar; and if $a + Mb = 0$, then $\mathbf{x}'\mathbf{R}\mathbf{x} = 0$ for any $\mathbf{x}$ of that form—note that any $\mathbf{x}$ of that form is nonnull. Thus, if $a + Mb < 0$, then $\mathbf{R}$ is not nonnegative definite (i.e., is neither positive definite nor positive semidefinite); and if $a + Mb = 0$, then $\mathbf{R}$ is not positive definite. Similarly, if $a < 0$, then $\mathbf{x}'\mathbf{R}\mathbf{x} < 0$ for any $\mathbf{x}$ with $\sum_{i=1}^{M} x_i = 0$ and with $x_i \neq x_j$ for some i and j; and if $a = 0$, then $\mathbf{x}'\mathbf{R}\mathbf{x} = 0$ for any such $\mathbf{x}$ (and any such $\mathbf{x}$ is nonnull). Thus, if $a < 0$, then $\mathbf{R}$ is not nonnegative definite; and if $a = 0$, then $\mathbf{R}$ is not positive definite. To complete the proof, it suffices to observe that if $a > 0$ and $a + Mb > 0$, then $\mathbf{x}'\mathbf{R}\mathbf{x} = 0$ only if $\mathbf{x}$ is such that both $\sum_{i=1}^{M}(x_i - \bar{x})^2 = 0$ and $M\bar{x}^2 = 0$ and hence only if $\mathbf{x} = \mathbf{0}$. Q.E.D.

Form of the variance-covariance matrix. Let us now return to the main development. Upon applying Lemma 4.4.1 with $a = 1 - \rho_k$ and $b = \rho_k$, we find that (when $N_k \geq 2$) the correlation matrix $\mathbf{R}_k$ of the vector $\mathbf{e}_k$ is nonnegative definite if and only if $1 - \rho_k \geq 0$ and $1 + (N_k - 1)\rho_k \geq 0$, or equivalently if and only if

$$-\frac{1}{N_k - 1} \leq \rho_k \leq 1, \tag{4.15}$$

and similarly that $\mathbf{R}_k$ is positive definite if and only if

$$-\frac{1}{N_k - 1} < \rho_k < 1. \tag{4.16}$$

The permissible values of ρ_k are those in the interval (4.15).

The variance-covariance matrix of the vector $\mathbf{e}_k$ is such that all of its diagonal elements (which are the variances) equal each other and all of its off-diagonal elements (which are the covariances) also equal each other. Such a variance-covariance matrix is said to be *compound symmetric* (e.g., Milliken and Johnson 2009, p. 536).

The variance-covariance matrix of the vector $\mathbf{e}$ of residual effects is positive definite if and only if all K of the matrices $\mathbf{R}_1, \mathbf{R}_2, \ldots, \mathbf{R}_K$ are positive definite—refer to result (4.13) and "recall" Lemma 2.13.31. Thus, $\text{var}(\mathbf{e})$ is positive definite if, for every k for which $N_k \geq 2$, ρ_k is in the interior (4.16) of interval (4.15). [If, for every k for which $N_k \geq 2$, ρ_k is in interval (4.15) but, for at least one such k, ρ_k equals an end point of interval (4.15), $\text{var}(\mathbf{e})$ is positive semidefinite.]

There are applications in which it may reasonably be assumed that the intraclass correlations are nonnegative (i.e., that $\rho_k \geq 0$ for every k for which $N_k \geq 2$). In some such applications, it is further assumed that all of the intraclass correlations are equal. Together, these assumptions are equivalent to the assumption that, for some scalar ρ in the interval $0 \leq \rho \leq 1$,

$$\rho_k = \rho \quad (\text{for every } k \text{ for which } N_k \geq 2). \tag{4.17}$$

Now, suppose that assumption (4.17) is adopted. If ρ were taken to be 0, $\text{var}(\mathbf{e})$ would be of the form considered in Part 2 of Subsection a. When ρ as well as $\sigma_1, \sigma_2, \ldots, \sigma_K$ are regarded as unknown parameters, $\text{var}(\mathbf{e})$ is of the form $\mathbf{V}(\boldsymbol{\theta})$ of $\text{var}(\mathbf{e})$ in the general linear model—take $\boldsymbol{\theta}$ to be a $(K+1)$-dimensional (column) vector whose elements are $\sigma_1, \sigma_2, \ldots, \sigma_K$, and ρ.

In some applications, there may be a willingness to augment assumption (4.17) with the additional assumption that the residual effects are completely homoscedastic, that is, with the additional assumption that, for some strictly positive scalar σ,

$$\sigma_1 = \sigma_2 = \cdots = \sigma_K = \sigma. \tag{4.18}$$

Then,

$$\text{var}(\mathbf{e}) = \sigma^2 \,\text{diag}(\mathbf{R}_1, \mathbf{R}_2, \ldots, \mathbf{R}_K), \tag{4.19}$$

and when σ is regarded as an unknown parameter and ρ is taken to be known, $\text{var}(\mathbf{e})$ is of the form

$\sigma^2\mathbf{H}$ of var($\mathbf{e}$) in the Aitken generalization of the G–M model. When both σ and ρ are regarded as unknown parameters, var($\mathbf{e}$) is of the form $\mathbf{V}(\boldsymbol{\theta})$ of var($\mathbf{e}$) in the general linear model—take $\boldsymbol{\theta}$ to be the 2-dimensional (column) vector whose elements are σ and ρ. Assumption (4.18) leads to a model that is less flexible but more parsimonious than that obtained by allowing the variances of the residual effects to differ from class to class.

Decomposition of the residual effects. A supposition that an intraclass correlation is nonnegative, but strictly less than 1, is the equivalent of a supposition that each residual effect can be regarded as the sum of two uncorrelated components, one of which is specific to that residual effect and the other of which is shared by all of the residual effects in the same class. Let us consider this equivalence in some detail. Accordingly, take a_k $(k = 1, 2, \ldots, K)$ and r_{ks} $(k = 1, 2, \ldots, K;\, s = 1, 2, \ldots, N_k)$ to be uncorrelated random variables, each with mean 0, such that $\text{var}(a_k) = \tau_k^2$ for some nonnegative scalar τ_k and $\text{var}(r_{ks}) = \phi_k^2$ $(s = 1, 2, \ldots, N_k)$ for some strictly positive scalar ϕ_k. And suppose that, for $k = 1, 2, \ldots, K$ and $s = 1, 2, \ldots, N_k$,

$$e_{ks} = a_k + r_{ks}. \tag{4.20}$$

Here, a_k is the component of the sth residual effect in the kth class that is shared by all of the residual effects in the kth class and r_{ks} is the component that is specific to e_{ks}.

We find that the variance σ_k^2 of the residual effects in the kth class is expressible as

$$\sigma_k^2 = \phi_k^2 + \tau_k^2. \tag{4.21}$$

Further, upon observing that (for $t \neq s = 1, 2, \ldots, N_k$) $\text{cov}(e_{ks}, e_{kt}) = \tau_k^2$, we find that the correlation of any two residual effects in the kth class (the intraclass correlation) is expressible as

$$\rho_k = \frac{\tau_k^2}{\sigma_k^2} = \frac{\tau_k^2}{\phi_k^2 + \tau_k^2} = \frac{(\tau_k/\phi_k)^2}{1 + (\tau_k/\phi_k)^2} \tag{4.22}$$

—this expression is well-defined even if $N_k = 1$. And, in addition to representations (4.10) and (4.11) for the correlation matrix $\mathbf{R}_k$ of the vector $\mathbf{e}_k$ and representation (4.12) for var($\mathbf{e}_k$), we have the representations

$$\mathbf{R}_k = \frac{\phi_k^2}{\phi_k^2 + \tau_k^2}\mathbf{I}_{N_k} + \frac{\tau_k^2}{\phi_k^2 + \tau_k^2}\mathbf{1}_{N_k}\mathbf{1}'_{N_k} \tag{4.23}$$

and

$$\text{var}(\mathbf{e}_k) = \phi_k^2\mathbf{I}_{N_k} + \tau_k^2\mathbf{1}_{N_k}\mathbf{1}'_{N_k} \tag{4.24}$$

$$= \phi_k^2[\mathbf{I}_{N_k} + (\tau_k/\phi_k)^2\mathbf{1}_{N_k}\mathbf{1}'_{N_k}]. \tag{4.25}$$

Note that result (4.22) implies that the intraclass correlation is nonnegative, but strictly less than 1. Note also that an assumption that ρ_k does not vary with k can be restated as an assumption that τ_k^2/σ_k^2 does not vary with k and also as an assumption that τ_k/ϕ_k (or τ_k^2/ϕ_k^2) does not vary with k.

The effects of intraclass competition. In some applications, the residual effects in each class may be those for data on entities among which there is competition. For example, the entities might consist of individual animals that are kept in the same pen and that may compete for space and for feed. Or they might consist of individual plants that are in close proximity and that may compete for water, nutrients, and light. In the presence of such competition, the residual effects in each class may tend to be less alike than would otherwise be the case.

The decomposition of the residual effects considered in the previous part can be modified so as to reflect the effects of competition. Let us suppose that the N_k residual effects in the kth class are those for data on N_k of a possibly larger number N_k^* of entities among which there is competition. Define a_k $(k = 1, 2, \ldots, K)$ and r_{ks} $(k = 1, 2, \ldots, K;\; s = 1, 2, \ldots, N_k)$ as in the previous part. And take d_{ks} $(k = 1, 2, \ldots, K; s = 1, 2, \ldots, N_k^*)$ to be random variables, each with mean 0, that are uncorrelated with the a_k's and the r_{ks}'s, and suppose that $\text{cov}(d_{ks}, d_{jt}) = 0$ for $j \neq k = 1, 2, \ldots, K$

(and for all s and t). Suppose further that (for $k = 1, 2, \ldots, K$) $\sum_{s=1}^{N_k^*} d_{ks} = 0$, that $\text{var}(d_{ks}) = \omega_k^2$ ($s = 1, 2, \ldots, N_k^*$) for some nonnegative scalar ω_k, and that $\text{cov}(d_{ks}, d_{kt})$ has the same value for all $t \neq s = 1, 2, \ldots, N_k^*$. Finally, suppose that, for $k = 1, 2, \ldots, K$ and $s = 1, 2, \ldots, N_k$,

$$e_{ks} = a_k + d_{ks} + r_{ks}. \tag{4.26}$$

Decomposition (4.26) can be regarded as a modification of decomposition (4.20) in which the term r_{ks} is replaced by the sum $d_{ks} + r_{ks}$. This modification is for the purpose of accounting for the possibility of intraclass competition.

In light of the various suppositions, we have that (for an arbitrary s and $t \neq s$)

$$\text{var}\Bigl(\sum_{s'=1}^{N_k^*} d_{ks'}\Bigr) = N_k^* \omega_k^2 + N_k^*(N_k^* - 1)\,\text{cov}(d_{ks}, d_{kt}) \tag{4.27}$$

and that

$$\text{var}\Bigl(\sum_{s'=1}^{N_k^*} d_{ks'}\Bigr) = \text{var}(0) = 0. \tag{4.28}$$

Together, results (4.27) and (4.28) imply that (for $t \neq s = 1, 2, \ldots, N_k^*$)

$$\text{cov}(d_{ks}, d_{kt}) = -\frac{1}{N_k^* - 1}\omega_k^2. \tag{4.29}$$

Decomposition (4.26) is such that the variance σ_k^2 of the residual effects in the kth class is expressible as

$$\sigma_k^2 = \phi_k^2 + \tau_k^2 + \omega_k^2. \tag{4.30}$$

Further, for $t \neq s = 1, 2, \ldots, N_k$,

$$\text{cov}(e_{ks}, e_{kt}) = \tau_k^2 + \text{cov}(d_{ks}, d_{kt}) = \tau_k^2 - \frac{1}{N_k^* - 1}\omega_k^2. \tag{4.31}$$

Thus, the correlation of any two residual effects in the kth class (the intraclass correlation) is expressible as

$$\rho_k = \frac{\tau_k^2 - \frac{1}{N_k^*-1}\omega_k^2}{\sigma_k^2} = \frac{\tau_k^2 - \frac{1}{N_k^*-1}\omega_k^2}{\phi_k^2 + \tau_k^2 + \omega_k^2} = \frac{(\tau_k/\phi_k)^2 - \frac{1}{N_k^*-1}(\omega_k/\phi_k)^2}{1 + (\tau_k/\phi_k)^2 + (\omega_k/\phi_k)^2}. \tag{4.32}$$

In light of expression (4.32), the permissible values of ρ_k are those in the interval

$$-\frac{1}{N_k^* - 1} < \rho_k < 1. \tag{4.33}$$

The intraclass correlation ρ_k approaches the upper end point of this interval as $\tau_k/\phi_k \to \infty$ (for fixed ω_k/ϕ_k) and approaches the lower end point as $\omega_k/\phi_k \to \infty$ (for fixed τ_k/ϕ_k).

Expression (4.32) "simplifies" to a considerable extent if neither τ_k/ϕ_k nor ω_k/ϕ_k varies with k (or, equivalently, if neither τ_k/σ_k nor ω_k/σ_k varies with k). However, even then, ρ_k may depend on N_k^* and hence may vary with k.

c. Example: corn-milling data

Littell, Milliken, Stroup, Wolfinger, and Schabenberger (2006, sec. 16.2) reported the results of an experimental study of how the milling of corn is affected by the moisture content of the corn and by the operating characteristics of the grinding mill. The operating characteristics were those associated with three variables: roll gap, screen size, and roller speed. Three equally spaced settings of each of these three variables were represented in the study. The experimental material consisted of ten

TABLE 4.4. Amount of grits obtained from each of 30 1-minute runs of a grinding mill and the batch, moisture content, roll gap, screen size, and roller speed for each run (Littell, Milliken, Stroup, Wolfinger, and Schabenberger 2006, sec. 16.2).

		Transformed setting (+1, high; 0, med.; −1, low)			
Amount of grits	Batch	Moisture content	Roll gap	Screen size	Roller speed
505	1	+1	+1	+1	+1
493	1	+1	−1	−1	−1
491	1	+1	−1	+1	−1
498	2	+1	+1	−1	0
504	2	+1	+1	−1	−1
500	2	+1	−1	+1	0
494	3	−1	0	−1	−1
498	3	−1	0	+1	0
498	3	−1	−1	0	+1
496	4	0	−1	−1	0
503	4	0	0	+1	+1
496	4	0	−1	0	−1
503	5	−1	−1	+1	+1
495	5	−1	+1	+1	−1
494	5	−1	−1	+1	−1
486	6	0	0	0	0
501	6	0	+1	+1	−1
490	6	0	+1	−1	+1
494	7	−1	+1	0	0
497	7	−1	+1	+1	+1
492	7	−1	−1	+1	−1
503	8	+1	−1	+1	+1
499	8	+1	0	0	−1
493	8	+1	0	−1	+1
505	9	+1	+1	+1	−1
500	9	+1	+1	0	+1
490	9	+1	−1	−1	+1
494	10	−1	−1	−1	+1
497	10	−1	+1	−1	−1
495	10	−1	−1	−1	−1

30-kilogram "batches" of corn. Each batch was tempered so that its moisture content conformed to a specified setting (one of three equally spaced settings selected for inclusion in the study).

Following its preparation, each batch was split into three equal (10 kg) parts. And for each part of each batch, settings were specified for the roll gap, screen size, and roller speed, the grinding mill was configured (to conform to the specified settings), the processing of the corn was undertaken, and the amount of grits obtained from a one-minute run was determined. The moisture content, roll gap, screen size, and roller speed were recorded on a transformed scale chosen so that the values of the high, medium, and low settings were $+1$, 0, and -1, respectively. The results of the 30 experimental runs are reproduced in Table 4.4.

The corn-milling experiment was conducted in accordance with what is known as a split-plot design (e.g., Hinkelmann and Kempthorne 2008, chap. 13). The ten batches are the so-called whole plots, and the three moisture-content settings constitute the so-called whole-plot treatments. Further, the 30 parts (obtained by splitting the 10 batches into 3 parts each) are the so-called split plots, and the various combinations of settings for roll gap, screen size, and roller speed constitute the so-called split-plot treatments—21 of a possible 27 combinations were included in the experiment.

The data from the corn-milling experiment might be suitable for the application of a general linear model. We could take $N = 30$, take the observed value of y_i to be the amount of grits obtained on the ith experimental run ($i = 1, 2, \ldots, 30$), take $C = 4$, and take u_1, u_2, u_3, and u_4 to be the moisture content, roll gap, screen size, and roller speed, respectively (with each being expressed on the transformed scale). And we could consider taking $\delta(\mathbf{u})$ to be a polynomial of degree 2 in u_1, u_2, u_3, and u_4. The nature of the application is such that $\delta(\mathbf{u})$ defines what is commonly referred to as a "response surface."

The batches in the corn-milling experiment define classes within which the residual effects are likely to be correlated. Moreover, this correlation is likely to be compound symmetric, that is, to be the same for every two residual effects in the same class. In the simplest case, the intraclass correlation would be regarded as having the same value, say ρ, for every class. Then, assuming that the data points (and the corresponding residual effects) are numbered $1, 2, \ldots, 30$ in the order in which the data points are listed in Table 4.4 and that the residual effects have a common variance σ^2, the variance-covariance matrix of the vector $\mathbf{e}$ of residual effects would be

$$\text{var}(\mathbf{e}) = \sigma^2 \,\text{diag}(\mathbf{R}, \mathbf{R}, \ldots, \mathbf{R}),$$

where $\mathbf{R} = \begin{pmatrix} 1 & \rho & \rho \\ \rho & 1 & \rho \\ \rho & \rho & 1 \end{pmatrix}$.

The batches can be expected to differ from one another in ways that go beyond any of the specified differences in moisture content. To the extent that any such differences are reflected in the amounts of grits obtained from the various one-minute runs, they contribute positively to the intraclass correlation. However, their influence could be offset to at least some small extent by the effects of splitting the batches into parts. If the splitting is such that (by the very nature of the process) some parts are favored at the expense of others (in ways that may affect the amount of grits), then the splitting may contribute negatively to the intraclass correlation. In effect, the splitting may introduce what could be regarded as a form of intraclass competition.

d. Example: shear-strength data

Khuri (1992, sec. 4) reported the results of an experimental study of how the effectiveness of an 11-component adhesive is influenced by the temperature and the curing time employed in its manufacture. The measure of effectiveness was taken to be the shear strength of the bond between galvanized steel bars created through application of the adhesive. Results were obtained for three temperatures (375°F, 400°F, and 450°F) and three curing times (30, 35, and 40 seconds) in all (nine) possible combinations. The steel used in assessing the shear strength consisted of "aliquots" obtained by sampling (at random) from the supply on hand in a warehouse; a sample was taken on each of 12 dates between July 11th and October 10th, inclusive.

The experiment was conducted in accordance with what could be regarded as a randomized complete block design (e.g., Hinkelmann and Kempthorne 2008, chap. 9). There are 12 "blocks" corresponding to the 12 dates on which samples were taken and 9 "treatments" corresponding to the 9 possible combinations of the 3 temperatures and the 3 curing times. The basic design was augmented by including (in four of the blocks) some "replicates" of one of the treatments (the one comprising a temperature of 400°F and a curing time of 35 seconds).

The data are reproduced in Table 4.5. These data might be suitable for the application of a general linear model. We could take $N = (12 \times 9) + 3 + 3 + 2 + 2 = 118$, and take the observed value of $\mathbf{y}$ to be the (118-dimensional) column vector formed from the 12 columns of shear strengths in Table 4.5 by listing them successively one under the other. And we could take $C = 2$, take u_1 and

TABLE 4.5. Shear strength (psi) of the bond between galvanized steel bars created through application of an adhesive: data for each of nine combinations of temperature and curing time (those employed in the manufacture of the adhesive) obtained using steel aliquots selected at random from those on hand on each of twelve dates (Khuri 1992, sec. 4).

Temp.	Time	Date (month/day)											
(°F)	(sec.)	07/11	07/16	07/20	08/07	08/08	08/14	08/20	08/22	09/11	09/24	10/03	10/10
375	30	1,226	1,075	1,172	1,213	1,282	1,142	1,281	1,305	1,091	1,281	1,305	1,207
400	30	1,898	1,790	1,804	1,961	1,940	1,699	1,833	1,774	1,588	1,992	2,011	1,742
450	30	2,142	1,843	2,061	2,184	2,095	1,935	2,116	2,133	1,913	2,213	2,192	1,995
375	35	1,472	1,121	1,506	1,606	1,572	1,608	1,502	1,580	1,343	1,691	1,584	1,486
400	35	2,010	2,175	2,279	2,450	2,291	2,374	2,417	2,393	2,205	2,142	2,052	2,339
400	35	1,882			2,355					2,268		2,032	
400	35	1,915			2,420					2,103		2,190	
400	35	2,106			2,240								
450	35	2,352	2,274	2,168	2,298	2,147	2,413	2,430	2,440	2,093	2,208	2,201	2,216
375	40	1,491	1,691	1,707	1,882	1,741	1,846	1,645	1,688	1,582	1,692	1,744	1,751
400	40	2,078	2,513	2,392	2,531	2,366	2,392	2,392	2,413	2,392	2,488	2,392	2,390
450	40	2,531	2,588	2,617	2,609	2,431	2,408	2,517	2,604	2,477	2,601	2,588	2,572

u_2 to be the temperature and curing time (employed in the manufacture of the adhesive), and take $\delta(\mathbf{u})$ to be a polynomial of degree 2 in u_1 and u_2. As in the previous example (that of Subsection c), the nature of the application is such that $\delta(\mathbf{u})$ defines a "response surface."

Steel aliquots chosen at random from those on hand on any particular date may tend to resemble each other more closely than ones chosen at random from those on hand on different dates. Accordingly, it may be prudent to regard the 12 blocks as "classes" and to allow for the possibility of an intraclass correlation. If (in doing so) it is assumed that the intraclass correlation and the variance of the random effects have values, say ρ and σ^2, respectively, that do not vary from block to block, then the variance-covariance matrix of the vector $\mathbf{e}$ of residual effects is

$$\text{var}(\mathbf{e}) = \sigma^2 \,\text{diag}(\mathbf{R}_1, \mathbf{R}_2, \ldots, \mathbf{R}_{12}), \tag{4.34}$$

where (for $k = 1, 2, \ldots, 12$)

$$\mathbf{R}_k = (1-\rho)\mathbf{I}_{N_k} + \rho \mathbf{1}_{N_k}\mathbf{1}'_{N_k}$$

with

$$N_k = \begin{cases} 9 & \text{if } k = 2, 3, 5, 6, 7, 8, 10, \text{ or } 12, \\ 11 & \text{if } k = 9 \text{ or } 11, \\ 12 & \text{if } k = 1 \text{ or } 4. \end{cases}$$

In arriving at expression (4.34), it was implicitly assumed that the residual effects associated with the data points in any one block are uncorrelated with those associated with the data points in any other block. It is conceivable that steel aliquots chosen at random from those on hand on different dates may tend to be more alike when the intervening time (between dates) is short than when it is long. If we wished to account for any such tendency, we would need to allow for the possibility that the residual effects associated with the data points in different blocks may be correlated to an extent that diminishes with the separation (in time) between the blocks. That would seem to call for taking $\text{var}(\mathbf{e})$ to be of a different and more complex form than the block-diagonal form (4.34).

e. Longitudinal data

There are situations where the data are obtained by recording the value of what is essentially the same variate for each of a number of "observational units" on each of a number of occasions (corresponding to different points in time). The observational units might be people, animals, plants, laboratory specimens, households, experimental plots (of land), or other such entities. For example, in a clinical trial of drugs for lowering blood pressure, each drug might be administered to a different group of people, with a placebo being administered to an additional group, and each person's blood pressure might be recorded periodically, including at least once prior to the administration of the drug or placebo—in this example, each person constitutes an observational unit. Data of this kind are referred to as *longitudinal data*.

Suppose that the observed values of the random variables $y_1, y_2, \ldots, y_N$ in the general linear model are longitudinal data. Further, let t represent time, and denote by $t_1, t_2, \ldots, t_N$ the values of t corresponding to $y_1, y_2, \ldots, y_N$, respectively. And assume (as can be done essentially without loss of generality) that the vector $\mathbf{u}$ of explanatory variables $u_1, u_2, \ldots, u_C$ is such that t is one of the explanatory variables or, more generally, is expressible as a function of $\mathbf{u}$, so that (for $i = 1, 2, \ldots, N$) the ith value t_i of t is determinable from the corresponding (ith) value $\mathbf{u}_i$ of $\mathbf{u}$.

Denote by K the number of observational units represented in the data, suppose that the observational units are numbered $1, 2, \ldots, K$, and define N_k to be the number of data points pertaining to the kth observational unit (so that $\sum_{k=1}^{K} N_k = N$). Assume that the numbering (from 1 through N) of the N data points is such that they are ordered by observational unit and by time within observational unit (so that if the ith data point pertains to the kth observational unit and the i'th data point to the k'th observational unit where $i' > i$, then either $k' > k$ or $k' = k$ and $t_{i'} \ge t_i$)—it is always possible to number the data points in such a way. The setting is one in which it is customary and convenient to use two subscripts, rather than one, to identify the random variables $y_1, y_2, \ldots, y_N$, residual effects $e_1, e_2, \ldots, e_N$, and times $t_1, t_2, \ldots, t_N$. Accordingly, let us write e_{ks} for the residual effect corresponding to the sth of those data points that pertain to the kth observational unit, and t_{ks} for the time corresponding to that data point.

It is often possible to account for the more "systematic" effects of time through the choice of the form of the function $\delta(\mathbf{u})$. However, even then, it is seldom appropriate to assume (as in the G–M model) that all N of the residual effects are uncorrelated with each other.

The vector $\mathbf{e}$ of residual effects is such that

$$\mathbf{e}' = (\mathbf{e}_1', \mathbf{e}_2', \ldots, \mathbf{e}_K'),$$

where (for $k = 1, 2, \ldots, K$) $\mathbf{e}_k = (e_{k1}, e_{k2}, \ldots, e_{kN_k})'$. It is assumed that $\text{cov}(\mathbf{e}_k, \mathbf{e}_j) = \mathbf{0}$ for $j \ne k = 1, 2, \ldots, K$, so that

$$\text{var}(\mathbf{e}) = \text{diag}[\text{var}(\mathbf{e}_1),\ \text{var}(\mathbf{e}_2),\ \ldots,\ \text{var}(\mathbf{e}_K)]. \tag{4.35}$$

For $k = 1, 2, \ldots, K$, $\text{var}(\mathbf{e}_k)$ is the $N_k \times N_k$ matrix with rth diagonal element $\text{var}(e_{kr})$ and rsth (where $s \ne r$) off-diagonal element $\text{cov}(e_{kr}, e_{ks})$. It is to be expected that e_{kr} and e_{ks} will be positively correlated, with the extent of their correlation depending on $|t_{ks} - t_{kr}|$; typically, the correlation of e_{kr} and e_{ks} is a strictly decreasing function of $|t_{kr} - t_{ks}|$. There are various kinds of stochastic processes that exhibit that kind of correlation structure. Among the simplest and most prominent of them is the following.

Stationary first-order autoregressive processes. Let x_1 represent a random variable having mean 0 (and a strictly positive variance), and let $x_2, x_3, \ldots$ represent a possibly infinite sequence of random variables generated successively (starting with x_1) in accordance with the following relationship:

$$x_{i+1} = \rho x_i + d_{i+1}. \tag{4.36}$$

Here, ρ is a (nonrandom) scalar in the interval $0 < \rho < 1$, and $d_2, d_3, \ldots$ are random variables, each with mean 0 (and a finite variance), that are uncorrelated with x_1 and with each other. The sequence of random variables $x_1, x_2, x_3, \ldots$ represents a stochastic process that is characterized as first-order autoregressive.

Note that

$$\mathrm{E}(x_i) = 0 \qquad \text{(for all } i\text{)}. \tag{4.37}$$

Note also that for $r = 1, 2, \ldots, i$, $\operatorname{cov}(x_r, d_{i+1}) = 0$, as can be readily verified by mathematical induction—because $x_r = \rho x_{r-1} + d_r$, $\operatorname{cov}(x_{r-1}, d_{i+1}) = 0$ implies that $\operatorname{cov}(x_r, d_{i+1}) = 0$. In particular, $\operatorname{cov}(x_i, d_{i+1}) = 0$. Thus,

$$\operatorname{var}(x_{i+1}) = \rho^2 \operatorname{var}(x_i) + \operatorname{var}(d_{i+1}). \tag{4.38}$$

Let us determine the conditions under which the sequence of random variables $x_1, x_2, x_3, \ldots$ is stationary in the sense that

$$\operatorname{var}(x_1) = \operatorname{var}(x_2) = \operatorname{var}(x_3) = \cdots. \tag{4.39}$$

In light of equality (4.38),

$$\operatorname{var}(x_{i+1}) = \operatorname{var}(x_i) \quad \Leftrightarrow \quad \operatorname{var}(d_{i+1}) = (1 - \rho^2) \operatorname{var}(x_i).$$

Accordingly, the sequence $x_1, x_2, x_3, \ldots$ satisfies condition (4.39) if (and only if)

$$\operatorname{var}(d_{i+1}) = (1 - \rho^2) \operatorname{var}(x_1) \qquad \text{(for all } i\text{)}. \tag{4.40}$$

By making repeated use of the defining relationship (4.36), we find that, for an arbitrary positive integer s,

$$x_{i+s} = \rho^s x_i + \sum_{j=0}^{s-1} \rho^j d_{i+s-j}. \tag{4.41}$$

And it follows that

$$\operatorname{cov}(x_i, x_{i+s}) = \operatorname{cov}(x_i, \rho^s x_i) = \rho^s \operatorname{var}(x_i). \tag{4.42}$$

Thus, in the special case where the sequence $x_1, x_2, x_3, \ldots$ satisfies condition (4.39), we have that

$$\operatorname{corr}(x_i, x_{i+s}) = \rho^s \tag{4.43}$$

or, equivalently, that (for $r \neq i = 1, 2, 3, \ldots$)

$$\operatorname{corr}(x_i, x_r) = \rho^{|r-i|}. \tag{4.44}$$

In summary, we have that when $d_2, d_3, d_4, \ldots$ satisfy condition (4.40), the sequence of random variables $x_1, x_2, x_3, \ldots$ satisfies condition (4.39) and condition (4.43) or (4.44). Accordingly, when $d_2, d_3, d_4, \ldots$ satisfy condition (4.40), the sequence $x_1, x_2, x_3, \ldots$ is of a kind that is sometimes referred to as stationary in the wide sense (e.g., Parzen 1960, chap. 10).

The entries in the sequence $x_1, x_2, x_3, \ldots$ may represent the state of some phenomenon at a succession of times, say times $t_1, t_2, t_3, \ldots$. The coefficient of x_i in expression (4.36) is ρ, which does not vary with i and hence does not vary with the "elapsed times" $t_2 - t_1, t_3 - t_2, t_4 - t_3, \ldots$. So, for the sequence $x_1, x_2, x_3, \ldots$ to be a suitable reflection of the evolution of the phenomenon over time, it would seem to be necessary that $t_1, t_2, t_3, \ldots$ be equally spaced.

A sequence that may be suitable even if $t_1, t_2, t_3, \ldots$ are not equally spaced can be achieved by introducing a modified version of the defining relationship (4.36). The requisite modification can be discerned from result (4.41) by thinking of the implications of that result as applied to a situation where the successive differences in time ($t_2 - t_1, t_3 - t_2, t_4 - t_3, \ldots$) are equal and arbitrarily small. Specifically, what is needed is to replace relationship (4.36) with the relationship

$$x_{i+1} = \lambda^{t_{i+1} - t_i} x_i + d_{i+1}, \tag{4.45}$$

where λ is a (nonrandom) scalar in the interval $0 < \lambda < 1$.

Suppose that this replacement is made. Then, in lieu of result (4.38), we have that

$$\operatorname{var}(x_{i+1}) = \lambda^{2(t_{i+1} - t_i)} \operatorname{var}(x_i) + \operatorname{var}(d_{i+1}).$$

And by employing essentially the same reasoning as before, we find that the sequence $x_1, x_2, x_3, \ldots$ satisfies condition (4.39) [i.e., the condition that the sequence is stationary in the sense that $\text{var}(x_1) = \text{var}(x_2) = \text{var}(x_3) = \cdots$] if (and only if)

$$\text{var}(d_{i+1}) = [1 - \lambda^{2(t_{i+1}-t_i)}]\,\text{var}(x_1) \qquad \text{(for all } i\text{)}. \tag{4.46}$$

Further, in lieu of result (4.42), we have that

$$\text{cov}(x_i, x_{i+s}) = \lambda^{t_{i+s}-t_i}\,\text{var}(x_i).$$

Thus, in the special case where the sequence $x_1, x_2, x_3, \ldots$ satisfies condition (4.39), we have that

$$\text{corr}(x_i, x_{i+s}) = \lambda^{t_{i+s}-t_i}$$

or, equivalently, that (for $r \neq i = 1, 2, 3, \ldots$)

$$\text{corr}(x_i, x_r) = \lambda^{|t_r - t_i|} \tag{4.47}$$

—these two results take the place of results (4.43) and (4.44), respectively.

In connection with result (4.47), it is worth noting that [in the special case where the sequence $x_1, x_2, x_3, \ldots$ satisfies condition (4.39)] $\text{corr}(x_i, x_r) \to 1$ as $|t_r - t_i| \to 0$ and $\text{corr}(x_i, x_r) \to 0$ as $|t_r - t_i| \to \infty$. And (in that special case) the quantity λ represents the correlation of any two of the x_i's that are separated from each other by a single unit of time.

In what follows, it is the stochastic process defined by relationship (4.45) that is referred to as a first-order autoregressive process. Moreover, in the special case where the stochastic process defined by relationship (4.45) satisfies condition (4.39), it is referred to as a stationary first-order autoregressive process.

Variance-covariance matrix of a subvector of residual effects. Let us now return to the main development, and consider further the form of the matrices $\text{var}(\mathbf{e}_1), \text{var}(\mathbf{e}_2), \ldots, \text{var}(\mathbf{e}_K)$. We could consider taking (for an "arbitrary" k) $\text{var}(\mathbf{e}_k)$ to be of the form that would result from regarding $e_{k1}, e_{k2}, \ldots, e_{kN_k}$ as N_k successive members of a stationary first-order autoregressive process, in which case we would have (based on the results of Part 1) that (for some strictly positive scalar σ and some scalar λ in the interval $0 < \lambda < 1$) $\text{var}(e_{k1}) = \text{var}(e_{k2}) = \cdots = \text{var}(e_{kN_k}) = \sigma^2$ and $\text{corr}(e_{ks}, e_{kj}) = \lambda^{|t_{kj}-t_{ks}|}$ ($j \neq s = 1, 2, \ldots, N_k$). However, for most applications, taking $\text{var}(\mathbf{e}_k)$ to be of that form would be inappropriate. It would imply that residual effects corresponding to "replicate" data points, that is, data points pertaining to the same observational unit and to the same time, are perfectly correlated. It would also imply that residual effects corresponding to data points that pertain to the same observational unit, but that are widely separated in time, are essentially uncorrelated. Neither characteristic conforms to what is found in many applications. There may be "measurement error," in which case the residual effects corresponding to replicate data points are expected to differ from one another. And all of the data points that pertain to the same observational unit (including ones that are widely separated in time) may be subject to some common influences, in which case every two of the residual effects corresponding to those data points may be correlated to at least some minimal extent.

Results that are better suited for most applications can be obtained by adopting a somewhat more elaborate approach. This approach builds on the approach introduced in Part 3 of Subsection b in connection with compound symmetry. Take a_k ($k = 1, 2, \ldots, K$) and r_{ks} ($k = 1, 2, \ldots, K$; $s = 1, 2, \ldots, N_k$) to be uncorrelated random variables, each with mean 0, such that $\text{var}(a_k) = \tau_k^2$ for some nonnegative scalar τ_k and $\text{var}(r_{ks}) = \phi_k^2$ ($s = 1, 2, \ldots, N_k$) for some strictly positive scalar ϕ_k. Further, take f_{ks} ($k = 1, 2, \ldots, K$; $s = 1, 2, \ldots, N_k$) to be random variables, each with mean 0, that are uncorrelated with the a_k's and the r_{ks}'s; take the elements of each of the K sets $\{f_{k1}, f_{k2}, \ldots, f_{kN_k}\}$ ($k = 1, 2, \ldots, K$) to be uncorrelated with those of each of the others; and take the variances and covariances of $f_{k1}, f_{k2}, \ldots, f_{kN_k}$ to be those obtained by regarding these random variables as N_k successive members of a stationary first-order autoregressive process, so that (for some strictly positive scalar η_k and some scalar λ_k in the interval $0 < \lambda_k < 1$) $\text{var}(f_{k1}) = \text{var}(f_{k2}) = \cdots = \text{var}(f_{kN_k}) = \eta_k^2$ and $\text{corr}(f_{ks}, f_{kj}) = \lambda_k^{|t_{kj}-t_{ks}|}$ $(j \neq s = 1, 2, \ldots, N_k)$.

Now, suppose that, for $k = 1, 2, \ldots, K$ and $s = 1, 2, \ldots, N_k$,

$$e_{ks} = a_k + f_{ks} + r_{ks} \tag{4.48}$$

—clearly, this supposition is compatible with the assumptions that $\text{E}(\mathbf{e}) = \mathbf{0}$ and that $\text{var}(\mathbf{e})$ is of the block-diagonal form (4.35). Equality (4.48) decomposes e_{ks} into three uncorrelated components; it implies that $\text{var}(e_{ks})$ equals the quantity σ_k^2 defined by

$$\sigma_k^2 = \tau_k^2 + \eta_k^2 + \phi_k^2. \tag{4.49}$$

Further, for $j \neq s$,

$$\text{cov}(e_{ks}, e_{kj}) = \tau_k^2 + \eta_k^2 \lambda_k^{|t_{kj}-t_{ks}|} \tag{4.50}$$

and hence

$$\text{corr}(e_{ks}, e_{kj}) = \frac{\tau_k^2 + \eta_k^2 \lambda_k^{|t_{kj}-t_{ks}|}}{\sigma_k^2} = \frac{\tau_k^2 + \eta_k^2 \lambda_k^{|t_{kj}-t_{ks}|}}{\phi_k^2 + \tau_k^2 + \eta_k^2} = \frac{(\tau_k/\phi_k)^2 + (\eta_k/\phi_k)^2 \lambda_k^{|t_{kj}-t_{ks}|}}{1 + (\tau_k/\phi_k)^2 + (\eta_k/\phi_k)^2}.$$

The correlation $\text{corr}(e_{ks}, e_{kj})$ can be regarded as a function of the elapsed time $|t_{kj} - t_{ks}|$. Clearly, if the τ_k's, η_k's, ϕ_k's, and λ_k's are such that τ_k/ϕ_k, η_k/ϕ_k, and λ_k do not vary with k, then this function does not vary with k.

In connection with supposition (4.48), it may be advisable to extend the parameter space of η_k by appending the value 0, which corresponds to allowing for the possibility that $\text{var}(f_{k1}) = \text{var}(f_{k2}) = \cdots = \text{var}(f_{kN_k}) = 0$ or, equivalently, that $f_{k1} = f_{k2} = \cdots = f_{kN_k} = 0$ with probability 1. When $\eta_k = 0$, $\text{var}(\mathbf{e}_k)$ is of the compound-symmetric form considered in Subsection b.

In practice, it is common to make the simplifying assumption that the τ_k's, η_k's, ϕ_k's, and λ_k's do not vary with k, so that $\tau_1 = \tau_2 = \cdots = \tau_K = \tau$, $\eta_1 = \eta_2 = \cdots = \eta_K = \eta$, $\phi_1 = \phi_2 = \cdots = \phi_K = \phi$, and $\lambda_1 = \lambda_2 = \cdots = \lambda_K = \lambda$ for some strictly positive scalar ϕ, for some nonnegative scalar τ, for some strictly positive (or alternatively nonnegative) scalar η, and for some scalar λ in the interval $0 < \lambda < 1$. Under that assumption, neither $\text{var}(e_{ks})$ nor $\text{corr}(e_{ks}, e_{kj})$ varies with k. Moreover, if λ and the ratios τ/ϕ and η/ϕ are known, then $\text{var}(\mathbf{e})$ is of the form $\sigma^2\mathbf{H}$ of $\text{var}(\mathbf{e})$ in the Aitken generalization of the G–M model. When τ, η, ϕ, and λ are regarded as unknown parameters and are taken to be the elements of the vector $\boldsymbol{\theta}$, $\text{var}(\mathbf{e})$ is of the form $\mathbf{V}(\boldsymbol{\theta})$ of $\text{var}(\mathbf{e})$ in the general linear model.

f. Example: dental data

Potthoff and Roy (1964) reported the results of a study involving youngsters (carried out at the University of North Carolina Dental School) of how the distance from the center of the pituitary to the pteryomaxillary fissure changes with age. A measurement of this distance was obtained for each of 27 youngsters (11 girls and 16 boys) at each of 4 ages (8, 10, 12, and 14 years). The resultant data are reproduced in Table 4.6.

These data may be suitable for application of a general linear model. We can take $N = 27 \times 4 = 108$. And, upon taking $\mathbf{y}_1, \mathbf{y}_2, \ldots, \mathbf{y}_{27}$ to be the 4-dimensional subvectors of $\mathbf{y}$ defined implicitly by the partitioning $\mathbf{y}' = (\mathbf{y}_1', \mathbf{y}_2', \ldots, \mathbf{y}_{27}')$, we can (for $k = 1, 2, \ldots, 27$) regard the measurements obtained on the kth youngster at ages 8, 10, 12, and 14 as the observed values of the elements of $\mathbf{y}_k$. Further, we can take $C = 2$, take u_1 to be the youngster's age at the time of the measurement, and take u_2 to be a variable that has one value, say 0, if the youngster is a girl, and a second (different) value, say 1, if the youngster is a boy.

TABLE 4.6. Data on 27 youngsters (11 girls and 16 boys) at 4 different ages: each data point is a measurement of the distance (in millimeters) from the center of the pituitary to the pteryomaxillary fissure (Potthoff and Roy 1964).

Girls					Boys				
	Age (in years)					Age (in years)			
Youngster	8	10	12	14	Youngster	8	10	12	14
1	21	20	21.5	23	12	26	25	29	31
2	21	21.5	24	25.5	13	21.5	22.5	23	26.5
3	20.5	24	24.5	26	14	23	22.5	24	27.5
4	23.5	24.5	25	26.5	15	25.5	27.5	26.5	27
5	21.5	23	22.5	23.5	16	20	23.5	22.5	26
6	20	21	21	22.5	17	24.5	25.5	27	28.5
7	21.5	22.5	23	25	18	22	22	24.5	26.5
8	23	23	23.5	24	19	24	21.5	24.5	25.5
9	20	21	22	21.5	20	23	20.5	31	26
10	16.5	19	19	19.5	21	27.5	28	31	31.5
11	24.5	25	28	28	22	23	23	23.5	25
					23	21.5	23.5	24	28
					24	17	24.5	26	29.5
					25	22.5	25.5	25.5	26
					26	23	24.5	26	30
					27	22	21.5	23.5	25

We might consider taking

$$\delta(\mathbf{u}) = \begin{cases} r(u_1) & \text{if } u_2 = 0, \\ s(u_1) & \text{if } u_2 = 1, \end{cases} \tag{4.51}$$

where $r(u_1)$ and $s(u_1)$ are both polynomials in u_1 (the coefficients of which are unknown parameters). For example, $r(u_1)$ and $s(u_1)$ might both be polynomials of degree 3 in u_1; that is, $r(u_1)$ and $s(u_1)$ might be of the form

$$r(u_1) = \beta_1 + \beta_2 u_1 + \beta_3 u_1^2 + \beta_4 u_1^3 \tag{4.52}$$

and

$$s(u_1) = \beta_5 + \beta_6 u_1 + \beta_7 u_1^2 + \beta_8 u_1^3, \tag{4.53}$$

where $\beta_1, \beta_2, \ldots, \beta_8$ are unknown parameters. If $\delta(\mathbf{u})$ is taken to be of the form (4.51) for polynomials $r(u_1)$ and $s(u_1)$ of the form (4.52) and (4.53), then the vector $\boldsymbol{\beta}$ (in the general linear model) is of dimension 8 (with elements $\beta_1, \beta_2, \ldots, \beta_8$) and the model matrix is

$$\mathbf{X} = \begin{pmatrix} \mathbf{X}_* & \mathbf{0} \\ \mathbf{X}_* & \mathbf{0} \\ \vdots & \vdots \\ \mathbf{X}_* & \mathbf{0} \\ \mathbf{0} & \mathbf{X}_* \\ \mathbf{0} & \mathbf{X}_* \\ \vdots & \vdots \\ \mathbf{0} & \mathbf{X}_* \end{pmatrix}, \quad \text{where } \mathbf{X}_* = \begin{pmatrix} 1 & 8 & 64 & 512 \\ 1 & 10 & 100 & 1000 \\ 1 & 12 & 144 & 1728 \\ 1 & 14 & 196 & 2744 \end{pmatrix}.$$

Taking $\delta(\mathbf{u})$ to be of the form (4.51) allows for the possibility that the relationship between distance and age may be markedly different for boys than for girls.

The distance measurements can be regarded as longitudinal data. The 27 youngsters (11 girls and 16 boys) form $K = 27$ observational units, each of which contributes 4 data points. And age plays the role of time.

Partition the (column) vector $\mathbf{e}$ of residual effects (in the general linear model) into 4-dimensional subvectors $\mathbf{e}_1, \mathbf{e}_2, \ldots, \mathbf{e}_{27}$ [in such a way that $\mathbf{e}' = (\mathbf{e}_1', \mathbf{e}_2', \ldots, \mathbf{e}_{27}')$]. (This partitioning corresponds to the partitioning of $\mathbf{y}$ into the subvectors $\mathbf{y}_1, \mathbf{y}_2, \ldots, \mathbf{y}_{27}$.) And (for $k = 1, 2, \ldots, 27$) denote by e_{k1}, e_{k2}, e_{k3}, and e_{k4} the elements of the subvector $\mathbf{e}_k$—these are the residual effects that correspond to the distance measurements on the kth youngster at ages 8, 10, 12, and 14. Then, proceeding as in Subsection e, we could take var($\mathbf{e}$) to be of the form

$$\text{var}(\mathbf{e}) = \text{diag}[\text{var}(\mathbf{e}_1),\ \text{var}(\mathbf{e}_2),\ \ldots,\ \text{var}(\mathbf{e}_K)].$$

Further, for $k = 1, 2, \ldots, 27$, we could take var($\mathbf{e}_k$) to be of the form associated with the supposition that (for $s = 1, 2, 3, 4$) e_{ks} is expressible in the form of the decomposition (4.48). That is, we could take the diagonal elements of var($\mathbf{e}_k$) to be of the form (4.49), and the off-diagonal elements to be of the form (4.50). Expressions (4.49) and (4.50) determine var($\mathbf{e}_k$) up to the values of the four parameters τ_k, η_k, ϕ_k, and λ_k. It is assumed that, for $k = 1, 2, \ldots, 11$ (corresponding to the 11 girls), the values of τ_k, η_k, ϕ_k, and λ_k do not vary with k, and similarly that, for $k = 12, 13, \ldots, 27$ (corresponding to the 16 boys), the values of τ_k, η_k, ϕ_k, and λ_k do not vary with k. Under that assumption, var($\mathbf{e}$) would depend on 8 parameters. Presumably, those parameters would be unknown, in which case the vector $\boldsymbol{\theta}$ of unknown parameters in the variance-covariance matrix $\mathbf{V}(\boldsymbol{\theta})$ (of the vector $\mathbf{e}$ of residual effects in the general linear model) would be of dimension 8.

g. Spatial data

There are situations where each of the N data points is associated with a specific location (in 1-, 2-, or 3-dimensional space). For example, the data points might be measurements of the hardness of samples of water obtained from different wells. Data of this kind are among those referred to as *spatial data*.

Suppose that the observed value of each of the random variables $y_1, y_2, \ldots, y_N$ in the general linear model is associated with a specific location in D-dimensional space. Further, let us represent an arbitrary location in D-dimensional space by a D-dimensional column vector $\mathbf{s}$ of "coordinates," denote by $\mathbf{s}_1, \mathbf{s}_2, \ldots, \mathbf{s}_N$ the values of $\mathbf{s}$ corresponding to $\mathbf{y}_1, \mathbf{y}_2, \ldots, \mathbf{y}_N$, respectively, and take S to be a finite or infinite set of values of $\mathbf{s}$ that includes $\mathbf{s}_1, \mathbf{s}_2, \ldots, \mathbf{s}_N$ (and perhaps other values of $\mathbf{s}$ that may be of interest). Assume (as can be done essentially without loss of generality) that the vector $\mathbf{u}$ of explanatory variables includes the vector $\mathbf{s}$ as a subvector or, more generally, that the elements of $\mathbf{s}$ are expressible as functions of $\mathbf{u}$, so that (for $i = 1, 2, \ldots, N$) the ith value $\mathbf{s}_i$ of $\mathbf{s}$ is determinable from the ith value $\mathbf{u}_i$ of $\mathbf{u}$.

Typically, data points associated with locations that are in close proximity tend to be more alike than those associated with locations that are farther apart. This phenomenon may be due in part to "systematic forces" that manifest themselves as "trends" or "gradients" in the surface defined by the function $\delta(\mathbf{u})$—whatever part may be due to these systematic forces is sometimes referred to as large-scale variation. However, typically not all of this phenomenon is attributable to systematic forces or is reflected in the surface defined by $\delta(\mathbf{u})$. There is generally a nonsystematic component. The nonsystematic component takes the form of what is sometimes called small-scale variation and is reflected in the correlation matrix of the residual effects $e_1, e_2, \ldots, e_N$. The residual effects may be positively correlated, with the correlation being greatest among residual effects corresponding to locations that are in close proximity.

It is supposed that the residual effects $e_1, e_2, \ldots, e_N$ are expressible as follows:

$$e_i = a_i + r_i \qquad (i = 1, 2, \ldots, N). \tag{4.54}$$

Here, $a_1, a_2, \ldots, a_N$ and $r_1, r_2, \ldots, r_N$ are random variables, each with an expected value of 0. Moreover, $r_1, r_2, \ldots, r_N$ are assumed to be uncorrelated with each other and with $a_1, a_2, \ldots, a_N$. And it is sometimes assumed that

$$\text{var}(a_i) = \tau^2 \quad \text{and} \quad \text{var}(r_i) = \phi^2 \qquad (i = 1, 2, \ldots, N) \tag{4.55}$$

for some nonnegative scalar τ and some strictly positive scalar ϕ, in which case

$$\text{var}(e_i) = \sigma^2 \qquad (i = 1, 2, \ldots, N), \tag{4.56}$$

where $\sigma^2 = \tau^2 + \phi^2 > 0$.

The random variables $a_1, a_2, \ldots, a_N$ may be correlated with the magnitude of the correlation between any two of them, say a_i and a_j, depending on the corresponding locations $\mathbf{s}_i$ and $\mathbf{s}_j$. Of particular interest is the case where

$$\text{cov}(a_i, a_j) = \sqrt{\text{var}\, a_i}\sqrt{\text{var}\, a_j}\, K(\mathbf{s}_i - \mathbf{s}_j) \qquad (i, j = 1, 2, \ldots, N). \tag{4.57}$$

Here, $K(\cdot)$ is a function whose domain is the set

$$H = \{\mathbf{h} \in \mathbb{R}^D \;:\; \mathbf{h} = \mathbf{s} - \mathbf{t} \text{ for } \mathbf{s}, \mathbf{t} \in S\}$$

and that has the following three properties: (1) $K(\mathbf{0}) = 1$; (2) $K(-\mathbf{h}) = K(\mathbf{h})$ for $\mathbf{h} \in H$; and (3) $\sum_{i=1}^{M}\sum_{j=1}^{M} x_i x_j K(\mathbf{t}_i - \mathbf{t}_j) \geq 0$ for every positive integer M, for all not-necessarily-distinct vectors $\mathbf{t}_1, \mathbf{t}_2, \ldots, \mathbf{t}_M$ in S, and for all scalars $x_1, x_2, \ldots, x_M$. Note that the third property of $K(\cdot)$, in combination with the first two properties, establishes that the $M \times M$ matrix with ijth element $K(\mathbf{t}_i - \mathbf{t}_j)$ can serve as a correlation matrix—it is symmetric and nonnegative definite and its diagonal elements equal 1. It establishes, in particular, that the $N \times N$ matrix with ijth element $K(\mathbf{s}_i - \mathbf{s}_j)$ can serve as a correlation matrix. The function $K(\cdot)$ is sometimes referred to as an *autocorrelation function* or a *correlogram*.

Suppose that the variances of the random variables $r_1, r_2, \ldots, r_N$ and the variances and covariances of the random variables $a_1, a_2, \ldots, a_N$ are of the form specified in conditions (4.55) and (4.57). Then, the variances of the residual effects $e_1, e_2, \ldots, e_N$ are of the form (4.56), and the covariances of the residual effects are of the form

$$\text{cov}(e_i, e_j) = \text{cov}(a_i, a_j) = \tau^2 K(\mathbf{s}_i - \mathbf{s}_j) \qquad (j \neq i = 1, 2, \ldots, N). \tag{4.58}$$

Further, the correlation between the ith and jth residual effects is of the form

$$\text{corr}(e_i, e_j) = \frac{\tau^2}{\sigma^2} K(\mathbf{s}_i - \mathbf{s}_j) = \frac{\tau^2}{\tau^2 + \phi^2} K(\mathbf{s}_i - \mathbf{s}_j) = \frac{(\tau/\phi)^2}{1 + (\tau/\phi)^2} K(\mathbf{s}_i - \mathbf{s}_j).$$

Accordingly, the distribution of the residual effects is weakly stationary in the sense that all of the residual effects have the same variance and in the sense that the covariance and correlation between any two residual effects depend on the corresponding locations only through the difference between the locations.

The variance σ^2 of the residual effects is the sum of two components: (1) the variance τ^2 of the a_i's and (2) the variance ϕ^2 of the r_i's. The first component τ^2 accounts for a part of whatever variability is spatial in origin, that is, a part of whatever variability is related to differences in location—it accounts for the so-called small-scale variation. The second component ϕ^2 accounts for the remaining variability, including any variability that may be attributable to measurement error.

Depending on the application, some of the N locations $\mathbf{s}_1, \mathbf{s}_2, \ldots, \mathbf{s}_N$ may be identical. For example, the N data points may represent the scores achieved on a standardized test by N different students and (for $i = 1, 2, \ldots, N$) $\mathbf{s}_i$ may represent the location of the school system in which the the ith student is enrolled. In such a circumstance, the variation accounted for by ϕ^2 would include the variation among those residual effects for which the corresponding locations are the same. If there were L distinct values of $\mathbf{s}$ represented among the N locations $\mathbf{s}_1, \mathbf{s}_2, \ldots, \mathbf{s}_N$ and the residual effects $e_1, e_2, \ldots, e_N$ were divided into L classes in such a way that the locations corresponding to the residual effects in any particular class were identical, then the ratio $\tau^2/(\tau^2 + \phi^2)$ could be regarded as the intraclass correlation—refer to Subsection b.

There may be spatial variation that is so localized in nature that it does not contribute to the covariances among the residual effects. This kind of spatial variation is sometimes referred to as microscale variation (e.g., Cressie 1993). The range of its influence is less than the distance between any two of the locations $\mathbf{s}_1, \mathbf{s}_2, \ldots, \mathbf{s}_N$. Its contribution, which has come to be known as the "nugget effect," is reflected in the component ϕ^2.

To complete the specification of the form (4.58) of the covariances of the residual effects $e_1, e_2, \ldots, e_N$, it remains to specify the form of the function $K(\cdot)$. In that regard, it suffices to take [for an arbitrary D-dimensional column vector $\mathbf{h}$ in the domain H of $K(\cdot)$]

$$K(\mathbf{h}) = \mathrm{E}(\cos \mathbf{h}'\mathbf{w}), \tag{4.59}$$

where $\mathbf{w}$ is a D-dimensional random column vector (the distribution of which may depend on unknown parameters).

Let us confirm this claim; that is, let us confirm that when $K(\cdot)$ is taken to be of the form (4.59), it has (even if S and hence H comprise all of $\mathbb{R}^D$) the three properties required of an autocorrelation function. Recall that $\cos 0 = 1$; that for any real number x, $\cos(-x) = \cos x$; and that for any real numbers x and z, $\cos(x-z) = (\cos x)(\cos z) + (\sin x)(\sin z)$. When $K(\cdot)$ is taken to be of the form (4.59), we have (in light of the properties of the cosine operator) that (1)

$$K(\mathbf{0}) = \mathrm{E}(\cos \mathbf{0}'\mathbf{w}) = \mathrm{E}(\cos 0) = \mathrm{E}(1) = 1;$$

that (2) for $\mathbf{h} \in H$,

$$K(-\mathbf{h}) = \mathrm{E}[\cos(-\mathbf{h})'\mathbf{w}] = \mathrm{E}[\cos(-\mathbf{h}'\mathbf{w})] = \mathrm{E}(\cos \mathbf{h}'\mathbf{w}) = K(\mathbf{h});$$

and that (3) for every positive integer M, any M vectors $\mathbf{t}_1, \mathbf{t}_2, \ldots, \mathbf{t}_M$ in S, and any M scalars $x_1, x_2, \ldots, x_M$,

$$\begin{aligned}
\sum_{i=1}^{M}\sum_{j=1}^{M} x_i x_j K(\mathbf{t}_i - \mathbf{t}_j) &= \sum_{i=1}^{M}\sum_{j=1}^{M} x_i x_j \,\mathrm{E}[\cos(\mathbf{t}_i'\mathbf{w} - \mathbf{t}_j'\mathbf{w})] \\
&= \mathrm{E}\Big[\sum_{i=1}^{M}\sum_{j=1}^{M} x_i x_j \cos(\mathbf{t}_i'\mathbf{w} - \mathbf{t}_j'\mathbf{w})\Big] \\
&= \mathrm{E}\Big[\sum_{i=1}^{M}\sum_{j=1}^{M} x_i x_j (\cos \mathbf{t}_i'\mathbf{w})(\cos \mathbf{t}_j'\mathbf{w}) + \sum_{i=1}^{M}\sum_{j=1}^{M} x_i x_j (\sin \mathbf{t}_i'\mathbf{w})(\sin \mathbf{t}_j'\mathbf{w})\Big] \\
&= \mathrm{E}\Big[\Big(\sum_{i=1}^{M} x_i \cos \mathbf{t}_i'\mathbf{w}\Big)^2 + \Big(\sum_{i=1}^{M} x_i \sin \mathbf{t}_i'\mathbf{w}\Big)^2\Big] \\
&\geq 0.
\end{aligned}$$

Thus, when $K(\cdot)$ is taken to be of the form (4.59), it has (even if S and H comprise all of $\mathbb{R}^D$) the requisite three properties.

Expression (4.59) depends on the distribution of $\mathbf{w}$. The evaluation of this expression for a distribution of any particular form is closely related to the evaluation of the characteristic function of a distribution of that form. The characteristic function of the distribution of $\mathbf{w}$ is the function $c(\cdot)$ defined (for an arbitrary D-dimensional column vector $\mathbf{h}$) by $c(\mathbf{h}) = \mathrm{E}(e^{i\mathbf{h}'\mathbf{w}})$ (where $i = \sqrt{-1}$). The characteristic function of the distribution of $\mathbf{w}$ can be expressed in the form

$$c(\mathbf{h}) = \mathrm{E}(\cos \mathbf{h}'\mathbf{w}) + i\,\mathrm{E}(\sin \mathbf{h}'\mathbf{w}), \tag{4.60}$$

the real component of which is identical to expression (4.59). It is worth mentioning that if the distribution of $\mathbf{w}$ has a moment generating function, say $m(\cdot)$, then

$$c(\mathbf{h}) = m(i\mathbf{h}) \tag{4.61}$$

(e.g., Grimmett and Welsh 1986, p. 117).

In the special case where $\mathbf{w}$ is distributed symmetrically about $\mathbf{0}$ (i.e., where $-\mathbf{w} \sim \mathbf{w}$), expression (4.60) simplifies to

$$c(\mathbf{h}) = \mathrm{E}(\cos \mathbf{h}'\mathbf{w}),$$

as is evident upon recalling that (for any real number x) $\sin(-x) = -\sin x$ and then observing that $\mathrm{E}(\sin \mathbf{h}'\mathbf{w}) = \mathrm{E}[\sin \mathbf{h}'(-\mathbf{w})] = \mathrm{E}[\sin(-\mathbf{h}'\mathbf{w})] = -\mathrm{E}(\sin \mathbf{h}'\mathbf{w})$ [which implies that $\mathrm{E}(\sin \mathbf{h}'\mathbf{w}) = 0$]. Thus, in the special case where $\mathbf{w}$ is distributed symmetrically about $\mathbf{0}$, taking $K(\cdot)$ to be of the form (4.59) is equivalent to taking $K(\cdot)$ to be of the form

$$K(\mathbf{h}) = c(\mathbf{h}).$$

Suppose, for example, that $\mathbf{w} \sim N(\mathbf{0}, \boldsymbol{\Gamma})$, where $\boldsymbol{\Gamma} = \{\gamma_{ij}\}$ is a symmetric nonnegative definite matrix. Then, it follows from result (3.5.47) that the moment generating function $m(\cdot)$ of the distribution of $\mathbf{w}$ is

$$m(\mathbf{h}) = \exp(\tfrac{1}{2}\mathbf{h}'\boldsymbol{\Gamma}\mathbf{h}),$$

implying [in light of result (4.61)] that the characteristic function $c(\cdot)$ of the distribution of $\mathbf{w}$ is

$$c(\mathbf{h}) = \exp(-\tfrac{1}{2}\mathbf{h}'\boldsymbol{\Gamma}\mathbf{h}).$$

Thus, the choices for the form of the function $K(\cdot)$ include

$$K(\mathbf{h}) = \exp(-\tfrac{1}{2}\mathbf{h}'\boldsymbol{\Gamma}\mathbf{h}). \tag{4.62}$$

Autocorrelation functions of the form (4.62) are referred to as Gaussian.

When $K(\cdot)$ is taken to be of the form (4.62), the matrix $\mathbf{V}(\boldsymbol{\theta})$ (the variance-covariance matrix of the residual effects in the general linear model) is the $N \times N$ matrix whose diagonal elements are $\sigma^2 = \tau^2 + \phi^2$ and whose ijth off-diagonal element equals $\tau^2 \exp[-\frac{1}{2}(\mathbf{s}_i - \mathbf{s}_j)'\boldsymbol{\Gamma}(\mathbf{s}_i - \mathbf{s}_j)]$. Assuming that $\boldsymbol{\Gamma}$ is to be regarded as unknown, the vector $\boldsymbol{\theta}$ (on which the elements of the matrix $\mathbf{V}(\boldsymbol{\theta})$ are functionally dependent) could be taken to be the $[2 + D(D+1)/2]$-dimensional column vector whose elements are ϕ, τ, and γ_{ij} $(j > i = 1, 2, \ldots, D)$.

If in addition to satisfying the three properties required of an autocorrelation function, the function $K(\cdot)$ is such that $K(\mathbf{h})$ depends on $\mathbf{h}$ only through the value of $\|\mathbf{h}\|$ (the usual norm of the vector $\mathbf{h}$), the function is said to be *isotropic* (e.g., Cressie 1993). For example, an autocorrelation function of the form (4.62) would be isotropic if the matrix $\boldsymbol{\Gamma}$ were restricted to $D \times D$ matrices of the form $\gamma\mathbf{I}$, where γ is a nonnegative scalar. If $\boldsymbol{\Gamma}$ were restricted in that way (and the scalar γ were regarded as an unknown parameter), then (in connection with the variance-covariance matrix $\mathbf{V}(\boldsymbol{\theta})$ of the residual effects in the general linear model) the vector $\boldsymbol{\theta}$ could be taken to be the 3-dimensional (column) vector with elements ϕ, τ, and γ. Isotropic autocorrelation functions are quite popular. For a list of some of the more widely employed isotropic autocorrelation functions, refer, for example, to Littell et al. (2006, sec. 11.3).

h. Example: tree-height data

Zhang, Bi, Cheng, and Davis (2004) considered a set of measurements of the diameters (at breast height overbark) and heights of trees in a circular plot of radius 40 m. The plot is located in the regrowth *Eucalyptus fastigata* forests in Glenbog State Forest in New South Wales, Australia. While *E. fastigata* was the dominant species, other eucalypts were also present as was a species of smaller trees or shrubs. Double stems, coppicing (i.e., the formation of trees or shrubs from shoots or root

TABLE 4.7. The diameters and heights and the coordinates of the locations of 101 trees in a circular region of radius 40 m—the locations are relative to the center of the region (Zhang, Bi, Cheng, and Davis 2004).

Tree	Diam. (cm)	Hgt. (m)	Coordinates 1st (m)	Coordinates 2nd (m)
1	17.5	14.0	0.73	1.31
2	18.3	15.1	1.99	−0.24
3	10.1	7.3	2.96	−3.65
4	49.0	26.5	32.90	−15.34
5	32.9	22.9	18.28	−11.87
6	50.6	16.3	22.31	−1.95
7	66.2	26.7	27.69	−3.40
8	73.5	31.0	29.30	−0.51
9	39.7	26.5	37.65	−1.97
10	69.8	32.5	39.75	−3.47
11	67.4	28.3	15.78	0.83
12	68.6	31.1	33.45	9.60
13	27.0	15.2	34.43	10.53
14	44.8	25.2	36.40	12.54
15	44.4	29.2	32.00	15.61
16	44.5	32.4	29.53	17.05
17	72.0	28.8	9.92	6.20
18	94.2	33.0	12.70	9.57
19	63.8	29.0	29.39	22.97
20	12.7	12.9	20.61	22.89
21	40.6	23.9	14.65	22.56
22	31.5	13.2	13.90	23.14
23	40.2	26.7	16.37	26.21
24	38.9	22.6	16.70	34.25
25	20.0	15.0	16.14	34.62
26	17.7	9.1	10.60	29.13
27	24.3	18.0	11.18	29.13
28	10.4	10.1	−0.34	3.89
29	52.2	28.9	−0.71	20.39
30	17.7	13.1	0.00	21.90
31	19.9	13.6	−0.75	21.49
32	56.4	29.0	−1.68	24.04
33	27.0	16.6	−2.55	36.51
34	37.0	28.5	−2.73	19.41
35	55.2	24.1	−3.43	32.62
36	81.2	30.0	−4.59	32.68
37	41.3	27.4	−4.37	8.21
38	16.9	13.8	−19.75	34.21
39	15.2	12.3	−19.61	33.94
40	11.6	12.4	−17.09	17.69
41	12.0	10.5	−19.64	21.06
42	11.6	13.6	−22.89	32.68
43	10.6	10.9	−23.09	18.04
44	36.7	20.0	−26.04	22.63
45	31.5	15.9	−25.49	22.95
46	65.8	33.0	−10.57	6.86
47	51.1	27.4	−14.41	6.72
48	19.4	10.5	−14.74	7.19
49	30.0	21.9	−18.61	9.07
50	40.2	23.3	−18.71	9.53
51	136.0	32.5	−27.23	13.28
52	43.7	31.5	−32.13	18.55
53	23.5	22.0	−31.55	18.95
54	53.1	25.1	−19.79	4.93
55	18.4	12.1	−21.30	6.92
56	38.8	24.7	−24.06	7.82
57	66.7	35.3	−32.98	12.00
58	61.7	31.5	−36.32	9.73
59	15.9	11.2	−35.84	9.60
60	14.5	11.5	−22.11	4.70
61	18.8	14.2	−1.98	−0.31
62	19.4	14.8	−2.26	−0.44
63	16.2	12.7	−8.25	−0.87
64	55.9	26.0	−13.10	−2.31
65	55.9	25.6	−13.10	−2.31
66	39.6	24.3	−28.44	−2.99
67	12.7	11.4	−29.43	−2.06
68	10.2	10.6	−31.08	−2.72
69	10.2	9.2	−31.18	−2.73
70	10.3	7.2	−31.88	−2.79
71	15.3	13.2	−33.65	−1.77
72	13.5	8.4	−36.08	−1.26
73	10.8	8.3	−34.33	−11.16
74	55.8	27.9	−28.09	−11.35
75	55.8	27.5	−28.09	−11.35
76	41.4	25.5	−25.72	−14.85
77	70.4	25.0	−18.45	−11.99
78	79.0	27.5	−25.29	−19.77
79	12.0	10.6	−8.19	−13.63
80	14.7	14.5	−28.34	−27.37
81	12.4	10.2	−16.44	−24.38
82	19.4	16.8	−18.33	−29.34
83	120.0	34.0	−2.41	−17.13
84	10.2	10.0	−13.75	−30.88
85	28.7	21.3	−14.37	−33.88
86	36.5	23.9	−14.49	−34.15
87	30.4	21.0	−9.79	−32.04
88	18.8	12.1	−9.76	−31.94
89	59.0	26.6	−6.62	−24.73
90	38.7	19.7	−5.88	−23.58
91	15.0	14.1	−8.03	−34.79
92	58.0	30.8	3.24	−13.00
93	34.7	26.6	3.54	−33.71
94	38.3	29.2	3.65	−34.71
95	44.4	29.2	5.51	−25.92
96	48.3	25.7	6.36	−25.52
97	32.8	15.7	7.73	−23.78
98	83.7	26.7	14.39	−15.43
99	39.0	25.1	19.46	−16.32
100	49.0	25.4	28.67	−21.60
101	14.4	11.4	3.61	−29.38

suckers rather than seed), and double leaders occurred with some frequency. The diameters and heights of the trees are listed in Table 4.7, along with the location of each tree. Only those trees with a diameter greater than 10 cm were considered; there were 101 such trees.

Interest centered on ascertaining how tree height relates to tree diameter. Refer to Zhang et al. (2004) for some informative graphical displays that bear on this relationship and that indicate how the trees are distributed within the plot.

The setting is one that might be suitable for the application of a general linear model. We could take $N = 101$, take the observed value of y_i to be the logarithm of the height of the ith tree ($i = 1, 2, \ldots, 101$), take $C = 3$, take u_1 to be the diameter of the tree, and take u_2 and u_3 to be the first and second coordinates of the location of the tree. And we could consider taking $\delta(\mathbf{u})$ to be of the simple form

$$\delta(\mathbf{u}) = \beta_1 + \beta_2 \log u_1, \tag{4.63}$$

where β_1 and β_2 are unknown parameters; expression (4.63) is a first-degree polynomial in $\log u_1$ with coefficients β_1 and β_2. Or, following Zhang et al. (2004), we could consider taking $\delta(\mathbf{u})$ to be a variation on the first-degree polynomial (4.63) in which the coefficients of the polynomial are allowed to differ from location to location (i.e., allowed to vary with u_2 and u_3).

The logarithms of the heights of the 101 trees constitute spatial data. The logarithm of the height of each tree is associated with a specific location in 2-dimensional space; this location is that represented by the value of the 2-dimensional vector $\mathbf{s}$ whose elements (in this particular setting) are u_2 and u_3. Accordingly, the residual effects $e_1, e_2, \ldots, e_N$ are likely to be correlated to an extent that depends on the relative locations of the trees with which they are associated. Trees that are in close proximity may tend to be subject to similar conditions, and consequently the residual effects identified with those trees may tend to be similar. It is worth noting, however, that the influence on the residual effects of any such similarity in conditions could be at least partially offset by the influence of competition for resources among neighboring trees.

In light of the spatial nature of the data, we might wish to take the variance-covariance matrix $\mathbf{V}(\boldsymbol{\theta})$ of the vector $\mathbf{e}$ of residual effects in the general linear model to be that whose diagonal and off-diagonal elements are of the form (4.56) and (4.58). Among the choices for the form of the autocorrelation function $K(\cdot)$ is that specified by expression (4.62). It might or might not be realistic to restrict the matrix $\boldsymbol{\Gamma}$ in expression (4.62) to be of the form $\boldsymbol{\Gamma} = \gamma\mathbf{I}$ (where γ is a nonnegative scalar). When $\boldsymbol{\Gamma}$ is restricted in that way, the autocorrelation function $K(\cdot)$ is isotropic, and the vector $\boldsymbol{\theta}$ could be taken to be the 3-dimensional (column) vector with elements ϕ, τ, and γ (assuming that each of these 3 scalars is to be regarded as an unknown parameter).

4.5 Multivariate Data

There are situations where the data are multivariate in nature. That is, there are situations where the data consist of possibly multiple observations on each of a number of "observational units" and where the multiple observations represent the observed values of different "response" variables. For example, the observational units might be individual people, and the response variables might be a person's height, weight, and blood pressure.

There is a similarity to longitudinal data; longitudinal data were discussed earlier (in Section 4.4e). In both cases, there are possibly multiple observations on each of a number of observational units. However, in the case of multivariate data, the multiple observations represent the observed values of different response variables, while in the case of longitudinal data, the multiple observations represent the values obtained at possibly different points in time for what is essentially the same variable. Actually, longitudinal data can be regarded as a special kind of multivariate data—think of the different points in time as defining different response variables. Nevertheless, there is a very meaningful distinction between longitudinal data and the sort of "unstructured" multivariate data that is the subject of the present section. Longitudinal data exhibit a structure that can be exploited for modeling purposes. Models (like those considered in Section 4.4e) that exploit that structure are not suitable for use with unstructured multivariate data.

a. Application of the general linear model to multivariate data

Let us consider the application of the general linear model under circumstances where the observed values of the random variables $y_1, y_2, \dots, y_N$ represent multivariate data. Suppose that there are R observational units, numbered $1, 2, \dots, R$, and S response variables, numbered $1, 2, \dots, S$. In general, the data on any particular observational unit may be incomplete; that is, the observed value of one or more of the S response variables may not be available for that observational unit. Let R_s represent the number of observational units for which the observed value of the sth response variable is available ($s = 1, 2, \dots, S$); and observe that $\sum_{s=1}^{S} R_s = N$ (the total number of data points). Further, let $r_{s1}, r_{s2}, \dots, r_{sR_s}$ represent the subsequence of the sequence $1, 2, \dots, R$ such that the integer r is in the subsequence if the sth response variable is available on the rth observational unit.

Let us assume that the random variables $y_1, y_2, \dots, y_N$ have been numbered in such a way that

$$\mathbf{y}' = (\mathbf{y}_1', \mathbf{y}_2', \dots, \mathbf{y}_S'),$$

where (for $s = 1, 2, \dots, S$) $\mathbf{y}_s$ is the R_s-dimensional column vector, the kth element of which is the random variable whose observed value is the value obtained for the sth response variable on the r_{sk}th observational unit ($k = 1, 2, \dots, R_s$). The setting is such that it is convenient to use two subscripts instead of one to distinguish among $y_1, y_2, \dots, y_N$ and also to distinguish among the residual effects $e_1, e_2, \dots, e_N$. Let us use the first subscript to identify the response variable and the second to identify the observational unit. Accordingly, let us write $y_{sr_{sk}}$ for the kth element of $\mathbf{y}_s$, so that, by definition,

$$\mathbf{y}_s = (y_{sr_{s1}}, y_{sr_{s2}}, \dots, y_{sr_{sR_s}})'.$$

And let us write $e_{sr_{sk}}$ for the residual effect corresponding to $y_{sr_{sk}}$.

In a typical application, the model is taken to be such that the model matrix is of the block-diagonal form

$$\mathbf{X} = \begin{pmatrix} \mathbf{X}_1 & \mathbf{0} & \dots & \mathbf{0} \\ \mathbf{0} & \mathbf{X}_2 & & \mathbf{0} \\ \vdots & & \ddots & \\ \mathbf{0} & \mathbf{0} & & \mathbf{X}_S \end{pmatrix}, \tag{5.1}$$

where (for $s = 1, 2, \dots, S$) $\mathbf{X}_s$ is of dimensions $R_s \times P_s$ (and where $P_1, P_2, \dots, P_S$ are positive integers that sum to P). Now, suppose that $\mathbf{X}$ is of the form (5.1), and partition the parameter vector $\boldsymbol{\beta}$ into S subvectors $\boldsymbol{\beta}_1, \boldsymbol{\beta}_2, \dots, \boldsymbol{\beta}_S$ of dimensions $P_1, P_2, \dots, P_S$, respectively, so that

$$\boldsymbol{\beta}' = (\boldsymbol{\beta}_1', \boldsymbol{\beta}_2', \dots, \boldsymbol{\beta}_S').$$

Further, partition the vector $\mathbf{e}$ of residual effects into S subvectors $\mathbf{e}_1, \mathbf{e}_2, \dots, \mathbf{e}_S$ of dimensions $R_1, R_2, \dots, R_S$, respectively, so that

$$\mathbf{e}' = (\mathbf{e}_1', \mathbf{e}_2', \dots, \mathbf{e}_S'),$$

where (for $s = 1, 2, \dots, S$) $\mathbf{e}_s' = (e_{sr_{s1}}, e_{sr_{s2}}, \dots, e_{sr_{sR_s}})'$—the partitioning of $\mathbf{e}$ is conformal to the partitioning of $\mathbf{y}$. Then, the model equation $\mathbf{y} = \mathbf{X}\boldsymbol{\beta} + \mathbf{e}$ is reexpressible as

$$\mathbf{y}_s = \mathbf{X}_s\boldsymbol{\beta}_s + \mathbf{e}_s \quad (s = 1, 2, \dots, S). \tag{5.2}$$

Note that the model equation for the vector $\mathbf{y}_s$ of observations on the sth response variable depends on only P_s of the elements of $\boldsymbol{\beta}$, namely, those P_s elements that are members of the subvector $\boldsymbol{\beta}_s$. In practice, it is often the case that $P_1 = P_2 = \cdots = P_S = P_*$ (where $P_* = P/S$) and that there is a matrix $\mathbf{X}_*$ of dimensions $R \times P_*$ such that (for $s = 1, 2, \ldots, S$) the (first through R_sth) rows of $\mathbf{X}_s$ are respectively the $r_{s1}, r_{s2}, \ldots, r_{sR_s}$th rows of $\mathbf{X}_*$.

An important special case is that where there is complete information on every observational unit; that is, where every one of the S response variables is observed on every one of the R observational units. Then, $R_1 = R_2 = \cdots = R_S = R$. And, commonly, $P_1 = P_2 = \cdots = P_S = P_*$ $(= P/S)$ and $\mathbf{X}_1 = \mathbf{X}_2 = \cdots = \mathbf{X}_S = \mathbf{X}_*$ (for some $R \times P_*$ matrix $\mathbf{X}_*$). Under those conditions, it is possible to reexpress the model equation $\mathbf{y} = \mathbf{X}\boldsymbol{\beta} + \mathbf{e}$ as

$$\mathbf{Y} = \mathbf{X}_*\mathbf{B} + \mathbf{E}, \tag{5.3}$$

where $\mathbf{Y} = (\mathbf{y}_1, \mathbf{y}_2, \ldots, \mathbf{y}_S)$, $\mathbf{B} = (\boldsymbol{\beta}_1, \boldsymbol{\beta}_2, \ldots, \boldsymbol{\beta}_S)$, and $\mathbf{E} = (\mathbf{e}_1, \mathbf{e}_2, \ldots, \mathbf{e}_S)$.

The N residual effects $e_{sr_{sk}}$ $(s = 1, 2, \ldots, S;\ k = 1, 2, \ldots, R_s)$ can be regarded as a subset of a set of RS random variables e_{sr} $(s = 1, 2, \ldots, S;\ r = 1, 2, \ldots, R)$ having expected values of 0—think of these RS random variables as the residual effects for the special case where there is complete information on every observational unit. It is assumed that the distribution of the random variables e_{sr} $(s = 1, 2, \ldots, S;\ r = 1, 2, \ldots, R)$ is such that the R vectors $(e_{1r}, e_{2r}, \ldots, e_{Sr})'$ $(r = 1, 2, \ldots, R)$ are uncorrelated and each has the same variance-covariance matrix $\boldsymbol{\Sigma} = \{\sigma_{ij}\}$. Then, in the special case where there is complete information on every observational unit, the variance-covariance matrix of the vector $\mathbf{e}$ of residual effects is

$$\begin{pmatrix} \sigma_{11}\mathbf{I}_R & \sigma_{12}\mathbf{I}_R & \cdots & \sigma_{1S}\mathbf{I}_R \\ \sigma_{12}\mathbf{I}_R & \sigma_{22}\mathbf{I}_R & \cdots & \sigma_{2S}\mathbf{I}_R \\ \vdots & \vdots & \ddots & \vdots \\ \sigma_{1S}\mathbf{I}_R & \sigma_{2S}\mathbf{I}_R & \cdots & \sigma_{SS}\mathbf{I}_R \end{pmatrix}. \tag{5.4}$$

Moreover, in the general case (where the information on some or all of the observational units is incomplete), the variance-covariance matrix of $\mathbf{e}$ is a submatrix of the matrix (5.4). Specifically, it is the submatrix obtained by replacing the stth block of matrix (5.4) with the $R_s \times R_t$ submatrix formed from that block by striking out all of the rows and columns of the block save the $r_{s1}, r_{s2}, \ldots,$ r_{sR_s}th rows and the $r_{t1}, r_{t2}, \ldots, r_{tR_t}$th columns $(s, t = 1, 2, \ldots, S)$.

Typically, the matrix $\boldsymbol{\Sigma}$ (which is inherently symmetric and nonnegative definite) is assumed to be positive definite, and its $S(S+1)/2$ distinct elements, say σ_{ij} $(j \ge i = 1, 2, \ldots, S)$, are regarded as unknown parameters. The situation is such that (even in the absence of the assumption that $\boldsymbol{\Sigma}$ is positive definite) var$(\mathbf{e})$ is of the form $\mathbf{V}(\boldsymbol{\theta})$ of var$(\mathbf{e})$ in the general linear model—the parameter vector $\boldsymbol{\theta}$ can be taken to be the $[S(S+1)/2]$-dimensional (column) vector with elements σ_{ij} $(j \ge i = 1, 2, \ldots, S)$.

b. Example: data on four characteristics of whey-protein gels

Schmidt, Illingworth, Deng, and Cornell (1979) reported the results of a study of the effects of the use of various levels of two reagents on the formation of whey-protein gels. The reagents are cysteine and $CaCl_2$ (calcium chloride). The effects of interest are those on the texture and the water-retention capacity of the gels. Various textural characteristics are reflected in the values of three variables known as hardness, cohesiveness, and springiness, and water-retention capacity can be assessed on the basis of a variable that is referred to as compressible H_2O.

Data on these variables were obtained from an experiment conducted in accordance with a response-surface design known as a central composite design (one with added center points). The experiment consisted of 13 trials, each of which took the form of five replications conducted at pre-specified levels of cysteine and $CaCl_2$. The five replicate values obtained in each trial for each of the four response variables (hardness, cohesiveness, springiness, and compressible H_2O) were

TABLE 4.8. Data on the textural characteristics (hardness, cohesiveness, and springiness) and water-retention capacity (compressible H_2O) of whey-protein gels at various levels of two reagents (cysteine and $CaCl_2$) used in their formation (Schmidt, Illingworth, Deng, and Cornell 1979).

	Reagent levels		Characteristics of gel			
Trial	Cysteine (mM)	$CaCl_2$ (mM)	Hardness (kg)	Cohesiveness	Springiness (mm)	Compressible H_2O (g)
1	8.0	6.5	2.48	0.55	1.95	0.22
2	34.0	6.5	0.91	0.52	1.37	0.67
3	8.0	25.9	0.71	0.67	1.74	0.57
4	34.0	25.9	0.41	0.36	1.20	0.69
5	2.6	16.2	2.28	0.59	1.75	0.33
6	39.4	16.2	0.35	0.31	1.13	0.67
7	21.0	2.5	2.14	0.54	1.68	0.42
8	21.0	29.9	0.78	0.51	1.51	0.57
9	21.0	16.2	1.50	0.66	1.80	0.44
10	21.0	16.2	1.66	0.66	1.79	0.50
11	21.0	16.2	1.48	0.66	1.79	0.50
12	21.0	16.2	1.41	0.66	1.77	0.43
13	21.0	16.2	1.58	0.66	1.73	0.47

averaged. Accordingly, each trial resulted in four data points, one for each response variable. The data from the 13 trials are reproduced in Table 4.8.

These data are multivariate in nature. The trials constitute the observational units; there are $R = 13$ of them. And there are $S = 4$ response variables: hardness, cohesiveness, springiness, and compressible H_2O. Moreover, there is complete information on every observational unit; every one of the four response variables was observed on every one of the 13 trials.

The setting is one that might be suitable for the application of a general linear model. More specifically, it might be suitable for the application of a general linear model of the form considered in Subsection a. In such an application, we might take $C = 3$, take u_1 to be the level of cysteine, take u_2 to be the level of $CaCl_2$, and take the value of u_3 to be 1, 2, 3, or 4 depending on whether the response variable is hardness, cohesiveness, springiness, or compressible H_2O.

Further, following Schmidt et al. (1979) and adopting the notation of Subsection a, we might take (for $s = 1, 2, 3, 4$ and $r = 1, 2, \ldots, 13$)

$$y_{sr} = \beta_{s1} + \beta_{s2}u_{1r} + \beta_{s3}u_{2r} + \beta_{s4}u_{1r}^2 + \beta_{s5}u_{2r}^2 + \beta_{s6}u_{1r}u_{2r} + e_{sr}, \tag{5.5}$$

where u_{1r} and u_{2r} are the values of u_1 and u_2 for the rth observational unit and where $\beta_{s1}, \beta_{s2}, \ldots, \beta_{s6}$ are unknown parameters. Taking (for $s = 1, 2, 3, 4$ and $r = 1, 2, \ldots, 13$) y_{sr} to be of the form (5.5) is equivalent to taking (for $s = 1, 2, 3, 4$) the vector $\mathbf{y}_s$ (with elements $y_{s1}, y_{s2}, \ldots, y_{sR}$) to be of the form (5.2) and to taking $\mathbf{X}_1 = \mathbf{X}_2 = \mathbf{X}_3 = \mathbf{X}_4 = \mathbf{X}_*$, where $\mathbf{X}_*$ is the 13×6 matrix whose rth row is $(1, u_{1r}, u_{2r}, u_{1r}^2, u_{2r}^2, u_{1r}u_{2r})$.

Exercises

Exercise 1. Verify formula (2.3).

Exercise 2. Write out the elements of the vector $\boldsymbol{\beta}$, of the observed value of the vector $\mathbf{y}$, and of the matrix $\mathbf{X}$ (in the model equation $\mathbf{y} = \mathbf{X}\boldsymbol{\beta} + \mathbf{e}$) in an application of the G–M model to the cement data of Section 4.2d. In doing so, regard the measurements of the heat that evolves during hardening as the data points, take $C = 4$, take u_1, u_2, u_3, and u_4 to be the respective amounts of tricalcium aluminate, tricalcium silicate, tetracalcium aluminoferrite, and β-dicalcium silicate, and take $\delta(\mathbf{u})$ to be of the form (2.11).

Exercise 3. Write out the elements of the vector $\boldsymbol{\beta}$, of the observed value of the vector $\mathbf{y}$, and of the matrix $\mathbf{X}$ (in the model equation $\mathbf{y} = \mathbf{X}\boldsymbol{\beta} + \mathbf{e}$) in an application of the G–M model to the lettuce data of Section 4.2e. In doing so, regard the yields of lettuce as the data points, take $C = 3$, take u_1, u_2, and u_3 to be the transformed amounts of Cu, Mo, and Fe, respectively, and take $\delta(\mathbf{u})$ to be of the form (2.14).

Exercise 4. Let y represent a random variable and $\mathbf{u}$ a C-dimensional random column vector such that the joint distribution of y and $\mathbf{u}$ is MVN (with a nonsingular variance-covariance matrix). And take $\mathbf{z} = \{z_j\}$ to be a transformation (of $\mathbf{u}$) of the form

$$\mathbf{z} = \mathbf{R}'[\mathbf{u} - \mathrm{E}(\mathbf{u})],$$

where $\mathbf{R}$ is a nonsingular (nonrandom) matrix such that $\mathrm{var}(\mathbf{z}) = \mathbf{I}$—the existence of such a matrix follows from the results of Section 3.3b. Show that

$$\frac{\mathrm{E}(y \mid \mathbf{u}) - \mathrm{E}(y)}{\sqrt{\mathrm{var}\, y}} = \sum_{j=1}^{C} \mathrm{corr}(y, z_j)\, z_j.$$

Exercise 5. Let y represent a random variable and $\mathbf{u} = (u_1, u_2, \ldots, u_C)'$ a C-dimensional random column vector, assume that $\mathrm{var}(\mathbf{u})$ is nonsingular, and suppose that $\mathrm{E}(y \mid \mathbf{u})$ is expressible in the form

$$\mathrm{E}(y \mid \mathbf{u}) = \beta_1 + \beta_2(u_1 - a_1) + \beta_3(u_2 - a_2) + \cdots + \beta_{C+1}(u_C - a_C),$$

where $a_1, a_2, \ldots, a_C$ and $\beta_1, \beta_2, \beta_3, \ldots, \beta_{C+1}$ are nonrandom scalars.

(a) Using the results of Section 3.4 (or otherwise), show that

$$(\beta_2, \beta_3, \ldots, \beta_{C+1}) = \mathrm{cov}(y, \mathbf{u})(\mathrm{var}\, \mathbf{u})^{-1},$$

and that

$$\beta_1 = \mathrm{E}(y) + \sum_{j=2}^{C+1} \beta_j[a_{j-1} - \mathrm{E}(u_{j-1})],$$

in agreement with the results obtained in Section 4.3 (under the assumption that the joint distribution of y and $\mathbf{u}$ is MVN).

(b) Show that

$$\mathrm{E}[\mathrm{var}(y \mid \mathbf{u})] = \mathrm{var}(y) - \mathrm{cov}(y, \mathbf{u})(\mathrm{var}\, \mathbf{u})^{-1}\mathrm{cov}(\mathbf{u}, y).$$

Exercise 6. Suppose that (in conformance with the development in Section 4.4b) the residual effects in the general linear model have been partitioned into K mutually exclusive and exhaustive subsets or classes numbered $1, 2, \ldots, K$. And for $k = 1, 2, \ldots, K$, write $e_{k1}, e_{k2}, \ldots, e_{kN_k}$ for the residual effects in the kth class. Take a_k^* $(k = 1, 2, \ldots, K)$ and r_{ks}^* $(k = 1, 2, \ldots, K;\ s = 1, 2, \ldots, N_k)$ to be uncorrelated random variables, each with mean 0, such that $\mathrm{var}(a_k^*) = \tau_k^{*2}$ for some nonnegative scalar τ_k^* and $\mathrm{var}(r_{ks}^*) = \phi_k^{*2}$ $(s = 1, 2, \ldots, N_k)$ for some strictly positive scalar ϕ_k^*. Consider the effect of taking the residual effects to be of the form

$$e_{ks} = a_k^* + r_{ks}^*, \tag{E.1}$$

rather than of the form (4.26). Are there values of τ_k^{*2} and ϕ_k^{*2} for which the value of $\mathrm{var}(\mathbf{e})$ is the same when the residual effects are taken to be of the form (E.1) as when they are taken to be of the form (4.26)? If so, what are those values; if not, why not?

Exercise 7. Develop a correlation structure for the residual effects in the general linear model that, in the application of the model to the shear-strength data (of Section 4.4d), would allow for the possibility that steel aliquots chosen at random from those on hand on different dates may tend to be more alike when the intervening time is short than when it is long. Do so by making use of the results (in Section 4.4e) on stationary first-order autoregressive processes.

Exercise 8. Suppose (as in Section 4.4g) that the residual effects $e_1, e_2, \ldots, e_N$ in the general linear model correspond to locations in D-dimensional space, that these locations are represented by D-dimensional column vectors $\mathbf{s}_1, \mathbf{s}_2, \ldots, \mathbf{s}_N$ of coordinates, and that S is a finite or infinite set of D-dimensional column vectors that includes $\mathbf{s}_1, \mathbf{s}_2, \ldots, \mathbf{s}_N$. Suppose further that $e_1, e_2, \ldots, e_N$ are expressible in the form (4.54) and that conditions (4.55) and (4.57) are applicable. And take $\psi(\cdot)$ to be the function defined on the set $H = \{\mathbf{h} \in \mathbb{R}^D : \mathbf{h} = \mathbf{s} - \mathbf{t} \text{ for } \mathbf{s}, \mathbf{t} \in S\}$ by

$$\psi(\mathbf{h}) = \phi^2 + \tau^2[1 - K(\mathbf{h})].$$

(a) Show that, for $j \neq i = 1, 2, \ldots, N$,

$$\tfrac{1}{2}\operatorname{var}(e_i - e_j) = \psi(\mathbf{s}_i - \mathbf{s}_j)$$

—this result serves to establish the function $\psi^*(\cdot)$, defined by

$$\psi^*(\mathbf{h}) = \begin{cases} \psi(\mathbf{h}) & \text{if } \mathbf{h} \neq \mathbf{0}, \\ 0 & \text{if } \mathbf{h} = \mathbf{0}, \end{cases}$$

as what in spatial statistics is known as a semivariogram (e.g., Cressie 1993).

(b) Show that (1) $\psi(\mathbf{0}) = \phi^2$; that (2) $\psi(-\mathbf{h}) = \psi(\mathbf{h})$ for $\mathbf{h} \in H$; and that (3) $\sum_{i=1}^{M}\sum_{j=1}^{M} x_i x_j \psi(\mathbf{t}_i - \mathbf{t}_j) \leq 0$ for every positive integer M, for all not-necessarily-distinct vectors $\mathbf{t}_1, \mathbf{t}_2, \ldots, \mathbf{t}_M$ in S, and for all scalars $x_1, x_2, \ldots, x_M$ such that $\sum_{i=1}^{M} x_i = 0$.

Exercise 9. Suppose that the general linear model is applied to the example of Section 4.5b (in the way described in Section 4.5b). What is the form of the function $\delta(\mathbf{u})$?

Bibliographic and Supplementary Notes

§1. What is herein called the model matrix is sometimes referred to as the design matrix. The term model matrix seems preferable in that the data may not have come from a designed experiment. Moreover, as discussed by Kempthorne (1980), different models (with different model matrices) can be contemplated even in the case of a designed experiment.

§2e. The data on yield reported by Hader, Harward, Mason, and Moore (1957) and presented herein are part of a larger collection of data reported by Moore, Harward, Mason, Hader, Lott, and Jackson (1957). In addition to the data on yield, data were obtained on the Cu content and the Fe content of the lettuce plants. And the experiment was one of four similar experiments; these experiments differed from each other in regard to the source of Fe and/or the source of nitrogen.

§3. For an illuminating discussion of regression from a historical perspective (that includes a detailed account of the contibutions of Sir Francis Galton), refer to Stigler (1986, chap. 8; 1999, chap. 9).

§4a. For additional discussion of the ways in which the variability of the residual effects can depend on the explanatory variables, refer to Pinheiro and Bates (2000, sec. 5.2), Carroll and Ruppert (1988, chap. 3), and/or Davidian and Giltinan (1995, sec. 2.2). They consider the choice of a nonnegative function $v(\mathbf{u})$ of $\mathbf{u}$ such that (for $i = 1, 2, \ldots, N$) $\operatorname{var}(e_i) = v(\mathbf{u}_i)$. They do so in a broad framework that includes choices for $v(\mathbf{u})$ in which the dependence on $\mathbf{u}$ can be wholly or partly through the value of $\delta(\mathbf{u})$. Such choices result in models for $y_1, y_2, \ldots, y_N$ of a form not covered by the general linear model (as defined herein)—they result in models in which the variances of $y_1, y_2, \ldots, y_N$ are related to their expected values.

§4b. In some presentations, the intraclass correlation is taken to be the same for every class, and the permissible values of the intraclass correlation are taken to be all of the values for which the variance-covariance matrix of the residual effects is nonnegative definite. Then, assuming that there are K classes of sizes $N_1, N_2, \ldots, N_K$ and denoting the intraclass correlation by ρ, the permissible values would be those in the interval

$$-\min_k \frac{1}{N_k - 1} \leq \rho \leq 1.$$

One could question whether the intraclass correlation's being the same for every class is compatible with its being negative. Assuming that a negative intraclass correlation is indicative of competition among some number ($\geq$ the class size) of entities, it would seem that the correlation would depend on the number of entities—presumably, the pairwise competition would be less intense and the correlation less affected if the number of entities were relatively large.

§4e. The development in Part 2 is based on taking the correlation structure of the sequence $f_{k1}, f_{k2}, \ldots, f_{kN_k}$ to be that of a stationary first-order autoregressive process. There are other possible choices for this correlation structure; see, for example, Diggle, Heagerty, Liang, and Zeger (2002, secs. 4.2.2 and 5.2) and Laird (2004, sec. 1.3).

§4g. For extensive (book-length) treatises on spatial statistics, refer, e.g., to Cressie (1993) and Schabenberger and Gotway (2005). Gaussian autocorrelation functions may be regarded as "artificial" and their use discouraged; they have certain characteristics that are considered by Schabenberger and Gotway—refer to their Section 4.3—and by many others to be inconsistent with the characteristics of real physical and biological processes.

§5a. By transposing both of its sides, model equation (5.3) can be reexpressed in the form of the equation

$$\mathbf{Y}' = \mathbf{B}'\mathbf{X}_*' + \mathbf{E}',$$

each side of which is an $S \times R$ matrix whose rows correspond to the response variables and whose columns correspond to the observational units. In many publications, the model equation is presented in this alternative form rather than in the form (5.3). As pointed out, for example, by Arnold (1981, p. 348), the form (5.3) has the appealing property that, in the special case of univariate data (i.e., the special case where $S = 1$), each side of the equation reduces to a column vector (rather than a row vector), in conformance with the usual representation for that case.

5

Estimation and Prediction: Classical Approach

Models of the form of the general linear model, and in particular those of the form of the Gauss–Markov or Aitken model, are often used to obtain point estimates of the unobservable quantities represented by various parametric functions. In many cases, the parametric functions are ones that are expressible in the form $\boldsymbol{\lambda}'\boldsymbol{\beta}$, where $\boldsymbol{\lambda} = (\lambda_1, \lambda_2, \ldots, \lambda_P)'$ is a P-dimensional column vector of constants, or equivalently ones that are expressible in the form $\sum_{j=1}^{P} \lambda_j \beta_j$. Models of the form of the G–M, Aitken, or general linear model may also be used to obtain predictions for future quantities; these would be future quantities that are represented by unobservable random variables with expected values of the form $\boldsymbol{\lambda}'\boldsymbol{\beta}$. The emphasis in this chapter is on the G–M model (in which the only parameter other than $\beta_1, \beta_2, \ldots, \beta_P$ is the standard deviation σ) and on what might be regarded as a classical approach to estimation and prediction.

5.1 Linearity and Unbiasedness

Suppose that $\mathbf{y}$ is an $N \times 1$ observable random vector that follows the G–M, Aitken, or general linear model, and consider the estimation of a parametric function of the form $\boldsymbol{\lambda}'\boldsymbol{\beta} = \sum_{j=1}^{P} \lambda_j \beta_j$, where $\boldsymbol{\lambda} = (\lambda_1, \lambda_2, \ldots, \lambda_P)'$. Is it "possible" to estimate $\boldsymbol{\lambda}'\boldsymbol{\beta}$ from the available information, and if so, which estimator is best and in what sense is it best? One way to judge the "goodness" of an estimator is on the basis of its mean squared error (MSE) or its root mean squared error (root MSE)—the root MSE is the square root of the MSE. When a function $t(\mathbf{y})$ of $\mathbf{y}$ is regarded as an estimator of $\boldsymbol{\lambda}'\boldsymbol{\beta}$, its MSE is (by definition) $\mathrm{E}\{[t(\mathbf{y}) - \boldsymbol{\lambda}'\boldsymbol{\beta}]^2\}$.

The information about the distribution of $\mathbf{y}$ provided by the G–M, Aitken, or general linear model is limited; it is confined to information about $\mathrm{E}(\mathbf{y})$ and $\mathrm{var}(\mathbf{y})$. The evaluation and comparison of potential estimators are greatly facilitated by restricting attention to estimators that are of a relatively simple form or that satisfy certain criteria and/or by making assumptions about the distribution of $\mathbf{y}$ that go beyond those inherent in the G–M, Aitken, or general linear model. If the evaluations and comparisons are to be meaningful, the restrictions need to be ones that have appeal in their own right, and the assumptions need to be realistic.

An estimator of $\boldsymbol{\lambda}'\boldsymbol{\beta}$, say an estimator $t(\mathbf{y})$, is said to be *linear* if it is expressible in the form

$$t(\mathbf{y}) = c + \sum_{i=1}^{N} a_i y_i,$$

where c and $a_1, a_2, \ldots, a_N$ are constants, or equivalently if it is expressible in the form

$$t(\mathbf{y}) = c + \mathbf{a}'\mathbf{y},$$

where c is a constant and $\mathbf{a} = (a_1, a_2, \ldots, a_N)'$ is an N-dimensional column vector of constants. Linear estimators of $\boldsymbol{\lambda}'\boldsymbol{\beta}$ are of a relatively simple form, which makes them readily amenable to evaluation, comparison, and interpretation. Accordingly, it is convenient and of some interest to obtain results on the estimation of $\boldsymbol{\lambda}'\boldsymbol{\beta}$ in the special case where consideration is restricted to linear estimators.

Attention is sometimes restricted to estimators that are unbiased. By definition, an estimator $t(\mathbf{y})$ of $\boldsymbol{\lambda}'\boldsymbol{\beta}$ is unbiased if $\mathrm{E}[t(\mathbf{y})] = \boldsymbol{\lambda}'\boldsymbol{\beta}$. If $t(\mathbf{y})$ is an unbiased estimator of $\boldsymbol{\lambda}'\boldsymbol{\beta}$, then

$$\mathrm{E}\{[t(\mathbf{y}) - \boldsymbol{\lambda}'\boldsymbol{\beta}]^2\} = \mathrm{var}[t(\mathbf{y})], \tag{1.1}$$

that is, its MSE equals its variance.

In the case of a linear estimator $c + \mathbf{a}'\mathbf{y}$, the expected value of the estimator is

$$\mathrm{E}(c + \mathbf{a}'\mathbf{y}) = c + \mathbf{a}'\mathrm{E}(\mathbf{y}) = c + \mathbf{a}'\mathbf{X}\boldsymbol{\beta}. \tag{1.2}$$

Accordingly, $c + \mathbf{a}'\mathbf{y}$ is an unbiased estimator of $\boldsymbol{\lambda}'\boldsymbol{\beta}$ if and only if, for every P-dimensional column vector $\underline{\boldsymbol{\beta}}$,

$$c + \mathbf{a}'\mathbf{X}\underline{\boldsymbol{\beta}} = \boldsymbol{\lambda}'\underline{\boldsymbol{\beta}}. \tag{1.3}$$

Clearly, a sufficient condition for the unbiasedness of the linear estimator $c + \mathbf{a}'\mathbf{y}$ is

$$c = 0 \quad \text{and} \quad \mathbf{a}'\mathbf{X} = \boldsymbol{\lambda}' \tag{1.4}$$

or, equivalently,

$$c = 0 \quad \text{and} \quad \mathbf{X}'\mathbf{a} = \boldsymbol{\lambda}. \tag{1.5}$$

This condition is also a necessary condition for the unbiasedness of $c + \mathbf{a}'\mathbf{y}$ as is evident upon observing that if equality (1.3) holds for every column vector $\underline{\boldsymbol{\beta}}$ in $\mathbb{R}^P$, then it holds in particular when $\underline{\boldsymbol{\beta}}$ is taken to be the $P \times 1$ null vector $\mathbf{0}$ (so that $c = 0$) and when (for each integer j between 1 and P, inclusive) $\underline{\boldsymbol{\beta}}$ is taken to be the jth column of $\mathbf{I}_P$ (so that the jth element of $\mathbf{a}'\mathbf{X}$ equals the jth element of $\boldsymbol{\lambda}'$).

In the special case of a linear unbiased estimator $\mathbf{a}'\mathbf{y}$, expression (1.1) for the MSE of an unbiased estimator of $\boldsymbol{\lambda}'\boldsymbol{\beta}$ simplifies to

$$\mathrm{E}[(\mathbf{a}'\mathbf{y} - \boldsymbol{\lambda}'\boldsymbol{\beta})^2] = \mathbf{a}'\mathrm{var}(\mathbf{y})\mathbf{a}. \tag{1.6}$$

5.2 Translation Equivariance

Suppose (as in Section 5.1) that $\mathbf{y}$ is an $N \times 1$ observable random vector that follows the Gauss–Markov, Aitken, or general linear model and that we wish to estimate a parametric function of the form $\boldsymbol{\lambda}'\boldsymbol{\beta}$. Attention is sometimes restricted to estimators that are unbiased. However, unbiasedness is not the only criterion that could be used to restrict the quantity of estimators under consideration. Another possible criterion is *translation equivariance* (also known as location equivariance).

Let $\mathbf{k}$ represent a P-dimensional column vector of known constants, and define $\mathbf{z} = \mathbf{y} + \mathbf{X}\mathbf{k}$. The vector $\mathbf{z}$, like the vector $\mathbf{y}$, is an N-dimensional observable random vector. Moreover,

$$\mathbf{z} = \mathbf{X}\boldsymbol{\tau} + \mathbf{e}, \tag{2.1}$$

where $\boldsymbol{\tau} = \boldsymbol{\beta} + \mathbf{k}$. Accordingly, $\mathbf{z}$ follows a G–M, Aitken, or general linear model that is identical in all respects to the model followed by $\mathbf{y}$, except that the role of the parameter vector $\boldsymbol{\beta}$ is played by a vector [represented by $\boldsymbol{\tau}$ in equality (2.1)] that has a different interpretation.

It can be argued that an estimator, say $t(\mathbf{y})$, of $\boldsymbol{\lambda}'\boldsymbol{\beta}$ should be such that the results obtained in using $t(\mathbf{y})$ to estimate $\boldsymbol{\lambda}'\boldsymbol{\beta}$ are consistent with those obtained in using $t(\mathbf{z})$ to estimate the corresponding parametric function ($\boldsymbol{\lambda}'\boldsymbol{\tau} = \boldsymbol{\lambda}'\boldsymbol{\beta} + \boldsymbol{\lambda}'\mathbf{k}$). Here, the consistency is in the sense that

$$t(\mathbf{y}) + \boldsymbol{\lambda}'\mathbf{k} = t(\mathbf{z})$$

or, equivalently, that

$$t(\mathbf{y}) + \boldsymbol{\lambda}'\mathbf{k} = t(\mathbf{y} + \mathbf{X}\mathbf{k}). \tag{2.2}$$

When applied to a linear estimator $c + \mathbf{a}'\mathbf{y}$, condition (2.2) becomes

$$c + \mathbf{a}'\mathbf{y} + \boldsymbol{\lambda}'\mathbf{k} = c + \mathbf{a}'(\mathbf{y} + \mathbf{X}\mathbf{k}),$$

which (after some simplification) can be restated as

$$\mathbf{a}'\mathbf{X}\mathbf{k} = \boldsymbol{\lambda}'\mathbf{k}. \tag{2.3}$$

The estimator $t(\mathbf{y})$ is said to be *translation equivariant* if it is such that condition (2.2) is satisfied for every $\mathbf{k} \in \mathcal{R}^P$ (and for every value of $\mathbf{y}$). Accordingly, the linear estimator $c + \mathbf{a}'\mathbf{y}$ is translation equivariant if and only if condition (2.3) is satisfied for every $\mathbf{k} \in \mathcal{R}^P$ or, equivalently, if and only if

$$\mathbf{a}'\mathbf{X} = \boldsymbol{\lambda}'. \tag{2.4}$$

Observe (in light of the results of Section 5.1) that condition (2.4) is identical to one of the conditions needed for unbiasedness—for unbiasedness, we also need the condition $c = 0$. Thus, the motivation for requiring that the coefficient vector $\mathbf{a}'$ in the linear estimator $c + \mathbf{a}'\mathbf{y}$ satisfy the condition $\mathbf{a}'\mathbf{X} = \boldsymbol{\lambda}'$ can come from a desire to achieve unbiasedness or translation equivariance or both.

5.3 Estimability

Suppose (as in Sections 5.1 and 5.2) that $\mathbf{y}$ is an $N \times 1$ observable random vector that follows the G–M, Aitken, or general linear model, and consider the estimation of a parametric function that is expressible in the form $\boldsymbol{\lambda}'\boldsymbol{\beta}$ or $\sum_{j=1}^{P} \lambda_j \beta_j$, where $\boldsymbol{\lambda} = (\lambda_1, \lambda_2, \ldots, \lambda_P)'$ is a $P \times 1$ vector of coefficients. If there exists a linear unbiased estimator of $\boldsymbol{\lambda}'\boldsymbol{\beta}$ [i.e., if there exists a constant c and an $N \times 1$ vector of constants $\mathbf{a}$ such that $\mathrm{E}(c + \mathbf{a}'\mathbf{y}) = \boldsymbol{\lambda}'\boldsymbol{\beta}$], then $\boldsymbol{\lambda}'\boldsymbol{\beta}$ is said to be *estimable*. Otherwise (if no such estimator exists), $\boldsymbol{\lambda}'\boldsymbol{\beta}$ is said to be *nonestimable*.

If $\boldsymbol{\lambda}'\boldsymbol{\beta}$ is estimable, then the data provide at least some information about $\boldsymbol{\lambda}'\boldsymbol{\beta}$. Estimability can be of critical importance in the design of an experiment. If the data from the experiment are to be regarded as having originated from a G–M, Aitken, or general linear model and if the quantities of interest are to be formulated as parametric functions of the form $\boldsymbol{\lambda}'\boldsymbol{\beta}$ (as is common practice), then it is imperative that every one of the relevant functions be estimable.

It follows immediately from the results of Section 5.1 that $\boldsymbol{\lambda}'\boldsymbol{\beta}$ is estimable if and only if there exists an $N \times 1$ vector $\mathbf{a}$ such that

$$\boldsymbol{\lambda}' = \mathbf{a}'\mathbf{X} \tag{3.1}$$

or, equivalently, such that

$$\boldsymbol{\lambda} = \mathbf{X}'\mathbf{a}. \tag{3.2}$$

Thus, for $\boldsymbol{\lambda}'\boldsymbol{\beta}$ to be estimable (under the G–M, Aitken, or general linear model), it is necessary and sufficient that

$$\boldsymbol{\lambda}' \in \mathcal{R}(\mathbf{X}) \tag{3.3}$$

or, equivalently, that

$$\boldsymbol{\lambda} \in \mathcal{C}(\mathbf{X}'). \tag{3.4}$$

Note that it follows from the very definition of estimability [as well as from condition (3.1)] that if $\boldsymbol{\lambda}'\boldsymbol{\beta}$ is estimable, then there exists an $N \times 1$ vector $\mathbf{a}$ such that

$$\boldsymbol{\lambda}'\boldsymbol{\beta} = \mathbf{a}'\,\mathrm{E}(\mathbf{y}). \tag{3.5}$$

Thus, if $\boldsymbol{\lambda}'\boldsymbol{\beta}$ is estimable, it is interpretable in terms of the expected values of $y_1, y_2, \ldots, y_N$. In fact, if $\boldsymbol{\lambda}'\boldsymbol{\beta}$ is estimable, it may be expressible in the form (3.5) for each of a number of different choices of **a** and, consequently, it may have multiple interpretations in terms of the expected values of $y_1, y_2, \ldots, y_N$.

Two basic and readily verifiable observations about linear combinations of parametric functions of the form $\boldsymbol{\lambda}'\boldsymbol{\beta}$ are as follows:

(1) linear combinations of estimable functions are estimable; and

(2) linear combinations of nonestimable functions are not necessarily nonestimable.

How many "essentially different" estimable functions are there? Let $\boldsymbol{\lambda}_1'\boldsymbol{\beta}, \boldsymbol{\lambda}_2'\boldsymbol{\beta}, \ldots, \boldsymbol{\lambda}_K'\boldsymbol{\beta}$ represent K (where K is an arbitrary positive integer) linear combinations of the elements of $\boldsymbol{\beta}$. These linear combinations are said to be linearly independent if their coefficient vectors $\boldsymbol{\lambda}_1', \boldsymbol{\lambda}_2', \ldots, \boldsymbol{\lambda}_K'$ are linearly independent vectors. A question as to whether $\boldsymbol{\lambda}_1'\boldsymbol{\beta}, \boldsymbol{\lambda}_2'\boldsymbol{\beta}, \ldots, \boldsymbol{\lambda}_K'\boldsymbol{\beta}$ are essentially different can be made precise by taking *essentially different* to mean *linearly independent*.

Letting $R = \text{rank}(\mathbf{X})$, some basic and readily verifiable observations about linearly independent parametric functions of the form $\boldsymbol{\lambda}'\boldsymbol{\beta}$ and about their estimability or nonestimability are as follows:

(1) there exists a set of R linearly independent estimable functions;

(2) no set of estimable functions contains more than R linearly independent estimable functions; and

(3) if the model is not of full rank (i.e., if $R < P$), then at least one and, in fact, at least $P - R$ of the individual parameters $\beta_1, \beta_2, \ldots, \beta_P$ are nonestimable.

When the model matrix $\mathbf{X}$ has full column rank P, the model is said to be of *full rank*. In the special case of a full-rank model, $\mathcal{R}(\mathbf{X}) = \mathcal{R}^P$, and every parametric function of the form $\boldsymbol{\lambda}'\boldsymbol{\beta}$ is estimable.

Note that the existence of an $N \times 1$ vector **a** that satisfies equality (3.2) is equivalent to the consistency of a linear system (in an $N \times 1$ vector **a** of unknowns), namely, the linear system with coefficient matrix $\mathbf{X}'$ (which is of dimensions $P \times N$) and with right side $\boldsymbol{\lambda}$. The significance of this equivalence is that any result on the consistency of a linear system can be readily translated into a result on the estimability of the parametric function $\boldsymbol{\lambda}'\boldsymbol{\beta}$. Consider, in particular, Theorem 2.11.1. Upon applying this theorem [and observing that $(\mathbf{X}^-)'$ is a generalized inverse of $\mathbf{X}'$], we find that for $\boldsymbol{\lambda}'\boldsymbol{\beta}$ to be estimable, it is necessary and sufficient that

$$\boldsymbol{\lambda}'\mathbf{X}^-\mathbf{X} = \boldsymbol{\lambda}' \tag{3.6}$$

or, equivalently, that

$$\boldsymbol{\lambda}'(\mathbf{I} - \mathbf{X}^-\mathbf{X}) = \mathbf{0}. \tag{3.7}$$

If $\text{rank}(\mathbf{X}) = P$, then (in light of Lemma 2.10.3) $\mathbf{X}^-\mathbf{X} = \mathbf{I}$. Thus, in the special case of a full-rank model, conditions (3.6) and (3.7) are vacuous.

a. A result on the consistency of a linear system

The following result on the consistency of a linear system can be used to obtain additional results on the estimability of a parametric function of the form $\boldsymbol{\lambda}'\boldsymbol{\beta}$.

Theorem 5.3.1. A linear system $\mathbf{AX} = \mathbf{B}$ (in $\mathbf{X}$) is consistent if and only if $\mathbf{k}'\mathbf{B} = \mathbf{0}$ for every column vector **k** (of compatible dimension) such that $\mathbf{k}'\mathbf{A} = \mathbf{0}$.

Proof. Denote by M the number of rows in **A** (and in **B**), let **k** represent an M-dimensional column vector, and observe (in light of Corollary 2.11.4 and Lemma 2.10.10) that

$$\begin{aligned}
\mathbf{k}'\mathbf{A} = \mathbf{0} \quad &\Leftrightarrow \quad \mathbf{A}'\mathbf{k} = \mathbf{0} \\
&\Leftrightarrow \quad \mathbf{k} \in \mathcal{N}(\mathbf{A}') \\
&\Leftrightarrow \quad \mathbf{k} \in \mathcal{C}[\mathbf{I} - (\mathbf{A}^-)'\mathbf{A}'] \\
&\Leftrightarrow \quad \mathbf{k} = [\mathbf{I} - (\mathbf{A}^-)'\mathbf{A}']\mathbf{r} \text{ for some } M \times 1 \text{ vector } \mathbf{r} \\
&\Leftrightarrow \quad \mathbf{k}' = \mathbf{r}'(\mathbf{I} - \mathbf{A}\mathbf{A}^-) \text{ for some } M \times 1 \text{ vector } \mathbf{r}.
\end{aligned}$$

Thus, recalling Lemma 2.2.2, we find that

$$\begin{aligned}\mathbf{k}'\mathbf{B} = \mathbf{0} \text{ for every } \mathbf{k} \text{ such that } \mathbf{k}'\mathbf{A} = \mathbf{0} \\ \Leftrightarrow \quad \mathbf{r}'(\mathbf{I} - \mathbf{A}\mathbf{A}^-)\mathbf{B} = \mathbf{0} \text{ for every } M \times 1 \text{ vector } \mathbf{r} \\ \Leftrightarrow \quad (\mathbf{I} - \mathbf{A}\mathbf{A}^-)\mathbf{B} = \mathbf{0}.\end{aligned}$$

And based on Theorem 2.11.1, we conclude that the linear system $\mathbf{AX} = \mathbf{B}$ is consistent if and only if $\mathbf{k}'\mathbf{B} = \mathbf{0}$ for every $\mathbf{k}$ such that $\mathbf{k}'\mathbf{A} = \mathbf{0}$. Q.E.D.

Theorem 5.3.1 establishes that the consistency of the linear system $\mathbf{AX} = \mathbf{B}$ is equivalent to a condition that is sometimes referred to as compatibility; the linear system $\mathbf{AX} = \mathbf{B}$ is said to be compatible if every linear relationship that exists among the rows of the coefficient matrix $\mathbf{A}$ also exists among the rows of the right side $\mathbf{B}$ (in the sense that $\mathbf{k}'\mathbf{A}=\mathbf{0} \Rightarrow \mathbf{k}'\mathbf{B}=\mathbf{0}$). The proof presented herein differs from that presented in *Matrix Algebra from a Statistician's Perspective* (Harville 1997, sec. 7.3); it makes use of results on generalized inverses.

b. Some alternative necessary and sufficient conditions for estimability

Let us now consider further the estimability of a parametric function of the form $\boldsymbol{\lambda}'\boldsymbol{\beta}$ (under the G–M, Aitken, or general linear model). As noted earlier, $\boldsymbol{\lambda}'\boldsymbol{\beta}$ is estimable if and only if the linear system $\mathbf{X}'\mathbf{a} = \boldsymbol{\lambda}$ (in $\mathbf{a}$) is consistent. Accordingly, it follows from Theorem 5.3.1 that for $\boldsymbol{\lambda}'\boldsymbol{\beta}$ to be estimable, it is necessary and sufficient that

$$\mathbf{k}'\boldsymbol{\lambda} = 0 \text{ for every } P \times 1 \text{ vector } \mathbf{k} \text{ such that } \mathbf{k}'\mathbf{X}' = \mathbf{0} \tag{3.8}$$

or, equivalently, that

$$\mathbf{k}'\boldsymbol{\lambda} = 0 \text{ for every } P \times 1 \text{ vector } \mathbf{k} \text{ in } \mathcal{N}(\mathbf{X}). \tag{3.9}$$

Let $S = \dim[\mathcal{N}(\mathbf{X})]$. And observe (in light of Lemma 2.11.5) that

$$S = P - \text{rank}(\mathbf{X}).$$

Unless the model is of full rank [in which case $S = 0$, $\mathcal{N}(\mathbf{X}) = \{\mathbf{0}\}$, and conditions (3.8) and (3.9) are vacuous], condition (3.9) comprises an infinite number of equalities—there is one equality for each vector $\mathbf{k}$ in the S-dimensional linear space $\mathcal{N}(\mathbf{X})$. Fortunately, all but S of the equalities that form condition (3.9) can be eliminated without affecting the necessity or sufficiency of the condition.

Let $\mathbf{k}_1, \mathbf{k}_2, \ldots, \mathbf{k}_S$ represent any S linearly independent vectors in $\mathcal{N}(\mathbf{X})$, that is, any S linearly independent (P-dimensional) column vectors such that $\mathbf{X}\mathbf{k}_1 = \mathbf{X}\mathbf{k}_2 = \cdots = \mathbf{X}\mathbf{k}_S = \mathbf{0}$. Then, for $\boldsymbol{\lambda}'\boldsymbol{\beta}$ to be estimable, it is necessary and sufficient that

$$\mathbf{k}_1'\boldsymbol{\lambda} = \mathbf{k}_2'\boldsymbol{\lambda} = \cdots = \mathbf{k}_S'\boldsymbol{\lambda} = 0. \tag{3.10}$$

To verify the necessity and sufficiency of condition (3.10), it suffices to establish that condition (3.10) is equivalent to condition (3.9). In fact, it is enough to establish that condition (3.10) implies condition (3.9)—that condition (3.9) implies condition (3.10) is obvious. Accordingly, let $\mathbf{k}$ represent an arbitrary member of $\mathcal{N}(\mathbf{X})$. And observe (in light of Theorem 2.4.11) that the set $\{\mathbf{k}_1, \mathbf{k}_2, \ldots, \mathbf{k}_S\}$ is a basis for $\mathcal{N}(\mathbf{X})$, implying the existence of scalars $a_1, a_2, \ldots, a_S$ such that

$$\mathbf{k} = a_1\mathbf{k}_1 + a_2\mathbf{k}_2 + \cdots + a_S\mathbf{k}_S$$

and hence such that

$$\mathbf{k}'\boldsymbol{\lambda} = a_1\mathbf{k}_1'\boldsymbol{\lambda} + a_2\mathbf{k}_2'\boldsymbol{\lambda} + \cdots + a_S\mathbf{k}_S'\boldsymbol{\lambda}.$$

Thus, if $\mathbf{k}_1'\boldsymbol{\lambda} = \mathbf{k}_2'\boldsymbol{\lambda} = \cdots = \mathbf{k}_S'\boldsymbol{\lambda} = 0$, then $\mathbf{k}'\boldsymbol{\lambda} = 0$, leading to the conclusion that condition (3.10) implies condition (3.9).

Condition (3.10) comprises only S of the infinite number of equalities that form condition (3.9), making it much easier to administer than condition (3.9).

c. A related concept: identifiability

Let us continue to suppose that $\mathbf{y}$ is an $N \times 1$ observable random vector that follows a G–M, Aitken, or general linear model. Recall that $\mathrm{E}(\mathbf{y}) = \mathbf{X}\boldsymbol{\beta}$. Does a parametric function of the form $\boldsymbol{\lambda}'\boldsymbol{\beta}$ have a fixed value for each value of $\mathrm{E}(\mathbf{y})$? This question can be restated more formally as follows: is $\boldsymbol{\lambda}'\boldsymbol{\beta}_1 = \boldsymbol{\lambda}'\boldsymbol{\beta}_2$ for every pair of P-dimensional column vectors $\boldsymbol{\beta}_1$ and $\boldsymbol{\beta}_2$ such that $\mathbf{X}\boldsymbol{\beta}_1 = \mathbf{X}\boldsymbol{\beta}_2$? Or, equivalently, is $\mathbf{X}\boldsymbol{\beta}_1 \neq \mathbf{X}\boldsymbol{\beta}_2$ for every pair of P-dimensional column vectors $\boldsymbol{\beta}_1$ and $\boldsymbol{\beta}_2$ such that $\boldsymbol{\lambda}'\boldsymbol{\beta}_1 \neq \boldsymbol{\lambda}'\boldsymbol{\beta}_2$? Unless the model is of full rank, the answer depends on the coefficient vector $\boldsymbol{\lambda}'$. When the answer is yes, the parametric function $\boldsymbol{\lambda}'\boldsymbol{\beta}$ is said to be *identifiable*—this terminology is consistent with that of Hinkelmann and Kempthorne (2008, sec. 4.4).

The parametric function $\boldsymbol{\lambda}'\boldsymbol{\beta}$ is identifiable if and only if it is estimable. To see this, suppose that $\boldsymbol{\lambda}'\boldsymbol{\beta}$ is estimable. Then, $\boldsymbol{\lambda}' = \mathbf{a}'\mathbf{X}$ for some column vector $\mathbf{a}$. And for any P-dimensional column vectors $\boldsymbol{\beta}_1$ and $\boldsymbol{\beta}_2$ such that $\mathbf{X}\boldsymbol{\beta}_1 = \mathbf{X}\boldsymbol{\beta}_2$,

$$\boldsymbol{\lambda}'\boldsymbol{\beta}_1 = \mathbf{a}'\mathbf{X}\boldsymbol{\beta}_1 = \mathbf{a}'\mathbf{X}\boldsymbol{\beta}_2 = \boldsymbol{\lambda}'\boldsymbol{\beta}_2.$$

Accordingly, $\boldsymbol{\lambda}'\boldsymbol{\beta}$ is identifiable.

Conversely, suppose that $\boldsymbol{\lambda}'\boldsymbol{\beta}$ is identifiable. Then, $\boldsymbol{\lambda}'\boldsymbol{\beta}_1 = \boldsymbol{\lambda}'\mathbf{0}$ for every P-dimensional column vector $\boldsymbol{\beta}_1$ such that $\mathbf{X}\boldsymbol{\beta}_1 = \mathbf{X}\mathbf{0}$, or equivalently $\boldsymbol{\lambda}'\mathbf{k} = 0$ for every vector $\mathbf{k}$ in $\mathcal{N}(\mathbf{X})$. And based on the results on estimability established in Subsection b (and on the observation that $\boldsymbol{\lambda}'\mathbf{k} = \mathbf{k}'\boldsymbol{\lambda}$), we conclude that $\boldsymbol{\lambda}'\boldsymbol{\beta}$ is estimable.

d. Polynomials (in 1 variable)

Suppose that $\mathbf{y}$ is an N-dimensional observable random vector that follows a G–M, Aitken, or general linear model. Suppose further that there is a single explanatory variable, so that $C = 1$ and $\mathbf{u} = (u_1)$. And (for the sake of simplicity) let us write u for u_1 or (depending on the context) for $\mathbf{u}$.

Let us consider the case (considered initially in Section 4.2a) where $\delta(u)$ is a polynomial. Specifically, let us consider the case where

$$\delta(u) = \beta_1 + \beta_2 u + \beta_3 u^2 + \cdots + \beta_P u^{P-1}. \tag{3.11}$$

Under what circumstances are all P of the coefficients $\beta_1, \beta_2, \ldots, \beta_P$ estimable? Or, equivalently, under what circumstances is the model of full rank? The answer to this question can be established with the help of a result on a kind of matrix known as a Vandermonde matrix.

Vandermonde matrices. A Vandermonde matrix is a square matrix $\mathbf{A}$ of the general form

$$\mathbf{A} = \begin{pmatrix} 1 & t_1 & t_1^2 & \ldots & t_1^{K-1} \\ 1 & t_2 & t_2^2 & \ldots & t_2^{K-1} \\ 1 & t_3 & t_3^2 & \ldots & t_3^{K-1} \\ \vdots & \vdots & \vdots & \ddots & \vdots \\ 1 & t_K & t_K^2 & \ldots & t_K^{K-1} \end{pmatrix},$$

where $t_1, t_2, t_3, \ldots, t_K$ are arbitrary scalars. The determinant of a Vandermonde matrix is obtainable from the formula

$$|\mathbf{A}| = \prod_{\substack{i,j \\ (j<i)}} (t_i - t_j). \tag{3.12}$$

For a derivation of this formula, refer, for example, to Harville (1997, sec. 13.6).

Denote by R the number of distinct values represented among $t_1, t_2, t_3, \ldots, t_K$. Then, it follows from result (3.12) that

$$R = K \quad \Leftrightarrow \quad |\mathbf{A}| \neq 0$$

and hence (in light of Theorem 2.14.20) that

$$R = K \quad \Leftrightarrow \quad \mathbf{A} \text{ is nonsingular.} \tag{3.13}$$

Rank of the model matrix. We are now in a position to determine the rank of the model matrix $\mathbf{X}$ (of a G–M, Aitken, or general linear model) in the special case where $C = 1$ and where $\delta(u)$ is the polynomial (3.11) (which is of degree $P - 1$ in the lone explanatory variable u). Denote by D the number of distinct values of u represented among the N values of u corresponding to the N observable random variables $y_1, y_2, \ldots, y_N$. And take $i_1, i_2, \ldots, i_D$ ($i_1 < i_2 < \cdots < i_D$) to be integers between 1 and N, inclusive, such that the values of u corresponding to $y_{i_1}, y_{i_2}, \ldots, y_{i_D}$ are distinct. Each of the N rows of $\mathbf{X}$ is either among its $i_1, i_2, \ldots, i_D$th rows or is a duplicate of one of those rows. Thus, $\mathcal{R}(\mathbf{X})$ is spanned by the $i_1, i_2, \ldots, i_D$th rows of $\mathbf{X}$, and it follows that $\text{rank}(\mathbf{X}) \le D$ and hence [since $\text{rank}(\mathbf{X}) \le P$] that $\text{rank}(\mathbf{X}) \le M$, where $M = \min(D, P)$. Moreover, it follows from result (3.13) that the $M \times M$ submatrix of $\mathbf{X}$ formed from its $i_1, i_2, \ldots, i_M$th rows and its first M columns is nonsingular, implying (in light of Theorem 2.4.19) that $\text{rank}(\mathbf{X}) \ge M$. And we conclude that

$$\text{rank}(\mathbf{X}) = \min(D, P). \tag{3.14}$$

In light of result (3.14), it is evident that [in the special case where $\delta(u)$ is the $(P-1)$-degree polynomial (3.11)] the model is of full rank if and only if $D \ge P$, that is, if and only if at least P of the N values of u (the N values of u corresponding to $y_1, y_2, \ldots, y_N$) are distinct. When the model is of full rank, all P of the coefficients $\beta_1, \beta_2, \ldots, \beta_P$ [in the $(P-1)$-degree polynomial] are estimable.

In the application to the ouabain data (of Section 4.2b), there are 4 distinct values of the explanatory variable, representing the 4 different rates of injection (or perhaps the logarithms of the 4 different rates). Accordingly, if $\delta(u)$ were taken to be a polynomial of the form (3.11), the model would be of full rank if and only if the degree of the polynomial were taken to be 3 or less.

e. An illustration: mixture data

Consider an application of the G–M, Aitken, or general linear model in which the C explanatory variables $u_1, u_2, \ldots, u_C$ represent the proportionate amounts of the ingredients in a mixture. By their very nature, the explanatory variables are such that

$$\sum_{j=1}^{C} u_j = 1. \tag{3.15}$$

Now, suppose that

$$\delta(\mathbf{u}) = \beta_1 + \beta_2 u_1 + \beta_3 u_2 + \cdots + \beta_{C+1} u_C \tag{3.16}$$

(in which case, $P = C + 1$). And let $\mathbf{x}_1, \mathbf{x}_2, \ldots, \mathbf{x}_{C+1}$ represent the first through last columns of the model matrix $\mathbf{X}$. Then,

$$\mathbf{x}_1 = \sum_{j=2}^{C+1} \mathbf{x}_j \tag{3.17}$$

or, equivalently,

$$\mathbf{X}\begin{pmatrix} -1 \\ \mathbf{1}_C \end{pmatrix} = \mathbf{0}. \tag{3.18}$$

Suppose, for example, that the mixtures are fruit juices, each of which is a blend of watermelon juice, orange juice, pineapple juice, and grapefruit juice (in which case, the data points might be flavor scores). Suppose further that $N = 6$ and that the data are those obtained for the following 6 blends:

u_1	u_2	u_3	u_4
0.8	0.2	0	0
0.2	0	0.8	0
0.4	0	0	0.6
0.5	0.1	0.4	0
0.6	0.1	0	0.3
0.3	0	0.4	0.3

What parametric functions of the form $\boldsymbol{\lambda}'\boldsymbol{\beta}$ are estimable? In light of equality (3.18), $\text{rank}(\mathbf{X}) \leq C$, and it follows from the results of Subsection b that for $\boldsymbol{\lambda}'\boldsymbol{\beta}$ to be estimable, it is necessary that

$$\begin{pmatrix} -1 \\ \mathbf{1}_C \end{pmatrix}' \boldsymbol{\lambda} = 0 \tag{3.19}$$

or, equivalently, that

$$\sum_{j=2}^{C+1} \lambda_j = \lambda_1. \tag{3.20}$$

Is this condition sufficient as well as necessary? The answer to this question depends on $\text{rank}(\mathbf{X})$. It follows from the results of Subsection b that if $\text{rank}(\mathbf{X}) = C$, then condition (3.20) is sufficient (as well as necessary) for the estimability of $\boldsymbol{\lambda}'\boldsymbol{\beta}$; however, if $\text{rank}(\mathbf{X}) < C$, then condition (3.20) is not (in and of itself) sufficient.

In the case of the 6 blends of the 4 juices,

$$\text{rank}(\mathbf{X}) = 3 = C - 1 < C.$$

To see this, observe that the last 3 columns of $\mathbf{X}$ (which contain the values of u_2, u_3, and u_4, respectively) are linearly independent, so that $\text{rank}(\mathbf{X}) \geq 3$. Observe also that each of the 6 blends of the 4 juices is such that

$$u_4 = 1.5u_1 - 6u_2 - 0.375u_3,$$

so that (in the case of the 6 blends of the 4 juices)

$$\mathbf{x}_5 = 1.5\mathbf{x}_2 - 6\mathbf{x}_3 - 0.375\mathbf{x}_4, \tag{3.21}$$

which [together with result (3.17)] implies that the first and last columns of $\mathbf{X}$ are expressible as linear combinations of the other 3 columns and hence that $\text{rank}(\mathbf{X}) \leq 3$. Thus, $\text{rank}(\mathbf{X}) = 3$.

To obtain conditions that are both necessary and sufficient for the estimability of $\boldsymbol{\lambda}'\boldsymbol{\beta}$ (from the information provided by the data on the 6 blends of the 4 juices), observe that equality (3.21) can be reexpressed in the form

$$\mathbf{X} \begin{pmatrix} 0 \\ 1.5 \\ -6 \\ -0.375 \\ -1 \end{pmatrix} = \mathbf{0}.$$

And it follows from the results of Subsection b that for $\boldsymbol{\lambda}'\boldsymbol{\beta}$ to be estimable, it is necessary that

$$\begin{pmatrix} 0 \\ 1.5 \\ -6 \\ -0.375 \\ -1 \end{pmatrix}' \boldsymbol{\lambda} = 0 \tag{3.22}$$

or, equivalently, that

$$\lambda_5 = 1.5\lambda_2 - 6\lambda_3 - 0.375\lambda_4. \tag{3.23}$$

Moreover, together, the two conditions (3.19) and (3.22) or, equivalently, the two conditions (3.20) and (3.23) are sufficient (as well as necessary) for the estimability of $\boldsymbol{\lambda}'\boldsymbol{\beta}$.

The example provided by the 6 blends of the 4 juices is one in which $\text{rank}(\mathbf{X}) = C-1\ (= P-2)$. It is easy to construct examples in which $\text{rank}(\mathbf{X}) = C\ (= P-1)$ and to do so for any $N\ (\geq C)$—in light of result (3.17) or (3.18), $\text{rank}(\mathbf{X})$ cannot be larger than C. In what is perhaps the simplest way to construct such an example, we can take any C of the blends corresponding to the N data points to be the C pure blends, each of which consists entirely of one ingredient. This approach results in a model matrix $\mathbf{X}$ whose N rows include the vectors $(1, 1, 0, 0, \ldots, 0, 0)$, $(1, 0, 1, 0, \ldots, 0, 0)$, ..., $(1, 0, 0, 0, \ldots, 0, 1)$. Clearly, these C vectors are linearly independent, implying that $\text{rank}(\mathbf{X}) \geq C$ and hence [since $\text{rank}(\mathbf{X}) \leq C$] that $\text{rank}(\mathbf{X}) = C$.

When $\text{rank}(\mathbf{X}) = C$, the condition $\begin{pmatrix} -1 \\ \mathbf{1}_C \end{pmatrix}' \boldsymbol{\lambda} = 0$, or equivalently the condition $\sum_{j=2}^{C+1} \lambda_j = \lambda_1$, is sufficient (as well as necessary) for the estimability of $\boldsymbol{\lambda}'\boldsymbol{\beta}$. It is worth noting that this condition is not satisfied by any of the individual parameters $\beta_1, \beta_2, \ldots, \beta_{C+1}$ and, consequently, none of these parameters is estimable. Thus, we have established (by means of an example) that not only is it possible for all P of the individual parameters $\beta_1, \beta_2, \ldots, \beta_P$ of a G–M, Aitken, or general linear model to be nonestimable, but it is possible even if $\text{rank}(\mathbf{X}) = P - 1$.

f. Inherent versus noninherent restrictions on estimability

Let us continue to consider the estimability of parametric functions of the form $\boldsymbol{\lambda}'\boldsymbol{\beta}$ (under the G–M, Aitken, or general linear model). Unless the model is of full rank, estimability is restricted to a proper subset of these parametric functions. The restriction is in the form of restrictions on the coefficient vector $\boldsymbol{\lambda}'$. As discussed in Subsection b, a necessary and sufficient condition for $\boldsymbol{\lambda}'\boldsymbol{\beta}$ to be estimable is that

$$\mathbf{k}'\boldsymbol{\lambda} = 0 \tag{3.24}$$

for every $P \times 1$ vector $\mathbf{k}$ in $\mathcal{N}(\mathbf{X})$.

It can be helpful to think of each of the restrictions of the form (3.24) as being either an inherent restriction or a noninherent restriction. Let U represent a set consisting of C-dimensional column vectors that are considered to be "feasible" values of the vector $\mathbf{u}$ of explanatory variables—a value might be deemed infeasible either because of the nature of the explanatory variables or because of the "limitations" of the model. The set U is assumed to include the values $\mathbf{u}_1, \mathbf{u}_2, \ldots, \mathbf{u}_N$ of $\mathbf{u}$ that correspond to the N data points.

A restriction of the form (3.24) is an *inherent restriction* if it would be applicable regardless of the number of data points and regardless of the values of $\mathbf{u}$ (in U) corresponding to the data points. Otherwise, the restriction is a *noninherent restriction*. A parametric function of the form $\boldsymbol{\lambda}'\boldsymbol{\beta}$ that fails to satisfy an inherent restriction tends not to be conceptually meaningful. Accordingly, the nonestimability of such a function tends not to be of concern.

A parametric function of the form $\boldsymbol{\lambda}'\boldsymbol{\beta}$ that is nonestimable but that would be estimable if it were not for the presence of noninherent restrictions tends to be conceptually meaningful. Its nonestimability may or may not be of concern, depending on whether or not it represents a quantity of interest.

It is informative to consider these concepts in the context of the illustrative example introduced and discussed in Subsection e. Accordingly, suppose that the C explanatory variables $u_1, u_2, \ldots, u_C$ represent the proportionate amounts of the ingredients in a mixture, in which case they are subject to the constraint $\sum_{j=1}^{C} u_j = 1$. And suppose that $\delta(\mathbf{u})$ is of the form (3.16). Then, as discussed in Subsection e, not all parametric functions of the form $\boldsymbol{\lambda}'\boldsymbol{\beta}$ are estimable; for $\boldsymbol{\lambda}'\boldsymbol{\beta}$ to be estimable, it is necessary that

$$\lambda_1 - \sum_{j=2}^{C+1} \lambda_j = 0 \tag{3.25}$$

or, equivalently, that

$$\sum_{j=2}^{C+1} \lambda_j = \lambda_1. \tag{3.26}$$

In this setting, the set U of feasible values of $\mathbf{u}$ would be the set

$$\{\mathbf{u} : u_j \geq 0 \text{ (for } j = 1, 2, \ldots, C) \text{ and } \textstyle\sum_{j=1}^{C} u_j = 1\} \tag{3.27}$$

(or, perhaps, a nondegenerate subset of that set). Geometrically, the form of the set (3.27) is that of a $(C-1)$-dimensional simplex. Regardless of the number of data points and regardless of which values of $\mathbf{u}$ correspond to the data points, the model matrix $\mathbf{X}$ would be such that $\mathbf{X}\begin{pmatrix} -1 \\ \mathbf{1}_C \end{pmatrix} = \mathbf{0}$ and, consequently, condition (3.25), or equivalently condition (3.26), would be a necessary condition for the estimability of $\boldsymbol{\lambda}'\boldsymbol{\beta}$. Accordingly, condition (3.26) constitutes an inherent restriction on the estimability of $\boldsymbol{\lambda}'\boldsymbol{\beta}$.

As discussed in Subsection e, the parametric functions for which condition (3.26) is not satisfied include all $C+1$ of the individual parameters $\beta_1, \beta_2, \ldots, \beta_{C+1}$. If the explanatory variables $u_1, u_2, \ldots, u_C$ were unrestricted, the individual parameters would have meaningful interpretations, emanating from the observation that $\beta_1 = \delta(\mathbf{0})$ and that (for $j = 1, 2, \ldots, C$) β_{j+1} equals the change in $\delta(\mathbf{u})$ effected by a unit change in the jth explanatory variable u_j when the other $C-1$ explanatory variables are held constant. However, those interpretations are rendered meaningless by the restriction of the vector $\mathbf{u}$ of explanatory variables to the set (3.27). The interpretation of β_1 emanating from the observation that $\beta_1 = \delta(\mathbf{0})$ is meaningless because there is no mixture for which $\mathbf{u} = \mathbf{0}$. The interpretations of $\beta_2, \beta_3, \ldots, \beta_{C+1}$ in terms of the change in $\delta(\mathbf{u})$ effected by changing one of the explanatory variables while holding the others constant are also meaningless. By their very nature, the explanatory variables $u_1, u_2, \ldots, u_C$ are such that $\sum_{j=1}^{C} u_j = 1$, making it impossible to change one of the explanatory variables without changing any of the others.

Subsection e includes an example in which $N = 6$ and $C = 4$ and in which the mixtures consist of blends of juices. In that example, $\mathbf{X}\begin{pmatrix} 0 \\ 1.5 \\ -6 \\ -0.375 \\ -1 \end{pmatrix} = \mathbf{0}$, so that for a parametric function of the form $\boldsymbol{\lambda}'\boldsymbol{\beta}$ to be estimable, it is necessary that

$$1.5\lambda_2 - 6\lambda_3 - 0.375\lambda_4 - \lambda_5 = 0 \tag{3.28}$$

or, equivalently, that

$$\lambda_5 = 1.5\lambda_2 - 6\lambda_3 - 0.375\lambda_4. \tag{3.29}$$

Condition (3.29) constitutes a noninherent restriction on the estimability of $\boldsymbol{\lambda}'\boldsymbol{\beta}$. It would not be applicable if, for instance, the value of $\mathbf{u}$ corresponding to the fourth of the example's 6 data points were $(0.5, 0.25, 0.25, 0)'$ rather than $(0.5, 0.1, 0.4, 0)'$ [in which case, the first 4 rows of the model matrix $\mathbf{X}$ would be linearly independent and rank($\mathbf{X}$) would be 4 rather than 3]. Parametric functions of the form $\boldsymbol{\lambda}'\boldsymbol{\beta}$ that satisfy restriction (3.26), but not restriction (3.29), are conceptually meaningful. Because of restriction (3.29), the only mixtures for which the "response function" $\delta(\mathbf{u})$ would be estimable from the 6 data points in the example are those for which

$$u_4 = 1.5u_1 - 6u_2 - 0.375u_3.$$

If it had been the case that the only restriction on estimability were that determined by the inherent restriction (3.26), then the value of $\delta(\mathbf{u})$ would have been estimable for every mixture [i.e., for every $\mathbf{u}$ in the set (3.27)].

Noninherent restrictions on the estimability of parametric functions of the form $\boldsymbol{\lambda}'\boldsymbol{\beta}$ may be encountered in cases where the data are "observational" in nature. They may also be encountered in cases where the data come from a designed experiment or a sample survey. The extent to which their presence is of concern would seem to depend on which parametric functions are rendered nonestimable and on the extent to which those functions are of interest.

In the case of data from a designed experiment or a sample survey, the presence of noninherent restrictions may be either inadvertent or by intent. Their presence may be attributable to problems in execution or design. Or when the affected parametric functions are ones that are considered to be of little importance and/or of negligible size, the presence of noninherent restrictions may be viewed as an acceptable consequence of an attempt to make the best possible use of limited resources.

5.4 The Method of Least Squares

Let $\underline{y}_1, \underline{y}_2, \ldots, \underline{y}_N$ represent N data points, and let $\mathbf{u}_1, \mathbf{u}_2, \ldots, \mathbf{u}_N$ represent the corresponding values of a C-dimensional column vector $\mathbf{u}$ of explanatory variables. Consider the problem of choosing a function $\underline{\delta}(\mathbf{u})$ of $\mathbf{u}$ on the basis of how well the values $\underline{\delta}(\mathbf{u}_1), \underline{\delta}(\mathbf{u}_2), \ldots, \underline{\delta}(\mathbf{u}_N)$ approximate $\underline{y}_1, \underline{y}_2, \ldots, \underline{y}_N$. Let Δ represent the set of candidates from which the function $\underline{\delta}(\cdot)$ is to be chosen.

Which member of Δ provides the best approximation? One way of addressing this question is through the minimization [for $\underline{\delta}(\cdot) \in \Delta$] of the norm of the N-dimensional column vector with elements $\underline{y}_1 - \underline{\delta}(\mathbf{u}_1), \underline{y}_2 - \underline{\delta}(\mathbf{u}_2), \ldots, \underline{y}_N - \underline{\delta}(\mathbf{u}_N)$. When the norm is taken to be the usual norm, this method is equivalent to minimizing $\sum_{i=1}^{N} [\underline{y}_i - \underline{\delta}(\mathbf{u}_i)]^2$, and is referred to as the *method of least squares* or simply as *least squares*.

The origins of the method of least squares are a matter of some dispute, but date back at least to an 1805 publication by Adrien Marie Legendre. For discussions of the history of the method, refer, for example, to Plackett (1972) and Stigler (1986, chap. 1).

The focus herein is on the method of least squares as applied to settings in which the data points are regarded as the values of observable random variables that follow a G–M, Aitken, or general linear model. In such a setting, the method of least squares can be used to obtain estimates of estimable functions (of the model's parameters) of the form $\boldsymbol{\lambda}'\boldsymbol{\beta}$. In the special case of the G–M model, least squares estimators have certain optimal properties.

In what follows, consideration of the method of least squares is restricted to the special case where Δ consists of those functions (of $\mathbf{u}$) that are expressible as linear combinations of P specified functions $\delta_1(\cdot), \delta_2(\cdot), \ldots, \delta_P(\cdot)$. Thus, $\underline{\delta}(\cdot)$ is a member of Δ if $\underline{\delta}(\mathbf{u})$ is expressible in the form

$$\underline{\delta}(\mathbf{u}) = b_1\delta_1(\mathbf{u}) + b_2\delta_2(\mathbf{u}) + \cdots + b_P\delta_P(\mathbf{u}), \tag{4.1}$$

where $b_1, b_2, \ldots, b_P$ are arbitrary scalars. When $\underline{\delta}(\cdot)$ is such that $\underline{\delta}(\mathbf{u})$ is expressible in the form (4.1),

$$\sum_{i=1}^{N} [\underline{y}_i - \underline{\delta}(\mathbf{u}_i)]^2 = \sum_{i=1}^{N} \Bigl(\underline{y}_i - \sum_{j=1}^{P} x_{ij}b_j\Bigr)^2,$$

where (for $i = 1, 2, \ldots, N$ and $j = 1, 2, \ldots, P$) $x_{ij} = \delta_j(\mathbf{u}_i)$. Accordingly, in the special case under consideration, the minimization of $\sum_{i=1}^{N} [\underline{y}_i - \underline{\delta}(\mathbf{u}_i)]^2$ with respect to $\underline{\delta}(\cdot)$ [for $\underline{\delta}(\cdot) \in \Delta$] is equivalent to the minimization of $\sum_{i=1}^{N} \bigl(\underline{y}_i - \sum_{j=1}^{P} x_{ij}b_j\bigr)^2$ with respect to $b_1, b_2, \ldots, b_P$. Moreover, upon letting $\underline{\mathbf{y}} = (\underline{y}_1, \underline{y}_2, \ldots, \underline{y}_N)'$ and $\mathbf{b} = (b_1, b_2, \ldots, b_P)'$ and taking $\mathbf{X}$ to be the $N \times P$ matrix with ijth element x_{ij},

$$\sum_{i=1}^{N} \Bigl(\underline{y}_i - \sum_{j=1}^{P} x_{ij}b_j\Bigr)^2 = (\underline{\mathbf{y}} - \mathbf{X}\mathbf{b})'(\underline{\mathbf{y}} - \mathbf{X}\mathbf{b}), \tag{4.2}$$

so that the minimization of $\sum_{i=1}^{N}\big(\underset{\sim}{y}_i - \sum_{j=1}^{P} x_{ij} b_j\big)^2$ with respect to $b_1, b_2, \ldots, b_P$ is equivalent to the minimization of $(\underset{\sim}{\mathbf{y}} - \mathbf{X}\mathbf{b})'(\underset{\sim}{\mathbf{y}} - \mathbf{X}\mathbf{b})$ with respect to the P-dimensional vector $\mathbf{b}$ (where $\mathbf{b}$ is an arbitrary member of $\mathbb{R}^P$).

We wish to obtain a solution to the problem of minimizing the quantity $\sum_{i=1}^{N}\big(\underset{\sim}{y}_i - \sum_{j=1}^{P} x_{ij} b_j\big)^2$ with respect to $b_1, b_2, \ldots, b_P$. Conditions that are necessary for this quantity to attain a minimum value can be obtained by differentiating with respect to $b_1, b_2, \ldots, b_P$ and by equating the resultant partial derivatives to 0. Or in what can be regarded as an appealing variation on this approach, we can reformulate the minimization problem in matrix notation [as the problem of minimizing $(\underset{\sim}{\mathbf{y}} - \mathbf{X}\mathbf{b})'(\underset{\sim}{\mathbf{y}} - \mathbf{X}\mathbf{b})$ with respect to $\mathbf{b}$] and take advantage of some basic results on vector differentiation.

a. Some basic results on vector differentiation

Let $\mathbf{x} = (x_1, x_2, \ldots, x_M)'$ represent an M-dimensional column vector of variables. And let $f(\mathbf{x})$ represent a function of $\mathbf{x}$. Write $\frac{\partial f(\mathbf{x})}{\partial \mathbf{x}}$, or simply $\frac{\partial f}{\partial \mathbf{x}}$, for the M-dimensional column vector whose jth element is the (first-order) partial derivative $\frac{\partial f(\mathbf{x})}{\partial x_j}$ of f with respect to x_j; this vector may be referred to as the *derivative of* $f(\mathbf{x})$ *with respect to* $\mathbf{x}$ and is sometimes called the *gradient vector* of f. And write $\frac{\partial f(\mathbf{x})}{\partial \mathbf{x}'}$, or $\frac{\partial f}{\partial \mathbf{x}'}$, for $\left[\frac{\partial f(\mathbf{x})}{\partial \mathbf{x}}\right]'$; this vector may be referred to as the *derivative of* $f(\mathbf{x})$ *with respect to* $\mathbf{x}'$. Further, write $\frac{\partial^2 f(\mathbf{x})}{\partial \mathbf{x}\,\partial \mathbf{x}'}$, or $\frac{\partial^2 f}{\partial \mathbf{x}\,\partial \mathbf{x}'}$, for the $M \times M$ matrix whose ijth element is the second-order partial derivative $\frac{\partial^2 f(\mathbf{x})}{\partial x_i\,\partial x_j}$—or, when $j = i$, $\frac{\partial^2 f(\mathbf{x})}{\partial x_i^2}$—of f with respect to x_i and x_j, and refer to this matrix as the *Hessian matrix* of f.

The formulas for obtaining the partial derivatives of a linear combination of functions, a product of functions, and a ratio of two functions with respect to a single variable extend in an altogether straightforward way to vector differentiation. In particular, in the case of the product of two functions $f(\mathbf{x})$ and $g(\mathbf{x})$ of $\mathbf{x}$,

$$\frac{\partial f(\mathbf{x})g(\mathbf{x})}{\partial \mathbf{x}} = f(\mathbf{x})\frac{\partial g(\mathbf{x})}{\partial \mathbf{x}} + g(\mathbf{x})\frac{\partial f(\mathbf{x})}{\partial \mathbf{x}}. \tag{4.3}$$

Refer, for example, to Harville (1997, sec. 15.2) for further particulars.

Denote by $\mathbf{a} = (a_1, a_2, \ldots, a_M)'$ an M-dimensional column vector of constants and by $\mathbf{A} = \{a_{ik}\}$ an $M \times M$ matrix of constants. And consider the differentiation of the linear form $\mathbf{a}'\mathbf{x} = \sum_i a_i x_i$ and of the quadratic form $\mathbf{x}'\mathbf{A}\mathbf{x} = \sum_{i,k} a_{ik} x_i x_k$. The first-order partial derivatives of $\mathbf{a}'\mathbf{x}$ and the first- and second-order partial derivatives of $\mathbf{x}'\mathbf{A}\mathbf{x}$ are

$$\frac{\partial\, \mathbf{a}'\mathbf{x}}{\partial x_j} = a_j \qquad (j = 1, 2, \ldots, M), \tag{4.4}$$

$$\frac{\partial\, \mathbf{x}'\mathbf{A}\mathbf{x}}{\partial x_j} = \sum_i a_{ij} x_i + \sum_k a_{jk} x_k \qquad (j = 1, 2, \ldots, M), \tag{4.5}$$

and

$$\frac{\partial^2\, \mathbf{x}'\mathbf{A}\mathbf{x}}{\partial x_s\,\partial x_j} = a_{sj} + a_{js} \qquad (s, j = 1, 2, \ldots, M), \tag{4.6}$$

as can be verified via a relatively simple exercise—refer, e.g., to Harville (1997, sec. 15.3) for the details.

In light of result (4.4), the gradient vector of $\mathbf{a}'\mathbf{x}$ is

$$\frac{\partial\, \mathbf{a}'\mathbf{x}}{\partial \mathbf{x}} = \mathbf{a}. \tag{4.7}$$

And in light of result (4.5), the gradient vector of $\mathbf{x}'\mathbf{A}\mathbf{x}$ is

$$\frac{\partial\, \mathbf{x}'\mathbf{A}\mathbf{x}}{\partial \mathbf{x}} = (\mathbf{A} + \mathbf{A}')\mathbf{x} \tag{4.8}$$

(as is evident upon observing that $\sum_k a_{jk}x_k$ is the jth element of the column vector $\mathbf{Ax}$ and $\sum_i a_{ij}x_i$ is the jth element of $\mathbf{A}'\mathbf{x}$). Further, in light of result (4.6), the Hessian matrix of $\mathbf{x}'\mathbf{A}\mathbf{x}$ is

$$\frac{\partial^2\, \mathbf{x}'\mathbf{A}\mathbf{x}}{\partial \mathbf{x}\,\partial \mathbf{x}'} = \mathbf{A} + \mathbf{A}'. \tag{4.9}$$

In the special case where $\mathbf{A}$ is symmetric, results (4.8) and (4.9) simplify to

$$\frac{\partial\, \mathbf{x}'\mathbf{A}\mathbf{x}}{\partial \mathbf{x}} = 2\mathbf{A}\mathbf{x} \tag{4.10}$$

and

$$\frac{\partial^2\, \mathbf{x}'\mathbf{A}\mathbf{x}}{\partial \mathbf{x}\,\partial \mathbf{x}'} = 2\mathbf{A}. \tag{4.11}$$

Let $\mathbf{f}(\mathbf{x})$ represent a $(P \times 1)$-dimensional vector-valued function of the M-dimensional (column) vector $\mathbf{x} = (x_1, x_2, \ldots, x_M)'$, and denote by $f_1(\mathbf{x}),\ f_2(\mathbf{x}),\ \ldots,\ f_P(\mathbf{x})$ the P functions of $\mathbf{x}$ that constitute the first, second, ..., Pth elements of $\mathbf{f}(\mathbf{x})$. Let us write $\dfrac{\partial \mathbf{f}(\mathbf{x})}{\partial \mathbf{x}'}$, or simply $\dfrac{\partial \mathbf{f}}{\partial \mathbf{x}'}$, for the $P \times M$ matrix whose sjth element is $\dfrac{\partial f_s}{\partial x_j}$; this matrix may be referred to as the *derivative of* $\mathbf{f}(\mathbf{x})$ *with respect to* $\mathbf{x}'$ and is sometimes called the *Jacobian matrix* of $\mathbf{f}$. And let us write $\dfrac{\partial \mathbf{f}'(\mathbf{x})}{\partial \mathbf{x}}$, or simply $\dfrac{\partial \mathbf{f}'}{\partial \mathbf{x}}$, for the $M \times P$ matrix whose jsth element is $\dfrac{\partial f_s}{\partial x_j}$ or, equivalently, whose sth column is $\dfrac{\partial f_s}{\partial \mathbf{x}}$; this matrix may be referred to as the *derivative of* $\mathbf{f}'(\mathbf{x})$ *with respect to* $\mathbf{x}$ and is sometimes called the *gradient matrix* of $\mathbf{f}$. Note that for any $P \times M$ matrix of constants $\mathbf{B}$,

$$\frac{\partial\, (\mathbf{B}\mathbf{x})}{\partial \mathbf{x}'} = \mathbf{B} \quad \text{and} \quad \frac{\partial\, (\mathbf{B}\mathbf{x})'}{\partial \mathbf{x}} = \mathbf{B}'. \tag{4.12}$$

b. Solution to the least squares minimization problem

Let us now consider the implications of the results of Subsection a as applied to the minimization (with respect to the $P \times 1$ vector $\mathbf{b} = \{b_j\}$) of $(\underline{\mathbf{y}} - \mathbf{X}\mathbf{b})'(\underline{\mathbf{y}} - \mathbf{X}\mathbf{b})$, where $\underline{\mathbf{y}} = \{\underline{y}_i\}$ is an $N \times 1$ data vector and $\mathbf{X} = \{x_{ij}\}$ an $N \times P$ matrix. Take $q(\cdot)$ to be the function (of $\mathbf{b}$) defined by $q(\mathbf{b}) = (\underline{\mathbf{y}} - \mathbf{X}\mathbf{b})'(\underline{\mathbf{y}} - \mathbf{X}\mathbf{b})$. And observe that

$$q(\mathbf{b}) = \underline{\mathbf{y}}'\underline{\mathbf{y}} - 2(\mathbf{X}'\underline{\mathbf{y}})'\mathbf{b} + \mathbf{b}'\mathbf{X}'\mathbf{X}\mathbf{b}.$$

Then, as an immediate application of formulas (4.7), (4.10), and (4.11), we have that

$$\frac{\partial\, q(\mathbf{b})}{\partial \mathbf{b}} = -2\mathbf{X}'\underline{\mathbf{y}} + 2\mathbf{X}'\mathbf{X}\mathbf{b} \tag{4.13}$$

and

$$\frac{\partial^2 q(\mathbf{b})}{\partial \mathbf{b}\,\partial \mathbf{b}'} = 2\mathbf{X}'\mathbf{X}. \tag{4.14}$$

It follows from basic results on unconstrained minimization (e.g., Luenberger and Ye 2016, sec. 7.1) that a necessary condition for $q(\mathbf{b})$ to attain a minimum value at a point $\tilde{\mathbf{b}}$ is that $\dfrac{\partial\, q(\mathbf{b})}{\partial \mathbf{b}} = \mathbf{0}$ at $\mathbf{b} = \tilde{\mathbf{b}}$. Clearly,

$$\frac{\partial q(\mathbf{b})}{\partial \mathbf{b}} = \mathbf{0} \quad \Leftrightarrow \quad \mathbf{X}'\mathbf{X}\mathbf{b} = \mathbf{X}'\underset{\sim}{\mathbf{y}}.$$

Thus, a necessary condition for $q(\mathbf{b})$ to attain a minimum value at a point $\tilde{\mathbf{b}}$ is that $\tilde{\mathbf{b}}$ constitute a solution to the linear system

$$\mathbf{X}'\mathbf{X}\mathbf{b} = \mathbf{X}'\underset{\sim}{\mathbf{y}} \tag{4.15}$$

(in the $P \times 1$ vector $\mathbf{b}$). Linear system (4.15) consists of P equations; collectively, these equations are known as the *normal equations*.

The normal equations are consistent, as can be verified by, for example, observing [in light of equality (2.12.2)] that $\mathbf{X}'\mathbf{X}(\mathbf{X}'\mathbf{X})^{-}\mathbf{X}'\underset{\sim}{\mathbf{y}} = \mathbf{X}'\underset{\sim}{\mathbf{y}}$ [which implies that $(\mathbf{X}'\mathbf{X})^{-}\mathbf{X}'\underset{\sim}{\mathbf{y}}$ is a solution to the normal equations]. Moreover, for $q(\mathbf{b})$ to attain a minimum value at a point $\tilde{\mathbf{b}}$, it is sufficient, as well as necessary, that $\tilde{\mathbf{b}}$ constitute a solution to the normal equations. That is, the set of points at which $q(\mathbf{b})$ attains a minimum value equals the solution set of linear system (4.15).

Let us verify that every solution to the normal equations is a point at which $q(\mathbf{b})$ attains a minimum value. There are at least two different ways of accomplishing the verification. We could start with the observation that $\frac{\partial^2 q(\mathbf{b})}{\partial \mathbf{b} \partial \mathbf{b}'}$ is a nonnegative definite matrix and that, as a consequence, $q(\mathbf{b})$ is a convex function (e.g., Luenberger and Ye 2016, sec. 7.4). We could then take advantage of a general result on the minimization of convex functions (e.g., Luenberger and Ye 2016, sec. 7.5) to conclude that for any $P \times 1$ vector $\tilde{\mathbf{b}}$ such that $\frac{\partial q(\mathbf{b})}{\partial \mathbf{b}} = \mathbf{0}$ at $\mathbf{b} = \tilde{\mathbf{b}}$ or, equivalently, for any solution $\tilde{\mathbf{b}}$ to the normal equations, $q(\mathbf{b})$ attains a minimum value at $\mathbf{b} = \tilde{\mathbf{b}}$.

An alternative way of accomplishing the verification is to do so directly (without making any demands on the reader's knowledge of the literature on optimization). For any solution $\tilde{\mathbf{b}}$ to the normal equations,

$$q(\mathbf{b}) = q(\tilde{\mathbf{b}}) + [\mathbf{X}(\mathbf{b} - \tilde{\mathbf{b}})]'\mathbf{X}(\mathbf{b} - \tilde{\mathbf{b}}) \geq q(\tilde{\mathbf{b}}), \tag{4.16}$$

as is evident upon observing that

$$q(\mathbf{b}) = [\underset{\sim}{\mathbf{y}} - \mathbf{X}\tilde{\mathbf{b}} - \mathbf{X}(\mathbf{b} - \tilde{\mathbf{b}})]'[\underset{\sim}{\mathbf{y}} - \mathbf{X}\tilde{\mathbf{b}} - \mathbf{X}(\mathbf{b} - \tilde{\mathbf{b}})],$$

that

$$[\mathbf{X}(\mathbf{b} - \tilde{\mathbf{b}})]'(\underset{\sim}{\mathbf{y}} - \mathbf{X}\tilde{\mathbf{b}}) = (\mathbf{b} - \tilde{\mathbf{b}})'(\mathbf{X}'\underset{\sim}{\mathbf{y}} - \mathbf{X}'\mathbf{X}\tilde{\mathbf{b}}) = 0,$$

that

$$(\underset{\sim}{\mathbf{y}} - \mathbf{X}\tilde{\mathbf{b}})'\mathbf{X}(\mathbf{b} - \tilde{\mathbf{b}}) = [(\underset{\sim}{\mathbf{y}} - \mathbf{X}\tilde{\mathbf{b}})'\mathbf{X}(\mathbf{b} - \tilde{\mathbf{b}})]' = [\mathbf{X}(\mathbf{b} - \tilde{\mathbf{b}})]'(\underset{\sim}{\mathbf{y}} - \mathbf{X}\tilde{\mathbf{b}}),$$

and that $[\mathbf{X}(\mathbf{b} - \tilde{\mathbf{b}})]'\mathbf{X}(\mathbf{b} - \tilde{\mathbf{b}})$ is a sum of squares [of the elements of $\mathbf{X}(\mathbf{b} - \tilde{\mathbf{b}})$]. And it follows immediately from result (4.16) that $q(\mathbf{b})$ attains a minimum value at $\mathbf{b} = \tilde{\mathbf{b}}$.

Result (4.16) also serves to confirm that any point at which $q(\mathbf{b})$ attains a minimum value is a solution to the normal equations (or, equivalently, a point at which $\frac{\partial q(\mathbf{b})}{\partial \mathbf{b}} = \mathbf{0}$). To see this, observe that

$$\begin{aligned} q(\mathbf{b}) = q(\tilde{\mathbf{b}}) \quad &\Rightarrow \quad [\mathbf{X}(\mathbf{b} - \tilde{\mathbf{b}})]'\mathbf{X}(\mathbf{b} - \tilde{\mathbf{b}}) = 0 \\ &\Rightarrow \quad \mathbf{X}(\mathbf{b} - \tilde{\mathbf{b}}) = \mathbf{0} \\ &\Rightarrow \quad \mathbf{X}\mathbf{b} = \mathbf{X}\tilde{\mathbf{b}} \\ &\Rightarrow \quad \mathbf{X}'\mathbf{X}\mathbf{b} = \mathbf{X}'\mathbf{X}\tilde{\mathbf{b}} = \mathbf{X}'\underset{\sim}{\mathbf{y}}. \end{aligned} \tag{4.17}$$

The value of the vector $\mathbf{X}\tilde{\mathbf{b}}$ is the same for any solution $\tilde{\mathbf{b}}$ to the normal equations or, equivalently, for any vector $\tilde{\mathbf{b}}$ at which $q(\mathbf{b})$ attains its minimum value. That is, for any two solutions $\tilde{\mathbf{b}}_1$ and $\tilde{\mathbf{b}}_2$ to the normal equations [or any two vectors $\tilde{\mathbf{b}}_1$ and $\tilde{\mathbf{b}}_2$ at which $q(\mathbf{b})$ attains its minimum value],

$$\mathbf{X}\tilde{\mathbf{b}}_1 = \mathbf{X}\tilde{\mathbf{b}}_2. \tag{4.18}$$

Equality (4.18) can be established by applying result (4.17) or, alternatively, by observing that $\mathbf{X}'\mathbf{X}\tilde{\mathbf{b}}_1 = \mathbf{X}'\underset{\sim}{\mathbf{y}} = \mathbf{X}'\mathbf{X}\tilde{\mathbf{b}}_2$ and then observing (in light of Corollary 2.3.4) that $\mathbf{X}'\mathbf{X}\tilde{\mathbf{b}}_1 = \mathbf{X}'\mathbf{X}\tilde{\mathbf{b}}_2 \Rightarrow$

$\mathbf{X}\tilde{\mathbf{b}}_1 = \mathbf{X}\tilde{\mathbf{b}}_2$. The vector $(\mathbf{X}'\mathbf{X})^-\mathbf{X}'\underset{\sim}{\mathbf{y}}$ is a solution to the normal equations. Thus, as a variation on result (4.18), we have that for any solution $\tilde{\mathbf{b}}$ to the normal equations or, equivalently, for any vector $\tilde{\mathbf{b}}$ at which $q(\mathbf{b})$ attains its minimum value,

$$\mathbf{X}\tilde{\mathbf{b}} = \mathbf{X}(\mathbf{X}'\mathbf{X})^-\mathbf{X}'\underset{\sim}{\mathbf{y}} = \mathbf{P}_{\mathbf{X}}\underset{\sim}{\mathbf{y}} \tag{4.19}$$

[where $\mathbf{P}_{\mathbf{X}} = \mathbf{X}(\mathbf{X}'\mathbf{X})^-\mathbf{X}'$].

Let $\tilde{\mathbf{b}}$ represent any solution to the normal equations. Then, the minimum value of $q(\mathbf{b})$ is

$$q(\tilde{\mathbf{b}}) = (\underset{\sim}{\mathbf{y}} - \mathbf{X}\tilde{\mathbf{b}})'(\underset{\sim}{\mathbf{y}} - \mathbf{X}\tilde{\mathbf{b}}).$$

This value is reexpressible in the form

$$q(\tilde{\mathbf{b}}) = \underset{\sim}{\mathbf{y}}'(\underset{\sim}{\mathbf{y}} - \mathbf{X}\tilde{\mathbf{b}}) \tag{4.20}$$

and in the form

$$q(\tilde{\mathbf{b}}) = \underset{\sim}{\mathbf{y}}'\underset{\sim}{\mathbf{y}} - \tilde{\mathbf{b}}'\mathbf{X}'\underset{\sim}{\mathbf{y}}, \tag{4.21}$$

as is evident upon observing that

$$(\mathbf{X}\tilde{\mathbf{b}})'(\underset{\sim}{\mathbf{y}} - \mathbf{X}\tilde{\mathbf{b}}) = \tilde{\mathbf{b}}'(\mathbf{X}'\underset{\sim}{\mathbf{y}} - \mathbf{X}'\mathbf{X}\tilde{\mathbf{b}}) = \tilde{\mathbf{b}}'\mathbf{0} = 0$$

[and that $\underset{\sim}{\mathbf{y}}'\mathbf{X}\tilde{\mathbf{b}} = (\underset{\sim}{\mathbf{y}}'\mathbf{X}\tilde{\mathbf{b}})' = \tilde{\mathbf{b}}'\mathbf{X}'\underset{\sim}{\mathbf{y}}$]. Moreover, in light of results (4.19) and (4.20), the minimum value of $q(\mathbf{b})$ is also expressible as

$$q(\tilde{\mathbf{b}}) = \underset{\sim}{\mathbf{y}}'(\mathbf{I} - \mathbf{P}_{\mathbf{X}})\underset{\sim}{\mathbf{y}}. \tag{4.22}$$

Note that expression (4.22) is a quadratic form (in $\underset{\sim}{\mathbf{y}}$), the matrix of which is $\mathbf{I} - \mathbf{P}_{\mathbf{X}}$.

In summary, we have established the following:

(1) The function $q(\mathbf{b}) = (\underset{\sim}{\mathbf{y}} - \mathbf{X}\mathbf{b})'(\underset{\sim}{\mathbf{y}} - \mathbf{X}\mathbf{b})$ attains a minimum value at a point $\tilde{\mathbf{b}}$ if and only if $\tilde{\mathbf{b}}$ is a solution to the linear system $\mathbf{X}'\mathbf{X}\mathbf{b} = \mathbf{X}'\underset{\sim}{\mathbf{y}}$ (comprising the normal equations).

(2) The linear system $\mathbf{X}'\mathbf{X}\mathbf{b} = \mathbf{X}'\underset{\sim}{\mathbf{y}}$ is consistent.

(3) $\mathbf{X}\tilde{\mathbf{b}}_1 = \mathbf{X}\tilde{\mathbf{b}}_2$ for any two solutions $\tilde{\mathbf{b}}_1$ and $\tilde{\mathbf{b}}_2$ to the linear system $\mathbf{X}'\mathbf{X}\mathbf{b} = \mathbf{X}'\underset{\sim}{\mathbf{y}}$ or, equivalently, for any two vectors $\tilde{\mathbf{b}}_1$ amd $\tilde{\mathbf{b}}_2$ at which $q(\mathbf{b})$ attains its minimum value.

(3′) $\mathbf{X}\tilde{\mathbf{b}} = \mathbf{P}_{\mathbf{X}}\underset{\sim}{\mathbf{y}}$ for any solution $\tilde{\mathbf{b}}$ to the linear system $\mathbf{X}'\mathbf{X}\mathbf{b} = \mathbf{X}'\underset{\sim}{\mathbf{y}}$ or, equivalently, for any vector $\tilde{\mathbf{b}}$ at which $q(\mathbf{b})$ attains its minimum value.

(4) For any solution $\tilde{\mathbf{b}}$ to the linear system $\mathbf{X}'\mathbf{X}\mathbf{b} = \mathbf{X}'\underset{\sim}{\mathbf{y}}$,

$$\min_{\mathbf{b}} q(\mathbf{b}) = q(\tilde{\mathbf{b}}) = \underset{\sim}{\mathbf{y}}'(\underset{\sim}{\mathbf{y}} - \mathbf{X}\tilde{\mathbf{b}}) = \underset{\sim}{\mathbf{y}}'\underset{\sim}{\mathbf{y}} - \tilde{\mathbf{b}}'\mathbf{X}'\underset{\sim}{\mathbf{y}} = \underset{\sim}{\mathbf{y}}'(\mathbf{I} - \mathbf{P}_{\mathbf{X}})\underset{\sim}{\mathbf{y}}.$$

c. Least squares estimators of estimable functions

Suppose the setting is one in which N data points are regarded as the respective values of observable random variables $y_1, y_2, \ldots, y_N$ that follow a G–M, Aitken, or general linear model. And consider the method of least squares as applied to the estimation of an estimable parametric function of the form $\boldsymbol{\lambda}'\boldsymbol{\beta}$. Take the functions $\delta_1(\cdot), \delta_2(\cdot), \ldots, \delta_P(\cdot)$ (and the number of such functions P) to be those for which (under the G–M, Aitken, or general linear model)

$$\mathrm{E}(y_i) = \beta_1\delta_1(\mathbf{u}_i) + \beta_2\delta_2(\mathbf{u}_i) + \cdots + \beta_P\delta_P(\mathbf{u}_i) \qquad (i = 1, 2, \ldots, N).$$

Further, let $\mathbf{y} = (y_1, y_2, \ldots, y_N)'$, and continue to take $\mathbf{X}$ to be the $N \times P$ matrix with ijth element $x_{ij} = \delta_j(\mathbf{u}_i)$.

By definition, the *least squares estimator* of an estimable function $\boldsymbol{\lambda}'\boldsymbol{\beta}$ is the function, say $\ell(\mathbf{y})$, of $\mathbf{y}$ whose value at $\mathbf{y} = \underset{\sim}{\mathbf{y}}$ (an arbitrary $N \times 1$ vector) is taken to be $\boldsymbol{\lambda}'\tilde{\mathbf{b}}$, where $\tilde{\mathbf{b}}$ is any solution to the linear system

$$\mathbf{X}'\mathbf{X}\mathbf{b} = \mathbf{X}'\underset{\sim}{\mathbf{y}} \tag{4.23}$$

(in the $P \times 1$ vector $\mathbf{b}$), comprising the normal equations. Unless $\text{rank}(\mathbf{X}) = P$, there are an infinite number of solutions to the normal equations and hence an infinite number of choices for $\tilde{\mathbf{b}}$. Nevertheless, $\boldsymbol{\lambda}'\tilde{\mathbf{b}}$ is uniquely defined; that is, $\boldsymbol{\lambda}'\tilde{\mathbf{b}}$ is invariant to the choice of $\tilde{\mathbf{b}}$. To see this, let $\tilde{\mathbf{b}}_1$ and $\tilde{\mathbf{b}}_2$ represent any two solutions to linear system (4.23), and observe (in light of the results of Subsection b and Section 5.3) that $\mathbf{X}\tilde{\mathbf{b}}_1 = \mathbf{X}\tilde{\mathbf{b}}_2$ and that (because of the estimability of $\boldsymbol{\lambda}'\boldsymbol{\beta}$) $\boldsymbol{\lambda}' = \mathbf{a}'\mathbf{X}$ for some $N \times 1$ vector $\mathbf{a}$, so that

$$\boldsymbol{\lambda}'\tilde{\mathbf{b}}_1 = \mathbf{a}'\mathbf{X}\tilde{\mathbf{b}}_1 = \mathbf{a}'\mathbf{X}\tilde{\mathbf{b}}_2 = \boldsymbol{\lambda}'\tilde{\mathbf{b}}_2.$$

The solutions to linear system (4.23) include the vector $(\mathbf{X}'\mathbf{X})^-\mathbf{X}'\underset{\sim}{\mathbf{y}}$. Thus, among the representations for the least squares estimator $\ell(\mathbf{y})$ of an estimable function $\boldsymbol{\lambda}'\boldsymbol{\beta}$ is the representation

$$\ell(\mathbf{y}) = \boldsymbol{\lambda}'(\mathbf{X}'\mathbf{X})^-\mathbf{X}'\mathbf{y}, \tag{4.24}$$

so the least squares estimator is a linear estimator. In the special case where $\mathbf{X}$ is of full column rank P, linear system (4.23) has the unique solution $(\mathbf{X}'\mathbf{X})^{-1}\mathbf{X}'\mathbf{y}$. And in that special case, expression (4.24) becomes

$$\ell(\mathbf{y}) = \boldsymbol{\lambda}'(\mathbf{X}'\mathbf{X})^{-1}\mathbf{X}'\mathbf{y}.$$

Some further results on estimability. Let us continue to suppose that $\mathbf{y}$ is a column vector of observable random variables $y_1, y_2, \ldots, y_N$ that follow a G–M, Aitken, or general linear model. And let us consider further the subject of Section 5.3, namely, the estimability of a parametric function of the form $\boldsymbol{\lambda}'\boldsymbol{\beta}$ $(= \sum_{j=1}^{P} \lambda_j \beta_j)$.

In Section 5.3, the estimability of $\boldsymbol{\lambda}'\boldsymbol{\beta}$ was related to various characteristics of the model matrix $\mathbf{X}$. A number of conditions were set forth, each of which is necessary and sufficient for estimability. Those conditions can be restated in terms of various characteristics of the $P \times P$ matrix $\mathbf{X}'\mathbf{X}$, which is the coefficient matrix of the normal equations. Their restatement is based on the following results:

$$\mathcal{R}(\mathbf{X}'\mathbf{X}) = \mathcal{R}(\mathbf{X}); \tag{4.25}$$

$$\text{rank}(\mathbf{X}'\mathbf{X}) = \text{rank}(\mathbf{X}); \tag{4.26}$$

$$\mathbf{X}[(\mathbf{X}'\mathbf{X})^-\mathbf{X}']\mathbf{X} = \mathbf{X} \tag{4.27}$$

[i.e., $(\mathbf{X}'\mathbf{X})^-\mathbf{X}'$ is a generalized inverse of $\mathbf{X}$]; and

$$\mathbf{k}'\mathbf{X}'\mathbf{X} = \mathbf{0} \quad \Leftrightarrow \quad \mathbf{k}'\mathbf{X}' = \mathbf{0} \tag{4.28}$$

(where $\mathbf{k}$ is an arbitrary $P \times 1$ vector) or, equivalently,

$$\mathbf{X}'\mathbf{X}\mathbf{k} = \mathbf{0} \quad \Leftrightarrow \quad \mathbf{X}\mathbf{k} = \mathbf{0}$$

or, also equivalently,

$$\mathcal{N}(\mathbf{X}'\mathbf{X}) = \mathcal{N}(\mathbf{X})$$

—results (4.25), (4.26), and (4.27) were established in Section 2.12, and result (4.28) is a consequence of Corollary 2.3.4.

In light of results (4.25), (4.26), (4.27), and (4.28), it follows from the results of Section 5.3 that each of the following conditions is necessary and sufficient for the estimability of $\boldsymbol{\lambda}'\boldsymbol{\beta}$:

(1) $\boldsymbol{\lambda}' \in \mathcal{R}(\mathbf{X}'\mathbf{X})$;

(2) $\boldsymbol{\lambda}' = \mathbf{r}'\mathbf{X}'\mathbf{X}$ for some $P \times 1$ vector $\mathbf{r}$;

(3) $\boldsymbol{\lambda}'(\mathbf{X}'\mathbf{X})^-\mathbf{X}'\mathbf{X} = \boldsymbol{\lambda}'$ or, equivalently,

(3′) $\boldsymbol{\lambda}'[\mathbf{I} - (\mathbf{X}'\mathbf{X})^-\mathbf{X}'\mathbf{X}] = \mathbf{0}$;

(4) $\mathbf{k}'\boldsymbol{\lambda} = 0$ for every $P \times 1$ vector $\mathbf{k}$ such that $\mathbf{k}'\mathbf{X}'\mathbf{X} = \mathbf{0}$ or, equivalently,

(4′) $$\mathbf{k}'\boldsymbol{\lambda} = 0 \text{ for every } P \times 1 \text{ vector } \mathbf{k} \text{ in } \mathcal{N}(\mathbf{X}'\mathbf{X});$$

(5) $$\mathbf{k}_1'\boldsymbol{\lambda} = \mathbf{k}_2'\boldsymbol{\lambda} = \cdots = \mathbf{k}_S'\boldsymbol{\lambda} = 0,$$

where $S = N - \text{rank}(\mathbf{X}'\mathbf{X})$ and where $\mathbf{k}_1, \mathbf{k}_2, \ldots, \mathbf{k}_S$ are any S linearly independent vectors in $\mathcal{N}(\mathbf{X}'\mathbf{X})$ [i.e., any S linearly independent (P-dimensional) column vectors such that $\mathbf{X}'\mathbf{X}\mathbf{k}_1 = \mathbf{X}'\mathbf{X}\mathbf{k}_2 = \cdots = \mathbf{X}'\mathbf{X}\mathbf{k}_S = \mathbf{0}$].

Conditions (1), (2), (3), and (3′) are stated in terms of the row vector $\boldsymbol{\lambda}'$. Alternative versions of what are essentially the same conditions can be obtained by restating the conditions in terms of the column vector $\boldsymbol{\lambda}$ (and by observing that $[(\mathbf{X}'\mathbf{X})^-]'$, like $(\mathbf{X}'\mathbf{X})^-$ itself, is a generalized inverse of $\mathbf{X}'\mathbf{X}$). The alternative versions are as follows:

(1) $$\boldsymbol{\lambda} \in \mathcal{C}(\mathbf{X}'\mathbf{X});$$

(2) $$\boldsymbol{\lambda} = \mathbf{X}'\mathbf{X}\mathbf{r} \;\; \text{for some } P \times 1 \text{ vector } \mathbf{r};$$

(3) $$\mathbf{X}'\mathbf{X}(\mathbf{X}'\mathbf{X})^-\boldsymbol{\lambda} = \boldsymbol{\lambda};$$

(3′) $$[\mathbf{I} - \mathbf{X}'\mathbf{X}(\mathbf{X}'\mathbf{X})^-]\boldsymbol{\lambda} = \mathbf{0}.$$

As in the case of the original versions, each of these conditions is necessary and sufficient for $\boldsymbol{\lambda}'\boldsymbol{\beta}$ to be estimable.

Note that the P-dimensional random column vector $\mathbf{X}'\mathbf{y}$, defined by the right side of the normal equations, is such that

$$\text{E}(\mathbf{X}'\mathbf{y}) = \mathbf{X}'\mathbf{X}\boldsymbol{\beta}. \tag{4.29}$$

And observe that if $\boldsymbol{\lambda}'\boldsymbol{\beta}$ is estimable, then (in light of our results on estimability) there exists a $P \times 1$ vector $\mathbf{r}$ such that

$$\boldsymbol{\lambda}'\boldsymbol{\beta} = \mathbf{r}'\text{E}(\mathbf{X}'\mathbf{y}). \tag{4.30}$$

Thus, an estimable function is interpretable in terms of the expected values of the P elements of $\mathbf{X}'\mathbf{y}$. In fact, unless $\text{rank}(\mathbf{X}) = P$, an estimable function $\boldsymbol{\lambda}'\boldsymbol{\beta}$ will have multiple representations of the form (4.30) and hence multiple interpretations in terms of the expected values of the elements of $\mathbf{X}'\mathbf{y}$.

As a further implication of result (4.29), we have that, for any $P \times 1$ vector $\mathbf{r}$ of constants,

$$\text{E}(\mathbf{r}'\mathbf{X}'\mathbf{y}) = \mathbf{r}'\mathbf{X}'\mathbf{X}\boldsymbol{\beta}.$$

And upon observing that the least squares estimator of $\mathbf{r}'\mathbf{X}'\mathbf{X}\boldsymbol{\beta}$ is $\mathbf{r}'\mathbf{X}'\mathbf{y}$, it follows that any linear combination of the elements of the vector $\mathbf{X}'\mathbf{y}$ (defined by the right side of the normal equations) is the least squares estimator of its expected value.

Conjugate normal equations. Let us resume our discussion of least squares estimation, taking the setting to that in which N data points are regarded as the respective values of the elements of an $N \times 1$ observable random vector $\mathbf{y}$ that follows a G–M, Aitken, or general linear model. Corresponding to any linear combination $\boldsymbol{\lambda}'\boldsymbol{\beta}$ (of the elements of the vector $\boldsymbol{\beta}$) is the linear system

$$\mathbf{X}'\mathbf{X}\mathbf{r} = \boldsymbol{\lambda}, \tag{4.31}$$

comprising P equations in a $P \times 1$ vector $\mathbf{r}$ of unknowns. The coefficient matrix $\mathbf{X}'\mathbf{X}$ of this linear system is the same as that of the linear system

$$\mathbf{X}'\mathbf{X}\mathbf{b} = \mathbf{X}'\underset{\sim}{\mathbf{y}}, \tag{4.32}$$

comprising the normal equations; however, the right side of linear system (4.31) is the coefficient vector $\boldsymbol{\lambda}$, while that of linear system (4.32) is $\mathbf{X}'\underset{\sim}{\mathbf{y}}$. The P equations that form linear system (4.31) are sometimes referred to (collectively) as the *conjugate normal equations*.

It follows from the results of Part 1 of the present subsection that $\boldsymbol{\lambda}'\boldsymbol{\beta}$ is estimable if and only if the conjugate normal equations are consistent. Now, suppose that $\boldsymbol{\lambda}'\boldsymbol{\beta}$ is estimable, and consider the least squares estimator $\ell(\mathbf{y})$ of $\boldsymbol{\lambda}'\boldsymbol{\beta}$. The value $\ell(\underset{\sim}{\mathbf{y}})$ of $\ell(\mathbf{y})$ at $\mathbf{y} = \underset{\sim}{\mathbf{y}}$ is expressible in terms of any solution to the normal equations; for any solution $\tilde{\mathbf{b}}$ to linear system (4.32),

$$\ell(\underset{\sim}{\mathbf{y}}) = \boldsymbol{\lambda}'\tilde{\mathbf{b}}. \tag{4.33}$$

The value of $\ell(\mathbf{y})$ is also expressible in terms of any solution to the conjugate normal equations. For any solution $\tilde{\mathbf{r}}$ to linear system (4.31) [and any solution $\tilde{\mathbf{b}}$ to linear system (4.32)], we find that

$$\ell(\underset{\sim}{\mathbf{y}}) = \boldsymbol{\lambda}'\tilde{\mathbf{b}} = (\mathbf{X}'\mathbf{X}\tilde{\mathbf{r}})'\tilde{\mathbf{b}} = \tilde{\mathbf{r}}'\mathbf{X}'\mathbf{X}\tilde{\mathbf{b}} = \tilde{\mathbf{r}}'\mathbf{X}'\underset{\sim}{\mathbf{y}}. \tag{4.34}$$

The upshot of result (4.34) is that the roles of the normal equations and the conjugate normal equations are (in a certain sense) interchangeable. The least squares estimate $\ell(\underset{\sim}{\mathbf{y}})$ can be obtained by forming the (usual) inner product $\boldsymbol{\lambda}'\tilde{\mathbf{b}}$ of a solution $\tilde{\mathbf{b}}$ to the normal equations and of the right side $\boldsymbol{\lambda}$ of the conjugate normal equations. Or, alternatively, it can be obtained by forming the inner product $\tilde{\mathbf{r}}'\mathbf{X}'\underset{\sim}{\mathbf{y}}$ of a solution $\tilde{\mathbf{r}}$ to the conjugate normal equations and of the right side $\mathbf{X}'\underset{\sim}{\mathbf{y}}$ of the normal equations.

The general form and expected values, variances, and covariances of least squares estimators. Let us continue to take $\mathbf{y}$ to be an N-dimensional observable random (column) vector that follows a G–M, Aitken, or general linear model. And let us consider the general form and expected values, variances, and covariances of least squares estimators of estimable linear combinations of the elements of $\boldsymbol{\beta}$.

Suppose that $\boldsymbol{\lambda}'\boldsymbol{\beta}$ is an estimable linear combination. Then, in light of the results of Part 2 of the present subsection, the least squares estimator $\ell(\mathbf{y})$ of $\boldsymbol{\lambda}'\boldsymbol{\beta}$ is expressible in the form

$$\ell(\mathbf{y}) = \tilde{\mathbf{r}}'\mathbf{X}'\mathbf{y}, \tag{4.35}$$

where $\tilde{\mathbf{r}}$ is any solution to the conjugate normal equations $\mathbf{X}'\mathbf{X}\mathbf{r} = \boldsymbol{\lambda}$. It follows immediately that the least squares estimator is a linear estimator, in confirmation of what was established earlier (in the introductory part of the present subsection) via a different approach. Moreover,

$$\mathrm{E}(\tilde{\mathbf{r}}'\mathbf{X}'\mathbf{y}) = \tilde{\mathbf{r}}'\mathbf{X}'\mathrm{E}(\mathbf{y}) = \tilde{\mathbf{r}}'\mathbf{X}'\mathbf{X}\boldsymbol{\beta} = (\mathbf{X}'\mathbf{X}\tilde{\mathbf{r}})'\boldsymbol{\beta} = \boldsymbol{\lambda}'\boldsymbol{\beta}. \tag{4.36}$$

Thus, the least squares estimator is a linear unbiased estimator.

The vector $\mathbf{X}\tilde{\mathbf{r}}$ is invariant to the choice of the solution $\tilde{\mathbf{r}}$ to the conjugate normal equations (as is evident, e.g., from Corollary 2.3.4). The solutions to the conjugate normal equations include the vector $(\mathbf{X}'\mathbf{X})^-\boldsymbol{\lambda}$, and {because $[(\mathbf{X}'\mathbf{X})^-]'$, like $(\mathbf{X}'\mathbf{X})^-$ itself, is a generalized inverse of $\mathbf{X}'\mathbf{X}$}they also include the vector $[(\mathbf{X}'\mathbf{X})^-]'\boldsymbol{\lambda}$. Accordingly,

$$\mathbf{X}\tilde{\mathbf{r}} = \mathbf{X}(\mathbf{X}'\mathbf{X})^-\boldsymbol{\lambda} = \mathbf{X}[(\mathbf{X}'\mathbf{X})^-]'\boldsymbol{\lambda}. \tag{4.37}$$

Under the general linear model, the variance of the least squares estimator of $\boldsymbol{\lambda}'\boldsymbol{\beta}$ is [in light of result (4.37) and the equality $\tilde{\mathbf{r}}'\mathbf{X}' = (\mathbf{X}\tilde{\mathbf{r}})'$] expressible as

$$\mathrm{var}(\tilde{\mathbf{r}}'\mathbf{X}'\mathbf{y}) = \tilde{\mathbf{r}}'\mathbf{X}'\mathbf{V}(\boldsymbol{\theta})\mathbf{X}\tilde{\mathbf{r}} = \boldsymbol{\lambda}'(\mathbf{X}'\mathbf{X})^-\mathbf{X}'\mathbf{V}(\boldsymbol{\theta})\mathbf{X}(\mathbf{X}'\mathbf{X})^-\boldsymbol{\lambda}. \tag{4.38}$$

Result (4.38) can be extended. Suppose that $\boldsymbol{\lambda}_1'\boldsymbol{\beta}$ and $\boldsymbol{\lambda}_2'\boldsymbol{\beta}$ are two estimable linear combinations of the elements of $\boldsymbol{\beta}$. Then, the least squares estimator of $\boldsymbol{\lambda}_1'\boldsymbol{\beta}$ equals $\tilde{\mathbf{r}}_1'\mathbf{X}'\mathbf{y}$ and that of $\boldsymbol{\lambda}_2'\boldsymbol{\beta}$ equals $\tilde{\mathbf{r}}_2'\mathbf{X}'\mathbf{y}$. Here, $\tilde{\mathbf{r}}_1$ is any solution to the linear system $\mathbf{X}'\mathbf{X}\mathbf{r}_1 = \boldsymbol{\lambda}_1$ (in $\mathbf{r}_1$) and $\tilde{\mathbf{r}}_2$ any solution to the linear system $\mathbf{X}'\mathbf{X}\mathbf{r}_2 = \boldsymbol{\lambda}_2$ (in $\mathbf{r}_2$). And under the general linear model,

$$\mathrm{cov}(\tilde{\mathbf{r}}_1'\mathbf{X}'\mathbf{y}, \tilde{\mathbf{r}}_2'\mathbf{X}'\mathbf{y}) = \tilde{\mathbf{r}}_1'\mathbf{X}'\mathbf{V}(\boldsymbol{\theta})\mathbf{X}\tilde{\mathbf{r}}_2 = \boldsymbol{\lambda}_1'(\mathbf{X}'\mathbf{X})^-\mathbf{X}'\mathbf{V}(\boldsymbol{\theta})\mathbf{X}(\mathbf{X}'\mathbf{X})^-\boldsymbol{\lambda}_2. \tag{4.39}$$

In the special case of the Aitken model, result (4.38) "simplifies" to

$$\mathrm{var}(\tilde{\mathbf{r}}'\mathbf{X}'\mathbf{y}) = \sigma^2\tilde{\mathbf{r}}'\mathbf{X}'\mathbf{H}\mathbf{X}\tilde{\mathbf{r}} = \sigma^2\boldsymbol{\lambda}'(\mathbf{X}'\mathbf{X})^-\mathbf{X}'\mathbf{H}\mathbf{X}(\mathbf{X}'\mathbf{X})^-\boldsymbol{\lambda}, \tag{4.40}$$

and result (4.39) to

$$\operatorname{cov}(\tilde{\mathbf{r}}_1'\mathbf{X}'\mathbf{y},\ \tilde{\mathbf{r}}_2'\mathbf{X}'\mathbf{y}) = \sigma^2\tilde{\mathbf{r}}_1'\mathbf{X}'\mathbf{H}\mathbf{X}\tilde{\mathbf{r}}_2 = \sigma^2\boldsymbol{\lambda}_1'(\mathbf{X}'\mathbf{X})^-\mathbf{X}'\mathbf{H}\mathbf{X}(\mathbf{X}'\mathbf{X})^-\boldsymbol{\lambda}_2. \tag{4.41}$$

Under the G–M model, considerable further simplification is possible, and various additional representations are obtainable. Specifically, we find that (under the G–M model)

$$\operatorname{var}(\tilde{\mathbf{r}}'\mathbf{X}'\mathbf{y}) = \sigma^2\tilde{\mathbf{r}}'\mathbf{X}'\mathbf{X}\tilde{\mathbf{r}} = \sigma^2\tilde{\mathbf{r}}'\boldsymbol{\lambda} = \sigma^2\boldsymbol{\lambda}'\tilde{\mathbf{r}} \tag{4.42}$$

$$= \sigma^2\tilde{\mathbf{r}}'\mathbf{X}'\mathbf{X}(\mathbf{X}'\mathbf{X})^-\mathbf{X}'\mathbf{X}\tilde{\mathbf{r}} = \sigma^2\boldsymbol{\lambda}'(\mathbf{X}'\mathbf{X})^-\boldsymbol{\lambda}, \tag{4.43}$$

and, similarly,

$$\operatorname{cov}(\tilde{\mathbf{r}}_1'\mathbf{X}'\mathbf{y},\ \tilde{\mathbf{r}}_2'\mathbf{X}'\mathbf{y}) = \sigma^2\tilde{\mathbf{r}}_1'\mathbf{X}'\mathbf{X}\tilde{\mathbf{r}}_2 = \sigma^2\tilde{\mathbf{r}}_1'\boldsymbol{\lambda}_2 = \sigma^2\boldsymbol{\lambda}_1'\tilde{\mathbf{r}}_2 \tag{4.44}$$

$$= \sigma^2\tilde{\mathbf{r}}_1'\mathbf{X}'\mathbf{X}(\mathbf{X}'\mathbf{X})^-\mathbf{X}'\mathbf{X}\tilde{\mathbf{r}}_2 = \sigma^2\boldsymbol{\lambda}_1'(\mathbf{X}'\mathbf{X})^-\boldsymbol{\lambda}_2. \tag{4.45}$$

d. The geometry of least squares

It can be informative to consider the method of least squares from a geometrical perspective. As a preliminary to doing so, let us extend some of the basic definitions of plane and solid geometry from $\mathcal{R}^2$ and $\mathcal{R}^3$, where they can be interpreted visually, to $\mathcal{R}^M$ (where M may be greater than 3).

Geometrically-related definitions. The *inner* (or dot) *product* of two M-dimensional column vectors, say $\mathbf{x} = \{x_i\}$ and $\mathbf{y} = \{y_i\}$, is denoted by the symbol $\mathbf{x} \bullet \mathbf{y}$. For a general definition of the inner product of two M-dimensional column vectors (or two $M \times N$ matrices), refer to Section 2.4f. The usual inner product is that for which

$$\mathbf{x} \bullet \mathbf{y} = \mathbf{y}'\mathbf{x} = \mathbf{x}'\mathbf{y} = \sum_{i=1}^{M} x_i y_i. \tag{4.46}$$

The usual inner product is (in the special cases $M = 2$ and $M = 3$) the inner product customarily employed in plane and solid geometry.

The definition of an inner product underlies the definitions of various other quantities. Consider, in particular, the *norm* (also known as the length or magnitude) of an M-dimensional column vector $\mathbf{x} = \{x_i\}$. The norm of $\mathbf{x}$ is denoted by the symbol $\|\mathbf{x}\|$. By definition,

$$\|\mathbf{x}\| = (\mathbf{x} \bullet \mathbf{x})^{1/2}.$$

When the inner product is taken to be the usual inner product,

$$\|\mathbf{x}\| = (\mathbf{x}'\mathbf{x})^{1/2} = \Big(\sum_{i=1}^{M} x_i^2\Big)^{1/2}, \tag{4.47}$$

and the norm is referred to as the usual norm.

The *distance* between two M-dimensional column vectors $\mathbf{x} = \{x_i\}$ and $\mathbf{y} = \{y_i\}$ is defined to be the norm $\|\mathbf{x}-\mathbf{y}\|$ of the difference $\mathbf{x} - \mathbf{y}$ between $\mathbf{x}$ and $\mathbf{y}$. In the case of the usual inner product,

$$\|\mathbf{x}-\mathbf{y}\| = [(\mathbf{x}-\mathbf{y})'(\mathbf{x}-\mathbf{y})]^{1/2} = \Big[\sum_{i=1}^{M} (x_i - y_i)^2\Big]^{1/2}. \tag{4.48}$$

The *angle* between two nonnull M-dimensional column vectors $\mathbf{x} = \{x_i\}$ and $\mathbf{y} = \{y_i\}$ is defined indirectly in terms of its cosine. Specifically, the angle between $\mathbf{x}$ and $\mathbf{y}$ is the angle θ $(0 \le \theta \le \pi)$ defined by

$$\cos\theta = \frac{\mathbf{x} \bullet \mathbf{y}}{\|\mathbf{x}\|\,\|\mathbf{y}\|} \tag{4.49}$$

—it follows from Theorem 2.4.21 (the Cauchy–Schwarz inequality) that $-1 \le \dfrac{\mathbf{x} \bullet \mathbf{y}}{\|\mathbf{x}\|\,\|\mathbf{y}\|} \le 1$. In the case of the usual inner product (and usual norm), equality (4.49) can be reexpressed in the form

$$\cos\theta = \frac{\mathbf{x}'\mathbf{y}}{(\mathbf{x}'\mathbf{x})^{1/2}(\mathbf{y}'\mathbf{y})^{1/2}} = \frac{\sum_{i=1}^{M} x_i y_i}{\left(\sum_{i=1}^{M} x_i^2\right)^{1/2}\left(\sum_{i=1}^{M} y_i^2\right)^{1/2}}. \tag{4.50}$$

By definition, two M-dimensional column vectors $\mathbf{x} = \{x_i\}$ and $\mathbf{y} = \{y_i\}$ are *orthogonal* (or perpendicular) to each other if $\mathbf{x} \bullet \mathbf{y} = 0$. Thus, when the inner product is taken to be the usual inner product, $\mathbf{x}$ and $\mathbf{y}$ are orthogonal to each other if $\mathbf{x}'\mathbf{y} = 0$ or, equivalently, if $\sum_{i=1}^{M} x_i y_i = 0$. The statement that $\mathbf{x}$ and $\mathbf{y}$ are orthogonal to each other is sometimes abbreviated to the statement that $\mathbf{x}$ and $\mathbf{y}$ are orthogonal. Clearly, two nonnull vectors are orthogonal if and only if the angle between them is $\pi/2$ (90°) or, equivalently, the cosine of that angle is 0.

If an M-dimensional column vector $\mathbf{x}$ is orthogonal to every vector in a subspace $\mathcal{U}$ of M-dimensional column vectors, $\mathbf{x}$ is said to be orthogonal to $\mathcal{U}$. The set consisting of all M-dimensional column vectors that are orthogonal to the subspace $\mathcal{U}$ is called the *orthogonal complement* of $\mathcal{U}$ and is denoted by the symbol $\mathcal{U}^{\perp}$. The set $\mathcal{U}^{\perp}$ is a linear space (as can be readily verified). When $\mathcal{U} = \mathcal{C}(\mathbf{X})$ (where $\mathbf{X}$ is a matrix), we may write $\mathcal{C}^{\perp}(\mathbf{X})$ for $\mathcal{U}^{\perp}$.

Least squares revisited: the projection and decomposition of the data vector. Denote by $\mathbf{y} = (y_1, y_2, \ldots, y_N)'$ an N-dimensional column vector of data points—this notation differs somewhat from that employed earlier in the section (which included an underline). Further, suppose that $y_1, y_2, \ldots, y_N$ are accompanied by the corresponding values $\mathbf{u}_1, \mathbf{u}_2, \ldots, \mathbf{u}_N$ of a C-dimensional column vector $\mathbf{u}$ of explanatory variables. Let us consider the approximation of $y_1, y_2, \ldots, y_N$ by $\underline{\delta}(\mathbf{u}_1), \underline{\delta}(\mathbf{u}_2), \ldots, \underline{\delta}(\mathbf{u}_N)$, where $\underline{\delta}(\mathbf{u})$ is a function of $\mathbf{u}$. Which of the possible choices for the function $\underline{\delta}(\cdot)$ results in the "best" approximation (and in what sense)? In particular, which results in the best approximation when the choice for $\underline{\delta}(\cdot)$ is restricted to functions (of $\mathbf{u}$) that are expressible as linear combinations of P specified functions $\delta_1(\cdot), \delta_2(\cdot), \ldots, \delta_P(\cdot)$; that is, when the choice is restricted to those functions that are expressible in the form

$$\underline{\delta}(\mathbf{u}) = b_1\delta_1(\mathbf{u}) + b_2\delta_2(\mathbf{u}) + \cdots + b_P\delta_P(\mathbf{u}), \tag{4.51}$$

where $b_1, b_2, \ldots, b_P$ are arbitrary scalars.

In the method of least squares, the function $\underline{\delta}(\cdot)$ is chosen so as to minimize the quantity $\{\sum_{i=1}^{N}[y_i - \underline{\delta}(\mathbf{u}_i)]^2\}^{1/2}$. This quantity is the (usual) norm of the N-dimensional vector whose elements are the individual errors of approximation $y_1 - \underline{\delta}(\mathbf{u}_1), y_2 - \underline{\delta}(\mathbf{u}_2), \ldots, y_N - \underline{\delta}(\mathbf{u}_N)$. It is interpretable as the (ordinary) distance between the N-dimensional data vector $\mathbf{y}$ and the N-dimensional vector whose elements are the approximations $\underline{\delta}(\mathbf{u}_1), \underline{\delta}(\mathbf{u}_2), \ldots, \underline{\delta}(\mathbf{u}_N)$.

Suppose now that $\underline{\delta}(\cdot)$ is taken to be of the form (4.51). And (for $i = 1, 2, \ldots, N$ and $j = 1, 2, \ldots, P$) define $x_{ij} = \delta_j(\mathbf{u}_i)$. Further, let $\mathbf{X}$ represent the $N \times P$ matrix with ijth element x_{ij}, and (for $j = 1, 2, \ldots, P$) take $\mathbf{x}_j = (x_{1j}, x_{2j}, \ldots, x_{Nj})'$ (so that $\mathbf{x}_j$ is the jth column of $\mathbf{X}$). Then, as discussed earlier (in the introductory part of the present section),

$$\sum_{i=1}^{N}[y_i - \underline{\delta}(\mathbf{u}_i)]^2 = \sum_{i=1}^{N}\Bigl(y_i - \sum_{j=1}^{P} x_{ij}b_j\Bigr)^2 = (\mathbf{y} - \mathbf{X}\mathbf{b})'(\mathbf{y} - \mathbf{X}\mathbf{b}), \tag{4.52}$$

where $\mathbf{b} = (b_1, b_2, \ldots, b_P)'$. Note [in connection with result (4.52)] that

$$(\mathbf{y} - \mathbf{X}\mathbf{b})'(\mathbf{y} - \mathbf{X}\mathbf{b}) = \Bigl(\mathbf{y} - \sum_{j=1}^{P} b_j\mathbf{x}_j\Bigr)'\Bigl(\mathbf{y} - \sum_{j=1}^{P} b_j\mathbf{x}_j\Bigr).$$

In light of result (4.52), the minimization problem that gives rise to the method of least squares can be regarded as that of minimizing $(\mathbf{y} - \mathbf{X}\mathbf{b})'(\mathbf{y} - \mathbf{X}\mathbf{b})$ [or, equivalently, that of minimizing the (usual) norm of $\mathbf{y} - \mathbf{X}\mathbf{b}$] with respect to the $P \times 1$ vector $\mathbf{b}$. As previously indicated (in Subsection b), $(\mathbf{y} - \mathbf{X}\mathbf{b})'(\mathbf{y} - \mathbf{X}\mathbf{b})$ attains a minimum value at a $P \times 1$ vector $\tilde{\mathbf{b}}$ if and only if $\tilde{\mathbf{b}}$ is a solution to the normal equations $\mathbf{X}'\mathbf{X}\mathbf{b} = \mathbf{X}'\mathbf{y}$.

Some further insights into the method of least squares can be obtained by transforming the underlying minimization problem into a more geometrically meaningful form. Let $\mathcal{U} = \mathcal{C}(\mathbf{X})$, and observe that an N-dimensional column vector $\mathbf{w}$ is a member of $\mathcal{U}$ if and only if $\mathbf{w} = \mathbf{X}\mathbf{b}$ for some $\mathbf{b}$, in which case the elements of $\mathbf{b}$ are the "coordinates" of $\mathbf{w}$ with respect to the spanning set $\{\mathbf{x}_1, \mathbf{x}_2, \ldots, \mathbf{x}_P\}$. Accordingly, the problem of minimizing $(\mathbf{y} - \mathbf{X}\mathbf{b})'(\mathbf{y} - \mathbf{X}\mathbf{b})$ with respect to $\mathbf{b}$ can be reformulated as the "coordinate-free" problem of minimizing $(\mathbf{y} - \mathbf{w})'(\mathbf{y} - \mathbf{w})$ with respect to $\mathbf{w}$, where $\mathbf{w}$ is an arbitrary member of the linear space $\mathcal{U}$. The latter problem depends on the matrix $\mathbf{X}$ only through its column space. From a geometrical perspective, the problem is that of finding the vector in the subspace $\mathcal{U}$ (of $\mathcal{R}^N$) that is "closest" to the data vector $\mathbf{y}$.

It follows from the results of Subsection b that $(\mathbf{y} - \mathbf{w})'(\mathbf{y} - \mathbf{w})$ attains a minimum value over the subspace $\mathcal{U}$, doing so at a unique point $\mathbf{z}$ that is expressible as

$$\mathbf{z} = \mathbf{X}\tilde{\mathbf{b}}, \tag{4.53}$$

where $\tilde{\mathbf{b}}$ is any solution to the normal equations $\mathbf{X}'\mathbf{X}\mathbf{b} = \mathbf{X}'\mathbf{y}$, and also as

$$\mathbf{z} = \mathbf{P}_{\mathbf{X}}\mathbf{y}. \tag{4.54}$$

Taking (here and in the remainder of the present subsection) the inner product to be the usual inner product, the vector $\mathbf{z}$ is such that $\mathbf{y} - \mathbf{z} \in \mathcal{U}^{\perp}$; that is, the difference between $\mathbf{y}$ and $\mathbf{z}$ is orthogonal to every vector in $\mathcal{U}$. To see this, let $\mathbf{a}$ represent an arbitrary member of $\mathcal{U}$ [$= \mathcal{C}(\mathbf{X})$], and observe that $\mathbf{a} = \mathbf{X}\mathbf{r}$ for some $P \times 1$ vector $\mathbf{r}$ and hence that

$$\mathbf{a}'(\mathbf{y} - \mathbf{z}) = \mathbf{r}'\mathbf{X}'(\mathbf{y} - \mathbf{X}\tilde{\mathbf{b}}) = \mathbf{r}'(\mathbf{X}'\mathbf{y} - \mathbf{X}'\mathbf{X}\tilde{\mathbf{b}}) = \mathbf{r}'(\mathbf{X}'\mathbf{y} - \mathbf{X}'\mathbf{y}) = \mathbf{r}'\mathbf{0} = 0.$$

Moreover, there is no member $\mathbf{w}$ of $\mathcal{U}$ other than $\mathbf{z}$ for which $\mathbf{y} - \mathbf{w} \in \mathcal{U}^{\perp}$, as is evident upon observing that if $\mathbf{w} \in \mathcal{U}$ and $\mathbf{y} - \mathbf{w} \in \mathcal{U}^{\perp}$, then $\mathbf{w} - \mathbf{z} \in \mathcal{U}$ and

$$\mathbf{w} - \mathbf{z} \;=\; (\mathbf{y} - \mathbf{z}) - (\mathbf{y} - \mathbf{w}) \;\in\; \mathcal{U}^{\perp}$$

(so that the vector $\mathbf{w} - \mathbf{z}$ is orthogonal to itself), implying that

$$(\mathbf{w} - \mathbf{z})'(\mathbf{w} - \mathbf{z}) = 0$$

and hence that $\mathbf{w} - \mathbf{z} = \mathbf{0}$ or, equivalently, that

$$\mathbf{w} = \mathbf{z}.$$

In summary, there is a unique vector $\mathbf{w}$ in $\mathcal{U}$ such that $\mathbf{y} - \mathbf{w} \in \mathcal{U}^{\perp}$, namely, the vector $\mathbf{z}$. This vector is referred to as the *orthogonal projection* of $\mathbf{y}$ on $\mathcal{U}$ or simply as the *projection* of $\mathbf{y}$ on $\mathcal{U}$. As previously indicated, the matrix $\mathbf{P}_{\mathbf{X}}$ is referred to as a projection matrix; the reason why is apparent from expression (4.54).

Conceptually, the point $\mathbf{z}$ in $\mathcal{R}^N$ at which $(\mathbf{y} - \mathbf{w})'(\mathbf{y} - \mathbf{w})$ attains its minimum value for $\mathbf{w} \in \mathcal{U}$ is obtainable by "projecting" the point $\mathbf{y}$ onto the surface $\mathcal{U}$. The point in $\mathcal{R}^N$ located by this operation is such that the "line" formed by joining that point with the point $\mathbf{y}$ is orthogonal (perpendicular) to the surface $\mathcal{U}$.

Corresponding to the projection $\mathbf{z}$ of $\mathbf{y}$ on $\mathcal{U}$ is the decomposition

$$\mathbf{y} = \mathbf{z} + \mathbf{d}, \tag{4.55}$$

where $\mathbf{d} = \mathbf{y} - \mathbf{z}$. The first component of this decomposition is a member of the linear space $\mathcal{U}$ [$= \mathcal{C}(\mathbf{X})$], and the second component is a member of the orthogonal complement $\mathcal{U}^{\perp}$ [$= \mathcal{C}^{\perp}(\mathbf{X})$] of

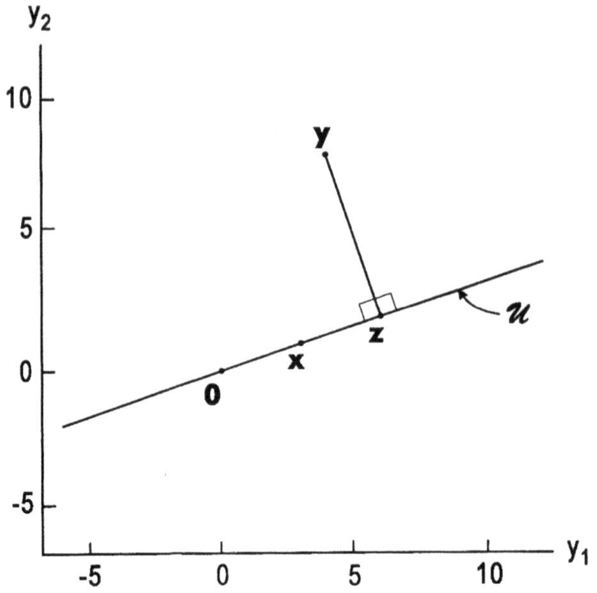

FIGURE 5.1. The projection $\mathbf{z}$ of the 2-dimensional data vector $\mathbf{y} = (4,\ 8)'$ on the 1-dimensional linear space $\mathcal{U} = \mathcal{C}(\mathbf{X})$, where $\mathbf{X} = \mathbf{x} = (3,\ 1)'$.

$\mathcal{U}$. In this context, the linear space $\mathcal{U}$ is sometimes referred to as the *estimation space*—logically, it could also be referred to as the approximation space—and the linear space $\mathcal{U}^{\perp}$ is sometimes referred to as the *error space*. Decomposition (4.55) is unique; if $\mathbf{y}$ is expressed as the sum of two components, the first of which is in the estimation space $\mathcal{U}$ and the second of which is in the error space $\mathcal{U}^{\perp}$, then necessarily the first component equals $\mathbf{z}$ and the second equals $\mathbf{d}$ $(= \mathbf{y} - \mathbf{z})$.

Example: N **= 2.** Suppose that $N = 2$, that $\mathbf{y} = (4,\ 8)'$, and that $\mathbf{X} = \mathbf{x}$, where $\mathbf{x}$ is the 2-dimensional column vector $\mathbf{x} = (3,\ 1)'$ (in which case $P = 1$). Then, the linear system $\mathbf{X}'\mathbf{X}\mathbf{b} = \mathbf{X}'\mathbf{y}$ becomes $(10)\mathbf{b} = (20)$, which has the unique solution $\mathbf{b} = (2)$. Thus, the projection of $\mathbf{y}$ on the linear space $\mathcal{U}$ $[= \mathcal{C}(\mathbf{X})]$ is the vector

$$\mathbf{z} = \begin{pmatrix} 3 \\ 1 \end{pmatrix} (2) = \begin{pmatrix} 6 \\ 2 \end{pmatrix},$$

as depicted in Figure 5.1.

Example: N **= 3.** Suppose that $N = 3$, that $\mathbf{y} = (3,\ -38/5,\ 74/5)'$, and that $\mathbf{X} = (\mathbf{x}_1,\ \mathbf{x}_2,\ \mathbf{x}_3)$, where

$$\mathbf{x}_1 = \begin{pmatrix} 0 \\ 3 \\ 6 \end{pmatrix}, \quad \mathbf{x}_2 = \begin{pmatrix} -2 \\ 2 \\ 4 \end{pmatrix}, \quad \text{and} \quad \mathbf{x}_3 = \begin{pmatrix} -2 \\ 1 \\ 2 \end{pmatrix}.$$

Clearly, $\mathbf{x}_1$ and $\mathbf{x}_2$ are linearly independent, and $\mathbf{x}_3 = \mathbf{x}_2 - (1/3)\mathbf{x}_1$. Thus, the linear space $\mathcal{U}$ $[= \mathcal{C}(\mathbf{X})]$ is of dimension 2.

The normal equations $\mathbf{X}'\mathbf{X}\mathbf{b} = \mathbf{X}'\mathbf{y}$ become

$$\begin{pmatrix} 45 & 30 & 15 \\ 30 & 24 & 14 \\ 15 & 14 & 9 \end{pmatrix} \mathbf{b} = \begin{pmatrix} 66 \\ 38 \\ 16 \end{pmatrix}.$$

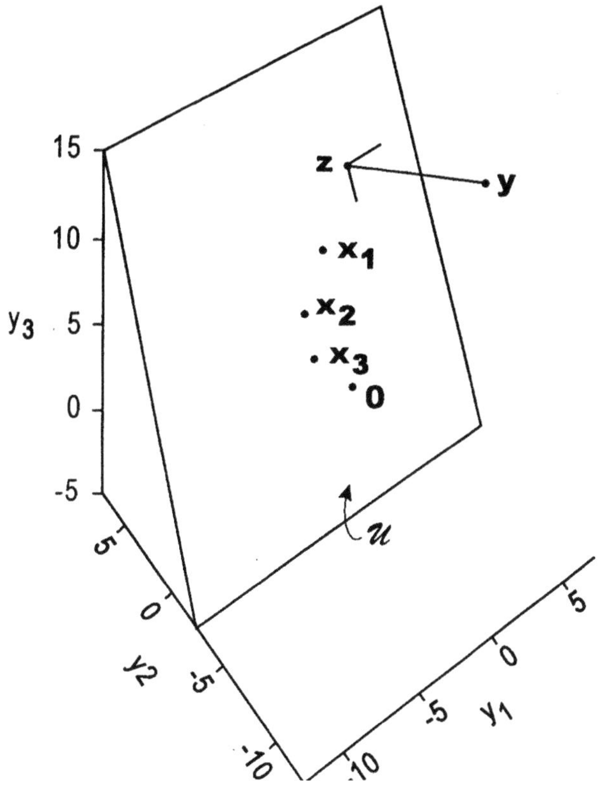

FIGURE 5.2. The projection $\mathbf{z}$ of the 3-dimensional data vector $\mathbf{y} = (3, -38/5, 74/5)'$ on the 2-dimensional linear space $\mathcal{U} = \mathcal{C}(\mathbf{X})$, where $\mathbf{X} = (\mathbf{x}_1, \mathbf{x}_2, \mathbf{x}_3)$, with $\mathbf{x}_1 = (0, 3, 6)'$, $\mathbf{x}_2 = (-2, 2, 4)'$, and $\mathbf{x}_3 = (-2, 1, 2)'$.

One solution to these equations is the vector $(32/15, -1/2, -1)'$. Thus, the projection of $\mathbf{y}$ on the linear space $\mathcal{U}$ [$= \mathcal{C}(\mathbf{X})$] is

$$\mathbf{z} = \begin{pmatrix} 0 & -2 & -2 \\ 3 & 2 & 1 \\ 6 & 4 & 2 \end{pmatrix} \begin{pmatrix} 32/15 \\ -1/2 \\ -1 \end{pmatrix} = \begin{pmatrix} 3 \\ 22/5 \\ 44/5 \end{pmatrix},$$

as depicted in Figure 5.2.

e. Least squares computations

Let us continue to take $\mathbf{y} = (y_1, y_2, \ldots, y_N)'$ to be an N-dimensional vector of data points and to suppose that $y_1, y_2, \ldots, y_N$ are accompanied by the values $\mathbf{u}_1, \mathbf{u}_2, \ldots, \mathbf{u}_N$ of a C-dimensional (column) vector $\mathbf{u}$ of explanatory variables. And let us continue to consider the approximation of $y_1, y_2, \ldots, y_N$ by $\underline{\delta}(\mathbf{u}_1), \underline{\delta}(\mathbf{u}_2), \ldots, \underline{\delta}(\mathbf{u}_N)$, where $\underline{\delta}(\cdot)$ is a function (of $\mathbf{u}$) that is expressible in the form of a linear combination

$$\underline{\delta}(\mathbf{u}) = b_1\delta_1(\mathbf{u}) + b_2\delta_2(\mathbf{u}) + \cdots + b_P\delta_P(\mathbf{u})$$

of P specified functions $\delta_1(\cdot), \delta_2(\cdot), \ldots, \delta_P(\cdot)$. Further, take $\mathbf{b} = (b_1, b_2, \ldots, b_P)'$ to be the $P \times 1$ vector of coefficients, and take $\mathbf{X}$ to be the $N \times P$ matrix whose ijth element x_{ij} is defined by $x_{ij} = \delta_j(\mathbf{u}_i)$. In the method of least squares, the value of $\mathbf{b}$ is taken to be a value at which the quantity $(\mathbf{y} - \mathbf{X}\mathbf{b})'(\mathbf{y} - \mathbf{X}\mathbf{b})$ attains a minimum value.

As discussed in Subsection b, $(\mathbf{y} - \mathbf{X}\mathbf{b})'(\mathbf{y} - \mathbf{X}\mathbf{b})$ attains a minimum value at a $P \times 1$ vector $\tilde{\mathbf{b}}$ if and only if $\tilde{\mathbf{b}}$ is a solution to the normal equations $\mathbf{X}'\mathbf{X}\mathbf{b} = \mathbf{X}'\mathbf{y}$. Accordingly, the least squares computations can be carried out by forming and solving the normal equations. Alternatively, by making use of various results on the decomposition of a matrix (as applied to the matrix $\mathbf{X}$), they can be carried out in a way that does not require the formation of the normal equations. The alternative approach can be advantageous from the standpoint of numerical accuracy, though any such advantage typically comes at the expense of greater demands on computing resources. In many implementations of the alternative approach, the underlying decomposition of the matrix $\mathbf{X}$ is taken to be a decomposition that is known as the QR decomposition.

QR decomposition of a matrix. Any $M \times N$ matrix $\mathbf{A}$ of full column rank N is expressible in the form

$$\mathbf{A} = \mathbf{Q}_1\mathbf{R}_1, \tag{4.56}$$

where $\mathbf{Q}_1$ is an $M \times N$ matrix with orthonormal columns and $\mathbf{R}_1$ is an upper triangular matrix with (strictly) positive diagonal elements. Moreover, the matrices $\mathbf{Q}_1$ and $\mathbf{R}_1$ are unique. The existence of a decomposition of the form (4.56) can be established by, for example, applying Gram–Schmidt orthogonalization. For a proof of the existence and uniqueness of a decomposition of the form (4.56), refer, for example, to Harville (1997, sec. 6.4).

As a variation on expression (4.56), we have the expression

$$\mathbf{A} = \mathbf{Q}\mathbf{R}, \tag{4.57}$$

where $\mathbf{Q} = (\mathbf{Q}_1, \mathbf{Q}_2)$ is an $M \times M$ orthogonal matrix and where $\mathbf{R} = \begin{pmatrix} \mathbf{R}_1 \\ \mathbf{0} \end{pmatrix}$. The columns of the $M \times (M-N)$ submatrix $\mathbf{Q}_2$ are any $M-N$ M-dimensional column vectors that together with the N columns of $\mathbf{Q}_1$ form an orthonormal basis for $\mathcal{R}^M$.

Either of the two decompositions (4.56) and (4.57) might be referred to as a *QR decomposition*.

QR decomposition as a basis for least squares computations. Let us now resume our discussion of the computational aspects of the minimization of $(\mathbf{y}-\mathbf{X}\mathbf{b})'(\mathbf{y}-\mathbf{X}\mathbf{b})$. Assume that the $N \times P$ matrix $\mathbf{X}$ is of full column rank P—discussion of the general case where $\operatorname{rank}(\mathbf{X})$ may be less than P is deferred until the final part of the present subsection. Consider the QR decomposition of $\mathbf{X}$. That is, consider a decomposition of $\mathbf{X}$ of the form

$$\mathbf{X} = \mathbf{Q}_1\mathbf{R}_1, \tag{4.58}$$

where $\mathbf{Q}_1$ is an $N \times P$ matrix with orthonormal columns and $\mathbf{R}_1$ is an upper triangular matrix with (strictly) positive diagonal elements, or of the related form

$$\mathbf{X} = \mathbf{Q}\mathbf{R}, \tag{4.59}$$

where $\mathbf{Q} = (\mathbf{Q}_1, \mathbf{Q}_2)$ is an $N \times N$ orthogonal matrix and where $\mathbf{R} = \begin{pmatrix} \mathbf{R}_1 \\ \mathbf{0} \end{pmatrix}$.

Let $\mathbf{z} = \mathbf{Q}'\mathbf{y}$, and partition $\mathbf{z}$ as $\mathbf{z} = \begin{pmatrix} \mathbf{z}_1 \\ \mathbf{z}_2 \end{pmatrix}$, where $\mathbf{z}_1 = \mathbf{Q}_1'\mathbf{y}$ and $\mathbf{z}_2 = \mathbf{Q}_2'\mathbf{y}$. Then,

$$\begin{aligned} \mathbf{y} - \mathbf{X}\mathbf{b} &= \mathbf{Q}(\mathbf{z} - \mathbf{R}\mathbf{b}) \\ &= \mathbf{Q}_1(\mathbf{z}_1 - \mathbf{R}_1\mathbf{b}) + \mathbf{Q}_2\mathbf{z}_2. \end{aligned} \tag{4.60}$$

And

$$\begin{aligned} (\mathbf{y} - \mathbf{X}\mathbf{b})'(\mathbf{y} - \mathbf{X}\mathbf{b}) &= (\mathbf{z} - \mathbf{R}\mathbf{b})'\mathbf{Q}'\mathbf{Q}(\mathbf{z} - \mathbf{R}\mathbf{b}) \\ &= (\mathbf{z} - \mathbf{R}\mathbf{b})'(\mathbf{z} - \mathbf{R}\mathbf{b}) \\ &= (\mathbf{z}_1 - \mathbf{R}_1\mathbf{b})'(\mathbf{z}_1 - \mathbf{R}_1\mathbf{b}) + \mathbf{z}_2'\mathbf{z}_2. \end{aligned} \tag{4.61}$$

Now, consider the linear system

$$\mathbf{R}_1\mathbf{b} = \mathbf{z}_1 \tag{4.62}$$

(in the vector $\mathbf{b}$), whose coefficient matrix is $\mathbf{R}_1$ and right side is $\mathbf{z}_1$. It follows from elementary results on triangular matrices (e.g., Harville 1997, corollary 8.5.6) that $\mathbf{R}_1$ is nonsingular. Thus, linear system (4.62) has a unique solution, say $\tilde{\mathbf{b}}$. Clearly, the first term of expression (4.61) equals 0 if $\mathbf{b} = \tilde{\mathbf{b}}$, and is greater than 0 otherwise (i.e., if $\mathbf{b} \neq \tilde{\mathbf{b}}$). And because the second term of expression (4.61) does not depend on $\mathbf{b}$, we conclude that $(\mathbf{y} - \mathbf{X}\mathbf{b})'(\mathbf{y} - \mathbf{X}\mathbf{b})$ attains a minimum value of $\mathbf{z}_2'\mathbf{z}_2$ and does so uniquely at the point $\tilde{\mathbf{b}}$. Moreover, in light of result (4.60),

$$\mathbf{y} - \mathbf{X}\tilde{\mathbf{b}} = \mathbf{Q}_2\mathbf{z}_2. \tag{4.63}$$

These results serve as the basis for an alternative approach to the least squares computations (not requiring the formation of the normal equations). In the alternative approach, the formation of the matrix $\mathbf{R}_1$ and the vector $\mathbf{z}_1$ are at the heart of the computations. Their formation can be accomplished through the use of Householder transformations (reflections) or Givens transformations (rotations) or through the use of a modified Gram–Schmidt procedure—refer, for example, to Golub and Van Loan (2013, chap. 5) for a detailed discussion. The value $\tilde{\mathbf{b}}$ at which $(\mathbf{y} - \mathbf{X}\mathbf{b})'(\mathbf{y} - \mathbf{X}\mathbf{b})$ attains its minimum value is determined from $\mathbf{R}_1$ and $\mathbf{z}_1$ by solving linear system (4.62), doing so in a way that exploits the triangularity of $\mathbf{R}_1$—refer, for example, to Harville (1997, sec. 11.8) for a discussion of the solution of a linear system with a triangular coefficient matrix.

Our results on the alternative approach to the least squares computations can be extended to the general case where the matrix $\mathbf{X}$ is not necessarily of full column rank. The extension requires some familiarity with a type of matrix called a permutation matrix.

Permutation matrices. A *permutation matrix* is a square matrix whose columns can be obtained by permuting (rearranging) the columns of an identity matrix. Thus, letting $\mathbf{u}_1, \mathbf{u}_2, \ldots, \mathbf{u}_N$ represent the first, second, ..., Nth columns, respectively, of $\mathbf{I}_N$, an $N \times N$ permutation matrix is a matrix of the general form $(\mathbf{u}_{k_1}, \mathbf{u}_{k_2}, \ldots, \mathbf{u}_{k_N})$, where $k_1, k_2, \ldots, k_N$ is an arbitrary permutation of the first N positive integers $1, 2, \ldots, N$. For example, one permutation matrix of order $N = 3$ is the 3×3 matrix

$$(\mathbf{u}_3,\ \mathbf{u}_1,\ \mathbf{u}_2) = \begin{pmatrix} 0 & 1 & 0 \\ 0 & 0 & 1 \\ 1 & 0 & 0 \end{pmatrix},$$

whose columns are the third, first, and second columns, respectively, of $\mathbf{I}_3$. Clearly, the columns of any permutation matrix form an orthonormal (with respect to the usual inner product) set, and hence any permutation matrix is an orthogonal matrix.

The jth element of the k_jth row of the $N \times N$ permutation matrix $(\mathbf{u}_{k_1}, \mathbf{u}_{k_2}, \ldots, \mathbf{u}_{k_N})$ is 1, and its other $N-1$ elements are 0. That is, the jth row $\mathbf{u}_j'$ of $\mathbf{I}_N$ is the k_jth row of $(\mathbf{u}_{k_1}, \mathbf{u}_{k_2}, \ldots, \mathbf{u}_{k_N})$ or, equivalently, the jth column $\mathbf{u}_j$ of $\mathbf{I}_N$ is the k_jth column of $(\mathbf{u}_{k_1}, \mathbf{u}_{k_2}, \ldots, \mathbf{u}_{k_N})'$. Thus, the transpose of any permutation matrix is itself a permutation matrix. Further, the rows of any permutation matrix are a permutation of the rows of an identity matrix and, conversely, any matrix whose rows can be obtained by permuting the rows of an identity matrix is a permutation matrix.

The effect of postmultiplying an $M \times N$ matrix $\mathbf{A}$ by an $N \times N$ permutation matrix $\mathbf{P}$ is to permute the columns of $\mathbf{A}$ in the same way that the columns of $\mathbf{I}_N$ were permuted in forming $\mathbf{P}$. Thus, if $\mathbf{a}_1, \mathbf{a}_2, \ldots, \mathbf{a}_N$ are the first, second, ..., Nth columns of $\mathbf{A}$, the first, second, ..., Nth columns of the product $\mathbf{A}(\mathbf{u}_{k_1}, \mathbf{u}_{k_2}, \ldots, \mathbf{u}_{k_N})$ of $\mathbf{A}$ and the $N \times N$ permutation matrix $(\mathbf{u}_{k_1}, \mathbf{u}_{k_2}, \ldots, \mathbf{u}_{k_N})$ are $\mathbf{a}_{k_1}, \mathbf{a}_{k_2}, \ldots, \mathbf{a}_{k_N}$, respectively. Further, the first, second ..., Nth columns $\mathbf{a}_1, \mathbf{a}_2, \ldots, \mathbf{a}_N$ of $\mathbf{A}$ are the $k_1, k_2, \ldots, k_N$th columns, respectively, of the product $\mathbf{A}(\mathbf{u}_{k_1}, \mathbf{u}_{k_2}, \ldots, \mathbf{u}_{k_N})'$ of $\mathbf{A}$ and the permutation matrix $(\mathbf{u}_{k_1}, \mathbf{u}_{k_2}, \ldots, \mathbf{u}_{k_N})'$. When $N = 3$, we have, for example, that

$$\mathbf{A}(\mathbf{u}_3, \mathbf{u}_1, \mathbf{u}_2) = (\mathbf{a}_1, \mathbf{a}_2, \mathbf{a}_3)\begin{pmatrix} 0 & 1 & 0 \\ 0 & 0 & 1 \\ 1 & 0 & 0 \end{pmatrix} = (\mathbf{a}_3, \mathbf{a}_1, \mathbf{a}_2)$$

and

$$\mathbf{A}(\mathbf{u}_3, \mathbf{u}_1, \mathbf{u}_2)' = (\mathbf{a}_1, \mathbf{a}_2, \mathbf{a}_3)\begin{pmatrix} 0 & 0 & 1 \\ 1 & 0 & 0 \\ 0 & 1 & 0 \end{pmatrix} = (\mathbf{a}_2, \mathbf{a}_3, \mathbf{a}_1).$$

Similarly, the effect of premultiplying an $N \times M$ matrix $\mathbf{A}$ by an $N \times N$ permutation matrix is to permute the rows of $\mathbf{A}$. If the first, second, ..., Nth rows of $\mathbf{A}$ are $\mathbf{a}_1', \mathbf{a}_2', \ldots, \mathbf{a}_N'$, respectively, then the first, second, ..., Nth rows of the product $(\mathbf{u}_{k_1}, \mathbf{u}_{k_2}, \ldots, \mathbf{u}_{k_N})'\mathbf{A}$ of the permutation matrix $(\mathbf{u}_{k_1}, \mathbf{u}_{k_2}, \ldots, \mathbf{u}_{k_N})'$ and $\mathbf{A}$ are $\mathbf{a}_{k_1}', \mathbf{a}_{k_2}', \ldots, \mathbf{a}_{k_N}'$, respectively, and $\mathbf{a}_1', \mathbf{a}_2', \ldots, \mathbf{a}_N'$ are the $k_1, k_2, \ldots, k_N$th rows, respectively, of $(\mathbf{u}_{k_1}, \mathbf{u}_{k_2}, \ldots, \mathbf{u}_{k_N})\mathbf{A}$. When $N = 3$, we have, for example, that

$$(\mathbf{u}_3, \mathbf{u}_1, \mathbf{u}_2)'\mathbf{A} = \begin{pmatrix} 0 & 0 & 1 \\ 1 & 0 & 0 \\ 0 & 1 & 0 \end{pmatrix}\begin{pmatrix} \mathbf{a}_1' \\ \mathbf{a}_2' \\ \mathbf{a}_3' \end{pmatrix} = \begin{pmatrix} \mathbf{a}_3' \\ \mathbf{a}_1' \\ \mathbf{a}_2' \end{pmatrix}$$

and

$$(\mathbf{u}_3, \mathbf{u}_1, \mathbf{u}_2)\mathbf{A} = \begin{pmatrix} 0 & 1 & 0 \\ 0 & 0 & 1 \\ 1 & 0 & 0 \end{pmatrix}\begin{pmatrix} \mathbf{a}_1' \\ \mathbf{a}_2' \\ \mathbf{a}_3' \end{pmatrix} = \begin{pmatrix} \mathbf{a}_2' \\ \mathbf{a}_3' \\ \mathbf{a}_1' \end{pmatrix}.$$

Alternative approach to least squares computations: general case. Let us now extend our initial results on the alternative approach to the least squares computations. Accordingly, suppose that we wish to minimize the quantity $(\mathbf{y} - \mathbf{Xb})'(\mathbf{y} - \mathbf{Xb})$ and that $\text{rank}(\mathbf{X}) = K$, where K is possibly less than P—our initial results (the results of Part 2) were obtained under the simplifying assumption that the $N \times P$ matrix $\mathbf{X}$ is of full column rank P.

Let $\mathbf{L}$ represent any $P \times P$ permutation matrix such that the first K columns of the $N \times P$ matrix $\mathbf{XL}$ are linearly independent, and partition $\mathbf{L}$ as $\mathbf{L} = (\mathbf{L}_1, \mathbf{L}_2)$, where $\mathbf{L}_1$ is of dimensions $P \times K$. Then,

$$\mathbf{XL} = (\mathbf{XL}_1, \mathbf{XL}_2),$$

and $\mathbf{XL}_1$ is of full column rank K. Decompose $\mathbf{XL}_1$ as

$$\mathbf{XL}_1 = \mathbf{Q}_1\mathbf{R}_1, \tag{4.64}$$

where $\mathbf{Q}_1$ is an $N \times K$ matrix with orthonormal columns and $\mathbf{R}_1$ is an upper triangular matrix with (strictly) positive diagonal elements. And observe that the columns of $\mathbf{Q}_1$ form a basis for $\mathcal{C}(\mathbf{XL})$ [$= \mathcal{C}(\mathbf{XL}_1)$], so that

$$\mathbf{XL}_2 = \mathbf{Q}_1\mathbf{R}_2, \tag{4.65}$$

for some matrix $\mathbf{R}_2$. Together, results (4.64) and (4.65) imply that

$$\mathbf{XL} = \mathbf{Q}_1(\mathbf{R}_1, \mathbf{R}_2)$$

and also that

$$\mathbf{XL} = \mathbf{QR},$$

where $\mathbf{Q} = (\mathbf{Q}_1, \mathbf{Q}_2)$ is an $N \times N$ orthogonal matrix and where $\mathbf{R} = \begin{pmatrix} \mathbf{R}_1 & \mathbf{R}_2 \\ \mathbf{0} & \mathbf{0} \end{pmatrix}$. Or, equivalently,

$$\mathbf{X} = \mathbf{Q}_1(\mathbf{R}_1, \mathbf{R}_2)\mathbf{L}' = \mathbf{Q}_1\mathbf{R}_1\mathbf{L}_1' + \mathbf{Q}_1\mathbf{R}_2\mathbf{L}_2'$$

and

$$\mathbf{X} = \mathbf{QRL}'. \tag{4.66}$$

Let $\mathbf{h} = \mathbf{L}'\mathbf{b}$, and partition $\mathbf{h}$ as $\mathbf{h} = \begin{pmatrix} \mathbf{h}_1 \\ \mathbf{h}_2 \end{pmatrix}$, where $\mathbf{h}_1 = \mathbf{L}_1'\mathbf{b}$ and $\mathbf{h}_2 = \mathbf{L}_2'\mathbf{b}$. Further, let

$\mathbf{z} = \mathbf{Q}'\mathbf{y}$, and partition $\mathbf{z}$ as $\mathbf{z} = \begin{pmatrix} \mathbf{z}_1 \\ \mathbf{z}_2 \end{pmatrix}$, where $\mathbf{z}_1 = \mathbf{Q}_1'\mathbf{y}$ and $\mathbf{z}_2 = \mathbf{Q}_2'\mathbf{y}$. Then,

$$\begin{aligned} \mathbf{y} - \mathbf{X}\mathbf{b} &= \mathbf{Q}(\mathbf{z} - \mathbf{R}\mathbf{h}) \\ &= \mathbf{Q}_1(\mathbf{z}_1 - \mathbf{R}_1\mathbf{h}_1 - \mathbf{R}_2\mathbf{h}_2) + \mathbf{Q}_2\mathbf{z}_2, \end{aligned} \tag{4.67}$$

which is a generalization of result (4.60). And

$$\begin{aligned} (\mathbf{y} - \mathbf{X}\mathbf{b})'(\mathbf{y} - \mathbf{X}\mathbf{b}) &= (\mathbf{z} - \mathbf{R}\mathbf{h})'\mathbf{Q}'\mathbf{Q}(\mathbf{z} - \mathbf{R}\mathbf{h}) \\ &= (\mathbf{z} - \mathbf{R}\mathbf{h})'(\mathbf{z} - \mathbf{R}\mathbf{h}) \\ &= (\mathbf{z}_1 - \mathbf{R}_1\mathbf{h}_1 - \mathbf{R}_2\mathbf{h}_2)'(\mathbf{z}_1 - \mathbf{R}_1\mathbf{h}_1 - \mathbf{R}_2\mathbf{h}_2) + \mathbf{z}_2'\mathbf{z}_2, \end{aligned} \tag{4.68}$$

which is a generalization of result (4.61).

Now, consider the minimization of $(\mathbf{y} - \mathbf{X}\mathbf{b})'(\mathbf{y} - \mathbf{X}\mathbf{b})$ with respect to the transformed vector $\mathbf{h}$ $(= \mathbf{L}'\mathbf{b})$. It follows from result (4.68) that $(\mathbf{y} - \mathbf{X}\mathbf{b})'(\mathbf{y} - \mathbf{X}\mathbf{b})$ attains a minimum value of $\mathbf{z}_2'\mathbf{z}_2$ and that it does so at those values of $\mathbf{h}$ for which the first term of expression (4.68) equals 0 or, equivalently, at those values for which $\mathbf{z}_1 - \mathbf{R}_1\mathbf{h}_1 - \mathbf{R}_2\mathbf{h}_2 = \mathbf{0}$. Accordingly, $(\mathbf{y} - \mathbf{X}\mathbf{b})'(\mathbf{y} - \mathbf{X}\mathbf{b})$ attains a minimum value of $\mathbf{z}_2'\mathbf{z}_2$ at values $\tilde{\mathbf{h}}_1$ and $\tilde{\mathbf{h}}_2$ of $\mathbf{h}_1$ and $\mathbf{h}_2$, respectively, if and only if $\mathbf{R}_1\tilde{\mathbf{h}}_1 = \mathbf{z}_1 - \mathbf{R}_2\tilde{\mathbf{h}}_2$ or, equivalently, if and only if $\tilde{\mathbf{h}}_1$ is the solution to the linear system

$$\mathbf{R}_1\mathbf{h}_1 = \mathbf{z}_1 - \mathbf{R}_2\tilde{\mathbf{h}}_2 \tag{4.69}$$

(in the vector $\mathbf{h}_1$). Thus, an arbitrary one of the values of $\mathbf{h}$ at which $(\mathbf{y} - \mathbf{X}\mathbf{b})'(\mathbf{y} - \mathbf{X}\mathbf{b})$ attains a minimum value is obtained by assigning $\mathbf{h}_2$ an arbitrary value $\tilde{\mathbf{h}}_2$ and by then taking the value $\tilde{\mathbf{h}}_1$ of $\mathbf{h}_1$ to be the solution to linear system (4.69)—the matrix $\mathbf{R}_1$ is nonsingular, so that $\tilde{\mathbf{h}}_1$ is uniquely determined by $\tilde{\mathbf{h}}_2$. In particular, we could take the value of $\mathbf{h}_2$ to be $\mathbf{0}$, and take the value of $\mathbf{h}_1$ to be the (unique) solution to the linear system $\mathbf{R}_1\mathbf{h}_1 = \mathbf{z}_1$.

We conclude that $(\mathbf{y} - \mathbf{X}\mathbf{b})'(\mathbf{y} - \mathbf{X}\mathbf{b})$ attains a minimum value of $\mathbf{z}_2'\mathbf{z}_2$ and that it does so at a value $\tilde{\mathbf{b}}$ of $\mathbf{b}$ if and only if for some $(P-K) \times 1$ vector $\tilde{\mathbf{h}}_2$, $\tilde{\mathbf{b}} = \mathbf{L}_1\tilde{\mathbf{h}}_1 + \mathbf{L}_2\tilde{\mathbf{h}}_2$, where $\tilde{\mathbf{h}}_1$ is the solution to linear system (4.69). Note [in light of result (4.67)] that for any such (minimizing) value $\tilde{\mathbf{b}}$ of $\mathbf{b}$,

$$\mathbf{y} - \mathbf{X}\tilde{\mathbf{b}} = \mathbf{Q}_2\mathbf{z}_2. \tag{4.70}$$

These results generalize the results obtained earlier (in Part 2) for the special case where the rank K of the $N \times P$ matrix $\mathbf{X}$ equals P. They provide a basis for extending the alternative approach to the least squares computations to the general case (where K may be less than P). As in the special case, the formation of the matrix $\mathbf{R}_1$ and the vector $\mathbf{z}_1$ are at the heart of the computations. (And, as in the special case, the formation of $\mathbf{R}_1$ and $\mathbf{z}_1$ can be accomplished via any of several procedures devised for that purpose.) In the general case, there is also a need to determine the permutation matrix $\mathbf{L}$ (i.e., to identify K linearly independent columns of $\mathbf{X}$) and possibly the matrix $\mathbf{R}_2$—if the value of $\mathbf{h}_2$ is taken to be $\mathbf{0}$, then $\mathbf{R}_2$ is not needed. A value $\tilde{\mathbf{b}}$ at which $(\mathbf{y} - \mathbf{X}\mathbf{b})'(\mathbf{y} - \mathbf{X}\mathbf{b})$ attains its minimum value is determined from $\mathbf{R}_1$, $\mathbf{z}_1$, $\mathbf{L}$, and possibly $\mathbf{R}_2$ by taking $\tilde{\mathbf{h}}_2$ to be any $(P-K) \times 1$ vector, by computing the solution $\tilde{\mathbf{h}}_1$ to linear system (4.69), and by setting $\tilde{\mathbf{b}} = \mathbf{L}_1\tilde{\mathbf{h}}_1 + \mathbf{L}_2\tilde{\mathbf{h}}_2$.

5.5 Best Linear Unbiased or Translation-Equivariant Estimation of Estimable Functions (under the G–M Model)

Suppose that $\mathbf{y}$ is an $N \times 1$ observable random vector that follows a G–M, Aitken, or general linear model. And consider the least squares estimator $\ell(\mathbf{y})$ of an estimable linear combination $\boldsymbol{\lambda}'\boldsymbol{\beta}$ of the elements of the parametric vector $\boldsymbol{\beta}$. The least squares estimator is a linear estimator, as was demonstrated in Section 5.4c—refer to representation (4.24) or (4.35). Moreover, the least squares estimator is an unbiased estimator. Its unbiasedness can be established directly by verifying that $\mathrm{E}[\ell(\mathbf{y})] = \boldsymbol{\lambda}'\boldsymbol{\beta}$, as was done in Section 5.4c—refer to result (4.36). Alternatively, its unbiasedness can be established by applying the following result (from Section 5.1) on the unbiasedness of linear estimators: for an estimator of the form $c + \mathbf{a}'\mathbf{y}$ to be an unbiased estimator of $\boldsymbol{\lambda}'\boldsymbol{\beta}$, it is necessary and sufficient that

$$c = 0 \quad \text{and} \quad \mathbf{X}'\mathbf{a} = \boldsymbol{\lambda}. \tag{5.1}$$

Upon observing [in light of result (4.35)] that $\ell(\mathbf{y}) = (\mathbf{X}\tilde{\mathbf{r}})'\mathbf{y}$, where $\tilde{\mathbf{r}}$ is any solution to the conjugate normal equations $\mathbf{X}'\mathbf{X}\mathbf{r} = \boldsymbol{\lambda}$, it follows immediately from the sufficiency of condition (5.1) that $\ell(\mathbf{y})$ is an unbiased estimator of $\boldsymbol{\lambda}'\boldsymbol{\beta}$.

The least squares estimator of $\boldsymbol{\lambda}'\boldsymbol{\beta}$ is translation equivariant as well as unbiased. To see this, recall (from Section 5.2) that for an estimator of the form $c + \mathbf{a}'\mathbf{y}$ to be a translation-equivariant estimator of $\boldsymbol{\lambda}'\boldsymbol{\beta}$, it is necessary and sufficient that $\mathbf{a}'\mathbf{X} = \boldsymbol{\lambda}'$ or, equivalently, that $\mathbf{X}'\mathbf{a} = \boldsymbol{\lambda}$. The translation equivariance of the least squares estimator $\ell(\mathbf{y})$ [which is expressible in the form $\ell(\mathbf{y}) = (\mathbf{X}\tilde{\mathbf{r}})'\mathbf{y}$] follows from the sufficiency of the condition $\mathbf{X}'\mathbf{a} = \boldsymbol{\lambda}$ in much the same way that its unbiasedness follows from the sufficiency of condition (5.1).

When $\mathbf{y}$ follows a G–M model, the least squares estimator of $\boldsymbol{\lambda}'\boldsymbol{\beta}$ is superior to other linear unbiased or translation-equivariant estimators in a sense that is to be discussed in Subsections a and b. More generally (when $\mathbf{y}$ follows an Aitken or general linear model), this superiority is confined to special cases. These special cases include, of course, G–M models, but also a limited number of other models.

a. Gauss–Markov theorem

In the special case where $\mathbf{y}$ is an $N \times 1$ observable random vector that follows a G–M model, the least squares estimator of an estimable linear combination $\boldsymbol{\lambda}'\boldsymbol{\beta}$ of the elements of the parametric vector $\boldsymbol{\beta}$ is the best linear unbiased estimator in the sense described in the following theorem.

Theorem 5.5.1 (Gauss–Markov theorem). Suppose that $\mathbf{y}$ is an $N \times 1$ observable random vector that follows a G–M, Aitken, or general linear model, and suppose that $\boldsymbol{\lambda}'\boldsymbol{\beta}$ is an estimable linear combination of the elements of the parametric vector $\boldsymbol{\beta}$. Then, the least squares estimator of $\boldsymbol{\lambda}'\boldsymbol{\beta}$ is a linear unbiased estimator. Moreover, in the special case where $\mathbf{y}$ follows a G–M model, the variance (and hence the mean squared error) of the least squares estimator is uniformly smaller than that of any other linear unbiased estimator.

Proof. That the least squares estimator is a linear unbiased estimator was established earlier (in the introductory part of the present section). Now, take $c + \mathbf{a}'\mathbf{y}$ to be an arbitrary linear unbiased estimator of the estimable linear combination $\boldsymbol{\lambda}'\boldsymbol{\beta}$, in which case

$$c = 0 \quad \text{and} \quad \mathbf{X}'\mathbf{a} = \boldsymbol{\lambda}$$

(as noted earlier). And recall that the least squares estimator of $\boldsymbol{\lambda}'\boldsymbol{\beta}$ is expressible in the form $(\mathbf{X}\tilde{\mathbf{r}})'\mathbf{y}$, where $\tilde{\mathbf{r}}$ is any solution to the conjugate normal equations $\mathbf{X}'\mathbf{X}\mathbf{r} = \boldsymbol{\lambda}$.

In the special case where $\mathbf{y}$ follows a G–M model, we find that

$$\begin{aligned} \operatorname{cov}[(\mathbf{X}\tilde{\mathbf{r}})'\mathbf{y},\, c + \mathbf{a}'\mathbf{y} - (\mathbf{X}\tilde{\mathbf{r}})'\mathbf{y}] &= \tilde{\mathbf{r}}'\mathbf{X}'(\sigma^2\mathbf{I})(\mathbf{a} - \mathbf{X}\tilde{\mathbf{r}}) \\ &= \sigma^2\tilde{\mathbf{r}}'(\mathbf{X}'\mathbf{a} - \mathbf{X}'\mathbf{X}\tilde{\mathbf{r}}) \\ &= \sigma^2\tilde{\mathbf{r}}'(\boldsymbol{\lambda} - \boldsymbol{\lambda}) = 0. \end{aligned}$$

Thus, in that special case,

$$\begin{aligned}\operatorname{var}(c+\mathbf{a}'\mathbf{y}) &= \operatorname{var}[(\mathbf{X}\tilde{\mathbf{r}})'\mathbf{y} + c + \mathbf{a}'\mathbf{y} - (\mathbf{X}\tilde{\mathbf{r}})'\mathbf{y}] \\ &= \operatorname{var}[(\mathbf{X}\tilde{\mathbf{r}})'\mathbf{y}] + \operatorname{var}[c + \mathbf{a}'\mathbf{y} - (\mathbf{X}\tilde{\mathbf{r}})'\mathbf{y}] \\ &\qquad + 2\operatorname{cov}[(\mathbf{X}\tilde{\mathbf{r}})'\mathbf{y},\, c + \mathbf{a}'\mathbf{y} - (\mathbf{X}\tilde{\mathbf{r}})'\mathbf{y}] \\ &= \operatorname{var}[(\mathbf{X}\tilde{\mathbf{r}})'\mathbf{y}] + \operatorname{var}[c + \mathbf{a}'\mathbf{y} - (\mathbf{X}\tilde{\mathbf{r}})'\mathbf{y}] \\ &\geq \operatorname{var}[(\mathbf{X}\tilde{\mathbf{r}})'\mathbf{y}], \end{aligned} \tag{5.2}$$

with equality holding if and only if $\operatorname{var}[c + \mathbf{a}'\mathbf{y} - (\mathbf{X}\tilde{\mathbf{r}})'\mathbf{y}] = 0$. Moreover, in the special case of the G–M model,

$$\begin{aligned}\operatorname{var}[c + \mathbf{a}'\mathbf{y} - (\mathbf{X}\tilde{\mathbf{r}})'\mathbf{y}] &= (\mathbf{a} - \mathbf{X}\tilde{\mathbf{r}})'(\sigma^2\mathbf{I})(\mathbf{a} - \mathbf{X}\tilde{\mathbf{r}}) \\ &= \sigma^2(\mathbf{a} - \mathbf{X}\tilde{\mathbf{r}})'(\mathbf{a} - \mathbf{X}\tilde{\mathbf{r}}),\end{aligned}$$

so that equality holds in inequality (5.2) if and only if $\mathbf{a} - \mathbf{X}\tilde{\mathbf{r}} = \mathbf{0}$ or, equivalently, if and only if $\mathbf{a} = \mathbf{X}\tilde{\mathbf{r}}$. We conclude that in the special case of the G–M model, the variance of the least squares estimator is uniformly smaller than that of any other linear unbiased estimator. Q.E.D.

Theorem 5.5.1 (in one form or another) has come to be known as the Gauss–Markov theorem (in honor of the contributions of Carl Friedrich Gauss and Andrei Andreevich Markov). It is one of the most famous theoretical results in all of statistics. Seal (1967, sec. 3) considered this result from a historical perspective. That Gauss's name has come to be attached to the result of Theorem 5.5.1 seems altogether appropriate. The case for the attachment of Markov's name appears to be much weaker.

It is customary (both in the present setting and in general) to refer to a linear unbiased estimator that has minimum variance among all linear unbiased estimators as a BLUE (an acronym for best linear unbiased estimator or estimation). If $\mathbf{y}$ is an $N \times 1$ observable random vector that follows a G–M model, then (according to the Gauss–Markov theorem) the least squares estimator of an estimable linear combination $\boldsymbol{\lambda}'\boldsymbol{\beta}$ of the elements of the parametric vector $\boldsymbol{\beta}$ is the unique BLUE of $\boldsymbol{\lambda}'\boldsymbol{\beta}$. Albert (1972, sec. 6.1), in a comment he characterized as jocular, suggested that the least squares estimator of an estimable linear combination could be referred to as a TRUE (an acronym for tiniest residual unbiased estimator). Accordingly, when the least squares estimator is a BLUE, it could be referred to as a TRUE-BLUE—someone who is unswervingly loyal or faithful is said to be true-blue.

b. A corollary

Suppose that $\mathbf{y}$ is an $N \times 1$ observable random vector that follows a G–M, Aitken, or general linear model, and take $\boldsymbol{\lambda}'\boldsymbol{\beta}$ to be an estimable linear combination of the elements of the parametric vector $\boldsymbol{\beta}$. Further, let $\ell(\mathbf{y}) = (\mathbf{X}\tilde{\mathbf{r}})'\mathbf{y}$, where $\tilde{\mathbf{r}}$ is any solution to the conjugate normal equations $\mathbf{X}'\mathbf{X}\mathbf{r} = \boldsymbol{\lambda}$ [so that $\ell(\mathbf{y})$ is the least squares estimator of $\boldsymbol{\lambda}'\boldsymbol{\beta}$]. And let $c + \mathbf{a}'\mathbf{y}$ represent an arbitrary linear translation-equivariant estimator of $\boldsymbol{\lambda}'\boldsymbol{\beta}$ or, equivalently, any estimator of the form $c + \mathbf{a}'\mathbf{y}$ that satisfies the condition $\mathbf{a}'\mathbf{X} = \boldsymbol{\lambda}'$; and recall (from the introductory part of the present section) that the least squares estimator is a linear translation-equivariant estimator.

Clearly, $\mathrm{E}(\mathbf{a}'\mathbf{y}) = \boldsymbol{\lambda}'\boldsymbol{\beta}$, that is, $\mathbf{a}'\mathbf{y}$ is an unbiased estimator of $\boldsymbol{\lambda}'\boldsymbol{\beta}$. And, as a consequence, the MSE (mean squared error) of $c + \mathbf{a}'\mathbf{y}$ is

$$\begin{aligned}\mathrm{E}[(c + \mathbf{a}'\mathbf{y} - \boldsymbol{\lambda}'\boldsymbol{\beta})^2] &= c^2 + \mathrm{E}[(\mathbf{a}'\mathbf{y} - \boldsymbol{\lambda}'\boldsymbol{\beta})^2] + 2c\,\mathrm{E}(\mathbf{a}'\mathbf{y} - \boldsymbol{\lambda}'\boldsymbol{\beta}) \\ &= c^2 + \operatorname{var}(\mathbf{a}'\mathbf{y}) \\ &\geq \operatorname{var}(\mathbf{a}'\mathbf{y}),\end{aligned}$$

with equality holding if and only if $c = 0$ and hence if and only if $c + \mathbf{a}'\mathbf{y} = \mathbf{a}'\mathbf{y}$. Moreover, in the special case where $\mathbf{y}$ follows a G–M model, it follows from the Gauss–Markov theorem that

$$\text{var}(\mathbf{a}'\mathbf{y}) \geq \text{var}[\ell(\mathbf{y})],$$

with equality holding if and only if $\mathbf{a}'\mathbf{y} = \ell(\mathbf{y})$, that is, if and only if $\mathbf{a}'\mathbf{y}$ is the least squares estimator. Accordingly, in that special case, the MSE of the least squares estimator is uniformly smaller than the MSE of any other linear translation-equivariant estimator.

In summary, we have the following result, the main part of which can be regarded as a corollary of the Gauss–Markov theorem.

Corollary 5.5.2. Suppose that $\mathbf{y}$ is an $N \times 1$ observable random vector that follows a G–M, Aitken, or general linear model, and suppose that $\boldsymbol{\lambda}'\boldsymbol{\beta}$ is an estimable linear combination of the elements of the parametric vector $\boldsymbol{\beta}$. Then, the least squares estimator of $\boldsymbol{\lambda}'\boldsymbol{\beta}$ is a linear translation-equivariant estimator. Moreover, in the special case where $\mathbf{y}$ follows a G–M model, the mean squared error of the least squares estimator is uniformly smaller than that of any other linear translation-equivariant estimator.

5.6 Simultaneous Estimation

Suppose that $\mathbf{y}$ is an $N \times 1$ observable random vector that follows a G–M, Aitken, or general linear model. And consider the estimation of estimable linear combinations of the elements of the parametric vector $\boldsymbol{\beta}$. Specifically, suppose that we wish to estimate a finite number M of such linear combinations, say $\boldsymbol{\lambda}_1'\boldsymbol{\beta}, \boldsymbol{\lambda}_2'\boldsymbol{\beta}, \ldots, \boldsymbol{\lambda}_M'\boldsymbol{\beta}$ (and perhaps some or all linear combinations of $\boldsymbol{\lambda}_1'\boldsymbol{\beta}, \boldsymbol{\lambda}_2'\boldsymbol{\beta}, \ldots, \boldsymbol{\lambda}_M'\boldsymbol{\beta}$). The Gauss–Markov theorem is relevant to the estimation of these linear combinations when the linear combinations are considered individually. However, that each of these linear combinations is to be estimated simultaneously with $M-1$ or more other linear combinations is not reflected in the criterion employed in the Gauss–Markov theorem. In this section, we obtain some results that account explicitly for the simultaneous estimation of the various linear combinations.

Let $\boldsymbol{\Lambda} = (\boldsymbol{\lambda}_1, \boldsymbol{\lambda}_2, \ldots, \boldsymbol{\lambda}_M)$, so that $\boldsymbol{\Lambda}'\boldsymbol{\beta}$ is the M-dimensional column vector whose elements are the M linear combinations $\boldsymbol{\lambda}_1'\boldsymbol{\beta}, \boldsymbol{\lambda}_2'\boldsymbol{\beta}, \ldots, \boldsymbol{\lambda}_M'\boldsymbol{\beta}$—when all M linear combinations $\boldsymbol{\lambda}_1'\boldsymbol{\beta}, \boldsymbol{\lambda}_2'\boldsymbol{\beta}, \ldots, \boldsymbol{\lambda}_M'\boldsymbol{\beta}$ are estimable (as is being assumed), the vector $\boldsymbol{\Lambda}'\boldsymbol{\beta}$ is said to be estimable. By definition,the least squares estimator of $\boldsymbol{\Lambda}'\boldsymbol{\beta}$ is the $M \times 1$ vector $\boldsymbol{\ell}(\mathbf{y}) = [\ell_1(\mathbf{y}), \ell_2(\mathbf{y}), \ldots, \ell_M(\mathbf{y})]'$, where $\ell_1(\mathbf{y}), \ell_2(\mathbf{y}), \ldots, \ell_M(\mathbf{y})$ are the least squares estimators of $\boldsymbol{\lambda}_1'\boldsymbol{\beta}, \boldsymbol{\lambda}_2'\boldsymbol{\beta}, \ldots, \boldsymbol{\lambda}_M'\boldsymbol{\beta}$, respectively. In light of result (4.24), the least squares estimator is expressible as

$$\boldsymbol{\ell}(\mathbf{y}) = \boldsymbol{\Lambda}'(\mathbf{X}'\mathbf{X})^{-}\mathbf{X}'\mathbf{y}. \tag{6.1}$$

And, in light of result (4.35), it is also expressible as

$$\boldsymbol{\ell}(\mathbf{y}) = \tilde{\mathbf{R}}'\mathbf{X}'\mathbf{y}, \tag{6.2}$$

where $\tilde{\mathbf{R}}$ is any solution to the linear system $\mathbf{X}'\mathbf{X}\mathbf{R} = \boldsymbol{\Lambda}$ (in the $P \times M$ matrix $\mathbf{R}$).

We have that

$$\text{E}[\boldsymbol{\ell}(\mathbf{y})] = \boldsymbol{\Lambda}'\boldsymbol{\beta}, \tag{6.3}$$

as is evident upon taking the expected value of expression (6.2) or, alternatively, upon using result (4.36) to establish that each element of $\boldsymbol{\Lambda}'\boldsymbol{\beta}$ equals the corresponding element of $\text{E}[\boldsymbol{\ell}(\mathbf{y})]$.

The least squares estimators $\ell_1(\mathbf{y}), \ell_2(\mathbf{y}), \ldots, \ell_M(\mathbf{y})$ of $\boldsymbol{\lambda}_1'\boldsymbol{\beta}, \boldsymbol{\lambda}_2'\boldsymbol{\beta}, \ldots, \boldsymbol{\lambda}_M'\boldsymbol{\beta}$ have the following basic property: the least squares estimator of $\sum_{j=1}^{M} k_j \boldsymbol{\lambda}_j'\boldsymbol{\beta}$ (where $k_1, k_2, \ldots, k_M$ are arbitrary constants) is $\sum_{j=1}^{M} k_j \ell_j(\mathbf{y})$—recall (from Section 5.3) that linear combinations of estimable functions are estimable. Upon letting $\mathbf{k} = (k_1, k_2, \ldots, k_M)'$, this property can be restated (in matrix notation) as follows: the least squares estimator of $\mathbf{k}'\boldsymbol{\Lambda}'\boldsymbol{\beta}$ $[= (\boldsymbol{\Lambda}\mathbf{k})'\boldsymbol{\beta}]$ is $\mathbf{k}'\boldsymbol{\ell}(\mathbf{y})$. In light of results (4.24) and (6.1), this property can be readily verified by observing that the least squares estimator of $(\boldsymbol{\Lambda}\mathbf{k})'\boldsymbol{\beta}$ is

$$(\boldsymbol{\Lambda}\mathbf{k})'(\mathbf{X}'\mathbf{X})^{-}\mathbf{X}'\mathbf{y} = \mathbf{k}'\boldsymbol{\Lambda}'(\mathbf{X}'\mathbf{X})^{-}\mathbf{X}'\mathbf{y} = \mathbf{k}'\boldsymbol{\ell}(\mathbf{y}).$$

Alternatively, in light of results (4.35) and (6.2), it can be verified by observing that (for any solution $\tilde{\mathbf{R}}$ to $\mathbf{X}'\mathbf{X}\mathbf{R} = \boldsymbol{\Lambda}$) $\tilde{\mathbf{R}}\mathbf{k}$ is a solution to the linear system $\mathbf{X}'\mathbf{X}\mathbf{r} = \boldsymbol{\Lambda}\mathbf{k}$ and hence that the least squares estimator of $(\boldsymbol{\Lambda}\mathbf{k})'\boldsymbol{\beta}$ is

$$(\tilde{\mathbf{R}}\mathbf{k})'\mathbf{X}'\mathbf{y} = \mathbf{k}'\tilde{\mathbf{R}}'\mathbf{X}'\mathbf{y} = \mathbf{k}'\boldsymbol{\ell}(\mathbf{y}).$$

Under the general linear model, the variance-covariance matrix of the least squares estimator of $\boldsymbol{\Lambda}'\boldsymbol{\beta}$ is expressible as

$$\text{var}[\boldsymbol{\ell}(\mathbf{y})] = \tilde{\mathbf{R}}'\mathbf{X}'\mathbf{V}(\boldsymbol{\theta})\mathbf{X}\tilde{\mathbf{R}} = \boldsymbol{\Lambda}'(\mathbf{X}'\mathbf{X})^{-}\mathbf{X}'\mathbf{V}(\boldsymbol{\theta})\mathbf{X}(\mathbf{X}'\mathbf{X})^{-}\boldsymbol{\Lambda} \tag{6.4}$$

(where $\tilde{\mathbf{R}}$ is any solution to $\mathbf{X}'\mathbf{X}\mathbf{R} = \boldsymbol{\Lambda}$). Result (6.4) can be deduced from result (4.39): start with the expressions for $\text{var}[\ell_i(\mathbf{y}),\ \ell_j(\mathbf{y})]$ obtained by applying result (4.39), and then observe that these expressions are essentially the same as the ijth elements of the expressions for $\text{var}[\boldsymbol{\ell}(\mathbf{y})]$ given by result (6.4) $(i, j = 1, 2, \ldots, M)$. In the special case of the Aitken model, result (6.4) "simplifies" to

$$\text{var}[\boldsymbol{\ell}(\mathbf{y})] = \sigma^2\tilde{\mathbf{R}}'\mathbf{X}'\mathbf{H}\mathbf{X}\tilde{\mathbf{R}} = \sigma^2\boldsymbol{\Lambda}'(\mathbf{X}'\mathbf{X})^{-}\mathbf{X}'\mathbf{H}\mathbf{X}(\mathbf{X}'\mathbf{X})^{-}\boldsymbol{\Lambda}. \tag{6.5}$$

And in the further special case of the G–M model, we find that

$$\text{var}[\boldsymbol{\ell}(\mathbf{y})] = \sigma^2\tilde{\mathbf{R}}'\mathbf{X}'\mathbf{X}\tilde{\mathbf{R}} = \sigma^2\tilde{\mathbf{R}}'\boldsymbol{\Lambda} = \sigma^2\boldsymbol{\Lambda}'\tilde{\mathbf{R}} \tag{6.6}$$

$$= \sigma^2\tilde{\mathbf{R}}'\mathbf{X}'\mathbf{X}(\mathbf{X}'\mathbf{X})^{-}\mathbf{X}'\mathbf{X}\tilde{\mathbf{R}} = \sigma^2\boldsymbol{\Lambda}'(\mathbf{X}'\mathbf{X})^{-}\boldsymbol{\Lambda}. \tag{6.7}$$

a. Best linear unbiased estimation

Let us consider further the estimation of $\boldsymbol{\Lambda}'\boldsymbol{\beta}$ (based on an $N \times 1$ observable random vector $\mathbf{y}$ that follows a G–M, Aitken, or general linear model). An estimator of $\boldsymbol{\Lambda}'\boldsymbol{\beta}$ is said to be a linear estimator if each of its elements is a linear estimator (of the corresponding element of $\boldsymbol{\Lambda}'\boldsymbol{\beta}$) or, equivalently, if it is expressible in the form $\mathbf{c} + \mathbf{A}'\mathbf{y}$ (where $\mathbf{c}$ is an $M \times 1$ vector of constants and $\mathbf{A}$ an $N \times M$ matrix of constants). An estimator $\mathbf{t}(\mathbf{y})$ of $\boldsymbol{\Lambda}'\boldsymbol{\beta}$ is said to be unbiased if each of its elements is an unbiased estimator of the corresponding element of $\boldsymbol{\Lambda}'\boldsymbol{\beta}$ or, equivalently, if $\text{E}[\mathbf{t}(\mathbf{y})] = \boldsymbol{\Lambda}'\boldsymbol{\beta}$. It follows from the results of Section 5.1 that for a linear estimator $\mathbf{c} + \mathbf{A}'\mathbf{y}$ of $\boldsymbol{\Lambda}'\boldsymbol{\beta}$ to be an unbiased estimator, it is necessary and sufficient that

$$\mathbf{c} = \mathbf{0} \quad \text{and} \quad \mathbf{A}'\mathbf{X} = \boldsymbol{\Lambda}'. \tag{6.8}$$

Moreover, it follows from what was established earlier (e.g., in Section 5.5) that the least squares estimator of $\boldsymbol{\Lambda}'\boldsymbol{\beta}$ is linear and unbiased.

How does the variance-covariance matrix of the least squares estimator of $\boldsymbol{\Lambda}'\boldsymbol{\beta}$ compare with the variance-covariance matrix of other linear unbiased estimators of $\boldsymbol{\Lambda}'\boldsymbol{\beta}$? The Gauss–Markov theorem implies that in the special case where $\mathbf{y}$ follows a G–M model, at least some of the diagonal elements of the variance-covariance matrix of the least squares estimator are (strictly) less than (and the others are equal to) the corresponding diagonal elements of the variance-covariance matrix of any other linear unbiased estimator. The following theorem makes a stronger statement.

Theorem 5.6.1. Suppose that $\mathbf{y}$ is an $N \times 1$ observable random vector that follows a G–M, Aitken, or general linear model, take $\boldsymbol{\Lambda}'\boldsymbol{\beta}$ to be any $M \times 1$ vector of estimable linear combinations of the elements of the parametric vector $\boldsymbol{\beta}$, denote by $\boldsymbol{\ell}(\mathbf{y})$ the least squares estimator of $\boldsymbol{\Lambda}'\boldsymbol{\beta}$, and let $\mathbf{c} + \mathbf{A}'\mathbf{y}$ represent an arbitrary linear unbiased estimator of $\boldsymbol{\Lambda}'\boldsymbol{\beta}$ (or, equivalently, any estimator of the form

$\mathbf{c}+\mathbf{A}'\mathbf{y}$ such that $\mathbf{c} = \mathbf{0}$ and $\mathbf{A}'\mathbf{X} = \mathbf{\Lambda}'$). Then, (the least squares estimator) $\boldsymbol{\ell}(\mathbf{y})$ is a linear unbiased estimator. Moreover, in the special case where $\mathbf{y}$ follows a G–M model, $\text{var}(\mathbf{c} + \mathbf{A}'\mathbf{y}) - \text{var}[\boldsymbol{\ell}(\mathbf{y})]$ is a nonnegative definite matrix, and $\text{var}(\mathbf{c} + \mathbf{A}'\mathbf{y}) - \text{var}[\boldsymbol{\ell}(\mathbf{y})] = \mathbf{0}$ or, equivalently, $\text{var}(\mathbf{c} + \mathbf{A}'\mathbf{y}) = \text{var}[\boldsymbol{\ell}(\mathbf{y})]$ if and only if $\mathbf{c} + \mathbf{A}'\mathbf{y} = \boldsymbol{\ell}(\mathbf{y})$.

Proof. That $\boldsymbol{\ell}(\mathbf{y})$ is a linear unbiased estimator of $\mathbf{\Lambda}'\boldsymbol{\beta}$ follows from what was established in Section 5.5 (as was noted previously). Now, suppose that $\mathbf{y}$ follows a G–M model, and consider the quadratic form

$$\mathbf{k}'\{\text{var}(\mathbf{c} + \mathbf{A}'\mathbf{y}) - \text{var}[\boldsymbol{\ell}(\mathbf{y})]\}\mathbf{k} \tag{6.9}$$

(in an M-dimensional column vector $\mathbf{k}$). Clearly, the quadratic form (6.9) is reexpressible as

$$\mathbf{k}'\{\text{var}(\mathbf{c} + \mathbf{A}'\mathbf{y}) - \text{var}[\boldsymbol{\ell}(\mathbf{y})]\}\mathbf{k} = \text{var}[\mathbf{k}'\mathbf{c} + (\mathbf{A}\mathbf{k})'\mathbf{y}] - \text{var}[\mathbf{k}'\boldsymbol{\ell}(\mathbf{y})]. \tag{6.10}$$

Moreover, $\mathbf{k}'\mathbf{c} + (\mathbf{A}\mathbf{k})'\mathbf{y}$ is a linear unbiased estimator of $(\mathbf{\Lambda}\mathbf{k})'\boldsymbol{\beta}$; the unbiasedness of which can be verified simply by observing that $\text{E}[\mathbf{k}'\mathbf{c} + (\mathbf{A}\mathbf{k})'\mathbf{y}] = \mathbf{k}'\,\text{E}(\mathbf{c} + \mathbf{A}'\mathbf{y}) = \mathbf{k}'\mathbf{\Lambda}'\boldsymbol{\beta} = (\mathbf{\Lambda}\mathbf{k})'\boldsymbol{\beta}$ or, alternatively, by observing [in light of the sufficiency of condition (1.4)] that $\mathbf{k}'\mathbf{c} = \mathbf{k}'\mathbf{0} = 0$ and that $(\mathbf{A}\mathbf{k})'\mathbf{X} = \mathbf{k}'\mathbf{A}'\mathbf{X} = \mathbf{k}'\mathbf{\Lambda}' = (\mathbf{\Lambda}\mathbf{k})'$. And as discussed in the introductory part of the present section, $\mathbf{k}'\boldsymbol{\ell}(\mathbf{y})$ is the least squares estimator of $(\mathbf{\Lambda}\mathbf{k})'\boldsymbol{\beta}$. Thus, it follows from the Gauss–Markov theorem that $\text{var}[\mathbf{k}'\boldsymbol{\ell}(\mathbf{y})] \le \text{var}[\mathbf{k}'\mathbf{c} + (\mathbf{A}\mathbf{k})'\mathbf{y}]$ or, equivalently, that

$$\text{var}[\mathbf{k}'\mathbf{c} + (\mathbf{A}\mathbf{k})'\mathbf{y}] - \text{var}[\mathbf{k}'\boldsymbol{\ell}(\mathbf{y})] \ge 0. \tag{6.11}$$

Together, results (6.10) and (6.11) imply that the quadratic form $\mathbf{k}'\{\text{var}(\mathbf{c} + \mathbf{A}'\mathbf{y}) - \text{var}[\boldsymbol{\ell}(\mathbf{y})]\}\mathbf{k}$ is nonnegative definite and hence that the matrix $\text{var}(\mathbf{c} + \mathbf{A}'\mathbf{y}) - \text{var}[\boldsymbol{\ell}(\mathbf{y})]$ is nonnegative definite.

As a further implication of the Gauss–Markov theorem, we have that $\text{var}[\mathbf{k}'\mathbf{c} + (\mathbf{A}\mathbf{k})'\mathbf{y}] = \text{var}[\mathbf{k}'\boldsymbol{\ell}(\mathbf{y})]$ or, equivalently, that equality holds in inequality (6.11) if and only if $\mathbf{k}'\mathbf{c} + (\mathbf{A}\mathbf{k})'\mathbf{y} = \mathbf{k}'\boldsymbol{\ell}(\mathbf{y})$, leading [in light of equality (6.10) and Corollary 2.13.4] to the conclusion that $\text{var}(\mathbf{c} + \mathbf{A}'\mathbf{y}) - \text{var}[\boldsymbol{\ell}(\mathbf{y})] = \mathbf{0}$ if and only if $\mathbf{k}'\mathbf{c} + (\mathbf{A}\mathbf{k})'\mathbf{y} = \mathbf{k}'\boldsymbol{\ell}(\mathbf{y})$ for every $\mathbf{k}$ and hence if and only if $\mathbf{c} + \mathbf{A}'\mathbf{y} = \boldsymbol{\ell}(\mathbf{y})$. Q.E.D.

Suppose (in connection with Theorem 5.6.1) that $\mathbf{y}$ follows a G–M model, in which case $\text{var}(\mathbf{c} + \mathbf{A}'\mathbf{y}) - \text{var}[\boldsymbol{\ell}(\mathbf{y})]$ is nonnegative definite. Then, $\text{var}(\mathbf{c} + \mathbf{A}'\mathbf{y}) - \text{var}[\boldsymbol{\ell}(\mathbf{y})] = \mathbf{R}'\mathbf{R}$ for some matrix $\mathbf{R}$ (as is evident from Corollary 2.13.25). And upon recalling Lemma 2.3.2 and observing that $\mathbf{R}'\mathbf{R}$ equals $\mathbf{0}$ if and only if all M of its diagonal elements equal 0 and upon letting (for $j = 1, 2, \ldots, M$) c_j represent the jth element of $\mathbf{c}$, $\mathbf{a}_j$ the jth column of $\mathbf{A}$, and $\ell_j(\mathbf{y})$ the jth element of $\boldsymbol{\ell}(\mathbf{y})$, it follows that

$$\begin{aligned}\text{var}(\mathbf{c} + \mathbf{A}'\mathbf{y}) - \text{var}[\boldsymbol{\ell}(\mathbf{y})] = \mathbf{0}&\\ &\Leftrightarrow\ \text{var}(c_j + \mathbf{a}_j'\mathbf{y}) - \text{var}[\ell_j(\mathbf{y})] = 0 \ \ (j = 1, 2, \ldots, M)\\ &\Leftrightarrow\ \text{tr}\{\text{var}(\mathbf{c} + \mathbf{A}'\mathbf{y}) - \text{var}[\boldsymbol{\ell}(\mathbf{y})]\} = 0\end{aligned}$$

or, equivalently, that

$$\begin{aligned}\text{var}(\mathbf{c} + \mathbf{A}'\mathbf{y}) = \text{var}[\boldsymbol{\ell}(\mathbf{y})] \ &\Leftrightarrow\ \text{var}(c_j + \mathbf{a}_j'\mathbf{y}) = \text{var}[\ell_j(\mathbf{y})] \ \ (j = 1, 2, \ldots, M)\\ &\Leftrightarrow\ \text{tr}[\text{var}(\mathbf{c} + \mathbf{A}'\mathbf{y})] = \text{tr}\{\text{var}[\boldsymbol{\ell}(\mathbf{y})]\}.\end{aligned}$$

Because the diagonal elements of a nonnegative definite matrix are inherently nonnegative (as evidenced by Corollary 2.13.14), the following result can be regarded as a corollary of Theorem 5.6.1.

Corollary 5.6.2. Suppose that $\mathbf{y}$ is an $N \times 1$ observable random vector that follows a G–M model, take $\mathbf{\Lambda}'\boldsymbol{\beta}$ to be any $M \times 1$ vector of estimable linear combinations of the elements of the parametric vector $\boldsymbol{\beta}$, denote by $\boldsymbol{\ell}(\mathbf{y})$ the least squares estimator of $\mathbf{\Lambda}'\boldsymbol{\beta}$, and let $\mathbf{c} + \mathbf{A}'\mathbf{y}$ represent an arbitrary linear unbiased estimator of $\mathbf{\Lambda}'\boldsymbol{\beta}$. Then,

$$\text{tr}\{\text{var}[\boldsymbol{\ell}(\mathbf{y})]\} \le \text{tr}[\text{var}(\mathbf{c} + \mathbf{A}'\mathbf{y})],$$

with equality holding if and only if $\mathbf{c} + \mathbf{A}'\mathbf{y} = \boldsymbol{\ell}(\mathbf{y})$.

Alternatively, Corollary 5.6.2 can be established as an almost immediate consequence of the Gauss–Markov theorem. A more substantial implication of Theorem 5.6.1 is provided by the following corollary.

Corollary 5.6.3. Suppose that $\mathbf{y}$ is an $N \times 1$ observable random vector that follows a G–M model, take $\boldsymbol{\Lambda}'\boldsymbol{\beta}$ to be any $M \times 1$ vector of estimable linear combinations of the elements of the parametric vector $\boldsymbol{\beta}$, denote by $\boldsymbol{\ell}(\mathbf{y})$ the least squares estimator of $\boldsymbol{\Lambda}'\boldsymbol{\beta}$, and let $\mathbf{c} + \mathbf{A}'\mathbf{y}$ represent an arbitrary linear unbiased estimator of $\boldsymbol{\Lambda}'\boldsymbol{\beta}$. Then,

$$\det\{\mathrm{var}[\boldsymbol{\ell}(\mathbf{y})]\} \le \det[\mathrm{var}(\mathbf{c} + \mathbf{A}'\mathbf{y})],$$

with equality holding if and only if $\mathrm{rank}(\boldsymbol{\Lambda}) < M$ (in which case both sides of the inequality equal 0) or $\mathbf{c} + \mathbf{A}'\mathbf{y} = \boldsymbol{\ell}(\mathbf{y})$.

Corollary 5.6.3 can be derived from Theorem 5.6.1 by applying the following result on determinants: for any $M \times M$ symmetric nonnegative definite matrix $\mathbf{B}$ and for any $M \times M$ symmetric matrix $\mathbf{C}$ such that $\mathbf{C} - \mathbf{B}$ is nonnegative definite, $|\mathbf{C}| \ge |\mathbf{B}|$, with equality holding if and only if $\mathbf{C}$ is singular or $\mathbf{C} = \mathbf{B}$—for a proof of this result, refer, e.g., to Harville (1997, corollary 18.1.8). Specifically, the application is that obtained by setting $\mathbf{B} = \mathrm{var}[\boldsymbol{\ell}(\mathbf{y})]$ and $\mathbf{C} = \mathrm{var}(\mathbf{c} + \mathbf{A}'\mathbf{y})$.

The determinant of the variance-covariance matrix of a vector-valued estimator is sometimes referred to as the generalized variance of the estimator. Accordingly, the result of Corollary 5.6.3 implies that (under the G–M model) the least squares estimator of the M-dimensional vector $\boldsymbol{\Lambda}'\boldsymbol{\beta}$ is a best linear unbiased estimator in the sense that its generalized variance is less than or equal to that of any other linear unbiased estimator—if $\mathrm{rank}(\boldsymbol{\Lambda}) = M$, the generalized variance of the least squares estimator is (strictly) less than that of any other linear unbiased estimator.

b. Best linear translation-equivariant estimation

Let us continue to consider the estimation of $\boldsymbol{\Lambda}'\boldsymbol{\beta}$ (based on an $N \times 1$ observable random vector $\mathbf{y}$ that follows a G–M, Aitken, or general linear model). An estimator, say $\mathbf{t}(\mathbf{y})$, of $\boldsymbol{\Lambda}'\boldsymbol{\beta}$ is said to be translation equivariant if the elements of $\mathbf{t}(\mathbf{y})$ are translation-equivariant estimators of the corresponding elements of $\boldsymbol{\Lambda}'\boldsymbol{\beta}$. Thus, in light of the discussion of Section 5.2, a necessary and sufficient condition for $\mathbf{t}(\mathbf{y})$ to be a translation-equivariant estimator of $\boldsymbol{\Lambda}'\boldsymbol{\beta}$ is that

$$\mathbf{t}(\mathbf{y}) + \boldsymbol{\Lambda}'\mathbf{k} = \mathbf{t}(\mathbf{y} + \mathbf{X}\mathbf{k}) \tag{6.12}$$

for every $P \times 1$ vector $\mathbf{k}$ (and for every value of $\mathbf{y}$). Further, for a linear estimator $\mathbf{c} + \mathbf{A}'\mathbf{y}$ to be a translation-equivariant estimator of $\boldsymbol{\Lambda}'\boldsymbol{\beta}$, it is necessary and sufficient that

$$\mathbf{A}'\mathbf{X} = \boldsymbol{\Lambda}'. \tag{6.13}$$

Moreover, it follows from what was established earlier (in Section 5.5) that the least squares estimator of $\boldsymbol{\Lambda}'\boldsymbol{\beta}$ is translation equivariant.

As what can be regarded as an additional corollary of Theorem 5.6.1, we have the following result.

Corollary 5.6.4. Suppose that $\mathbf{y}$ is an $N \times 1$ observable random vector that follows a G–M, Aitken, or general linear model, take $\boldsymbol{\Lambda}'\boldsymbol{\beta}$ to be any $M \times 1$ vector of estimable linear combinations of the elements of the parametric vector $\boldsymbol{\beta}$, denote by $\boldsymbol{\ell}(\mathbf{y})$ the least squares estimator of $\boldsymbol{\Lambda}'\boldsymbol{\beta}$, and let $\mathbf{c} + \mathbf{A}'\mathbf{y}$ represent an arbitrary linear translation-equivariant estimator of $\boldsymbol{\Lambda}'\boldsymbol{\beta}$ (or, equivalently, any estimator of the form $\mathbf{c} + \mathbf{A}'\mathbf{y}$ such that $\mathbf{A}'\mathbf{X} = \boldsymbol{\Lambda}'$). Then, (the least squares estimator) $\boldsymbol{\ell}(\mathbf{y})$ is a linear translation-equivariant estimator. Moreover, in the special case where $\mathbf{y}$ follows a G–M model, the difference

$$\mathrm{E}[(\mathbf{c}+\mathbf{A}'\mathbf{y}-\boldsymbol{\Lambda}'\boldsymbol{\beta})(\mathbf{c}+\mathbf{A}'\mathbf{y}-\boldsymbol{\Lambda}'\boldsymbol{\beta})'] - \mathrm{E}\{[\boldsymbol{\ell}(\mathbf{y})-\boldsymbol{\Lambda}'\boldsymbol{\beta}][\boldsymbol{\ell}(\mathbf{y})-\boldsymbol{\Lambda}'\boldsymbol{\beta}]'\} \tag{6.14}$$

between the mean-squared-error matrices of $\mathbf{c} + \mathbf{A}'\mathbf{y}$ and $\boldsymbol{\ell}(\mathbf{y})$ is a nonnegative definite matrix, and this difference equals $\mathbf{0}$ if and only if $\mathbf{c} + \mathbf{A}'\mathbf{y} = \boldsymbol{\ell}(\mathbf{y})$.

The nonnegative definiteness of the matrix (6.14) and the condition $[\mathbf{c} + \mathbf{A}'\mathbf{y} = \boldsymbol{\ell}(\mathbf{y})]$ under which it equals $\mathbf{0}$ follow from Theorem 5.6.1 in much the same way that the main part of Corollary 5.5.2 follows from the Gauss–Markov theorem.

5.7 Estimation of Variability and Covariability

Suppose that $\mathbf{y} = (y_1, y_2, \ldots, y_N)'$ is an $N \times 1$ observable random vector that follows a G–M, Aitken, or general linear model. Then, the N diagonal elements $\text{var}(y_1), \text{var}(y_2), \ldots, \text{var}(y_N)$ of the matrix $\text{var}(\mathbf{y})$ represent the underlying variability and the $N(N-1)$ off-diagonal elements $\text{cov}(y_i, y_j)$ $(j \neq i = 1, 2, \ldots, N)$ represent the underlying covariability. In the case of the G–M model, $\text{var}(\mathbf{y}) = \sigma^2\mathbf{I}$, so that $y_1, y_2, \ldots, y_N$ are uncorrelated with a common (strictly) positive variance σ^2 (of unknown value). In the case of the Aitken model, $\text{var}(\mathbf{y}) = \sigma^2\mathbf{H}$ (where $\mathbf{H}$ is a known symmetric nonnegative definite matrix), so that the variances and covariances of $y_1, y_2, \ldots, y_N$ are known up to the (unknown) value of a (strictly) positive scalar multiple σ^2. And, in the case of the general linear model, $\text{var}(\mathbf{y}) = \mathbf{V}(\boldsymbol{\theta})$ [where $\mathbf{V}(\boldsymbol{\theta})$ is a symmetric nonnegative definite matrix whose elements are known functions of a $T \times 1$ parameter vector $\boldsymbol{\theta}$], so that the variances and covariances of $y_1, y_2, \ldots, y_N$ are known up to the (unknown) value of $\boldsymbol{\theta}$.

The matrix $\text{var}(\mathbf{y})$ is of interest because its value determines the variances and covariances of the least squares estimators of estimable linear combinations of the elements of the parametric vector $\boldsymbol{\beta}$. Moreover, the underlying variability and covariability [represented by the diagonal and off-diagonal elements of $\text{var}(\mathbf{y})$] may be of interest in their own right.

In the present section, some initial results are obtained on the estimation of variability and covariability. The emphasis is on results that are specific to the estimation of σ^2 under the G–M model. As a preliminary, formulas are derived for the expected values and variances and covariances of quadratic forms (in a random vector) and for the covariance of a quadratic form and a linear form. And, prior to that, some matrix operations that enter in various of those formulas are introduced and briefly discussed.

a. Some matrix operations

Vec of a matrix. Let $\mathbf{A}$ represent an $M \times N$ matrix. It is sometimes convenient to rearrange the elements of $\mathbf{A}$ in the form of an MN-dimensional column vector. The conventional way of doing so is to successively place the first, second, ..., Nth columns $\mathbf{a}_1, \mathbf{a}_2, \ldots, \mathbf{a}_N$ of $\mathbf{A}$ one under the other, giving the column vector

$$\begin{pmatrix} \mathbf{a}_1 \\ \mathbf{a}_2 \\ \vdots \\ \mathbf{a}_N \end{pmatrix}. \tag{7.1}$$

The vector (7.1) is referred to as the *vec* of $\mathbf{A}$, and is denoted by the symbol $\text{vec}(\mathbf{A})$ or (when the parentheses are not needed for clarity) by $\text{vec}\,\mathbf{A}$. By definition, the ijth element of $\mathbf{A}$ is the $[(j-1)M+i]$th element of $\text{vec}\,\mathbf{A}$.

Vech of a symmetric matrix. Let $\mathbf{A} = \{a_{ij}\}$ represent an $N \times N$ symmetric matrix. The values of all N^2 elements of $\mathbf{A}$ can be determined from the values of those $N(N+1)/2$ elements that are on or below the diagonal [or, alternatively, from those $N(N+1)/2$ elements that are on or above the diagonal]. Accordingly, in rearranging the elements of $\mathbf{A}$ in the form of a vector (as in forming the vec), we may wish to exclude the $N(N-1)/2$ "duplicate" elements. Thus, as an alternative to the vec of $\mathbf{A}$, we may wish to consider the $N(N+1)/2$-dimensional column vector

$$\begin{pmatrix} \mathbf{a}_1^* \\ \mathbf{a}_2^* \\ \vdots \\ \mathbf{a}_N^* \end{pmatrix}, \tag{7.2}$$

where (for $i = 1, 2, \ldots, N$) $\mathbf{a}_i^* = (a_{ii}, a_{i+1,i}, \ldots, a_{Ni})'$ is the subvector of the ith column of $\mathbf{A}$ obtained by striking out its first $i - 1$ elements. The vector (7.2) is referred to as the *vech* of $\mathbf{A}$ and is denoted by the symbol vech($\mathbf{A}$) or vech $\mathbf{A}$. For $N = 1$, $N = 2$, and $N = 3$,

$$\text{vech}\,\mathbf{A} = (a_{11}), \quad \text{vech}\,\mathbf{A} = \begin{pmatrix} a_{11} \\ a_{21} \\ a_{22} \end{pmatrix}, \text{ and vech}\,\mathbf{A} = \begin{pmatrix} a_{11} \\ a_{21} \\ a_{31} \\ a_{22} \\ a_{32} \\ a_{33} \end{pmatrix}, \text{ respectively.}$$

Every element of $\mathbf{A}$, and hence every element of vec $\mathbf{A}$, is either an element of vech $\mathbf{A}$ or a "duplicate" of an element of vech $\mathbf{A}$. Thus, there exists a unique $[N^2 \times N(N+1)/2]$-dimensional matrix, to be denoted by the symbol $\mathbf{G}_N$, such that (for every $N \times N$ symmetric matrix $\mathbf{A}$)

$$\text{vec}\,\mathbf{A} = \mathbf{G}_N\,\text{vech}\,\mathbf{A}.$$

This matrix is called the *duplication matrix*. Clearly,

$$\mathbf{G}_1 = (1), \quad \mathbf{G}_2 = \begin{pmatrix} 1 & 0 & 0 \\ 0 & 1 & 0 \\ 0 & 1 & 0 \\ 0 & 0 & 1 \end{pmatrix}, \text{ and } \mathbf{G}_3 = \begin{pmatrix} 1 & 0 & 0 & 0 & 0 & 0 \\ 0 & 1 & 0 & 0 & 0 & 0 \\ 0 & 0 & 1 & 0 & 0 & 0 \\ 0 & 1 & 0 & 0 & 0 & 0 \\ 0 & 0 & 0 & 1 & 0 & 0 \\ 0 & 0 & 0 & 0 & 1 & 0 \\ 0 & 0 & 1 & 0 & 0 & 0 \\ 0 & 0 & 0 & 0 & 1 & 0 \\ 0 & 0 & 0 & 0 & 0 & 1 \end{pmatrix}.$$

Note that

$$\text{rank}\,\mathbf{G}_N = N(N+1)/2 \tag{7.3}$$

(so that $\mathbf{G}_N$ is of full column rank), as is evident upon observing that every row of $\mathbf{I}_{N(N+1)/2}$ is a row of $\mathbf{G}_N$ and hence that $\mathbf{G}_N$ contains $N(N+1)/2$ linearly independent rows.

Kronecker product. The *Kronecker product* of two matrices, say an $M \times N$ matrix $\mathbf{A} = \{a_{ij}\}$ and a $P \times Q$ matrix $\mathbf{B} = \{b_{ij}\}$, is denoted by the symbol $\mathbf{A} \otimes \mathbf{B}$ and is defined to be the $MP \times NQ$ matrix

$$\mathbf{A} \otimes \mathbf{B} = \begin{pmatrix} a_{11}\mathbf{B} & a_{12}\mathbf{B} & \ldots & a_{1N}\mathbf{B} \\ a_{21}\mathbf{B} & a_{22}\mathbf{B} & \ldots & a_{2N}\mathbf{B} \\ \vdots & \vdots & & \vdots \\ a_{M1}\mathbf{B} & a_{M2}\mathbf{B} & \ldots & a_{MN}\mathbf{B} \end{pmatrix}$$

obtained by replacing (for $i = 1, 2, \ldots, M$ and $j = 1, 2, \ldots, N$) the ij element of $\mathbf{A}$ with the $P \times Q$ matrix $a_{ij}\mathbf{B}$. Thus, the Kronecker product of $\mathbf{A}$ and $\mathbf{B}$ can be regarded as a partitioned matrix, comprising M rows and N columns of $(P \times Q)$-dimensional blocks, the ijth of which is $a_{ij}\mathbf{B}$.

Among the various properties of the Kronecker product operation is the following: for any matrices $\mathbf{A}$ and $\mathbf{B}$,

$$(\mathbf{A} \otimes \mathbf{B})' = \mathbf{A}' \otimes \mathbf{B}' \tag{7.4}$$

—for a verification of equality (7.4), refer, e.g., to Harville (1997, sec. 16.1).

Two formulas. There are two formulas that will be convenient to have at our disposal; one of these is for the vec of a product of three matrices and the other is for the trace of the product of four matrices. The two formulas are as follows. For any $M \times N$ matrix $\mathbf{A}$, $N \times P$ matrix $\mathbf{B}$, and $P \times Q$ matrix $\mathbf{C}$,

$$\text{vec}\, \mathbf{ABC} = (\mathbf{C}' \otimes \mathbf{A})\, \text{vec}\, \mathbf{B}. \tag{7.5}$$

And for any $M \times N$ matrix $\mathbf{A}$, $M \times P$ matrix $\mathbf{B}$, $P \times Q$ matrix $\mathbf{C}$, and $N \times Q$ matrix $\mathbf{D}$,

$$\text{tr}(\mathbf{A}'\mathbf{BCD}') = (\text{vec}\, \mathbf{A})'(\mathbf{D} \otimes \mathbf{B})\, \text{vec}\, \mathbf{C}. \tag{7.6}$$

For a derivation of formulas (7.5) and (7.6), refer, for example, to Harville (1997, sec. 16.2).

b. Expected values and variances of quadratic forms (and their covariances with each other and with linear forms)

Suppose that $\mathbf{x}$ is an N-dimensional random column vector. Then, it is customary to refer to a linear combination, say $\mathbf{a}'\mathbf{x}$, of the elements of $\mathbf{x}$ (where $\mathbf{a}$ is an $N \times 1$ vector of constants) as a linear form (in $\mathbf{x}$).

Formulas for the expected values and the variances and covariances of linear forms are available from the results of Sections 3.1 and 3.2. If the random vector $\mathbf{x}$ has a mean vector $\boldsymbol{\mu}$, then the expected value of a linear form $\mathbf{a}'\mathbf{x}$ (in $\mathbf{x}$) is expressible as

$$\text{E}(\mathbf{a}'\mathbf{x}) = \mathbf{a}'\boldsymbol{\mu}. \tag{7.7}$$

And if, in addition, $\mathbf{x}$ has a variance-covariance matrix $\boldsymbol{\Sigma}$, then the variance of $\mathbf{a}'\mathbf{x}$ is expressible as

$$\text{var}(\mathbf{a}'\mathbf{x}) = \mathbf{a}'\boldsymbol{\Sigma}\mathbf{a}, \tag{7.8}$$

and, more generally, the covariance of $\mathbf{a}'\mathbf{x}$ and $\mathbf{b}'\mathbf{x}$ (where $\mathbf{b}'\mathbf{x}$ is a second linear form in $\mathbf{x}$) is expressible as

$$\text{cov}(\mathbf{a}'\mathbf{x}, \mathbf{b}'\mathbf{x}) = \mathbf{a}'\boldsymbol{\Sigma}\mathbf{b}. \tag{7.9}$$

In what follows, these results are extended by obtaining formulas for the expected values and variances of quadratic forms (in a random column vector) and formulas for the covariances of the quadratic forms with each other and with linear forms.

Main results. The main results are presented in a series of three theorems.

Theorem 5.7.1. Let $\mathbf{x}$ represent an N-dimensional random column vector having mean vector $\boldsymbol{\mu} = \{\mu_i\}$ and variance-covariance matrix $\boldsymbol{\Sigma} = \{\sigma_{ij}\}$, and take $\mathbf{A} = \{a_{ij}\}$ to be an $N \times N$ matrix of constants. Then,

$$\text{E}(\mathbf{x}'\mathbf{A}\mathbf{x}) = \sum_{i,j} a_{ij}(\sigma_{ij} + \mu_i\mu_j) \tag{7.10}$$

$$= \text{tr}(\mathbf{A}\boldsymbol{\Sigma}) + \boldsymbol{\mu}'\mathbf{A}\boldsymbol{\mu}. \tag{7.11}$$

Proof. Letting x_i represent the ith element of $\mathbf{x}$ $(i = 1, 2, \ldots, N)$, we find that

$$\begin{aligned}
\mathrm{E}(\mathbf{x}'\mathbf{A}\mathbf{x}) = \mathrm{E}\Big(\sum_{i,j} a_{ij} x_i x_j\Big) &= \sum_{i,j} a_{ij}\,\mathrm{E}(x_i x_j) \\
&= \sum_{i,j} a_{ij}(\sigma_{ij} + \mu_i \mu_j) \\
&= \sum_i \Big(\sum_j a_{ij}\sigma_{ji}\Big) + \sum_{i,j} a_{ij}\mu_i\mu_j \\
&= \mathrm{tr}(\mathbf{A}\boldsymbol{\Sigma}) + \boldsymbol{\mu}'\mathbf{A}\boldsymbol{\mu}.
\end{aligned}$$

Q.E.D.

Theorem 5.7.2. Let $\mathbf{x}$ represent an N-dimensional random column vector having mean vector $\boldsymbol{\mu} = \{\mu_i\}$, variance-covariance matrix $\boldsymbol{\Sigma} = \{\sigma_{ij}\}$, and third central moments $\lambda_{ijk} = \mathrm{E}[(x_i - \mu_i)(x_j - \mu_j)(x_k - \mu_k)]$ $(i, j, k = 1, 2, \ldots, N)$, and take $\mathbf{b} = \{b_i\}$ to be an N-dimensional column vector of constants and $\mathbf{A} = \{a_{ij}\}$ to be an $N \times N$ symmetric matrix of constants. Then,

$$\mathrm{cov}(\mathbf{b}'\mathbf{x},\, \mathbf{x}'\mathbf{A}\mathbf{x}) = \sum_{i,j,k} b_i a_{jk}(\lambda_{ijk} + 2\mu_j\sigma_{ik}) \tag{7.12}$$

$$= \mathbf{b}'\boldsymbol{\Lambda}\,\mathrm{vec}\,\mathbf{A} + 2\mathbf{b}'\boldsymbol{\Sigma}\mathbf{A}\boldsymbol{\mu}, \tag{7.13}$$

where $\boldsymbol{\Lambda}$ is an $N \times N^2$ matrix whose entry for the ith row and jkth column [i.e., column $(j-1)N+k$] is λ_{ijk}.

Proof. Letting $\mathbf{z} = \{z_i\} = \mathbf{x} - \boldsymbol{\mu}$ (in which case $\mathbf{x} = \mathbf{z} + \boldsymbol{\mu}$) and using Theorem 5.7.1 [and observing that $\mathbf{z}'\mathbf{A}\boldsymbol{\mu} = (\mathbf{z}'\mathbf{A}\boldsymbol{\mu})' = \boldsymbol{\mu}'\mathbf{A}\mathbf{z}$ and that $\mathbf{b}'\mathbf{z} = \mathbf{z}'\mathbf{b}$], we find that

$$\begin{aligned}
\mathrm{cov}(\mathbf{b}'\mathbf{x},\, \mathbf{x}'\mathbf{A}\mathbf{x}) &= \mathrm{cov}(\mathbf{b}'\mathbf{z},\, \mathbf{x}'\mathbf{A}\mathbf{x}) = \mathrm{E}[(\mathbf{b}'\mathbf{z})\mathbf{x}'\mathbf{A}\mathbf{x}] \\
&= \mathrm{E}[(\mathbf{b}'\mathbf{z})(\mathbf{z}'\mathbf{A}\mathbf{z} + 2\boldsymbol{\mu}'\mathbf{A}\mathbf{z} + \boldsymbol{\mu}'\mathbf{A}\boldsymbol{\mu})] \\
&= \mathrm{E}\Big[\Big(\sum_i b_i z_i\Big)\Big(\sum_{j,k} a_{jk} z_j z_k\Big)\Big] + 2\,\mathrm{E}[\mathbf{z}'\mathbf{b}\boldsymbol{\mu}'\mathbf{A}\mathbf{z}] + 0 \\
&= \mathrm{E}\Big(\sum_{i,j,k} b_i a_{jk} z_i z_j z_k\Big) + 2\sum_{i,k} b_i\Big(\sum_j \mu_j a_{jk}\Big)\sigma_{ik} \\
&= \sum_{i,j,k} b_i a_{jk}(\lambda_{ijk} + 2\mu_j\sigma_{ik}) \\
&= \sum_{j,k}\Big(\sum_i b_i\lambda_{ijk}\Big)a_{kj} + 2\sum_k\Big(\sum_i b_i\sigma_{ik}\Big)\Big(\sum_j a_{kj}\mu_j\Big) \\
&= \mathbf{b}'\boldsymbol{\Lambda}\,\mathrm{vec}\,\mathbf{A} + 2\mathbf{b}'\boldsymbol{\Sigma}\mathbf{A}\boldsymbol{\mu}.
\end{aligned}$$

Q.E.D.

If the distribution of $\mathbf{x}$ is MVN, then $\boldsymbol{\Lambda} = \mathbf{0}$ (as is evident from the results of Section 3.5n). More generally, if the distribution of $\mathbf{x}$ is symmetric [in the sense that $-(\mathbf{x}-\boldsymbol{\mu}) \sim \mathbf{x} - \boldsymbol{\mu}$], then $\boldsymbol{\Lambda} = \mathbf{0}$ [as is evident upon observing that if the distribution of $\mathbf{x}$ is symmetric, then (for all i, j, and k) $-\lambda_{ijk} = \lambda_{ijk}$]. When $\boldsymbol{\Lambda} = \mathbf{0}$, formula (7.13) simplifies to

$$\mathrm{cov}(\mathbf{b}'\mathbf{x},\, \mathbf{x}'\mathbf{A}\mathbf{x}) = 2\mathbf{b}'\boldsymbol{\Sigma}\mathbf{A}\boldsymbol{\mu}. \tag{7.14}$$

Thus, if the distribution of $\mathbf{x}$ is symmetric and its mean vector is null, then any linear form in $\mathbf{x}$ is uncorrelated with any quadratic form.

Theorem 5.7.3. Let $\mathbf{x}$ represent an N-dimensional random column vector having mean vector $\boldsymbol{\mu} = \{\mu_i\}$, variance-covariance matrix $\boldsymbol{\Sigma} = \{\sigma_{ij}\}$, third central moments $\lambda_{ijk} = \mathrm{E}[(x_i - \mu_i)$

$(x_j-\mu_j)(x_k-\mu_k)]$ $(i,j,k=1,2,\ldots,N)$, and fourth central moments $\gamma_{ijkm}=\mathrm{E}[(x_i-\mu_i)(x_j-\mu_j)(x_k-\mu_k)(x_m-\mu_m)]$ $(i,j,k,m=1,2,\ldots,N)$; and take $\mathbf{A}=\{a_{ij}\}$ and $\mathbf{H}=\{h_{ij}\}$ to be $N\times N$ symmetric matrices of constants. Then,

$$
\begin{aligned}
\operatorname{cov}(\mathbf{x}'\mathbf{A}\mathbf{x},\ \mathbf{x}'\mathbf{H}\mathbf{x})
&= \sum_{i,j,k,m} a_{ij}h_{km}[(\gamma_{ijkm}-\sigma_{ij}\sigma_{km}-\sigma_{ik}\sigma_{jm}-\sigma_{im}\sigma_{jk}) \\
&\qquad + 2\mu_k\lambda_{ijm} + 2\mu_i\lambda_{jkm} + 2\sigma_{ik}\sigma_{jm} + 4\mu_j\mu_k\sigma_{im}] \qquad (7.15)\\
&= (\operatorname{vec}\mathbf{A})'\boldsymbol{\Omega}\operatorname{vec}\mathbf{H} + 2\boldsymbol{\mu}'\mathbf{H}\boldsymbol{\Lambda}\operatorname{vec}\mathbf{A} + 2\boldsymbol{\mu}'\mathbf{A}\boldsymbol{\Lambda}\operatorname{vec}\mathbf{H} \\
&\qquad + 2\operatorname{tr}(\mathbf{A}\boldsymbol{\Sigma}\mathbf{H}\boldsymbol{\Sigma}) + 4\boldsymbol{\mu}'\mathbf{A}\boldsymbol{\Sigma}\mathbf{H}\boldsymbol{\mu}, \qquad (7.16)
\end{aligned}
$$

where $\boldsymbol{\Omega}$ is an $N^2\times N^2$ matrix whose entry for the ijth row [row $(i-1)N+j$] and kmth column [column $(k-1)N+m$] is $\gamma_{ijkm}-\sigma_{ij}\sigma_{km}-\sigma_{ik}\sigma_{jm}-\sigma_{im}\sigma_{jk}$ and where $\boldsymbol{\Lambda}$ is an $N\times N^2$ matrix whose entry for the jth row and kmth column [column $(k-1)N+m$] is λ_{jkm}.

Proof. Letting $\mathbf{z}=\{z_i\}=\mathbf{x}-\boldsymbol{\mu}$ (in which case $\mathbf{x}=\mathbf{z}+\boldsymbol{\mu}$) and using Theorems 5.7.1 and 5.7.2 [and observing that $\mathbf{z}'\mathbf{A}\boldsymbol{\mu}=(\mathbf{z}'\mathbf{A}\boldsymbol{\mu})'=\boldsymbol{\mu}'\mathbf{A}\mathbf{z}$ and similarly that $\mathbf{z}'\mathbf{H}\boldsymbol{\mu}=\boldsymbol{\mu}'\mathbf{H}\mathbf{z}$), we find that

$$
\begin{aligned}
&\operatorname{cov}(\mathbf{x}'\mathbf{A}\mathbf{x},\ \mathbf{x}'\mathbf{H}\mathbf{x}) \\
&\quad = \mathrm{E}[(\mathbf{x}'\mathbf{A}\mathbf{x})(\mathbf{x}'\mathbf{H}\mathbf{x})] - \mathrm{E}(\mathbf{x}'\mathbf{A}\mathbf{x})\,\mathrm{E}(\mathbf{x}'\mathbf{H}\mathbf{x}) \\
&\quad = \mathrm{E}[(\mathbf{z}+\boldsymbol{\mu})'\mathbf{A}(\mathbf{z}+\boldsymbol{\mu})(\mathbf{z}+\boldsymbol{\mu})'\mathbf{H}(\mathbf{z}+\boldsymbol{\mu})] \\
&\qquad\qquad - [\mathrm{E}(\mathbf{z}'\mathbf{A}\mathbf{z}) + \boldsymbol{\mu}'\mathbf{A}\boldsymbol{\mu}][\mathrm{E}(\mathbf{z}'\mathbf{H}\mathbf{z}) + \boldsymbol{\mu}'\mathbf{H}\boldsymbol{\mu}] \\
&\quad = \mathrm{E}[(\mathbf{z}'\mathbf{A}\mathbf{z})(\mathbf{z}'\mathbf{H}\mathbf{z})] + 2\,\mathrm{E}[(\mathbf{z}'\mathbf{A}\mathbf{z})(\boldsymbol{\mu}'\mathbf{H}\mathbf{z})] + 2\,\mathrm{E}[(\boldsymbol{\mu}'\mathbf{A}\mathbf{z})(\mathbf{z}'\mathbf{H}\mathbf{z})] \\
&\qquad + 4\,\mathrm{E}[(\mathbf{z}'\mathbf{A}\boldsymbol{\mu})(\boldsymbol{\mu}'\mathbf{H}\mathbf{z})] + 2\,\mathrm{E}[(\mathbf{z}'\mathbf{A}\boldsymbol{\mu})(\boldsymbol{\mu}'\mathbf{H}\boldsymbol{\mu})] + 2\,\mathrm{E}[(\boldsymbol{\mu}'\mathbf{A}\boldsymbol{\mu})(\boldsymbol{\mu}'\mathbf{H}\mathbf{z})] \\
&\qquad - \mathrm{E}(\mathbf{z}'\mathbf{A}\mathbf{z})\,\mathrm{E}(\mathbf{z}'\mathbf{H}\mathbf{z}) \\
&\quad = \mathrm{E}[(\mathbf{z}'\mathbf{A}\mathbf{z})(\mathbf{z}'\mathbf{H}\mathbf{z})] + 2\operatorname{cov}(\boldsymbol{\mu}'\mathbf{H}\mathbf{z},\ \mathbf{z}'\mathbf{A}\mathbf{z}) + 2\operatorname{cov}(\boldsymbol{\mu}'\mathbf{A}\mathbf{z},\ \mathbf{z}'\mathbf{H}\mathbf{z}) \\
&\qquad\qquad + 4\,\mathrm{E}(\mathbf{z}'\mathbf{A}\boldsymbol{\mu}\boldsymbol{\mu}'\mathbf{H}\mathbf{z}) + 0 + 0 - \mathrm{E}(\mathbf{z}'\mathbf{A}\mathbf{z})\,\mathrm{E}(\mathbf{z}'\mathbf{H}\mathbf{z}) \\
&\quad = \mathrm{E}\Big(\sum_{i,j,k,m} a_{ij}h_{km}z_iz_jz_kz_m\Big) \\
&\qquad + 2\sum_{i,j,m} a_{ij}\Big(\sum_k \mu_kh_{km}\Big)\lambda_{ijm} + 2\sum_{j,k,m} h_{km}\Big(\sum_i \mu_ia_{ij}\Big)\lambda_{jkm} \\
&\qquad + 4\sum_{i,m}\Big(\sum_j a_{ij}\mu_j\Big)\Big(\sum_k h_{km}\mu_k\Big)\sigma_{im} - \sum_{i,j} a_{ij}\sigma_{ij}\sum_{k,m} h_{km}\sigma_{km} \\
&\quad = \sum_{i,j,k,m} a_{ij}h_{km}[\gamma_{ijkm} + 2\mu_k\lambda_{ijm} + 2\mu_i\lambda_{jkm} + 4\mu_j\mu_k\sigma_{im} - \sigma_{ij}\sigma_{km}] \\
&\quad = \sum_{i,j,k,m} a_{ij}h_{km}[(\gamma_{ijkm}-\sigma_{ij}\sigma_{km}-\sigma_{ik}\sigma_{jm}-\sigma_{im}\sigma_{jk}) \\
&\qquad\qquad + 2\mu_k\lambda_{ijm} + 2\mu_i\lambda_{jkm} + 2\sigma_{ik}\sigma_{jm} + 4\mu_j\mu_k\sigma_{im}] \\
&\quad = \sum_{i,j,k,m} a_{ji}h_{mk}(\gamma_{ijkm}-\sigma_{ij}\sigma_{km}-\sigma_{ik}\sigma_{jm}-\sigma_{im}\sigma_{jk}) \\
&\qquad + 2\sum_{i,j}\Big[\sum_m\Big(\sum_k \mu_kh_{km}\Big)\lambda_{mij}\Big]a_{ji} + 2\sum_{k,m}\Big[\sum_j\Big(\sum_i \mu_ia_{ij}\Big)\lambda_{jkm}\Big]h_{mk} \\
&\qquad\qquad + 2\sum_i\sum_m\Big(\sum_j a_{ij}\sigma_{jm}\Big)\Big(\sum_k h_{mk}\sigma_{ki}\Big) \\
&\qquad\qquad + 4\sum_{i,m}\Big(\sum_j \mu_ja_{ji}\Big)\Big(\sum_k h_{mk}\mu_k\Big)\sigma_{im}
\end{aligned}
$$

$$= (\text{vec}\,\mathbf{A})'\boldsymbol{\Omega}\,\text{vec}\,\mathbf{H} + 2\boldsymbol{\mu}'\mathbf{H}\boldsymbol{\Lambda}\,\text{vec}\,\mathbf{A} + 2\boldsymbol{\mu}'\mathbf{A}\boldsymbol{\Lambda}\,\text{vec}\,\mathbf{H} + 2\,\text{tr}(\mathbf{A}\boldsymbol{\Sigma}\mathbf{H}\boldsymbol{\Sigma}) + 4\boldsymbol{\mu}'\mathbf{A}\boldsymbol{\Sigma}\mathbf{H}\boldsymbol{\mu}.$$

Q.E.D.

As a special case of formula (7.16) (that where $\mathbf{H} = \mathbf{A}$), we have that

$$\text{var}(\mathbf{x}'\mathbf{A}\mathbf{x}) = (\text{vec}\,\mathbf{A})'\boldsymbol{\Omega}\,\text{vec}\,\mathbf{A} + 4\boldsymbol{\mu}'\mathbf{A}\boldsymbol{\Lambda}\,\text{vec}\,\mathbf{A} + 2\,\text{tr}[(\mathbf{A}\boldsymbol{\Sigma})^2] + 4\boldsymbol{\mu}'\mathbf{A}\boldsymbol{\Sigma}\mathbf{A}\boldsymbol{\mu}. \tag{7.17}$$

If the distribution of $\mathbf{x}$ is symmetric, then formula (7.16) simplifies to

$$\text{cov}(\mathbf{x}'\mathbf{A}\mathbf{x},\ \mathbf{x}'\mathbf{H}\mathbf{x}) = (\text{vec}\,\mathbf{A})'\boldsymbol{\Omega}\,\text{vec}\,\mathbf{H} + 2\,\text{tr}(\mathbf{A}\boldsymbol{\Sigma}\mathbf{H}\boldsymbol{\Sigma}) + 4\boldsymbol{\mu}'\mathbf{A}\boldsymbol{\Sigma}\mathbf{H}\boldsymbol{\mu}. \tag{7.18}$$

If the distribution of $\mathbf{x}$ is MVN, then it is symmetric and, in addition, it is such that $\boldsymbol{\Omega} = \mathbf{0}$ (as is evident from the results of Section 3.5n), in which case there is a further simplification to

$$\text{cov}(\mathbf{x}'\mathbf{A}\mathbf{x},\ \mathbf{x}'\mathbf{H}\mathbf{x}) = 2\,\text{tr}(\mathbf{A}\boldsymbol{\Sigma}\mathbf{H}\boldsymbol{\Sigma}) + 4\boldsymbol{\mu}'\mathbf{A}\boldsymbol{\Sigma}\mathbf{H}\boldsymbol{\mu}, \tag{7.19}$$

or, in the special case where $\mathbf{H} = \mathbf{A}$, to

$$\text{var}(\mathbf{x}'\mathbf{A}\mathbf{x}) = 2\,\text{tr}[(\mathbf{A}\boldsymbol{\Sigma})^2] + 4\boldsymbol{\mu}'\mathbf{A}\boldsymbol{\Sigma}\mathbf{A}\boldsymbol{\mu}. \tag{7.20}$$

The formulas of Theorems 5.7.2 and 5.7.3 were derived under the assumption that the matrix $\mathbf{A}$ of the quadratic form $\mathbf{x}'\mathbf{A}\mathbf{x}$ is symmetric and (in the case of Theorem 5.7.3) the assumption that the matrix $\mathbf{H}$ of the quadratic form $\mathbf{x}'\mathbf{H}\mathbf{x}$ is symmetric. Note that whether or not $\mathbf{A}$ and/or $\mathbf{H}$ are symmetric, it would be the case that $\mathbf{x}'\mathbf{A}\mathbf{x} = \frac{1}{2}\mathbf{x}'(\mathbf{A}+\mathbf{A}')\mathbf{x}$ and that $\mathbf{x}'\mathbf{H}\mathbf{x} = \frac{1}{2}\mathbf{x}'(\mathbf{H}+\mathbf{H}')\mathbf{x}$. Thus, the formulas of Theorems 5.7.2 and 5.7.3 could be extended to the case where the matrices of the quadratic forms are possibly nonsymmetric simply by substituting $\frac{1}{2}(\mathbf{A}+\mathbf{A}')$ for $\mathbf{A}$ and (in the case of Theorem 5.7.3) $\frac{1}{2}(\mathbf{H}+\mathbf{H}')$ for $\mathbf{H}$.

Some alternative representations. By making use of the vec and vech operations, the expressions provided by the formulas of Theorems 5.7.1, 5.7.2, and 5.7.3 can be recast in ways that are informative about the nature of the dependence of the expressions on the elements of the matrices of the quadratic forms.

An alternative to the matrix expression (7.11) provided by Theorem 5.7.1 for the expected value of the quadratic form $\mathbf{x}'\mathbf{A}\mathbf{x}$ is as follows:

$$\text{E}(\mathbf{x}'\mathbf{A}\mathbf{x}) = [\text{vec}(\boldsymbol{\Sigma}) + (\boldsymbol{\mu} \otimes \boldsymbol{\mu})]'\,\text{vec}\,\mathbf{A} \tag{7.21}$$

[as can be readily verified from expression (7.10)]. Expression (7.21) is a linear form in vec $\mathbf{A}$; that is, it is a linear combination of the elements of vec $\mathbf{A}$ (which are the elements of $\mathbf{A}$). If $\mathbf{A}$ is symmetric, then expression (7.21) can be restated as follows:

$$\text{E}(\mathbf{x}'\mathbf{A}\mathbf{x}) = [\text{vec}(\boldsymbol{\Sigma}) + (\boldsymbol{\mu} \otimes \boldsymbol{\mu})]'\mathbf{G}_N\,\text{vech}\,\mathbf{A} \tag{7.22}$$

(where $\mathbf{G}_N$ is the duplication matrix). Expression (7.22) is a linear form in vech $\mathbf{A}$, the elements of which are $N(N+1)/2$ nonredundant elements of $\mathbf{A}$—if $\mathbf{A}$ is symmetric, $N(N-1)/2$ of its elements are redundant.

Now, consider the matrix expression (7.13) provided by Theorem 5.7.2 for the covariance of the linear form $\mathbf{b}'\mathbf{x}$ and the quadratic form $\mathbf{x}'\mathbf{A}\mathbf{x}$. Making use of result (7.6), we find that the second term of expression (7.13) can be reexpressed as follows:

$$2\mathbf{b}'\boldsymbol{\Sigma}\mathbf{A}\boldsymbol{\mu} = 2\,\text{tr}(\mathbf{b}'\boldsymbol{\Sigma}\mathbf{A}\boldsymbol{\mu}) = 2\,\text{tr}[\mathbf{b}'\boldsymbol{\Sigma}\mathbf{A}(\boldsymbol{\mu}')'] = 2(\text{vec}\,\mathbf{b})'(\boldsymbol{\mu}' \otimes \boldsymbol{\Sigma})\,\text{vec}\,\mathbf{A} = 2\mathbf{b}'(\boldsymbol{\mu}' \otimes \boldsymbol{\Sigma})\,\text{vec}\,\mathbf{A}.$$

Thus, formula (7.13) can be restated as follows:

$$\text{cov}(\mathbf{b}'\mathbf{x},\ \mathbf{x}'\mathbf{A}\mathbf{x}) = \mathbf{b}'[\boldsymbol{\Lambda} + 2(\boldsymbol{\mu}' \otimes \boldsymbol{\Sigma})]\,\text{vec}\,\mathbf{A} = \mathbf{b}'[\boldsymbol{\Lambda} + 2(\boldsymbol{\mu}' \otimes \boldsymbol{\Sigma})]\mathbf{G}_N\,\text{vech}\,\mathbf{A}. \tag{7.23}$$

Expression (7.23) is a bilinear form in the N-dimensional column vector $\mathbf{b}$ and the $N(N+1)/2$-dimensional column vector vech $\mathbf{A}$, that is, for any particular value of $\mathbf{b}$, it is a linear form in vech $\mathbf{A}$, and for any particular value of vech $\mathbf{A}$, it is a linear form in $\mathbf{b}$.

Further, consider the matrix expression (7.16) provided by Theorem 5.7.3 for the covariance of the two quadratic forms $\mathbf{x}'\mathbf{A}\mathbf{x}$ and $\mathbf{x}'\mathbf{H}\mathbf{x}$. Making use of results (7.5) and (7.4), we find that the second and third terms of expression (7.16) can be reexpressed as follows:

$$\begin{aligned} 2\boldsymbol{\mu}'\mathbf{H}\boldsymbol{\Lambda}\operatorname{vec}\mathbf{A} &= 2[\boldsymbol{\mu}'\mathbf{H}\boldsymbol{\Lambda}\operatorname{vec}\mathbf{A}]' \\ &= 2(\operatorname{vec}\mathbf{A})'\boldsymbol{\Lambda}'\mathbf{H}\boldsymbol{\mu} \\ &= 2(\operatorname{vec}\mathbf{A})'\operatorname{vec}(\boldsymbol{\Lambda}'\mathbf{H}\boldsymbol{\mu}) \\ &= 2(\operatorname{vec}\mathbf{A})'(\boldsymbol{\mu}'\otimes\boldsymbol{\Lambda}')\operatorname{vec}\mathbf{H} \\ &= 2(\operatorname{vec}\mathbf{A})'(\boldsymbol{\mu}\otimes\boldsymbol{\Lambda})'\operatorname{vec}\mathbf{H} \end{aligned} \tag{7.24}$$

and

$$\begin{aligned} 2\boldsymbol{\mu}'\mathbf{A}\boldsymbol{\Lambda}\operatorname{vec}\mathbf{H} &= 2(\boldsymbol{\Lambda}'\mathbf{A}\boldsymbol{\mu})'\operatorname{vec}\mathbf{H} \\ &= 2[\operatorname{vec}(\boldsymbol{\Lambda}'\mathbf{A}\boldsymbol{\mu})]'\operatorname{vec}\mathbf{H} \\ &= 2[(\boldsymbol{\mu}'\otimes\boldsymbol{\Lambda}')\operatorname{vec}\mathbf{A}]'\operatorname{vec}\mathbf{H} \\ &= 2(\operatorname{vec}\mathbf{A})'(\boldsymbol{\mu}\otimes\boldsymbol{\Lambda})\operatorname{vec}\mathbf{H}. \end{aligned} \tag{7.25}$$

Moreover, making use of result (7.6) (and Lemma 2.3.1), the fourth and fifth terms of expression (7.16) are reexpressible as

$$2\operatorname{tr}(\mathbf{A}\boldsymbol{\Sigma}\mathbf{H}\boldsymbol{\Sigma}) = 2\operatorname{tr}(\mathbf{A}'\boldsymbol{\Sigma}\mathbf{H}\boldsymbol{\Sigma}') = 2(\operatorname{vec}\mathbf{A})'(\boldsymbol{\Sigma}\otimes\boldsymbol{\Sigma})\operatorname{vec}\mathbf{H} \tag{7.26}$$

and

$$\begin{aligned} 4\boldsymbol{\mu}'\mathbf{A}\boldsymbol{\Sigma}\mathbf{H}\boldsymbol{\mu} &= 4\operatorname{tr}(\boldsymbol{\mu}'\mathbf{A}\boldsymbol{\Sigma}\mathbf{H}\boldsymbol{\mu}) \\ &= 4\operatorname{tr}(\mathbf{A}\boldsymbol{\Sigma}\mathbf{H}\boldsymbol{\mu}\boldsymbol{\mu}') \\ &= 4\operatorname{tr}[\mathbf{A}'\boldsymbol{\Sigma}\mathbf{H}(\boldsymbol{\mu}\boldsymbol{\mu}')'] \\ &= 4(\operatorname{vec}\mathbf{A})'[(\boldsymbol{\mu}\boldsymbol{\mu}')\otimes\boldsymbol{\Sigma}]\operatorname{vec}\mathbf{H}. \end{aligned} \tag{7.27}$$

And based on results (7.24), (7.25), (7.26), and (7.27), formula (7.16) can be restated as follows:

$$\begin{aligned} \operatorname{cov}(\mathbf{x}'\mathbf{A}\mathbf{x},\, \mathbf{x}'\mathbf{H}\mathbf{x}) &= (\operatorname{vec}\mathbf{A})'\{\boldsymbol{\Omega} + 2(\boldsymbol{\mu}\otimes\boldsymbol{\Lambda})' + 2(\boldsymbol{\mu}\otimes\boldsymbol{\Lambda}) \\ &\qquad + 2(\boldsymbol{\Sigma}\otimes\boldsymbol{\Sigma}) + 4[(\boldsymbol{\mu}\boldsymbol{\mu}')\otimes\boldsymbol{\Sigma}]\}\operatorname{vec}\mathbf{H} \qquad (7.28) \\ &= (\operatorname{vech}\mathbf{A})'\mathbf{G}_N'\{\boldsymbol{\Omega} + 2(\boldsymbol{\mu}\otimes\boldsymbol{\Lambda})' + 2(\boldsymbol{\mu}\otimes\boldsymbol{\Lambda}) \\ &\qquad + 2(\boldsymbol{\Sigma}\otimes\boldsymbol{\Sigma}) + 4[(\boldsymbol{\mu}\boldsymbol{\mu}')\otimes\boldsymbol{\Sigma}]\}\mathbf{G}_N\operatorname{vech}\mathbf{H}. \qquad (7.29) \end{aligned}$$

In result (7.29), $\operatorname{cov}(\mathbf{x}'\mathbf{A}\mathbf{x},\, \mathbf{x}'\mathbf{H}\mathbf{x})$ is expressed as a bilinear form in the $N(N+1)/2$-dimensional column vectors vech $\mathbf{A}$ and vech $\mathbf{H}$. As a special case of result (7.29) (that where $\mathbf{H} = \mathbf{A}$), we have the result

$$\begin{aligned} \operatorname{var}(\mathbf{x}'\mathbf{A}\mathbf{x}) &= (\operatorname{vech}\mathbf{A})'\mathbf{G}_N'\{\boldsymbol{\Omega} + 2(\boldsymbol{\mu}\otimes\boldsymbol{\Lambda})' + 2(\boldsymbol{\mu}\otimes\boldsymbol{\Lambda}) \\ &\qquad + 2(\boldsymbol{\Sigma}\otimes\boldsymbol{\Sigma}) + 4[(\boldsymbol{\mu}\boldsymbol{\mu}')\otimes\boldsymbol{\Sigma}]\}\mathbf{G}_N\operatorname{vech}\mathbf{A}, \end{aligned} \tag{7.30}$$

in which $\operatorname{var}(\mathbf{x}'\mathbf{A}\mathbf{x})$ is expressed as a quadratic form in the $N(N+1)/2$-dimensional column vector vech $\mathbf{A}$.

c. Estimation of σ^2 (under the G–M model)

Let us add to our earlier discussion of the method of least squares by introducing some notation, terminology, and results that are relevant to making inferences about variability and covariability.

Suppose that $\mathbf{y} = (y_1, y_2, \ldots, y_N)'$ is an $N \times 1$ observable random vector that follows a G–M, Aitken, or general linear model. And let

$$\tilde{\mathbf{e}} = \mathbf{y} - \mathbf{P_X y} \quad [= (\mathbf{I} - \mathbf{P_X})\mathbf{y}]$$

[where $\mathbf{P_X} = \mathbf{X}(\mathbf{X}'\mathbf{X})^{-}\mathbf{X}'$], so that (for $i = 1, 2, \ldots, N$) the ith element of $\tilde{\mathbf{e}}$ is the difference between y_i and the least squares estimator of $\mathrm{E}(y_i)$. It is customary to refer to the elements of $\tilde{\mathbf{e}}$ (or to their observed values) as *least squares residuals*, or simply as *residuals*, and to refer to $\tilde{\mathbf{e}}$ itself (or to its observed value) as the *residual vector*.

Upon applying formulas (3.1.7) and (3.2.47) and observing (in light of Theorem 2.12.2) that $\mathbf{P_X X} = \mathbf{X}$ and that $\mathbf{P_X}$ is symmetric and idempotent, we find that

$$\mathrm{E}(\tilde{\mathbf{e}}) = (\mathbf{I} - \mathbf{P_X})\mathbf{X}\boldsymbol{\beta} = (\mathbf{X} - \mathbf{X})\boldsymbol{\beta} = \mathbf{0} \tag{7.31}$$

and that, in the special case where $\mathbf{y}$ follows a G–M model,

$$\mathrm{var}(\tilde{\mathbf{e}}) = (\mathbf{I} - \mathbf{P_X})(\sigma^2\mathbf{I})(\mathbf{I} - \mathbf{P_X})' = \sigma^2(\mathbf{I} - \mathbf{P_X}). \tag{7.32}$$

Corresponding to the vector $\tilde{\mathbf{e}}$ is the quantity $\tilde{\mathbf{e}}'\tilde{\mathbf{e}}$, which is customarily referred to as the *residual sum of squares*. It follows from the results of Section 5.4b (on least squares minimization) that, for every value of $\mathbf{y}$,

$$\tilde{\mathbf{e}}'\tilde{\mathbf{e}} = \min_{\mathbf{b}} (\mathbf{y} - \mathbf{Xb})'(\mathbf{y} - \mathbf{Xb}). \tag{7.33}$$

Moreover,

$$\tilde{\mathbf{e}}'\tilde{\mathbf{e}} = \mathbf{y}'(\mathbf{I} - \mathbf{P_X})\mathbf{y}, \tag{7.34}$$

as is evident from the results of Section 5.4b or upon observing (in light of the symmetry and idempotency of $\mathbf{P_X}$) that $(\mathbf{I} - \mathbf{P_X})'(\mathbf{I} - \mathbf{P_X}) = \mathbf{I} - \mathbf{P_X}$.

In what follows (i.e., in the remainder of Subsection c), it is supposed that $\mathbf{y}$ follows a G–M model, and the emphasis is on the estimation of the parameter σ^2.

An unbiased estimator. The expected value of the residual sum of squares can be derived by applying formula (7.11) (for the expected value of a quadratic form) to expression (7.34) (which is a quadratic form in the random vector $\mathbf{y}$). Recalling that $\mathbf{P_X X} = \mathbf{X}$, we find that

$$\begin{aligned}
\mathrm{E}(\tilde{\mathbf{e}}'\tilde{\mathbf{e}}) &= \mathrm{E}[\mathbf{y}'(\mathbf{I} - \mathbf{P_X})\mathbf{y}] \\
&= \mathrm{tr}[(\mathbf{I} - \mathbf{P_X})(\sigma^2\mathbf{I})] + (\mathbf{X}\boldsymbol{\beta})'(\mathbf{I} - \mathbf{P_X})\mathbf{X}\boldsymbol{\beta} \\
&= \sigma^2\,\mathrm{tr}(\mathbf{I} - \mathbf{P_X}) + 0 \\
&= \sigma^2[N - \mathrm{tr}(\mathbf{P_X})].
\end{aligned}$$

Moreover, because $\mathbf{P_X}$ is idempotent and because $(\mathbf{X}'\mathbf{X})^{-}\mathbf{X}'$ is a generalized inverse of $\mathbf{X}$—refer to Part (1) of Theorem 2.12.2—we have (in light of Corollary 2.8.3 and Lemma 2.10.13) that

$$\mathrm{tr}(\mathbf{P_X}) = \mathrm{rank}(\mathbf{P_X}) = \mathrm{rank}[\mathbf{X}(\mathbf{X}'\mathbf{X})^{-}\mathbf{X}'] = \mathrm{rank}\,\mathbf{X}. \tag{7.35}$$

Thus, the expected value of the residual sum of squares is

$$\mathrm{E}(\tilde{\mathbf{e}}'\tilde{\mathbf{e}}) = \sigma^2(N - \mathrm{rank}\,\mathbf{X}). \tag{7.36}$$

Assume that the rank of the model matrix $\mathbf{X}$ is (strictly) less than N. Then, upon dividing the residual sum of squares by $N - \mathrm{rank}\,\mathbf{X}$, we obtain the quantity

$$\hat{\sigma}^2 = \frac{\tilde{\mathbf{e}}'\tilde{\mathbf{e}}}{N - \mathrm{rank}\,\mathbf{X}}.$$

Clearly,

$$\mathrm{E}(\hat{\sigma}^2) = \sigma^2, \tag{7.37}$$

that is, the quantity $\hat{\sigma}^2$ obtained by dividing the residual sum of squares by $N - \mathrm{rank}\,\mathbf{X}$ is an unbiased estimator of the parameter σ^2.

Let us find the variance of the estimator $\hat{\sigma}^2$. Suppose that the fourth-order moments of the distribution of the vector $\mathbf{e} = (e_1, e_2, \ldots, e_N)'$ of residual effects are such that (for $i, j, k, m = 1, 2, \ldots, N$)

$$\mathrm{E}(e_i e_j e_k e_m) = \begin{cases} 3\sigma^4 & \text{if } m = k = j = i, \\ \sigma^4 & \text{if } j = i \text{ and } m = k \neq i, \text{ if } k = i \text{ and } m = j \neq i, \\ & \quad \text{or if } m = i \text{ and } k = j \neq i, \\ 0 & \text{otherwise} \end{cases} \tag{7.38}$$

(as would be the case if the distribution of $\mathbf{e}$ were MVN). Further, let $\boldsymbol{\Lambda}$ represent the $N \times N^2$ matrix whose entry for the jth row and kmth column [column $(k-1)N + m$] is $\mathrm{E}(e_j e_k e_m)$. Then, upon applying formula (7.17) and once again making use of the properties of the $\mathbf{P_X}$ matrix (set forth in Theorem 2.12.2), we find that

$$\begin{aligned} \mathrm{var}(\tilde{\mathbf{e}}'\tilde{\mathbf{e}}) &= \mathrm{var}[\mathbf{y}'(\mathbf{I} - \mathbf{P_X})\mathbf{y}] \\ &= 4\boldsymbol{\beta}'\mathbf{X}'(\mathbf{I} - \mathbf{P_X})\boldsymbol{\Lambda}\,\mathrm{vec}(\mathbf{I} - \mathbf{P_X}) + 2\,\mathrm{tr}[(\mathbf{I} - \mathbf{P_X})(\sigma^2\mathbf{I})(\mathbf{I} - \mathbf{P_X})(\sigma^2\mathbf{I})] \\ &\qquad + 4\boldsymbol{\beta}'\mathbf{X}'(\mathbf{I} - \mathbf{P_X})(\sigma^2\mathbf{I})(\mathbf{I} - \mathbf{P_X})\mathbf{X}\boldsymbol{\beta} \\ &= 0 + 2\sigma^4\,\mathrm{tr}(\mathbf{I} - \mathbf{P_X}) + 0 \\ &= 2\sigma^4[N - \mathrm{tr}(\mathbf{P_X})]. \end{aligned}$$

And upon applying result (7.35), we conclude that

$$\mathrm{var}(\tilde{\mathbf{e}}'\tilde{\mathbf{e}}) = 2\sigma^4(N - \mathrm{rank}\,\mathbf{X}). \tag{7.39}$$

Moreover, as a particular implication of result (7.39), we have that

$$\mathrm{var}(\hat{\sigma}^2) = \frac{\mathrm{var}(\tilde{\mathbf{e}}'\tilde{\mathbf{e}})}{(N - \mathrm{rank}\,\mathbf{X})^2} = \frac{2\sigma^4}{N - \mathrm{rank}\,\mathbf{X}}. \tag{7.40}$$

The Hodges–Lehmann estimator. The estimator $\hat{\sigma}^2$ is of the general form

$$\frac{\tilde{\mathbf{e}}'\tilde{\mathbf{e}}}{k}, \tag{7.41}$$

where k is a (strictly) positive constant. It is the estimator of the form (7.41) obtained by taking $k = N - \mathrm{rank}\,\mathbf{X}$. Taking $k = N - \mathrm{rank}\,\mathbf{X}$ achieves unbiasedness. Nevertheless, it can be of interest to consider other choices for k.

Let us derive the MSE (mean squared error) of the estimator (7.41). And, in doing so, let us continue to suppose that the fourth-order moments of the distribution of the vector $\mathbf{e} = (e_1, e_2, \ldots, e_N)'$ of residual effects are such that (for $i, j, k, m = 1, 2, \ldots, N$) $\mathrm{E}(e_i e_j e_k e_m)$ satisfies condition (7.38) (as would be the case if the distribution of $\mathbf{e}$ were MVN).

The MSE of the estimator (7.41) can be regarded as a function, say $m(k)$, of the scalar k. Making use of results (7.36) and (7.39), we find that

$$m(k) = \text{var}\left(\frac{\tilde{\mathbf{e}}'\tilde{\mathbf{e}}}{k}\right) + \left[\mathrm{E}\left(\frac{\tilde{\mathbf{e}}'\tilde{\mathbf{e}}}{k}\right) - \sigma^2\right]^2$$
$$= \frac{1}{k^2}\left\{\text{var}(\tilde{\mathbf{e}}'\tilde{\mathbf{e}}) + [\mathrm{E}(\tilde{\mathbf{e}}'\tilde{\mathbf{e}}) - k\sigma^2]^2\right\}$$
$$= \frac{\sigma^4}{k^2}\left\{2(N - \text{rank}\,\mathbf{X}) + [N - \text{rank}(\mathbf{X}) - k]^2\right\} \tag{7.42}$$
$$= \sigma^4\left\{\frac{(N - \text{rank}\,\mathbf{X})[N - \text{rank}(\mathbf{X}) + 2 - 2k]}{k^2} + 1\right\}. \tag{7.43}$$

For what choice of k does $m(k)$ attain its minimum value? Upon differentiating $m(k)$ and engaging in some algebraic simplification, we find that

$$\frac{dm(k)}{dk} = \frac{-2\sigma^4(N - \text{rank}\,\mathbf{X})[N - \text{rank}(\mathbf{X}) + 2 - k]}{k^3},$$

so that $\frac{dm(k)}{dk} < 0$ if $k < N - \text{rank}(\mathbf{X}) + 2$, $\frac{dm(k)}{dk} = 0$ if $k = N - \text{rank}(\mathbf{X}) + 2$, and $\frac{dm(k)}{dk} > 0$ if $k > N - \text{rank}(\mathbf{X}) + 2$. Thus, $m(k)$ is a decreasing function of k over the interval $0 < k \le N - \text{rank}(\mathbf{X}) + 2$, is an increasing function over the interval $k \ge N - \text{rank}(\mathbf{X}) + 2$, and attains its minimum value at $k = N - \text{rank}(\mathbf{X}) + 2$.

We conclude that among estimators of σ^2 of the form (7.41), the estimator

$$\frac{\tilde{\mathbf{e}}'\tilde{\mathbf{e}}}{N - \text{rank}(\mathbf{X}) + 2} \tag{7.44}$$

has minimum MSE. The estimator (7.44) is sometimes referred to as the *Hodges–Lehmann estimator*. In light of results (7.36) and (7.42), it has a bias of

$$\mathrm{E}\left[\frac{\tilde{\mathbf{e}}'\tilde{\mathbf{e}}}{N - \text{rank}(\mathbf{X}) + 2}\right] - \sigma^2 = \sigma^2\left[\frac{N - \text{rank}\,\mathbf{X}}{N - \text{rank}(\mathbf{X}) + 2} - 1\right]$$
$$= \frac{-2\sigma^2}{N - \text{rank}(\mathbf{X}) + 2} \tag{7.45}$$

and an MSE of

$$\frac{\sigma^4[2(N - \text{rank}\,\mathbf{X}) + (-2)^2]}{[N - \text{rank}(\mathbf{X}) + 2]^2} = \frac{2\sigma^4}{N - \text{rank}(\mathbf{X}) + 2}. \tag{7.46}$$

By way of comparison, the unbiased estimator $\hat{\sigma}^2$ (obtained by taking $k = N - \text{rank}\,\mathbf{X}$) has an MSE of $2\sigma^4/(N - \text{rank}\,\mathbf{X})$.

Statistical independence. Let us conclude the present subsection (Subsection c) with some results pertaining to least squares estimators of estimable linear combinations of the elements of the parametric vector $\boldsymbol{\beta}$. The least squares estimator of any such linear combination is expressible as $\mathbf{r}'\mathbf{X}'\mathbf{y}$ for some $P \times 1$ vector $\mathbf{r}$ of constants; more generally, the M-dimensional column vector whose elements are the least squares estimators of M such linear combinations is expressible as $\mathbf{R}'\mathbf{X}'\mathbf{y}$ for some $P \times M$ matrix $\mathbf{R}$ of constants. Making use of formula (3.2.46) and of Parts (4) and (2) of Theorem 2.12.2, we find that

$$\text{cov}(\mathbf{r}'\mathbf{X}'\mathbf{y}, \tilde{\mathbf{e}}) = \mathbf{r}'\mathbf{X}'(\sigma^2\mathbf{I})(\mathbf{I} - \mathbf{P}_{\mathbf{X}})' = \sigma^2\mathbf{r}'\mathbf{X}'(\mathbf{I} - \mathbf{P}_{\mathbf{X}}) = \mathbf{0} \tag{7.47}$$

and, similarly (and more generally), that

$$\text{cov}(\mathbf{R}'\mathbf{X}'\mathbf{y},\ \tilde{\mathbf{e}}) = \mathbf{0}. \tag{7.48}$$

Thus, the least squares estimator $\mathbf{r}'\mathbf{X}'\mathbf{y}$ and the residual vector $\tilde{\mathbf{e}}$ are uncorrelated. And, more generally, the vector $\mathbf{R}'\mathbf{X}'\mathbf{y}$ of least squares estimators and the residual vector $\tilde{\mathbf{e}}$ are uncorrelated.

Is the least squares estimator $\mathbf{r}'\mathbf{X}'\mathbf{y}$ uncorrelated with the residual sum of squares $\tilde{\mathbf{e}}'\tilde{\mathbf{e}}$? Or, equivalently, is $\mathbf{r}'\mathbf{X}'\mathbf{y}$ uncorrelated with an estimator of σ^2 of the form (7.41), including the unbiased estimator $\hat{\sigma}^2$ (and the Hodges–Lehmann estimator)? Assuming the model is such that the distribution of the vector $\mathbf{e} = (e_1, e_2, \ldots, e_N)'$ of residual effects has third-order moments $\lambda_{ijk} = \text{E}(e_i e_j e_k)$ $(i, j, k = 1, 2, \ldots, N)$ and making use of formula (7.13), we find that

$$\text{cov}(\mathbf{r}'\mathbf{X}'\mathbf{y},\ \tilde{\mathbf{e}}'\tilde{\mathbf{e}}) = \mathbf{r}'\mathbf{X}'\boldsymbol{\Lambda}\ \text{vec}(\mathbf{I} - \mathbf{P}_{\mathbf{X}}) + 2\mathbf{r}'\mathbf{X}'(\sigma^2\mathbf{I})(\mathbf{I} - \mathbf{P}_{\mathbf{X}})\mathbf{X}\boldsymbol{\beta}, \tag{7.49}$$

where $\boldsymbol{\Lambda}$ is an $N \times N^2$ matrix whose entry for the ith row and jkth column [column $(j-1)N + k$] is λ_{ijk}. The second term of expression (7.49) equals 0, as is evident upon recalling that $\mathbf{P}_{\mathbf{X}}\mathbf{X} = \mathbf{X}$, and the first term equals 0 if $\boldsymbol{\Lambda} = \mathbf{0}$, as would be the case if the distribution of $\mathbf{e}$ were MVN or, more generally, if the distribution of $\mathbf{e}$ were symmetric. Thus, if the distribution of $\mathbf{e}$ is symmetric, then

$$\text{cov}(\mathbf{r}'\mathbf{X}'\mathbf{y},\ \tilde{\mathbf{e}}'\tilde{\mathbf{e}}) = 0 \tag{7.50}$$

and, more generally,

$$\text{cov}(\mathbf{R}'\mathbf{X}'\mathbf{y},\ \tilde{\mathbf{e}}'\tilde{\mathbf{e}}) = \mathbf{0}. \tag{7.51}$$

Accordingly, if the distribution of $\mathbf{e}$ is symmetric, $\mathbf{r}'\mathbf{X}'\mathbf{y}$ and $\mathbf{R}'\mathbf{X}'\mathbf{y}$ are uncorrelated with any estimator of σ^2 of the form (7.41), including the unbiased estimator $\hat{\sigma}^2$ (and the Hodges–Lehmann estimator).

Are the vector $\mathbf{R}'\mathbf{X}'\mathbf{y}$ of least squares estimators and the residual vector $\tilde{\mathbf{e}}$ statistically independent (as well as uncorrelated)? If the model is such that the distribution of $\mathbf{e}$ is MVN (in which case the distribution of $\mathbf{y}$ is also MVN), then it follows from Corollary 3.5.6 that the answer is yes. That is, if the model is such that the distribution of $\mathbf{e}$ is MVN, then $\tilde{\mathbf{e}}$ is distributed independently of $\mathbf{R}'\mathbf{X}'\mathbf{y}$ (and, in particular, $\tilde{\mathbf{e}}$ is distributed independently of $\mathbf{r}'\mathbf{X}'\mathbf{y}$). Moreover, $\tilde{\mathbf{e}}$ being distributed independently of $\mathbf{R}'\mathbf{X}'\mathbf{y}$ implies that "any" function of $\tilde{\mathbf{e}}$ is distributed independently of $\mathbf{R}'\mathbf{X}'\mathbf{y}$—refer, e.g., to Casella and Berger (2002, theorem 4.6.12). Accordingly, if the distribution of $\mathbf{e}$ is MVN, then the residual sum of squares $\tilde{\mathbf{e}}'\tilde{\mathbf{e}}$ is distributed independently of $\mathbf{R}'\mathbf{X}'\mathbf{y}$ and any estimator of σ^2 of the form (7.41) (including the unbiased estimator $\hat{\sigma}^2$ and the Hodges–Lehmann estimator) is distributed independently of $\mathbf{R}'\mathbf{X}'\mathbf{y}$.

d. Translation invariance

Suppose that $\mathbf{y}$ is an $N \times 1$ observable random vector that follows a G–M, Aitken, or general linear model. And suppose that we wish to make inferences about σ^2 (in the case of a G–M or Aitken model) or $\boldsymbol{\theta}$ (in the case of a general linear model) or about various functions of σ^2 or $\boldsymbol{\theta}$. In making such inferences, it is common practice to restrict attention to procedures that depend on the value of $\mathbf{y}$ only through the value of a (possibly vector-valued) statistic having a property known as translation invariance (or location invariance).

Proceeding as in Section 5.2 (in discussing the translation-equivariant estimation of a parametric function of the form $\boldsymbol{\lambda}'\boldsymbol{\beta}$), let $\mathbf{k}$ represent a P-dimensional column vector of known constants, and define $\mathbf{z} = \mathbf{y} + \mathbf{X}\mathbf{k}$. Then, $\mathbf{z} = \mathbf{X}\boldsymbol{\tau} + \mathbf{e}$, where $\boldsymbol{\tau} = \boldsymbol{\beta} + \mathbf{k}$. And $\mathbf{z}$ can be regarded as an $N \times 1$ observable random vector that follows a G–M, Aitken, or general linear model that is identical in all respects to the model followed by $\mathbf{y}$, except that the role of the parametric vector $\boldsymbol{\beta}$ is played by a vector (represented by $\boldsymbol{\tau}$) having a different interpretation. It can be argued that inferences about σ^2 or $\boldsymbol{\theta}$, or about functions of σ^2 or $\boldsymbol{\theta}$, should be made on the basis of a statistical procedure that depends on the value of $\mathbf{y}$ only through the value of a (possibly vector-valued) statistic $\mathbf{h}(\mathbf{y})$ that, for every $\mathbf{k} \in \mathcal{R}^P$ (and for every value of $\mathbf{y}$), satisfies the condition

$$\mathbf{h}(\mathbf{y}) = \mathbf{h}(\mathbf{z})$$

or, equivalently, the condition

$$\mathbf{h}(\mathbf{y}) = \mathbf{h}(\mathbf{y} + \mathbf{X}\mathbf{k}). \tag{7.52}$$

Any statistic $\mathbf{h}(\mathbf{y})$ that satisfies condition (7.52) and that does so for every $\mathbf{k} \in \mathcal{R}^P$ (and for every value of $\mathbf{y}$) is said to be *translation invariant*.

If the statistic $\mathbf{h}(\mathbf{y})$ is translation invariant, then

$$\mathbf{h}(\mathbf{y}) = \mathbf{h}[\mathbf{y} + \mathbf{X}(-\boldsymbol{\beta})] = \mathbf{h}(\mathbf{y} - \mathbf{X}\boldsymbol{\beta}) = \mathbf{h}(\mathbf{e}). \tag{7.53}$$

Thus, the statistical properties of a statistical procedure that depends on the value of $\mathbf{y}$ only through the value of a translation-invariant statistic $\mathbf{h}(\mathbf{y})$ are completely determined by the distribution of the vector $\mathbf{e}$ of residual effects. They do not depend on the vector $\boldsymbol{\beta}$.

Let us now consider condition (7.52) in the special case where $\mathbf{h}(\mathbf{y})$ is a scalar-valued statistic $h(\mathbf{y})$ of the form

$$h(\mathbf{y}) = \mathbf{y}'\mathbf{A}\mathbf{y},$$

where $\mathbf{A}$ is a symmetric matrix of constants. In this special case,

$$\begin{aligned} \mathbf{h}(\mathbf{y}) = \mathbf{h}(\mathbf{y} + \mathbf{X}\mathbf{k}) \quad &\Leftrightarrow \quad \mathbf{y}'\mathbf{A}\mathbf{X}\mathbf{k} + \mathbf{k}'\mathbf{X}'\mathbf{A}\mathbf{y} + \mathbf{k}'\mathbf{X}'\mathbf{A}\mathbf{X}\mathbf{k} = 0 \\ &\Leftrightarrow \quad 2\mathbf{y}'\mathbf{A}\mathbf{X}\mathbf{k} = -\mathbf{k}'\mathbf{X}'\mathbf{A}\mathbf{X}\mathbf{k}. \end{aligned} \tag{7.54}$$

For condition (7.54) to be satisfied for every $\mathbf{k} \in \mathcal{R}^P$ (and for every value of $\mathbf{y}$), it is sufficient that $\mathbf{A}\mathbf{X} = \mathbf{0}$. It is also necessary. To see this, suppose that condition (7.54) is satisfied for every $\mathbf{k} \in \mathcal{R}^P$ (and for every value of $\mathbf{y}$). Then, upon setting $\mathbf{y} = \mathbf{0}$ in condition (7.54), we find that $\mathbf{k}'\mathbf{X}'\mathbf{A}\mathbf{X}\mathbf{k} = 0$ for every $\mathbf{k} \in \mathcal{R}^P$, implying (in light of Corollary 2.13.4) that $\mathbf{X}'\mathbf{A}\mathbf{X} = \mathbf{0}$. Thus, $\mathbf{y}'\mathbf{A}\mathbf{X}\mathbf{k} = 0$ for every $\mathbf{k} \in \mathcal{R}^P$ (and every value of $\mathbf{y}$), implying that every element of $\mathbf{A}\mathbf{X}$ equals 0 and hence that $\mathbf{A}\mathbf{X} = \mathbf{0}$.

In summary, we have established that the quadratic form $\mathbf{y}'\mathbf{A}\mathbf{y}$ (where $\mathbf{A}$ is a symmetric matrix of constants) is a translation-invariant statistic if and only if the matrix $\mathbf{A}$ of the quadratic form satisfies the condition

$$\mathbf{A}\mathbf{X} = \mathbf{0}. \tag{7.55}$$

Adopting the same notation and terminology as in Subsection c, consider the concept of translation invariance as applied to the residual vector $\tilde{\mathbf{e}}$ and to the residual sum of squares $\tilde{\mathbf{e}}'\tilde{\mathbf{e}}$. Recall that $\tilde{\mathbf{e}}$ is expressible as $\tilde{\mathbf{e}} = (\mathbf{I} - \mathbf{P_X})\mathbf{y}$ and $\tilde{\mathbf{e}}'\tilde{\mathbf{e}}$ as $\tilde{\mathbf{e}}'\tilde{\mathbf{e}} = \mathbf{y}'(\mathbf{I} - \mathbf{P_X})\mathbf{y}$. Recall also that $\mathbf{P_X}\mathbf{X} = \mathbf{X}$ and hence that $(\mathbf{I} - \mathbf{P_X})\mathbf{X} = \mathbf{0}$. Thus, for any $P \times 1$ vector $\mathbf{k}$ (and for any value of $\mathbf{y}$),

$$(\mathbf{I} - \mathbf{P_X})(\mathbf{y} + \mathbf{X}\mathbf{k}) = (\mathbf{I} - \mathbf{P_X})\mathbf{y}.$$

And it follows that $\tilde{\mathbf{e}}$ is translation invariant. Moreover, $\tilde{\mathbf{e}}'\tilde{\mathbf{e}}$ is also translation invariant, as is evident upon observing that it depends on $\mathbf{y}$ only through the value of $\tilde{\mathbf{e}}$ or, alternatively, upon applying condition (7.55) (with $\mathbf{A} = \mathbf{I} - \mathbf{P_X}$)—that condition (7.55) is applicable is evident upon recalling that $\mathbf{P_X}$ is symmetric and hence that the matrix $\mathbf{I} - \mathbf{P_X}$ of the quadratic form $\mathbf{y}'(\mathbf{I} - \mathbf{P_X})\mathbf{y}$ is symmetric.

Let us now specialize by supposing that $\mathbf{y}$ follows a G–M model, and let us add to the results obtained in Subsection c (on the estimation of σ^2) by obtaining some results on translation-invariant estimation. Since the residual sum of squares $\tilde{\mathbf{e}}'\tilde{\mathbf{e}}$ is translation invariant, any estimator of σ^2 of the form (7.41) is translation invariant. In particular, the unbiased estimator $\hat{\sigma}^2$ is translation invariant (and the Hodges–Lehmann estimator is translation invariant).

A quadratic form $\mathbf{y}'\mathbf{A}\mathbf{y}$ in the observable random vector $\mathbf{y}$ (where $\mathbf{A}$ is a symmetric matrix of constants) is an unbiased estimator of σ^2 and is translation invariant if and only if

$$\mathrm{E}(\mathbf{y}'\mathbf{A}\mathbf{y}) = \sigma^2 \quad \text{and} \quad \mathbf{A}\mathbf{X} = \mathbf{0} \tag{7.56}$$

(in which case the quadratic form is referred to as a quadratic unbiased translation-invariant estimator). As an application of formula (7.11), we have that

$$\mathrm{E}(\mathbf{y}'\mathbf{A}\mathbf{y}) = \mathrm{tr}[\mathbf{A}(\sigma^2\mathbf{I})] + \boldsymbol{\beta}'\mathbf{X}'\mathbf{A}\mathbf{X}\boldsymbol{\beta} = \sigma^2\,\mathrm{tr}(\mathbf{A}) + \boldsymbol{\beta}'\mathbf{X}'\mathbf{A}\mathbf{X}\boldsymbol{\beta}. \tag{7.57}$$

In light of result (7.57), condition (7.56) is equivalent to the condition

$$\mathrm{tr}(\mathbf{A}) = 1 \quad \text{and} \quad \mathbf{A}\mathbf{X} = \mathbf{0}. \tag{7.58}$$

Thus, the quadratic form $\mathbf{y}'\mathbf{A}\mathbf{y}$ is a quadratic unbiased translation-invariant estimator of σ^2 if and only if the matrix $\mathbf{A}$ of the quadratic form satisfies condition (7.58).

Clearly, the estimator $\hat{\sigma}^2$ [which is expressible in the form $\hat{\sigma}^2 = \mathbf{y}'(\mathbf{I} - \mathbf{P_X})\mathbf{y}$] is a quadratic unbiased translation-invariant estimator of σ^2. In fact, if the fourth-order moments of the distribution of the vector $\mathbf{e} = (e_1, e_2, \ldots, e_N)'$ of residual effects are such that (for $i, j, k, m = 1, 2, \ldots, N$) $\mathrm{E}(e_i e_j e_k e_m)$ satisfies condition (7.38) (as would be the case if the distribution of $\mathbf{e}$ were MVN), then the estimator $\hat{\sigma}^2$ has minimum variance (and hence minimum MSE) among all quadratic unbiased translation-invariant estimators of σ^2, as we now proceed to show.

Suppose that the fourth-order moments of the distribution of the vector $\mathbf{e} = (e_1, e_2, \ldots, e_N)'$ are such that (for $i, j, k, m = 1, 2, \ldots, N$) $\mathrm{E}(e_i e_j e_k e_m)$ satisfies condition (7.38). And denote by $\boldsymbol{\Lambda}$ the $N \times N^2$ matrix whose entry for the jth row and kmth column [column $(k-1)N + m$] is $\mathrm{E}(e_j e_k e_m)$. Then, for any quadratic unbiased translation-invariant estimator $\mathbf{y}'\mathbf{A}\mathbf{y}$ of σ^2 (where $\mathbf{A}$ is symmetric), we find [upon applying formula (7.17) and observing that $\mathbf{A}\mathbf{X} = \mathbf{0}$] that

$$\begin{aligned}\mathrm{var}(\mathbf{y}'\mathbf{A}\mathbf{y}) &= 4\boldsymbol{\beta}'(\mathbf{A}\mathbf{X})'\boldsymbol{\Lambda}\,\mathrm{vec}\,\mathbf{A} + 2\sigma^4\,\mathrm{tr}(\mathbf{A}^2) + 4\sigma^2\boldsymbol{\beta}'\mathbf{X}'\mathbf{A}\mathbf{A}\mathbf{X}\boldsymbol{\beta} \\ &= 0 + 2\sigma^4\,\mathrm{tr}(\mathbf{A}^2) + 0 = 2\sigma^4\,\mathrm{tr}(\mathbf{A}^2).\end{aligned} \tag{7.59}$$

Let $\mathbf{R} = \mathbf{A} - \dfrac{1}{N - \mathrm{rank}\,\mathbf{X}}(\mathbf{I} - \mathbf{P_X})$, so that

$$\mathbf{A} = \frac{1}{N - \mathrm{rank}\,\mathbf{X}}(\mathbf{I} - \mathbf{P_X}) + \mathbf{R}. \tag{7.60}$$

Further, observe that (since $\mathbf{P_X}$ is symmetric) $\mathbf{R}' = \mathbf{R}$, that (since $\mathbf{A}\mathbf{X} = \mathbf{0}$ and $\mathbf{P_X}\mathbf{X} = \mathbf{X}$)

$$\mathbf{R}\mathbf{X} = \mathbf{A}\mathbf{X} - \frac{1}{N - \mathrm{rank}\,\mathbf{X}}(\mathbf{I} - \mathbf{P_X})\mathbf{X} = \mathbf{0} - \mathbf{0} = \mathbf{0},$$

and that

$$\mathbf{X}'\mathbf{R} = \mathbf{X}'\mathbf{R}' = (\mathbf{R}\mathbf{X})' = \mathbf{0}' = \mathbf{0}.$$

Accordingly, upon substituting expression (7.60) for $\mathbf{A}$ (and recalling that $\mathbf{P_X}$ is idempotent), we find that

$$\mathbf{A}^2 = \frac{1}{(N - \mathrm{rank}\,\mathbf{X})^2}(\mathbf{I} - \mathbf{P_X}) + \frac{2}{N - \mathrm{rank}\,\mathbf{X}}\mathbf{R} + \mathbf{R}'\mathbf{R}. \tag{7.61}$$

Moreover, because $\mathrm{tr}(\mathbf{A}) = 1$, we have [in light of result (7.35)] that

$$\mathrm{tr}(\mathbf{R}) = \mathrm{tr}(\mathbf{A}) - \frac{1}{N - \mathrm{rank}\,\mathbf{X}}\,\mathrm{tr}(\mathbf{I} - \mathbf{P_X}) = 1 - 1 = 0. \tag{7.62}$$

And upon substituting expression (7.61) for $\mathbf{A}^2$ in expression (7.59) and making use of results (7.35) and (7.62), we find that

$$\begin{aligned}\mathrm{var}(\mathbf{y}'\mathbf{A}\mathbf{y}) &= 2\sigma^4\left[\frac{1}{(N - \mathrm{rank}\,\mathbf{X})^2}\,\mathrm{tr}(\mathbf{I} - \mathbf{P_X}) + \frac{2}{N - \mathrm{rank}\,\mathbf{X}}\,\mathrm{tr}(\mathbf{R}) + \mathrm{tr}(\mathbf{R}'\mathbf{R})\right] \\ &= 2\sigma^4\left[\frac{1}{N - \mathrm{rank}\,\mathbf{X}} + \mathrm{tr}(\mathbf{R}'\mathbf{R})\right].\end{aligned} \tag{7.63}$$

Finally, upon observing that $\mathrm{tr}(\mathbf{R}'\mathbf{R}) = \sum_{i,j} r_{ij}^2$, where (for $i, j = 1, 2, \ldots, N$) r_{ij} is the ijth element of $\mathbf{R}$, we conclude that $\mathrm{var}(\mathbf{y}'\mathbf{A}\mathbf{y})$ attains a minimum value of $2\sigma^4/(N - \mathrm{rank}\,\mathbf{X})$ and does so uniquely when $\mathbf{R} = \mathbf{0}$ or, equivalently, when $\mathbf{A} = \dfrac{1}{N - \mathrm{rank}\,\mathbf{X}}(\mathbf{I} - \mathbf{P_X})$ (i.e., when $\mathbf{y}'\mathbf{A}\mathbf{y} = \hat{\sigma}^2$).

5.8 Best (Minimum-Variance) Unbiased Estimation

Take $\mathbf{y}$ to be an $N \times 1$ observable random vector that follows a G–M model, and consider the estimation of an estimable linear combination $\boldsymbol{\lambda}'\boldsymbol{\beta}$ of the elements of the parametric vector $\boldsymbol{\beta}$ and consider also the estimation of the parameter σ^2. In Section 5.5a, it was determined that the least squares estimator of $\boldsymbol{\lambda}'\boldsymbol{\beta}$ has minimum variance among all linear unbiased estimators. And in Section 5.7d, it was determined that the estimator $\hat{\sigma}^2 = \mathbf{y}'(\mathbf{I} - \mathbf{P_X})\mathbf{y}/(N - \operatorname{rank} \mathbf{X})$ has minimum variance among all quadratic unbiased translation-invariant estimators of σ^2 [provided that the fourth-order moments of the distribution of the vector $\mathbf{e} = (e_1, e_2, \ldots, e_N)'$ of residual effects are such that (for $i, j, k, m = 1, 2, \ldots, N$) $\mathrm{E}(e_i e_j e_k e_m)$ satisfies condition (7.38)].

If the distribution of $\mathbf{e}$ is assumed to be MVN, something more can be said. It can be shown that under the assumption of multivariate normality, $\mathbf{X}'\mathbf{y}$ and $\mathbf{y}'(\mathbf{I} - \mathbf{P_X})\mathbf{y}$ form a complete sufficient statistic—refer, e.g., to Casella and Berger (2002, def. 6.2.21) or to Schervish (1995, def. 2.34) for the definition of completeness—in which case "any" function, say $t[\mathbf{X}'\mathbf{y}, \mathbf{y}'(\mathbf{I} - \mathbf{P_X})\mathbf{y}]$, of $\mathbf{X}'\mathbf{y}$ and $\mathbf{y}'(\mathbf{I} - \mathbf{P_X})\mathbf{y}$ is a best (minimum-variance) unbiased estimator of $\mathrm{E}\{t[\mathbf{X}'\mathbf{y}, \mathbf{y}'(\mathbf{I} - \mathbf{P_X})\mathbf{y}]\}$ (e.g., Schervish 1995, theorem 5.5; Casella and Berger 2002, theorem 7.3.23). It follows, in particular, that under the assumption of multivariate normality, the least squares estimator of $\boldsymbol{\lambda}'\boldsymbol{\beta}$ has minimum variance among all unbiased estimators (linear or not) and the estimator $\hat{\sigma}^2$ has minimum variance among all unbiased estimators of σ^2 (quadratic and/or translation invariant or not).

Let us assume that the distribution of $\mathbf{e}$ is MVN, and verify that (under the assumption of multivariate normality) $\mathbf{X}'\mathbf{y}$ and $\mathbf{y}'(\mathbf{I} - \mathbf{P_X})\mathbf{y}$ form a complete sufficient statistic. Let us begin by introducing a transformation of $\mathbf{X}'\mathbf{y}$ and $\mathbf{y}'(\mathbf{I} - \mathbf{P_X})\mathbf{y}$ that facilitates the verification.

Define $K = \operatorname{rank} \mathbf{X}$. And observe that there exists an $N \times K$ matrix, say $\mathbf{W}$, whose columns form a basis for $\mathcal{C}(\mathbf{X})$. Observe also that $\mathbf{W} = \mathbf{XR}$ for some matrix $\mathbf{R}$ and that $\mathbf{X} = \mathbf{WS}$ for some $(K \times P)$ matrix $\mathbf{S}$ (of rank K). Moreover, $\mathbf{X}'\mathbf{y}$ and $\mathbf{y}'(\mathbf{I} - \mathbf{P_X})\mathbf{y}$ are expressible in terms of the $(K \times 1)$ vector $\mathbf{W}'\mathbf{y}$ and the sum of squares $\mathbf{y}'\mathbf{y}$; we have that

$$\mathbf{X}'\mathbf{y} = \mathbf{S}'\mathbf{W}'\mathbf{y} \quad \text{and} \quad \mathbf{y}'(\mathbf{I} - \mathbf{P_X})\mathbf{y} = \mathbf{y}'\mathbf{y} - (\mathbf{W}'\mathbf{y})'\mathbf{S}(\mathbf{X}'\mathbf{X})^{-}\mathbf{S}'\mathbf{W}'\mathbf{y}. \tag{8.1}$$

Conversely, $\mathbf{W}'\mathbf{y}$ and $\mathbf{y}'\mathbf{y}$ are expressible in terms of $\mathbf{X}'\mathbf{y}$ and $\mathbf{y}'(\mathbf{I} - \mathbf{P_X})\mathbf{y}$; we have that

$$\mathbf{W}'\mathbf{y} = \mathbf{R}'\mathbf{X}'\mathbf{y} \quad \text{and} \quad \mathbf{y}'\mathbf{y} = \mathbf{y}'(\mathbf{I} - \mathbf{P_X})\mathbf{y} + (\mathbf{X}'\mathbf{y})'(\mathbf{X}'\mathbf{X})^{-}\mathbf{X}'\mathbf{y}. \tag{8.2}$$

Thus, corresponding to any function $g[\mathbf{X}'\mathbf{y}, \mathbf{y}'(\mathbf{I} - \mathbf{P_X})\mathbf{y}]$ of $\mathbf{X}'\mathbf{y}$ and $\mathbf{y}'(\mathbf{I} - \mathbf{P_X})\mathbf{y}$, there is a function, say $g_*(\mathbf{W}'\mathbf{y}, \mathbf{y}'\mathbf{y})$, of $\mathbf{W}'\mathbf{y}$ and $\mathbf{y}'\mathbf{y}$ such that $g_*(\mathbf{W}'\mathbf{y}, \mathbf{y}'\mathbf{y}) = g[\mathbf{X}'\mathbf{y}, \mathbf{y}'(\mathbf{I} - \mathbf{P_X})\mathbf{y}]$ for every value of $\mathbf{y}$; namely, the function $g_*(\mathbf{W}'\mathbf{y}, \mathbf{y}'\mathbf{y})$ defined by

$$g_*(\mathbf{W}'\mathbf{y}, \mathbf{y}'\mathbf{y}) = g[\mathbf{S}'\mathbf{W}'\mathbf{y}, \mathbf{y}'\mathbf{y} - (\mathbf{W}'\mathbf{y})'\mathbf{S}(\mathbf{X}'\mathbf{X})^{-}\mathbf{S}'\mathbf{W}'\mathbf{y}].$$

Similarly, corresponding to any function $h(\mathbf{W}'\mathbf{y}, \mathbf{y}'\mathbf{y})$, of $\mathbf{W}'\mathbf{y}$ and $\mathbf{y}'\mathbf{y}$, there is a function, say $h_*[\mathbf{X}'\mathbf{y}, \mathbf{y}'(\mathbf{I} - \mathbf{P_X})\mathbf{y}]$, of $\mathbf{X}'\mathbf{y}$ and $\mathbf{y}'(\mathbf{I} - \mathbf{P_X})\mathbf{y}$ such that $h_*[\mathbf{X}'\mathbf{y}, \mathbf{y}'(\mathbf{I} - \mathbf{P_X})\mathbf{y}] = h(\mathbf{W}'\mathbf{y}, \mathbf{y}'\mathbf{y})$ for every value of $\mathbf{y}$; namely, the function $h_*[\mathbf{X}'\mathbf{y}, \mathbf{y}'(\mathbf{I} - \mathbf{P_X})\mathbf{y}]$ defined by

$$h_*[\mathbf{X}'\mathbf{y}, \mathbf{y}'(\mathbf{I} - \mathbf{P_X})\mathbf{y}] = h[\mathbf{R}'\mathbf{X}'\mathbf{y}, \mathbf{y}'(\mathbf{I} - \mathbf{P_X})\mathbf{y} + (\mathbf{X}'\mathbf{y})'(\mathbf{X}'\mathbf{X})^{-}\mathbf{X}'\mathbf{y}].$$

Now, suppose that $\mathbf{W}'\mathbf{y}$ and $\mathbf{y}'\mathbf{y}$ form a complete sufficient statistic. Then, it follows from result (8.2) that $\mathbf{X}'\mathbf{y}$ and $\mathbf{y}'(\mathbf{I} - \mathbf{P_X})\mathbf{y}$ form a sufficient statistic. Moreover, if $\mathrm{E}\{g[\mathbf{X}'\mathbf{y}, \mathbf{y}'(\mathbf{I} - \mathbf{P_X})\mathbf{y}]\} = 0$, then $\mathrm{E}[g_*(\mathbf{W}'\mathbf{y}, \mathbf{y}'\mathbf{y})] = 0$, implying that $\Pr[g_*(\mathbf{W}'\mathbf{y}, \mathbf{y}'\mathbf{y}) = 0] = 1$ and hence that $\Pr\{g[\mathbf{X}'\mathbf{y}, \mathbf{y}'(\mathbf{I} - \mathbf{P_X})\mathbf{y}] = 0\} = 1$. Thus, $\mathbf{X}'\mathbf{y}$ and $\mathbf{y}'(\mathbf{I} - \mathbf{P_X})\mathbf{y}$ form a complete statistic.

Conversely, suppose that $\mathbf{X}'\mathbf{y}$ and $\mathbf{y}'(\mathbf{I} - \mathbf{P_X})\mathbf{y}$ form a complete sufficient statistic. Then, it follows from result (8.1) that $\mathbf{W}'\mathbf{y}$ and $\mathbf{y}'\mathbf{y}$ form a sufficient statistic. Moreover, if $\mathrm{E}[h(\mathbf{W}'\mathbf{y}, \mathbf{y}'\mathbf{y})] = 0$, then $\mathrm{E}\{h_*[\mathbf{X}'\mathbf{y}, \mathbf{y}'(\mathbf{I} - \mathbf{P_X})\mathbf{y}]\} = 0$, implying that $\Pr\{h_*[\mathbf{X}'\mathbf{y}, \mathbf{y}'(\mathbf{I} - \mathbf{P_X})\mathbf{y}] = 0\} = 1$ and hence that $\Pr[h(\mathbf{W}'\mathbf{y}, \mathbf{y}'\mathbf{y}) = 0] = 1$. Thus, $\mathbf{W}'\mathbf{y}$ and $\mathbf{y}'\mathbf{y}$ form a complete statistic.

At this point, we have established that $\mathbf{X}'\mathbf{y}$ and $\mathbf{y}'(\mathbf{I} - \mathbf{P_X})\mathbf{y}$ form a complete sufficient statistic if and only if $\mathbf{W}'\mathbf{y}$ and $\mathbf{y}'\mathbf{y}$ form a complete sufficient statistic. Thus, for purposes of verifying that $\mathbf{X}'\mathbf{y}$ and $\mathbf{y}'(\mathbf{I} - \mathbf{P_X})\mathbf{y}$ form a complete sufficient statistic, it suffices to consider the sufficiency and the completeness of the statistic formed by $\mathbf{W}'\mathbf{y}$ and $\mathbf{y}'\mathbf{y}$. In that regard, the probability density function of $\mathbf{y}$, say $f(\cdot)$, is expressible as follows:

$$\begin{aligned} f(\mathbf{y}) &= \frac{1}{(2\pi\sigma^2)^{N/2}} \exp\Big[-\frac{1}{2\sigma^2}(\mathbf{y} - \mathbf{X}\boldsymbol{\beta})'(\mathbf{y} - \mathbf{X}\boldsymbol{\beta})\Big] \\ &= \frac{1}{(2\pi\sigma^2)^{N/2}} \exp\Big[-\frac{1}{2\sigma^2}(\mathbf{y}'\mathbf{y} - 2\boldsymbol{\beta}'\mathbf{X}'\mathbf{y} + \boldsymbol{\beta}'\mathbf{X}'\mathbf{X}\boldsymbol{\beta})\Big] \\ &= \frac{1}{(2\pi\sigma^2)^{N/2}} \exp\Big[-\frac{1}{2\sigma^2}\boldsymbol{\beta}'\mathbf{X}'\mathbf{X}\boldsymbol{\beta}\Big] \exp\Big[-\frac{1}{2\sigma^2}\mathbf{y}'\mathbf{y} + \Big(\frac{1}{\sigma^2}\mathbf{S}\boldsymbol{\beta}\Big)'\mathbf{W}'\mathbf{y}\Big]. \end{aligned} \tag{8.3}$$

Based on a well-known result on complete sufficient statistics for exponential families of distributions [a result that is theorem 2.74 in Schervish's (1995) book], it follows from result (8.3) that $\mathbf{W}'\mathbf{y}$ and $\mathbf{y}'\mathbf{y}$ form a complete sufficient statistic—to establish that the result on complete sufficient statistics for exponential families is applicable, it suffices to observe [in connection with expression (8.3)] that the parametric function $-1/(2\sigma^2)$ and the $(K \times 1)$ vector $(1/\sigma^2)\mathbf{S}\boldsymbol{\beta}$ of parametric functions are such that, for any (strictly) negative scalar c and any $K \times 1$ vector $\mathbf{d}$, $-1/(2\sigma^2) = c$ and $(1/\sigma^2)\mathbf{S}\boldsymbol{\beta} = \mathbf{d}$ for some value of σ^2 and some value of $\boldsymbol{\beta}$ (as is evident upon noting that $\mathbf{S}$ contains K linearly independent columns). It remains only to observe that since $\mathbf{W}'\mathbf{y}$ and $\mathbf{y}'\mathbf{y}$ form a complete sufficient statistic, $\mathbf{X}'\mathbf{y}$ and $\mathbf{y}'(\mathbf{I} - \mathbf{P_X})\mathbf{y}$ form a complete sufficient statistic.

5.9 Likelihood-Based Methods

A likelihood-based method, known as maximum likelihood (ML), can be used to estimate functions of the parameters (σ and the elements of $\boldsymbol{\beta}$) of the G–M or Aitken model. More generally, it can be used to estimate the parameters (the elements of $\boldsymbol{\beta}$ and of $\boldsymbol{\theta}$) of the general linear model. The use of this method requires an assumption that the distribution of the vector $\mathbf{e}$ of residual effects is known up to the value of σ (in the special case of a G–M or Aitken model) or up to the value of $\boldsymbol{\theta}$ (in the case of a general linear model). Typically, the distribution of $\mathbf{e}$ is taken to be MVN (multivariate normal).

a. (Ordinary) maximum likelihood estimation

It is convenient and instructive to begin by considering ML estimation in the relatively simple case of a G–M model.

G–M model. Suppose that $\mathbf{y}$ is an $N \times 1$ observable random vector that follows a G–M model. Suppose further that the distribution of the vector $\mathbf{e}$ of residual effects is MVN. Then, $\mathbf{y} \sim N(\mathbf{X}\boldsymbol{\beta}, \sigma^2\mathbf{I})$.

Let $f(\cdot\,; \boldsymbol{\beta}, \sigma)$ represent the probability density function (pdf) of the distribution of $\mathbf{y}$, and denote by $\underline{\mathbf{y}}$ the observed value of $\mathbf{y}$. Then, by definition, the likelihood function is the function, say $L(\boldsymbol{\beta}, \sigma; \underline{\mathbf{y}})$, of the parameters (which consist of σ and the elements of $\boldsymbol{\beta}$) defined (for $\boldsymbol{\beta} \in \mathcal{R}^P$ and $\sigma > 0$) by $L(\boldsymbol{\beta}, \sigma; \underline{\mathbf{y}}) = f(\underline{\mathbf{y}}; \boldsymbol{\beta}, \sigma)$. Accordingly,

$$L(\boldsymbol{\beta}, \sigma; \underline{\mathbf{y}}) = \frac{1}{(2\pi\sigma^2)^{N/2}} \exp\Big[-\frac{1}{2\sigma^2}(\underline{\mathbf{y}} - \mathbf{X}\boldsymbol{\beta})'(\underline{\mathbf{y}} - \mathbf{X}\boldsymbol{\beta})\Big]. \tag{9.1}$$

And the log-likelihood function, say $\ell(\boldsymbol{\beta}, \sigma; \underline{\mathbf{y}})$ [which, by definition, is the function obtained by equating $\ell(\boldsymbol{\beta}, \sigma; \underline{\mathbf{y}})$ to the logarithm of the likelihood function, i.e., to $\log L(\boldsymbol{\beta}, \sigma; \underline{\mathbf{y}})$], is expressible as

$$\ell(\boldsymbol{\beta}, \sigma; \underline{\mathbf{y}}) = -\frac{N}{2}\log(2\pi) - \frac{N}{2}\log\sigma^2 - \frac{1}{2\sigma^2}(\underline{\mathbf{y}} - \mathbf{X}\boldsymbol{\beta})'(\underline{\mathbf{y}} - \mathbf{X}\boldsymbol{\beta}). \tag{9.2}$$

Now, consider the maximization of the likelihood function $L(\boldsymbol{\beta}, \sigma; \underline{\mathbf{y}})$ or, equivalently, of the log-likelihood function $\ell(\boldsymbol{\beta}, \sigma; \underline{\mathbf{y}})$. Irrespective of the value of σ, $\ell(\boldsymbol{\beta}, \sigma; \underline{\mathbf{y}})$ attains its maximum value with respect to $\boldsymbol{\beta}$ at any value of $\boldsymbol{\beta}$ that minimizes $(\mathbf{y} - \mathbf{X}\boldsymbol{\beta})'(\mathbf{y} - \mathbf{X}\boldsymbol{\beta})$. Thus, in light of the results of Section 5.4b (on least squares minimization), $\ell(\boldsymbol{\beta}, \sigma; \underline{\mathbf{y}})$ attains its maximum value with respect to $\boldsymbol{\beta}$ at a point $\tilde{\boldsymbol{\beta}}$ if and only if

$$\mathbf{X}'\mathbf{X}\tilde{\boldsymbol{\beta}} = \mathbf{X}'\underline{\mathbf{y}}, \tag{9.3}$$

that is, if and only if $\tilde{\boldsymbol{\beta}}$ is a solution to the normal equations.

Letting $\tilde{\boldsymbol{\beta}}$ represent any $P \times 1$ vector that satisfies condition (9.3), it remains to consider the maximization of $\ell(\tilde{\boldsymbol{\beta}}, \sigma; \underline{\mathbf{y}})$ with respect to σ. In that regard, take $g(\sigma)$ to be a function of σ of the form

$$g(\sigma) = a - \frac{K}{2}\log\sigma^2 - \frac{c}{2\sigma^2}, \tag{9.4}$$

where a is a constant, c is a (strictly) positive constant, and K is a (strictly) positive integer. And observe that, unless $\underline{\mathbf{y}} - \mathbf{X}\tilde{\boldsymbol{\beta}} = \mathbf{0}$ (which is an event of probability 0), $\ell(\tilde{\boldsymbol{\beta}}, \sigma; \underline{\mathbf{y}})$ is of the form (9.4); in the special case where $a = -(N/2)\log(2\pi)$, $c = (\underline{\mathbf{y}} - \mathbf{X}\tilde{\boldsymbol{\beta}})'(\underline{\mathbf{y}} - \mathbf{X}\tilde{\boldsymbol{\beta}})$, and $K = N$, $g(\sigma) = \ell(\tilde{\boldsymbol{\beta}}, \sigma; \underline{\mathbf{y}})$. Clearly,

$$\frac{dg(\sigma)}{d\sigma} = -\frac{K}{\sigma} + \frac{c}{\sigma^3} = -\frac{K}{\sigma^3}\left(\sigma^2 - \frac{c}{K}\right).$$

Thus, $dg(\sigma)/d\sigma > 0$ if $\sigma^2 < c/K$, $dg(\sigma)/d\sigma = 0$ if $\sigma^2 = c/K$, and $dg(\sigma)/d\sigma < 0$ if $\sigma^2 > c/K$, so that $g(\sigma)$ is an increasing function of σ for $\sigma < \sqrt{c/K}$, is a decreasing function for $\sigma > \sqrt{c/K}$, and attains its maximum value at $\sigma = \sqrt{c/K}$.

Unless the model is of full rank (i.e., unless $\operatorname{rank}\mathbf{X} = P$), there are an infinite number of solutions to the normal equations and hence an infinite number of values of $\boldsymbol{\beta}$ that maximize $\ell(\boldsymbol{\beta}, \sigma; \underline{\mathbf{y}})$. However, the value of an estimable linear combination $\boldsymbol{\lambda}'\boldsymbol{\beta}$ of the elements of $\boldsymbol{\beta}$ is the same for every value of $\boldsymbol{\beta}$ that maximizes $\ell(\boldsymbol{\beta}, \sigma; \underline{\mathbf{y}})$—recall (from the results of Section 5.4 on the method of least squares) that $\boldsymbol{\lambda}'\tilde{\mathbf{b}}$ has the same value for every solution $\tilde{\mathbf{b}}$ to the normal equations.

In effect, we have established that the least squares estimator of any estimable linear combination of the elements of $\boldsymbol{\beta}$ is also the ML estimator. Moreover, since condition (9.3) can be satisfied by taking $\tilde{\boldsymbol{\beta}} = (\mathbf{X}'\mathbf{X})^{-}\mathbf{X}'\underline{\mathbf{y}}$, the ML estimator of σ^2 (the square root of which is the ML estimator of σ) is the estimator

$$\frac{\tilde{\mathbf{e}}'\tilde{\mathbf{e}}}{N}, \tag{9.5}$$

where $\tilde{\mathbf{e}} = \mathbf{y} - \mathbf{P}_{\mathbf{X}}\mathbf{y}$. Like the unbiased estimator $\tilde{\mathbf{e}}'\tilde{\mathbf{e}}/(N - \operatorname{rank}\mathbf{X})$ and the Hodges–Lehmann estimator $\tilde{\mathbf{e}}'\tilde{\mathbf{e}}/[N - \operatorname{rank}(\mathbf{X}) + 2]$, the ML estimator of σ^2 is of the form (7.41).

A result on minimization and some results on matrices. As a preliminary to considering ML estimation as applied to a general linear model (or an Aitken model), it is convenient to establish the following result on minimization.

Theorem 5.9.1. Let $\mathbf{b}$ represent a $P \times 1$ vector of (unconstrained) variables, and define $f(\mathbf{b}) = (\mathbf{y} - \mathbf{X}\mathbf{b})'\mathbf{W}(\mathbf{y} - \mathbf{X}\mathbf{b})$, where $\mathbf{W}$ is an $N \times N$ symmetric nonnegative definite matrix, $\mathbf{X}$ is an $N \times P$ matrix, and $\mathbf{y}$ is an $N \times 1$ vector. Then, the linear system $\mathbf{X}'\mathbf{W}\mathbf{X}\mathbf{b} = \mathbf{X}'\mathbf{W}\mathbf{y}$ (in $\mathbf{b}$) is consistent. Further, $f(\mathbf{b})$ attains its minimum value at a point $\tilde{\mathbf{b}}$ if and only if $\tilde{\mathbf{b}}$ is a solution to $\mathbf{X}'\mathbf{W}\mathbf{X}\mathbf{b} = \mathbf{X}'\mathbf{W}\mathbf{y}$, in which case $f(\tilde{\mathbf{b}}) = \mathbf{y}'\mathbf{W}\mathbf{y} - \tilde{\mathbf{b}}'\mathbf{X}'\mathbf{W}\mathbf{y}$.

Proof. Let $\mathbf{R}$ represent a matrix such that $\mathbf{W} = \mathbf{R}'\mathbf{R}$—the existence of such a matrix is guaranteed by Corollary 2.13.25. Then, upon letting $\mathbf{t} = \mathbf{Ry}$ and $\mathbf{U} = \mathbf{RX}$, $f(\mathbf{b})$ is expressible as $f(\mathbf{b}) = (\mathbf{t} - \mathbf{Ub})'(\mathbf{t} - \mathbf{Ub})$. Moreover, it follows from the results of Section 5.4b (on least squares minimization) that the linear system $\mathbf{U}'\mathbf{Ub} = \mathbf{U}'\mathbf{t}$ (in $\mathbf{b}$) is consistent and that $(\mathbf{t} - \mathbf{Ub})'(\mathbf{t} - \mathbf{Ub})$ attains its minimum value at a point $\tilde{\mathbf{b}}$ if and only if $\tilde{\mathbf{b}}$ is a solution to $\mathbf{U}'\mathbf{Ub} = \mathbf{U}'\mathbf{t}$, in which case

$$(\mathbf{t} - \mathbf{U}\tilde{\mathbf{b}})'(\mathbf{t} - \mathbf{U}\tilde{\mathbf{b}}) = \mathbf{t}'\mathbf{t} - \tilde{\mathbf{b}}'\mathbf{U}'\mathbf{t}.$$

It remains only to observe that $\mathbf{U}'\mathbf{U} = \mathbf{X}'\mathbf{WX}$, that $\mathbf{U}'\mathbf{t} = \mathbf{X}'\mathbf{Wy}$, and that $\mathbf{t}'\mathbf{t} = \mathbf{y}'\mathbf{Wy}$. Q.E.D.

In addition to Theorem 5.9.1, it is convenient to have at our disposal the following lemma, which can be regarded as a generalization of Lemma 2.12.1.

Lemma 5.9.2. For any $N \times P$ matrix $\mathbf{X}$ and any $N \times N$ symmetric nonnegative definite matrix $\mathbf{W}$,

$$\mathcal{R}(\mathbf{X}'\mathbf{WX}) = \mathcal{R}(\mathbf{WX}), \quad \mathcal{C}(\mathbf{X}'\mathbf{WX}) = \mathcal{C}(\mathbf{X}'\mathbf{W}), \quad \text{and} \quad \text{rank}(\mathbf{X}'\mathbf{WX}) = \text{rank}(\mathbf{WX}).$$

Proof. In light of Corollary 2.13.25, $\mathbf{W} = \mathbf{R}'\mathbf{R}$ for some matrix $\mathbf{R}$. And upon observing that $\mathbf{X}'\mathbf{WX} = (\mathbf{RX})'\mathbf{RX}$ and making use of Corollary 2.4.4 and Lemma 2.12.1, we find that

$$\mathcal{R}(\mathbf{WX}) = \mathcal{R}(\mathbf{R}'\mathbf{RX}) \subset \mathcal{R}(\mathbf{RX}) = \mathcal{R}[(\mathbf{RX})'\mathbf{RX}] = \mathcal{R}(\mathbf{X}'\mathbf{WX}) \subset \mathcal{R}(\mathbf{WX})$$

and hence that $\mathcal{R}(\mathbf{X}'\mathbf{WX}) = \mathcal{R}(\mathbf{WX})$. Moreover, that $\mathcal{R}(\mathbf{X}'\mathbf{WX}) = \mathcal{R}(\mathbf{WX})$ implies that $\text{rank}(\mathbf{X}'\mathbf{WX}) = \text{rank}(\mathbf{WX})$ and, in light of Lemma 2.4.6, that $\mathcal{C}(\mathbf{X}'\mathbf{WX}) = \mathcal{C}(\mathbf{X}'\mathbf{W})$. Q.E.D.

In the special case of Lemma 5.9.2 where $\mathbf{W}$ is a (symmetric) positive definite matrix (and hence is nonsingular), it follows from Corollary 2.5.6 that $\mathcal{R}(\mathbf{WX}) = \mathcal{R}(\mathbf{X})$, $\mathcal{C}(\mathbf{X}'\mathbf{W}) = \mathcal{C}(\mathbf{X}')$, and $\text{rank}(\mathbf{WX}) = \text{rank}(\mathbf{X})$. Thus, we have the following corollary, which (like Lemma 5.9.2 itself) can be regarded as a generalization of Lemma 2.12.1.

Corollary 5.9.3. For any $N \times P$ matrix $\mathbf{X}$ and any $N \times N$ symmetric positive definite matrix $\mathbf{W}$,

$$\mathcal{R}(\mathbf{X}'\mathbf{WX}) = \mathcal{R}(\mathbf{X}), \quad \mathcal{C}(\mathbf{X}'\mathbf{WX}) = \mathcal{C}(\mathbf{X}'), \quad \text{and} \quad \text{rank}(\mathbf{X}'\mathbf{WX}) = \text{rank}(\mathbf{X}).$$

As an additional corollary of Lemma 5.9.2, we have the following result.

Corollary 5.9.4. For any $N \times P$ matrix $\mathbf{X}$ and any $N \times N$ symmetric nonnegative definite matrix $\mathbf{W}$,

$$\mathbf{WX}(\mathbf{X}'\mathbf{WX})^-\mathbf{X}'\mathbf{WX} = \mathbf{WX} \quad \text{and} \quad \mathbf{X}'\mathbf{WX}(\mathbf{X}'\mathbf{WX})^-\mathbf{X}'\mathbf{W} = \mathbf{X}'\mathbf{W}.$$

Proof. In light of Lemmas 5.9.2 and 2.4.3, $\mathbf{WX} = \mathbf{L}'\mathbf{X}'\mathbf{WX}$ for some $P \times N$ matrix $\mathbf{L}$. Thus,

$$\mathbf{WX}(\mathbf{X}'\mathbf{WX})^-\mathbf{X}'\mathbf{WX} = \mathbf{L}'\mathbf{X}'\mathbf{WX}(\mathbf{X}'\mathbf{WX})^-\mathbf{X}'\mathbf{WX} = \mathbf{L}'\mathbf{X}'\mathbf{WX} = \mathbf{WX}$$

and [since $\mathbf{X}'\mathbf{W} = (\mathbf{WX})' = (\mathbf{L}'\mathbf{X}'\mathbf{WX})' = \mathbf{X}'\mathbf{WXL}$]

$$\mathbf{X}'\mathbf{WX}(\mathbf{X}'\mathbf{WX})^-\mathbf{X}'\mathbf{W} = \mathbf{X}'\mathbf{WX}(\mathbf{X}'\mathbf{WX})^-\mathbf{X}'\mathbf{WXL} = \mathbf{X}'\mathbf{WXL} = \mathbf{X}'\mathbf{W}.$$

Q.E.D.

General linear model. Suppose that $\mathbf{y}$ is an $N \times 1$ observable random vector that follows a general linear model. Suppose further that the distribution of the vector $\mathbf{e}$ of residual effects is MVN, so that $\mathbf{y} \sim N[\mathbf{X}\boldsymbol{\beta}, \mathbf{V}(\boldsymbol{\theta})]$. And suppose that $\mathbf{V}(\boldsymbol{\theta})$ is of rank N (for every $\boldsymbol{\theta} \in \Theta$).

Let us consider the ML estimation of functions of the model's parameters (which consist of the elements $\beta_1, \beta_2, \ldots, \beta_P$ of the vector $\boldsymbol{\beta}$ and the elements $\theta_1, \theta_2, \ldots, \theta_T$ of the vector $\boldsymbol{\theta}$). Let $f(\cdot\,; \boldsymbol{\beta}, \boldsymbol{\theta})$ represent the pdf of the distribution of $\mathbf{y}$, and denote by $\underline{\mathbf{y}}$ the observed value of $\mathbf{y}$. Then, the likelihood function is the function, say $L(\boldsymbol{\beta}, \boldsymbol{\theta}; \underline{\mathbf{y}})$, of $\boldsymbol{\beta}$ and $\boldsymbol{\theta}$ defined (for $\boldsymbol{\beta} \in \mathbb{R}^P$ and $\boldsymbol{\theta} \in \Theta$) by $L(\boldsymbol{\beta}, \boldsymbol{\theta}; \underline{\mathbf{y}}) = f(\underline{\mathbf{y}}; \boldsymbol{\beta}, \boldsymbol{\theta})$. Accordingly,

$$L(\boldsymbol{\beta}, \boldsymbol{\theta}; \underline{\mathbf{y}}) = \frac{1}{(2\pi)^{N/2}|\mathbf{V}(\boldsymbol{\theta})|^{1/2}} \exp\left\{-\frac{1}{2}(\underline{\mathbf{y}} - \mathbf{X}\boldsymbol{\beta})'[\mathbf{V}(\boldsymbol{\theta})]^{-1}(\underline{\mathbf{y}} - \mathbf{X}\boldsymbol{\beta})\right\}. \tag{9.6}$$

And the log-likelihood function, say $\ell(\boldsymbol{\beta}, \boldsymbol{\theta}; \underline{\mathbf{y}})$, is expressible as

$$\ell(\boldsymbol{\beta},\boldsymbol{\theta};\underline{\mathbf{y}}) = -\frac{N}{2}\log(2\pi) - \frac{1}{2}\log|\mathbf{V}(\boldsymbol{\theta})| - \frac{1}{2}(\underline{\mathbf{y}} - \mathbf{X}\boldsymbol{\beta})'[\mathbf{V}(\boldsymbol{\theta})]^{-1}(\underline{\mathbf{y}} - \mathbf{X}\boldsymbol{\beta}). \tag{9.7}$$

Maximum likelihood estimates are obtained by maximizing $L(\boldsymbol{\beta},\boldsymbol{\theta};\underline{\mathbf{y}})$ or, equivalently, $\ell(\boldsymbol{\beta},\boldsymbol{\theta};\underline{\mathbf{y}})$ with respect to $\boldsymbol{\beta}$ and $\boldsymbol{\theta}$: if $L(\boldsymbol{\beta},\boldsymbol{\theta};\underline{\mathbf{y}})$ or $\ell(\boldsymbol{\beta},\boldsymbol{\theta};\underline{\mathbf{y}})$ attains its maximum value at values $\tilde{\boldsymbol{\beta}}$ and $\tilde{\boldsymbol{\theta}}$ (of $\boldsymbol{\beta}$ and $\boldsymbol{\theta}$, respectively), then an ML estimate of a function, say $h(\boldsymbol{\beta},\boldsymbol{\theta})$, of $\boldsymbol{\beta}$ and/or $\boldsymbol{\theta}$ is provided by the quantity $h(\tilde{\boldsymbol{\beta}},\tilde{\boldsymbol{\theta}})$ obtained by substituting $\tilde{\boldsymbol{\beta}}$ and $\tilde{\boldsymbol{\theta}}$ for $\boldsymbol{\beta}$ and $\boldsymbol{\theta}$. In considering the maximization of the likelihood or log-likelihood function, it is helpful to begin by regarding the value of $\boldsymbol{\theta}$ as "fixed" and considering the maximization of the likelihood or log-likelihood function with respect to $\boldsymbol{\beta}$ alone.

Observe [in light of result (4.5.5) and Corollary 2.13.12] that (regardless of the value of $\boldsymbol{\theta}$) $[\mathbf{V}(\boldsymbol{\theta})]^{-1}$ is a symmetric positive definite matrix. Accordingly, it follows from Theorem 5.9.1 that for any particular value of $\boldsymbol{\theta}$, the linear system

$$\mathbf{X}'[\mathbf{V}(\boldsymbol{\theta})]^{-1}\mathbf{X}\mathbf{b} = \mathbf{X}'[\mathbf{V}(\boldsymbol{\theta})]^{-1}\underline{\mathbf{y}} \tag{9.8}$$

(in the $P \times 1$ vector $\mathbf{b}$) is consistent. Further, $(\underline{\mathbf{y}}-\mathbf{X}\boldsymbol{\beta})'[\mathbf{V}(\boldsymbol{\theta})]^{-1}(\underline{\mathbf{y}}-\mathbf{X}\boldsymbol{\beta})$ attains its minimum value, or equivalently $-(1/2)(\underline{\mathbf{y}}-\mathbf{X}\boldsymbol{\beta})'[\mathbf{V}(\boldsymbol{\theta})]^{-1}(\underline{\mathbf{y}}-\mathbf{X}\boldsymbol{\beta})$ attains its maximum value, at a value $\tilde{\boldsymbol{\beta}}(\boldsymbol{\theta})$ of $\boldsymbol{\beta}$ if and only if $\tilde{\boldsymbol{\beta}}(\boldsymbol{\theta})$ is a solution to linear system (9.8), that is, if and only if

$$\mathbf{X}'[\mathbf{V}(\boldsymbol{\theta})]^{-1}\mathbf{X}\tilde{\boldsymbol{\beta}}(\boldsymbol{\theta}) = \mathbf{X}'[\mathbf{V}(\boldsymbol{\theta})]^{-1}\underline{\mathbf{y}}, \tag{9.9}$$

in which case

$$\begin{aligned}\max_{\boldsymbol{\beta}\in\mathbb{R}^P} -\frac{1}{2}(\underline{\mathbf{y}}-\mathbf{X}\boldsymbol{\beta})'[\mathbf{V}(\boldsymbol{\theta})]^{-1}(\underline{\mathbf{y}}-\mathbf{X}\boldsymbol{\beta}) &\\ = -\frac{1}{2}[\underline{\mathbf{y}}-\mathbf{X}\tilde{\boldsymbol{\beta}}(\boldsymbol{\theta})]'[\mathbf{V}(\boldsymbol{\theta})]^{-1}[\underline{\mathbf{y}}-\mathbf{X}\tilde{\boldsymbol{\beta}}(\boldsymbol{\theta})] &\\ = -\frac{1}{2}\{\underline{\mathbf{y}}'[\mathbf{V}(\boldsymbol{\theta})]^{-1}\underline{\mathbf{y}} - [\tilde{\boldsymbol{\beta}}(\boldsymbol{\theta})]'\mathbf{X}'[\mathbf{V}(\boldsymbol{\theta})]^{-1}\underline{\mathbf{y}}\}. &\end{aligned} \tag{9.10}$$

Now, suppose that (for $\boldsymbol{\theta} \in \Theta$) $\tilde{\boldsymbol{\beta}}(\boldsymbol{\theta})$ satisfies condition (9.9). Then, for any matrix $\mathbf{A}$ such that $\mathcal{R}(\mathbf{A}) \subset \mathcal{R}(\mathbf{X})$, the value of $\mathbf{A}\tilde{\boldsymbol{\beta}}(\boldsymbol{\theta})$ (at any particular value of $\boldsymbol{\theta}$) does not depend on the choice of $\tilde{\boldsymbol{\beta}}(\boldsymbol{\theta})$, as is evident upon observing (in light of Corollary 5.9.3) that $\mathbf{A} = \mathbf{T}(\boldsymbol{\theta})\mathbf{X}'[\mathbf{V}(\boldsymbol{\theta})]^{-1}\mathbf{X}$ for some matrix-valued function $\mathbf{T}(\boldsymbol{\theta})$ of $\boldsymbol{\theta}$ and hence that

$$\mathbf{A}\tilde{\boldsymbol{\beta}}(\boldsymbol{\theta}) = \mathbf{T}(\boldsymbol{\theta})\mathbf{X}'[\mathbf{V}(\boldsymbol{\theta})]^{-1}\mathbf{X}\tilde{\boldsymbol{\beta}}(\boldsymbol{\theta}) = \mathbf{T}(\boldsymbol{\theta})\mathbf{X}'[\mathbf{V}(\boldsymbol{\theta})]^{-1}\underline{\mathbf{y}}.$$

Thus, $\mathbf{X}\tilde{\boldsymbol{\beta}}(\boldsymbol{\theta})$ does not depend on the choice of $\tilde{\boldsymbol{\beta}}(\boldsymbol{\theta})$, and for any estimable linear combination $\boldsymbol{\lambda}'\boldsymbol{\beta}$ of the elements of $\boldsymbol{\beta}$, $\boldsymbol{\lambda}'\tilde{\boldsymbol{\beta}}(\boldsymbol{\theta})$ does not depend on the choice of $\tilde{\boldsymbol{\beta}}(\boldsymbol{\theta})$. Among the possible choices for $\tilde{\boldsymbol{\beta}}(\boldsymbol{\theta})$ are the vector $\{\mathbf{X}'[\mathbf{V}(\boldsymbol{\theta})]^{-1}\mathbf{X}\}^-\mathbf{X}'[\mathbf{V}(\boldsymbol{\theta})]^{-1}\underline{\mathbf{y}}$ and the vector $(\{\mathbf{X}'[\mathbf{V}(\boldsymbol{\theta})]^{-1}\mathbf{X}\}^-)'\mathbf{X}'[\mathbf{V}(\boldsymbol{\theta})]^{-1}\underline{\mathbf{y}}$.

Define

$$L_*(\boldsymbol{\theta};\underline{\mathbf{y}}) = L[\tilde{\boldsymbol{\beta}}(\boldsymbol{\theta}),\boldsymbol{\theta};\underline{\mathbf{y}}] \quad \text{and} \quad \ell_*(\boldsymbol{\theta};\underline{\mathbf{y}}) = \ell[\tilde{\boldsymbol{\beta}}(\boldsymbol{\theta}),\boldsymbol{\theta};\underline{\mathbf{y}}] \;[=\log L_*(\boldsymbol{\theta};\underline{\mathbf{y}})]. \tag{9.11}$$

Then,

$$L_*(\boldsymbol{\theta};\underline{\mathbf{y}}) = \max_{\boldsymbol{\beta}\in\mathbb{R}^P} L(\boldsymbol{\beta},\boldsymbol{\theta};\underline{\mathbf{y}}) \quad \text{and} \quad \ell_*(\boldsymbol{\theta};\underline{\mathbf{y}}) = \max_{\boldsymbol{\beta}\in\mathbb{R}^P} \ell(\boldsymbol{\beta},\boldsymbol{\theta};\underline{\mathbf{y}}), \tag{9.12}$$

so that $L_*(\boldsymbol{\theta};\underline{\mathbf{y}})$ is a profile likelihood function and $\ell_*(\boldsymbol{\theta};\underline{\mathbf{y}})$ is a profile log-likelihood function—refer, e.g., to Severini (2000, sec 4.6) for the definition of a profile likelihood or profile log-likelihood function. Moreover,

$$\ell_*(\boldsymbol{\theta};\underline{\mathbf{y}}) = -\frac{N}{2}\log(2\pi) - \frac{1}{2}\log|\mathbf{V}(\boldsymbol{\theta})| - \frac{1}{2}[\underline{\mathbf{y}}-\mathbf{X}\tilde{\boldsymbol{\beta}}(\boldsymbol{\theta})]'[\mathbf{V}(\boldsymbol{\theta})]^{-1}[\underline{\mathbf{y}}-\mathbf{X}\tilde{\boldsymbol{\beta}}(\boldsymbol{\theta})] \tag{9.13}$$

$$= -\frac{N}{2}\log(2\pi) - \frac{1}{2}\log|\mathbf{V}(\boldsymbol{\theta})| - \frac{1}{2}\{\underline{\mathbf{y}}'[\mathbf{V}(\boldsymbol{\theta})]^{-1}\underline{\mathbf{y}} - [\tilde{\boldsymbol{\beta}}(\boldsymbol{\theta})]'\mathbf{X}'[\mathbf{V}(\boldsymbol{\theta})]^{-1}\underline{\mathbf{y}}\} \tag{9.14}$$

$$\begin{aligned}&= -\frac{N}{2}\log(2\pi) - \frac{1}{2}\log|\mathbf{V}(\boldsymbol{\theta})| \\ &\quad -\frac{1}{2}\underline{\mathbf{y}}'\big([\mathbf{V}(\boldsymbol{\theta})]^{-1} - [\mathbf{V}(\boldsymbol{\theta})]^{-1}\mathbf{X}\{\mathbf{X}'[\mathbf{V}(\boldsymbol{\theta})]^{-1}\mathbf{X}\}^-\mathbf{X}'[\mathbf{V}(\boldsymbol{\theta})]^{-1}\big)\underline{\mathbf{y}}.\end{aligned} \tag{9.15}$$

Result (9.12) is significant from a computational standpoint. It "reduces" the problem of maximizing $L(\boldsymbol{\beta}, \boldsymbol{\theta}; \underset{\sim}{y})$ or $\ell(\boldsymbol{\beta}, \boldsymbol{\theta}; \underset{\sim}{y})$ with respect to $\boldsymbol{\beta}$ and $\boldsymbol{\theta}$ to that of maximizing $L_*(\boldsymbol{\theta}; \underset{\sim}{y})$ or $\ell_*(\boldsymbol{\theta}; \underset{\sim}{y})$ with respect to $\boldsymbol{\theta}$ alone. Values of $\boldsymbol{\beta}$ and $\boldsymbol{\theta}$ at which $L(\boldsymbol{\beta}, \boldsymbol{\theta}; \underset{\sim}{y})$ or $\ell(\boldsymbol{\beta}, \boldsymbol{\theta}; \underset{\sim}{y})$ attains its maximum value can be obtained by taking the value of $\boldsymbol{\theta}$ to be a value, say $\tilde{\boldsymbol{\theta}}$, at which $L_*(\boldsymbol{\theta}; \underset{\sim}{y})$ or $\ell_*(\boldsymbol{\theta}; \underset{\sim}{y})$ attains its maximum value and by then taking the value of $\boldsymbol{\beta}$ to be a solution $\tilde{\boldsymbol{\beta}}(\tilde{\boldsymbol{\theta}})$ to the linear system

$$\mathbf{X}'[\mathbf{V}(\tilde{\boldsymbol{\theta}})]^{-1}\mathbf{X}\mathbf{b} = \mathbf{X}'[\mathbf{V}(\tilde{\boldsymbol{\theta}})]^{-1}\underset{\sim}{y}$$

(in the $P \times 1$ vector $\mathbf{b}$).

In general, a solution to the problem of maximizing $\ell_*(\boldsymbol{\theta}; \underset{\sim}{y})$ is not obtainable in "closed form"; rather, the maximization must be accomplished numerically via an iterative procedure—the discussion of such procedures is deferred until later in the book. Nevertheless, there are special cases where the maximization of $\ell_*(\boldsymbol{\theta}; \underset{\sim}{y})$, and hence that of $\ell(\boldsymbol{\beta}, \boldsymbol{\theta}; \underset{\sim}{y})$, can be accomplished without resort to indirect (iterative) numerical methods. Indirect numerical methods are not needed in the special case where $\mathbf{y}$ follows a G–M model; that special case was discussed in Part 1 of the present subsection. More generally, indirect numerical methods are not needed in the special case where $\mathbf{y}$ follows an Aitken model, as is to be demonstrated in what follows.

Aitken model. Suppose that $\mathbf{y}$ follows an Aitken model (and that $\mathbf{H}$ is nonsingular and that the distribution of $\mathbf{e}$ is MVN). And regard the Aitken model as the special case of the general linear model where $T = 1$ (i.e., where $\boldsymbol{\theta}$ has only 1 element), where $\theta_1 = \sigma$, and where $\mathbf{V}(\boldsymbol{\theta}) = \sigma^2\mathbf{H}$. In that special case, linear system (9.8) is equivalent to the linear system

$$\mathbf{X}'\mathbf{H}^{-1}\mathbf{X}\mathbf{b} = \mathbf{X}'\mathbf{H}^{-1}\underset{\sim}{y} \tag{9.16}$$

—the equivalence is in the sense that both linear systems have the same set of solutions. The equations comprising linear system (9.16) are known as the Aitken equations. When $\mathbf{H} = \mathbf{I}$ (i.e., when the model is a G–M model), the linear system (9.16) of Aitken equations simplifies to the linear system $\mathbf{X}'\mathbf{X}\mathbf{b} = \mathbf{X}'\underset{\sim}{y}$ of normal equations.

In this setting, we are free to choose the vector $\tilde{\beta}(\boldsymbol{\theta})$ in such a way that it has the same value for every value of $\boldsymbol{\theta}$. Accordingly, for every value of $\boldsymbol{\theta}$, take $\tilde{\beta}(\boldsymbol{\theta})$ to be $\tilde{\beta}$, where $\tilde{\beta}$ is any solution to the Aitken equations. Then, writing σ for $\boldsymbol{\theta}$, the profile log-likelihood function $\ell_*(\sigma; \underset{\sim}{y})$ is expressible as

$$\ell_*(\sigma; \underset{\sim}{y}) = \ell(\tilde{\beta}, \sigma; \underset{\sim}{y}) = -\frac{N}{2}\log(2\pi) - \frac{N}{2}\log\sigma^2 - \frac{1}{2}\log|\mathbf{H}| - \frac{1}{2\sigma^2}(\underset{\sim}{y} - \mathbf{X}\tilde{\boldsymbol{\beta}})'\mathbf{H}^{-1}(\underset{\sim}{y} - \mathbf{X}\tilde{\boldsymbol{\beta}}).$$

Unless $\underset{\sim}{y} - \mathbf{X}\tilde{\boldsymbol{\beta}} = \mathbf{0}$ (which is an event of probability 0), $\ell_*(\sigma; \underset{\sim}{y})$ is of the form of the function $g(\sigma)$ defined (in Part 1 of the present subsection) by equality (9.4); upon setting $a = -(N/2)\log(2\pi) - (1/2)\log|\mathbf{H}|$, $c = (\underset{\sim}{y} - \mathbf{X}\tilde{\boldsymbol{\beta}})'\mathbf{H}^{-1}(\underset{\sim}{y} - \mathbf{X}\tilde{\boldsymbol{\beta}})$, and $K = N$, $g(\sigma) = \ell_*(\sigma; \underset{\sim}{y})$. Thus, it follows from the results of Part 1 that $\ell_*(\sigma; \underset{\sim}{y})$ attains its maximum value when σ^2 equals

$$\frac{(\underset{\sim}{y} - \mathbf{X}\tilde{\boldsymbol{\beta}})'\mathbf{H}^{-1}(\underset{\sim}{y} - \mathbf{X}\tilde{\boldsymbol{\beta}})}{N}. \tag{9.17}$$

And we conclude that $\ell(\boldsymbol{\beta}, \sigma; \underset{\sim}{y})$ attains its maximum value when $\boldsymbol{\beta}$ equals $\tilde{\boldsymbol{\beta}}$ and when σ^2 equals the quantity (9.17). This conclusion serves to generalize the conclusion reached in Part 1, where it was determined that in the special case of the G–M model, the log-likelihood function attains its maximum value when β equals a solution, say $\tilde{\boldsymbol{\beta}}$, to the normal equations (i.e., to the linear system $\mathbf{X}'\mathbf{X}\mathbf{b} = \mathbf{X}'\underset{\sim}{y}$) and when σ^2 equals $(\underset{\sim}{y} - \mathbf{X}\tilde{\boldsymbol{\beta}})'(\underset{\sim}{y} - \mathbf{X}\tilde{\boldsymbol{\beta}})/N$.

b. Restricted or residual maximum likelihood estimation (REML estimation)

Suppose that $\mathbf{y}$ is an $N \times 1$ observable random vector that follows a general linear model. Suppose further that the distribution of the vector $\mathbf{e}$ of residual effects is MVN, so that $\mathbf{y} \sim N[\mathbf{X}\boldsymbol{\beta}, \mathbf{V}(\boldsymbol{\theta})]$.

And let $\ell(\boldsymbol{\beta}, \boldsymbol{\theta}; \underline{\mathbf{y}})$ represent the log-likelihood function [where $\underline{\mathbf{y}}$ is the observed value of $\mathbf{y}$ and where $\mathbf{V}(\boldsymbol{\theta})$ is assumed to be of rank N (for every $\boldsymbol{\theta} \in \Theta$)]. This function has the representation (9.7).

Suppose that $\tilde{\boldsymbol{\beta}}$ and $\tilde{\boldsymbol{\theta}}$ are values of $\boldsymbol{\beta}$ and $\boldsymbol{\theta}$ at which $\ell(\boldsymbol{\beta}, \boldsymbol{\theta}; \underline{\mathbf{y}})$ attains its maximum value. And observe that $\tilde{\boldsymbol{\theta}}$ is a value of $\boldsymbol{\theta}$ at which $\ell(\tilde{\boldsymbol{\beta}}, \boldsymbol{\theta}; \underline{\mathbf{y}})$ attains its maximum value. There is an implication that $\tilde{\boldsymbol{\theta}}$ is identical to the value of $\boldsymbol{\theta}$ that would be obtained from maximizing the likelihood function under a supposition that $\boldsymbol{\beta}$ is a known $(P \times 1)$ vector (rather than a vector of unknown parameters) and under the further supposition that $\boldsymbol{\beta}$ equals $\tilde{\boldsymbol{\beta}}$ (or, perhaps more precisely, $\mathbf{X}\boldsymbol{\beta}$ equals $\mathbf{X}\tilde{\boldsymbol{\beta}}$). Thus, in a certain sense, maximum likelihood estimators of functions of $\boldsymbol{\theta}$ fail to account for the estimation of $\boldsymbol{\beta}$. This failure can be disconcerting and can have undesirable consequences.

It is informative to consider the manifestation of this phenomenon in the relatively simple special case of a G–M model. In that special case, the use of maximum likelihood estimation results in σ^2 being estimated by the quantity (9.5), in which the residual sum of squares is divided by N rather than by $N - \operatorname{rank} \mathbf{X}$ as in the case of the unbiased estimator [or by $N - \operatorname{rank}(\mathbf{X}) + 2$ as in the case of the Hodges–Lehmann estimator].

The failure of ML estimators of functions of $\boldsymbol{\theta}$ to account for the estimation of $\boldsymbol{\beta}$ has led to the widespread use of a variant of maximum likelihood that has come to be known by the acronym REML (which is regarded by some as standing for restricted maximum likelihood and by others as standing for residual maximum likelihood). In REML, inferences about functions of $\boldsymbol{\theta}$ are based on the likelihood function associated with a vector of what are sometimes called error contrasts.

An *error contrast* is a linear unbiased estimator of 0, that is, a linear combination, say $\mathbf{r}'\mathbf{y}$, of the elements of $\mathbf{y}$ such that $E(\mathbf{r}'\mathbf{y}) = 0$ or, equivalently, such that $\mathbf{X}'\mathbf{r} = \mathbf{0}$. Thus, $\mathbf{r}'\mathbf{y}$ is an error contrast if and only if $\mathbf{r} \in \mathcal{N}(\mathbf{X}')$. Moreover, in light of Lemma 2.11.5,

$$\dim[\mathcal{N}(\mathbf{X}'] = N - \operatorname{rank}(\mathbf{X}') = N - \operatorname{rank} \mathbf{X}.$$

And it follows that there exists a set of $N - \operatorname{rank} \mathbf{X}$ linearly independent error contrasts and that no set of error contrasts contains more than $N - \operatorname{rank} \mathbf{X}$ linearly independent error contrasts.

Accordingly, let $\mathbf{R}$ represent an $N \times (N - \operatorname{rank} \mathbf{X})$ matrix (of constants) of full column rank $N - \operatorname{rank} \mathbf{X}$ such that $\mathbf{X}'\mathbf{R} = \mathbf{0}$ [or, equivalently, an $N \times (N - \operatorname{rank} \mathbf{X})$ matrix whose columns are linearly independent members of the null space $\mathcal{N}(\mathbf{X}')$ of $\mathbf{X}'$]. And take $\mathbf{z}$ to be the $(N - \operatorname{rank} \mathbf{X}) \times 1$ vector defined by $\mathbf{z} = \mathbf{R}'\mathbf{y}$ (so that the elements of $\mathbf{z}$ are $N - \operatorname{rank} \mathbf{X}$ linearly independent error contrasts). Then, $\mathbf{z} \sim N[\mathbf{0}, \mathbf{R}'\mathbf{V}(\boldsymbol{\theta})\mathbf{R}]$, and [in light of the assumption that $\mathbf{V}(\boldsymbol{\theta})$ is nonsingular and in light of Theorem 2.13.10] $\mathbf{R}'\mathbf{V}(\boldsymbol{\theta})\mathbf{R}$ is nonsingular. Further, let $f(\cdot\,; \boldsymbol{\theta})$ represent the pdf of the distribution of $\mathbf{z}$, and take $L(\boldsymbol{\theta}; \mathbf{R}'\underline{\mathbf{y}})$ to be the function of $\boldsymbol{\theta}$ defined (for $\boldsymbol{\theta} \in \Theta$) by $L(\boldsymbol{\theta}; \mathbf{R}'\underline{\mathbf{y}}) = f(\mathbf{R}'\underline{\mathbf{y}}; \boldsymbol{\theta})$. The function $L(\boldsymbol{\theta}; \mathbf{R}'\underline{\mathbf{y}})$ is a likelihood function; it is the likelihood function obtained by regarding the observed value of $\mathbf{z}$ as the data vector. In REML, the inferences about functions of $\boldsymbol{\theta}$ are based on the likelihood function $L(\boldsymbol{\theta}; \mathbf{R}'\underline{\mathbf{y}})$ [or on a likelihood function that is equivalent to $L(\boldsymbol{\theta}; \mathbf{R}'\underline{\mathbf{y}})$ in the sense that it differs from $L(\boldsymbol{\theta}; \mathbf{R}'\underline{\mathbf{y}})$ by no more than a multiplicative constant].

It is worth noting that the use of REML results in the same inferences regardless of the choice of the matrix $\mathbf{R}$. To see that REML has this property, let $\mathbf{R}_1$ and $\mathbf{R}_2$ represent any two choices for $\mathbf{R}$, that is, take $\mathbf{R}_1$ and $\mathbf{R}_2$ to be any two $N \times (N - \operatorname{rank} \mathbf{X})$ matrices of full column rank such that $\mathbf{X}'\mathbf{R}_1 = \mathbf{X}'\mathbf{R}_2 = \mathbf{0}$. Further, define $\mathbf{z}_1 = \mathbf{R}_1'\mathbf{y}$ and $\mathbf{z}_2 = \mathbf{R}_2'\mathbf{y}$. And let $f_1(\cdot\,; \boldsymbol{\theta})$ represent the pdf of the distribution of $\mathbf{z}_1$ and $f_2(\cdot\,; \boldsymbol{\theta})$ the pdf of the distribution of $\mathbf{z}_2$; and take $L_1(\boldsymbol{\theta}; \mathbf{R}_1'\underline{\mathbf{y}})$ and $L_2(\boldsymbol{\theta}; \mathbf{R}_2'\underline{\mathbf{y}})$ to be the functions of $\boldsymbol{\theta}$ defined by $L_1(\boldsymbol{\theta}; \mathbf{R}_1'\underline{\mathbf{y}}) = f_1(\mathbf{R}_1'\underline{\mathbf{y}}; \boldsymbol{\theta})$ and $L_2(\boldsymbol{\theta}; \mathbf{R}_2'\underline{\mathbf{y}}) = f_2(\mathbf{R}_2'\underline{\mathbf{y}}; \boldsymbol{\theta})$.

There exists an $(N - \operatorname{rank} \mathbf{X}) \times (N - \operatorname{rank} \mathbf{X})$ matrix $\mathbf{A}$ such that $\mathbf{R}_2 = \mathbf{R}_1\mathbf{A}$, as is evident upon observing that the columns of each of the two matrices $\mathbf{R}_1$ and $\mathbf{R}_2$ form a basis for the $(N - \operatorname{rank} \mathbf{X})$-dimensional linear space $\mathcal{N}(\mathbf{X}')$; necessarily, $\mathbf{A}$ is nonsingular. Moreover, the pdf's of the distributions of $\mathbf{z}_1$ and $\mathbf{z}_2$ are such that (for every value of $\mathbf{z}_1$)

$$f_1(\mathbf{z}_1) = |\det \mathbf{A}|\, f_2(\mathbf{A}'\mathbf{z}_1)$$

—this relationship can be verified directly from formula (3.5.32) for the pdf of an MVN distribution or simply by observing that $\mathbf{z}_2 = \mathbf{A}'\mathbf{z}_1$ and making use of standard results (e.g., Bickel and Doksum 2001, sec. B.2) on a change of variables. Thus,

$$L_2(\boldsymbol{\theta}; \mathbf{R}_2'\underline{\mathbf{y}}) = f_2(\mathbf{R}_2'\underline{\mathbf{y}}; \boldsymbol{\theta}) = f_2(\mathbf{A}'\mathbf{R}_1'\underline{\mathbf{y}}; \boldsymbol{\theta}) = |\det \mathbf{A}|^{-1} f_1(\mathbf{R}_1'\underline{\mathbf{y}}; \boldsymbol{\theta}) = |\det \mathbf{A}|^{-1} L_1(\boldsymbol{\theta}; \mathbf{R}_1'\underline{\mathbf{y}}).$$

We conclude that the two likelihood functions $L_1(\boldsymbol{\theta}; \mathbf{R}_1'\underline{\mathbf{y}})$ and $L_2(\boldsymbol{\theta}; \mathbf{R}_2'\underline{\mathbf{y}})$ differ from each other by no more than a multiplicative constant and hence that they are equivalent.

The $(N - \operatorname{rank} \mathbf{X})$-dimensional vector $\mathbf{z} = \mathbf{R}'\mathbf{y}$ of error contrasts is translation invariant, as is evident upon observing that for every $P \times 1$ vector $\mathbf{k}$ (and every value of $\mathbf{y}$),

$$\mathbf{R}'(\mathbf{y} + \mathbf{X}\mathbf{k}) = \mathbf{R}'\mathbf{y} + (\mathbf{X}'\mathbf{R})'\mathbf{k} = \mathbf{R}'\mathbf{y} + \mathbf{0}\mathbf{k} = \mathbf{R}'\mathbf{y}.$$

In fact, $\mathbf{z}$ is a maximal invariant: in the present context, a (possibly vector-valued) statistic $\mathbf{h}(\mathbf{y})$ is said to be a *maximal invariant* if it is invariant and if corresponding to each pair of values $\mathbf{y}_1$ and $\mathbf{y}_2$ of $\mathbf{y}$ such that $\mathbf{h}(\mathbf{y}_2) = \mathbf{h}(\mathbf{y}_1)$, there exists a $P \times 1$ vector $\mathbf{k}$ such that $\mathbf{y}_2 = \mathbf{y}_1 + \mathbf{X}\mathbf{k}$—refer, e.g., to Lehmann and Romano (2005b, sec 6.2) for a general definition (of a maximal invariant).

To confirm that $\mathbf{z}$ is a maximal invariant, take $\mathbf{y}_1$ and $\mathbf{y}_2$ to be any pair of values of $\mathbf{y}$ such that $\mathbf{R}'\mathbf{y}_2 = \mathbf{R}'\mathbf{y}_1$. And observe that $\mathbf{y}_2 = \mathbf{y}_1 + (\mathbf{y}_2 - \mathbf{y}_1)$ and that $\mathbf{y}_2 - \mathbf{y}_1 \in \mathcal{N}(\mathbf{R}')$. Observe also (in light of Lemma 2.11.5) that $\dim[\mathcal{N}(\mathbf{R}')] = \operatorname{rank} \mathbf{X}$. Moreover, $\mathbf{R}'\mathbf{X} = (\mathbf{X}'\mathbf{R})' = \mathbf{0}$, implying (in light of Lemma 2.4.2) that $\mathcal{C}(\mathbf{X}) \subset \mathcal{N}(\mathbf{R}')$ and hence (in light of Theorem 2.4.10) that $\mathcal{C}(\mathbf{X}) = \mathcal{N}(\mathbf{R}')$. Thus, the linear space $\mathcal{N}(\mathbf{R}')$ is spanned by the columns of $\mathbf{X}$, leading to the conclusion that there exists a $P \times 1$ vector $\mathbf{k}$ such that $\mathbf{y}_2 - \mathbf{y}_1 = \mathbf{X}\mathbf{k}$ and hence such that $\mathbf{y}_2 = \mathbf{y}_1 + \mathbf{X}\mathbf{k}$.

That $\mathbf{z} = \mathbf{R}'\mathbf{y}$ is a maximal invariant is of interest because any maximal invariant, say $\mathbf{h}(\mathbf{y})$, has (in the present context) the following property: a (possibly vector-valued) statistic, say $\mathbf{g}(\mathbf{y})$, is translation invariant if and only if $\mathbf{g}(\mathbf{y})$ depends on the value of $\mathbf{y}$ only through $\mathbf{h}(\mathbf{y})$, that is, if and only if there exists a function $\mathbf{s}(\cdot)$ such that $\mathbf{g}(\mathbf{y}) = \mathbf{s}[\mathbf{h}(\mathbf{y})]$ (for every value of $\mathbf{y}$). To see that $\mathbf{h}(\mathbf{y})$ has this property, observe that if [for some function $\mathbf{s}(\cdot)$] $\mathbf{g}(\mathbf{y}) = \mathbf{s}[\mathbf{h}(\mathbf{y})]$ (for every value of $\mathbf{y}$), then (for every $P \times 1$ vector $\mathbf{k}$)

$$\mathbf{g}(\mathbf{y} + \mathbf{X}\mathbf{k}) = \mathbf{s}[\mathbf{h}(\mathbf{y} + \mathbf{X}\mathbf{k})] = \mathbf{s}[\mathbf{h}(\mathbf{y})] = \mathbf{g}(\mathbf{y}),$$

so that $\mathbf{g}(\mathbf{y})$ is translation invariant. Conversely, if $\mathbf{g}(\mathbf{y})$ is translation invariant and if $\mathbf{y}_1$ and $\mathbf{y}_2$ are any pair of values of $\mathbf{y}$ such that $\mathbf{h}(\mathbf{y}_2) = \mathbf{h}(\mathbf{y}_1)$, then $\mathbf{y}_2 = \mathbf{y}_1 + \mathbf{X}\mathbf{k}$ for some vector $\mathbf{k}$ and, consequently, $\mathbf{g}(\mathbf{y}_2) = \mathbf{g}(\mathbf{y}_1 + \mathbf{X}\mathbf{k}) = \mathbf{g}(\mathbf{y}_1)$.

The vector $\mathbf{z}$ consists of $N - \operatorname{rank} \mathbf{X}$ linearly independent linear combinations of the elements of the $N \times 1$ vector $\mathbf{y}$. Suppose that we introduce an additional $\operatorname{rank} \mathbf{X}$ linear combinations in the form of the $(\operatorname{rank} \mathbf{X}) \times 1$ vector $\mathbf{u}$ defined by $\mathbf{u} = \mathbf{X}_*'\mathbf{y}$, where $\mathbf{X}_*$ is any $N \times (\operatorname{rank} \mathbf{X})$ matrix (of constants) whose columns are linearly independent columns of $\mathbf{X}$ or, more generally, whose columns form a basis for $\mathcal{C}(\mathbf{X})$. Then,

$$\begin{pmatrix} \mathbf{u} \\ \mathbf{z} \end{pmatrix} = (\mathbf{X}_*, \mathbf{R})'\mathbf{y}.$$

And (since $\mathbf{X}_* = \mathbf{X}\mathbf{A}$ for some matrix $\mathbf{A}$)

$$\begin{aligned} \operatorname{rank}(\mathbf{X}_*, \mathbf{R}) &= \operatorname{rank}[(\mathbf{X}_*, \mathbf{R})'(\mathbf{X}_*, \mathbf{R})] \\ &= \operatorname{rank} \operatorname{diag}(\mathbf{X}_*'\mathbf{X}_*, \mathbf{R}'\mathbf{R}) \\ &= \operatorname{rank}(\mathbf{X}_*'\mathbf{X}_*) + \operatorname{rank}(\mathbf{R}'\mathbf{R}) \\ &= \operatorname{rank} \mathbf{X}_* + \operatorname{rank} \mathbf{R} \\ &= \operatorname{rank}(\mathbf{X}) + N - \operatorname{rank}(\mathbf{X}) = N. \end{aligned} \tag{9.18}$$

Accordingly, the likelihood function that would result from regarding the observed value $(\mathbf{X}_*, \mathbf{R})'\underline{\mathbf{y}}$ of $\begin{pmatrix} \mathbf{u} \\ \mathbf{z} \end{pmatrix}$ as the data vector differs by no more than a multiplicative constant from that obtained by regarding the observed value $\underset{\sim}{y}$ of $\mathbf{y}$ as the data vector (as can be readily verified). When viewed in this context, the likelihood function that is employed in REML can be regarded as what is known as a marginal likelihood—refer, e.g., to Pawitan (2001, sec. 10.3) for the definition of a marginal likelihood.

The vector $\tilde{\mathbf{e}} = (\mathbf{I} - \mathbf{P_X})\mathbf{y}$ [where $\mathbf{P_X} = \mathbf{X}(\mathbf{X}'\mathbf{X})^-\mathbf{X}'$] is the vector of (least squares) residuals. Observe [in light of Theorem 2.12.2 and Lemma 2.8.4] that $\mathbf{X}'(\mathbf{I} - \mathbf{P_X}) = \mathbf{0}$ and that

$$\text{rank}(\mathbf{I} - \mathbf{P_X}) = N - \text{rank}\,\mathbf{P_X} = N - \text{rank}\,\mathbf{X}. \tag{9.19}$$

Thus, among the choices for the $N \times (N - \text{rank}\,\mathbf{X})$ matrix $\mathbf{R}$ (of full column rank $N - \text{rank}\,\mathbf{X}$ such that $\mathbf{X}'\mathbf{R} = \mathbf{0}$) is any $N \times (N - \text{rank}\,\mathbf{X})$ matrix whose columns are a linearly independent subset of the columns of the (symmetric) matrix $\mathbf{I} - \mathbf{P_X}$. For any such choice of $\mathbf{R}$, the elements of the $(N - \text{rank}\,\mathbf{X}) \times 1$ vector $\mathbf{z} = \mathbf{R}'\mathbf{y}$ consist of linearly independent (least squares) residuals.

The letters R and E in the acronym REML can be regarded as representing either restricted or residual. REML is restricted ML in the sense that in the formation of the likelihood function, the data are restricted to those inherent in the values of the $N - \text{rank}\,\mathbf{X}$ linearly independent error contrasts. REML is residual ML in the sense that the $N - \text{rank}\,\mathbf{X}$ linearly independent error contrasts can be taken to be (least squares) residuals.

It might seem as though the use of REML would result in the loss of some information about functions of $\boldsymbol{\theta}$. However, in at least one regard, there is no loss of information. Consider the profile likelihood function $L_*(\cdot\,;\underset{\sim}{y})$ or profile log-likelihood function $\ell_*(\cdot\,;\underset{\sim}{y})$ of definition (9.11)—the (ordinary) ML estimate of a function of $\boldsymbol{\theta}$ is obtained from a value of $\boldsymbol{\theta}$ at which $L_*(\boldsymbol{\theta};\underset{\sim}{y})$ or $\ell_*(\boldsymbol{\theta};\underset{\sim}{y})$ attains its maximum value. The identity of the function $L_*(\cdot\,;\underset{\sim}{y})$ or, equivalently, that of the function $\ell_*(\cdot\,;\underset{\sim}{y})$ can be determined solely from knowledge of the observed value $\mathbf{R}'\underset{\sim}{y}$ of the vector $\mathbf{z}$ of error contrasts; complete knowledge of the observed value $\underset{\sim}{y}$ of $\mathbf{y}$ is not required. Thus, the (ordinary) ML estimator of a function of $\boldsymbol{\theta}$ (like the REML estimator) depends on the value of $\mathbf{y}$ only through the value of the vector of error contrasts.

Let us verify that the identity of the function $\ell_*(\cdot\,;\underset{\sim}{y})$ is determinable solely from knowledge of $\mathbf{R}'\underset{\sim}{y}$. Let $\underset{\sim}{\tilde{e}} = (\mathbf{I} - \mathbf{P_X})\underset{\sim}{y}$, and observe (in light of Theorem 2.12.2) that $\mathbf{X}'(\mathbf{I} - \mathbf{P_X})' = \mathbf{0}$, implying [since the columns of $\mathbf{R}$ form a basis for $\mathcal{N}(\mathbf{X}')$] that $(\mathbf{I} - \mathbf{P_X})' = \mathbf{RK}$ for some matrix $\mathbf{K}$ and hence that

$$\underset{\sim}{\tilde{e}} = (\mathbf{RK})'\underset{\sim}{y} = \mathbf{K}'\mathbf{R}'\underset{\sim}{y} \tag{9.20}$$

—$\underset{\sim}{\tilde{e}}$ is the observed value of the vector $\tilde{\mathbf{e}} = (\mathbf{I} - \mathbf{P_X})\mathbf{y}$. Moreover, upon observing [in light of result (2.5.5) and Corollary 2.13.12] that $[\mathbf{V}(\boldsymbol{\theta})]^{-1}$ is a symmetric positive definite matrix, it follows from Corollary 5.9.4 that

$$\big([\mathbf{V}(\boldsymbol{\theta})]^{-1} - [\mathbf{V}(\boldsymbol{\theta})]^{-1}\mathbf{X}\{\mathbf{X}'[\mathbf{V}(\boldsymbol{\theta})]^{-1}\mathbf{X}\}^-\mathbf{X}'[\mathbf{V}(\boldsymbol{\theta})]^{-1}\big)\mathbf{X} = \mathbf{0}$$

and that

$$\mathbf{X}'\big([\mathbf{V}(\boldsymbol{\theta})]^{-1} - [\mathbf{V}(\boldsymbol{\theta})]^{-1}\mathbf{X}\{\mathbf{X}'[\mathbf{V}(\boldsymbol{\theta})]^{-1}\mathbf{X}\}^-\mathbf{X}'[\mathbf{V}(\boldsymbol{\theta})]^{-1}\big) = \mathbf{0}.$$

And as a consequence, formula (9.15) for $\ell_*(\boldsymbol{\theta};\underset{\sim}{y})$ can be reexpressed as follows:

$$\begin{aligned}\ell_*(\boldsymbol{\theta};\underset{\sim}{y}) = -\frac{N}{2}\log(2\pi) - \frac{1}{2}\log|\mathbf{V}(\boldsymbol{\theta})| \\ -\frac{1}{2}\underset{\sim}{\tilde{e}}'\big([\mathbf{V}(\boldsymbol{\theta})]^{-1} - [\mathbf{V}(\boldsymbol{\theta})]^{-1}\mathbf{X}\{\mathbf{X}'[\mathbf{V}(\boldsymbol{\theta})]^{-1}\mathbf{X}\}^-\mathbf{X}'[\mathbf{V}(\boldsymbol{\theta})]^{-1}\big)\underset{\sim}{\tilde{e}}.\end{aligned} \tag{9.21}$$

Together, results (9.21) and (9.20) imply that the identity of the function $\ell_*(\cdot\,;\underset{\sim}{y})$ is determinable solely from knowledge of $\mathbf{R}'\underset{\sim}{y}$.

Some results on symmetric idempotent matrices and on null spaces. As a preliminary to considering REML in the special case of a G–M model, it is helpful to establish the following three results on symmetric idempotent matrices and on null spaces.

Theorem 5.9.5. Every symmetric idempotent matrix is nonnegative definite. Moreover, if $\mathbf{A}$ is an $N \times N$ symmetric idempotent matrix of rank $R > 0$, then there exists an $N \times R$ matrix $\mathbf{Q}$ such that $\mathbf{A} = \mathbf{QQ}'$, and, for any such $N \times R$ matrix $\mathbf{Q}$, $\operatorname{rank} \mathbf{Q} = R$ and $\mathbf{Q}'\mathbf{Q} = \mathbf{I}$. And, conversely, for any $N \times R$ matrix $\mathbf{Q}$ such that $\mathbf{Q}'\mathbf{Q} = \mathbf{I}$, $\mathbf{QQ}'$ is an $N \times N$ symmetric idempotent matrix of rank R.

Proof. Suppose that $\mathbf{A}$ is an $N \times N$ symmetric idempotent matrix of rank R (≥ 0). Then, $\mathbf{A} = \mathbf{A}^2 = \mathbf{A}'\mathbf{A}$, and it follows from Corollary 2.13.15 that $\mathbf{A}$ is nonnegative definite. Moreover, assuming that $R > 0$, it follows from Corollary 2.13.23 that there exists an $N \times R$ matrix $\mathbf{Q}$ such that $\mathbf{A} = \mathbf{QQ}'$. And for any such $N \times R$ matrix $\mathbf{Q}$, we find, in light of Lemma 2.12.1 [and result (2.4.1)], that $\operatorname{rank} \mathbf{Q} = R$ and that $\mathbf{Q}'\mathbf{Q}$ is nonsingular and, in addition, we find that

$$\mathbf{Q}'\mathbf{QQ}'\mathbf{QQ}'\mathbf{Q} = \mathbf{Q}'\mathbf{A}^2\mathbf{Q} = \mathbf{Q}'\mathbf{AQ} = \mathbf{Q}'\mathbf{QQ}'\mathbf{Q} \tag{9.22}$$

and hence [upon premultiplying and postmultiplying both sides of equality (9.22) by $(\mathbf{Q}'\mathbf{Q})^{-1}$] that $\mathbf{Q}'\mathbf{Q} = \mathbf{I}$.

Conversely, suppose that $\mathbf{Q}$ is an $N \times R$ matrix such that $\mathbf{Q}'\mathbf{Q} = \mathbf{I}$. Then, upon observing that $\mathbf{QQ}' = \mathbf{P_Q}$ and (in light of Lemma 2.12.1) that

$$\operatorname{rank}(\mathbf{Q}) = \operatorname{rank}(\mathbf{Q}'\mathbf{Q}) = \operatorname{rank}(\mathbf{I}_R) = R,$$

it follows from Theorem 2.12.2 that $\mathbf{QQ}'$ is a symmetric idempotent matrix of rank R. Q.E.D.

Theorem 5.9.6. Let $\mathbf{X}$ represent an $N \times P$ matrix of rank R ($< N$). Then, $\mathbf{I} - \mathbf{P_X}$ is an $N \times N$ symmetric idempotent matrix of rank $N - R$, and there exists an $N \times (N - R)$ matrix $\mathbf{Q}$ such that $\mathbf{I} - \mathbf{P_X} = \mathbf{QQ}'$. Moreover, for any $N \times (N - R)$ matrix $\mathbf{Q}$, $\mathbf{I} - \mathbf{P_X} = \mathbf{QQ}'$ if and only if $\mathbf{X}'\mathbf{Q} = \mathbf{0}$ and $\mathbf{Q}'\mathbf{Q} = \mathbf{I}$ (in which case $\mathbf{Q}$ is of full column rank $N - R$).

Proof. In light of Theorem 2.12.2 and Lemmas 2.8.1 and 2.8.4, it is clear that $\mathbf{I} - \mathbf{P_X}$ is a symmetric idempotent matrix of rank $N - R$. And in light of Theorem 5.9.5, there exists an $N \times (N - R)$ matrix $\mathbf{Q}$ such that $\mathbf{I} - \mathbf{P_X} = \mathbf{QQ}'$.

Now, suppose that $\mathbf{Q}$ is any $N \times (N - R)$ matrix such that $\mathbf{X}'\mathbf{Q} = \mathbf{0}$ and $\mathbf{Q}'\mathbf{Q} = \mathbf{I}$. Then, $\mathbf{QQ}'$ is a symmetric idempotent matrix of rank $N - R$ (as is evident from Theorem 5.9.5), and $\mathbf{P_X Q} = \mathbf{0}$ and $\mathbf{Q}'\mathbf{P_X} = \mathbf{0}$. And, consequently, $\mathbf{I} - \mathbf{P_X} - \mathbf{QQ}'$ is a symmetric idempotent matrix. Further, making use of Corollary 2.8.3 and Lemma 2.12.1, we find that

$$\begin{aligned}
\operatorname{rank}(\mathbf{I} - \mathbf{P_X} - \mathbf{QQ}') &= \operatorname{tr}(\mathbf{I} - \mathbf{P_X} - \mathbf{QQ}') \\
&= \operatorname{tr}(\mathbf{I} - \mathbf{P_X}) - \operatorname{tr}(\mathbf{QQ}') \\
&= \operatorname{rank}(\mathbf{I} - \mathbf{P_X}) - \operatorname{rank}(\mathbf{QQ}') \\
&= N - R - (N - R) = 0,
\end{aligned}$$

implying that $\mathbf{I} - \mathbf{P_X} - \mathbf{QQ}' = \mathbf{0}$ and hence that $\mathbf{I} - \mathbf{P_X} = \mathbf{QQ}'$.

Conversely, suppose that $\mathbf{Q}$ is any $N \times (N - R)$ matrix such that $\mathbf{I} - \mathbf{P_X} = \mathbf{QQ}'$. Then, according to Theorem 5.9.5, $\mathbf{Q}'\mathbf{Q} = \mathbf{I}$. Moreover, making use of Theorem 2.12.2, we find that

$$\mathbf{X}'\mathbf{QQ}' = \mathbf{X}'(\mathbf{I} - \mathbf{P_X}) = \mathbf{0},$$

implying (in light of Corollary 2.3.4) that $\mathbf{X}'\mathbf{Q} = \mathbf{0}$. Q.E.D.

Lemma 5.9.7. Let $\mathbf{X}$ represent an $N \times P$ matrix of rank R ($< N$). Then, for any $N \times (N - R)$ matrix $\mathbf{Q}$, $\mathbf{X}'\mathbf{Q} = \mathbf{0}$ and $\mathbf{Q}'\mathbf{Q} = \mathbf{I}$ if and only if the columns of $\mathbf{Q}$ form an orthonormal basis for $\mathcal{N}(\mathbf{X}')$.

Proof. If the columns of $\mathbf{Q}$ form an orthonormal basis for $\mathcal{N}(\mathbf{X}')$, then clearly $\mathbf{X}'\mathbf{Q} = \mathbf{0}$ and $\mathbf{Q}'\mathbf{Q} = \mathbf{I}$. Conversely, suppose that $\mathbf{X}'\mathbf{Q} = \mathbf{0}$ and $\mathbf{Q}'\mathbf{Q} = \mathbf{I}$. Then, clearly, the $N - R$ columns of $\mathbf{Q}$ are orthonormal, and each of them is contained in $\mathcal{N}(\mathbf{X}')$. And since orthonormal vectors are linearly independent (as is evident from Lemma 2.4.22) and since (according to Lemma 2.11.5)

$\dim[\mathcal{N}(\mathbf{X}')] = N - R$, it follows from Theorem 2.4.11 that the columns of $\mathbf{Q}$ form a basis for $\mathcal{N}(\mathbf{X}')$. Q.E.D.

REML in the special case of a G–M model. Let us consider REML in the special case where the $N \times 1$ observable random vector $\mathbf{y}$ follows a G–M model. And in doing so, let us continue to suppose that the distribution of the vector $\mathbf{e}$ of residual effects is MVN. Then, $\mathbf{y} \sim N(\mathbf{X}\boldsymbol{\beta}, \sigma^2\mathbf{I})$.

What is the REML estimator of σ^2, and how does it compare with other estimators of σ^2, including the (ordinary) ML estimator (which was derived in Subsection a)? These questions can be readily answered by making a judicious choice for the $N \times (N - \operatorname{rank}\mathbf{X})$ matrix $\mathbf{R}$ (of full column rank $N - \operatorname{rank}\mathbf{X}$) such that $\mathbf{X}'\mathbf{R} = \mathbf{0}$.

Let $\mathbf{Q}$ represent an $N \times (N - \operatorname{rank}\mathbf{X})$ matrix whose columns form an orthonormal basis for $\mathcal{N}(\mathbf{X}')$. Or, equivalently (in light of Lemma 5.9.7), take $\mathbf{Q}$ to be an $N \times (N - \operatorname{rank}\mathbf{X})$ matrix such that $\mathbf{X}'\mathbf{Q} = \mathbf{0}$ and $\mathbf{Q}'\mathbf{Q} = \mathbf{I}$. And observe (in light of Theorem 5.9.6) that

$$\mathbf{I} - \mathbf{P}_{\mathbf{X}} = \mathbf{Q}\mathbf{Q}'$$

(and that $\mathbf{Q}$ is of full column rank).

Suppose that in implementing REML, we set $\mathbf{R} = \mathbf{Q}$—clearly, that is a legitimate choice for $\mathbf{R}$. Then, $\mathbf{z} = \mathbf{Q}'\mathbf{y} \sim N(\mathbf{0}, \sigma^2\mathbf{I})$. And, letting $\underline{\mathbf{y}}$ represent the observed value of $\mathbf{y}$, the log-likelihood function that results from regarding the observed value $\mathbf{Q}'\underline{\mathbf{y}}$ of $\mathbf{z}$ as the data vector is the function $\ell(\sigma, \mathbf{Q}'\underline{\mathbf{y}})$ of σ given by

$$\begin{aligned} \ell(\sigma, \mathbf{Q}'\underline{\mathbf{y}}) &= -\frac{N - \operatorname{rank}\mathbf{X}}{2}\log(2\pi) - \frac{1}{2}\log\left|\sigma^2\mathbf{I}_{N-\operatorname{rank}\mathbf{X}}\right| - \frac{1}{2}\underline{\mathbf{y}}'\mathbf{Q}(\sigma^2\mathbf{I})^{-1}\mathbf{Q}'\underline{\mathbf{y}} \\ &= -\frac{N - \operatorname{rank}\mathbf{X}}{2}\log(2\pi) - \frac{N - \operatorname{rank}\mathbf{X}}{2}\log\sigma^2 - \frac{1}{2\sigma^2}\underline{\mathbf{y}}'(\mathbf{I} - \mathbf{P}_{\mathbf{X}})\underline{\mathbf{y}} \\ &= -\frac{N - \operatorname{rank}\mathbf{X}}{2}\log(2\pi) - \frac{N - \operatorname{rank}\mathbf{X}}{2}\log\sigma^2 - \frac{1}{2\sigma^2}[(\mathbf{I} - \mathbf{P}_{\mathbf{X}})\underline{\mathbf{y}}]'(\mathbf{I} - \mathbf{P}_{\mathbf{X}})\underline{\mathbf{y}}. \end{aligned} \tag{9.23}$$

Unless $(\mathbf{I} - \mathbf{P}_{\mathbf{X}})\underline{\mathbf{y}} = \mathbf{0}$ (which is an event of probability 0), $\ell(\sigma, \mathbf{Q}'\underline{\mathbf{y}})$ is of the form of the function $g(\sigma)$ defined by equality (9.4); upon setting $a = -[(N - \operatorname{rank}\mathbf{X})/2]\log(2\pi)$, $c = [(\mathbf{I} - \mathbf{P}_{\mathbf{X}})\underline{\mathbf{y}}]'(\mathbf{I} - \mathbf{P}_{\mathbf{X}})\underline{\mathbf{y}}$, and $K = N - \operatorname{rank}\mathbf{X}$, $g(\sigma) = \ell(\sigma, \mathbf{Q}'\underline{\mathbf{y}})$. Accordingly, it follows from the results of Part 1 of Subsection a that $\ell(\sigma, \mathbf{Q}'\underline{\mathbf{y}})$ attains its maximum value when σ^2 equals

$$\frac{[(\mathbf{I} - \mathbf{P}_{\mathbf{X}})\underline{\mathbf{y}}]'(\mathbf{I} - \mathbf{P}_{\mathbf{X}})\underline{\mathbf{y}}}{N - \operatorname{rank}\mathbf{X}}.$$

Thus, the REML estimator of σ^2 is the estimator

$$\frac{\tilde{\mathbf{e}}'\tilde{\mathbf{e}}}{N - \operatorname{rank}\mathbf{X}}, \tag{9.24}$$

where $\tilde{\mathbf{e}} = \mathbf{y} - \mathbf{P}_{\mathbf{X}}\mathbf{y}$.

The REML estimator (9.24) is of the form (7.41) considered in Section 5.7c; it is the estimator of the form (7.41) that is unbiased. Unlike the (ordinary) ML estimator $\tilde{\mathbf{e}}'\tilde{\mathbf{e}}/N$ [which was derived in Part 1 of Subsection a and is also of the form (7.41)], it "accounts for the estimation of $\boldsymbol{\beta}$"; in the REML estimation of σ^2, the residual sum of squares $\tilde{\mathbf{e}}'\tilde{\mathbf{e}}$ is divided by $N - \operatorname{rank}\mathbf{X}$ rather than by N.

A matrix lemma. Preliminary to the further discussion of REML, it is convenient to establish the following lemma.

Lemma 5.9.8. Let $\mathbf{A}$ represent a $Q \times S$ matrix. Then, for any $K \times Q$ matrix $\mathbf{C}$ of full column rank Q and any $S \times T$ matrix $\mathbf{B}$ of full row rank S, $\mathbf{B}(\mathbf{CAB})^-\mathbf{C}$ is a generalized inverse of $\mathbf{A}$.

Proof. In light of Lemma 2.5.1, $\mathbf{C}$ has a left inverse, say $\mathbf{L}$, and $\mathbf{B}$ has a right inverse, say $\mathbf{R}$. And it follows that

$$\mathbf{AB}(\mathbf{CAB})^-\mathbf{CA} = \mathbf{IAB}(\mathbf{CAB})^-\mathbf{CAI} = \mathbf{LCAB}(\mathbf{CAB})^-\mathbf{CABR} = \mathbf{LCABR} = \mathbf{IAI} = \mathbf{A}.$$

Thus, $\mathbf{B}(\mathbf{CAB})^{-}\mathbf{C}$ is a generalized inverse of $\mathbf{A}$. Q.E.D.

Note that in the special case where $\mathbf{A}$ is nonsingular (i.e., the special case where $\mathbf{A}$ is a $Q \times Q$ matrix of rank Q), the result of Lemma 5.9.8 can be restated as follows:

$$\mathbf{A}^{-1} = \mathbf{B}(\mathbf{CAB})^{-}\mathbf{C}. \tag{9.25}$$

An informative and computationally useful expression for the REML log-likelihood function. Suppose that $\mathbf{y}$ is an $N \times 1$ observable random vector that follows a general linear model. Suppose further that the distribution of the vector $\mathbf{e}$ of residual effects is MVN and that the variance-covariance matrix $\mathbf{V}(\boldsymbol{\theta})$ of $\mathbf{e}$ is nonsingular (for every $\boldsymbol{\theta} \in \Theta$). And let $\mathbf{z} = \mathbf{R}'\mathbf{y}$, where $\mathbf{R}$ is an $N \times (N - \text{rank}\,\mathbf{X})$ matrix of full column rank $N - \text{rank}\,\mathbf{X}$ such that $\mathbf{X}'\mathbf{R} = \mathbf{0}$, and denote by $\underline{\mathbf{y}}$ the observed value of $\mathbf{y}$.

In REML, inferences about functions of $\boldsymbol{\theta}$ are based on the likelihood function $L(\boldsymbol{\theta};\, \mathbf{R}'\underline{\mathbf{y}})$ obtained by regarding the observed value $\mathbf{R}'\underline{\mathbf{y}}$ of $\mathbf{z}$ as the data vector. Corresponding to $L(\boldsymbol{\theta};\, \mathbf{R}'\underline{\mathbf{y}})$ is the log-likelihood function $\ell(\boldsymbol{\theta};\, \mathbf{R}'\underline{\mathbf{y}}) = \log L(\boldsymbol{\theta};\, \mathbf{R}'\underline{\mathbf{y}})$. We have that $\mathbf{z} \sim N[\mathbf{0},\, \mathbf{R}'\mathbf{V}(\boldsymbol{\theta})\mathbf{R}]$, and it follows that

$$\ell(\boldsymbol{\theta};\, \mathbf{R}'\underline{\mathbf{y}}) = -\frac{N - \text{rank}\,\mathbf{X}}{2}\log(2\pi) - \frac{1}{2}\log|\mathbf{R}'\mathbf{V}(\boldsymbol{\theta})\mathbf{R}| - \frac{1}{2}\underline{\mathbf{y}}'\mathbf{R}[\mathbf{R}'\mathbf{V}(\boldsymbol{\theta})\mathbf{R}]^{-1}\mathbf{R}'\underline{\mathbf{y}} \tag{9.26}$$

—recall that $\mathbf{R}'\mathbf{V}(\boldsymbol{\theta})\mathbf{R}$ is nonsingular.

REML estimates of functions of $\boldsymbol{\theta}$ are obtained from a value, say $\hat{\boldsymbol{\theta}}$, of $\boldsymbol{\theta}$ at which $L(\boldsymbol{\theta};\, \mathbf{R}'\underline{\mathbf{y}})$ or, equivalently, $\ell(\boldsymbol{\theta};\, \mathbf{R}'\underline{\mathbf{y}})$ attains its maximum value. By way of comparison, (ordinary) ML estimates of such functions are obtained from a value, say $\tilde{\boldsymbol{\theta}}$, at which the profile likelihood function $L_*(\boldsymbol{\theta};\, \underline{\mathbf{y}})$ or profile log-likelihood function $\ell_*(\boldsymbol{\theta};\, \underline{\mathbf{y}})$ attains its maximum value; the (ordinary) ML estimate of a function $h(\boldsymbol{\theta})$ of $\boldsymbol{\theta}$ is $h(\tilde{\boldsymbol{\theta}})$, whereas the REML estimate is $h(\hat{\boldsymbol{\theta}})$. It is of potential interest to compare $\ell(\boldsymbol{\theta};\, \mathbf{R}'\underline{\mathbf{y}})$ with $\ell_*(\boldsymbol{\theta};\, \underline{\mathbf{y}})$. Expressions for $\ell_*(\boldsymbol{\theta};\, \underline{\mathbf{y}})$ are given by results (9.13), (9.14), and (9.15). However, expression (9.26) [for $\ell(\boldsymbol{\theta};\, \mathbf{R}'\underline{\mathbf{y}})$] is not of a form that facilitates meaningful comparisons with any of those expressions. Moreover, depending on the nature of the variance-covariance matrix $\mathbf{V}(\boldsymbol{\theta})$ (and on the choice of the matrix $\mathbf{R}$), expression (9.26) may not be well-suited for computational purposes [such as in computing the values of $\ell(\boldsymbol{\theta};\, \mathbf{R}'\underline{\mathbf{y}})$ corresponding to various values of $\boldsymbol{\theta}$].

For purposes of obtaining a more useful expression for $\ell(\boldsymbol{\theta};\, \mathbf{R}'\underline{\mathbf{y}})$, take $\mathbf{S}$ to be any matrix (with N rows) whose columns span $\mathcal{C}(\mathbf{X})$, that is, any matrix such that $\mathcal{C}(\mathbf{S}) = \mathcal{C}(\mathbf{X})$ (in which case, $\mathbf{S} = \mathbf{XA}$ for some matrix $\mathbf{A}$). And, temporarily (for the sake of simplicity) writing $\mathbf{V}$ for $\mathbf{V}(\boldsymbol{\theta})$, observe that

$$(\mathbf{V}^{-1}\mathbf{S},\ \mathbf{R})'\mathbf{V}(\mathbf{V}^{-1}\mathbf{S},\ \mathbf{R}) = \text{diag}(\mathbf{S}'\mathbf{V}^{-1}\mathbf{S},\ \mathbf{R}'\mathbf{V}\mathbf{R}) \tag{9.27}$$

and [in light of result (2.5.5), Corollary 2.13.12, and Corollary 5.9.3] that

$$\begin{aligned} \text{rank}[(\mathbf{V}^{-1}\mathbf{S},\ \mathbf{R})'\mathbf{V}(\mathbf{V}^{-1}\mathbf{S},\ \mathbf{R})] &= \text{rank}[\text{diag}(\mathbf{S}'\mathbf{V}^{-1}\mathbf{S},\ \mathbf{R}'\mathbf{V}\mathbf{R})] \\ &= \text{rank}(\mathbf{S}'\mathbf{V}^{-1}\mathbf{S}) + \text{rank}(\mathbf{R}'\mathbf{V}\mathbf{R}) \\ &= \text{rank}(\mathbf{S}) + \text{rank}(\mathbf{R}) \\ &= \text{rank}(\mathbf{X}) + N - \text{rank}(\mathbf{X}) = N. \end{aligned} \tag{9.28}$$

Result (9.28) implies (in light of Corollary 5.9.3) that

$$\text{rank}\,(\mathbf{V}^{-1}\mathbf{S},\ \mathbf{R}) = N \tag{9.29}$$

or, equivalently, that $(\mathbf{V}^{-1}\mathbf{S},\ \mathbf{R})$ is of full row rank. Thus, upon applying formula (9.25), it follows from result (9.27) that

$$\begin{aligned} \mathbf{V}^{-1} &= (\mathbf{V}^{-1}\mathbf{S},\ \mathbf{R})\,\text{diag}[(\mathbf{S}'\mathbf{V}^{-1}\mathbf{S})^{-},\ (\mathbf{R}'\mathbf{V}\mathbf{R})^{-1}](\mathbf{V}^{-1}\mathbf{S},\ \mathbf{R})' \\ &= \mathbf{V}^{-1}\mathbf{S}(\mathbf{S}'\mathbf{V}^{-1}\mathbf{S})^{-}\mathbf{S}'\mathbf{V}^{-1} + \mathbf{R}(\mathbf{R}'\mathbf{V}\mathbf{R})^{-1}\mathbf{R}' \end{aligned} \tag{9.30}$$

and hence that

$$\mathbf{R}(\mathbf{R}'\mathbf{V}\mathbf{R})^{-1}\mathbf{R}' = \mathbf{V}^{-1} - \mathbf{V}^{-1}\mathbf{S}(\mathbf{S}'\mathbf{V}^{-1}\mathbf{S})^{-}\mathbf{S}'\mathbf{V}^{-1}. \tag{9.31}$$

Moreover, as a special case of equality (9.31) (that where $\mathbf{S} = \mathbf{X}$), we obtain the following expression for $\mathbf{R}(\mathbf{R}'\mathbf{V}\mathbf{R})^{-1}\mathbf{R}'$ [a quantity which appears in the 3rd term of expression (9.26) for $\ell(\boldsymbol{\theta};\, \mathbf{R}'\underline{\mathbf{y}})$]:

$$\mathbf{R}(\mathbf{R}'\mathbf{V}\mathbf{R})^{-1}\mathbf{R}' = \mathbf{V}^{-1} - \mathbf{V}^{-1}\mathbf{X}(\mathbf{X}'\mathbf{V}^{-1}\mathbf{X})^{-}\mathbf{X}'\mathbf{V}^{-1}. \tag{9.32}$$

Now, consider the quantity $|\mathbf{R}'\mathbf{V}\mathbf{R}|$ [which appears in the 2nd term of expression (9.26)]. Take $\mathbf{X}_*$ to be any $N \times (\operatorname{rank}\mathbf{X})$ matrix whose columns are linearly independent columns of $\mathbf{X}$ or, more generally, whose columns form a basis for $\mathcal{C}(\mathbf{X})$ (in which case, $\mathbf{X}_* = \mathbf{X}\mathbf{A}$ for some matrix $\mathbf{A}$). Observing that

$$(\mathbf{X}_*,\, \mathbf{R})'(\mathbf{X}_*,\, \mathbf{R}) = \operatorname{diag}(\mathbf{X}_*'\mathbf{X}_*,\, \mathbf{R}'\mathbf{R})$$

and making use of basic properties of determinants, we find that

$$\begin{aligned} |(\mathbf{X}_*,\, \mathbf{R})'\mathbf{V}(\mathbf{X}_*,\, \mathbf{R})| &= |(\mathbf{X}_*,\, \mathbf{R})'|\,|(\mathbf{X}_*,\, \mathbf{R})|\,|\mathbf{V}| \\ &= |(\mathbf{X}_*,\, \mathbf{R})'(\mathbf{X}_*,\, \mathbf{R})|\,|\mathbf{V}| \\ &= |\operatorname{diag}(\mathbf{X}_*'\mathbf{X}_*,\, \mathbf{R}'\mathbf{R})|\,|\mathbf{V}| \\ &= |\mathbf{X}_*'\mathbf{X}_*|\,|\mathbf{R}'\mathbf{R}|\,|\mathbf{V}|. \end{aligned} \tag{9.33}$$

And making use of formula (2.14.29) for the determinant of a partitioned matrix, we find that

$$\begin{aligned} |(\mathbf{X}_*,\, \mathbf{R})'\mathbf{V}(\mathbf{X}_*,\, \mathbf{R})| &= \begin{vmatrix} \mathbf{X}_*'\mathbf{V}\mathbf{X}_* & \mathbf{X}_*'\mathbf{V}\mathbf{R} \\ \mathbf{R}'\mathbf{V}\mathbf{X}_* & \mathbf{R}'\mathbf{V}\mathbf{R} \end{vmatrix} \\ &= |\mathbf{R}'\mathbf{V}\mathbf{R}|\,|\mathbf{X}_*'\mathbf{V}\mathbf{X}_* - \mathbf{X}_*'\mathbf{V}\mathbf{R}(\mathbf{R}'\mathbf{V}\mathbf{R})^{-1}\mathbf{R}'\mathbf{V}\mathbf{X}_*| \\ &= |\mathbf{R}'\mathbf{V}\mathbf{R}|\,|\mathbf{X}_*'[\mathbf{V} - \mathbf{V}\mathbf{R}(\mathbf{R}'\mathbf{V}\mathbf{R})^{-1}\mathbf{R}'\mathbf{V}]\mathbf{X}_*|. \end{aligned} \tag{9.34}$$

Moreover, as a special case of equality (9.30) (that where $\mathbf{S} = \mathbf{X}_*$), we have (since, in light of Corollary 5.9.3, $\mathbf{X}_*'\mathbf{V}^{-1}\mathbf{X}_*$ is nonsingular) that

$$\mathbf{V}^{-1} = \mathbf{V}^{-1}\mathbf{X}_*(\mathbf{X}_*'\mathbf{V}^{-1}\mathbf{X}_*)^{-1}\mathbf{X}_*'\mathbf{V}^{-1} + \mathbf{R}(\mathbf{R}'\mathbf{V}\mathbf{R})^{-1}\mathbf{R}'$$

and (upon premultiplying and postmultiplying by $\mathbf{V}$ and rearranging terms) that

$$\mathbf{V} - \mathbf{V}\mathbf{R}(\mathbf{R}'\mathbf{V}\mathbf{R})^{-1}\mathbf{R}'\mathbf{V} = \mathbf{X}_*(\mathbf{X}_*'\mathbf{V}^{-1}\mathbf{X}_*)^{-1}\mathbf{X}_*'. \tag{9.35}$$

Upon replacing $\mathbf{V} - \mathbf{V}\mathbf{R}(\mathbf{R}'\mathbf{V}\mathbf{R})^{-1}\mathbf{R}'\mathbf{V}$ with expression (9.35), result (9.34) simplifies as follows:

$$\begin{aligned} |(\mathbf{X}_*,\, \mathbf{R})'\mathbf{V}(\mathbf{X}_*,\, \mathbf{R})| &= |\mathbf{R}'\mathbf{V}\mathbf{R}|\,|\mathbf{X}_*'\mathbf{X}_*(\mathbf{X}_*'\mathbf{V}^{-1}\mathbf{X}_*)^{-1}\mathbf{X}_*'\mathbf{X}_*| \\ &= |\mathbf{R}'\mathbf{V}\mathbf{R}|\,|\mathbf{X}_*'\mathbf{X}_*|^2/|\mathbf{X}_*'\mathbf{V}^{-1}\mathbf{X}_*|. \end{aligned} \tag{9.36}$$

It remains to equate expressions (9.33) and (9.36); doing so leads to the following expression for $|\mathbf{R}'\mathbf{V}\mathbf{R}|$:

$$|\mathbf{R}'\mathbf{V}\mathbf{R}| = |\mathbf{R}'\mathbf{R}|\,|\mathbf{V}|\,|\mathbf{X}_*'\mathbf{V}^{-1}\mathbf{X}_*|/|\mathbf{X}_*'\mathbf{X}_*|. \tag{9.37}$$

Upon substituting expressions (9.32) and (9.37) [for $\mathbf{R}(\mathbf{R}'\mathbf{V}\mathbf{R})^{-1}\mathbf{R}'$ and $|\mathbf{R}'\mathbf{V}\mathbf{R}|$] into expression (9.26), we find that the REML log-likelihood function $\ell(\boldsymbol{\theta};\, \mathbf{R}'\underline{\mathbf{y}})$ is reexpressible as follows:

$$\begin{aligned} \ell(\boldsymbol{\theta};\mathbf{R}'\underline{\mathbf{y}}) = &-\frac{N - \operatorname{rank}\mathbf{X}}{2}\log(2\pi) - \frac{1}{2}\log|\mathbf{R}'\mathbf{R}| + \frac{1}{2}\log|\mathbf{X}_*'\mathbf{X}_*| \\ &- \frac{1}{2}\log|\mathbf{V}(\boldsymbol{\theta})| - \frac{1}{2}\log|\mathbf{X}_*'[\mathbf{V}(\boldsymbol{\theta})]^{-1}\mathbf{X}_*| \\ &- \frac{1}{2}\underline{\mathbf{y}}'\big([\mathbf{V}(\boldsymbol{\theta})]^{-1} - [\mathbf{V}(\boldsymbol{\theta})]^{-1}\mathbf{X}\{\mathbf{X}'[\mathbf{V}(\boldsymbol{\theta})]^{-1}\mathbf{X}\}^{-}\mathbf{X}'[\mathbf{V}(\boldsymbol{\theta})]^{-1}\big)\underline{\mathbf{y}}. \end{aligned} \tag{9.38}$$

If $\mathbf{R}$ is taken to be a matrix whose columns form an orthonormal basis for $\mathcal{N}(\mathbf{X}')$, then the second term of expression (9.38) equals 0; similarly, if $\mathbf{X}_*$ is taken to be a matrix whose columns form an orthonormal basis for $\mathcal{C}(\mathbf{X})$, then the third term of expression (9.38) equals 0. However, what is more important is that the choice of $\mathbf{R}$ affects expression (9.38) only through its second term, which is a constant (i.e., does not involve $\boldsymbol{\theta}$). And for any two choices of $\mathbf{X}_*$, say $\mathbf{X}_1$ and $\mathbf{X}_2$, $\mathbf{X}_2 = \mathbf{X}_1\mathbf{B}$ for some matrix $\mathbf{B}$ (which is necessarily nonsingular), implying that

$$-\frac{1}{2}\log|\mathbf{X}_2'[\mathbf{V}(\boldsymbol{\theta})]^{-1}\mathbf{X}_2| = -\frac{1}{2}\log|\mathbf{B}'\mathbf{X}_1'[\mathbf{V}(\boldsymbol{\theta})]^{-1}\mathbf{X}_1\mathbf{B}| = -\log|\det\mathbf{B}| - \frac{1}{2}\log|\mathbf{X}_1'[\mathbf{V}(\boldsymbol{\theta})]^{-1}\mathbf{X}_1|$$

and, similarly, that

$$\frac{1}{2}\log|\mathbf{X}_2'\mathbf{X}_2| = \log|\det\mathbf{B}| + \frac{1}{2}\log|\mathbf{X}_1'\mathbf{X}_1|,$$

so that the only effect on expression (9.38) of a change in the choice of $\mathbf{X}_*$ from $\mathbf{X}_1$ to $\mathbf{X}_2$ is to add a constant to the third term and to subtract the same constant from the fifth term. Thus, the choice of $\mathbf{R}$ and the choice of $\mathbf{X}_*$ are immaterial.

The last term of expression (9.38) can be reexpressed in terms of an arbitrary solution, say $\tilde{\boldsymbol{\beta}}(\boldsymbol{\theta})$, to the linear system

$$\mathbf{X}'[\mathbf{V}(\boldsymbol{\theta})]^{-1}\mathbf{X}\mathbf{b} = \mathbf{X}'[\mathbf{V}(\boldsymbol{\theta})]^{-1}\underset{\sim}{\mathbf{y}} \tag{9.39}$$

(in the $P \times 1$ vector $\mathbf{b}$)—recall (from Subsection a) that this linear system is consistent, that $\mathbf{X}\tilde{\boldsymbol{\beta}}(\boldsymbol{\theta})$ does not depend on the choice of $\tilde{\boldsymbol{\beta}}(\boldsymbol{\theta})$, and that the choices for $\tilde{\boldsymbol{\beta}}(\boldsymbol{\theta})$ include the vector $(\{\mathbf{X}'[\mathbf{V}(\boldsymbol{\theta})]^{-1}\mathbf{X}\}^-)'\mathbf{X}'[\mathbf{V}(\boldsymbol{\theta})]^{-1}\underset{\sim}{\mathbf{y}}$. We find that

$$\begin{aligned}\underset{\sim}{\mathbf{y}}'([\mathbf{V}(\boldsymbol{\theta})]^{-1} &- [\mathbf{V}(\boldsymbol{\theta})]^{-1}\mathbf{X}\{\mathbf{X}'[\mathbf{V}(\boldsymbol{\theta})]^{-1}\mathbf{X}\}^-\mathbf{X}'[\mathbf{V}(\boldsymbol{\theta})]^{-1})\underset{\sim}{\mathbf{y}} \\ &= \underset{\sim}{\mathbf{y}}'[\mathbf{V}(\boldsymbol{\theta})]^{-1}\underset{\sim}{\mathbf{y}} - [\tilde{\boldsymbol{\beta}}(\boldsymbol{\theta})]'\mathbf{X}'[\mathbf{V}(\boldsymbol{\theta})]^{-1}\underset{\sim}{\mathbf{y}} && (9.40)\\ &= [\underset{\sim}{\mathbf{y}} - \mathbf{X}\tilde{\boldsymbol{\beta}}(\boldsymbol{\theta})]'[\mathbf{V}(\boldsymbol{\theta})]^{-1}[\underset{\sim}{\mathbf{y}} - \mathbf{X}\tilde{\boldsymbol{\beta}}(\boldsymbol{\theta})]. && (9.41)\end{aligned}$$

It is informative to compare expression (9.38) for $\ell(\boldsymbol{\theta};\, \mathbf{R}'\underset{\sim}{\mathbf{y}})$ with expression (9.15) for the profile log-likelihood function $\ell_*(\boldsymbol{\theta};\underset{\sim}{\mathbf{y}})$. Aside from the terms that do not depend on $\boldsymbol{\theta}$ [the 1st term of expression (9.15) and the first 3 terms of expression (9.38)], the only difference between the two expressions is the inclusion in expression (9.38) of the term $-\frac{1}{2}\log|\mathbf{X}_*'[\mathbf{V}(\boldsymbol{\theta})]^{-1}\mathbf{X}_*|$. This term depends on $\boldsymbol{\theta}$, but not on $\underset{\sim}{\mathbf{y}}$. Its inclusion serves to adjust the profile log-likelihood function $\ell_*(\boldsymbol{\theta};\underset{\sim}{\mathbf{y}})$ so as to compensate for the failure of ordinary ML (in estimating functions of $\boldsymbol{\theta}$) to account for the estimation of $\boldsymbol{\beta}$. Unlike the profile log-likelihood function, $\ell(\boldsymbol{\theta};\, \mathbf{R}'\underset{\sim}{\mathbf{y}})$ is the logarithm of an actual likelihood function and, consequently, has the properties thereof—it is the logarithm of the likelihood function $L(\boldsymbol{\theta};\, \mathbf{R}'\underset{\sim}{\mathbf{y}})$ obtained by regarding the observed value $\mathbf{R}'\underset{\sim}{\mathbf{y}}$ of $\mathbf{z}$ as the data vector.

If the form of the $N \times N$ matrix $\mathbf{V}(\boldsymbol{\theta})$ is such that $\mathbf{V}(\boldsymbol{\theta})$ is relatively easy to invert (as is often the case in practice), then expression (9.38) for $\ell(\boldsymbol{\theta};\, \mathbf{R}'\underset{\sim}{\mathbf{y}})$ is likely to be much more useful for computational purposes than expression (9.26). Expression (9.38) [along with expression (9.40) or (9.41)] serves to relate the numerical evaluation of $\ell(\boldsymbol{\theta};\, \mathbf{R}'\underset{\sim}{\mathbf{y}})$ for any particular value of $\boldsymbol{\theta}$ to the solution of the linear system (9.39), comprising P equations in P "unknowns."

Special case: Aitken model. Let us now specialize to the case where $\mathbf{y}$ follows an Aitken model (and where $\mathbf{H}$ is nonsingular). As in Subsection a, this case is to be regarded as the special case of the general linear model where $T = 1$, where $\boldsymbol{\theta} = (\sigma)$, and where $\mathbf{V}(\boldsymbol{\theta}) = \sigma^2\mathbf{H}$. In this special case, linear system (9.39) is equivalent to (i.e., has the same solutions as) the linear system

$$\mathbf{X}'\mathbf{H}^{-1}\mathbf{X}\mathbf{b} = \mathbf{X}'\mathbf{H}^{-1}\underset{\sim}{\mathbf{y}}, \tag{9.42}$$

comprising the Aitken equations. And taking $\tilde{\boldsymbol{\beta}}$ to be any solution to linear system (9.42), we find [in light of results (9.38) and (9.41)] that the log-likelihood function $\ell(\boldsymbol{\theta};\, \mathbf{R}'\underset{\sim}{\mathbf{y}})$ is expressible as

$$\begin{aligned}\ell(\sigma;\, \mathbf{R}'\underset{\sim}{\mathbf{y}}) = &-\frac{N-\operatorname{rank}\mathbf{X}}{2}\log(2\pi) - \frac{1}{2}\log|\mathbf{R}'\mathbf{R}| + \frac{1}{2}\log|\mathbf{X}_*'\mathbf{X}_*| \\ &- \frac{1}{2}\log|\mathbf{H}| - \frac{1}{2}\log|\mathbf{X}_*'\mathbf{H}^{-1}\mathbf{X}_*| - \frac{N-\operatorname{rank}\mathbf{X}}{2}\log\sigma^2 \\ &- \frac{1}{2\sigma^2}(\underset{\sim}{\mathbf{y}}-\mathbf{X}\tilde{\boldsymbol{\beta}})'\mathbf{H}^{-1}(\underset{\sim}{\mathbf{y}}-\mathbf{X}\tilde{\boldsymbol{\beta}}). \qquad (9.43)\end{aligned}$$

Unless $\underset{\sim}{\mathbf{y}} - \mathbf{X}\tilde{\boldsymbol{\beta}} = \mathbf{0}$ (which is an event of probability 0), $\ell(\sigma;\mathbf{R}'\underset{\sim}{\mathbf{y}})$ is of the form of the function $g(\sigma)$ defined (in Part 1 of Subsection a) by equality (9.4); upon setting $a = -[(N - \operatorname{rank}\mathbf{X})/2]\log(2\pi) - (1/2)\log|\mathbf{R}'\mathbf{R}| + (1/2)\log|\mathbf{X}_*'\mathbf{X}_*| - (1/2)\log|\mathbf{H}| - (1/2)\log|\mathbf{X}_*'\mathbf{H}^{-1}\mathbf{X}_*|$, $c = (\underset{\sim}{\mathbf{y}}-\mathbf{X}\tilde{\boldsymbol{\beta}})'\mathbf{H}^{-1}(\underset{\sim}{\mathbf{y}}-\mathbf{X}\tilde{\boldsymbol{\beta}})$, and $K = N - \operatorname{rank}\mathbf{X}$, $g(\sigma) = \ell(\sigma;\mathbf{R}'\underset{\sim}{\mathbf{y}})$. Accordingly, it follows from the results of Part 1 of Subsection a that $\ell(\sigma;\mathbf{R}'\underset{\sim}{\mathbf{y}})$ attains its maximum value when σ^2 equals

$$\frac{(\underset{\sim}{\mathbf{y}}-\mathbf{X}\tilde{\boldsymbol{\beta}})'\mathbf{H}^{-1}(\underset{\sim}{\mathbf{y}}-\mathbf{X}\tilde{\boldsymbol{\beta}})}{N-\operatorname{rank}\mathbf{X}}. \qquad (9.44)$$

The quantity (9.44) is the REML estimate of σ^2; it is the estimate obtained by dividing $(\underset{\sim}{\mathbf{y}} - \mathbf{X}\tilde{\boldsymbol{\beta}})'\mathbf{H}^{-1}(\underset{\sim}{\mathbf{y}} - \mathbf{X}\tilde{\boldsymbol{\beta}})$ by $N - \operatorname{rank}\mathbf{X}$. It differs from the (ordinary) ML estimate of σ^2; which (as is evident from the results of Subsection a) is obtained by dividing $(\underset{\sim}{\mathbf{y}}-\mathbf{X}\tilde{\boldsymbol{\beta}})'\mathbf{H}^{-1}(\underset{\sim}{\mathbf{y}}-\mathbf{X}\tilde{\boldsymbol{\beta}})$ by N. Note that in the further special case of the G–M model (i.e., the further special case where $\mathbf{H} = \mathbf{I}$), the Aitken equations simplify to the normal equations $\mathbf{X}'\mathbf{X}\mathbf{b} = \mathbf{X}'\underset{\sim}{\mathbf{y}}$ and expression (9.44) (for the REML estimate) is (upon setting $\tilde{\boldsymbol{\beta}} = (\mathbf{X}'\mathbf{X})^-\mathbf{X}'\underset{\sim}{\mathbf{y}}$) reexpressible as $[(\mathbf{I} - \mathbf{P}_{\mathbf{X}})\underset{\sim}{\mathbf{y}}]'(\mathbf{I} - \mathbf{P}_{\mathbf{X}})\underset{\sim}{\mathbf{y}}/(N - \operatorname{rank}\mathbf{X})$, in agreement with the expression for the REML estimator [expression (9.24)] derived in a previous part of the present subsection.

c. Elliptical distributions

The results of Subsections a and b (on the ML and REML estimation of functions of the parameters of a G–M, Aitken, or general linear model) were obtained under the assumption that the distribution of the vector $\mathbf{e}$ of residual effects is MVN. Some of the properties of the MVN distribution extend (in a relatively straightforward way) to a broader class of distributions called elliptical distributions (or elliptically contoured or elliptically symmetric distributions). Elliptical distributions are introduced (and some of their basic properties described) in the present subsection—this follows the presentation (in Part 1 of the present subsection) of a useful result on orthogonal matrices. Then, in Subsection d, the results of Subsections a and b are revisited with the intent of obtaining extensions suitable for G–M, Aitken, or general linear models when the form of the distribution of the vector $\mathbf{e}$ of residual effects is taken to be that of an elliptical distribution other than a multivariate normal distribution.

A matrix lemma.

Lemma 5.9.9. For any two M-dimensional column vectors $\mathbf{x}_1$ and $\mathbf{x}_2$, $\mathbf{x}_2'\mathbf{x}_2 = \mathbf{x}_1'\mathbf{x}_1$ if and only if there exists an $M \times M$ orthogonal matrix $\mathbf{O}$ such that $\mathbf{x}_2 = \mathbf{O}\mathbf{x}_1$.

Proof. If there exists an orthogonal matrix $\mathbf{O}$ such that $\mathbf{x}_2 = \mathbf{O}\mathbf{x}_1$, then, clearly,

$$\mathbf{x}_2'\mathbf{x}_2 = (\mathbf{O}\mathbf{x}_1)'\mathbf{O}\mathbf{x}_1 = \mathbf{x}_1'\mathbf{O}'\mathbf{O}\mathbf{x}_1 = \mathbf{x}_1'\mathbf{x}_1.$$

For purposes of establishing the converse, take $\mathbf{u} = (1, 0, 0, \ldots, 0)'$ to be the first column of $\mathbf{I}_M$, and assume that both $\mathbf{x}_1$ and $\mathbf{x}_2$ are nonnull—if $\mathbf{x}_2'\mathbf{x}_2 = \mathbf{x}_1'\mathbf{x}_1$ and either $\mathbf{x}_1$ or $\mathbf{x}_2$ is null, then both $\mathbf{x}_1$ and $\mathbf{x}_2$ are null, in which case $\mathbf{x}_2 = \mathbf{O}\mathbf{x}_1$ for any $M \times M$ orthogonal matrix $\mathbf{O}$. And for $i = 1, 2$, define

$$\mathbf{P}_i = \mathbf{I} - 2(\mathbf{v}_i'\mathbf{v}_i)^{-1}\mathbf{v}_i\mathbf{v}_i',$$

where $\mathbf{v}_i = \mathbf{x}_i - (\mathbf{x}_i'\mathbf{x}_i)^{1/2}\mathbf{u}$—if $\mathbf{v}_i = \mathbf{0}$, take $\mathbf{P}_i = \mathbf{I}$. The two matrices $\mathbf{P}_1$ and $\mathbf{P}_2$ are Householder matrices; they are orthogonal and are such that, for $i = 1, 2$, $\mathbf{P}_i\mathbf{x}_i = (\mathbf{x}_i'\mathbf{x}_i)^{1/2}\mathbf{u}$—refer, e.g., to Golub and Van Loan (2013, sec. 5.1.2). Thus, if $\mathbf{x}_2'\mathbf{x}_2 = \mathbf{x}_1'\mathbf{x}_1$, then

$$\mathbf{P}_2\mathbf{x}_2 = (\mathbf{x}_2'\mathbf{x}_2)^{1/2}\mathbf{u} = (\mathbf{x}_1'\mathbf{x}_1)^{1/2}\mathbf{u} = \mathbf{P}_1\mathbf{x}_1,$$

implying that

$$\mathbf{x}_2 = \mathbf{P}_2'\mathbf{P}_1\mathbf{x}_1$$

and hence (since $\mathbf{P}_2'\mathbf{P}_1$ is orthogonal) that there exists an orthogonal matrix $\mathbf{O}$ such that $\mathbf{x}_2 = \mathbf{O}\mathbf{x}_1$. Q.E.D.

Spherical distributions. Elliptical distributions are defined in terms of spherical distributions (which are themselves elliptical distributions, albeit of a relatively simple kind). An $M \times 1$ random vector $\mathbf{z}$ is said to have a *spherical* (or spherically symmetric) *distribution* if, for every $M \times M$ orthogonal matrix $\mathbf{O}$, the distribution of $\mathbf{O}\mathbf{z}$ is the same as that of $\mathbf{z}$. For example, the $N(\mathbf{0}, \sigma^2\mathbf{I})$ distribution (where σ is any nonnegative scalar) is a spherical distribution.

Suppose that the distribution of the M-dimensional random vector $\mathbf{z} = (z_1, z_2, \ldots, z_M)'$ is spherical. Then, upon observing that $-\mathbf{I}_M$ is an orthogonal matrix, we find that

$$-\mathbf{z} = -\mathbf{I}_M\mathbf{z} \sim \mathbf{z}. \tag{9.45}$$

Thus, a spherical distribution is symmetric. And, it follows, in particular, that if $\mathrm{E}(\mathbf{z})$ exists, then

$$\mathrm{E}(\mathbf{z}) = \mathbf{0}. \tag{9.46}$$

Further, if the second-order moments of the distribution of $\mathbf{z}$ exist, then

$$\mathrm{var}(\mathbf{z}) = c\mathbf{I} \tag{9.47}$$

for some nonnegative scalar c.

To verify result (9.47), take $\mathbf{O}_i$ to be the $M \times M$ orthogonal matrix obtained by interchanging the first and ith rows of $\mathbf{I}_M$, and take $\mathbf{P}_i$ to be the $M \times M$ orthogonal matrix obtained by multiplying the ith row of $\mathbf{I}_M$ by -1. Then, upon observing that $\mathbf{O}_i\mathbf{z} \sim \mathbf{z}$ and that z_i is the first element of $\mathbf{O}_i\mathbf{z}$, we find that

$$z_i \sim z_1. \tag{9.48}$$

And upon observing that $\mathbf{P}_i\mathbf{z} \sim \mathbf{z}$ and that $-z_i$ is the ith element of $\mathbf{P}_i\mathbf{z}$, we find that (for $j > i$)

$$\begin{pmatrix} -z_i \\ z_j \end{pmatrix} \sim \begin{pmatrix} z_i \\ z_j \end{pmatrix}$$

and hence that

$$-z_i z_j \sim z_i z_j. \tag{9.49}$$

It follows from equality (9.48) that the diagonal elements of $\mathrm{var}(\mathbf{z})$ have a common value c and from equality (9.49) that the off-diagonal elements of $\mathrm{var}(\mathbf{z})$ [the ijth of which equals $\mathrm{E}(z_i z_j)$] are 0.

According to result (9.47), the M elements $z_1, z_2, \ldots, z_M$ of the spherically distributed random vector $\mathbf{z}$ are uncorrelated. However, it is only in the special case where the distribution of $\mathbf{z}$ is MVN that $z_1, z_2, \ldots, z_M$ are statistically independent—refer, e.g., to Kollo and von Rosen (2005, sec. 2.3) or to Fang, Kotz, and Ng (1990, sec. 4.3) for a proof.

The variance-covariance matrix of the spherically distributed random vector $\mathbf{z}$ is a scalar multiple $c\mathbf{I}$ of $\mathbf{I}$. Note that [aside from the degenerate special case where $\mathrm{var}(\mathbf{z}) = \mathbf{0}$ or, equivalently, where $\mathbf{z} = \mathbf{0}$ with probability 1] the elements of $\mathbf{z}$ can be rescaled by dividing each of them by $\sqrt{c}$, the effect of which is to transform $\mathbf{z}$ into the vector $c^{-1/2}\mathbf{z}$ whose variance-covariance matrix is $\mathbf{I}$. Note also that, like $\mathbf{z}$, the transformed vector $c^{-1/2}\mathbf{z}$ has a spherical distribution.

Pdf of a spherical distribution. Take $\mathbf{z} = (z_1, z_2, \ldots, z_M)'$ to be an M-dimensional random (column) vector that has an absolutely continuous distribution with pdf $f(\cdot)$. Clearly, whether or not this distribution is spherical depends on the nature of the pdf.

Define $\mathbf{u} = \mathbf{O}\mathbf{z}$, where $\mathbf{O}$ is an arbitrary $M \times M$ orthogonal matrix, and denote by u_i the ith element of $\mathbf{u}$. Then, the distribution of $\mathbf{u}$ has as a pdf the function $h(\cdot)$ obtained by taking (for every value of $\mathbf{u}$)

$$h(\mathbf{u}) = |\det \mathbf{J}|\, f(\mathbf{O}'\mathbf{u}),$$

where $\mathbf{J}$ is the $M \times M$ matrix with ijth element $\partial z_i / \partial u_j$ (e.g., Bickel and Doksum 2001, sec. B.2). Moreover, $\mathbf{J} = \mathbf{O}'$, implying (in light of Corollary 2.14.19) that $\det \mathbf{J} = \pm 1$. Thus,

$$h(\mathbf{u}) = f(\mathbf{O}'\mathbf{u}) \quad \text{or, equivalently,} \quad h(\mathbf{O}\mathbf{z}) = f(\mathbf{z}).$$

And upon observing [in light of the fundamental theorem of (integral) calculus (e.g., Billingsley 1995)] that $\mathbf{u} \sim \mathbf{z}$ if and only if $h(\mathbf{O}\mathbf{z}) = f(\mathbf{O}\mathbf{z})$ (with probability 1), it follows that $\mathbf{u} \sim \mathbf{z}$ if and only if

$$f(\mathbf{O}\mathbf{z}) = f(\mathbf{z}) \quad \text{(with probability 1).} \tag{9.50}$$

In effect, we have established that $\mathbf{z}$ has a spherical distribution if and only if, for every orthogonal matrix $\mathbf{O}$, the pdf $f(\cdot)$ satisfies condition (9.50). Now, suppose that $f(\mathbf{z})$ depends on the value of $\mathbf{z}$ only through $\mathbf{z}'\mathbf{z}$ or, equivalently, that there exists a (nonnegative) function $g(\cdot)$ (of a single nonnegative variable) such that

$$f(\mathbf{z}) = g(\mathbf{z}'\mathbf{z}) \quad \text{(for every value of } \mathbf{z}\text{).} \tag{9.51}$$

Clearly, if $f(\cdot)$ is of the form (9.51), then, for every orthogonal matrix $\mathbf{O}$, $f(\cdot)$ satisfies condition (9.50) and, in fact, satisfies the more stringent condition

$$f(\mathbf{O}\mathbf{z}) = f(\mathbf{z}) \quad \text{(for every value of } \mathbf{z}\text{).} \tag{9.52}$$

Thus, if $f(\cdot)$ is of the form (9.51), then the distribution of $\mathbf{z}$ is spherical.

Consider the converse. Suppose that the distribution of $\mathbf{z}$ is spherical and hence that, for every orthogonal matrix $\mathbf{O}$, $f(\cdot)$ satisfies condition (9.50). Is $f(\cdot)$ necessarily of the form (9.51)? If for every orthogonal matrix $\mathbf{O}$, $f(\cdot)$ satisfies condition (9.52), then the answer is yes.

To see this, suppose that (for every orthogonal matrix $\mathbf{O}$) $f(\cdot)$ satisfies condition (9.52). Then, for any $M \times 1$ vectors $\mathbf{z}_1$ and $\mathbf{z}_2$ such that $\mathbf{z}_2'\mathbf{z}_2 = \mathbf{z}_1'\mathbf{z}_1$, we find [upon observing (in light of Lemma 5.9.9) that $\mathbf{z}_2 = \mathbf{O}\mathbf{z}_1$ for some orthogonal matrix $\mathbf{O}$] that $f(\mathbf{z}_2) = f(\mathbf{z}_1)$. Thus, for any particular (nonnegative) constant c, $f(\mathbf{z})$ has the same value for every $\mathbf{z}$ for which $\mathbf{z}'\mathbf{z} = c$. And it follows that there exists a function $g(\cdot)$ for which $f(\cdot)$ is expressible in the form (9.51).

Subsequently, there will be occasion to refer to the distribution of an M-dimensional random column vector that is absolutely continuous with a pdf $f(\cdot)$ of the form (9.51). Accordingly, as a matter of convenience, let us interpret any reference to an absolutely continuous spherical distribution as a reference to a distribution with those characteristics.

Let $g(\cdot)$ represent a nonnegative function whose domain is the interval $[0, \infty)$. And let $\mathbf{z} = (z_1, z_2, \ldots, z_M)'$ represent an $M \times 1$ vector of (unrestricted) variables, and suppose that

$$0 < \int_{\mathbb{R}^M} g(\mathbf{z}'\mathbf{z})\, d\mathbf{z} < \infty. \tag{9.53}$$

Further, take $f(\mathbf{z})$ to be the (nonnegative) function of $\mathbf{z}$ defined by

$$f(\mathbf{z}) = c^{-1} g(\mathbf{z}'\mathbf{z}), \tag{9.54}$$

where $c = \int_{\mathbb{R}^M} g(\mathbf{z}'\mathbf{z})\, d\mathbf{z}$ (and observe that $\int_{\mathbb{R}^M} f(\mathbf{z})\, d\mathbf{z} = 1$). Then, there is an absolutely continuous distribution (of an $M \times 1$ random vector) having $f(\cdot)$ as a pdf, and [since $f(\cdot)$ is of the form (9.51)] that distribution is spherical.

The M-dimensional integral $\int_{\mathbb{R}^M} g(\mathbf{z}'\mathbf{z})\, d\mathbf{z}$ can be simplified. Clearly,

$$\int_{\mathbb{R}^M} g(\mathbf{z}'\mathbf{z})\, d\mathbf{z} = 2^M \int_0^\infty \int_0^\infty \cdots \int_0^\infty g\big(\textstyle\sum_{i=1}^M z_i^2\big)\, dz_1 dz_2 \cdots dz_M.$$

Upon making the change of variables $u_i = z_i^2$ $(i = 1, 2, \ldots, M)$ and observing that $\partial z_i/\partial u_i = (1/2)u_i^{-1/2}$, we find that

$$\begin{aligned}\int_{\mathbb{R}^M} g(\mathbf{z}'\mathbf{z})\, d\mathbf{z} &= 2^M \int_0^\infty \int_0^\infty \cdots \int_0^\infty g\big(\textstyle\sum_{i=1}^M u_i\big) \big(\tfrac{1}{2}\big)^M \prod_{i=1}^M u_i^{-1/2}\, du_1 du_2 \cdots du_M \\ &= \int_0^\infty \int_0^\infty \cdots \int_0^\infty g\big(\textstyle\sum_{i=1}^M u_i\big) \prod_{i=1}^M u_i^{-1/2}\, du_1 du_2 \cdots du_M.\end{aligned}$$

And upon making the further change of variables $y_i = u_i$ $(i = 1, 2, \ldots, M-1)$, $y_M = \sum_{i=1}^M u_i$ and observing that the $M \times M$ matrix with ijth element $\partial u_i/\partial y_j$ equals $\begin{pmatrix} \mathbf{I} & \mathbf{0} \\ -\mathbf{1}' & 1 \end{pmatrix}$ (the determinant of which equals 1), we find that

$$\int_{\mathbb{R}^M} g(\mathbf{z}'\mathbf{z})\, d\mathbf{z} = \int_D g(y_M) \prod_{i=1}^{M-1} y_i^{-1/2} \big(y_M - \textstyle\sum_{i=1}^{M-1} y_i\big)^{-1/2}\, dy_1 dy_2 \cdots dy_M,$$

where $D = \{y_1, y_2, \ldots, y_M \;:\; y_i \geq 0 \;(i = 1, 2, \ldots, M-1),\; y_M \geq \sum_{i=1}^{M-1} y_i\}$. Moreover, upon making yet another change of variables $w_i = y_i/y_M$ $(i = 1, 2, \ldots, M-1)$, $w_M = y_M$ and observing that the $M \times M$ matrix with ijth element $\partial y_i/\partial w_j$ equals $\begin{bmatrix} w_M \mathbf{I} & (w_1, w_2, \ldots, w_{M-1})' \\ \mathbf{0} & 1 \end{bmatrix}$ (the determinant of which equals w_M^{M-1}), we find that

$$\begin{aligned}\int_{\mathbb{R}^M} g(\mathbf{z}'\mathbf{z})\, d\mathbf{z} = \int_{D_*} \prod_{i=1}^{M-1} w_i^{-1/2} \big(1 - \textstyle\sum_{i=1}^{M-1} w_i\big)^{-1/2}\, dw_1 dw_2 \cdots dw_{M-1} \\ \times \int_0^\infty w_M^{(M/2)-1} g(w_M)\, dw_M,\end{aligned} \tag{9.55}$$

where $D_* = \{w_1, w_2, \ldots, w_{M-1} \;:\; w_i \geq 0 \;(i = 1, 2, \ldots, M-1),\; \sum_{i=1}^{M-1} w_i \leq 1\}$.

According to a basic result on the normalizing constant for the pdf of a Dirichlet distribution—the Dirichlet distribution is the subject of Section 6.1e—

$$\int_{D_*} \prod_{i=1}^{M-1} w_i^{-1/2} \big(1 - \textstyle\sum_{i=1}^{M-1} w_i\big)^{-1/2}\, dw_1 dw_2 \cdots dw_{M-1} = \frac{[\Gamma(1/2)]^M}{\Gamma(M/2)} = \frac{\pi^{M/2}}{\Gamma(M/2)}.$$

Thus,

$$\int_{\mathbb{R}^M} g(\mathbf{z}'\mathbf{z})\, d\mathbf{z} = \frac{\pi^{M/2}}{\Gamma(M/2)} \int_0^\infty w_M^{(M/2)-1} g(w_M)\, dw_M; \tag{9.56}$$

and upon introducing the change of variable $s = w_M^{1/2}$ and observing that $dw_M/ds = 2s$, we find that

$$\int_{\mathbb{R}^M} g(\mathbf{z}'\mathbf{z})\, d\mathbf{z} = \frac{2\pi^{M/2}}{\Gamma(M/2)} \int_0^\infty s^{M-1} g(s^2)\, ds. \tag{9.57}$$

In light of result (9.57), the function $g(\cdot)$ satisfies condition (9.53) if and only if

$$0 < \int_0^\infty s^{M-1} g(s^2)\, ds < \infty,$$

in which case the constant c in expression (9.54) is expressible in the form (9.57).

Moment generating function of a spherical distribution. Spherical distributions can be characterized in terms of their moment generating functions (or, more generally, their characteristic functions) as well as in terms of their pdfs. Take $\mathbf{z} = (z_1, z_2, \ldots, z_M)'$ to be an M-dimensional random (column) vector, and suppose that the distribution of $\mathbf{z}$ has a moment generating function, say $\psi(\cdot)$. Then, for the distribution of $\mathbf{z}$ to be spherical, it is necessary and sufficient that

$$\psi(\mathbf{Ot}) = \psi(\mathbf{t}) \text{ for every } M \times M \text{ orthogonal matrix } \mathbf{O} \text{ (and for every } M \times 1 \text{ vector } \mathbf{t} \text{ in a neighborhood of } \mathbf{0}). \quad (9.58)$$

To see this, let $\mathbf{O}$ represent an arbitrary $M \times M$ matrix, and observe that (for any $M \times 1$ vector $\mathbf{t}$)

$$\psi(\mathbf{Ot}) = \mathrm{E}\big[e^{(\mathbf{Ot})'\mathbf{z}}\big] = \mathrm{E}\big[e^{\mathbf{t}'(\mathbf{O}'\mathbf{z})}\big]$$

and hence that $\psi(\mathbf{Ot}) = \psi(\mathbf{t})$ (for every $M \times 1$ vector $\mathbf{t}$ in a neighborhood of $\mathbf{0}$) if and only if $\psi(\cdot)$ is the moment generating function of the distribution of $\mathbf{O}'\mathbf{z}$ (as well as that of the distribution of $\mathbf{z}$), or equivalently—refer, e.g., to Casella and Berger (2002, p. 65)—if and only if $\mathbf{O}'\mathbf{z}$ and $\mathbf{z}$ have the same distribution.

For the distribution of $\mathbf{z}$ to be spherical, it is necessary and sufficient that $\psi(\mathbf{t})$ depend on the $M \times 1$ vector $\mathbf{t}$ only through the value of $\mathbf{t}'\mathbf{t}$ or, equivalently, that there exists a function $\phi(\cdot)$ (of a single nonnegative variable) such that

$$\psi(\mathbf{t}) = \phi(\mathbf{t}'\mathbf{t}) \text{ (for every } M \times 1 \text{ vector } \mathbf{t} \text{ in a neighborhood of } \mathbf{0}). \quad (9.59)$$

Let us verify the necessity and sufficiency of the existence of a function $\phi(\cdot)$ that satisfies condition (9.59). If there exists a function $\phi(\cdot)$ that satisfies condition (9.59), then for every $M \times M$ orthogonal matrix $\mathbf{O}$ (and for every $M \times 1$ vector $\mathbf{t}$ in a neighborhood of $\mathbf{0}$),

$$\psi(\mathbf{Ot}) = \phi[(\mathbf{Ot})'\mathbf{Ot}] = \phi(\mathbf{t}'\mathbf{O}'\mathbf{Ot}) = \phi(\mathbf{t}'\mathbf{t}) = \psi(\mathbf{t}),$$

so that condition (9.58) is satisfied and, consequently, the distribution of $\mathbf{z}$ is spherical.

Conversely, suppose that the distribution of $\mathbf{z}$ is spherical and hence that condition (9.58) is satisfied. Then, for "any" $M \times 1$ vectors $\mathbf{t}_1$ and $\mathbf{t}_2$ such that $\mathbf{t}_2'\mathbf{t}_2 = \mathbf{t}_1'\mathbf{t}_1$, we find [upon observing (in light of Lemma 5.9.9) that $\mathbf{t}_2 = \mathbf{Ot}_1$ for some orthogonal matrix $\mathbf{O}$] that $\psi(\mathbf{t}_2) = \psi(\mathbf{t}_1)$. Thus, for any sufficiently small nonnegative constant c, $\psi(\mathbf{t})$ has the same value for every $M \times 1$ vector $\mathbf{t}$ for which $\mathbf{t}'\mathbf{t} = c$. And it follows that there exists a function $\phi(\cdot)$ that satisfies condition (9.59).

What can be said about the nature of the function $\phi(\cdot)$? Clearly, $\phi(0) = 1$. Moreover, $\phi(\cdot)$ is a strictly increasing function. To see this, take $\mathbf{t}$ to be any $M \times 1$ vector (of constants) such that $\mathbf{t}'\mathbf{t} = 1$, and observe that for any nonnegative scalar k,

$$\phi(k) = \phi(k\mathbf{t}'\mathbf{t}) = \tfrac{1}{2}\phi(k\mathbf{t}'\mathbf{t}) + \tfrac{1}{2}\phi[k(-\mathbf{t})'(-\mathbf{t})] = \tfrac{1}{2}\,\mathrm{E}\Big[e^{\sqrt{k}\mathbf{t}'\mathbf{z}} + e^{-\sqrt{k}\mathbf{t}'\mathbf{z}}\Big].$$

Observe also that (for $k > 0$)

$$\begin{aligned}\frac{d\big[e^{\sqrt{k}\mathbf{t}'\mathbf{z}} + e^{-\sqrt{k}\mathbf{t}'\mathbf{z}}\big]}{dk} &= (1/2)k^{-1/2}\mathbf{t}'\mathbf{z}\Big(e^{\sqrt{k}\mathbf{t}'\mathbf{z}} - e^{-\sqrt{k}\mathbf{t}'\mathbf{z}}\Big)\\ &> 0 \quad \text{if } \mathbf{t}'\mathbf{z} \neq 0.\end{aligned}$$

Thus (for $k > 0$)

$$\frac{d\phi(k)}{dk} = \tfrac{1}{2}\,\mathrm{E}\left\{\frac{d\big[e^{\sqrt{k}\mathbf{t}'\mathbf{z}} + e^{-\sqrt{k}\mathbf{t}'\mathbf{z}}\big]}{dk}\right\} > 0,$$

which confirms that $\phi(\cdot)$ is a strictly increasing function.

Linear transformation of a spherically distributed random vector. Let M and N represent arbitrary positive integers. And define

$$\mathbf{x} = \boldsymbol{\mu} + \boldsymbol{\Gamma}'\mathbf{z}, \quad (9.60)$$

where $\boldsymbol{\mu}$ is an arbitrary M-dimensional nonrandom column vector, $\boldsymbol{\Gamma}$ is an arbitrary $N \times M$ nonrandom matrix, and $\mathbf{z}$ is an N-dimensional spherically distributed random column vector. Further, let $\boldsymbol{\Sigma} = \boldsymbol{\Gamma}'\boldsymbol{\Gamma}$.

If E(**z**) exists, then E(**x**) exists and [in light of result (9.46)]

$$\mathrm{E}(\mathbf{x}) = \boldsymbol{\mu}. \tag{9.61}$$

And if the second-order moments of the distribution of **z** exist, then so do those of the distribution of **x** and [in light of result (9.47)]

$$\mathrm{var}(\mathbf{x}) = c\boldsymbol{\Sigma}, \tag{9.62}$$

where c is the variance of any element of **z**—every element of **z** has the same variance.

If the distribution of **z** has a moment generating function, say $\omega(\cdot)$, then there exists a (nonnegative) function $\phi(\cdot)$ (of a single nonnegative variable) such that (for every $N \times 1$ vector **s** in a neighborhood of **0**) $\omega(\mathbf{s}) = \phi(\mathbf{s}'\mathbf{s})$, and the distribution of **x** has the moment generating function $\psi(\cdot)$, where (for every $M \times 1$ vector **t** in a neighborhood of **0**)

$$\psi(\mathbf{t}) = \mathrm{E}\big(e^{\mathbf{t}'\mathbf{x}}\big) = \mathrm{E}\big[e^{\mathbf{t}'(\boldsymbol{\mu}+\boldsymbol{\Gamma}'\mathbf{z})}\big] = e^{\mathbf{t}'\boldsymbol{\mu}}\,\mathrm{E}\big[e^{(\boldsymbol{\Gamma}\mathbf{t})'\mathbf{z}}\big] = e^{\mathbf{t}'\boldsymbol{\mu}}\omega(\boldsymbol{\Gamma}\mathbf{t}) = e^{\mathbf{t}'\boldsymbol{\mu}}\phi(\mathbf{t}'\boldsymbol{\Sigma}\mathbf{t}). \tag{9.63}$$

Note that the moment generating function of the distribution of **x** and hence the distribution itself depend on the value of the $N \times M$ matrix $\boldsymbol{\Gamma}$ only through the value of the $M \times M$ matrix $\boldsymbol{\Sigma}$.

Marginal distributions (of spherically distributed random vectors). Let **z** represent an N-dimensional spherically distributed random column vector. And take $\mathbf{z}_*$ to be an M-dimensional subvector of **z** (where $M < N$), say the subvector obtained by striking out all of the elements of **z** save the $i_1, i_2, \ldots, i_M$th elements.

Suppose that the distribution of **z** has a moment generating function, say $\psi(\cdot)$. Then, necessarily, there exists a (nonnegative) function $\phi(\cdot)$ (of a single nonnegative variable) such that $\psi(\mathbf{s}) = \phi(\mathbf{s}'\mathbf{s})$ (for every $N \times 1$ vector **s** in a neighborhood of **0**). Clearly, the subvector $\mathbf{z}_*$ can be regarded as a special case of the random column vector **x** defined by expression (9.60); it is the special case obtained by setting $\boldsymbol{\mu} = \mathbf{0}$ and taking $\boldsymbol{\Gamma}$ to be the $N \times M$ matrix whose first, second, ..., Mth columns are, respectively, the $i_1, i_2, \ldots, i_M$th columns of $\mathbf{I}_N$. And (in light of the results of the preceding part of the present subsection) it follows that the distribution of $\mathbf{z}_*$ has a moment generating function, say $\psi_*(\cdot)$, and that (for every $M \times 1$ vector **t** in some neighborhood of **0**)

$$\psi_*(\mathbf{t}) = \phi(\mathbf{t}'\mathbf{t}). \tag{9.64}$$

Thus, the moment generating function of the distribution of the subvector $\mathbf{z}_*$ is characterized by the same function $\phi(\cdot)$ as that of the distribution of **z** itself.

Suppose now that **u** is an M-dimensional random column vector whose distribution has a moment generating function, say $\omega(\cdot)$, and that (for every $M \times 1$ vector **t** in a neighborhood of **0**) $\omega(\mathbf{t}) = \phi(\mathbf{t}'\mathbf{t})$. Then, the distribution of **u** is spherical. Moreover, it has the same moment generating function as the distribution of $\mathbf{z}_*$ (and, consequently, $\mathbf{u} \sim \mathbf{z}_*$). There is an implication that the elements of **u**, like those of $\mathbf{z}_*$, have the same variance as the elements of **z**.

The moment generating function of a marginal distribution of **z** (i.e., of the distribution of a subvector of **z**) is characterized by the same function $\phi(\cdot)$ as that of the distribution of **z** itself. In the case of pdfs, the relationship is more complex.

Suppose that the distribution of the N-dimensional spherically distributed random column vector **z** is an absolutely continuous spherical distribution. Then, the distribution of **z** has a pdf $f(\cdot)$, where $f(\mathbf{z}) = g(\mathbf{z}'\mathbf{z})$ for some (nonnegative) function $g(\cdot)$ of a single nonnegative variable (and for every value of **z**). Accordingly, the distribution of the M-dimensional subvector $\mathbf{z}_*$ is the absolutely continuous distribution with pdf $f_*(\cdot)$ defined (for every value of $\mathbf{z}_*$) by

$$f_*(\mathbf{z}_*) = \int_{\mathcal{R}^{N-M}} g(\mathbf{z}_*'\mathbf{z}_* + \bar{\mathbf{z}}_*'\bar{\mathbf{z}}_*)\, d\bar{\mathbf{z}}_*,$$

where $\bar{\mathbf{z}}_*$ is the $(N-M)$-dimensional subvector of **z** obtained by striking out the $i_1, i_2, \ldots, i_M$th elements. And upon regarding $g(\mathbf{z}_*'\mathbf{z}_* + w)$ as a function of a nonnegative variable w and applying result (9.57), we find that (for every value of $\mathbf{z}_*$)

$$f_*(\mathbf{z}_*) = \frac{2\pi^{(N-M)/2}}{\Gamma[(N-M)/2]} \int_0^\infty s^{N-M-1} g(\mathbf{z}_*'\mathbf{z}_* + s^2)\, ds. \tag{9.65}$$

Clearly, $f(\mathbf{z}_*)$ depends on the value of $\mathbf{z}_*$ only through $\mathbf{z}_*'\mathbf{z}_*$, so that (as could have been anticipated from our results on the moment generating function of the distribution of a subvector of a spherically distributed random vector) the distribution of $\mathbf{z}_*$ is spherical. Further, upon introducing the changes of variable $w = s^2$ and $u = \mathbf{z}_*'\mathbf{z}_* + w$, we obtain the following variations on expression (9.65):

$$f_*(\mathbf{z}_*) = \frac{\pi^{(N-M)/2}}{\Gamma[(N-M)/2]} \int_0^\infty w^{[(N-M)/2]-1} g(\mathbf{z}_*'\mathbf{z}_* + w)\, dw \tag{9.66}$$

$$= \frac{\pi^{(N-M)/2}}{\Gamma[(N-M)/2]} \int_{\mathbf{z}_*'\mathbf{z}_*}^\infty (u - \mathbf{z}_*'\mathbf{z}_*)^{[(N-M)/2]-1} g(u)\, du. \tag{9.67}$$

Elliptical distributions: definition. The distribution of a random column vector of the form of the vector $\mathbf{x}$ of equality (9.60) is said to be *elliptical*. And a random column vector whose distribution is that of the vector $\mathbf{x}$ of equality (9.60) may be referred to as being distributed elliptically about $\boldsymbol{\mu}$ or, in the special case where $\boldsymbol{\Gamma} = \mathbf{I}$ (or where $\boldsymbol{\Gamma}$ is orthogonal), as being distributed spherically about $\boldsymbol{\mu}$. Clearly, a random column vector $\mathbf{x}$ is distributed elliptically about $\boldsymbol{\mu}$ if and only if $\mathbf{x} - \boldsymbol{\mu}$ is distributed elliptically about $\mathbf{0}$ and is distributed spherically about $\boldsymbol{\mu}$ if and only if $\mathbf{x} - \boldsymbol{\mu}$ is distributed spherically about $\mathbf{0}$. Let us consider the definition of an elliptical distribution as applied to distributions whose second-order moments exist.

For any $M \times 1$ vector $\boldsymbol{\mu}$ and any $M \times M$ nonnegative definite matrix $\boldsymbol{\Sigma}$, an $M \times 1$ random vector $\mathbf{x}$ has an elliptical distribution with mean vector $\boldsymbol{\mu}$ and variance-covariance matrix $\boldsymbol{\Sigma}$ if and only if

$$\mathbf{x} \sim \boldsymbol{\mu} + \boldsymbol{\Gamma}'\mathbf{z} \tag{9.68}$$

for some matrix $\boldsymbol{\Gamma}$ such that $\boldsymbol{\Sigma} = \boldsymbol{\Gamma}'\boldsymbol{\Gamma}$ and some random (column) vector $\mathbf{z}$ (of compatible dimension) having a spherical distribution with variance-covariance matrix $\mathbf{I}$—recall that if a random column vector $\mathbf{z}$ has a spherical distribution with a variance-covariance matrix that is a nonzero scalar multiple $c\mathbf{I}$ of $\mathbf{I}$, then the rescaled vector $c^{-1/2}\mathbf{z}$ has a spherical distribution with variance-covariance matrix $\mathbf{I}$. In connection with condition (9.68), define $K = \operatorname{rank} \boldsymbol{\Sigma}$, and denote by N the number of rows in the matrix $\boldsymbol{\Gamma}$ (or, equivalently, the number of elements in $\mathbf{z}$)—necessarily, $N \geq K$. For any particular N, the distribution of $\boldsymbol{\mu} + \boldsymbol{\Gamma}'\mathbf{z}$ does not depend on the choice of $\boldsymbol{\Gamma}$ [as is evident (for the case where the distribution of $\mathbf{z}$ has a moment generating function) from result (9.63)]; rather, it depends only on $\boldsymbol{\mu}$, $\boldsymbol{\Sigma}$, and the distribution of $\mathbf{z}$.

Now, consider the distribution of $\boldsymbol{\mu} + \boldsymbol{\Gamma}'\mathbf{z}$ for different choices of N. Assume that $K \geq 1$—if $K = 0$, then (for any choice of N) $\boldsymbol{\Gamma} = \mathbf{0}$ and hence $\boldsymbol{\mu} + \boldsymbol{\Gamma}'\mathbf{z} = \boldsymbol{\mu}$. And take $\boldsymbol{\Gamma}_*$ to be a $K \times M$ matrix such that $\boldsymbol{\Sigma} = \boldsymbol{\Gamma}_*'\boldsymbol{\Gamma}_*$, and take $\mathbf{z}_*$ to be a $K \times 1$ random vector having a spherical distribution with variance-covariance matrix $\mathbf{I}_K$.

Suppose that the distribution of $\mathbf{z}_*$ has a moment generating function, say $\omega_*(\cdot)$. Then, because the distribution of $\mathbf{z}_*$ is spherical, there exists a (nonnegative) function $\phi(\cdot)$ (of a single nonnegative variable) such that (for every $K \times 1$ vector $\mathbf{t}_*$ in a neighborhood of $\mathbf{0}$) $\omega_*(\mathbf{t}_*) = \phi(\mathbf{t}_*'\mathbf{t}_*)$.

Take $\omega(\mathbf{t})$ to be the function of an $N \times 1$ vector $\mathbf{t}$ defined (for every value of $\mathbf{t}$ in some neighborhood of $\mathbf{0}$) by $\omega(\mathbf{t}) = \phi(\mathbf{t}'\mathbf{t})$. There may or may not exist an (N-dimensional) distribution having $\omega(\cdot)$ as a moment generating function. If such a distribution exists, then that distribution is spherical, and for any random vector, say $\mathbf{w}$, having that distribution, the distribution of $\mathbf{z}_*$ is a marginal distribution of $\mathbf{w}$ and $\operatorname{var}(\mathbf{w}) = \mathbf{I}_N$. Accordingly, if there exists a distribution having $\omega(\cdot)$ as a moment generating function, then the distribution of the random vector $\mathbf{z}$ [in expression (9.68)] could be taken to be that distribution, in which case the distribution of $\boldsymbol{\mu} + \boldsymbol{\Gamma}'\mathbf{z}$ would have the same moment generating function as the distribution of $\boldsymbol{\mu} + \boldsymbol{\Gamma}_*'\mathbf{z}_*$ [as is evident from result (9.63)] and it would follow that $\boldsymbol{\mu} + \boldsymbol{\Gamma}'\mathbf{z} \sim \boldsymbol{\mu} + \boldsymbol{\Gamma}_*'\mathbf{z}_*$.

Thus, as long as there exists an (N-dimensional) distribution having $\omega(\cdot)$ as a moment generating function [where $\omega(\mathbf{t}) = \phi(\mathbf{t}'\mathbf{t})$] and as long as the distribution of $\mathbf{z}$ is taken to be that distribution, the distribution of $\boldsymbol{\mu} + \boldsymbol{\Gamma}'\mathbf{z}$ is invariant to the choice of N. This invariance extends to every N for which there exists an (N-dimensional) distribution having $\omega(\cdot)$ as a moment generating function.

Let us refer to the function $\phi(\cdot)$ as the *mgf generator* of the distribution of the M-dimensional random vector $\boldsymbol{\mu} + \boldsymbol{\Gamma}_*'\mathbf{z}_*$ (with mgf being regarded as an acronym for moment generating function). The moment generating function of the distribution of $\boldsymbol{\mu} + \boldsymbol{\Gamma}_*'\mathbf{z}_*$ is the function $\psi(\cdot)$ defined (for every $M \times 1$ vector $\mathbf{t}$ in a neighborhood of $\mathbf{0}$) by

$$\psi(\mathbf{t}) = e^{\mathbf{t}'\boldsymbol{\mu}}\phi(\mathbf{t}'\boldsymbol{\Sigma}\mathbf{t}) \tag{9.69}$$

[as is evident from result (9.63)]. The distribution of $\boldsymbol{\mu} + \boldsymbol{\Gamma}_*'\mathbf{z}_*$ is completely determined by the mean vector $\boldsymbol{\mu}$, the variance-covariance matrix $\boldsymbol{\Sigma}$, and the mgf generator $\phi(\cdot)$. Accordingly, we may refer to this distribution as an (M-dimensional) elliptical distribution with mean $\boldsymbol{\mu}$, variance-covariance matrix $\boldsymbol{\Sigma}$, and mgf generator $\phi(\cdot)$. The mgf generator $\phi(\cdot)$ serves to identify the applicable distribution of $\mathbf{z}_*$; alternatively, some other characteristic of the distribution of $\mathbf{z}_*$ could be used for that purpose (e.g., the pdf). Note that the $N(\boldsymbol{\mu}, \boldsymbol{\Sigma})$ distribution is an elliptical distribution with mean $\boldsymbol{\mu}$, variance-covariance matrix $\boldsymbol{\Sigma}$, and mgf generator $\phi_*(\cdot)$, where (for every nonnegative scalar u) $\phi_*(u) = \exp(u/2)$.

Pdf of an elliptical distribution. Let $\mathbf{x} = (x_1, x_2, \ldots, x_M)'$ represent an $M \times 1$ random vector, and suppose that for some $M \times 1$ (nonrandom) vector $\boldsymbol{\mu}$ and some $M \times M$ (nonrandom) positive definite matrix $\boldsymbol{\Sigma}$,

$$\mathbf{x} = \boldsymbol{\mu} + \boldsymbol{\Gamma}'\mathbf{z},$$

where $\boldsymbol{\Gamma}$ is an $M \times M$ (nonsingular) matrix such that $\boldsymbol{\Sigma} = \boldsymbol{\Gamma}'\boldsymbol{\Gamma}$ and where $\mathbf{z} = (z_1, z_2, \ldots, z_M)'$ is an $M \times 1$ spherically distributed random vector with variance-covariance matrix $\mathbf{I}$. Then, $\mathbf{x}$ has an elliptical distribution with mean vector $\boldsymbol{\mu}$ and variance-covariance matrix $\boldsymbol{\Sigma}$.

Now, suppose that the distribution of $\mathbf{z}$ is an absolutely continuous spherical distribution. Then, the distribution of $\mathbf{z}$ is absolutely continuous with a pdf $h(\cdot)$ defined as follows in terms of some (nonnegative) function $g(\cdot)$ (of a single nonnegative variable) for which $\int_0^\infty s^{M-1} g(s^2)\, ds < \infty$:

$$h(\mathbf{z}) = c^{-1} g(\mathbf{z}'\mathbf{z}),$$

where $c = [2\pi^{M/2}/\Gamma(M/2)] \int_0^\infty s^{M-1} g(s^2)\, ds$. And the distribution of $\mathbf{x}$ is absolutely continuous with a pdf, say $f(\cdot)$, that is derivable from the pdf of the distribution of $\mathbf{z}$.

Let us derive an expression for $f(\mathbf{x})$. Clearly, $\mathbf{z} = (\boldsymbol{\Gamma}')^{-1}(\mathbf{x} - \boldsymbol{\mu})$, and the $M \times M$ matrix with ijth element $\partial z_i/\partial x_j$ equals $(\boldsymbol{\Gamma}')^{-1}$. Moreover,

$$\begin{aligned} |\det(\boldsymbol{\Gamma}')^{-1}| = |\det \boldsymbol{\Gamma}'|^{-1} = [(\det \boldsymbol{\Gamma}')^2]^{-1/2} &= [(\det \boldsymbol{\Gamma}') \det \boldsymbol{\Gamma}]^{-1/2} \\ &= [\det(\boldsymbol{\Gamma}'\boldsymbol{\Gamma})]^{-1/2} = (\det \boldsymbol{\Sigma})^{-1/2}. \end{aligned}$$

Thus, making use of standard results on a change of variables (e.g., Bickel and Doksum 2001, sec. B.2) and observing that $\boldsymbol{\Sigma}^{-1} = \boldsymbol{\Gamma}^{-1}(\boldsymbol{\Gamma}')^{-1} = [(\boldsymbol{\Gamma}')^{-1}]'(\boldsymbol{\Gamma}')^{-1}$, we find that

$$f(\mathbf{x}) = c^{-1}|\boldsymbol{\Sigma}|^{-1/2} g[(\mathbf{x} - \boldsymbol{\mu})'\boldsymbol{\Sigma}^{-1}(\mathbf{x} - \boldsymbol{\mu})]. \tag{9.70}$$

Linear transformation of an elliptically distributed random vector. Let $\mathbf{x}$ represent an $N \times 1$ random vector that has an (N-dimensional) elliptical distribution with mean $\boldsymbol{\mu}$, variance-covariance matrix $\boldsymbol{\Sigma}$, and (if $\boldsymbol{\Sigma} \neq \mathbf{0}$) mgf generator $\phi(\cdot)$. And take $\mathbf{y}$ to be the $M \times 1$ random vector obtained by transforming $\mathbf{x}$ as follows:

$$\mathbf{y} = \mathbf{c} + \mathbf{A}\mathbf{x}, \tag{9.71}$$

where $\mathbf{c}$ is an $M \times 1$ (nonrandom) vector and $\mathbf{A}$ an $M \times N$ (nonrandom) matrix. Then, $\mathbf{y}$ has an (M-dimensional) elliptical distribution with mean $\mathbf{c} + \mathbf{A}\boldsymbol{\mu}$, variance-covariance matrix $\mathbf{A}\boldsymbol{\Sigma}\mathbf{A}'$, and (if $\mathbf{A}\boldsymbol{\Sigma}\mathbf{A}' \neq \mathbf{0}$) mgf generator $\phi(\cdot)$ (identical to the mgf generator of the distribution of $\mathbf{x}$).

Let us verify that $\mathbf{y}$ has this distribution. Define $K = \operatorname{rank} \boldsymbol{\Sigma}$. And suppose that $K > 0$ (or, equivalently, that $\boldsymbol{\Sigma} \neq \mathbf{0}$), in which case

$$\mathbf{x} \sim \boldsymbol{\mu} + \boldsymbol{\Gamma}'\mathbf{z},$$

where $\boldsymbol{\Gamma}$ is any $K \times N$ (nonrandom) matrix such that $\boldsymbol{\Sigma} = \boldsymbol{\Gamma}'\boldsymbol{\Gamma}$ and where $\mathbf{z}$ is a $K \times 1$ random vector having a spherical distribution with a moment generating function $\omega(\cdot)$ defined (for every $K \times 1$ vector $\mathbf{s}$ in a neighborhood of $\mathbf{0}$) by $\omega(\mathbf{s}) = \phi(\mathbf{s}'\mathbf{s})$. Then,

$$\mathbf{y} \sim \mathbf{c} + \mathbf{A}(\boldsymbol{\mu} + \boldsymbol{\Gamma}'\mathbf{z}) = \mathbf{c} + \mathbf{A}\boldsymbol{\mu} + (\boldsymbol{\Gamma}\mathbf{A}')'\mathbf{z}. \tag{9.72}$$

Now, let $K_* = \operatorname{rank}(\mathbf{A}\boldsymbol{\Sigma}\mathbf{A}')$, and observe that $K_* \leq K$ and that $\mathbf{A}\boldsymbol{\Sigma}\mathbf{A}' = (\boldsymbol{\Gamma}\mathbf{A}')'\boldsymbol{\Gamma}\mathbf{A}'$. Further, suppose that $K_* > 0$ (or, equivalently, that $\mathbf{A}\boldsymbol{\Sigma}\mathbf{A}' \neq \mathbf{0}$), take $\boldsymbol{\Gamma}_*$ to be any $K_* \times M$ (nonrandom) matrix such that $\mathbf{A}\boldsymbol{\Sigma}\mathbf{A}' = \boldsymbol{\Gamma}_*'\boldsymbol{\Gamma}_*$, and take $\mathbf{z}_*$ to be a $K_* \times 1$ random vector having a distribution that is a marginal distribution of $\mathbf{z}$ and that, consequently, has a moment generating function $\omega_*(\cdot)$ defined (for every $K_* \times 1$ vector $\mathbf{s}_*$ in a neighborhood of $\mathbf{0}$) by $\omega_*(\mathbf{s}_*) = \phi(\mathbf{s}_*'\mathbf{s}_*)$. Then, it follows from what was established earlier (in defining elliptical distributions) that

$$\mathbf{c} + \mathbf{A}\boldsymbol{\mu} + (\boldsymbol{\Gamma}\mathbf{A}')'\mathbf{z} \sim \mathbf{c} + \mathbf{A}\boldsymbol{\mu} + \boldsymbol{\Gamma}_*'\mathbf{z}_*,$$

which [in combination with result (9.72)] implies that $\mathbf{y}$ has an elliptical distribution with mean $\mathbf{c} + \mathbf{A}\boldsymbol{\mu}$, variance-covariance matrix $\mathbf{A}\boldsymbol{\Sigma}\mathbf{A}'$, and mgf generator $\phi(\cdot)$. It remains only to observe that even in the "degenerate" case where $\boldsymbol{\Sigma} = \mathbf{0}$ or, more generally, where $\mathbf{A}\boldsymbol{\Sigma}\mathbf{A}' = \mathbf{0}$, $\mathrm{E}(\mathbf{y}) = \mathbf{c} + \mathbf{A}\boldsymbol{\mu}$ and $\operatorname{var}(\mathbf{y}) = \mathbf{A}\boldsymbol{\Sigma}\mathbf{A}'$ (and to observe that the distribution of a random vector whose variance-covariance matrix equals a null matrix qualifies as an elliptical distribution).

Marginal distributions (of elliptically distributed random vectors). Let $\mathbf{x}$ represent an $N \times 1$ random vector that has an (N-dimensional) elliptical distribution with mean $\boldsymbol{\mu}$, nonnull variance-covariance matrix $\boldsymbol{\Sigma}$, and mgf generator $\phi(\cdot)$. And take $\mathbf{x}_*$ to be an M-dimensional subvector of $\mathbf{x}$ (where $M < N$), say the subvector obtained by striking out all of the elements of $\mathbf{x}$ save the $i_1, i_2, \ldots, i_M$th elements. Further, take $\boldsymbol{\mu}_*$ to be the M-dimensional subvector of $\boldsymbol{\mu}$ obtained by striking out all of the elements of $\boldsymbol{\mu}$ save the $i_1, i_2, \ldots, i_M$th elements and $\boldsymbol{\Sigma}_*$ to be the $M \times M$ submatrix of $\boldsymbol{\Sigma}$ obtained by striking out all of the rows and columns of $\boldsymbol{\Sigma}$ save the $i_1, i_2, \ldots, i_M$th rows and columns.

Consider the distribution of $\mathbf{x}_*$. Clearly, $\mathbf{x}_* = \mathbf{A}\mathbf{x}$, where $\mathbf{A}$ is the $M \times N$ matrix whose first, second, ..., Mth rows are, respectively, the $i_1, i_2, \ldots, i_M$th rows of $\mathbf{I}_N$. Thus, upon observing that $\mathbf{A}\boldsymbol{\mu} = \boldsymbol{\mu}_*$ and that $\mathbf{A}\boldsymbol{\Sigma}\mathbf{A}' = \boldsymbol{\Sigma}_*$, it follows from the result of the preceding subsection (the subsection pertaining to linear transformation of elliptically distributed random vectors) that $\mathbf{x}_*$ has an elliptical distribution with mean $\boldsymbol{\mu}_*$, variance-covariance matrix $\boldsymbol{\Sigma}_*$, and (if $\boldsymbol{\Sigma}_* \neq \mathbf{0}$) mgf generator $\phi(\cdot)$ (identical to the mgf generator of $\mathbf{x}$).

d. Maximum likelihood as applied to elliptical distributions (besides the MVN distribution)

Suppose that $\mathbf{y}$ is an $N \times 1$ observable random vector that follows a general linear model. Suppose further that the variance-covariance matrix $\mathbf{V}(\boldsymbol{\theta})$ of the vector $\mathbf{e}$ of residual effects is nonsingular and that

$$\mathbf{e} \sim [\boldsymbol{\Gamma}(\boldsymbol{\theta})]'\mathbf{u}, \tag{9.73}$$

where $\boldsymbol{\Gamma}(\boldsymbol{\theta})$ is an $N \times N$ (nonsingular) matrix (whose elements may be functionally dependent on $\boldsymbol{\theta}$) such that $\mathbf{V}(\boldsymbol{\theta}) = [\boldsymbol{\Gamma}(\boldsymbol{\theta})]'\boldsymbol{\Gamma}(\boldsymbol{\theta})$ and where $\mathbf{u}$ is an $N \times 1$ random vector having an absolutely continuous spherical distribution with variance-covariance matrix $\mathbf{I}$. The distribution of $\mathbf{u}$ has a pdf $h(\cdot)$, where (for every value of $\mathbf{u}$)

$$h(\mathbf{u}) = c^{-1} g(\mathbf{u}'\mathbf{u}).$$

Here, $g(\cdot)$ is a nonnegative function (of a single nonnegative variable) such that $\int_0^\infty s^{N-1} g(s^2)\, ds < \infty$, and $c = [2\pi^{N/2}/\Gamma(N/2)] \int_0^\infty s^{N-1} g(s^2)\, ds$. As a consequence of supposition (9.73), $\mathbf{y}$ has an elliptical distribution.

Let us consider the ML estimation of functions of the parameters of the general linear model (i.e., functions of the elements $\beta_1, \beta_2, \ldots, \beta_P$ of the vector $\boldsymbol{\beta}$ and the elements $\theta_1, \theta_2, \ldots, \theta_T$ of the vector $\boldsymbol{\theta}$). That topic was considered earlier (in Subsection a) in the special case where $\mathbf{e} \sim N[\mathbf{0}, \mathbf{V}(\boldsymbol{\theta})]$—when $g(s^2) = \exp(-s^2/2)$, $h(\mathbf{u}) = (2\pi)^{-N/2} \exp(-\frac{1}{2}\mathbf{u}'\mathbf{u})$, which is the pdf of the $N(\mathbf{0}, \mathbf{I}_N)$ distribution.

Let $f(\cdot\,; \boldsymbol{\beta}, \boldsymbol{\theta})$ represent the pdf of the distribution of $\mathbf{y}$, and denote by $\underline{\mathbf{y}}$ the observed value of $\mathbf{y}$. Then, the likelihood function is the function, say $L(\boldsymbol{\beta}, \boldsymbol{\theta}; \underline{\mathbf{y}})$ of $\boldsymbol{\beta}$ and $\boldsymbol{\theta}$ defined by $L(\boldsymbol{\beta}, \boldsymbol{\theta}; \underline{\mathbf{y}}) = f(\underline{\mathbf{y}}; \boldsymbol{\beta}, \boldsymbol{\theta})$. Accordingly, it follows from result (9.70) that

$$L(\boldsymbol{\beta}, \boldsymbol{\theta}; \underline{\mathbf{y}}) = c^{-1} |\mathbf{V}(\boldsymbol{\theta})|^{-1/2} g\{(\underline{\mathbf{y}} - \mathbf{X}\boldsymbol{\beta})'[\mathbf{V}(\boldsymbol{\theta})]^{-1}(\underline{\mathbf{y}} - \mathbf{X}\boldsymbol{\beta})\}. \tag{9.74}$$

Maximum likelihood estimates of functions of $\boldsymbol{\beta}$ and/or $\boldsymbol{\theta}$ are obtained from values, say $\tilde{\boldsymbol{\beta}}$ and $\tilde{\boldsymbol{\theta}}$, of $\boldsymbol{\beta}$ and $\boldsymbol{\theta}$ at which $L(\boldsymbol{\beta}, \boldsymbol{\theta}; \underline{\mathbf{y}})$ attains its maximum value: a maximum likelihood estimate of a function, say $r(\boldsymbol{\beta}, \boldsymbol{\theta})$, of $\boldsymbol{\beta}$ and/or $\boldsymbol{\theta}$ is provided by the quantity $r(\tilde{\boldsymbol{\beta}}, \tilde{\boldsymbol{\theta}})$ obtained by substituting $\tilde{\boldsymbol{\beta}}$ and $\tilde{\boldsymbol{\theta}}$ for $\boldsymbol{\beta}$ and $\boldsymbol{\theta}$.

Profile likelihood function. Now, suppose that the function $g(\cdot)$ is a strictly decreasing function (as in the special case where the distribution of $\mathbf{e}$ is MVN). Then, for any particular value of $\boldsymbol{\theta}$, the maximization of $L(\boldsymbol{\beta}, \boldsymbol{\theta}; \underline{\mathbf{y}})$ with respect to $\boldsymbol{\beta}$ is equivalent to the minimization of $(\underline{\mathbf{y}} - \mathbf{X}\boldsymbol{\beta})'[\mathbf{V}(\boldsymbol{\theta})]^{-1}(\underline{\mathbf{y}} - \mathbf{X}\boldsymbol{\beta})$ with respect to $\boldsymbol{\beta}$. Thus, upon regarding the value of $\boldsymbol{\theta}$ as "fixed," upon recalling (from Part 3 of Subsection a) that the linear system

$$\mathbf{X}'[\mathbf{V}(\boldsymbol{\theta})]^{-1}\mathbf{X}\mathbf{b} = \mathbf{X}'[\mathbf{V}(\boldsymbol{\theta})]^{-1}\underline{\mathbf{y}} \tag{9.75}$$

(in the $P \times 1$ vector $\mathbf{b}$) is consistent, and upon employing the same line of reasoning as in Part 3 of Subsection a, we find that $L(\boldsymbol{\beta}, \boldsymbol{\theta}; \underline{\mathbf{y}})$ attains its maximum value at a value $\tilde{\boldsymbol{\beta}}(\boldsymbol{\theta})$ of $\boldsymbol{\beta}$ if and only if $\tilde{\boldsymbol{\beta}}(\boldsymbol{\theta})$ is a solution to linear system (9.75) or, equivalently, if and only if

$$\mathbf{X}'[\mathbf{V}(\boldsymbol{\theta})]^{-1}\mathbf{X}\tilde{\boldsymbol{\beta}}(\boldsymbol{\theta}) = \mathbf{X}'[\mathbf{V}(\boldsymbol{\theta})]^{-1}\underline{\mathbf{y}},$$

in which case

$$L[\tilde{\boldsymbol{\beta}}(\boldsymbol{\theta}), \boldsymbol{\theta}; \underline{\mathbf{y}}] = c^{-1} |\mathbf{V}(\boldsymbol{\theta})|^{-1/2} g\{[\underline{\mathbf{y}} - \mathbf{X}\tilde{\boldsymbol{\beta}}(\boldsymbol{\theta})]'[\mathbf{V}(\boldsymbol{\theta})]^{-1}[\underline{\mathbf{y}} - \mathbf{X}\tilde{\boldsymbol{\beta}}(\boldsymbol{\theta})]\} \tag{9.76}$$

$$= c^{-1} |\mathbf{V}(\boldsymbol{\theta})|^{-1/2} g\{\underline{\mathbf{y}}'[\mathbf{V}(\boldsymbol{\theta})]^{-1}\underline{\mathbf{y}} - [\tilde{\boldsymbol{\beta}}(\boldsymbol{\theta})]'\mathbf{X}'[\mathbf{V}(\boldsymbol{\theta})]^{-1}\underline{\mathbf{y}}\} \tag{9.77}$$

$$= c^{-1} |\mathbf{V}(\boldsymbol{\theta})|^{-1/2} g\{\underline{\mathbf{y}}'([\mathbf{V}(\boldsymbol{\theta})]^{-1} - [\mathbf{V}(\boldsymbol{\theta})]^{-1}\mathbf{X}\{\mathbf{X}'[\mathbf{V}(\boldsymbol{\theta})]^{-1}\mathbf{X}\}^{-}\mathbf{X}'[\mathbf{V}(\boldsymbol{\theta})]^{-1})\underline{\mathbf{y}}\}. \tag{9.78}$$

Accordingly, the function $L_*(\boldsymbol{\theta}; \underline{\mathbf{y}})$ of $\boldsymbol{\theta}$ defined by $L_*(\boldsymbol{\theta}; \underline{\mathbf{y}}) = L[\tilde{\boldsymbol{\beta}}(\boldsymbol{\theta}), \boldsymbol{\theta}; \underline{\mathbf{y}}]$ is a profile likelihood function.

Values, say $\tilde{\boldsymbol{\beta}}$ and $\tilde{\boldsymbol{\theta}}$ (of $\boldsymbol{\beta}$ and $\boldsymbol{\theta}$, respectively), at which $L(\boldsymbol{\beta}, \boldsymbol{\theta}; \underline{\mathbf{y}})$ attains its maximum value can be obtained by taking $\tilde{\boldsymbol{\theta}}$ to be a value at which the profile likelihood function $L_*(\boldsymbol{\theta}; \underline{\mathbf{y}})$ attains its maximum value and by then taking $\tilde{\boldsymbol{\beta}}$ to be a solution to the linear system

$$\mathbf{X}'[\mathbf{V}(\tilde{\boldsymbol{\theta}})]^{-1}\mathbf{X}\mathbf{b} = \mathbf{X}'[\mathbf{V}(\tilde{\boldsymbol{\theta}})]^{-1}\underline{\mathbf{y}}.$$

Except for relatively simple special cases, the maximization of $L_*(\boldsymbol{\theta}; \underline{\mathbf{y}})$ must be accomplished numerically via an iterative procedure.

REML variant. REML is a variant of ML in which inferences about functions of $\boldsymbol{\theta}$ are based on the likelihood function associated with a vector of so-called error contrasts. REML was introduced and discussed in an earlier subsection (Subsection b) under the assumption that the distribution of $\mathbf{e}$ is MVN. Let us consider REML in the present, more general context (where the distribution of $\mathbf{e}$ is taken to be elliptical).

Let $\mathbf{R}$ represent an $N \times (N - \operatorname{rank} \mathbf{X})$ matrix (of constants) of full column rank $N - \operatorname{rank} \mathbf{X}$ such that $\mathbf{X}'\mathbf{R} = \mathbf{0}$, and take $\mathbf{z}$ to be the $(N - \operatorname{rank} \mathbf{X}) \times 1$ vector defined by $\mathbf{z} = \mathbf{R}'\mathbf{y}$. Note that $\mathbf{z} = \mathbf{R}'\mathbf{e}$ and hence that the distribution of $\mathbf{z}$ does not depend on $\boldsymbol{\beta}$. Further, let $k(\cdot\,; \boldsymbol{\theta})$ represent the pdf of the distribution of $\mathbf{z}$, and take $L(\boldsymbol{\theta};\, \mathbf{R}'\underline{\mathbf{y}})$ to be the function of $\boldsymbol{\theta}$ defined (for $\boldsymbol{\theta} \in \Theta$) by $L(\boldsymbol{\theta};\, \mathbf{R}'\underline{\mathbf{y}}) = k(\mathbf{R}'\underline{\mathbf{y}};\, \boldsymbol{\theta})$. The function $L(\boldsymbol{\theta};\, \mathbf{R}'\underline{\mathbf{y}})$ is a likelihood function; it is the likelihood function obtained by regarding the value of $\mathbf{z}$ as the data vector.

Now, suppose that the (N-dimensional spherical) distribution of the random vector $\mathbf{u}$ [in expression (9.73)] has a moment generating function, say $\psi(\cdot)$. Then, necessarily, there exists a (nonnegative) function $\phi(\cdot)$ (of a single nonnegative variable) such that (for every $N \times 1$ vector $\mathbf{t}$ in a neighborhood of $\mathbf{0}$) $\psi(\mathbf{t}) = \phi(\mathbf{t}'\mathbf{t})$. And in light of the results of Subsection c, it follows that $\mathbf{z}$ has an $[(N - \operatorname{rank} \mathbf{X})$-dimensional$]$ elliptical distribution with mean $\mathbf{0}$, variance-covariance matrix $\mathbf{R}'\mathbf{V}(\boldsymbol{\theta})\mathbf{R}$, and mgf generator $\phi(\cdot)$. Further,

$$\mathbf{z} \sim [\boldsymbol{\Gamma}_*(\boldsymbol{\theta})]'\mathbf{u}_*,$$

where $\boldsymbol{\Gamma}_*(\boldsymbol{\theta})$ is any $(N - \operatorname{rank} \mathbf{X}) \times (N - \operatorname{rank} \mathbf{X})$ matrix such that $\mathbf{R}'\mathbf{V}(\boldsymbol{\theta})\mathbf{R} = [\boldsymbol{\Gamma}_*(\boldsymbol{\theta})]'\boldsymbol{\Gamma}_*(\boldsymbol{\theta})$ and where $\mathbf{u}_*$ is an $(N - \operatorname{rank} \mathbf{X}) \times 1$ random vector whose distribution is spherical with variance-covariance matrix $\mathbf{I}$ and with moment generating function $\psi_*(\cdot)$ defined [for every $(N - \operatorname{rank} \mathbf{X}) \times 1$ vector $\mathbf{t}_*$ in a neighborhood of $\mathbf{0}$] by $\psi_*(\mathbf{t}_*) = \phi(\mathbf{t}_*'\mathbf{t}_*)$—the distribution of $\mathbf{u}_*$ is a marginal distribution of $\mathbf{u}$.

The distribution of $\mathbf{u}_*$ is absolutely continuous with a pdf $h_*(\cdot)$ that (at least in principle) is determinable from the pdf of the distribution of $\mathbf{u}$ and that is expressible in the form

$$h_*(\mathbf{u}_*) = c_*^{-1} g_*(\mathbf{u}_*'\mathbf{u}_*),$$

where $g_*(\cdot)$ is a nonnegative function (of a single nonnegative variable) such that $\int_0^\infty s^{N-\operatorname{rank}(\mathbf{X})-1} g_*(s^2)\, ds < \infty$ and where c_* is a strictly positive constant. Necessarily,

$$c_* = \frac{2\pi^{(N-\operatorname{rank}\mathbf{X})/2}}{\Gamma[(N - \operatorname{rank}\mathbf{X})/2]} \int_0^\infty s^{N-\operatorname{rank}(\mathbf{X})-1} g_*(s^2)\, ds.$$

Thus, in light of result (9.70), the pdf of the distribution of $\mathbf{z}$ is absolutely continuous with a pdf $k(\cdot\,; \boldsymbol{\theta})$ that is expressible as

$$k(\mathbf{z}; \boldsymbol{\theta}) = c_*^{-1} |\mathbf{R}'\mathbf{V}(\boldsymbol{\theta})\mathbf{R}|^{-1/2} g_*\{\mathbf{z}'[\mathbf{R}'\mathbf{V}(\boldsymbol{\theta})\mathbf{R}]^{-1}\mathbf{z}\}.$$

And it follows that the REML likelihood function is expressible as

$$L(\boldsymbol{\theta};\, \mathbf{R}'\underline{\mathbf{y}}) = c_*^{-1} |\mathbf{R}'\mathbf{V}(\boldsymbol{\theta})\mathbf{R}|^{-1/2} g_*\{\underline{\mathbf{y}}'\mathbf{R}[\mathbf{R}'\mathbf{V}(\boldsymbol{\theta})\mathbf{R}]^{-1}\mathbf{R}'\underline{\mathbf{y}}\}. \tag{9.79}$$

As in the special case where the distribution of $\mathbf{e}$ is MVN, an alternative expression for $L(\boldsymbol{\theta};\, \mathbf{R}'\underline{\mathbf{y}})$ can be obtained by taking advantage of identities (9.32) and (9.37). Taking $\mathbf{X}_*$ to be any $N \times (\operatorname{rank} \mathbf{X})$ matrix whose columns form a basis for $\mathcal{C}(\mathbf{X})$, we find that

$$\begin{aligned} L(\boldsymbol{\theta};\, \mathbf{R}'\underline{\mathbf{y}}) = {} & c_*^{-1} |\mathbf{R}'\mathbf{R}|^{-1/2} |\mathbf{X}_*'\mathbf{X}_*|^{1/2} |\mathbf{V}(\boldsymbol{\theta})|^{-1/2} |\mathbf{X}_*'[\mathbf{V}(\boldsymbol{\theta})]^{-1}\mathbf{X}_*|^{-1/2} \\ & \times g_*\{\underline{\mathbf{y}}'([\mathbf{V}(\boldsymbol{\theta})]^{-1} - [\mathbf{V}(\boldsymbol{\theta})]^{-1}\mathbf{X}\{\mathbf{X}'[\mathbf{V}(\boldsymbol{\theta})]^{-1}\mathbf{X}\}^{-}\mathbf{X}'[\mathbf{V}(\boldsymbol{\theta})]^{-1})\underline{\mathbf{y}}\}. \end{aligned} \tag{9.80}$$

Alternative versions of this expression can be obtained by replacing the argument of the function $g_*(\cdot)$ with expression (9.40) or expression (9.41).

As in the special case where the distribution of $\mathbf{e}$ is MVN, $L(\boldsymbol{\theta};\, \mathbf{R}'\underline{\mathbf{y}})$ depends on the choice of the matrix $\mathbf{R}$ only through the multiplicative constant $|\mathbf{R}'\mathbf{R}|^{-1/2}$. In some special cases including that where the distribution of $\mathbf{e}$ is MVN, the function $g_*(\cdot)$ differs from the function $g(\cdot)$ by no more than a multiplicative constant. However, in general, the relationship between $g_*(\cdot)$ and $g(\cdot)$ is more complex.

5.10 Prediction

a. Some general results

Let $\mathbf{y}$ represent an $N \times 1$ observable random vector. And consider the use of $\mathbf{y}$ in predicting an unobservable random variable or, more generally, an unobservable random vector, say an $M \times 1$ unobservable random vector $\mathbf{w} = (w_1, w_2, \ldots, w_M)'$. That is, consider the use of the observed value of $\mathbf{y}$ (the so-called data vector) in making inferences about an unobservable quantity that can be regarded as a realization (i.e., sample value) of $\mathbf{w}$. Here, an unobservable quantity is a quantity that is unobservable at the time the inferences are to be made; it may become observable at some future time (as suggested by the use of the word prediction). In the present section, the focus is on obtaining a point estimate of the unobservable quantity; that is, on what might be deemed a point prediction.

Suppose that the second-order moments of the joint distribution of $\mathbf{w}$ and $\mathbf{y}$ exist. And adopt the following notation: $\boldsymbol{\mu}_y = \mathrm{E}(\mathbf{y})$, $\boldsymbol{\mu}_w = \mathrm{E}(\mathbf{w})$, $\mathbf{V}_y = \mathrm{var}(\mathbf{y})$, $\mathbf{V}_{yw} = \mathrm{cov}(\mathbf{y}, \mathbf{w})$, and $\mathbf{V}_w = \mathrm{var}(\mathbf{w})$. Further, in considering the special case $M = 1$, let us write w, μ_w, $\mathbf{v}_{yw}$, and v_w for $\mathbf{w}$, $\boldsymbol{\mu}_w$, $\mathbf{V}_{yw}$, and $\mathbf{V}_w$, respectively.

It is informative to consider the prediction of $\mathbf{w}$ under each of the following states of knowledge: (1) the joint distribution of $\mathbf{y}$ and $\mathbf{w}$ is known; (2) only $\boldsymbol{\mu}_y$, $\boldsymbol{\mu}_w$, $\mathbf{V}_y$, $\mathbf{V}_{yw}$, and $\mathbf{V}_w$ are known; and (3) only $\mathbf{V}_y$, $\mathbf{V}_{yw}$, and $\mathbf{V}_w$ are known.

Let $\tilde{\mathbf{w}}(\mathbf{y})$ represent an $(M \times 1)$-dimensional vector-valued function of $\mathbf{y}$ that qualifies as a (point) predictor of $\mathbf{w}$—in considering the special case where $M = 1$, let us write $\tilde{w}(\mathbf{y})$ for $\tilde{\mathbf{w}}(\mathbf{y})$. That $\tilde{\mathbf{w}}(\mathbf{y})$ qualifies as a predictor implies that the vector-valued function $\tilde{\mathbf{w}}(\cdot)$ depends on the joint distribution of $\mathbf{y}$ and $\mathbf{w}$ (if at all) only through characteristics of the joint distribution that are known.

The difference $\tilde{\mathbf{w}}(\mathbf{y}) - \mathbf{w}$ is referred to as the *prediction error*. The predictor $\tilde{\mathbf{w}}(\mathbf{y})$ is said to be *unbiased* if $\mathrm{E}[\tilde{\mathbf{w}}(\mathbf{y}) - \mathbf{w}] = \mathbf{0}$, that is, if the expected value of the prediction error equals $\mathbf{0}$, or, equivalently, if $\mathrm{E}[\tilde{\mathbf{w}}(\mathbf{y})] = \boldsymbol{\mu}_w$, that is, if the expected value of the predictor is the same as that of the random vector $\mathbf{w}$ whose realization is being predicted.

Attention is sometimes restricted to linear predictors. An $(M \times 1)$-dimensional vector-valued function $\mathbf{t}(\mathbf{y})$ of $\mathbf{y}$ is said to be *linear* if it is expressible in the form

$$\mathbf{t}(\mathbf{y}) = \mathbf{c} + \mathbf{A}'\mathbf{y}, \tag{10.1}$$

where $\mathbf{c}$ is an $M \times 1$ vector of constants and $\mathbf{A}$ is an $M \times N$ matrix of constants. A vector-valued function $\mathbf{t}(\mathbf{y})$ that is expressible in the form (10.1) is regarded as linear even if the vector $\mathbf{c}$ and the matrix $\mathbf{A}$ depend on the joint distribution of $\mathbf{y}$ and $\mathbf{w}$—the linearity reflects the nature of the dependence on the value of $\mathbf{y}$, not the nature of any dependence on the joint distribution. And it qualifies as a predictor if any dependence on the joint distribution of $\mathbf{y}$ and $\mathbf{w}$ is confined to characteristics of the joint distribution that are known.

The $M \times M$ matrix $\mathrm{E}\{[\tilde{\mathbf{w}}(\mathbf{y}) - \mathbf{w}][\tilde{\mathbf{w}}(\mathbf{y}) - \mathbf{w}]'\}$ is referred to as the *mean-squared-error* (MSE) *matrix* of the predictor $\tilde{\mathbf{w}}(\mathbf{y})$. If $\tilde{\mathbf{w}}(\mathbf{y})$ is an unbiased predictor (of $\mathbf{w}$), then

$$\mathrm{E}\{[\tilde{\mathbf{w}}(\mathbf{y}) - \mathbf{w}][\tilde{\mathbf{w}}(\mathbf{y}) - \mathbf{w}]'\} = \mathrm{var}[\tilde{\mathbf{w}}(\mathbf{y}) - \mathbf{w}].$$

That is, the MSE matrix of an unbiased predictor equals the variance-covariance matrix of its prediction error (not the variance-covariance matrix of the predictor itself). Note that in the special case where $M = 1$, the MSE matrix has only one element, which is expressible as $\mathrm{E}\{[\tilde{w}(\mathbf{y}) - w]^2\}$ and which is referred to simply as the *mean squared error* (MSE) of the (scalar-valued) predictor $\tilde{w}(\mathbf{y})$.

State (1): joint distribution known. Suppose that the joint distribution of $\mathbf{y}$ and $\mathbf{w}$ is known or that, at the very least, enough is known about the joint distribution to determine the conditional expected value $\mathrm{E}(\mathbf{w} \mid \mathbf{y})$ of $\mathbf{w}$ given $\mathbf{y}$. And observe that

$$\mathrm{E}[\mathrm{E}(\mathbf{w} \mid \mathbf{y}) - \mathbf{w} \mid \mathbf{y}] = \mathbf{0} \quad \text{(with probability 1).} \tag{10.2}$$

Observe also that, for "any" column vector $\mathbf{h}(\mathbf{y})$ of functions of $\mathbf{y}$,

$$\mathrm{E}\{\mathbf{h}(\mathbf{y})[\mathrm{E}(\mathbf{w} \mid \mathbf{y}) - \mathbf{w}]' \mid \mathbf{y}\} = \mathbf{0} \quad \text{(with probability 1).} \tag{10.3}$$

Now, let $\mathbf{t}(\mathbf{y})$ represent "any" $(M \times 1)$-dimensional vector-valued function of $\mathbf{y}$—in the special case where $M = 1$, let us write $t(\mathbf{y})$ for $\mathbf{t}(\mathbf{y})$. Then, upon observing that

$$\begin{aligned}[\mathbf{t}(\mathbf{y}) - \mathbf{w}][\mathbf{t}(\mathbf{y}) - \mathbf{w}]' &= \{\mathbf{t}(\mathbf{y}) - \mathrm{E}(\mathbf{w} \mid \mathbf{y}) + [\mathrm{E}(\mathbf{w} \mid \mathbf{y}) - \mathbf{w}]\}\{\mathbf{t}(\mathbf{y}) - \mathrm{E}(\mathbf{w} \mid \mathbf{y}) + [\mathrm{E}(\mathbf{w} \mid \mathbf{y}) - \mathbf{w}]\}' \\ &= [\mathbf{t}(\mathbf{y}) - \mathrm{E}(\mathbf{w} \mid \mathbf{y})][\mathbf{t}(\mathbf{y}) - \mathrm{E}(\mathbf{w} \mid \mathbf{y})]' + [\mathrm{E}(\mathbf{w} \mid \mathbf{y}) - \mathbf{w}][\mathrm{E}(\mathbf{w} \mid \mathbf{y}) - \mathbf{w}]' \\ &\quad + [\mathbf{t}(\mathbf{y}) - \mathrm{E}(\mathbf{w} \mid \mathbf{y})][\mathrm{E}(\mathbf{w} \mid \mathbf{y}) - \mathbf{w}]' + \{[\mathbf{t}(\mathbf{y}) - \mathrm{E}(\mathbf{w} \mid \mathbf{y})][\mathrm{E}(\mathbf{w} \mid \mathbf{y}) - \mathbf{w}]'\}'\end{aligned}$$

and [in light of result (10.3)] that

$$\mathrm{E}\{[\mathbf{t}(\mathbf{y}) - \mathrm{E}(\mathbf{w} \mid \mathbf{y})][\mathrm{E}(\mathbf{w} \mid \mathbf{y}) - \mathbf{w}]' \mid \mathbf{y}\} = \mathbf{0} \quad \text{(with probability 1),} \tag{10.4}$$

we find that

$$\begin{aligned}&\mathrm{E}\{[\mathbf{t}(\mathbf{y}) - \mathbf{w}][\mathbf{t}(\mathbf{y}) - \mathbf{w}]' \mid \mathbf{y}\} \\ &\qquad = [\mathbf{t}(\mathbf{y}) - \mathrm{E}(\mathbf{w} \mid \mathbf{y})][\mathbf{t}(\mathbf{y}) - \mathrm{E}(\mathbf{w} \mid \mathbf{y})]' \\ &\qquad\quad + \mathrm{E}\{[\mathrm{E}(\mathbf{w} \mid \mathbf{y}) - \mathbf{w}][\mathrm{E}(\mathbf{w} \mid \mathbf{y}) - \mathbf{w}]' \mid \mathbf{y}\} \quad \text{(with probability 1).}\end{aligned} \tag{10.5}$$

Result (10.5) implies that $\mathrm{E}(\mathbf{w} \mid \mathbf{y})$ is an optimal predictor of $\mathbf{w}$. It is optimal in the sense that the difference $\mathrm{E}\{[\mathbf{t}(\mathbf{y})-\mathbf{w}][\mathbf{t}(\mathbf{y})-\mathbf{w}]' \mid \mathbf{y}\} - \mathrm{E}\{[\mathrm{E}(\mathbf{w} \mid \mathbf{y})-\mathbf{w}][\mathrm{E}(\mathbf{w} \mid \mathbf{y})-\mathbf{w}]' \mid \mathbf{y}\}$ between the conditional (given $\mathbf{y}$) MSE matrix of an arbitrary predictor $\mathbf{t}(\mathbf{y})$ and that of $\mathrm{E}(\mathbf{w} \mid \mathbf{y})$ equals (with probability 1) the matrix $[\mathbf{t}(\mathbf{y}) - \mathrm{E}(\mathbf{w} \mid \mathbf{y})][\mathbf{t}(\mathbf{y}) - \mathrm{E}(\mathbf{w} \mid \mathbf{y})]'$, which is nonnegative definite and which equals $\mathbf{0}$ if and only if $\mathbf{t}(\mathbf{y}) - \mathrm{E}(\mathbf{w} \mid \mathbf{y}) = \mathbf{0}$ or, equivalently, if and only if $\mathbf{t}(\mathbf{y}) = \mathrm{E}(\mathbf{w} \mid \mathbf{y})$. In the special case where $M = 1$, we have that [for an arbitrary predictor $t(\mathbf{y})$]

$$\mathrm{E}\{[t(\mathbf{y}) - w]^2 \mid \mathbf{y}\} \geq \mathrm{E}\{[\mathrm{E}(w \mid \mathbf{y}) - w]^2 \mid \mathbf{y}\} \quad \text{(with probability 1).}$$

It is worth noting that

$$\mathrm{E}\{[\mathbf{t}(\mathbf{y}) - \mathbf{w}][\mathbf{t}(\mathbf{y}) - \mathbf{w}]'\} = \mathrm{E}\big(\mathrm{E}\{[\mathbf{t}(\mathbf{y}) - \mathbf{w}][\mathbf{t}(\mathbf{y}) - \mathbf{w}]' \mid \mathbf{y}\}\big),$$

so that $\mathrm{E}(\mathbf{w} \mid \mathbf{y})$ is optimal when the various predictors are compared on the basis of their unconditional MSE matrices as well as when they are compared on the basis of their conditional MSE matrices. The conditional MSE matrix of the optimal predictor $\mathrm{E}(\mathbf{w} \mid \mathbf{y})$ is

$$\mathrm{E}\{[\mathrm{E}(\mathbf{w} \mid \mathbf{y}) - \mathbf{w}][\mathrm{E}(\mathbf{w} \mid \mathbf{y}) - \mathbf{w}]' \mid \mathbf{y}\} = \mathrm{var}(\mathbf{w} \mid \mathbf{y}),$$

and the (unconditional) MSE matrix of $\mathrm{E}(\mathbf{w} \mid \mathbf{y})$ or, equivalently, the (unconditional) variance-covariance matrix of $\mathrm{E}(\mathbf{w} \mid \mathbf{y}) - \mathbf{w}$ is

$$\mathrm{var}[\mathrm{E}(\mathbf{w} \mid \mathbf{y}) - \mathbf{w}] = \mathrm{E}\{[\mathrm{E}(\mathbf{w} \mid \mathbf{y}) - \mathbf{w}][\mathrm{E}(\mathbf{w} \mid \mathbf{y}) - \mathbf{w}]'\} = \mathrm{E}[\mathrm{var}(\mathbf{w} \mid \mathbf{y})].$$

Clearly, $\mathrm{E}(\mathbf{w} \mid \mathbf{y})$ is an unbiased predictor; in fact, the expected value of its prediction error equals $\mathbf{0}$ conditionally on $\mathbf{y}$ (albeit with probability 1) as well as unconditionally [as is evident from result (10.2)]. Whether or not $\mathrm{E}(\mathbf{w} \mid \mathbf{y})$ is a linear predictor (or, more generally, equal to a linear predictor with probability 1) depends on the form of the joint distribution of $\mathbf{y}$ and $\mathbf{w}$; a sufficient (but not a necessary) condition for $\mathrm{E}(\mathbf{w} \mid \mathbf{y})$ to be linear (or, at least, "linear with probability 1") is that the joint distribution of $\mathbf{y}$ and $\mathbf{w}$ be MVN.

State (2): only the means and the variances and covariances are known. Suppose that $\boldsymbol{\mu}_y$, $\boldsymbol{\mu}_w$, $\mathbf{V}_y$, $\mathbf{V}_{yw}$, and $\mathbf{V}_w$ are known, but that nothing else is known about the joint distribution of $\mathbf{y}$ and $\mathbf{w}$. Then, $\mathrm{E}(\mathbf{w} \mid \mathbf{y})$ is not determinable from what is known, forcing us to look elsewhere for a predictor of $\mathbf{w}$.

Assume (for the sake of simplicity) that $\mathbf{V}_y$ is nonsingular. And consider the predictor

$$\boldsymbol{\eta}(\mathbf{y}) = \boldsymbol{\mu}_w + \mathbf{V}_{yw}'\mathbf{V}_y^{-1}(\mathbf{y} - \boldsymbol{\mu}_y) = \boldsymbol{\tau} + \mathbf{V}_{yw}'\mathbf{V}_y^{-1}\mathbf{y},$$

where $\boldsymbol{\tau} = \boldsymbol{\mu}_w - \mathbf{V}_{yw}'\mathbf{V}_y^{-1}\boldsymbol{\mu}_y$—in the special case where $M = 1$, let us write $\eta(\mathbf{y})$ for $\boldsymbol{\eta}(\mathbf{y})$.

Clearly, $\boldsymbol{\eta}(\mathbf{y})$ is linear; it is also unbiased. Now, consider its MSE matrix $\mathrm{E}\{[\boldsymbol{\eta}(\mathbf{y})-\mathbf{w}][\boldsymbol{\eta}(\mathbf{y})-\mathbf{w}]'\}$ or, equivalently, the variance-covariance matrix $\mathrm{var}[\boldsymbol{\eta}(\mathbf{y}) - \mathbf{w}]$ of its prediction error. Let us compare the MSE matrix of $\boldsymbol{\eta}(\mathbf{y})$ with the MSE matrices of other linear predictors.

Let $\mathbf{t}(\mathbf{y})$ represent an $(M\times 1)$-dimensional vector-valued function of $\mathbf{y}$ of the form $\mathbf{t}(\mathbf{y}) = \mathbf{c}+\mathbf{A}'\mathbf{y}$, where $\mathbf{c}$ is an $M \times 1$ vector of constants and $\mathbf{A}$ an $N \times M$ matrix of constants—in the special case where $M = 1$, let us write $t(\mathbf{y})$ for $\mathbf{t}(\mathbf{y})$. Further, decompose the difference between $\mathbf{t}(\mathbf{y})$ and $\mathbf{w}$ into two components as follows:

$$\mathbf{t}(\mathbf{y}) - \mathbf{w} = [\mathbf{t}(\mathbf{y}) - \boldsymbol{\eta}(\mathbf{y})] + [\boldsymbol{\eta}(\mathbf{y}) - \mathbf{w}]. \tag{10.6}$$

And observe that

$$\mathrm{cov}[\mathbf{y},\, \boldsymbol{\eta}(\mathbf{y}) - \mathbf{w}] = \mathrm{cov}(\mathbf{y},\, \mathbf{V}_{yw}'\mathbf{V}_y^{-1}\mathbf{y} - \mathbf{w}) = \mathbf{V}_y[\mathbf{V}_{yw}'\mathbf{V}_y^{-1}]' - \mathbf{V}_{yw} = \mathbf{0}. \tag{10.7}$$

Then, because $\mathrm{E}[\boldsymbol{\eta}(\mathbf{y}) - \mathbf{w}] = \mathbf{0}$ and because $\mathbf{t}(\mathbf{y}) - \boldsymbol{\eta}(\mathbf{y}) = \mathbf{c} - \boldsymbol{\tau} + (\mathbf{A}' - \mathbf{V}_{yw}'\mathbf{V}_y^{-1})\mathbf{y}$, it follows that

$$\begin{aligned}\mathrm{E}\{[\mathbf{t}(\mathbf{y}) - \boldsymbol{\eta}(\mathbf{y})][\boldsymbol{\eta}(\mathbf{y}) - \mathbf{w}]'\} &= \mathrm{cov}[\mathbf{t}(\mathbf{y}) - \boldsymbol{\eta}(\mathbf{y}),\, \boldsymbol{\eta}(\mathbf{y}) - \mathbf{w}] \\ &= (\mathbf{A}' - \mathbf{V}_{yw}'\mathbf{V}_y^{-1})\,\mathrm{cov}[\mathbf{y},\, \boldsymbol{\eta}(\mathbf{y}) - \mathbf{w}] = \mathbf{0}. \end{aligned} \tag{10.8}$$

Thus,

$$\begin{aligned}&\mathrm{E}\{[\mathbf{t}(\mathbf{y}) - \mathbf{w}][\mathbf{t}(\mathbf{y}) - \mathbf{w}]'\} \\ &\quad = \mathrm{E}\big(\{[\mathbf{t}(\mathbf{y}) - \boldsymbol{\eta}(\mathbf{y})] + [\boldsymbol{\eta}(\mathbf{y}) - \mathbf{w}]\}\{[\mathbf{t}(\mathbf{y}) - \boldsymbol{\eta}(\mathbf{y})]' + [\boldsymbol{\eta}(\mathbf{y}) - \mathbf{w}]'\}\big) \\ &\quad = \mathrm{E}\{[\mathbf{t}(\mathbf{y}) - \boldsymbol{\eta}(\mathbf{y})][\mathbf{t}(\mathbf{y}) - \boldsymbol{\eta}(\mathbf{y})]'\} + \mathrm{var}[\boldsymbol{\eta}(\mathbf{y}) - \mathbf{w}]. \end{aligned} \tag{10.9}$$

Any linear predictor of $\mathbf{w}$ is expressible in the form $[\mathbf{t}(\mathbf{y}) = \mathbf{c} + \mathbf{A}'\mathbf{y}]$ of the vector-valued function $\mathbf{t}(\mathbf{y})$. Accordingly, result (10.9) implies that $\boldsymbol{\eta}(\mathbf{y})$ is the best linear predictor of $\mathbf{w}$. It is the best linear predictor in the sense that the difference between the MSE matrix $\mathrm{E}\{[\mathbf{t}(\mathbf{y})-\mathbf{w}][\mathbf{t}(\mathbf{y})-\mathbf{w}]'\}$ of an arbitrary linear predictor $\mathbf{t}(\mathbf{y})$ and the matrix $\mathrm{var}[\boldsymbol{\eta}(\mathbf{y}) - \mathbf{w}]$ [which is the MSE matrix of $\boldsymbol{\eta}(\mathbf{y})$] equals the matrix $\mathrm{E}\{[\mathbf{t}(\mathbf{y}) - \boldsymbol{\eta}(\mathbf{y})][\mathbf{t}(\mathbf{y}) - \boldsymbol{\eta}(\mathbf{y})]'\}$, which is nonnegative definite and which equals $\mathbf{0}$ if and only if $\mathbf{t}(\mathbf{y}) - \boldsymbol{\eta}(\mathbf{y}) = \mathbf{0}$ or, equivalently, if and only if $\mathbf{t}(\mathbf{y}) = \boldsymbol{\eta}(\mathbf{y})$. (To see that $\mathrm{E}\{[\mathbf{t}(\mathbf{y})-\boldsymbol{\eta}(\mathbf{y})][\mathbf{t}(\mathbf{y})-\boldsymbol{\eta}(\mathbf{y})]'\} = \mathbf{0}$ implies that $\mathbf{t}(\mathbf{y})-\boldsymbol{\eta}(\mathbf{y}) = \mathbf{0}$, observe that (for $j = 1, 2, \ldots, M$) the jth element of $\mathbf{t}(\mathbf{y}) - \boldsymbol{\eta}(\mathbf{y})$ equals $k_j + \boldsymbol{\ell}_j'\mathbf{y}$, where k_j is the jth element of $\mathbf{c} - \boldsymbol{\tau}$ and $\boldsymbol{\ell}_j$ the jth column of $\mathbf{A} - \mathbf{V}_y^{-1}\mathbf{V}_{yw}$, that the jth diagonal element of $\mathrm{E}\{[\mathbf{t}(\mathbf{y}) - \boldsymbol{\eta}(\mathbf{y})][\mathbf{t}(\mathbf{y}) - \boldsymbol{\eta}(\mathbf{y})]'\}$ equals $\mathrm{E}[(k_j + \boldsymbol{\ell}_j'\mathbf{y})^2]$, and that $\mathrm{E}[(k_j + \boldsymbol{\ell}_j'\mathbf{y})^2] = 0$ implies that $\mathrm{E}(k_j + \boldsymbol{\ell}_j'\mathbf{y}) = 0$ and $\mathrm{var}(k_j + \boldsymbol{\ell}_j'\mathbf{y}) = 0$ and hence that $\boldsymbol{\ell}_j = \mathbf{0}$ and $k_j = 0$.) In the special case where $M = 1$, we have [for an arbitrary linear predictor $t(\mathbf{y})$] that

$$\mathrm{E}\{[t(\mathbf{y}) - w]^2\} \geq \mathrm{var}[\eta(\mathbf{y}) - w] \quad (= \mathrm{E}\{[\eta(\mathbf{y}) - w]^2\}), \tag{10.10}$$

with equality holding in inequality (10.10) if and only if $t(\mathbf{y}) = \eta(\mathbf{y})$.

The prediction error of the best linear predictor $\boldsymbol{\eta}(\mathbf{y})$ can be decomposed into two components on the basis of the following identity:

$$\boldsymbol{\eta}(\mathbf{y}) - \mathbf{w} = [\boldsymbol{\eta}(\mathbf{y}) - \mathrm{E}(\mathbf{w} \mid \mathbf{y})] + [\mathrm{E}(\mathbf{w} \mid \mathbf{y}) - \mathbf{w}]. \tag{10.11}$$

The second component $\mathrm{E}(\mathbf{w} \mid \mathbf{y}) - \mathbf{w}$ of this decomposition has an expected value of $\mathbf{0}$ [conditionally on $\mathbf{y}$ (albeit with probability 1) as well as unconditionally], and because $\mathrm{E}[\boldsymbol{\eta}(\mathbf{y}) - \mathbf{w}] = \mathbf{0}$, the first component $\boldsymbol{\eta}(\mathbf{y}) - E(\mathbf{w} \mid \mathbf{y})$ also has an expected value of $\mathbf{0}$. Moreover, it follows from result (10.3) that

$$E\{[\boldsymbol{\eta}(\mathbf{y}) - E(\mathbf{w} \mid \mathbf{y})][E(\mathbf{w} \mid \mathbf{y}) - \mathbf{w}]' \mid \mathbf{y}\} = \mathbf{0} \quad \text{(with probability 1)}, \tag{10.12}$$

implying that

$$E\{[\boldsymbol{\eta}(\mathbf{y}) - E(\mathbf{w} \mid \mathbf{y})][E(\mathbf{w} \mid \mathbf{y}) - \mathbf{w}]'\} = \mathbf{0} \tag{10.13}$$

and hence that the two components $\boldsymbol{\eta}(\mathbf{y}) - E(\mathbf{w} \mid \mathbf{y})$ and $E(\mathbf{w} \mid \mathbf{y}) - \mathbf{w}$ of decomposition (10.11) are uncorrelated. And upon applying result (10.5) [with $\mathbf{t}(\mathbf{y}) = \boldsymbol{\eta}(\mathbf{y})$], we find that

$$\begin{aligned} E\{[\boldsymbol{\eta}(\mathbf{y}) - \mathbf{w}][\boldsymbol{\eta}(\mathbf{y}) - \mathbf{w}]' \mid \mathbf{y}\} = [\boldsymbol{\eta}(\mathbf{y}) &- E(\mathbf{w} \mid \mathbf{y})][\boldsymbol{\eta}(\mathbf{y}) - E(\mathbf{w} \mid \mathbf{y})]' \\ &+ \text{var}(\mathbf{w} \mid \mathbf{y}) \quad \text{(with probability 1)}. \end{aligned} \tag{10.14}$$

In the special case where $M = 1$, result (10.14) is reexpressible as

$$E\{[\eta(\mathbf{y}) - w]^2 \mid \mathbf{y}\} = [\eta(\mathbf{y}) - E(w \mid \mathbf{y})]^2 + \text{var}(w \mid \mathbf{y}) \quad \text{(with probability 1)}.$$

Equality (10.14) serves to decompose the conditional (on $\mathbf{y}$) MSE matrix of the best linear predictor $\boldsymbol{\eta}(\mathbf{y})$ into two components, corresponding to the two components of the decomposition (10.11) of the prediction error of $\boldsymbol{\eta}(\mathbf{y})$. The (unconditional) MSE matrix of $\boldsymbol{\eta}(\mathbf{y})$ or, equivalently, the (unconditional) variance-covariance matrix of $\boldsymbol{\eta}(\mathbf{y}) - \mathbf{w}$ lends itself to a similar decomposition. We find that

$$\begin{aligned} \text{var}[\boldsymbol{\eta}(\mathbf{y}) - \mathbf{w}] &= \text{var}[\boldsymbol{\eta}(\mathbf{y}) - E(\mathbf{w} \mid \mathbf{y})] + \text{var}[E(\mathbf{w} \mid \mathbf{y}) - \mathbf{w}] \\ &= \text{var}[\boldsymbol{\eta}(\mathbf{y}) - E(\mathbf{w} \mid \mathbf{y})] + E[\text{var}(\mathbf{w} \mid \mathbf{y})]. \end{aligned} \tag{10.15}$$

Of the two components of the prediction error of $\boldsymbol{\eta}(\mathbf{y})$, the second component $E(\mathbf{w} \mid \mathbf{y}) - \mathbf{w}$ can be regarded as an "inherent" component. It is inherent in the sense that it is an error that would be incurred even if enough were known about the joint distribution of $\mathbf{y}$ and $\mathbf{w}$ that $E(\mathbf{w} \mid \mathbf{y})$ were determinable and were employed as the predictor. The first component $\boldsymbol{\eta}(\mathbf{y}) - E(\mathbf{w} \mid \mathbf{y})$ of the prediction error can be regarded as a "nonlinearity" component; it equals $\mathbf{0}$ if and only if $E(\mathbf{w} \mid \mathbf{y}) = \mathbf{c} + \mathbf{A}'\mathbf{y}$ for some vector $\mathbf{c}$ of constants and some matrix $\mathbf{A}$ of constants.

The variance-covariance matrix of the prediction error of $\boldsymbol{\eta}(\mathbf{y})$ is expressible as

$$\begin{aligned} \text{var}[\boldsymbol{\eta}(\mathbf{y}) - \mathbf{w}] &= \text{var}(\mathbf{V}_{yw}'\mathbf{V}_y^{-1}\mathbf{y} - \mathbf{w}) \\ &= \mathbf{V}_w + \mathbf{V}_{yw}'\mathbf{V}_y^{-1}\mathbf{V}_y(\mathbf{V}_{yw}'\mathbf{V}_y^{-1})' - \mathbf{V}_{yw}'\mathbf{V}_y^{-1}\mathbf{V}_{yw} - [\mathbf{V}_{yw}'\mathbf{V}_y^{-1}\mathbf{V}_{yw}]' \\ &= \mathbf{V}_w - \mathbf{V}_{yw}'\mathbf{V}_y^{-1}\mathbf{V}_{yw}. \end{aligned} \tag{10.16}$$

It differs from the variance-covariance matrix of $\boldsymbol{\eta}(\mathbf{y})$; the latter variance-covariance matrix is expressible as

$$\text{var}[\boldsymbol{\eta}(\mathbf{y})] = \text{var}(\mathbf{V}_{yw}'\mathbf{V}_y^{-1}\mathbf{y}) = \mathbf{V}_{yw}'\mathbf{V}_y^{-1}\mathbf{V}_y[\mathbf{V}_{yw}'\mathbf{V}_y^{-1}]' = \mathbf{V}_{yw}'\mathbf{V}_y^{-1}\mathbf{V}_{yw}.$$

In fact, $\text{var}[\boldsymbol{\eta}(\mathbf{y}) - \mathbf{w}]$ and $\text{var}[\boldsymbol{\eta}(\mathbf{y})]$ are the first and second components in the following decomposition of $\text{var}(\mathbf{w})$:

$$\text{var}(\mathbf{w}) = \text{var}[\boldsymbol{\eta}(\mathbf{y}) - \mathbf{w}] + \text{var}[\boldsymbol{\eta}(\mathbf{y})].$$

The best linear predictor $\boldsymbol{\eta}(\mathbf{y})$ can be regarded as an approximation to $E(\mathbf{w} \mid \mathbf{y})$. The expected value $E[\boldsymbol{\eta}(\mathbf{y}) - E(\mathbf{w} \mid \mathbf{y})]$ of the error of this approximation equals $\mathbf{0}$. Note that $\text{var}[\boldsymbol{\eta}(\mathbf{y}) - E(\mathbf{w} \mid \mathbf{y})] = E\{[\boldsymbol{\eta}(\mathbf{y}) - E(\mathbf{w} \mid \mathbf{y})][\boldsymbol{\eta}(\mathbf{y}) - E(\mathbf{w} \mid \mathbf{y})]'\}$. Further, $\boldsymbol{\eta}(\mathbf{y})$ is the best linear approximation to $E(\mathbf{w} \mid \mathbf{y})$ in the sense that, for any $(M \times 1)$-dimensional vector-valued function $\mathbf{t}(\mathbf{y})$ of the form $\mathbf{t}(\mathbf{y}) = \mathbf{c} + \mathbf{A}'\mathbf{y}$, the difference between the matrix $E\{[\mathbf{t}(\mathbf{y}) - E(\mathbf{w} \mid \mathbf{y})][\mathbf{t}(\mathbf{y}) - E(\mathbf{w} \mid \mathbf{y})]'\}$ and the matrix $\text{var}[\boldsymbol{\eta}(\mathbf{y}) - E(\mathbf{w} \mid \mathbf{y})]$ equals the matrix $E\{[\mathbf{t}(\mathbf{y}) - \boldsymbol{\eta}(\mathbf{y})][\mathbf{t}(\mathbf{y}) - \boldsymbol{\eta}(\mathbf{y})]'\}$, which is nonnegative definite and which equals $\mathbf{0}$ if and only if $\mathbf{t}(\mathbf{y}) = \boldsymbol{\eta}(\mathbf{y})$. This result follows from what has already been established (in regard to the best linear prediction of $\mathbf{w}$) upon observing [in light of result (10.5)] that

$$E\{[\mathbf{t}(\mathbf{y}) - \mathbf{w}][\mathbf{t}(\mathbf{y}) - \mathbf{w}]'\} = E\{[\mathbf{t}(\mathbf{y}) - E(\mathbf{w} \mid \mathbf{y})][\mathbf{t}(\mathbf{y}) - E(\mathbf{w} \mid \mathbf{y})]'\} + E[\text{var}(\mathbf{w} \mid \mathbf{y})],$$

which in combination with result (10.15) implies that the difference between the two matrices $E\{[\mathbf{t}(\mathbf{y}) - E(\mathbf{w} \mid \mathbf{y})][\mathbf{t}(\mathbf{y}) - E(\mathbf{w} \mid \mathbf{y})]'\}$ and $\text{var}[\boldsymbol{\eta}(\mathbf{y}) - E(\mathbf{w} \mid \mathbf{y})]$ is the same as that between the two matrices $E\{[\mathbf{t}(\mathbf{y}) - \mathbf{w}][\mathbf{t}(\mathbf{y}) - \mathbf{w}]'\}$ and $\text{var}[\boldsymbol{\eta}(\mathbf{y}) - \mathbf{w}]$. In the special case where $M = 1$, we have [for any function $t(\mathbf{y})$ of $\mathbf{y}$ of the form $t(\mathbf{y}) = c + \mathbf{a}'\mathbf{y}$ (where c is a constant and $\mathbf{a}$ an $N \times 1$ vector of constants)] that

$$E\{[t(\mathbf{y}) - E(w \mid \mathbf{y})]^2\} \geq \text{var}[\eta(\mathbf{y}) - E(w \mid \mathbf{y})] \quad \left(= E\{[\eta(\mathbf{y}) - E(w \mid \mathbf{y})]^2\}\right), \tag{10.17}$$

with equality holding in inequality (10.17) if and only if $t(\mathbf{y}) = \eta(\mathbf{y})$.

Hartigan (1969) refers to $\boldsymbol{\eta}(\mathbf{y})$ as the *linear expectation of* $\mathbf{w}$ *given* $\mathbf{y}$. And in the special case where $M = 1$, he refers to $\text{var}[\mathbf{w} - \boldsymbol{\eta}(\mathbf{y})]$ $(= \mathbf{V}_w - \mathbf{V}_{yw}'\mathbf{V}_y^{-1}\mathbf{V}_{yw})$ as the *linear variance of* $\mathbf{w}$ *given* $\mathbf{y}$—in the general case (where M can exceed 1), this quantity could be referred to as the *linear variance-covariance matrix of* $\mathbf{w}$ *given* $\mathbf{y}$. It is only in special cases, such as that where the joint distribution of $\mathbf{y}$ and $\mathbf{w}$ is MVN, that the linear expectation and linear variance-covariance matrix of $\mathbf{w}$ given $\mathbf{y}$ coincide with the conditional expectation $E(\mathbf{w} \mid \mathbf{y})$ and conditional variance-covariance matrix $\text{var}(\mathbf{w} \mid \mathbf{y})$ of $\mathbf{w}$ given $\mathbf{y}$.

Note that for the vector-valued function $\boldsymbol{\eta}(\cdot)$ to be determinable from what is known about the joint distribution of $\mathbf{y}$ and $\mathbf{w}$, the supposition that $\boldsymbol{\mu}_y$, $\boldsymbol{\mu}_w$, $\mathbf{V}_y$, $\mathbf{V}_{yw}$, and $\mathbf{V}_w$ are known is stronger than necessary. It suffices to know the vector $\boldsymbol{\tau} = \boldsymbol{\mu}_w - \mathbf{V}_{yw}'\mathbf{V}_y^{-1}\boldsymbol{\mu}_y$ and the matrix $\mathbf{V}_{yw}'\mathbf{V}_y^{-1}$.

State (3): only the variances and covariances are known. Suppose that $\mathbf{V}_y$, $\mathbf{V}_{yw}$, and $\mathbf{V}_w$ are known (and that $\mathbf{V}_y$ is nonsingular), but that nothing else is known about the joint distribution of $\mathbf{y}$ and $\mathbf{w}$. Then, $\boldsymbol{\eta}(\cdot)$ is not determinable from what is known, and consequently $\boldsymbol{\eta}(\mathbf{y})$ does not qualify as a predictor. Thus, we are forced to look elsewhere for a predictor of $\mathbf{w}$.

Corresponding to any estimator $\tilde{\boldsymbol{\tau}}(\mathbf{y})$ of the vector $\boldsymbol{\tau}$ $(= \boldsymbol{\mu}_w - \mathbf{V}_{yw}'\mathbf{V}_y^{-1}\boldsymbol{\mu}_y)$ is the predictor $\tilde{\boldsymbol{\eta}}(\mathbf{y})$ of $\mathbf{w}$ obtained from $\boldsymbol{\eta}(\mathbf{y})$ by substituting $\tilde{\boldsymbol{\tau}}(\mathbf{y})$ for $\boldsymbol{\tau}$. That is, corresponding to $\tilde{\boldsymbol{\tau}}(\mathbf{y})$ is the predictor $\tilde{\boldsymbol{\eta}}(\mathbf{y})$ defined as follows:

$$\tilde{\boldsymbol{\eta}}(\mathbf{y}) = \tilde{\boldsymbol{\tau}}(\mathbf{y}) + \mathbf{V}_{yw}'\mathbf{V}_y^{-1}\mathbf{y}. \tag{10.18}$$

Equality (10.18) serves to establish a one-to-one correspondence between estimators of $\boldsymbol{\tau}$ and predictors of $\mathbf{w}$—corresponding to any predictor $\tilde{\boldsymbol{\eta}}(\mathbf{y})$ of $\mathbf{w}$ is a unique estimator $\tilde{\boldsymbol{\tau}}(\mathbf{y})$ of $\boldsymbol{\tau}$ that satisfies equality (10.18), namely, the estimator $\tilde{\boldsymbol{\tau}}(\mathbf{y})$ defined by $\tilde{\boldsymbol{\tau}}(\mathbf{y}) = \tilde{\boldsymbol{\eta}}(\mathbf{y}) - \mathbf{V}_{yw}'\mathbf{V}_y^{-1}\mathbf{y}$.

Clearly, the predictor $\tilde{\boldsymbol{\eta}}(\mathbf{y})$ is linear if and only if the corresponding estimator $\tilde{\boldsymbol{\tau}}(\mathbf{y})$ is linear. Moreover,

$$E[\tilde{\boldsymbol{\eta}}(\mathbf{y})] = E[\tilde{\boldsymbol{\tau}}(\mathbf{y})] + \mathbf{V}_{yw}'\mathbf{V}_y^{-1}\boldsymbol{\mu}_y \tag{10.19}$$

and, consequently,

$$E[\tilde{\boldsymbol{\eta}}(\mathbf{y})] = \boldsymbol{\mu}_w \quad \Leftrightarrow \quad E[\tilde{\boldsymbol{\tau}}(\mathbf{y})] = \boldsymbol{\tau}. \tag{10.20}$$

Thus, $\tilde{\boldsymbol{\eta}}(\mathbf{y})$ is an unbiased predictor of $\mathbf{w}$ if and only if the corresponding estimator $\tilde{\boldsymbol{\tau}}(\mathbf{y})$ is an unbiased estimator of $\boldsymbol{\tau}$.

The following identity serves to decompose the prediction error of the predictor $\tilde{\boldsymbol{\eta}}(\mathbf{y})$ into two components:

$$\tilde{\boldsymbol{\eta}}(\mathbf{y}) - \mathbf{w} = [\tilde{\boldsymbol{\eta}}(\mathbf{y}) - \boldsymbol{\eta}(\mathbf{y})] + [\boldsymbol{\eta}(\mathbf{y}) - \mathbf{w}]. \tag{10.21}$$

Clearly,

$$\tilde{\boldsymbol{\eta}}(\mathbf{y}) - \boldsymbol{\eta}(\mathbf{y}) = \tilde{\boldsymbol{\tau}}(\mathbf{y}) - \boldsymbol{\tau}. \tag{10.22}$$

Thus, decomposition (10.21) can be reexpressed as follows:

$$\tilde{\boldsymbol{\eta}}(\mathbf{y}) - \mathbf{w} = [\tilde{\boldsymbol{\tau}}(\mathbf{y}) - \boldsymbol{\tau}] + [\boldsymbol{\eta}(\mathbf{y}) - \mathbf{w}]. \tag{10.23}$$

Let us now specialize to linear predictors. Let us write $\tilde{\eta}_L(\mathbf{y})$ for a linear predictor of $\mathbf{w}$ and $\tilde{\boldsymbol{\tau}}_L(\mathbf{y})$ for the corresponding estimator of $\boldsymbol{\tau}$ [which, like $\tilde{\eta}_L(\mathbf{y})$, is linear]. Then, in light of results (10.22) and (10.8),

$$E\{[\tilde{\boldsymbol{\tau}}_L(\mathbf{y}) - \boldsymbol{\tau}][\boldsymbol{\eta}(\mathbf{y}) - \mathbf{w}]'\} = E\{[\tilde{\boldsymbol{\eta}}_L(\mathbf{y}) - \boldsymbol{\eta}(\mathbf{y})][\boldsymbol{\eta}(\mathbf{y}) - \mathbf{w}]'\} = \mathbf{0}. \tag{10.24}$$

And making use of results (10.23) and (10.16), it follows that

$$\begin{aligned} E\{[\tilde{\boldsymbol{\eta}}_L(\mathbf{y}) - \mathbf{w}][\tilde{\boldsymbol{\eta}}_L(\mathbf{y}) - \mathbf{w}]'\} \\ &= E\big(\{[\tilde{\boldsymbol{\tau}}_L(\mathbf{y}) - \boldsymbol{\tau}] + [\boldsymbol{\eta}(\mathbf{y}) - \mathbf{w}]\}\{[\tilde{\boldsymbol{\tau}}_L(\mathbf{y}) - \boldsymbol{\tau}]' + [\boldsymbol{\eta}(\mathbf{y}) - \mathbf{w}]'\}\big) \\ &= E\{[\tilde{\boldsymbol{\tau}}_L(\mathbf{y}) - \boldsymbol{\tau}][\tilde{\boldsymbol{\tau}}_L(\mathbf{y}) - \boldsymbol{\tau}]'\} + \text{var}[\boldsymbol{\eta}(\mathbf{y}) - \mathbf{w}] \\ &= E\{[\tilde{\boldsymbol{\tau}}_L(\mathbf{y}) - \boldsymbol{\tau}][\tilde{\boldsymbol{\tau}}_L(\mathbf{y}) - \boldsymbol{\tau}]'\} + \mathbf{V}_w - \mathbf{V}_{yw}'\mathbf{V}_y^{-1}\mathbf{V}_{yw}. \end{aligned} \tag{10.25}$$

Let $\mathcal{L}_p$ represent a collection of linear predictors of $\mathbf{w}$. And let $\mathcal{L}_e$ represent the collection of (linear) estimators of $\boldsymbol{\tau}$ that correspond to the predictors in $\mathcal{L}_p$. Then, for a predictor, say $\hat{\boldsymbol{\eta}}_L(\mathbf{y})$, in the collection $\mathcal{L}_p$ to be best in the sense that, for every predictor $\tilde{\boldsymbol{\eta}}_L(\mathbf{y})$ in $\mathcal{L}_p$, the matrix $E\{[\tilde{\boldsymbol{\eta}}_L(\mathbf{y}) - \mathbf{w}][\tilde{\boldsymbol{\eta}}_L(\mathbf{y}) - \mathbf{w}]'\} - E\{[\hat{\boldsymbol{\eta}}_L(\mathbf{y}) - \mathbf{w}][\hat{\boldsymbol{\eta}}_L(\mathbf{y}) - \mathbf{w}]'\}$ is nonnegative definite, it is necessary and sufficient that

$$\hat{\boldsymbol{\eta}}_L(\mathbf{y}) = \hat{\boldsymbol{\tau}}_L(\mathbf{y}) + \mathbf{V}_{yw}'\mathbf{V}_y^{-1}\mathbf{y}$$

for some estimator $\hat{\boldsymbol{\tau}}_L(\mathbf{y})$ in $\mathcal{L}_e$ that is best in the sense that, for every estimator $\tilde{\boldsymbol{\tau}}_L(\mathbf{y})$ in $\mathcal{L}_e$, the matrix $E\{[\tilde{\boldsymbol{\tau}}_L(\mathbf{y}) - \boldsymbol{\tau}][\tilde{\boldsymbol{\tau}}_L(\mathbf{y}) - \boldsymbol{\tau}]'\} - E\{[\hat{\boldsymbol{\tau}}_L(\mathbf{y}) - \boldsymbol{\tau}][\hat{\boldsymbol{\tau}}_L(\mathbf{y}) - \boldsymbol{\tau}]'\}$ is nonnegative definite. In general, there may or may not be an estimator that is best in such a sense; the existence of such an estimator depends on the nature of the collection $\mathcal{L}_e$ and on any assumptions that may be made about $\boldsymbol{\mu}_y$ and $\boldsymbol{\mu}_w$.

If $\mathcal{L}_p$ is the collection of all linear unbiased predictors of $\mathbf{w}$, then $\mathcal{L}_e$ is the collection of all linear unbiased estimators of $\boldsymbol{\tau}$. As previously indicated (in Section 5.5a), it is customary to refer to an estimator that is best among linear unbiased estimators as a BLUE (an acronym for best linear unbiased estimator or estimation). Similarly, a predictor that is best among linear unbiased predictors is customarily referred to as a BLUP (an acronym for best linear unbiased predictor or prediction).

The prediction error of the predictor $\tilde{\boldsymbol{\eta}}(\mathbf{y})$ can be decomposed into three components by starting with decomposition (10.23) and by expanding the component $\boldsymbol{\eta}(\mathbf{y}) - \mathbf{w}$ into two components on the basis of decomposition (10.11). As specialized to the linear predictor $\tilde{\boldsymbol{\eta}}_L(\mathbf{y})$, the resultant decomposition is

$$\tilde{\boldsymbol{\eta}}_L(\mathbf{y}) - \mathbf{w} = [\tilde{\boldsymbol{\tau}}_L(\mathbf{y}) - \boldsymbol{\tau}] + [\boldsymbol{\eta}(\mathbf{y}) - E(\mathbf{w} \mid \mathbf{y})] + [E(\mathbf{w} \mid \mathbf{y}) - \mathbf{w}]. \tag{10.26}$$

Recall (from the preceding part of the present subsection) that $\boldsymbol{\eta}(\mathbf{y}) - E(\mathbf{w} \mid \mathbf{y})$ and $E(\mathbf{w} \mid \mathbf{y}) - \mathbf{w}$ [which are the 2nd and 3rd components of decomposition (10.26)] are uncorrelated and that each has an expected value of $\mathbf{0}$. Moreover, it follows from result (10.3) that $\tilde{\boldsymbol{\tau}}_L(\mathbf{y}) - \boldsymbol{\tau}$ is uncorrelated with $E(\mathbf{w} \mid \mathbf{y}) - \mathbf{w}$ and from result (10.7) that it is uncorrelated with $\boldsymbol{\eta}(\mathbf{y}) - \mathbf{w}$ and hence uncorrelated with $\boldsymbol{\eta}(\mathbf{y}) - E(\mathbf{w} \mid \mathbf{y})$ [which is expressible as the difference between $\boldsymbol{\eta}(\mathbf{y}) - \mathbf{w}$ and $E(\mathbf{w} \mid \mathbf{y}) - \mathbf{w}$]. Thus, all three components of decomposition (10.26) are uncorrelated. Expanding on the terminology introduced in the preceding part of the present subsection, the first, second, and third components of decomposition (10.26) can be regarded, respectively, as an "unknown-means" component, a "nonlinearity" component, and an "inherent" component.

Corresponding to decomposition (10.26) of the prediction error of $\tilde{\boldsymbol{\eta}}_L(\mathbf{y})$ is the following decomposition of the MSE matrix of $\tilde{\boldsymbol{\eta}}_L(\mathbf{y})$:

$$\begin{aligned} E\{[\tilde{\boldsymbol{\eta}}_L(\mathbf{y}) - \mathbf{w}][\tilde{\boldsymbol{\eta}}_L(\mathbf{y}) - \mathbf{w}]'\} \\ &= E\{[\tilde{\boldsymbol{\tau}}_L(\mathbf{y}) - \boldsymbol{\tau}][\tilde{\boldsymbol{\tau}}_L(\mathbf{y}) - \boldsymbol{\tau}]'\} + \text{var}[\boldsymbol{\eta}(\mathbf{y}) - E(\mathbf{w} \mid \mathbf{y})] + \text{var}[E(\mathbf{w} \mid \mathbf{y}) - \mathbf{w}]. \end{aligned} \tag{10.27}$$

In the special case where $M = 1$, this decomposition can [upon writing $\tilde{\eta}_L(\mathbf{y})$ for $\tilde{\boldsymbol{\eta}}_L(\mathbf{y})$, $\tilde{\tau}_L(\mathbf{y})$ for $\tilde{\boldsymbol{\tau}}_L(\mathbf{y})$, and τ for $\boldsymbol{\tau}$, as well as $\eta(\mathbf{y})$ for $\boldsymbol{\eta}(\mathbf{y})$ and w for $\mathbf{w}$] be reexpressed as follows:

$$\mathrm{E}\{[\tilde{\eta}_L(\mathbf{y}) - w]^2\} = \mathrm{E}\{[\tilde{\tau}_L(\mathbf{y}) - \tau]^2\} + \mathrm{var}[\eta(\mathbf{y}) - \mathrm{E}(w \mid \mathbf{y})] + \mathrm{var}[\mathrm{E}(w \mid \mathbf{y}) - w].$$

In taking $\tilde{\boldsymbol{\tau}}(\mathbf{y})$ to be an estimator of $\boldsymbol{\tau}$ and regarding $\tilde{\boldsymbol{\eta}}(\mathbf{y})$ as a predictor of $\mathbf{w}$, it is implicitly assumed that the functions $\tilde{\boldsymbol{\tau}}(\cdot)$ and $\tilde{\boldsymbol{\eta}}(\cdot)$ depend on the joint distribution of $\mathbf{y}$ and $\mathbf{w}$ only through $\mathbf{V}_y$, $\mathbf{V}_{yw}$, and $\mathbf{V}_w$. In practice, the dependence may only be through the elements of the matrix $\mathbf{V}_{yw}'\mathbf{V}_y^{-1}$ and through various functions of the elements of $\mathbf{V}_y^{-1}$, in which case $\tilde{\boldsymbol{\tau}}(\mathbf{y})$ may qualify as an estimator and $\tilde{\boldsymbol{\eta}}(\mathbf{y})$ as a predictor even in the absence of complete knowledge of $\mathbf{V}_y$, $\mathbf{V}_{yw}$, and $\mathbf{V}_w$.

b. Prediction on the basis of a G–M, Aitken, or general linear model

Suppose that the value of an $N \times 1$ observable random vector $\mathbf{y}$ is to be used to predict the realization of an $M \times 1$ unobservable random vector $\mathbf{w} = (w_1, w_2, \ldots, w_M)'$. How might we proceed? As is evident from the results of Subsection a, the answer depends on what is "known" about the joint distribution of $\mathbf{y}$ and $\mathbf{w}$.

We could refer to whatever assumptions are made about the joint distribution of $\mathbf{y}$ and $\mathbf{w}$ as a (statistical) model. However, while doing so might be logical, it would be unconventional and hence potentially confusing. It is customary to restrict the use of the word model to the assumptions made about the distribution of the observable random vector $\mathbf{y}$.

Irrespective of the terminology, the assumptions made about the distribution of $\mathbf{y}$ do not in and of themselves provide an adequate basis for prediction. The prediction of $\mathbf{w}$ requires the larger set of assumptions that apply to the joint distribution of $\mathbf{y}$ and $\mathbf{w}$. It is this larger set of assumptions that establishes a statistical relationship between $\mathbf{y}$ and $\mathbf{w}$.

Now, assume that $\mathbf{y}$ follows a general linear model. And for purposes of predicting the realization of $\mathbf{w}$ from the value of $\mathbf{y}$, let us augment that assumption with an assumption that

$$\mathrm{E}(\mathbf{w}) = \boldsymbol{\Lambda}'\boldsymbol{\beta} \tag{10.28}$$

for some $(P \times M)$ matrix $\boldsymbol{\Lambda}$ of (known) constants and an assumption that

$$\mathrm{cov}(\mathbf{y}, \mathbf{w}) = \mathbf{V}_{yw}(\boldsymbol{\theta}) \qquad \text{and} \qquad \mathrm{var}(\mathbf{w}) = \mathbf{V}_w(\boldsymbol{\theta}) \tag{10.29}$$

for some matrices $\mathbf{V}_{yw}(\boldsymbol{\theta})$ and $\mathbf{V}_w(\boldsymbol{\theta})$ whose elements are known functions of the parametric vector $\boldsymbol{\theta}$—it is assumed that cov($\mathbf{y}$, $\mathbf{w}$) and var($\mathbf{w}$), like var($\mathbf{y}$), do not depend on $\boldsymbol{\beta}$. Further, let us (in the present context) write $\mathbf{V}_y(\boldsymbol{\theta})$ for $\mathbf{V}(\boldsymbol{\theta})$. Note that the $(N+M) \times (N+M)$ matrix $\begin{bmatrix} \mathbf{V}_y(\boldsymbol{\theta}) & \mathbf{V}_{yw}(\boldsymbol{\theta}) \\ [\mathbf{V}_{yw}(\boldsymbol{\theta})]' & \mathbf{V}_w(\boldsymbol{\theta}) \end{bmatrix}$—which is the variance-covariance matrix of the $(N+M)$-dimensional vector $\begin{pmatrix} \mathbf{y} \\ \mathbf{w} \end{pmatrix}$—is inherently nonnegative definite.

Note that the assumption that $\mathbf{w}$ satisfies condition (10.28) is consistent with taking $\mathbf{w}$ to be of the form

$$\mathbf{w} = \boldsymbol{\Lambda}'\boldsymbol{\beta} + \mathbf{d}, \tag{10.30}$$

where $\mathbf{d}$ is an M-dimensional random column vector with $\mathrm{E}(\mathbf{d}) = \mathbf{0}$. The vector $\mathbf{d}$ can be regarded as the counterpart of the vector $\mathbf{e}$ of residual effects in the model equation (1.14). Upon taking $\mathbf{w}$ to be of the form (10.30), assumption (10.29) is reexpressible as

$$\mathrm{cov}(\mathbf{e}, \mathbf{d}) = \mathbf{V}_{yw}(\boldsymbol{\theta}) \qquad \text{and} \qquad \mathrm{var}(\mathbf{d}) = \mathbf{V}_w(\boldsymbol{\theta}). \tag{10.31}$$

In the present context, a predictor, say $\tilde{\mathbf{w}}(\mathbf{y})$, of $\mathbf{w}$ is unbiased if and only if $\mathrm{E}[\tilde{\mathbf{w}}(\mathbf{y})] = \boldsymbol{\Lambda}'\boldsymbol{\beta}$. If there exists a linear unbiased predictor of $\mathbf{w}$, let us refer to $\mathbf{w}$ as *predictable*; otherwise, let us refer to $\mathbf{w}$ as *unpredictable*. Clearly, $\tilde{\mathbf{w}}(\mathbf{y})$ is an unbiased predictor of $\mathbf{w}$ if and only if it is an unbiased estimator of $\boldsymbol{\Lambda}'\boldsymbol{\beta}$. And $\mathbf{w}$ is predictable if and only if $\boldsymbol{\Lambda}'\boldsymbol{\beta}$ is estimable, that is, if and only if all M of the elements of the vector $\boldsymbol{\Lambda}'\boldsymbol{\beta}$ are estimable linear combinations of the P elements of the parametric vector $\boldsymbol{\beta}$.

As defined and discussed in Sections 5.2 and 5.6b, translation equivariance is a criterion that is applicable to estimators of a linear combination of the elements of $\boldsymbol{\beta}$ or, more generally, to estimators of a vector of such linear combinations. This criterion can also be applied to predictors. A predictor $\tilde{\mathbf{w}}(\mathbf{y})$ of the random vector $\mathbf{w}$ (the expected value of which is $\boldsymbol{\Lambda}'\boldsymbol{\beta}$) is said to be *translation equivariant* if $\tilde{\mathbf{w}}(\mathbf{y}+\mathbf{X}\mathbf{k}) = \tilde{\mathbf{w}}(\mathbf{y}) + \boldsymbol{\Lambda}'\mathbf{k}$ for every $P \times 1$ vector $\mathbf{k}$ (and for every value of $\mathbf{y}$). Clearly, $\tilde{\mathbf{w}}(\mathbf{y})$ is a translation-equivariant predictor of $\mathbf{w}$ if and only if it is a translation-equivariant estimator of the expected value $\boldsymbol{\Lambda}'\boldsymbol{\beta}$ of $\mathbf{w}$.

Special case: Aitken and G–M models. Let us now specialize to the case where $\mathbf{y}$ follows an Aitken model. Under the Aitken model, $\text{var}(\mathbf{y})$ is an unknown scalar multiple of a known (nonnegative definite) matrix $\mathbf{H}$. It is convenient (and potentially useful) to consider the prediction of $\mathbf{w}$ under the assumption that $\text{var}\begin{pmatrix}\mathbf{y}\\ \mathbf{w}\end{pmatrix}$ is also an unknown scalar multiple of a known (nonnegative definite) matrix. Accordingly, it is supposed that $\text{cov}(\mathbf{y},\,\mathbf{w})$ and $\text{var}(\mathbf{w})$ are of the form

$$\text{cov}(\mathbf{y},\,\mathbf{w}) = \sigma^2\mathbf{H}_{yw} \qquad \text{and} \qquad \text{var}(\mathbf{w}) = \sigma^2\mathbf{H}_w,$$

where $\mathbf{H}_{yw}$ and $\mathbf{H}_w$ are known matrices. Thus, writing $\mathbf{H}_y$ for $\mathbf{H}$, the setup is such that

$$\text{var}\begin{pmatrix}\mathbf{y}\\ \mathbf{w}\end{pmatrix} = \sigma^2\begin{pmatrix}\mathbf{H}_y & \mathbf{H}_{yw}\\ \mathbf{H}'_{yw} & \mathbf{H}_w\end{pmatrix}.$$

As in the general case, it is supposed that

$$\text{E}(\mathbf{w}) = \boldsymbol{\Lambda}'\boldsymbol{\beta}$$

(where $\boldsymbol{\Lambda}$ is a known matrix).

The setup can be regarded as a special case of the more general setup where $\mathbf{y}$ follows a general linear model and where $\text{E}(\mathbf{w})$ is of the form (10.28) and $\text{cov}(\mathbf{y},\,\mathbf{w})$ and $\text{var}(\mathbf{w})$ of the form (10.29). Specifically, it can be regarded as the special case where $\boldsymbol{\theta}$ is the one-dimensional vector whose only element is $\theta_1 = \sigma$, where $\Theta = \{\boldsymbol{\theta} \mid \theta_1 > 0\}$, and where $\mathbf{V}_y(\boldsymbol{\theta}) = \theta_1^2\mathbf{H}_y$, $\mathbf{V}_{yw}(\boldsymbol{\theta}) = \theta_1^2\mathbf{H}_{yw}$, and $\mathbf{V}_w(\boldsymbol{\theta}) = \theta_1^2\mathbf{H}_w$. Clearly, in this special case, $\text{var}\begin{pmatrix}\mathbf{y}\\ \mathbf{w}\end{pmatrix}$ is known up to the value of the unknown scalar multiple $\theta_1^2 = \sigma^2$.

In the further special case where $\mathbf{y}$ follows a G–M model [i.e., where $\text{var}(\mathbf{y}) = \sigma^2\mathbf{I}$], $\mathbf{H}_y = \mathbf{I}$. When $\mathbf{H}_y = \mathbf{I}$, the case where $\mathbf{H}_{yw} = \mathbf{0}$ and $\mathbf{H}_w = \mathbf{I}$ is often singled out for special attention. The case where $\mathbf{H}_y = \mathbf{I}$, $\mathbf{H}_{yw} = \mathbf{0}$, and $\mathbf{H}_w = \mathbf{I}$ is encountered in applications where the realization of $\mathbf{w}$ corresponds to a vector of future data points and where the augmented vector $\begin{pmatrix}\mathbf{y}\\ \mathbf{w}\end{pmatrix}$ is assumed to follow a G–M model, the model matrix of which is $\begin{pmatrix}\mathbf{X}\\ \boldsymbol{\Lambda}'\end{pmatrix}$.

Best linear unbiased prediction (under a G–M model). Suppose that the $N \times 1$ observable random vector $\mathbf{y}$ follows a G–M model, in which case $\text{E}(\mathbf{y}) = \mathbf{X}\boldsymbol{\beta}$ and $\text{var}(\mathbf{y}) = \sigma^2\mathbf{I}$. And consider the prediction of the $M \times 1$ unobservable random vector $\mathbf{w}$ whose expected value is of the form

$$\text{E}(\mathbf{w}) = \boldsymbol{\Lambda}'\boldsymbol{\beta} \tag{10.32}$$

(where $\boldsymbol{\Lambda}$ is a matrix of known constants). Assume that $\text{cov}(\mathbf{y},\,\mathbf{w})$ and $\text{var}(\mathbf{w})$ are of the form

$$\text{cov}(\mathbf{y},\,\mathbf{w}) = \sigma^2\mathbf{H}_{yw} \qquad \text{and} \qquad \text{var}(\mathbf{w}) = \sigma^2\mathbf{H}_w \tag{10.33}$$

(where $\mathbf{H}_{yw}$ and $\mathbf{H}_w$ are known matrices). Assume also that $\mathbf{w}$ is predictable or, equivalently, that $\boldsymbol{\Lambda}'\boldsymbol{\beta}$ is estimable.

For purposes of applying the results of the final part of the preceding subsection (Subsection a), take $\boldsymbol{\tau}$ to be the $M \times 1$ vector (of linear combinations of the elements of $\boldsymbol{\beta}$) defined as follows:

$$\boldsymbol{\tau} = \text{E}(\mathbf{w}) - [\text{cov}(\mathbf{y},\,\mathbf{w})]'[\text{var}(\mathbf{y})]^{-1}\,\text{E}(\mathbf{y}) = \boldsymbol{\Lambda}'\boldsymbol{\beta} - \mathbf{H}'_{yw}\mathbf{X}\boldsymbol{\beta} = (\boldsymbol{\Lambda}' - \mathbf{H}'_{yw}\mathbf{X})\boldsymbol{\beta}. \tag{10.34}$$

Clearly, $\boldsymbol{\tau}$ is estimable, and its least squares estimator is the vector $\hat{\boldsymbol{\tau}}_L(\mathbf{y})$ defined as follows:

$$\hat{\boldsymbol{\tau}}_L(\mathbf{y}) = (\boldsymbol{\Lambda}' - \mathbf{H}'_{yw}\mathbf{X})(\mathbf{X}'\mathbf{X})^-\mathbf{X}'\mathbf{y} = \boldsymbol{\Lambda}'(\mathbf{X}'\mathbf{X})^-\mathbf{X}'\mathbf{y} - \mathbf{H}'_{yw}\mathbf{P}_{\mathbf{X}}\mathbf{y} \tag{10.35}$$

[where $\mathbf{P}_{\mathbf{X}} = \mathbf{X}(\mathbf{X}'\mathbf{X})^-\mathbf{X}'$]. Moreover, according to Theorem 5.6.1, $\hat{\boldsymbol{\tau}}_L(\mathbf{y})$ is a linear unbiased estimator of $\boldsymbol{\tau}$ and, in fact, is the BLUE (best linear unbiased estimator) of $\boldsymbol{\tau}$. It is the BLUE in the sense that the difference between the MSE matrix $\mathrm{E}\{[\tilde{\boldsymbol{\tau}}_L(\mathbf{y}) - \boldsymbol{\tau}][\tilde{\boldsymbol{\tau}}_L(\mathbf{y}) - \boldsymbol{\tau}]'\} = \mathrm{var}[\tilde{\boldsymbol{\tau}}_L(\mathbf{y})]$ of an arbitrary linear unbiased estimator $\tilde{\boldsymbol{\tau}}_L(\mathbf{y})$ of $\boldsymbol{\tau}$ and the MSE matrix $\mathrm{E}\{[\hat{\boldsymbol{\tau}}_L(\mathbf{y}) - \boldsymbol{\tau}][\hat{\boldsymbol{\tau}}_L(\mathbf{y}) - \boldsymbol{\tau}]'\} = \mathrm{var}[\hat{\boldsymbol{\tau}}_L(\mathbf{y})]$ of the least squares estimator $\hat{\boldsymbol{\tau}}_L(\mathbf{y})$ is nonnegative definite [and is equal to $\mathbf{0}$ if and only if $\tilde{\boldsymbol{\tau}}_L(\mathbf{y}) = \hat{\boldsymbol{\tau}}_L(\mathbf{y})$].

Now, let

$$\hat{\mathbf{w}}_L(\mathbf{y}) = \hat{\boldsymbol{\tau}}_L(\mathbf{y}) + [\mathrm{cov}(\mathbf{y}, \mathbf{w})]'[\mathrm{var}(\mathbf{y})]^{-1}\mathbf{y} = \boldsymbol{\Lambda}'(\mathbf{X}'\mathbf{X})^-\mathbf{X}'\mathbf{y} + \mathbf{H}'_{yw}(\mathbf{I} - \mathbf{P}_{\mathbf{X}})\mathbf{y}. \tag{10.36}$$

Then, it follows from the results of the final part of Subsection a that $\hat{\mathbf{w}}_L(\mathbf{y})$ is a linear unbiased predictor of $\mathbf{w}$ and, in fact, is the BLUP (best linear unbiased predictor) of $\mathbf{w}$. It is the BLUP in the sense that the difference between the MSE matrix $\mathrm{E}\{[\tilde{\mathbf{w}}_L(\mathbf{y})-\mathbf{w}][\tilde{\mathbf{w}}_L(\mathbf{y})-\mathbf{w}]'\} = \mathrm{var}[\tilde{\mathbf{w}}_L(\mathbf{y})-\mathbf{w}]$ of an arbitrary linear unbiased predictor $\tilde{\mathbf{w}}_L(\mathbf{y})$ of $\mathbf{w}$ and the MSE matrix $\mathrm{E}\{[\hat{\mathbf{w}}_L(\mathbf{y})-\mathbf{w}][\hat{\mathbf{w}}_L(\mathbf{y})-\mathbf{w}]'\} = \mathrm{var}[\hat{\mathbf{w}}_L(\mathbf{y}) - \mathbf{w}]$ of $\hat{\mathbf{w}}_L(\mathbf{y})$ is nonnegative definite [and is equal to $\mathbf{0}$ if and only if $\tilde{\mathbf{w}}_L(\mathbf{y}) = \hat{\mathbf{w}}_L(\mathbf{y})$]. In the special case where $M = 1$, the sense in which $\hat{\mathbf{w}}_L(\mathbf{y})$ is the BLUP can [upon writing $\hat{w}_L(\mathbf{y})$ for $\hat{\mathbf{w}}_L(\mathbf{y})$ and w for $\mathbf{w}$] be restated as follows: the MSE of $\hat{w}_L(\mathbf{y})$ [or, equivalently, the variance of the prediction error of $\hat{w}_L(\mathbf{y})$] is smaller than that of any other linear unbiased predictor of w.

In light of result (6.7), the variance-covariance matrix of the least squares estimator of $\boldsymbol{\tau}$ is

$$\mathrm{var}[\hat{\boldsymbol{\tau}}_L(\mathbf{y})] = \sigma^2(\boldsymbol{\Lambda}' - \mathbf{H}'_{yw}\mathbf{X})(\mathbf{X}'\mathbf{X})^-(\boldsymbol{\Lambda} - \mathbf{X}'\mathbf{H}_{yw}). \tag{10.37}$$

Accordingly, it follows from result (10.25) that the MSE matrix of the BLUP of $\mathbf{w}$ or, equivalently, the variance-covariance matrix of the prediction error of the BLUP is

$$\mathrm{var}[\hat{\mathbf{w}}_L(\mathbf{y}) - \mathbf{w}] = \sigma^2(\boldsymbol{\Lambda}' - \mathbf{H}'_{yw}\mathbf{X})(\mathbf{X}'\mathbf{X})^-(\boldsymbol{\Lambda} - \mathbf{X}'\mathbf{H}_{yw}) + \sigma^2(\mathbf{H}_w - \mathbf{H}'_{yw}\mathbf{H}_{yw}). \tag{10.38}$$

In the special case where $\mathbf{H}_{yw} = \mathbf{0}$, we find that $\boldsymbol{\tau} = \boldsymbol{\Lambda}'\boldsymbol{\beta}$, $\hat{\mathbf{w}}_L(\mathbf{y}) = \hat{\boldsymbol{\tau}}_L(\mathbf{y})$, $\mathrm{var}[\hat{\boldsymbol{\tau}}_L(\mathbf{y})] = \sigma^2\boldsymbol{\Lambda}'(\mathbf{X}'\mathbf{X})^-\boldsymbol{\Lambda}$, and

$$\mathrm{var}[\hat{\mathbf{w}}_L(\mathbf{y}) - \mathbf{w}] = \sigma^2\boldsymbol{\Lambda}'(\mathbf{X}'\mathbf{X})^-\boldsymbol{\Lambda} + \sigma^2\mathbf{H}_w.$$

Note that even in this special case [where the BLUP of $\mathbf{w}$ equals the BLUE of $\boldsymbol{\tau}$ and where $\boldsymbol{\tau} = \mathrm{E}(\mathbf{w})$], the MSE matrix of the BLUP typically differs from that of the BLUE. The difference between the two MSE matrices [$\sigma^2\mathbf{H}_w$ in the special case and $\sigma^2(\mathbf{H}_w - \mathbf{H}'_{yw}\mathbf{H}_{yw})$ in the general case] is nonnegative definite. This difference is attributable to the variability of $\mathbf{w}$, which contributes to the variability of the prediction error $\hat{\mathbf{w}}(\mathbf{y}) - \mathbf{w}$ but not to the variability of $\hat{\boldsymbol{\tau}}(\mathbf{y}) - \boldsymbol{\tau}$.

Best linear translation-equivariant prediction (under a G–M model). Let us continue to consider the prediction of the $M \times 1$ unobservable random vector $\mathbf{w}$ on the basis of the $N \times 1$ observable random vector $\mathbf{y}$, doing so under the same conditions as in the preceding part of the present subsection. Thus, it is supposed that $\mathbf{y}$ follows a G–M model, that $\mathrm{E}(\mathbf{w})$ is of the form (10.32), that $\mathrm{cov}(\mathbf{y}, \mathbf{w})$ and $\mathrm{var}(\mathbf{w})$ are of the form (10.33), and that $\mathbf{w}$ is predictable. Further, define $\boldsymbol{\tau}$, $\hat{\boldsymbol{\tau}}_L(\mathbf{y})$, and $\hat{\mathbf{w}}_L(\mathbf{y})$ as in equations (10.34), (10.35), and (10.36) [so that $\hat{\boldsymbol{\tau}}_L(\mathbf{y})$ is the BLUE of $\boldsymbol{\tau}$ and $\hat{\mathbf{w}}_L(\mathbf{y})$ the BLUP of $\mathbf{w}$].

Let us consider the translation-equivariant prediction of $\mathbf{w}$. Denote by $\tilde{\mathbf{w}}(\mathbf{y})$ an arbitrary predictor of $\mathbf{w}$, and take

$$\tilde{\boldsymbol{\tau}}(\mathbf{y}) = \tilde{\mathbf{w}}(\mathbf{y}) - [\mathrm{cov}(\mathbf{y}, \mathbf{w})]'[\mathrm{var}(\mathbf{y})]^{-1}\mathbf{y} = \tilde{\mathbf{w}}(\mathbf{y}) - \mathbf{H}'_{yw}\mathbf{y}$$

to be the corresponding estimator of $\boldsymbol{\tau}$—refer to the final part of Subsection a. Then, $\tilde{\mathbf{w}}(\mathbf{y})$ is a translation-equivariant predictor (of $\mathbf{w}$) if and only if $\tilde{\boldsymbol{\tau}}(\mathbf{y})$ is a translation-equivariant estimator (of $\boldsymbol{\tau}$), as can be readily verified. Further, $\tilde{\mathbf{w}}(\mathbf{y})$ is a linear translation-equivariant predictor (of $\mathbf{w}$) if and only if $\tilde{\boldsymbol{\tau}}(\mathbf{y})$ is a linear translation-equivariant estimator (of $\boldsymbol{\tau}$).

In light of Corollary 5.6.4, the estimator $\hat{\boldsymbol{\tau}}_L(\mathbf{y})$ is a linear translation-equivariant estimator of $\boldsymbol{\tau}$ and, in fact, is the best linear translation-equivariant estimator of $\boldsymbol{\tau}$. It is the best linear translation-equivariant estimator in the sense that the difference between the MSE matrix $\mathrm{E}\{[\tilde{\boldsymbol{\tau}}_L(\mathbf{y})-\boldsymbol{\tau}][\tilde{\boldsymbol{\tau}}_L(\mathbf{y})-\boldsymbol{\tau}]'\}$ of an arbitrary linear translation-equivariant estimator $\tilde{\boldsymbol{\tau}}_L(\mathbf{y})$ of $\boldsymbol{\tau}$ and the MSE matrix $\mathrm{E}\{[\hat{\boldsymbol{\tau}}_L(\mathbf{y})-\boldsymbol{\tau}][\hat{\boldsymbol{\tau}}_L(\mathbf{y})-\boldsymbol{\tau}]'\}$ of $\hat{\boldsymbol{\tau}}_L(\mathbf{y})$ is nonnegative definite [and is equal to $\mathbf{0}$ if and only if $\tilde{\boldsymbol{\tau}}_L(\mathbf{y}) = \hat{\boldsymbol{\tau}}_L(\mathbf{y})$]. And upon recalling the results of the final part of Subsection a, it follows that the predictor $\hat{\mathbf{w}}_L(\mathbf{y})$ is a linear translation-equivariant predictor of $\mathbf{w}$ and, in fact, is the best linear translation-equivariant predictor of $\mathbf{w}$. It is the best linear translation-equivariant predictor in the sense that the difference between the MSE matrix $\mathrm{E}\{[\tilde{\mathbf{w}}_L(\mathbf{y})-\mathbf{w}][\tilde{\mathbf{w}}_L(\mathbf{y})-\mathbf{w}]'\}$ of an arbitrary linear translation-equivariant predictor $\tilde{\mathbf{w}}_L(\mathbf{y})$ of $\mathbf{w}$ and the MSE matrix $\mathrm{E}\{[\hat{\mathbf{w}}_L(\mathbf{y})-\mathbf{w}][\hat{\mathbf{w}}_L(\mathbf{y})-\mathbf{w}]'\} = \mathrm{var}[\hat{\mathbf{w}}_L(\mathbf{y})-\mathbf{w}]$ of $\hat{\mathbf{w}}_L(\mathbf{y})$ is nonnegative definite [and is equal to $\mathbf{0}$ if and only if $\tilde{\mathbf{w}}_L(\mathbf{y}) = \hat{\mathbf{w}}_L(\mathbf{y})$]. In the special case where $M = 1$, the sense in which $\hat{\mathbf{w}}_L(\mathbf{y})$ is the best linear translation-equivariant predictor can [upon writing $\hat{w}_L(\mathbf{y})$ for $\hat{\mathbf{w}}_L(\mathbf{y})$ and w for $\mathbf{w}$] be restated as follows: the MSE of $\hat{w}_L(\mathbf{y})$ is smaller than that of any other linear translation-equivariant predictor of w.

c. Conditional expected values: elliptical distributions

Let $\mathbf{w}$ represent an $M \times 1$ random vector and $\mathbf{y}$ an $N \times 1$ random vector. Suppose that the second-order moments of the joint distribution of $\mathbf{w}$ and $\mathbf{y}$ exist. And adopt the following notation: $\boldsymbol{\mu}_y = \mathrm{E}(\mathbf{y})$, $\boldsymbol{\mu}_w = \mathrm{E}(\mathbf{w})$, $\mathbf{V}_y = \mathrm{var}(\mathbf{y})$, $\mathbf{V}_{yw} = \mathrm{cov}(\mathbf{y}, \mathbf{w})$, and $\mathbf{V}_w = \mathrm{var}(\mathbf{w})$. Further, suppose that $\mathbf{V}_y$ is nonsingular.

Let $\boldsymbol{\eta}(\mathbf{y}) = \boldsymbol{\mu}_w + \mathbf{V}_{yw}'\mathbf{V}_y^{-1}(\mathbf{y}-\boldsymbol{\mu}_y)$. If $\mathbf{y}$ is observable but $\mathbf{w}$ is unobservable, we might wish to use the value of $\mathbf{y}$ to predict the realization of $\mathbf{w}$. If $\boldsymbol{\eta}(\cdot)$ is determinable from what is known about the joint distribution of $\mathbf{w}$ and $\mathbf{y}$, we could use $\boldsymbol{\eta}(\mathbf{y})$ to make the prediction; it would be the best linear predictor in the sense described in Part 2 of Subsection a. If enough more is known about the joint distribution of $\mathbf{w}$ and $\mathbf{y}$ that $\mathrm{E}(\mathbf{w} \mid \mathbf{y})$ is determinable, we might prefer to use $\mathrm{E}(\mathbf{w} \mid \mathbf{y})$ to make the prediction; it would be the best predictor in the sense described in Part 1 of Subsection a.

Under what circumstances is $\mathrm{E}(\mathbf{w} \mid \mathbf{y})$ equal to $\boldsymbol{\eta}(\mathbf{y})$ (at least with probability 1) or, equivalently, under what circumstances is $\mathrm{E}(\mathbf{w} \mid \mathbf{y})$ linear (or at least "linear with probability 1"). As previously indicated (in Part 1 of Subsection a), one such circumstance is that where the joint distribution of $\mathbf{w}$ and $\mathbf{y}$ is MVN. More generally, $\mathrm{E}(\mathbf{w} \mid \mathbf{y})$ equals $\boldsymbol{\eta}(\mathbf{y})$ (at least with probability 1) if the joint distribution of $\mathbf{w}$ and $\mathbf{y}$ is elliptical, as will now be shown.

Let $\mathbf{e} = \mathbf{w} - \boldsymbol{\eta}(\mathbf{y})$, and observe that

$$\begin{pmatrix}\mathbf{e}\\ \mathbf{y}\end{pmatrix} = \begin{pmatrix}-\boldsymbol{\mu}_w + \mathbf{V}_{yw}'\mathbf{V}_y^{-1}\boldsymbol{\mu}_y\\ \mathbf{0}\end{pmatrix} + \begin{pmatrix}\mathbf{I} & -\mathbf{V}_{yw}'\mathbf{V}_y^{-1}\\ \mathbf{0} & \mathbf{I}\end{pmatrix}\begin{pmatrix}\mathbf{w}\\ \mathbf{y}\end{pmatrix}.$$

Observe also [in light of Part (2) of Theorem 3.5.9] that

$$\mathrm{E}(\mathbf{e}) = \mathbf{0}, \quad \mathrm{var}(\mathbf{e}) = \mathbf{V}_w - \mathbf{V}_{yw}'\mathbf{V}_y^{-1}\mathbf{V}_{yw}, \quad \text{and} \quad \mathrm{cov}(\mathbf{e}, \mathbf{y}) = \mathbf{0}.$$

Now, suppose that the distribution of the vector $\begin{pmatrix}\mathbf{w}\\ \mathbf{y}\end{pmatrix}$ is elliptical with mgf generator $\phi(\cdot)$. Then, it follows from the next-to-last part of Section 5.9c that the vector $\begin{pmatrix}\mathbf{e}\\ \mathbf{y}\end{pmatrix}$ has an elliptical distribution with mean $\begin{pmatrix}\mathbf{0}\\ \boldsymbol{\mu}_y\end{pmatrix}$, variance-covariance matrix $\begin{pmatrix}\mathbf{V}_w - \mathbf{V}_{yw}'\mathbf{V}_y^{-1}\mathbf{V}_{yw} & \mathbf{0}\\ \mathbf{0} & \mathbf{V}_y\end{pmatrix}$, and mgf generator $\phi(\cdot)$ and that the vector $\begin{pmatrix}-\mathbf{e}\\ \mathbf{y}\end{pmatrix}$ has this same distribution. Thus, the conditional distribution of $-\mathbf{e}$ given $\mathbf{y}$ is the same as that of $\mathbf{e}$ given $\mathbf{y}$, so that the conditional distribution of $\mathbf{e}$ is symmetrical about $\mathbf{0}$ and hence $\mathrm{E}(\mathbf{e} \mid \mathbf{y}) = \mathbf{0}$ (with probability 1). And since $\mathbf{w} = \boldsymbol{\eta}(\mathbf{y}) + \mathbf{e}$, we conclude that $\mathrm{E}(\mathbf{w} \mid \mathbf{y}) = \boldsymbol{\eta}(\mathbf{y})$ (with probability 1).

Exercises

Exercise 1. Take the context to be that of estimating parametric functions of the form $\boldsymbol{\lambda}'\boldsymbol{\beta}$ from an $N \times 1$ observable random vector $\mathbf{y}$ that follows a G–M, Aitken, or general linear model. Verify (1) that linear combinations of estimable functions are estimable and (2) that linear combinations of nonestimable functions are not necessarily nonestimable.

Exercise 2. Take the context to be that of estimating parametric functions of the form $\boldsymbol{\lambda}'\boldsymbol{\beta}$ from an $N \times 1$ observable random vector $\mathbf{y}$ that follows a G–M, Aitken, or general linear model. And let $R = \text{rank}(\mathbf{X})$.

(a) Verify (1) that there exists a set of R linearly independent estimable functions; (2) that no set of estimable functions contains more than R linearly independent estimable functions; and (3) that if the model is not of full rank (i.e., if $R < P$), then at least one and, in fact, at least $P - R$ of the individual parameters $\beta_1, \beta_2, \ldots, \beta_P$ are nonestimable.

(b) Show that the jth of the individual parameters $\beta_1, \beta_2, \ldots, \beta_P$ is estimable if and only if the jth element of every vector in $\mathcal{N}(\mathbf{X})$ equals 0 $(j = 1, 2, \ldots, P)$.

Exercise 3. Show that for a parametric function of the form $\boldsymbol{\lambda}'\boldsymbol{\beta}$ to be estimable from an $N \times 1$ observable random vector $\mathbf{y}$ that follows a G–M, Aitken, or general linear model, it is necessary and sufficient that

$$\text{rank}(\mathbf{X}', \boldsymbol{\lambda}) = \text{rank}(\mathbf{X}).$$

Exercise 4. Suppose that $\mathbf{y}$ is an $N \times 1$ observable random vector that follows a G–M, Aitken, or general linear model. Further, take $\underline{\mathbf{y}}$ to be any value of $\mathbf{y}$, and consider the quantity $\boldsymbol{\lambda}'\tilde{\mathbf{b}}$, where $\boldsymbol{\lambda}$ is an arbitrary $P \times 1$ vector of constants and $\tilde{\mathbf{b}}$ is any solution to the linear system $\mathbf{X}'\mathbf{X}\mathbf{b} = \mathbf{X}'\underline{\mathbf{y}}$ (in the $P \times 1$ vector $\mathbf{b}$). Show that if $\boldsymbol{\lambda}'\tilde{\mathbf{b}}$ is invariant to the choice of the solution $\tilde{\mathbf{b}}$, then $\boldsymbol{\lambda}'\boldsymbol{\beta}$ is an estimable function. And discuss the implications of this result.

Exercise 5. Suppose that $\mathbf{y}$ is an $N \times 1$ observable random vector that follows a G–M, Aitken, or general linear model. And let $\mathbf{a}$ represent an arbitrary $N \times 1$ vector of constants. Show that $\mathbf{a}'\mathbf{y}$ is the least squares estimator of its expected value $\text{E}(\mathbf{a}'\mathbf{y})$ (i.e., of the parametric function $\mathbf{a}'\mathbf{X}\boldsymbol{\beta}$) if and only if $\mathbf{a} \in \mathcal{C}(\mathbf{X})$.

Exercise 6. Let $\mathcal{U}$ represent a subspace of the linear space $\mathcal{R}^M$ of all M-dimensional column vectors. Verify that the set $\mathcal{U}^{\perp}$ (comprising all M-dimensional column vectors that are orthogonal to $\mathcal{U}$) is a linear space.

Exercise 7. Let $\mathbf{X}$ represent an $N \times P$ matrix. A $P \times N$ matrix $\mathbf{G}$ is said to be a *least squares generalized inverse* of $\mathbf{X}$ if it is a generalized inverse of $\mathbf{X}$ (i.e., if $\mathbf{XGX} = \mathbf{X}$) and if, in addition, $(\mathbf{XG})' = \mathbf{XG}$ (i.e., $\mathbf{XG}$ is symmetric).

(a) Show that $\mathbf{G}$ is a least squares generalized inverse of $\mathbf{X}$ if and only if $\mathbf{X}'\mathbf{X}\mathbf{G} = \mathbf{X}'$.

(b) Using Part (a) (or otherwise), establish the existence of a least squares generalized inverse of $\mathbf{X}$.

(c) Show that if $\mathbf{G}$ is a least squares generalized inverse of $\mathbf{X}$, then, for any $N \times Q$ matrix $\mathbf{Y}$, the matrix $\mathbf{GY}$ is a solution to the linear system $\mathbf{X}'\mathbf{X}\mathbf{B} = \mathbf{X}'\mathbf{Y}$ (in the $P \times Q$ matrix $\mathbf{B}$).

Exercise 8. Let $\mathbf{A}$ represent an $M \times N$ matrix. An $N \times M$ matrix $\mathbf{H}$ is said to be a *minimum norm generalized inverse* of $\mathbf{A}$ if it is a generalized inverse of $\mathbf{A}$ (i.e., if $\mathbf{AHA} = \mathbf{A}$) and if, in addition, $(\mathbf{HA})' = \mathbf{HA}$ (i.e., $\mathbf{HA}$ is symmetric).

(a) Show that $\mathbf{H}$ is a minimum norm generalized inverse of $\mathbf{A}$ if and only if $\mathbf{H}'$ is a least squares generalized inverse of $\mathbf{A}'$ (where least squares generalized inverse is as defined in Exercise 7).

(b) Using the results of Exercise 7 (or otherwise), establish the existence of a minimum norm generalized inverse of $\mathbf{A}$.

(c) Show that if $\mathbf{H}$ is a minimum norm generalized inverse of $\mathbf{A}$, then, for any vector $\mathbf{b} \in \mathcal{C}(\mathbf{A})$, $\|\mathbf{x}\|$ attains its minimum value over the set $\{\mathbf{x} : \mathbf{A}\mathbf{x} = \mathbf{b}\}$ [comprising all solutions to the linear system $\mathbf{A}\mathbf{x} = \mathbf{b}$ (in $\mathbf{x}$)] uniquely at $\mathbf{x} = \mathbf{H}\mathbf{b}$ (where $\|\cdot\|$ denotes the usual norm).

Exercise 9. Let $\mathbf{X}$ represent an $N \times P$ matrix, and let $\mathbf{G}$ represent a $P \times N$ matrix that is subject to the following four conditions: (1) $\mathbf{XGX} = \mathbf{X}$; (2) $\mathbf{GXG} = \mathbf{G}$; (3) $(\mathbf{XG})' = \mathbf{XG}$; and (4) $(\mathbf{GX})' = \mathbf{GX}$.

(a) Show that if a $P \times P$ matrix $\mathbf{H}$ is a minimum norm generalized inverse of $\mathbf{X}'\mathbf{X}$, then conditions (1)–(4) can be satisfied by taking $\mathbf{G} = \mathbf{H}\mathbf{X}'$.

(b) Use Part (a) and the result of Part (b) of Exercise 8 (or other means) to establish the existence of a $P \times N$ matrix $\mathbf{G}$ that satisfies conditions (1)–(4) and show that there is only one such matrix.

(c) Let $\mathbf{X}^+$ represent the unique $P \times N$ matrix $\mathbf{G}$ that satisfies conditions (1)–(4)—this matrix is customarily referred to as the *Moore–Penrose inverse*, and conditions (1)–(4) are customarily referred to as the *Moore–Penrose conditions*. Using Parts (a) and (b) and the results of Part (c) of Exercise 7 and Part (c) of Exercise 8 (or otherwise), show that $\mathbf{X}^+\mathbf{y}$ is a solution to the linear system $\mathbf{X}'\mathbf{X}\mathbf{b} = \mathbf{X}'\mathbf{y}$ (in $\mathbf{b}$) and that $\|\mathbf{b}\|$ attains its minimum value over the set $\{\mathbf{b} : \mathbf{X}'\mathbf{X}\mathbf{b} = \mathbf{X}'\mathbf{y}\}$ (comprising all solutions to the linear system) uniquely at $\mathbf{b} = \mathbf{X}^+\mathbf{y}$ (where $\|\cdot\|$ denotes the usual norm).

Exercise 10. Consider further the alternative approach to the least squares computations, taking the formulation and the notation to be those of the final part of Section 5.4e.

(a) Let $\tilde{\mathbf{b}} = \mathbf{L}_1\tilde{\mathbf{h}}_1 + \mathbf{L}_2\tilde{\mathbf{h}}_2$, where $\tilde{\mathbf{h}}_2$ is an arbitrary $(P-K)$-dimensional column vector and $\tilde{\mathbf{h}}_1$ is the solution to the linear system $\mathbf{R}_1\mathbf{h}_1 = \mathbf{z}_1 - \mathbf{R}_2\tilde{\mathbf{h}}_2$. Show that $\|\tilde{\mathbf{b}}\|$ is minimized by taking

$$\tilde{\mathbf{h}}_2 = [\mathbf{I} + (\mathbf{R}_1^{-1}\mathbf{R}_2)'\mathbf{R}_1^{-1}\mathbf{R}_2]^{-1}(\mathbf{R}_1^{-1}\mathbf{R}_2)'\mathbf{R}_1^{-1}\mathbf{z}_1.$$

Do so by formulating this minimization problem as a least squares problem in which the role of $\mathbf{y}$ is played by the vector $\begin{pmatrix}\mathbf{R}_1^{-1}\mathbf{z}_1\\ \mathbf{0}\end{pmatrix}$, the role of $\mathbf{X}$ is played by the matrix $\begin{pmatrix}\mathbf{R}_1^{-1}\mathbf{R}_2\\ \mathbf{I}\end{pmatrix}$, and the role of $\mathbf{b}$ is played by $\tilde{\mathbf{h}}_2$.

(b) Let $\mathbf{O}_1$ represent a $P \times K$ matrix with orthonormal columns and $\mathbf{T}_1$ a $K \times K$ upper triangular matrix such that $\begin{pmatrix}\mathbf{R}_1'\\ \mathbf{R}_2'\end{pmatrix} = \mathbf{O}_1\mathbf{T}_1'$—the existence of a decomposition of this form can be established in much the same way as the existence of the QR decomposition (in which $\mathbf{T}_1$ would be lower triangular rather than upper triangular). Further, take $\mathbf{O}_2$ to be any $P \times (P-K)$ matrix such that the $P \times P$ matrix $\mathbf{O}$ defined by $\mathbf{O} = (\mathbf{O}_1, \mathbf{O}_2)$ is orthogonal.

(1) Show that $\mathbf{X} = \mathbf{Q}\mathbf{T}(\mathbf{L}\mathbf{O})'$, where $\mathbf{T} = \begin{pmatrix}\mathbf{T}_1 & \mathbf{0}\\ \mathbf{0} & \mathbf{0}\end{pmatrix}$.

(2) Show that $\mathbf{y} - \mathbf{X}\mathbf{b} = \mathbf{Q}_1(\mathbf{z}_1 - \mathbf{T}_1\mathbf{d}_1) + \mathbf{Q}_2\mathbf{z}_2$, where $\mathbf{d} = (\mathbf{L}\mathbf{O})'\mathbf{b}$ and $\mathbf{d}$ is partitioned as $\mathbf{d} = \begin{pmatrix}\mathbf{d}_1\\ \mathbf{d}_2\end{pmatrix}$.

(3) Show that $(\mathbf{y} - \mathbf{X}\mathbf{b})'(\mathbf{y} - \mathbf{X}\mathbf{b}) = (\mathbf{z}_1 - \mathbf{T}_1\mathbf{d}_1)'(\mathbf{z}_1 - \mathbf{T}_1\mathbf{d}_1) + \mathbf{z}_2'\mathbf{z}_2$.

(4) Taking $\tilde{\mathbf{d}}_1$ to be the solution to the linear system $\mathbf{T}_1\mathbf{d}_1 = \mathbf{z}_1$ (in $\mathbf{d}_1$), show that $(\mathbf{y} -$

$\mathbf{Xb})'(\mathbf{y} - \mathbf{Xb})$ attains a minimum value of $\mathbf{z}_2'\mathbf{z}_2$ and that it does so at a value $\tilde{\mathbf{b}}$ of $\mathbf{b}$ if and only if $\tilde{\mathbf{b}} = \mathbf{LO}\begin{pmatrix}\tilde{\mathbf{d}}_1 \\ \tilde{\mathbf{d}}_2\end{pmatrix}$ for some $(P-K) \times 1$ vector $\tilde{\mathbf{d}}_2$.

(5) Letting $\tilde{\mathbf{b}}$ represent an arbitrary one of the values of $\mathbf{b}$ at which $(\mathbf{y} - \mathbf{Xb})'(\mathbf{y} - \mathbf{Xb})$ attains a minimum value [and, as in Part (4), taking $\tilde{\mathbf{d}}_1$ to be the solution to $\mathbf{T}_1\mathbf{d}_1 = \mathbf{z}_1$], show that $\|\tilde{\mathbf{b}}\|^2$ (where $\|\cdot\|$ denotes the usual norm) attains a minimum value of $\tilde{\mathbf{d}}_1'\tilde{\mathbf{d}}_1$ and that it does so uniquely at $\tilde{\mathbf{b}} = \mathbf{LO}\begin{pmatrix}\tilde{\mathbf{d}}_1 \\ \mathbf{0}\end{pmatrix}$.

Exercise 11. Verify that the difference (6.14) is a nonnegative definite matrix and that it equals $\mathbf{0}$ if and only if $\mathbf{c} + \mathbf{A}'\mathbf{y} = \boldsymbol{\ell}(\mathbf{y})$.

Exercise 12. Suppose that $\mathbf{y}$ is an $N \times 1$ observable random vector that follows a G–M, Aitken, or general linear model. And let $s(\mathbf{y})$ represent any particular translation-equivariant estimator of an estimable linear combination $\boldsymbol{\lambda}'\boldsymbol{\beta}$ of the elements of the parametric vector $\boldsymbol{\beta}$—e.g., $s(\mathbf{y})$ could be the least squares estimator of $\boldsymbol{\lambda}'\boldsymbol{\beta}$. Show that an estimator $t(\mathbf{y})$ of $\boldsymbol{\lambda}'\boldsymbol{\beta}$ is translation equivariant if and only if

$$t(\mathbf{y}) = s(\mathbf{y}) + d(\mathbf{y})$$

for some translation-invariant statistic $d(\mathbf{y})$.

Exercise 13. Suppose that $\mathbf{y}$ is an $N \times 1$ observable random vector that follows a G–M model. And let $\mathbf{y}'\mathbf{A}\mathbf{y}$ represent a quadratic unbiased nonnegative-definite estimator of σ^2, that is, a quadratic form in $\mathbf{y}$ whose matrix $\mathbf{A}$ is a symmetric nonnegative definite matrix of constants and whose expected value is σ^2.

(a) Show that $\mathbf{y}'\mathbf{A}\mathbf{y}$ is translation invariant.

(b) Suppose that the fourth-order moments of the distribution of the vector $\mathbf{e} = (e_1, e_2, \ldots, e_N)'$ are such that (for $i, j, k, m = 1, 2, \ldots, N$) $\mathrm{E}(e_i e_j e_k e_m)$ satisfies condition (7.38). For what choice of $\mathbf{A}$ is the variance of the quadratic unbiased nonnegative-definite estimator $\mathbf{y}'\mathbf{A}\mathbf{y}$ a minimum? Describe your reasoning.

Exercise 14. Suppose that $\mathbf{y}$ is an $N \times 1$ observable random vector that follows a G–M model. Suppose further that the distribution of the vector $\mathbf{e} = (e_1, e_2, \ldots, e_N)'$ has third-order moments $\lambda_{jkm} = \mathrm{E}(e_j e_k e_m)$ $(j, k, m = 1, 2, \ldots, N)$ and fourth-order moments $\gamma_{ijkm} = \mathrm{E}(e_i e_j e_k e_m)$ $(i, j, k, m = 1, 2, \ldots, N)$. And let $\mathbf{A} = \{a_{ij}\}$ represent an $N \times N$ symmetric matrix of constants.

(a) Show that in the special case where the elements $e_1, e_2, \ldots, e_N$ of $\mathbf{e}$ are statistically independent,

$$\mathrm{var}(\mathbf{y}'\mathbf{A}\mathbf{y}) = \mathbf{a}'\boldsymbol{\Omega}^*\mathbf{a} + 4\boldsymbol{\beta}'\mathbf{X}'\mathbf{A}\boldsymbol{\Lambda}^*\mathbf{a} + 2\sigma^4\,\mathrm{tr}(\mathbf{A}^2) + 4\sigma^2\boldsymbol{\beta}'\mathbf{X}'\mathbf{A}^2\mathbf{X}\boldsymbol{\beta}, \qquad \text{(E.1)}$$

where $\boldsymbol{\Omega}^*$ is the $N \times N$ diagonal matrix whose ith diagonal element is $\gamma_{iiii} - 3\sigma^4$, where $\boldsymbol{\Lambda}^*$ is the $N \times N$ diagonal matrix whose ith diagonal element is λ_{iii}, and where $\mathbf{a}$ is the $N \times 1$ vector whose elements are the diagonal elements $a_{11}, a_{22}, \ldots, a_{NN}$ of $\mathbf{A}$.

(b) Suppose that the elements $e_1, e_2, \ldots, e_N$ of $\mathbf{e}$ are statistically independent, that (for $i = 1, 2, \ldots, N$) $\gamma_{iiii} = \gamma$ (for some scalar γ), and that all N of the diagonal elements of the $\mathbf{P_X}$ matrix are equal to each other. Show that the estimator $\hat{\sigma}^2 = \tilde{\mathbf{e}}'\tilde{\mathbf{e}}/(N - \mathrm{rank}\,\mathbf{X})$ [where $\tilde{\mathbf{e}} = (\mathbf{I} - \mathbf{P_X})\mathbf{y}$] has minimum variance among all quadratic unbiased translation-invariant estimators of σ^2.

Exercise 15. Suppose that $\mathbf{y}$ is an $N \times 1$ observable random vector that follows a G–M model, and assume that the distribution of the vector $\mathbf{e}$ of residual effects is MVN.

(a) Letting $\boldsymbol{\lambda}'\boldsymbol{\beta}$ represent an estimable linear combination of the elements of the parametric vector $\boldsymbol{\beta}$, find a minimum-variance unbiased estimator of $(\boldsymbol{\lambda}'\boldsymbol{\beta})^2$.

(b) Find a minimum-variance unbiased estimator of σ^4.

Exercise 16. Suppose that $\mathbf{y}$ is an $N \times 1$ observable random vector that follows a G–M model, and assume that the distribution of the vector $\mathbf{e}$ of residual effects is MVN. Show that if σ^2 were known, $\mathbf{X}'\mathbf{y}$ would be a complete sufficient statistic.

Exercise 17. Suppose that $\mathbf{y}$ is an $N \times 1$ observable random vector that follows a general linear model. Suppose further that the distribution of the vector $\mathbf{e}$ of residual effects is MVN or, more generally, that the distribution of $\mathbf{e}$ is known up to the value of the vector $\boldsymbol{\theta}$. And take $\mathbf{h}(\mathbf{y})$ to be any (possibly vector-valued) translation-invariant statistic.

(a) Show that if $\boldsymbol{\theta}$ were known, $\mathbf{h}(\mathbf{y})$ would be an ancillary statistic—for a definition of ancillarity, refer, e.g., to Casella and Berger (2002, def. 6.2.16) or to Lehmann and Casella (1998, p. 41).

(b) Suppose that $\mathbf{X}'\mathbf{y}$ would be a complete sufficient statistic if $\boldsymbol{\theta}$ were known. Show (1) that the least squares estimator of any estimable linear combination $\boldsymbol{\lambda}'\boldsymbol{\beta}$ of the elements of the parametric vector $\boldsymbol{\beta}$ has minimum variance among all unbiased estimators, (2) that any vector of least squares estimators of estimable linear combinations (of the elements of $\boldsymbol{\beta}$) is distributed independently of $\mathbf{h}(\mathbf{y})$, and (3) (using the result of Exercise 12 or otherwise) that the least squares estimator of any estimable linear combination $\boldsymbol{\lambda}'\boldsymbol{\beta}$ has minimum mean squared error among all translation-equivariant estimators. {*Hint* [for Part (2)]. Make use of Basu's theorem—refer, e.g., to Lehmann and Casella (1998, p. 42) for a statement of Basu's theorem.}

Exercise 18. Suppose that $\mathbf{y}$ is an $N \times 1$ observable random vector that follows a G–M model. Suppose further that the distribution of the vector $\mathbf{e}$ of residual effects is MVN. And, letting $\tilde{\mathbf{e}} = \mathbf{y} - \mathbf{P_X}\mathbf{y}$, take $\tilde{\sigma}^2 = \tilde{\mathbf{e}}'\tilde{\mathbf{e}}/N$ to be the ML estimator of σ^2 and $\hat{\sigma}^2 = \tilde{\mathbf{e}}'\tilde{\mathbf{e}}/(N - \operatorname{rank} \mathbf{X})$ to be the unbiased estimator.

(a) Find the bias and the MSE of the ML estimator $\tilde{\sigma}^2$.

(b) Compare the MSE of the ML estimator $\tilde{\sigma}^2$ with that of the unbiased estimator $\hat{\sigma}^2$: for which values of N and of $\operatorname{rank} \mathbf{X}$ is the MSE of the ML estimator smaller than that of the unbiased estimator and for which values is it larger?

Exercise 19. Suppose that $\mathbf{y}$ is an $N \times 1$ observable random vector that follows a general linear model, that the distribution of the vector $\mathbf{e}$ of residual effects is MVN, and that the variance-covariance matrix $\mathbf{V}(\boldsymbol{\theta})$ of $\mathbf{e}$ is nonsingular (for all $\boldsymbol{\theta} \in \Theta$). And, letting $K = N - \operatorname{rank} \mathbf{X}$, take $\mathbf{R}$ to be any $N \times K$ matrix (of constants) of full column rank K such that $\mathbf{X}'\mathbf{R} = \mathbf{0}$, and (as in Section 5.9b) define $\mathbf{z} = \mathbf{R}'\mathbf{y}$. Further, let $\mathbf{w} = \mathbf{s}(\mathbf{z})$, where $\mathbf{s}(\cdot)$ is a $K \times 1$ vector of real-valued functions that defines a one-to-one mapping of $\mathbb{R}^K$ onto some set $\mathcal{W}$.

(a) Show that $\mathbf{w}$ is a maximal invariant.

(b) Let $f_1(\cdot\,;\boldsymbol{\theta})$ represent the pdf of the distribution of $\mathbf{z}$, and assume that $\mathbf{s}(\cdot)$ is such that the distribution of $\mathbf{w}$ has a pdf, say $f_2(\cdot\,;\boldsymbol{\theta})$, that is obtainable from $f_1(\cdot\,;\boldsymbol{\theta})$ via an application of the basic formula (e.g., Bickel and Doksum 2001, sec. B.2) for a change of variables. And, taking $L_1(\boldsymbol{\theta};\, \mathbf{R}'\underline{\mathbf{y}})$ and $L_2[\boldsymbol{\theta};\, \mathbf{s}(\mathbf{R}'\underline{\mathbf{y}})]$ (where $\underline{\mathbf{y}}$ denotes the observed value of $\mathbf{y}$) to be the likelihood functions defined by $L_1(\boldsymbol{\theta};\, \mathbf{R}'\underline{\mathbf{y}}) = f_1(\mathbf{R}'\underline{\mathbf{y}};\, \boldsymbol{\theta})$ and $L_2[\boldsymbol{\theta};\, \mathbf{s}(\mathbf{R}'\underline{\mathbf{y}})] = f_2[\mathbf{s}(\mathbf{R}'\underline{\mathbf{y}});\, \boldsymbol{\theta}]$, show that $L_1(\boldsymbol{\theta};\, \mathbf{R}'\underline{\mathbf{y}})$ and $L_2[\boldsymbol{\theta};\, \mathbf{s}(\mathbf{R}'\underline{\mathbf{y}})]$ differ from each other by no more than a multiplicative constant.

Exercise 20. Suppose that $\mathbf{y}$ is an $N \times 1$ observable random vector that follows a general linear model, that the distribution of the vector $\mathbf{e}$ of residual effects is MVN, and that the variance-covariance matrix $\mathbf{V}(\boldsymbol{\theta})$ of $\mathbf{e}$ is nonsingular (for all $\boldsymbol{\theta} \in \Theta$). Further, let $\mathbf{z} = \mathbf{R}'\mathbf{y}$, where $\mathbf{R}$ is any $N \times (N - \operatorname{rank} \mathbf{X})$ matrix (of constants) of full column rank $N - \operatorname{rank} \mathbf{X}$ such that $\mathbf{X}'\mathbf{R} = \mathbf{0}$; and let $\mathbf{u} = \mathbf{X}_*'\mathbf{y}$, where

$\mathbf{X}_*$ is any $N \times (\text{rank}\,\mathbf{X})$ matrix (of constants) whose columns form a basis for $\mathcal{C}(\mathbf{X})$. And denote by $\underset{\sim}{\mathbf{y}}$ the observed value of $\mathbf{y}$.

(a) Verify that the likelihood function that would result from regarding the observed value $(\mathbf{X}_*, \mathbf{R})'\underset{\sim}{\mathbf{y}}$ of $\begin{pmatrix}\mathbf{u}\\ \mathbf{z}\end{pmatrix}$ as the data vector differs by no more than a multiplicative constant from that obtained by regarding the observed value $\underset{\sim}{\mathbf{y}}$ of $\mathbf{y}$ as the data vector.

(b) Let $f_0(\cdot\,|\,\cdot\,;\boldsymbol{\beta},\boldsymbol{\theta})$ represent the pdf of the conditional distribution of $\mathbf{u}$ given $\mathbf{z}$. And take $L_0[\boldsymbol{\beta},\boldsymbol{\theta};\,(\mathbf{X}_*, \mathbf{R})'\underset{\sim}{\mathbf{y}}]$ to be the function of $\boldsymbol{\beta}$ and $\boldsymbol{\theta}$ defined by $L_0[\boldsymbol{\beta},\boldsymbol{\theta};\,(\mathbf{X}_*, \mathbf{R})'\underset{\sim}{\mathbf{y}}] = f_0(\mathbf{X}_*'\underset{\sim}{\mathbf{y}}\,|\,\mathbf{R}'\underset{\sim}{\mathbf{y}};\boldsymbol{\beta},\boldsymbol{\theta})$. Show that

$$L_0[\boldsymbol{\beta},\boldsymbol{\theta};\,(\mathbf{X}_*, \mathbf{R})'\underset{\sim}{\mathbf{y}}] = (2\pi)^{-(\text{rank}\,\mathbf{X})/2}\,|\mathbf{X}_*'\mathbf{X}_*|^{-1}\,|\mathbf{X}_*'[\mathbf{V}(\boldsymbol{\theta})]^{-1}\mathbf{X}_*|^{1/2} \times \exp\{-\tfrac{1}{2}[\tilde{\boldsymbol{\beta}}(\boldsymbol{\theta})-\boldsymbol{\beta}]'\mathbf{X}'[\mathbf{V}(\boldsymbol{\theta})]^{-1}\mathbf{X}[\tilde{\boldsymbol{\beta}}(\boldsymbol{\theta})-\boldsymbol{\beta}]\},$$

where $\tilde{\boldsymbol{\beta}}(\boldsymbol{\theta})$ is any solution to the linear system $\mathbf{X}'[\mathbf{V}(\boldsymbol{\theta})]^{-1}\mathbf{X}\mathbf{b} = \mathbf{X}'[\mathbf{V}(\boldsymbol{\theta})]^{-1}\underset{\sim}{\mathbf{y}}$ (in the $P \times 1$ vector $\mathbf{b}$).

(c) In connection with Part (b), show (1) that

$$[\tilde{\boldsymbol{\beta}}(\boldsymbol{\theta})-\boldsymbol{\beta}]'\mathbf{X}'[\mathbf{V}(\boldsymbol{\theta})]^{-1}\mathbf{X}[\tilde{\boldsymbol{\beta}}(\boldsymbol{\theta})-\boldsymbol{\beta}] = (\underset{\sim}{\mathbf{y}} - \mathbf{X}\boldsymbol{\beta})'[\mathbf{V}(\boldsymbol{\theta})]^{-1}\mathbf{X}\{\mathbf{X}'[\mathbf{V}(\boldsymbol{\theta})]^{-1}\mathbf{X}\}^{-}\mathbf{X}'[\mathbf{V}(\boldsymbol{\theta})]^{-1}(\underset{\sim}{\mathbf{y}} - \mathbf{X}\boldsymbol{\beta})$$

and (2) that the distribution of the random variable s defined by

$$s = (\mathbf{y} - \mathbf{X}\boldsymbol{\beta})'[\mathbf{V}(\boldsymbol{\theta})]^{-1}\mathbf{X}\{\mathbf{X}'[\mathbf{V}(\boldsymbol{\theta})]^{-1}\mathbf{X}\}^{-}\mathbf{X}'[\mathbf{V}(\boldsymbol{\theta})]^{-1}(\mathbf{y} - \mathbf{X}\boldsymbol{\beta})$$

does not depend on $\boldsymbol{\beta}$.

Exercise 21. Suppose that $\mathbf{z}$ is an $S \times 1$ observable random vector and that $\mathbf{z} \sim N(\mathbf{0}, \sigma^2\mathbf{I})$, where σ is a (strictly) positive unknown parameter.

(a) Show that $\mathbf{z}'\mathbf{z}$ is a complete sufficient statistic.

(b) Take $\mathbf{w}(\mathbf{z})$ to be the S-dimensional vector-valued statistic defined by $\mathbf{w}(\mathbf{z}) = (\mathbf{z}'\mathbf{z})^{-1/2}\mathbf{z}$—$\mathbf{w}(\mathbf{z})$ is defined for $\mathbf{z} \neq \mathbf{0}$ and hence with probability 1. Show that $\mathbf{z}'\mathbf{z}$ and $\mathbf{w}(\mathbf{z})$ are statistically independent. (*Hint.* Make use of Basu's theorem.)

(c) Show that any estimator of σ^2 of the form $\mathbf{z}'\mathbf{z}/k$ (where k is a nonzero constant) is scale equivariant—an estimator, say $t(\mathbf{z})$, of σ^2 is to be regarded as *scale equivariant* if for every (strictly) positive scalar c (and for every nonnull value of $\mathbf{z}$) $t(c\mathbf{z}) = c^2 t(\mathbf{z})$.

(d) Let $t_0(\mathbf{z})$ represent any particular scale-equivariant estimator of σ^2 such that $t_0(\mathbf{z}) \neq 0$ for $\mathbf{z} \neq \mathbf{0}$. Show that an estimator $t(\mathbf{z})$ of σ^2 is scale equivariant if and only if, for some function $u(\cdot)$ such that $u(c\mathbf{z}) = u(\mathbf{z})$ (for every strictly positive constant c and every nonnull value of $\mathbf{z}$),

$$t(\mathbf{z}) = u(\mathbf{z})t_0(\mathbf{z}) \quad \text{for} \quad \mathbf{z} \neq \mathbf{0}. \tag{E.2}$$

(e) Show that a function $u(\mathbf{z})$ of $\mathbf{z}$ is such that $u(c\mathbf{z}) = u(\mathbf{z})$ (for every strictly positive constant c and every nonnull value of $\mathbf{z}$) if and only if $u(\mathbf{z})$ depends on the value of $\mathbf{z}$ only through $\mathbf{w}(\mathbf{z})$ [where $\mathbf{w}(\mathbf{z})$ is as defined in Part (b)].

(f) Show that the estimator $\mathbf{z}'\mathbf{z}/(S+2)$ has minimum MSE among all scale-equivariant estimators of σ^2.

Exercise 22. Suppose that $\mathbf{y}$ is an $N \times 1$ observable random vector that follows a G–M model and that the distribution of the vector $\mathbf{e}$ of residual effects is MVN. Using the result of Part (f) of Exercise 21 (or otherwise), show that the Hodges–Lehmann estimator $\mathbf{y}'(\mathbf{I}-\mathbf{P_X})\mathbf{y}/[N-\text{rank}(\mathbf{X})+2]$ has minimum MSE among all translation-invariant estimators of σ^2 that are scale equivariant—a translation-invariant estimator, say $t(\mathbf{y})$, of σ^2 is to be regarded as *scale equivariant* if $t(c\mathbf{y}) = c^2 t(\mathbf{y})$ for every (strictly) positive scalar c and for every nonnull value of $\mathbf{y}$ in $\mathfrak{N}(\mathbf{X}')$.

Exercise 23. Let $\mathbf{z} = (z_1, z_2, \ldots, z_M)'$ represent an M-dimensional random (column) vector that has an absolutely continuous distribution with a pdf $f(\cdot)$. And suppose that for some (nonnegative) function $g(\cdot)$ (of a single nonnegative variable), $f(\mathbf{z}) \propto g(\mathbf{z}'\mathbf{z})$ (in which case the distribution of $\mathbf{z}$ is spherical). Show (for $i = 1, 2, \ldots, M$) that $\mathrm{E}(z_i^2)$ exists if and only if $\int_0^\infty s^{M+1} g(s^2)\, ds < \infty$, in which case

$$\mathrm{var}(z_i) = \mathrm{E}(z_i^2) = \frac{1}{M} \frac{\int_0^\infty s^{M+1} g(s^2)\, ds}{\int_0^\infty s^{M-1} g(s^2)\, ds}.$$

Exercise 24. Let $\mathbf{z}$ represent an N-dimensional random column vector, and let $\mathbf{z}_*$ represent an M-dimensional subvector of $\mathbf{z}$ (where $M < N$). And suppose that the distributions of $\mathbf{z}$ and $\mathbf{z}_*$ are absolutely continuous with pdfs $f(\cdot)$ and $f_*(\cdot)$, respectively. Suppose also that there exist (nonnegative) functions $g(\cdot)$ and $g_*(\cdot)$ (of a single nonnegative variable) such that (for every value of $\mathbf{z}$) $f(\mathbf{z}) = g(\mathbf{z}'\mathbf{z})$ and (for every value of $\mathbf{z}_*$) $f_*(\mathbf{z}_*) = g_*(\mathbf{z}_*'\mathbf{z}_*)$ (in which case the distributions of $\mathbf{z}$ and $\mathbf{z}_*$ are spherical).

(a) Show that (for $v \geq 0$)

$$g_*(v) = \frac{\pi^{(N-M)/2}}{\Gamma[(N-M)/2]} \int_v^\infty (u - v)^{[(N-M)/2]-1} g(u)\, du.$$

(b) Show that if $N - M = 2$, then (for $v > 0$)

$$g(v) = -\frac{1}{\pi} g_*'(v),$$

where $g_*'(\cdot)$ is the derivative of $g_*(\cdot)$.

Exercise 25. Let $\mathbf{y}$ represent an $N \times 1$ random vector and $\mathbf{w}$ an $M \times 1$ random vector. Suppose that the second-order moments of the joint distribution of $\mathbf{y}$ and $\mathbf{w}$ exist, and adopt the following notation: $\boldsymbol{\mu}_y = \mathrm{E}(\mathbf{y})$, $\boldsymbol{\mu}_w = \mathrm{E}(\mathbf{w})$, $\mathbf{V}_y = \mathrm{var}(\mathbf{y})$, $\mathbf{V}_{yw} = \mathrm{cov}(\mathbf{y}, \mathbf{w})$, and $\mathbf{V}_w = \mathrm{var}(\mathbf{w})$. Further, assume that $\mathbf{V}_y$ is nonsingular.

(a) Show that the matrix $\mathbf{V}_w - \mathbf{V}_{yw}'\mathbf{V}_y^{-1}\mathbf{V}_{yw} - \mathrm{E}[\mathrm{var}(\mathbf{w} \mid \mathbf{y})]$ is nonnegative definite and that it equals $\mathbf{0}$ if and only if (for some nonrandom vector $\mathbf{c}$ and some nonrandom matrix $\mathbf{A}$) $\mathrm{E}(\mathbf{w} \mid \mathbf{y}) = \mathbf{c} + \mathbf{A}'\mathbf{y}$ (with probability 1).

(b) Show that the matrix $\mathrm{var}[\mathrm{E}(\mathbf{w} \mid \mathbf{y})] - \mathbf{V}_{yw}'\mathbf{V}_y^{-1}\mathbf{V}_{yw}$ is nonnegative definite and that it equals $\mathbf{0}$ if and only if (for some nonrandom vector $\mathbf{c}$ and some nonrandom matrix $\mathbf{A}$) $\mathrm{E}(\mathbf{w} \mid \mathbf{y}) = \mathbf{c} + \mathbf{A}'\mathbf{y}$ (with probability 1).

Exercise 26. Let $\mathbf{y}$ represent an $N \times 1$ observable random vector and $\mathbf{w}$ an $M \times 1$ unobservable random vector. Suppose that the second-order moments of the joint distribution of $\mathbf{y}$ and $\mathbf{w}$ exist, and adopt the following notation: $\boldsymbol{\mu}_y = \mathrm{E}(\mathbf{y})$, $\boldsymbol{\mu}_w = \mathrm{E}(\mathbf{w})$, $\mathbf{V}_y = \mathrm{var}(\mathbf{y})$, $\mathbf{V}_{yw} = \mathrm{cov}(\mathbf{y}, \mathbf{w})$, and $\mathbf{V}_w = \mathrm{var}(\mathbf{w})$. Assume that $\boldsymbol{\mu}_y$, $\boldsymbol{\mu}_w$, $\mathbf{V}_y$, $\mathbf{V}_{yw}$, and $\mathbf{V}_w$ are known. Further, define $\boldsymbol{\eta}(\mathbf{y}) = \boldsymbol{\mu}_w + \mathbf{V}_{yw}'\mathbf{V}_y^{-}(\mathbf{y} - \boldsymbol{\mu}_y)$, and take $\mathbf{t}(\mathbf{y})$ to be an $(M \times 1)$-dimensional vector-valued function of the form $\mathbf{t}(\mathbf{y}) = \mathbf{c} + \mathbf{A}'\mathbf{y}$, where $\mathbf{c}$ is a vector of constants and $\mathbf{A}$ is an $N \times M$ matrix of constants. Extend various of the results of Section 5.10a (to the case where $\mathbf{V}_y$ may be singular) by using Theorem 3.5.11 to show (1) that $\boldsymbol{\eta}(\mathbf{y})$ is the best linear predictor of $\mathbf{w}$ in the sense that the difference between the matrix $\mathrm{E}\{[\mathbf{t}(\mathbf{y}) - \mathbf{w}][\mathbf{t}(\mathbf{y}) - \mathbf{w}]'\}$ and the matrix $\mathrm{var}[\boldsymbol{\eta}(\mathbf{y}) - \mathbf{w}]$ [which is the MSE matrix of $\boldsymbol{\eta}(\mathbf{y})$] equals the matrix $\mathrm{E}\{[\mathbf{t}(\mathbf{y}) - \boldsymbol{\eta}(\mathbf{y})][\mathbf{t}(\mathbf{y}) - \boldsymbol{\eta}(\mathbf{y})]'\}$, which is nonnegative definite and which equals $\mathbf{0}$ if and only if $\mathbf{t}(\mathbf{y}) = \boldsymbol{\eta}(\mathbf{y})$ for every value of $\mathbf{y}$ such that $\mathbf{y} - \boldsymbol{\mu}_y \in \mathcal{C}(\mathbf{V}_y)$, (2) that $\Pr[\mathbf{y} - \boldsymbol{\mu}_y \in \mathcal{C}(\mathbf{V}_y)] = 1$, and (3) that $\mathrm{var}[\boldsymbol{\eta}(\mathbf{y}) - \mathbf{w}] = \mathbf{V}_w - \mathbf{V}_{yw}'\mathbf{V}_y^{-}\mathbf{V}_{yw}$.

Exercise 27. Suppose that $\mathbf{y}$ is an $N \times 1$ observable random vector that follows a G–M model, and take $\mathbf{w}$ to be an $M \times 1$ unobservable random vector whose value is to be predicted. Suppose further that $E(\mathbf{w})$ is of the form $E(\mathbf{w}) = \boldsymbol{\Lambda}'\boldsymbol{\beta}$ (where $\boldsymbol{\Lambda}$ is a matrix of known constants) and that $\mathrm{cov}(\mathbf{y}, \mathbf{w})$ is of the form $\mathrm{cov}(\mathbf{y}, \mathbf{w}) = \sigma^2 \mathbf{H}_{yw}$ (where $\mathbf{H}_{yw}$ is a known matrix). Let $\boldsymbol{\tau} = (\boldsymbol{\Lambda}' - \mathbf{H}_{yw}'\mathbf{X})\boldsymbol{\beta}$, denote by $\tilde{\mathbf{w}}(\mathbf{y})$ an arbitrary predictor (of $\mathbf{w}$), and define $\tilde{\boldsymbol{\tau}}(\mathbf{y}) = \tilde{\mathbf{w}}(\mathbf{y}) - \mathbf{H}_{yw}'\mathbf{y}$. Verify that $\tilde{\mathbf{w}}(\mathbf{y})$ is a translation-equivariant predictor (of $\mathbf{w}$) if and only if $\tilde{\boldsymbol{\tau}}(\mathbf{y})$ is a translation-equivariant estimator of $\boldsymbol{\tau}$.

Bibliographic and Supplementary Notes

§2. In some presentations, the use of the term translation (or location) invariance is extended to include what is herein referred to as translation equivariance.

§3e. For an extensive (book-length) discussion of mixture data, refer to Cornell (2002).

§4c. My acquaintance with the term conjugate normal equations came through some class notes authored by Oscar Kempthorne.

§4d. For a discussion of projections that is considerably more extensive and at a somewhat more general level than that provided herein, refer, for example, to Harville (1997, chaps. 12 and 17).

§7a. For a relatively extensive discussion of the vec and vech operations and of Kronecker products, refer, for example, to Chapter 16 of Harville's (1997) book and to the references cited therein.

§7c. Justification for referring to the estimator (7.44) as the Hodges–Lehmann estimator is provided by results presented by Hodges and Lehmann in their 1951 paper. Refer to the expository note by David (2009) for some discussion of a historical nature that relates to the statistical independence of the residual sum of squares and a least squares estimator.

§7d. The results in this subsection that pertain to the minimum-variance quadratic unbiased translation-invariant estimation of σ^2 (and the related results that are the subject of Exercise 14) are variations on the results of Atiqullah (1962) on the minimum-variance quadratic unbiased nonnegative-definite estimation of σ^2, which are covered by Ravishanker and Dey (2002) in their Section 4.4—Exercise 13 serves to relate the minimum-variance quadratic unbiased nonnegative-definite estimation of σ^2 to the minimum-variance quadratic unbiased translation-invariant estimation of σ^2.

§9b. REML originated with the work of Patterson and R. Thompson (1971)—while related ideas can be found in earlier work by others (e.g., W. A. Thompson, Jr. 1962), it was Patterson and R. Thompson who provided the kind of substantive development that was needed for REML to become a viable alternative to ordinary ML. The discussion of maximal invariants is based on results presented (in a more general context) in Section 6.2 of Lehmann and Romano's (2005b) book. Refer to Verbyla (1990) and to LaMotte (2007) for discussion of various matters pertaining to the derivation of expression (9.38) (and related expressions) for the log-likelihood function $\ell(\boldsymbol{\theta};\, \mathbf{R}'\underline{\mathbf{y}})$ employed in REML in making inferences about functions of $\boldsymbol{\theta}$.

§9c. Refer, e.g., to Kollo and von Rosen (2005, table 2.3.1) or Fang, Kotz, and Ng (1990, table 3.1) for a table [originating with Jensen (1985)] that characterizes (in terms of the pdf or the characteristic function) various subclasses of multidimensional spherical distributions.

§10a. The approach (to point prediction) taken in this subsection is essentially the same as that taken in Harville's (1985) paper.

Exercises 7, 8, and 9. For a relatively extensive discussion of generalized inverses that satisfy one or more of Moore–Penrose conditions (2)–(4) [as well as Moore–Penrose condition (1)], including least squares generalized inverses, minimum norm generalized inverses, and the Moore–Penrose inverse itself, refer, e.g., to Harville (1997, chap. 20).

Exercise 20. For some general discussion bearing on the implications of Parts (b) and (c) of Exercise 20, refer to Sprott (1975).

Exercise 25. Exercise 25 is based on results presented by Harville (2003a).

6

Some Relevant Distributions and Their Properties

The multivariate normal distribution was introduced and was discussed extensively in Section 3.5. A broader class of multivariate distributions, comprising so-called elliptical distributions, was considered in Section 5.9c. Numerous results on the first- and second-order moments of linear and quadratic forms (in random vectors) were presented in Chapter 3 and in Section 5.7b.

Knowledge of the multivariate normal distribution and of other elliptical distributions and knowledge of results on the first- and second-order moments of linear and quadratic forms provide a more-or-less adequate background for the discussion of the classical approach to point estimation and prediction, which was the subject of Chapter 5. However, when it comes to extending the results on point estimation and prediction to the construction and evaluation of confidence regions and of test procedures, this knowledge, while still relevant, is far from adequate. It needs to be augmented with a knowledge of the distributions of certain functions of normally distributed random vectors and a knowledge of various related distributions and with a knowledge of the properties of such distributions. It is these distributions and their properties that form the subject matter of the present chapter.

6.1 Chi-Square, Gamma, Beta, and Dirichlet Distributions

Let $\mathbf{z} = (z_1, z_2, \ldots, z_N)'$ represent an $N \times 1$ random vector whose distribution is $N(\mathbf{0}, \mathbf{I})$ or, equivalently, whose N elements are distributed independently and identically as $N(0, 1)$. The sum of squares $\mathbf{z}'\mathbf{z} = \sum_{i=1}^{N} z_i^2$ of the elements of $\mathbf{z}$ has a distribution that is known as a chi-square (or chi-squared) distribution—this distribution depends on the number N, which is referred to as the degrees of freedom of the distribution. Chi-square distributions (and various related distributions) play an important role in making statistical inferences on the basis of linear statistical models. Chi-square distributions belong to a broader class of distributions known as gamma distributions. Gamma distributions, including chi-square distributions, give rise to other important distributions known as beta distributions or, more generally, Dirichlet distributions. Some relevant background is provided in what follows.

a. Gamma distribution

For strictly positive scalars α and β, let $f(\cdot)$ represent the function defined (on the real line) by

$$f(x) = \begin{cases} \dfrac{1}{\Gamma(\alpha)\beta^{\alpha}}\, x^{\alpha-1} e^{-x/\beta}, & \text{for } 0 < x < \infty, \\ 0, & \text{elsewhere.} \end{cases} \tag{1.1}$$

Clearly, $f(x) \geq 0$ for $-\infty < x < \infty$. And $\int_{-\infty}^{\infty} f(x)\, dx = 1$, as is evident upon introducing the change of variable $w = x/\beta$ and recalling the definition—refer to expression (3.5.5)—of the gamma function. Thus, the function $f(\cdot)$ qualifies as a pdf (probability density function). The distribution determined by this pdf is known as the *gamma distribution* (with parameters α and β). Let us use the symbol $Ga(\alpha, \beta)$ to denote this distribution.

If a random variable x has a $Ga(\alpha, \beta)$ distribution, then for any (strictly) positive constant c,

$$cx \sim Ga(\alpha, c\beta), \tag{1.2}$$

as can be readily verified.

Suppose that two random variables w_1 and w_2 are distributed independently as $Ga(\alpha_1, \beta)$ and $Ga(\alpha_2, \beta)$, respectively. Let

$$w = w_1 + w_2 \quad \text{and} \quad s = \frac{w_1}{w_1 + w_2}. \tag{1.3}$$

And note that equalities (1.3) define a one-to-one transformation from the rectangular region $\{w_1, w_2 : w_1 > 0,\ w_2 > 0\}$ onto the rectangular region

$$\{w, s : w > 0,\ 0 < s < 1\}. \tag{1.4}$$

Note also that the inverse transformation is defined by

$$w_1 = sw \quad \text{and} \quad w_2 = (1 - s)w.$$

We find that

$$\det\begin{pmatrix} \partial w_1/\partial w & \partial w_1/\partial s \\ \partial w_2/\partial w & \partial w_2/\partial s \end{pmatrix} = \det\begin{pmatrix} s & w \\ 1-s & -w \end{pmatrix} = -w.$$

Let $f(\cdot\,,\cdot)$ represent the pdf of the joint distribution of w and s, and $f_1(\cdot)$ and $f_2(\cdot)$ the pdfs of the distributions of w_1 and w_2, respectively. Then, for values of w and s in the region (1.4),

$$\begin{aligned} f(w, s) &= f_1(sw)\, f_2[(1 - s)w]\, |-w| \\ &= \frac{1}{\Gamma(\alpha_1)\Gamma(\alpha_2)\beta^{\alpha_1+\alpha_2}}\, w^{\alpha_1+\alpha_2-1} s^{\alpha_1-1}(1 - s)^{\alpha_2-1} e^{-w/\beta} \end{aligned}$$

—for values of w and s outside region (1.4), $f(w, s) = 0$.

Clearly,

$$f(w, s) = g(w)\, h(s) \quad \text{(for all values of } w \text{ amd } s\text{)}, \tag{1.5}$$

where $g(\cdot)$ and $h(\cdot)$ are functions defined as follows:

$$g(w) = \begin{cases} \dfrac{1}{\Gamma(\alpha_1 + \alpha_2)\beta^{\alpha_1+\alpha_2}}\, w^{\alpha_1+\alpha_2-1} e^{-w/\beta}, & \text{for } w > 0, \\ 0, & \text{elsewhere,} \end{cases} \tag{1.6}$$

and

$$h(s) = \begin{cases} \dfrac{\Gamma(\alpha_1 + \alpha_2)}{\Gamma(\alpha_1)\Gamma(\alpha_2)}\, s^{\alpha_1-1}(1 - s)^{\alpha_2-1}, & \text{for } 0 < s < 1, \\ 0, & \text{elsewhere.} \end{cases} \tag{1.7}$$

The function $g(\cdot)$ is seen to be the pdf of the $Ga(\alpha_1 + \alpha_2, \beta)$ distribution. And because $h(s) \geq 0$ (for every value of s) and because

$$\int_{-\infty}^{\infty} h(s)\, ds = \int_{-\infty}^{\infty} h(s)\, ds \int_{-\infty}^{\infty} g(w)\, dw = \int_{-\infty}^{\infty}\int_{-\infty}^{\infty} f(w, s)\, dw\, ds = 1, \tag{1.8}$$

$h(\cdot)$, like $g(\cdot)$, is a pdf; it is the pdf of the distribution of the random variable $s = w_1/(w_1 + w_2)$. Moreover, the random variables w and s are distributed independently.

Based on what has been established, we have the following result.

Theorem 6.1.1. If two random variables w_1 and w_2 are distributed independently as $Ga(\alpha_1, \beta)$ and $Ga(\alpha_2, \beta)$, respectively, then (1) $w_1 + w_2$ is distributed as $Ga(\alpha_1 + \alpha_2, \beta)$ and (2) $w_1 + w_2$ is distributed independently of $w_1/(w_1 + w_2)$.

By employing a simple induction argument, we can establish the following generalization of the first part of Theorem 6.1.1.

Theorem 6.1.2. If N random variables $w_1, w_2, \ldots, w_N$ are distributed independently as $Ga(\alpha_1, \beta), Ga(\alpha_2, \beta), \ldots, Ga(\alpha_N, \beta)$, respectively, then $w_1 + w_2 + \cdots + w_N$ is distributed as $Ga(\alpha_1 + \alpha_2 + \cdots + \alpha_N, \beta)$.

b. Beta distribution and function

The distribution with pdf $h(\cdot)$ defined by expression (1.7) is known as the *beta distribution* (with parameters α_1 and α_2) and is of interest in its own right. Let us use the symbol $Be(\alpha_1, \alpha_2)$ to denote this distribution.

Take $B(\cdot\,, \cdot)$ to be the function (whose domain consists of the coordinates of the points in the first quadrant of the plane) defined by

$$B(y, z) = \int_0^1 t^{y-1}(1-t)^{z-1}\, dt \qquad (y > 0,\ z > 0). \tag{1.9}$$

This function is known as the *beta function*. The beta function is expressible in terms of the gamma function:

$$B(y, z) = \frac{\Gamma(y)\Gamma(z)}{\Gamma(y+z)}, \tag{1.10}$$

as is evident from result (1.8).

Note that the pdf $h(\cdot)$ of the $Be(\alpha_1, \alpha_2)$ distribution can be reexpressed in terms of the beta function. We have that

$$h(s) = \begin{cases} \dfrac{1}{B(\alpha_1, \alpha_2)} s^{\alpha_1 - 1}(1-s)^{\alpha_2 - 1}, & \text{for } 0 < s < 1, \\ 0, & \text{elsewhere.} \end{cases} \tag{1.11}$$

For $0 \le x \le 1$ and for $y > 0$ and $z > 0$, define

$$I_x(y, z) = \frac{1}{B(y, z)} \int_0^x t^{y-1}(1-t)^{z-1}\, dt. \tag{1.12}$$

The function $I_x(\cdot\,, \cdot)$ is known as the *incomplete beta function ratio*. Clearly, the function $F(\cdot)$ defined (in terms of the incomplete beta function ratio) by

$$F(x) = I_x(\alpha_1, \alpha_2) \qquad (0 \le x \le 1) \tag{1.13}$$

coincides with the cdf (cumulative distribution function) of the $Be(\alpha_1, \alpha_2)$ distribution over the interval $0 \le x \le 1$.

Upon making the change of variable $r = 1 - t$ in expression (1.12), we find that

$$\begin{aligned} I_x(y, z) &= \frac{1}{B(z, y)} \int_{1-x}^1 r^{z-1}(1-r)^{y-1}\, dr \\ &= \frac{1}{B(z, y)} \left[B(z, y) - \int_0^{1-x} r^{z-1}(1-r)^{y-1}\, dr \right]. \end{aligned}$$

Thus,

$$I_x(y, z) = 1 - I_{1-x}(z, y). \tag{1.14}$$

c. Chi-square distribution: definition and relationship to the gamma distribution

Let $\mathbf{z} = (z_1, z_2, \ldots, z_N)'$ represent an N-dimensional random (column) vector whose distribution is $N(\mathbf{0}, \mathbf{I})$ or, equivalently, whose elements are distributed independently and identically as $N(0, 1)$. The distribution of the sum of squares $\mathbf{z}'\mathbf{z} = \sum_{i=1}^{N} z_i^2$ is known as the *chi-square distribution*. The positive integer N is regarded as a parameter of the chi-square distribution—there is a family of chi-square distributions, one for each value of N. It is customary to refer to the parameter N as the *degrees of freedom* of the distribution.

Let us find the pdf of the distribution of $\mathbf{z}'\mathbf{z}$. It is fruitful to begin by finding the pdf of the distribution of z^2, where z is a random variable whose distribution is $N(0, 1)$. Define $u = z^2$. And denote by $f(\cdot)$ the pdf of the $N(0, 1)$ distribution, and by $h(\cdot)$ the pdf of the distribution of u. Note that there are two values of z that give rise to each nonzero value of u, namely, $z = \pm\sqrt{u}$. Note also that $d\sqrt{u}/du = \bigl(2\sqrt{u}\bigr)^{-1}$. Accordingly, using standard techniques for a change of variable (e.g., Casella and Berger 2002, sec. 2.1), we find that, for $u > 0$,

$$h(u) = f\bigl(-\sqrt{u}\bigr)\left|-\bigl(2\sqrt{u}\bigr)^{-1}\right| + f\bigl(\sqrt{u}\bigr)\bigl(2\sqrt{u}\bigr)^{-1} = \frac{1}{\sqrt{2\pi u}}\,e^{-u/2}$$

and that, for $u \le 0$, $h(u) = 0$.

Upon recalling that $\Gamma\bigl(\tfrac{1}{2}\bigr) = \sqrt{\pi}$, we find that the pdf $h(\cdot)$ is identical to the pdf of the $Ga\bigl(\tfrac{1}{2}, 2\bigr)$ distribution. Thus, $z^2 \sim Ga\bigl(\tfrac{1}{2}, 2\bigr)$, so that $z_1^2, z_2^2, \ldots, z_N^2$ are distributed independently—functions of statistically independent random variables are statistically independent—and identically as $Ga\bigl(\tfrac{1}{2}, 2\bigr)$. And upon applying Theorem 6.1.2, it follows that

$$\mathbf{z}'\mathbf{z} = \sum_{i=1}^{N} z_i^2 \sim Ga\bigl(\tfrac{N}{2}, 2\bigr). \tag{1.15}$$

Instead of defining the chi-square distribution (with N degrees of freedom) in terms of the N-variate standard normal distribution (as was done herein where it was defined to be the distribution of $\mathbf{z}'\mathbf{z}$), some authors define it directly in terms of the gamma distribution. Specifically, they define the chi-square distribution with N degrees of freedom to be the special case of the $Ga(\alpha, \beta)$ distribution where $\alpha = N/2$ and $\beta = 2$. In their approach, the derivation of the distribution of $\mathbf{z}'\mathbf{z}$ serves a different purpose; it serves to establish what is no longer true by definition, namely, that $\mathbf{z}'\mathbf{z}$ has a chi-square distribution with N degrees of freedom. The two approaches (their approach and the one taken herein) can be regarded as different means to the same end. In either case, the chi-square distribution with N degrees of freedom is the distribution with a pdf, say $g(\cdot)$, that is the pdf of the $Ga\bigl(\tfrac{N}{2}, 2\bigr)$ distribution and that is expressible as follows:

$$g(x) = \begin{cases} \dfrac{1}{\Gamma(N/2)2^{N/2}}\,x^{(N/2)-1}e^{-x/2}, & \text{for } 0 < x < \infty, \\ 0, & \text{elsewhere.} \end{cases} \tag{1.16}$$

Let us write $\chi^2(N)$ for a chi-square distribution with N degrees of freedom. Here, N is assumed to be an integer. Reference is sometimes made to a chi-square distribution with noninteger (but strictly positive) degrees of freedom N. Unless otherwise indicated, such a reference is to be interpreted as a reference to the $Ga\bigl(\tfrac{N}{2}, 2\bigr)$ distribution; this interpretation is that obtained when the relationship between the $\chi^2(N)$ and $Ga\bigl(\tfrac{N}{2}, 2\bigr)$ distributions is regarded as extending to noninteger values of N.

In light of the relationship of the chi-square distribution to the gamma distribution, various results on the gamma distribution can be translated into results on the chi-square distribution. In particular, if a random variable x has a $\chi^2(N)$ distribution and if c is a (strictly) positive constant, then it follows from result (1.2) that

$$cx \sim Ga\bigl(\tfrac{N}{2}, 2c\bigr). \tag{1.17}$$

The following result is an immediate consequence of Theorem 6.1.2.

Theorem 6.1.3. If K random variables $u_1, u_2, \ldots, u_K$ are distributed independently as $\chi^2(N_1), \chi^2(N_2), \ldots, \chi^2(N_K)$, respectively, then $u_1 + u_2 + \cdots + u_K$ is distributed as $\chi^2(N_1 + N_2 + \cdots + N_K)$.

d. Moment generating function, moments, and cumulants of the gamma and chi-square distributions

Let w represent a random variable whose distribution is $Ga(\alpha, \beta)$. And denote by $f(\cdot)$ the pdf of the $Ga(\alpha, \beta)$ distribution. Further, let u represent a random variable whose distribution is $\chi^2(N)$.

For $t < 1/\beta$,

$$\begin{aligned} \mathrm{E}(e^{tw}) &= \int_0^\infty e^{tx} f(x)\, dx \\ &= \int_0^\infty \frac{1}{\Gamma(\alpha)\beta^\alpha} x^{\alpha-1} e^{-x(1-\beta t)/\beta}\, dx \\ &= \frac{1}{(1-\beta t)^\alpha} \int_0^\infty \frac{1}{\Gamma(\alpha)\xi^\alpha} x^{\alpha-1} e^{-x/\xi}\, dx, \end{aligned} \tag{1.18}$$

where $\xi = \beta/(1-\beta t)$. The integrand of the integral in expression (1.18) equals $g(x)$, where $g(\cdot)$ is the pdf of the $Ga(\alpha, \xi)$ distribution, so that the integral equals 1. Thus, the mgf (moment generating function), say $m(\cdot)$, of the $Ga(\alpha, \beta)$ distribution is

$$m(t) = (1-\beta t)^{-\alpha} \qquad (t < 1/\beta). \tag{1.19}$$

As a special case of result (1.19), we have that the mgf, say $m(\cdot)$, of the $\chi^2(N)$ distribution is

$$m(t) = (1-2t)^{-N/2} \qquad (t < 1/2). \tag{1.20}$$

For $r > -\alpha$,

$$\begin{aligned} \mathrm{E}(w^r) &= \int_0^\infty x^r f(x)\, dx \\ &= \int_0^\infty \frac{1}{\Gamma(\alpha)\beta^\alpha} x^{\alpha+r-1} e^{-x/\beta}\, dx \\ &= \frac{\beta^r \Gamma(\alpha+r)}{\Gamma(\alpha)} \int_0^\infty \frac{1}{\Gamma(\alpha+r)\beta^{\alpha+r}} x^{\alpha+r-1} e^{-x/\beta}\, dx. \end{aligned} \tag{1.21}$$

The integrand of the integral in expression (1.21) equals $g(x)$, where $g(\cdot)$ is the pdf of the $Ga(\alpha+r, \beta)$ distribution, so that the integral equals 1. Thus, for $r > -\alpha$,

$$\mathrm{E}(w^r) = \beta^r \frac{\Gamma(\alpha+r)}{\Gamma(\alpha)}. \tag{1.22}$$

The gamma function $\Gamma(\cdot)$ is such that, for $x > 0$ and for any positive integer r,

$$\Gamma(x+r) = (x+r-1)\cdots(x+2)(x+1)x\Gamma(x), \tag{1.23}$$

as is evident upon the repeated application of result (3.5.6). In light of result (1.23), it follows from formula (1.22) that the rth (positive, integer-valued) moment of the $Ga(\alpha, \beta)$ distribution is

$$\mathrm{E}(w^r) = \beta^r \alpha(\alpha+1)(\alpha+2)\cdots(\alpha+r-1). \tag{1.24}$$

Thus, the mean and variance of the $Ga(\alpha, \beta)$ distribution are

$$\mathrm{E}(w) = \alpha\beta \tag{1.25}$$

and

$$\text{var}(w) = \beta^2\alpha(\alpha + 1) - (\alpha\beta)^2 = \alpha\beta^2. \tag{1.26}$$

Upon setting $\alpha = N/2$ and $\beta = 2$ in expression (1.22), we find that (for $r > -N/2$)

$$\text{E}(u^r) = 2^r \frac{\Gamma[(N/2) + r]}{\Gamma(N/2)}. \tag{1.27}$$

Further, the rth (positive, integer-valued) moment of the $\chi^2(N)$ distribution is

$$\begin{aligned}\text{E}(u^r) &= 2^r(N/2)[(N/2) + 1][(N/2) + 2]\cdots[(N/2) + r - 1] \\ &= N(N + 2)(N + 4)\cdots[N + 2(r - 1)]. \end{aligned} \tag{1.28}$$

And the mean and variance of the $\chi^2(N)$ distribution are

$$\text{E}(u) = N \tag{1.29}$$

and

$$\text{var}(u) = 2N. \tag{1.30}$$

Upon applying formula (1.27) [and making use of result (1.23)], we find that the rth moment of the reciprocal of a chi-square random variable (with N degrees of freedom) is (for $r = 1, 2, \ldots < N/2$)

$$\begin{aligned}\text{E}(u^{-r}) &= 2^{-r} \frac{\Gamma[(N/2) - r]}{\Gamma(N/2)} \\ &= \{2^r[(N/2) - 1][(N/2) - 2]\cdots[(N/2) - r]\}^{-1} \\ &= [(N - 2)(N - 4)\cdots(N - 2r)]^{-1}. \end{aligned} \tag{1.31}$$

In particular, we find that (for $N > 2$)

$$\text{E}\left(\frac{1}{u}\right) = \frac{1}{N - 2}. \tag{1.32}$$

Upon taking the logarithm of the mgf $m(\cdot)$ of the $Ga(\alpha, \beta)$ distribution, we obtain the cumulant generating function, say $c(\cdot)$, of this distribution—refer, e.g., to Bickel and Doksum (2001, sec. A.12) for an introduction to cumulants and cumulant generating functions. In light of result (1.19),

$$c(t) = -\alpha \log(1 - \beta t) \qquad (t < 1/\beta). \tag{1.33}$$

And, for $-1/\beta \le t < 1/\beta$,

$$c(t) = \alpha \sum_{r=1}^{\infty} (t\beta)^r/r = \sum_{r=1}^{\infty} \alpha\beta^r(r - 1)!\, t^r/r!.$$

Thus, the rth cumulant of the $Ga(\alpha, \beta)$ distribution is

$$\alpha\beta^r(r - 1)!. \tag{1.34}$$

As a special case, we have that the rth cumulant of the $\chi^2(N)$ distribution is

$$N2^{r-1}(r - 1)!. \tag{1.35}$$

e. Dirichlet distribution

Let $w_1, w_2, \ldots, w_K$, and w_{K+1} represent $K+1$ random variables that are distributed independently as $Ga(\alpha_1, \beta), Ga(\alpha_2, \beta), \ldots, Ga(\alpha_K, \beta)$, and $Ga(\alpha_{K+1}, \beta)$, respectively. And consider the joint distribution of the $K+1$ random variables w and $s_1, s_2, \ldots, s_K$ defined as follows:

$$w = \sum_{k=1}^{K+1} w_k \qquad \text{and} \qquad s_k = \frac{w_k}{\sum_{k'=1}^{K+1} w_{k'}} \quad (k = 1, 2, \ldots, K). \tag{1.36}$$

A derivation of the pdf of the joint distribution of w and $s_1, s_2, \ldots, s_K$ was presented for the special case where $K = 1$ in Subsection a. Let us extend that derivation to the general case (where K is an arbitrary positive integer).

The $K+1$ equalities (1.36) define a one-to-one transformation from the rectangular region $\{w_1, w_2, \ldots, w_K, w_{K+1} : w_k > 0 \ (k = 1, 2, \ldots, K, K+1)\}$ onto the region

$$\{w, s_1, s_2, \ldots, s_K : w > 0,\ s_k > 0\ (k = 1, 2, \ldots, K),\ \textstyle\sum_{k=1}^{K} s_k < 1\}. \tag{1.37}$$

The inverse transformation is defined by

$$w_k = s_k w \quad (k = 1, 2, \ldots, K) \qquad \text{and} \qquad w_{K+1} = \big(1 - \textstyle\sum_{k=1}^{K} s_k\big) w.$$

For $j, k = 1, 2, \ldots, K$,

$$\frac{\partial w_j}{\partial s_k} = \begin{cases} w, & \text{if } k = j, \\ 0, & \text{if } k \neq j. \end{cases}$$

Further, $\dfrac{\partial w_j}{\partial w} = s_j \ (j = 1, 2, \ldots, K)$, $\dfrac{\partial w_{K+1}}{\partial s_k} = -w \ (k = 1, 2, \ldots, K)$, and $\dfrac{\partial w_{K+1}}{\partial w} = 1 - \sum_{k=1}^{K} s_k$. Thus, letting $\mathbf{s} = (s_1, s_2, \ldots, s_K)'$ and making use of Theorem 2.14.9, formula (2.14.29), and Corollary 2.14.2, we find that

$$\begin{aligned}
\begin{vmatrix}
\frac{\partial w_1}{\partial w} & \frac{\partial w_1}{\partial s_1} & \frac{\partial w_1}{\partial s_2} & \cdots & \frac{\partial w_1}{\partial s_K} \\
\frac{\partial w_2}{\partial w} & \frac{\partial w_2}{\partial s_1} & \frac{\partial w_2}{\partial s_2} & \cdots & \frac{\partial w_2}{\partial s_K} \\
\vdots & \vdots & \vdots & \ddots & \vdots \\
\frac{\partial w_K}{\partial w} & \frac{\partial w_K}{\partial s_1} & \frac{\partial w_K}{\partial s_2} & \cdots & \frac{\partial w_K}{\partial s_K} \\
\frac{\partial w_{K+1}}{\partial w} & \frac{\partial w_{K+1}}{\partial s_1} & \frac{\partial w_{K+1}}{\partial s_2} & \cdots & \frac{\partial w_{K+1}}{\partial s_K}
\end{vmatrix}
&= \begin{vmatrix} \mathbf{s} & w\mathbf{I} \\ 1 - \sum_{k=1}^{K} s_k & -w\mathbf{1}' \end{vmatrix} \\
&= (-1)^K \begin{vmatrix} w\mathbf{I} & \mathbf{s} \\ -w\mathbf{1}' & 1 - \sum_{k=1}^{K} s_k \end{vmatrix} \\
&= (-1)^K w^K \big(1 - \textstyle\sum_{k=1}^{K} s_k + \sum_{k=1}^{K} s_k\big) \\
&= (-w)^K.
\end{aligned}$$

Now, let $f(\cdot, \cdot, \cdot, \ldots, \cdot)$ represent the pdf of the joint distribution of w and $s_1, s_2, \ldots, s_K$, define $\alpha = \sum_{k=1}^{K+1} \alpha_k$, and, for $k = 1, 2, \ldots, K$, denote by $f_k(\cdot)$ the pdf of the distribution of w_k. Then, for $w, s_1, s_2, \ldots, s_K$ in the region (1.37),

$$\begin{aligned}
&f(w, s_1, s_2, \ldots, s_K) \\
&\quad = f_1(s_1 w) f_2(s_2 w) \cdots f_K(s_K w) f_{K+1}\big[\big(1 - \textstyle\sum_{k=1}^{K} s_k\big) w\big] \, |(-w)^K| \\
&\quad = \frac{1}{\Gamma(\alpha_1)\Gamma(\alpha_2)\cdots\Gamma(\alpha_K)\Gamma(\alpha_{K+1})\beta^{\alpha}} \\
&\qquad\qquad \times w^{\alpha-1} s_1^{\alpha_1-1} s_2^{\alpha_2-1} \cdots s_K^{\alpha_K-1} \big(1 - \textstyle\sum_{k=1}^{K} s_k\big)^{\alpha_{K+1}-1} e^{-w/\beta}
\end{aligned}$$

—for $w, s_1, s_2, \ldots, s_K$ outside region (1.37), $f(w, s_1, s_2, \ldots, s_K) = 0$.

Clearly,

$$f(w, s_1, s_2, \ldots, s_K) = g(w)h(s_1, s_2, \ldots, s_K) \quad \text{(for all } w \text{ and } s_1, s_2, \ldots, s_K\text{)},$$

where

$$g(w) = \begin{cases} \dfrac{1}{\Gamma(\alpha)\beta^{\alpha}}\, w^{\alpha-1}e^{-w/\beta}, & \text{for } 0 < w < \infty, \\ 0, & \text{elsewhere,} \end{cases}$$

and, letting $S = \{(s_1, s_2, \ldots, s_K) : s_k > 0\ (k = 1, 2, \ldots, K),\ \sum_{k=1}^{K} s_k < 1\}$,

$$h(s_1, s_2, \ldots, s_K) = \begin{cases} \dfrac{\Gamma(\alpha_1+\alpha_2+\cdots+\alpha_K+\alpha_{K+1})}{\Gamma(\alpha_1)\Gamma(\alpha_2)\cdots\Gamma(\alpha_K)\Gamma(\alpha_{K+1})} \\ \qquad \times\, s_1^{\alpha_1-1}s_2^{\alpha_2-1}\cdots s_K^{\alpha_K-1}\left(1-\sum_{k=1}^{K} s_k\right)^{\alpha_{K+1}-1}, \\ \qquad\qquad \text{for } (s_1, s_2, \ldots, s_K) \in S, \\ 0, \qquad\qquad \text{elsewhere.} \end{cases} \tag{1.38}$$

The function $g(\cdot)$ is seen to be the pdf of the $Ga(\sum_{k=1}^{K+1} \alpha_k,\ \beta)$ distribution. And because $h(s_1, s_2, \ldots, s_K) \geq 0$ for all $s_1, s_2, \ldots, s_K$ and because

$$\begin{aligned} &\int_{-\infty}^{\infty}\int_{-\infty}^{\infty}\cdots\int_{-\infty}^{\infty} h(s_1, s_2, \ldots, s_K)\, ds_1 ds_2 \ldots ds_K \\ &\quad = \int_{-\infty}^{\infty}\int_{-\infty}^{\infty}\cdots\int_{-\infty}^{\infty} h(s_1, s_2, \ldots, s_K)\, ds_1 ds_2 \ldots ds_K \int_{-\infty}^{\infty} g(w)\, dw \\ &\quad = \int_{-\infty}^{\infty}\int_{-\infty}^{\infty}\int_{-\infty}^{\infty}\cdots\int_{-\infty}^{\infty} f(w, s_1, s_2, \ldots, s_K)\, dw ds_1 ds_2 \ldots ds_K = 1, \end{aligned}$$

$h(\cdot, \cdot, \ldots, \cdot)$, like $g(\cdot)$, is a pdf; it is the pdf of the joint distribution of the K random variables $s_k = w_k / \sum_{k'=1}^{K+1} w_{k'}$ $(k = 1, 2, \ldots, K)$. Moreover, w is distributed independently of $s_1, s_2, \ldots, s_K$.

Based on what has been established, we have the following generalization of Theorem 6.1.1.

Theorem 6.1.4. If $K+1$ random variables $w_1, w_2, \ldots, w_K, w_{K+1}$ are distributed independently as $Ga(\alpha_1,\ \beta), Ga(\alpha_2,\ \beta), \ldots, Ga(\alpha_K,\ \beta), Ga(\alpha_{K+1},\ \beta)$, respectively, then (1) $\sum_{k=1}^{K+1} w_k$ is distributed as $Ga(\sum_{k=1}^{K+1} \alpha_k,\ \beta)$ and (2) $\sum_{k=1}^{K+1} w_k$ is distributed independently of $w_1/\sum_{k=1}^{K+1} w_k,\ w_2/\sum_{k=1}^{K+1} w_k, \ldots,\ w_K/\sum_{k=1}^{K+1} w_k$.

Note that Part (1) of Theorem 6.1.4 is essentially a restatement of a result established earlier (in the form of Theorem 6.1.2) via a mathematical induction argument.

The joint distribution of the K random variables $s_1, s_2, \ldots, s_K$, the pdf of which is the function $h(\cdot, \cdot, \ldots, \cdot)$ defined by expression (1.38), is known as the *Dirichlet distribution* (with parameters $\alpha_1, \alpha_2, \ldots, \alpha_K$, and α_{K+1}). Let us denote this distribution by the symbol $Di(\alpha_1, \alpha_2, \ldots, \alpha_K, \alpha_{K+1}; K)$. The beta distribution is a special case of the Dirichlet distribution; specifically, the $Be(\alpha_1, \alpha_2)$ distribution is identical to the $Di(\alpha_1, \alpha_2; 2)$ distribution.

Some results on the Dirichlet distribution. Some results on the Dirichlet distribution are stated in the form of the following theorem.

Theorem 6.1.5. Take $s_1, s_2, \ldots, s_K$ to be K random variables whose joint distribution is $Di(\alpha_1, \alpha_2, \ldots, \alpha_K, \alpha_{K+1}; K)$, and define $s_{K+1} = 1 - \sum_{k=1}^{K} s_k$. Further, partition the integers $1, \ldots, K, K+1$ into $I+1$ (nonempty) mutually exclusive and exhaustive subsets, say $B_1, \ldots, B_I, B_{I+1}$, of sizes $K_1+1, \ldots, K_I+1, K_{I+1}+1$, respectively, and denote by $\{i_1, i_2, \ldots, i_P\}$ the subset of $\{1, \ldots, I, I+1\}$ consisting of every integer i between 1 and $I+1$, inclusive, for which $K_i \geq 1$. And for $i = 1, \ldots, I, I+1$, let $s_i^* = \sum_{k \in B_i} s_k$ and $\alpha_i^* = \sum_{k \in B_i} \alpha_k$; and for $p = 1, 2, \ldots, P$, let $\mathbf{u}_p$ represent the $K_{i_p} \times 1$ vector whose elements are the first K_{i_p} of the $K_{i_p} + 1$ quantities $s_k / \sum_{k' \in B_{i_p}} s_{k'}$ $(k \in B_{i_p})$. Then,

(1) the $P + 1$ random vectors $\mathbf{u}_1, \mathbf{u}_2, \ldots, \mathbf{u}_P$, and $(s_1^*, \ldots, s_I^*, s_{I+1}^*)'$ are statistically independent;

(2) the joint distribution of $s_1^*, s_2^*, \ldots, s_I^*$ is $Di(\alpha_1^*, \alpha_2^*, \ldots, \alpha_I^*, \alpha_{I+1}^*; I)$; and

(3) for $p = 1, 2, \ldots, P$, the joint distribution of the K_{i_p} elements of $\mathbf{u}_p$ is Dirichlet with parameters α_k $(k \in B_{i_p})$.

Proof. Let w_{ij} $(i = 1, \ldots, I, I+1;\ j = 1, \ldots, K_i, K_i + 1)$ represent statistically independent random variables, and suppose that (for all i and j) $w_{ij} \sim Ga(\alpha_{ij}, \beta)$, where α_{ij} is the jth of the $K_i + 1$ parameters α_k $(k \in B_i)$. Further, let $\mathbf{w}_i = (w_{i1}, \ldots, w_{iK_i}, w_{i,K_i+1})'$. And observe that in light of the very definition of the Dirichlet distribution, it suffices (for purposes of the proof) to set the $K_i + 1$ random variables s_k $(k \in B_i)$ equal to $w_{ij} / \sum_{i'=1}^{I+1} \sum_{j'=1}^{K_{i'}+1} w_{i'j'}$ $(j = 1, \ldots, K_i, K_i + 1)$, respectively $(i = 1, \ldots, I, I+1)$. Upon doing so, we find that

$$s_i^* = \sum_{j=1}^{K_i+1} w_{ij} \Big/ \sum_{i'=1}^{I+1} \sum_{j'=1}^{K_{i'}+1} w_{i'j'} \qquad (i = 1, \ldots, I, I+1) \tag{1.39}$$

and that (for $p = 1, 2, \ldots, P$) the K_{i_p} elements of the vector $\mathbf{u}_p$ are $u_{p1}, u_{p2}, \ldots, u_{pK_{i_p}}$, where (for $j = 1, 2, \ldots, K_{i_p}$)

$$u_{pj} = \frac{w_{i_p j} \Big/ \sum_{i'=1}^{I+1} \sum_{j'=1}^{K_{i'}+1} w_{i'j'}}{\sum_{j'=1}^{K_{i_p}+1} w_{i_p j'} \Big/ \sum_{i'=1}^{I+1} \sum_{j'=1}^{K_{i'}+1} w_{i'j'}} = \frac{w_{i_p j}}{\sum_{j'=1}^{K_{i_p}+1} w_{i_p j'}}. \tag{1.40}$$

Part (3) of Theorem 6.1.5 follows immediately from result (1.40). And upon observing that the I+1 sums $\sum_{j=1}^{K_i+1} w_{ij}$ $(i = 1, \ldots, I, I$+1) are statistically independent and observing also [in light of Theorem 6.1.2 or Part (1) of Theorem 6.1.4] that (for $i = 1, \ldots, I, I$+1) $\sum_{j=1}^{K_i+1} w_{ij} \sim Ga(\alpha_i^*, \beta)$, Part (2) follows from result (1.39).

It remains to verify Part (1). Let $\mathbf{y} = (y_1, \ldots, y_I, y_{I+1})'$, where (for $i = 1, \ldots, I, I+1$) $y_i = \sum_{j=1}^{K_i+1} w_{ij}$, and consider a change of variables from the elements of the $I + 1$ (statistically independent) random vectors $\mathbf{w}_1, \ldots, \mathbf{w}_I, \mathbf{w}_{I+1}$ to the elements of the $P + 1$ random vectors $\mathbf{u}_1, \mathbf{u}_2, \ldots, \mathbf{u}_P$, and $\mathbf{y}$. We find that the pdf, say $f(\cdot, \cdot, \ldots, \cdot, \cdot)$ of $\mathbf{u}_1, \mathbf{u}_2, \ldots, \mathbf{u}_P$, and $\mathbf{y}$ is expressible as

$$f(\mathbf{u}_1, \mathbf{u}_2, \ldots, \mathbf{u}_P, \mathbf{y}) = \prod_{i=1}^{I+1} g_i(y_i) \prod_{p=1}^{P} h_p(\mathbf{u}_p), \tag{1.41}$$

where $g_i(\cdot)$ is the pdf of the $Ga(\alpha_i^*, \beta)$ distribution and $h_p(\cdot)$ is the pdf of the $Di(\alpha_{i_p 1}, \ldots, \alpha_{i_p K_{i_p}}, \alpha_{i_p, K_{i_p}+1}; K_{i_p})$ distribution. And upon observing that $s_1^*, \ldots, s_I^*, s_{I+1}^*$ are expressible as functions of $\mathbf{y}$, it follows from result (1.41) that $\mathbf{u}_1, \mathbf{u}_2, \ldots, \mathbf{u}_P$, and $(s_1^*, \ldots, s_I^*, s_{I+1}^*)'$ are statistically independent. Q.E.D.

Marginal distributions. Define $s_1, s_2, \ldots, s_K$, and s_{K+1} as in Theorem 6.1.5; that is, take $s_1, s_2, \ldots, s_K$ to be K random variables whose joint distribution is $Di(\alpha_1, \alpha_2, \ldots, \alpha_K, \alpha_{K+1}; K)$, and let $s_{K+1} = 1 - \sum_{k=1}^{K} s_k$. And, taking I to be an arbitrary integer between 1 and K, inclusive, consider the joint distribution of any I elements of the set $\{s_1, \ldots, s_K, s_{K+1}\}$, say the $k_1, k_2, \ldots, k_I$th elements $s_{k_1}, s_{k_2}, \ldots, s_{k_I}$.

The joint distribution of $s_{k_1}, s_{k_2}, \ldots, s_{k_I}$ can be readily determined by applying Part (2) of Theorem 6.1.5 (in the special case where $B_1 = \{k_1\}$, $B_2 = \{k_2\}, \ldots, B_I = \{k_I\}$ and where B_{I+1} is the $(K+1-I)$-dimensional subset of $\{1, \ldots, K, K+1\}$ obtained by striking out $k_1, k_2, \ldots, k_I$). The joint distribution is $Di(\alpha_{k_1}, \alpha_{k_2}, \ldots, \alpha_{k_I}, \alpha_{I+1}^*; I)$, where $\alpha_{I+1}^* = \sum_{k \in B_{I+1}} \alpha_k$. In particular, for an arbitrary one of the integers $1, \ldots, K, K+1$, say the integer k, the (marginal) distribution of s_k is $Di(\alpha_k, \sum_{k'=1\,(k' \neq k)}^{K+1} \alpha_{k'}; 1)$ or, equivalently, $Be(\alpha_k, \sum_{k'=1\,(k' \neq k)}^{K+1} \alpha_{k'})$.

f. Applications to the multivariate standard normal distribution

Let $\mathbf{z} = (z_1, \ldots, z_K, z_{K+1})'$ represent a $(K+1)$-dimensional random (column) vector whose distribution is $(K+1)$-variate standard normal, that is, whose distribution is $N(\mathbf{0}, \mathbf{I}_{K+1})$ or, equivalently, whose elements are distributed independently and identically as $N(0, 1)$. Then, $z_1^2, \ldots, z_K^2$, and z_{K+1}^2 are distributed independently, and each has a $\chi^2(1)$ distribution or, equivalently, a $Ga\big(\frac{1}{2}, 2\big)$ distribution.

Consider the distribution of the $K+1$ random variables $z_1^2/\sum_{k=1}^{K+1} z_k^2, \ldots, z_K^2/\sum_{k=1}^{K+1} z_k^2$, and $\sum_{k=1}^{K+1} z_k^2$. As a special case of Part (2) of Theorem 6.1.4, we have that $\sum_{k=1}^{K+1} z_k^2$ is distributed independently of $z_1^2/\sum_{k=1}^{K+1} z_k^2, \ldots, z_K^2/\sum_{k=1}^{K+1} z_k^2$. And as a special case of Theorem 6.1.3 [or of Part (1) of Theorem 6.1.4], we have that $\sum_{k=1}^{K+1} z_k^2 \sim \chi^2(K+1)$. Moreover, $z_1^2/\sum_{k=1}^{K+1} z_k^2, \ldots, z_K^2/\sum_{k=1}^{K+1} z_k^2$ have a $Di\big(\frac{1}{2}, \ldots, \frac{1}{2}, \frac{1}{2}; K\big)$ distribution, and, more generally, any K' (where $1 \le K' \le K$) of the random variables $z_1^2/\sum_{k=1}^{K+1} z_k^2, \ldots, z_K^2/\sum_{k=1}^{K+1} z_k^2, z_{K+1}^2/\sum_{k=1}^{K+1} z_k^2$ have a $Di\big(\frac{1}{2}, \ldots, \frac{1}{2}, \frac{K+1-K'}{2}; K'\big)$ distribution.

Now, let

$$u = \sum_{k=1}^{K+1} z_k^2 \quad \text{and (for } k = 1, \ldots, K, K+1) \quad y_k = \frac{z_k}{\big(\sum_{j=1}^{K+1} z_j^2\big)^{1/2}},$$

and consider the joint distribution of u and any K' (where $1 \le K' \le K$) of the random variables $y_1, \ldots, y_K, y_{K+1}$ (which for notational convenience and without any essential loss of generality are taken to be the first K' of these random variables). Let us reexpress u and $y_1, y_2, \ldots, y_{K'}$ as

$$u = v + \sum_{k=1}^{K'} z_k^2 \quad \text{and (for } k = 1, 2, \ldots, K') \quad y_k = \frac{z_k}{\big(v + \sum_{j=1}^{K'} z_j^2\big)^{1/2}}, \tag{1.42}$$

where $v = \sum_{k=K'+1}^{K+1} z_k^2$. Clearly, v is distributed independently of $z_1, z_2, \ldots, z_{K'}$ as $\chi^2(K-K'+1)$.

Define $\mathbf{y}_* = (y_1, y_2, \ldots, y_{K'})'$. And observe that the $K'+1$ equalities (1.42) define a one-to-one transformation from the $(K'+1)$-dimensional region defined by the $K'+1$ inequalities $0 < v < \infty$ and $-\infty < z_k < \infty$ $(k = 1, 2, \ldots, K')$ onto the region

$$\{u, \mathbf{y}_* \;:\; 0 < u < \infty, \; \mathbf{y}_* \in D^*\},$$

where $D^* = \{\mathbf{y}_* \;:\; \sum_{k=1}^{K'} y_k^2 < 1\}$. Observe also that the inverse of this transformation is the transformation defined by the $K'+1$ equalities

$$v = u\big(1 - \textstyle\sum_{k=1}^{K'} y_k^2\big) \qquad \text{and (for } k = 1, 2, \ldots, K') \qquad z_k = u^{1/2} y_k.$$

Further, letting $\mathbf{A}$ represent the $(K'+1) \times (K'+1)$ matrix whose ijth element is the partial derivative of the ith element of the vector $(v, z_1, z_2, \ldots, z_{K'})$ with respect to the jth element of the vector $(u, y_1, y_2, \ldots, y_{K'})$ and recalling Theorem 2.14.22, we find that

$$|\mathbf{A}| = \begin{vmatrix} 1 - \sum_{j=1}^{K'} y_j^2 & -2u\mathbf{y}_*' \\ (1/2)u^{-1/2}\mathbf{y}_* & u^{1/2}\mathbf{I} \end{vmatrix} = u^{K'/2}.$$

Thus, denoting by $d(\cdot)$ the pdf of the $\chi^2(K-K'+1)$ distribution and by $b(\cdot)$ the pdf of the $N(\mathbf{0}, \mathbf{I}_{K'})$ distribution and making use of standard results on a change of variables, the joint distribution of $u, y_1, y_2, \ldots, y_{K'}$ has as a pdf the function $q(\cdot, \cdot, \cdot, \ldots, \cdot)$ (of K' variables) obtained by taking (for

$0 < u < \infty$ and $\mathbf{y}_* \in D^*$)

$$
\begin{aligned}
q(u, y_1, y_2, \ldots, y_{K'}) \\
&= d\big[u\big(1 - \textstyle\sum_{k=1}^{K'} y_k^2\big)\big]\, b\big(u^{1/2}\mathbf{y}_*\big)\, u^{K'/2} \\
&= \frac{1}{2^{(K+1)/2}\Gamma[(K-K'+1)/2]\pi^{K'/2}} u^{[(K+1)/2]-1} \\
&\qquad \times \big(1-\textstyle\sum_{k=1}^{K'} y_k^2\big)^{[(K-K'+1)/2]-1} e^{-u/2} \\
&= \frac{1}{2^{(K+1)/2}\Gamma[(K+1)/2]} u^{[(K+1)/2]-1} e^{-u/2} \\
&\qquad \times \frac{\Gamma[(K+1)/2]}{\Gamma[(K-K'+1)/2]\pi^{K'/2}} \big(1-\textstyle\sum_{k=1}^{K'} y_k^2\big)^{[(K-K'+1)/2]-1} \qquad (1.43)
\end{aligned}
$$

—for u and $\mathbf{y}_*$ such that $-\infty < u \le 0$ or $\mathbf{y}_* \notin D^*$, $q(u, y_1, y_2, \ldots, y_{K'}) = 0$. Accordingly, we conclude that (for all u and $y_1, y_2, \ldots, y_{K'}$)

$$q(u, y_1, y_2, \ldots, y_{K'}) = r(u)\, h^*(y_1, y_2, \ldots, y_{K'}), \qquad (1.44)$$

where $r(\cdot)$ is the pdf of the $\chi^2(K+1)$ distribution and $h^*(\cdot, \cdot, \ldots, \cdot)$ is the function (of K' variables) defined (for all $y_1, y_2, \ldots, y_{K'}$) as follows:

$$
h^*(y_1, y_2, \ldots, y_{K'}) = \begin{cases} \dfrac{\Gamma[(K+1)/2]}{\Gamma[(K-K'+1)/2]\,\pi^{K'/2}} \big(1-\sum_{k=1}^{K'} y_k^2\big)^{[(K-K'+1)/2]-1}, & \text{if } \sum_{k=1}^{K'} y_k^2 < 1, \\ 0, & \text{otherwise.} \end{cases} \qquad (1.45)
$$

In effect, we have established that $y_1, y_2, \ldots, y_{K'}$ are statistically independent of $\sum_{k=1}^{K+1} z_k^2$ and that the distribution of $y_1, y_2, \ldots, y_{K'}$ has as a pdf the function $h^*(\cdot, \cdot, \ldots, \cdot)$ defined by expression (1.45). In the special case where $K' = K$, the function $h^*(\cdot, \cdot, \ldots, \cdot)$ is the pdf of the joint distribution of $y_1, y_2, \ldots, y_K$. When $K' = K$, $h^*(\cdot, \cdot, \ldots, \cdot)$ is reexpressible as follows: for all $y_1, y_2, \ldots, y_K$,

$$
h^*(y_1, y_2, \ldots, y_K) = \begin{cases} \dfrac{\Gamma[(K+1)/2]}{\pi^{(K+1)/2}} \big(1-\sum_{k=1}^{K} y_k^2\big)^{-1/2}, & \text{if } \sum_{k=1}^{K} y_k^2 < 1, \\ 0, & \text{otherwise.} \end{cases} \qquad (1.46)
$$

Denote by $i(\cdot)$ the function of a single variable, say z, defined as follows:

$$
i(z) = \begin{cases} 1, & \text{for } z > 0, \\ 0, & \text{for } z = 0, \\ -1, & \text{for } z < 0. \end{cases}
$$

Clearly,

$$y_{K+1} = i_{K+1}\big(1 - \textstyle\sum_{k=1}^{K} y_k^2\big)^{1/2},$$

where $i_{K+1} = i(z_{K+1})$. Moreover, $\Pr(i_{K+1} = 0) = 0$. And the joint distribution of $z_1, z_2, \ldots, z_K, -z_{K+1}$ is the same as that of $z_1, z_2, \ldots, z_K, z_{K+1}$, implying that the joint distribution of $u, y_1, y_2, \ldots, y_K, -i_{K+1}$ is the same as that of $u, y_1, y_2, \ldots, y_K, i_{K+1}$ and hence that $\Pr(i_{K+1} = -1) = \Pr(i_{K+1} = 1) = \frac{1}{2}$, both unconditionally and conditionally on $u, y_1, y_2, \ldots, y_K$ or $y_1, y_2, \ldots, y_K$. Thus, conditionally on $u, y_1, y_2, \ldots, y_K$ or $y_1, y_2, \ldots, y_K$,

$$
y_{K+1} = \begin{cases} \big(1 - \sum_{k=1}^{K} y_k^2\big)^{1/2}, & \text{with probability } \frac{1}{2}, \\ -\big(1 - \sum_{k=1}^{K} y_k^2\big)^{1/2}, & \text{with probability } \frac{1}{2}. \end{cases}
$$

Random variables, say $x_1, \ldots, x_K, x_{K+1}$, whose joint distribution is that of the random variables $y_1, \ldots, y_K, y_{K+1}$ are said to be *distributed uniformly on the surface of a* $(K+1)$*-dimensional unit ball*—refer, e.g., to definition 1.1 of Gupta and Song (1997). More generally, random variables $x_1, \ldots, x_K, x_{K+1}$ whose joint distribution is that of the random variables $ry_1, \ldots, ry_K, ry_{K+1}$ are said to be *distributed uniformly on the surface of a* $(K+1)$*-dimensional ball of radius* r. Note that if $x_1, \ldots, x_K, x_{K+1}$ are distributed uniformly on the surface of a $(K+1)$-dimensional unit ball, then $x_1^2, \ldots, x_K^2$ have a $Di\big(\frac{1}{2}, \ldots, \frac{1}{2}, \frac{1}{2}; K\big)$ distribution.

The $(K+1)$-dimensional random vector $\mathbf{z}$ [which has a $(K+1)$-dimensional standard normal distribution] is expressible in terms of the random variable u [which has a $\chi^2(K+1)$ distribution] and the $K+1$ random variables $y_1, \ldots, y_K, y_{K+1}$ [which are distributed uniformly on the surface of a $(K+1)$-dimensional unit ball independently of u]. Clearly,

$$\mathbf{z} = \sqrt{u}\,\mathbf{y}, \tag{1.47}$$

where $\mathbf{y} = (y_1, \ldots, y_K, y_{K+1})'$.

The distribution of the (positive) square root of a chi-square random variable, say a chi-square random variable with N degrees of freedom, is sometimes referred to as a *chi distribution* (with N degrees of freedom). This distribution has a pdf $b(\cdot)$ that is expressible as

$$b(x) = \begin{cases} \dfrac{1}{\Gamma(N/2)\, 2^{(N/2)-1}}\, x^{N-1} e^{-x^2/2}, & \text{for } 0 < x < \infty, \\ 0, & \text{elsewhere,} \end{cases} \tag{1.48}$$

as can be readily verified. Accordingly, the random variable $\sqrt{u}$, which appears in expression (1.47), has a chi distribution with $K+1$ degrees of freedom, the pdf of which is obtainable from expression (1.48) (upon setting $N = K+1$).

g. Extensions to spherical distributions

A transformation from a spherical distribution to a Dirichlet distribution. Let $z_1, z_2, \ldots, z_M$ represent M random variables whose joint distribution is absolutely continuous with a pdf $f(\cdot, \cdot, \ldots, \cdot)$. And suppose that for some (nonnegative) function $g(\cdot)$ (of a single nonnegative variable),

$$f(z_1, z_2, \ldots, z_M) = g\big(\textstyle\sum_{i=1}^M z_i^2\big) \qquad \text{(for all values of } z_1, z_2, \ldots, z_M\text{).} \tag{1.49}$$

Then, as discussed in Section 5.9c, the joint distribution of $z_1, z_2, \ldots, z_M$ is spherical.

Define $u_i = z_i^2$ $(i = 1, 2, \ldots, M)$, and denote by $q(\cdot, \cdot, \ldots, \cdot)$ the pdf of the joint distribution of $u_1, u_2, \ldots, u_M$. Then, upon observing that $\partial z_i / \partial u_i = \pm(1/2) u_i^{-1/2}$ $(i = 1, 2, \ldots, M)$ and making use of standard results on a change of variables (e.g., Casella and Berger 2002, sec. 4.6), we find that (for $u_1 > 0, u_2 > 0, \ldots, u_M > 0$)

$$\begin{aligned} q(u_1, u_2, \ldots, u_M) &= 2^M g\big(\textstyle\sum_{i=1}^M u_i\big) \big(\frac{1}{2}\big)^M \prod_{i=1}^M u_i^{-1/2} \\ &= g\big(\textstyle\sum_{i=1}^M u_i\big) \prod_{i=1}^M u_i^{-1/2} \end{aligned}$$

—for $u_1, u_2, \ldots, u_M$ such that $u_i \le 0$ for some i, $q(u_1, u_2, \ldots, u_M) = 0$.

Now, let $w_i = z_i^2 / \sum_{j=1}^M z_j^2$ $(i = 1, \ldots, M-1)$ and $w_M = \sum_{i=1}^M z_i^2$, and denote by $p(\cdot, \ldots, \cdot, \cdot)$ the pdf of the joint distribution of $w_1, \ldots, w_{M-1}, w_M$. Further, take D_* to be the set

$$\big\{(w_1, \ldots, w_{M-1}) : w_i > 0 \ (i = 1, \ldots, M-1),\ \textstyle\sum_{i=1}^{M-1} w_i < 1\big\}.$$

And take $y_i = u_i$ $(i = 1, \ldots, M-1)$ and $y_M = \sum_{i=1}^M u_i$, and observe that $w_i = y_i / y_M$

$(i = 1, \ldots, M-1)$ and $w_M = y_M$. Then, proceeding in essentially the same way as in the derivation of result (5.9.55), we find that for $(w_1, \ldots, w_{M-1}) \in D_*$ and for $w_M > 0$,

$$\begin{aligned} p(w_1, \ldots, w_{M-1}, w_M) &= \prod_{i=1}^{M-1} w_i^{-1/2} \bigl(1 - \sum_{i=1}^{M-1} w_i\bigr)^{-1/2} w_M^{(M/2)-1} g(w_M) \\ &= \frac{\Gamma(M/2)}{\pi^{M/2}} \prod_{i=1}^{M-1} w_i^{-1/2} \bigl(1 - \sum_{i=1}^{M-1} w_i\bigr)^{-1/2} \frac{\pi^{M/2}}{\Gamma(M/2)} w_M^{(M/2)-1} g(w_M) \end{aligned}$$

—if $(w_1, \ldots, w_{M-1}) \notin D_*$ or if $w_M \leq 0$, then $p(w_1, \ldots, w_{M-1}, w_M) = 0$. Thus, for all values of $w_1, \ldots, w_{M-1}, w_M$,

$$p(w_1, \ldots, w_{M-1}, w_M) = r(w_M) h(w_1, \ldots, w_{M-1}), \tag{1.50}$$

where $r(\cdot)$ and $h(\cdot, \ldots, \cdot)$ are functions defined as follows:

$$r(w_M) = \begin{cases} \dfrac{\pi^{M/2}}{\Gamma(M/2)} w_M^{(M/2)-1} g(w_M), & \text{for } 0 < w_M < \infty, \\ 0, & \text{for } -\infty < w_M \leq 0, \end{cases} \tag{1.51}$$

and

$$h(w_1, \ldots, w_{M-1}) = \begin{cases} \dfrac{\Gamma(M/2)}{\pi^{M/2}} \prod_{i=1}^{M-1} w_i^{-1/2} \bigl(1 - \sum_{i=1}^{M-1} w_i\bigr)^{-1/2}, \\ \qquad\qquad \text{for } (w_1, \ldots, w_{M-1}) \in D_*, \\ 0, \qquad\qquad \text{elsewhere.} \end{cases} \tag{1.52}$$

The function $h(\cdot, \ldots, \cdot)$ is the pdf of a $Di\bigl(\frac{1}{2}, \ldots, \frac{1}{2}, \frac{1}{2}; M-1\bigr)$ distribution. And the function $r(\cdot)$ is nonnegative and [in light of result (5.9.56)] $\int_{-\infty}^{\infty} r(w_M)\, dw_M = 1$, so that $r(\cdot)$, like $h(\cdot, \ldots, \cdot)$, is a pdf. Accordingly, we conclude that the random variable $\sum_{i=1}^{M} z_i^2$ is distributed independently of the random variables $z_1^2 / \sum_{i=1}^{M} z_i^2, \ldots, z_{M-1}^2 / \sum_{i=1}^{M} z_i^2$, that the distribution of $\sum_{i=1}^{M} z_i^2$ is the distribution with pdf $r(\cdot)$, and that $z_1^2 / \sum_{i=1}^{M} z_i^2, \ldots, z_{M-1}^2 / \sum_{i=1}^{M} z_i^2$ have an $(M-1)$-dimensional Dirichlet distribution (with parameters $\frac{1}{2}, \ldots, \frac{1}{2}, \frac{1}{2}$). Further, the distribution of the random variable $\bigl(\sum_{i=1}^{M} z_i^2\bigr)^{1/2}$ is the distribution with pdf $b(\cdot)$, where

$$b(s) = \begin{cases} \dfrac{2\pi^{M/2}}{\Gamma(M/2)} s^{M-1} g(s^2), & \text{for } 0 < s < \infty, \\ 0, & \text{elsewhere,} \end{cases} \tag{1.53}$$

as can be readily verified.

In the special case where the joint distribution of $z_1, z_2, \ldots, z_M$ is $N(\mathbf{0}, \mathbf{I})$ (which is the special case considered in the preceding subsection, i.e., in Subsection f) the function $g(\cdot)$ is that for which $g(x) = (2\pi)^{-M/2} e^{-x/2}$ (for every scalar x). In that special case, $r(\cdot)$ is the pdf of the $\chi^2(M)$ distribution, and $b(\cdot)$ is the pdf of the chi distribution (with M degrees of freedom).

Decomposition of spherically distributed random variables. Let $z_1, \ldots, z_K, z_{K+1}$ represent $K+1$ random variables whose joint distribution is absolutely continuous with a pdf $f(\cdot, \ldots, \cdot, \cdot)$. And suppose that (for arbitrary values of $z_1, \ldots, z_K, z_{K+1}$)

$$f(z_1, \ldots, z_K, z_{K+1}) = g\bigl(\sum_{k=1}^{K+1} z_k^2\bigr), \tag{1.54}$$

where $g(\cdot)$ is a (nonnegative) function of a single nonnegative variable (in which case the joint distribution of $z_1, \ldots, z_K, z_{K+1}$ is spherical). Further, let

$$u = \sum_{k=1}^{K+1} z_k^2 \quad \text{and (for } k = 1, \ldots, K, K+1) \quad y_k = \frac{z_k}{\bigl(\sum_{j=1}^{K+1} z_j^2\bigr)^{1/2}}.$$

And consider the joint distribution of $u, y_1, y_2, \ldots, y_K$.

Clearly,

$$u = v + \sum_{k=1}^{K} z_k^2 \quad \text{and (for } k = 1, \ldots, K, K+1\text{)} \quad y_k = \frac{z_k}{\bigl(v + \sum_{j=1}^{K} z_j^2\bigr)^{1/2}},$$

where $v = z_{K+1}^2$. Moreover, upon applying standard results on a change of variables (e.g., Casella and Berger 2002, sec. 4.6), we find that the joint distribution of the random variables $v, z_1, z_2, \ldots, z_K$ has as a pdf the function $f^*(\cdot, \cdot, \cdot, \ldots, \cdot)$ (of $K+1$ variables) obtained by taking (for $0 < v < \infty$ and for all $z_1, z_2, \ldots, z_K$)

$$f^*(v, z_1, z_2, \ldots, z_K) = 2g\bigl(v + \textstyle\sum_{k=1}^{K} z_k^2\bigr)(1/2)v^{-1/2} = g\bigl(v + \sum_{k=1}^{K} z_k^2\bigr)v^{-1/2}$$

—if $-\infty < v \le 0$, take $f^*(v, z_1, z_2, \ldots, z_K) = 0$.

Let $\mathbf{y}_* = (y_1, y_2, \ldots, y_K)'$, and define $D^* = \{\mathbf{y}_* : \sum_{k=1}^{K} y_k^2 < 1\}$. Then, proceeding in essentially the same way as in the derivation of result (1.43), we find that the joint distribution of $u, y_1, y_2, \ldots, y_K$ has as a pdf the function $q(\cdot, \cdot, \cdot, \ldots, \cdot)$ (of $K+1$ variables) obtained by taking (for $0 < u < \infty$ and $\mathbf{y}_* \in D^*$)

$$\begin{aligned} q(u, y_1, y_2, \ldots, y_K) &= f^*\bigl[u\bigl(1 - \textstyle\sum_{k=1}^{K} y_k^2\bigr), u^{1/2}y_1, u^{1/2}y_2, \ldots, u^{1/2}y_K\bigr]\, u^{K/2} \\ &= g(u)\, u^{-1/2}\bigl(1 - \textstyle\sum_{k=1}^{K} y_k^2\bigr)^{-1/2} u^{K/2} \\ &= \frac{\pi^{(K+1)/2}}{\Gamma[(K+1)/2]}\, u^{[(K+1)/2]-1} g(u)\, \frac{\Gamma[(K+1)/2]}{\pi^{(K+1)/2}} \bigl(1 - \textstyle\sum_{k=1}^{K} y_k^2\bigr)^{-1/2} \end{aligned}$$

—for u and $\mathbf{y}_*$ such that $-\infty < u \le 0$ or $\mathbf{y}_* \notin D^*$, take $q(u, y_1, y_2, \ldots, y_K) = 0$. Thus, for all u and $y_1, y_2, \ldots, y_K$,

$$q(u, y_1, y_2, \ldots, y_K) = r(u)\, h^*(y_1, y_2, \ldots, y_K), \tag{1.55}$$

where $r(\cdot)$ is the function (of a single variable) defined (for all u) and $h^*(\cdot, \cdot, \ldots, \cdot)$ the function (of K variables) defined (for all $y_1, y_2, \ldots, y_K$) as follows:

$$r(u) = \begin{cases} \dfrac{\pi^{(K+1)/2}}{\Gamma[(K+1)/2]}\, u^{[(K+1)/2]-1} g(u), & \text{for } 0 < u < \infty, \\ 0, & \text{for } -\infty < u \le 0, \end{cases} \tag{1.56}$$

and

$$h^*(y_1, y_2, \ldots, y_K) = \begin{cases} \dfrac{\Gamma[(K+1)/2]}{\pi^{(K+1)/2}} \bigl(1 - \sum_{k=1}^{K} y_k^2\bigr)^{-1/2}, & \text{if } \sum_{k=1}^{K} y_k^2 < 1, \\ 0, & \text{otherwise.} \end{cases} \tag{1.57}$$

As is evident from the results of Part 1 of the present subsection, the function $r(\cdot)$ is a pdf; it is the pdf of the distribution of $\sum_{k=1}^{K+1} z_k^2$. Further, $y_1, y_2, \ldots, y_K$ are statistically independent of $\sum_{k=1}^{K+1} z_k^2$, and the distribution of $y_1, y_2, \ldots, y_K$ has as a pdf the function $h^*(\cdot, \cdot, \ldots, \cdot)$ defined by expression (1.57). And, conditionally on $u, y_1, y_2, \ldots, y_K$ or $y_1, y_2, \ldots, y_K$,

$$y_{K+1} = \begin{cases} \bigl(1 - \sum_{k=1}^{K} y_k^2\bigr)^{1/2}, & \text{with probability } \frac{1}{2}, \\ -\bigl(1 - \sum_{k=1}^{K} y_k^2\bigr)^{1/2}, & \text{with probability } \frac{1}{2}, \end{cases}$$

as can be established in the same way as in Subsection f [where it was assumed that the joint distribution of $z_1, \ldots, z_K, z_{K+1}$ is $N(\mathbf{0}, \mathbf{I})$]. Accordingly, $y_1, \ldots, y_K, y_{K+1}$ are distributed uniformly on the surface of a $(K+1)$-dimensional unit ball.

Let $\mathbf{z} = (z_1, \ldots, z_K, z_{K+1})'$, and consider the decomposition of the vector $\mathbf{z}$ defined by the identity

$$\mathbf{z} = \sqrt{u}\,\mathbf{y}, \tag{1.58}$$

where $\mathbf{y} = (y_1, \ldots, y_K, y_{K+1})'$. This decomposition was considered previously (in Subsection f) in the special case where $\mathbf{z} \sim N(\mathbf{0}, \mathbf{I})$. As in the special case, $\mathbf{y}$ is distributed uniformly on the surface of a $(K+1)$-dimensional unit ball (and is distributed independently of u or $\sqrt{u}$). In the present (general) case of an arbitrary absolutely continuous spherical distribution [i.e., where the distribution of $\mathbf{z}$ is any absolutely continuous distribution with a pdf of the form (1.54)], the distribution of u is the distribution with the pdf $r(\cdot)$ given by expression (1.56) and (recalling the results of Part 1 of the present subsection) the distribution of $\sqrt{u}$ is the distribution with the pdf $b(\cdot)$ given by the expression

$$b(x) = \begin{cases} \dfrac{2\pi^{(K+1)/2}}{\Gamma[(K+1)/2]}\, x^K g(x^2), & \text{for } 0 < x < \infty, \\ 0, & \text{elsewhere.} \end{cases} \tag{1.59}$$

In the special case where $\mathbf{z} \sim N(\mathbf{0}, \mathbf{I})$, $u \sim \chi^2(K+1)$, and $\sqrt{u}$ has a chi distribution (with $K+1$ degrees of freedom).

6.2 Noncentral Chi-Square Distribution

Chi-square distributions were considered in Section 6.1. Those distributions form a subclass of a larger class of distributions known as noncentral chi-square distributions. Preliminary to defining and discussing noncentral chi-square distributions, it is convenient to introduce an orthogonal matrix known as a Helmert matrix.

a. Helmert matrix

Let $\mathbf{a} = (a_1, a_2, \ldots, a_N)'$ represent an N-dimensional nonnull (column) vector. Does there exist an $N \times N$ orthogonal matrix, one of whose rows, say the first row, is proportional to $\mathbf{a}'$? Or, equivalently, does there exist an orthonormal basis for $\mathcal{R}^N$ that includes the vector $\mathbf{a}$? In what follows, the answer is shown to be yes. The approach taken is to describe a particular $N \times N$ orthogonal matrix whose first row is proportional to $\mathbf{a}'$. Other approaches are possible—refer, e.g., to Harville (1997, sec 6.4).

Let us begin by considering a special case. Suppose that $a_i \neq 0$ for $i = 1, 2, \ldots, N$. And consider the $N \times N$ matrix $\mathbf{P}$, whose first through Nth rows, say $\mathbf{p}_1', \mathbf{p}_2', \ldots, \mathbf{p}_N'$, are each of norm 1 and are further defined as follows: take $\mathbf{p}_1'$ proportional to $\mathbf{a}'$, take $\mathbf{p}_2'$ proportional to

$$(a_1,\ -a_1^2/a_2,\ 0,\ 0, \ldots,\ 0);$$

take $\mathbf{p}_3'$ proportional to

$$[a_1,\ a_2,\ -(a_1^2+a_2^2)/a_3,\ 0,\ 0, \ldots,\ 0];$$

and, more generally, take the second through Nth rows proportional to

$$(a_1,\ a_2, \ldots,\ a_{k-1},\ -\textstyle\sum_{i=1}^{k-1} a_i^2/a_k,\ 0,\ 0, \ldots,\ 0) \qquad (k = 2, 3, \ldots, N), \tag{2.1}$$

respectively.

Clearly, the $N-1$ vectors (2.1) are orthogonal to each other and to the vector $\mathbf{a}'$. Thus, $\mathbf{P}$ is an orthogonal matrix. Moreover, upon "normalizing" $\mathbf{a}'$ and the vectors (2.1), we find that

$$\mathbf{p}_1' = \left(\textstyle\sum_{i=1}^{N} a_i^2\right)^{-1/2} (a_1, a_2, \ldots, a_N) \tag{2.2}$$

and that, for $k = 2, 3, \ldots, N$,

$$\mathbf{p}_k' = \left[\frac{a_k^2}{(\sum_{i=1}^{k-1} a_i^2)(\sum_{i=1}^{k} a_i^2)}\right]^{1/2} (a_1, a_2, \ldots, a_{k-1}, -\textstyle\sum_{i=1}^{k-1} a_i^2/a_k, 0, 0, \ldots, 0). \tag{2.3}$$

When $\mathbf{a}' = (1, 1, \ldots, 1)'$, formulas (2.2) and (2.3) simplify to

$$\mathbf{p}_1' = N^{-1/2}(1, 1, \ldots, 1)$$

and

$$\mathbf{p}_k' = [k(k-1)]^{-1/2}(1, 1, \ldots, 1, 1-k, 0, 0, \ldots, 0) \qquad (k = 2, 3, \ldots, N),$$

and $\mathbf{P}$ reduces to a matrix known as the *Helmert matrix* (of order N). (In some presentations, it is the transpose of this matrix that is called the Helmert matrix.)

We have established that in the special case where $a_i \neq 0$ for $i = 1, 2, \ldots, N$, there exists an $N \times N$ orthogonal matrix whose first row is proportional to $\mathbf{a}'$; the matrix $\mathbf{P}$, whose N rows $\mathbf{p}_1', \mathbf{p}_2', \ldots, \mathbf{p}_N'$ are determinable from formulas (2.2) and (2.3), is such a matrix. Now, consider the general case, in which as many as $N-1$ elements of $\mathbf{a}$ may equal 0.

Suppose that K of the elements of $\mathbf{a}$ are nonzero (where $1 \le K \le N$), say the $j_1, j_2, \ldots, j_K$th elements (and that the other $N-K$ elements of $\mathbf{a}$ equal 0). Then, it follows from what has already been established that there exists a $K \times K$ orthogonal matrix, say $\mathbf{Q}$, whose first row is proportional to the vector $(a_{j_1}, a_{j_2}, \ldots, a_{j_K})$. And, denoting by $\mathbf{q}_1, \mathbf{q}_2, \ldots, \mathbf{q}_K$ the columns of $\mathbf{Q}$, an $N \times N$ orthogonal matrix, say $\mathbf{P}$, whose first row is proportional to $\mathbf{a}'$ is obtainable as follows: take

$$\mathbf{P} = \begin{pmatrix} \mathbf{P}_1 \\ \mathbf{P}_2 \end{pmatrix},$$

where $\mathbf{P}_1$ is a $K \times N$ matrix whose $j_1, j_2, \ldots, j_K$th columns are $\mathbf{q}_1, \mathbf{q}_2, \ldots, \mathbf{q}_K$, respectively, and whose other $N-K$ columns equal $\mathbf{0}$ and where $\mathbf{P}_2$ is an $(N-K) \times N$ matrix whose $j_1, j_2, \ldots, j_K$th columns equal $\mathbf{0}$ and whose other $N-K$ columns are the columns of $\mathbf{I}_{N-K}$—that $\mathbf{P}$ is orthogonal is evident upon observing that its columns are orthonormal. Thus, as in the special case where $\mathbf{a}_i \neq 0$ for $i = 1, 2, \ldots, N$, there exists an $N \times N$ orthogonal matrix whose first row is proportional to $\mathbf{a}'$.

b. Noncentral chi-square distribution: definition

Let $\mathbf{z} = (z_1, z_2, \ldots, z_N)'$ represent an N-dimensional random (column) vector whose distribution is N-variate normal with mean vector $\boldsymbol{\mu} = (\mu_1, \mu_2, \ldots, \mu_N)'$ and variance-covariance matrix $\mathbf{I}$ or, equivalently, whose elements are distributed independently as $N(\mu_1, 1), N(\mu_2, 1), \ldots, N(\mu_N, 1)$. Define

$$w = \mathbf{z}'\mathbf{z} \quad \text{or, equivalently,} \quad w = \textstyle\sum_{i=1}^N z_i^2.$$

When $\boldsymbol{\mu} = \mathbf{0}$, the distribution of w is, by definition, a chi-square distribution (with N degrees of freedom)—the chi-square distribution was introduced and discussed in Section 6.1. Let us now consider the distribution of w in the general case where $\boldsymbol{\mu}$ is not necessarily null.

Let

$$\lambda = \boldsymbol{\mu}'\boldsymbol{\mu} = \textstyle\sum_{i=1}^N \mu_i^2,$$

and let

$$\tau = \sqrt{\lambda} = \sqrt{\boldsymbol{\mu}'\boldsymbol{\mu}} = \sqrt{\textstyle\sum_{i=1}^N \mu_i^2}.$$

If $\tau \neq 0$ (or, equivalently, if $\boldsymbol{\mu} \neq \mathbf{0}$), define $\mathbf{a}_1 = (1/\tau)\boldsymbol{\mu}$; otherwise (if $\tau = 0$ or, equivalently, if $\boldsymbol{\mu} = \mathbf{0}$), take $\mathbf{a}_1$ to be any $N \times 1$ (nonrandom) vector such that $\mathbf{a}_1'\mathbf{a}_1 = 1$. Further, take $\mathbf{A}_2$ to be any $N \times (N-1)$ (nonrandom) matrix such that the matrix $\mathbf{A}$ defined by $\mathbf{A} = (\mathbf{a}_1, \mathbf{A}_2)$ is orthogonal—the existence of such a matrix follows from the results of Subsection a. And define

$$x_1 = \mathbf{a}_1'\mathbf{z} \quad \text{and} \quad \mathbf{x}_2 = \mathbf{A}_2'\mathbf{z}.$$

Then, clearly,

$$\begin{pmatrix} x_1 \\ \mathbf{x}_2 \end{pmatrix} \sim N\left[\begin{pmatrix} \tau \\ \mathbf{0} \end{pmatrix}, \begin{pmatrix} 1 & \mathbf{0} \\ \mathbf{0} & \mathbf{I}_{N-1} \end{pmatrix}\right], \tag{2.4}$$

and

$$w = \mathbf{z}'\mathbf{z} = \mathbf{z}'\mathbf{I}\mathbf{z} = \mathbf{z}'\mathbf{A}\mathbf{A}'\mathbf{z} = (\mathbf{A}'\mathbf{z})'\mathbf{A}'\mathbf{z} = \begin{pmatrix} x_1 \\ \mathbf{x}_2 \end{pmatrix}' \begin{pmatrix} x_1 \\ \mathbf{x}_2 \end{pmatrix} = x_1^2 + \mathbf{x}_2'\mathbf{x}_2. \tag{2.5}$$

The distribution of the random variable w is called the *noncentral chi-square distribution*. As is evident from results (2.4) and (2.5), this distribution depends on the value of $\boldsymbol{\mu}$ only through τ or, equivalently, only through λ. In the special case where $\lambda = 0$, the noncentral chi-square distribution is identical to the distribution that we have been referring to as the chi-square distribution. In this special case, the distribution is sometimes (for the sake of clarity and/or emphasis) referred to as the *central* chi-square distribution.

The noncentral chi-square distribution depends on two parameters: in addition to the degrees of freedom N, it depends on the quantity λ, which is referred to as the *noncentrality parameter*. Let us use the symbol $\chi^2(N, \lambda)$ to represent a noncentral chi-square distribution with degrees of freedom N and noncentrality parameter λ.

c. Pdf of the noncentral chi-square distribution

As a first step in deriving the pdf of the $\chi^2(N, \lambda)$ distribution (for arbitrary degrees of freedom N), let us derive the pdf of the $\chi^2(1, \lambda)$ distribution.

Pdf of the $\chi^2(1, \lambda)$ distribution. The derivation of the pdf of the $\chi^2(1, \lambda)$ distribution parallels the derivation (in Section 6.1c) of the pdf of the $\chi^2(1)$ distribution.

Let λ represent an arbitrary nonnegative scalar, take z to be a random variable whose distribution is $N(\sqrt{\lambda},\ 1)$, and define $u = z^2$. Then, by definition, $u \sim \chi^2(1, \lambda)$. And a pdf, say $h(\cdot)$, of the distribution of u is obtainable from the pdf, say $f(\cdot)$, of the $N(\sqrt{\lambda},\ 1)$ distribution: for $u > 0$, take

$$\begin{aligned} h(u) &= f(-\sqrt{u})\,\big|-(2\sqrt{u})^{-1}\big| + f(\sqrt{u})\,(2\sqrt{u})^{-1} \\ &= (2\sqrt{2\pi u})^{-1}\big[e^{-(\sqrt{u}+\sqrt{\lambda})^2/2} + e^{-(\sqrt{u}-\sqrt{\lambda})^2/2}\big] \\ &= (2\sqrt{2\pi u})^{-1} e^{-(u+\lambda)/2}\big(e^{\sqrt{\lambda u}} + e^{-\sqrt{\lambda u}}\big) \end{aligned}$$

—for $u \le 0$, take $h(u) = 0$.

The pdf $h(\cdot)$ is reexpressible in terms of the hyperbolic cosine function $\cosh(\cdot)$. By definition,

$$\cosh(x) = \tfrac{1}{2}(e^x + e^{-x}) \qquad \text{(for every scalar } x\text{)}.$$

Thus,

$$h(u) = \begin{cases} \dfrac{1}{\sqrt{2\pi u}} e^{-(u+\lambda)/2} \cosh(\sqrt{\lambda u}), & \text{for } u > 0, \\ 0, & \text{for } u \le 0. \end{cases} \tag{2.6}$$

The pdf $h(\cdot)$ can be further reexpressed by making use of the power-series representation

$$\cosh(x) = \sum_{r=0}^{\infty} \frac{x^{2r}}{(2r)!} \qquad (-\infty < x < \infty)$$

for the hyperbolic cosine function and by making use of result (3.5.11). We find that, for $u > 0$,

$$\begin{aligned} h(u) &= \frac{1}{\sqrt{2\pi u}} e^{-(u+\lambda)/2} \sum_{r=0}^{\infty} \frac{(\lambda u)^r}{(2\pi)^{-1/2}\, 2^{2r+(1/2)}\, r!\, \Gamma\big(r+\frac{1}{2}\big)} \\ &= \sum_{r=0}^{\infty} \frac{(\lambda/2)^r e^{-\lambda/2}}{r!} \frac{1}{\Gamma[(2r+1)/2]\, 2^{(2r+1)/2}}\, u^{[(2r+1)/2]-1} e^{-u/2}. \end{aligned} \tag{2.7}$$

And letting (for $j = 1, 2, 3, \ldots$) $g_j(\cdot)$ represent the pdf of a central chi-square distribution with j degrees of freedom, it follows that (for all u)

$$h(u) = \sum_{r=0}^{\infty} \frac{(\lambda/2)^r e^{-\lambda/2}}{r!}\, g_{2r+1}(u). \tag{2.8}$$

The coefficients $(\lambda/2)^r e^{-\lambda/2}/r!$ $(r = 0, 1, 2, \ldots)$ of the quantities $g_{2r+1}(u)$ $(r = 0, 1, 2, \ldots)$ in the sum (2.8) can be regarded as the values $p(r)$ $(r = 0, 1, 2, \ldots)$ of a function $p(\cdot)$; this function is the probability mass function of a Poisson distribution with parameter $\lambda/2$. Thus, the pdf $h(\cdot)$ of the $\chi^2(1, \lambda)$ distribution is a weighted average of the pdfs of central chi-square distributions. A distribution whose pdf (or cumulative distribution function) is expressible as a weighted average of the pdfs (or cumulative distribution functions) of other distributions is known as a *mixture distribution*.

Extension to the general case (of arbitrary degrees of freedom). According to result (2.5), the $\chi^2(N, \lambda)$ distribution can be regarded as the sum of two independently distributed random variables, the first of which has a $\chi^2(1, \lambda)$ distribution and the second of which has a $\chi^2(N-1)$ distribution. Moreover, in Part 1 (of the present subsection), the pdf of the $\chi^2(1, \lambda)$ distribution was determined to be that given by expression (2.8), which is a weighted average of pdfs of central chi-square distributions. Accordingly, take w_1 to be a random variable whose distribution is that with a pdf $h(\cdot)$ of the form

$$h(w_1) = \sum_{r=0}^{\infty} p_r g_{N_r}(w_1), \tag{2.9}$$

where $p_0, p_1, p_2, \ldots$ are nonnegative constants such that $\sum_{r=0}^{\infty} p_r = 1$, where $N_0, N_1, N_2, \ldots$ are (strictly) positive integers, and where (as in Part 1) $g_j(\cdot)$ denotes (for any strictly positive integer j) the pdf of a central chi-square distribution with j degrees of freedom. And for an arbitrary (strictly) positive integer K, take w_2 to be a random variable that is distributed independently of w_1 as $\chi^2(K)$, and define

$$w = w_1 + w_2.$$

Let us determine the pdf of the distribution of w. Denote by $b(\cdot)$ the pdf of the $\chi^2(K)$ distribution, and define

$$s = \frac{w_1}{w_1 + w_2}.$$

Then, proceeding in essentially the same way as in Section 6.1a (in arriving at Theorem 6.1.1), we find that a pdf, say $f(\cdot, \cdot)$, of the joint distribution of w and s is obtained by taking, for $w > 0$ and $0 < s < 1$,

$$\begin{aligned} f(w, s) &= h(sw)\, b[(1-s)w]\, |-w| \\ &= w \sum_{r=0}^{\infty} p_r\, g_{N_r}(sw)\, b[(1-s)w] \\ &= \sum_{r=0}^{\infty} p_r \frac{1}{\Gamma[(N_r+K)/2]\, 2^{(N_r+K)/2}}\, w^{[(N_r+K)/2]-1} e^{-w/2} \\ &\qquad \times \frac{\Gamma[(N_r+K)/2]}{\Gamma(N_r/2)\Gamma(K/2)}\, s^{(N_r/2)-1}(1-s)^{(K/2)-1} \end{aligned} \tag{2.10}$$

—for $w \le 0$ and for s such that $s \le 0$ or $s \ge 1$, $f(w, s) = 0$. And letting $d_{\alpha_1,\alpha_2}(\cdot)$ represent the pdf of a $Be(\alpha_1, \alpha_2)$ distribution (for arbitrary values of the parameters α_1 and α_2), it follows that (for all w and s)

$$f(w, s) = \sum_{r=0}^{\infty} p_r\, g_{N_r+K}(w)\, d_{N_r/2,\, K/2}(s). \tag{2.11}$$

Thus, as a pdf of the (marginal) distribution of w, we have the function $q(\cdot)$ obtained by taking (for all w)

$$q(w) = \int_0^1 f(w, s)\, ds = \sum_{r=0}^{\infty} p_r\, g_{N_r+K}(w) \int_0^1 d_{N_r/2,\, K/2}(s)\, ds = \sum_{r=0}^{\infty} p_r\, g_{N_r+K}(w). \tag{2.12}$$

The distribution of w, like that of w_1, is a mixture distribution. As in the case of the pdf of the distribution of w_1, the pdf of the distribution of w is a weighted average of the pdfs of central chi-square distributions. Moreover, the sequence $p_0, p_1, p_2, \ldots$ of weights is the same in the case of the pdf of the distribution of w as in the case of the pdf of the distribution of w_1. And the central chi-square distributions represented in one of these weighted averages are related in a simple way to those represented in the other; each of the central chi-square distributions represented in the weighted average (2.12) have an additional K degrees of freedom.

In light of results (2.5) and (2.8), the pdf of the $\chi^2(N, \lambda)$ distribution is obtainable as a special case of the pdf (2.12). Specifically, upon setting $K = N-1$ and setting (for $r = 0, 1, 2, \ldots$) $N_r = 2r+1$ and $p_r = p(r)$ [where, as in Part 1 of the present subsection, $p(r) = (\lambda/2)^r e^{-\lambda/2}/r!$], we obtain the pdf of the $\chi^2(N, \lambda)$ distribution as a special case of the pdf (2.12). Accordingly, the pdf of the $\chi^2(N, \lambda)$ distribution is the function $q(\cdot)$ that is expressible as follows:

$$q(w) = \sum_{r=0}^{\infty} p(r)\, g_{2r+N}(w) \tag{2.13}$$

$$= \sum_{r=0}^{\infty} \frac{(\lambda/2)^r e^{-\lambda/2}}{r!}\, g_{2r+N}(w) \tag{2.14}$$

$$= \begin{cases} \displaystyle\sum_{r=0}^{\infty} \frac{(\lambda/2)^r e^{-\lambda/2}}{r!} \frac{1}{\Gamma[(2r+N)/2]\, 2^{(2r+N)/2}}\, w^{[(2r+N)/2]-1} e^{-w/2}, & \text{for } w > 0, \\ 0, & \text{for } w \le 0. \end{cases} \tag{2.15}$$

d. Distribution of a sum of noncentral chi-square random variables

Suppose that two random variables, say w_1 and w_2, are distributed independently as $\chi^2(N_1, \lambda_1)$ and $\chi^2(N_2, \lambda_2)$, respectively. And consider the distribution of the sum $w = w_1 + w_2$.

Let $\mathbf{x}$ represent an $(N_1+N_2) \times 1$ random vector whose distribution is

$$N\left[\begin{pmatrix} \sqrt{\lambda_1/N_1}\, \mathbf{1}_{N_1} \\ \sqrt{\lambda_2/N_2}\, \mathbf{1}_{N_2} \end{pmatrix}, \begin{pmatrix} \mathbf{I}_{N_1} & \mathbf{0} \\ \mathbf{0} & \mathbf{I}_{N_2} \end{pmatrix}\right],$$

and partition $\mathbf{x}$ as $\mathbf{x} = \begin{pmatrix} \mathbf{x}_1 \\ \mathbf{x}_2 \end{pmatrix}$ (where $\mathbf{x}_1$ is of dimensions $N_1 \times 1$). And observe (in light of Theorem 3.5.3 and Corollary 3.5.5) that $\mathbf{x}_1$ and $\mathbf{x}_2$ are distributed independently as $N\left(\sqrt{\lambda_1/N_1}\, \mathbf{1}_{N_1},\, \mathbf{I}_{N_1}\right)$ and $N\left(\sqrt{\lambda_2/N_2}\, \mathbf{1}_{N_2},\, \mathbf{I}_{N_2}\right)$, respectively. Observe also (in light of the very definition of the noncentral chi-square distribution) that the joint distribution of $\mathbf{x}_1'\mathbf{x}_1$ and $\mathbf{x}_2'\mathbf{x}_2$ is the same as that of w_1 and w_2. Thus,

$$\mathbf{x}'\mathbf{x} = \mathbf{x}_1'\mathbf{x}_1 + \mathbf{x}_2'\mathbf{x}_2 \sim w_1 + w_2 = w.$$

Moreover, $\mathbf{x}'\mathbf{x}$ has a noncentral chi-square distribution with N_1+N_2 degrees of freedom and noncentrality parameter

$$\begin{pmatrix} \sqrt{\lambda_1/N_1}\, \mathbf{1}_{N_1} \\ \sqrt{\lambda_2/N_2}\, \mathbf{1}_{N_2} \end{pmatrix}' \begin{pmatrix} \sqrt{\lambda_1/N_1}\, \mathbf{1}_{N_1} \\ \sqrt{\lambda_2/N_2}\, \mathbf{1}_{N_2} \end{pmatrix} = (\lambda_1/N_1)\mathbf{1}_{N_1}'\mathbf{1}_{N_1} + (\lambda_2/N_2)\mathbf{1}_{N_2}'\mathbf{1}_{N_2} = \lambda_1 + \lambda_2,$$

leading to the conclusion that

$$w \sim \chi^2(N_1+N_2,\, \lambda_1+\lambda_2). \tag{2.16}$$

Upon employing a simple mathematical-induction argument, we arrive at the following generalization of result (2.16).

Theorem 6.2.1. If K random variables $w_1, w_2, \ldots, w_K$ are distributed independently as $\chi^2(N_1, \lambda_1), \chi^2(N_2, \lambda_2), \ldots, \chi^2(N_K, \lambda_K)$, respectively, then $w_1 + w_2 + \cdots + w_K$ is distributed as $\chi^2(N_1 + N_2 + \cdots + N_K, \lambda_1 + \lambda_2 + \cdots + \lambda_K)$.

Note that in the special case where $\lambda_1 = \lambda_2 = \cdots = \lambda_K = 0$, Theorem 6.2.1 reduces to Theorem 6.1.3.

e. Moment generating function and cumulants

Let z represent a random variable whose distribution is $N(\mu, 1)$. Further, let $f(\cdot)$ represent the pdf of the $N(\mu, 1)$ distribution. Then, for $t < 1/2$,

$$\begin{aligned}
\mathrm{E}\big(e^{tz^2}\big) &= \int_{-\infty}^{\infty} e^{tx^2} f(x)\, dx \\
&= \int_{-\infty}^{\infty} \frac{1}{\sqrt{2\pi}} \exp\!\left[-\frac{(1-2t)}{2} x^2 + \mu x - \frac{\mu^2}{2}\right] dx \\
&= \int_{-\infty}^{\infty} \frac{1}{\sqrt{2\pi}} \exp\!\left[-\frac{(1-2t)}{2} \left(x - \frac{\mu}{1-2t}\right)^2 + \frac{t\mu^2}{1-2t}\right] dx \\
&= (1-2t)^{-1/2} \exp\!\left(\frac{t\mu^2}{1-2t}\right) \int_{-\infty}^{\infty} h_t(x)\, dx,
\end{aligned}$$

where $h_t(\cdot)$ is the pdf of the $N[\mu/(1-2t),\ (1-2t)^{-1}]$ distribution. Thus, for $t < 1/2$,

$$\mathrm{E}\big(e^{tz^2}\big) = (1-2t)^{-1/2} \exp\!\left(\frac{t\mu^2}{1-2t}\right).$$

And it follows that the moment generating function, say $m(\cdot)$, of the $\chi^2(1, \lambda)$ distribution is

$$m(t) = (1-2t)^{-1/2} \exp\!\left(\frac{\lambda t}{1-2t}\right) \qquad (t < 1/2). \tag{2.17}$$

To obtain an expression for the moment generating function of the $\chi^2(N, \lambda)$ distribution (where N is any strictly positive integer), it suffices (in light of Theorem 6.2.1) to find the moment generating function of the distribution of the sum $w = w_1 + w_2$ of two random variables w_1 and w_2 that are distributed independently as $\chi^2(N{-}1)$ and $\chi^2(1, \lambda)$, respectively. Letting $m_1(\cdot)$ represent the moment generating function of the $\chi^2(N-1)$ distribution and $m_2(\cdot)$ the moment generating function of the $\chi^2(1, \lambda)$ distribution and making use of results (1.20) and (2.17), we find that, for $t < 1/2$,

$$\begin{aligned}
\mathrm{E}\big(e^{tw}\big) &= \mathrm{E}\big(e^{tw_1}\big)\, \mathrm{E}\big(e^{tw_2}\big) = m_1(t)\, m_2(t) \\
&= (1-2t)^{-(N-1)/2} (1-2t)^{-1/2} \exp\!\left(\frac{\lambda t}{1-2t}\right) \\
&= (1-2t)^{-N/2} \exp\!\left(\frac{\lambda t}{1-2t}\right).
\end{aligned}$$

Thus, the moment generating function, say $m(\cdot)$, of the $\chi^2(N, \lambda)$ distribution is

$$m(t) = (1-2t)^{-N/2} \exp\!\left(\frac{\lambda t}{1-2t}\right) \qquad (t < 1/2). \tag{2.18}$$

Or, upon reexpressing $\exp[\lambda t/(1-2t)]$ as

$$\exp\!\left(\frac{\lambda t}{1-2t}\right) = \exp(-\lambda/2) \exp\!\left(\frac{\lambda/2}{1-2t}\right)$$

and replacing $\exp[(\lambda/2)/(1-2t)]$ with its power-series expansion

$$\exp\left(\frac{\lambda/2}{1-2t}\right) = \sum_{r=0}^{\infty} \frac{(\lambda/2)^r}{(1-2t)^r\, r!}, \tag{2.19}$$

we obtain the alternative representation

$$m(t) = \sum_{r=0}^{\infty} \frac{(\lambda/2)^r e^{-\lambda/2}}{r!} (1-2t)^{-(2r+N)/2} \qquad (t < 1/2). \tag{2.20}$$

Alternatively, expression (2.20) for the moment generating function of the $\chi^2(N, \lambda)$ distribution can be derived from the pdf (2.14). Letting (as in Subsection c) $g_j(\cdot)$ represent (for an arbitrary strictly positive integer j) the pdf of a $\chi^2(j)$ distribution, the alternative approach gives

$$\begin{aligned} m(t) &= \int_0^{\infty} e^{tw} \sum_{r=0}^{\infty} \frac{(\lambda/2)^r e^{-\lambda/2}}{r!}\, g_{2r+N}(w)\, dw \\ &= \sum_{r=0}^{\infty} \frac{(\lambda/2)^r e^{-\lambda/2}}{r!} \int_0^{\infty} e^{tw} g_{2r+N}(w)\, dw. \end{aligned} \tag{2.21}$$

If we use formula (1.20) to evaluate (for each r) the integral $\int_0^{\infty} e^{tw} g_{2r+N}(w)\, dw$, we arrive immediately at expression (2.20).

The cumulant generating function, say $c(\cdot)$, of the $\chi^2(N, \lambda)$ distribution is [in light of result (2.18)]

$$c(t) = \log m(t) = -(N/2)\log(1-2t) + \lambda t(1-2t)^{-1} \qquad (t < 1/2). \tag{2.22}$$

Upon expanding $c(t)$ in a power series (about 0), we find that (for $-1/2 < t < 1/2$)

$$\begin{aligned} c(t) &= (N/2)\sum_{r=1}^{\infty} (2t)^r/r + \lambda t \sum_{r=1}^{\infty} (2t)^{r-1} \\ &= \sum_{r=1}^{\infty} 2^{r-1} t^r (N+\lambda r)/r \\ &= \sum_{r=1}^{\infty} (N+\lambda r)\, 2^{r-1}(r-1)!\, t^r/r!. \end{aligned} \tag{2.23}$$

Thus, the rth cumulant of the $\chi^2(N, \lambda)$ distribution is

$$(N+\lambda r)\, 2^{r-1}(r-1)!. \tag{2.24}$$

f. Moments

Mean and variance. Let w represent a random variable whose distribution is $\chi^2(N, \lambda)$. And [for purposes of determining $\mathrm{E}(w)$, $\mathrm{var}(w)$, and $\mathrm{E}(w^2)$] let $x = \sqrt{\lambda} + z$, where z is a random variable that has a standard normal distribution, so that $x \sim N(\sqrt{\lambda},\, 1)$ and hence $x^2 \sim \chi^2(1, \lambda)$. Then, in light of Theorem 6.2.1,

$$w \sim x^2 + u,$$

where u is a random variable that is distributed independently of z (and hence distributed independently of x and x^2) as $\chi^2(N-1)$.

Clearly,

$$\mathrm{E}(x^2) = \mathrm{var}(x) + [\mathrm{E}(x)]^2 = 1 + \lambda. \tag{2.25}$$

And, making use of results (3.5.19) and (3.5.20), we find that

$$\begin{aligned}\mathrm{E}(x^4) &= \mathrm{E}\big[(\sqrt{\lambda} + z)^4\big] \\ &= \mathrm{E}\big(z^4 + 4z^3\sqrt{\lambda} + 6z^2\lambda + 4z\lambda^{3/2} + \lambda^2\big) \\ &= 3 + 0 + 6\lambda + 0 + \lambda^2 = \lambda^2 + 6\lambda + 3, \end{aligned} \tag{2.26}$$

which [in combination with result (2.25)] implies that

$$\mathrm{var}(x^2) = \mathrm{E}(x^4) - [\mathrm{E}(x^2)]^2 = 4\lambda + 2 = 2(2\lambda + 1). \tag{2.27}$$

Thus, upon recalling results (1.29) and (1.30), it follows that

$$\mathrm{E}(w) = \mathrm{E}(x^2) + \mathrm{E}(u) = 1 + \lambda + N - 1 = N + \lambda, \tag{2.28}$$

$$\mathrm{var}(w) = \mathrm{var}(x^2) + \mathrm{var}(u) = 4\lambda + 2 + 2(N-1) = 2(N + 2\lambda), \tag{2.29}$$

and

$$\mathrm{E}(w^2) = \mathrm{var}(w) + [\mathrm{E}(w)]^2 = (N+2)(N+2\lambda) + \lambda^2. \tag{2.30}$$

Higher-order moments. Let (as in Part 1) w represent a random variable whose distribution is $\chi^2(N, \lambda)$. Further, take (for an arbitrary strictly positive integer k) $g_k(\cdot)$ to be the pdf of a $\chi^2(k)$ distribution, and recall [from result (2.14)] that the pdf, say $q(\cdot)$, of the $\chi^2(N, \lambda)$ distribution is expressible (for all w) as

$$q(w) = \sum_{j=0}^{\infty} \frac{(\lambda/2)^j e^{-\lambda/2}}{j!} g_{2j+N}(w).$$

Then, using result (1.27), we find that (for $r > -N/2$)

$$\begin{aligned}\mathrm{E}(w^r) &= \int_0^{\infty} w^r q(w)\, dw \\ &= \sum_{j=0}^{\infty} \frac{(\lambda/2)^j e^{-\lambda/2}}{j!} \int_0^{\infty} w^r g_{2j+N}(w)\, dw \\ &= \sum_{j=0}^{\infty} \frac{(\lambda/2)^j e^{-\lambda/2}}{j!} 2^r \frac{\Gamma[(N/2) + j + r]}{\Gamma[(N/2) + j]} \\ &= 2^r e^{-\lambda/2} \sum_{j=0}^{\infty} \frac{(\lambda/2)^j \Gamma[(N/2) + j + r]}{j!\, \Gamma[(N/2) + j]}. \end{aligned} \tag{2.31}$$

Now, define $m_r = \mathrm{E}(w^r)$, and regard m_r as a function of the noncentrality parameter λ. And observe that (for $r > -N/2$)

$$\frac{dm_r}{d\lambda} = -(1/2)m_r + 2^{r-1}e^{-\lambda/2} \sum_{j=0}^{\infty} \frac{j\,(\lambda/2)^{j-1}\Gamma[(N/2) + j + r]}{j!\, \Gamma[(N/2) + j]}$$

and hence that

$$\lambda\Big(m_r + 2\frac{dm_r}{d\lambda}\Big) = 2^{r+1}e^{-\lambda/2} \sum_{j=0}^{\infty} \frac{j\,(\lambda/2)^{j}\Gamma[(N/2) + j + r]}{j!\, \Gamma[(N/2) + j]}.$$

Thus, for $r > -N/2$,

$$\begin{aligned} m_{r+1} &= 2^{r+1}e^{-\lambda/2}\sum_{j=0}^{\infty}\frac{(\lambda/2)^j\,\Gamma[(N/2)+j+r+1]}{j!\,\Gamma[(N/2)+j]} \\ &= 2^{r+1}e^{-\lambda/2}\sum_{j=0}^{\infty}\frac{(\lambda/2)^j\,[(N/2)+j+r]\,\Gamma[(N/2)+j+r]}{j!\,\Gamma[(N/2)+j]} \\ &= 2[(N/2)+r]m_r+\lambda\Big(m_r+2\frac{dm_r}{d\lambda}\Big) \\ &= (N+2r+\lambda)m_r+2\lambda\frac{dm_r}{d\lambda}. \end{aligned} \tag{2.32}$$

Formula (2.32) relates $E(w^{r+1})$ to $E(w^r)$ and to the derivative of $E(w^r)$ (with respect to λ). Since clearly $E(w^0) = 1$, formula (2.32) can be used to determine the moments $E(w^1)$, $E(w^2)$, $E(w^3)$, … [of the $\chi^2(N,\lambda)$ distribution] recursively. In particular,

$$\begin{aligned} E(w) = m_1 &= [N+2(0)+\lambda]m_0+2\lambda\frac{dm_0}{d\lambda} \\ &= (N+\lambda)1+2\lambda(0) = N+\lambda \end{aligned} \tag{2.33}$$

and

$$\begin{aligned} E(w^2) = m_2 &= [N+2(1)+\lambda]m_1+2\lambda\frac{dm_1}{d\lambda} \\ &= (N+\lambda+2)(N+\lambda)+2\lambda(1) \\ &= (N+2)(N+2\lambda)+\lambda^2, \end{aligned} \tag{2.34}$$

in agreement with results (2.28) and (2.30) from Part 1.

Explicit expressions for the moments of the $\chi^2(N,\lambda)$ distribution are provided by the following result: for $r = 0, 1, 2, \ldots,$

$$E(w^r) = 2^r\,\Gamma[(N/2)+r]\sum_{j=0}^{r}\binom{r}{j}\frac{(\lambda/2)^j}{\Gamma[(N/2)+j]} \tag{2.35}$$

—interpret 0^0 as 1. This result, whose verification is the subject of the final part of the present subsection, provides a representation for the rth moment of the $\chi^2(N,\lambda)$ distribution that consists of a sum of $r+1$ terms. In contrast, the representation provided by formula (2.31) consists of a sum of an infinite number of terms. Moreover, by making use of result (1.23), result (2.35) can be reexpressed in the following, "simplified" form: for $r = 0, 1, 2, \ldots,$

$$E(w^r) = \lambda^r + \sum_{j=0}^{r-1}(N+2j)[N+2(j+1)][N+2(j+2)]\cdots[N+2(r-1)]\binom{r}{j}\lambda^j. \tag{2.36}$$

Verification of formula (2.35). The verification of formula (2.35) is by mathematical induction. The formula is valid for $r = 0$—according to the formula, $E(w^0) = 1$. Now, suppose that the formula is valid for $r = k$ (where k is an arbitrary nonnegative integer), that is, suppose that

$$m_k = 2^k\,\Gamma[(N/2)+k]\sum_{j=0}^{k}\binom{k}{j}\frac{(\lambda/2)^j}{\Gamma[(N/2)+j]} \tag{2.37}$$

—as before, $m_j = E(w^j)$ (for $j > -N/2$). We wish to show that formula (2.35) is valid for $r = k+1$, that is, to show that

$$m_{k+1} = 2^{k+1}\,\Gamma[(N/2)+k+1]\sum_{j=0}^{k+1}\binom{k+1}{j}\frac{(\lambda/2)^j}{\Gamma[(N/2)+j]}. \tag{2.38}$$

From result (2.32), we have that

$$m_{k+1} = (N + 2k)m_k + \lambda m_k + 2\lambda \frac{dm_k}{d\lambda}. \tag{2.39}$$

And it follows from supposition (2.37) that

$$\begin{aligned} \lambda m_k &= 2^{k+1}\Gamma[(N/2) + k] \sum_{s=0}^{k} \binom{k}{s} \frac{(\lambda/2)^{s+1}}{\Gamma[(N/2) + s]} \\ &= 2^{k+1}\Gamma[(N/2) + k] \sum_{j=1}^{k+1} \binom{k}{j-1} \frac{(\lambda/2)^{j}}{\Gamma[(N/2) + j - 1]} \\ &= 2^{k+1}\Gamma[(N/2) + k] \sum_{j=1}^{k+1} \binom{k}{j-1} \frac{[(N/2) + j - 1](\lambda/2)^{j}}{\Gamma[(N/2) + j]} \end{aligned} \tag{2.40}$$

and that

$$2\lambda \frac{dm_k}{d\lambda} = 2^{k+1}\Gamma[(N/2) + k] \sum_{j=1}^{k} \binom{k}{j} \frac{j(\lambda/2)^{j}}{\Gamma[(N/2) + j]}. \tag{2.41}$$

Upon starting with expression (2.39) and substituting expressions (2.37), (2.40), and (2.41) for m_k, λm_k, and $2\lambda(dm_k/d\lambda)$, respectively, we find that

$$\begin{aligned} m_{k+1} = 2^{k+1}\Gamma[(N/2) + k] \Bigg\{ &[(N/2) + k] \sum_{j=0}^{k} \binom{k}{j} \frac{(\lambda/2)^{j}}{\Gamma[(N/2) + j]} \\ &+ \sum_{j=1}^{k+1} \binom{k}{j-1} \frac{[(N/2) + j - 1](\lambda/2)^{j}}{\Gamma[(N/2) + j]} \\ &+ \sum_{j=1}^{k} \binom{k}{j} \frac{j(\lambda/2)^{j}}{\Gamma[(N/2) + j]} \Bigg\}. \end{aligned} \tag{2.42}$$

Expressions (2.38) and (2.42) are both polynomials of degree $k+1$ in $\lambda/2$. Thus, to establish equality (2.38), it suffices to establish that the coefficient of $(\lambda/2)^j$ is the same for each of these two polynomials ($j = 0, 1, \ldots, k+1$). In the case of the polynomial (2.42), the coefficient of $(\lambda/2)^0$ is

$$2^{k+1}\Gamma[(N/2)+k][(N/2)+k]/\Gamma(N/2) = 2^{k+1}\Gamma[(N/2)+k+1]/\Gamma(N/2), \tag{2.43}$$

the coefficient of $(\lambda/2)^{k+1}$ is

$$2^{k+1}\Gamma[(N/2) + k][(N/2) + k]/\Gamma[(N/2) + k + 1] = 2^{k+1}, \tag{2.44}$$

and, for $j = 1, 2, \ldots, k$, the coefficient of $(\lambda/2)^j$ is

$$\begin{aligned} &\frac{2^{k+1}\Gamma[(N/2)+k]}{\Gamma[(N/2)+j]} \left\{ [(N/2)+k]\binom{k}{j} + [(N/2)-1]\binom{k}{j-1} + j\left[\binom{k}{j-1} + \binom{k}{j}\right] \right\} \\ &\quad = \frac{2^{k+1}\Gamma[(N/2)+k]}{\Gamma[(N/2)+j]} \left\{ [(N/2)+k]\binom{k}{j} + [(N/2)-1]\binom{k}{j-1} + j\binom{k+1}{j} \right\} \\ &\quad = \frac{2^{k+1}\Gamma[(N/2)+k]}{\Gamma[(N/2)+j]} \binom{k+1}{j} \left\{ [(N/2)+k]\frac{k-j+1}{k+1} + [(N/2)-1]\frac{j}{k+1} + j \right\} \end{aligned}$$

$$= \frac{2^{k+1}\Gamma[(N/2)+k]}{\Gamma[(N/2)+j]}\binom{k+1}{j}[(N/2)+k]$$

$$= \frac{2^{k+1}\Gamma[(N/2)+k+1]}{\Gamma[(N/2)+j]}\binom{k+1}{j}. \tag{2.45}$$

Clearly, the coefficients (2.43), (2.44), and (2.45) are identical to the coefficients of the polynomial (2.38). Accordingly, equality (2.38 is established, and the mathematical-induction argument is complete.

g. An extension: the "noncentral gamma distribution"

For strictly positive parameters α and β and for a nonnegative parameter δ, let $f(\cdot)$ represent the function defined (on the real line) by

$$f(x) = \sum_{r=0}^{\infty} \frac{\delta^r e^{-\delta}}{r!} h_r(x), \tag{2.46}$$

where $h_r(\cdot)$ is the pdf of a $Ga(\alpha+r,\ \beta)$ distribution. The function $f(\cdot)$ is the pdf of a distribution. More specifically, it is the pdf of a mixture distribution; it is a weighted average of the pdfs of gamma distributions, where the weights $\delta^r e^{-\delta}/r!$ $(r = 0, 1, 2, \ldots)$ are the values assumed by the probability mass function of a Poisson distribution with parameter δ.

The noncentral chi-square distribution is related to the distribution with pdf (2.46) in much the same way that the (central) chi-square distribution is related to the gamma distribution. In the special case where $\alpha = N/2$, $\beta = 2$, and $\delta = \lambda/2$, the pdf (2.46) is identical to the pdf of the $\chi^2(N, \lambda)$ distribution. Accordingly, the pdf (2.46) provides a basis for extending the definition of the noncentral chi-square distribution to noninteger degrees of freedom.

Let us denote the distribution with pdf (2.46) by the symbol $Ga(\alpha, \beta, \delta)$. And let us write $m(\cdot)$ for the moment generating function of that distribution. Then, proceeding in much the same way as in arriving at expression (2.21) and recalling result (1.19), we find that (for $t < 1/\beta$)

$$m(t) = \sum_{r=0}^{\infty} \frac{\delta^r e^{-\delta}}{r!} \int_0^{\infty} e^{tx} h_r(x)\,dx$$

$$= \sum_{r=0}^{\infty} \frac{\delta^r e^{-\delta}}{r!} (1-\beta t)^{-(\alpha+r)}. \tag{2.47}$$

Moreover, analogous to expression (2.19), we have the power-series representation

$$\exp\left(\frac{\delta}{1-\beta t}\right) = \sum_{r=0}^{\infty} \frac{\delta^r}{(1-\beta t)^r\, r!},$$

so that the moment generating function is reexpressible in the form

$$m(t) = (1-\beta t)^{-\alpha} \exp\left(\frac{\delta\beta t}{1-\beta t}\right) \qquad (t < 1/\beta). \tag{2.48}$$

The cumulants of the $Ga(\alpha, \beta, \delta)$ distribution can be determined from expression (2.48) in essentially the same way that the cumulants of the $\chi^2(N, \lambda)$ distribution were determined from expression (2.18): for $r = 1, 2, 3, \ldots$, the rth cumulant is

$$(\alpha + \delta r)\,\beta^r\,(r-1)!. \tag{2.49}$$

Theorem 6.2.1 (on the distribution of a sum of noncentral chi-square random variables) can be generalized. Suppose that K random variables $w_1, w_2, \ldots, w_K$ are distributed independently as $Ga(\alpha_1, \beta, \delta_1)$, $Ga(\alpha_2, \beta, \delta_2)$, ..., $Ga(\alpha_K, \beta, \delta_K)$, respectively. Then, the moment generating function, say $m(\cdot)$, of the sum $\sum_{k=1}^{K} w_k$ of $w_1, w_2, \ldots, w_K$ is given by the formula

$$m(t) = (1-\beta t)^{-(\alpha_1+\alpha_2+\cdots+\alpha_K)} \exp\left[\frac{(\delta_1+\delta_2+\cdots+\delta_K)\beta t}{1-\beta t}\right] \qquad (t < 1/\beta),$$

as is evident from result (2.48) upon observing that

$$\mathrm{E}\big(e^{t\sum_{k=1}^{K} w_k}\big) = \mathrm{E}\Big(\prod_{k=1}^{K} e^{t w_k}\Big) = \prod_{k=1}^{K} \mathrm{E}(e^{t w_k}).$$

Thus, $m(\cdot)$ is the moment generating function of a $Ga\big(\sum_{k=1}^{K}\alpha_k,\ \beta,\ \sum_{k=1}^{K}\delta_k\big)$ distribution, and it follows that $\sum_{k=1}^{K} w_k \sim Ga\big(\sum_{k=1}^{K}\alpha_k,\ \beta,\ \sum_{k=1}^{K}\delta_k\big)$. In conclusion, we have the following theorem, which can be regarded as a generalization of Theorem 6.1.2 as well as of Theorem 6.2.1.

Theorem 6.2.2. If K random variables $w_1, w_2, \ldots, w_K$ are distributed independently as $Ga(\alpha_1, \beta, \delta_1)$, $Ga(\alpha_2, \beta, \delta_2)$, ..., $Ga(\alpha_K, \beta, \delta_K)$, respectively, then $w_1 + w_2 + \cdots + w_K \sim Ga\big(\sum_{k=1}^{K}\alpha_k,\ \beta,\ \sum_{k=1}^{K}\delta_k\big)$.

It remains to extend the results of Subsection f (on the moments of the noncentral chi-square distribution) to the moments of the distribution with pdf (2.46). Let w represent a random variable whose distribution is $Ga(\alpha, \beta, \delta)$. Then, as a readily derivable generalization of formula (2.31), we have that (for $r > -\alpha$)

$$\mathrm{E}(w^r) = \beta^r e^{-\delta} \sum_{j=0}^{\infty} \frac{\delta^j \Gamma(\alpha + j + r)}{j!\,\Gamma(\alpha + j)}. \tag{2.50}$$

Further, defining $m_r = \mathrm{E}(w^r)$, regarding m_r as a function of δ, and proceeding as in the derivation of result (2.32), we find that

$$m_{r+1} = \beta\Big[(\alpha + r + \delta)m_r + \delta\frac{dm_r}{d\delta}\Big]. \tag{2.51}$$

The recursive relationship (2.51) can be used to derive the first two moments of the $Ga(\alpha, \beta, \delta)$ distribution in much the same way that the recursive relationship (2.32) was used to derive the first two moments of the $\chi^2(N, \lambda)$ distribution. Upon observing that $\mathrm{E}(w^0) = 1$, we find that

$$\begin{aligned}\mathrm{E}(w) = m_1 &= \beta\Big[(\alpha + 0 + \delta)m_0 + \delta\frac{dm_0}{d\delta}\Big] \\ &= \beta(\alpha + \delta)\end{aligned} \tag{2.52}$$

and that

$$\begin{aligned}\mathrm{E}(w^2) = m_2 &= \beta\Big[(\alpha + 1 + \delta)m_1 + \delta\frac{dm_1}{d\delta}\Big] \\ &= \beta\Big[(\alpha + 1 + \delta)\beta(\alpha+\delta) + \delta\beta\Big] \\ &= \beta^2[(\alpha+1)(\alpha+2\delta) + \delta^2].\end{aligned} \tag{2.53}$$

Further,

$$\mathrm{var}(w) = \mathrm{E}(w^2) - [\mathrm{E}(w)]^2 = \beta^2(\alpha + 2\delta). \tag{2.54}$$

Finally, as a generalization of formula (2.35), we have that (for $r = 0, 1, 2, \ldots$)

$$\mathrm{E}(w^r) = \beta^r \Gamma(\alpha + r) \sum_{j=0}^{r} \binom{r}{j} \frac{\delta^j}{\Gamma(\alpha + j)}. \tag{2.55}$$

Formula (2.55) can be verified via a mathematical-induction argument akin to that employed in Part 3 of Subsection f, and, in what is essentially a generalization of result (2.36), is [in light of result (1.23)] reexpressible in the "simplified" form

$$\mathrm{E}(w^r) = \beta^r \Big[\delta^r + \sum_{j=0}^{r-1} (\alpha+j)(\alpha+j+1)(\alpha+j+2)\cdots(\alpha+r-1)\binom{r}{j}\delta^j. \tag{2.56}$$

h. An extension: distribution of the sum of the squared elements of a random vector that is distributed spherically about a nonnull vector of constants

Let $\mathbf{z} = (z_1, z_2, \ldots, z_N)'$ represent an N-dimensional random (column) vector that has any particular spherical distribution. Further, let $\boldsymbol{\mu} = (\mu_1, \mu_2, \ldots, \mu_N)'$ represent any N-dimensional nonrandom (column) vector. And consider the distribution of the quantity $\mathbf{x}'\mathbf{x} = \sum_{i=1}^N x_i^2$, where $\mathbf{x} = (x_1, x_2, \ldots, x_N)'$ is the N-dimensional random (column) vector defined as follows: $\mathbf{x} = \boldsymbol{\mu} + \mathbf{z}$ or, equivalently, $x_i = \mu_i + z_i$ $(i = 1, 2, \ldots, N)$ (so that $\mathbf{x}$ is distributed spherically about $\boldsymbol{\mu}$).

Define $\lambda = \boldsymbol{\mu}'\boldsymbol{\mu} = \sum_{i=1}^N \mu_i^2$ and $\tau = \sqrt{\lambda} = \sqrt{\boldsymbol{\mu}'\boldsymbol{\mu}} = \sqrt{\sum_{i=1}^N \mu_i^2}$. It follows from what was established earlier (in Subsection b) that in the special case where $\mathbf{z} \sim N(\mathbf{0}, \mathbf{I})$ and hence where $\mathbf{x} \sim N(\boldsymbol{\mu}, \mathbf{I})$, the distribution of $\mathbf{x}'\mathbf{x}$ depends on the value of $\boldsymbol{\mu}$ only through τ or, equivalently, only through λ—in that special case, the distribution of $\mathbf{x}'\mathbf{x}$ is, by definition, the noncentral chi-square distribution with parameters N and λ. In fact, the distribution of $\mathbf{x}'\mathbf{x}$ has this property (i.e., the property of depending on the value of $\boldsymbol{\mu}$ only through τ or, equivalently, only through λ) not only in the special case where $\mathbf{z} \sim N(\mathbf{0}, \mathbf{I})$, but also in the general case (where the distribution of $\mathbf{z}$ is any particular spherical distribution).

To see this, take (as in Subsection b) $\mathbf{A}$ to be any $N \times N$ orthogonal matrix whose first column is $(1/\tau)\boldsymbol{\mu}$—if $\tau = 0$ (or, equivalently, if $\boldsymbol{\mu} = \mathbf{0}$), take $\mathbf{A}$ to be an arbitrary $N \times N$ orthogonal matrix. And observe that

$$\mathbf{x}'\mathbf{x} = \mathbf{x}'\mathbf{I}\mathbf{x} = \mathbf{x}'\mathbf{A}\mathbf{A}'\mathbf{x} = (\mathbf{A}'\mathbf{x})'\mathbf{A}'\mathbf{x} \tag{2.57}$$

and that

$$\mathbf{A}'\mathbf{x} = \mathbf{A}'\boldsymbol{\mu} + \mathbf{w}, \tag{2.58}$$

where $\mathbf{w} = \mathbf{A}'\mathbf{z}$. Observe also that

$$\mathbf{A}'\boldsymbol{\mu} = \begin{pmatrix} \tau \\ \mathbf{0} \end{pmatrix}, \tag{2.59}$$

that

$$\mathbf{w} \sim \mathbf{z}, \tag{2.60}$$

and, upon partitioning $\mathbf{w}$ as $\mathbf{w} = \begin{pmatrix} w_1 \\ \mathbf{w}_2 \end{pmatrix}$ (where w_1 is a scalar), that

$$\mathbf{x}'\mathbf{x} = (\mathbf{A}'\boldsymbol{\mu} + \mathbf{w})'(\mathbf{A}'\boldsymbol{\mu} + \mathbf{w}) = (\tau + w_1)^2 + \mathbf{w}_2'\mathbf{w}_2 = \lambda + 2\tau w_1 + \mathbf{w}'\mathbf{w}. \tag{2.61}$$

It is now clear [in light of results (2.60) and (2.61)] that the distribution of $\mathbf{x}'\mathbf{x}$ depends on the value of $\boldsymbol{\mu}$ only through τ or, equivalently, only through λ.

Consider now the special case of an absolutely continuous spherical distribution where the distribution of $\mathbf{z}$ is absolutely continuous with a pdf $f(\cdot)$ such that

$$f(\mathbf{z}) = g(\mathbf{z}'\mathbf{z}) \qquad (\mathbf{z} \in \mathcal{R}^N) \tag{2.62}$$

for some (nonnegative) function $g(\cdot)$ (of a single nonnegative variable). Letting $u = \mathbf{w}'\mathbf{w}$ and $v = w_1/(\mathbf{w}'\mathbf{w})^{1/2}$, we find [in light of result (2.61)] that (for $\mathbf{w} \neq \mathbf{0}$) $\mathbf{x}'\mathbf{x}$ is expressible in the form

$$\mathbf{x}'\mathbf{x} = \lambda + 2\tau v u^{1/2} + u. \tag{2.63}$$

Moreover, in light of result (2.60), the joint distribution of u and v is the same as that of $\mathbf{z}'\mathbf{z}$ and $z_1/(\mathbf{z}'\mathbf{z})^{1/2}$, implying (in light of the results of Section 6.1g) that u and v are distributed independently, that the distribution of u is the distribution with pdf $r(\cdot)$, where

$$r(u) = \begin{cases} \dfrac{\pi^{N/2}}{\Gamma(N/2)} u^{(N/2)-1} g(u), & \text{for } 0 < u < \infty, \\ 0, & \text{for } -\infty < u \le 0, \end{cases} \tag{2.64}$$

and that the distribution of v is the distribution with pdf $h^*(\cdot)$, where

$$h^*(v) = \begin{cases} \dfrac{\Gamma(N/2)}{\Gamma[(N-1)/2]\pi^{1/2}} (1-v^2)^{(N-3)/2}, & \text{if } -1 < v < 1, \\ 0, & \text{otherwise.} \end{cases} \tag{2.65}$$

Define $y = \mathbf{x}'\mathbf{x}$. In the special case where $\mathbf{z} \sim N(\mathbf{0}, \mathbf{I})$ [and hence where $\mathbf{x} \sim N(\boldsymbol{\mu}, \mathbf{I})$], y has (by definition) a $\chi^2(N, \lambda)$ distribution, the pdf of which is a function $q(\cdot)$ that is expressible in the form (2.15). Let us obtain an expression for the pdf of the distribution of y in the general case [where the distribution of $\mathbf{z}$ is any (spherical) distribution that is absolutely continuous with a pdf $f(\cdot)$ of the form (2.62)].

Denote by $d(\cdot\,,\cdot)$ the pdf of the joint distribution of u and v, so that (for all u and v)

$$d(u, v) = r(u)h^*(v).$$

Now, introduce a change of variables from u and v to the random variables y and s, where

$$s = \frac{\tau + w_1}{y^{1/2}} = \frac{\tau + vu^{1/2}}{(\lambda + 2\tau v u^{1/2} + u)^{1/2}}$$

(and where $y = \lambda + 2\tau v u^{1/2} + u$). And observe that

$$\frac{\partial y}{\partial u} = 1 + \tau v u^{-1/2}, \qquad \frac{\partial y}{\partial v} = 2\tau u^{1/2},$$

$$\frac{\partial s}{\partial u} = (1/y)\big[(1/2)y^{1/2} v u^{-1/2} - (1/2)y^{-1/2}(1 + \tau v u^{-1/2})(\tau + vu^{1/2})\big], \quad \text{and}$$

$$\frac{\partial s}{\partial v} = (1/y)\big[y^{1/2}u^{1/2} - y^{-1/2}\tau u^{1/2}(\tau + vu^{1/2})\big];$$

accordingly,

$$\begin{vmatrix} \partial y/\partial u & \partial y/\partial v \\ \partial s/\partial u & \partial s/\partial v \end{vmatrix} = u^{1/2} y^{-1/2},$$

as can be readily verified. Observe also that $vu^{1/2} = sy^{1/2} - \tau$ and hence that

$$u = y - \lambda - 2\tau v u^{1/2} = y - 2\tau s y^{1/2} + \lambda$$

and

$$1 - v^2 = \frac{u - (sy^{1/2} - \tau)^2}{u} = \frac{y - s^2 y}{u} = (1 - s^2)(y/u).$$

Further, denoting by $b(\cdot\,,\cdot)$ the pdf of the joint distribution of y and s and making use of standard results on a change of variables (e.g., Bickel and Doksum 2001, sec. B.2.1), we find that, for $0 < y < \infty$ and for $-1 < s < 1$,

$$b(y, s) = \frac{\pi^{(N-1)/2}}{\Gamma[(N-1)/2]} y^{(N/2)-1} (1 - s^2)^{(N-3)/2} g(y - 2\tau s y^{1/2} + \lambda) \tag{2.66}$$

—for $-\infty < y \le 0$ and for s such that $s \le -1$ or $s \ge 1$, $b(y, s) = 0$.

We conclude that a pdf, say $q(\cdot)$, of the (marginal) distribution of the random variable y ($= \mathbf{x}'\mathbf{x}$) is obtainable by taking, for $0 < y < \infty$,

$$q(y) = \frac{\pi^{(N-1)/2}}{\Gamma[(N-1)/2]} y^{(N/2)-1} \int_{-1}^{1} (1 - s^2)^{(N-3)/2} g(y - 2\tau s y^{1/2} + \lambda)\, ds \tag{2.67}$$

[and by taking, for $-\infty < y \le 0$, $q(y) = 0$]. In the special case where $\lambda = 0$ (and hence where $\tau = 0$), expression (2.67) simplifies to

$$q(y) = \frac{\pi^{N/2}}{\Gamma(N/2)} y^{(N/2)-1} g(y),$$

in agreement with the pdf (1.56) derived earlier (in Section 6.1g) for that special case.

6.3 Central and Noncentral *F* Distributions

Let

$$F = \frac{u/M}{v/N} \quad \left[= (N/M)\frac{u}{v}\right], \tag{3.1}$$

where u and v are random variables that are distributed independently as $\chi^2(M)$ and $\chi^2(N)$, respectively. The distribution of the random variable F plays an important role in the use of linear models to make statistical inferences. This distribution is known as *Snedecor's F distribution* or simply as the *F distribution*. It is also sometimes referred to as Fisher's variance-ratio distribution.

The distribution of the random variable F depends on two parameters: M, which is referred to as the *numerator degrees of freedom*. and N, which is referred to as the *denominator degrees of freedom*. Let us denote the F distribution with M numerator degrees of freedom and N denominator degrees of freedom by the symbol $SF(M, N)$.

As a generalization of the F distribution, we have the noncentral F distribution. The *noncentral F distribution* is the distribution of the random variable

$$F^* = \frac{u^*/M}{v/N} \quad \left[= (N/M)\frac{u^*}{v}\right] \tag{3.2}$$

obtained from expression (3.1) for the random variable F upon replacing the random variable u (which has a central chi-square distribution) with a random variable u^* that has the noncentral chi-square distribution $\chi^2(M, \lambda)$ (and that, like u, is distributed independently of v). The distribution of the random variable F^* has three parameters: the *numerator degrees of freedom* M, the *denominator degrees of freedom* N, and the *noncentrality parameter* λ. Let us denote the noncentral F distribution with parameters M, N, and λ by the symbol $SF(M, N, \lambda)$.

Clearly, the F distribution $SF(M, N)$ can be regarded as the special case $SF(M, N, 0)$ of the noncentral F distribution obtained upon setting the noncentrality parameter λ equal to 0. For the sake of clarity (i.e., to distinguish it from the noncentral F distribution), the "ordinary" F distribution may be referred to as the *central F distribution*.

a. (Central) *F* distribution

Relationship to the beta distribution. Let u and v represent random variables that are distributed independently as $\chi^2(M)$ and $\chi^2(N)$, respectively. Further, define

$$w = \frac{u/M}{v/N} \quad \text{and} \quad x = \frac{u}{u+v}.$$

Then,

$$w = \frac{Nx}{M(1-x)} \quad \text{and} \quad x = \frac{Mw}{N+Mw} \quad \left[= \frac{(M/N)w}{1+(M/N)w}\right], \tag{3.3}$$

as can be readily verified.

By definition, $w \sim SF(M, N)$. And in light of the discussion of Sections 6.1a, 6.1b, and 6.1c, $x \sim Be(M/2,\ N/2)$. In effect, we have established the following result: if x is a random variable that is distributed as $Be(M/2, N/2)$, then

$$\frac{Nx}{M(1-x)} \sim SF(M, N); \tag{3.4}$$

and if w is a random variable that is distributed as $SF(M, N)$, then

$$\frac{Mw}{N+Mw} \sim Be(M/2,\ N/2). \tag{3.5}$$

The cdf (cumulative distribution function) of an F distribution can be reexpressed in terms of an incomplete beta function ratio (which coincides with the cdf of a beta distribution)—the incomplete beta function ratio was introduced and discussed in Section 6.1b. Denote by $F(\cdot)$ the cdf of the $SF(M, N)$ distribution. Then, letting x represent a random variable that is distributed as $Be(M/2, N/2)$, we find [in light of result (3.4)] that (for any nonnegative scalar c)

$$F(c) = \Pr\biggl[\frac{Nx}{M(1-x)} \le c\biggr] = \Pr\biggl[x \le \frac{Mc}{N+Mc}\biggr] = I_{Mc/(N+Mc)}(M/2,\ N/2) \tag{3.6}$$

—for $c < 0$, $F(c) = 0$. Moreover, in light of result (1.14) on the incomplete beta function ratio, the cdf of the $SF(M, N)$ distribution can also be expressed (for $c \ge 0$) as

$$F(c) = 1 - I_{N/(N+Mc)}(N/2,\ M/2). \tag{3.7}$$

Distribution of the reciprocal. Let w represent a random variable that has an $SF(M, N)$ distribution. Then, clearly,

$$\frac{1}{w} \sim SF(N, M). \tag{3.8}$$

Now, let $F(\cdot)$ represent the cdf (cumulative distribution function) of the $SF(M, N)$ distribution and $G(\cdot)$ the cdf of the $SF(N, M)$ distribution. Then, for any strictly positive scalar c,

$$F(c) = 1 - G(1/c), \tag{3.9}$$

as is evident upon observing that

$$\Pr(w \le c) = \Pr(1/w \ge 1/c) = \Pr(1/w > 1/c) = 1 - \Pr(1/w \le 1/c)$$

[and as could also be ascertained from results (3.6) and (3.7)]. Moreover, for $0 < \alpha < 1$, the upper $100\alpha\%$ point, say $\bar{F}_\alpha(M, N)$, of the $SF(M, N)$ distribution is related to the upper $100(1-\alpha)\%$ point, say $\bar{F}_{1-\alpha}(N, M)$, of the $SF(N, M)$ distribution as follows:

$$\bar{F}_\alpha(M, N) = 1/\bar{F}_{1-\alpha}(N, M) \tag{3.10}$$

—this relationship can be readily verified by applying result (3.9) [with $c = 1/\bar{F}_{1-\alpha}(N, M)$] or simply by observing that

$$\Pr[w > 1/\bar{F}_{1-\alpha}(N, M)] = \Pr[1/w < \bar{F}_{1-\alpha}(N, M)] = \Pr[1/w \le \bar{F}_{1-\alpha}(N, M)] = \alpha.$$

Joint distribution. As in Part 1 of the present subsection, take u and v to be statistically independent random variables that are distributed as $\chi^2(M)$ and $\chi^2(N)$, respectively, and define $w = (u/M)/(v/N)$ and $x = u/(u+v)$. Let us consider the joint distribution of w and the random variable y defined as follows:

$$y = u + v.$$

By definition, $w \sim SF(M, N)$. And in light of Theorem 6.1.3, $y \sim \chi^2(M+N)$. Moreover, upon observing (in light of the results of Section 6.1c) that the $\chi^2(M)$ distribution is identical to the $Ga(M/2,\ 2)$ distribution and that the $\chi^2(N)$ distribution is identical to the $Ga(N/2,\ 2)$ distribution, it follows from Theorem 6.1.1 that y is distributed independently of x and hence [since, in light of result (3.3), w is expressible as a function of x] that y is distributed independently of w.

Probability density function (pdf). Let x represent a random variable whose distribution is $Be(M/2,\ N/2)$ (where M and N are arbitrary strictly positive integers). As is evident from the results of Sections 6.1a and 6.1b, the $Be(M/2, N/2)$ distribution has as a pdf the function $h(\cdot)$ given by the formula

$$h(x) = \begin{cases} \dfrac{\Gamma[(M+N)/2]}{\Gamma(M/2)\Gamma(N/2)}\, x^{(M/2)-1}(1-x)^{(N/2)-1}, & \text{for } 0 < x < 1, \\ 0, & \text{elsewhere.} \end{cases}$$

Consider the random variable w defined as follows:

$$w = \frac{Nx}{M(1-x)}. \tag{3.11}$$

According to result (3.4), $w \sim SF(M, N)$. Moreover, equality (3.11) defines a one-to-one transformation from the interval $0 < x < 1$ onto the interval $0 < w < \infty$; the inverse transformation is that defined by the equality

$$x = (M/N)w[1 + (M/N)w]^{-1}.$$

Thus, a pdf, say $f(\cdot)$, of the $SF(M, N)$ distribution is obtainable from the pdf of the $Be(M/2,\ N/2)$ distribution. Upon observing that

$$\frac{dx}{dw} = (M/N)[1 + (M/N)w]^{-2} \quad \text{and} \quad 1 - x = [1 + (M/N)w]^{-1}$$

and making use of standard results on a change of variable, we find that, for $0 < w < \infty$,

$$\begin{aligned} f(w) &= h\{(M/N)w[1 + (M/N)w]^{-1}\}(M/N)[1 + (M/N)w]^{-2} \\ &= \frac{\Gamma[(M+N)/2]}{\Gamma(M/2)\Gamma(N/2)}(M/N)^{M/2} w^{(M/2)-1}[1 + (M/N)w]^{-(M+N)/2} \end{aligned} \tag{3.12}$$

—for $-\infty < w \le 0,\ f(w) = 0$.

Moments. Let w represent a random variable that has an $SF(M, N)$ distribution. Then, by definition, $w \sim (N/M)(u/v)$, where u and v are random variables that are distributed independently as $\chi^2(M)$ and $\chi^2(N)$, respectively. And making use of result (1.27), we find that, for $-M/2 < r < N/2$,

$$\mathrm{E}(w^r) = (N/M)^r\, \mathrm{E}(u^r)\,\mathrm{E}(v^{-r}) = (N/M)^r \frac{\Gamma[(M/2)+r]}{\Gamma(M/2)} \frac{\Gamma[(N/2)-r]}{\Gamma(N/2)}. \tag{3.13}$$

Further, it follows from results (1.28) and (1.31) that the rth (integer) moment of the $SF(M, N)$ distribution is expressible as

$$\mathrm{E}(w^r) = (N/M)^r \frac{M(M+2)(M+4)\cdots[M+2(r-1)]}{(N-2)(N-4)\cdots(N-2r)} \tag{3.14}$$

($r = 1, 2, \ldots < N/2$). For $r \ge N/2$, the rth moment of the $SF(M, N)$ distribution does not exist. (And, as a consequence, the F distribution does not have a moment generating function.)

The mean of the $SF(M, N)$ distribution is (if $N > 2$)

$$\mathrm{E}(w) = N/(N-2). \tag{3.15}$$

And the second moment and the variance are (if $N > 4$)

$$\mathrm{E}(w^2) = (N/M)^2 \frac{M(M+2)}{(N-2)(N-4)} = \frac{M+2}{M} \frac{N^2}{(N-2)(N-4)} \tag{3.16}$$

and

$$\mathrm{var}(w) = \mathrm{E}(w^2) - [\mathrm{E}(w)]^2 = \frac{2N^2(M+N-2)}{M(N-2)^2(N-4)}. \tag{3.17}$$

Noninteger degrees of freedom. The definition of the F distribution can be extended to noninteger degrees of freedom in much the same way as the definition of the chi-square distribution. For arbitrary strictly positive numbers M and N, take u and v to be random variables that are distributed independently as $Ga\big(\frac{M}{2}, 2\big)$ and $Ga\big(\frac{N}{2}, 2\big)$, respectively, and define

$$w = \frac{u/M}{v/N}. \tag{3.18}$$

Let us regard the distribution of the random variable w as an F distribution with M (possibly noninteger) numerator degrees of freedom and N (possibly noninteger) denominator degrees of freedom. In the special case where M and N are (strictly positive) integers, the $Ga\big(\frac{M}{2}, 2\big)$ distribution is identical to the $\chi^2(M)$ distribution and the $Ga\big(\frac{N}{2}, 2\big)$ distribution is identical to the $\chi^2(N)$ distribution (as is evident from the results of Section 6.1c), so that this usage of the term F distribution is consistent with our previous usage of this term.

Note that in the definition (3.18) of the random variable w, we could have taken u and v to be random variables that are distributed independently as $Ga\big(\frac{M}{2}, \beta\big)$ and $Ga\big(\frac{N}{2}, \beta\big)$, respectively (where β is an arbitrary strictly positive scalar). The distribution of w is unaffected by the choice of β. To see this, observe (as in Part 1 of the present subsection) that $w = Nx/[M(1-x)]$, where $x = u/(u+v)$, and that (irrespective of the value of β) $u/(u+v) \sim Be\big(\frac{M}{2}, \frac{N}{2}\big)$.

A function of random variables that are distributed independently and identically as $N(0,1)$. Let $\mathbf{z} = (z_1, z_2, \dots, z_N)'$ (where $N \geq 2$) represent an (N-dimensional) random (column) vector that has an N-variate standard normal distribution or, equivalently, whose elements are distributed independently and identically as $N(0,1)$. And let K represent any integer between 1 and $N-1$, inclusive. Then,

$$\frac{\sum_{i=1}^{K} z_i^2/K}{\sum_{i=K+1}^{N} z_i^2/(N-K)} \sim SF(K, N-K), \tag{3.19}$$

as is evident from the very definitions of the chi-square and F distributions.

A multivariate version. Let $u_1, u_2, \dots, u_J$, and v represent random variables that are distributed independently as $\chi^2(M_1), \chi^2(M_2), \dots, \chi^2(M_J)$, and $\chi^2(N)$, respectively. Further, define (for $j = 1, 2, \dots, J$)

$$w_j = \frac{u_j/M_j}{v/N}.$$

Then, the (marginal) distribution of w_j is $SF(M_j, N)$. And the (marginal) distribution of w_j is related to a beta distribution; as is evident from the results of Part 1 of the present subsection (and as can be readily verified),

$$w_j = \frac{Nx_j}{M_j(1-x_j)},$$

where $x_j = u_j/(u_j+v) \sim Be\big(\frac{M_j}{2}, \frac{N}{2}\big)$.

Now, consider the joint distribution of the random variables $w_1, w_2, \dots, w_J$. The numerators $u_1/M_1, u_2/M_2, \dots, u_J/M_J$ of $w_1, w_2, \dots, w_J$ are statistically independent, however they have a common denominator v/N.

For $j = 1, 2, \ldots, J$, let
$$s_j = \frac{u_j}{v + \sum_{j'=1}^{J} u_{j'}}.$$
Then, $1 - \sum_{j'=1}^{J} s_{j'} = v \big/ \big(v + \sum_{j'=1}^{J} u_{j'}\big)$ and, consequently,
$$w_j = \frac{N s_j}{M_j \big(1 - \sum_{j'=1}^{J} s_{j'}\big)}.$$
Moreover, the joint distribution of $s_1, s_2, \ldots, s_J$ is the Dirichlet distribution $Di\big(\frac{M_1}{2}, \frac{M_2}{2}, \ldots, \frac{M_J}{2}, \frac{N}{2}; J\big)$. Thus, the joint distribution of $w_1, w_2, \ldots, w_J$ is expressible in terms of a Dirichlet distribution.

More specifically, suppose that $\mathbf{z}_1, \mathbf{z}_2, \ldots, \mathbf{z}_J, \mathbf{z}_{J+1}$ are random column vectors of dimensions $N_1, N_2, \ldots, N_J, N_{J+1}$, respectively, the joint distribution of which is $\big(\sum_{j=1}^{J+1} N_j\big)$-variate standard normal $N(\mathbf{0}, \mathbf{I})$. And, for $j = 1, 2, \ldots, J, J+1$, denote by z_{jk} the kth element of $\mathbf{z}_j$. Further, define
$$w_j = \frac{\sum_{k=1}^{N_j} z_{jk}^2 / N_j}{\sum_{k=1}^{N_{J+1}} z_{J+1,k}^2 / N_{J+1}} \qquad (j = 1, 2, \ldots, J),$$
and observe that the $J+1$ sums of squares $\sum_{k=1}^{N_1} z_{1k}^2, \sum_{k=1}^{N_2} z_{2k}^2, \ldots, \sum_{k=1}^{N_J} z_{Jk}^2$, and $\sum_{k=1}^{N_{J+1}} z_{J+1,k}^2$ are distributed independently as $\chi^2(N_1), \chi^2(N_2), \ldots, \chi^2(N_J)$, and $\chi^2(N_{J+1})$, respectively. Then, for $j = 1, 2, \ldots, J$, the (marginal) distribution of w_j is $SF(N_j, N_{J+1})$ and is related to a beta distribution; for $j = 1, 2, \ldots, J$,
$$w_j = \frac{N_{J+1} x_j}{N_j (1 - x_j)},$$
where $x_j = \sum_{k=1}^{N_j} z_{jk}^2 \big/ \big(\sum_{k=1}^{N_j} z_{jk}^2 + \sum_{k=1}^{N_{J+1}} z_{J+1,k}^2\big) \sim Be\big(\frac{N_j}{2}, \frac{N_{J+1}}{2}\big)$. The joint distribution of $w_1, w_2, \ldots, w_J$ is related to a Dirichlet distribution; clearly,
$$w_j = \frac{N_{J+1} s_j}{N_j \big(1 - \sum_{j'=1}^{J} s_{j'}\big)}, \quad \text{where} \quad s_j = \frac{\sum_{k=1}^{N_j} z_{jk}^2}{\sum_{j'=1}^{J+1} \sum_{k=1}^{N_{j'}} z_{j'k}^2} \qquad (j = 1, 2, \ldots, J),$$
and $s_1, s_2, \ldots, s_J$ are jointly distributed as $Di\big(\frac{N_1}{2}, \frac{N_2}{2}, \ldots, \frac{N_J}{2}, \frac{N_{J+1}}{2}; J\big)$.

b. The wider applicability of various results derived under an assumption of normality

Let $\mathbf{z} = (z_1, \ldots, z_K, z_{K+1})'$ represent any $(K+1)$-dimensional random (column) vector having an absolutely continuous spherical distribution. And define $w_0 = \sum_{k=1}^{K+1} z_k^2$ and (for $k = 1, \ldots, K, K+1$)
$$y_k = \frac{z_k}{\big(\sum_{j=1}^{K+1} z_j^2\big)^{1/2}} \quad \text{and} \quad w_k = y_k^2 = \frac{z_k^2}{\sum_{j=1}^{K+1} z_j^2},$$
in which case $w_{K+1} = 1 - \sum_{k=1}^{K} w_k$. As is evident from the results of Part 1 of Section 6.1g, $w_1, w_2, \ldots, w_K$ are statistically independent of w_0 and have a $Di\big(\frac{1}{2}, \ldots, \frac{1}{2}, \frac{1}{2}; K\big)$ distribution; and as is evident from the results of Part 2, $y_1, \ldots, y_K, y_{K+1}$ are statistically independent of w_0 and are distributed uniformly on the surface of a $(K+1)$-dimensional unit ball. There is an implication that the distribution of $w_1, \ldots, w_K, w_{K+1}$ and the distribution of $y_1, \ldots, y_K, y_{K+1}$ are the same in the general case (of an arbitrary absolutely continuous spherical distribution) as in the special case where $\mathbf{z} \sim N(\mathbf{0}, \mathbf{I}_{K+1})$. Thus, we have the following theorem.

Theorem 6.3.1. Let $\mathbf{z} = (z_1, \ldots, z_K, z_{K+1})'$ represent any $(K+1)$-dimensional random (column) vector having an absolutely continuous spherical distribution. And define $w_0 = \sum_{k=1}^{K+1} z_k^2$ and (for $k = 1, \ldots, K, K+1$)

$$y_k = \frac{z_k}{\left(\sum_{j=1}^{K+1} z_j^2\right)^{1/2}} \quad \text{and} \quad w_k = y_k^2 = \frac{z_k^2}{\sum_{j=1}^{K+1} z_j^2}.$$

(1) For "any" function $g(\cdot)$ (defined on $\mathcal{R}^{K+1}$) such that $g(\mathbf{z})$ depends on the value of $\mathbf{z}$ only through $w_1, \dots, w_K, w_{K+1}$ or, more generally, only through $y_1, \dots, y_K, y_{K+1}$, the random variable $g(\mathbf{z})$ is statistically independent of w_0 and has the same distribution in the general case (of an arbitrary absolutely continuous spherical distribution) as in the special case where $\mathbf{z} \sim N(\mathbf{0}, \mathbf{I}_{K+1})$. (2) For "any" P functions $g_1(\cdot), g_2(\cdot), \dots, g_P(\cdot)$ (defined on $\mathcal{R}^{K+1}$) such that (for $j = 1, 2, \dots, P$) $g_j(\mathbf{z})$ depends on the value of $\mathbf{z}$ only through $w_1, \dots, w_K, w_{K+1}$ or, more generally, only through $y_1, \dots, y_K, y_{K+1}$, the random variables $g_1(\mathbf{z}), g_2(\mathbf{z}), \dots, g_P(\mathbf{z})$ are statistically independent of w_0 and have the same (joint) distribution in the general case (of an arbitrary absolutely continuous spherical distribution) as in the special case where $\mathbf{z} \sim N(\mathbf{0}, \mathbf{I}_{K+1})$.

Application to the F distribution. Let $\mathbf{z} = (z_1, z_2, \dots, z_N)'$ (where $N \geq 2$) represent an N-dimensional random (column) vector. In the next-to-last part of Subsection a, it was established that if $\mathbf{z} \sim N(\mathbf{0}, \mathbf{I}_N)$, then (for any integer K between 1 and $N-1$, inclusive)

$$\frac{\sum_{i=1}^{K} z_i^2/K}{\sum_{i=K+1}^{N} z_i^2/(N-K)} \sim SF(K, N-K). \tag{3.20}$$

Clearly,

$$\frac{\sum_{i=1}^{K} z_i^2/K}{\sum_{i=K+1}^{N} z_i^2/(N-K)} = \frac{\sum_{i=1}^{K} w_i/K}{\sum_{i=K+1}^{N} w_i/(N-K)}, \tag{3.21}$$

where (for $i = 1, 2, \dots, N$) $w_i = z_i^2 / \sum_{i'=1}^{N} z_{i'}^2$. And we conclude [on the basis of Part (1) of Theorem 6.3.1] that the random variable (3.21) is distributed (independently of $\sum_{i=1}^{N} z_i^2$) as $SF(K, N-K)$ provided only that $\mathbf{z}$ has an absolutely continuous spherical distribution—that $\mathbf{z} \sim N(\mathbf{0}, \mathbf{I}_N)$ is sufficient, but it is not necessary.

A "multivariate" version of the application. Let $\mathbf{z}_1, \mathbf{z}_2, \dots, \mathbf{z}_J, \mathbf{z}_{J+1}$ represent random column vectors of dimensions $N_1, N_2, \dots, N_J, N_{J+1}$, respectively. And, for $j = 1, 2, \dots, J, J+1$, denote by z_{jk} the kth element of $\mathbf{z}_j$. Further, define

$$w_j = \frac{\sum_{k=1}^{N_j} z_{jk}^2/N_j}{\sum_{k=1}^{N_{J+1}} z_{J+1,k}^2/N_{J+1}} \quad (j = 1, 2, \dots, J).$$

Then, as observed earlier (in the last part of Subsection a) in connection with the special case where the joint distribution of $\mathbf{z}_1, \mathbf{z}_2, \dots, \mathbf{z}_J, \mathbf{z}_{J+1}$ is $\left(\sum_{j=1}^{J+1} N_j\right)$-variate standard normal $N(\mathbf{0}, \mathbf{I})$ (and as can be readily verified),

$$w_j = \frac{N_{J+1} s_j}{N_j\left(1 - \sum_{j'=1}^{J} s_{j'}\right)}, \quad \text{where} \quad s_j = \frac{\sum_{k=1}^{N_j} z_{jk}^2}{\sum_{j'=1}^{J+1} \sum_{k=1}^{N_{j'}} z_{j'k}^2} \quad (j = 1, 2, \dots, J).$$

Clearly, $s_1, s_2, \dots, s_J$ and hence $w_1, w_2, \dots, w_J$ depend on the values of the $\sum_{j=1}^{J+1} N_j$ quantities $z_{jk}^2 / \sum_{j'=1}^{J+1} \sum_{k'=1}^{N_{j'}} z_{j'k'}^2$ ($j = 1, 2, \dots, J+1$; $k = 1, 2, \dots, N_j$). Thus, it follows from Theorem 6.3.1 that if the joint distribution of $\mathbf{z}_1, \mathbf{z}_2, \dots, \mathbf{z}_J, \mathbf{z}_{J+1}$ is an absolutely continuous spherical distribution, then $s_1, s_2, \dots, s_J$ and $w_1, w_2, \dots, w_J$ are distributed independently of $\sum_{j=1}^{J+1} \sum_{k=1}^{N_j} z_{jk}^2$, and the joint distribution of $s_1, s_2, \dots, s_J$ and the joint distribution of $w_1, w_2, \dots, w_J$ are the same as in the special case where the joint distribution of $\mathbf{z}_1, \mathbf{z}_2, \dots, \mathbf{z}_J, \mathbf{z}_{J+1}$ is $N(\mathbf{0}, \mathbf{I})$. And in light of the results obtained earlier (in the last part of Subsection a) for that special case, we are able to infer that if the joint distribution of $\mathbf{z}_1, \mathbf{z}_2, \dots, \mathbf{z}_J, \mathbf{z}_{J+1}$ is an absolutely continuous spherical distribution, then the joint distribution of $s_1, s_2, \dots, s_J$ is $Di\left(\frac{N_1}{2}, \frac{N_2}{2}, \dots, \frac{N_J}{2}, \frac{N_{J+1}}{2}; J\right)$. [The same inference could be made "directly" by using the results of Section 6.1g to establish that if the joint distribution of $\mathbf{z}_1, \mathbf{z}_2, \ldots, \mathbf{z}_J, \mathbf{z}_{J+1}$ is an absolutely continuous spherical distribution, then the joint distribution of the $\sum_{j=1}^{J+1} N_j$ random variables $z_{jk}^2 / \sum_{j'=1}^{J+1} \sum_{k'=1}^{N_{j'}} z_{j'k'}^2$ is $Di\left(\frac{1}{2}, \frac{1}{2}, \ldots, \frac{1}{2}, \frac{1}{2}; \sum_{j=1}^{J+1} N_j - 1\right)$ and by then applying Theorem 6.1.5.]

c. Noncentral F distribution

A related distribution: the noncentral beta distribution. Let u^* and v represent random variables that are distributed independently as $\chi^2(M, \lambda)$ and $\chi^2(N)$, respectively. Further, define

$$w = \frac{u^*/M}{v/N} \quad \text{and} \quad x = \frac{u^*}{u^* + v}.$$

Then, as in the case of result (3.3),

$$w = \frac{Nx}{M(1-x)} \quad \text{and} \quad x = \frac{Mw}{N + Mw} \left[= \frac{(M/N)w}{1 + (M/N)w} \right]. \tag{3.22}$$

By definition, $w \sim SF(M, N, \lambda)$. And the distribution of x is a generalization of the $Be\left(\frac{M}{2}, \frac{N}{2}\right)$ distribution that is referred to as a *noncentral beta distribution* with parameters $M/2$, $N/2$, and $\lambda/2$ and that is to be denoted by the symbol $Be\left(\frac{M}{2}, \frac{N}{2}, \frac{\lambda}{2}\right)$.

Let $y = u^* + v$. Then, in light of results (2.11) and (2.13), the joint distribution of y and x has as a pdf the function $f(\cdot\,,\cdot)$ obtained by taking (for all y and x)

$$f(y, x) = \sum_{r=0}^{\infty} p(r)\, g_{M+N+2r}(y)\, d_{(M+2r)/2,\, N/2}(x), \tag{3.23}$$

where (for $r = 0, 1, 2, \ldots$) $p(r) = (\lambda/2)^r e^{-\lambda/2}/r!$, where (for any strictly positive integer j) $g_j(\cdot)$ denotes the pdf of a $\chi^2(j)$ distribution, and where (for any values of the parameters α_1 and α_2) $d_{\alpha_1,\alpha_2}(\cdot)$ denotes the pdf of a $Be(\alpha_1, \alpha_2)$ distribution. Thus, a pdf, say $h(\cdot)$, of the (marginal) distribution of x is obtained by taking (for all x)

$$\begin{aligned} h(x) = \int_0^{\infty} f(y, x)\, dy &= \sum_{r=0}^{\infty} p(r)\, d_{(M+2r)/2,\, N/2}(x) \int_0^{\infty} g_{M+N+2r}(y)\, dy \\ &= \sum_{r=0}^{\infty} p(r)\, d_{(M+2r)/2,\, N/2}(x). \end{aligned} \tag{3.24}$$

An extension. The definition of the noncentral beta distribution can be extended. Take $x = u^*/(u^* + v)$, where u^* and v are random variables that are distributed independently as $Ga(\alpha_1, \beta, \delta)$ and $Ga(\alpha_2, \beta)$, respectively—here, α_1, α_2, and β are arbitrary strictly positive scalars and δ is an arbitrary nonnegative scalar. Letting $y = u^* + v$ and proceeding in essentially the same way as in the derivation of result (2.11), we find that the joint distribution of y and x has as a pdf the function $f(\cdot\,,\cdot)$ obtained by taking (for all y and x)

$$f(y, x) = \sum_{r=0}^{\infty} \frac{\delta^r e^{-\delta}}{r!}\, g_r^*(y)\, d_{\alpha_1+r,\, \alpha_2}(x), \tag{3.25}$$

where (for $r = 0, 1, 2, \ldots$) $g_r^*(\cdot)$ represents the pdf of a $Ga(\alpha_1 + \alpha_2 + r, \beta)$ distribution and $d_{\alpha_1+r,\alpha_2}(\cdot)$ represents the pdf of a $Be(\alpha_1 + r, \alpha_2)$ distribution. Accordingly, the pdf of the (marginal) distribution of x is the function $h(\cdot)$ obtained by taking (for all x)

$$h(x) = \int_0^{\infty} f(y, x)\, dy = \sum_{r=0}^{\infty} \frac{\delta^r e^{-\delta}}{r!}\, d_{\alpha_1+r,\, \alpha_2}(x). \tag{3.26}$$

Formulas (3.25) and (3.26) can be regarded as extensions of formulas (3.23) and (3.24); formulas (3.23) and (3.24) are for the special case where $\alpha_1 = M/2$, $\beta = 2$, and $\alpha_2 = N/2$ (and where δ is expressed in the form $\lambda/2$).

Take the *noncentral beta distribution* with parameters α_1, α_2, and δ (where $\alpha_1 > 0, \alpha_2 > 0$, and $\delta \geq 0$) to be the distribution of the random variable x or, equivalently, the distribution with pdf $h(\cdot)$ given by expression (3.26)—the distribution of x does not depend on the parameter β. And denote this distribution by the symbol $Be(\alpha_1, \alpha_2, \delta)$.

Probability density function (of the noncentral F distribution). Earlier (in Subsection a), the pdf (3.12) of the (central) F distribution was derived from the pdf of a beta distribution. By taking a similar approach, the pdf of the noncentral F distribution can be derived from the pdf of a noncentral beta distribution.

Let x represent a random variable that has a noncentral beta distribution with parameters $M/2$, $N/2$, and $\lambda/2$ (where M and N are arbitrary strictly positive integers and λ is an arbitrary nonnegative scalar). And define

$$w = \frac{Nx}{M(1-x)}. \tag{3.27}$$

Then, in light of what was established earlier (in Part 1 of the present subsection),

$$w \sim SF(M, N, \lambda).$$

Moreover, equality (3.27) defines a one-to-one transformation from the interval $0 < x < 1$ onto the interval $0 < w < \infty$; the inverse transformation is that defined by the equality

$$x = (M/N)w[1 + (M/N)w]^{-1}.$$

Thus, a pdf, say $f(\cdot)$, of the $SF(M, N, \lambda)$ distribution is obtainable from a pdf of the distribution of x.

For $r = 0, 1, 2, \ldots$, let $p(r) = (\lambda/2)^r e^{-\lambda/2}/r!$, and, for arbitrary strictly positive scalars α_1 and α_2, denote by $d_{\alpha_1,\alpha_2}(\cdot)$ the pdf of a $Be(\alpha_1, \alpha_2)$ distribution. Then, according to result (3.24), a pdf, say $h(\cdot)$, of the distribution of x is obtained by taking (for all x)

$$h(x) = \sum_{r=0}^{\infty} p(r)\, d_{(M+2r)/2,\, N/2}(x).$$

And upon observing that

$$\frac{dx}{dw} = (M/N)[1 + (M/N)w]^{-2} \quad \text{and} \quad 1 - x = [1 + (M/N)w]^{-1}$$

and making use of standard results on a change of variable, we find that, for $0 < w < \infty$,

$$\begin{aligned} f(w) &= h\{(M/N)w[1 + (M/N)w]^{-1}\}(M/N)[1 + (M/N)w]^{-2} \\ &= \sum_{r=0}^{\infty} p(r) \frac{\Gamma[(M+N+2r)/2]}{\Gamma[(M+2r)/2]\Gamma(N/2)} (M/N)^{(M+2r)/2} \\ &\qquad\qquad \times w^{[(M+2r)/2]-1}[1 + (M/N)w]^{-(M+N+2r)/2} \\ &= \sum_{r=0}^{\infty} p(r)[M/(M+2r)]\, g_r\{[M/(M+2r)]w\}, \end{aligned} \tag{3.28}$$

where (for $r = 0, 1, 2, \ldots$) $g_r(\cdot)$ denotes the pdf of the $SF(M+2r,\ N)$ distribution—for $-\infty < w \leq 0,\ f(w) = 0$.

Moments. Let w represent a random variable that has an $SF(M, N, \lambda)$ distribution. Then, by definition, $w \sim (N/M)(u/v)$, where u and v are random variables that are distributed independently as $\chi^2(M, \lambda)$ and $\chi^2(N)$, respectively. And making use of results (1.27) and (2.31), we find that, for $-M/2 < r < N/2$,

$$\begin{aligned}
\mathrm{E}(w^r) &= (N/M)^r\,\mathrm{E}(u^r)\,\mathrm{E}(v^{-r}) \\
&= (N/M)^r\,\frac{\Gamma[(N/2)-r]}{2^r\,\Gamma(N/2)}\,\mathrm{E}(u^r) \qquad (3.29)\\
&= (N/M)^r\,\frac{\Gamma[(N/2)-r]}{\Gamma(N/2)}\,e^{-\lambda/2}\sum_{j=0}^{\infty}\frac{(\lambda/2)^j\,\Gamma[(M/2)+j+r]}{j!\,\Gamma[(M/2)+j]}. \qquad (3.30)
\end{aligned}$$

For $r \geq N/2$, the rth moment of the $SF(M, N, \lambda)$ distribution, like that of the central F distribution, does not exist. And (as in the case of the central F distribution) the noncentral F distribution does not have a moment generating function.

Upon recalling results (2.28) and (2.30) [or (2.33) and (2.34)] and result (1.23) and applying formula (3.29), we find that the mean of the $SF(M, N, \lambda)$ distribution is (if $N > 2$)

$$\mathrm{E}(w) = \frac{N}{N-2}\left(1+\frac{\lambda}{M}\right) \qquad (3.31)$$

and the second moment and variance are (if $N > 4$)

$$\begin{aligned}
\mathrm{E}(w^2) &= (N/M)^2\,\frac{(M+2)(M+2\lambda)+\lambda^2}{(N-2)(N-4)} \\
&= \frac{N^2}{(N-2)(N-4)}\left[\left(1+\frac{\lambda}{M}\right)^2+\frac{2}{M}\left(1+\frac{2\lambda}{M}\right)\right] \qquad (3.32)
\end{aligned}$$

and

$$\operatorname{var}(w) = \mathrm{E}(w^2)-[\mathrm{E}(w)]^2 = \frac{2N^2\{M[1+(\lambda/M)]^2+(N-2)[1+2(\lambda/M)]\}}{M(N-2)^2(N-4)}. \qquad (3.33)$$

Moreover, results (3.31) and (3.32) can be regarded as special cases of a more general result: in light of results (1.23) and (2.36), it follows from result (3.29) that (for $r = 1, 2, \ldots < N/2$) the rth moment of the $SF(M, N, \lambda)$ distribution is

$$\begin{aligned}
\mathrm{E}(w^r) = \frac{N^r}{(N-2)(N-4)\cdots(N-2r)}&\left\{\left(\frac{\lambda}{M}\right)^r\right.\\
&\left.+\sum_{j=0}^{r-1}\frac{(M+2j)[M+2(j+1)][M+2(j+2)]\cdots[M+2(r-1)]}{M^{r-j}}\binom{r}{j}\left(\frac{\lambda}{M}\right)^j\right\}. \qquad (3.34)
\end{aligned}$$

Noninteger degrees of freedom. The definition of the noncentral F distribution can be extended to noninteger degrees of freedom in much the same way as the definitions of the (central) F distribution and the central and noncentral chi-square distributions. For arbitrary strictly positive numbers M and N (and an arbitrary nonnegative number λ), take u^* and v to be random variables that are distributed independently as $Ga(M/2,\ N/2,\ \lambda/2)$ and $Ga(N/2,\ 2)$, respectively. Further, define

$$w = \frac{u^*/M}{v/N}.$$

Let us regard the distribution of the random variable w as a noncentral F distribution with M (possibly noninteger) numerator degrees of freedom and N (possibly noninteger) denominator degrees of freedom (and with noncentrality parameter λ). When M is an integer, the $Ga(M/2,\ 2,\ \lambda/2)$

distribution is identical to the $\chi^2(M, \lambda)$ distribution, and when N is an integer, the $Ga(N/2, 2)$ distribution is identical to the $\chi^2(N)$ distribution, so that this usage of the term noncentral F distribution is consistent with our previous usage.

Let $x = u^*/(u^* + v)$. As in the special case where M and N are integers, w and x are related to each other as follows:

$$w = \frac{Nx}{M(1-x)} \quad \text{and} \quad x = \frac{(M/N)w}{1+(M/N)w}.$$

By definition, x has a $Be(M/2,\ N/2,\ \lambda/2)$ distribution—refer to Part 2 of the present subsection. Accordingly, the distribution of w is related to the $Be(M/2,\ N/2,\ \lambda/2)$ distribution in the same way as in the special case where M and N are integers. Further, the distribution of x and hence that of w would be unaffected if the distributions of the (statistically independent) random variables u^* and v were taken to be $Ga(M/2,\ \beta,\ \lambda/2)$ and $Ga(N/2,\ \beta)$, respectively, where β is an arbitrary strictly positive number (not necessarily equal to 2).

6.4 Central, Noncentral, and Multivariate *t* Distributions

Let

$$t = \frac{z}{\sqrt{v/N}}, \tag{4.1}$$

where z and v are random variables that are statistically independent with $z \sim N(0, 1)$ and $v \sim \chi^2(N)$. The distribution of the random variable t is known as *Student's t distribution* or simply as the *t distribution*. Like the F distribution, it plays an important role in the use of linear models to make statistical inferences.

The distribution of the random variable t depends on one parameter; this parameter is the quantity N, which [as in the case of the $\chi^2(N)$ distribution] is referred to as the *degrees of freedom*. Let us denote the t distribution with N degrees of freedom by the symbol $St(N)$.

As a generalization of the t distribution, we have the noncentral t distribution. The *noncentral t distribution* is the distribution of the random variable

$$t^* = \frac{x}{\sqrt{v/N}},$$

obtained from expression (4.1) for the random variable t upon replacing the random variable z [which has an $N(0, 1)$ distribution] with a random variable x that has an $N(\mu, 1)$ distribution (where μ is an arbitrary scalar and where x, like z, is statistically independent of v). The distribution of the random variable t^* depends on two parameters: the *degrees of freedom* N and the scalar μ, which is referred to as the *noncentrality parameter*. Let us denote the noncentral t distribution with parameters N and μ by the symbol $St(N, \mu)$.

Clearly, the t distribution $St(N)$ can be regarded as the special case $St(N, 0)$ of the noncentral t distribution obtained upon setting the noncentrality parameter μ equal to 0. For the sake of clarity (i.e., to distinguish it from the noncentral t distribution), the "ordinary" t distribution may be referred to as the *central t distribution*.

There is a multivariate version of the t distribution. Let us continue to take v to be a random variable that is distributed as $\chi^2(N)$, and let us take $\mathbf{z}$ to be an M-dimensional random column vector that is distributed independently of v as $N(\mathbf{0}, \mathbf{R})$, where $\mathbf{R} = \{r_{ij}\}$ is an arbitrary $(M \times M)$ correlation matrix. Then, the distribution of the M-dimensional random column vector

$$\mathbf{t} = \frac{1}{\sqrt{v/N}}\mathbf{z}$$

is referred to as the *M-variate t distribution* (or when the dimension of $\mathbf{t}$ is unspecified or is clear from the context) as the *multivariate t distribution*. The parameters of this distribution consist of the degrees of freedom N and the $M(M-1)/2$ correlations r_{ij} $(i > j = 1, 2, \ldots, M)$—the diagonal elements of $\mathbf{R}$ equal 1 and (because $\mathbf{R}$ is symmetric) only $M(M-1)/2$ of its off-diagonal elements are distinct. Let us denote the multivariate t distribution with N degrees of freedom and with correlation matrix $\mathbf{R}$ by the symbol $MVt(N, \mathbf{R})$—the number M of variables is discernible from the dimensions of $\mathbf{R}$. The ordinary (univariate) t distribution $St(N)$ can be regarded as a special case of the multivariate t distribution; it is the special case $MVt(N, 1)$ where the correlation matrix is the 1×1 matrix whose only element equals 1.

a. (Central) *t* distribution

Related distributions. The t distribution is closely related to the F distribution (as is apparent from the very definitions of the t and F distributions). More specifically, the $St(N)$ distribution is related to the $SF(1, N)$ distribution.

Let t represent a random variable that is distributed as $St(N)$, and F a random variable that is distributed as $SF(1, N)$. Then,

$$t^2 \sim F, \tag{4.2}$$

or, equivalently,

$$|t| \sim \sqrt{F}. \tag{4.3}$$

The $St(N)$ distribution (i.e., the t distribution with N degrees of freedom) is also closely related to the distribution of the random variable y defined as follows:

$$y = \frac{z}{(v + z^2)^{1/2}},$$

where z and v are as defined in the introduction to the present section [i.e., where z and v are random variables that are distributed independently as $N(0, 1)$ and $\chi^2(N)$, respectively]. Now, let $t = z/\sqrt{v/N}$, in which case $t \sim St(N)$ (as is evident from the very definition of the t distribution). Then, t and y are related as follows:

$$t = \frac{\sqrt{N}\,y}{(1 - y^2)^{1/2}} \quad \text{and} \quad y = \frac{t}{(N + t^2)^{1/2}}. \tag{4.4}$$

Probability density function (pdf). Let us continue to take z and v to be random variables that are distributed independently as $N(0, 1)$ and $\chi^2(N)$, respectively, and to take $y = z/(v + z^2)^{1/2}$ and $t = z/\sqrt{v/N}$. Further, define $u = v + z^2$.

Let us determine the pdf of the joint distribution of u and y and the pdf of the (marginal) distribution of y—in light of the relationships (4.4), the pdf of the distribution of t is determinable from the pdf of the distribution of y. The equalities $u = v + z^2$ and $y = z/(v + z^2)^{1/2}$ define a one-to-one transformation from the region defined by the inequalities $0 < v < \infty$ and $-\infty < z < \infty$ onto the region defined by the inequalities $0 < u < \infty$ and $-1 < y < 1$. The inverse of this transformation is the transformation defined by the equalities

$$v = u(1 - y^2) \quad \text{and} \quad z = u^{1/2} y.$$

Further,

$$\begin{vmatrix} \partial v/\partial u & \partial v/\partial y \\ \partial z/\partial u & \partial z/\partial y \end{vmatrix} = \begin{vmatrix} 1 - y^2 & -2uy \\ (1/2)u^{-1/2}y & u^{1/2} \end{vmatrix} = u^{1/2}.$$

Thus, denoting by $d(\cdot)$ the pdf of the $\chi^2(N)$ distribution and by $b(\cdot)$ the pdf of the $N(0, 1)$ distribution and making use of standard results on a change of variables, the joint distribution of u and y has as a pdf the function $q(\cdot\,,\cdot)$ (of 2 variables) obtained by taking (for $0 < u < \infty$ and $-1 < y < 1$)

$$\begin{aligned} q(u, y) &= d[u(1-y^2)]\, b(u^{1/2}y)\, u^{1/2} \\ &= \frac{1}{\Gamma(N/2)\, 2^{(N+1)/2} \pi^{1/2}}\, u^{[(N+1)/2]-1} e^{-u/2} (1-y^2)^{(N/2)-1} \\ &= \frac{1}{\Gamma[(N+1)/2]\, 2^{(N+1)/2}}\, u^{[(N+1)/2]-1} e^{-u/2}\, \frac{\Gamma[(N+1)/2]}{\Gamma(N/2)\,\pi^{1/2}} (1-y^2)^{(N/2)-1} \end{aligned} \tag{4.5}$$

—for u and y such that $-\infty < u \le 0$ or $1 \le |y| < \infty$, $q(u, y) = 0$. The derivation of expression (4.5) is more or less the same as the derivation (in Section 6.1f) of expression (1.43).

The quantity $q(u, y)$ is reexpressible (for all u and y) in the form

$$q(u, y) = g(u)\, h^*(y), \tag{4.6}$$

where $g(\cdot)$ is the pdf of the $\chi^2(N+1)$ distribution and where $h^*(\cdot)$ is the function (of a single variable) defined as follows:

$$h^*(y) = \begin{cases} \dfrac{\Gamma[(N+1)/2]}{\Gamma(N/2)\,\pi^{1/2}} (1-y^2)^{(N/2)-1}, & \text{for } -1 < y < 1, \\ 0, & \text{elsewhere.} \end{cases} \tag{4.7}$$

Accordingly, we conclude that $h^*(\cdot)$ is a pdf; it is the pdf of the distribution of y. Moreover, y is distributed independently of u.

Now, upon making a change of variable from y to t [based on the relationships (4.4)] and observing that

$$\frac{d\,t(N+t^2)^{-1/2}}{dt} = N(N+t^2)^{-3/2},$$

we find that the distribution of t is the distribution with pdf $f(\cdot)$ defined (for all t) by

$$f(t) = h^*[t(N+t^2)^{-1/2}]\, N(N+t^2)^{-3/2} = \frac{\Gamma[(N+1)/2]}{\Gamma(N/2)\,\pi^{1/2}} N^{-1/2} \left(1 + \frac{t^2}{N}\right)^{-(N+1)/2}. \tag{4.8}$$

And t (like y) is distributed independently of u (i.e., independently of $v + z^2$).

In the special case where $N = 1$ (i.e., in the special case of a t distribution with 1 degree of freedom), expression (4.8) simplifies to

$$f(t) = \frac{1}{\pi} \frac{1}{1+t^2}. \tag{4.9}$$

The distribution with pdf (4.9) is known as the (standard) *Cauchy distribution*. Thus, the t distribution with 1 degree of freedom is identical to the Cauchy distribution.

As $N \to \infty$, the pdf of the $St(N)$ distribution converges (on all of $\mathcal{R}^1$) to the pdf of the $N(0, 1)$ distribution—refer, e.g., to Casella and Berger (2002, exercise 5.18). And, accordingly, t converges in distribution to z.

The pdf of the $St(5)$ distribution is displayed in Figure 6.1 along with the pdf of the $St(1)$ (Cauchy) distribution and the pdf of the $N(0, 1)$ (standard normal) distribution.

Symmetry and (absolute, odd, and even) moments. The absolute moments of the t distribution are determinable from the results of Section 6.3a on the F distribution. Let t represent a random variable that is distributed as $St(N)$ and w a random variable that is distributed as $SF(1, N)$. Then, as an implication of the relationship (4.3), we have (for an arbitrary scalar r) that $\text{E}(|t|^r)$ exists if and only if $\text{E}(w^{r/2})$ exists, in which case

$$\text{E}(|t|^r) = \text{E}(w^{r/2}). \tag{4.10}$$

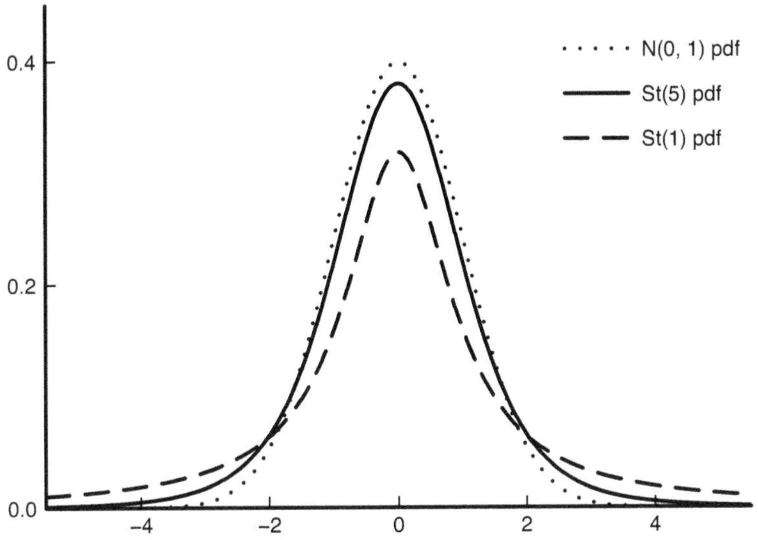

FIGURE 6.1. The probability density functions of the $N(0, 1)$ (standard normal), $St(5)$, and $St(1)$ (Cauchy) distributions.

And upon applying result (3.13), we find that, for $-1 < r < N$,

$$\mathrm{E}(|t|^r) = N^{r/2} \frac{\Gamma[(r+1)/2]}{\Gamma(1/2)} \frac{\Gamma[(N-r)/2]}{\Gamma(N/2)}. \tag{4.11}$$

For any even positive integer r, $|t|^r = t^r$. Thus, upon applying result (3.14), we find [in light of result (4.10)] that, for $r = 2, 4, 6, \ldots < N$, the rth moment of the $St(N)$ distribution exists and is expressible as

$$\mathrm{E}(t^r) = N^{r/2} \frac{(r-1)(r-3)\cdots(3)(1)}{(N-2)(N-4)\cdots(N-r)}. \tag{4.12}$$

For $r \geq N$, the rth moment of the $St(N)$ distribution does not exist (and, as a consequence, the t distribution, like the F distribution, does not have a moment generating function).

The $St(N)$ distribution is symmetric (about 0), that is,

$$-t \sim t \tag{4.13}$$

(as is evident from the very definition of the t distribution). And upon observing that, for any odd positive integer r, $-t^r = (-t)^r$, we find that, for $r = 1, 3, 5, \ldots < N$,

$$-\mathrm{E}(t^r) = \mathrm{E}(-t^r) = \mathrm{E}[(-t)^r] = \mathrm{E}(t^r)$$

and hence that (for $r = 1, 3, 5, \ldots < N$)

$$\mathrm{E}(t^r) = 0. \tag{4.14}$$

Thus, those odd moments of the $St(N)$ distribution that exist (which are those of order less than N) are all equal to 0. In particular, for $N > 1$,

$$\mathrm{E}(t) = 0. \tag{4.15}$$

Note that none of the moments of the $St(1)$ (Cauchy) distribution exist, not even the mean. And the $St(2)$ distribution has a mean (which equals 0), but does not have a second moment (or any other moments of order greater than 1) and hence does not have a variance.

For $N > 2$, we have [upon applying result (4.12)] that

$$\mathrm{var}(t) = \mathrm{E}(t^2) = \frac{N}{N-2}. \tag{4.16}$$

And, for $N > 4$,

$$E(t^4) = \frac{3N^2}{(N-2)(N-4)} = 3\left(\frac{N}{N-2}\right)^2 \frac{N-2}{N-4}, \tag{4.17}$$

which in combination with result (4.16) implies that (for $N > 4$)

$$\frac{E(t^4)}{[\mathrm{var}(t)]^2} = 3\,\frac{N-2}{N-4}. \tag{4.18}$$

[Expression (4.18) is an expression for a quantity that is sometimes referred to as the *kurtosis*, though in some presentations it is the difference between this quantity and the number 3 that is referred to as the kurtosis.]

Standardized version. Let t represent a random variable that is distributed as $St(N)$. And suppose that $N > 2$. Then, a standardized version, say s, of the random variable t can be created by taking

$$s = t/a, \tag{4.19}$$

where $a = \sqrt{\mathrm{var}(t)} = \sqrt{N/(N-2)}$.

The distribution of s has mean 0 and variance 1. Accordingly, the mean and variance of the distribution of s are identical to the mean and variance of the $N(0, 1)$ distribution. And if $N > 3$, the third moment $E(s^3)$ of the distribution of s equals 0, which is identical to the third moment of the $N(0, 1)$ distribution. Further, if $N > 4$, the fourth moment of the distribution of s is

$$E(s^4) = \frac{E(t^4)}{[\mathrm{var}(t)]^2} = 3\,\frac{N-2}{N-4} \tag{4.20}$$

[as is evident from result (4.18)]; by way of comparison, the fourth moment of the $N(0, 1)$ distribution is 3.

Let $f(\cdot)$ represent the pdf of the distribution of t, that is, the pdf of the $St(N)$ distribution. Then, making use of expression (4.8) and of standard results on a change of variable, we find that the distribution of s has as a pdf the function $f^*(\cdot)$ obtained by taking (for all s)

$$f^*(s) = af(as) = \frac{\Gamma[(N+1)/2]}{\Gamma(N/2)\,\pi^{1/2}}\,\frac{1}{(N-2)^{1/2}}\left(1 + \frac{s^2}{N-2}\right)^{-(N+1)/2}. \tag{4.21}$$

The pdf $f^*(\cdot)$ is displayed in Figure 6.2 for the case where $N = 5$ and for the case where $N = 3$; for purposes of comparison, the pdf of the $N(0, 1)$ (standard normal) distribution is also displayed.

Noninteger degrees of freedom. The definition of the t distribution can be extended to noninteger degrees of freedom by proceeding along the same lines as in Sections 6.1c and 6.3a in extending the definitions of the chi-square and F distributions. For any (strictly) positive number N, take the t distribution with N degrees of freedom to be the distribution of the random variable $t = z/\sqrt{v/N}$, where z and v are random variables that are statistically independent with $z \sim N(0, 1)$ and $v \sim Ga(\frac{N}{2}, 2)$. In the special case where N is a (strictly positive) integer, the $Ga(\frac{N}{2}, 2)$ distribution is identical to the $\chi^2(N)$ distribution, so that this usage of the term t distribution is consistent with our previous usage of this term.

Percentage points. Let t represent a random variable that is distributed as $St(N)$. Further, for $0 < \alpha < 1$, denote by $\bar{t}_\alpha(N)$ the upper $100\alpha\%$ point of the $St(N)$ distribution, that is, the point c such that $\Pr(t > c) = \alpha$. And observe (in light of the symmetry and absolute continuity of the t distribution) that, for any (nonrandom) scalar c,

$$\Pr(t \le -c) = \Pr(t < -c) = \Pr(-t < -c) = \Pr(t > c) \tag{4.22}$$

and that, for $c \ge 0$,

$$\Pr(|t| > c) = \Pr(t > c) + \Pr(t < -c) = 2\Pr(t > c). \tag{4.23}$$

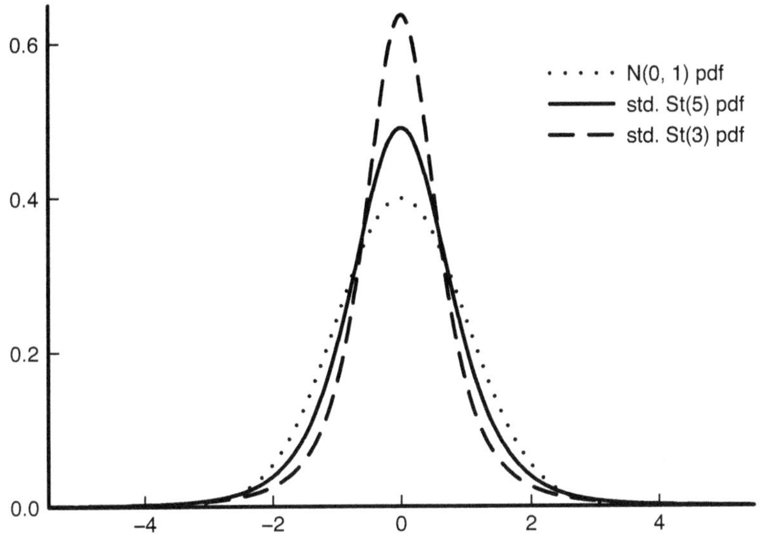

FIGURE 6.2. The probability density functions of the $N(0, 1)$ (standard normal) distribution and of the distributions of the standardized versions of a random variable having an $St(5)$ distribution and a random variable having an $St(3)$ distribution.

In light of result (4.22), we find that (for $0 < \alpha < 1$)

$$\Pr[t \le -\bar{t}_{1-\alpha}(N)] = \Pr[t > \bar{t}_{1-\alpha}(N)] = 1 - \alpha = 1 - \Pr[t > \bar{t}_\alpha(N)] = \Pr[t \le \bar{t}_\alpha(N)],$$

implying that

$$\bar{t}_\alpha(N) = -\bar{t}_{1-\alpha}(N). \tag{4.24}$$

And in light of result (4.23), we find that

$$\Pr[|t| > \bar{t}_{\alpha/2}(N)] = 2\Pr[t > \bar{t}_{\alpha/2}(N)] = 2(\alpha/2) = \alpha, \tag{4.25}$$

so that the upper $100\alpha\%$ point of the distribution of $|t|$ equals the upper $100(\alpha/2)\%$ point $\bar{t}_{\alpha/2}(N)$ of the distribution of t [i.e., of the $St(N)$ distribution]. Moreover, in light of relationship (4.3),

$$\bar{t}_{\alpha/2}(N) = \sqrt{\bar{F}_\alpha(1, N)}, \tag{4.26}$$

where $\bar{F}_\alpha(1, N)$ is the upper $100\alpha\%$ point of the $SF(1, N)$ distribution.

The t distribution as the distribution of a function of a random vector having a multivariate standard normal distribution or a spherical distribution. Let $\mathbf{z} = (z_1, z_2, \ldots, z_{N+1})'$ represent an $(N+1)$-dimensional random (column) vector. And let

$$t = \frac{z_{N+1}}{\sqrt{(1/N)\sum_{i=1}^{N} z_i^2}}.$$

Suppose that $\mathbf{z} \sim N(\mathbf{0}, \mathbf{I}_{N+1})$. Then, z_{N+1} and $\sum_{i=1}^{N} z_i^2$ are statistically independent random variables with $z_{N+1} \sim N(0, 1)$ and $\sum_{i=1}^{N} z_i^2 \sim \chi^2(N)$. Thus,

$$t \sim St(N). \tag{4.27}$$

Moreover, it follows from the results of Part 2 (of the present subsection) that t is distributed independently of $\sum_{i=1}^{N+1} z_i^2$.

More generally, suppose that $\mathbf{z}$ has an absolutely continuous spherical distribution. And [recalling result (4.4)] observe that

$$t = \frac{\sqrt{N}\,y}{(1-y^2)^{1/2}}, \qquad \text{where } y = \frac{z_{N+1}}{\left(\sum_{i=1}^{N+1} z_i^2\right)^{1/2}}.$$

Then, it follows from Theorem 6.3.1 that [as in the special case where $z \sim N(\mathbf{0}, \mathbf{I})$]

$$t \sim St(N), \tag{4.28}$$

and t is distributed independently of $\sum_{i=1}^{N+1} z_i^2$.

b. Noncentral *t* distribution

Related distributions. Let x and v represent random variables that are distributed independently as $N(\mu, 1)$ and $\chi^2(N)$, respectively. And observe that (by definition) $x/\sqrt{v/N} \sim St(N, \mu)$. Observe also that x^2 is distributed independently of v as $\chi^2(1, \mu^2)$ and hence that $x^2/(v/N) \sim SF(1, N, \mu^2)$. Thus, if t is a random variable that is distributed as $St(N, \mu)$, then

$$t^2 \sim SF(1, N, \mu^2) \tag{4.29}$$

or, equivalently,

$$|t| \sim \sqrt{F}, \tag{4.30}$$

where F is a random variable that is distributed as $SF(1, N, \mu^2)$.

Now, let $t = x/\sqrt{v/N}$, in which case $t \sim St(N, \mu)$, and define

$$y = \frac{x}{\sqrt{v + x^2}}.$$

Then, as in the case of result (4.4), t and y are related as follows:

$$t = \frac{\sqrt{N}\,y}{(1-y^2)^{1/2}} \qquad \text{and} \qquad y = \frac{t}{(N+t^2)^{1/2}}. \tag{4.31}$$

Probability density function (pdf). Let us derive an expression for the pdf of the noncentral t distribution. Let us do so by following an approach analogous to the one taken in Subsection a in deriving expression (4.8) for the pdf of the central t distribution.

Take x and v to be random variables that are distributed independently as $N(\mu, 1)$ and $\chi^2(N)$, respectively. Further, take $t = x/\sqrt{v/N}$, in which case $t \sim St(N, \mu)$, and define $u = v + x^2$ and $y = x/(v+x^2)^{1/2}$, in which case t and y are related by equalities (4.31). Then, denoting by $d(\cdot)$ the pdf of the $\chi^2(N)$ distribution and by $b(\cdot)$ the pdf of the $N(\mu, 1)$ distribution and proceeding in the same way as in arriving at expression (4.5), we find that the joint distribution of u and y has as a pdf the function $q(\cdot\,,\cdot)$ obtained by taking (for $0 < u < \infty$ and $-1 < y < 1$)

$$\begin{aligned} q(u, y) &= d[u(1-y^2)]\,b(u^{1/2}y)\,u^{1/2} \\ &= \frac{1}{\Gamma(N/2)\,2^{(N+1)/2}\pi^{1/2}}\,(1-y^2)^{(N/2)-1}u^{(N-1)/2}e^{-u/2}e^{u^{1/2}y\mu}e^{-\mu^2/2} \end{aligned} \tag{4.32}$$

—for u and y such that $-\infty < u \le 0$ or $1 \le |y| < \infty$, $q(u, y) = 0$. And upon replacing the quantity $e^{u^{1/2}y\mu}$ in expression (4.32) with its power-series representation

$$e^{u^{1/2}y\mu} = \sum_{r=0}^{\infty} \frac{(u^{1/2}y\mu)^r}{r!},$$

we find that (for $0 < u < \infty$ and $-1 < y < 1$)

$$q(u, y) = \frac{\Gamma[(N+1)/2]}{\Gamma(N/2)\,\pi^{1/2}} (1-y^2)^{(N/2)-1} e^{-\mu^2/2} \sum_{r=0}^{\infty} \frac{\Gamma[(N+r+1)/2]}{\Gamma[(N+1)/2]\, r!} (\sqrt{2}\,\mu y)^r g_r(u), \quad (4.33)$$

where (for $r = 0, 1, 2, \ldots$) $g_r(\cdot)$ is the pdf of the $\chi^2(N+r+1)$ distribution.

As a consequence of result (4.33), we have that (for $-1 < y < 1$)

$$\int_0^{\infty} q(u, y)\, du = \frac{\Gamma[(N+1)/2]}{\Gamma(N/2)\,\pi^{1/2}} (1-y^2)^{(N/2)-1} e^{-\mu^2/2} \sum_{r=0}^{\infty} \frac{\Gamma[(N+r+1)/2]}{\Gamma[(N+1)/2]\, r!} (\sqrt{2}\,\mu y)^r.$$

Thus, the (marginal) distribution of y has as a pdf the function $h^*(\cdot)$ defined as follows:

$$h^*(y) = \begin{cases} \dfrac{\Gamma[(N+1)/2]}{\Gamma(N/2)\,\pi^{1/2}} (1-y^2)^{(N/2)-1} e^{-\mu^2/2} \\ \qquad\qquad \times \displaystyle\sum_{r=0}^{\infty} \frac{\Gamma[(N+r+1)/2]}{\Gamma[(N+1)/2]\, r!} (\sqrt{2}\,\mu y)^r, & \text{for } -1 < y < 1, \\ 0, & \text{elsewhere.} \end{cases}$$

Finally, making a change of variable from y to t [based on the relationship (4.31)] and proceeding in the same way as in arriving at expression (4.8), we find that the distribution of t has as a pdf the function $f(\cdot)$ obtained by taking (for all t)

$$\begin{aligned} f(t) &= h^*[t(N+t^2)^{-1/2}]\, N(N+t^2)^{-3/2} \\ &= \frac{\Gamma[(N+1)/2]}{\Gamma(N/2)\,\pi^{1/2}} N^{-1/2} e^{-\mu^2/2} \sum_{r=0}^{\infty} \frac{\Gamma[(N+r+1)/2]}{\Gamma[(N+1)/2]\, r!} \left(\frac{\sqrt{2}\,\mu t}{\sqrt{N}}\right)^r \left(1+\frac{t^2}{N}\right)^{-(N+r+1)/2}. \end{aligned} \quad (4.34)$$

In the special case where $\mu = 0$, expression (4.34) simplifies to expression (4.8) for the pdf of the central t distribution $St(N)$.

Moments: relationship to the moments of the $N(\mu, 1)$ distribution. Let t represent a random variable that has an $St(N, \mu)$ distribution. Then, by definition, $t \sim x/\sqrt{v/N}$, where x and v are random variables that are distributed independently as $N(\mu, 1)$ and $\chi^2(N)$, respectively. And, for $r = 1, 2, \ldots < N$, the rth moment of the $St(N, \mu)$ distribution exists and is expressible as

$$\mathrm{E}(t^r) = N^{r/2}\, \mathrm{E}(v^{-r/2})\, \mathrm{E}(x^r) \quad (4.35)$$

or [in light of results (1.27) and (1.31)] as

$$\mathrm{E}(t^r) = (N/2)^{r/2} \frac{\Gamma[(N-r)/2]}{\Gamma(N/2)}\, \mathrm{E}(x^r) \quad (4.36)$$

or, in the special case where r is an even number,

$$\mathrm{E}(t^r) = \frac{N^{r/2}}{(N-2)(N-4)\cdots(N-r)}\, \mathrm{E}(x^r). \quad (4.37)$$

Like the $St(N)$ distribution, the $St(N, \mu)$ distribution does not have moments of order N or greater and, accordingly, does not have a moment generating function.

In light of results (4.36) and (4.37), we find that the mean of the $St(N, \mu)$ distribution is (if $N > 1$)

$$\mathrm{E}(t) = \sqrt{N/2}\, \frac{\Gamma[(N-1)/2]}{\Gamma(N/2)}\, \mu \quad (4.38)$$

and the second moment is (if $N > 2$)

$$\mathrm{E}(t^2) = \frac{N}{N-2}(1+\mu^2). \tag{4.39}$$

Noninteger degrees of freedom. The definition of the noncentral t distribution can be extended to noninteger degrees of freedom in essentially the same way as the definition of the central t distribution. For any (strictly) positive scalar N (and any scalar μ), take the noncentral t distribution with N degrees of freedom and noncentrality parameter μ to be the distribution of the random variable $t = x/\sqrt{v/N}$, where x and v are random variables that are statistically independent with $x \sim N(\mu, 1)$ and $v \sim Ga\left(\frac{N}{2}, 2\right)$. In the special case where N is a (strictly positive) integer, the $Ga\left(\frac{N}{2}, 2\right)$ distribution is identical to the $\chi^2(N)$ distribution, so that this usage of the term noncental t distribution is consistent with our previous usage of this term.

Some relationships. Let t represent a random variable that has an $ST(N, \mu)$ distribution and t^* a random variable that has an $St(N, -\mu)$ distribution. Then, clearly,

$$t^* \sim -t. \tag{4.40}$$

And for any (nonrandom) scalar c, we have [as a generalization of result (4.22)] that

$$\Pr(t^* \leq -c) = \Pr(t^* < -c) = \Pr(-t < -c) = \Pr(t > c), \tag{4.41}$$

implying in particular that

$$\Pr(t^* > -c) = 1 - \Pr(t > c) \tag{4.42}$$

and that

$$\Pr(t^* \leq -c) = 1 - \Pr(t \leq c). \tag{4.43}$$

In light of relationships (4.42) and (4.43), it suffices in evaluating $\Pr(t > c)$ or $\Pr(t \leq c)$ to restrict attention to nonnegative values of μ or, alternatively, to nonnegative values of c. Further, for $c \geq 0$, we have [as a generalization of result (4.23)] that

$$\begin{aligned}\Pr(|t| > c) = \Pr(t > c) + \Pr(t < -c) &= \Pr(t > c) + \Pr(-t > c) \\ &= \Pr(t > c) + \Pr(t^* > c).\end{aligned} \tag{4.44}$$

c. A result on determinants of matrices of the form R + STU

As a preliminary to deriving (in Subsection d) some results on the multivariate t distribution, it is convenient to introduce the following result on determinants.

Theorem 6.4.1. Let $\mathbf{R}$ represent an $N \times N$ matrix, $\mathbf{S}$ an $N \times M$ matrix, $\mathbf{T}$ an $M \times M$ matrix, and $\mathbf{U}$ an $M \times N$ matrix. If $\mathbf{R}$ and $\mathbf{T}$ are nonsingular, then

$$|\mathbf{R} + \mathbf{STU}| = |\mathbf{R}||\mathbf{T}||\mathbf{T}^{-1} + \mathbf{UR}^{-1}\mathbf{S}| \tag{4.45}$$

Proof. Suppose that $\mathbf{R}$ and $\mathbf{T}$ are nonsingular. Then, making use of Theorem 2.14.22, we find that

$$\begin{vmatrix} \mathbf{R} & -\mathbf{S} \\ \mathbf{U} & \mathbf{T}^{-1} \end{vmatrix} = |\mathbf{T}^{-1}||\mathbf{R} - (-\mathbf{S})(\mathbf{T}^{-1})^{-1}\mathbf{U}| = |\mathbf{T}^{-1}||\mathbf{R} + \mathbf{STU}|$$

and also that

$$\begin{vmatrix} \mathbf{R} & -\mathbf{S} \\ \mathbf{U} & \mathbf{T}^{-1} \end{vmatrix} = |\mathbf{R}||\mathbf{T}^{-1} - \mathbf{UR}^{-1}(-\mathbf{S})| = |\mathbf{R}||\mathbf{T}^{-1} + \mathbf{UR}^{-1}\mathbf{S}|.$$

Thus,

$$|\mathbf{T}^{-1}||\mathbf{R} + \mathbf{STU}| = |\mathbf{R}||\mathbf{T}^{-1} + \mathbf{UR}^{-1}\mathbf{S}|$$

or, equivalently (since $|\mathbf{T}^{-1}| = 1/|\mathbf{T}|$),

$$|\mathbf{R} + \mathbf{STU}| = |\mathbf{R}||\mathbf{T}||\mathbf{T}^{-1} + \mathbf{UR}^{-1}\mathbf{S}|.$$

Q.E.D.

In the special case where $\mathbf{R} = \mathbf{I}_N$ and $\mathbf{T} = \mathbf{I}_M$, Theorem 6.4.1 simplifies to the following result.

Corollary 6.4.2. For any $N \times M$ matrix $\mathbf{S}$ and any $M \times N$ matrix $\mathbf{U}$,

$$|\mathbf{I}_N + \mathbf{SU}| = |\mathbf{I}_M + \mathbf{US}|. \tag{4.46}$$

In the special case where $M = 1$, Corollary 6.4.2 can be restated as the following corollary.

Corollary 6.4.3. For any N-dimensional column vectors $\mathbf{s} = \{s_i\}$ and $\mathbf{u} = \{u_i\}$,

$$|\mathbf{I}_N + \mathbf{su}'| = 1 + \mathbf{u}'\mathbf{s} = 1 + \mathbf{s}'\mathbf{u} = 1 + \sum_{i=1}^{N} s_i u_i. \tag{4.47}$$

d. Multivariate *t* distribution

Related distributions. Let $\mathbf{t} = (t_1, t_2, \ldots, t_M)'$ represent a random (column) vector that has an $MVt(N, \mathbf{R})$ distribution, that is, an M-variate t distribution with N degrees of freedom and correlation matrix $\mathbf{R}$. If $\mathbf{t}_*$ is any subvector of $\mathbf{t}$, say the M^*-dimensional subvector $(t_{i_1}, t_{i_2}, \ldots, t_{i_{M^*}})'$ consisting of the $i_1, i_2, \ldots, i_{M^*}$th elements, then, clearly,

$$\mathbf{t}_* \sim MVt(N, \mathbf{R}_*), \tag{4.48}$$

where $\mathbf{R}_*$ is the $M^* \times M^*$ submatrix of $\mathbf{R}$ formed by striking out all of the rows and columns of $\mathbf{R}$ save its $i_1, i_2, \ldots, i_{M^*}$th rows and columns. And, for $i = 1, 2, \ldots, M$,

$$t_i \sim St(N), \tag{4.49}$$

that is, the (marginal) distribution of each of the elements of $\mathbf{t}$ is a (univariate) t distribution with the same number of degrees of freedom (N) as the distribution of $\mathbf{t}$.

In the special case where $\mathbf{R} = \mathbf{I}$,

$$M^{-1}\mathbf{t}'\mathbf{t} \sim SF(M, N). \tag{4.50}$$

More generally, if $\mathbf{t}$ is partitioned into some number of subvectors, say K subvectors $\mathbf{t}_1, \mathbf{t}_2, \ldots, \mathbf{t}_K$ of dimensions $M_1, M_2, \ldots, M_K$, respectively, then, letting $u_1, u_2, \ldots, u_K$, and v represent random variables that are distributed independently as $\chi^2(M_1), \chi^2(M_2), \ldots, \chi^2(M_K)$, and $\chi^2(N)$, respectively, we find that, in the special case where $\mathbf{R} = \mathbf{I}$, the joint distribution of the K quantities $M_1^{-1}\mathbf{t}_1'\mathbf{t}_1$, $M_2^{-1}\mathbf{t}_2'\mathbf{t}_2$, $\ldots$, $M_K^{-1}\mathbf{t}_K'\mathbf{t}_K$ is identical to the joint distribution of the K ratios $(u_1/M_1)/(v/N)$, $(u_2/M_2)/(v/N)$, $\ldots$, $(u_K/M_K)/(v/N)$ [the marginal distributions of which are $SF(M_1, N), SF(M_2, N), \ldots, SF(M_K, N)$, respectively, and which have a common denominator v/N].

The multivariate t distribution is related asymptotically to the multivariate normal distribution. It can be shown that as $N \to \infty$, the $MVt(N, \mathbf{R})$ distribution converges to the $N(\mathbf{0}, \mathbf{R})$ distribution.

Now, take $\mathbf{z}$ to be an M-dimensional random column vector and v a random variable that are statistically independent with $\mathbf{z} \sim N(\mathbf{0}, \mathbf{R})$ and $v \sim \chi^2(N)$. And take $\mathbf{t} = (v/N)^{-1/2}\mathbf{z}$, in which case $\mathbf{t} \sim MVt(N, \mathbf{R})$, and define $\mathbf{y} = (v + \mathbf{z}'\mathbf{R}^{-1}\mathbf{z})^{-1/2}\mathbf{z}$. Then, $\mathbf{t}$ and $\mathbf{y}$ are related as follows:

$$\mathbf{t} = [N/(1 - \mathbf{y}'\mathbf{R}^{-1}\mathbf{y})]^{1/2}\mathbf{y} \quad \text{and} \quad \mathbf{y} = (N + \mathbf{t}'\mathbf{R}^{-1}\mathbf{t})^{-1/2}\mathbf{t}. \tag{4.51}$$

Probability density function (pdf). Let us continue to take $\mathbf{z}$ to be an M-dimensional random column vector and v a random variable that are statistically independent with $\mathbf{z} \sim N(\mathbf{0}, \mathbf{R})$ and $v \sim \chi^2(N)$. And let us continue to take $\mathbf{t} = (v/N)^{-1/2}\mathbf{z}$ and to define $\mathbf{y} = (v + \mathbf{z}'\mathbf{R}^{-1}\mathbf{z})^{-1/2}\mathbf{z}$. Further, define $u = v + \mathbf{z}'\mathbf{R}^{-1}\mathbf{z}$.

Consider the joint distribution of the random variable u and the random vector $\mathbf{y}$. The equalities $u = v + \mathbf{z}'\mathbf{R}^{-1}\mathbf{z}$ and $\mathbf{y} = (v + \mathbf{z}'\mathbf{R}^{-1}\mathbf{z})^{-1/2}\mathbf{z}$ define a one-to-one transformation from the region $\{v, \mathbf{z} : 0 < v < \infty, \mathbf{z} \in \mathcal{R}^M\}$ onto the region $\{u, \mathbf{y} : 0 < u < \infty, \mathbf{y}'\mathbf{R}^{-1}\mathbf{y} < 1\}$. The inverse of this transformation is the transformation defined by the equalities

$$v = u(1 - \mathbf{y}'\mathbf{R}^{-1}\mathbf{y}) \quad \text{and} \quad \mathbf{z} = u^{1/2}\mathbf{y}.$$

Further, letting $\mathbf{J}$ represent the $(M+1) \times (M+1)$ matrix whose ijth element is the partial derivative of the ith element of the vector $\begin{pmatrix} v \\ \mathbf{z} \end{pmatrix}$ with respect to the jth element of the vector $\begin{pmatrix} u \\ \mathbf{y} \end{pmatrix}$ and making use of results (5.4.10), (2.14.29), (2.14.11), and (2.14.9), we find that

$$|\mathbf{J}| = \begin{vmatrix} \partial v/\partial u & \partial v/\partial \mathbf{y}' \\ \partial \mathbf{z}/\partial u & \partial \mathbf{z}/\partial \mathbf{y}' \end{vmatrix} = \begin{vmatrix} 1-\mathbf{y}'\mathbf{R}^{-1}\mathbf{y} & -2u\mathbf{y}'\mathbf{R}^{-1} \\ (1/2)u^{-1/2}\mathbf{y} & u^{1/2}\mathbf{I} \end{vmatrix} = u^{M/2}.$$

Thus, denoting by $d(\cdot)$ the pdf of the $\chi^2(N)$ distribution and by $b(\cdot)$ the pdf of the $N(\mathbf{0}, \mathbf{R})$ distribution and making use of standard results on a change of variables, the joint distribution of u and $\mathbf{y}$ has as a pdf the function $q(\cdot\,,\cdot)$ obtained by taking (for u and $\mathbf{y}$ such that $0 < u < \infty$ and $\mathbf{y}'\mathbf{R}^{-1}\mathbf{y} < 1$)

$$\begin{aligned} q(u,\mathbf{y}) &= d[u(1-\mathbf{y}'\mathbf{R}^{-1}\mathbf{y})]\, b(u^{1/2}\mathbf{y})\, u^{M/2} \\ &= \frac{1}{\Gamma(N/2)\, 2^{(N+M)/2} \pi^{M/2} |\mathbf{R}|^{1/2}}\, u^{[(N+M)/2]-1} e^{-u/2} (1-\mathbf{y}'\mathbf{R}^{-1}\mathbf{y})^{(N/2)-1} \\ &= \frac{1}{\Gamma[(N+M)/2]\, 2^{(N+M)/2}}\, u^{[(N+M)/2]-1} e^{-u/2} \\ &\qquad\qquad \times \frac{\Gamma[(N+M)/2]}{\Gamma(N/2)\, \pi^{M/2} |\mathbf{R}|^{1/2}} (1-\mathbf{y}'\mathbf{R}^{-1}\mathbf{y})^{(N/2)-1} \end{aligned} \tag{4.52}$$

—for u and $\mathbf{y}$ such that $-\infty < u \le 0$ or $\mathbf{y}'\mathbf{R}^{-1}\mathbf{y} \ge 1$, $q(u,\mathbf{y}) = 0$. The derivation of expression (4.52) parallels the derivation (in Part 2 of Subsection a) of expression (4.5) and is very similar to the derivation (in Section 6.1f) of expression (1.43).

The quantity $q(u,\mathbf{y})$ is reexpressible (for all u and $\mathbf{y}$) in the form

$$q(u,\mathbf{y}) = g(u)\, h^*(\mathbf{y}), \tag{4.53}$$

where $g(\cdot)$ is the pdf of the $\chi^2(N+M)$ distribution and where $h^*(\cdot)$ is the function (of an $M \times 1$ vector) defined as follows: for all $\mathbf{y}$,

$$h^*(\mathbf{y}) = \begin{cases} \dfrac{\Gamma[(N+M)/2]}{\Gamma(N/2)\, \pi^{M/2} |\mathbf{R}|^{1/2}} (1 - \mathbf{y}'\mathbf{R}^{-1}\mathbf{y})^{(N/2)-1}, & \text{if } \mathbf{y}'\mathbf{R}^{-1}\mathbf{y} < 1, \\ 0, & \text{otherwise.} \end{cases} \tag{4.54}$$

Accordingly, we conclude that $h^*(\cdot)$ is a pdf; it is the pdf of the distribution of $\mathbf{y}$. Moreover, $\mathbf{y}$ is distributed independently of u.

Now, for $j = 1, 2, \ldots, M$, let y_j represent the jth element of $\mathbf{y}$, t_j the jth element of $\mathbf{t}$, and $\mathbf{e}_j$

the jth column of $\mathbf{I}_M$, and observe [in light of relationship (4.51) and result (5.4.10)] that

$$\begin{aligned}\frac{\partial y_j}{\partial \mathbf{t}} &= \frac{\partial (N+\mathbf{t}'\mathbf{R}^{-1}\mathbf{t})^{-1/2} t_j}{\partial \mathbf{t}} \\ &= (N+\mathbf{t}'\mathbf{R}^{-1}\mathbf{t})^{-1/2} \frac{\partial t_j}{\partial \mathbf{t}} + t_j \frac{\partial (N+\mathbf{t}'\mathbf{R}^{-1}\mathbf{t})^{-1/2}}{\partial \mathbf{t}} \\ &= (N+\mathbf{t}'\mathbf{R}^{-1}\mathbf{t})^{-1/2} \mathbf{e}_j + t_j(-1/2)(N+\mathbf{t}'\mathbf{R}^{-1}\mathbf{t})^{-3/2} \frac{\partial \mathbf{t}'\mathbf{R}^{-1}\mathbf{t}}{\partial \mathbf{t}} \\ &= (N+\mathbf{t}'\mathbf{R}^{-1}\mathbf{t})^{-1/2} \mathbf{e}_j - (N+\mathbf{t}'\mathbf{R}^{-1}\mathbf{t})^{-3/2} t_j \mathbf{R}^{-1}\mathbf{t}.\end{aligned}$$

Then,

$$\begin{aligned}\frac{\partial \mathbf{y}'}{\partial \mathbf{t}} = \left(\frac{\partial y_1}{\partial \mathbf{t}}, \frac{\partial y_2}{\partial \mathbf{t}}, \ldots, \frac{\partial y_M}{\partial \mathbf{t}}\right) &= (N+\mathbf{t}'\mathbf{R}^{-1}\mathbf{t})^{-1/2}\mathbf{I} - (N+\mathbf{t}'\mathbf{R}^{-1}\mathbf{t})^{-3/2}\mathbf{R}^{-1}\mathbf{t}\mathbf{t}' \\ &= (N+\mathbf{t}'\mathbf{R}^{-1}\mathbf{t})^{-1/2}[\mathbf{I} - (N+\mathbf{t}'\mathbf{R}^{-1}\mathbf{t})^{-1}\mathbf{R}^{-1}\mathbf{t}\mathbf{t}'],\end{aligned}$$

implying [in light of Lemma 2.14.3 and Corollaries 2.14.6 and 6.4.3] that

$$\left|\frac{\partial \mathbf{y}}{\partial \mathbf{t}'}\right| = \left|\frac{\partial \mathbf{y}'}{\partial \mathbf{t}}\right| = (N+\mathbf{t}'\mathbf{R}^{-1}\mathbf{t})^{-M/2}[1 - (N+\mathbf{t}'\mathbf{R}^{-1}\mathbf{t})^{-1}\mathbf{t}'\mathbf{R}^{-1}\mathbf{t}] = N(N+\mathbf{t}'\mathbf{R}^{-1}\mathbf{t})^{-(M/2)-1}.$$

Thus, upon making a change of variables from the elements of $\mathbf{y}$ to the elements of $\mathbf{t}$, we find that the distribution of $\mathbf{t}$ (which is the M-variate t distribution with N degrees of freedom) has as a pdf the function $f(\cdot)$ obtained by taking (for all $\mathbf{t}$)

$$\begin{aligned}f(\mathbf{t}) &= h^*[(N+\mathbf{t}'\mathbf{R}^{-1}\mathbf{t})^{-1/2}\mathbf{t}]\, N(N+\mathbf{t}'\mathbf{R}^{-1}\mathbf{t})^{-(M/2)-1} \\ &= \frac{\Gamma[(N+M)/2]}{\Gamma(N/2)\,\pi^{M/2}|\mathbf{R}|^{1/2}} N^{N/2}(N+\mathbf{t}'\mathbf{R}^{-1}\mathbf{t})^{-(N+M)/2} \\ &= \frac{\Gamma[(N+M)/2]}{\Gamma(N/2)\,\pi^{M/2}|\mathbf{R}|^{1/2}} N^{-M/2}\left(1 + \frac{\mathbf{t}'\mathbf{R}^{-1}\mathbf{t}}{N}\right)^{-(N+M)/2}. \qquad (4.55)\end{aligned}$$

And $\mathbf{t}$ (like $\mathbf{y}$) is distributed independently of u (i.e., independently of $v + \mathbf{z}'\mathbf{R}^{-1}\mathbf{z}$).

In the special case where $M = 1$, expression (4.55) simplifies to expression (4.8) for the pdf of the (univariate) t distribution with N degrees of freedom.

Moments. Let $\mathbf{t} = (t_1, t_2, \ldots, t_M)'$ represent an M-dimensional random (column) vector that has an $MVt(N, \mathbf{R})$ distribution. Further, for $i, j = 1, 2, \ldots, M$, denote by r_{ij} the ijth element of the correlation matrix $\mathbf{R}$. And denote by k an arbitrary (strictly) positive integer, and by $k_1, k_2, \ldots, k_M$ any nonnegative integers such that $k = \sum_{i=1}^{M} k_i$.

By definition, $\mathbf{t} \sim (v/N)^{-1/2}\mathbf{z}$, where $\mathbf{z} = (z_1, z_2, \ldots, z_M)'$ is an M-dimensional random (column) vector and v a random variable that are statistically independent with $\mathbf{z} \sim N(\mathbf{0}, \mathbf{R})$ and $v \sim \chi^2(N)$. For $k < N$, $\mathrm{E}(v^{-k/2})$ exists, and the kth-order moment $\mathrm{E}(t_1^{k_1} t_2^{k_2} \cdots t_M^{k_M})$ of the $MVt(N, \mathbf{R})$ distribution is expressible as follows:

$$\begin{aligned}\mathrm{E}(t_1^{k_1} t_2^{k_2} \cdots t_M^{k_M}) &= \mathrm{E}[(v/N)^{-k/2} z_1^{k_1} z_2^{k_2} \cdots z_M^{k_M}] \\ &= N^{k/2}\mathrm{E}(v^{-k/2})\mathrm{E}(z_1^{k_1} z_2^{k_2} \cdots z_M^{k_M}). \qquad (4.56)\end{aligned}$$

Moreover, $-\mathbf{z} \sim \mathbf{z}$, so that

$$\mathrm{E}(z_1^{k_1} z_2^{k_2} \cdots z_M^{k_M}) = \mathrm{E}[(-z_1)^{k_1}(-z_2)^{k_2} \cdots (-z_M)^{k_M}] = (-1)^k\, \mathrm{E}(z_1^{k_1} z_2^{k_2} \cdots z_M^{k_M}).$$

Thus, if k is an odd number, then $\mathrm{E}(z_1^{k_1} z_2^{k_2} \cdots z_M^{k_M}) = 0$ and hence (if k is an odd number smaller than N)

$$\mathrm{E}(t_1^{k_1} t_2^{k_2} \cdots t_M^{k_M}) = 0. \qquad (4.57)$$

Alternatively, if k is an even number (smaller than N), then [in light of result (1.31)]

$$\mathrm{E}\big(t_1^{k_1} t_2^{k_2} \cdots t_M^{k_M}\big) = \frac{N^{k/2}}{(N-2)(N-4)\cdots(N-k)}\, \mathrm{E}\big(z_1^{k_1} z_2^{k_2} \cdots z_M^{k_M}\big). \tag{4.58}$$

We conclude that (for $k < N$) each kth-order moment of the $MVt(N,\, \mathbf{R})$ distribution is either 0 or is obtainable from the corresponding kth-order moment of the $N(\mathbf{0},\, \mathbf{R})$ distribution (depending on whether k is odd or even). In particular, for $i = 1, 2, \ldots, M$, we find that (if $N > 1$)

$$\mathrm{E}(t_i) = 0 \tag{4.59}$$

and that (if $N > 2$)

$$\mathrm{var}(t_i) = \mathrm{E}\big(t_i^2\big) = \frac{N}{N-2} r_{ii} = \frac{N}{N-2}, \tag{4.60}$$

in agreement with results (4.15) and (4.16). And, for $j \neq i = 1, 2, \ldots, M$, we find that (if $N > 2$)

$$\mathrm{cov}(t_i,\, t_j) = \mathrm{E}(t_i t_j) = \frac{N}{N-2} r_{ij} \tag{4.61}$$

and that (if $N > 2$)

$$\mathrm{corr}(t_i,\, t_j) = r_{ij} \;[= \mathrm{corr}(z_i,\, z_j)]. \tag{4.62}$$

In matrix notation, we have that (if $N > 1$)

$$\mathrm{E}(\mathbf{t}) = \mathbf{0} \tag{4.63}$$

and that (if $N > 2$)

$$\mathrm{var}(\mathbf{t}) = \frac{N}{N-2} \mathbf{R}. \tag{4.64}$$

Noninteger degrees of freedom. The definition of the multivariate t distribution can be extended to noninteger degrees of freedom in essentially the same way as the definition of the (univariate) t distribution. For any (strictly) positive number N, take the M-variate t distribution with degrees of freedom N and correlation matrix $\mathbf{R}$ to be the distribution of the M-variate random (column) vector $\mathbf{t} = (v/N)^{-1/2}\mathbf{z}$, where $\mathbf{z}$ is an M-dimensional random column vector and v a random variable that are statistically independent with $\mathbf{z} \sim N(\mathbf{0},\, \mathbf{R})$ and $v \sim Ga\big(\frac{N}{2},\, 2\big)$. In the special case where N is a (strictly positive) integer, the $Ga\big(\frac{N}{2},\, 2\big)$ distribution is identical to the $\chi^2(N)$ distribution, so that this usage of the term M-variate t distribution is consistent with our previous usage of this term.

Sphericity and ellipticity. The $MVt(N,\, \mathbf{I}_M)$ distribution is spherical. To see this, take $\mathbf{z}$ to be an M-dimensional random column vector and v a random variable that are statistically independent with $\mathbf{z} \sim N(\mathbf{0},\, \mathbf{I}_M)$ and $v \sim \chi^2(N)$, and take $\mathbf{O}$ to be any $M \times M$ orthogonal matrix of constants. Further, let $\mathbf{t} = (v/N)^{-1/2}\mathbf{z}$, in which case $\mathbf{t} \sim MVt(N,\, \mathbf{I}_M)$, and observe that the $M \times 1$ vector $\mathbf{Oz}$, like $\mathbf{z}$ itself, is distributed independently of v as $N(\mathbf{0}, \mathbf{I}_M)$. Thus,

$$\mathbf{Ot} = (v/N)^{-1/2}(\mathbf{Oz}) \sim (v/N)^{-1/2}\mathbf{z} = \mathbf{t}.$$

And we conclude that the distribution of $\mathbf{t}$ [the $MVt(N,\, \mathbf{I}_M)$ distribution] is spherical.

That the $MVt(N,\, \mathbf{I}_M)$ distribution is spherical can also be inferred from the form of its pdf. As is evident from result (4.55), the $MVt(N,\, \mathbf{I}_M)$ distribution has a pdf $f(\cdot)$ that is expressible (for all $\mathbf{t}$) in the form

$$f(\mathbf{t}) = g(\mathbf{t}'\mathbf{t}), \tag{4.65}$$

where $g(\cdot)$ is the following (nonnegative) function of a single nonnegative variable, say w:

$$g(w) = \frac{\Gamma[(N+M)/2]}{\Gamma(N/2)\,\pi^{M/2}}\, N^{-M/2}\Big(1 + \frac{w}{N}\Big)^{-(N+M)/2}.$$

Now, let $\mathbf{R}$ represent an arbitrary $M \times M$ correlation matrix. Then, in light of Corollary 2.13.24, there exists an $M \times M$ matrix $\mathbf{S}$ such that $\mathbf{R} = \mathbf{S}'\mathbf{S}$. And since the distribution of $\mathbf{t}$ is spherical, the distribution of $\mathbf{S}'\mathbf{t}$ is (by definition) elliptical; $\mathbf{S}'\mathbf{t}$ is distributed elliptically about $\mathbf{0}$. Moreover,

$$\mathbf{S}'\mathbf{t} = (v/N)^{-1/2}(\mathbf{S}'\mathbf{z}),$$

and $\mathbf{S}'\mathbf{z}$ is distributed independently of v as $N(\mathbf{0}, \mathbf{R})$, so that $\mathbf{S}'\mathbf{t} \sim MVt(N, \mathbf{R})$. Thus, the $MVt(N, \mathbf{R})$ distribution is elliptical.

The $MVt(N, \mathbf{I}_M)$ distribution as the distribution of a vector-valued function of a random vector having a standard normal distribution or a spherical distribution. Let $\mathbf{z} = (z_1, z_2, \ldots, z_N, z_{N+1}, z_{N+2}, \ldots, z_{N+M})'$ represent an $(N+M)$-dimensional random (column) vector. Further, let

$$\mathbf{t} = [(1/N)\textstyle\sum_{i=1}^{N} z_i^2]^{-1/2}\mathbf{z}_*,$$

where $\mathbf{z}_* = (z_{N+1}, z_{N+2}, \ldots, z_{N+M})'$.

Suppose that $\mathbf{z} \sim N(\mathbf{0}, \mathbf{I}_{N+M})$. Then, $\mathbf{z}_*$ and $\sum_{i=1}^{N} z_i^2$ are distributed independently as $N(\mathbf{0}, \mathbf{I}_M)$ and $\chi^2(N)$, respectively. Thus,

$$\mathbf{t} \sim MVt(N, \mathbf{I}_M). \tag{4.66}$$

Moreover, it follows from the results of Part 2 (of the present subsection) that $\mathbf{t}$ is distributed independently of $\sum_{i=1}^{N+M} z_i^2$.

More generally, suppose that $\mathbf{z}$ has an absolutely continuous spherical distribution. And [recalling result (4.51)] observe that

$$\mathbf{t} = [N/(1-\mathbf{y}'\mathbf{y})]^{1/2}\mathbf{y}, \qquad \text{where } \mathbf{y} = \left(\textstyle\sum_{i=1}^{N+M} z_i^2\right)^{-1/2}\mathbf{z}_*.$$

Then, it follows from Theorem 6.3.1 that [as in the special case where $\mathbf{z} \sim N(\mathbf{0}, \mathbf{I})$]

$$\mathbf{t} \sim MVt(N, \mathbf{I}_M), \tag{4.67}$$

and $\mathbf{t}$ is statistically independent of $\sum_{i=1}^{N+M} z_i^2$.

6.5 Moment Generating Function of the Distribution of One or More Quadratic Forms or Second-Degree Polynomials (in a Normally Distributed Random Vector)

a. Some preliminary results

As a preliminary to deriving the moment generating function of the distribution of one or more quadratic forms or second-degree polynomials (in a normally distributed random vector), it is convenient to introduce some basic results on the positive definiteness of linear combinations of matrices. The linear combinations of immediate interest are those of the form $\mathbf{I}_M - t\mathbf{A}$, where $\mathbf{A}$ is an $M \times M$ symmetric matrix and t is a scalar, or, more generally, those of the form $\mathbf{I}_M - \sum_{i=1}^{K} t_i\mathbf{A}_i$, where $\mathbf{A}_1, \mathbf{A}_2, \ldots, \mathbf{A}_K$ are $M \times M$ symmetric matrices and $t_1, t_2, \ldots, t_K$ are scalars. For what values of t is $\mathbf{I}_M - t\mathbf{A}$ positive definite? Or, more generally, for what values of $t_1, t_2, \ldots, t_K$ is $\mathbf{I}_M - \sum_{i=1}^{K} t_i\mathbf{A}_i$ positive definite?

Existence of a neighborhood within which a linear combination is positive definite.

Lemma 6.5.1. Corresponding to any $M \times M$ symmetric matrix $\mathbf{A}$, there exists a (strictly) positive scalar c such that $\mathbf{I}_M - t\mathbf{A}$ is positive definite for every scalar t in the interval $-c < t < c$.

Lemma 6.5.1 is essentially a special case of the following lemma.

Lemma 6.5.2. Let $\mathbf{A}_1, \mathbf{A}_2, \ldots, \mathbf{A}_K$ represent $M \times M$ symmetric matrices, and let $\mathbf{t} = (t_1, t_2, \ldots, t_K)'$ represent a K-dimensional (column) vector. Then, there exists a neighborhood N of the $K \times 1$ null vector $\mathbf{0}$ such that $\mathbf{I}_M - \sum_{i=1}^{K} t_i\mathbf{A}_i$ is positive definite for $\mathbf{t} \in N$.

Proof (of Lemma 6.5.2). For $p = 1, 2, \ldots, M$, let $\mathbf{A}_1^{(p)}, \mathbf{A}_2^{(p)}, \ldots, \mathbf{A}_K^{(p)}$ represent the leading principal submatrices of order p of $\mathbf{A}_1, \mathbf{A}_2, \ldots, \mathbf{A}_K$, respectively. Then, when regarded as a function of the vector $\mathbf{t}$, $|\mathbf{I}_p - \sum_{i=1}^K t_i \mathbf{A}_i^{(p)}|$ is continuous at $\mathbf{0}$ (and at every other point in $\mathbb{R}^K$), as is evident from the very definition of a determinant. Moreover,

$$\lim_{\mathbf{t}\to\mathbf{0}} |\mathbf{I}_p - \sum_{i=1}^{K} t_i \mathbf{A}_i^{(p)}| = |\mathbf{I}_p - \sum_{i=1}^{K} 0\mathbf{A}_i^{(p)}| = |\mathbf{I}_p| = 1.$$

Thus, there exists a neighborhood, say N_p, of $\mathbf{0}$ such that $|\mathbf{I}_p - \sum_{i=1}^K t_i \mathbf{A}_i^{(p)}| > 0$ for $\mathbf{t} \in N_p$.

Now, take N to be the smallest of the M neighborhoods $N_1, N_2, \ldots, N_M$. Then, for $\mathbf{t} \in N$, $|\mathbf{I}_p - \sum_{i=1}^K t_i \mathbf{A}_i^{(p)}| > 0$ $(p = 1, 2, \ldots, M)$. And upon observing that the matrices $\mathbf{I}_p - \sum_{i=1}^K t_i \mathbf{A}_i^{(p)}$ $(p = 1, 2, \ldots, M)$ are the leading principal submatrices of the matrix $\mathbf{I}_M - \sum_{i=1}^K t_i \mathbf{A}_i$, it follows from Theorem 2.14.23 that $\mathbf{I}_M - \sum_{i=1}^K t_i \mathbf{A}_i$ is positive definite for $\mathbf{t} \in N$. Q.E.D.

A more specific result. Let $\mathbf{A}$ represent an $M \times M$ symmetric matrix, and let t represent an arbitrary scalar. And take S to be the subset of $\mathbb{R}$ defined as follows:

$$S = \{t \ : \ \mathbf{I}_M - t\mathbf{A} \text{ is positive definite}\}.$$

According to Lemma 6.5.1, there exists a (strictly) positive scalar c such that the interval $(-c,\ c)$ is contained in the set S. Let us investigate the nature of the set S more thoroughly.

Let $\mathbf{x}$ represent an M-dimensional column vector of variables. Further, take $q(\cdot)$ to be the function defined (on $\mathbb{R}^M$) as follows: $q(\mathbf{x}) = \mathbf{x}'\mathbf{A}\mathbf{x}$. Then, $q(\mathbf{x})$ attains a maximum value and a minimum value over the set $\{\mathbf{x} \ : \ \mathbf{x}'\mathbf{x} = 1\}$, as is evident upon observing that the function $f(\cdot)$ is continuous and that the set $\{\mathbf{x} \ : \ \mathbf{x}'\mathbf{x} = 1\}$ is closed and bounded—"recall" that any continuous function attains a maximum value and a minimum value over any closed and bounded set (e.g., Bartle 1976, secs. 11 and 22; Bartle and Sherbert 2011).

Accordingly, define

$$d_0 = \min_{\mathbf{x}\,:\,\mathbf{x}'\mathbf{x}=1} \mathbf{x}'\mathbf{A}\mathbf{x} \quad \text{and} \quad d_1 = \max_{\mathbf{x}\,:\,\mathbf{x}'\mathbf{x}=1} \mathbf{x}'\mathbf{A}\mathbf{x}.$$

[The scalars d_0 and d_1 are eigenvalues of the matrix $\mathbf{A}$; in fact, they are respectively the smallest and largest eigenvalues of $\mathbf{A}$, as can be ascertained from results to be presented subsequently (in Section 6.7a).] And observe that

$$\begin{aligned}
\mathbf{I} - t\mathbf{A} \text{ positive definite} &\Leftrightarrow \mathbf{x}'(\mathbf{I} - t\mathbf{A})\mathbf{x} > 0 \text{ for } \mathbf{x} \text{ such that } \mathbf{x} \neq \mathbf{0} \\
&\Leftrightarrow \mathbf{x}'(\mathbf{I} - t\mathbf{A})\mathbf{x} > 0 \text{ for } \mathbf{x} \text{ such that } \mathbf{x}'\mathbf{x} = 1 \\
&\Leftrightarrow t\mathbf{x}'\mathbf{A}\mathbf{x} < 1 \text{ for } \mathbf{x} \text{ such that } \mathbf{x}'\mathbf{x} = 1 \\
&\Leftrightarrow \begin{cases} 1/d_0 < t < 1/d_1, & \text{if } d_0 < 0 \text{ and } d_1 > 0, \\ 1/d_0 < t, & \text{if } d_0 < 0 \text{ and } d_1 \le 0, \\ t < 1/d_1, & \text{if } d_0 \ge 0 \text{ and } d_1 > 0, \\ 0t < 1, & \text{if } d_0 = d_1 = 0. \end{cases}
\end{aligned}$$

Thus,

$$S = \begin{cases} (1/d_0,\ 1/d_1), & \text{if } d_0 < 0 \text{ and } d_1 > 0, \\ (1/d_0,\ \infty), & \text{if } d_0 < 0 \text{ and } d_1 \le 0, \\ (-\infty,\ 1/d_1), & \text{if } d_0 \ge 0 \text{ and } d_1 > 0, \\ (-\infty,\ \infty), & \text{if } d_0 = d_1 = 0. \end{cases} \tag{5.1}$$

Extended applicability of conditions. Let $\mathbf{A}$ represent an $M \times M$ symmetric matrix, t an arbitrary scalar, and $\mathbf{V}$ an $M \times M$ symmetric positive definite matrix. Consider the extension of the conditions under which a matrix of the form $\mathbf{I}_M - t\mathbf{A}$ is positive definite to matrices of the more general form $\mathbf{V} - t\mathbf{A}$.

According to Corollary 2.13.29, there exists an $M \times M$ nonsingular matrix $\mathbf{Q}$ such that $\mathbf{V} = \mathbf{Q}'\mathbf{Q}$. And upon observing that

$$\mathbf{V} - t\mathbf{A} = \mathbf{Q}'[\mathbf{I} - t(\mathbf{Q}^{-1})'\mathbf{A}\mathbf{Q}^{-1}]\mathbf{Q} \quad \text{and} \quad \mathbf{I} - t(\mathbf{Q}^{-1})'\mathbf{A}\mathbf{Q}^{-1} = (\mathbf{Q}^{-1})'(\mathbf{V} - t\mathbf{A})\mathbf{Q}^{-1},$$

it becomes clear (in light of Corollary 2.13.11) that $\mathbf{V} - t\mathbf{A}$ is positive definite if and only if the matrix $\mathbf{I}_M - t(\mathbf{Q}^{-1})'\mathbf{A}\mathbf{Q}^{-1}$ is positive definite. Thus, the applicability of the conditions under which a matrix of the form $\mathbf{I}_M - t\mathbf{A}$ is positive definite can be readily extended to a matrix of the more general form $\mathbf{V} - t\mathbf{A}$; it is a simple matter of applying those conditions with $(\mathbf{Q}^{-1})'\mathbf{A}\mathbf{Q}^{-1}$ in place of $\mathbf{A}$. More generally, conditions under which a matrix of the form $\mathbf{I}_M - \sum_{i=1}^K t_i\mathbf{A}_i$ (where $\mathbf{A}_1, \mathbf{A}_2, \ldots, \mathbf{A}_K$ are $M \times M$ symmetric matrices and $t_1, t_2, \ldots, t_K$ arbitrary scalars) is positive definite can be translated into conditions under which a matrix of the form $\mathbf{V} - \sum_{i=1}^K t_i\mathbf{A}_i$ is positive definite by replacing $\mathbf{A}_1, \mathbf{A}_2, \ldots, \mathbf{A}_K$ with $(\mathbf{Q}^{-1})'\mathbf{A}_1\mathbf{Q}^{-1}$, $(\mathbf{Q}^{-1})'\mathbf{A}_2\mathbf{Q}^{-1}$, $\ldots$, $(\mathbf{Q}^{-1})'\mathbf{A}_K\mathbf{Q}^{-1}$, respectively.

Note that conditions under which $\mathbf{I} - t\mathbf{A}$ or $\mathbf{I} - \sum_{i=1}^K t_i\mathbf{A}_i$ (or $\mathbf{V} - t\mathbf{A}$ or $\mathbf{V} - \sum_{i=1}^K t_i\mathbf{A}_i$) is positive definite can be easily translated into conditions under which $\mathbf{I} + t\mathbf{A}$ or $\mathbf{I} + \sum_{i=1}^K t_i\mathbf{A}_i$ (or $\mathbf{V} + t\mathbf{A}$ or $\mathbf{V} + \sum_{i=1}^K t_i\mathbf{A}_i$) is positive definite.

b. Main results

Let us derive the moment generating function of the distribution of a quadratic form $\mathbf{x}'\mathbf{A}\mathbf{x}$, or more generally of the distribution of a second-degree polynomial $c + \mathbf{b}'\mathbf{x} + \mathbf{x}'\mathbf{A}\mathbf{x}$, in a random column vector $\mathbf{x}$, where $\mathbf{x} \sim N(\boldsymbol{\mu}, \boldsymbol{\Sigma})$. And let us derive the moment generating function of the joint distribution of two or more quadratic forms or second-degree polynomials. Let us do so by establishing and exploiting the following theorem.

Theorem 6.5.3. Let $\mathbf{z}$ represent an M-dimensional random column vector that has an M-variate standard normal distribution $N(\mathbf{0}, \mathbf{I}_M)$. Then, for any constant c and any M-dimensional column vector $\mathbf{b}$ (of constants) and for any $M \times M$ symmetric matrix $\mathbf{A}$ (of constants) such that $\mathbf{I} - 2\mathbf{A}$ is positive definite,

$$\mathrm{E}\big(e^{c+\mathbf{b}'\mathbf{z}+\mathbf{z}'\mathbf{A}\mathbf{z}}\big) = |\mathbf{I} - 2\mathbf{A}|^{-1/2} e^{c+(1/2)\mathbf{b}'(\mathbf{I}-2\mathbf{A})^{-1}\mathbf{b}}. \tag{5.2}$$

Proof. Let $f(\cdot)$ represent the pdf of the $N(\mathbf{0}, \mathbf{I}_M)$ distribution and $g(\cdot)$ the pdf of the $N[(\mathbf{I} - 2\mathbf{A})^{-1}\mathbf{b},\ (\mathbf{I}-2\mathbf{A})^{-1}]$ distribution. Then, for all $\mathbf{z}$,

$$e^{c+\mathbf{b}'\mathbf{z}+\mathbf{z}'\mathbf{A}\mathbf{z}} f(\mathbf{z}) = |\mathbf{I} - 2\mathbf{A}|^{-1/2} e^{c+(1/2)\mathbf{b}'(\mathbf{I}-2\mathbf{A})^{-1}\mathbf{b}} g(\mathbf{z}),$$

as can be readily verified. And it follows that

$$\begin{aligned}\mathrm{E}\big(e^{c+\mathbf{b}'\mathbf{z}+\mathbf{z}'\mathbf{A}\mathbf{z}}\big) &= \int_{\mathbb{R}^M} e^{c+\mathbf{b}'\mathbf{z}+\mathbf{z}'\mathbf{A}\mathbf{z}} f(\mathbf{z})\, d\mathbf{z} \\ &= |\mathbf{I} - 2\mathbf{A}|^{-1/2} e^{c+(1/2)\mathbf{b}'(\mathbf{I}-2\mathbf{A})^{-1}\mathbf{b}} \int_{\mathbb{R}^M} g(\mathbf{z})\, d\mathbf{z} \\ &= |\mathbf{I} - 2\mathbf{A}|^{-1/2} e^{c+(1/2)\mathbf{b}'(\mathbf{I}-2\mathbf{A})^{-1}\mathbf{b}}.\end{aligned}$$

Q.E.D.

Moment generating function of the distribution of a single quadratic form or second-degree polynomial. Let $\mathbf{x}$ represent an M-dimensional random column vector that has an $N(\boldsymbol{\mu}, \boldsymbol{\Sigma})$ distribution (where the rank of $\boldsymbol{\Sigma}$ is possibly less than M). Further, take $\boldsymbol{\Gamma}$ to be any matrix such that $\boldsymbol{\Sigma} = \boldsymbol{\Gamma}'\boldsymbol{\Gamma}$—the existence of such a matrix follows from Corollary 2.13.25—and denote by R the number of rows in $\boldsymbol{\Gamma}$. And observe that

$$\mathbf{x} \sim \boldsymbol{\mu} + \boldsymbol{\Gamma}'\mathbf{z}, \tag{5.3}$$

where $\mathbf{z}$ is an R-dimensional random (column) vector that has an $N(\mathbf{0}, \mathbf{I}_R)$ distribution.

Let c represent a constant, $\mathbf{b}$ an M-dimensional column vector of constants, and $\mathbf{A}$ an $M \times M$ symmetric matrix of constants. And denote by $m(\cdot)$ the moment generating function of the second-degree polynomial $c + \mathbf{b}'\mathbf{x} + \mathbf{x}'\mathbf{A}\mathbf{x}$ (in the random vector $\mathbf{x}$), and let t represent an arbitrary scalar. Further, take S to be the subset of $\mathcal{R}$ defined as follows:

$$S = \{t \; : \; \text{the matrix } \mathbf{I} - 2t\mathbf{\Gamma}\mathbf{A}\mathbf{\Gamma}' \text{ is positive definite}\}.$$

As is evident from Lemma 6.5.1, this subset includes a neighborhood of 0, and, letting $d_0 = 2\min_{\mathbf{z}\,:\,\mathbf{z}'\mathbf{z}=1} \mathbf{z}'\mathbf{\Gamma}\mathbf{A}\mathbf{\Gamma}'\mathbf{z}$ and $d_1 = 2\max_{\mathbf{z}\,:\,\mathbf{z}'\mathbf{z}=1} \mathbf{z}'\mathbf{\Gamma}\mathbf{A}\mathbf{\Gamma}'\mathbf{z}$, it is [in light of result (5.1)] expressible in the form

$$S = \begin{cases} (1/d_0,\ 1/d_1), & \text{if } d_0 < 0 \text{ and } d_1 > 0, \\ (1/d_0,\ \infty), & \text{if } d_0 < 0 \text{ and } d_1 \le 0, \\ (-\infty,\ 1/d_1), & \text{if } d_0 \ge 0 \text{ and } d_1 > 0, \\ (-\infty,\ \infty), & \text{if } d_0 = d_1 = 0. \end{cases}$$

Upon observing [in light of result (5.3)] that

$$\begin{aligned} c + \mathbf{b}'\mathbf{x} + \mathbf{x}'\mathbf{A}\mathbf{x} &\sim c + \mathbf{b}'(\boldsymbol{\mu} + \mathbf{\Gamma}'\mathbf{z}) + (\boldsymbol{\mu} + \mathbf{\Gamma}'\mathbf{z})'\mathbf{A}(\boldsymbol{\mu} + \mathbf{\Gamma}'\mathbf{z}) \\ &= c + \mathbf{b}'\boldsymbol{\mu} + \boldsymbol{\mu}'\mathbf{A}\boldsymbol{\mu} + [\mathbf{\Gamma}(\mathbf{b} + 2\mathbf{A}\boldsymbol{\mu})]'\mathbf{z} + \mathbf{z}'\mathbf{\Gamma}\mathbf{A}\mathbf{\Gamma}'\mathbf{z} \end{aligned} \tag{5.4}$$

and upon applying result (5.2), we find that, for $t \in S$,

$$\begin{aligned} m(t) &= \mathrm{E}\big[e^{t(c + \mathbf{b}'\mathbf{x} + \mathbf{x}'\mathbf{A}\mathbf{x})}\big] \\ &= \mathrm{E}\big\{e^{t(c + \mathbf{b}'\boldsymbol{\mu} + \boldsymbol{\mu}'\mathbf{A}\boldsymbol{\mu}) + [t\mathbf{\Gamma}(\mathbf{b} + 2\mathbf{A}\boldsymbol{\mu})]'\mathbf{z} + \mathbf{z}'(t\mathbf{\Gamma}\mathbf{A}\mathbf{\Gamma}')\mathbf{z}}\big\} \\ &= |\mathbf{I} - 2t\mathbf{\Gamma}\mathbf{A}\mathbf{\Gamma}'|^{-1/2} \exp[t(c + \mathbf{b}'\boldsymbol{\mu} + \boldsymbol{\mu}'\mathbf{A}\boldsymbol{\mu})] \\ &\qquad \times \exp[(1/2)t^2(\mathbf{b} + 2\mathbf{A}\boldsymbol{\mu})'\mathbf{\Gamma}'(\mathbf{I} - 2t\mathbf{\Gamma}\mathbf{A}\mathbf{\Gamma}')^{-1}\mathbf{\Gamma}(\mathbf{b} + 2\mathbf{A}\boldsymbol{\mu})]. \end{aligned} \tag{5.5}$$

The dependence of expression (5.5) on the variance-covariance matrix $\mathbf{\Sigma}$ is through the "intermediary" $\mathbf{\Gamma}$. The moment generating function can be reexpressed in terms of $\mathbf{\Sigma}$ itself. In light of Corollary 6.4.2,

$$|\mathbf{I} - 2t\mathbf{\Gamma}\mathbf{A}\mathbf{\Gamma}'| = |\mathbf{I} - \mathbf{\Gamma}(2t\mathbf{A})\mathbf{\Gamma}'| = |\mathbf{I} - 2t\mathbf{A}\mathbf{\Gamma}'\mathbf{\Gamma}| = |\mathbf{I} - 2t\mathbf{A}\mathbf{\Sigma}|, \tag{5.6}$$

implying that

$$|\mathbf{I} - 2t\mathbf{A}\mathbf{\Sigma}| > 0 \quad \text{for } t \in S \tag{5.7}$$

and hence that $\mathbf{I} - 2t\mathbf{A}\mathbf{\Sigma}$ is nonsingular for $t \in S$. Moreover,

$$(\mathbf{I} - 2t\mathbf{\Gamma}\mathbf{A}\mathbf{\Gamma}')^{-1}\mathbf{\Gamma} = \mathbf{\Gamma}(\mathbf{I} - 2t\mathbf{A}\mathbf{\Sigma})^{-1} \quad \text{for } t \in S, \tag{5.8}$$

as is evident upon observing that

$$\mathbf{\Gamma}(\mathbf{I} - 2t\mathbf{A}\mathbf{\Sigma}) = (\mathbf{I} - 2t\mathbf{\Gamma}\mathbf{A}\mathbf{\Gamma}')\mathbf{\Gamma}$$

and upon premultiplying both sides of this equality by $(\mathbf{I} - 2t\mathbf{\Gamma}\mathbf{A}\mathbf{\Gamma}')^{-1}$ and postmultiplying both sides by $(\mathbf{I} - 2t\mathbf{A}\mathbf{\Sigma})^{-1}$.

Results (5.6) and (5.8) can be used to reexpress expression (5.5) (for the moment generating function) as follows: for $t \in S$,

$$\begin{aligned} m(t) &= |\mathbf{I} - 2t\mathbf{A}\mathbf{\Sigma}|^{-1/2} \exp[t(c + \mathbf{b}'\boldsymbol{\mu} + \boldsymbol{\mu}'\mathbf{A}\boldsymbol{\mu})] \\ &\qquad \times \exp[(1/2)t^2(\mathbf{b} + 2\mathbf{A}\boldsymbol{\mu})'\mathbf{\Sigma}(\mathbf{I} - 2t\mathbf{A}\mathbf{\Sigma})^{-1}(\mathbf{b} + 2\mathbf{A}\boldsymbol{\mu})]. \end{aligned} \tag{5.9}$$

In the special case where $c = 0$ and $\mathbf{b} = \mathbf{0}$ [i.e., where $m(\cdot)$ is the moment generating function of the quadratic form $\mathbf{x}'\mathbf{A}\mathbf{x}$], expression (5.9) simplifies as follows: for $t \in S$,

$$m(t) = |\mathbf{I} - 2t\mathbf{A}\boldsymbol{\Sigma}|^{-1/2}\exp\{t\boldsymbol{\mu}'[\mathbf{I} + 2t\mathbf{A}\boldsymbol{\Sigma}(\mathbf{I} - 2t\mathbf{A}\boldsymbol{\Sigma})^{-1}]\mathbf{A}\boldsymbol{\mu}\}$$
$$= |\mathbf{I} - 2t\mathbf{A}\boldsymbol{\Sigma}|^{-1/2}\exp[t\boldsymbol{\mu}'(\mathbf{I} - 2t\mathbf{A}\boldsymbol{\Sigma})^{-1}\mathbf{A}\boldsymbol{\mu}]. \tag{5.10}$$

And in the further special case where (in addition to $c = 0$ and $\mathbf{b} = \mathbf{0}$) $\boldsymbol{\Sigma}$ is nonsingular, the moment generating function (of the distribution of $\mathbf{x}'\mathbf{A}\mathbf{x}$) is also expressible as follows: for $t \in S$,

$$m(t) = |\mathbf{I} - 2t\mathbf{A}\boldsymbol{\Sigma}|^{-1/2}\exp\{-(1/2)\boldsymbol{\mu}'[\mathbf{I} - (\mathbf{I} - 2t\mathbf{A}\boldsymbol{\Sigma})^{-1}]\boldsymbol{\Sigma}^{-1}\boldsymbol{\mu}\}, \tag{5.11}$$

as is evident upon observing that

$$t(\mathbf{I} - 2t\mathbf{A}\boldsymbol{\Sigma})^{-1}\mathbf{A} = -(1/2)(\mathbf{I} - 2t\mathbf{A}\boldsymbol{\Sigma})^{-1}(-2t\mathbf{A}\boldsymbol{\Sigma})\boldsymbol{\Sigma}^{-1} = -(1/2)[\mathbf{I} - (\mathbf{I} - 2t\mathbf{A}\boldsymbol{\Sigma})^{-1}]\boldsymbol{\Sigma}^{-1}.$$

Moment generating function of the joint distribution of multiple quadratic forms or second-degree polynomials. Let us continue to take $\mathbf{x}$ to be an M-dimensional random column vector that has an $N(\boldsymbol{\mu}, \boldsymbol{\Sigma})$ distribution, to take $\boldsymbol{\Gamma}$ to be any matrix such that $\boldsymbol{\Sigma} = \boldsymbol{\Gamma}'\boldsymbol{\Gamma}$, to denote by R the number of rows in $\boldsymbol{\Gamma}$, and to take $\mathbf{z}$ to be an R-dimensional random column vector that has an $N(\mathbf{0}, \mathbf{I}_R)$ distribution.

For $i = 1, 2, \ldots, K$ (where K is a strictly positive integer), let c_i represent a constant, $\mathbf{b}_i$ an M-dimensional column vector of constants, and $\mathbf{A}_i$ an $M \times M$ symmetric matrix of constants. And denote by $m(\cdot)$ the moment generating function of the distribution of the K-dimensional random column vector whose ith element is the second-degree polynomial $c_i + \mathbf{b}_i'\mathbf{x} + \mathbf{x}'\mathbf{A}_i\mathbf{x}$ (in the random vector $\mathbf{x}$), and let $\mathbf{t} = (t_1, t_2, \ldots, t_K)'$ represent an arbitrary K-dimensional column vector. Further, take S to be the subset of $\mathbb{R}^K$ defined as follows:

$$S = \{(t_1, t_2, \ldots, t_K)' \ : \ \mathbf{I} - 2\textstyle\sum_{i=1}^K t_i\boldsymbol{\Gamma}\mathbf{A}_i\boldsymbol{\Gamma}' \text{ is positive definite}\}.$$

As indicated by Lemma 6.5.2, this subset includes a neighborhood of $\mathbf{0}$.

Now, recalling (from Part 1) that $\mathbf{x} \sim \boldsymbol{\mu} + \boldsymbol{\Gamma}'\mathbf{z}$, we have [analogous to result (5.4)] that

$$c_i + \mathbf{b}_i'\mathbf{x} + \mathbf{x}'\mathbf{A}_i\mathbf{x} \sim c_i + \mathbf{b}_i'\boldsymbol{\mu} + \boldsymbol{\mu}'\mathbf{A}_i\boldsymbol{\mu} + [\boldsymbol{\Gamma}(\mathbf{b}_i + 2\mathbf{A}_i\boldsymbol{\mu})]'\mathbf{z} + \mathbf{z}'\boldsymbol{\Gamma}\mathbf{A}_i\boldsymbol{\Gamma}'\mathbf{z}$$

($i = 1, 2, \ldots, K$). And upon applying result (5.2), we obtain the following generalization of result (5.5): for $\mathbf{t} \in S$,

$$m(\mathbf{t}) = \mathrm{E}\big[e^{\sum_i t_i(c_i + \mathbf{b}_i'\mathbf{x} + \mathbf{x}'\mathbf{A}_i\mathbf{x})}\big]$$
$$= \mathrm{E}\big\{e^{\sum_i t_i(c_i + \mathbf{b}_i'\boldsymbol{\mu} + \boldsymbol{\mu}'\mathbf{A}_i\boldsymbol{\mu}) + [\boldsymbol{\Gamma}\sum_i t_i(\mathbf{b}_i + 2\mathbf{A}_i\boldsymbol{\mu})]'\mathbf{z} + \mathbf{z}'(\sum_i t_i\boldsymbol{\Gamma}\mathbf{A}_i\boldsymbol{\Gamma}')\mathbf{z}}\big\}$$
$$= \big|\mathbf{I} - 2\textstyle\sum_i t_i\boldsymbol{\Gamma}\mathbf{A}_i\boldsymbol{\Gamma}'\big|^{-1/2}\exp\big[\sum_i t_i(c_i + \mathbf{b}_i'\boldsymbol{\mu} + \boldsymbol{\mu}'\mathbf{A}_i\boldsymbol{\mu})\big]$$
$$\times \exp\big\{(1/2)\big[\textstyle\sum_i t_i(\mathbf{b}_i + 2\mathbf{A}_i\boldsymbol{\mu})\big]'\boldsymbol{\Gamma}'\big(\mathbf{I} - 2\sum_i t_i\boldsymbol{\Gamma}\mathbf{A}_i\boldsymbol{\Gamma}'\big)^{-1}\boldsymbol{\Gamma}\sum_i t_i(\mathbf{b}_i + 2\mathbf{A}_i\boldsymbol{\mu})\big\}. \tag{5.12}$$

As a straightforward generalization of result (5.6), we have that

$$\big|\mathbf{I} - 2\textstyle\sum_{i=1}^K t_i\boldsymbol{\Gamma}\mathbf{A}_i\boldsymbol{\Gamma}'\big| = \big|\mathbf{I} - 2\sum_{i=1}^K t_i\mathbf{A}_i\boldsymbol{\Sigma}\big|, \tag{5.13}$$

implying that

$$\big|\mathbf{I} - 2\textstyle\sum_{i=1}^K t_i\mathbf{A}_i\boldsymbol{\Sigma}\big| > 0 \quad \text{for } \mathbf{t} \in S \tag{5.14}$$

and hence that $\mathbf{I} - 2\sum_{i=1}^K t_i\mathbf{A}_i\boldsymbol{\Sigma}$ is nonsingular for $\mathbf{t} \in S$. And as a straightforward generalization of result (5.8), we have that

$$\big(\mathbf{I} - 2\textstyle\sum_{i=1}^K t_i\boldsymbol{\Gamma}\mathbf{A}_i\boldsymbol{\Gamma}'\big)^{-1}\boldsymbol{\Gamma} = \boldsymbol{\Gamma}\big(\mathbf{I} - 2\sum_{i=1}^K t_i\mathbf{A}_i\boldsymbol{\Sigma}\big)^{-1} \quad \text{for } \mathbf{t} \in S. \tag{5.15}$$

Based on results (5.13) and (5.15), we obtain, as a variation on expression (5.12) for the moment generating function, the following generalization of expression (5.9): for $\mathbf{t} \in S$,

$$m(\mathbf{t}) = \left|\mathbf{I}-2\textstyle\sum_i t_i \mathbf{A}_i \boldsymbol{\Sigma}\right|^{-1/2} \exp\left[\textstyle\sum_i t_i (c_i + \mathbf{b}_i'\boldsymbol{\mu} + \boldsymbol{\mu}'\mathbf{A}_i\boldsymbol{\mu})\right]$$
$$\times \exp\left\{(1/2)\left[\textstyle\sum_i t_i (\mathbf{b}_i + 2\mathbf{A}_i\boldsymbol{\mu})\right]' \boldsymbol{\Sigma}\left(\mathbf{I}-2\textstyle\sum_i t_i \mathbf{A}_i \boldsymbol{\Sigma}\right)^{-1} \textstyle\sum_i t_i (\mathbf{b}_i + 2\mathbf{A}_i\boldsymbol{\mu})\right\}. \quad (5.16)$$

In the special case where $c_1 = c_2 = \cdots = c_K = 0$ and $\mathbf{b}_1 = \mathbf{b}_2 = \cdots = \mathbf{b}_K = \mathbf{0}$ [i.e., where $m(\cdot)$ is the moment generating function of the joint distribution of the quadratic forms $\mathbf{x}'\mathbf{A}_1\mathbf{x}, \mathbf{x}'\mathbf{A}_2\mathbf{x}, \ldots, \mathbf{x}'\mathbf{A}_K\mathbf{x}$], expression (5.16) simplifies to the following generalization of expression (5.10): for $\mathbf{t} \in S$,

$$m(\mathbf{t}) = \left|\mathbf{I}-2\textstyle\sum_i t_i \mathbf{A}_i \boldsymbol{\Sigma}\right|^{-1/2} \exp\left[\boldsymbol{\mu}'\left(\mathbf{I}-2\textstyle\sum_i t_i \mathbf{A}_i \boldsymbol{\Sigma}\right)^{-1} \textstyle\sum_i t_i \mathbf{A}_i \boldsymbol{\mu}\right]. \quad (5.17)$$

And in the further special case where (in addition to $c_1 = c_2 = \cdots = c_K = 0$ and $\mathbf{b}_1 = \mathbf{b}_2 = \cdots = \mathbf{b}_K = \mathbf{0}$) $\boldsymbol{\Sigma}$ is nonsingular, the moment generating function (of the joint distribution of $\mathbf{x}'\mathbf{A}_1\mathbf{x}, \mathbf{x}'\mathbf{A}_2\mathbf{x}, \ldots, \mathbf{x}'\mathbf{A}_K\mathbf{x}$) is alternatively expressible as the following generalization of expression (5.11): for $\mathbf{t} \in S$,

$$m(\mathbf{t}) = \left|\mathbf{I}-2\textstyle\sum_i t_i \mathbf{A}_i \boldsymbol{\Sigma}\right|^{-1/2} \exp\left\{-(1/2)\boldsymbol{\mu}'\left[\mathbf{I} - \left(\mathbf{I}-2\textstyle\sum_i t_i \mathbf{A}_i \boldsymbol{\Sigma}\right)^{-1}\right]\boldsymbol{\Sigma}^{-1}\boldsymbol{\mu}\right\}. \quad (5.18)$$

6.6 Distribution of Quadratic Forms or Second-Degree Polynomials (in a Normally Distributed Random Vector): Chi-Squareness

Suppose that $\mathbf{x}$ is an $M \times 1$ random column vector that has an $N(\boldsymbol{\mu}, \boldsymbol{\Sigma})$ distribution. Under what conditions does the quadratic form $\mathbf{x}'\mathbf{A}\mathbf{x}$ (where $\mathbf{A}$ is an $M \times M$ symmetric matrix of constants) have a (possibly noncentral) chi-square distribution? And, more generally, under what conditions does the second-degree polynomial $c + \mathbf{b}'\mathbf{x} + \mathbf{x}'\mathbf{A}\mathbf{x}$ (where c is a constant and $\mathbf{b}$ an $M \times 1$ vector of constants) have a (possibly noncentral) chi-square distribution? In answering these questions, it is convenient to initially restrict attention to the special case where $\boldsymbol{\mu} = \mathbf{0}$ and $\boldsymbol{\Sigma} = \mathbf{I}$ (i.e., the special case where $\mathbf{x}$ has an M-variate standard normal distribution).

a. Special case: quadratic form or second-degree polynomial in a random vector that has a multivariate standard normal distribution

The following theorem gives conditions that are necessary and sufficient for a second-degree polynomial (in a random vector that has a multivariate standard normal distribution) to have a noncentral chi-square distribution.

Theorem 6.6.1. Let $\mathbf{z}$ represent an M-dimensional random column vector that has an $N(\mathbf{0}, \mathbf{I}_M)$ distribution, and take $q = c + \mathbf{b}'\mathbf{z} + \mathbf{z}'\mathbf{A}\mathbf{z}$, where c is a constant, $\mathbf{b}$ an M-dimensional column vector of constants, and $\mathbf{A}$ an $M \times M$ (nonnull) symmetric matrix of constants. If

$$\mathbf{A}^2 = \mathbf{A}, \quad (6.1)$$

$$\mathbf{b} = \mathbf{A}\mathbf{b}, \quad (6.2)$$

and

$$c = \tfrac{1}{4}\mathbf{b}'\mathbf{b}, \quad (6.3)$$

then $q \sim \chi^2(R, c)$, where $R = \operatorname{rank} \mathbf{A} = \operatorname{tr}(\mathbf{A})$. Conversely, if $q \sim \chi^2(R, \lambda)$ (for some strictly positive integer R), then $\mathbf{A}$, $\mathbf{b}$, and c satisfy conditions (6.1), (6.2), and (6.3), $R = \operatorname{rank} \mathbf{A} = \operatorname{tr}(\mathbf{A})$, and $\lambda = c$.

In connection with Theorem 6.6.1, it is worth noting that if $\mathbf{A}$, $\mathbf{b}$, and c satisfy conditions (6.2) and (6.3), then the second-degree polynomial q is reexpressible as a quadratic form

$$q = \left(\mathbf{z} + \tfrac{1}{2}\mathbf{b}\right)'\mathbf{A}\left(\mathbf{z} + \tfrac{1}{2}\mathbf{b}\right) \tag{6.4}$$

[in the vector $\mathbf{z} + \frac{1}{2}\mathbf{b}$, the distribution of which is $N\left(\frac{1}{2}\mathbf{b},\, \mathbf{I}_M\right)$]. Moreover, if $\mathbf{A}$, $\mathbf{b}$, and c satisfy all three of conditions (6.1), (6.2), and (6.3), then q is reexpressible as a sum of squares

$$q = \left(\mathbf{A}\mathbf{z} + \tfrac{1}{2}\mathbf{b}\right)'\left(\mathbf{A}\mathbf{z} + \tfrac{1}{2}\mathbf{b}\right) \tag{6.5}$$

[of the elements of the vector $\mathbf{A}\mathbf{z} + \frac{1}{2}\mathbf{b}$, the distribution of which is $N\left(\frac{1}{2}\mathbf{b},\, \mathbf{A}\right)$].

Theorem 6.6.1 asserts that conditions (6.1), (6.2), and (6.3) are necessary and sufficient for the second-degree polynomial q to have a noncentral chi-square distribution. In proving Theorem 6.6.1, it is convenient to devote our initial efforts to establishing the sufficiency of these conditions. The proof of sufficiency is considerably simpler than that of necessity. And, perhaps fortuitously, it is the sufficiency that is of the most importance; it is typically the sufficiency of the conditions that is invoked in an application of the theorem rather than their necessity.

Proof (of Theorem 6.6.1): sufficiency. Suppose that the symmetric matrix $\mathbf{A}$, the column vector $\mathbf{b}$, and the scalar c satisfy conditions (6.1), (6.2), and (6.3). Then, in conformance with the earlier observation (6.4),

$$q = \left(\mathbf{z} + \tfrac{1}{2}\mathbf{b}\right)'\mathbf{A}\left(\mathbf{z} + \tfrac{1}{2}\mathbf{b}\right).$$

Moreover, it follows from Theorem 5.9.5 that there exists a matrix $\mathbf{O}$ of dimensions $M \times R$, where $R = \operatorname{rank} \mathbf{A}$, such that $\mathbf{A} = \mathbf{O}\mathbf{O}'$ and that necessarily this matrix is such that $\mathbf{O}'\mathbf{O} = \mathbf{I}_R$. Thus,

$$q = \left(\mathbf{z} + \tfrac{1}{2}\mathbf{b}\right)'\mathbf{O}\mathbf{O}'\left(\mathbf{z} + \tfrac{1}{2}\mathbf{b}\right) = \mathbf{x}'\mathbf{x},$$

where $\mathbf{x} = \mathbf{O}'\mathbf{z} + \frac{1}{2}\mathbf{O}'\mathbf{b}$. And upon observing that

$$\mathbf{x} \sim N\left(\tfrac{1}{2}\mathbf{O}'\mathbf{b},\, \mathbf{I}_R\right),$$

we conclude (on the basis of the very definition of the noncentral chi-square distribution) that

$$q \sim \chi^2\left[R,\, \left(\tfrac{1}{2}\mathbf{O}'\mathbf{b}\right)'\left(\tfrac{1}{2}\mathbf{O}'\mathbf{b}\right)\right].$$

It remains only to observe that (because $\mathbf{A}$ is idempotent) $\operatorname{rank} \mathbf{A} = \operatorname{tr}(\mathbf{A})$—refer to Corollary 2.8.3—and that

$$\left(\tfrac{1}{2}\mathbf{O}'\mathbf{b}\right)'\left(\tfrac{1}{2}\mathbf{O}'\mathbf{b}\right) = \tfrac{1}{4}\mathbf{b}'\mathbf{O}\mathbf{O}'\mathbf{b} = \tfrac{1}{4}\mathbf{b}'\mathbf{A}\mathbf{b} = \tfrac{1}{4}\mathbf{b}'\mathbf{b} = c.$$

Q.E.D.

The proof of the "necessity part" of Theorem 6.6.1 is deferred until Section 6.7, subsequent to a discussion of the spectral decomposition of a symmetric matrix and of the introduction of some results on polynomials.

b. Extension to quadratic forms or second-degree polynomials in a random vector having an arbitrary multivariate normal distribution

Theorem 6.6.1 can be generalized as follows.

Theorem 6.6.2. Let $\mathbf{x}$ represent an M-dimensional random column vector that has an $N(\boldsymbol{\mu}, \boldsymbol{\Sigma})$ distribution, and take $q = c + \mathbf{b}'\mathbf{x} + \mathbf{x}'\mathbf{A}\mathbf{x}$, where c is a constant, $\mathbf{b}$ an M-dimensional column vector of constants, and $\mathbf{A}$ an $M \times M$ symmetric matrix of constants (such that $\boldsymbol{\Sigma}\mathbf{A}\boldsymbol{\Sigma} \neq \mathbf{0}$). If

$$\boldsymbol{\Sigma}\mathbf{A}\boldsymbol{\Sigma}\mathbf{A}\boldsymbol{\Sigma} = \boldsymbol{\Sigma}\mathbf{A}\boldsymbol{\Sigma}, \tag{6.6}$$

$$\boldsymbol{\Sigma}(\mathbf{b} + 2\mathbf{A}\boldsymbol{\mu}) = \boldsymbol{\Sigma}\mathbf{A}\boldsymbol{\Sigma}(\mathbf{b} + 2\mathbf{A}\boldsymbol{\mu}), \tag{6.7}$$

and

$$c + \mathbf{b}'\boldsymbol{\mu} + \boldsymbol{\mu}'\mathbf{A}\boldsymbol{\mu} = \tfrac{1}{4}(\mathbf{b} + 2\mathbf{A}\boldsymbol{\mu})'\boldsymbol{\Sigma}(\mathbf{b} + 2\mathbf{A}\boldsymbol{\mu}), \tag{6.8}$$

then $q \sim \chi^2(R,\, c+\mathbf{b}'\boldsymbol{\mu}+\boldsymbol{\mu}'\mathbf{A}\boldsymbol{\mu})$, where $R = \text{rank}(\boldsymbol{\Sigma}\mathbf{A}\boldsymbol{\Sigma}) = \text{tr}(\mathbf{A}\boldsymbol{\Sigma})$. Conversely, if $q \sim \chi^2(R, \lambda)$ (for some strictly positive integer R), then $\mathbf{A}$, $\mathbf{b}$, and c (and $\boldsymbol{\Sigma}$ and $\boldsymbol{\mu}$) satisfy conditions (6.6), (6.7), and (6.8), $R = \text{rank}(\boldsymbol{\Sigma}\mathbf{A}\boldsymbol{\Sigma}) = \text{tr}(\mathbf{A}\boldsymbol{\Sigma})$, and $\lambda = c + \mathbf{b}'\boldsymbol{\mu} + \boldsymbol{\mu}'\mathbf{A}\boldsymbol{\mu}$.

Proof. Let $\mathbf{d} = \mathbf{b} + 2\mathbf{A}\boldsymbol{\mu}$, take $\boldsymbol{\Gamma}$ to be any matrix (with M columns) such that $\boldsymbol{\Sigma} = \boldsymbol{\Gamma}'\boldsymbol{\Gamma}$ (the existence of which follows from Corollary 2.13.25), and denote by P the number of rows in $\boldsymbol{\Gamma}$. Further, take $\mathbf{z}$ to be a P-dimensional random column vector that is distributed as $N(\mathbf{0}, \mathbf{I}_P)$. And observe that $\mathbf{x} \sim \boldsymbol{\mu} + \boldsymbol{\Gamma}'\mathbf{z}$ and hence that

$$q \sim c + \mathbf{b}'(\boldsymbol{\mu} + \boldsymbol{\Gamma}'\mathbf{z}) + (\boldsymbol{\mu} + \boldsymbol{\Gamma}'\mathbf{z})'\mathbf{A}(\boldsymbol{\mu} + \boldsymbol{\Gamma}'\mathbf{z}) = c + \mathbf{b}'\boldsymbol{\mu} + \boldsymbol{\mu}'\mathbf{A}\boldsymbol{\mu} + (\boldsymbol{\Gamma}\mathbf{d})'\mathbf{z} + \mathbf{z}'\boldsymbol{\Gamma}\mathbf{A}\boldsymbol{\Gamma}'\mathbf{z}.$$

Observe also (in light of Corollary 2.3.4) that

$$\boldsymbol{\Sigma}\mathbf{A}\boldsymbol{\Sigma} \neq \mathbf{0} \quad \Leftrightarrow \quad \boldsymbol{\Sigma}\mathbf{A}\boldsymbol{\Gamma}' \neq \mathbf{0} \quad \Leftrightarrow \quad \boldsymbol{\Gamma}\mathbf{A}\boldsymbol{\Gamma}' \neq \mathbf{0}.$$

Accordingly, it follows from Theorem 6.6.1 that if

$$\boldsymbol{\Gamma}\mathbf{A}\boldsymbol{\Gamma}'\boldsymbol{\Gamma}\mathbf{A}\boldsymbol{\Gamma}' = \boldsymbol{\Gamma}\mathbf{A}\boldsymbol{\Gamma}', \tag{6.9}$$

$$\boldsymbol{\Gamma}\mathbf{d} = \boldsymbol{\Gamma}\mathbf{A}\boldsymbol{\Gamma}'\boldsymbol{\Gamma}\mathbf{d}, \tag{6.10}$$

and

$$c + \mathbf{b}'\boldsymbol{\mu} + \boldsymbol{\mu}'\mathbf{A}\boldsymbol{\mu} = \tfrac{1}{4}\mathbf{d}'\boldsymbol{\Gamma}'\boldsymbol{\Gamma}\mathbf{d}, \tag{6.11}$$

then $q \sim \chi^2(R,\, c+\mathbf{b}'\boldsymbol{\mu}+\boldsymbol{\mu}'\mathbf{A}\boldsymbol{\mu})$, where $R = \text{rank}(\boldsymbol{\Gamma}\mathbf{A}\boldsymbol{\Gamma}') = \text{tr}(\boldsymbol{\Gamma}\mathbf{A}\boldsymbol{\Gamma}')$; and, conversely, if $q \sim \chi^2(R, \lambda)$ (for some strictly positive integer R), then $\mathbf{A}$, $\mathbf{b}$, and c (and $\boldsymbol{\Gamma}$ and $\boldsymbol{\mu}$) satisfy conditions (6.9), (6.10), and (6.11), $R = \text{rank}(\boldsymbol{\Gamma}\mathbf{A}\boldsymbol{\Gamma}') = \text{tr}(\boldsymbol{\Gamma}\mathbf{A}\boldsymbol{\Gamma}')$, and $\lambda = c + \mathbf{b}'\boldsymbol{\mu} + \boldsymbol{\mu}'\mathbf{A}\boldsymbol{\mu}$. Moreover, in light of Lemma 2.12.3,

$$\text{rank}(\boldsymbol{\Gamma}\mathbf{A}\boldsymbol{\Gamma}') = \text{rank}(\boldsymbol{\Sigma}\mathbf{A}\boldsymbol{\Gamma}') = \text{rank}(\boldsymbol{\Sigma}\mathbf{A}\boldsymbol{\Sigma}),$$

and in light of Lemma 2.3.1, $\text{tr}(\boldsymbol{\Gamma}\mathbf{A}\boldsymbol{\Gamma}') = \text{tr}(\mathbf{A}\boldsymbol{\Sigma})$. Since $\mathbf{d}'\boldsymbol{\Gamma}'\boldsymbol{\Gamma}\mathbf{d} = (\mathbf{b} + 2\mathbf{A}\boldsymbol{\mu})'\boldsymbol{\Sigma}(\mathbf{b} + 2\mathbf{A}\boldsymbol{\mu})$, it remains only to observe (in light of Corollary 2.3.4) that

$$\boldsymbol{\Gamma}\mathbf{A}\boldsymbol{\Gamma}'\boldsymbol{\Gamma}\mathbf{A}\boldsymbol{\Gamma}' = \boldsymbol{\Gamma}\mathbf{A}\boldsymbol{\Gamma}' \quad \Leftrightarrow \quad \boldsymbol{\Gamma}'\boldsymbol{\Gamma}\mathbf{A}\boldsymbol{\Gamma}'\boldsymbol{\Gamma}\mathbf{A}\boldsymbol{\Gamma}' = \boldsymbol{\Gamma}'\boldsymbol{\Gamma}\mathbf{A}\boldsymbol{\Gamma}' \quad \Leftrightarrow \quad \boldsymbol{\Gamma}'\boldsymbol{\Gamma}\mathbf{A}\boldsymbol{\Gamma}'\boldsymbol{\Gamma}\mathbf{A}\boldsymbol{\Gamma}'\boldsymbol{\Gamma} = \boldsymbol{\Gamma}'\boldsymbol{\Gamma}\mathbf{A}\boldsymbol{\Gamma}'\boldsymbol{\Gamma}$$

[so that conditions (6.6) and (6.9) are equivalent] and that

$$\boldsymbol{\Gamma}\mathbf{d} = \boldsymbol{\Gamma}\mathbf{A}\boldsymbol{\Gamma}'\boldsymbol{\Gamma}\mathbf{d} \quad \Leftrightarrow \quad \boldsymbol{\Gamma}'\boldsymbol{\Gamma}\mathbf{d} = \boldsymbol{\Gamma}'\boldsymbol{\Gamma}\mathbf{A}\boldsymbol{\Gamma}'\boldsymbol{\Gamma}\mathbf{d}$$

[so that conditions (6.7) and (6.10) are equivalent]. Q.E.D.

Note that condition (6.6) is satisfied if $\mathbf{A}$ is such that

$$(\mathbf{A}\boldsymbol{\Sigma})^2 = \mathbf{A}\boldsymbol{\Sigma} \quad [\text{or, equivalently, } (\boldsymbol{\Sigma}\mathbf{A})^2 = \boldsymbol{\Sigma}\mathbf{A}],$$

that is, if $\mathbf{A}\boldsymbol{\Sigma}$ is idempotent (or, equivalently, $\boldsymbol{\Sigma}\mathbf{A}$ is idempotent), in which case

$$\text{tr}(\mathbf{A}\boldsymbol{\Sigma}) = \text{rank}(\mathbf{A}\boldsymbol{\Sigma}) \qquad [= \text{rank}(\boldsymbol{\Sigma}\mathbf{A}\boldsymbol{\Sigma})].$$

And note that conditions (6.6) and (6.7) are both satisfied if $\mathbf{A}$ and $\mathbf{b}$ are such that

$$(\mathbf{A}\boldsymbol{\Sigma})^2 = \mathbf{A}\boldsymbol{\Sigma} \qquad \text{and} \qquad \mathbf{b} \in \mathcal{C}(\mathbf{A}).$$

Note also that all three of conditions (6.6), (6.7), and (6.8) are satisfied if $\mathbf{A}$, $\mathbf{b}$, and c are such that

$$\mathbf{A}\boldsymbol{\Sigma}\mathbf{A} = \mathbf{A}, \qquad \mathbf{b} \in \mathcal{C}(\mathbf{A}), \qquad \text{and} \qquad c = \tfrac{1}{4}\mathbf{b}'\boldsymbol{\Sigma}\mathbf{b}. \tag{6.12}$$

Finally, note that (by definition) $\mathbf{A}\boldsymbol{\Sigma}\mathbf{A} = \mathbf{A}$ if and only if $\boldsymbol{\Sigma}$ is a generalized inverse of $\mathbf{A}$.

In the special case where $\boldsymbol{\Sigma}$ is nonsingular, condition (6.12) is a necessary condition for $\mathbf{A}$, $\mathbf{b}$, and c to satisfy all three of conditions (6.6), (6.7), and (6.8) (as well as a sufficient condition), as can be readily verified. Moreover, if $\boldsymbol{\Sigma}$ is nonsingular, then $\text{rank}(\boldsymbol{\Sigma}\mathbf{A}\boldsymbol{\Sigma}) = \text{rank}(\mathbf{A})$. Thus, as a corollary of Theorem 6.6.2, we have the following result.

Corollary 6.6.3. Let $\mathbf{x}$ represent an M-dimensional random column vector that has an $N(\boldsymbol{\mu}, \boldsymbol{\Sigma})$ distribution, where $\boldsymbol{\Sigma}$ is nonsingular. And take $q = c + \mathbf{b}'\mathbf{x} + \mathbf{x}'\mathbf{A}\mathbf{x}$, where c is a constant, $\mathbf{b}$ an M-dimensional column vector of constants, and $\mathbf{A}$ an $M \times M$ (nonnull) symmetric matrix of constants. If

$$\mathbf{A}\boldsymbol{\Sigma}\mathbf{A} = \mathbf{A}, \qquad \mathbf{b} \in \mathcal{C}(\mathbf{A}), \qquad \text{and} \qquad c = \tfrac{1}{4}\mathbf{b}'\boldsymbol{\Sigma}\mathbf{b}, \tag{6.13}$$

then $q \sim \chi^2(\operatorname{rank}\mathbf{A},\ c+\mathbf{b}'\boldsymbol{\mu}+\boldsymbol{\mu}'\mathbf{A}\boldsymbol{\mu})$. Conversely, if $q \sim \chi^2(R, \lambda)$ (for some strictly positive integer R), then $\mathbf{A}$, $\mathbf{b}$, and c (and $\boldsymbol{\Sigma}$) satisfy condition (6.13), $R = \operatorname{rank}\mathbf{A}$, and $\lambda = c + \mathbf{b}'\boldsymbol{\mu} + \boldsymbol{\mu}'\mathbf{A}\boldsymbol{\mu}$.

In the special case where q is a quadratic form (i.e., where $c = 0$ and $\mathbf{b} = \mathbf{0}$), Corollary 6.6.3 simplifies to the following result.

Corollary 6.6.4. Let $\mathbf{x}$ represent an M-dimensional random column vector that has an $N(\boldsymbol{\mu}, \boldsymbol{\Sigma})$ distribution, where $\boldsymbol{\Sigma}$ is nonsingular. And take $\mathbf{A}$ to be an $M \times M$ (nonnull) symmetric matrix of constants. If $\mathbf{A}\boldsymbol{\Sigma}\mathbf{A} = \mathbf{A}$, then $\mathbf{x}'\mathbf{A}\mathbf{x} \sim \chi^2(\operatorname{rank}\mathbf{A},\ \boldsymbol{\mu}'\mathbf{A}\boldsymbol{\mu})$. Conversely, if $\mathbf{x}'\mathbf{A}\mathbf{x} \sim \chi^2(R, \lambda)$ (for some strictly positive integer R), then $\mathbf{A}\boldsymbol{\Sigma}\mathbf{A} = \mathbf{A}$, $R = \operatorname{rank}\mathbf{A}$, and $\lambda = \boldsymbol{\mu}'\mathbf{A}\boldsymbol{\mu}$.

In connection with Corollaries 6.6.3 and 6.6.4, note that if $\boldsymbol{\Sigma}$ is nonsingular, then

$$\mathbf{A}\boldsymbol{\Sigma}\mathbf{A} = \mathbf{A} \quad \Leftrightarrow \quad (\mathbf{A}\boldsymbol{\Sigma})^2 = \mathbf{A}\boldsymbol{\Sigma} \text{ (i.e., } \mathbf{A}\boldsymbol{\Sigma} \text{ is idempotent).}$$

Moreover, upon taking $\mathbf{k}$ to be an M-dimensional column vector of constants and upon applying Corollary 6.6.3 (with $\mathbf{A} = \boldsymbol{\Sigma}^{-1}$, $\mathbf{b} = -2\boldsymbol{\Sigma}^{-1}\mathbf{k}$, and $c = \mathbf{k}'\boldsymbol{\Sigma}^{-1}\mathbf{k}$), we find that [for an M-dimensional random column vector $\mathbf{x}$ that has an $N(\boldsymbol{\mu}, \boldsymbol{\Sigma})$ distribution, where $\boldsymbol{\Sigma}$ is nonsingular]

$$(\mathbf{x}-\mathbf{k})'\boldsymbol{\Sigma}^{-1}(\mathbf{x}-\mathbf{k}) \sim \chi^2[M,\ (\boldsymbol{\mu}-\mathbf{k})'\boldsymbol{\Sigma}^{-1}(\boldsymbol{\mu}-\mathbf{k})]. \tag{6.14}$$

In the special case where $\mathbf{k} = \mathbf{0}$, result (6.14) simplifies to the following result:

$$\mathbf{x}'\boldsymbol{\Sigma}^{-1}\mathbf{x} \sim \chi^2(M,\ \boldsymbol{\mu}'\boldsymbol{\Sigma}^{-1}\boldsymbol{\mu}). \tag{6.15}$$

Alternatively, result (6.15) is obtainable as an application of Corollary 6.6.4 (that where $\mathbf{A} = \boldsymbol{\Sigma}^{-1}$).

c. Some results on linear spaces (of M-dimensional row or column vectors or, more generally, of $M \times N$ matrices)

At this point in the discussion of the distribution of quadratic forms, it is helpful to introduce some additional results on linear spaces. According to Theorem 2.4.7, every linear space (of $M \times N$ matrices) has a basis. And according to Theorem 2.4.11, any set of R linearly independent matrices in an R-dimensional linear space $\mathcal{V}$ (of $M \times N$ matrices) is a basis for $\mathcal{V}$. A useful generalization of these results is provided by the following theorem.

Theorem 6.6.5. For any set S of R linearly independent matrices in a K-dimensional linear space $\mathcal{V}$ (of $M \times N$ matrices), there exists a basis for $\mathcal{V}$ that includes all R of the matrices in S (and $K-R$ additional matrices).

For a proof of the result set forth in Theorem 6.6.5, refer, for example, to Harville (1997, sec. 4.3g).

Not only does every linear space (of $M \times N$ matrices) have a basis (as asserted by Theorem 2.4.7), but (according to Theorem 2.4.23) every linear space (of $M \times N$ matrices) has an orthonormal basis. A useful generalization of this result is provided by the following variation on Theorem 6.6.5.

Theorem 6.6.6. For any orthonormal set S of R matrices in a K-dimensional linear space $\mathcal{V}$ (of $M \times N$ matrices), there exists an orthonormal basis for $\mathcal{V}$ that includes all R of the matrices in S (and $K-R$ additional matrices).

Theorem 6.6.6 can be derived from Theorem 6.6.5 in much the same way that Theorem 2.4.23 can be derived from Theorem 2.4.7—refer, e.g., to Harville (1997, sec. 6.4c) for some specifics.

d. Some variations on the results of Subsections a and b

Suppose that $\mathbf{z}$ is an M-dimensional random column vector that is distributed as $N(\mathbf{0}, \mathbf{I}_M)$ and that $\mathbf{A}$ is an $M \times M$ (nonnull) symmetric matrix of constants. As a special case of Theorem 6.6.1, we have the following result: if $\mathbf{A}^2 = \mathbf{A}$, then $\mathbf{z}'\mathbf{A}\mathbf{z} \sim \chi^2(R)$, where $R = \operatorname{rank} \mathbf{A} = \operatorname{tr}(\mathbf{A})$; and, conversely, if $\mathbf{z}'\mathbf{A}\mathbf{z} \sim \chi^2(R)$ (for some strictly positive integer R), then $\mathbf{A}^2 = \mathbf{A}$ and $R = \operatorname{rank} \mathbf{A} = \operatorname{tr}(\mathbf{A})$. A variation on this result is as follows.

Theorem 6.6.7. Let $\mathbf{z}$ represent an M-dimensional random column vector that has an $N(\mathbf{0}, \mathbf{I}_M)$ distribution, take $y_1, y_2, \ldots, y_M$ to be statistically independent random variables that are distributed identically as $N(0, 1)$, and denote by $\mathbf{A}$ an $M \times M$ (nonnull) symmetric matrix of constants. If $\mathbf{A}^2 = \mathbf{A}$, then

$$\frac{\mathbf{z}'\mathbf{A}\mathbf{z}}{\mathbf{z}'\mathbf{z}} \sim \frac{\sum_{i=1}^{R} y_i^2}{\sum_{i=1}^{M} y_i^2},$$

where $R = \operatorname{rank} \mathbf{A} = \operatorname{tr}(\mathbf{A})$; and, conversely, if $\mathbf{z}'\mathbf{A}\mathbf{z}/\mathbf{z}'\mathbf{z} \sim \sum_{i=1}^{R} y_i^2 / \sum_{i=1}^{M} y_i^2$ for some integer R between 1 and M, inclusive, then $\mathbf{A}^2 = \mathbf{A}$ and $R = \operatorname{rank} \mathbf{A} = \operatorname{tr}(\mathbf{A})$.

In connection with Theorem 6.6.7, note that if $\mathbf{z}$ is an M-dimensional random column vector that has an $N(\mathbf{0}, \mathbf{I}_M)$ distribution and if $y_1, y_2, \ldots, y_M$ are statistically independent random variables that are distributed identically as $N(0, 1)$, then for any integer R between 1 and $M-1$, inclusive,

$$\frac{\mathbf{z}'\mathbf{A}\mathbf{z}}{\mathbf{z}'\mathbf{z}} \sim \frac{\sum_{i=1}^{R} y_i^2}{\sum_{i=1}^{M} y_i^2} \quad \Leftrightarrow \quad \frac{\mathbf{z}'\mathbf{A}\mathbf{z}}{\mathbf{z}'\mathbf{z}} \sim Be\left(\tfrac{R}{2}, \tfrac{M-R}{2}\right) \tag{6.16}$$

—for $R = M$, $\sum_{i=1}^{R} y_i^2 / \sum_{i=1}^{M} y_i^2 = 1$.

Proof (of Theorem 6.6.7). Suppose that $\mathbf{A}^2 = \mathbf{A}$ [in which case $\operatorname{rank} \mathbf{A} = \operatorname{tr}(\mathbf{A})$]. Then, according to Theorem 5.9.5, there exists a matrix $\mathbf{Q}_1$ of dimensions $M \times R$, where $R = \operatorname{rank} \mathbf{A}$, such that $\mathbf{A} = \mathbf{Q}_1\mathbf{Q}_1'$, and, necessarily, this matrix is such that $\mathbf{Q}_1'\mathbf{Q}_1 = \mathbf{I}_R$. Now, take $\mathbf{Q}$ to be the $M \times M$ matrix defined as follows: if $R = M$, take $\mathbf{Q} = \mathbf{Q}_1$; if $R < M$, take $\mathbf{Q} = (\mathbf{Q}_1, \mathbf{Q}_2)$, where $\mathbf{Q}_2$ is an $M \times (M-R)$ matrix whose columns consist of any $M-R$ vectors that, together with the R columns of $\mathbf{Q}_1$, form (when the inner product is taken to be the usual inner product) an orthonormal basis for $\mathbb{R}^M$—the existence of such vectors follows from Theorem 6.6.6. Further, define $\mathbf{y}_1 = \mathbf{Q}_1'\mathbf{z}$ and $\mathbf{y} = \mathbf{Q}'\mathbf{z}$, and observe that $\mathbf{Q}$ is orthogonal. And upon observing that $\mathbf{y} \sim N(\mathbf{0}, \mathbf{I}_M)$, that

$$\frac{\mathbf{z}'\mathbf{A}\mathbf{z}}{\mathbf{z}'\mathbf{z}} = \frac{\mathbf{z}'\mathbf{Q}_1\mathbf{Q}_1'\mathbf{z}}{\mathbf{z}'\mathbf{Q}\mathbf{Q}'\mathbf{z}} = \frac{\mathbf{y}_1'\mathbf{y}_1}{\mathbf{y}'\mathbf{y}},$$

and that the elements of $\mathbf{y}_1$ are the first R elements of $\mathbf{y}$, we conclude that

$$\frac{\mathbf{z}'\mathbf{A}\mathbf{z}}{\mathbf{z}'\mathbf{z}} \sim \frac{\sum_{i=1}^{R} y_i^2}{\sum_{i=1}^{M} y_i^2}.$$

Conversely, suppose that $\mathbf{z}'\mathbf{A}\mathbf{z}/\mathbf{z}'\mathbf{z} \sim \sum_{i=1}^{R} y_i^2 / \sum_{i=1}^{M} y_i^2$ for some integer R between 1 and M, inclusive. Then, letting $z_1, z_2, \ldots, z_M$ represent the elements of $\mathbf{z}$ (and observing that the joint distribution of $z_1, z_2, \ldots, z_M$ is identical to that of $y_1, y_2, \ldots, y_M$),

$$\frac{\mathbf{z}'\mathbf{A}\mathbf{z}}{\mathbf{z}'\mathbf{z}} \sim \frac{\sum_{i=1}^{R} z_i^2}{\sum_{i=1}^{M} z_i^2}.$$

Moreover, each of the quantities $\mathbf{z}'\mathbf{A}\mathbf{z}/\mathbf{z}'\mathbf{z}$ and $\sum_{i=1}^{R} z_i^2 / \sum_{i=1}^{M} z_i^2$ depends on the value of $\mathbf{z}$ only through $(\mathbf{z}'\mathbf{z})^{-1/2}\mathbf{z}$, and, consequently, it follows from the results of Section 6.1f that each of these quantities is distributed independently of $\mathbf{z}'\mathbf{z}$ $(= \sum_{i=1}^{M} z_i^2)$. Thus,

$$\mathbf{z}'\mathbf{A}\mathbf{z} = \mathbf{z}'\mathbf{z}\frac{\mathbf{z}'\mathbf{A}\mathbf{z}}{\mathbf{z}'\mathbf{z}} \sim \mathbf{z}'\mathbf{z}\frac{\sum_{i=1}^{R} z_i^2}{\sum_{i=1}^{M} z_i^2} = \sum_{i=1}^{R} z_i^2.$$

It is now clear that $\mathbf{z}'\mathbf{A}\mathbf{z} \sim \chi^2(R)$ and hence, upon applying the "necessity part" of Theorem 6.6.1, that $\mathbf{A}^2 = \mathbf{A}$ and that $R = \text{rank}\,\mathbf{A} = \text{tr}(\mathbf{A})$. Q.E.D.

The result of Theorem 6.6.7 can be generalized. Suppose that $\mathbf{x}$ is an M-dimensional random column vector that is distributed as $N(\mathbf{0}, \mathbf{\Sigma})$, let $P = \text{rank}\,\mathbf{\Sigma}$, suppose that $P > 0$, and take $\mathbf{A}$ to be an $M \times M$ symmetric matrix of constants (such that $\mathbf{\Sigma A \Sigma} \neq \mathbf{0}$). Further, take $\mathbf{\Gamma}$ to be a matrix of dimensions $P \times M$ such that $\mathbf{\Sigma} = \mathbf{\Gamma}'\mathbf{\Gamma}$ (the existence of which follows from Corollary 2.13.23), and take $\mathbf{z}$ to be a P-dimensional random column vector that has an $N(\mathbf{0}, \mathbf{I}_P)$ distribution.

The matrix $\mathbf{\Gamma}$ has full row rank P (as is evident from Corollary 2.13.23), implying (in light of Lemma 2.5.1) that it has a right inverse, say $\mathbf{\Lambda}$. Accordingly, it follows from Theorem 2.10.5 that $\mathbf{\Lambda\Lambda}'$ is a generalized inverse of $\mathbf{\Sigma}$. Further, upon observing that $\mathbf{x} \sim \mathbf{\Gamma}'\mathbf{z}$ and (in light of Theorem 2.12.2) that $\mathbf{\Gamma\Sigma}^-\mathbf{\Gamma}'$ is invariant to the choice of the generalized inverse $\mathbf{\Sigma}^-$, we find that

$$\frac{\mathbf{x}'\mathbf{A}\mathbf{x}}{\mathbf{x}'\mathbf{\Sigma}^-\mathbf{x}} \sim \frac{\mathbf{z}'\mathbf{\Gamma A \Gamma}'\mathbf{z}}{\mathbf{z}'\mathbf{\Gamma\Sigma}^-\mathbf{\Gamma}'\mathbf{z}} = \frac{\mathbf{z}'\mathbf{\Gamma A \Gamma}'\mathbf{z}}{\mathbf{z}'\mathbf{\Gamma\Lambda\Lambda}'\mathbf{\Gamma}'\mathbf{z}} = \frac{\mathbf{z}'\mathbf{\Gamma A \Gamma}'\mathbf{z}}{\mathbf{z}'\mathbf{z}}. \tag{6.17}$$

And upon applying Theorem 6.6.7 [and taking $y_1, y_2, \ldots, y_P$ to be statistically independent random variables that are distributed identically as $N(0, 1)$], we conclude that if $\mathbf{\Gamma A \Sigma A \Gamma}' = \mathbf{\Gamma A \Gamma}'$, then

$$\frac{\mathbf{x}'\mathbf{A}\mathbf{x}}{\mathbf{x}'\mathbf{\Sigma}^-\mathbf{x}} \sim \frac{\sum_{i=1}^{R} y_i^2}{\sum_{i=1}^{P} y_i^2},$$

where $R = \text{rank}(\mathbf{\Gamma A \Gamma}') = \text{tr}(\mathbf{\Gamma A \Gamma}')$; and, conversely, if $\dfrac{\mathbf{x}'\mathbf{A}\mathbf{x}}{\mathbf{x}'\mathbf{\Sigma}^-\mathbf{x}} \sim \dfrac{\sum_{i=1}^{R} y_i^2}{\sum_{i=1}^{P} y_i^2}$ for some integer R between 1 and P, inclusive, then $\mathbf{\Gamma A \Sigma A \Gamma}' = \mathbf{\Gamma A \Gamma}'$ and $R = \text{rank}(\mathbf{\Gamma A \Gamma}') = \text{tr}(\mathbf{\Gamma A \Gamma}')$—note (in light of Corollary 2.3.4) that $\mathbf{\Sigma A \Sigma} \neq \mathbf{0} \Leftrightarrow \mathbf{\Gamma A \Gamma}' \neq \mathbf{0}$.

The condition $\mathbf{\Gamma A \Sigma A \Gamma}' = \mathbf{\Gamma A \Gamma}'$ can be restated in terms that do not involve $\mathbf{\Gamma}$. Upon applying Corollary 2.3.4, we find that

$$\mathbf{\Gamma A \Sigma A \Gamma}' = \mathbf{\Gamma A \Gamma}' \quad \Leftrightarrow \quad \mathbf{\Sigma A \Sigma A \Sigma} = \mathbf{\Sigma A \Sigma}.$$

Moreover, as observed earlier (in the proof of Theorem 6.6.2),

$$\text{rank}(\mathbf{\Gamma A \Gamma}') = \text{rank}(\mathbf{\Sigma A \Sigma}) \quad \text{and} \quad \text{tr}(\mathbf{\Gamma A \Gamma}') = \text{tr}(\mathbf{A\Sigma}).$$

In summary, we have the following theorem, which generalizes Theorem 6.6.7 and which relates to Theorem 6.6.2 in the same way that Theorem 6.6.7 relates to Theorem 6.6.1.

Theorem 6.6.8. Let $\mathbf{x}$ represent an M-dimensional random column vector that has an $N(\mathbf{0}, \mathbf{\Sigma})$ distribution, let $P = \text{rank}\,\mathbf{\Sigma}$, suppose that $P > 0$, take $y_1, y_2, \ldots, y_P$ to be statistically independent random variables that are distributed identically as $N(0, 1)$, and denote by $\mathbf{A}$ an $M \times M$ symmetric matrix of constants (such that $\mathbf{\Sigma A \Sigma} \neq \mathbf{0}$). If $\mathbf{\Sigma A \Sigma A \Sigma} = \mathbf{\Sigma A \Sigma}$, then

$$\frac{\mathbf{x}'\mathbf{A}\mathbf{x}}{\mathbf{x}'\mathbf{\Sigma}^-\mathbf{x}} \sim \frac{\sum_{i=1}^{R} y_i^2}{\sum_{i=1}^{P} y_i^2},$$

where $R = \text{rank}(\mathbf{\Sigma A \Sigma}) = \text{tr}(\mathbf{A\Sigma})$; and, conversely, if $\dfrac{\mathbf{x}'\mathbf{A}\mathbf{x}}{\mathbf{x}'\mathbf{\Sigma}^-\mathbf{x}} \sim \dfrac{\sum_{i=1}^{R} y_i^2}{\sum_{i=1}^{P} y_i^2}$ for some integer R between 1 and P, inclusive, then $\mathbf{\Sigma A \Sigma A \Sigma} = \mathbf{\Sigma A \Sigma}$ and $R = \text{rank}(\mathbf{\Sigma A \Sigma}) = \text{tr}(\mathbf{A\Sigma})$.

In connection with Theorem 6.6.8, note that (for $1 \leq R \leq P-1$) the condition $\dfrac{\mathbf{x}'\mathbf{A}\mathbf{x}}{\mathbf{x}'\mathbf{\Sigma}^-\mathbf{x}} \sim \dfrac{\sum_{i=1}^{R} y_i^2}{\sum_{i=1}^{P} y_i^2}$ is equivalent to the condition

$$\frac{\mathbf{x}'\mathbf{A}\mathbf{x}}{\mathbf{x}'\boldsymbol{\Sigma}^{-}\mathbf{x}} \sim Be\left(\tfrac{R}{2}, \tfrac{P-R}{2}\right). \tag{6.18}$$

Note also that the condition $\boldsymbol{\Sigma}\mathbf{A}\boldsymbol{\Sigma}\mathbf{A}\boldsymbol{\Sigma} = \boldsymbol{\Sigma}\mathbf{A}\boldsymbol{\Sigma}$ is satisfied if, in particular, $(\mathbf{A}\boldsymbol{\Sigma})^2 = \mathbf{A}\boldsymbol{\Sigma}$, in which case

$$\text{tr}(\mathbf{A}\boldsymbol{\Sigma}) = \text{rank}(\mathbf{A}\boldsymbol{\Sigma}) \qquad [= \text{rank}(\boldsymbol{\Sigma}\mathbf{A}\boldsymbol{\Sigma})].$$

Finally, note that if $\boldsymbol{\Sigma}$ is nonsingular, then the condition $\boldsymbol{\Sigma}\mathbf{A}\boldsymbol{\Sigma}\mathbf{A}\boldsymbol{\Sigma} = \boldsymbol{\Sigma}\mathbf{A}\boldsymbol{\Sigma}$ is equivalent to the condition $\mathbf{A}\boldsymbol{\Sigma}\mathbf{A} = \mathbf{A}$, and $\text{rank}(\boldsymbol{\Sigma}\mathbf{A}\boldsymbol{\Sigma}) = \text{rank}\,\mathbf{A}$.

e. Extensions to spherically or elliptically distributed random vectors

The results of Theorem 6.6.7 [and result (6.16)] pertain to the distribution of $\mathbf{z}'\mathbf{A}\mathbf{z}/\mathbf{z}'\mathbf{z}$, where $\mathbf{z}$ is an M-dimensional random column vector that has an $N(\mathbf{0}, \mathbf{I}_M)$ distribution (and where $\mathbf{A}$ is an $M \times M$ nonnull symmetric matrix of constants). The validity of these results is not limited to the case where the distribution of the M-dimensional random column vector $\mathbf{z}$ is $N(\mathbf{0}, \mathbf{I}_M)$; it extends to the more general case where the distribution of $\mathbf{z}$ is an absolutely continuous spherical distribution. To see this, suppose that the distribution of $\mathbf{z}$ is an absolutely continuous spherical distribution, and observe that

$$\frac{\mathbf{z}'\mathbf{A}\mathbf{z}}{\mathbf{z}'\mathbf{z}} = [(\mathbf{z}'\mathbf{z})^{-1/2}\mathbf{z}]'\mathbf{A}[(\mathbf{z}'\mathbf{z})^{-1/2}\mathbf{z}]$$

and that the normalized vector $(\mathbf{z}'\mathbf{z})^{-1/2}\mathbf{z}$ has the same distribution as in the special case where the distribution of $\mathbf{z}$ is $N(\mathbf{0}, \mathbf{I}_M)$; as in the special case, $(\mathbf{z}'\mathbf{z})^{-1/2}\mathbf{z}$ is distributed uniformly on the surface of an M-dimensional unit ball—refer to the results of Sections 6.1f and 6.1g. Thus, the distribution of $\mathbf{z}'\mathbf{A}\mathbf{z}/\mathbf{z}'\mathbf{z}$ is the same in the general case where the distribution of $\mathbf{z}$ is an absolutely continuous spherical distribution as in the special case where $\mathbf{z} \sim N(\mathbf{0}, \mathbf{I}_M)$.

Now, consider the results summarized in Theorem 6.6.8 [and result (6.18)]; these results pertain to the distribution of $\mathbf{x}'\mathbf{A}\mathbf{x}/\mathbf{x}'\boldsymbol{\Sigma}^{-}\mathbf{x}$, where $\mathbf{x}$ is an M-dimensional random column vector that has an $N(\mathbf{0}, \boldsymbol{\Sigma})$ distribution [and where $\boldsymbol{\Sigma}$ is an $M \times M$ symmetric nonnegative definite matrix of rank $P\ (> 0)$ and where $\mathbf{A}$ is an $M \times M$ symmetric matrix of constants (such that $\boldsymbol{\Sigma}\mathbf{A}\boldsymbol{\Sigma} \neq \mathbf{0}$)]. The validity of these results is not limited to the case where the distribution of the M-dimensional random column vector $\mathbf{x}$ is $N(\mathbf{0}, \boldsymbol{\Sigma})$; it extends to the more general case where the distribution of $\mathbf{x}$ is that of the vector $\boldsymbol{\Gamma}'\mathbf{z}$, where $\boldsymbol{\Gamma}$ is a $P \times M$ matrix such that $\boldsymbol{\Sigma} = \boldsymbol{\Gamma}'\boldsymbol{\Gamma}$ and where $\mathbf{z}$ is a P-dimensional random column vector that has an absolutely continuous spherical distribution—in the more general case, $\mathbf{x}$ is distributed elliptically about $\mathbf{0}$.

To see this, suppose that $\mathbf{x} \sim \boldsymbol{\Gamma}'\mathbf{z}$ (where $\boldsymbol{\Gamma}$ is a $P \times M$ matrix such that $\boldsymbol{\Sigma} = \boldsymbol{\Gamma}'\boldsymbol{\Gamma}$ and where $\mathbf{z}$ is a P-dimensional random column vector that has an absolutely continuous spherical distribution). And observe that, as in the special case of result (6.17) [where $\mathbf{x} \sim N(\mathbf{0}, \boldsymbol{\Sigma})$ and $\mathbf{z} \sim N(\mathbf{0}, \mathbf{I}_P)$],

$$\frac{\mathbf{x}'\mathbf{A}\mathbf{x}}{\mathbf{x}'\boldsymbol{\Sigma}^{-}\mathbf{x}} \sim \frac{\mathbf{z}'\boldsymbol{\Gamma}\mathbf{A}\boldsymbol{\Gamma}'\mathbf{z}}{\mathbf{z}'\mathbf{z}},$$

that

$$\frac{\mathbf{z}'\boldsymbol{\Gamma}\mathbf{A}\boldsymbol{\Gamma}'\mathbf{z}}{\mathbf{z}'\mathbf{z}} = [(\mathbf{z}'\mathbf{z})^{-1/2}\mathbf{z}]'\boldsymbol{\Gamma}\mathbf{A}\boldsymbol{\Gamma}'[(\mathbf{z}'\mathbf{z})^{-1/2}\mathbf{z}],$$

and that the normalized vector $(\mathbf{z}'\mathbf{z})^{-1/2}\mathbf{z}$ has the same distribution as in the special case where $\mathbf{z} \sim N(\mathbf{0}, \mathbf{I}_P)$. Accordingly, the distribution of $\mathbf{x}'\mathbf{A}\mathbf{x}/\mathbf{x}'\boldsymbol{\Sigma}^{-}\mathbf{x}$ is the same in the general case where (for a $P \times M$ matrix $\boldsymbol{\Gamma}$ such that $\boldsymbol{\Sigma} = \boldsymbol{\Gamma}'\boldsymbol{\Gamma}$ and a P-dimensional random column vector $\mathbf{z}$ that has an absolutely continuous spherical distribution) $\mathbf{x} \sim \boldsymbol{\Gamma}'\mathbf{z}$ as in the special case where $\mathbf{x} \sim N(\mathbf{0}, \boldsymbol{\Sigma})$.

6.7 The Spectral Decomposition, with Application to the Distribution of Quadratic Forms

The existence of a decomposition (of a symmetric matrix) known as the spectral decomposition can be extremely useful, and in some cases indispensable, in establishing various results on the distribution of quadratic forms. There is an intimate relationship between the spectral decomposition (of a symmetric matrix) and the so-called eigenvalues and eigenvectors of the matrix.

a. Eigenvalues, eigenvectors, and the spectral decomposition

Let $\mathbf{A} = \{a_{ij}\}$ represent an $N \times N$ matrix. A scalar (real number) λ is said to be an *eigenvalue* of $\mathbf{A}$ if there exists an N-dimensional nonnull column vector $\mathbf{x}$ such that

$$\mathbf{Ax} = \lambda\mathbf{x}$$

or, equivalently, such that

$$(\mathbf{A} - \lambda\mathbf{I}_N)\mathbf{x} = \mathbf{0}.$$

Consider the function $p(\lambda)$ of a single variable λ defined (for all λ) as follows:

$$p(\lambda) = |\mathbf{A} - \lambda\mathbf{I}_N|.$$

It follows from the very definition of a determinant that $p(\lambda)$ is a polynomial (in λ) of degree N; this polynomial is referred to as the *characteristic polynomial* of the matrix $\mathbf{A}$. Upon equating $p(\lambda)$ to 0, we obtain the equality

$$p(\lambda) = 0,$$

which can be regarded as an equation (in λ) and (when so regarded) is referred to as the *characteristic equation*. Clearly, a scalar is an eigenvalue of $\mathbf{A}$ if and only if it is a root of the characteristic polynomial or, equivalently, is a solution to the characteristic equation.

An N-dimensional nonnull column vector $\mathbf{x}$ is said to be an *eigenvector* of the $N \times N$ matrix $\mathbf{A}$ if there exists a scalar (real number) λ such that $\mathbf{Ax} = \lambda\mathbf{x}$, in which case λ is (by definition) an eigenvalue of $\mathbf{A}$. For any particular eigenvector $\mathbf{x}$ (of $\mathbf{A}$), there is only one eigenvalue λ such that $\mathbf{Ax} = \lambda\mathbf{x}$, which (since $\mathbf{Ax} = \lambda\mathbf{x} \Rightarrow \mathbf{x}'\mathbf{Ax} = \lambda\mathbf{x}'\mathbf{x}$) is

$$\lambda = \frac{\mathbf{x}'\mathbf{Ax}}{\mathbf{x}'\mathbf{x}}.$$

The eigenvector $\mathbf{x}$ is said to correspond to (or belong to) this eigenvalue.

Note that if $\mathbf{x}$ is an eigenvector of $\mathbf{A}$ corresponding to an eigenvalue λ, then for any nonzero scalar c, the scalar multiple $c\mathbf{x}$ is also an eigenvector of $\mathbf{A}$, and $c\mathbf{x}$ corresponds to the same eigenvalue as $\mathbf{x}$. In particular, if $\mathbf{x}$ is an eigenvector of $\mathbf{A}$ corresponding to an eigenvalue λ, then the vector $(\mathbf{x}'\mathbf{x})^{-1/2}\mathbf{x}$, which is the scalar multiple of $\mathbf{x}$ having a norm of 1, is also an eigenvector of $\mathbf{A}$ corresponding to λ.

Existence of eigenvalues. Does an $N \times N$ matrix necessarily have an eigenvalue? The corollary of the following theorem indicates that in the case of a symmetric matrix, the answer is yes.

Theorem 6.7.1. Let $\mathbf{A}$ represent an $N \times N$ matrix. Then, there exist N-dimensional nonnull column vectors $\mathbf{x}_0$ and $\mathbf{x}_1$ such that

$$\frac{\mathbf{x}_0'\mathbf{Ax}_0}{\mathbf{x}_0'\mathbf{x}_0} \le \frac{\mathbf{x}'\mathbf{Ax}}{\mathbf{x}'\mathbf{x}} \le \frac{\mathbf{x}_1'\mathbf{Ax}_1}{\mathbf{x}_1'\mathbf{x}_1}$$

for every nonnull column vector $\mathbf{x}$ in $\mathcal{R}^N$ (or, equivalently, such that

$$\frac{\mathbf{x}_0'\mathbf{Ax}_0}{\mathbf{x}_0'\mathbf{x}_0} = \min_{\mathbf{x}\neq\mathbf{0}} \frac{\mathbf{x}'\mathbf{Ax}}{\mathbf{x}'\mathbf{x}} = \min_{\mathbf{x}\,:\,\mathbf{x}'\mathbf{x}=1} \mathbf{x}'\mathbf{Ax}$$

and

$$\frac{\mathbf{x}_1'\mathbf{A}\mathbf{x}_1}{\mathbf{x}_1'\mathbf{x}_1} = \max_{\mathbf{x}\neq\mathbf{0}} \frac{\mathbf{x}'\mathbf{A}\mathbf{x}}{\mathbf{x}'\mathbf{x}} = \max_{\mathbf{x}\,:\,\mathbf{x}'\mathbf{x}=1} \mathbf{x}'\mathbf{A}\mathbf{x}).$$

Moreover, if $\mathbf{A}$ is symmetric, then $\frac{\mathbf{x}_0'\mathbf{A}\mathbf{x}_0}{\mathbf{x}_0'\mathbf{x}_0}$ and $\frac{\mathbf{x}_1'\mathbf{A}\mathbf{x}_1}{\mathbf{x}_1'\mathbf{x}_1}$ are eigenvalues of $\mathbf{A}$—they are respectively the smallest and largest eigenvalues of $\mathbf{A}$—and $\mathbf{x}_0$ and $\mathbf{x}_1$ are eigenvectors corresponding to $\frac{\mathbf{x}_0'\mathbf{A}\mathbf{x}_0}{\mathbf{x}_0'\mathbf{x}_0}$ and $\frac{\mathbf{x}_1'\mathbf{A}\mathbf{x}_1}{\mathbf{x}_1'\mathbf{x}_1}$, respectively.

Proof. Let $\mathbf{x}$ represent an N-dimensional column vector of (unconstrained) variables, and take $f(\cdot)$ to be the function defined (on $\mathcal{R}^N$) as follows: $f(\mathbf{x}) = \mathbf{x}'\mathbf{A}\mathbf{x}$. Further, define $S = \{\mathbf{x} : \mathbf{x}'\mathbf{x} = 1\}$. And observe that the function $f(\cdot)$ is continuous and that the set S is closed and bounded. Then, upon recalling (as in Section 6.5a) that a continuous function attains a minimum value and a maximum value over any closed and bounded set, it follows that S contains vectors $\mathbf{x}_0$ and $\mathbf{x}_1$ such that, for $\mathbf{x} \in S$,

$$\mathbf{x}_0'\mathbf{A}\mathbf{x}_0 \leq \mathbf{x}'\mathbf{A}\mathbf{x} \leq \mathbf{x}_1'\mathbf{A}\mathbf{x}_1.$$

Thus, for $\mathbf{x} \neq \mathbf{0}$,

$$\frac{\mathbf{x}_0'\mathbf{A}\mathbf{x}_0}{\mathbf{x}_0'\mathbf{x}_0} = \mathbf{x}_0'\mathbf{A}\mathbf{x}_0 \leq \frac{\mathbf{x}'\mathbf{A}\mathbf{x}}{\mathbf{x}'\mathbf{x}} \leq \mathbf{x}_1'\mathbf{A}\mathbf{x}_1 = \frac{\mathbf{x}_1'\mathbf{A}\mathbf{x}_1}{\mathbf{x}_1'\mathbf{x}_1}.$$

Now, suppose that $\mathbf{A}$ is symmetric, and take $\mathbf{x}_0$ and $\mathbf{x}_1$ to be any N-dimensional nonnull column vectors such that

$$\frac{\mathbf{x}_0'\mathbf{A}\mathbf{x}_0}{\mathbf{x}_0'\mathbf{x}_0} \leq \frac{\mathbf{x}'\mathbf{A}\mathbf{x}}{\mathbf{x}'\mathbf{x}} \leq \frac{\mathbf{x}_1'\mathbf{A}\mathbf{x}_1}{\mathbf{x}_1'\mathbf{x}_1}$$

for every nonnull column vector $\mathbf{x}$ in $\mathcal{R}^N$. And observe that, for $\mathbf{x} \neq \mathbf{0}$,

$$\frac{1}{\mathbf{x}'\mathbf{x}}\mathbf{x}'[\mathbf{A} - (\mathbf{x}_0'\mathbf{A}\mathbf{x}_0/\mathbf{x}_0'\mathbf{x}_0)\mathbf{I}_N]\mathbf{x} \geq 0$$

or, equivalently,

$$\mathbf{x}'[\mathbf{A} - (\mathbf{x}_0'\mathbf{A}\mathbf{x}_0/\mathbf{x}_0'\mathbf{x}_0)\mathbf{I}_N]\mathbf{x} \geq 0.$$

Thus, $\mathbf{A} - (\mathbf{x}_0'\mathbf{A}\mathbf{x}_0/\mathbf{x}_0'\mathbf{x}_0)\mathbf{I}_N$ is a symmetric nonnegative definite matrix, and upon observing that

$$\mathbf{x}_0'[\mathbf{A} - (\mathbf{x}_0'\mathbf{A}\mathbf{x}_0/\mathbf{x}_0'\mathbf{x}_0)\mathbf{I}_N]\mathbf{x}_0 = 0,$$

it follows from Corollary 2.13.27 that

$$[\mathbf{A} - (\mathbf{x}_0'\mathbf{A}\mathbf{x}_0/\mathbf{x}_0'\mathbf{x}_0)\mathbf{I}_N]\mathbf{x}_0 = \mathbf{0}.$$

It is now clear that $\mathbf{x}_0'\mathbf{A}\mathbf{x}_0/\mathbf{x}_0'\mathbf{x}_0$ is an eigenvalue of $\mathbf{A}$, that $\mathbf{x}_0$ is an eigenvector of $\mathbf{A}$ corresponding to $\mathbf{x}_0'\mathbf{A}\mathbf{x}_0/\mathbf{x}_0'\mathbf{x}_0$, and (since if λ is an eigenvalue of $\mathbf{A}$, $\boldsymbol{\lambda} = \mathbf{x}'\mathbf{A}\mathbf{x}/\mathbf{x}'\mathbf{x}$ for some nonnull vector $\mathbf{x}$) that $\mathbf{x}_0'\mathbf{A}\mathbf{x}_0/\mathbf{x}_0'\mathbf{x}_0$ is the smallest eigenvalue of $\mathbf{A}$. It follows from a similar argument that $\mathbf{x}_1'\mathbf{A}\mathbf{x}_1/\mathbf{x}_1'\mathbf{x}_1$ is an eigenvalue of $\mathbf{A}$, that $\mathbf{x}_1$ is an eigenvector of $\mathbf{A}$ corresponding to $\mathbf{x}_1'\mathbf{A}\mathbf{x}_1/\mathbf{x}_1'\mathbf{x}_1$, and that $\mathbf{x}_1'\mathbf{A}\mathbf{x}_1/\mathbf{x}_1'\mathbf{x}_1$ is the largest eigenvalue of $\mathbf{A}$. Q.E.D.

As an immediate consequence of Theorem 6.7.1, we have the following corollary.

Corollary 6.7.2. Every symmetric matrix has an eigenvalue.

Does the result of Corollary 6.7.2 extend to $(N \times N)$ nonsymmetric matrices? That is, does an $(N \times N)$ nonsymmetric matrix necessarily have an eigenvalue? The answer to this question depends on whether an eigenvalue is required to be a real number (and an eigenvector a vector of real numbers), as is the case herein, or whether the definition of an eigenvalue (and the definition of an eigenvector) are extended, as is done in many presentations on the subject, so that a complex number can qualify as an eigenvalue (and a vector of complex numbers as an eigenvector).

Consider, for example, the 2×2 matrix $\begin{pmatrix} 0 & 1 \\ -1 & 0 \end{pmatrix}$. The characteristic polynomial of this matrix is

$$p(\lambda) = \begin{vmatrix} -\lambda & 1 \\ -1 & -\lambda \end{vmatrix} = \lambda^2 + 1,$$

which has no real roots but 2 imaginary roots (namely, $\lambda = i$ and $\lambda = -i$). In fact, the fundamental theorem of algebra—the proof of which involves some higher-level mathematics—guarantees that any $N \times N$ matrix (symmetric or not) has a possibly complex root and hence a "possibly complex eigenvalue."

Orthogonality of the eigenvectors of a symmetric matrix. Any two eigenvectors of a symmetric matrix that correspond to different eigenvalues have the basic property described in the following lemma.

Lemma 6.7.3. Suppose that $\mathbf{A}$ is an $N \times N$ symmetric matrix that has an eigenvector $\mathbf{x}_1$ corresponding to an eigenvalue λ_1 and an eigenvector $\mathbf{x}_2$ corresponding to an eigenvalue λ_2. If $\lambda_1 \neq \lambda_2$, then $\mathbf{x}_1$ and $\mathbf{x}_2$ are orthogonal (with respect to the usual inner product).

Proof. By definition, $\mathbf{A}\mathbf{x}_1 = \lambda_1\mathbf{x}_1$ and $\mathbf{A}\mathbf{x}_2 = \lambda_2\mathbf{x}_2$. Further, upon premultiplying both sides of the first equality by $\mathbf{x}_2'$ and both sides of the second by $\mathbf{x}_1'$, we find that

$$\mathbf{x}_2'\mathbf{A}\mathbf{x}_1 = \lambda_1\mathbf{x}_2'\mathbf{x}_1 \quad \text{and} \quad \mathbf{x}_1'\mathbf{A}\mathbf{x}_2 = \lambda_2\mathbf{x}_1'\mathbf{x}_2.$$

And since $\mathbf{A}$ is symmetric, it follows that

$$\lambda_1\mathbf{x}_2'\mathbf{x}_1 = \mathbf{x}_2'\mathbf{A}\mathbf{x}_1 = (\mathbf{x}_1'\mathbf{A}\mathbf{x}_2)' = (\lambda_2\mathbf{x}_1'\mathbf{x}_2)' = \lambda_2\mathbf{x}_2'\mathbf{x}_1,$$

implying that $(\lambda_1 - \lambda_2)\mathbf{x}_2'\mathbf{x}_1 = 0$ and hence if $\lambda_1 \neq \lambda_2$ that $\mathbf{x}_2'\mathbf{x}_1 = 0$. Thus, if $\lambda_1 \neq \lambda_2$, then $\mathbf{x}_1$ and $\mathbf{x}_2$ are orthogonal. Q.E.D.

Diagonalization. An $N \times N$ matrix $\mathbf{A}$ is said to be *diagonalizable* (or *diagonable*) if there exists an $N \times N$ nonsingular matrix $\mathbf{Q}$ such that $\mathbf{Q}^{-1}\mathbf{A}\mathbf{Q} = \mathbf{D}$ for some diagonal matrix $\mathbf{D}$, in which case $\mathbf{Q}$ is said to *diagonalize* $\mathbf{A}$ (or $\mathbf{A}$ is said to be *diagonalized* by $\mathbf{Q}$). Note (in connection with this definition) that

$$\mathbf{Q}^{-1}\mathbf{A}\mathbf{Q} = \mathbf{D} \quad \Leftrightarrow \quad \mathbf{A}\mathbf{Q} = \mathbf{Q}\mathbf{D} \quad \Leftrightarrow \quad \mathbf{A} = \mathbf{Q}\mathbf{D}\mathbf{Q}^{-1}. \tag{7.1}$$

An $N \times N$ matrix $\mathbf{A}$ is said to be *orthogonally diagonalizable* if it can be diagonalized by an orthogonal matrix. Thus, an $N \times N$ matrix $\mathbf{A}$ is orthogonally diagonalizable if there exists an $N \times N$ orthogonal matrix $\mathbf{Q}$ such that $\mathbf{Q}'\mathbf{A}\mathbf{Q} = \mathbf{D}$ for some diagonal matrix $\mathbf{D}$, in which case

$$\mathbf{A} = \mathbf{Q}\mathbf{D}\mathbf{Q}'. \tag{7.2}$$

Since $\mathbf{Q}\mathbf{D}\mathbf{Q}'$ is symmetric, it is clear from equality (7.2) that a necessary condition for an $N \times N$ matrix $\mathbf{A}$ to be orthogonally diagonalizable is that $\mathbf{A}$ be symmetric—certain nonsymmetric matrices are diagonalizable, however they are not orthogonally diagonalizable. This condition is also sufficient, as indicated by the following theorem.

Theorem 6.7.4. Every symmetric matrix is orthogonally diagonalizable.

Proof. The proof is by mathematical induction. Clearly, every 1×1 matrix is orthogonally diagonalizable. Now, suppose that every $(N-1) \times (N-1)$ symmetric matrix is orthogonally diagonalizable (where $N \geq 2$). Then, it suffices to show that every $N \times N$ symmetric matrix is orthogonally diagonalizable.

Let $\mathbf{A}$ represent an $N \times N$ symmetric matrix. And let λ represent an eigenvalue of $\mathbf{A}$ (the existence of which is guaranteed by Corollary 6.7.2), and take $\mathbf{u}$ to be an eigenvector (of $\mathbf{A}$) with (usual) norm 1 that corresponds to λ. Further, take $\mathbf{V}$ to be any $N \times (N-1)$ matrix such that the N vectors consisting of $\mathbf{u}$ and the $N-1$ columns of $\mathbf{V}$ form an orthonormal basis for $\mathbb{R}^N$—the existence of such a matrix follows from Theorem 6.6.6—or, equivalently, such that $(\mathbf{u}, \mathbf{V})$ is an $N \times N$ orthogonal matrix. Then, $\mathbf{Au} = \lambda\mathbf{u}$, $\mathbf{u}'\mathbf{u} = 1$, and $\mathbf{V}'\mathbf{u} = \mathbf{0}$, and, consequently,

$$(\mathbf{u}, \mathbf{V})'\mathbf{A}(\mathbf{u}, \mathbf{V}) = \begin{bmatrix} \mathbf{u}'\mathbf{A}\mathbf{u} & (\mathbf{V}'\mathbf{A}\mathbf{u})' \\ \mathbf{V}'\mathbf{A}\mathbf{u} & \mathbf{V}'\mathbf{A}\mathbf{V} \end{bmatrix} = \begin{pmatrix} \lambda & \mathbf{0}' \\ \mathbf{0} & \mathbf{V}'\mathbf{A}\mathbf{V} \end{pmatrix}.$$

Clearly, $\mathbf{V}'\mathbf{A}\mathbf{V}$ is a symmetric matrix of order $N-1$, so that (by supposition) there exists an $(N-1) \times (N-1)$ orthogonal matrix $\mathbf{R}$ such that $\mathbf{R}'\mathbf{V}'\mathbf{A}\mathbf{V}\mathbf{R} = \mathbf{F}$ for some diagonal matrix $\mathbf{F}$ (of order $N-1$). Define $\mathbf{S} = \mathrm{diag}(1, \mathbf{R})$, and let $\mathbf{P} = (\mathbf{u}, \mathbf{V})\mathbf{S}$. Then,

$$\mathbf{S}'\mathbf{S} = \mathrm{diag}(1, \mathbf{R}'\mathbf{R}) = \mathrm{diag}(1, \mathbf{I}_{N-1}) = \mathbf{I}_N,$$

so that $\mathbf{S}$ is orthogonal and hence (according to Lemma 2.7.1) $\mathbf{P}$ is orthogonal. Further,

$$\mathbf{P}'\mathbf{A}\mathbf{P} = \mathbf{S}'(\mathbf{u}, \mathbf{V})'\mathbf{A}(\mathbf{u}, \mathbf{V})\mathbf{S} = \mathbf{S}'\,\mathrm{diag}(\lambda, \mathbf{V}'\mathbf{A}\mathbf{V})\mathbf{S} = \mathrm{diag}(\lambda, \mathbf{R}'\mathbf{V}'\mathbf{A}\mathbf{V}\mathbf{R}) = \mathrm{diag}(\lambda, \mathbf{F}),$$

so that $\mathbf{P}'\mathbf{A}\mathbf{P}$ is a diagonal matrix. Thus, $\mathbf{A}$ is orthogonally diagonalizable. Q.E.D.

Spectral decomposition: definition and some basic properties. Let $\mathbf{A}$ represent an $N \times N$ symmetric matrix. Further, let $\mathbf{Q}$ represent an $N \times N$ orthogonal matrix and $\mathbf{D}$ an $N \times N$ diagonal matrix such that

$$\mathbf{Q}'\mathbf{A}\mathbf{Q} = \mathbf{D} \tag{7.3}$$

or, equivalently, such that

$$\mathbf{A} = \mathbf{Q}\mathbf{D}\mathbf{Q}' \tag{7.4}$$

—the existence of an orthogonal matrix $\mathbf{Q}$ and a diagonal matrix $\mathbf{D}$ that satisfy condition (7.3) follows from Theorem 6.7.4. And observe that equality (7.4) is also expressible in the form

$$\mathbf{A} = \sum_{i=1}^{N} d_i \mathbf{q}_i \mathbf{q}_i', \tag{7.5}$$

where (for $i = 1, 2, \ldots, N$) d_i represents the ith diagonal element of $\mathbf{D}$ and $\mathbf{q}_i$ the ith column of $\mathbf{Q}$. Expression (7.4) or expression (7.5) is sometimes referred to as the *spectral decomposition* or *spectral representation* of the matrix $\mathbf{A}$.

The characteristic polynomial $p(\cdot)$ (of $\mathbf{A}$) can be reexpressed in terms related to the spectral decomposition (7.4) or (7.5); specifically, it can be reexpressed in terms of the diagonal elements $d_1, d_2, \ldots, d_N$ of the diagonal matrix $\mathbf{D}$. For $\lambda \in \mathbb{R}$,

$$p(\lambda) = |\mathbf{A} - \lambda\mathbf{I}_N| = |\mathbf{Q}'(\mathbf{D} - \lambda\mathbf{I}_N)\mathbf{Q}| = |\mathbf{Q}|^2|\mathbf{D} - \lambda\mathbf{I}_N| = |\mathbf{D} - \lambda\mathbf{I}_N|$$

and hence

$$p(\lambda) = (-1)^N \textstyle\prod_{i=1}^{N} (\lambda - d_i) \tag{7.6}$$

$$= (-1)^N \textstyle\prod_{j=1}^{K} (\lambda - \lambda_j)^{N_j}, \tag{7.7}$$

where $\{\lambda_1, \lambda_2, \ldots, \lambda_K\}$ is a set whose elements consist of the distinct values represented among the N scalars $d_1, d_2, \ldots, d_N$ and where (for $j = 1, 2, \ldots, K$) N_j represents the number of values of the integer i (between 1 and N, inclusive) for which $d_i = \lambda_j$.

In light of expression (7.6) or (7.7), it is clear that a scalar λ is an eigenvalue of $\mathbf{A}$ if and only if $\lambda = d_i$ for some integer i (between 1 and N, inclusive) or, equivalently, if and only if λ is contained in the set $\{\lambda_1, \lambda_2, \ldots, \lambda_K\}$. Accordingly, $\lambda_1, \lambda_2, \ldots, \lambda_K$ may be referred to as the *distinct eigenvalues* of $\mathbf{A}$. And, collectively, $d_1, d_2, \ldots, d_N$ may be referred to as the *not-necessarily-distinct eigenvalues* of $\mathbf{A}$. Further, the set $\{\lambda_1, \lambda_2, \ldots, \lambda_K\}$ is sometimes referred to as the *spectrum* of $\mathbf{A}$, and (for $j = 1, 2, \ldots, K$) N_j is sometimes referred to as the *multiplicity* of λ_j.

Clearly,

$$\mathbf{AQ} = \mathbf{QD}, \tag{7.8}$$

or, equivalently,

$$\mathbf{A}\mathbf{q}_i = d_i \mathbf{q}_i \quad (i = 1, 2, \ldots, N). \tag{7.9}$$

Thus, the N orthonormal vectors $\mathbf{q}_1, \mathbf{q}_2, \ldots, \mathbf{q}_N$ are eigenvectors of $\mathbf{A}$, the ith of which corresponds to the eigenvalue d_i. Note that result (7.8) or (7.9) is also expressible in the form

$$\mathbf{A}\mathbf{Q}_j = \lambda_j \mathbf{Q}_j \quad (j = 1, 2, \ldots, K), \tag{7.10}$$

where $\mathbf{Q}_j$ is the $N \times N_j$ matrix whose columns consist of those of the eigenvectors $\mathbf{q}_1, \mathbf{q}_2, \ldots, \mathbf{q}_N$ for which the corresponding eigenvalue equals λ_j. Note also that the K equalities in the collection (7.10) are reexpressible as

$$(\mathbf{A} - \lambda_j \mathbf{I}_N)\mathbf{Q}_j = \mathbf{0} \quad (j = 1, 2, \ldots, K). \tag{7.11}$$

It is clear from result (7.11) that the columns of $\mathbf{Q}_j$ are members of $\mathcal{N}(\mathbf{A} - \lambda_j \mathbf{I}_N)$. In fact, they form a basis (an orthonormal basis) for $\mathcal{N}(\mathbf{A} - \lambda_j \mathbf{I}_N)$, as is evident upon observing (in light of Lemma 2.11.5) that

$$\begin{aligned} \dim[\mathcal{N}(\mathbf{A}-\lambda_j \mathbf{I}_N)] &= N - \text{rank}(\mathbf{A}-\lambda_j \mathbf{I}_N) \\ &= N - \text{rank}[\mathbf{Q}'(\mathbf{A}-\lambda_j \mathbf{I}_N)\mathbf{Q}] \\ &= N - \text{rank}(\mathbf{D}-\lambda_j \mathbf{I}_N) = N - (N - N_j) = N_j. \end{aligned} \tag{7.12}$$

In general, a distinction needs to be made between the algebraic multiplicity and the geometric multiplicity of an eigenvalue λ_j; algebraic multiplicity refers to the multiplicity of $\lambda - \lambda_j$ as a factor of the characteristic polynomial $p(\lambda)$, whereas geometric multiplicity refers to the dimension of the linear space $\mathcal{N}(\mathbf{A}-\lambda_j \mathbf{I}_N)$. However, in the present context (where the eigenvalue is that of a symmetic matrix $\mathbf{A}$), the algebraic and geometric multiplicities are equal, so that no distinction is necessary.

To what extent is the spectral decomposition of $\mathbf{A}$ unique? The distinct eigenvalues $\lambda_1, \lambda_2, \ldots, \lambda_K$ are (aside from order) unique and their multiplicities $N_1, N_2, \ldots, N_K$ are unique, as is evident from result (7.7). Moreover, for j (an integer between 1 and K, inclusive) such that $N_j = 1$, $\mathbf{Q}_j$ is unique, as is evident from result (7.11) upon observing [in light of result (7.12)] that (if $N_j = 1$) $\dim[\mathcal{N}(\mathbf{A}-\lambda_j \mathbf{I}_N)] = 1$. For j such that $N_j > 1$, $\mathbf{Q}_j$ is not uniquely determined; $\mathbf{Q}_j$ can be taken to be any $N \times N_j$ matrix whose columns are orthonormal eigenvectors (of $\mathbf{A}$) corresponding to the eigenvalue λ_j or, equivalently, whose columns form an orthonormal basis for $\mathcal{N}(\mathbf{A}-\lambda_j \mathbf{I}_N)$—refer to Lemma 6.7.3. However, even for j such that $N_j > 1$, $\mathbf{Q}_j \mathbf{Q}_j'$ is uniquely determined—refer, e.g., to Harville (1997, sec. 21.5) for a proof. Accordingly, a decomposition of $\mathbf{A}$ that is unique (aside from the order of the terms) is obtained upon reexpressing decomposition (7.5) in the form

$$\mathbf{A} = \textstyle\sum_{j=1}^{K} \lambda_j \mathbf{E}_j, \tag{7.13}$$

where $\mathbf{E}_j = \mathbf{Q}_j \mathbf{Q}_j'$. Decomposition (7.13), like decompositions (7.4) and 7.5), is sometimes referred to as the spectral decomposition.

***Rank, trace, and determinant of a symmetric matrix*.** The following theorem provides expressions for the rank, trace, and determinant of a symmetric matrix (in terms of its eigenvalues).

Theorem 6.7.5. Let $\mathbf{A}$ represent an $N \times N$ symmetric matrix with not-necessarily-distinct eigenvalues $d_1, d_2, \ldots, d_N$ and with distinct eigenvalues $\lambda_1, \lambda_2, \ldots, \lambda_K$ of multiplicities $N_1, N_2, \ldots, N_K$, respectively. Then,

(1) $\text{rank}\,\mathbf{A} = N - N_0$, where $N_0 = N_j$ if $\lambda_j = 0$ $(1 \le j \le K)$ and where $N_0 = 0$ if $0 \notin \{\lambda_1, \lambda_2, \ldots, \lambda_K\}$ (i.e., where N_0 equals the multiplicity of the eigenvalue 0 if 0 is an eigenvalue of $\mathbf{A}$ and equals 0 otherwise);

(2) $\text{tr}(\mathbf{A}) = \sum_{i=1}^{N} d_i = \sum_{j=1}^{K} N_j\lambda_j$; and

(3) $\det(\mathbf{A}) = \prod_{i=1}^{N} d_i = \prod_{j=1}^{K} \lambda_j^{N_j}$.

Proof. Let $\mathbf{Q}$ represent an $N \times N$ orthogonal matrix such that $\mathbf{A} = \mathbf{QDQ}'$, where $\mathbf{D} = \text{diag}(d_1, d_2, \ldots, d_N)$—the existence of such a matrix follows from the results of the preceding part of the present subsection (i.e., the part pertaining to the spectral decomposition).

(1) Clearly, rank $\mathbf{A}$ equals rank $\mathbf{D}$, and rank $\mathbf{D}$ equals the number of diagonal elements of $\mathbf{D}$ that are nonzero. Thus, rank $\mathbf{A} = N - N_0$.

(2) Making use of Lemma 2.3.1, we find that

$$\text{tr}(\mathbf{A}) = \text{tr}(\mathbf{QDQ}') = \text{tr}(\mathbf{DQ}'\mathbf{Q}) = \text{tr}(\mathbf{DI}) = \text{tr}(\mathbf{D}) = \sum_{i=1}^{N} d_i = \sum_{j=1}^{K} N_j\lambda_j.$$

(3) Making use of result (2.14.25), Lemma 2.14.3, and Corollary 2.14.19, we find that

$$|\mathbf{A}| = |\mathbf{QDQ}'| = |\mathbf{Q}||\mathbf{D}||\mathbf{Q}'| = |\mathbf{Q}|^2|\mathbf{D}| = |\mathbf{D}| = \prod_{i=1}^{N} d_i = \prod_{j=1}^{K} \lambda_j^{N_j}.$$

Q.E.D.

When is a symmetric matrix nonnegative definite, positive definite, or positive semidefinite? Let $\mathbf{A}$ represent an $N \times N$ symmetric matrix. And take $\mathbf{Q}$ to be an $N \times N$ orthogonal matrix and $\mathbf{D}$ an $N \times N$ diagonal matrix such that $\mathbf{A} = \mathbf{QDQ}'$—the existence of such an orthogonal matrix and such a diagonal matrix follows from Theorem 6.7.4. Then, upon recalling (from the discussion of the spectral decomposition) that the diagonal elements of $\mathbf{D}$ constitute the (not-necessarily-distinct) eigenvalues of $\mathbf{A}$ and upon applying Corollary 2.13.16, we arrive at the following result.

Theorem 6.7.6. Let $\mathbf{A}$ represent an $N \times N$ symmetric matrix with not-necessarily-distinct eigenvalues $d_1, d_2, \ldots, d_N$. Then, (1) $\mathbf{A}$ is nonnegative definite if and only if $d_1, d_2, \ldots, d_N$ are nonnegative; (2) $\mathbf{A}$ is positive definite if and only if $d_1, d_2, \ldots, d_N$ are (strictly) positive; and (3) $\mathbf{A}$ is positive semidefinite if and only if $d_i \geq 0$ for $i = 1, 2, \ldots, N$ with equality holding for one or more values of i.

When is a symmetric matrix idempotent? The following theorem characterizes the idempotency of a symmetric matrix in terms of its eigenvalues.

Theorem 6.7.7. An $N \times N$ symmetric matrix is idempotent if (and only if) it has no eigenvalues other than 0 or 1.

Proof. Let $\mathbf{A}$ represent an $N \times N$ symmetric matrix, and denote by $d_1, d_2, \ldots, d_N$ its not-necessarily-distinct eigenvalues. And observe (in light of the discussion of the spectral decomposition) that there exists an $N \times N$ orthogonal matrix $\mathbf{Q}$ such that $\mathbf{A} = \mathbf{QDQ}'$, where $\mathbf{D} = \text{diag}(d_1, d_2, \ldots, d_N)$. Observe also that

$$\mathbf{A}^2 = \mathbf{QDQ}'\mathbf{QDQ}' = \mathbf{QD}^2\mathbf{Q}'.$$

Thus,

$$\mathbf{A}^2 = \mathbf{A} \quad \Leftrightarrow \quad \mathbf{D}^2 = \mathbf{D} \quad \Leftrightarrow \quad d_i^2 = d_i \ (i = 1, 2, \ldots, N).$$

Moreover, $d_i^2 = d_i$ if and only if either $d_i = 0$ or $d_i = 1$. It is now clear that $\mathbf{A}$ is idempotent if (and only if) it has no eigenvalues other than 0 or 1. Q.E.D.

b. Reexpression of a quadratic form (in a normally distributed random vector) as a linear combination of independently distributed random variables

Let $\mathbf{z}$ represent an N-dimensional random column vector that has an $N(\mathbf{0}, \mathbf{I}_N)$ distribution. And consider the distribution of a second-degree polynomial (in $\mathbf{z}$), say the second-degree polynomial $q = c + \mathbf{b}'\mathbf{z} + \mathbf{z}'\mathbf{Az}$, where c is a constant, $\mathbf{b}$ an N-dimensional column vector of constants, and $\mathbf{A}$ an $N \times N$ symmetric matrix of constants. Further, take $\mathbf{O}$ to be an $N \times N$ orthogonal matrix and $\mathbf{D}$ an $N \times N$ diagonal matrix such that

$$\mathbf{A} = \mathbf{ODO}' \tag{7.14}$$

—the existence of such an orthogonal matrix and such a diagonal matrix follows from Theorem 6.7.4. As previously indicated (in Subsection a), the representation (7.14) is sometimes referred to as the spectral decomposition or representation (of the matrix **A**).

The second-degree polynomial q can be reexpressed in terms related to the representation (7.14). Let $\mathbf{r} = \mathbf{O}'\mathbf{b}$ and $\mathbf{u} = \mathbf{O}'\mathbf{z}$. Then, clearly,

$$q = c + \mathbf{r}'\mathbf{u} + \mathbf{u}'\mathbf{D}\mathbf{u} = \sum_{i=1}^{N} (c_i + r_i u_i + d_i u_i^2), \tag{7.15}$$

where $c_1, c_2, \ldots, c_N$ represent any constants such that $\sum_{i=1}^{N} c_i = c$ and where (for $i = 1, 2, \ldots, N$) r_i represents the ith element of $\mathbf{r}$, u_i the ith element of $\mathbf{u}$, and d_i the ith diagonal element of $\mathbf{D}$. Moreover, $\mathbf{u} \sim N(\mathbf{0}, \mathbf{I}_N)$ or, equivalently, $u_1, u_2, \ldots, u_N$ are distributed independently and identically as $N(0, 1)$.

Now, consider the distribution of q in the special case where $c = 0$ and $\mathbf{b} = \mathbf{0}$, that is, the special case where q is the quadratic form $\mathbf{z}'\mathbf{A}\mathbf{z}$. Based on result (7.15), we have that

$$\mathbf{z}'\mathbf{A}\mathbf{z} = \mathbf{u}'\mathbf{D}\mathbf{u} = \sum_{i=1}^{N} d_i u_i^2. \tag{7.16}$$

And it follows that

$$\mathbf{z}'\mathbf{A}\mathbf{z} \sim \sum_{i=1}^{N} d_i w_i, \tag{7.17}$$

where $w_1, w_2, \ldots, w_N$ are random variables that are distributed independently and identically as $\chi^2(1)$.

According to result (7.17), $\mathbf{z}'\mathbf{A}\mathbf{z}$ is distributed as a linear combination of the random variables $w_1, w_2, \ldots, w_N$. Upon observing that the coefficients in this linear combination are $d_1, d_2, \ldots, d_N$ and upon recalling (from Subsection a) that $d_1, d_2, \ldots, d_N$ are the not-necessarily-distinct eigenvalues of the matrix **A**, we obtain the further result that

$$\mathbf{z}'\mathbf{A}\mathbf{z} \sim \sum_{j=1}^{K} \lambda_j v_j, \tag{7.18}$$

where $\lambda_1, \lambda_2, \ldots, \lambda_K$ are the distinct eigenvalues of **A** with multiplicities $N_1, N_2, \ldots, N_K$, respectively, and where $v_1, v_2, \ldots, v_K$ are random variables that are distributed independently as $\chi^2(N_1), \chi^2(N_2), \ldots, \chi^2(N_K)$, respectively.

The moment generating function of the second-degree polynomial q can be expressed in terms of c, the elements of $\mathbf{r}$, and the not-necessarily-distinct eigenvalues $d_1, d_2, \ldots, d_N$ of the matrix **A**. Let $m(\cdot)$ represent the moment generating function of q, define $t_0 = 2\min_i d_i$ and $t_1 = 2\max_i d_i$, and take

$$S = \{t \in \mathcal{R} : \mathbf{I}_N - 2t\mathbf{D} \text{ is positive definite}\}$$

or, equivalently,

$$S = \begin{cases} (1/t_0,\ 1/t_1), & \text{if } t_0 < 0 \text{ and } t_1 > 0, \\ (1/t_0,\ \infty), & \text{if } t_0 < 0 \text{ and } t_1 \le 0, \\ (-\infty,\ 1/t_1), & \text{if } t_0 \ge 0 \text{ and } t_1 > 0, \\ (-\infty,\ \infty), & \text{if } t_0 = t_1 = 0. \end{cases}$$

Then, upon regarding q as a second-degree polynomial in the random vector $\mathbf{u}$ (rather than in the random vector $\mathbf{z}$) and applying formula (5.9), we find that, for $t \in S$,

$$m(t) = |\mathbf{I} - 2t\mathbf{D}|^{-1/2} \exp[tc + (1/2)t^2\mathbf{r}'(\mathbf{I} - 2t\mathbf{D})^{-1}\mathbf{r}] \tag{7.19}$$

$$= \textstyle\prod_{i=1}^{N} (1 - 2td_i)^{-1/2} \exp[tc + (1/2)t^2 \sum_{i=1}^{N} (1 - 2td_i)^{-1} r_i^2]. \tag{7.20}$$

In the special case where $c = 0$ and $\mathbf{b} = \mathbf{0}$, that is, the special case where $q = \mathbf{z}'\mathbf{A}\mathbf{z}$, we have that, for $t \in S$,

$$m(t) = |\mathbf{I} - 2t\mathbf{D}|^{-1/2} = \prod_{i=1}^{N} (1 - 2td_i)^{-1/2}. \tag{7.21}$$

An extension. The results on the distribution of the second-degree polynomial in an N-dimensional random column vector that has an $N(\mathbf{0}, \mathbf{I}_N)$ distribution can be readily extended to the distribution of a second-degree polynomial in an N-dimensional random column vector that has an N-variate normal distribution with an arbitrary mean vector and arbitrary variance-covariance matrix. Let $\mathbf{x}$ represent an N-dimensional random column vector that has an $N(\boldsymbol{\mu}, \boldsymbol{\Sigma})$ distribution. And consider the distribution of the second-degree polynomial $q = c + \mathbf{b}'\mathbf{x} + \mathbf{x}'\mathbf{A}\mathbf{x}$ (where, as before, c is a constant, $\mathbf{b}$ an N-dimensional column vector of constants, and $\mathbf{A}$ a symmetric matrix of constants).

Take $\boldsymbol{\Gamma}$ to be any matrix (with N columns) such that $\boldsymbol{\Sigma} = \boldsymbol{\Gamma}'\boldsymbol{\Gamma}$, and denote by P the number of rows in $\boldsymbol{\Gamma}$. Further, take $\mathbf{z}$ to be a P-dimensional random column vector that has an $N(\mathbf{0}, \mathbf{I}_P)$ distribution. Then, $\mathbf{x} \sim \boldsymbol{\mu} + \boldsymbol{\Gamma}'\mathbf{z}$, and as observed earlier [in result (5.4)],

$$q \sim c + \mathbf{b}'\boldsymbol{\mu} + \boldsymbol{\mu}'\mathbf{A}\boldsymbol{\mu} + [\boldsymbol{\Gamma}(\mathbf{b}+2\mathbf{A}\boldsymbol{\mu})]'\mathbf{z} + \mathbf{z}'\boldsymbol{\Gamma}\mathbf{A}\boldsymbol{\Gamma}'\mathbf{z}. \tag{7.22}$$

Thus, upon applying result (7.15) with $c + \mathbf{b}'\boldsymbol{\mu} + \boldsymbol{\mu}'\mathbf{A}\boldsymbol{\mu}$, $\boldsymbol{\Gamma}(\mathbf{b}+2\mathbf{A}\boldsymbol{\mu})$, and $\boldsymbol{\Gamma}\mathbf{A}\boldsymbol{\Gamma}'$ in place of c, $\mathbf{b}$, and $\mathbf{A}$, respectively, and taking $\mathbf{O}$ to be a $P \times P$ orthogonal matrix and $\mathbf{D} = \{d_i\}$ a $P \times P$ diagonal matrix such that $\boldsymbol{\Gamma}\mathbf{A}\boldsymbol{\Gamma}' = \mathbf{O}\mathbf{D}\mathbf{O}'$, we find that

$$q \sim k + \mathbf{r}'\mathbf{u} + \mathbf{u}'\mathbf{D}\mathbf{u} = \sum_{i=1}^{N} (k_i + r_i u_i + d_i u_i^2), \tag{7.23}$$

where $k = c + \mathbf{b}'\boldsymbol{\mu} + \boldsymbol{\mu}'\mathbf{A}\boldsymbol{\mu}$ and $k_1, k_2, \ldots, k_P$ are any constants such that $\sum_{i=1}^{N} k_i = k$, where $\mathbf{r} = \{r_i\} = \mathbf{O}'\boldsymbol{\Gamma}(\mathbf{b}+2\mathbf{A}\boldsymbol{\mu})$, and where $\mathbf{u} = \{u_i\}$ is a P-dimensional column vector that has an $N(\mathbf{0}, \mathbf{I}_P)$ distribution [or, equivalently, whose elements $u_1, u_2, \ldots, u_P$ are distributed independently and identically as $N(0, 1)$].

Note that the same approach that led to result (7.23) could be used to obtain a formula for the moment generating function of q that could be regarded as an extension of formula (7.19) or (7.20). It is simply a matter of applying formula (7.19) or (7.20) to the second-degree polynomial in $\mathbf{z}$ that has the same distribution as q—refer to result (7.22). That is, it is simply a matter of applying formula (7.19) or (7.20) with $c + \mathbf{b}'\boldsymbol{\mu} + \boldsymbol{\mu}'\mathbf{A}\boldsymbol{\mu}$, $\boldsymbol{\Gamma}(\mathbf{b}+2\mathbf{A}\boldsymbol{\mu})$, and $\boldsymbol{\Gamma}\mathbf{A}\boldsymbol{\Gamma}'$ in place of c, $\mathbf{b}$, and $\mathbf{A}$, respectively.

A variation. Let $\mathbf{z}$ represent an N-dimensional random column vector that has an $N(\mathbf{0}, \mathbf{I}_N)$ distribution. And consider the distribution of the ratio $\mathbf{z}'\mathbf{A}\mathbf{z}/\mathbf{z}'\mathbf{z}$, where $\mathbf{A}$ is an $N \times N$ symmetric matrix of constants. Note that $\mathbf{z}'\mathbf{A}\mathbf{z}/\mathbf{z}'\mathbf{z}$ is reexpressible as a quadratic form $[(\mathbf{z}'\mathbf{z})^{-1/2}\mathbf{z}]'\mathbf{A}[(\mathbf{z}'\mathbf{z})^{-1/2}\mathbf{z}]$ in the normalized vector $(\mathbf{z}'\mathbf{z})^{-1/2}\mathbf{z}$, the elements of which are distributed uniformly on the surface of an N-dimensional unit ball.

Take $\mathbf{O}$ to be an $N \times N$ orthogonal matrix and $\mathbf{D}$ an $N \times N$ diagonal matrix such that $\mathbf{A} = \mathbf{O}\mathbf{D}\mathbf{O}'$ (i.e., such that $\mathbf{A} = \mathbf{O}\mathbf{D}\mathbf{O}'$ is the spectral decomposition of $\mathbf{A}$). And define $\mathbf{u} = \mathbf{O}'\mathbf{z}$, and (for $i = 1, 2, \ldots, N$) denote by u_i the ith element of $\mathbf{u}$ and by d_i the ith diagonal element of $\mathbf{D}$ (so that d_i is the ith of the not-necessarily-distinct eigenvalues of $\mathbf{A}$). Then, $\mathbf{u} \sim N(\mathbf{0}, \mathbf{I}_N)$, and as was established earlier—refer to result (7.16)—

$$\mathbf{z}'\mathbf{A}\mathbf{z} = \sum_{i=1}^{N} d_i u_i^2.$$

Further,

$$\mathbf{z}'\mathbf{z} = \mathbf{z}'\mathbf{I}\mathbf{z} = \mathbf{z}'\mathbf{O}\mathbf{O}'\mathbf{z} = \mathbf{u}'\mathbf{u} = \sum_{i=1}^{N} u_i^2.$$

Thus,

$$\frac{\mathbf{z}'\mathbf{A}\mathbf{z}}{\mathbf{z}'\mathbf{z}} = \frac{\sum_{i=1}^{N} d_i u_i^2}{\sum_{i=1}^{N} u_i^2} \sim \frac{\sum_{j=1}^{K} \lambda_j v_j}{\sum_{j=1}^{K} v_j},$$

where $\lambda_1, \lambda_2, \ldots, \lambda_K$ are the distinct eigenvalues of $\mathbf{A}$ with multiplicities $N_1, N_2, \ldots, N_K$, respectively, and $v_1, v_2, \ldots, v_K$ are random variables that are distributed independently as $\chi^2(N_1), \chi^2(N_2), \ldots, \chi^2(N_K)$, respectively. Upon recalling the definition of the Dirichlet distribution (and the relationship between the chi-square distribution and the gamma distribution), we conclude that

$$\frac{\mathbf{z}'\mathbf{A}\mathbf{z}}{\mathbf{z}'\mathbf{z}} \sim \sum_{j=1}^{K} \lambda_j w_j, \tag{7.24}$$

where $w_1, w_2, \ldots, w_{K-1}$ are random variables that are jointly distributed as $Di\left(\frac{N_1}{2}, \frac{N_2}{2}, \ldots, \frac{N_{K-1}}{2}, \frac{N_K}{2}; K-1\right)$ and where $w_K = 1 - \sum_{j=1}^{K-1} w_j$—in the "degenerate" case where $K = 1$, $w_1 = 1$ and $\mathbf{z}'\mathbf{A}\mathbf{z}/\mathbf{z}'\mathbf{z} = \lambda_1$.

Result (7.24) can be generalized. Suppose that $\mathbf{x}$ is an N-dimensional random column vector that is distributed as $N(\mathbf{0}, \boldsymbol{\Sigma})$ (where $\boldsymbol{\Sigma} \neq \mathbf{0}$), let $P = \operatorname{rank} \boldsymbol{\Sigma}$, and take $\mathbf{A}$ to be an $N \times N$ symmetric matrix of constants. Further, take $\boldsymbol{\Gamma}$ to be a matrix of dimensions $P \times N$ such that $\boldsymbol{\Sigma} = \boldsymbol{\Gamma}'\boldsymbol{\Gamma}$ (the existence of which follows from Corollary 2.13.23), and take $\mathbf{z}$ to be a P-dimensional random column vector that has an $N(\mathbf{0}, \mathbf{I}_P)$ distribution. Then, as in the case of result (6.17), we have that

$$\frac{\mathbf{x}'\mathbf{A}\mathbf{x}}{\mathbf{x}'\boldsymbol{\Sigma}^{-}\mathbf{x}} \sim \frac{\mathbf{z}'\boldsymbol{\Gamma}\mathbf{A}\boldsymbol{\Gamma}'\mathbf{z}}{\mathbf{z}'\mathbf{z}}.$$

And upon applying result (7.24) (with the $P \times P$ matrix $\boldsymbol{\Gamma}\mathbf{A}\boldsymbol{\Gamma}'$ in place of the $N \times N$ matrix $\mathbf{A}$), we find that

$$\frac{\mathbf{x}'\mathbf{A}\mathbf{x}}{\mathbf{x}'\boldsymbol{\Sigma}^{-}\mathbf{x}} \sim \sum_{j=1}^{K} \lambda_j w_j, \tag{7.25}$$

where $\lambda_1, \lambda_2, \ldots, \lambda_K$ are the distinct eigenvalues of $\boldsymbol{\Gamma}\mathbf{A}\boldsymbol{\Gamma}'$ with multiplicities $N_1, N_2, \ldots, N_K$, respectively, where $w_1, w_2, \ldots, w_{K-1}$ are random variables that are jointly distributed as $Di\left(\frac{N_1}{2}, \frac{N_2}{2}, \ldots, \frac{N_{K-1}}{2}, \frac{N_K}{2}; K-1\right)$, and where $w_K = 1 - \sum_{j=1}^{K-1} w_j$—in the "degenerate" case where $K = 1$, $w_1 = 1$ and $\mathbf{x}'\mathbf{A}\mathbf{x}/\mathbf{x}'\boldsymbol{\Sigma}^{-}\mathbf{x} = \lambda_1$ (with probability 1).

Applicability to spherically or elliptically distributed random vectors. Result (7.24) on the distribution of the ratio $\mathbf{z}'\mathbf{A}\mathbf{z}/\mathbf{z}'\mathbf{z}$ (where $\mathbf{z}$ is an N-dimensional random column vector and $\mathbf{A}$ an $N \times N$ symmetric matrix of constants) was derived under the assumption that $\mathbf{z} \sim N(\mathbf{0}, \mathbf{I}_N)$. However, as discussed earlier (in Section 6.6e), $\mathbf{z}'\mathbf{A}\mathbf{z}/\mathbf{z}'\mathbf{z}$ has the same distribution when the distribution of $\mathbf{z}$ is taken to be an arbitrary absolutely continuous spherical distribution as it does when $\mathbf{z} \sim N(\mathbf{0}, \mathbf{I}_N)$. Thus, result (7.24) is applicable even if the distribution of $\mathbf{z}$ is an absolutely continuous distribution other than the $N(\mathbf{0}, \mathbf{I}_N)$ distribution.

Now, consider result (7.25) on the distribution of the ratio $\mathbf{x}'\mathbf{A}\mathbf{x}/\mathbf{x}'\boldsymbol{\Sigma}^{-}\mathbf{x}$, where $\mathbf{x}$ is an N-dimensional random column vector that has an $N(\mathbf{0}, \boldsymbol{\Sigma})$ distribution (and where $\boldsymbol{\Sigma} \neq \mathbf{0}$ and where $\mathbf{A}$ is an $N \times N$ symmetric matrix of constants). Let $P = \operatorname{rank} \boldsymbol{\Sigma}$, take $\boldsymbol{\Gamma}$ to be a $P \times N$ matrix such that $\boldsymbol{\Sigma} = \boldsymbol{\Gamma}'\boldsymbol{\Gamma}$, and take $\mathbf{z}$ to be a P-dimensional random column vector that has an absolutely continuous spherical distribution. Then, $\boldsymbol{\Gamma}'\mathbf{z}$ is distributed elliptically about $\mathbf{0}$. Moreover,

$$\frac{(\boldsymbol{\Gamma}'\mathbf{z})'\mathbf{A}\boldsymbol{\Gamma}'\mathbf{z}}{(\boldsymbol{\Gamma}'\mathbf{z})'\boldsymbol{\Sigma}^{-}\boldsymbol{\Gamma}'\mathbf{z}} \sim \frac{\mathbf{x}'\mathbf{A}\mathbf{x}}{\mathbf{x}'\boldsymbol{\Sigma}^{-}\mathbf{x}}.$$

—refer to Section 6.6e. Thus, result (7.25) is applicable when $\mathbf{x}$ is taken to be an N-dimensional random column vector whose distribution is that of the elliptically distributed random vector $\boldsymbol{\Gamma}'\mathbf{z}$ as well as when $\mathbf{x} \sim N(\mathbf{0}, \boldsymbol{\Sigma})$—note that $\boldsymbol{\Gamma}'\mathbf{z} \sim N(\mathbf{0}, \boldsymbol{\Sigma})$ in the special case where $\mathbf{z} \sim N(\mathbf{0}, \mathbf{I}_P)$.

c. Some properties of polynomials (in a single variable)

Let us consider some properties of polynomials (in a single variable). The immediate objective in doing so is to set the stage for proving (in Subsection d) the "necessity part" of Theorem 6.6.1.

Let x represent a real variable. Then, a function of x, say $p(x)$, that is expressible in the form

$$p(x) = a_0 + a_1 x + a_2 x^2 + \cdots + a_N x^N,$$

where N is a nonnegative integer and where the coefficients $a_0, a_1, a_2, \ldots, a_N$ are real numbers, is referred to as a *polynomial* (in x). The polynomial $p(x)$ is said to be *nonzero* if one or more of the coefficients $a_0, a_1, a_2, \ldots, a_N$ are nonzero, in which case the largest nonnegative integer k such that $a_k \neq 0$ is referred to as the *degree* of $p(x)$ and is denoted by the symbol $\deg[p(x)]$. When it causes no confusion, $p(x)$ may be abbreviated to p {and $\deg[p(x)]$ to $\deg(p)$}.

A polynomial q is said to be a *factor* of a polynomial p if there exists a polynomial r such that $p \equiv qr$. And a real number c is said to be a *root* (or a *zero*) of a polynomial p if $p(c) = 0$.

A basic property of polynomials is as follows.

Theorem 6.7.8. Let $p(x)$ and $q(x)$ represent polynomials (in a variable x). And suppose that $p(x) = q(x)$ for all x in some nondegenerate interval. Or, more generally, taking N to be a nonnegative integer such that $N > \deg(p)$ (if p is nonzero) and $N > \deg(q)$ (if q is nonzero), suppose that $p(x) = q(x)$ for N distinct values of x. Then, $p(x) = q(x)$ for all x.

Proof. Suppose that $N > 0$, and take $x_1, x_2, \ldots, x_N$ to be N distinct values of x such that $p(x_i) = q(x_i)$ for $i = 1, 2, \ldots, N$—if $N = 0$, then neither p nor q is nonzero, in which case $p(x) = 0 = q(x)$ for all x. And observe that there exist real numbers $a_0, a_1, a_2, \ldots, a_{N-1}$ and $b_0, b_1, b_2, \ldots, b_{N-1}$ such that

$$p(x) = a_0 + a_1 x + a_2 x^2 + \cdots + a_{N-1} x^{N-1}$$

and

$$q(x) = b_0 + b_1 x + b_2 x^2 + \cdots + b_{N-1} x^{N-1}.$$

Further, let $\mathbf{p} = [p(x_1), p(x_2), \ldots, p(x_N)]'$ and $\mathbf{q} = [q(x_1), q(x_2), \ldots, q(x_N)]'$, define $\mathbf{a} = (a_0, a_1, a_2, \ldots, a_{N-1})'$ and $\mathbf{b} = (b_0, b_1, b_2, \ldots, b_{N-1})'$, and take $\mathbf{H}$ to be the $N \times N$ matrix with ijth element x_i^{j-1}—by convention, $0^0 = 1$. Then, clearly,

$$\mathbf{H}(\mathbf{a} - \mathbf{b}) = \mathbf{Ha} - \mathbf{Hb} = \mathbf{p} - \mathbf{q} = \mathbf{0}.$$

Moreover, the matrix $\mathbf{H}$ is a Vandermonde matrix, and upon applying result (5.3.13), it follows that $\mathbf{H}$ is nonsingular. Thus, $\mathbf{a} - \mathbf{b} = \mathbf{H}^{-1}\mathbf{0} = \mathbf{0}$, implying that $\mathbf{a} = \mathbf{b}$ and hence that $p(\mathbf{x}) = q(\mathbf{x})$ for all $\mathbf{x}$. Q.E.D.

Some additional basic properties of polynomials are described in the following four theorems.

Theorem 6.7.9 (the division algorithm)**.** Let p and q represent polynomials. Suppose that q is nonzero. Then, there exist unique polynomials b and r such that

$$p \equiv bq + r,$$

where either $r \equiv 0$ or else $\deg(r) < \deg(q)$.

Theorem 6.7.10. Let $p(x)$ represent a nonzero polynomial (in x) of degree N. Then, for any real number c, $p(x)$ has a unique representation of the form

$$p(x) = b_0 + b_1(x-c) + b_2(x-c)^2 + \cdots + b_N(x-c)^N,$$

where $b_0, b_1, b_2, \ldots, b_N$ are real numbers.

Theorem 6.7.11 (the factor theorem)**.** A real number c is a root of a polynomial $p(x)$ (in x) if and only if the polynomial $x - c$ is a factor of $p(x)$.

Theorem 6.7.12. Let $p(x)$, $q(x)$, and $r(x)$ represent polynomials (in x). And suppose that

$$p(x)q(x) = (x-c)^M r(x),$$

where M is a positive integer and where c is a real number that is not a root of $p(x)$. Then, $(x-c)^M$ is a factor of $q(x)$; that is, there exists a polynomial $s(x)$ such that

$$q(x) = (x-c)^M s(x).$$

Refer, for example, to Beaumont and Pierce (1963) for proofs of Theorems 6.7.9, 6.7.10, and 6.7.11, which are equivalent to their theorems 9-3.3, 9-3.5, and 9-7.5, and to Harville (1997, appendix to chap. 21) for a proof of Theorem 6.7.12.

The various basic properties of polynomials can be used to establish the following result.

Theorem 6.7.13. Let $r_1(x), s_1(x)$ and $s_2(x)$ represent polynomials in a real variable x. And take $r_2(x)$ to be a function of x defined as follows:

$$r_2(x) = \gamma(x-\lambda_1)^{M_1}(x-\lambda_2)^{M_2}\cdots(x-\lambda_K)^{M_K},$$

where K is a nonnegative integer, where $M_1, M_2, \ldots, M_K$ are (strictly) positive integers, where γ is a nonzero real number, and where $\lambda_1, \lambda_2, \ldots, \lambda_K$ are real numbers—when $K = 0$, $r_2(x) = \gamma$. Suppose that

$$\log\left[\frac{s_1(x)}{s_2(x)}\right] = \frac{r_1(x)}{r_2(x)}$$

for all x in some nondegenerate interval [that does not include $\lambda_1, \lambda_2, \ldots, \lambda_K$ or any roots of $s_2(x)$]. Then, there exists a real number α such that $r_1(x) = \alpha r_2(x)$ for all x. Further, $s_1(x) = e^{\alpha}s_2(x)$ for all x.

For a proof of Theorem 6.7.13, refer to Harville (1997, sec. 21.16) or to Harville and Kempthorne (1997, sec. 3).

d. Proof of the "necessity part" of Theorem 6.6.1

Let $\mathbf{z}$ represent an M-dimensional random column vector that has an $N(\mathbf{0}, \mathbf{I}_M)$ distribution, and take $q = c + \mathbf{b}'\mathbf{z} + \mathbf{z}'\mathbf{A}\mathbf{z}$, where c is a constant, $\mathbf{b}$ an M-dimensional column vector of constants, and $\mathbf{A}$ an $M \times M$ (nonnull) symmetric matrix of constants. Let us complete the proof of Theorem 6.6.1 by showing that if $q \sim \chi^2(R, \lambda)$ (for some strictly positive integer R), then $\mathbf{A}$, $\mathbf{b}$, and c satisfy the conditions

$$\mathbf{A}^2 = \mathbf{A}, \quad \mathbf{b} = \mathbf{A}\mathbf{b}, \quad \text{and} \quad c = \tfrac{1}{4}\mathbf{b}'\mathbf{b}$$

and R and λ are such that

$$R = \operatorname{rank}\mathbf{A} = \operatorname{tr}(\mathbf{A}) \quad \text{and} \quad \lambda = c.$$

And let us do so in a way that takes advantage of the results of Subsections a, b, and c.

Define $K = \operatorname{rank}\mathbf{A}$. And take $\mathbf{O}$ to be an $M \times M$ orthogonal matrix such that

$$\mathbf{O}'\mathbf{A}\mathbf{O} = \operatorname{diag}(d_1, d_2, \ldots, d_M)$$

for some scalars $d_1, d_2, \ldots, d_M$—the existence of such an $M \times M$ orthogonal matrix follows from Theorem 6.7.4—and define $\mathbf{u} = \mathbf{O}'\mathbf{b}$. In light of Theorem 6.7.5, K of the scalars $d_1, d_2, \ldots, d_M$ (which are the not-necessarily-distinct eigenvalues of $\mathbf{A}$) are nonzero, and the rest of them equal 0. Assume (without loss of generality) that it is the first K of the scalars $d_1, d_2, \ldots, d_M$ that are nonzero, so that $d_{K+1} = d_{K+2} = \cdots = d_M = 0$—this assumption can always be satisfied by reordering $d_1, d_2, \ldots, d_M$ and the corresponding columns of $\mathbf{O}$ (as necessary). Further, letting $\mathbf{D}_1 = \operatorname{diag}(d_1, d_2, \ldots, d_K)$ and partitioning $\mathbf{O}$ as $\mathbf{O} = (\mathbf{O}_1, \mathbf{O}_2)$ (where the dimensions of $\mathbf{O}_1$ are $M \times K$), observe that

$$\mathbf{A} = \mathbf{O}_1\mathbf{D}_1\mathbf{O}_1'.$$

Let $m(\cdot)$ represent the moment generating function of q. Then, upon applying result (7.20), we find that, for every scalar t in some nondegenerate interval that includes 0,

$$m(t) = \prod_{i=1}^{K}(1-2td_i)^{-1/2}\exp\left[tc + \frac{t^2}{2}\left(\sum_{i=1}^{K}\frac{u_i^2}{1-2td_i} + \sum_{i=K+1}^{M}u_i^2\right)\right], \tag{7.26}$$

where $u_1, u_2, \ldots, u_M$ represent the elements of $\mathbf{u}$.

Now, suppose that $q \sim \chi^2(R, \lambda)$. Then, it follows from result (2.18) that, for every scalar t such that $t < 1/2$,

$$m(t) = (1-2t)^{-R/2} \exp[t\lambda/(1-2t)]. \tag{7.27}$$

And upon equating expressions (7.26) and (7.27) and squaring both sides of the resultant equality, we find that, for every scalar t in some interval I that includes 0,

$$\frac{\prod_{i=1}^{K}(1-2td_i)}{(1-2t)^R} = \exp\left[2tc + t^2\left(\sum_{i=1}^{K}\frac{u_i^2}{1-2td_i} + \sum_{i=K+1}^{M} u_i^2\right) - \frac{2t\lambda}{1-2t}\right]. \tag{7.28}$$

Upon applying Theorem 6.7.13 to result (7.28) [and observing that, at $t = 0$, $\prod_{i=1}^{K}(1-2td_i) = (1-2t)^R$], we obtain the following result:

$$\prod_{i=1}^{K}(1-2td_i) = (1-2t)^R \quad \text{(for all } t\text{)}. \tag{7.29}$$

Thus, $R = K$ [since, otherwise, the polynomials forming the left and right sides of equality (7.29) would be of different degrees]. And, for $i = 1, 2, \ldots, K$, $d_i = 1$ [since the left side of equality (7.29) has a root at $t = 1/(2d_i)$, while, if $d_i \neq 1$, the right side does not]. We conclude (on the basis of Theorem 6.7.7) that $\mathbf{A}^2 = \mathbf{A}$ and (in light of Corollary 2.8.3) that $K = \text{tr}(\mathbf{A})$.

It remains to show that $\mathbf{b} = \mathbf{Ab}$ and that $\lambda = c = \frac{1}{4}\mathbf{b}'\mathbf{b}$. Since $R = K$ and $d_1 = d_2 = \cdots = d_K = 1$, it follows from result (7.28) that (for $t \in I$)

$$2tc + t^2\left(\sum_{i=1}^{K}\frac{u_i^2}{1-2t} + \sum_{i=K+1}^{M} u_i^2\right) - \frac{2t\lambda}{1-2t} = 0. \tag{7.30}$$

And upon multiplying both sides of equality (7.30) by $1-2t$, we obtain the equality

$$2(c-\lambda)t - 4\left[c - \tfrac{1}{4}\textstyle\sum_{i=1}^{M} u_i^2\right]t^2 - 2t^3\textstyle\sum_{i=K+1}^{M} u_i^2 = 0$$

[which, since both sides are polynomials in t, holds for all t]. Thus,

$$c - \lambda = c - \tfrac{1}{4}\textstyle\sum_{i=1}^{M} u_i^2 = \textstyle\sum_{i=K+1}^{M} u_i^2 = 0.$$

Moreover,

$$\textstyle\sum_{i=1}^{M} u_i^2 = \mathbf{u}'\mathbf{u} = (\mathbf{O}'\mathbf{b})'\mathbf{O}'\mathbf{b} = \mathbf{b}'\mathbf{b},$$

and

$$\textstyle\sum_{i=K+1}^{M} u_i^2 = (u_{K+1}, u_{K+2}, \ldots, u_M)(u_{K+1}, u_{K+2}, \ldots, u_M)' = (\mathbf{O}_2'\mathbf{b})'\mathbf{O}_2'\mathbf{b}.$$

We conclude that

$$\lambda = c = \tfrac{1}{4}\mathbf{b}'\mathbf{b}$$

and that $\mathbf{O}_2'\mathbf{b} = \mathbf{0}$ and hence that

$$\mathbf{b} = \mathbf{OO}'\mathbf{b} = (\mathbf{O}_1\mathbf{O}_1' + \mathbf{O}_2\mathbf{O}_2')\mathbf{b} = \mathbf{O}_1\mathbf{O}_1'\mathbf{b} = \mathbf{O}_1\mathbf{I}\mathbf{O}_1'\mathbf{b} = \mathbf{O}_1\mathbf{D}_1\mathbf{O}_1'\mathbf{b} = \mathbf{Ab}.$$

6.8 More on the Distribution of Quadratic Forms or Second-Degree Polynomials (in a Normally Distributed Random Vector)

a. Statistical independence of quadratic forms or second-degree polynomials (in a normally distributed random vector)

The following theorem can be used to determine whether or not two or more quadratic forms or second-degree polynomials (in a normally distributed random vector) are statistically independent.

Theorem 6.8.1. Let $\mathbf{x}$ represent an M-dimensional random column vector that has an $N(\boldsymbol{\mu}, \boldsymbol{\Sigma})$ distribution. And, for $i = 1, 2, \ldots, K$, take $q_i = c_i + \mathbf{b}_i'\mathbf{x} + \mathbf{x}'\mathbf{A}_i\mathbf{x}$, where c_i is a constant, $\mathbf{b}_i$ an M-dimensional column vector of constants, and $\mathbf{A}_i$ an $M \times M$ symmetric matrix of constants. Then, $q_1, q_2, \ldots, q_K$ are distributed independently if and only if, for $j \neq i = 1, 2, \ldots, K$,

$$\boldsymbol{\Sigma}\mathbf{A}_i\boldsymbol{\Sigma}\mathbf{A}_j\boldsymbol{\Sigma} = \mathbf{0}, \tag{8.1}$$

$$\boldsymbol{\Sigma}\mathbf{A}_i\boldsymbol{\Sigma}(\mathbf{b}_j + 2\mathbf{A}_j\boldsymbol{\mu}) = \mathbf{0}, \tag{8.2}$$

and

$$(\mathbf{b}_i + 2\mathbf{A}_i\boldsymbol{\mu})'\boldsymbol{\Sigma}(\mathbf{b}_j + 2\mathbf{A}_j\boldsymbol{\mu}) = 0. \tag{8.3}$$

In connection with conditions (8.1) and (8.3), it is worth noting that

$$\boldsymbol{\Sigma}\mathbf{A}_i\boldsymbol{\Sigma}\mathbf{A}_j\boldsymbol{\Sigma} = \mathbf{0} \quad \Leftrightarrow \quad \boldsymbol{\Sigma}\mathbf{A}_j\boldsymbol{\Sigma}\mathbf{A}_i\boldsymbol{\Sigma} = \mathbf{0}$$

and

$$(\mathbf{b}_i + 2\mathbf{A}_i\boldsymbol{\mu})'\boldsymbol{\Sigma}(\mathbf{b}_j + 2\mathbf{A}_j\boldsymbol{\mu}) = 0 \quad \Leftrightarrow \quad (\mathbf{b}_j + 2\mathbf{A}_j\boldsymbol{\mu})'\boldsymbol{\Sigma}(\mathbf{b}_i + 2\mathbf{A}_i\boldsymbol{\mu}) = 0,$$

as is evident upon "taking transposes."

In the special case of second-degree polynomials in a normally distributed random vector that has a nonsingular variance-covariance matrix, we obtain the following corollary of Theorem 6.8.1.

Corollary 6.8.2. Let $\mathbf{x}$ represent an M-dimensional random column vector that has an $N(\boldsymbol{\mu}, \boldsymbol{\Sigma})$ distribution, where $\boldsymbol{\Sigma}$ is nonsingular. And, for $i = 1, 2, \ldots, K$, take $q_i = c_i + \mathbf{b}_i'\mathbf{x} + \mathbf{x}'\mathbf{A}_i\mathbf{x}$, where c_i is a constant, $\mathbf{b}_i$ an M-dimensional column vector of constants, and $\mathbf{A}_i$ an $M \times M$ symmetric matrix of constants. Then, $q_1, q_2, \ldots, q_K$ are distributed independently if and only if, for $j \neq i = 1, 2, \ldots, K$,

$$\mathbf{A}_i\boldsymbol{\Sigma}\mathbf{A}_j = \mathbf{0}, \tag{8.4}$$

$$\mathbf{A}_i\boldsymbol{\Sigma}\mathbf{b}_j = \mathbf{0}, \tag{8.5}$$

and

$$\mathbf{b}_i'\boldsymbol{\Sigma}\mathbf{b}_j = 0. \tag{8.6}$$

Note that in the special case of Corollary 6.8.2 where $q_1, q_2, \ldots, q_K$ are quadratic forms (i.e, the special case where $c_1 = c_2 = \cdots = c_K = 0$ and $\mathbf{b}_1 = \mathbf{b}_2 = \cdots = \mathbf{b}_K = \mathbf{0}$), conditions (8.5) and (8.6) are vacuous; in this special case, only condition (8.4) is "operative." Note also that in the special case of Corollary 6.8.2 where $\boldsymbol{\Sigma} = \mathbf{I}$, Corollary 6.8.2 reduces to the following result.

Theorem 6.8.3. Let $\mathbf{x}$ represent an M-dimensional random column vector that has an $N(\boldsymbol{\mu}, \mathbf{I}_M)$ distribution. And, for $i = 1, 2, \ldots, K$, take $q_i = c_i + \mathbf{b}_i'\mathbf{x} + \mathbf{x}'\mathbf{A}_i\mathbf{x}$, where c_i is a constant, $\mathbf{b}_i$ an M-dimensional column vector of constants, and $\mathbf{A}_i$ an $M \times M$ symmetric matrix of constants. Then, $q_1, q_2, \ldots, q_K$ are distributed independently if and only if, for $j \neq i = 1, 2, \ldots, K$,

$$\mathbf{A}_i\mathbf{A}_j = \mathbf{0}, \quad \mathbf{A}_i\mathbf{b}_j = \mathbf{0}, \quad \text{and} \quad \mathbf{b}_i'\mathbf{b}_j = 0. \tag{8.7}$$

Verification of theorems. To prove Theorem 6.8.1, it suffices to prove Theorem 6.8.3. In fact, it suffices to prove the special case of Theorem 6.8.3 where $\boldsymbol{\mu} = \mathbf{0}$. To see this, define $\mathbf{x}$ and $q_1, q_2, \ldots, q_K$ as in Theorem 6.8.1. That is, take $\mathbf{x}$ to be an M-dimensional random column vector that has an $N(\boldsymbol{\mu}, \boldsymbol{\Sigma})$ distribution. And, for $i = 1, 2, \ldots, K$, take $q_i = c_i + \mathbf{b}_i'\mathbf{x} + \mathbf{x}'\mathbf{A}_i\mathbf{x}$, where c_i is a constant, $\mathbf{b}_i$ an M-dimensional column vector of constants, and $\mathbf{A}_i$ an $M \times M$ symmetric matrix of constants. Further, take $\boldsymbol{\Gamma}$ to be any matrix (with M columns) such that $\boldsymbol{\Sigma} = \boldsymbol{\Gamma}'\boldsymbol{\Gamma}$, denote by P the number of rows in $\boldsymbol{\Gamma}$, and define $\mathbf{z}$ to be a P-dimensional random column vector that has an $N(\mathbf{0}, \mathbf{I}_P)$ distribution. Then, $\mathbf{x} \sim \boldsymbol{\mu} + \boldsymbol{\Gamma}'\mathbf{z}$, and hence the joint distribution of $q_1, q_2, \ldots, q_K$ is identical to that of $q_1^*, q_2^*, \ldots, q_K^*$, where (for $i = 1, 2, \ldots, K$)

$$q_i^* = c_i + \mathbf{b}_i'(\boldsymbol{\mu} + \boldsymbol{\Gamma}'\mathbf{z}) + (\boldsymbol{\mu} + \boldsymbol{\Gamma}'\mathbf{z})'\mathbf{A}_i(\boldsymbol{\mu} + \boldsymbol{\Gamma}'\mathbf{z}).$$

For $i = 1, 2, \ldots, K$, q_i^* is reexpressible in the form

$$q_i^* = c_i + \mathbf{b}_i'\boldsymbol{\mu} + \boldsymbol{\mu}'\mathbf{A}_i\boldsymbol{\mu} + [\boldsymbol{\Gamma}(\mathbf{b}_i + 2\mathbf{A}_i\boldsymbol{\mu})]'\mathbf{z} + \mathbf{z}'\boldsymbol{\Gamma}\mathbf{A}_i\boldsymbol{\Gamma}'\mathbf{z},$$

which is a second-degree polynomial in $\mathbf{z}$. Thus, it follows from Theorem 6.8.3 (upon applying the theorem to $q_1^*, q_2^*, \ldots, q_K^*$) that $q_1^*, q_2^*, \ldots, q_K^*$ (and hence $q_1, q_2, \ldots, q_K$) are statistically independent if and only if, for $i = 1, 2, \ldots, K$,

$$\boldsymbol{\Gamma}\mathbf{A}_i\boldsymbol{\Gamma}'\boldsymbol{\Gamma}\mathbf{A}_j\boldsymbol{\Gamma}' = \mathbf{0}, \tag{8.8}$$

$$\boldsymbol{\Gamma}\mathbf{A}_i\boldsymbol{\Gamma}'\boldsymbol{\Gamma}(\mathbf{b}_j + 2\mathbf{A}_j\boldsymbol{\mu}) = \mathbf{0}, \tag{8.9}$$

and

$$(\mathbf{b}_i + 2\mathbf{A}_i\boldsymbol{\mu})'\boldsymbol{\Gamma}'\boldsymbol{\Gamma}(\mathbf{b}_j + 2\mathbf{A}_j\boldsymbol{\mu}) = 0. \tag{8.10}$$

Moreover, in light of Corollary 2.3.4,

$$\boldsymbol{\Gamma}\mathbf{A}_i\boldsymbol{\Gamma}'\boldsymbol{\Gamma}\mathbf{A}_j\boldsymbol{\Gamma}' = \mathbf{0} \quad \Leftrightarrow \quad \boldsymbol{\Gamma}'\boldsymbol{\Gamma}\mathbf{A}_i\boldsymbol{\Gamma}'\boldsymbol{\Gamma}\mathbf{A}_j\boldsymbol{\Gamma}' = \mathbf{0} \quad \Leftrightarrow \quad \boldsymbol{\Gamma}'\boldsymbol{\Gamma}\mathbf{A}_i\boldsymbol{\Gamma}'\boldsymbol{\Gamma}\mathbf{A}_j\boldsymbol{\Gamma}'\boldsymbol{\Gamma} = \mathbf{0}$$

and

$$\boldsymbol{\Gamma}\mathbf{A}_i\boldsymbol{\Gamma}'\boldsymbol{\Gamma}(\mathbf{b}_j + 2\mathbf{A}_j\boldsymbol{\mu}) = \mathbf{0} \quad \Leftrightarrow \quad \boldsymbol{\Gamma}'\boldsymbol{\Gamma}\mathbf{A}_i\boldsymbol{\Gamma}'\boldsymbol{\Gamma}(\mathbf{b}_j + 2\mathbf{A}_j\boldsymbol{\mu}) = \mathbf{0}.$$

And we conclude that the conditions [conditions (8.8), (8.9), and (8.10)] derived from the application of Theorem 6.8.3 to $q_1^*, q_2^*, \ldots, q_K^*$ are equivalent to conditions (8.1), (8.2), and (8.3) (of Theorem 6.8.1).

Theorem 6.8.3, like Theorem 6.6.1, pertains to second-degree polynomials in a normally distributed random column vector. Theorem 6.8.3 gives conditions that are necessary and sufficient for such second-degree polynomials to be statistically independent, whereas Theorem 6.6.1 gives conditions that are necessary and sufficient for such a second-degree polynomial to have a noncentral chi-square distribution. In the case of Theorem 6.8.3, as in the case of Theorem 6.6.1, it is much easier to prove sufficiency than necessity, and the sufficiency is more important than the necessity (in the sense that it is typically the sufficiency that is invoked in an application). Accordingly, the following proof of Theorem 6.8.3 is a proof of sufficiency; the proof of necessity is deferred until a subsequent subsection (Subsection e).

Proof (of Theorem 6.8.3): sufficiency. For $i = 1, 2, \ldots, K$, q_i is reexpressible in the form

$$q_i = c_i + \mathbf{b}_i'\mathbf{x} + (\mathbf{A}_i\mathbf{x})'\mathbf{A}_i^-(\mathbf{A}_i\mathbf{x}),$$

and hence q_i depends on the value of $\mathbf{x}$ only through the $(M+1)$-dimensional column vector $(\mathbf{b}_i,\ \mathbf{A}_i)'\mathbf{x}$. Moreover, for $j \neq i = 1, 2, \ldots, K$,

$$\text{cov}[(\mathbf{b}_i,\ \mathbf{A}_i)'\mathbf{x},\ (\mathbf{b}_j,\ \mathbf{A}_j)'\mathbf{x}] = (\mathbf{b}_i,\ \mathbf{A}_i)'(\mathbf{b}_j,\ \mathbf{A}_j) = \begin{bmatrix} \mathbf{b}_i'\mathbf{b}_j & (\mathbf{A}_j\mathbf{b}_i)' \\ \mathbf{A}_i\mathbf{b}_j & \mathbf{A}_i\mathbf{A}_j \end{bmatrix}. \tag{8.11}$$

And the joint distribution of the K vectors $(\mathbf{b}_1,\ \mathbf{A}_1)'\mathbf{x},\ (\mathbf{b}_2,\ \mathbf{A}_2)'\mathbf{x}, \ldots,\ (\mathbf{b}_K,\ \mathbf{A}_K)'\mathbf{x}$ is multivariate normal.

Now, suppose that, for $j \neq i = 1, 2, \ldots, K$, condition (8.7) is satisfied. Then, in light of result (8.11), $(\mathbf{b}_1,\ \mathbf{A}_1)'\mathbf{x},\ (\mathbf{b}_2,\ \mathbf{A}_2)'\mathbf{x}, \ldots,\ (\mathbf{b}_K,\ \mathbf{A}_K)'\mathbf{x}$ are uncorrelated and hence (since their joint distribution is multivariate normal) statistically independent, leading to the conclusion that the second-degree polynomials $q_1, q_2, \ldots, q_K$ [each of which depends on a different one of the vectors $(\mathbf{b}_1,\ \mathbf{A}_1)'\mathbf{x},\ (\mathbf{b}_2,\ \mathbf{A}_2)'\mathbf{x}, \ldots,\ (\mathbf{b}_K,\ \mathbf{A}_K)'\mathbf{x}$] are statistically independent. Q.E.D.

An extension. The coverage of Theorem 6.8.1 includes the special case where one or more of the quantities $q_1, q_2, \ldots, q_K$ (whose statistical independence is in question) are linear forms. In the following generalization, the coverage is extended to include vectors of linear forms.

Theorem 6.8.4. Let $\mathbf{x}$ represent an M-dimensional random column vector that has an $N(\boldsymbol{\mu}, \boldsymbol{\Sigma})$ distribution. And, for $i = 1, 2, \ldots, K$, take $q_i = c_i + \mathbf{b}_i'\mathbf{x} + \mathbf{x}'\mathbf{A}_i\mathbf{x}$, where c_i is a constant, $\mathbf{b}_i$ an M-dimensional column vector of constants, and $\mathbf{A}_i$ an $M \times M$ symmetric matrix of constants. Further, for $s = 1, 2, \ldots, R$, denote by $\mathbf{d}_s$ an N_s-dimensional column vector of constants and by $\mathbf{L}_s$ an $M \times N_s$ matrix of constants. Then, $q_1, q_2, \ldots, q_K,\ \mathbf{d}_1 + \mathbf{L}_1'\mathbf{x},\ \mathbf{d}_2 + \mathbf{L}_2'\mathbf{x}, \ldots,\ \mathbf{d}_R + \mathbf{L}_R'\mathbf{x}$ are distributed independently if and only if, for $j \neq i = 1, 2, \ldots, K$,

$$\boldsymbol{\Sigma}\mathbf{A}_i\boldsymbol{\Sigma}\mathbf{A}_j\boldsymbol{\Sigma} = \mathbf{0}, \tag{8.12}$$

$$\boldsymbol{\Sigma}\mathbf{A}_i\boldsymbol{\Sigma}(\mathbf{b}_j + 2\mathbf{A}_j\boldsymbol{\mu}) = \mathbf{0}, \tag{8.13}$$

and

$$(\mathbf{b}_i + 2\mathbf{A}_i \boldsymbol{\mu})' \boldsymbol{\Sigma} (\mathbf{b}_j + 2\mathbf{A}_j \boldsymbol{\mu}) = 0, \tag{8.14}$$

for $i = 1, 2, \ldots, K$ and $s = 1, 2, \ldots, R$,

$$\boldsymbol{\Sigma} \mathbf{A}_i \boldsymbol{\Sigma} \mathbf{L}_s = \mathbf{0} \tag{8.15}$$

and

$$(\mathbf{b}_i + 2\mathbf{A}_i \boldsymbol{\mu})' \boldsymbol{\Sigma} \mathbf{L}_s = \mathbf{0}, \tag{8.16}$$

and, for $t \neq s = 1, 2, \ldots, R$,

$$\mathbf{L}_t' \boldsymbol{\Sigma} \mathbf{L}_s = \mathbf{0}. \tag{8.17}$$

Note that in the special case where $\boldsymbol{\Sigma}$ is nonsingular, conditions (8.15) and (8.16) are (collectively) equivalent to the condition

$$\mathbf{A}_i \boldsymbol{\Sigma} \mathbf{L}_s = \mathbf{0} \quad \text{and} \quad \mathbf{b}_i' \boldsymbol{\Sigma} \mathbf{L}_s = \mathbf{0}. \tag{8.18}$$

Accordingly, in the further special case where $\boldsymbol{\Sigma} = \mathbf{I}$, the result of Theorem 6.8.4 can be restated in the form of the following generalization of Theorem 6.8.3.

Theorem 6.8.5. Let $\mathbf{x}$ represent an M-dimensional random column vector that has an $N(\boldsymbol{\mu}, \mathbf{I}_M)$ distribution. And, for $i = 1, 2, \ldots, K$, take $q_i = c_i + \mathbf{b}_i'\mathbf{x} + \mathbf{x}'\mathbf{A}_i\mathbf{x}$, where c_i is a constant, $\mathbf{b}_i$ an M-dimensional column vector of constants, and $\mathbf{A}_i$ an $M \times M$ symmetric matrix of constants. Further, for $s = 1, 2, \ldots, R$, denote by $\mathbf{d}_s$ an N_s-dimensional column vector of constants and by $\mathbf{L}_s$ an $M \times N_s$ matrix of constants. Then, $q_1, q_2, \ldots, q_K$, $\mathbf{d}_1 + \mathbf{L}_1'\mathbf{x}$, $\mathbf{d}_2 + \mathbf{L}_2'\mathbf{x}, \ldots,$ $\mathbf{d}_R + \mathbf{L}_R'\mathbf{x}$ are distributed independently if and only if, for $j \neq i = 1, 2, \ldots, K$,

$$\mathbf{A}_i \mathbf{A}_j = \mathbf{0}, \quad \mathbf{A}_i \mathbf{b}_j = \mathbf{0}, \quad \text{and} \quad \mathbf{b}_i' \mathbf{b}_j = 0, \tag{8.19}$$

for $i = 1, 2, \ldots, K$ and $s = 1, 2, \ldots, R$,

$$\mathbf{A}_i \mathbf{L}_s = \mathbf{0} \quad \text{and} \quad \mathbf{b}_i' \mathbf{L}_s = \mathbf{0}, \tag{8.20}$$

and, for $t \neq s = 1, 2, \ldots, R$,

$$\mathbf{L}_t' \mathbf{L}_s = \mathbf{0}. \tag{8.21}$$

To prove Theorem 6.8.4, it suffices to prove Theorem 6.8.5, as can be established via a straightforward extension of the argument used in establishing that to prove Theorem 6.8.1, it suffices to prove Theorem 6.8.3. Moreover, the "sufficiency part" of Theorem 6.8.5 can be established via a straightforward extension of the argument used to establish the sufficiency part of Theorem 6.8.3.

Now, consider the "necessity part" of Theorem 6.8.5. Suppose that $q_1, q_2, \ldots, q_K$, $\mathbf{d}_1 + \mathbf{L}_1'\mathbf{x}$, $\mathbf{d}_2 + \mathbf{L}_2'\mathbf{x}, \ldots,$ $\mathbf{d}_R + \mathbf{L}_R'\mathbf{x}$ are statistically independent. Then, for arbitrary column vectors $\mathbf{h}_1, \mathbf{h}_2, \ldots, \mathbf{h}_R$ of constants (of dimensions $N_1, N_2, \ldots, N_R$, respectively) $q_1, q_2, \ldots, q_K$, $\mathbf{h}_1'\mathbf{d}_1 + (\mathbf{L}_1\mathbf{h}_1)'\mathbf{x}$, $\mathbf{h}_2'\mathbf{d}_2 + (\mathbf{L}_2\mathbf{h}_2)'\mathbf{x}, \ldots,$ $\mathbf{h}_R'\mathbf{d}_R + (\mathbf{L}_R\mathbf{h}_R)'\mathbf{x}$ are statistically independent. And it follows from the necessity part of Theorem 6.8.3 that, for $j \neq i = 1, 2, \ldots, K$,

$$\mathbf{A}_i \mathbf{A}_j = \mathbf{0}, \quad \mathbf{A}_i \mathbf{b}_j = \mathbf{0}, \quad \text{and} \quad \mathbf{b}_i' \mathbf{b}_j = 0,$$

for $i = 1, 2, \ldots, K$ and $s = 1, 2, \ldots, R$,

$$\mathbf{A}_i \mathbf{L}_s \mathbf{h}_s = \mathbf{0} \quad \text{and} \quad \mathbf{b}_i' \mathbf{L}_s \mathbf{h}_s = 0, \tag{8.22}$$

and, for $t \neq s = 1, 2, \ldots, R$,

$$\mathbf{h}_t' \mathbf{L}_t' \mathbf{L}_s \mathbf{h}_s = 0. \tag{8.23}$$

Moreover, since the vectors $\mathbf{h}_1, \mathbf{h}_2, \ldots, \mathbf{h}_R$ are arbitrary, results (8.22) and (8.23) imply that, for $i = 1, 2, \ldots, K$ and $s = 1, 2, \ldots, R$,

$$\mathbf{A}_i \mathbf{L}_s = \mathbf{0} \quad \text{and} \quad \mathbf{b}_i' \mathbf{L}_s = \mathbf{0}$$

and, for $t \neq s = 1, 2, \ldots, R$,

$$\mathbf{L}_t' \mathbf{L}_s = \mathbf{0}.$$

Thus, the necessity part of Theorem 6.8.5 follows from the necessity part of Theorem 6.8.3.

Statistical independence versus zero correlation. Let $\mathbf{x}$ represent an M-dimensional random column vector that has an $N(\boldsymbol{\mu}, \boldsymbol{\Sigma})$ distribution. For an $M \times N_1$ matrix of constants $\mathbf{L}_1$ and an $M \times N_2$ matrix of constants $\mathbf{L}_2$,

$$\text{cov}(\mathbf{L}_1'\mathbf{x}, \mathbf{L}_2'\mathbf{x}) = \mathbf{L}_1'\boldsymbol{\Sigma}\mathbf{L}_2.$$

Thus, two vectors of linear forms in a normally distributed random column vector are statistically independent if and only if they are uncorrelated.

For an $M \times N$ matrix of constants $\mathbf{L}$ and an $M \times M$ symmetric matrix of constants $\mathbf{A}$,

$$\text{cov}(\mathbf{L}'\mathbf{x}, \mathbf{x}'\mathbf{A}\mathbf{x}) = 2\mathbf{L}'\boldsymbol{\Sigma}\mathbf{A}\boldsymbol{\mu}$$

—refer to result (5.7.14)—so that the vector $\mathbf{L}'\mathbf{x}$ of linear forms and the quadratic form $\mathbf{x}'\mathbf{A}\mathbf{x}$ (in the normally distributed random vector $\mathbf{x}$) are uncorrelated if and only if $\mathbf{L}'\boldsymbol{\Sigma}\mathbf{A}\boldsymbol{\mu} = \mathbf{0}$, but are statistically independent if and only if $\boldsymbol{\Sigma}\mathbf{A}\boldsymbol{\Sigma}\mathbf{L} = \mathbf{0}$ and $\boldsymbol{\mu}'\mathbf{A}\boldsymbol{\Sigma}\mathbf{L} = \mathbf{0}$ (or, equivalently, if and only if $\mathbf{L}'\boldsymbol{\Sigma}\mathbf{A}\boldsymbol{\Sigma} = \mathbf{0}$ and $\mathbf{L}'\boldsymbol{\Sigma}\mathbf{A}\boldsymbol{\mu} = \mathbf{0}$). And, for two $M \times M$ symmetric matrices of constants $\mathbf{A}_1$ and $\mathbf{A}_2$,

$$\text{cov}(\mathbf{x}'\mathbf{A}_1\mathbf{x}, \mathbf{x}'\mathbf{A}_2\mathbf{x}) = 2\,\text{tr}(\mathbf{A}_1\boldsymbol{\Sigma}\mathbf{A}_2\boldsymbol{\Sigma}) + 4\boldsymbol{\mu}'\mathbf{A}_1\boldsymbol{\Sigma}\mathbf{A}_2\boldsymbol{\mu}$$

—refer to result (5.7.19)—so that the two quadratic forms $\mathbf{x}'\mathbf{A}_1\mathbf{x}$ and $\mathbf{x}'\mathbf{A}_2\mathbf{x}$ (in the normally distributed random vector $\mathbf{x}$) are uncorrelated if and only if

$$\text{tr}(\mathbf{A}_1\boldsymbol{\Sigma}\mathbf{A}_2\boldsymbol{\Sigma}) + 2\boldsymbol{\mu}'\mathbf{A}_1\boldsymbol{\Sigma}\mathbf{A}_2\boldsymbol{\mu} = 0, \tag{8.24}$$

but are statistically independent if and only if

$$\boldsymbol{\Sigma}\mathbf{A}_1\boldsymbol{\Sigma}\mathbf{A}_2\boldsymbol{\Sigma} = \mathbf{0},\ \boldsymbol{\Sigma}\mathbf{A}_1\boldsymbol{\Sigma}\mathbf{A}_2\boldsymbol{\mu} = \mathbf{0},\ \boldsymbol{\Sigma}\mathbf{A}_2\boldsymbol{\Sigma}\mathbf{A}_1\boldsymbol{\mu} = \mathbf{0}, \text{ and } \boldsymbol{\mu}'\mathbf{A}_1\boldsymbol{\Sigma}\mathbf{A}_2\boldsymbol{\mu} = 0.$$

b. Cochran's theorem

Theorem 6.8.1 can be used to determine whether two or more second-degree polynomials (in a normally distributed random vector) are statistically independent. The following theorem can be used to determine whether the second-degree polynomials are not only statistically independent but whether, in addition, they have noncentral chi-square distributions.

Theorem 6.8.6. Let $\mathbf{x}$ represent an M-dimensional random column vector that has an $N(\boldsymbol{\mu}, \boldsymbol{\Sigma})$ distribution. And, for $i = 1, 2, \ldots, K$, take $q_i = c_i + \mathbf{b}_i'\mathbf{x} + \mathbf{x}'\mathbf{A}_i\mathbf{x}$, where c_i is a constant, $\mathbf{b}_i$ an M-dimensional column vector of constants, and $\mathbf{A}_i$ an $M \times M$ symmetric matrix of constants (such that $\boldsymbol{\Sigma}\mathbf{A}_i\boldsymbol{\Sigma} \neq \mathbf{0}$). Further, define $\mathbf{A} = \mathbf{A}_1 + \mathbf{A}_2 + \cdots + \mathbf{A}_K$. If

$$\boldsymbol{\Sigma}\mathbf{A}\boldsymbol{\Sigma}\mathbf{A}\boldsymbol{\Sigma} = \boldsymbol{\Sigma}\mathbf{A}\boldsymbol{\Sigma}, \tag{8.25}$$

$$\text{rank}(\boldsymbol{\Sigma}\mathbf{A}_1\boldsymbol{\Sigma}) + \text{rank}(\boldsymbol{\Sigma}\mathbf{A}_2\boldsymbol{\Sigma}) + \cdots + \text{rank}(\boldsymbol{\Sigma}\mathbf{A}_K\boldsymbol{\Sigma}) = \text{rank}(\boldsymbol{\Sigma}\mathbf{A}\boldsymbol{\Sigma}), \tag{8.26}$$

$$\boldsymbol{\Sigma}(\mathbf{b}_i + 2\mathbf{A}_i\boldsymbol{\mu}) = \boldsymbol{\Sigma}\mathbf{A}\boldsymbol{\Sigma}(\mathbf{b}_i + 2\mathbf{A}_i\boldsymbol{\mu}) \quad (i = 1, 2, \ldots, K) \tag{8.27}$$

and

$$c_i + \mathbf{b}_i'\boldsymbol{\mu} + \boldsymbol{\mu}'\mathbf{A}_i\boldsymbol{\mu} = \tfrac{1}{4}(\mathbf{b}_i + 2\mathbf{A}_i\boldsymbol{\mu})'\boldsymbol{\Sigma}(\mathbf{b}_i + 2\mathbf{A}_i\boldsymbol{\mu}), \quad (i = 1, 2, \ldots, K), \tag{8.28}$$

then $q_1, q_2, \ldots, q_K$ are statistically independent and (for $i = 1, 2, \ldots, K$) $q_i \sim \chi^2(R_i, c_i + \mathbf{b}_i'\boldsymbol{\mu} + \boldsymbol{\mu}'\mathbf{A}_i\boldsymbol{\mu})$, where $R_i = \text{rank}(\boldsymbol{\Sigma}\mathbf{A}_i\boldsymbol{\Sigma}) = \text{tr}(\mathbf{A}_i\boldsymbol{\Sigma})$. Conversely, if $q_1, q_2, \ldots, q_K$ are statistically independent and (for $i = 1, 2, \ldots, K$) $q_i \sim \chi^2(R_i, \lambda_i)$ (where R_i is a strictly positive integer), then conditions (8.25), (8.26), (8.27), and (8.28) are satisfied and (for $i = 1, 2, \ldots, K$) $R_i = \text{rank}(\boldsymbol{\Sigma}\mathbf{A}_i\boldsymbol{\Sigma}) = \text{tr}(\mathbf{A}_i\boldsymbol{\Sigma})$ and $\lambda_i = c_i + \mathbf{b}_i'\boldsymbol{\mu} + \boldsymbol{\mu}'\mathbf{A}_i\boldsymbol{\mu}$.

Some results on matrices. The proof of Theorem 6.8.6 makes use of certain properties of matrices. These properties are presented in the form of a generalization of the following theorem.

Theorem 6.8.7. Let $\mathbf{A}_1, \mathbf{A}_2, \ldots, \mathbf{A}_K$ represent $N \times N$ matrices, and define $\mathbf{A} = \mathbf{A}_1 + \mathbf{A}_2 + \cdots + \mathbf{A}_K$. Suppose that $\mathbf{A}$ is idempotent. Then, *each* of the following conditions implies the other two:

(1) $\mathbf{A}_i\mathbf{A}_j = \mathbf{0}$ (for $j \neq i = 1, 2, \ldots, K$) and $\text{rank}(\mathbf{A}_i^2) = \text{rank}(\mathbf{A}_i)$ (for $i = 1, 2, \ldots, K$);
(2) $\mathbf{A}_i^2 = \mathbf{A}_i$ (for $i = 1, 2, \ldots, K$);
(3) $\text{rank}(\mathbf{A}_1) + \text{rank}(\mathbf{A}_2) + \cdots + \text{rank}(\mathbf{A}_K) = \text{rank}(\mathbf{A})$.

As a preliminary to proving Theorem 6.8.7, let us establish the following two lemmas on idempotent matrices.

Lemma 6.8.8. An $N \times N$ matrix $\mathbf{A}$ is idempotent if and only if $\text{rank}(\mathbf{A}) + \text{rank}(\mathbf{I} - \mathbf{A}) = N$.

Proof (of Lemma 6.8.8). That $\text{rank}(\mathbf{A}) + \text{rank}(\mathbf{I} - \mathbf{A}) = N$ if $\mathbf{A}$ is idempotent is an immediate consequence of Lemma 2.8.4. Now, for purposes of establishing the converse, observe that $\mathcal{N}(\mathbf{A}) \subset \mathcal{C}(\mathbf{I} - \mathbf{A})$—if $\mathbf{x} \in \mathcal{N}(\mathbf{A})$, then $\mathbf{Ax} = \mathbf{0}$, in which case $\mathbf{x} = (\mathbf{I} - \mathbf{A})\mathbf{x} \in \mathcal{C}(\mathbf{I} - \mathbf{A})$. Observe also (in light of Lemma 2.11.5) that $\dim[\mathcal{N}(\mathbf{A})] = N - \text{rank}(\mathbf{A})$. Thus, if $\text{rank}(\mathbf{A}) + \text{rank}(\mathbf{I} - \mathbf{A}) = N$, then

$$\dim[\mathcal{N}(\mathbf{A})] = \text{rank}(\mathbf{I} - \mathbf{A}) = \dim[\mathcal{C}(\mathbf{I} - \mathbf{A})],$$

implying (in light of Theorem 2.4.10) that $\mathcal{N}(\mathbf{A}) = \mathcal{C}(\mathbf{I} - \mathbf{A})$ and hence that every column of $\mathbf{I} - \mathbf{A}$ is a member of $\mathcal{N}(\mathbf{A})$, in which case $\mathbf{A}(\mathbf{I} - \mathbf{A}) = \mathbf{0}$ or, equivalently, $\mathbf{A}^2 = \mathbf{A}$. Q.E.D.

Lemma 6.8.9. For any two idempotent matrices $\mathbf{A}$ and $\mathbf{B}$ (of the same size), $\mathbf{A}+\mathbf{B}$ is idempotent if and only if $\mathbf{BA} = \mathbf{AB} = \mathbf{0}$.

Proof (of Lemma 6.8.9). Clearly,

$$(\mathbf{A} + \mathbf{B})^2 = \mathbf{A}^2 + \mathbf{B}^2 + \mathbf{AB} + \mathbf{BA} = \mathbf{A} + \mathbf{B} + \mathbf{AB} + \mathbf{BA}.$$

Thus, $\mathbf{A} + \mathbf{B}$ is idempotent if and only if $\mathbf{AB} + \mathbf{BA} = \mathbf{0}$.

Now, suppose that $\mathbf{BA} = \mathbf{AB} = \mathbf{0}$. Then, obviously, $\mathbf{AB} + \mathbf{BA} = \mathbf{0}$, and, consequently, $\mathbf{A} + \mathbf{B}$ is idempotent.

Conversely, suppose that $\mathbf{AB} + \mathbf{BA} = \mathbf{0}$ (as would be the case if $\mathbf{A} + \mathbf{B}$ were idempotent). Then,

$$\mathbf{AB} + \mathbf{ABA} = \mathbf{A}^2\mathbf{B} + \mathbf{ABA} = \mathbf{A}(\mathbf{AB} + \mathbf{BA}) = \mathbf{0},$$

and

$$\mathbf{ABA} + \mathbf{BA} = \mathbf{ABA} + \mathbf{BA}^2 = (\mathbf{AB} + \mathbf{BA})\mathbf{A} = \mathbf{0},$$

implying that

$$\mathbf{AB} - \mathbf{BA} = \mathbf{AB} + \mathbf{ABA} - (\mathbf{ABA} + \mathbf{BA}) = \mathbf{0} - \mathbf{0} = \mathbf{0}$$

and hence that

$$\mathbf{BA} = \mathbf{AB}.$$

Moreover,

$$\mathbf{AB} = \tfrac{1}{2}(\mathbf{AB} + \mathbf{AB}) = \tfrac{1}{2}(\mathbf{AB} + \mathbf{BA}) = \mathbf{0}.$$

Q.E.D.

Proof (of Theorem 6.8.7). The proof consists of successively showing that Condition (1) implies Condition (2), that Condition (2) implies Condition (3), and that Condition (3) implies Condition (1).

(1) $\Rightarrow$ (2). Suppose that Condition (1) is satisfied. Then, for an arbitrary integer i between 1 and K, inclusive,

$$\mathbf{A}_i^2 = \mathbf{A}_i\mathbf{A} = \mathbf{A}_i\mathbf{A}\mathbf{A} = \mathbf{A}_i^2\mathbf{A} = \mathbf{A}_i^3.$$

Moreover, since $\text{rank}(\mathbf{A}_i^2) = \text{rank}(\mathbf{A}_i)$, it follows from Corollary 2.4.17 that $\mathcal{C}(\mathbf{A}_i^2) = \mathcal{C}(\mathbf{A}_i)$, so that $\mathbf{A}_i = \mathbf{A}_i^2\mathbf{L}_i$ for some matrix $\mathbf{L}_i$. Thus,

$$\mathbf{A}_i = \mathbf{A}_i^2\mathbf{L}_i = \mathbf{A}_i^3\mathbf{L}_i = \mathbf{A}_i(\mathbf{A}_i^2\mathbf{L}_i) = \mathbf{A}_i\mathbf{A}_i = \mathbf{A}_i^2.$$

(2) $\Rightarrow$ (3). Suppose that Condition (2) is satisfied. Then, making use of Corollary 2.8.3, we find that

$$\sum_{i=1}^{K} \text{rank}(\mathbf{A}_i) = \sum_{i=1}^{K} \text{tr}(\mathbf{A}_i) = \text{tr}\Big(\sum_{i=1}^{K} \mathbf{A}_i\Big) = \text{tr}(\mathbf{A}) = \text{rank}(\mathbf{A}).$$

(3) $\Rightarrow$ (1). Suppose that Condition (3) is satisfied. And define $\mathbf{A}_0 = \mathbf{I}_N - \mathbf{A}$. Then, $\sum_{i=0}^{K} \mathbf{A}_i = \mathbf{I}$.

Moreover, it follows from Lemma 2.8.4 (and also from Lemma 6.8.8) that $\text{rank}(\mathbf{A}_0) = N - \text{rank}(\mathbf{A})$ and hence that $\sum_{i=0}^{K} \text{rank}(\mathbf{A}_i) = N$.

Now, making use of inequality (2.4.24), we find (for an arbitrary integer i between 1 and K, inclusive) that

$$\text{rank}(\mathbf{I} - \mathbf{A}_i) = \text{rank}\Bigl(\sum_{s=0\ (s\neq i)}^{K} \mathbf{A}_s\Bigr) \leq \sum_{s=0\ (s\neq i)}^{K} \text{rank}(\mathbf{A}_s) = N - \text{rank}(\mathbf{A}_i),$$

so that $\text{rank}(\mathbf{A}_i) + \text{rank}(\mathbf{I} - \mathbf{A}_i) \leq N$, implying [since $\text{rank}(\mathbf{A}_i) + \text{rank}(\mathbf{I} - \mathbf{A}_i) \geq \text{rank}(\mathbf{A}_i + \mathbf{I} - \mathbf{A}_i) = \text{rank}(\mathbf{I}_N) = N$] that

$$\text{rank}(\mathbf{A}_i) + \text{rank}(\mathbf{I} - \mathbf{A}_i) = N.$$

And upon applying Lemma 6.8.8, it follows that $\mathbf{A}_i$ is idempotent [and that $\text{rank}(\mathbf{A}_i^2) = \text{rank}(\mathbf{A}_i)$].

Further, upon again making use of inequality (2.4.24), we find (for $j \neq i = 1, 2, \ldots, K$) that

$$\begin{aligned}\text{rank}(\mathbf{I} - \mathbf{A}_i - \mathbf{A}_j) &= \text{rank}\Bigl(\sum_{s=0\ (s\neq i,\, j)}^{K} \mathbf{A}_s\Bigr)\\ &\leq \sum_{s=0\ (s\neq i,\, j)}^{K} \text{rank}(\mathbf{A}_s) = N - [\text{rank}(\mathbf{A}_i) + \text{rank}(\mathbf{A}_j)]\\ &\leq N - \text{rank}(\mathbf{A}_i + \mathbf{A}_j),\end{aligned}$$

so that $\text{rank}(\mathbf{A}_i + \mathbf{A}_j) + \text{rank}(\mathbf{I} - \mathbf{A}_i - \mathbf{A}_j) \leq N$, implying [since $\text{rank}(\mathbf{A}_i + \mathbf{A}_j) + \text{rank}(\mathbf{I} - \mathbf{A}_i - \mathbf{A}_j) \geq \text{rank}(\mathbf{A}_i + \mathbf{A}_j + \mathbf{I} - \mathbf{A}_i - \mathbf{A}_j) = \text{rank}(\mathbf{I}_N) = N$] that

$$\text{rank}(\mathbf{A}_i + \mathbf{A}_j) + \text{rank}[\mathbf{I} - (\mathbf{A}_i + \mathbf{A}_j)] = N$$

and hence (in light of Lemma 6.8.8) that $\mathbf{A}_i + \mathbf{A}_j$ is idempotent and leading (in light of Lemma 6.8.9) to the conclusion that $\mathbf{A}_i\mathbf{A}_j = \mathbf{0}$. Q.E.D.

When the matrices $\mathbf{A}_1, \mathbf{A}_2, \ldots, \mathbf{A}_K$ are symmetric, Condition (1) of Theorem 6.8.7 reduces to the condition

$$\mathbf{A}_i\mathbf{A}_j = \mathbf{0} \quad (\text{for } j \neq i = 1, 2, \ldots, K).$$

To see this, observe that if $\mathbf{A}_i$ is symmetric, then $\mathbf{A}_i^2 = \mathbf{A}_i'\mathbf{A}_i$, implying (in light of Lemma 2.12.1) that $\text{rank}(\mathbf{A}_i^2) = \text{rank}(\mathbf{A}_i)$ $(i = 1, 2, \ldots, K)$. As what can be regarded as a generalization of the special case of Theorem 6.8.7 where $\mathbf{A}_1, \mathbf{A}_2, \ldots, \mathbf{A}_K$ are symmetric, we have the following result.

Theorem 6.8.10. Let $\mathbf{A}_1, \mathbf{A}_2, \ldots, \mathbf{A}_K$ represent $N \times N$ symmetric matrices, define $\mathbf{A} = \mathbf{A}_1 + \mathbf{A}_2 + \cdots + \mathbf{A}_K$, and take $\boldsymbol{\Sigma}$ to be an $N \times N$ symmetric nonnegative definite matrix. Suppose that $\boldsymbol{\Sigma}\mathbf{A}\boldsymbol{\Sigma}\mathbf{A}\boldsymbol{\Sigma} = \boldsymbol{\Sigma}\mathbf{A}\boldsymbol{\Sigma}$. Then, *each* of the following conditions implies the other two:

(1) $\boldsymbol{\Sigma}\mathbf{A}_i\boldsymbol{\Sigma}\mathbf{A}_j\boldsymbol{\Sigma} = \mathbf{0}$ (for $j \neq i = 1, 2, \ldots, K$);

(2) $\boldsymbol{\Sigma}\mathbf{A}_i\boldsymbol{\Sigma}\mathbf{A}_i\boldsymbol{\Sigma} = \boldsymbol{\Sigma}\mathbf{A}_i\boldsymbol{\Sigma}$ (for $i = 1, 2, \ldots, K$);

(3) $\text{rank}(\boldsymbol{\Sigma}\mathbf{A}_1\boldsymbol{\Sigma}) + \text{rank}(\boldsymbol{\Sigma}\mathbf{A}_2\boldsymbol{\Sigma}) + \cdots + \text{rank}(\boldsymbol{\Sigma}\mathbf{A}_K\boldsymbol{\Sigma}) = \text{rank}(\boldsymbol{\Sigma}\mathbf{A}\boldsymbol{\Sigma})$.

Proof (of Theorem 6.8.10). Take $\boldsymbol{\Gamma}$ to be any matrix (with N columns) such that $\boldsymbol{\Sigma} = \boldsymbol{\Gamma}'\boldsymbol{\Gamma}$. Clearly,

$$\boldsymbol{\Gamma}\mathbf{A}\boldsymbol{\Gamma}' = \boldsymbol{\Gamma}\mathbf{A}_1\boldsymbol{\Gamma}' + \boldsymbol{\Gamma}\mathbf{A}_2\boldsymbol{\Gamma}' + \cdots + \boldsymbol{\Gamma}\mathbf{A}_K\boldsymbol{\Gamma}'.$$

And Corollary 2.3.4 can be used to show that the condition $\boldsymbol{\Sigma}\mathbf{A}\boldsymbol{\Sigma}\mathbf{A}\boldsymbol{\Sigma} = \boldsymbol{\Sigma}\mathbf{A}\boldsymbol{\Sigma}$ is equivalent to the condition that $\boldsymbol{\Gamma}\mathbf{A}\boldsymbol{\Gamma}'$ be idempotent, to show that Condition (1) is equivalent to the condition

$$(\boldsymbol{\Gamma}\mathbf{A}_i\boldsymbol{\Gamma}')(\boldsymbol{\Gamma}\mathbf{A}_j\boldsymbol{\Gamma}') = \mathbf{0} \quad (\text{for } j \neq i = 1, 2, \ldots, K),$$

and to show that Condition (2) is equivalent to the condition

$$(\boldsymbol{\Gamma}\mathbf{A}_i\boldsymbol{\Gamma}')^2 = \boldsymbol{\Gamma}\mathbf{A}_i\boldsymbol{\Gamma}' \quad (\text{for } i = 1, 2, \ldots, K).$$

Moreover, Lemma 2.12.3 can be used to show that Condition (3) is equivalent to the condition

$$\text{rank}(\boldsymbol{\Gamma}\mathbf{A}_1\boldsymbol{\Gamma}') + \text{rank}(\boldsymbol{\Gamma}\mathbf{A}_2\boldsymbol{\Gamma}') + \cdots + \text{rank}(\boldsymbol{\Gamma}\mathbf{A}_K\boldsymbol{\Gamma}') = \text{rank}(\boldsymbol{\Gamma}\mathbf{A}\boldsymbol{\Gamma}').$$

Thus, Theorem 6.8.10 can be established by applying Theorem 6.8.7 to the symmetric matrices $\boldsymbol{\Gamma}\mathbf{A}_1\boldsymbol{\Gamma}', \boldsymbol{\Gamma}\mathbf{A}_2\boldsymbol{\Gamma}', \ldots, \boldsymbol{\Gamma}\mathbf{A}_K\boldsymbol{\Gamma}'$. Q.E.D.

A result related to Theorem 6.8.7 is that if $\mathbf{A}_1, \mathbf{A}_2, \ldots, \mathbf{A}_K$ are square matrices such that $\mathbf{A}_i^2 = \mathbf{A}_i$ and $\mathbf{A}_i\mathbf{A}_j = \mathbf{0}$ for $j \neq i = 1, 2, \ldots, K$, then $\mathbf{A}_1 + \mathbf{A}_2 + \cdots \mathbf{A}_K$ is idempotent. Similarly, a result related to Theorem 6.8.10 is that if $\mathbf{A}_1, \mathbf{A}_2, \ldots, \mathbf{A}_K$ are square matrices such that $\boldsymbol{\Sigma}\mathbf{A}_i\boldsymbol{\Sigma}\mathbf{A}_i\boldsymbol{\Sigma} = \boldsymbol{\Sigma}\mathbf{A}_i\boldsymbol{\Sigma}$ and $\boldsymbol{\Sigma}\mathbf{A}_i\boldsymbol{\Sigma}\mathbf{A}_j\boldsymbol{\Sigma} = \mathbf{0}$ for $j \neq i = 1, 2, \ldots, K$, then

$$\boldsymbol{\Sigma}\big(\textstyle\sum_{i=1}^K \mathbf{A}_i\big)\boldsymbol{\Sigma}\big(\sum_{i=1}^K \mathbf{A}_i\big)\boldsymbol{\Sigma} = \boldsymbol{\Sigma}\big(\sum_{i=1}^K \mathbf{A}_i\big)\boldsymbol{\Sigma}.$$

Proof of Theorem 6.8.6. Let us prove Theorem 6.8.6, doing so by taking advantage of Theorem 6.8.10. Suppose that conditions (8.25), (8.26), (8.27), and (8.28) are satisfied. In light of Theorem 6.8.10, it follows from conditions (8.25) and (8.26) that

$$\boldsymbol{\Sigma}\mathbf{A}_i\boldsymbol{\Sigma}\mathbf{A}_j\boldsymbol{\Sigma} = \mathbf{0} \quad (j \neq i = 1, 2, \ldots, K) \tag{8.29}$$

and that

$$\boldsymbol{\Sigma}\mathbf{A}_i\boldsymbol{\Sigma}\mathbf{A}_i\boldsymbol{\Sigma} = \boldsymbol{\Sigma}\mathbf{A}_i\boldsymbol{\Sigma} \quad (i = 1, 2, \ldots, K). \tag{8.30}$$

Moreover, for $j \neq i = 1, 2, \ldots, K$, condition (8.27) [in combination with result (8.29)] implies that

$$\boldsymbol{\Sigma}\mathbf{A}_i\boldsymbol{\Sigma}(\mathbf{b}_j + 2\mathbf{A}_j\boldsymbol{\mu}) = \boldsymbol{\Sigma}\mathbf{A}_i\boldsymbol{\Sigma}\mathbf{A}_j\boldsymbol{\Sigma}(\mathbf{b}_j + 2\mathbf{A}_j\boldsymbol{\mu}) = \mathbf{0}$$

and that

$$\begin{aligned}(\mathbf{b}_i + 2\mathbf{A}_i\boldsymbol{\mu})'\boldsymbol{\Sigma}(\mathbf{b}_j + 2\mathbf{A}_j\boldsymbol{\mu}) &= [\boldsymbol{\Sigma}(\mathbf{b}_i + 2\mathbf{A}_i\boldsymbol{\mu})]'(\mathbf{b}_j + 2\mathbf{A}_j\boldsymbol{\mu}) \\ &= [\boldsymbol{\Sigma}\mathbf{A}_i\boldsymbol{\Sigma}(\mathbf{b}_i + 2\mathbf{A}_i\boldsymbol{\mu})]'(\mathbf{b}_j + 2\mathbf{A}_j\boldsymbol{\mu}) \\ &= (\mathbf{b}_i + 2\mathbf{A}_i\boldsymbol{\mu})'\boldsymbol{\Sigma}\mathbf{A}_i\boldsymbol{\Sigma}(\mathbf{b}_j + 2\mathbf{A}_j\boldsymbol{\mu}) = 0.\end{aligned}$$

And upon applying Theorems 6.8.1 and 6.6.2, we conclude that $q_1, q_2, \ldots, q_K$ are statistically independent and that (for $i = 1, 2, \ldots, K$) $q_i \sim \chi^2(R_i, c_i + \mathbf{b}_i'\boldsymbol{\mu} + \boldsymbol{\mu}'\mathbf{A}_i\boldsymbol{\mu})$, where $R_i = \text{rank}(\boldsymbol{\Sigma}\mathbf{A}_i\boldsymbol{\Sigma}) = \text{tr}(\mathbf{A}_i\boldsymbol{\Sigma})$.

Conversely, suppose that $q_1, q_2, \ldots, q_K$ are statistically independent and that (for $i = 1, 2, \ldots, K$) $q_i \sim \chi^2(R_i, \lambda_i)$ (where R_i is a strictly positive integer). Then, from Theorem 6.6.2, we have that conditions (8.27) and (8.28) are satisfied, that (for $i = 1, 2, \ldots, K$) $R_i = \text{rank}(\boldsymbol{\Sigma}\mathbf{A}_i\boldsymbol{\Sigma}) = \text{tr}(\mathbf{A}_i\boldsymbol{\Sigma})$ and $\lambda_i = c_i + \mathbf{b}_i'\boldsymbol{\mu} + \boldsymbol{\mu}'\mathbf{A}_i\boldsymbol{\mu}$, and that

$$\boldsymbol{\Sigma}\mathbf{A}_i\boldsymbol{\Sigma}\mathbf{A}_i\boldsymbol{\Sigma} = \boldsymbol{\Sigma}\mathbf{A}_i\boldsymbol{\Sigma} \qquad (i = 1, 2, \ldots, K). \tag{8.31}$$

And from Theorem 6.8.1, we have that

$$\boldsymbol{\Sigma}\mathbf{A}_i\boldsymbol{\Sigma}\mathbf{A}_j\boldsymbol{\Sigma} = \mathbf{0} \qquad (j \neq i = 1, 2, \ldots, K). \tag{8.32}$$

Together, results (8.31) and (8.32) imply that condition (8.25) is satisfied. That condition (8.26) is also satisfied can be inferred from Theorem 6.8.10. Q.E.D.

Corollaries of Theorem 6.8.6. In light of Theorem 6.8.10, alternative versions of Theorem 6.8.6 can be obtained by replacing condition (8.26) with the condition

$$\boldsymbol{\Sigma}\mathbf{A}_i\boldsymbol{\Sigma}\mathbf{A}_j\boldsymbol{\Sigma} = \mathbf{0} \qquad (j \neq i = 1, 2, \ldots, K)$$

or with the condition

$$\boldsymbol{\Sigma}\mathbf{A}_i\boldsymbol{\Sigma}\mathbf{A}_i\boldsymbol{\Sigma} = \boldsymbol{\Sigma}\mathbf{A}_i\boldsymbol{\Sigma} \qquad (i = 1, 2, \ldots, K).$$

In either case, the replacement results in what can be regarded as a corollary of Theorem 6.8.6. The following result can also be regarded as a corollary of Theorem 6.8.6.

Corollary 6.8.11. Let $\mathbf{x}$ represent an M-dimensional random column vector that has an $N(\boldsymbol{\mu}, \mathbf{I}_M)$ distribution. And take $\mathbf{A}_1, \mathbf{A}_2, \ldots, \mathbf{A}_K$ to be $M \times M$ (nonnull) symmetric matrices of constants, and define $\mathbf{A} = \mathbf{A}_1 + \mathbf{A}_2 + \cdots + \mathbf{A}_K$. If $\mathbf{A}$ is idempotent and

$$\operatorname{rank}(\mathbf{A}_1) + \operatorname{rank}(\mathbf{A}_2) + \cdots + \operatorname{rank}(\mathbf{A}_K) = \operatorname{rank}(\mathbf{A}), \tag{8.33}$$

then the quadratic forms $\mathbf{x}'\mathbf{A}_1\mathbf{x}, \mathbf{x}'\mathbf{A}_2\mathbf{x}, \ldots, \mathbf{x}'\mathbf{A}_K\mathbf{x}$ are statistically independent and (for $i = 1, 2, \ldots, K$) $\mathbf{x}'\mathbf{A}_i\mathbf{x} \sim \chi^2(R_i, \boldsymbol{\mu}'\mathbf{A}_i\boldsymbol{\mu})$, where $R_i = \operatorname{rank}(\mathbf{A}_i) = \operatorname{tr}(\mathbf{A}_i)$. Conversely, if $\mathbf{x}'\mathbf{A}_1\mathbf{x}, \mathbf{x}'\mathbf{A}_2\mathbf{x}, \ldots, \mathbf{x}'\mathbf{A}_K\mathbf{x}$ are statistically independent and if (for $i = 1, 2, \ldots, K$) $\mathbf{x}'\mathbf{A}_i\mathbf{x} \sim \chi^2(R_i, \lambda_i)$ (where R_i is a strictly positive integer), then $\mathbf{A}$ is idempotent, condition (8.33) is satisfied, and (for $i = 1, 2, \ldots, K$) $R_i = \operatorname{rank}(\mathbf{A}_i) = \operatorname{tr}(\mathbf{A}_i)$ and $\lambda_i = \boldsymbol{\mu}'\mathbf{A}_i\boldsymbol{\mu}$.

Corollary 6.8.11 can be deduced from Theorem 6.8.6 by making use of Theorem 6.8.7. The special case of Corollary 6.8.11 where $\boldsymbol{\mu} = \mathbf{0}$ and where $\mathbf{A}_1, \mathbf{A}_2, \ldots, \mathbf{A}_K$ are such that $\mathbf{A} = \mathbf{I}$ was formulated and proved by Cochran (1934) and is known as Cochran's theorem. Cochran's theorem is one of the most famous theoretical results in all of statistics. Note that in light of Theorem 6.8.7, alternative versions of Corollary 6.8.11 can be obtained by replacing condition (8.33) with the condition

$$\mathbf{A}_i\mathbf{A}_j = \mathbf{0} \qquad (j \neq i = 1, 2, \ldots, K)$$

or with the condition

$$\mathbf{A}_i^2 = \mathbf{A}_i \qquad (i = 1, 2, \ldots, K).$$

Another result that can be regarded as a corollary of Theorem 6.8.6 is as follows.

Corollary 6.8.12. Let $\mathbf{x}$ represent an M-dimensional random column vector that has an $N(\boldsymbol{\mu}, \boldsymbol{\Sigma})$ distribution. And, for $i = 1, 2, \ldots, K$, take $\mathbf{A}_i$ to be an $M \times M$ symmetric matrix of constants (such that $\mathbf{A}_i\boldsymbol{\Sigma} \neq \mathbf{0}$). Further, define $\mathbf{A} = \mathbf{A}_1 + \mathbf{A}_2 + \cdots + \mathbf{A}_K$. If

$$\mathbf{A}\boldsymbol{\Sigma}\mathbf{A} = \mathbf{A} \tag{8.34}$$

and

$$\mathbf{A}_i\boldsymbol{\Sigma}\mathbf{A}_i = \mathbf{A}_i \qquad (i = 1, 2, \ldots, K), \tag{8.35}$$

then the quadratic forms $\mathbf{x}'\mathbf{A}_1\mathbf{x}, \mathbf{x}'\mathbf{A}_2\mathbf{x}, \ldots, \mathbf{x}'\mathbf{A}_K\mathbf{x}$ are statistically independent and (for $i = 1, 2, \ldots, K$) $\mathbf{x}'\mathbf{A}_i\mathbf{x} \sim \chi^2(R_i, \boldsymbol{\mu}'\mathbf{A}_i\boldsymbol{\mu})$, where $R_i = \operatorname{rank}(\mathbf{A}_i\boldsymbol{\Sigma}) = \operatorname{tr}(\mathbf{A}_i\boldsymbol{\Sigma})$. Conversely, if $\mathbf{x}'\mathbf{A}_1\mathbf{x}, \mathbf{x}'\mathbf{A}_2\mathbf{x}, \ldots, \mathbf{x}'\mathbf{A}_K\mathbf{x}$ are statistically independent and if (for $i = 1, 2, \ldots, K$) $\mathbf{x}'\mathbf{A}_i\mathbf{x} \sim \chi^2(R_i, \lambda_i)$ (where R_i is a strictly positive integer), then in the special case where $\boldsymbol{\Sigma}$ is nonsingular, conditions (8.34) and (8.35) are satisfied and (for $i = 1, 2, \ldots, K$) $R_i = \operatorname{rank}(\mathbf{A}_i\boldsymbol{\Sigma}) = \operatorname{tr}(\mathbf{A}_i\boldsymbol{\Sigma})$ and $\lambda_i = \boldsymbol{\mu}'\mathbf{A}_i\boldsymbol{\mu}$.

Proof. Condition (8.34) implies condition (8.25), and in the special case where $\boldsymbol{\Sigma}$ is nonsingular, conditions (8.34) and (8.25) are equivalent. And as noted earlier, condition (8.26) of Theorem 6.8.6 can be replaced by the condition

$$\boldsymbol{\Sigma}\mathbf{A}_i\boldsymbol{\Sigma}\mathbf{A}_i\boldsymbol{\Sigma} = \boldsymbol{\Sigma}\mathbf{A}_i\boldsymbol{\Sigma} \qquad (i = 1, 2, \ldots, K). \tag{8.36}$$

Moreover, condition (8.35) implies condition (8.36) and that (for $i = 1, 2, \ldots, K$) $\boldsymbol{\Sigma}\mathbf{A}_i\boldsymbol{\mu} = \boldsymbol{\Sigma}\mathbf{A}_i\boldsymbol{\Sigma}\mathbf{A}_i\boldsymbol{\mu}$ and $\boldsymbol{\mu}'\mathbf{A}_i\boldsymbol{\mu} = \boldsymbol{\mu}'\mathbf{A}_i\boldsymbol{\Sigma}\mathbf{A}_i\boldsymbol{\mu}$; in the special case where $\boldsymbol{\Sigma}$ is nonsingular, conditions (8.35) and (8.36) are equivalent. Condition (8.35) also implies that (for $i = 1, 2, \ldots, K$) $\mathbf{A}_i\boldsymbol{\Sigma}$ is idempotent and hence that $\operatorname{rank}(\mathbf{A}_i\boldsymbol{\Sigma}) = \operatorname{tr}(\mathbf{A}_i\boldsymbol{\Sigma})$. Accordingly, Corollary 6.8.12 is obtainable as an "application" of Theorem 6.8.6. Q.E.D.

If in Corollary 6.8.12, we substitute $(1/\gamma_i)\mathbf{A}_i$ for $\mathbf{A}_i$, where γ_i is a nonzero scalar, we obtain an additional corollary as follows.

Corollary 6.8.13. Let $\mathbf{x}$ represent an M-dimensional random column vector that has an $N(\boldsymbol{\mu}, \boldsymbol{\Sigma})$ distribution. And, for $i = 1, 2, \ldots, K$, take $\mathbf{A}_i$ to be an $M \times M$ symmetric matrix of constants (such that $\mathbf{A}_i\boldsymbol{\Sigma} \neq \mathbf{0}$), and take γ_i to be a nonzero constant. Further, define $\mathbf{B} = (1/\gamma_1)\mathbf{A}_1 + (1/\gamma_2)\mathbf{A}_2 + \cdots + (1/\gamma_K)\mathbf{A}_K$. If

$$\mathbf{B}\boldsymbol{\Sigma}\mathbf{B} = \mathbf{B} \tag{8.37}$$

and

$$\mathbf{A}_i \boldsymbol{\Sigma} \mathbf{A}_i = \gamma_i \mathbf{A}_i \qquad (i = 1, 2, \ldots, K), \tag{8.38}$$

then the quadratic forms $\mathbf{x}'\mathbf{A}_1\mathbf{x}, \mathbf{x}'\mathbf{A}_2\mathbf{x}, \ldots, \mathbf{x}'\mathbf{A}_K\mathbf{x}$ are statistically independent and (for $i = 1, 2, \ldots, K$) $\mathbf{x}'\mathbf{A}_i\mathbf{x}/\gamma_i \sim \chi^2(R_i, \boldsymbol{\mu}'\mathbf{A}_i\boldsymbol{\mu}/\gamma_i)$, where $R_i = \text{rank}(\mathbf{A}_i\boldsymbol{\Sigma}) = (1/\gamma_i)\,\text{tr}(\mathbf{A}_i\boldsymbol{\Sigma})$. Conversely, if $\mathbf{x}'\mathbf{A}_1\mathbf{x}, \mathbf{x}'\mathbf{A}_2\mathbf{x}, \ldots, \mathbf{x}'\mathbf{A}_K\mathbf{x}$ are statistically independent and if (for $i = 1, 2, \ldots, K$) $\mathbf{x}'\mathbf{A}_i\mathbf{x}/\gamma_i \sim \chi^2(R_i, \lambda_i)$ (where R_i is a strictly positive integer), then in the special case where $\boldsymbol{\Sigma}$ is nonsingular, conditions (8.37) and (8.38) are satisfied and (for $i = 1, 2, \ldots, K$) $R_i = \text{rank}(\mathbf{A}_i\boldsymbol{\Sigma}) = (1/\gamma_i)\,\text{tr}(\mathbf{A}_i\boldsymbol{\Sigma})$ and $\lambda_i = \boldsymbol{\mu}'\mathbf{A}_i\boldsymbol{\mu}/\gamma_i$.

As yet another corollary of Theorem 6.8.6, we have the following generalization of a result attributable to Albert (1976).

Corollary 6.8.14. Let $\mathbf{x}$ represent an M-dimensional random column vector that has an $N(\boldsymbol{\mu}, \boldsymbol{\Sigma})$ distribution. And, for $i = 1, 2, \ldots, K$, take $\mathbf{A}_i$ to be an $M \times M$ (nonnull) symmetric matrix of constants, and take γ_i to be a nonzero constant. Further, define $\mathbf{A} = \mathbf{A}_1 + \mathbf{A}_2 + \cdots \mathbf{A}_K$, and suppose that $\mathbf{A}$ is idempotent. If

$$\mathbf{A}_i^2 = \mathbf{A}_i \qquad (i = 1, 2, \ldots, K) \tag{8.39}$$

and

$$\mathbf{A}\boldsymbol{\Sigma}\mathbf{A}_i = \gamma_i \mathbf{A}_i \qquad (i = 1, 2, \ldots, K), \tag{8.40}$$

then the quadratic forms $\mathbf{x}'\mathbf{A}_1\mathbf{x}, \mathbf{x}'\mathbf{A}_2\mathbf{x}, \ldots, \mathbf{x}'\mathbf{A}_K\mathbf{x}$ are statistically independent and (for $i = 1, 2, \ldots, K$) $\gamma_i = \text{tr}(\mathbf{A}_i\boldsymbol{\Sigma})/\,\text{tr}(\mathbf{A}_i) > 0$ and $\mathbf{x}'\mathbf{A}_i\mathbf{x}/\gamma_i \sim \chi^2(R_i, \boldsymbol{\mu}'\mathbf{A}_i\boldsymbol{\mu}/\gamma_i)$, where $R_i = \text{rank}\,\mathbf{A}_i = \text{tr}(\mathbf{A}_i)$. Conversely, if $\mathbf{x}'\mathbf{A}_1\mathbf{x}, \mathbf{x}'\mathbf{A}_2\mathbf{x}, \ldots, \mathbf{x}'\mathbf{A}_K\mathbf{x}$ are statistically independent and if (for $i = 1, 2, \ldots, K$) $\mathbf{x}'\mathbf{A}_i\mathbf{x}/\gamma_i \sim \chi^2(R_i, \lambda_i)$ (where R_i is a strictly positive integer), then in the special case where $\boldsymbol{\Sigma}$ is nonsingular, conditions (8.39) and (8.40) are satisfied and (for $i = 1, 2, \ldots, K$) $\gamma_i = \text{tr}(\mathbf{A}_i\boldsymbol{\Sigma})/\,\text{tr}(\mathbf{A}_i) > 0$, $R_i = \text{rank}\,\mathbf{A}_i = \text{tr}(\mathbf{A}_i)$, and $\lambda_i = \boldsymbol{\mu}'\mathbf{A}_i\boldsymbol{\mu}/\gamma_i$.

Proof (of Corollary 6.8.14). Define $\mathbf{B} = (1/\gamma_1)\mathbf{A}_1 + (1/\gamma_2)\mathbf{A}_2 + \cdots + (1/\gamma_K)\mathbf{A}_K$.

Suppose that conditions (8.39) and (8.40) are satisfied. Then, for $j \neq i = 1, 2, \ldots, K$, $\mathbf{A}_i\mathbf{A}_j = \mathbf{0}$ (as is evident from Theorem 6.8.7). Thus, for $i = 1, 2, \ldots, K$,

$$\mathbf{A}_i\boldsymbol{\Sigma}\mathbf{A}_i = \mathbf{A}_i\mathbf{A}\boldsymbol{\Sigma}\mathbf{A}_i = \gamma_i\mathbf{A}_i^2 = \gamma_i\mathbf{A}_i, \tag{8.41}$$

implying in particular that

$$\gamma_i\,\text{tr}(\mathbf{A}_i) = \text{tr}(\mathbf{A}_i\boldsymbol{\Sigma}\mathbf{A}_i) = \text{tr}(\mathbf{A}_i^2\boldsymbol{\Sigma}) = \text{tr}(\mathbf{A}_i\boldsymbol{\Sigma})$$

and hence [since $\text{tr}(\mathbf{A}_i) = \text{tr}(\mathbf{A}_i^2) = \text{tr}(\mathbf{A}_i'\mathbf{A}_i) > 0$ and since $\mathbf{A}_i\boldsymbol{\Sigma}\mathbf{A}_i = \mathbf{A}_i'\boldsymbol{\Sigma}\mathbf{A}_i$ is a symmetric nonnegative definite matrix] that

$$\gamma_i = \text{tr}(\mathbf{A}_i\boldsymbol{\Sigma})/\,\text{tr}(\mathbf{A}_i) > 0. \tag{8.42}$$

Moreover, for $j \neq i = 1, 2, \ldots, K$,

$$\mathbf{A}_i\boldsymbol{\Sigma}\mathbf{A}_j = \mathbf{A}_i\mathbf{A}\boldsymbol{\Sigma}\mathbf{A}_j = \gamma_j\mathbf{A}_i\mathbf{A}_j = \mathbf{0},$$

so that

$$\mathbf{B}\boldsymbol{\Sigma}\mathbf{B} = \mathbf{B}.$$

And upon observing [in light of result (8.41) or (8.42)] that (for $i = 1, 2, \ldots, K$) $\mathbf{A}_i\boldsymbol{\Sigma} \neq \mathbf{0}$, it follows from Corollary 6.8.13 that $\mathbf{x}'\mathbf{A}_1\mathbf{x}, \mathbf{x}'\mathbf{A}_2\mathbf{x}, \ldots, \mathbf{x}'\mathbf{A}_K\mathbf{x}$ are statistically independent and that (for $i = 1, 2, \ldots, K$) $\mathbf{x}'\mathbf{A}_i\mathbf{x}/\gamma_i \sim \chi^2(R_i, \boldsymbol{\mu}'\mathbf{A}_i\boldsymbol{\mu}/\gamma_i)$, where $R_i = (1/\gamma_i)\,\text{tr}(\mathbf{A}_i\boldsymbol{\Sigma}) = \text{tr}(\mathbf{A}_i) = \text{rank}\,\mathbf{A}_i$.

Conversely, suppose that $\mathbf{x}'\mathbf{A}_1\mathbf{x}, \mathbf{x}'\mathbf{A}_2\mathbf{x}, \ldots, \mathbf{x}'\mathbf{A}_K\mathbf{x}$ are statistically independent and that (for $i = 1, 2, \ldots, K$) $\mathbf{x}'\mathbf{A}_i\mathbf{x}/\gamma_i \sim \chi^2(R_i, \lambda_i)$ (where R_i is a strictly positive integer). And suppose that $\boldsymbol{\Sigma}$ is nonsingular (in which case $\mathbf{A}_i\boldsymbol{\Sigma} \neq \mathbf{0}$ for $i = 1, 2, \ldots, K$). Then, from Corollary 6.8.13, we have that $\mathbf{B}\boldsymbol{\Sigma}\mathbf{B} = \mathbf{B}$ and (for $i = 1, 2, \ldots, K$) that

$$\mathbf{A}_i\boldsymbol{\Sigma}\mathbf{A}_i = \gamma_i\mathbf{A}_i$$

and also that $R_i = (1/\gamma_i)\,\text{tr}(\mathbf{A}_i\boldsymbol{\Sigma})$ and $\lambda_i = \boldsymbol{\mu}'\mathbf{A}_i\boldsymbol{\mu}/\gamma_i$. Further, $\boldsymbol{\Sigma}\mathbf{B}\boldsymbol{\Sigma}\mathbf{B}\boldsymbol{\Sigma} = \boldsymbol{\Sigma}\mathbf{B}\boldsymbol{\Sigma}$ and (for $i = 1, 2, \ldots, K$) $\boldsymbol{\Sigma}(\gamma_i^{-1}\mathbf{A}_i)\boldsymbol{\Sigma}(\gamma_i^{-1}\mathbf{A}_i)\boldsymbol{\Sigma} = \boldsymbol{\Sigma}(\gamma_i^{-1}\mathbf{A}_i)\boldsymbol{\Sigma}$, so that (in light of Theorem 6.8.10) we have that

$$(\gamma_i\gamma_j)^{-1}\mathbf{\Sigma}\mathbf{A}_i\mathbf{\Sigma}\mathbf{A}_j\mathbf{\Sigma} = \mathbf{\Sigma}(\gamma_i^{-1}\mathbf{A}_i)\mathbf{\Sigma}(\gamma_j^{-1}\mathbf{A}_j)\mathbf{\Sigma} = \mathbf{0}$$

or, equivalently,

$$\mathbf{A}_i\mathbf{\Sigma}\mathbf{A}_j = \mathbf{0}.$$

Thus, for $i = 1, 2, \ldots, K$, we find that

$$\mathbf{A}\mathbf{\Sigma}\mathbf{A}_i = \mathbf{A}_i\mathbf{\Sigma}\mathbf{A}_i = \gamma_i\mathbf{A}_i$$

and (since $\mathbf{A}$ is idempotent) that

$$\mathbf{A}_i^2 = \mathbf{A}_i'\mathbf{A}_i = \gamma_i^{-2}\mathbf{A}_i\mathbf{\Sigma}\mathbf{A}^2\mathbf{\Sigma}\mathbf{A}_i = \gamma_i^{-2}\mathbf{A}_i\mathbf{\Sigma}\mathbf{A}\mathbf{\Sigma}\mathbf{A}_i = \gamma_i^{-1}\mathbf{A}_i\mathbf{\Sigma}\mathbf{A}_i = \mathbf{A}_i,$$

leading to the conclusion that conditions (8.39) and (8.40) are satisfied and by implication—via the same argument that gave rise to result (8.42)—that (for $i = 1, 2, \ldots, K$) $\gamma_i = \operatorname{tr}(\mathbf{A}_i\mathbf{\Sigma})/\operatorname{tr}(\mathbf{A}_i) > 0$ [in which case $R_i = (1/\gamma_i)\operatorname{tr}(\mathbf{A}_i\mathbf{\Sigma}) = \operatorname{tr}(\mathbf{A}_i) = \operatorname{rank}\mathbf{A}_i$]. Q.E.D.

c. Some connections to the Dirichlet distribution

As a variation on Cochran's theorem—recall that in the special case where $\boldsymbol{\mu} = \mathbf{0}$ and $\mathbf{A} = \mathbf{I}$, Corollary 6.8.11 is Cochran's theorem—we have the following result.

Theorem 6.8.15. Let $\mathbf{z}$ represent an M-dimensional random column vector that has an $N(\mathbf{0}, \mathbf{I}_M)$ distribution, where $M > 1$. And take $\mathbf{A}_1, \mathbf{A}_2, \ldots, \mathbf{A}_K$ to be $M \times M$ (nonnull) symmetric matrices of constants, and define $\mathbf{A} = \mathbf{A}_1 + \mathbf{A}_2 + \cdots + \mathbf{A}_K$. If $\mathbf{A}$ is idempotent and

$$\operatorname{rank}(\mathbf{A}_1) + \operatorname{rank}(\mathbf{A}_2) + \cdots + \operatorname{rank}(\mathbf{A}_K) = \operatorname{rank}(\mathbf{A}) < M, \tag{8.43}$$

then $\mathbf{z}'\mathbf{A}_1\mathbf{z}/\mathbf{z}'\mathbf{z}, \mathbf{z}'\mathbf{A}_2\mathbf{z}/\mathbf{z}'\mathbf{z}, \ldots, \mathbf{z}'\mathbf{A}_K\mathbf{z}/\mathbf{z}'\mathbf{z}$ have a $Di\big[R_1/2,\ R_2/2, \ldots,\ R_K/2, \big(M - \sum_{i=1}^K R_i\big)/2;\ K\big]$ distribution, where (for $i = 1, 2, \ldots, K$) $R_i = \operatorname{rank}(\mathbf{A}_i) = \operatorname{tr}(\mathbf{A}_i)$. Conversely, if $\mathbf{z}'\mathbf{A}_1\mathbf{z}/\mathbf{z}'\mathbf{z}, \mathbf{z}'\mathbf{A}_2\mathbf{z}/\mathbf{z}'\mathbf{z}, \ldots, \mathbf{z}'\mathbf{A}_K\mathbf{z}/\mathbf{z}'\mathbf{z}$ have a $Di\big[R_1/2,\ R_2/2,\ \ldots,\ R_K/2,\ \big(M - \sum_{i=1}^K R_i\big)/2;\ K\big]$ distribution (where $R_1, R_2, \ldots, R_K$ are strictly positive integers such that $\sum_{i=1}^K R_i < M$), then $\mathbf{A}$ is idempotent, condition (8.43) is satisfied, and (for $i = 1, 2, \ldots, K$) $R_i = \operatorname{rank}(\mathbf{A}_i) = \operatorname{tr}(\mathbf{A}_i)$.

Proof. Suppose that $\mathbf{A}$ is idempotent and that condition (8.43) is satisfied. Further, let $\mathbf{A}_{K+1} = \mathbf{I} - \mathbf{A}$, so that $\sum_{i=1}^{K+1}\mathbf{A}_i = \mathbf{I}$. And observe (in light of Lemma 2.8.4) that $\operatorname{rank}(\mathbf{A}_{K+1}) = M - \operatorname{rank}(\mathbf{A})$ and hence that $\mathbf{A}_{K+1}$ is nonnull and that $\sum_{i=1}^{K+1}\operatorname{rank}(\mathbf{A}_i) = M$. Then, as an application of Corollary 6.8.11 (one where $\boldsymbol{\mu} = \mathbf{0}$ and $\mathbf{A} = \mathbf{I}$), we have that the $K+1$ quadratic forms $\mathbf{z}'\mathbf{A}_1\mathbf{z}, \mathbf{z}'\mathbf{A}_2\mathbf{z}, \ldots, \mathbf{z}'\mathbf{A}_K\mathbf{z}, \mathbf{z}'\mathbf{A}_{K+1}\mathbf{z}$ are statistically independent and that (for $i = 1, 2, \ldots, K, K+1$) $\mathbf{z}'\mathbf{A}_i\mathbf{z} \sim \chi^2(R_i)$, where $R_i = \operatorname{rank}(\mathbf{A}_i) = \operatorname{tr}(\mathbf{A}_i)$. Clearly, $\mathbf{z}'\mathbf{z} = \sum_{i=1}^{K+1}\mathbf{z}'\mathbf{A}_i\mathbf{z}$, and $R_{K+1} = M - \sum_{i=1}^K R_i$. Accordingly, we conclude that $\mathbf{z}'\mathbf{A}_1\mathbf{z}/\mathbf{z}'\mathbf{z}, \mathbf{z}'\mathbf{A}_2\mathbf{z}/\mathbf{z}'\mathbf{z}, \ldots, \mathbf{z}'\mathbf{A}_K\mathbf{z}/\mathbf{z}'\mathbf{z}$ have a $Di\big[R_1/2,\ R_2/2, \ldots,\ R_K/2, \big(M - \sum_{i=1}^K R_i\big)/2;\ K\big]$ distribution.

Conversely, suppose that $\mathbf{z}'\mathbf{A}_1\mathbf{z}/\mathbf{z}'\mathbf{z}, \mathbf{z}'\mathbf{A}_2\mathbf{z}/\mathbf{z}'\mathbf{z}, \ldots, \mathbf{z}'\mathbf{A}_K\mathbf{z}/\mathbf{z}'\mathbf{z}$ have a $Di\big[R_1/2,\ R_2/2,\ \ldots,\ R_K/2,\ \big(M - \sum_{i=1}^K R_i\big)/2;\ K\big]$ distribution (where $R_1, R_2, \ldots, R_K$ are strictly positive integers such that $\sum_{i=1}^K R_i < M$). And partition $\mathbf{z}$ into subvectors $\mathbf{z}_1, \mathbf{z}_2, \ldots, \mathbf{z}_K, \mathbf{z}_{K+1}$ of dimensions $R_1, R_2, \ldots, R_K, M - \sum_{i=1}^K R_i$, respectively, so that $\mathbf{z}' = (\mathbf{z}_1', \mathbf{z}_2', \ldots, \mathbf{z}_K', \mathbf{z}_{K+1}')$. Then, the (joint) distribution of $\mathbf{z}'\mathbf{A}_1\mathbf{z}/\mathbf{z}'\mathbf{z}, \mathbf{z}'\mathbf{A}_2\mathbf{z}/\mathbf{z}'\mathbf{z}, \ldots, \mathbf{z}'\mathbf{A}_K\mathbf{z}/\mathbf{z}'\mathbf{z}$ is identical to that of the quantities $\mathbf{z}_1'\mathbf{z}_1/\mathbf{z}'\mathbf{z}$, $\mathbf{z}_2'\mathbf{z}_2/\mathbf{z}'\mathbf{z}, \ldots, \mathbf{z}_K'\mathbf{z}_K/\mathbf{z}'\mathbf{z}$. Moreover, the quantities $\mathbf{z}'\mathbf{A}_1\mathbf{z}/\mathbf{z}'\mathbf{z}, \mathbf{z}'\mathbf{A}_2\mathbf{z}/\mathbf{z}'\mathbf{z}, \ldots, \mathbf{z}'\mathbf{A}_K\mathbf{z}/\mathbf{z}'\mathbf{z}$ and the quantities $\mathbf{z}_1'\mathbf{z}_1/\mathbf{z}'\mathbf{z}, \mathbf{z}_2'\mathbf{z}_2/\mathbf{z}'\mathbf{z}, \ldots, \mathbf{z}_K'\mathbf{z}_K/\mathbf{z}'\mathbf{z}$ are both distributed independently of $\mathbf{z}'\mathbf{z}$, as is evident from the results of Section 6.1f upon observing that (for $i = 1, 2, \ldots, K$) $\mathbf{z}'\mathbf{A}_i\mathbf{z}/\mathbf{z}'\mathbf{z}$ and $\mathbf{z}_i'\mathbf{z}_i/\mathbf{z}'\mathbf{z}$ depend on the value of $\mathbf{z}$ only through $(\mathbf{z}'\mathbf{z})^{-1/2}\mathbf{z}$. Thus,

$$\begin{pmatrix} \mathbf{z}'\mathbf{A}_1\mathbf{z} \\ \mathbf{z}'\mathbf{A}_2\mathbf{z} \\ \vdots \\ \mathbf{z}'\mathbf{A}_K\mathbf{z} \end{pmatrix} = \mathbf{z}'\mathbf{z}\begin{pmatrix} \mathbf{z}'\mathbf{A}_1\mathbf{z}/\mathbf{z}'\mathbf{z} \\ \mathbf{z}'\mathbf{A}_2\mathbf{z}/\mathbf{z}'\mathbf{z} \\ \vdots \\ \mathbf{z}'\mathbf{A}_K\mathbf{z}/\mathbf{z}'\mathbf{z} \end{pmatrix} \sim \mathbf{z}'\mathbf{z}\begin{pmatrix} \mathbf{z}_1'\mathbf{z}_1/\mathbf{z}'\mathbf{z} \\ \mathbf{z}_2'\mathbf{z}_2/\mathbf{z}'\mathbf{z} \\ \vdots \\ \mathbf{z}_K'\mathbf{z}_K/\mathbf{z}'\mathbf{z} \end{pmatrix} = \begin{pmatrix} \mathbf{z}_1'\mathbf{z}_1 \\ \mathbf{z}_2'\mathbf{z}_2 \\ \vdots \\ \mathbf{z}_K'\mathbf{z}_K \end{pmatrix},$$

which implies that the quadratic forms $\mathbf{z}'\mathbf{A}_1\mathbf{z}, \mathbf{z}'\mathbf{A}_2\mathbf{z}, \ldots, \mathbf{z}'\mathbf{A}_K\mathbf{z}$ are statistically independent and that (for $i = 1, 2, \ldots, K$) $\mathbf{z}'\mathbf{A}_i\mathbf{z} \sim \chi^2(R_i)$. Upon applying Corollary 6.8.11, it follows that $\mathbf{A}$ is idempotent, that $\sum_{i=1}^K \text{rank}(\mathbf{A}_i) = \text{rank}(\mathbf{A})$, and that (for $i = 1, 2, \ldots, K$) $R_i = \text{rank}(\mathbf{A}_i) = \text{tr}(\mathbf{A}_i)$ (in which case $\sum_{i=1}^K \text{rank}(\mathbf{A}_i) = \sum_{i=1}^K R_i < M$). Q.E.D.

The result of Theorem 6.8.15 can be generalized in much the same fashion as Theorem 6.6.7. Take $\mathbf{x}$ to be an M-dimensional random column vector that is distributed as $N(\mathbf{0}, \mathbf{\Sigma})$, let $P = \text{rank}\,\mathbf{\Sigma}$, suppose that $P > 1$, and take $\mathbf{A}_1, \mathbf{A}_2, \ldots, \mathbf{A}_K$ to be $M \times M$ symmetric matrices of constants (such that $\mathbf{\Sigma A}_1\mathbf{\Sigma}, \mathbf{\Sigma A}_2\mathbf{\Sigma}, \ldots, \mathbf{\Sigma A}_K\mathbf{\Sigma}$ are nonnull). Further, take $\mathbf{\Gamma}$ to be a matrix of dimensions $P \times M$ such that $\mathbf{\Sigma} = \mathbf{\Gamma}'\mathbf{\Gamma}$ (the existence of which follows from Corollary 2.13.23), take $\mathbf{z}$ to be a P-dimensional random column vector that has an $N(\mathbf{0}, \mathbf{I}_P)$ distribution, and define $\mathbf{A} = \mathbf{A}_1 + \mathbf{A}_2 + \cdots \mathbf{A}_K$. And observe that $\mathbf{x} \sim \mathbf{\Gamma}'\mathbf{z}$, that (for $i = 1, 2, \ldots, K$) $(\mathbf{\Gamma}'\mathbf{z})'\mathbf{A}_i\mathbf{\Gamma}'\mathbf{z} = \mathbf{z}'\mathbf{\Gamma A}_i\mathbf{\Gamma}'\mathbf{z}$, and (in light of the discourse in Section 6.6d) that $(\mathbf{\Gamma}'\mathbf{z})'\mathbf{\Sigma}^-\mathbf{\Gamma}'\mathbf{z} = \mathbf{z}'\mathbf{z}$. Observe also that $\mathbf{\Gamma A}_1\mathbf{\Gamma}', \mathbf{\Gamma A}_2\mathbf{\Gamma}', \ldots, \mathbf{\Gamma A}_K\mathbf{\Gamma}'$ are nonnull (as can be readily verified by making use of Corollary 2.3.4).

Now, suppose that $\mathbf{\Gamma A \Gamma}'$ is idempotent and that

$$\text{rank}(\mathbf{\Gamma A}_1\mathbf{\Gamma}') + \text{rank}(\mathbf{\Gamma A}_2\mathbf{\Gamma}') + \cdots + \text{rank}(\mathbf{\Gamma A}_K\mathbf{\Gamma}') = \text{rank}(\mathbf{\Gamma A \Gamma}') < P. \tag{8.44}$$

Then, upon applying Theorem 6.8.15 (with $\mathbf{\Gamma A}_1\mathbf{\Gamma}', \mathbf{\Gamma A}_2\mathbf{\Gamma}', \ldots, \mathbf{\Gamma A}_K\mathbf{\Gamma}'$ in place of $\mathbf{A}_1, \mathbf{A}_2, \ldots, \mathbf{A}_K$), we find that $\mathbf{z}'\mathbf{\Gamma A}_1\mathbf{\Gamma}'\mathbf{z}/\mathbf{z}'\mathbf{z}, \mathbf{z}'\mathbf{\Gamma A}_2\mathbf{\Gamma}'\mathbf{z}/\mathbf{z}'\mathbf{z}, \ldots, \mathbf{z}'\mathbf{\Gamma A}_K\mathbf{\Gamma}'\mathbf{z}/\mathbf{z}'\mathbf{z}$ have a $Di\big[R_1/2, R_2/2, \ldots, R_K/2, \big(P - \sum_{i=1}^K R_i\big)/2; K\big]$ distribution, where (for $i = 1, 2, \ldots, K$) $R_i = \text{rank}(\mathbf{\Gamma A}_i\mathbf{\Gamma}') = \text{tr}(\mathbf{\Gamma A}_i\mathbf{\Gamma}')$. And, conversely, if $\mathbf{z}'\mathbf{\Gamma A}_1\mathbf{\Gamma}'\mathbf{z}/\mathbf{z}'\mathbf{z}, \mathbf{z}'\mathbf{\Gamma A}_2\mathbf{\Gamma}'\mathbf{z}/\mathbf{z}'\mathbf{z}, \ldots, \mathbf{z}'\mathbf{\Gamma A}_K\mathbf{\Gamma}'\mathbf{z}/\mathbf{z}'\mathbf{z}$ have a $Di\big[R_1/2, R_2/2, \ldots, R_K/2, \big(P - \sum_{i=1}^K R_i\big)/2; K\big]$ distribution (where $R_1, R_2, \ldots, R_K$ are strictly positive integers such that $\sum_{i=1}^K R_i < P$), then $\mathbf{\Gamma A \Gamma}'$ is idempotent, condition (8.44) is satisfied, and (for $i = 1, 2, \ldots, K$) $R_i = \text{rank}(\mathbf{\Gamma A}_i\mathbf{\Gamma}') = \text{tr}(\mathbf{\Gamma A}_i\mathbf{\Gamma}')$.

Clearly, $\mathbf{x}'\mathbf{A}_1\mathbf{x}/\mathbf{x}'\mathbf{\Sigma}^-\mathbf{x}, \mathbf{x}'\mathbf{A}_2\mathbf{x}/\mathbf{x}'\mathbf{\Sigma}^-\mathbf{x}, \ldots, \mathbf{x}'\mathbf{A}_K\mathbf{x}/\mathbf{x}'\mathbf{\Sigma}^-\mathbf{x}$ have the same distribution as $\mathbf{z}'\mathbf{\Gamma A}_1\mathbf{\Gamma}'\mathbf{z}/\mathbf{z}'\mathbf{z}, \mathbf{z}'\mathbf{\Gamma A}_2\mathbf{\Gamma}'\mathbf{z}/\mathbf{z}'\mathbf{z}, \ldots, \mathbf{z}'\mathbf{\Gamma A}_K\mathbf{\Gamma}'\mathbf{z}/\mathbf{z}'\mathbf{z}$. And by employing the same line of reasoning as in the proof of Theorem 6.6.2, it can be shown that

$$\mathbf{\Gamma A \Gamma}'\mathbf{\Gamma A \Gamma}' = \mathbf{\Gamma A \Gamma}' \quad \Leftrightarrow \quad \mathbf{\Sigma A \Sigma A \Sigma} = \mathbf{\Sigma A \Sigma},$$

that $\text{rank}(\mathbf{\Gamma A \Gamma}') = \text{rank}(\mathbf{\Sigma A \Sigma})$, that

$$\text{rank}(\mathbf{\Gamma A}_i\mathbf{\Gamma}') = \text{rank}(\mathbf{\Sigma A}_i\mathbf{\Sigma}) \qquad (i = 1, 2, \ldots, K),$$

and that

$$\text{tr}(\mathbf{\Gamma A}_i\mathbf{\Gamma}') = \text{tr}(\mathbf{A}_i\mathbf{\Sigma}) \qquad (i = 1, 2, \ldots, K).$$

Thus, we have the following generalization of Theorem 6.8.15.

Theorem 6.8.16. Let $\mathbf{x}$ represent an M-dimensional random column vector that has an $N(\mathbf{0}, \mathbf{\Sigma})$ distribution, let $P = \text{rank}\,\mathbf{\Sigma}$, suppose that $P > 1$, take $\mathbf{A}_1, \mathbf{A}_2, \ldots, \mathbf{A}_K$ to be $M \times M$ symmetric matrices of constants (such that $\mathbf{\Sigma A}_1\mathbf{\Sigma}, \mathbf{\Sigma A}_2\mathbf{\Sigma}, \ldots, \mathbf{\Sigma A}_K\mathbf{\Sigma}$ are nonnull), and define $\mathbf{A} = \mathbf{A}_1 + \mathbf{A}_2 + \ldots + \mathbf{A}_K$. If $\mathbf{\Sigma A \Sigma A \Sigma} = \mathbf{\Sigma A \Sigma}$ and

$$\text{rank}(\mathbf{\Sigma A}_1\mathbf{\Sigma}) + \text{rank}(\mathbf{\Sigma A}_2\mathbf{\Sigma}) + \cdots + \text{rank}(\mathbf{\Sigma A}_K\mathbf{\Sigma}) = \text{rank}(\mathbf{\Sigma A \Sigma}) < P, \tag{8.45}$$

then $\mathbf{x}'\mathbf{A}_1\mathbf{x}/\mathbf{x}'\mathbf{\Sigma}^-\mathbf{x}, \mathbf{x}'\mathbf{A}_2\mathbf{x}/\mathbf{x}'\mathbf{\Sigma}^-\mathbf{x}, \ldots, \mathbf{x}'\mathbf{A}_K\mathbf{x}/\mathbf{x}'\mathbf{\Sigma}^-\mathbf{x}$ have a $Di\big[R_1/2, R_2/2, \ldots, R_K/2, \big(P - \sum_{i=1}^K R_i\big)/2; K\big]$ distribution, where (for $i = 1, 2, \ldots, K$) $R_i = \text{rank}(\mathbf{\Sigma A}_i\mathbf{\Sigma}) = \text{tr}(\mathbf{A}_i\mathbf{\Sigma})$. Conversely, if $\mathbf{x}'\mathbf{A}_1\mathbf{x}/\mathbf{x}'\mathbf{\Sigma}^-\mathbf{x}, \mathbf{x}'\mathbf{A}_2\mathbf{x}/\mathbf{x}'\mathbf{\Sigma}^-\mathbf{x}, \ldots, \mathbf{x}'\mathbf{A}_K\mathbf{x}/\mathbf{x}'\mathbf{\Sigma}^-\mathbf{x}$ have a $Di\big[R_1/2, R_2/2, \ldots, R_K/2, \big(P - \sum_{i=1}^K R_i\big)/2; K\big]$ distribution (where $R_1, R_2, \ldots, R_K$ are strictly positive integers such that $\sum_{i=1}^K R_i < P$), then $\mathbf{\Sigma A \Sigma A \Sigma} = \mathbf{\Sigma A \Sigma}$, condition (8.45) is satisfied, and (for $i = 1, 2, \ldots, K$) $R_i = \text{rank}(\mathbf{\Sigma A}_i\mathbf{\Sigma}) = \text{tr}(\mathbf{A}_i\mathbf{\Sigma})$.

The validity of the results of Theorem 6.8.15 is not limited to the case where the distribution of the M-dimensional random column vector $\mathbf{z}$ is $N(\mathbf{0}, \mathbf{I}_M)$. Similarly, the validity of the results of Theorem 6.8.16 is not limited to the case where the distribution of the M-dimensional random column vector $\mathbf{x}$ is $N(\mathbf{0}, \mathbf{\Sigma})$. The validity of these results can be extended to a broader class of distributions by employing an approach analogous to that described in Section 6.6e for extending the results of Theorems 6.6.7 and 6.6.8.

Specifically, the validity of the results of Theorem 6.8.15 extends to the case where the distribution of the M-dimensional random column vector $\mathbf{z}$ is an arbitrary absolutely continuous spherical distribution. Similarly, the validity of the results of Theorem 6.8.16 extends to the case where the distribution of the M-dimensional random column vector $\mathbf{x}$ is that of the vector $\mathbf{\Gamma}'\mathbf{z}$, where (with $P = \text{rank}\,\mathbf{\Sigma}$) $\mathbf{\Gamma}$ is a $P \times M$ matrix such that $\mathbf{\Sigma} = \mathbf{\Gamma}'\mathbf{\Gamma}$ and where $\mathbf{z}$ is a P-dimensional random column vector that has an absolutely continuous spherical distribution (i.e., the case where $\mathbf{x}$ is distributed elliptically about $\mathbf{0}$).

d. Some results on matrices

Let us introduce some additional results on matrices, thereby setting the stage for proving the necessity part of Theorem 6.8.3 (on the statistical independence of second-degree polynomials).

Differentiation of a matrix: some basic results. Suppose that, for $i = 1, 2, \ldots, P$ and $j = 1, 2, \ldots, Q$, $f_{ij}(t)$ is a function of a variable t. And define $\mathbf{F}(t)$ to be the $P \times Q$ matrix with ijth element $f_{ij}(t)$, so that $\mathbf{F}(t)$ is a matrix-valued function of t. Further, write $\frac{\partial f_{ij}(t)}{\partial t}$, or simply $\frac{\partial f_{ij}}{\partial t}$, for the derivative of $f_{ij}(t)$ at an arbitrary value of t (for which the derivative exists), and write $\frac{\partial \mathbf{F}(t)}{\partial t}$, or simply $\frac{\partial \mathbf{F}}{\partial t}$, for the $P \times Q$ matrix with ijth element $\frac{\partial f_{ij}(t)}{\partial t}$ or $\frac{\partial f_{ij}}{\partial t}$.

Certain of the basic properties of the derivatives of scalar-valued functions extend in a straightforward way to matrix-valued functions. In particular, if $\mathbf{F}(t)$ is a $P \times Q$ matrix-valued function of a variable t, then

$$\frac{\partial \mathbf{F}'}{\partial t} = \left(\frac{\partial \mathbf{F}}{\partial t}\right)' \tag{8.46}$$

and, for any $R \times P$ matrix of constants $\mathbf{A}$ and any $Q \times S$ matrix of constants $\mathbf{B}$,

$$\frac{\partial \mathbf{AF}}{\partial t} = \mathbf{A}\frac{\partial \mathbf{F}}{\partial t} \quad \text{and} \quad \frac{\partial \mathbf{FB}}{\partial t} = \frac{\partial \mathbf{F}}{\partial t}\mathbf{B}. \tag{8.47}$$

And if $\mathbf{F}(t)$ is a $P \times Q$ matrix-valued function and $\mathbf{G}(t)$ a $Q \times S$ matrix-valued function of a variable t, then

$$\frac{\partial \mathbf{FG}}{\partial t} = \mathbf{F}\frac{\partial \mathbf{G}}{\partial t} + \frac{\partial \mathbf{F}}{\partial t}\mathbf{G}. \tag{8.48}$$

Further, if $g(t)$ is a scalar-valued function and $\mathbf{F}(t)$ a matrix-valued function of a variable t, then [writing $\frac{\partial g}{\partial t}$ for the derivative of $g(t)$ at an arbitrary value of t]

$$\frac{\partial g\mathbf{F}}{\partial t} = \frac{\partial g}{\partial t}\mathbf{F} + g\frac{\partial \mathbf{F}}{\partial t}. \tag{8.49}$$

Refer, for example, to Harville (1997, sec.15.4) for additional discussion of basic results pertaining to the differentiation of matrix-valued functions of a single variable.

Some results on matrices of the form $\mathbf{I} - t\mathbf{A}$ (where t is a scalar and $\mathbf{A}$ is a symmetric matrix. Let $\mathbf{A}$ represent an $N \times N$ symmetric matrix. And regard the matrix $\mathbf{I} - t\mathbf{A}$ as a matrix-valued function of a variable t. Further, take $\mathbf{O}$ to be an $N \times N$ orthogonal matrix and $\mathbf{D}$ an $N \times N$ diagonal matrix such that

$$\mathbf{A} = \mathbf{ODO}', \tag{8.50}$$

and denote by $d_1, d_2, \ldots, d_N$ the diagonal elements of $\mathbf{D}$—the decomposition (8.50) is the spectral decomposition (of the matrix $\mathbf{A}$), the existence of which follows from Theorem 6.7.4.

Clearly,

$$\mathbf{I} - t\mathbf{A} = \mathbf{O}(\mathbf{I} - t\mathbf{D})\mathbf{O}' = \mathbf{O}\,\mathrm{diag}(1{-}td_1,\, 1{-}td_2,\, \ldots,\, 1{-}td_N)\,\mathbf{O}'. \tag{8.51}$$

And in light of result (2.14.25), Lemma 2.14.3, and Corollaries 2.14.19 and 2.14.2, it follows that

$$|\mathbf{I} - t\mathbf{A}| = \prod_{i=1}^{N} (1 - td_i), \tag{8.52}$$

and in light of results (2.5.11) and (2.6.5), that

$$(\mathbf{I} - t\mathbf{A})^{-1} = \mathbf{O}\,\mathrm{diag}\Big(\frac{1}{1{-}td_1}, \frac{1}{1{-}td_2}, \ldots, \frac{1}{1{-}td_N}\Big)\mathbf{O}'. \tag{8.53}$$

Thus,

$$\frac{\partial\,|\mathbf{I} - t\mathbf{A}|}{\partial t} = -\sum_{i=1}^{N} d_i \prod_{j=1\,(j\neq i)}^{N} (1 - td_j),$$

implying (in light of Theorem 6.7.5) that

$$\frac{\partial\,|\mathbf{I} - t\mathbf{A}|}{\partial t}\bigg|_{t=0} = -\,\mathrm{tr}(\mathbf{A}). \tag{8.54}$$

Further,

$$\begin{aligned} \frac{\partial(\mathbf{I} - t\mathbf{A})^{-1}}{\partial t} &= \mathbf{O}\,\mathrm{diag}\Big[\frac{d_1}{(1{-}td_1)^2}, \frac{d_2}{(1{-}td_2)^2}, \ldots, \frac{d_N}{(1{-}td_N)^2}\Big]\mathbf{O}' \\ &= (\mathbf{I} - t\mathbf{A})^{-1}\mathbf{A}(\mathbf{I} - t\mathbf{A})^{-1} \end{aligned} \tag{8.55}$$

and [denoting by $\frac{\partial^k \mathbf{F}(t)}{\partial t^k}$ or $\frac{\partial^k \mathbf{F}}{\partial t^k}$ the $N \times N$ matrix whose ijth element is the kth order derivative of the ijth element of an $N \times N$ matrix $\mathbf{F}(t)$ of functions of a variable t]

$$\begin{aligned} \frac{\partial^2(\mathbf{I} - t\mathbf{A})^{-1}}{\partial t^2} &= \frac{\partial(\mathbf{I} - t\mathbf{A})^{-1}}{\partial t}\mathbf{A}(\mathbf{I} - t\mathbf{A})^{-1} + (\mathbf{I} - t\mathbf{A})^{-1}\mathbf{A}\frac{\partial(\mathbf{I} - t\mathbf{A})^{-1}}{\partial t} \\ &= 2\,(\mathbf{I} - t\mathbf{A})^{-1}\mathbf{A}(\mathbf{I} - t\mathbf{A})^{-1}\mathbf{A}(\mathbf{I} - t\mathbf{A})^{-1}. \end{aligned} \tag{8.56}$$

The matrix $\mathbf{I} - t\mathbf{A}$ is singular if $t = 1/d_i$ for some i such that $d_i \neq 0$; otherwise, $\mathbf{I} - t\mathbf{A}$ is nonsingular. And formulas (8.53), (8.55), and (8.56) are valid for any t for which $\mathbf{I} - t\mathbf{A}$ is nonsingular.

Formulas (8.54), (8.55), and (8.56) can be generalized. Let $\mathbf{V}$ represent an $N \times N$ symmetric nonnegative definite matrix (and continue to take $\mathbf{A}$ to be an $N \times N$ symmetric matrix and to regard t as a scalar-valued variable). Then, $\mathbf{V} = \mathbf{R}'\mathbf{R}$ for some matrix $\mathbf{R}$ (having N columns). And upon observing (in light of Corollary 6.4.2) that

$$|\mathbf{I} - t\mathbf{A}\mathbf{V}| = |\mathbf{I} - t\mathbf{R}\mathbf{A}\mathbf{R}'|$$

and applying result (8.54) (with $\mathbf{R}\mathbf{A}\mathbf{R}'$ in place of $\mathbf{A}$), we find that

$$\frac{\partial\,|\mathbf{I} - t\mathbf{A}\mathbf{V}|}{\partial t}\bigg|_{t=0} = -\,\mathrm{tr}(\mathbf{R}\mathbf{A}\mathbf{R}') = -\,\mathrm{tr}(\mathbf{A}\mathbf{R}'\mathbf{R}) = -\,\mathrm{tr}(\mathbf{A}\mathbf{V}). \tag{8.57}$$

Now, suppose that the $N \times N$ symmetric nonnegative definite matrix $\mathbf{V}$ is positive definite, and take $\mathbf{R}$ to be a nonsingular matrix (of order N) such that $\mathbf{V} = \mathbf{R}'\mathbf{R}$. Then, upon observing that

$$\mathbf{V} - t\mathbf{A} = \mathbf{R}'[\mathbf{I} - t(\mathbf{R}^{-1})'\mathbf{A}\mathbf{R}^{-1}]\mathbf{R}$$

{so that $(\mathbf{V}-t\mathbf{A})^{-1} = \mathbf{R}^{-1}[\mathbf{I}-t(\mathbf{R}^{-1})'\mathbf{A}\mathbf{R}^{-1}]^{-1}(\mathbf{R}^{-1})'$} and applying results (8.55) and (8.56), we find that

$$\frac{\partial(\mathbf{V}-t\mathbf{A})^{-1}}{\partial t} = \mathbf{R}^{-1}[\mathbf{I}-t(\mathbf{R}^{-1})'\mathbf{A}\mathbf{R}^{-1}]^{-1}(\mathbf{R}^{-1})'\mathbf{A}\mathbf{R}^{-1}[\mathbf{I}-t(\mathbf{R}^{-1})'\mathbf{A}\mathbf{R}^{-1}]^{-1}(\mathbf{R}^{-1})'$$
$$= (\mathbf{V}-t\mathbf{A})^{-1}\mathbf{A}(\mathbf{V}-t\mathbf{A})^{-1} \tag{8.58}$$

and that

$$\frac{\partial^2(\mathbf{V}-t\mathbf{A})^{-1}}{\partial t^2} = 2(\mathbf{V}-t\mathbf{A})^{-1}\mathbf{A}(\mathbf{V}-t\mathbf{A})^{-1}\mathbf{A}(\mathbf{V}-t\mathbf{A})^{-1}. \tag{8.59}$$

Formulas (8.58) and (8.59) are valid for any t for which $\mathbf{V}-t\mathbf{A}$ is nonsingular.

Some results on determinants. We are now in a position to establish the following result.

Theorem 6.8.17. Let $\mathbf{A}$ and $\mathbf{B}$ represent $N \times N$ symmetric matrices, and let c and d represent (strictly) positive scalars. Then, a necessary and sufficient condition for

$$|\mathbf{I}-t\mathbf{A}-u\mathbf{B}| = |\mathbf{I}-t\mathbf{A}|\,|\mathbf{I}-u\mathbf{B}| \tag{8.60}$$

for all (scalars) t and u satisfying $|t| < c$ and $|u| < d$ (or, equivalently, for all t and u) is that $\mathbf{AB} = \mathbf{0}$.

Proof. The sufficiency of the condition $\mathbf{AB} = \mathbf{0}$ is evident upon observing that (for all t and u)

$$|\mathbf{I}-t\mathbf{A}|\,|\mathbf{I}-u\mathbf{B}| = |(\mathbf{I}-t\mathbf{A})(\mathbf{I}-u\mathbf{B})| = |\mathbf{I}-t\mathbf{A}-u\mathbf{B}+tu\mathbf{AB}|.$$

Now, for purposes of establishing the necessity of this condition, suppose that equality (8.50) holds for all t and u satisfying $|t| < c$ and $|u| < d$. And let c^* ($\le c$) represent a (strictly) positive scalar such that $\mathbf{I}-t\mathbf{A}$ is positive definite whenever $|t| < c^*$—the existence of such a scalar is guaranteed by Lemma 6.5.1—and (for t satisfying $|t| < c^*$) let

$$\mathbf{H}(t) = (\mathbf{I}-t\mathbf{A})^{-1}.$$

If $|t| < c^*$, then $|\mathbf{I}-u\mathbf{B}| = |\mathbf{I}-u\mathbf{B}\mathbf{H}(t)|$ and $|\mathbf{I}-(-u)\mathbf{B}| = |\mathbf{I}-(-u)\mathbf{B}\mathbf{H}(t)|$, implying that

$$|\mathbf{I}-u^2\mathbf{B}^2| = |\mathbf{I}-u^2[\mathbf{B}\mathbf{H}(t)]^2|. \tag{8.61}$$

Since each side of equality (8.61) is (for fixed t) a polynomial in u^2, we have that

$$|\mathbf{I}-r\mathbf{B}^2| = |\mathbf{I}-r[\mathbf{B}\mathbf{H}(t)]^2|$$

for every scalar r (and for t such that $|t| < c^*$)—refer to Theorem 6.7.8.

Upon observing that $[\mathbf{B}\mathbf{H}(t)]^2 = [\mathbf{B}\mathbf{H}(t)\mathbf{B}]\mathbf{H}(t)$, that $\mathbf{B}^2$ and $\mathbf{B}\mathbf{H}(t)\mathbf{B}$ are symmetric, and that $\mathbf{H}(t)$ is symmetric and nonnegative definite, and upon applying results (8.54) and (8.57), we find that, for every scalar t such that $|t| < c^*$,

$$\operatorname{tr}(\mathbf{B}^2) = -\frac{\partial\,|\mathbf{I}-r\mathbf{B}^2|}{\partial r}\bigg|_{r=0} = -\frac{\partial\,|\mathbf{I}-r[\mathbf{B}\mathbf{H}(t)]^2|}{\partial r}\bigg|_{r=0} = \operatorname{tr}\{[\mathbf{B}\mathbf{H}(t)]^2\}.$$

Thus,

$$\frac{\partial^2 \operatorname{tr}\{[\mathbf{B}\mathbf{H}(t)]^2\}}{\partial t^2} = \frac{\partial^2 \operatorname{tr}(\mathbf{B}^2)}{\partial t^2} = 0 \tag{8.62}$$

(for t such that $|t| < c^*$). Moreover,

$$\frac{\partial \operatorname{tr}\{[\mathbf{B}\mathbf{H}(t)]^2\}}{\partial t} = \operatorname{tr}\left\{\frac{\partial\,[\mathbf{B}\mathbf{H}(t)]^2}{\partial t}\right\} = \operatorname{tr}\left[\mathbf{B}\frac{\partial\mathbf{H}(t)}{\partial t}\mathbf{B}\mathbf{H}(t) + \mathbf{B}\mathbf{H}(t)\mathbf{B}\frac{\partial\mathbf{H}(t)}{\partial t}\right]$$
$$= 2\operatorname{tr}\left[\mathbf{B}\mathbf{H}(t)\mathbf{B}\frac{\partial\mathbf{H}(t)}{\partial t}\right],$$

implying [in light of results (8.55) and (8.56)] that

$$\begin{aligned} \frac{\partial^2 \operatorname{tr}\{[\mathbf{BH}(t)]^2\}}{\partial t^2} &= 2 \operatorname{tr}\left[\mathbf{B}\frac{\partial \mathbf{H}(t)}{\partial t}\mathbf{B}\frac{\partial \mathbf{H}(t)}{\partial t} + \mathbf{BH}(t)\mathbf{B}\frac{\partial^2 \mathbf{H}(t)}{\partial t^2}\right] \\ &= 2 \operatorname{tr}[\mathbf{BH}(t)\mathbf{AH}(t)\mathbf{BH}(t)\mathbf{AH}(t) + 2\mathbf{BH}(t)\mathbf{BH}(t)\mathbf{AH}(t)\mathbf{AH}(t)]. \qquad (8.63) \end{aligned}$$

Combining results (8.62) and (8.63) and setting $t = 0$ gives

$$\begin{aligned} 0 &= \operatorname{tr}[(\mathbf{BA})^2] + 2 \operatorname{tr}(\mathbf{B}^2\mathbf{A}^2) \\ &= \operatorname{tr}[(\mathbf{BA})^2 + \mathbf{B}^2\mathbf{A}^2] + \operatorname{tr}(\mathbf{B}^2\mathbf{A}^2) \\ &= \operatorname{tr}[\mathbf{B}(\mathbf{AB} + \mathbf{BA})\mathbf{A}] + \operatorname{tr}(\mathbf{B}^2\mathbf{A}^2) \\ &= \tfrac{1}{2} \operatorname{tr}[\mathbf{B}(\mathbf{AB} + \mathbf{BA})\mathbf{A}] + \tfrac{1}{2} \operatorname{tr}\{[\mathbf{B}(\mathbf{AB} + \mathbf{BA})\mathbf{A}]'\} + \operatorname{tr}(\mathbf{BA}^2\mathbf{B}) \\ &= \tfrac{1}{2} \operatorname{tr}[(\mathbf{AB} + \mathbf{BA})\mathbf{AB}] + \tfrac{1}{2} \operatorname{tr}[\mathbf{A}(\mathbf{AB} + \mathbf{BA})'\mathbf{B}] + \operatorname{tr}(\mathbf{BA}^2\mathbf{B}) \\ &= \tfrac{1}{2} \operatorname{tr}[(\mathbf{AB} + \mathbf{BA})'\mathbf{AB}] + \tfrac{1}{2} \operatorname{tr}[(\mathbf{AB} + \mathbf{BA})'\mathbf{BA}] + \operatorname{tr}(\mathbf{BA}^2\mathbf{B}) \\ &= \tfrac{1}{2} \operatorname{tr}[(\mathbf{AB} + \mathbf{BA})'(\mathbf{AB} + \mathbf{BA})] + \operatorname{tr}[(\mathbf{AB})'\mathbf{AB}]. \qquad (8.64) \end{aligned}$$

Both terms of expression (8.64) are nonnegative and hence equal to 0. Moreover, $\operatorname{tr}[(\mathbf{AB})'\mathbf{AB}] = 0$ implies that $\mathbf{AB} = \mathbf{0}$—refer to Lemma 2.3.2. Q.E.D.

In light of Lemma 6.5.2, we have the following variation on Theorem 6.8.17.

Corollary 6.8.18. Let $\mathbf{A}$ and $\mathbf{B}$ represent $N \times N$ symmetric matrices. Then, there exist (strictly) positive scalars c and d such that $\mathbf{I} - t\mathbf{A}$, $\mathbf{I} - u\mathbf{B}$, and $\mathbf{I} - t\mathbf{A} - u\mathbf{B}$ are positive definite for all (scalars) t and u satisfying $|t| < c$ and $|u| < d$. And a necessary and sufficient condition for

$$\log\left[\frac{|\mathbf{I} - t\mathbf{A} - u\mathbf{B}|}{|\mathbf{I} - t\mathbf{A}||\mathbf{I} - u\mathbf{B}|}\right] = 0$$

for all t and u satisfying $|t| < c$ and $|u| < d$ is that $\mathbf{AB} = \mathbf{0}$.

The following theorem can be regarded as a generalization of Corollary 6.8.18.

Theorem 6.8.19. Let $\mathbf{A}$ and $\mathbf{B}$ represent $N \times N$ symmetric matrices. Then, there exist (strictly) positive scalars c and d such that $\mathbf{I} - t\mathbf{A}$, $\mathbf{I} - u\mathbf{B}$, and $\mathbf{I} - t\mathbf{A} - u\mathbf{B}$ are positive definite for all (scalars) t and u satisfying $|t| < c$ and $|u| < d$. And letting $h(t, u)$ represent a polynomial (in t and u), necessary and sufficient conditions for

$$\log\left[\frac{|\mathbf{I} - t\mathbf{A} - u\mathbf{B}|}{|\mathbf{I} - t\mathbf{A}||\mathbf{I} - u\mathbf{B}|}\right] = \frac{h(t,u)}{|\mathbf{I} - t\mathbf{A}||\mathbf{I} - u\mathbf{B}||\mathbf{I} - t\mathbf{A} - u\mathbf{B}|} \qquad (8.65)$$

for all t and u satisfying $|t| < c$ and $|u| < d$ are that $\mathbf{AB} = \mathbf{0}$ and that, for all t and u satisfying $|t| < c$ and $|u| < d$ (or, equivalently, for all t and u), $h(t, u) = 0$.

Proof (of Theorem 6.8.19). The sufficiency of these conditions is an immediate consequence of Corollary 6.8.18 [as is the existence of positive scalars c and d such that $\mathbf{I} - t\mathbf{A}$, $\mathbf{I} - u\mathbf{B}$, and $\mathbf{I} - t\mathbf{A} - u\mathbf{B}$ are positive definite for all t and u satisfying $|t| < c$ and $|u| < d$].

For purposes of establishing their necessity, take u to be an arbitrary scalar satisfying $|u| < d$, and observe (in light of Corollaries 2.13.12 and 2.13.29) that there exists an $N \times N$ nonsingular matrix $\mathbf{S}$ such that $(\mathbf{I} - u\mathbf{B})^{-1} = \mathbf{S}'\mathbf{S}$. Observe also (in light of Theorem 6.7.4) that there exist $N \times N$ matrices $\mathbf{P}$ and $\mathbf{Q}$ such that $\mathbf{A} = \mathbf{P} \operatorname{diag}(d_1, d_2, \ldots, d_N)\mathbf{P}'$ and $\mathbf{SAS}' = \mathbf{Q} \operatorname{diag}(f_1, f_2, \ldots, f_N)\mathbf{Q}'$ for some scalars $d_1, d_2 \ldots, d_N$ and $f_1, f_2, \ldots, f_N$—$d_1, d_2 \ldots, d_N$ are the not-necessarily-distinct eigenvalues of $\mathbf{A}$ and $f_1, f_2, \ldots, f_N$ the not-necessarily-distinct eigenvalues of $\mathbf{SAS}'$. Moreover, letting $R = \operatorname{rank} \mathbf{A}$, it follows from Theorem 6.7.5 that exactly R of the scalars $d_1, d_2 \ldots, d_N$ and exactly R of the scalars $f_1, f_2, \ldots, f_N$ are nonzero; assume (without any essential loss of generality) that it is the first R of the scalars $d_1, d_2 \ldots, d_N$ and the first R of the scalars $f_1, f_2, \ldots, f_N$ that are nonzero. Then, in light of results (2.14.25) and (2. 14.10) and Corollary 2.14.19, we find that

$$\begin{aligned} |\mathbf{I} - t\mathbf{A}| &= |\mathbf{P}[\mathbf{I} - \operatorname{diag}(td_1, \ldots, td_{R-1}, td_R, 0, 0, \ldots, 0)]\mathbf{P}'| \\ &= |\operatorname{diag}(1-td_1, \ldots, 1-td_{R-1}, 1-td_R, 1, 1, \ldots, 1)| \\ &= \prod_{i=1}^{R} (1 - td_i) = \prod_{i=1}^{R} (-d_i)(t - d_i^{-1}) \end{aligned}$$

and that

$$\begin{aligned} |\mathbf{I} - t\mathbf{A} - u\mathbf{B}| &= |\mathbf{S}^{-1}(\mathbf{I} - t\mathbf{SAS}')(\mathbf{S}')^{-1}| \\ &= |\mathbf{S}|^{-2}|\mathbf{I} - t\mathbf{SAS}'| \\ &= |\mathbf{S}|^{-2}|\mathbf{Q}[\mathbf{I} - \operatorname{diag}(tf_1, \ldots, tf_{R-1}, tf_R, 0, 0, \ldots, 0)]\mathbf{Q}'| \\ &= |\mathbf{S}|^{-2}|\operatorname{diag}(1-tf_1, \ldots, 1-tf_{R-1}, 1-tf_R, 1, 1, \ldots, 1)| \\ &= |\mathbf{S}|^{-2}\prod_{i=1}^{R} (1 - tf_i) = |\mathbf{S}|^{-2}\prod_{i=1}^{R} (-f_i)(t - f_i^{-1}), \end{aligned}$$

so that (for fixed u) $|\mathbf{I} - t\mathbf{A} - u\mathbf{B}|$ and $|\mathbf{I} - t\mathbf{A}||\mathbf{I} - u\mathbf{B}|$ are polynomials in t. And

$$|\mathbf{I} - t\mathbf{A}||\mathbf{I} - u\mathbf{B}||\mathbf{I} - t\mathbf{A} - u\mathbf{B}| = |\mathbf{I} - u\mathbf{B}||\mathbf{S}|^{-2}\prod_{i=1}^{R} d_i f_i (t - d_i^{-1})(t - f_i^{-1}), \tag{8.66}$$

which (for fixed u) is a polynomial in t of degree $2R$ with roots $d_1^{-1}, d_2^{-1}, \ldots, d_R^{-1}, f_1^{-1}, f_2^{-1}, \ldots, f_R^{-1}$.

Now, regarding u as fixed, suppose that equality (8.65) holds for all t satisfying $|t| < c$. Then, in light of equality (8.66), it follows from Theorem 6.7.13 that there exists a real number $\alpha(u)$ such that, for all t,

$$h(t, u) = \alpha(u)|\mathbf{I} - t\mathbf{A}||\mathbf{I} - u\mathbf{B}||\mathbf{I} - t\mathbf{A} - u\mathbf{B}|$$

and

$$|\mathbf{I} - t\mathbf{A} - u\mathbf{B}| = e^{\alpha(u)}|\mathbf{I} - t\mathbf{A}||\mathbf{I} - u\mathbf{B}|. \tag{8.67}$$

[In applying Theorem 6.7.13, take $x = t$, $s_1(t) = |\mathbf{I} - t\mathbf{A} - u\mathbf{B}|$, $s_2(t) = |\mathbf{I} - t\mathbf{A}||\mathbf{I} - u\mathbf{B}|$, $r_1(t) = h(t, u)$, and $r_2(t) = |\mathbf{I}-t\mathbf{A}||\mathbf{I}-u\mathbf{B}||\mathbf{I}-t\mathbf{A}-u\mathbf{B}|$.] Moreover, upon setting $t = 0$ in equality (8.67), we find that

$$|\mathbf{I} - u\mathbf{B}| = e^{\alpha(u)}|\mathbf{I} - u\mathbf{B}|,$$

implying that $e^{\alpha(u)} = 1$ or, equivalently, that $\alpha(u) = 0$. Thus, for all t,

$$h(t, u) = 0 \quad \text{and} \quad |\mathbf{I} - t\mathbf{A} - u\mathbf{B}| = |\mathbf{I} - t\mathbf{A}||\mathbf{I} - u\mathbf{B}|.$$

We conclude that if equality (8.65) holds for all t and u satisfying $|t| < c$ and $|u| < d$, then $h(t, u) = 0$ and $|\mathbf{I} - t\mathbf{A} - u\mathbf{B}| = |\mathbf{I} - t\mathbf{A}||\mathbf{I} - u\mathbf{B}|$ for all t and u satisfying $|u| < d$, implying (in light of Theorem 6.7.8) that $h(t, u) = 0$ for all t and u and (in light of Theorem 6.8.17) that $\mathbf{AB} = \mathbf{0}$. Q.E.D.

The cofactors of a (square) matrix. Let $\mathbf{A} = \{a_{ij}\}$ represent an $N \times N$ matrix. And (for $i, j = 1, 2, \ldots, N$) let $\mathbf{A}_{ij}$ represent the $(N-1) \times (N-1)$ submatrix of $\mathbf{A}$ obtained by striking out the row and column that contain the element a_{ij}, that is, by striking out the ith row and the jth column. The determinant $|\mathbf{A}_{ij}|$ of this submatrix is called the *minor* of the element a_{ij}; the "signed" minor $(-1)^{i+j}|\mathbf{A}_{ij}|$ is called the *cofactor* of a_{ij}.

The determinant of an $N \times N$ matrix $\mathbf{A}$ can be expanded in terms of the cofactors of the N elements of any particular row or column of $\mathbf{A}$, as described in the following theorem.

Theorem 6.8.20. Let $\mathbf{A}$ represent an $N \times N$ matrix. And (for $i, j = 1, 2, \ldots, N$) let a_{ij} represent the ijth element of $\mathbf{A}$ and let α_{ij} represent the cofactor of a_{ij}. Then, for $i = 1, 2, \ldots, N$,

$$|\mathbf{A}| = \sum_{j=1}^{N} a_{ij}\alpha_{ij} = a_{i1}\alpha_{i1} + a_{i2}\alpha_{i2} + \cdots + a_{iN}\alpha_{iN} \tag{8.68}$$

$$= \sum_{j=1}^{N} a_{ji}\alpha_{ji} = a_{1i}\alpha_{1i} + a_{2i}\alpha_{2i} + \cdots + a_{Ni}\alpha_{Ni}. \tag{8.69}$$

For a proof of Theorem 6.8.20, refer, for example, to Harville (1997, sec 13.5). The following theorem adds to the results of Theorem 6.8.20.

Theorem 6.8.21. Let $\mathbf{A}$ represent an $N \times N$ matrix. And (for $i, j = 1, 2, \ldots, N$) let a_{ij} represent the ijth element of $\mathbf{A}$ and let α_{ij} represent the cofactor of a_{ij}. Then, for $i' \neq i = 1, \ldots, N$,

$$\sum_{j=1}^{N} a_{ij}\alpha_{i'j} = a_{i1}\alpha_{i'1} + a_{i2}\alpha_{i'2} + \cdots + a_{iN}\alpha_{i'N} = 0, \tag{8.70}$$

and

$$\sum_{j=1}^{N} a_{ji}\alpha_{ji'} = a_{1i}\alpha_{1i'} + a_{2i}\alpha_{2i'} + \cdots + a_{Ni}\alpha_{Ni'} = 0. \tag{8.71}$$

Proof (of Theorem 6.8.21). Consider result (8.70). Let $\mathbf{B}$ represent a matrix whose i'th row equals the ith row of $\mathbf{A}$ and whose first, second, ..., $(i'-1)$th, $(i'+1)$th, ..., $(N-1)$th, Nth rows are identical to those of $\mathbf{A}$ (where $i' \neq i$). Observe that the i'th row of $\mathbf{B}$ is a duplicate of its ith row and hence (in light of Lemma 2.14.10) that $|\mathbf{B}| = 0$.

Let b_{kj} represent the kjth element of $\mathbf{B}$ $(k, j = 1, 2, \ldots, N)$. Clearly, the cofactor of $b_{i'j}$ is the same as that of $a_{i'j}$ $(j = 1, 2, \ldots, N)$. Thus, making use of Theorem 6.8.20, we find that

$$\sum_{j=1}^{N} a_{ij}\alpha_{i'j} = \sum_{j=1}^{N} b_{i'j}\alpha_{i'j} = |\mathbf{B}| = 0,$$

which establishes result (8.70). Result (8.71) can be proved via an analogous argument. Q.E.D.

For any $N \times N$ matrix $\mathbf{A} = \{a_{ij}\}$, the $N \times N$ matrix whose ijth element is the cofactor α_{ij} of a_{ij} is called the *matrix of cofactors* (or *cofactor matrix*) of $\mathbf{A}$. The transpose of this matrix is called the *adjoint* or *adjoint matrix* of $\mathbf{A}$ and is denoted by the symbol adj $\mathbf{A}$ or adj($\mathbf{A}$).

There is a close relationship between the adjoint of a nonsingular matrix $\mathbf{A}$ and the inverse of $\mathbf{A}$, as is evident from the following theorem and as is made explicit in the corollary of this theorem.

Theorem 6.8.22. For any $N \times N$ matrix $\mathbf{A}$,

$$\mathbf{A}\,\mathrm{adj}(\mathbf{A}) = (\mathrm{adj}\,\mathbf{A})\mathbf{A} = |\mathbf{A}|\mathbf{I}_N.$$

Proof. Let a_{ij} represent the ijth element of $\mathbf{A}$ and let α_{ij} represent the cofactor of a_{ij} $(i, j = 1, 2, \ldots, N)$. Then, the ii'th element of the matrix product $\mathbf{A}\,\mathrm{adj}(\mathbf{A})$ is $\sum_{j=1}^{N} a_{ij}\alpha_{i'j}$ $(i, i' = 1, 2, \ldots, N)$. Moreover, according to Theorems 6.8.20 and 6.8.21,

$$\sum_{j=1}^{N} a_{ij}\alpha_{i'j} = \begin{cases} |\mathbf{A}|, & \text{if } i' = i, \\ 0, & \text{if } i' \neq i. \end{cases}$$

Thus, $\mathbf{A}\,\mathrm{adj}(\mathbf{A}) = |\mathbf{A}|\mathbf{I}$. That $(\mathrm{adj}\,\mathbf{A})\mathbf{A} = |\mathbf{A}|\mathbf{I}$ can be established via a similar argument. Q.E.D.

Corollary 6.8.23. If $\mathbf{A}$ is an $N \times N$ nonsingular matrix, then

$$\mathrm{adj}\,\mathbf{A} = |\mathbf{A}|\mathbf{A}^{-1} \tag{8.72}$$

or, equivalently,

$$\mathbf{A}^{-1} = (1/|\mathbf{A}|)\,\mathrm{adj}(\mathbf{A}). \tag{8.73}$$

e. Proof of the "necessity part" of Theorem 6.8.3

As in Theorem 6.8.3, let $\mathbf{x}$ represent an M-dimensional random column vector that has an $N(\boldsymbol{\mu}, \mathbf{I}_M)$ distribution. And, for $i = 1, 2, \ldots, K$, take $q_i = c_i + \mathbf{b}_i'\mathbf{x} + \mathbf{x}'\mathbf{A}_i\mathbf{x}$, where c_i is a constant, $\mathbf{b}_i$ an M-dimensional column vector of constants, and $\mathbf{A}_i$ an $M \times M$ symmetric matrix of constants.

Suppose that $q_1, q_2, \ldots, q_K$ are statistically independent. We wish to show that (for $j \neq i = 1, 2, \ldots, K$)

$$\mathbf{A}_i\mathbf{A}_j = \mathbf{0}, \quad \mathbf{A}_i\mathbf{b}_j = \mathbf{0}, \quad \text{and} \quad \mathbf{b}_i'\mathbf{b}_j = 0,$$

thereby proving the necessity part of Theorem 6.8.3.

For $i = 1, 2, \ldots, K$, let $\mathbf{d}_i = \mathbf{b}_i + 2\mathbf{A}_i\boldsymbol{\mu}$. Further, for $j \neq i = 1, 2, \ldots, K$, denote by $m_{ij}(\cdot, \cdot)$ the moment generating function of the joint distribution of q_i and q_j. And letting i and j represent arbitrary distinct integers between 1 and K, inclusive, observe (in light of the results of Section 6.5) that there exists a neighborhood N_{ij} of the 2×1 null vector $\mathbf{0}$ such that, for $(t, u)' \in N_{ij}$ (where t and u are scalars), $\mathbf{I} - 2t\mathbf{A}_i - 2u\mathbf{A}_j$ is positive definite and

$$\begin{aligned} m_{ij}(t,u) = |\mathbf{I} - 2t\mathbf{A}_i - 2u\mathbf{A}_j|^{-1/2} \exp[t(c_i + \mathbf{b}_i'\boldsymbol{\mu} + \boldsymbol{\mu}'\mathbf{A}_i\boldsymbol{\mu}) + u(c_j + \mathbf{b}_j'\boldsymbol{\mu} + \boldsymbol{\mu}'\mathbf{A}_j\boldsymbol{\mu})] \\ \times \exp\bigl[\tfrac{1}{2}(t\mathbf{d}_i + u\mathbf{d}_j)'(\mathbf{I} - 2t\mathbf{A}_i - 2u\mathbf{A}_j)^{-1}(t\mathbf{d}_i + u\mathbf{d}_j)\bigr]. \quad (8.74) \end{aligned}$$

Observe also (in light of the statistical independence of $q_1, q_2, \ldots, q_K$) that, for scalars t and u such that $(t, u)' \in N_{ij}$,

$$m_{ij}(t,u) = m_{ij}(t,0)\, m_{ij}(0,u). \quad (8.75)$$

Upon squaring both sides of equality (8.75) and making use of formula (8.74), we find that, for t and u such that $(t, u)' \in N_{ij}$,

$$\log\left[\frac{|\mathbf{I} - 2t\mathbf{A}_i - 2u\mathbf{A}_j|}{|\mathbf{I} - 2t\mathbf{A}_i|\,|\mathbf{I} - 2u\mathbf{A}_j|}\right] = r_{ij}(t,u), \quad (8.76)$$

where

$$\begin{aligned} r_{ij}(t,u) = (t\mathbf{d}_i + u\mathbf{d}_j)'(\mathbf{I} - 2t\mathbf{A}_i - 2u\mathbf{A}_j)^{-1}(t\mathbf{d}_i + u\mathbf{d}_j) \\ - t^2\mathbf{d}_i'(\mathbf{I} - 2t\mathbf{A}_i)^{-1}\mathbf{d}_i - u^2\mathbf{d}_j'(\mathbf{I} - 2u\mathbf{A}_j)^{-1}\mathbf{d}_j. \quad (8.77) \end{aligned}$$

In light of Corollary 6.8.23,

$$(\mathbf{I} - 2t\mathbf{A}_i - 2u\mathbf{A}_j)^{-1} = (1/|\mathbf{I} - 2t\mathbf{A}_i - 2u\mathbf{A}_j|)\,\mathrm{adj}(\mathbf{I} - 2t\mathbf{A}_i - 2u\mathbf{A}_j)$$

and, similarly,

$$(\mathbf{I} - 2t\mathbf{A}_i)^{-1} = (1/|\mathbf{I} - 2t\mathbf{A}_i|)\,\mathrm{adj}(\mathbf{I} - 2t\mathbf{A}_i) \quad \text{and} \quad (\mathbf{I} - 2u\mathbf{A}_j)^{-1} = (1/|\mathbf{I} - 2u\mathbf{A}_j|)\,\mathrm{adj}(\mathbf{I} - 2u\mathbf{A}_j).$$

Moreover, the elements of $\mathrm{adj}(\mathbf{I} - 2t\mathbf{A}_i - 2u\mathbf{A}_j)$, $\mathrm{adj}(\mathbf{I} - 2t\mathbf{A}_i)$, and $\mathrm{adj}(\mathbf{I} - 2u\mathbf{A}_j)$ are polynomials in t and/or u. And $|\mathbf{I} - 2t\mathbf{A}_i - 2u\mathbf{A}_j|$, $|\mathbf{I} - 2t\mathbf{A}_i|$, and $|\mathbf{I} - 2u\mathbf{A}_j|$ are also polynomials in t and/or u. Thus, $r_{ij}(t,u)$ is expressible in the form

$$r_{ij}(t,u) = \frac{h_{ij}(t,u)}{|\mathbf{I} - 2t\mathbf{A}_i|\,|\mathbf{I} - 2u\mathbf{A}_j|\,|\mathbf{I} - 2t\mathbf{A}_i - 2u\mathbf{A}_j|}, \quad (8.78)$$

where $h_{ij}(t,u)$ is a polynomial in t and u.

In light of results (8.76) and (8.78), it follows from Theorem 6.8.19 that $\mathbf{A}_i\mathbf{A}_j = \mathbf{0}$. It also follows that $h_{ij}(t,u) = 0$ for all scalars t and u and hence that $r_{ij}(t,u) = 0$ for all t and u such that $(t, u)' \in N_{ij}$. Moreover, by making use of various of the results of Section 6.8d on matrix differentiation, it can be shown (via a straightforward, though tedious exercise) that

$$\left.\frac{\partial^2 r_{ij}(t,u)}{\partial t\,\partial u}\right|_{t=u=0} = 2\mathbf{d}_i'\mathbf{d}_j \quad (8.79)$$

and that

$$\left.\frac{\partial^4 r_{ij}(t,u)}{\partial t^2\,\partial u^2}\right|_{t=u=0} = 16(\mathbf{A}_j\mathbf{d}_i)'\mathbf{A}_j\mathbf{d}_i + 16(\mathbf{A}_i\mathbf{d}_j)'\mathbf{A}_i\mathbf{d}_j. \quad (8.80)$$

And upon observing that the partial derivatives of $r_{ij}(t, u)$ evaluated at $t = u = 0$, equal 0, it follows from result (8.79) that $\mathbf{d}_i'\mathbf{d}_j = 0$ and from result (8.80) that $\mathbf{A}_j\mathbf{d}_i = \mathbf{0}$ and $\mathbf{A}_i\mathbf{d}_j = \mathbf{0}$. Thus,

$$\mathbf{A}_i\mathbf{b}_j = \mathbf{A}_i(\mathbf{d}_j - 2\mathbf{A}_j\boldsymbol{\mu}) = \mathbf{A}_i\mathbf{d}_j - 2\mathbf{A}_i\mathbf{A}_j\boldsymbol{\mu} = \mathbf{0}$$

and

$$\mathbf{b}_i'\mathbf{b}_j = (\mathbf{d}_i - 2\mathbf{A}_i\boldsymbol{\mu})'(\mathbf{d}_j - 2\mathbf{A}_j\boldsymbol{\mu}) = \mathbf{d}_i'\mathbf{d}_j - 2\boldsymbol{\mu}'\mathbf{A}_i\mathbf{d}_j - 2(\mathbf{A}_j\mathbf{d}_i)'\boldsymbol{\mu} + 4\boldsymbol{\mu}'\mathbf{A}_i\mathbf{A}_j\boldsymbol{\mu} = 0.$$

The proof of the necessity part of Theorem 6.8.3 is now complete.

Exercises

Exercise 1. Let x represent a random variable whose distribution is $Ga(\alpha, \beta)$, and let c represent a (strictly) positive constant. Show that $cx \sim Ga(\alpha, c\beta)$ [thereby verifying result (1.2)].

Exercise 2. Let w represent a random variable whose distribution is $Ga(\alpha, \beta)$, where α is a (strictly positive) integer. Show that (for any strictly positive scalar t)

$$\Pr(w \le t) = \Pr(u \ge \alpha),$$

where u is a random variable whose distribution is Poisson with parameter t/β [so that $\Pr(u = s) = e^{-t/\beta}(t/\beta)^s/s!$ for $s = 0, 1, 2, \ldots$].

Exercise 3. Let u and w represent random variables that are distributed independently as $Be(\alpha, \delta)$ and $Be(\alpha+\delta, \lambda)$, respectively. Show that $uw \sim Be(\alpha, \lambda+\delta)$.

Exercise 4. Let x represent a random variable whose distribution is $Be(\alpha, \lambda)$.

(a) Show that, for $r > -\alpha$,

$$\mathrm{E}(x^r) = \frac{\Gamma(\alpha + r)\Gamma(\alpha + \lambda)}{\Gamma(\alpha)\Gamma(\alpha + \lambda + r)}.$$

(b) Show that

$$\mathrm{E}(x) = \frac{\alpha}{\alpha + \lambda} \quad \text{and} \quad \mathrm{var}(x) = \frac{\alpha\lambda}{(\alpha + \lambda)^2(\alpha + \lambda + 1)}.$$

Exercise 5. Take $x_1, x_2, \ldots, x_K$ to be K random variables whose joint distribution is $Di(\alpha_1, \alpha_2, \ldots, \alpha_K, \alpha_{K+1}; K)$, define $x_{K+1} = 1 - \sum_{k=1}^{K} x_k$, and let $\alpha = \alpha_1 + \cdots + \alpha_K + \alpha_{K+1}$.

(a) Generalize the results of Part (a) of Exercise 4 by showing that, for $r_1 > -\alpha_1, \ldots, r_K > -\alpha_K$, and $r_{K+1} > -\alpha_{K+1}$,

$$\mathrm{E}\big(x_1^{r_1} \cdots x_K^{r_K} x_{K+1}^{r_{K+1}}\big) = \frac{\Gamma(\alpha)}{\Gamma\big(\alpha + \sum_{k=1}^{K+1} r_k\big)} \prod_{k=1}^{K+1} \frac{\Gamma(\alpha_k + r_k)}{\Gamma(\alpha_k)}.$$

(b) Generalize the results of Part (b) of Exercise 4 by showing that (for an arbitrary integer k between 1 and $K+1$, inclusive)

$$\mathrm{E}(x_k) = \frac{\alpha_k}{\alpha} \quad \text{and} \quad \mathrm{var}(x_k) = \frac{\alpha_k(\alpha - \alpha_k)}{\alpha^2(\alpha + 1)}$$

and that (for any 2 distinct integers j and k between 1 and $K+1$, inclusive)

$$\mathrm{cov}(x_j, x_k) = \frac{-\alpha_j\alpha_k}{\alpha^2(\alpha + 1)}.$$

Exercise 6. Verify that the function $b(\cdot)$ defined by expression (1.48) is a pdf of the chi distribution with N degrees of freedom.

Exercise 7. For strictly positive integers J and K, let $s_1, \ldots, s_J, s_{J+1}, \ldots, s_{J+K}$ represent $J+K$ random variables whose joint distribution is $Di(\alpha_1, \ldots, \alpha_J, \alpha_{J+1}, \ldots, \alpha_{J+K}, \alpha_{J+K+1}; J+K)$. Further, for $k = 1, 2, \ldots, K$, let $x_k = s_{J+k} / \bigl(1 - \sum_{j=1}^{J} s_j\bigr)$. Show that the conditional distribution of $x_1, x_2, \ldots, x_K$ given $s_1, s_2, \ldots, s_J$ is $Di(\alpha_{J+1}, \ldots, \alpha_{J+K}, \alpha_{J+K+1}; K)$.

Exercise 8. Let $z_1, z_2, \ldots, z_M$ represent random variables whose joint distribution is absolutely continuous with a pdf $f(\cdot, \cdot, \ldots, \cdot)$ of the form $f(z_1, z_2, \ldots, z_M) = g\bigl(\sum_{i=1}^{M} z_i^2\bigr)$ [where $g(\cdot)$ is a nonnegative function of a single nonnegative variable]. Verify that the function $b(\cdot)$ defined by expression (1.53) is a pdf of the distribution of the random variable $\bigl(\sum_{i=1}^{M} z_i^2\bigr)^{1/2}$. (*Note*. This exercise can be regarded as a more general version of Exercise 6.)

Exercise 9. Use the procedure described in Section 6.2a to construct a 6×6 orthogonal matrix whose first row is proportional to the vector $(0, -3,\ 4,\ 2,\ 0, -1)$.

Exercise 10. Let $\mathbf{x}_1$ and $\mathbf{x}_2$ represent M-dimensional column vectors.

(a) Use the results of Section 6.2a (pertaining to Helmert matrices) to show that if $\mathbf{x}_2'\mathbf{x}_2 = \mathbf{x}_1'\mathbf{x}_1$, then there exist orthogonal matrices $\mathbf{O}_1$ and $\mathbf{O}_2$ such that $\mathbf{O}_2\mathbf{x}_2 = \mathbf{O}_1\mathbf{x}_1$.

(b) Use the result of Part (a) to devise an alternative proof of the "only if" part of Lemma 5.9.9.

Exercise 11. Let w represent a random variable whose distribution is $\chi^2(N, \lambda)$. Verify that the expressions for $\mathrm{E}(w)$ and $\mathrm{E}(w^2)$ provided by formula (2.36) are in agreement with those provided by results (2.33) and (2.34) [or, equivalently, by results (2.28) and (2.30)].

Exercise 12. Let w_1 and w_2 represent random variables that are distributed independently as $Ga(\alpha_1,\ \beta,\ \delta_1)$ and $Ga(\alpha_2,\ \beta,\ \delta_2)$, respectively, and define $w = w_1 + w_2$. Derive the pdf of the distribution of w by starting with the pdf of the joint distribution of w_1 and w_2 and introducing a suitable change of variables. [*Note*. This derivation serves the purpose of verifying that $w \sim Ga(\alpha_1 + \alpha_2,\ \beta,\ \delta_1 + \delta_2)$ and (when coupled with a mathematical-induction argument) represents an alternative way of establishing Theorem 6.2.2 (and Theorem 6.2.1).]

Exercise 13. Let $\mathbf{x} = \boldsymbol{\mu} + \mathbf{z}$, where $\mathbf{z}$ is an N-dimensional random column vector that has an absolutely continuous spherical distribution and where $\boldsymbol{\mu}$ is an N-dimensional nonrandom column vector. Verify that in the special case where $\mathbf{z} \sim N(\mathbf{0}, \mathbf{I})$, the pdf $q(\cdot)$ derived in Section 6.2h for the distribution of $\mathbf{x}'\mathbf{x}$ "simplifies to" (i.e., is reexpressible in the form of) the expression (2.15) given in Section 6.2c for the pdf of the noncentral chi-square distribution [with N degrees of freedom and with noncentrality parameter λ $(= \boldsymbol{\mu}'\boldsymbol{\mu})$].

Exercise 14. Let u and v represent random variables that are distributed independently as $\chi^2(M)$ and $\chi^2(N)$, respectively. And define $w = (u/M)/(v/N)$. Devise an alternative derivation of the pdf of the $SF(M, N)$ distribution by (1) deriving the pdf of the joint distribution of w and v and by (2) determining the pdf of the marginal distribution of w from the pdf of the joint distribution of w and v.

Exercise 15. Let $t = z \big/ \sqrt{v/N}$, where z and v are random variables that are statistically independent with $z \sim N(0, 1)$ and $v \sim \chi^2(N)$ [in which case $t \sim St(N)$].

(a) Starting with the pdf of the joint distribution of z and v, derive the pdf of the joint distribution of t and v.

(b) Derive the pdf of the $St(N)$ distribution from the pdf of the joint distribution of t and v, thereby providing an alternative to the derivation given in Part 2 of Section 6.4a.

Exercise 16. Let $t = (x_1 + x_2)/|x_1 - x_2|$, where x_1 and x_2 are random variables that are distributed independently and identically as $N(\mu, \sigma^2)$ (with $\sigma > 0$). Show that t has a noncentral t distribution, and determine the values of the parameters (the degrees of freedom and the noncentrality parameter) of this distribution.

Exercise 17. Let t represent a random variable that has an $St(N, \mu)$ distribution. And take r to be an arbitrary one of the integers $1, 2, \ldots < N$. Generalize expressions (4.38) and (4.39) [for $\mathrm{E}(t^1)$ and $\mathrm{E}(t^2)$, respectively] by obtaining an expression for $\mathrm{E}(t^r)$ (in terms of μ). (*Note.* This exercise is closely related to Exercise 3.12.)

Exercise 18. Let $\mathbf{t}$ represent an M-dimensional random column vector that has an $MVt(N, \mathbf{I}_M)$ distribution. And let $w = \mathbf{t}'\mathbf{t}$. Derive the pdf of the distribution of w in each of the following two ways: (1) as a special case of the pdf (1.51) and (2) by making use of the relationship (4.50).

Exercise 19. Let $\mathbf{x}$ represent an M-dimensional random column vector whose distribution has as a pdf a function $f(\cdot)$ that is expressible in the following form: for all $\mathbf{x}$,

$$f(\mathbf{x}) = \int_0^\infty h(\mathbf{x} \mid u) g(u)\, du,$$

where $g(\cdot)$ is the pdf of the distribution of a strictly positive random variable u and where (for every u) $h(\cdot \mid u)$ is the pdf of the $N(\mathbf{0}, u^{-1}\mathbf{I}_M)$ distribution.

(a) Show that the distribution of $\mathbf{x}$ is spherical.

(b) Show that the distribution of u can be chosen in such a way that $f(\cdot)$ is the pdf of the $MVt(N, \mathbf{I}_M)$ distribution.

Exercise 20. Show that if condition (6.7) of Theorem 6.6.2 is replaced by the condition

$$\boldsymbol{\Sigma}(\mathbf{b} + 2\mathbf{A}\boldsymbol{\mu}) \in \mathcal{C}(\boldsymbol{\Sigma}\mathbf{A}\boldsymbol{\Sigma}),$$

the theorem is still valid.

Exercise 21. Let $\mathbf{x}$ represent an M-dimensional random column vector that has an $N(\boldsymbol{\mu}, \boldsymbol{\Sigma})$ distribution (where $\boldsymbol{\Sigma} \neq \mathbf{0}$), and take $\mathbf{G}$ to be a symmetric generalized inverse of $\boldsymbol{\Sigma}$. Show that

$$\mathbf{x}'\mathbf{G}\mathbf{x} \sim \chi^2(\operatorname{rank} \boldsymbol{\Sigma},\ \boldsymbol{\mu}'\mathbf{G}\boldsymbol{\mu})$$

if $\boldsymbol{\mu} \in \mathcal{C}(\boldsymbol{\Sigma})$ or $\mathbf{G}\boldsymbol{\Sigma}\mathbf{G} = \mathbf{G}$. [*Note.* A symmetric generalized inverse $\mathbf{G}$ is obtainable from a possibly nonsymmetric generalized inverse, say $\mathbf{H}$, by taking $\mathbf{G} = \frac{1}{2}\mathbf{H} + \frac{1}{2}\mathbf{H}'$; the condition $\mathbf{G}\boldsymbol{\Sigma}\mathbf{G} = \mathbf{G}$ is the second of the so-called Moore–Penrose conditions—refer, e.g., to Harville (1997, chap. 20) for a discussion of the Moore–Penrose conditions.]

Exercise 22. Let $\mathbf{z}$ represent an N-dimensional random column vector. And suppose that the distribution of $\mathbf{z}$ is an absolutely continuous spherical distribution, so that the distribution of $\mathbf{z}$ has as a pdf a function $f(\cdot)$ such that (for all $\mathbf{z}$) $f(\mathbf{z}) = g(\mathbf{z}'\mathbf{z})$, where $g(\cdot)$ is a (nonnegative) function of a single nonnegative variable. Further, take $\mathbf{z}_*$ to be an M-dimensional subvector of $\mathbf{z}$ (where $M < N$), and let $v = \mathbf{z}_*'\mathbf{z}_*$.

(a) Show that the distribution of v has as a pdf the function $h(\cdot)$ defined as follows: for $v > 0$,

$$h(v) = \frac{\pi^{N/2}}{\Gamma(M/2)\,\Gamma[(N-M)/2]}\, v^{(M/2)-1} \int_0^\infty w^{[(N-M)/2]-1} g(v+w)\, dw;$$

for $v \leq 0$, $h(v) = 0$.

(b) Verify that in the special case where $\mathbf{z} \sim N(\mathbf{0}, \mathbf{I}_N)$, $h(\cdot)$ simplifies to the pdf of the $\chi^2(M)$ distribution.

Exercise 23. Let $\mathbf{z} = (z_1, z_2, \ldots, z_M)'$ represent an M-dimensional random (column) vector that has a spherical distribution. And take $\mathbf{A}$ to be an $M \times M$ symmetric idempotent matrix of rank R (where $R \geq 1$).

(a) Starting from first principles (i.e., from the definition of a spherical distribution), use the results of Theorems 5.9.5 and 6.6.6 to show that (1) $\mathbf{z}'\mathbf{A}\mathbf{z} \sim \sum_{i=1}^{R} z_i^2$ and [assuming that $\Pr(\mathbf{z} \neq \mathbf{0}) = 1$] that (2) $\mathbf{z}'\mathbf{A}\mathbf{z}/\mathbf{z}'\mathbf{z} \sim \sum_{i=1}^{R} z_i^2 / \sum_{i=1}^{M} z_i^2$.

(b) Provide an alternative "derivation" of results (1) and (2) of Part (a); do so by showing that (when $\mathbf{z}$ has an absolutely continuous spherical distribution) these two results can be obtained by applying Theorem 6.6.7 (and by making use of the results of Sections 6.1f and 6.1g).

Exercise 24. Let $\mathbf{A}$ represent an $N \times N$ symmetric matrix. And take $\mathbf{Q}$ to be an $N \times N$ orthogonal matrix and $\mathbf{D}$ an $N \times N$ diagonal matrix such that $\mathbf{A} = \mathbf{Q}\mathbf{D}\mathbf{Q}'$—the decomposition $\mathbf{A} = \mathbf{Q}\mathbf{D}\mathbf{Q}'$ is the spectral decomposition, the existence and properties of which are established in Section 6.7a. Further, denote by $d_1, d_2, \ldots, d_N$ the diagonal elements of $\mathbf{D}$ (which are the not-necessarily-distinct eigenvalues of $\mathbf{A}$), and taking $\mathbf{D}^+$ to be the $N \times N$ diagonal matrix whose ith diagonal element is d_i^+, where $d_i^+ = 0$ if $d_i = 0$ and where $d_i^+ = 1/d_i$ if $d_i \neq 0$, define $\mathbf{A}^+ = \mathbf{Q}\mathbf{D}^+\mathbf{Q}'$. Show that (1) $\mathbf{A}\mathbf{A}^+\mathbf{A} = \mathbf{A}$ (i.e., $\mathbf{A}^+$ is a generalized inverse of $\mathbf{A}$) and also that (2) $\mathbf{A}^+\mathbf{A}\mathbf{A}^+ = \mathbf{A}^+$, (3) $\mathbf{A}\mathbf{A}^+$ is symmetric, and (4) $\mathbf{A}^+\mathbf{A}$ is symmetric—as discussed, e.g., by Harville (1997, chap. 20), these four conditions are known as the Moore–Penrose conditions and they serve to determine a unique matrix $\mathbf{A}^+$ that is known as the Moore–Penrose inverse.

Exercise 25. Let $\boldsymbol{\Sigma}$ represent an $N \times N$ symmetric nonnegative definite matrix, and take $\boldsymbol{\Gamma}_1$ to be a $P_1 \times N$ matrix and $\boldsymbol{\Gamma}_2$ a $P_2 \times N$ matrix such that $\boldsymbol{\Sigma} = \boldsymbol{\Gamma}_1'\boldsymbol{\Gamma}_1 = \boldsymbol{\Gamma}_2'\boldsymbol{\Gamma}_2$. Further, take $\mathbf{A}$ to be an $N \times N$ symmetric matrix. And assuming that $P_2 \geq P_1$ (as can be done without any essential loss of generality), show that the P_2 not-necessarily-distinct eigenvalues of the $P_2 \times P_2$ matrix $\boldsymbol{\Gamma}_2\mathbf{A}\boldsymbol{\Gamma}_2'$ consist of the P_1 not-necessarily-distinct eigenvalues of the $P_1 \times P_1$ matrix $\boldsymbol{\Gamma}_1\mathbf{A}\boldsymbol{\Gamma}_1'$ and of $P_2 - P_1$ zeroes. (*Hint.* Make use of Corollary 6.4.2.)

Exercise 26. Let $\mathbf{A}$ represent an $M \times M$ symmetric matrix and $\boldsymbol{\Sigma}$ an $M \times M$ symmetric nonnegative definite matrix. Show that the condition $\boldsymbol{\Sigma}\mathbf{A}\boldsymbol{\Sigma}\mathbf{A}\boldsymbol{\Sigma} = \boldsymbol{\Sigma}\mathbf{A}\boldsymbol{\Sigma}$ (which appears in Theorem 6.6.2) is equivalent to *each* of the following three conditions:

(1) $(\mathbf{A}\boldsymbol{\Sigma})^3 = (\mathbf{A}\boldsymbol{\Sigma})^2$;

(2) $\mathrm{tr}[(\mathbf{A}\boldsymbol{\Sigma})^2] = \mathrm{tr}[(\mathbf{A}\boldsymbol{\Sigma})^3] = \mathrm{tr}[(\mathbf{A}\boldsymbol{\Sigma})^4]$; and

(3) $\mathrm{tr}[(\mathbf{A}\boldsymbol{\Sigma})^2] = \mathrm{tr}(\mathbf{A}\boldsymbol{\Sigma}) = \mathrm{rank}(\boldsymbol{\Sigma}\mathbf{A}\boldsymbol{\Sigma})$.

Exercise 27. Let $\mathbf{z}$ represent an M-dimensional random column vector that has an $N(\mathbf{0}, \mathbf{I}_M)$ distribution, and take $q = c + \mathbf{b}'\mathbf{z} + \mathbf{z}'\mathbf{A}\mathbf{z}$, where c is a constant, $\mathbf{b}$ an M-dimensional column vector of constants, and $\mathbf{A}$ an $M \times M$ (nonnull) symmetric matrix of constants. Further, denote by $m(\cdot)$ the moment generating function of q. Provide an alternative derivation of the "sufficiency part" of Theorem 6.6.1 by showing that if $\mathbf{A}^2 = \mathbf{A}$, $\mathbf{b} = \mathbf{A}\mathbf{b}$, and $c = \frac{1}{4}\mathbf{b}'\mathbf{b}$, then, for every scalar t in some neighborhood of 0, $m(t) = m_*(t)$, where $m_*(\cdot)$ is the moment generating function of a $\chi^2(R, c)$ distribution and where $R = \mathrm{rank}\,\mathbf{A} = \mathrm{tr}(\mathbf{A})$.

Exercise 28. Let $\mathbf{x}$ represent an M-dimensional random column vector that has an $N(\mathbf{0}, \boldsymbol{\Sigma})$ distribution, and denote by $\mathbf{A}$ an $M \times M$ symmetric matrix of constants. Construct an example where $M = 3$ and where $\boldsymbol{\Sigma}$ and $\mathbf{A}$ are such that $\mathbf{A}\boldsymbol{\Sigma}$ is *not* idempotent but are nevertheless such that $\mathbf{x}'\mathbf{A}\mathbf{x}$ has a chi-square distribution.

Exercise 29. Let $\mathbf{x}$ represent an M-dimensional random column vector that has an $N(\mathbf{0}, \boldsymbol{\Sigma})$ distribution. Further, partition $\mathbf{x}$ and $\boldsymbol{\Sigma}$ as

$$\mathbf{x} = \begin{pmatrix} \mathbf{x}_1 \\ \mathbf{x}_2 \end{pmatrix} \quad \text{and} \quad \boldsymbol{\Sigma} = \begin{pmatrix} \boldsymbol{\Sigma}_{11} & \boldsymbol{\Sigma}_{12} \\ \boldsymbol{\Sigma}_{21} & \boldsymbol{\Sigma}_{22} \end{pmatrix}$$

(where the dimensions of $\boldsymbol{\Sigma}_{11}$ are the same as the dimension of $\mathbf{x}_1$). And take $\mathbf{G}_1$ to be a generalized inverse of $\boldsymbol{\Sigma}_{11}$ and $\mathbf{G}_2$ a generalized inverse of $\boldsymbol{\Sigma}_{22}$. Show that $\mathbf{x}_1'\mathbf{G}_1\mathbf{x}_1$ and $\mathbf{x}_2'\mathbf{G}_2\mathbf{x}_2$ are distributed independently if and only if $\boldsymbol{\Sigma}_{12} = \mathbf{0}$.

Exercise 30. Let $\mathbf{x}$ represent an M-dimensional random column vector that has an $N(\boldsymbol{\mu}, \mathbf{I}_M)$ distribution. And, for $i = 1, 2, \ldots, K$, take $q_i = c_i + \mathbf{b}_i'\mathbf{x} + \mathbf{x}'\mathbf{A}_i\mathbf{x}$, where c_i is a constant, $\mathbf{b}_i$ an M-dimensional column vector of constants, and $\mathbf{A}_i$ an $M \times M$ symmetric matrix of constants. Further, denote by $m(\cdot, \cdot, \ldots, \cdot)$ the moment generating function of the joint distribution of $q_1, q_2, \ldots, q_K$. Provide an alternative derivation of the "sufficiency part" of Theorem 6.8.3 by showing that if, for $j \neq i = 1, 2, \ldots, K$, $\mathbf{A}_i\mathbf{A}_j = \mathbf{0}$, $\mathbf{A}_i\mathbf{b}_j = \mathbf{0}$, and $\mathbf{b}_i'\mathbf{b}_j = 0$, then there exist (strictly) positive scalars $h_1, h_2, \ldots, h_K$ such that, for any scalars $t_1, t_2, \ldots, t_K$ for which $|t_1| < h_1$, $|t_2| < h_2$, $\ldots$, $|t_K| < h_K$,

$$m(t_1, t_2, \ldots, t_K) = m(t_1, 0, 0, \ldots, 0)\, m(0, t_2, 0, 0, \ldots, 0) \cdots m(0, \ldots, 0, 0, t_K).$$

Exercise 31. Let $\mathbf{x}$ represent an M-dimensional random column vector that has an $N(\mathbf{0}, \boldsymbol{\Sigma})$ distribution. And take $\mathbf{A}_1$ and $\mathbf{A}_2$ to be $M \times M$ symmetric *nonnegative definite* matrices of constants. Show that the two quadratic forms $\mathbf{x}'\mathbf{A}_1\mathbf{x}$ and $\mathbf{x}'\mathbf{A}_2\mathbf{x}$ are statistically independent if and only if they are uncorrelated.

Exercise 32. Let $\mathbf{x}$ represent an M-dimensional random column vector that has an $N(\boldsymbol{\mu}, \mathbf{I}_M)$ distribution. Show (by producing an example) that there exist quadratic forms $\mathbf{x}'\mathbf{A}_1\mathbf{x}$ and $\mathbf{x}'\mathbf{A}_2\mathbf{x}$ (where $\mathbf{A}_1$ and $\mathbf{A}_2$ are $M \times M$ symmetric matrices of constants) that are uncorrelated for *every* $\boldsymbol{\mu} \in \mathbb{R}^M$ but that are not statistically independent for *any* $\boldsymbol{\mu} \in \mathbb{R}^M$.

Bibliographic and Supplementary Notes

§1e. Theorem 6.1.5 is more or less identical to Theorem 1.4 of Fang, Kotz, and Ng (1990).

§2b. In some presentations (e.g., Ravishanker and Dey 2002, sec 5.3), the noncentrality parameter of the noncentral chi-square distribution is defined to be $\boldsymbol{\mu}'\boldsymbol{\mu}/2$ (instead of $\boldsymbol{\mu}'\boldsymbol{\mu}$).

§2h. Refer, for example, to Cacoullos and Koutras (1984) for a considerably more extensive discussion of the distribution of the random variable $\mathbf{x}'\mathbf{x}$, where $\mathbf{x} = \boldsymbol{\mu} + \mathbf{z}$ for some N-dimensional spherically distributed random column vector $\mathbf{z}$ and for some N-dimensional nonrandom column vector $\boldsymbol{\mu}$.

§3. The contributions of (George W.) Snedecor to the establishment of Snedecor's F distribution would seem to be more modest than might be inferred from the terminology. Snedecor initiated the practice of using F as a symbol for a random variable whose distribution is that of the ratio (3.1) and indicated that he had done so to honor the contributions of R. A. Fisher, though, subsequently, in a letter to H. W. Heckstall-Smith—refer, e.g., to page 319 of the volume edited by Bennett (1990)—Fisher dismissed Snedecor's gesture as an "afterthought." And Snedecor was among the first to present tables of the percentage points of the F distribution; the earliest tables, which were those of Fisher, were expressed in terms related to the distribution of the logarithm of the ratio (3.1).

§4. The t distribution is attributable to William Sealy Gosset (1876–1937). His work on this distribution took place while he was in the employ of the Guinness Brewery and was published under the pseudonym Student (Student 1908)—refer, e.g., to Zabell (2008).

§4c. This subsection is essentially a replicate of the first part of Section 18.1a of Harville (1997).

§5a. The results of Part 1 of this subsection more or less duplicate results presented in Section 18.3 of Harville (1997).

§6d and §6e. The inspiration for the content of these subsections (and for Exercises 22 and 23) came from results like those presented by Anderson and Fang (1987).

§7a. The polynomial $p(\lambda)$ and equation $p(\lambda) = 0$ referred to herein as the characteristic polynomial and characteristic equation differ by a factor of $(-1)^N$ from what some authors refer to as the characteristic polynomial and equation; those authors refer to the polynomial $q(\lambda)$ obtained by taking (for all λ) $q(\lambda) = |\lambda \mathbf{I}_N - \mathbf{A}|$ as the characteristic polynomial and/or to the equation $q(\lambda) = 0$ as the characteristic equation. Theorem 6.7.1 is essentially the same as Theorem 21.5.6 of Harville (1997), and the approach taken in devising a proof is a variation on the approach taken by Harville.

§7c. Theorem 6.7.13 can be regarded as a variant of a "lemma" on polynomials in 2 variables that was stated (without proof) by Laha (1956) and that has come to be identified (at least among statisticians) with Laha's name—essentially the same lemma appears (along with a proof) in Ogawa's (1950) paper. Various approaches to the proof of Laha's lemma are discussed by Driscoll and Gundberg (1986, sec. 3)—refer also to Driscoll and Krasnicka (1995, sec. 4).

§7d. The proof (of the "necessity part" of Theorem 6.6.1) presented in Section 6.7d makes use of Theorem 6.7.13 (on polynomials). Driscoll (1999) and Khatri (1999) introduced (in the context of Corollary 6.6.4) an alternative proof of necessity—refer also to Ravishanker and Dey (2002, sec. 5.4) and to Khuri (2010, sec. 5.2).

§8a (and §8e). The result presented herein as Corollary 6.8.2 includes as a special case a result that has come to be widely known as Craig's theorem (in recognition of the contributions of A. T. Craig) and that (to acknowledge the relevance of the work of H. Sakamoto and/or K. Matusita) is also sometimes referred to as the Craig–Sakamoto or Craig–Sakamoto–Matusita theorem. Craig's theorem has a long and tortuous history that includes numerous attempts at proofs of necessity, many of which have been judged to be incomplete or otherwise flawed or deficient. Accounts of this history are provided by, for example, Driscoll and Gundberg (1986), Reid and Driscoll (1988), and Driscoll and Krasnicka (1995) and, more recently, Ogawa and Olkin (2008).

§8c. For more on the kind of variations on Cochran's theorem that are the subject of this subsection, refer, e.g., to Anderson and Fang (1987).

§8d. The proof of Theorem 6.8.17 is based on a proof presented by Rao and Mitra (1971, pp. 170–171). For some discussion (of a historical nature and also of a more general nature) pertaining to alternative proofs of the result of Theorem 6.8.17, refer to Ogawa and Olkin (2008).

§8e. The proof (of the "necessity part" of Theorem 6.8.3) presented in Section 6.8e is based on Theorem 6.8.19, which was proved (in Section 6.8d) by making use of Theorem 6.7.13 (on polynomials). Reid and Driscoll (1988) and Driscoll and Krasnicka (1995) introduced an alternative proof of the necessity of the conditions under which two or more quadratic forms (in a normally distributed random vector) are distributed independently—refer also to Khuri (2010, sec. 5.3).

Exercise 7. The result of Exercise 7 is essentially the same as Theorem 1.6 of Fang, Kotz, and Ng (1990).

Exercise 26. Conditions (1) and (3) of Exercise 26 correspond to conditions given by Shanbhag (1968).

7

Confidence Intervals (or Sets) and Tests of Hypotheses

Suppose that $\mathbf{y}$ is an $N \times 1$ observable random vector that follows a G–M model. And suppose that we wish to make inferences about a parametric function of the form $\boldsymbol{\lambda}'\boldsymbol{\beta}$ (where $\boldsymbol{\lambda}$ is a $P \times 1$ vector of constants) or, more generally, about a vector of such parametric functions. Or suppose that we wish to make inferences about the realization of an unobservable random variable whose expected value is of the form $\boldsymbol{\lambda}'\boldsymbol{\beta}$ or about the realization of a vector of such random variables. Inferences that take the form of point estimation or prediction were considered in Chapter 5. The present chapter is devoted to inferences that take the form of an interval or set of values. More specifically, it is devoted to confidence intervals and sets (and to the corresponding tests of hypotheses).

7.1 "Setting the Stage": Response Surfaces in the Context of a Specific Application and in General

Recall (from Section 4.2e) the description and discussion of the experimental study of how the yield of lettuce plants is affected by the levels of the three trace minerals Cu, Mo, and Fe. Let $\mathbf{u} = (u_1, u_2, u_3)'$, where u_1, u_2, and u_3 represent the transformed amounts of Cu, Mo, and Fe, respectively. The data from the experimental study consisted of the 20 yields listed in column 1 of Table 4.3. The ith of these yields came from the plants in a container in which the value, say $\mathbf{u}_i = (u_{i1}, u_{i2}, u_{i3})'$, of $\mathbf{u}$ is the 3×1 vector whose elements u_{i1}, u_{i2}, and u_{i3} are the values of u_1, u_2, and u_3 (i.e., of the transformed amounts of Cu, Mo, and Fe) listed in the ith row of Table 4.3 ($i = 1, 2, \ldots, 20$). The 20 yields are regarded as the observed values of random variables $y_1, y_2, \ldots, y_{20}$, respectively. It is assumed that $\mathrm{E}(y_i) = \delta(\mathbf{u}_i)$ ($i = 1, 2, \ldots, 20$) for some function $\delta(\mathbf{u})$ of the vector $\mathbf{u}$. The function $\delta(\mathbf{u})$ is an example of what (in geometric terms) is customarily referred to as a response surface.

Among the choices for the function $\delta(\mathbf{u})$ is that of a polynomial. Whether or not such a choice is likely to be satisfactory depends in part on the degree of the polynomial and on the relevant domain (i.e., relevant set of $\mathbf{u}$-values). In the case of a first-, second-, or third-order polynomial,

$$\delta(\mathbf{u}) = \beta_1 + \beta_2 u_1 + \beta_3 u_2 + \beta_4 u_3, \tag{1.1}$$

$$\begin{aligned}\delta(\mathbf{u}) = \beta_1 + \beta_2 u_1 + \beta_3 u_2 + \beta_4 u_3 \\ + \beta_{11} u_1^2 + \beta_{12} u_1 u_2 + \beta_{13} u_1 u_3 + \beta_{22} u_2^2 + \beta_{23} u_2 u_3 + \beta_{33} u_3^2,\end{aligned} \tag{1.2}$$

or

$$\begin{aligned}\delta(\mathbf{u}) = \beta_1 + \beta_2 u_1 + \beta_3 u_2 + \beta_4 u_3 \\ + \beta_{11} u_1^2 + \beta_{12} u_1 u_2 + \beta_{13} u_1 u_3 + \beta_{22} u_2^2 + \beta_{23} u_2 u_3 + \beta_{33} u_3^2 \\ + \beta_{111} u_1^3 + \beta_{112} u_1^2 u_2 + \beta_{113} u_1^2 u_3 + \beta_{122} u_1 u_2^2 + \beta_{123} u_1 u_2 u_3 \\ + \beta_{133} u_1 u_3^2 + \beta_{222} u_2^3 + \beta_{223} u_2^2 u_3 + \beta_{233} u_2 u_3^2 + \beta_{333} u_3^3,\end{aligned} \tag{1.3}$$

respectively. Here, the coefficients β_j ($j = 1, 2, \ldots, 4$), β_{jk} ($j = 1, 2, 3;\ k = j, \ldots, 3$), and $\beta_{jk\ell}$

($j = 1, 2, 3$; $k = j, \ldots, 3$; $\ell = k, \ldots, 3$) are regarded as unknown (unconstrained) parameters. Upon taking the function $\delta(\mathbf{u})$ to be the first-, second-, or third-order polynomial (1.1), (1.2), or (1.3), we obtain a G–M model (with $P = 4$, 10, or 20, respectively).

Let us (in the present context) refer to the three G–M models corresponding to the three choices (1.1), (1.2), and (1.3) for $\delta(\mathbf{u})$ as the first-order, second-order, and third-order models, respectively. Each of these models is expressible in terms related to the general formulation (in Section 4.1) of the G–M model. In each case, N equals 20, and for the sake of consistency with the notation introduced in Section 4.1 (where $y_1, y_2, \ldots, y_N$ were taken to be the first through Nth elements of $\mathbf{y}$) take $\mathbf{y}$ to be the 20×1 random vector whose ith element is the random variable y_i (the observed value of which is the ith of the lettuce yields listed in column 1 of Table 4.3). Further, letting $\boldsymbol{\beta}_1 = (\beta_1, \beta_2, \beta_3, \beta_4)'$, $\boldsymbol{\beta}_2 = (\beta_{11}, \beta_{12}, \beta_{13}, \beta_{22}, \beta_{23}, \beta_{33})'$, and $\boldsymbol{\beta}_3 = (\beta_{111}, \beta_{112}, \beta_{113}, \beta_{122}, \beta_{123}, \beta_{133}, \beta_{222}, \beta_{223}, \beta_{233}, \beta_{333})'$, take $\boldsymbol{\beta}$ for the first-, second-, or third-order model to be $\boldsymbol{\beta} = \boldsymbol{\beta}_1$, $\boldsymbol{\beta} = \begin{pmatrix} \boldsymbol{\beta}_1 \\ \boldsymbol{\beta}_2 \end{pmatrix}$, or $\boldsymbol{\beta} = \begin{pmatrix} \boldsymbol{\beta}_1 \\ \boldsymbol{\beta}_2 \\ \boldsymbol{\beta}_3 \end{pmatrix}$, respectively (in which case $P = 4$, $P = 10$, or $P = 20$, respectively). Then, letting $\mathbf{X}_1$ represent the 20×4 matrix with ith row $(1, \mathbf{u}_i') = (1, u_{i1}, u_{i2}, u_{i3})$, $\mathbf{X}_2$ the 20×6 matrix with ith row $(u_{i1}^2, u_{i1}u_{i2}, u_{i1}u_{i3}, u_{i2}^2, u_{i2}u_{i3}, u_{i3}^2)$, and $\mathbf{X}_3$ the 20×10 matrix with ith row $(u_{i1}^3, u_{i1}^2 u_{i2}, u_{i1}^2 u_{i3}, u_{i1}u_{i2}^2, u_{i1}u_{i2}u_{i3}, u_{i1}u_{i3}^2, u_{i2}^3, u_{i2}^2 u_{i3}, u_{i2}u_{i3}^2, u_{i3}^3)$, the model matrix $\mathbf{X}$ for the first-, second-, or third-order model is $\mathbf{X} = \mathbf{X}_1$, $\mathbf{X} = (\mathbf{X}_1, \mathbf{X}_2)$, or $\mathbf{X} = (\mathbf{X}_1, \mathbf{X}_2, \mathbf{X}_3)$, respectively.

The lettuce-yield data were obtained from a designed experiment. What seemed to be of interest were inferences about the response surface over a targeted region and perhaps inferences about the yields of lettuce that might be obtained from plants grown in the future under conditions similar to those present in the experimental study. Of particular interest were various characteristics of the response surface; these included the presence and magnitude of any interactions among Cu, Mo, and Fe (in their effects on yield) and the location of the optimal combination of levels of Cu, Mo, and Fe (i.e., the combination that results in the largest expected yield).

The range of $\mathbf{u}$-values for which the data were to be used to make inferences about the response surface (or its characteristics) and the range of $\mathbf{u}$-values represented in the experiment were such that (in light of information available from earlier studies) a second-order model was adopted. The intended design was of a kind known as a rotatable central composite design (e.g., Myers, Montgomery, and Anderson-Cook 2016, chap. 8). However, due to an error, the experiment was carried out in such a way that the level of Fe (on the transformed scale) was -0.4965 in the containers for which the intended level was -1—there were 4 such containers.

Suppose that $\delta(\mathbf{u})$ is the second-order polynomial (1.2) in the elements of the vector $\mathbf{u} = (u_1, u_2, u_3)'$. Or, more generally, suppose that $\delta(\mathbf{u})$ is the second-order polynomial (4.2.14) in the elements of the vector $\mathbf{u} = (u_1, u_2, \ldots, u_C)'$. Then, $\delta(\mathbf{u})$ is reexpressible in matrix notation. Clearly,

$$\delta(\mathbf{u}) = \beta_1 + \mathbf{a}'\mathbf{u} + \mathbf{u}'\mathbf{A}\mathbf{u}, \tag{1.4}$$

where $\mathbf{a} = (\beta_2, \beta_3, \ldots, \beta_{C+1})'$ and where $\mathbf{A}$ is the $C \times C$ symmetric matrix with ith diagonal element β_{ii} and with (for $j > i$) ijth and jith off-diagonal elements $\beta_{ij}/2$.

An expression for the gradient vector of the second-order polynomial $\delta(\mathbf{u})$ is obtained upon applying results (5.4.7) and (5.4.10) on vector differentiation. We find that

$$\frac{\partial \delta(\mathbf{u})}{\partial \mathbf{u}} = \mathbf{a} + 2\mathbf{A}\mathbf{u}. \tag{1.5}$$

By definition, a point, say $\mathbf{u}_0$, is a stationary point of $\delta(\mathbf{u})$ if and only if $\left.\frac{\partial \delta(\mathbf{u})}{\partial \mathbf{u}}\right|_{\mathbf{u}=\mathbf{u}_0} = \mathbf{0}$ and hence if and only if $\mathbf{u} = \mathbf{u}_0$ is a solution to the following linear system (in the vector $\mathbf{u}$):

$$-2\mathbf{A}\mathbf{u} = \mathbf{a}, \tag{1.6}$$

in which case

$$\delta(\mathbf{u}_0) = \beta_1 + \mathbf{a}'\mathbf{u}_0 + (\mathbf{A}\mathbf{u}_0)'\mathbf{u}_0 = \beta_1 + (1/2)\mathbf{a}'\mathbf{u}_0. \tag{1.7}$$

The linear system (1.6) is consistent (i.e., has a solution) if the matrix $\mathbf{A}$ is nonsingular, in which case the linear system has a unique solution $\mathbf{u}_0$ that is expressible as

$$\mathbf{u}_0 = -(1/2)\mathbf{A}^{-1}\mathbf{a}. \tag{1.8}$$

More generally, linear system (1.6) is consistent if (and only if) $\mathbf{a} \in \mathcal{C}(\mathbf{A})$.

In light of result (5.4.11), the Hessian matrix of $\delta(\mathbf{u})$ is expressible as follows:

$$\frac{\partial^2 \delta(\mathbf{u})}{\partial \mathbf{u}\, \partial \mathbf{u}'} = 2\mathbf{A}. \tag{1.9}$$

If the matrix $-\mathbf{A}$ is nonnegative definite, then the stationary points of $\delta(\mathbf{u})$ are points at which $\delta(\mathbf{u})$ attains a maximum value—refer, e.g., to Harville (1997, sec. 19.1).

Let us consider the use of the data on lettuce yields in making inferences (on the basis of the second-order model) about the response surface and various of its characteristics. Specifically, let us consider the use of these data in making inferences about the value of the second-order polynomial $\delta(\mathbf{u})$ for each value of $\mathbf{u}$ in the relevant region. And let us consider the use of these data in making inferences about the parameters of the second-order model [and hence about the values of the first-order derivatives (at $\mathbf{u} = \mathbf{0}$) and second-order derivatives of the second-order polynomial $\delta(\mathbf{u})$] and in making inferences about the location of the stationary points of $\delta(\mathbf{u})$.

The (20×10) model matrix $\mathbf{X} = (\mathbf{X}_1, \mathbf{X}_2)$ is of full column rank 10—the model matrix is sufficiently simple that its rank can be determined without resort to numerical means. Thus, all ten of the parameters that form the elements of $\boldsymbol{\beta}$ are estimable (and every linear combination of these parameters is estimable).

Let $\hat{\boldsymbol{\beta}} = (\hat{\beta}_1, \hat{\beta}_2, \hat{\beta}_3, \hat{\beta}_4, \hat{\beta}_{11}, \hat{\beta}_{12}, \hat{\beta}_{13}, \hat{\beta}_{22}, \hat{\beta}_{23}, \hat{\beta}_{33})'$ represent the least squares estimator of $\boldsymbol{\beta}$, that is, the 10×1 vector whose elements are the least squares estimators of the corresponding elements of $\boldsymbol{\beta}$; let $\hat{\sigma}^2$ represent the usual unbiased estimator of σ^2; and denote by $\underset{\sim}{\mathbf{y}}$ the data vector (the elements of which are listed in column 1 of Table 4.3). The least squares estimate of $\boldsymbol{\beta}$ (which is the value of $\hat{\boldsymbol{\beta}}$ when $\mathbf{y} = \underset{\sim}{\mathbf{y}}$) is obtainable as the (unique) solution, say $\tilde{\mathbf{b}}$, to the linear system $\mathbf{X}'\mathbf{X}\mathbf{b} = \mathbf{X}'\underset{\sim}{\mathbf{y}}$ (in the vector $\mathbf{b}$) comprising the so-called normal equations. The residual sum of squares equals 108.9407; dividing this quantity by $N - P = 10$ (to obtain the value of $\hat{\sigma}^2$) gives 10.89 as an estimate of σ^2. Upon taking the square root of this estimate of σ^2, we obtain 3.30 as an estimate of σ. The variance-covariance matrix of $\hat{\boldsymbol{\beta}}$ is expressible as $\text{var}(\hat{\boldsymbol{\beta}}) = \sigma^2(\mathbf{X}'\mathbf{X})^{-1}$ and is estimated unbiasedly by the matrix $\hat{\sigma}^2(\mathbf{X}'\mathbf{X})^{-1}$. The standard errors of the elements of $\hat{\boldsymbol{\beta}}$ are given by the square roots of the diagonal elements of $\sigma^2(\mathbf{X}'\mathbf{X})^{-1}$, and the estimated standard errors (those corresponding to the estimator $\hat{\sigma}^2$) are given by the square roots of the values of the diagonal elements of $\hat{\sigma}^2(\mathbf{X}'\mathbf{X})^{-1}$.

The least squares estimates of the elements of $\boldsymbol{\beta}$ (i.e., of the parameters $\beta_1, \beta_2, \beta_3, \beta_4, \beta_{11}, \beta_{12}, \beta_{13}, \beta_{22}, \beta_{23}$, and β_{33}) are presented in Table 7.1 along with their standard errors and estimated standard errors. And letting $\mathbf{S}$ represent the diagonal matrix of order $P = 10$ whose first through Pth diagonal elements are respectively the square roots of the first through Pth diagonal elements of the $P \times P$ matrix $(\mathbf{X}'\mathbf{X})^{-1}$, the correlation matrix of the vector $\hat{\boldsymbol{\beta}}$ of least squares estimators of the elements of $\boldsymbol{\beta}$ is

$$\mathbf{S}^{-1}(\mathbf{X}'\mathbf{X})^{-1}\mathbf{S}^{-1} = \begin{pmatrix} 1.00 & 0.00 & 0.00 & 0.09 & -0.57 & 0.00 & 0.00 & -0.57 & 0.00 & -0.52 \\ 0.00 & 1.00 & 0.00 & 0.00 & 0.00 & 0.00 & -0.24 & 0.00 & 0.00 & 0.00 \\ 0.00 & 0.00 & 1.00 & 0.00 & 0.00 & 0.00 & 0.00 & 0.00 & -0.24 & 0.00 \\ 0.09 & 0.00 & 0.00 & 1.00 & -0.10 & 0.00 & 0.00 & -0.10 & 0.00 & -0.22 \\ -0.57 & 0.00 & 0.00 & -0.10 & 1.00 & 0.00 & 0.00 & 0.13 & 0.00 & 0.19 \\ 0.00 & 0.00 & 0.00 & 0.00 & 0.00 & 1.00 & 0.00 & 0.00 & 0.00 & 0.00 \\ 0.00 & -0.24 & 0.00 & 0.00 & 0.00 & 0.00 & 1.00 & 0.00 & 0.00 & 0.00 \\ -0.57 & 0.00 & 0.00 & -0.10 & 0.13 & 0.00 & 0.00 & 1.00 & 0.00 & 0.19 \\ 0.00 & 0.00 & -0.24 & 0.00 & 0.00 & 0.00 & 0.00 & 0.00 & 1.00 & 0.00 \\ -0.52 & 0.00 & 0.00 & -0.22 & 0.19 & 0.00 & 0.00 & 0.19 & 0.00 & 1.00 \end{pmatrix}.$$

TABLE 7.1. Least squares estimates (with standard errors and estimated standard errors) obtained from the lettuce-yield data for the regression coefficients in a second-order G–M model.

Coefficient of	Regression coefficient	Least squares estimate	Std. error of the estimator	Estimated std. error of the estimator
1	β_1	24.31	0.404σ	1.34
u_1	β_2	−4.57	0.279σ	0.92
u_2	β_3	−0.82	0.279σ	0.92
u_3	β_4	−0.03	0.318σ	1.05
u_1^2	β_{11}	−5.27	0.267σ	0.88
u_1u_2	β_{12}	−1.31	0.354σ	1.17
u_1u_3	β_{13}	−1.29	0.462σ	1.52
u_2^2	β_{22}	−0.73	0.267σ	0.88
u_2u_3	β_{23}	0.66	0.462σ	1.52
u_3^2	β_{33}	−5.34	0.276σ	0.91

The model matrix $\mathbf{X}$ has a relatively simple structure, as might be expected in the case of a designed experiment. That structure is reflected in the correlation matrix of $\hat{\boldsymbol{\beta}}$ and in the various other quantities that depend on the model matrix through the matrix $\mathbf{X}'\mathbf{X}$. Of course, that structure would have been even simpler had not the level of Fe in the four containers where it was supposed to have been −1 (on the transformed scale) been taken instead to be −0.4965.

Assuming that the distribution of the vector $\mathbf{e}$ of residual effects in the G–M model is $N(\mathbf{0}, \sigma^2\mathbf{I})$ or, more generally, that the fourth-order moments of its distribution are identical to those of the $N(\mathbf{0}, \sigma^2\mathbf{I})$ distribution,

$$\text{var}(\hat{\sigma}^2) = \frac{2\sigma^4}{N - \text{rank}\,\mathbf{X}} = \frac{\sigma^4}{5} \tag{1.10}$$

—refer to result (5.7.40). And (under the same assumption)

$$\text{E}(\hat{\sigma}^4) = \text{var}(\hat{\sigma}^2) + [\text{E}(\hat{\sigma}^2)]^2 = \frac{N - \text{rank}(\mathbf{X}) + 2}{N - \text{rank}(\mathbf{X})}\,\sigma^4$$

and, consequently, $\text{var}(\hat{\sigma}^2)$ is estimated unbiasedly by

$$\frac{2\hat{\sigma}^4}{N - \text{rank}(\mathbf{X}) + 2}. \tag{1.11}$$

Corresponding to expression (1.10) for $\text{var}(\hat{\sigma}^2)$ is the expression

$$\sqrt{\frac{2}{N - \text{rank}\,\mathbf{X}}}\,\sigma^2 = \frac{\sigma^2}{\sqrt{5}} \tag{1.12}$$

for the standard error of $\hat{\sigma}^2$, and corresponding to the estimator (1.11) of $\text{var}(\hat{\sigma}^2)$ is the estimator

$$\sqrt{\frac{2}{N - \text{rank}(\mathbf{X}) + 2}}\,\hat{\sigma}^2. \tag{1.13}$$

of the standard error of $\hat{\sigma}^2$. The estimated standard error of $\hat{\sigma}^2$ [i.e., the value of the estimator (1.13)] is

$$\sqrt{\frac{2}{12}}\,10.89407 = 4.45.$$

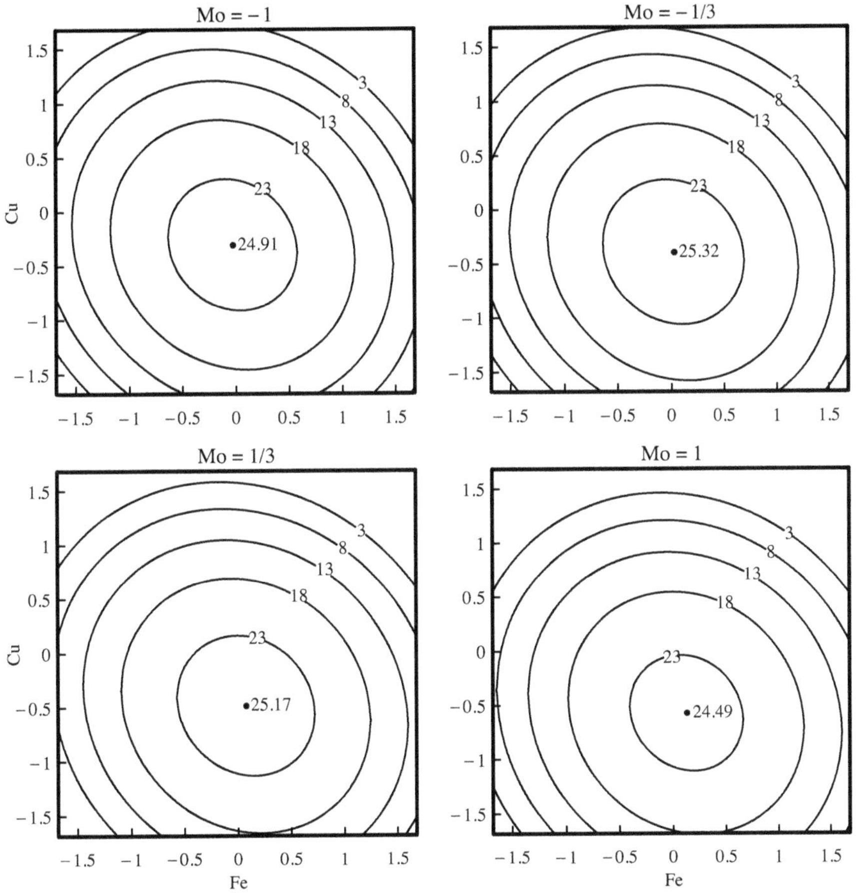

FIGURE 7.1. Contour plots of the estimated response surface obtained from the lettuce-yield data (on the basis of a second-order model). Each plot serves to relate the yield of lettuce plants to the levels of 2 trace minerals (Cu and Fe) at one of 4 levels of a third trace mineral (Mo).

Let

$$\hat{\mathbf{a}} = \begin{pmatrix} \hat{\beta}_2 \\ \hat{\beta}_3 \\ \hat{\beta}_4 \end{pmatrix} \quad \text{and} \quad \hat{\mathbf{A}} = \begin{pmatrix} \hat{\beta}_{11} & \frac{1}{2}\hat{\beta}_{12} & \frac{1}{2}\hat{\beta}_{13} \\ \frac{1}{2}\hat{\beta}_{12} & \hat{\beta}_{22} & \frac{1}{2}\hat{\beta}_{23} \\ \frac{1}{2}\hat{\beta}_{13} & \frac{1}{2}\hat{\beta}_{23} & \hat{\beta}_{33} \end{pmatrix}.$$

For any particular value of the vector $\mathbf{u} = (u_1, u_2, u_3)'$ (the elements of which represent the transformed levels of Cu, Mo, and Fe), the value of $\delta(\mathbf{u})$ is a linear combination of the ten regression coefficients β_1, β_2, β_3, β_4, β_{11}, β_{12}, β_{13}, β_{22}, β_{23}, and β_{33}. And upon taking the same linear combination of the least squares estimators of the regression coefficients, we obtain the least squares estimator of the value of $\delta(\mathbf{u})$. Accordingly, the least squares estimator of the value of $\delta(\mathbf{u})$ is the value of the function $\hat{\delta}(\mathbf{u})$ (of $\mathbf{u}$) defined as follows:

$$\hat{\delta}(\mathbf{u}) = \hat{\beta}_1 + \hat{\mathbf{a}}'\mathbf{u} + \mathbf{u}'\hat{\mathbf{A}}\mathbf{u}. \tag{1.14}$$

The estimated response surface defined by this function is depicted in Figure 7.1 in the form of four contour plots; each of these plots corresponds to a different one of four levels of Mo.

For any linear combination $\boldsymbol{\lambda}'\boldsymbol{\beta}$ of the regression coefficients β_1, β_2, β_3, β_4, β_{11}, β_{12}, β_{13}, β_{22}, β_{23}, and β_{33},

$$\text{var}(\boldsymbol{\lambda}'\hat{\boldsymbol{\beta}}) = \sigma^2 \boldsymbol{\lambda}'(\mathbf{X}'\mathbf{X})^{-1}\boldsymbol{\lambda}. \tag{1.15}$$

More generally, for any two linear combinations $\boldsymbol{\lambda}'\boldsymbol{\beta}$ and $\boldsymbol{\ell}'\boldsymbol{\beta}$,

$$\text{cov}(\boldsymbol{\lambda}'\hat{\boldsymbol{\beta}}, \boldsymbol{\ell}'\hat{\boldsymbol{\beta}}) = \sigma^2 \boldsymbol{\lambda}'(\mathbf{X}'\mathbf{X})^{-1}\boldsymbol{\ell}. \tag{1.16}$$

And upon replacing σ^2 in expression (1.15) or (1.16) with the unbiased estimator $\hat{\sigma}^2$, we obtain an unbiased estimator of $\text{var}(\boldsymbol{\lambda}'\hat{\boldsymbol{\beta}})$ or $\text{cov}(\boldsymbol{\lambda}'\hat{\boldsymbol{\beta}}, \boldsymbol{\ell}'\hat{\boldsymbol{\beta}})$. Further, upon taking the square root of $\text{var}(\boldsymbol{\lambda}'\hat{\boldsymbol{\beta}})$ and of its unbiased estimator, we obtain $\sigma\sqrt{\boldsymbol{\lambda}'(\mathbf{X}'\mathbf{X})^{-1}\boldsymbol{\lambda}}$ as an expression for the standard error of $\boldsymbol{\lambda}'\hat{\boldsymbol{\beta}}$ and $\hat{\sigma}\sqrt{\boldsymbol{\lambda}'(\mathbf{X}'\mathbf{X})^{-1}\boldsymbol{\lambda}}$ as an estimator of the standard error.

Note that these results (on the least squares estimators of the arbitrary linear combinations $\boldsymbol{\lambda}'\boldsymbol{\beta}$ and $\boldsymbol{\ell}'\boldsymbol{\beta}$) are applicable to $\hat{\delta}(\mathbf{u})$ and $\hat{\delta}(\mathbf{v})$, where $\mathbf{u} = (u_1, u_2, u_3)'$ and $\mathbf{v} = (v_1, v_2, v_3)'$ represent any particular points in 3-dimensional space. Upon setting $\boldsymbol{\lambda} = (1, u_1, u_2, u_3, u_1^2, u_1u_2, u_1u_3, u_2^2, u_2u_3, u_3^2)'$ and $\boldsymbol{\ell} = (1, v_1, v_2, v_3, v_1^2, v_1v_2, v_1v_3, v_2^2, v_2v_3, v_3^2)'$, we find that $\boldsymbol{\lambda}'\hat{\boldsymbol{\beta}} = \hat{\delta}(\mathbf{u})$ and $\boldsymbol{\ell}'\hat{\boldsymbol{\beta}} = \hat{\delta}(\mathbf{v})$. Clearly, $\hat{\delta}(\mathbf{0}) = \hat{\beta}_1$.

The results on the least squares estimators of the arbitrary linear combinations $\boldsymbol{\lambda}'\boldsymbol{\beta}$ and $\boldsymbol{\ell}'\boldsymbol{\beta}$ are also applicable to the elements of the vector $\hat{\mathbf{a}} + 2\hat{\mathbf{A}}\mathbf{u}$ and the elements of the matrix $2\hat{\mathbf{A}}$. These quantities are the least squares estimators of the elements of the vector $\mathbf{a} + 2\mathbf{A}\mathbf{u}$ and the elements of the matrix $2\mathbf{A}$—clearly, each element of $\mathbf{a} + 2\mathbf{A}\mathbf{u}$ and $2\mathbf{A}$ is expressible as a linear combination of the elements of $\boldsymbol{\beta}$. And (under the second-order G–M model) $\mathbf{a} + 2\mathbf{A}\mathbf{u}$ is the gradient vector and $2\mathbf{A}$ the Hessian matrix of the function $\delta(\mathbf{u})$. Note that, when $\mathbf{u} = \mathbf{0}$, the estimator $\hat{\mathbf{a}} + 2\hat{\mathbf{A}}\mathbf{u}$ of the gradient vector simplifies to the vector $\hat{\mathbf{a}}$, the elements of which are $\hat{\beta}_2$, $\hat{\beta}_3$, and $\hat{\beta}_4$.

Like the function $\delta(\mathbf{u})$, the function $\hat{\delta}(\mathbf{u})$ is a second-degree polynomial in the elements of the vector $\mathbf{u} = (u_1, u_2, u_3)'$. And analogous to results (1.5) and (1.9), we have that

$$\frac{\partial \hat{\delta}(\mathbf{u})}{\partial \mathbf{u}} = \hat{\mathbf{a}} + 2\hat{\mathbf{A}}\mathbf{u} \quad \text{and} \quad \frac{\partial^2 \hat{\delta}(\mathbf{u})}{\partial \mathbf{u}\, \partial \mathbf{u}'} = 2\hat{\mathbf{A}}. \tag{1.17}$$

Assume that $\hat{\mathbf{A}}$ is nonsingular with probability 1 (as would be the case if, e.g., the distribution of the vector $\mathbf{e}$ of residual effects in the second-order G–M model were MVN), and define

$$\hat{\mathbf{u}}_0 = -(1/2)\hat{\mathbf{A}}^{-1}\hat{\mathbf{a}}. \tag{1.18}$$

Then, with probability 1, the estimated response surface $\hat{\delta}(\mathbf{u})$ has a unique stationary point and that point equals $\hat{\mathbf{u}}_0$. Further,

$$\hat{\delta}(\hat{\mathbf{u}}_0) = \hat{\beta}_1 + (1/2)\hat{\mathbf{a}}'\hat{\mathbf{u}}_0; \tag{1.19}$$

and if $-\hat{\mathbf{A}}$ is positive definite, $\hat{\delta}(\mathbf{u})$ attains its maximum value (uniquely) at $\hat{\mathbf{u}}_0$. When the least squares estimates of β_1 and of the elements of $\mathbf{a}$ and $\mathbf{A}$ are those obtained from the lettuce-yield data, we find (upon, e.g., making use of Theorem 2.14.23) that $-\hat{\mathbf{A}}$ is positive definite and that

$$\hat{\mathbf{u}}_0 = (-0.42, -0.17, 0.04)' \quad \text{and} \quad \hat{\delta}(\hat{\mathbf{u}}_0) = 25.33.$$

Suppose that the second-order regression coefficients β_{11}, β_{12}, β_{13}, β_{22}, β_{23}, and β_{33} are such that the matrix $\mathbf{A}$ is nonsingular, in which case the function $\delta(\mathbf{u})$ has a unique stationary point $\mathbf{u}_0 = -(1/2)\mathbf{A}^{-1}\mathbf{a}$. Then, it seems natural to regard $\hat{\mathbf{u}}_0$ as an estimator of $\mathbf{u}_0$ (as suggested by the notation).

A more formal justification for regarding $\hat{\mathbf{u}}_0$ as an estimator of $\mathbf{u}_0$ is possible. Suppose that the distribution of the vector $\mathbf{e}$ of residual effects in the second-order (G–M) model is MVN. Then, it follows from the results of Section 5.9a that the elements of the matrix $\hat{\mathbf{A}}$ and the vector $\hat{\mathbf{a}}$ are the ML estimators of the corresponding elements of the matrix $\mathbf{A}$ and the vector $\mathbf{a}$ (and are the ML estimators even when the values of the second-order regression coefficients β_{11}, β_{12}, β_{13}, β_{22}, β_{23}, and β_{33} are restricted to those values for which $\mathbf{A}$ is nonsingular). And upon applying a well-known general result on the ML estimation of parametric functions [which is Theorem 5.1.1 of Zacks (1971)], we conclude that $\hat{\mathbf{u}}_0$ is the ML estimator of $\mathbf{u}_0$.

In making inferences from the results of the experimental study (of the yield of lettuce plants), there could be interest in predictive inferences as well as in inferences about the function $\delta(\mathbf{u})$ (and about various characteristics of this function). Specifically, there could be interest in making inferences about the yield to be obtained in the future from a container of lettuce plants, based on regarding the future yield as a realization of the random variable $\delta(\mathbf{u}) + d$, where d is a random variable that has an expected value of 0 and a variance of σ^2 (and that is uncorrelated with the vector $\mathbf{e}$). Then, with regard to point prediction, the BLUP of this quantity equals the least squares estimator $\hat{\delta}(\mathbf{u})$ of $\delta(\mathbf{u})$, and the variance of the prediction error (which equals the mean squared error of the BLUP) is $\sigma^2 + \text{var}[\hat{\delta}(\mathbf{u})]$. And the covariance of the prediction errors of the BLUPs of two future yields, one of which is modeled as the realization of $\delta(\mathbf{u}) + d$ and the other as the realization of $\delta(\mathbf{v}) + h$ (where h is a random variable that has an expected value of 0 and a variance of σ^2 and that is uncorrelated with $\mathbf{e}$ and with d), equals $\text{cov}[\hat{\delta}(\mathbf{u}),\, \hat{\delta}(\mathbf{v})]$.

Point estimation (or prediction) can be quite informative, especially when accompanied by standard errors or other quantities that reflect the magnitude of the underlying variability. However, it is generally desirable to augment any such inferences with inferences that take the form of intervals or sets. Following the presentation (in Section 7.2) of some results on "multi-part" G–M models, the emphasis (beginning in Section 7.3) in the present chapter is on confidence intervals and sets (and on the closely related topic of tests of hypotheses).

7.2 Augmented G–M Model

Section 7.1 was devoted to a discussion of the use of the data (from Section 4.2e) on the yield of lettuce plants in making statistical inferences on the basis of a G–M model. Three different versions of the G–M model were considered; these were referred to as the first-, second-, and third-order models. The second of these versions is obtainable from the first and the third from the second via the introduction of some additional terms. Let us consider the effects (both in general and in the context of the lettuce-yield application) of the inclusion of additional terms in a G–M model.

a. General results

Let $\mathbf{Z} = \{z_{ij}\}$ represent a matrix with N rows, and denote by Q the number of columns in $\mathbf{Z}$. And as an alternative to the original G–M model with model matrix $\mathbf{X}$, consider the following G–M model with model matrix $(\mathbf{X},\, \mathbf{Z})$:

$$\mathbf{y} = (\mathbf{X},\, \mathbf{Z})\begin{pmatrix}\boldsymbol{\beta}\\ \boldsymbol{\tau}\end{pmatrix} + \mathbf{e}, \tag{2.1}$$

where $\boldsymbol{\tau} = (\tau_1, \tau_2, \ldots, \tau_Q)'$ is a Q-dimensional (column) vector of additional parameters [and where the parameter space for the augmented parameter vector $\begin{pmatrix}\boldsymbol{\beta}\\ \boldsymbol{\tau}\end{pmatrix}$ is $\mathbb{R}^{P+Q}$]. To distinguish this model from the original G–M model, let us refer to it as the augmented G–M model. Note that the model equation (2.1) for the augmented G–M model is reexpressible in the form

$$\mathbf{y} = \mathbf{X}\boldsymbol{\beta} + \mathbf{Z}\boldsymbol{\tau} + \mathbf{e}. \tag{2.2}$$

Let $\boldsymbol{\Lambda}$ represent a matrix (with P rows) such that $\mathcal{R}(\boldsymbol{\Lambda}') = \mathcal{R}(\mathbf{X})$, and denote by M the number of columns in $\boldsymbol{\Lambda}$ (in which case $M \geq \text{rank}\,\boldsymbol{\Lambda} = \text{rank}\,\mathbf{X}$). Then, under the original G–M model $\mathbf{y} = \mathbf{X}\boldsymbol{\beta} + \mathbf{e}$, the M elements of the vector $\boldsymbol{\Lambda}'\boldsymbol{\beta}$ are estimable linear combinations of the elements of $\boldsymbol{\beta}$. Moreover, these estimable linear combinations include $\text{rank}(\mathbf{X})$ linearly independent estimable linear combinations [and no set of linearly independent estimable linear combinations can include more than $\text{rank}(\mathbf{X})$ linear combinations].

Under the original G–M model, the least squares estimators of the elements of the vector $\mathbf{\Lambda}'\boldsymbol{\beta}$ are the elements of the vector

$$\tilde{\mathbf{R}}'\mathbf{X}'\mathbf{y}, \tag{2.3}$$

where $\tilde{\mathbf{R}}$ is any solution to the linear system

$$\mathbf{X}'\mathbf{X}\mathbf{R} = \mathbf{\Lambda} \tag{2.4}$$

(in the $P \times M$ matrix $\mathbf{R}$)—refer to Section 5.4 or 5.6. Note that the matrix $\mathbf{X}\mathbf{R}$ and hence its transpose $\mathbf{R}'\mathbf{X}'$ do not vary with the choice of solution to linear system (2.4) (as is evident from, e.g., Corollary 2.3.4).

Under the augmented G–M model,

$$\text{E}(\tilde{\mathbf{R}}'\mathbf{X}'\mathbf{y}) = \tilde{\mathbf{R}}'\mathbf{X}'(\mathbf{X}\boldsymbol{\beta} + \mathbf{Z}\boldsymbol{\tau}) = \mathbf{\Lambda}'\boldsymbol{\beta} + \tilde{\mathbf{R}}'\mathbf{X}'\mathbf{Z}\boldsymbol{\tau}. \tag{2.5}$$

Thus, while $\tilde{\mathbf{R}}'\mathbf{X}'\mathbf{y}$ is an unbiased estimator of $\mathbf{\Lambda}'\boldsymbol{\beta}$ under the original G–M model (in fact, it is the best linear unbiased estimator of $\mathbf{\Lambda}'\boldsymbol{\beta}$ in the sense described in Section 5.6), it is not (in general) an unbiased estimator of $\mathbf{\Lambda}'\boldsymbol{\beta}$ under the augmented G–M model. In fact, under the augmented G–M model, the elements of $\tilde{\mathbf{R}}'\mathbf{X}'\mathbf{y}$ are the least squares estimators of the elements of $\mathbf{\Lambda}'\boldsymbol{\beta} + \tilde{\mathbf{R}}'\mathbf{X}'\mathbf{Z}\boldsymbol{\tau}$, as is evident upon observing that

$$(\mathbf{X},\, \mathbf{Z})'(\mathbf{X},\, \mathbf{Z}) = \begin{pmatrix} \mathbf{X}'\mathbf{X} & \mathbf{X}'\mathbf{Z} \\ \mathbf{Z}'\mathbf{X} & \mathbf{Z}'\mathbf{Z} \end{pmatrix},$$

$$\mathbf{\Lambda}'\boldsymbol{\beta} + \tilde{\mathbf{R}}'\mathbf{X}'\mathbf{Z}\boldsymbol{\tau} = (\mathbf{\Lambda}',\, \tilde{\mathbf{R}}'\mathbf{X}'\mathbf{Z})\begin{pmatrix} \boldsymbol{\beta} \\ \boldsymbol{\tau} \end{pmatrix} = \begin{pmatrix} \mathbf{\Lambda} \\ \mathbf{Z}'\mathbf{X}\tilde{\mathbf{R}} \end{pmatrix}'\begin{pmatrix} \boldsymbol{\beta} \\ \boldsymbol{\tau} \end{pmatrix}, \tag{2.6}$$

and

$$\begin{pmatrix} \mathbf{X}'\mathbf{X} & \mathbf{X}'\mathbf{Z} \\ \mathbf{Z}'\mathbf{X} & \mathbf{Z}'\mathbf{Z} \end{pmatrix}\begin{pmatrix} \tilde{\mathbf{R}} \\ \mathbf{0} \end{pmatrix} = \begin{pmatrix} \mathbf{\Lambda} \\ \mathbf{Z}'\mathbf{X}\tilde{\mathbf{R}} \end{pmatrix}$$

and as is also evident upon observing (in light of the results of Section 5.4) that any linear combination of the elements of the vector $(\mathbf{X},\, \mathbf{Z})'\mathbf{y}$ is the least squares estimator of its expected value.

Note that while in general $\text{E}(\tilde{\mathbf{R}}'\mathbf{X}'\mathbf{y})$ is affected by the augmentation of the G–M model, $\text{var}(\tilde{\mathbf{R}}'\mathbf{X}'\mathbf{y})$ is unaffected (in the sense that the same expressions continue to apply).

In regard to the coefficient matrix $(\mathbf{\Lambda}',\, \tilde{\mathbf{R}}'\mathbf{X}'\mathbf{Z}) = \begin{pmatrix} \mathbf{\Lambda} \\ \mathbf{Z}'\mathbf{X}\tilde{\mathbf{R}} \end{pmatrix}'$ of expression (2.6) for the vector $\mathbf{\Lambda}'\boldsymbol{\beta} + \tilde{\mathbf{R}}'\mathbf{X}'\mathbf{Z}\boldsymbol{\tau}$ of linear combinations of the elements of the parametric vector $\begin{pmatrix} \boldsymbol{\beta} \\ \boldsymbol{\tau} \end{pmatrix}$, note that

$$\text{rank}(\mathbf{\Lambda}',\, \tilde{\mathbf{R}}'\mathbf{X}'\mathbf{Z}) = \text{rank}\, \mathbf{\Lambda} \quad (= \text{rank}\, \mathbf{X}), \tag{2.7}$$

as is evident upon observing that $\mathbf{\Lambda}' = \tilde{\mathbf{R}}'\mathbf{X}'\mathbf{X}$ and

$$(\mathbf{\Lambda}',\, \tilde{\mathbf{R}}'\mathbf{X}'\mathbf{Z}) = \tilde{\mathbf{R}}'\mathbf{X}'(\mathbf{X},\, \mathbf{Z})$$

and hence (in light of Lemma 2.12.3 and Corollary 2.4.17) that

$$\text{rank}(\mathbf{\Lambda}') = \text{rank}(\tilde{\mathbf{R}}'\mathbf{X}') \geq \text{rank}(\mathbf{\Lambda}',\, \tilde{\mathbf{R}}'\mathbf{X}'\mathbf{Z}) \geq \text{rank}(\mathbf{\Lambda}').$$

Further, for any M-dimensional column vector $\boldsymbol{\ell}$, we find (upon observing that $\boldsymbol{\ell}'\mathbf{\Lambda}' = (\tilde{\mathbf{R}}\boldsymbol{\ell})'\mathbf{X}'\mathbf{X}$ and making use of Corollary 2.3.4) that

$$\boldsymbol{\ell}'\mathbf{\Lambda}' = \mathbf{0} \;\Leftrightarrow\; (\tilde{\mathbf{R}}\boldsymbol{\ell})'\mathbf{X}' = \mathbf{0} \;\Rightarrow\; (\tilde{\mathbf{R}}\boldsymbol{\ell})'\mathbf{X}'\mathbf{Z} = \mathbf{0} \;\Leftrightarrow\; \boldsymbol{\ell}'\tilde{\mathbf{R}}'\mathbf{X}'\mathbf{Z} = \mathbf{0} \tag{2.8}$$

and that

$$\boldsymbol{\ell}'\boldsymbol{\Lambda}' = \mathbf{0} \;\Leftrightarrow\; \boldsymbol{\ell}'(\boldsymbol{\Lambda}', \tilde{\mathbf{R}}'\mathbf{X}'\mathbf{Z}) = \mathbf{0}. \tag{2.9}$$

Thus, if any row or linear combination of rows of the matrix $\tilde{\mathbf{R}}'\mathbf{X}'\mathbf{Z}$ is nonnull, then the corresponding row or linear combination of rows of the matrix $\boldsymbol{\Lambda}'$ is also nonnull. And a subset of the rows of the matrix $(\boldsymbol{\Lambda}', \tilde{\mathbf{R}}'\mathbf{X}'\mathbf{Z})$ is linearly independent if and only if the corresponding subset of the rows of the matrix $\boldsymbol{\Lambda}'$ is linearly independent.

Let $\mathbf{S}$ represent a matrix (with N rows) such that $\mathcal{C}(\mathbf{S}) = \mathcal{N}(\mathbf{X}')$ (i.e., such that the columns of $\mathbf{S}$ span the null space of $\mathbf{X}'$) or equivalently {since (according to Lemma 2.11.5) $\dim[\mathcal{N}(\mathbf{X}')] = N - \text{rank}(\mathbf{X})$} such that $\mathbf{X}'\mathbf{S} = \mathbf{0}$ and $\text{rank}(\mathbf{S}) = N - \text{rank}(\mathbf{X})$. Further, denote by N_* the number of columns in $\mathbf{S}$—necessarily, $N_* \ge N - \text{rank}(\mathbf{X})$. And consider the N_*-dimensional column vector $\mathbf{S}'\mathbf{Z}\boldsymbol{\tau}$, the elements of which are linear combinations of the elements of the vector $\boldsymbol{\tau}$.

Under the augmented G–M model, the elements of the vector $\mathbf{S}'\mathbf{Z}\boldsymbol{\tau}$ (like those of the vector $\boldsymbol{\Lambda}'\boldsymbol{\beta} + \tilde{\mathbf{R}}'\mathbf{X}'\mathbf{Z}\boldsymbol{\tau}$) are estimable linear combinations of the elements of the vector $\begin{pmatrix}\boldsymbol{\beta}\\ \boldsymbol{\tau}\end{pmatrix}$, as is evident upon observing that

$$\mathbf{S}'\mathbf{X} = (\mathbf{X}'\mathbf{S})' = \mathbf{0}$$

and hence that

$$\mathbf{S}'\mathbf{Z}\boldsymbol{\tau} = (\mathbf{0}, \mathbf{S}'\mathbf{Z})\begin{pmatrix}\boldsymbol{\beta}\\ \boldsymbol{\tau}\end{pmatrix} = \mathbf{S}'(\mathbf{X}, \mathbf{Z})\begin{pmatrix}\boldsymbol{\beta}\\ \boldsymbol{\tau}\end{pmatrix}.$$

Clearly,

$$\begin{pmatrix}\boldsymbol{\Lambda}'\boldsymbol{\beta} + \tilde{\mathbf{R}}'\mathbf{X}'\mathbf{Z}\boldsymbol{\tau}\\ \mathbf{S}'\mathbf{Z}\boldsymbol{\tau}\end{pmatrix} = \begin{pmatrix}\boldsymbol{\Lambda}' & \tilde{\mathbf{R}}'\mathbf{X}'\mathbf{Z}\\ \mathbf{0} & \mathbf{S}'\mathbf{Z}\end{pmatrix}\begin{pmatrix}\boldsymbol{\beta}\\ \boldsymbol{\tau}\end{pmatrix}.$$

And upon observing that

$$\begin{pmatrix}\boldsymbol{\Lambda}' & \tilde{\mathbf{R}}'\mathbf{X}'\mathbf{Z}\\ \mathbf{0} & \mathbf{S}'\mathbf{Z}\end{pmatrix} = \begin{pmatrix}\tilde{\mathbf{R}}'\mathbf{X}'\\ \mathbf{S}'\end{pmatrix}(\mathbf{X}, \mathbf{Z})$$

and (in light of Lemmas 2.12.1, 2.6.1, and 2.12.3) that

$$\begin{aligned}
\text{rank}\begin{pmatrix}\tilde{\mathbf{R}}'\mathbf{X}'\\ \mathbf{S}'\end{pmatrix} = \text{rank}\left[\begin{pmatrix}\tilde{\mathbf{R}}'\mathbf{X}'\\ \mathbf{S}'\end{pmatrix}\begin{pmatrix}\tilde{\mathbf{R}}'\mathbf{X}'\\ \mathbf{S}'\end{pmatrix}'\right] &= \text{rank}\begin{bmatrix}(\mathbf{X}\tilde{\mathbf{R}})'\mathbf{X}\tilde{\mathbf{R}} & \mathbf{0}\\ \mathbf{0} & \mathbf{S}'\mathbf{S}\end{bmatrix}\\
&= \text{rank}[(\mathbf{X}\tilde{\mathbf{R}})'\mathbf{X}\tilde{\mathbf{R}}] + \text{rank}(\mathbf{S}'\mathbf{S})\\
&= \text{rank}(\mathbf{X}\tilde{\mathbf{R}}) + \text{rank}(\mathbf{S})\\
&= \text{rank}(\boldsymbol{\Lambda}) + \text{rank}(\mathbf{S})\\
&= \text{rank}(\mathbf{X}) + N - \text{rank}(\mathbf{X}) = N,
\end{aligned}$$

it follows from Lemma 2.5.5 that

$$\mathcal{R}\left[\begin{pmatrix}\boldsymbol{\Lambda}' & \tilde{\mathbf{R}}'\mathbf{X}'\mathbf{Z}\\ \mathbf{0} & \mathbf{S}'\mathbf{Z}\end{pmatrix}\right] = \mathcal{R}[(\mathbf{X}, \mathbf{Z})] \;\text{ and }\; \text{rank}\begin{pmatrix}\boldsymbol{\Lambda}' & \tilde{\mathbf{R}}'\mathbf{X}'\mathbf{Z}\\ \mathbf{0} & \mathbf{S}'\mathbf{Z}\end{pmatrix} = \text{rank}(\mathbf{X}, \mathbf{Z}). \tag{2.10}$$

Moreover,

$$\begin{pmatrix}\boldsymbol{\Lambda}' & \tilde{\mathbf{R}}'\mathbf{X}'\mathbf{Z}\\ \mathbf{0} & \mathbf{S}'\mathbf{Z}\end{pmatrix} = \begin{pmatrix}\boldsymbol{\Lambda}' & \boldsymbol{\Lambda}'\mathbf{L}\\ \mathbf{0} & \mathbf{S}'\mathbf{Z}\end{pmatrix} \tag{2.11}$$

for some matrix $\mathbf{L}$ [as is evident from Lemma 2.4.3 upon observing (in light of Corollary 2.4.4 and Lemma 2.12.3) that $\mathcal{C}(\tilde{\mathbf{R}}'\mathbf{X}'\mathbf{Z}) \subset \mathcal{C}(\tilde{\mathbf{R}}'\mathbf{X}') = \mathcal{C}(\tilde{\mathbf{R}}'\mathbf{X}'\mathbf{X}) = \mathcal{C}(\boldsymbol{\Lambda}')$], implying that

$$\begin{pmatrix}\boldsymbol{\Lambda}' & \tilde{\mathbf{R}}'\mathbf{X}'\mathbf{Z}\\ \mathbf{0} & \mathbf{S}'\mathbf{Z}\end{pmatrix} = \begin{pmatrix}\boldsymbol{\Lambda}' & \mathbf{0}\\ \mathbf{0} & \mathbf{S}'\mathbf{Z}\end{pmatrix}\begin{pmatrix}\mathbf{I}_P & \mathbf{L}\\ \mathbf{0} & \mathbf{I}_Q\end{pmatrix}$$

and hence [since (according to Lemma 2.6.2) $\begin{pmatrix}\mathbf{I} & \mathbf{L}\\ \mathbf{0} & \mathbf{I}\end{pmatrix}$ is nonsingular] that

$$\text{rank}\begin{pmatrix}\boldsymbol{\Lambda}' & \tilde{\mathbf{R}}'\mathbf{X}'\mathbf{Z}\\ \mathbf{0} & \mathbf{S}'\mathbf{Z}\end{pmatrix} = \text{rank}(\boldsymbol{\Lambda}) + \text{rank}(\mathbf{S}'\mathbf{Z}) = \text{rank}(\mathbf{X}) + \text{rank}(\mathbf{S}'\mathbf{Z}) \tag{2.12}$$

(as is evident from Corollary 2.5.6 and Lemma 2.6.1). Together with result (2.10), result (2.12) implies that

$$\operatorname{rank}(\mathbf{S}'\mathbf{Z}) = \operatorname{rank}(\mathbf{X},\ \mathbf{Z}) - \operatorname{rank}(\mathbf{X}). \tag{2.13}$$

For any M-dimensional column vector $\boldsymbol{\ell}_1$ and any N_*-dimensional column vector $\boldsymbol{\ell}_2$,

$$\begin{pmatrix}\boldsymbol{\ell}_1\\ \boldsymbol{\ell}_2\end{pmatrix}'\begin{pmatrix}\boldsymbol{\Lambda}' & \tilde{\mathbf{R}}'\mathbf{X}'\mathbf{Z}\\ \mathbf{0} & \mathbf{S}'\mathbf{Z}\end{pmatrix} = \mathbf{0} \quad\Leftrightarrow\quad \boldsymbol{\ell}_1'(\boldsymbol{\Lambda}',\ \tilde{\mathbf{R}}'\mathbf{X}'\mathbf{Z}) = \mathbf{0} \text{ and } \boldsymbol{\ell}_2'\mathbf{S}'\mathbf{Z} = \mathbf{0}, \tag{2.14}$$

as is evident from result (2.9). Thus, a subset of size $\operatorname{rank}(\mathbf{X},\ \mathbf{Z})$ of the $M+N_*$ rows of the matrix $\begin{pmatrix}\boldsymbol{\Lambda}' & \tilde{\mathbf{R}}'\mathbf{X}'\mathbf{Z}\\ \mathbf{0} & \mathbf{S}'\mathbf{Z}\end{pmatrix}$ is linearly independent if and only if the subset consists of $\operatorname{rank}(\mathbf{X})$ linearly independent rows of $(\boldsymbol{\Lambda}',\ \tilde{\mathbf{R}}'\mathbf{X}'\mathbf{Z})$ and $\operatorname{rank}(\mathbf{X},\ \mathbf{Z})-\operatorname{rank}(\mathbf{X})$ linearly independent rows of $(\mathbf{0},\ \mathbf{S}'\mathbf{Z})$. [In light of result (2.10), there exists a linearly independent subset of size $\operatorname{rank}(\mathbf{X},\ \mathbf{Z})$ of the rows of $\begin{pmatrix}\boldsymbol{\Lambda}' & \tilde{\mathbf{R}}'\mathbf{X}'\mathbf{Z}\\ \mathbf{0} & \mathbf{S}'\mathbf{Z}\end{pmatrix}$ and no linearly independent subset of a size larger than $\operatorname{rank}(\mathbf{X},\ \mathbf{Z})$.]

Under the augmented G–M model, a linear combination, say $\boldsymbol{\lambda}_1'\boldsymbol{\beta} + \boldsymbol{\lambda}_2'\boldsymbol{\tau}$, of the elements of $\boldsymbol{\beta}$ and $\boldsymbol{\tau}$ is [in light of result (2.10)] estimable if and only if

$$\boldsymbol{\lambda}_1' = \boldsymbol{\ell}_1'\boldsymbol{\Lambda}' \quad\text{and}\quad \boldsymbol{\lambda}_2' = \boldsymbol{\ell}_1'\tilde{\mathbf{R}}'\mathbf{X}'\mathbf{Z} + \boldsymbol{\ell}_2'\mathbf{S}'\mathbf{Z} \tag{2.15}$$

for some M-dimensional column vector $\boldsymbol{\ell}_1$ and some N_*-dimensional column vector $\boldsymbol{\ell}_2$. Moreover, in the special case where $\boldsymbol{\lambda}_1 = \mathbf{0}$ (i.e., in the special case of a linear combination of the elements of $\boldsymbol{\tau}$), this result can [in light of result (2.8)] be simplified as follows: under the augmented G–M model, $\boldsymbol{\lambda}_2'\boldsymbol{\tau}$ is estimable if and only if

$$\boldsymbol{\lambda}_2' = \boldsymbol{\ell}_2'\mathbf{S}'\mathbf{Z} \tag{2.16}$$

for some N_*-dimensional column vector $\boldsymbol{\ell}_2$.

Among the choices for the $N \times N_*$ matrix $\mathbf{S}$ is the $N \times N$ matrix $\mathbf{I}-\mathbf{P_X}$. Since (according to Theorem 2.12.2) $\mathbf{X}'\mathbf{P_X} = \mathbf{X}'$, $\mathbf{P_X}^2 = \mathbf{P_X}$, and $\operatorname{rank}(\mathbf{P_X}) = \operatorname{rank}(\mathbf{X})$,

$$\mathbf{X}'(\mathbf{I}-\mathbf{P_X}) = \mathbf{0}$$

and (in light of Lemma 2.8.4)

$$\operatorname{rank}(\mathbf{I}-\mathbf{P_X}) = N - \operatorname{rank}(\mathbf{X}).$$

Clearly, when $\mathbf{S} = \mathbf{I}-\mathbf{P_X}$, $N_* = N$.

Let $\tilde{\mathbf{T}}$ represent any solution to the following linear system [in the $(P+Q) \times N_*$ matrix $\mathbf{T}$]:

$$(\mathbf{X},\ \mathbf{Z})'(\mathbf{X},\ \mathbf{Z})\mathbf{T} = (\mathbf{0},\ \mathbf{S}'\mathbf{Z})' \tag{2.17}$$

—the existence of a solution follows (in light of the results of Section 5.4c) from the estimability (under the augmented G–M model) of the elements of the vector $(\mathbf{0},\ \mathbf{S}'\mathbf{Z})\begin{pmatrix}\boldsymbol{\beta}\\ \boldsymbol{\tau}\end{pmatrix}$. Further, partition $\tilde{\mathbf{T}}$ as $\tilde{\mathbf{T}} = \begin{pmatrix}\tilde{\mathbf{T}}_1\\ \tilde{\mathbf{T}}_2\end{pmatrix}$ (where $\tilde{\mathbf{T}}_1$ has P rows), and observe that

$$\mathbf{X}'\mathbf{X}\tilde{\mathbf{T}}_1 = -\mathbf{X}'\mathbf{Z}\tilde{\mathbf{T}}_2$$

and hence (in light of Theorem 2.12.2) that

$$\mathbf{X}\tilde{\mathbf{T}}_1 = \mathbf{P_X}\mathbf{X}\tilde{\mathbf{T}}_1 = \mathbf{X}(\mathbf{X}'\mathbf{X})^-\mathbf{X}'\mathbf{X}\tilde{\mathbf{T}}_1 = -\mathbf{X}(\mathbf{X}'\mathbf{X})^-\mathbf{X}'\mathbf{Z}\tilde{\mathbf{T}}_2 = -\mathbf{P_X}\mathbf{Z}\tilde{\mathbf{T}}_2. \tag{2.18}$$

Then, under the augmented G–M model, the least squares estimators of the elements of the vector $\mathbf{S}'\mathbf{Z}\boldsymbol{\tau}$ are [in light of result (5.4.35)] the (corresponding) elements of a vector that is expressible as follows:

$$\tilde{\mathbf{T}}'(\mathbf{X},\ \mathbf{Z})'\mathbf{y} = \tilde{\mathbf{T}}_1'\mathbf{X}'\mathbf{y}+\tilde{\mathbf{T}}_2'\mathbf{Z}'\mathbf{y} = (-\mathbf{P_X}\mathbf{Z}\tilde{\mathbf{T}}_2)'\mathbf{y}+\tilde{\mathbf{T}}_2'\mathbf{Z}'\mathbf{y} = \tilde{\mathbf{T}}_2'\mathbf{Z}'(\mathbf{I}-\mathbf{P_X})\mathbf{y}. \tag{2.19}$$

And (under the augmented G–M model)

$$\operatorname{cov}[\tilde{\mathbf{T}}_2'\mathbf{Z}'(\mathbf{I}-\mathbf{P_X})\mathbf{y},\ \tilde{\mathbf{R}}'\mathbf{X}'\mathbf{y}] = \sigma^2\tilde{\mathbf{T}}_2'\mathbf{Z}'(\mathbf{I}-\mathbf{P_X})\mathbf{X}\tilde{\mathbf{R}} = \mathbf{0} \tag{2.20}$$

(i.e., the least squares estimators of the elements of $\mathbf{S}'\mathbf{Z}\boldsymbol{\tau}$ are uncorrelated with those of the elements of $\boldsymbol{\Lambda}'\boldsymbol{\beta} + \tilde{\mathbf{R}}'\mathbf{X}'\mathbf{Z}\boldsymbol{\tau}$) and [in light of result (5.6.6)]

$$\mathrm{var}[\tilde{\mathbf{T}}_2'\mathbf{Z}'(\mathbf{I}-\mathbf{P}_{\mathbf{X}})\mathbf{y}] = \sigma^2\tilde{\mathbf{T}}'(\mathbf{0},\ \mathbf{S}'\mathbf{Z})' = \sigma^2\tilde{\mathbf{T}}_2'\mathbf{Z}'\mathbf{S}. \tag{2.21}$$

In connection with linear system (2.17), note [in light of result (2.18)] that

$$\mathbf{Z}'(\mathbf{I}-\mathbf{P}_{\mathbf{X}})\mathbf{Z}\tilde{\mathbf{T}}_2 = \mathbf{Z}'\mathbf{S} \quad [= \mathbf{Z}'(\mathbf{I}-\mathbf{P}_{\mathbf{X}})\mathbf{S}]$$

(and that $\mathbf{X}'\mathbf{X}\tilde{\mathbf{T}}_1 = -\mathbf{X}'\mathbf{Z}\tilde{\mathbf{T}}_2$). Conversely, if $\tilde{\mathbf{T}}_2$ is taken to be a solution to the linear system

$$\mathbf{Z}'(\mathbf{I}-\mathbf{P}_{\mathbf{X}})\mathbf{Z}\mathbf{T}_2 = \mathbf{Z}'\mathbf{S} \tag{2.22}$$

(in $\mathbf{T}_2$) and $\tilde{\mathbf{T}}_1$ a solution to the linear system $\mathbf{X}'\mathbf{X}\mathbf{T}_1 = -\mathbf{X}'\mathbf{Z}\tilde{\mathbf{T}}_2$ (in $\mathbf{T}_1$), then $\begin{pmatrix}\tilde{\mathbf{T}}_1\\ \tilde{\mathbf{T}}_2\end{pmatrix}$ is a solution to linear system (2.17).

b. Some results for a specific implementation

The results of Subsection a depend on the matrices $\boldsymbol{\Lambda}$ and $\mathbf{S}$. Among the choices for $\boldsymbol{\Lambda}$ and $\mathbf{S}$ are those derived from a particular decomposition of the model matrix $(\mathbf{X},\ \mathbf{Z})$ of the augmented G–M model; this decomposition is as follows:

$$(\mathbf{X},\ \mathbf{Z}) = \mathbf{O}\mathbf{U}, \tag{2.23}$$

where $\mathbf{O}$ is an $N \times \mathrm{rank}(\mathbf{X},\ \mathbf{Z})$ matrix with orthonormal columns and where, for some $\mathrm{rank}(\mathbf{X}) \times P$ matrix $\mathbf{U}_{11}$ (of full row rank), some $\mathrm{rank}(\mathbf{X}) \times Q$ matrix $\mathbf{U}_{12}$, and some $[\mathrm{rank}(\mathbf{X},\ \mathbf{Z}) - \mathrm{rank}(\mathbf{X})] \times Q$ matrix $\mathbf{U}_{22}$ (of full row rank),

$$\mathbf{U} = \begin{pmatrix}\mathbf{U}_{11} & \mathbf{U}_{12}\\ \mathbf{0} & \mathbf{U}_{22}\end{pmatrix}.$$

A decomposition of the form (2.23) can be constructed by, for example, applying Gram–Schmidt orthogonalization, or (in what would be preferable for numerical purposes) modified Gram–Schmidt orthogonalization, to the columns of the matrix $(\mathbf{X},\ \mathbf{Z})$. In fact, when this method of construction is employed, $\mathbf{U}$ is a submatrix of a $(P+Q) \times (P+Q)$ upper triangular matrix having $\mathrm{rank}(\mathbf{X},\ \mathbf{Z})$ positive diagonal elements and $P+Q-\mathrm{rank}(\mathbf{X},\ \mathbf{Z})$ null rows; it is the submatrix obtained by striking out the null rows—refer, e.g., to Harville (1997, chap. 6). When $\mathbf{U}$ is of this form, the decomposition (2.23) is what is known (at least in the special case of the decomposition of a matrix having full column rank) as the QR deomposition (or the "skinny" QR decomposition). The QR decomposition was encountered earlier; Section 5.4e included a discussion of the use of the QR decomposition in the computation of least squares estimates.

Partition the matrix $\mathbf{O}$ (conformally to the partitioning of $\mathbf{U}$) as $\mathbf{O} = (\mathbf{O}_1,\ \mathbf{O}_2)$, where $\mathbf{O}_1$ has $\mathrm{rank}(\mathbf{X})$ columns. And observe that

$$\mathbf{X} = \mathbf{O}_1\mathbf{U}_{11} \quad \text{and} \quad \mathbf{Z} = \mathbf{O}_2\mathbf{U}_{22} + \mathbf{O}_1\mathbf{U}_{12}. \tag{2.24}$$

Then, among the choices for $\boldsymbol{\Lambda}$ is that obtained by taking $\boldsymbol{\Lambda}' = \mathbf{U}_{11}$ (as is evident from Corollary 2.4.17). Further, the choices for $\mathbf{S}$ include the matrix $(\mathbf{O}_2,\ \mathbf{O}_3)$, where $\mathbf{O}_3$ is an $N \times [N - \mathrm{rank}(\mathbf{X},\ \mathbf{Z})]$ matrix whose columns form an orthonormal basis for $\mathcal{N}[(\mathbf{X},\mathbf{Z})]$ or, equivalently (in light of Corollary 2.4.17), $\mathcal{N}[(\mathbf{O}_1,\ \mathbf{O}_2)]$—the $N - \mathrm{rank}(\mathbf{X})$ columns of $(\mathbf{O}_2,\ \mathbf{O}_3)$ form an orthonormal basis for $\mathcal{N}(\mathbf{X})$.

In light of result (2.24), we have that

$$\mathbf{X}'\mathbf{X} = \mathbf{U}_{11}'\mathbf{U}_{11} \quad \text{and} \quad \mathbf{X}'\mathbf{Z} = \mathbf{U}_{11}'\mathbf{U}_{12}.$$

Moreover, when $\boldsymbol{\Lambda}' = \mathbf{U}_{11}$, we find that

$$\mathbf{X}'\mathbf{X}\mathbf{R} = \boldsymbol{\Lambda} \;\Leftrightarrow\; \mathbf{U}_{11}'\mathbf{U}_{11}\mathbf{R} = \mathbf{U}_{11}' \;\Leftrightarrow\; \mathbf{U}_{11}\mathbf{R} = \mathbf{I}$$

—that $\mathbf{U}_{11}'\mathbf{U}_{11}\mathbf{R} = \mathbf{U}_{11}' \Rightarrow \mathbf{U}_{11}\mathbf{R} = \mathbf{I}$ is clear upon, e.g., observing that $\mathbf{U}_{11}\mathbf{U}_{11}'$ is nonsingular and premultiplying both sides of the equation $\mathbf{U}_{11}'\mathbf{U}_{11}\mathbf{R} = \mathbf{U}_{11}'$ by $(\mathbf{U}_{11}\mathbf{U}_{11}')^{-1}\mathbf{U}_{11}$. Thus, when $\mathbf{\Lambda}' = \mathbf{U}_{11}$, the expected value $\mathbf{\Lambda}'\boldsymbol{\beta} + \tilde{\mathbf{R}}'\mathbf{X}'\mathbf{Z}\boldsymbol{\tau}$ (under the augmented G–M model) of the estimator $\tilde{\mathbf{R}}'\mathbf{X}'\mathbf{y}$ (where as in Subsection a, $\tilde{\mathbf{R}}$ represents an arbitrary solution to $\mathbf{X}'\mathbf{X}\mathbf{R} = \mathbf{\Lambda}$) is reexpressible as

$$\mathbf{\Lambda}'\boldsymbol{\beta} + \tilde{\mathbf{R}}'\mathbf{X}'\mathbf{Z}\boldsymbol{\tau} = \mathbf{U}_{11}\boldsymbol{\beta} + \tilde{\mathbf{R}}'\mathbf{U}_{11}'\mathbf{U}_{12}\boldsymbol{\tau} = \mathbf{U}_{11}\boldsymbol{\beta} + \mathbf{U}_{12}\boldsymbol{\tau}. \tag{2.25}$$

Moreover, $\tilde{\mathbf{R}}$ is a right inverse of $\mathbf{U}_{11}$, and

$$\text{var}(\tilde{\mathbf{R}}'\mathbf{X}'\mathbf{y}) = \sigma^2\tilde{\mathbf{R}}'\mathbf{X}'\mathbf{X}\tilde{\mathbf{R}} = \sigma^2\mathbf{\Lambda}'\tilde{\mathbf{R}} = \sigma^2\mathbf{U}_{11}\tilde{\mathbf{R}} = \sigma^2\mathbf{I}. \tag{2.26}$$

Now, suppose that $\mathbf{S} = (\mathbf{O}_2,\ \mathbf{O}_3)$, in which case

$$\mathbf{Z}'\mathbf{S} = (\mathbf{Z}'\mathbf{O}_2,\ \mathbf{Z}'\mathbf{O}_3) = (\mathbf{U}_{22}',\ \mathbf{0}),$$

so that

$$\mathbf{S}'\mathbf{Z}\boldsymbol{\tau} = \begin{pmatrix}\mathbf{U}_{22}\\ \mathbf{0}\end{pmatrix}\boldsymbol{\tau} = \begin{pmatrix}\mathbf{U}_{22}\boldsymbol{\tau}\\ \mathbf{0}\end{pmatrix}.$$

And observe (in light of Lemma 2.6.3) that

$$\text{rank}\,\mathbf{U} = \text{rank}(\mathbf{X},\ \mathbf{Z})$$

and hence that

$$\text{rank}(\mathbf{U}\mathbf{U}') = \text{rank}(\mathbf{X},\ \mathbf{Z}). \tag{2.27}$$

Observe also that $(\mathbf{O},\ \mathbf{O}_3)$ is an $(N \times N)$ orthogonal matrix and hence that

$$\mathbf{I} = (\mathbf{O},\ \mathbf{O}_3)(\mathbf{O},\ \mathbf{O}_3)' = \mathbf{O}\mathbf{O}' + \mathbf{O}_3\mathbf{O}_3'.$$

Then,

$$(\mathbf{X},\ \mathbf{Z})'(\mathbf{X},\ \mathbf{Z}) = (\mathbf{X},\ \mathbf{Z})'(\mathbf{O}\mathbf{O}' + \mathbf{O}_3\mathbf{O}_3')(\mathbf{X},\ \mathbf{Z}) = \mathbf{U}'\mathbf{U} \tag{2.28}$$

and

$$(\mathbf{X},\ \mathbf{Z})'\mathbf{y} = (\mathbf{X},\ \mathbf{Z})'(\mathbf{O}\mathbf{O}' + \mathbf{O}_3\mathbf{O}_3')\mathbf{y} = \mathbf{U}'\mathbf{O}'\mathbf{y}. \tag{2.29}$$

Moreover,

$$(\mathbf{0},\ \mathbf{S}'\mathbf{Z})' = \begin{pmatrix}\mathbf{0} & \mathbf{0}\\ \mathbf{U}_{22}' & \mathbf{0}\end{pmatrix} = \mathbf{U}'\begin{pmatrix}\mathbf{0} & \mathbf{0}\\ \mathbf{I} & \mathbf{0}\end{pmatrix}. \tag{2.30}$$

Thus, the solution $\tilde{\mathbf{T}}$ to the linear system $(\mathbf{X},\ \mathbf{Z})'(\mathbf{X},\ \mathbf{Z})\mathbf{T} = (\mathbf{0},\ \mathbf{S}'\mathbf{Z})'$ satisfies the equality

$$\mathbf{U}'\mathbf{U}\tilde{\mathbf{T}} = \mathbf{U}'\begin{pmatrix}\mathbf{0} & \mathbf{0}\\ \mathbf{I} & \mathbf{0}\end{pmatrix},$$

so that $\tilde{\mathbf{T}}$ also satisfies the equality

$$\mathbf{U}\mathbf{U}'\mathbf{U}\tilde{\mathbf{T}} = \mathbf{U}\mathbf{U}'\begin{pmatrix}\mathbf{0} & \mathbf{0}\\ \mathbf{I} & \mathbf{0}\end{pmatrix}$$

and hence [since according to result (2.27), $\mathbf{U}\mathbf{U}'$ is nonsingular] the equality

$$\mathbf{U}\tilde{\mathbf{T}} = \begin{pmatrix}\mathbf{0} & \mathbf{0}\\ \mathbf{I} & \mathbf{0}\end{pmatrix}. \tag{2.31}$$

Making use of equalities (2.29) and (2.31), we find that (under the augmented G–M model) the least squares estimators of the elements of the vector $\mathbf{S}'\mathbf{Z}\boldsymbol{\tau}\ \left[= \begin{pmatrix}\mathbf{U}_{22}\boldsymbol{\tau}\\ \mathbf{0}\end{pmatrix}\right]$ are the (corresponding) elements of the vector

$$\tilde{\mathbf{T}}'(\mathbf{X},\ \mathbf{Z})'\mathbf{y} = \tilde{\mathbf{T}}'\mathbf{U}'\mathbf{O}'\mathbf{y} = \begin{pmatrix}\mathbf{0} & \mathbf{0}\\ \mathbf{I} & \mathbf{0}\end{pmatrix}'\begin{pmatrix}\mathbf{O}_1'\mathbf{y}\\ \mathbf{O}_2'\mathbf{y}\end{pmatrix} = \begin{pmatrix}\mathbf{O}_2'\mathbf{y}\\ \mathbf{0}\end{pmatrix}. \tag{2.32}$$

Clearly, the variance-covariance matrix of this vector is

$$\sigma^2\begin{pmatrix}\mathbf{I} & \mathbf{0}\\ \mathbf{0} & \mathbf{0}\end{pmatrix}. \tag{2.33}$$

Note that $\mathbf{S}'\mathbf{Z}\boldsymbol{\tau}$, the estimator (2.32) of $\mathbf{S}'\mathbf{Z}\boldsymbol{\tau}$, and the variance-covariance matrix (2.33) of the estimator (2.32) do not depend on $\mathbf{O}_3$ (even though $\mathbf{S}$ itself depends on $\mathbf{O}_3$).

c. An illustration

Let us illustrate the results of Subsections a and b by using them to add to the results obtained earlier (in Section 7.1) for the lettuce-yield data. Accordingly, let us take $\mathbf{y}$ to be the 20×1 random vector whose observed value is the vector of lettuce yields. Further, let us adopt the terminology and notation introduced in Section 7.1 (along with those introduced in Subsections a and b of the present section).

Suppose that the original G–M model is the second-order G–M model, which is the model that was adopted in the analyses carried out in Section 7.1. And suppose that the augmented G–M model is the third-order G–M model. Then, $\mathbf{X} = (\mathbf{X}_1, \mathbf{X}_2)$ and $\mathbf{Z} = \mathbf{X}_3$ (where $\mathbf{X}_1$, $\mathbf{X}_2$, and $\mathbf{X}_3$ are as defined in Section 7.1). Further, $\boldsymbol{\beta} = (\beta_1, \beta_2, \beta_3, \beta_4, \beta_{11}, \beta_{12}, \beta_{13}, \beta_{22}, \beta_{23}, \beta_{33})'$ and $\boldsymbol{\tau} = (\beta_{111}, \beta_{112}, \beta_{113}, \beta_{122}, \beta_{123}, \beta_{133}, \beta_{222}, \beta_{223}, \beta_{233}, \beta_{333})'$.

Upon applying result (2.5) with $\tilde{\mathbf{R}} = (\mathbf{X}'\mathbf{X})^{-1}$ (corresponding to $\boldsymbol{\Lambda} = \mathbf{I}$), we find that under the augmented (third-order) model

$$\begin{aligned}
E(\hat{\beta}_1) &= \beta_1 - 0.064\beta_{113} - 0.064\beta_{223} + 0.148\beta_{333},\\
E(\hat{\beta}_2) &= \beta_2 + 1.805\beta_{111} + 0.560\beta_{122} + 0.278\beta_{133},\\
E(\hat{\beta}_3) &= \beta_3 + 0.560\beta_{112} + 1.805\beta_{222} + 0.278\beta_{233},\\
E(\hat{\beta}_4) &= \beta_4 + 0.427\beta_{113} + 0.427\beta_{223} + 1.958\beta_{333},\\
E(\hat{\beta}_{11}) &= \beta_{11} + 0.046\beta_{113} + 0.046\beta_{223} - 0.045\beta_{333},\\
E(\hat{\beta}_{12}) &= \beta_{12} + 0.252\beta_{123},\\
E(\hat{\beta}_{13}) &= \beta_{13} - 0.325\beta_{111} + 0.178\beta_{122} + 0.592\beta_{133},\\
E(\hat{\beta}_{22}) &= \beta_{22} + 0.046\beta_{113} + 0.046\beta_{223} - 0.045\beta_{333},\\
E(\hat{\beta}_{23}) &= \beta_{23} + 0.178\beta_{112} - 0.325\beta_{222} + 0.592\beta_{233}, \qquad \text{and}\\
E(\hat{\beta}_{33}) &= \beta_{33} + 0.110\beta_{113} + 0.110\beta_{223} - 0.204\beta_{333}.
\end{aligned}$$

All ten of the least squares estimators $\hat{\beta}_1$, $\hat{\beta}_2$, $\hat{\beta}_3$, $\hat{\beta}_4$, $\hat{\beta}_{11}$, $\hat{\beta}_{12}$, $\hat{\beta}_{13}$, $\hat{\beta}_{22}$, $\hat{\beta}_{23}$, and $\hat{\beta}_{33}$ of the elements of $\boldsymbol{\beta}$ are at least somewhat susceptible to biases occasioned by the exclusion from the model of third-order terms. The exposure to such biases appears to be greatest in the case of the estimators $\hat{\beta}_2$, $\hat{\beta}_3$, and $\hat{\beta}_4$ of the first-order regression coefficients β_2, β_3, and β_4. In fact, if the level of Fe in the first, second, third, and fifth containers had been -1 (on the transformed scale), which was the intended level, instead of -0.4965, the expected values of the other seven estimators ($\hat{\beta}_1$, $\hat{\beta}_{11}$, $\hat{\beta}_{12}$, $\hat{\beta}_{13}$, $\hat{\beta}_{22}$, $\hat{\beta}_{23}$, and $\hat{\beta}_{33}$) would have been the same under the third-order model as under the second-order model (i.e., would have equalled β_1, β_{11}, β_{12}, β_{13}, β_{22}, β_{23}, and β_{33}, respectively, under both models)—if the level of Fe in those containers had been -1, the expected values of $\hat{\beta}_2$, $\hat{\beta}_3$, and $\hat{\beta}_4$ under the third-order model would have been

$$\begin{aligned}
E(\hat{\beta}_2) &= \beta_2 + 1.757\beta_{111} + 0.586\beta_{122} + 0.586\beta_{133},\\
E(\hat{\beta}_3) &= \beta_3 + 0.586\beta_{112} + 1.757\beta_{222} + 0.586\beta_{233}, \qquad \text{and}\\
E(\hat{\beta}_4) &= \beta_4 + 0.586\beta_{113} + 0.586\beta_{223} + 1.757\beta_{333}.
\end{aligned}$$

The estimators $\hat{\beta}_1$, $\hat{\beta}_2$, $\hat{\beta}_3$, $\hat{\beta}_4$, $\hat{\beta}_{11}$, $\hat{\beta}_{12}$, $\hat{\beta}_{13}$, $\hat{\beta}_{22}$, $\hat{\beta}_{23}$, and $\hat{\beta}_{33}$ (which are the least squares estimators of the regression coefficients in the second-order model) have the "same" standard errors under the augmented (third-order) model as under the original (second-order) model; they are the same in the sense that the expressions given in Table 7.1 for the standard errors are still applicable. However, the "interpretation" of the parameter σ (which appears as a multiplicative factor in those expressions) differs. This difference manifests itself in the estimation of σ and σ^2. In the case of the augmented G–M model, the usual estimator $\hat{\sigma}^2$ of σ^2 is that represented by the quadratic form

$$\mathbf{y}'[\mathbf{I}-\mathbf{P}_{(\mathbf{X},\,\mathbf{Z})}]\mathbf{y}/[N-\text{rank}(\mathbf{X},\,\mathbf{Z})], \tag{2.34}$$

rather than (as in the case of the original G–M model) that represented by the quadratic form $\mathbf{y}'(\mathbf{I}-\mathbf{P}_{\mathbf{X}})\mathbf{y}/[N-\text{rank}(\mathbf{X})]$.

For the lettuce-yield data, $\mathbf{y}'[\mathbf{I}-\mathbf{P}_{(\mathbf{X},\,\mathbf{Z})}]\mathbf{y} = 52.6306$ and $\text{rank}(\mathbf{X},\,\mathbf{Z}) = 15$, so that when the estimator (2.34) is applied to the lettuce-yield data, we obtain $\hat{\sigma}^2 = 10.53$. Upon taking the square root of this value, we obtain $\hat{\sigma} = 3.24$, which is 1.70% smaller than the value (3.30) obtained for $\hat{\sigma}$ when the model was taken to be the original (second-order) model. Accordingly, when the model for the lettuce-yield data is taken to be the augmented (third-order) model, the estimated standard errors obtained for $\hat{\beta}_1$, $\hat{\beta}_2$, $\hat{\beta}_3$, $\hat{\beta}_4$, $\hat{\beta}_{11}$, $\hat{\beta}_{12}$, $\hat{\beta}_{13}$, $\hat{\beta}_{22}$, $\hat{\beta}_{23}$, and $\hat{\beta}_{33}$ are 1.70% smaller than those given in Table 7.1 (which were obtained on the basis of the second-order model).

Under the augmented G–M model, the $\text{rank}(\mathbf{X},\,\mathbf{Z})-\text{rank}(\mathbf{X})$ elements of the vector $\mathbf{U}_{22}\boldsymbol{\tau}$ [where $\mathbf{U}_{22}$ is defined in terms of the decomposition (2.23)] are linearly independent estimable linear combinations of the elements of $\boldsymbol{\tau}$ [and every estimable linear combination of the elements of $\boldsymbol{\tau}$ is expressible in terms of these $\text{rank}(\mathbf{X},\,\mathbf{Z})-\text{rank}(\mathbf{X})$ linear combinations]. In the lettuce-yield application [where the augmented G–M model is the third-order model and where $\text{rank}(\mathbf{X},\,\mathbf{Z})-\text{rank}(\mathbf{X}) = 5$], the elements of $\mathbf{U}_{22}\boldsymbol{\tau}$ [those obtained when the decomposition (2.23) is taken to be the QR decomposition] are the following linear combinations of the third-order regression coefficients:

$$3.253\beta_{111} - 1.779\beta_{122} - 0.883\beta_{133},$$
$$1.779\beta_{112} - 3.253\beta_{222} + 0.883\beta_{233},$$
$$1.554\beta_{113} + 1.554\beta_{223} - 3.168\beta_{333},$$
$$2.116\beta_{123},$$

and

$$0.471\beta_{333}.$$

The least squares estimates of these five linear combinations are −1.97, 4.22, −5.11, −2.05, and 2.07, respectively—the least squares estimators are uncorrelated, and each of them has a standard error of σ and an estimated standard error of 3.24.

7.3 The *F* Test (and Corresponding Confidence Set) and a Generalized *S* Method

Suppose that $\mathbf{y}$ is an $N \times 1$ observable random vector that follows the G–M model. And let $\tau = \boldsymbol{\lambda}'\boldsymbol{\beta}$, where $\boldsymbol{\lambda}$ is a $P \times 1$ vector of constants. Further, suppose that $\boldsymbol{\lambda}$ is nonnull and that $\boldsymbol{\lambda}' \in \mathcal{R}(\mathbf{X})$ (so that τ is a nontrivial estimable function).

In addition to the (point) estimation of τ—the estimation of such a function was considered earlier (in Chapter 5)—inferences about τ may take the form of a confidence interval (or set). They may also take the form of a test (of a specified size) of the null hypothesis $H_0: \tau = 0$ versus the alternative hypothesis $H_1: \tau \neq 0$ or, more generally, of $H_0: \tau = \tau^{(0)}$ (where $\tau^{(0)}$ is any hypothesized value) versus $H_1: \tau \neq \tau^{(0)}$.

In any particular application, there are likely to be a number of linear combinations of the elements of $\boldsymbol{\beta}$ that represent quantities of interest. For $i = 1, 2, \ldots, M$, let $\tau_i = \boldsymbol{\lambda}_i'\boldsymbol{\beta}$, where $\boldsymbol{\lambda}_1, \boldsymbol{\lambda}_2, \ldots, \boldsymbol{\lambda}_M$ are $P \times 1$ vectors of constants (one or more of which are nonnull). In "matrix notation," $\boldsymbol{\tau} = \boldsymbol{\Lambda}'\boldsymbol{\beta}$, where $\boldsymbol{\tau} = (\tau_1, \tau_2, \ldots, \tau_M)'$ and $\boldsymbol{\Lambda} = (\boldsymbol{\lambda}_1, \boldsymbol{\lambda}_2, \ldots, \boldsymbol{\lambda}_M)$.

Assume that $\tau_1, \tau_2, \ldots, \tau_M$ are estimable, and suppose that we wish to make inferences about

$\tau_1, \tau_2, \ldots, \tau_M$ and possibly about some or all linear combinations of $\tau_1, \tau_2, \ldots, \tau_M$. These inferences may take the form of a confidence set for the vector $\boldsymbol{\tau}$. Or they may take the form of individual confidence intervals (or sets) for $\tau_1, \tau_2, \ldots, \tau_M$ (and possibly for linear combinations of $\tau_1, \tau_2, \ldots, \tau_M$) that have a specified probability of simultaneous coverage. Alternatively, these inferences may take the form of a test of hypothesis. More specifically, they may take the form of a test (of a specified size) of the null hypothesis $H_0: \boldsymbol{\tau} = \mathbf{0}$ versus the alternative hypothesis $H_1: \boldsymbol{\tau} \neq \mathbf{0}$ or, more generally, of $H_0: \boldsymbol{\tau} = \boldsymbol{\tau}^{(0)}$ [where $\boldsymbol{\tau}^{(0)} = (\tau_1^{(0)}, \tau_2^{(0)}, \ldots, \tau_M^{(0)})'$ is any vector of hypothesized values] versus $H_1: \boldsymbol{\tau} \neq \boldsymbol{\tau}^{(0)}$. They may also take a form that consists of testing whether or not each of the M quantities $\tau_1, \tau_2, \ldots, \tau_M$ (and possibly each of various linear combinations of these M quantities) equals a hypothesized value (subject to some restriction on the probability of an "excessive overall number" of false rejections).

In testing $H_0: \boldsymbol{\tau} = \boldsymbol{\tau}^{(0)}$ (versus $H_1: \boldsymbol{\tau} \neq \boldsymbol{\tau}^{(0)}$) attention is restricted to what are called testable hypotheses. The null hypothesis $H_0: \boldsymbol{\tau} = \boldsymbol{\tau}^{(0)}$ is said to be *testable* if (in addition to $\tau_1, \tau_2, \ldots, \tau_M$ being estimable and $\boldsymbol{\Lambda}$ being nonnull) $\boldsymbol{\tau}^{(0)} \in \mathcal{C}(\boldsymbol{\Lambda}')$, that is, if $\boldsymbol{\tau}^{(0)} = \boldsymbol{\Lambda}'\boldsymbol{\beta}^{(0)}$ for some $P \times 1$ vector $\boldsymbol{\beta}^{(0)}$. Note that if $\boldsymbol{\tau}^{(0)} \notin \mathcal{C}(\boldsymbol{\Lambda}')$, there would not exist any values of $\boldsymbol{\beta}$ for which $\boldsymbol{\Lambda}'\boldsymbol{\beta}$ equals the hypothesized value $\boldsymbol{\tau}^{(0)}$ and, consequently, H_0 would be inherently false. It is worth noting that while the definition of testability adopted herein rules out the existence of any contradictions among the M equalities $\tau_i = \tau_i^{(0)}$ $(i = 1, 2, \ldots, M)$ that define H_0, it is sufficiently flexible to accommodate redundancies among these equalities—a more restrictive definition (one adopted by many authors) would be to require that $\text{rank}(\boldsymbol{\Lambda}) = M$.

Most of the results on confidence sets for $\boldsymbol{\tau}$ or its individual elements or for testing whether or not $\boldsymbol{\tau}$ or its individual elements equal hypothesized values are obtained under an assumption that the vector $\mathbf{e}$ of the residual effects in the G–M model has an $N(\mathbf{0}, \sigma^2\mathbf{I})$ distribution. However, for some of these results, this assumption is stronger than necessary; it suffices to assume that the distribution of $\mathbf{e}$ is spherically symmetric.

a. Canonical form (of the G–M model)

Let us continue to take $\mathbf{y}$ to be an $N \times 1$ observable random vector that follows the G–M model. And let us consider further inference about the $M \times 1$ vector $\boldsymbol{\tau}$ $(= \boldsymbol{\Lambda}'\boldsymbol{\beta})$.

The problem of making inferences about $\boldsymbol{\tau}$ can be reduced to its essence and considerable insight into this problem can be gained by introducing a suitable transformation. Accordingly, let $P_* = \text{rank}\,\mathbf{X}$ and $M_* = \text{rank}\,\boldsymbol{\Lambda}$, and assume that $\boldsymbol{\tau}$ is estimable and that $M_* \geq 1$. Further, let $\hat{\boldsymbol{\tau}} = (\hat{\tau}_1, \hat{\tau}_2, \ldots, \hat{\tau}_M)'$, where (for $j = 1, 2, \ldots, M$) $\hat{\tau}_j$ is the least squares estimator of τ_j, denote by $\tilde{\mathbf{R}}$ an arbitrary solution to the linear system $\mathbf{X}'\mathbf{X}\mathbf{R} = \boldsymbol{\Lambda}$ (in the $P \times M$ matrix $\mathbf{R}$), and recall (from Chapter 5) that

$$\hat{\boldsymbol{\tau}} = \tilde{\mathbf{R}}'\mathbf{X}'\mathbf{y} = \boldsymbol{\Lambda}'(\mathbf{X}'\mathbf{X})^-\mathbf{X}'\mathbf{y} \quad \text{and} \quad \text{var}(\hat{\boldsymbol{\tau}}) = \sigma^2\mathbf{C},$$

where

$$\mathbf{C} = \tilde{\mathbf{R}}'\mathbf{X}'\mathbf{X}\tilde{\mathbf{R}} = \boldsymbol{\Lambda}'\tilde{\mathbf{R}} = \tilde{\mathbf{R}}'\boldsymbol{\Lambda} = \boldsymbol{\Lambda}'(\mathbf{X}'\mathbf{X})^-\boldsymbol{\Lambda}.$$

Note that the matrix $\mathbf{C}$ is symmetric and nonnegative definite and (in light of Lemmas 2.12.1 and 2.12.3) that

$$\text{rank}\,\mathbf{C} = \text{rank}(\mathbf{X}\tilde{\mathbf{R}}) = \text{rank}(\mathbf{X}'\mathbf{X}\tilde{\mathbf{R}}) = \text{rank}\,\boldsymbol{\Lambda} = M_*, \tag{3.1}$$

implying in particular (in light of Corollary 2.13.23) that

$$\mathbf{C} = \mathbf{T}'\mathbf{T}$$

for some $M_* \times M$ matrix $\mathbf{T}$ of full row rank M_*. Now, take $\mathbf{S}$ to be any right inverse of $\mathbf{T}$—that $\mathbf{T}$ has a right inverse is evident from Lemma 2.5.1—or, more generally, take $\mathbf{S}$ to be any $M \times M_*$ matrix such that $\mathbf{TS}$ is orthogonal, in which case

$$\mathbf{S}'\mathbf{C}\mathbf{S} = (\mathbf{TS})'\mathbf{TS} = \mathbf{I} \tag{3.2}$$

—conversely, if $\mathbf{S}$ were taken to be any $M \times M_*$ matrix such that $\mathbf{S}'\mathbf{C}\mathbf{S} = \mathbf{I}$, $\mathbf{T}\mathbf{S}$ would be orthogonal. Then,

$$\text{rank}\,\mathbf{S} = \text{rank}(\mathbf{\Lambda}\mathbf{S}) = M_*, \tag{3.3}$$

as is evident upon observing that

$$M_* \geq \text{rank}\,\mathbf{S} \geq \text{rank}(\mathbf{\Lambda}\mathbf{S}) \geq \text{rank}(\mathbf{S}'\tilde{\mathbf{R}}'\mathbf{\Lambda}\mathbf{S}) = \text{rank}(\mathbf{S}'\mathbf{C}\mathbf{S}) = \text{rank}(\mathbf{I}_{M_*}) = M_*.$$

Further, let

$$\boldsymbol{\alpha} = \mathbf{S}'\boldsymbol{\tau} = \mathbf{S}'\mathbf{\Lambda}'\boldsymbol{\beta} = (\mathbf{\Lambda}\mathbf{S})'\boldsymbol{\beta},$$

so that $\boldsymbol{\alpha}$ is an $M_* \times 1$ vector whose elements are expressible as linearly independent linear combinations of the elements of either $\boldsymbol{\tau}$ or $\boldsymbol{\beta}$. And let $\hat{\boldsymbol{\alpha}}$ represent the least squares estimator of $\boldsymbol{\alpha}$. Then, clearly,

$$\hat{\boldsymbol{\alpha}} = (\tilde{\mathbf{R}}\mathbf{S})'\mathbf{X}'\mathbf{y} = \mathbf{S}'\hat{\boldsymbol{\tau}}$$

and

$$\text{var}(\hat{\boldsymbol{\alpha}}) = \sigma^2\mathbf{I}.$$

Inverse relationship. The transformation from the M-dimensional vector $\boldsymbol{\tau}$ to the M_*-dimensional vector $\boldsymbol{\alpha}$ is invertible. In light of result (3.3), the M_* columns of the matrix $\mathbf{\Lambda}\mathbf{S}$ form a basis for $\mathcal{C}(\mathbf{\Lambda})$, and, consequently, there exists a unique $M_* \times M$ matrix $\mathbf{W}$ (of full row rank M_*) such that

$$\mathbf{\Lambda}\mathbf{S}\mathbf{W} = \mathbf{\Lambda}. \tag{3.4}$$

And upon premultiplying both sides of equality (3.4) by $\mathbf{S}'\mathbf{\Lambda}'(\mathbf{X}'\mathbf{X})^-$ [and making use of result (3.2)], we find that

$$\mathbf{W} = (\mathbf{T}\mathbf{S})'\mathbf{T} \tag{3.5}$$

—in the special case where $\mathbf{S}$ is a right inverse of $\mathbf{T}$, $\mathbf{W} = \mathbf{T}$. Note that

$$\mathcal{C}(\mathbf{W}') = \mathcal{C}(\mathbf{\Lambda}') \quad \text{and} \quad \mathbf{W}\mathbf{S} = \mathbf{I}, \tag{3.6}$$

as is evident upon, for example, observing that $\mathcal{C}(\mathbf{\Lambda}') = \mathcal{C}[(\mathbf{\Lambda}\mathbf{S}\mathbf{W})'] \subset \mathcal{C}(\mathbf{W}')$ (and invoking Theorem 2.4.16) and upon observing [in light of result (3.5)] that $\mathbf{W}\mathbf{S} = (\mathbf{T}\mathbf{S})'\mathbf{T}\mathbf{S}$ [and applying result (3.2)]. Note also that

$$\mathbf{C} = \mathbf{\Lambda}'(\mathbf{X}'\mathbf{X})^-\mathbf{\Lambda} = (\mathbf{\Lambda}\mathbf{S}\mathbf{W})'(\mathbf{X}'\mathbf{X})^-\mathbf{\Lambda}\mathbf{S}\mathbf{W} = \mathbf{W}'\mathbf{S}'\mathbf{C}\mathbf{S}\mathbf{W} = \mathbf{W}'\mathbf{W}. \tag{3.7}$$

Clearly,

$$\boldsymbol{\tau} = \mathbf{\Lambda}'\boldsymbol{\beta} = (\mathbf{\Lambda}\mathbf{S}\mathbf{W})'\boldsymbol{\beta} = \mathbf{W}'\mathbf{S}'\boldsymbol{\tau} = \mathbf{W}'\boldsymbol{\alpha}. \tag{3.8}$$

Similarly,

$$\hat{\boldsymbol{\tau}} = \mathbf{\Lambda}'(\mathbf{X}'\mathbf{X})^-\mathbf{X}'\mathbf{y} = (\mathbf{\Lambda}\mathbf{S}\mathbf{W})'(\mathbf{X}'\mathbf{X})^-\mathbf{X}'\mathbf{y} = \mathbf{W}'\mathbf{S}'\hat{\boldsymbol{\tau}} = \mathbf{W}'\hat{\boldsymbol{\alpha}}. \tag{3.9}$$

A particular implementation. The vectors $\boldsymbol{\lambda}_1, \boldsymbol{\lambda}_2, \ldots, \boldsymbol{\lambda}_M$ that form the columns of the $P \times M$ matrix $\mathbf{\Lambda}$ (which is of rank M_*) include M_* linearly independent vectors. Let $j_1, j_2, \ldots, j_M$ (where $j_1 < j_2 < \ldots < j_{M_*}$ and $j_{M_*+1} < j_{M_*+2} < \ldots < j_M$) represent a permutation of the first M positive integers $1, 2, \ldots, M$ such that the $j_1, j_2, \ldots, j_{M_*}$th columns $\boldsymbol{\lambda}_{j_1}, \boldsymbol{\lambda}_{j_2}, \ldots, \boldsymbol{\lambda}_{j_{M_*}}$ of $\mathbf{\Lambda}$ are linearly independent; and denote by $\mathbf{\Lambda}_*$ the $P \times M_*$ submatrix of $\mathbf{\Lambda}$ whose first, second, ..., M_*th columns are $\boldsymbol{\lambda}_{j_1}, \boldsymbol{\lambda}_{j_2}, \ldots, \boldsymbol{\lambda}_{j_{M_*}}$, respectively. Then, $\boldsymbol{\lambda}_j = \mathbf{\Lambda}_*\mathbf{k}_j$ for some uniquely defined $M_* \times 1$ vector $\mathbf{k}_j$ ($j = 1, 2, \ldots, M$); $\mathbf{k}_{j_1}, \mathbf{k}_{j_2}, \ldots, \mathbf{k}_{j_{M_*}}$ are respectively the first, second, ..., M_*th columns of $\mathbf{I}_{M_*}$. Further, let $\mathbf{K} = (\mathbf{k}_1, \mathbf{k}_2, \ldots, \mathbf{k}_M)$, so that $\mathbf{\Lambda} = \mathbf{\Lambda}_*\mathbf{K}$. And take $\boldsymbol{\tau}_*$ and $\hat{\boldsymbol{\tau}}_*$ to be the $M_* \times 1$ subvectors of $\boldsymbol{\tau}$ and $\hat{\boldsymbol{\tau}}$, respectively, obtained by striking out their $j_{M_*+1}, j_{M_*+2}, \ldots, j_M$th elements and take $\mathbf{C}_*$ to be the $M_* \times M_*$ submatrix of $\mathbf{C}$ obtained by striking out its $j_{M_*+1}, j_{M_*+2}, \ldots, j_M$th rows and columns, so that

$$\boldsymbol{\tau}_* = \mathbf{\Lambda}_*'\boldsymbol{\beta}, \quad \hat{\boldsymbol{\tau}}_* = \mathbf{\Lambda}_*'(\mathbf{X}'\mathbf{X})^-\mathbf{X}'\mathbf{y}, \quad \text{and} \quad \mathbf{C}_* = \mathbf{\Lambda}_*'(\mathbf{X}'\mathbf{X})^-\mathbf{\Lambda}_*.$$

and observe that

$$\boldsymbol{\tau} = \mathbf{K}'\boldsymbol{\tau}_*, \quad \hat{\boldsymbol{\tau}} = \mathbf{K}'\hat{\boldsymbol{\tau}}_*, \quad \text{and} \quad \mathbf{C} = \mathbf{K}'\mathbf{C}_*\mathbf{K}.$$

The submatrix $\mathbf{C}_*$ is a symmetric positive definite matrix (as is evident from Corollary 2.13.28

upon observing that $\mathbf{C}_*$ is an $M_* \times M_*$ symmetric nonnegative matrix of rank M_*). Accordingly, let $\mathbf{T}_*$ represent an $M_* \times M_*$ nonsingular matrix such that $\mathbf{C}_* = \mathbf{T}_*'\mathbf{T}_*$—the existence of such a matrix is evident from Corollary 2.13.29. Further, let $\tilde{\mathbf{T}} = \mathbf{T}_*\mathbf{K}$, and define $\tilde{\mathbf{S}}$ to be the $M \times M_*$ matrix whose $j_1, j_2, \ldots, j_{M_*}$th rows are respectively the first, second, ..., M_*th rows of $\mathbf{T}_*^{-1}$ and whose remaining ($j_{M_*+1}, j_{M_*+2}, \ldots, j_M$th) rows are null vectors. Then,

$$\mathbf{C} = \tilde{\mathbf{T}}'\tilde{\mathbf{T}}$$

And

$$\tilde{\mathbf{T}}\tilde{\mathbf{S}} = \mathbf{T}_*\mathbf{T}_*^{-1} = \mathbf{I},$$

so that $\tilde{\mathbf{S}}$ is a right inverse of $\tilde{\mathbf{T}}$. Thus, among the choices for a matrix $\mathbf{T}$ such that $\mathbf{C} = \mathbf{T}'\mathbf{T}$ and for a matrix $\mathbf{S}$ such that $\mathbf{S}'\mathbf{C}\mathbf{S} = \mathbf{I}$ are $\mathbf{T} = \tilde{\mathbf{T}}$ and $\mathbf{S} = \tilde{\mathbf{S}}$; and when $\mathbf{T} = \tilde{\mathbf{T}}$ and $\mathbf{S} = \tilde{\mathbf{S}}$,

$$\mathbf{W} = \tilde{\mathbf{T}} = \mathbf{T}_*\mathbf{K},$$

in which case $\boldsymbol{\tau} = \mathbf{K}'\mathbf{T}_*'\boldsymbol{\alpha}$ and $\hat{\boldsymbol{\tau}} = \mathbf{K}'\mathbf{T}_*'\hat{\boldsymbol{\alpha}}$ or, equivalently,

$$\boldsymbol{\tau}_* = \mathbf{T}_*'\boldsymbol{\alpha} \quad \text{and} \quad \tau_{j_i} = \mathbf{k}_{j_i}'\boldsymbol{\tau}_* \ (i = M_*+1, M_*+2, \ldots, M),$$

and

$$\hat{\boldsymbol{\tau}}_* = \mathbf{T}_*'\hat{\boldsymbol{\alpha}} \quad \text{and} \quad \hat{\tau}_{j_i} = \mathbf{k}_{j_i}'\hat{\boldsymbol{\tau}}_* \ (i = M_*+1, M_*+2, \ldots, M).$$

An equivalent null hypothesis. Now, consider the problem of testing the null hypothesis $H_0: \boldsymbol{\tau} = \boldsymbol{\tau}^{(0)}$ versus the alternative hypothesis $H_1: \boldsymbol{\tau} \neq \boldsymbol{\tau}^{(0)}$. This problem can be reformulated in terms of the vector $\boldsymbol{\alpha}$. Assume that H_0 is testable, in which case $\boldsymbol{\tau}^{(0)} = \boldsymbol{\Lambda}'\boldsymbol{\beta}^{(0)}$ for some $P \times 1$ vector $\boldsymbol{\beta}^{(0)}$, let $\boldsymbol{\alpha}^{(0)} = \mathbf{S}'\boldsymbol{\tau}^{(0)}$, and consider the problem of testing the null hypothesis $\tilde{H}_0: \boldsymbol{\alpha} = \boldsymbol{\alpha}^{(0)}$ versus the alternative hypothesis $\tilde{H}_1: \boldsymbol{\alpha} \neq \boldsymbol{\alpha}^{(0)}$.

The problem of testing $\tilde{H}_0$ versus $\tilde{H}_1$ is equivalent to that of testing H_0 versus H_1; they are equivalent in the sense that any value of $\boldsymbol{\beta}$ that satisfies $\tilde{H}_0$ satisfies H_0 and vice versa. To see this, observe that a value of $\boldsymbol{\beta}$ satisfies $\tilde{H}_0$ if and only if it satisfies the equality $\mathbf{S}'\boldsymbol{\Lambda}'(\boldsymbol{\beta} - \boldsymbol{\beta}^{(0)}) = \mathbf{0}$, and it satisfies H_0 if and only if it satisfies the equality $\boldsymbol{\Lambda}'(\boldsymbol{\beta} - \boldsymbol{\beta}^{(0)}) = \mathbf{0}$. Observe also that $\mathcal{N}(\mathbf{S}'\boldsymbol{\Lambda}') \supset \mathcal{N}(\boldsymbol{\Lambda}')$ and [in light of Lemma 2.11.5 and result (3.3)] that

$$\dim[\mathcal{N}(\mathbf{S}'\boldsymbol{\Lambda}')] = P - \operatorname{rank}(\mathbf{S}'\boldsymbol{\Lambda}') = P - \operatorname{rank}(\boldsymbol{\Lambda}\mathbf{S}) = P - M_* = \dim[\mathcal{N}(\boldsymbol{\Lambda}')]$$

and hence (recalling Theorem 2.4.10) that $\mathcal{N}(\mathbf{S}'\boldsymbol{\Lambda}') = \mathcal{N}(\boldsymbol{\Lambda}')$.

A vector of error contrasts. Let $\mathbf{L}$ represent an $N \times (N - P_*)$ matrix whose columns form an orthonormal basis for $\mathcal{N}(\mathbf{X}')$—the existence of an orthonormal basis follows from Theorem 2.4.23 and as previously indicated (in Section 5.9b) and as is evident from Lemma 2.11.5, $\dim[\mathcal{N}(\mathbf{X}')] = N - P_*$. Further, let $\mathbf{d} = \mathbf{L}'\mathbf{y}$. And observe that

$$\operatorname{rank} \mathbf{L} = N - P_*, \quad \mathbf{X}'\mathbf{L} = \mathbf{0}, \quad \mathbf{L}'\mathbf{X} = \mathbf{0}, \quad \text{and} \quad \mathbf{L}'\mathbf{L} = \mathbf{I}, \tag{3.10}$$

that

$$\mathrm{E}(\mathbf{d}) = \mathbf{0} \quad \text{and} \quad \operatorname{var}(\mathbf{d}) = \sigma^2\mathbf{I},$$

and that $\operatorname{cov}(\mathbf{d}, \mathbf{X}'\mathbf{y}) = \mathbf{0}$ and hence that

$$\operatorname{cov}(\mathbf{d}, \hat{\boldsymbol{\alpha}}) = \mathbf{0}.$$

The $N - P_*$ elements of the vector $\mathbf{d}$ are error contrasts—error contrasts were discussed earlier (in Section 5.9b).

Complementary parametric functions and their least squares estimators. The least squares estimator $\hat{\boldsymbol{\alpha}}$ of the vector $\boldsymbol{\alpha}$ and the vector $\mathbf{d}$ of error contrasts can be combined into a single vector and expressed as follows:

$$\begin{pmatrix} \hat{\boldsymbol{\alpha}} \\ \mathbf{d} \end{pmatrix} = \begin{pmatrix} \mathbf{S}'\tilde{\mathbf{R}}'\mathbf{X}'\mathbf{y} \\ \mathbf{L}'\mathbf{y} \end{pmatrix} = (\mathbf{X}\tilde{\mathbf{R}}\mathbf{S}, \mathbf{L})'\mathbf{y}.$$

The columns of the $N \times (N - P_* + M_*)$ matrix $(\mathbf{X}\tilde{\mathbf{R}}\mathbf{S}, \ \mathbf{L})$ are orthonormal. And (in light of Lemma 2.11.5)

$$\dim\{\mathcal{N}[(\mathbf{X}\tilde{\mathbf{R}}\mathbf{S},\ \mathbf{L})']\} = N - (N - P_* + M_*) = P_* - M_*. \tag{3.11}$$

Further,

$$\mathcal{N}[(\mathbf{X}\tilde{\mathbf{R}}\mathbf{S},\ \mathbf{L})'] \subset \mathcal{N}(\mathbf{L}') = \mathcal{C}(\mathbf{X}) \tag{3.12}$$

—that $\mathcal{N}(\mathbf{L}') = \mathcal{C}(\mathbf{X})$ follows from Theorem 2.4.10 upon observing [in light of result (3.10) and Lemma 2.4.2] that $\mathcal{C}(\mathbf{X}) \subset \mathcal{N}(\mathbf{L}')$ and (in light of Lemma 2.11.5) that $\dim[\mathcal{N}(\mathbf{L}')] = N-(N-P_*) = P_* = \dim[\mathcal{C}(\mathbf{X})]$.

Let $\mathbf{U}$ represent a matrix whose columns form an orthonormal basis for $\mathcal{N}[(\mathbf{X}\tilde{\mathbf{R}}\mathbf{S},\ \mathbf{L})']$—the existence of such a matrix follows from Theorem 2.4.23. And observe [in light of result (3.11)] that $\mathbf{U}$ is of dimensions $N \times (P_* - M_*)$ and [in light of result (3.12)] that $\mathbf{U} = \mathbf{X}\mathbf{K}$ for some matrix $\mathbf{K}$ [of dimensions $P \times (P_* - M_*)$]. Observe also that

$$\mathbf{X}'\mathbf{X}\mathbf{K} = \mathbf{X}'\mathbf{U} = (\mathbf{U}'\mathbf{X})' \tag{3.13}$$

and (in light of Lemma 2.12.3) that

$$\operatorname{rank}(\mathbf{U}'\mathbf{X}) = \operatorname{rank}(\mathbf{X}'\mathbf{X}\mathbf{K}) = \operatorname{rank}(\mathbf{X}\mathbf{K}) = \operatorname{rank}(\mathbf{U}) = P_* - M_*. \tag{3.14}$$

Now, let $\boldsymbol{\eta} = \mathbf{U}'\mathbf{X}\boldsymbol{\beta}$ and $\hat{\boldsymbol{\eta}} = \mathbf{U}'\mathbf{y}$ $(= \mathbf{K}'\mathbf{X}'\mathbf{y})$. Then, in light of results (3.14) and (3.13), $\boldsymbol{\eta}$ is a vector of $P_* - M_*$ linearly independent estimable functions, and $\hat{\boldsymbol{\eta}}$ is the least squares estimator of $\boldsymbol{\eta}$. Further,

$$\operatorname{rank}\mathbf{U} = P_* - M_*, \qquad \mathbf{U}'\mathbf{L} = \mathbf{0}, \qquad \mathbf{U}'\mathbf{X}\tilde{\mathbf{R}}\mathbf{S} = \mathbf{0}, \qquad \text{and} \qquad \mathbf{U}'\mathbf{U} = \mathbf{I},$$

implying in particular that

$$\operatorname{cov}(\hat{\boldsymbol{\eta}},\ \mathbf{d}) = \mathbf{0}, \qquad \operatorname{cov}(\hat{\boldsymbol{\eta}},\ \hat{\boldsymbol{\alpha}}) = \mathbf{0}, \qquad \text{and} \qquad \operatorname{var}(\hat{\boldsymbol{\eta}}) = \sigma^2\mathbf{I}.$$

An orthogonal transformation. Let $\mathbf{z} = \mathbf{O}'\mathbf{y}$, where $\mathbf{O} = (\mathbf{X}\tilde{\mathbf{R}}\mathbf{S},\ \mathbf{U},\ \mathbf{L})$. Upon partitioning $\mathbf{z}$ (which is an N-dimensional column vector) into subvectors $\mathbf{z}_1$, $\mathbf{z}_2$, and $\mathbf{z}_3$ of dimensions M_*, $P_* - M_*$, and $N - P_*$, respectively, so that $\mathbf{z}' = (\mathbf{z}_1', \mathbf{z}_2', \mathbf{z}_3')$, we find that

$$\mathbf{z}_1 = \mathbf{S}'\tilde{\mathbf{R}}'\mathbf{X}'\mathbf{y} = \hat{\boldsymbol{\alpha}}, \qquad \mathbf{z}_2 = \mathbf{U}'\mathbf{y} = \hat{\boldsymbol{\eta}}, \qquad \text{and} \qquad \mathbf{z}_3 = \mathbf{L}'\mathbf{y} = \mathbf{d}.$$

And

$$\mathbf{O}'\mathbf{O} = \mathbf{I}, \tag{3.15}$$

that is, $\mathbf{O}$ is orthogonal, and

$$\mathbf{O}'\mathbf{X} = \begin{pmatrix} \mathbf{S}'\tilde{\mathbf{R}}'\mathbf{X}'\mathbf{X} \\ \mathbf{U}'\mathbf{X} \\ \mathbf{0} \end{pmatrix} = \begin{pmatrix} \mathbf{S}'\boldsymbol{\Lambda}' \\ \mathbf{U}'\mathbf{X} \\ \mathbf{0} \end{pmatrix}, \tag{3.16}$$

so that

$$\mathrm{E}(\mathbf{z}) = \mathbf{O}'\mathbf{X}\boldsymbol{\beta} = \begin{pmatrix} \boldsymbol{\alpha} \\ \boldsymbol{\eta} \\ \mathbf{0} \end{pmatrix} \qquad \text{and} \qquad \operatorname{var}(\mathbf{z}) = \mathbf{O}'(\sigma^2\mathbf{I})\mathbf{O} = \sigma^2\mathbf{I}. \tag{3.17}$$

Moreover, the distribution of $\mathbf{z}$ is determinable from that of $\mathbf{y}$ and vice versa. In particular, $\mathbf{z} \sim N\left[\begin{pmatrix} \boldsymbol{\alpha} \\ \boldsymbol{\eta} \\ \mathbf{0} \end{pmatrix},\ \sigma^2\mathbf{I}\right]$ if and only if $\mathbf{y} \sim N(\mathbf{X}\boldsymbol{\beta},\ \sigma^2\mathbf{I})$. Accordingly, in devising and evaluating procedures for making inferences about the vector $\boldsymbol{\tau}$, we can take advantage of the relatively simple and transparent form of the mean vector of $\mathbf{z}$ by working with $\mathbf{z}$ rather than $\mathbf{y}$.

Definition and "role" (of the canonical form). Since (by supposition) $\mathbf{y}$ follows the G–M model,

$$\mathbf{z} = \mathbf{O}'\mathbf{X}\boldsymbol{\beta} + \mathbf{O}'\mathbf{e}. \tag{3.18}$$

Thus,

$$\mathbf{z} = \begin{pmatrix} \mathbf{I}_{M_*} & \mathbf{0} \\ \mathbf{0} & \mathbf{I}_{P_*-M_*} \\ \mathbf{0} & \mathbf{0} \end{pmatrix} \begin{pmatrix} \boldsymbol{\alpha} \\ \boldsymbol{\eta} \end{pmatrix} + \mathbf{e}^*,$$

where $\mathbf{e}^* = \mathbf{O}'\mathbf{e}$. Clearly,

$$E(\mathbf{e}^*) = \mathbf{0} \quad \text{and} \quad \text{var}(\mathbf{e}^*) = \sigma^2\mathbf{I}.$$

Like $\mathbf{y}$, $\mathbf{z}$ is an observable random vector. Also like $\mathbf{y}$, it follows a G–M model. Specifically, it follows a G–M model in which the role of the $N \times P$ model matrix $\mathbf{X}$ is played by the $N \times P_*$ matrix $\begin{pmatrix} \mathbf{I}_{M_*} & \mathbf{0} \\ \mathbf{0} & \mathbf{I}_{P_*-M_*} \\ \mathbf{0} & \mathbf{0} \end{pmatrix}$ and the role of the $P \times 1$ parameter vector $\boldsymbol{\beta}$ is played by the $P_* \times 1$ vector $\begin{pmatrix} \boldsymbol{\alpha} \\ \boldsymbol{\eta} \end{pmatrix}$ and in which the M_*-dimensional vector $\boldsymbol{\alpha}$ and the (P_*-M_*)-dimensional vector $\boldsymbol{\eta}$ are regarded as vectors of unknown (unconstrained) parameters rather than as vectors of parametric functions. This model is referred to as the *canonical form* of the (G–M) model.

Suppose that the vector $\mathbf{z}$ is assumed to follow the canonical form of the G–M model (with parameterization $\boldsymbol{\alpha}$, $\boldsymbol{\eta}$, and σ^2). Suppose further that the distribution of the vector of residual effects in the canonical form is taken to be the same as that of the vector $\mathbf{e}$ of residual effects in the original form $\mathbf{y} = \mathbf{X}\boldsymbol{\beta} + \mathbf{e}$. And suppose that $\mathbf{O}'\mathbf{e} \sim \mathbf{e}$, as would be the case if $\mathbf{e} \sim N(\mathbf{0}, \sigma^2\mathbf{I})$ or, more generally, if the distribution of $\mathbf{e}$ is spherically symmetric. Then, the distribution of the vector $\mathbf{Oz}$ obtained (on the basis of the canonical form) upon setting the parameter vector $\boldsymbol{\alpha}$ equal to $\mathbf{S}'\boldsymbol{\Lambda}'\boldsymbol{\beta}$ and the parameter vector $\boldsymbol{\eta}$ equal to $\mathbf{U}'\mathbf{X}\boldsymbol{\beta}$ is identical to the distribution of $\mathbf{y}$ (i.e., to the distribution of $\mathbf{y}$ obtained from a direct application of the model $\mathbf{y} = \mathbf{X}\boldsymbol{\beta} + \mathbf{e}$). Note that (when it comes to making inferences about $\boldsymbol{\Lambda}'\boldsymbol{\beta}$) the elements of the parameter vector $\boldsymbol{\eta}$ of the canonical form can be viewed as "nuisance parameters."

Sufficiency and completeness. Suppose that $\mathbf{y} \sim N(\mathbf{X}\boldsymbol{\beta}, \sigma^2\mathbf{I})$, as is the case when the distribution of the vector $\mathbf{e}$ of residual effects in the G–M model is taken to be $N(\mathbf{0}, \sigma^2\mathbf{I})$ (which is a case where $\mathbf{O}'\mathbf{e} \sim \mathbf{e}$). Then, as was established earlier (in Section 5.8) by working directly with the pdf of the distribution of $\mathbf{y}$, $\mathbf{X}'\mathbf{y}$ and $\mathbf{y}'(\mathbf{I}-\mathbf{P_X})\mathbf{y}$ form a complete sufficient statistic. In establishing results of this kind, it can be advantageous to adopt an alternative approach that makes use of the canonical form of the model.

Suppose that the transformed vector $\mathbf{z}$ $(= \mathbf{O}'\mathbf{y})$ follows the canonical form of the G–M model. Suppose also that the distribution of the vector of residual effects in the canonical form is $N(\mathbf{0}, \sigma^2\mathbf{I})$, in which case $\mathbf{z} \sim N\left[\begin{pmatrix} \boldsymbol{\alpha} \\ \boldsymbol{\eta} \\ \mathbf{0} \end{pmatrix}, \sigma^2\mathbf{I}\right]$. Further, denote by $f(\cdot)$ the pdf of the distribution of $\mathbf{z}$. Then,

$$\begin{aligned} f(\mathbf{z}) &= \frac{1}{(2\pi\sigma^2)^{N/2}} \exp\left[-\frac{1}{2\sigma^2}\begin{pmatrix} \mathbf{z}_1-\boldsymbol{\alpha} \\ \mathbf{z}_2-\boldsymbol{\eta} \\ \mathbf{z}_3-\mathbf{0} \end{pmatrix}'\begin{pmatrix} \mathbf{z}_1-\boldsymbol{\alpha} \\ \mathbf{z}_2-\boldsymbol{\eta} \\ \mathbf{z}_3-\mathbf{0} \end{pmatrix}\right] \\ &= \frac{1}{(2\pi\sigma^2)^{N/2}} \exp\left\{-\frac{1}{2\sigma^2}\left[(\hat{\boldsymbol{\alpha}}-\boldsymbol{\alpha})'(\hat{\boldsymbol{\alpha}}-\boldsymbol{\alpha}) + (\hat{\boldsymbol{\eta}}-\boldsymbol{\eta})'(\hat{\boldsymbol{\eta}}-\boldsymbol{\eta}) + \mathbf{d}'\mathbf{d}\right]\right\} \\ &= \frac{1}{(2\pi\sigma^2)^{N/2}} \exp\left[-\frac{1}{2\sigma^2}(\boldsymbol{\alpha}'\boldsymbol{\alpha} + \boldsymbol{\eta}'\boldsymbol{\eta})\right] \\ &\qquad \times \exp\left[-\frac{1}{2\sigma^2}(\mathbf{d}'\mathbf{d} + \hat{\boldsymbol{\alpha}}'\hat{\boldsymbol{\alpha}} + \hat{\boldsymbol{\eta}}'\hat{\boldsymbol{\eta}}) + \frac{1}{\sigma^2}\boldsymbol{\alpha}'\hat{\boldsymbol{\alpha}} + \frac{1}{\sigma^2}\boldsymbol{\eta}'\hat{\boldsymbol{\eta}}\right]. \end{aligned} \tag{3.19}$$

And it follows from a standard result (e.g., Schervish 1995, theorem 2.74) on exponential families of distributions that $(\mathbf{d}'\mathbf{d} + \hat{\boldsymbol{\alpha}}'\hat{\boldsymbol{\alpha}} + \hat{\boldsymbol{\eta}}'\hat{\boldsymbol{\eta}})$, $\hat{\boldsymbol{\alpha}}$, and $\hat{\boldsymbol{\eta}}$ form a complete sufficient statistic.

The complete sufficient statistic can be reexpressed in various alternative forms. Upon observing that

$$\mathbf{d}'\mathbf{d} + \hat{\boldsymbol{\alpha}}'\hat{\boldsymbol{\alpha}} + \hat{\boldsymbol{\eta}}'\hat{\boldsymbol{\eta}} = \mathbf{z}'\mathbf{z} = \mathbf{z}'\mathbf{I}\mathbf{z} = \mathbf{z}'\mathbf{O}'\mathbf{O}\mathbf{z} = \mathbf{y}'\mathbf{y}, \tag{3.20}$$

[in light of Theorem 5.9.6 and result (3.10)] that

$$\mathbf{d}'\mathbf{d} = \mathbf{y}'\mathbf{L}\mathbf{L}'\mathbf{y} = \mathbf{y}'(\mathbf{I}-\mathbf{P_X})\mathbf{y}, \tag{3.21}$$

and [in light of result (3.16)] that

$$\mathbf{X}'\mathbf{y} = \mathbf{X}'\mathbf{I}\mathbf{y} = \mathbf{X}'\mathbf{O}\mathbf{O}'\mathbf{y} = (\mathbf{O}'\mathbf{X})'\mathbf{z} = \boldsymbol{\Lambda}\mathbf{S}\hat{\boldsymbol{\alpha}} + \mathbf{X}'\mathbf{U}\hat{\boldsymbol{\eta}} \tag{3.22}$$

and upon recalling result (3.9), we are able to conclude that each of the following combinations forms a complete sufficient statistic:

(1) $\mathbf{y}'\mathbf{y}$, $\hat{\boldsymbol{\alpha}}$, and $\hat{\boldsymbol{\eta}}$;
(2) $\mathbf{y}'\mathbf{y}$, $\hat{\boldsymbol{\tau}}$, and $\hat{\boldsymbol{\eta}}$;
(3) $\mathbf{y}'\mathbf{y}$ and $\mathbf{X}'\mathbf{y}$;
(4) $\mathbf{y}'(\mathbf{I}-\mathbf{P_X})\mathbf{y}$, $\hat{\boldsymbol{\alpha}}$, and $\hat{\boldsymbol{\eta}}$;
(5) $\mathbf{y}'(\mathbf{I}-\mathbf{P_X})\mathbf{y}$, $\hat{\boldsymbol{\tau}}$, and $\hat{\boldsymbol{\eta}}$; and
(6) $\mathbf{y}'(\mathbf{I}-\mathbf{P_X})\mathbf{y}$ and $\mathbf{X}'\mathbf{y}$.

An extension. Consider a generalization of the situation considered in the preceding part, where it was assumed that the vector $\mathbf{e}$ of residual effects in the G–M model is $N(\mathbf{0}, \sigma^2\mathbf{I})$. Continue to suppose that the distribution of the vector of residual effects in the canonical form is the same as that of the vector $\mathbf{e}$ in the G–M model. And suppose that the distribution of the vector $\sigma^{-1}\mathbf{e}$ of standardized residual effects is not necessarily $N(\mathbf{0}, \mathbf{I})$ but rather is an absolutely continuous spherically symmetric distribution with a pdf $h(\cdot)$ that (letting $\mathbf{u}$ represent an arbitrary $N \times 1$ vector) is of the form $h(\mathbf{u}) = c^{-1}g(\mathbf{u}'\mathbf{u})$, where $g(\cdot)$ is a known (nonnegative) function (of a single variable) such that $\int_0^\infty s^{N-1}g(s^2)\,ds < \infty$ and where $c = [2\pi^{N/2}/\Gamma(N/2)]\int_0^\infty s^{N-1}g(s^2)\,ds$. Then, $\mathbf{O}'\mathbf{e} \sim \mathbf{e}$, and in light of the discussion of Section 5.9—refer, in particular, to expression (5.9.70)—the distribution of $\mathbf{z}$ has a pdf $f(\cdot)$ of the form

$$f(\mathbf{z}) = c^{-1}(\sigma^2)^{-N/2} g\left[\frac{1}{\sigma^2}\begin{pmatrix}\mathbf{z}_1-\boldsymbol{\alpha}\\ \mathbf{z}_2-\boldsymbol{\eta}\\ \mathbf{z}_3-\mathbf{0}\end{pmatrix}'\begin{pmatrix}\mathbf{z}_1-\boldsymbol{\alpha}\\ \mathbf{z}_2-\boldsymbol{\eta}\\ \mathbf{z}_3-\mathbf{0}\end{pmatrix}\right].$$

And upon proceeding in essentially the same way as in arriving at expression (3.19) and upon applying the factorization theorem (e.g., Casella and Berger 2002, theorem 6.2.6), we arrive at the conclusion that the same quantities that form a complete sufficient statistic in the special case where $\mathbf{e}$ has an $N(\mathbf{0}, \sigma^2\mathbf{I})$ distribution form a sufficient (though not necessarily complete) statistic.

b. The test and confidence set and their basic properties

Let us continue to take $\mathbf{y}$ to be an $N \times 1$ observable random vector that follows the G–M model. And taking the results of Subsection a to be our starting point (and adopting the notation employed therein), let us consider further inferences about the vector $\boldsymbol{\tau}$ $(= \boldsymbol{\Lambda}'\boldsymbol{\beta})$.

Suppose that the distribution of the vector $\mathbf{e}$ of residual effects is MVN. Then, $\mathbf{y} \sim N(\mathbf{X}\boldsymbol{\beta}, \sigma^2\mathbf{I})$. And $\mathbf{z}$ $(= \mathbf{O}'\mathbf{y}) \sim N\left[\begin{pmatrix}\boldsymbol{\alpha}\\ \boldsymbol{\eta}\\ \mathbf{0}\end{pmatrix}, \sigma^2\mathbf{I}\right]$.

F statistic and pivotal quantity. Letting $\dot{\boldsymbol{\alpha}}$ represent an arbitrary value of $\boldsymbol{\alpha}$ (i.e., an arbitrary $M_* \times 1$ vector), consider the random variable $\tilde{F}(\dot{\boldsymbol{\alpha}})$ defined as follows:

$$\tilde{F}(\dot{\boldsymbol{\alpha}}) = \frac{(\hat{\boldsymbol{\alpha}}-\dot{\boldsymbol{\alpha}})'(\hat{\boldsymbol{\alpha}}-\dot{\boldsymbol{\alpha}})/M_*}{\mathbf{d}'\mathbf{d}/(N-P_*)}.$$

—if $\mathbf{d} = \mathbf{0}$ (which is an event of probability 0), interpret $\tilde{F}(\dot{\boldsymbol{\alpha}})$ as 0 or ∞, depending on whether $\hat{\boldsymbol{\alpha}} = \dot{\boldsymbol{\alpha}}$ or $\hat{\boldsymbol{\alpha}} \neq \dot{\boldsymbol{\alpha}}$. Observe (in light of the results of Subsection a) that

$$(1/\sigma)(\hat{\boldsymbol{\alpha}}-\dot{\boldsymbol{\alpha}}) \sim N[(1/\sigma)(\boldsymbol{\alpha}-\dot{\boldsymbol{\alpha}}), \mathbf{I}],$$

that

$$(1/\sigma)\mathbf{d} \sim N(\mathbf{0}, \mathbf{I}),$$

and that $\hat{\alpha}$ and $\mathbf{d}$ are statistically independent, implying that

$$(1/\sigma^2)(\hat{\boldsymbol{\alpha}}-\dot{\boldsymbol{\alpha}})'(\hat{\boldsymbol{\alpha}}-\dot{\boldsymbol{\alpha}}) \sim \chi^2[M_*,\ (1/\sigma^2)(\boldsymbol{\alpha}-\dot{\boldsymbol{\alpha}})'(\boldsymbol{\alpha}-\dot{\boldsymbol{\alpha}})],$$

that

$$(1/\sigma^2)\mathbf{d}'\mathbf{d} \sim \chi^2(N-P_*),$$

and that $(1/\sigma^2)(\hat{\boldsymbol{\alpha}}-\dot{\boldsymbol{\alpha}})'(\hat{\boldsymbol{\alpha}}-\dot{\boldsymbol{\alpha}})$ and $(1/\sigma^2)\mathbf{d}'\mathbf{d}$ are statistically independent. Since $\tilde{F}(\dot{\boldsymbol{\alpha}})$ is reexpressible as

$$\tilde{F}(\dot{\boldsymbol{\alpha}}) = \frac{(1/\sigma^2)(\hat{\boldsymbol{\alpha}}-\dot{\boldsymbol{\alpha}})'(\hat{\boldsymbol{\alpha}}-\dot{\boldsymbol{\alpha}})/M_*}{(1/\sigma^2)\mathbf{d}'\mathbf{d}/(N-P_*)}, \tag{3.23}$$

it follows that

$$\tilde{F}(\dot{\boldsymbol{\alpha}}) \sim SF[M_*,\ N-P_*,\ (1/\sigma^2)(\dot{\boldsymbol{\alpha}}-\boldsymbol{\alpha})'(\dot{\boldsymbol{\alpha}}-\boldsymbol{\alpha})]. \tag{3.24}$$

In particular (in the special case where $\dot{\boldsymbol{\alpha}} = \boldsymbol{\alpha}$),

$$\tilde{F}(\boldsymbol{\alpha}) \sim SF(M_*,\ N-P_*). \tag{3.25}$$

In the special case where $\dot{\boldsymbol{\alpha}} = \boldsymbol{\alpha}^{(0)}$, $\tilde{F}(\dot{\boldsymbol{\alpha}})$ can serve as a "test statistic"; it provides a basis for testing the null hypothesis $\tilde{H}_0 : \boldsymbol{\alpha} = \boldsymbol{\alpha}^{(0)}$ versus the alternative hypothesis $\tilde{H}_1 : \boldsymbol{\alpha} \neq \boldsymbol{\alpha}^{(0)}$ or, equivalently, $H_0 : \boldsymbol{\tau} = \boldsymbol{\tau}^{(0)}$ versus $H_1 : \boldsymbol{\tau} \neq \boldsymbol{\tau}^{(0)}$—it is being assumed that H_0 is testable. In the special case where $\dot{\boldsymbol{\alpha}} = \boldsymbol{\alpha}$, it can serve as a "pivotal quantity" for devising a confidence set for $\boldsymbol{\alpha}$ and ultimately for $\boldsymbol{\tau}$.

Some alternative expressions. As is evident from result (3.21), the "denominator" of $\tilde{F}(\dot{\boldsymbol{\alpha}})$ is reexpressible as follows:

$$\mathbf{d}'\mathbf{d}/(N-P_*) = \mathbf{y}'(\mathbf{I}-\mathbf{P_X})\mathbf{y}/(N-P_*) = \hat{\sigma}^2, \tag{3.26}$$

where $\hat{\sigma}^2$ is the customary unbiased estimator of σ^2 (discussed in Section 5.7c). Thus,

$$\tilde{F}(\dot{\boldsymbol{\alpha}}) = \frac{(\hat{\boldsymbol{\alpha}}-\dot{\boldsymbol{\alpha}})'(\hat{\boldsymbol{\alpha}}-\dot{\boldsymbol{\alpha}})/M_*}{\mathbf{y}'(\mathbf{I}-\mathbf{P_X})\mathbf{y}/(N-P_*)} = (\hat{\boldsymbol{\alpha}}-\dot{\boldsymbol{\alpha}})'(\hat{\boldsymbol{\alpha}}-\dot{\boldsymbol{\alpha}})/(M_*\hat{\sigma}^2). \tag{3.27}$$

Now, consider the "numerator" of $\tilde{F}(\dot{\boldsymbol{\alpha}})$. Let $\dot{\boldsymbol{\tau}}$ represent an arbitrary value of $\boldsymbol{\tau}$, that is, an arbitrary member of $\mathcal{C}(\boldsymbol{\Lambda}')$. Then, for $\dot{\boldsymbol{\alpha}} = \mathbf{S}'\dot{\boldsymbol{\tau}}$ (which is the "corresponding value" of $\boldsymbol{\alpha}$),

$$(\hat{\boldsymbol{\alpha}}-\dot{\boldsymbol{\alpha}})'(\hat{\boldsymbol{\alpha}}-\dot{\boldsymbol{\alpha}})/M_* = (\hat{\boldsymbol{\tau}}-\dot{\boldsymbol{\tau}})'\mathbf{C}^-(\hat{\boldsymbol{\tau}}-\dot{\boldsymbol{\tau}})/M_*. \tag{3.28}$$

To see this, observe that

$$(\hat{\boldsymbol{\alpha}}-\mathbf{S}'\dot{\boldsymbol{\tau}})'(\hat{\boldsymbol{\alpha}}-\mathbf{S}'\dot{\boldsymbol{\tau}}) = (\hat{\boldsymbol{\tau}}-\dot{\boldsymbol{\tau}})'\mathbf{S}\mathbf{S}'(\hat{\boldsymbol{\tau}}-\dot{\boldsymbol{\tau}})$$

and that

$$\mathbf{C}\mathbf{S}\mathbf{S}'\mathbf{C} = \mathbf{T}'\mathbf{T}\mathbf{S}(\mathbf{T}\mathbf{S})'\mathbf{T} = \mathbf{T}'\mathbf{I}\mathbf{T} = \mathbf{T}'\mathbf{T} = \mathbf{C} \tag{3.29}$$

(so that $\mathbf{S}\mathbf{S}'$ is a generalized inverse of $\mathbf{C}$). Recalling that $\hat{\boldsymbol{\tau}} = \boldsymbol{\Lambda}'(\mathbf{X}'\mathbf{X})^-\mathbf{X}'\mathbf{y}$, observe also that

$$\hat{\boldsymbol{\tau}}-\dot{\boldsymbol{\tau}} \in \mathcal{C}(\boldsymbol{\Lambda}') \tag{3.30}$$

and [in light of Corollary 2.4.17 and result (3.1)] that

$$\mathcal{C}(\boldsymbol{\Lambda}') = \mathcal{C}(\mathbf{C}), \tag{3.31}$$

implying that $\hat{\boldsymbol{\tau}}-\dot{\boldsymbol{\tau}} = \mathbf{C}\mathbf{r}$ for some vector $\mathbf{r}$ and hence that

$$(\hat{\boldsymbol{\tau}}-\dot{\boldsymbol{\tau}})'\mathbf{C}^-(\hat{\boldsymbol{\tau}}-\dot{\boldsymbol{\tau}}) = \mathbf{r}'\mathbf{C}\mathbf{C}^-\mathbf{C}\mathbf{r} = \mathbf{r}'\mathbf{C}\mathbf{r}$$

and leading to the conclusion that $(\hat{\boldsymbol{\tau}}-\dot{\boldsymbol{\tau}})'\mathbf{C}^-(\hat{\boldsymbol{\tau}}-\dot{\boldsymbol{\tau}})$ is invariant to the choice of the generalized inverse $\mathbf{C}^-$.

In light of result (3.28), $\tilde{F}(\dot{\boldsymbol{\alpha}})$ is reexpressible in terms of the quantity $F(\dot{\boldsymbol{\tau}})$ defined as follows:

$$F(\dot{\boldsymbol{\tau}}) = \frac{(\hat{\boldsymbol{\tau}}-\dot{\boldsymbol{\tau}})'\mathbf{C}^-(\hat{\boldsymbol{\tau}}-\dot{\boldsymbol{\tau}})/M_*}{\mathbf{y}'(\mathbf{I}-\mathbf{P_X})\mathbf{y}/(N-P_*)} = (\hat{\boldsymbol{\tau}}-\dot{\boldsymbol{\tau}})'\mathbf{C}^-(\hat{\boldsymbol{\tau}}-\dot{\boldsymbol{\tau}})/(M_*\hat{\sigma}^2).$$

—if $\mathbf{y}'(\mathbf{I}-\mathbf{P_X})\mathbf{y} = 0$ or, equivalently, $\hat{\sigma} = 0$ (which is an event of probability 0), interpret $F(\dot{\boldsymbol{\tau}})$ as 0 or ∞, depending on whether $\hat{\boldsymbol{\tau}} = \dot{\boldsymbol{\tau}}$ or $\hat{\boldsymbol{\tau}} \neq \dot{\boldsymbol{\tau}}$. For $\dot{\boldsymbol{\alpha}} = \mathbf{S}'\dot{\boldsymbol{\tau}}$ or, equivalently, for $\dot{\boldsymbol{\tau}} = \mathbf{W}'\dot{\boldsymbol{\alpha}}$,

$$\tilde{F}(\dot{\boldsymbol{\alpha}}) = F(\dot{\boldsymbol{\tau}}). \tag{3.32}$$

In particular [in the case of the test statistic $\tilde{F}(\boldsymbol{\alpha}^{(0)})$ and the pivotal quantity $\tilde{F}(\boldsymbol{\alpha})$],

$$\tilde{F}(\boldsymbol{\alpha}^{(0)}) = F(\boldsymbol{\tau}^{(0)}) \quad \text{and} \quad \tilde{F}(\boldsymbol{\alpha}) = F(\boldsymbol{\tau}). \tag{3.33}$$

An expression for the quantity $(1/\sigma^2)(\dot{\boldsymbol{\alpha}}-\boldsymbol{\alpha})'(\dot{\boldsymbol{\alpha}}-\boldsymbol{\alpha})$—this quantity appears in result (3.24)—can be obtained that is analogous to expression (3.28) for $(\hat{\boldsymbol{\alpha}}-\dot{\boldsymbol{\alpha}})'(\hat{\boldsymbol{\alpha}}-\dot{\boldsymbol{\alpha}})/M_*$ and that can be verified in essentially the same way. For $\dot{\boldsymbol{\alpha}} = \mathbf{S}'\dot{\boldsymbol{\tau}}$ [where $\dot{\boldsymbol{\tau}} \in \mathcal{C}(\boldsymbol{\Lambda}')$], we find that

$$(1/\sigma^2)(\dot{\boldsymbol{\alpha}}-\boldsymbol{\alpha})'(\dot{\boldsymbol{\alpha}}-\boldsymbol{\alpha}) = (1/\sigma^2)(\dot{\boldsymbol{\tau}}-\boldsymbol{\tau})'\mathbf{C}^-(\dot{\boldsymbol{\tau}}-\boldsymbol{\tau}). \tag{3.34}$$

A characterization of the members of $\mathcal{C}(\boldsymbol{\Lambda}')$. Let $\dot{\boldsymbol{\tau}} = (\dot{\tau}_1, \dot{\tau}_2, \ldots, \dot{\tau}_M)'$ represent an arbitrary $M \times 1$ vector. And (defining $j_1, j_2, \ldots, j_M$ as in Subsection a) let $\dot{\boldsymbol{\tau}}_* = (\dot{\tau}_{j_1}, \dot{\tau}_{j_2}, \ldots, \dot{\tau}_{j_{M_*}})'$. Then,

$$\dot{\boldsymbol{\tau}} \in \mathcal{C}(\boldsymbol{\Lambda}') \;\Leftrightarrow\; \dot{\boldsymbol{\tau}}_* \in \mathbb{R}^{M_*} \text{ and } \dot{\tau}_{j_i} = \mathbf{k}_{j_i}'\dot{\boldsymbol{\tau}}_* \; (i = M_*+1, M_*+2, \ldots, M) \tag{3.35}$$

(where $\mathbf{k}_{j_{M_*+1}}, \mathbf{k}_{j_{M_*+2}}, \ldots, \mathbf{k}_{j_M}$ are as defined in Subsection a), as becomes evident upon observing that $\dot{\boldsymbol{\tau}} \in \mathcal{C}(\boldsymbol{\Lambda}')$ if and only if $\dot{\boldsymbol{\tau}} = \boldsymbol{\Lambda}'\dot{\boldsymbol{\beta}}$ for some $P \times 1$ vector $\dot{\boldsymbol{\beta}}$ and that $\boldsymbol{\Lambda}' = \mathbf{K}'\boldsymbol{\Lambda}_*'$ (where $\mathbf{K}$ and $\boldsymbol{\Lambda}_*$ are as defined in Subsection a).

A particular choice for the generalized inverse of $\mathbf{C}$. Among the choices for the generalized inverse $\mathbf{C}^-$ is

$$\mathbf{C}^- = \tilde{\mathbf{S}}\tilde{\mathbf{S}}' \tag{3.36}$$

(where $\tilde{\mathbf{S}}$ is as defined in Subsection a). For $i,\ i' = 1, 2, \ldots, M_*$, the $j_i\,j_{i'}$th element of this particular generalized inverse equals the $i\,i'$th element of the ordinary inverse $\mathbf{C}_*^{-1}$ of the $M_* \times M_*$ nonsingular submatrix $\mathbf{C}_*$ of $\mathbf{C}$ (the submatrix obtained upon striking out the $j_{M_*+1}, j_{M_*+2}, \ldots, j_M$th rows and columns of $\mathbf{C}$); the remaining elements of this particular generalized inverse equal 0. Thus, upon setting $\mathbf{C}^-$ equal to this particular generalized inverse, we find that $(\hat{\boldsymbol{\tau}}-\dot{\boldsymbol{\tau}})'\mathbf{C}^-(\hat{\boldsymbol{\tau}}-\dot{\boldsymbol{\tau}})$ is expressible [for $\dot{\boldsymbol{\tau}} \in \mathcal{C}(\boldsymbol{\Lambda}')$] as follows:

$$(\hat{\boldsymbol{\tau}}-\dot{\boldsymbol{\tau}})'\mathbf{C}^-(\hat{\boldsymbol{\tau}}-\dot{\boldsymbol{\tau}}) = (\hat{\boldsymbol{\tau}}_*-\dot{\boldsymbol{\tau}}_*)'\mathbf{C}_*^{-1}(\hat{\boldsymbol{\tau}}_*-\dot{\boldsymbol{\tau}}_*) \tag{3.37}$$

(where, as in Subsection a, $\hat{\boldsymbol{\tau}}_*$ and $\dot{\boldsymbol{\tau}}_*$ represent the subvectors of $\hat{\boldsymbol{\tau}}$ and $\dot{\boldsymbol{\tau}}$, respectively, obtained by striking out their $j_{M_*+1}, j_{M_*+2}, \ldots, j_M$th elements).

Confidence set. Denote by $\tilde{A}_F$ a set of $\boldsymbol{\alpha}$-values defined as follows:

$$\tilde{A}_F = \{\dot{\boldsymbol{\alpha}} : \tilde{F}(\dot{\boldsymbol{\alpha}}) \le \bar{F}_{\dot{\gamma}}(M_*, N-P_*)\},$$

where $\bar{F}_{\dot{\gamma}}(M_*, N-P_*)$ is the upper $100\dot{\gamma}\%$ point of the $SF(M_*, N-P_*)$ distribution and $\dot{\gamma}$ is a scalar between 0 and 1. Since $\tilde{F}(\dot{\boldsymbol{\alpha}})$ varies with $\mathbf{z}$, the set $\tilde{A}_F$ also varies with $\mathbf{z}$. For purposes of making explicit the dependence of $\tilde{A}_F$ on $\mathbf{z}$, let us write $\tilde{A}_F(\mathbf{z})$, or alternatively $\tilde{A}_F(\hat{\boldsymbol{\alpha}}, \hat{\boldsymbol{\eta}}, \mathbf{d})$, for $\tilde{A}_F$.

On the basis of result (3.25), we have that

$$\Pr[\boldsymbol{\alpha} \in \tilde{A}_F(\mathbf{z})] = 1-\dot{\gamma}.$$

Thus, the set $\tilde{A}_F$ constitutes a $100(1-\dot{\gamma})\%$ confidence set for $\boldsymbol{\alpha}$. In light of result (3.27),

$$\tilde{A}_F = \{\dot{\boldsymbol{\alpha}} : (\dot{\boldsymbol{\alpha}}-\hat{\boldsymbol{\alpha}})'(\dot{\boldsymbol{\alpha}}-\hat{\boldsymbol{\alpha}}) \le M_*\hat{\sigma}^2\bar{F}_{\dot{\gamma}}(M_*, N-P_*)\}. \tag{3.38}$$

The geometrical form of the set $\tilde{A}_F$ is that of an M_*-dimensional closed ball centered at the point $\hat{\boldsymbol{\alpha}}$ and with radius $[M_*\hat{\sigma}^2\bar{F}_{\dot{\gamma}}(M_*, N-P_*)]^{1/2}$.

By exploiting the relationship $\boldsymbol{\tau} = \mathbf{W}'\boldsymbol{\alpha}$, a confidence set for $\boldsymbol{\tau}$ can be obtained from that for $\boldsymbol{\alpha}$. Define a set A_F (of $\boldsymbol{\tau}$-values) as follows:

$$A_F = \{\dot{\boldsymbol{\tau}} : \dot{\boldsymbol{\tau}} = \mathbf{W}'\dot{\boldsymbol{\alpha}},\ \dot{\boldsymbol{\alpha}} \in \tilde{A}_F\}. \tag{3.39}$$

Since $\tilde{A}_F$ depends on $\mathbf{z}$ and hence (since $\mathbf{z} = \mathbf{O}'\mathbf{y}$) on $\mathbf{y}$, the set A_F depends on $\mathbf{y}$. For purposes of making this dependence explicit, let us write $A_F(\mathbf{y})$ for A_F. Clearly,

$$\Pr[\boldsymbol{\tau} \in A_F(\mathbf{y})] = \Pr[\mathbf{W}'\boldsymbol{\alpha} \in A_F(\mathbf{y})] = \Pr[\boldsymbol{\alpha} \in \tilde{A}_F(\mathbf{z})] = 1-\dot{\gamma}.$$

Thus, the set A_F constitutes a $100(1-\dot{\gamma})\%$ confidence set for $\boldsymbol{\tau}$.

Making use of results (3.32) and (3.6), we find that

$$\begin{aligned} A_F &= \{\dot{\boldsymbol{\tau}} \in \mathcal{C}(\boldsymbol{\Lambda}') : F(\dot{\boldsymbol{\tau}}) \le \bar{F}_{\dot{\gamma}}(M_*, N-P_*)\} \\ &= \{\dot{\boldsymbol{\tau}} \in \mathcal{C}(\boldsymbol{\Lambda}') : (\dot{\boldsymbol{\tau}}-\hat{\boldsymbol{\tau}})'\mathbf{C}^-(\dot{\boldsymbol{\tau}}-\hat{\boldsymbol{\tau}}) \le M_*\hat{\sigma}^2\bar{F}_{\dot{\gamma}}(M_*, N-P_*)\}. \end{aligned} \tag{3.40}$$

In the special case where $M_* = M$, result (3.40) simplifies to

$$A_F = \{\dot{\boldsymbol{\tau}} : (\dot{\boldsymbol{\tau}}-\hat{\boldsymbol{\tau}})'\mathbf{C}^{-1}(\dot{\boldsymbol{\tau}}-\hat{\boldsymbol{\tau}}) \le M\hat{\sigma}^2\bar{F}_{\dot{\gamma}}(M, N-P_*)\}. \tag{3.41}$$

More generally,

$$\begin{aligned} A_F = \{\dot{\boldsymbol{\tau}} : (\dot{\boldsymbol{\tau}}_*-\hat{\boldsymbol{\tau}}_*)'\mathbf{C}_*^{-1}(\dot{\boldsymbol{\tau}}_*-\hat{\boldsymbol{\tau}}_*) &\le M_*\hat{\sigma}^2\bar{F}_{\dot{\gamma}}(M_*, N-P_*), \\ \dot{\tau}_{j_i} &= \mathbf{k}_{j_i}'\dot{\boldsymbol{\tau}}_*\ (i = M_*+1, M_*+2, \ldots, M)\} \end{aligned} \tag{3.42}$$

[where $\dot{\boldsymbol{\tau}}_* = (\dot{\tau}_{j_1}, \dot{\tau}_{j_2}, \ldots, \dot{\tau}_{j_{M_*}})'$, where $\dot{\tau}_1, \dot{\tau}_2, \ldots, \dot{\tau}_M$ represent the first, second, ..., Mth elements of $\dot{\boldsymbol{\tau}}$, and where $\mathbf{k}_{j_{M_*+1}}, \mathbf{k}_{j_{M_*+2}}, \ldots, \mathbf{k}_{j_M}$ and $j_1, j_2, \ldots, j_M$ are as defined in Subsection a]. Geometrically, the set $\{\dot{\boldsymbol{\tau}} : (\dot{\boldsymbol{\tau}}_*-\hat{\boldsymbol{\tau}}_*)'\mathbf{C}_*^{-1}(\dot{\boldsymbol{\tau}}_*-\hat{\boldsymbol{\tau}}_*) \le M_*\hat{\sigma}^2\bar{F}_{\dot{\gamma}}(M_*, N-P_*)\}$ is represented by the points in M_*-dimensional space enclosed by a surface that is "elliptical in nature."

F test. Define a set $\tilde{C}_F$ of $\mathbf{z}$-values as follows:

$$\begin{aligned} \tilde{C}_F &= \{\mathbf{z} : \tilde{F}(\boldsymbol{\alpha}^{(0)}) > \bar{F}_{\dot{\gamma}}(M_*, N-P_*)\} \\ &= \left\{\mathbf{z} : \frac{(\hat{\boldsymbol{\alpha}}-\boldsymbol{\alpha}^{(0)})'(\hat{\boldsymbol{\alpha}}-\boldsymbol{\alpha}^{(0)})/M_*}{\mathbf{d}'\mathbf{d}/(N-P_*)} > \bar{F}_{\dot{\gamma}}(M_*, N-P_*)\right\}. \end{aligned}$$

And take $\tilde{\phi}_F(\mathbf{z})$ to be the corresponding indicator function:

$$\tilde{\phi}_F(\mathbf{z}) = \begin{cases} 1, & \text{if } \mathbf{z} \in \tilde{C}_F, \\ 0, & \text{if } \mathbf{z} \notin \tilde{C}_F. \end{cases}$$

Then, under the null hypothesis $\tilde{H}_0 : \boldsymbol{\alpha} = \boldsymbol{\alpha}^{(0)}$ or, equivalently, $H_0 : \boldsymbol{\tau} = \boldsymbol{\tau}^{(0)}$,

$$\Pr(\mathbf{z} \in \tilde{C}_F) = \Pr[\tilde{\phi}_F(\mathbf{z}) = 1] = \dot{\gamma}.$$

Thus, as a size-$\dot{\gamma}$ test of $\tilde{H}_0$ or H_0, we have the test with critical (rejection) region $\tilde{C}_F$ and critical (test) function $\tilde{\phi}_F(\mathbf{z})$, that is, the test that rejects $\tilde{H}_0$ or H_0 if $\mathbf{z} \in \tilde{C}_F$ or $\tilde{\phi}_F(\mathbf{z}) = 1$ and accepts $\tilde{H}_0$ or H_0 otherwise. This test is referred to as the size-$\dot{\gamma}$ *F test*.

The critical region and critical function of the size-$\dot{\gamma}$ F test can be reexpressed in terms of $\mathbf{y}$. In light of result (3.33),

$$\mathbf{z} \in \tilde{C}_F \Leftrightarrow \mathbf{y} \in C_F \quad \text{and} \quad \tilde{\phi}_F(\mathbf{z}) = 1 \Leftrightarrow \phi_F(\mathbf{y}) = 1, \tag{3.43}$$

where

$$\begin{aligned} C_F &= \{\mathbf{y} : F(\boldsymbol{\tau}^{(0)}) > \bar{F}_{\dot{\gamma}}(M_*, N-P_*)\} \\ &= \left\{\mathbf{y} : \frac{(\hat{\boldsymbol{\tau}}-\boldsymbol{\tau}^{(0)})'\mathbf{C}^-(\hat{\boldsymbol{\tau}}-\boldsymbol{\tau}^{(0)})/M_*}{\mathbf{y}'(\mathbf{I}-\mathbf{P_X})\mathbf{y}/(N-P_*)} > \bar{F}_{\dot{\gamma}}(M_*, N-P_*)\right\} \end{aligned}$$

and

$$\phi_F(\mathbf{y}) = \begin{cases} 1, & \text{if } \mathbf{y} \in C_F, \\ 0, & \text{if } \mathbf{y} \notin C_F. \end{cases}$$

In connection with result (3.43), it is worth noting that [in light of result (3.37)]

$$(\hat{\boldsymbol{\tau}} - \boldsymbol{\tau}^{(0)})'\mathbf{C}^{-}(\hat{\boldsymbol{\tau}} - \boldsymbol{\tau}^{(0)}) = (\hat{\boldsymbol{\tau}}_* - \boldsymbol{\tau}_*^{(0)})'\mathbf{C}_*^{-1}(\hat{\boldsymbol{\tau}}_* - \boldsymbol{\tau}_*^{(0)}), \tag{3.44}$$

where $\boldsymbol{\tau}_*^{(0)}$ is the subvector of $\boldsymbol{\tau}^{(0)}$ obtained upon striking out its $j_{M_*+1}, j_{M_*+2}, \ldots, j_M$th elements—result (3.44) is consistent with the observation that the (testable) null hypothesis H_0 is equivalent to the null hypothesis that $\boldsymbol{\tau}_* = \boldsymbol{\tau}_*^{(0)}$.

A relationship. There is an intimate relationship between the F test of the null hypothesis $\tilde{H}_0$ or the equivalent null hypothesis H_0 and the confidence sets $\tilde{A}_F$ and A_F for $\boldsymbol{\alpha}$ and $\boldsymbol{\tau}$. Clearly,

$$\mathbf{z} \notin \tilde{C}_F \;\Leftrightarrow\; \boldsymbol{\alpha}^{(0)} \in \tilde{A}_F \qquad \text{and} \qquad \mathbf{y} \notin C_F \;\Leftrightarrow\; \boldsymbol{\tau}^{(0)} \in A_F. \tag{3.45}$$

Thus, the confidence set $\tilde{A}_F$ for $\boldsymbol{\alpha}$ consists of those values of $\boldsymbol{\alpha}^{(0)}$ for which the null hypothesis $\tilde{H}_0 : \boldsymbol{\alpha} = \boldsymbol{\alpha}^{(0)}$ is accepted, and the confidence set A_F for $\boldsymbol{\tau}$ consists of those values of $\boldsymbol{\tau}^{(0)}$ [$\in \mathcal{C}(\boldsymbol{\Lambda}')$] for which the equivalent null hypothesis $H_0 : \boldsymbol{\tau} = \boldsymbol{\tau}^{(0)}$ is accepted.

Power function and probability of false coverage. The probability of $\tilde{H}_0$ or H_0 being rejected by the F test (or any other test) depends on the model's parameters; when regarded as a function of the model's parameters, this probability is referred to as the *power function* of the test. The power function of the size-$\dot{\gamma}$ F test of $\tilde{H}_0$ or H_0 is expressible in terms of $\boldsymbol{\alpha}$, $\boldsymbol{\eta}$, and σ; specifically, it is expressible as the function $\tilde{\gamma}_F(\boldsymbol{\alpha}, \boldsymbol{\eta}, \sigma)$ defined as follows:

$$\tilde{\gamma}_F(\boldsymbol{\alpha}, \boldsymbol{\eta}, \sigma) = \Pr[\tilde{F}(\boldsymbol{\alpha}^{(0)}) > \bar{F}_{\dot{\gamma}}(M_*, \; N - P_*)].$$

And [as is evident from result (3.24)]

$$\tilde{F}(\boldsymbol{\alpha}^{(0)}) \sim SF[M_*, \; N - P_*, \; (1/\sigma^2)(\boldsymbol{\alpha} - \boldsymbol{\alpha}^{(0)})'(\boldsymbol{\alpha} - \boldsymbol{\alpha}^{(0)})]. \tag{3.46}$$

Thus, $\tilde{\gamma}_F(\boldsymbol{\alpha}, \boldsymbol{\eta}, \sigma)$ does not depend on $\boldsymbol{\eta}$, and it depends on $\boldsymbol{\alpha}$ and σ only through the quantity

$$(1/\sigma^2)(\boldsymbol{\alpha} - \boldsymbol{\alpha}^{(0)})'(\boldsymbol{\alpha} - \boldsymbol{\alpha}^{(0)}) = [(1/\sigma)(\boldsymbol{\alpha} - \boldsymbol{\alpha}^{(0)})]'[(1/\sigma)(\boldsymbol{\alpha} - \boldsymbol{\alpha}^{(0)})]. \tag{3.47}$$

This quantity is interpretable as the squared distance (in units of σ) between the true and hypothesized values of $\boldsymbol{\alpha}$. When $\boldsymbol{\alpha} = \boldsymbol{\alpha}^{(0)}$, $\tilde{\gamma}_F(\boldsymbol{\alpha}, \boldsymbol{\eta}, \sigma) = \dot{\gamma}$.

The power function can be reexpressed as a function, say $\gamma_F(\boldsymbol{\tau}, \boldsymbol{\eta}, \sigma)$, of $\boldsymbol{\tau}$, $\boldsymbol{\eta}$, and σ. Clearly,

$$\gamma_F(\boldsymbol{\tau}, \boldsymbol{\eta}, \sigma) = \tilde{\gamma}_F(\mathbf{S}'\boldsymbol{\tau}, \boldsymbol{\eta}, \sigma),$$

and, in light of result (3.34),

$$(1/\sigma^2)(\mathbf{S}'\boldsymbol{\tau} - \boldsymbol{\alpha}^{(0)})'(\mathbf{S}'\boldsymbol{\tau} - \boldsymbol{\alpha}^{(0)}) = (1/\sigma^2)(\boldsymbol{\tau} - \boldsymbol{\tau}^{(0)})'\mathbf{C}^{-}(\boldsymbol{\tau} - \boldsymbol{\tau}^{(0)}) \tag{3.48}$$

$$= (1/\sigma^2)(\boldsymbol{\tau}_* - \boldsymbol{\tau}_*^{(0)})'\mathbf{C}_*^{-1}(\boldsymbol{\tau}_* - \boldsymbol{\tau}_*^{(0)}). \tag{3.49}$$

For $\boldsymbol{\alpha} \neq \boldsymbol{\alpha}^{(0)}$ or equivalently $\boldsymbol{\tau} \neq \boldsymbol{\tau}^{(0)}$, $\tilde{\gamma}_F(\boldsymbol{\alpha}, \boldsymbol{\eta}, \sigma)$ or $\gamma_F(\boldsymbol{\tau}, \boldsymbol{\eta}, \sigma)$ represents the *power* of the size-$\dot{\gamma}$ F test, that is, the probability of rejecting $\tilde{H}_0$ or H_0 when $\tilde{H}_0$ or H_0 is false. The power of a size-$\dot{\gamma}$ test is a widely adopted criterion for assessing the test's effectiveness.

In the case of a $100(1-\dot{\gamma})\%$ confidence region for $\boldsymbol{\alpha}$ or $\boldsymbol{\tau}$, the assessment of its effectiveness might be based on the *probability of false coverage*, which (by definition) is the probability that the region will cover (i.e., include) a vector $\boldsymbol{\alpha}^{(0)}$ when $\boldsymbol{\alpha}^{(0)} \neq \boldsymbol{\alpha}$ or a vector $\boldsymbol{\tau}^{(0)}$ when $\boldsymbol{\tau}^{(0)} \neq \boldsymbol{\tau}$. In light of the relationships (3.43) and (3.45), the probability $\Pr(\boldsymbol{\tau}^{(0)} \in A_F)$ of A_F covering $\boldsymbol{\tau}^{(0)}$ [where $\boldsymbol{\tau}^{(0)} \in \mathcal{C}(\boldsymbol{\Lambda}')$] equals the probability $\Pr(\boldsymbol{\alpha}^{(0)} \in \tilde{A}_F)$ of $\tilde{A}_F$ covering $\boldsymbol{\alpha}^{(0)}$ ($= \mathbf{S}'\boldsymbol{\tau}^{(0)}$), and their probability of coverage equals $1 - \gamma_F(\boldsymbol{\tau}, \boldsymbol{\eta}, \sigma)$ or (in terms of $\boldsymbol{\alpha}$ and σ) $1 - \tilde{\gamma}_F(\boldsymbol{\alpha}, \boldsymbol{\eta}, \sigma)$.

A property of the noncentral F distribution. The power function $\tilde{\gamma}_F(\boldsymbol{\alpha}, \boldsymbol{\eta}, \sigma)$ or $\gamma_F(\boldsymbol{\tau}, \boldsymbol{\eta}, \sigma)$ of the F test of $\tilde{H}_0$ or H_0 depends on the model's parameters only through the noncentrality parameter of a noncentral F distribution with numerator degrees of freedom M_* and denominator degrees of freedom $N - P_*$. An important characteristic of this dependence is discernible from the following lemma.

Lemma 7.3.1. Let w represent a random variable that has an $SF(r, s, \lambda)$ distribution (where $0 < r < \infty$, $0 < s < \infty$, and $0 \le \lambda < \infty$). Then, for any (strictly) positive constant c, $\Pr(w > c)$ is a strictly increasing function of λ.

Preliminary to proving Lemma 7.3.1, it is convenient to establish (in the form of the following 2 lemmas) some results on central or noncentral chi-square distributions.

Lemma 7.3.2. Let u represent a random variable that has a $\chi^2(r)$ distribution (where $0 < r < \infty$). Then, for any (strictly) positive constant c, $\Pr(u > c)$ is a strictly increasing function of r.

Proof (of Lemma 7.3.2). Let v represent a random variable that is distributed independently of u as $\chi^2(s)$ (where $0 < s < \infty$). Then, $u+v \sim \chi^2(r+s)$, and it suffices to observe that

$$\Pr(u+v > c) = \Pr(u > c) + \Pr(u \le c,\ v > c-u) > \Pr(u > c).$$

Q.E.D.

Lemma 7.3.3. Let u represent a random variable that has a $\chi^2(r, \lambda)$ distribution (where $0 < r < \infty$ and $0 \le \lambda < \infty$). Then, for any (strictly) positive constant c, $\Pr(u > c)$ is a strictly increasing function of λ.

Proof (of Lemma 7.3.3). Let $h(\cdot)$ represent the pdf of the $\chi^2(r, \lambda)$ distribution. Further, for $j = 1, 2, 3, \ldots$, let $g_j(\cdot)$ represent the pdf of the $\chi^2(j)$ distribution, and let v_j represent a random variable that has a $\chi^2(j)$ distribution. Then, making use of expression (6.2.14), we find that (for $\lambda > 0$)

$$\begin{aligned}
\frac{d\,\Pr(u > c)}{d\lambda} &= \int_c^\infty \frac{dh(u)}{d\lambda}\,du \\
&= \int_c^\infty \sum_{k=0}^\infty \frac{d[(\lambda/2)^k e^{-\lambda/2}/k!]}{d\lambda}\, g_{2k+r}(u)\,du \\
&= \int_c^\infty \Big\{ -\tfrac{1}{2} e^{-\lambda/2} g_r(u) + \tfrac{1}{2} \sum_{k=1}^\infty \frac{1}{k!}\big[k(\lambda/2)^{k-1} e^{-\lambda/2} \\
&\qquad\qquad - (\lambda/2)^k e^{-\lambda/2}\big] g_{2k+r}(u) \Big\}\,du \\
&= \tfrac{1}{2} \int_c^\infty \Big\{ \sum_{j=1}^\infty \big[(\lambda/2)^{j-1} e^{-\lambda/2}/(j-1)!\big]\, g_{2j+r}(u) \\
&\qquad\qquad - \sum_{k=0}^\infty \big[(\lambda/2)^k e^{-\lambda/2}/k!\big]\, g_{2k+r}(u) \Big\}\,du \\
&= \tfrac{1}{2} \int_c^\infty \Big\{ \sum_{k=0}^\infty \big[(\lambda/2)^k e^{-\lambda/2}/k!\big]\, g_{2k+r+2}(u) \\
&\qquad\qquad - \sum_{k=0}^\infty \big[(\lambda/2)^k e^{-\lambda/2}/k!\big]\, g_{2k+r}(u) \Big\}\,du \\
&= \tfrac{1}{2} \sum_{k=0}^\infty \big[(\lambda/2)^k e^{-\lambda/2}/k!\big]\big[\Pr(v_{2k+r+2} > c) - \Pr(v_{2k+r} > c)\big].
\end{aligned}$$

And in light of Lemma 7.3.2, it follows that $\dfrac{d\,\Pr(u > c)}{d\lambda} > 0$ and hence that $\Pr(u > c)$ is a strictly increasing function of λ. Q.E.D.

Proof (of Lemma 7.3.1). Let u and v represent random variables that are distributed independently as $\chi^2(r, \lambda)$ and $\chi^2(s)$, respectively. Then, observing that $w \sim \dfrac{u/r}{v/s}$ and denoting by $g(\cdot)$ the pdf of the $\chi^2(s)$ distribution, we find that

$$\Pr(w > c) = \Pr\Big(\frac{u/r}{v/s} > c\Big) = \Pr\Big(u > \frac{crv}{s}\Big) = \int_c^{\infty} \Pr\Big(u > \frac{cr\dot{v}}{s}\Big) g(\dot{v})\, d\dot{v}.$$

And based on Lemma 7.3.3, we conclude that $\Pr(w > c)$ is a strictly increasing function of λ. Q.E.D.

As an application of Lemma 7.3.1, we have that the size-$\dot{\gamma}$ F test of $\tilde{H}_0$ or H_0 is such that the probability of rejection is a strictly increasing function of the quantity (3.47), (3.48), or (3.49). And in the case of the $100(1-\dot{\gamma})\%$ confidence region $\tilde{A}_F$ or A_F, there is an implication that the probability $\Pr(\boldsymbol{\alpha}^{(0)} \in \tilde{A}_F)$ or $\Pr(\boldsymbol{\tau}^{(0)} \in A_F)$ of $\tilde{A}_F$ or A_F covering the vector $\boldsymbol{\alpha}^{(0)}$ or $\boldsymbol{\tau}^{(0)}$ [where $\boldsymbol{\tau}^{(0)} \in \mathcal{C}(\boldsymbol{\Lambda}')$] is a strictly decreasing function of the quantity (3.47), (3.48), or (3.49).

Similarity and unbiasedness. The size-$\dot{\gamma}$ F test of the null hypothesis $\tilde{H}_0$ or H_0 is a *similar test* in the sense that the probability of rejection is the same (equal to $\dot{\gamma}$) for all values of the model's parameters for which the null hypothesis is satisfied. And the size-$\dot{\gamma}$ F test of $\tilde{H}_0$ or H_0 is an *unbiased test* in the sense that the probability of rejection is at least as great for all values of the model's parameters for which the alternative hypothesis is satisfied as for any values for which the null hypothesis is satisfied. In fact, as is evident from the results of the preceding part of the present subsection, it is *strictly unbiased* in the sense that the probability of rejection is (strictly) greater for all values of the model's parameters for which the alternative hypothesis is satisfied than for any values for which the null hypothesis is satisfied.

The $100(1-\dot{\gamma})\%$ confidence regions $\tilde{A}_F$ or A_F possess properties analogous to those of the size-$\dot{\gamma}$ test. The probability of coverage $\Pr(\boldsymbol{\alpha} \in \tilde{A}_F)$ or $\Pr(\boldsymbol{\tau} \in A_F)$ is the same (equal to $1-\dot{\gamma}$) for all values of the model's parameters. And the probability of $\tilde{A}_F$ or A_F covering a vector $\boldsymbol{\alpha}^{(0)}$ or $\boldsymbol{\tau}^{(0)}$ that differs from $\boldsymbol{\alpha}$ or $\boldsymbol{\tau}$, respectively, is (strictly) less than $1-\dot{\gamma}$.

A special case: $M = M_* = 1$. Suppose (for the time being) that $M = M_* = 1$. Then, each of the quantities $\boldsymbol{\tau}$, $\hat{\boldsymbol{\tau}}$, $\mathbf{C}$, $\boldsymbol{\alpha}$, and $\hat{\boldsymbol{\alpha}}$ has a single element; let us write τ, $\hat{\tau}$, c, α, or $\hat{\alpha}$, respectively, for this element. Each of the matrices $\mathbf{T}$ and $\mathbf{S}$ also has a single element. The element of $\mathbf{T}$ is $\pm\sqrt{c}$, and hence the element of $\mathbf{S}$ can be taken to be either $1/\sqrt{c}$ or $-1/\sqrt{c}$; let us take it to be $1/\sqrt{c}$, in which case

$$\alpha = \tau/\sqrt{c} \quad \text{and} \quad \hat{\alpha} = \hat{\tau}/\sqrt{c}.$$

We find that (for arbitrary values $\dot{\alpha}$ and $\dot{\tau}$ of α and τ, respectively)

$$\tilde{F}(\dot{\alpha}) = [\tilde{t}(\dot{\alpha})]^2 \quad \text{and} \quad F(\dot{\tau}) = [t(\dot{\tau})]^2, \tag{3.50}$$

where $\tilde{t}(\dot{\alpha}) = (\hat{\alpha} - \dot{\alpha})/\hat{\sigma}$ and $t(\dot{\tau}) = (\hat{\tau} - \dot{\tau})/(\sqrt{c}\hat{\sigma})$—if $\hat{\sigma} = 0$ or, equivalently, $\mathbf{d} = \mathbf{0}$, interpret $\tilde{t}(\dot{\alpha})$ as 0, ∞, or $-\infty$ depending on whether $\hat{\alpha} = \dot{\alpha}$, $\hat{\alpha} > \dot{\alpha}$, or $\hat{\alpha} < \dot{\alpha}$ and, similarly, interpret $t(\dot{\tau})$ as 0, ∞, or $-\infty$ depending on whether $\hat{\tau} = \dot{\tau}$, $\hat{\tau} > \dot{\tau}$, or $\hat{\tau} < \dot{\tau}$. Further,

$$\tilde{t}(\dot{\alpha}) = \frac{(1/\sigma)(\hat{\alpha} - \dot{\alpha})}{\sqrt{(1/\sigma^2)\mathbf{d}'\mathbf{d}/(N-P_*)}} \sim St[N-P_*,\, (\alpha-\dot{\alpha})/\sigma]. \tag{3.51}$$

And for $\dot{\alpha} = \dot{\tau}/\sqrt{c}$ or, equivalently, for $\dot{\tau} = \sqrt{c}\dot{\alpha}$,

$$\tilde{t}(\dot{\alpha}) = t(\dot{\tau}) \quad \text{and} \quad (\alpha-\dot{\alpha})/\sigma = (\tau-\dot{\tau})/(\sqrt{c}\sigma). \tag{3.52}$$

Let $\bar{t}_{\dot{\gamma}/2}(N-P_*)$ represent the upper $100(\dot{\gamma}/2)\%$ point of the $St(N-P_*)$ distribution, and observe [in light of result (6.4.26)] that $\bar{t}_{\dot{\gamma}/2}(N-P_*) = \sqrt{\bar{F}_{\dot{\gamma}}(1, N-P_*)}$. Then, writing $\tau^{(0)}$ for $\boldsymbol{\tau}^{(0)}$, the critical region C_F of the size-$\dot{\gamma}$ F test of the null hypothesis $H_0: \tau = \tau^{(0)}$ is reexpressible in the form

$$C_F = \{\mathbf{y} \,:\, |t(\tau^{(0)})| > \bar{t}_{\dot{\gamma}/2}(N-P_*)\} \tag{3.53}$$

$$= \{\mathbf{y} \,:\, t(\tau^{(0)}) > \bar{t}_{\dot{\gamma}/2}(N-P_*) \text{ or } t(\tau^{(0)}) < -\bar{t}_{\dot{\gamma}/2}(N-P_*)\}. \tag{3.54}$$

And the $100(1-\dot{\gamma})\%$ confidence set A_F for τ is reexpressible in the form

$$A_F = \{\dot{\tau} \in \mathcal{R}^1 \,:\, \hat{\tau} - \sqrt{c}\,\hat{\sigma}\bar{t}_{\dot{\gamma}/2}(N-P_*) \le \dot{\tau} \le \hat{\tau} + \sqrt{c}\,\hat{\sigma}\bar{t}_{\dot{\gamma}/2}(N-P_*)\}. \tag{3.55}$$

Accordingly, in the special case under consideration (that where $M = M_* = 1$), the F test of H_0 is equivalent to what is commonly referred to as the two-sided t test, and the corresponding confidence set takes the form of an interval whose end points are equidistant from $\hat{\tau}$—this interval is sometimes referred to as a t interval.

An extension. Let us resume our discussion of the general case (where M is an arbitrary positive integer and where M_* may be less than M). And let us consider the extent to which the "validity" of the F test of $\tilde{H}_0$ or H_0 and of the corresponding confidence set for $\boldsymbol{\alpha}$ or $\boldsymbol{\tau}$ depend on the assumption that the distribution of the vector $\mathbf{e}$ of residual effects is MVN.

Recalling result (3.18), observe that

$$\begin{pmatrix} \hat{\boldsymbol{\alpha}} - \boldsymbol{\alpha} \\ \hat{\boldsymbol{\eta}} - \boldsymbol{\eta} \\ \mathbf{d} \end{pmatrix} = \mathbf{O}'\mathbf{e}.$$

Observe also that if $\mathbf{e}$ has an absolutely continuous spherical distribution, then so does the transformed vector $\mathbf{O}'\mathbf{e}$. Moreover, if $\mathbf{O}'\mathbf{e}$ has an absolutely continuous spherical distribution, then so does the vector $\begin{pmatrix} \hat{\boldsymbol{\alpha}} - \boldsymbol{\alpha} \\ \mathbf{d} \end{pmatrix}$ (as is evident upon observing that this vector is a subvector of $\mathbf{O}'\mathbf{e}$ and upon recalling the discussion of Section 5.9c). And in light of expression (3.23), it follows from the discussion of Section 6.3b that $\tilde{F}(\boldsymbol{\alpha})$ has the same distribution when $\begin{pmatrix} \hat{\boldsymbol{\alpha}} - \boldsymbol{\alpha} \\ \mathbf{d} \end{pmatrix}$ has any absolutely continuous spherical distribution as it does in the special case where the distribution of this vector is $N(\mathbf{0}, \sigma^2\mathbf{I})$ [which is its distribution when $\mathbf{e} \sim N(\mathbf{0}, \sigma^2\mathbf{I})$]. It is now clear that the assumption that the distribution of $\mathbf{e}$ is MVN is stronger than needed to insure the validity of the size-$\dot{\gamma}$ F test of $\tilde{H}_0$ or H_0 and of the corresponding $100(1-\dot{\gamma})\%$ confidence set $\tilde{A}_F$ or A_F. When the distribution of $\mathbf{e}$ or, more generally, the distribution of $\begin{pmatrix} \hat{\boldsymbol{\alpha}} - \boldsymbol{\alpha} \\ \mathbf{d} \end{pmatrix}$, is an absolutely continuous spherical distribution, the size of this test and the probability of coverage of this confidence set are the same as in the special case where the distribution of $\mathbf{e}$ is $N(\mathbf{0}, \sigma^2\mathbf{I})$.

It is worth noting that while the size of the test and the probability of coverage of the confidence set are the same for all absolutely continuous spherical distributions, the power of the test and the probability of false coverage are not the same. For $\dot{\boldsymbol{\alpha}} = \boldsymbol{\alpha}$, the distribution of $\tilde{F}(\dot{\boldsymbol{\alpha}})$ is the same for all absolutely continuous spherical distributions; however, for $\dot{\boldsymbol{\alpha}} \neq \boldsymbol{\alpha}$, it differs from one absolutely continuous spherical distribution to another. If the distribution of the vector $\begin{pmatrix} \hat{\boldsymbol{\alpha}} - \boldsymbol{\alpha} \\ \mathbf{d} \end{pmatrix}$ is $N(\mathbf{0}, \sigma^2\mathbf{I})$ [as would be the case if $\mathbf{e} \sim N(\mathbf{0}, \sigma^2\mathbf{I})$], then the distribution of $\tilde{F}(\dot{\boldsymbol{\alpha}})$ is noncentral F with degrees of freedom M_* and $N - P_*$ and noncentrality parameter $(1/\sigma^2)(\dot{\boldsymbol{\alpha}} - \boldsymbol{\alpha})'(\dot{\boldsymbol{\alpha}} - \boldsymbol{\alpha})$. More generally, if the distribution of the vector $(1/\sigma)\begin{pmatrix} \hat{\boldsymbol{\alpha}} - \boldsymbol{\alpha} \\ \mathbf{d} \end{pmatrix}$ (which has expected value $\mathbf{0}$ and variance-covariance matrix $\mathbf{I}$) is an absolutely continuous spherical distribution (that does not depend on the model's parameters), then the distribution of $\tilde{F}(\dot{\boldsymbol{\alpha}})$ depends on the model's parameters only through the value of the quantity $(1/\sigma^2)(\dot{\boldsymbol{\alpha}} - \boldsymbol{\alpha})'(\dot{\boldsymbol{\alpha}} - \boldsymbol{\alpha})$ [as can be readily verified via a development similar to that leading up to result (6.2.61)]. However, only in special cases is this distribution a noncentral F distribution.

Invariance/equivariance. Let $A(\mathbf{y})$ (or simply A) represent an arbitrary $100(1-\dot{\gamma})\%$ confidence set for the vector $\boldsymbol{\tau}$ $(= \boldsymbol{\Lambda}'\boldsymbol{\beta})$. And let $\tilde{A}(\mathbf{z})$ (or simply $\tilde{A}$) represent the corresponding $100(1-\dot{\gamma})\%$ confidence set for the vector $\boldsymbol{\alpha}$ $(= \mathbf{S}'\boldsymbol{\tau})$, so that

$$\tilde{A}(\mathbf{z}) = \{\dot{\boldsymbol{\alpha}} \,:\, \dot{\boldsymbol{\alpha}} = \mathbf{S}'\dot{\boldsymbol{\tau}},\ \dot{\boldsymbol{\tau}} \in A(\mathbf{O}\mathbf{z})\}.$$

Further, take $\phi(\cdot)$ and $\tilde{\phi}(\cdot)$ to be functions (of $\mathbf{y}$ and $\mathbf{z}$, respectively, defined as follows:

$$\phi(\mathbf{y}) = \begin{cases} 1, & \text{if } \boldsymbol{\tau}^{(0)} \notin A(\mathbf{y}), \\ 0, & \text{if } \boldsymbol{\tau}^{(0)} \in A(\mathbf{y}), \end{cases} \quad \text{and} \quad \tilde{\phi}(\mathbf{z}) = \begin{cases} 1, & \text{if } \boldsymbol{\alpha}^{(0)} \notin \tilde{A}(\mathbf{z}), \\ 0, & \text{if } \boldsymbol{\alpha}^{(0)} \in \tilde{A}(\mathbf{z}), \end{cases} \tag{3.56}$$

in which case $\phi(\mathbf{y})$ is the critical function for a size-$\dot{\gamma}$ test of the null hypothesis $H_0\colon \boldsymbol{\tau} = \boldsymbol{\tau}^{(0)}$ (versus the alternative hypothesis $H_1 : \boldsymbol{\tau} \neq \boldsymbol{\tau}^{(0)}$) or, equivalently, of $\tilde{H}_0 : \boldsymbol{\alpha} = \boldsymbol{\alpha}^{(0)}$ (versus $\tilde{H}_1 : \boldsymbol{\alpha} \neq \boldsymbol{\alpha}^{(0)}$) and $\tilde{\phi}(\mathbf{z}) = \phi(\mathbf{Oz})$. Or, more generally, define $\phi(\mathbf{y})$ to be the critical function for an arbitrary size-$\dot{\gamma}$ test of H_0 or $\tilde{H}_0$ (versus H_1 or $\tilde{H}_1$), and define $\tilde{\phi}(\mathbf{z}) = \phi(\mathbf{Oz})$.

Taking $\mathbf{u}$ to be an $N \times 1$ unobservable random vector that has an $N(\mathbf{0}, \mathbf{I})$ distribution or some other absolutely continuous spherical distribution with mean $\mathbf{0}$ and variance-covariance matrix $\mathbf{I}$ and assuming that $\mathbf{y} \sim \mathbf{X}\boldsymbol{\beta} + \sigma\mathbf{u}$ and hence that $\mathbf{z} \sim \begin{pmatrix} \boldsymbol{\alpha} \\ \boldsymbol{\eta} \\ \mathbf{0} \end{pmatrix} + \sigma\mathbf{u}$, let $\tilde{T}(\cdot)$ represent a one-to-one (linear) transformation from $\mathbb{R}^N$ onto $\mathbb{R}^N$ for which there exist corresponding one-to-one transformations $\tilde{F}_1(\cdot)$ from $\mathbb{R}^{M_*}$ onto $\mathbb{R}^{M_*}$, $\tilde{F}_2(\cdot)$ from $\mathbb{R}^{P_*-M_*}$ onto $\mathbb{R}^{P_*-M_*}$, and $\tilde{F}_3(\cdot)$ from the interval $(0, \infty)$ onto $(0, \infty)$ [where $\tilde{T}(\cdot)$, $\tilde{F}_1(\cdot)$, $\tilde{F}_2(\cdot)$, and $\tilde{F}_3(\cdot)$ do not depend on $\boldsymbol{\beta}$ or σ] such that

$$\tilde{T}(\mathbf{z}) \sim \begin{bmatrix} \tilde{F}_1(\boldsymbol{\alpha}) \\ \tilde{F}_2(\boldsymbol{\eta}) \\ \mathbf{0} \end{bmatrix} + \tilde{F}_3(\sigma)\mathbf{u}, \tag{3.57}$$

so that the problem of making inferences about the vector $\tilde{F}_1(\boldsymbol{\alpha})$ on the basis of the transformed vector $\tilde{T}(\mathbf{z})$ is of the same general form as that of making inferences about $\boldsymbol{\alpha}$ on the basis of $\mathbf{z}$. Further, let $\tilde{G}$ represent a group of such transformations—refer, e.g., to Casella and Berger (2002, sec. 6.4) for the definition of a group. And let us write $\tilde{T}(\mathbf{z}_1, \mathbf{z}_2, \mathbf{z}_3)$ for $\tilde{T}(\mathbf{z})$ whenever it is convenient to do so.

In making a choice for the $100(1-\dot{\gamma})\%$ confidence set A (for $\boldsymbol{\tau}$), there would seem to be some appeal in restricting attention to choices for which the corresponding $100(1-\dot{\gamma})\%$ confidence set $\tilde{A}$ (for $\boldsymbol{\alpha}$) is such that, for every value of $\mathbf{z}$ and for every transformation $\tilde{T}(\cdot)$ in $\tilde{G}$,

$$\tilde{A}[\tilde{T}(\mathbf{z})] = \{\ddot{\boldsymbol{\alpha}} : \ddot{\boldsymbol{\alpha}} = \tilde{F}_1(\dot{\boldsymbol{\alpha}}),\ \dot{\boldsymbol{\alpha}} \in \tilde{A}(\mathbf{z})\}. \tag{3.58}$$

A choice for $\tilde{A}$ having this property is said to be *invariant* or *equivariant* with respect to $\tilde{G}$, with the term invariant being reserved for the special case where

$$\tilde{F}_1(\dot{\boldsymbol{\alpha}}) = \dot{\boldsymbol{\alpha}} \text{ for every } \dot{\boldsymbol{\alpha}} \in \mathbb{R}^{M_*} \text{ [and every } \tilde{T}(\cdot) \in \tilde{G}] \tag{3.59}$$

—in that special case, condition (3.58) simplifies to $\tilde{A}[\tilde{T}(\mathbf{z})] = \tilde{A}(\mathbf{z})$.

Clearly,

$$\boldsymbol{\alpha} = \boldsymbol{\alpha}^{(0)} \Leftrightarrow \tilde{F}_1(\boldsymbol{\alpha}) = \tilde{F}_1(\boldsymbol{\alpha}^{(0)}). \tag{3.60}$$

Suppose that

$$\tilde{F}_1(\boldsymbol{\alpha}^{(0)}) = \boldsymbol{\alpha}^{(0)} \quad [\text{for every } \tilde{T}(\cdot) \in \tilde{G}]. \tag{3.61}$$

Then, in making a choice for a size-$\dot{\gamma}$ test of the null hypothesis $\tilde{H}_0\colon \boldsymbol{\alpha} = \boldsymbol{\alpha}^{(0)}$ (versus the alternative hypothesis $\tilde{H}_1 : \boldsymbol{\alpha} \neq \boldsymbol{\alpha}^{(0)}$), there would seem to be some appeal in restricting attention to choices for which the critical function $\tilde{\phi}(\cdot)$ is such that, for every value of $\mathbf{z}$ and for every transformation $\tilde{T}(\cdot)$ in $\tilde{G}$,

$$\tilde{\phi}[\tilde{T}(\mathbf{z})] = \tilde{\phi}(\mathbf{z}). \tag{3.62}$$

—the appeal may be enhanced if condition (3.59) [which is more restrictive than condition (3.61)] is satisfied. A choice for a size-$\dot{\gamma}$ test (of $\tilde{H}_0$ versus $\tilde{H}_1$) having this property is said to be *invariant* with respect to $\tilde{G}$ (as is the critical function itself). In the special case where condition (3.59) is satisfied, a size-$\dot{\gamma}$ test with critical function of the form (3.56) is invariant with respect to $\tilde{G}$ for every hypothesized value $\boldsymbol{\alpha}^{(0)} \in \mathbb{R}^{M_*}$ if and only if the corresponding $100(1-\dot{\gamma})\%$ confidence set $\tilde{A}$ is invariant with respect to $\tilde{G}$.

The characterization of equivariance and invariance can be recast in terms of $\boldsymbol{\tau}$ and/or in terms of transformations of $\mathbf{y}$. Corresponding to the one-to-one transformation $\tilde{F}_1(\cdot)$ (from $\mathcal{R}^{M_*}$ onto $\mathcal{R}^{M_*}$) is the one-to-one transformation $F_1(\cdot)$ from $\mathcal{C}(\boldsymbol{\Lambda}')$ onto $\mathcal{C}(\boldsymbol{\Lambda}')$ defined as follows:

$$F_1(\dot{\boldsymbol{\tau}}) = \mathbf{W}'\tilde{F}_1(\mathbf{S}'\dot{\boldsymbol{\tau}}) \quad [\text{for every } \dot{\boldsymbol{\tau}} \in \mathcal{C}(\boldsymbol{\Lambda}')] \tag{3.63}$$

or, equivalently,

$$\tilde{F}_1(\dot{\boldsymbol{\alpha}}) = \mathbf{S}'F_1(\mathbf{W}'\dot{\boldsymbol{\alpha}}) \quad (\text{for every } \dot{\boldsymbol{\alpha}} \in \mathcal{R}^{M_*}). \tag{3.64}$$

And corresponding to the one-to-one transformation $\tilde{T}(\cdot)$ (from $\mathcal{R}^N$ onto $\mathcal{R}^N$) is the one-to-one transformation $T(\cdot)$ (also from $\mathcal{R}^N$ onto $\mathcal{R}^N$) defined as follows:

$$T(\mathbf{y}) = \mathbf{O}\tilde{T}(\mathbf{O}'\mathbf{y}) \quad \text{or, equivalently,} \quad \tilde{T}(\mathbf{z}) = \mathbf{O}'T(\mathbf{O}\mathbf{z}). \tag{3.65}$$

Clearly,

$$T(\mathbf{y}) = \mathbf{O}\tilde{T}(\mathbf{z}) \sim \mathbf{O}\begin{bmatrix} \tilde{F}_1(\boldsymbol{\alpha}) \\ \tilde{F}_2(\boldsymbol{\eta}) \\ \mathbf{0} \end{bmatrix} + \tilde{F}_3(\sigma)\mathbf{u}. \tag{3.66}$$

Moreover,

$$\mathbf{O}\begin{bmatrix} \tilde{F}_1(\boldsymbol{\alpha}) \\ \tilde{F}_2(\boldsymbol{\eta}) \\ \mathbf{0} \end{bmatrix} = \mathbf{X}\tilde{\mathbf{R}}\mathbf{S}\mathbf{S}'F_1(\boldsymbol{\tau}) + \mathbf{U}\tilde{F}_2(\boldsymbol{\eta}) = \mathbf{X}\tilde{\mathbf{R}}\mathbf{C}^{-}F_1(\boldsymbol{\tau}) + \mathbf{U}\tilde{F}_2(\boldsymbol{\eta}), \tag{3.67}$$

as can be readily verified.

Now, let G represent the group of transformations of $\mathbf{y}$ obtained upon reexpressing each of the transformations (of $\mathbf{z}$) in the group $\tilde{G}$ in terms of $\mathbf{y}$, so that $T(\cdot) \in G$ if and only if $\tilde{T}(\cdot) \in \tilde{G}$, where $\tilde{T}(\cdot)$ is the unique transformation (of $\mathbf{z}$) that corresponds to the transformation $T(\cdot)$ (of $\mathbf{y}$) in the sense determined by relationship (3.65). Then, condition (3.59) is satisfied if and only if

$$F_1(\dot{\boldsymbol{\tau}}) = \dot{\boldsymbol{\tau}} \text{ for every } \dot{\boldsymbol{\tau}} \in \mathcal{C}(\boldsymbol{\Lambda}') \text{ [and every } T(\cdot) \in G], \tag{3.68}$$

and condition (3.61) is satisfied if and only if

$$F_1(\boldsymbol{\tau}^{(0)}) = \boldsymbol{\tau}^{(0)} \quad [\text{for every } T(\cdot) \in G]. \tag{3.69}$$

—the equivalence of conditions (3.59) and (3.68) and the equivalence of conditions (3.61) and (3.69) can be readily verified.

Further, the $100(1-\dot{\gamma})\%$ confidence set $\tilde{A}(\mathbf{z})$ for $\boldsymbol{\alpha}$ is equivariant or invariant with respect to $\tilde{G}$ if and only if the corresponding $100(1-\dot{\gamma})\%$ confidence set $A(\mathbf{y})$ for $\boldsymbol{\tau}$ is respectively equivariant or, in the special case where condition (3.68) is satisfied, invariant in the sense that, for every value of $\mathbf{y}$ and for every transformation $T(\cdot)$ in G,

$$A[T(\mathbf{y})] = \{\ddot{\boldsymbol{\tau}} \,:\, \ddot{\boldsymbol{\tau}} = F_1(\dot{\boldsymbol{\tau}}),\ \dot{\boldsymbol{\tau}} \in A(\mathbf{y})\} \tag{3.70}$$

—in the special case where condition (3.68) is satisfied, condition (3.70) simplifies to $A[T(\mathbf{y})] = A(\mathbf{y})$. To see this, observe that [for $\dot{\boldsymbol{\tau}}$ and $\ddot{\boldsymbol{\tau}}$ in $\mathcal{C}(\boldsymbol{\Lambda}')$]

$$\dot{\boldsymbol{\tau}} \in A(\mathbf{y}) \;\Leftrightarrow\; \dot{\boldsymbol{\tau}} \in A(\mathbf{O}\mathbf{z}) \;\Leftrightarrow\; \mathbf{S}'\dot{\boldsymbol{\tau}} \in \tilde{A}(\mathbf{z}),$$

that

$$\ddot{\boldsymbol{\tau}} = F_1(\dot{\boldsymbol{\tau}}) \;\Leftrightarrow\; \ddot{\boldsymbol{\tau}} = \mathbf{W}'\tilde{F}_1(\mathbf{S}'\dot{\boldsymbol{\tau}}) \;\Leftrightarrow\; \mathbf{S}'\ddot{\boldsymbol{\tau}} = \tilde{F}_1(\mathbf{S}'\dot{\boldsymbol{\tau}}),$$

and that

$$\ddot{\boldsymbol{\tau}} \in A[T(\mathbf{y})] \;\Leftrightarrow\; \mathbf{S}'\ddot{\boldsymbol{\tau}} \in \tilde{A}[\tilde{T}(\mathbf{z})].$$

Now, consider the size-$\dot{\gamma}$ test of H_0 or $\tilde{H}_0$ (versus H_1 or $\tilde{H}_1$) with critical function $\tilde{\phi}(\mathbf{z})$. This test is identical to that with critical function $\phi(\mathbf{y})$ [$= \tilde{\phi}(\mathbf{O}'\mathbf{y})$]. And assuming that condition (3.61) or equivalently condition (3.69) is satisfied, the test and the critical function $\phi(\mathbf{z})$ are invariant with respect to $\tilde{G}$ [in the sense (3.62)] if and only if the test and the critical function $\phi(\mathbf{y})$ are invariant with respect to G in the sense that [for every value of $\mathbf{y}$ and for every transformation $T(\cdot)$ in G]

$$\phi[T(\mathbf{y})] = \phi(\mathbf{y}), \tag{3.71}$$

as is evident upon observing that

$$\phi[T(\mathbf{y})] = \phi(\mathbf{y}) \;\Leftrightarrow\; \phi[\mathbf{O}\tilde{T}(\mathbf{z})] = \phi(\mathbf{O}\mathbf{z}) \;\Leftrightarrow\; \tilde{\phi}[\tilde{T}(\mathbf{z})] = \tilde{\phi}(\mathbf{z}).$$

In the special case where condition (3.68) is satisfied, a size-$\dot{\gamma}$ test with critical function of the form (3.56) is invariant with respect to G for every hypothesized value $\boldsymbol{\tau}^{(0)} \in \mathcal{C}(\boldsymbol{\Lambda}')$ if and only if the corresponding $100(1-\dot{\gamma})\%$ confidence set A is invariant with respect to G.

Translation (location) invariance/equivariance. As special cases of the size-$\dot{\gamma}$ test (of $\tilde{H}_0$ or H_0) with critical function $\tilde{\phi}(\cdot)$ or $\phi(\cdot)$ and of the $100(1-\dot{\gamma})\%$ confidence sets $\tilde{A}$ and A (for $\boldsymbol{\alpha}$ and $\boldsymbol{\tau}$, respectively), we have the size-$\dot{\gamma}$ F test and the corresponding confidence sets. These are the special cases where $\tilde{\phi}(\cdot) = \tilde{\phi}_F(\cdot)$, $\phi(\cdot) = \phi_F(\cdot)$, $\tilde{A} = \tilde{A}_F$, and $A = A_F$. Are the size-$\dot{\gamma}$ F test and the corresponding confidence sets invariant or equivariant?

Let $\tilde{T}_0(\mathbf{z})$ and $\tilde{T}_1(\mathbf{z})$ represent one-to-one transformations from $\mathbb{R}^N$ onto $\mathbb{R}^N$ of the following form:

$$\tilde{T}_0(\mathbf{z}) = \begin{pmatrix} \mathbf{z}_1+\mathbf{a} \\ \mathbf{z}_2 \\ \mathbf{z}_3 \end{pmatrix} \quad \text{and} \quad \tilde{T}_1(\mathbf{z}) = \begin{pmatrix} \mathbf{z}_1 \\ \mathbf{z}_2+\mathbf{c} \\ \mathbf{z}_3 \end{pmatrix},$$

where $\mathbf{a}$ is an M_*-dimensional and $\mathbf{c}$ a (P_*-M_*)-dimensional vector of constants. And let $\tilde{T}_{01}(\mathbf{z})$ represent the transformation formed by composition from $\tilde{T}_0(\mathbf{z})$ and $\tilde{T}_1(\mathbf{z})$, so that

$$\tilde{T}_{01}(\mathbf{z}) = \tilde{T}_0[\tilde{T}_1(\mathbf{z})] = \tilde{T}_1[\tilde{T}_0(\mathbf{z})] = \begin{pmatrix} \mathbf{z}_1+\mathbf{a} \\ \mathbf{z}_2+\mathbf{c} \\ \mathbf{z}_3 \end{pmatrix}.$$

Clearly, $\tilde{T}_0(\mathbf{z})$, $\tilde{T}_1(\mathbf{z})$, and $\tilde{T}_{01}(\mathbf{z})$ are special cases of the transformation $\tilde{T}(\mathbf{z})$, which is characterized by property (3.57)—$\tilde{F}_1(\boldsymbol{\alpha}) = \boldsymbol{\alpha}+\mathbf{a}$ for $\tilde{T}_0(\mathbf{z})$ and $\tilde{T}_{01}(\mathbf{z})$ and $= \boldsymbol{\alpha}$ for $\tilde{T}_1(\mathbf{z})$, $\tilde{F}_2(\boldsymbol{\eta}) = \boldsymbol{\eta}+\mathbf{c}$ for $\tilde{T}_1(\mathbf{z})$ and $\tilde{T}_{01}(\mathbf{z})$ and $= \boldsymbol{\eta}$ for $\tilde{T}_0(\mathbf{z})$, and $\tilde{F}_3(\sigma) = \sigma$ for $\tilde{T}_0(\mathbf{z})$, $\tilde{T}_1(\mathbf{z})$, and $\tilde{T}_{01}(\mathbf{z})$. Further, corresponding to $\tilde{T}_0(\mathbf{z})$, $\tilde{T}_1(\mathbf{z})$, and $\tilde{T}_{01}(\mathbf{z})$ are groups $\tilde{G}_0$, $\tilde{G}_1$, and $\tilde{G}_{01}$ of transformations [of the form $\tilde{T}(\mathbf{z})$] defined as follows: $\tilde{T}(\cdot) \in \tilde{G}_0$ if, for some $\mathbf{a}$, $\tilde{T}(\cdot) = \tilde{T}_0(\cdot)$; $\tilde{T}(\cdot) \in \tilde{G}_1$ if, for some $\mathbf{c}$, $\tilde{T}(\cdot) = \tilde{T}_1(\cdot)$; and $\tilde{T}(\cdot) \in \tilde{G}_{01}$ if, for some $\mathbf{a}$ and $\mathbf{c}$, $\tilde{T}(\cdot) = \tilde{T}_{01}(\cdot)$.

Corresponding to the transformations $\tilde{T}_0(\mathbf{z})$, $\tilde{T}_1(\mathbf{z})$, and $\tilde{T}_{01}(\mathbf{z})$ [of the form $\tilde{T}(\mathbf{z})$] are the transformations $T_0(\mathbf{y})$, $T_1(\mathbf{y})$, and $T_{01}(\mathbf{y})$, respectively [of the form $T(\mathbf{y})$] defined as follows:

$$T_0(\mathbf{y}) = \mathbf{y}+\mathbf{v}_0, \quad T_1(\mathbf{y}) = \mathbf{y}+\mathbf{v}_1, \quad \text{and} \quad T_{01}(\mathbf{y}) = \mathbf{y}+\mathbf{v}, \tag{3.72}$$

where $\mathbf{v}_0 \in \mathcal{C}(\mathbf{X}\tilde{\mathbf{R}}\mathbf{S})$, $\mathbf{v}_1 \in \mathcal{C}(\mathbf{U})$, and $\mathbf{v} \in \mathcal{C}(\mathbf{X})$, or, equivalently,

$$T_0(\mathbf{y}) = \mathbf{y}+\mathbf{X}\mathbf{b}_0, \quad T_1(\mathbf{y}) = \mathbf{y}+\mathbf{X}\mathbf{b}_1, \quad \text{and} \quad T_{01}(\mathbf{y}) = \mathbf{y}+\mathbf{X}\mathbf{b}, \tag{3.73}$$

where $\mathbf{b}_0 \in \mathcal{N}(\mathbf{U}'\mathbf{X})$, $\mathbf{b}_1 \in \mathcal{N}(\mathbf{S}'\boldsymbol{\Lambda}')$, and $\mathbf{b} \in \mathbb{R}^P$. And corresponding to $T_0(\mathbf{y})$, $T_1(\mathbf{y})$, and $T_{01}(\mathbf{y})$ are groups G_0, G_1, and G_{01} of transformations [of the form $T(\mathbf{y})$] defined as follows: $T(\cdot) \in G_0$ if, for some $\mathbf{v}_0$ or $\mathbf{b}_0$, $T(\cdot) = T_0(\cdot)$; $T(\cdot) \in G_1$ if, for some $\mathbf{v}_1$ or $\mathbf{b}_1$, $T(\cdot) = T_1(\cdot)$; and $T(\cdot) \in G_{01}$ if, for some $\mathbf{v}$ or $\mathbf{b}$, $T(\cdot) = T_{01}(\cdot)$. The groups G_0, G_1, and G_{01} do not vary with the choice of the matrices $\mathbf{S}$, $\mathbf{U}$, and $\mathbf{L}$, as can be readily verified.

Clearly, the F test (of $\tilde{H}_0$ or H_0) and the corresponding confidence sets $\tilde{A}_F(\mathbf{z})$ and $A_F(\mathbf{y})$ are invariant with respect to the group $\tilde{G}_1$ or G_1 of transformations of the form $\tilde{T}_1(\cdot)$ or $T_1(\cdot)$. And the confidence sets $\tilde{A}_F(\mathbf{z})$ and $A_F(\mathbf{y})$ are equivariant with respect to the group $\tilde{G}_0$ or G_0 of transformations of the form $\tilde{T}_0(\cdot)$ or $T_0(\cdot)$ and with respect to the group $\tilde{G}_{01}$ or G_{01} of transformations of the form $\tilde{T}_{01}(\cdot)$ or $T_{01}(\cdot)$.

Scale invariance/equivariance. Let us consider further the invariance or equivariance of the F test of $\tilde{H}_0$ or H_0 and of the corresponding confidence sets for $\boldsymbol{\alpha}$ and $\boldsymbol{\tau}$. For $\boldsymbol{\alpha}^{(0)} \in \mathbb{R}^{M_*}$, let $\tilde{T}_2(\mathbf{z}; \boldsymbol{\alpha}^{(0)})$ represent a one-to-one transformation from $\mathbb{R}^N$ onto $\mathbb{R}^N$ of the following form:

$$\tilde{T}_2(\mathbf{z}; \boldsymbol{\alpha}^{(0)}) = \begin{pmatrix} \boldsymbol{\alpha}^{(0)} + k(\mathbf{z}_1 - \boldsymbol{\alpha}^{(0)}) \\ k\mathbf{z}_2 \\ k\mathbf{z}_3 \end{pmatrix}, \tag{3.74}$$

where k is a strictly positive scalar. Note that in the special case where $\boldsymbol{\alpha}^{(0)} = \mathbf{0}$, equality (3.74) simplifies to

$$\tilde{T}_2(\mathbf{z}; \mathbf{0}) = k\mathbf{z}. \tag{3.75}$$

The transformation $\tilde{T}_2(\mathbf{z}; \boldsymbol{\alpha}^{(0)})$ is a special case of the transformation $\tilde{T}(\mathbf{z})$. In this special case,

$$\tilde{F}_1(\boldsymbol{\alpha}) = \boldsymbol{\alpha}^{(0)} + k(\boldsymbol{\alpha} - \boldsymbol{\alpha}^{(0)}), \quad \tilde{F}_2(\boldsymbol{\eta}) = k\boldsymbol{\eta}, \quad \text{and} \quad \tilde{F}_3(\sigma) = k\sigma.$$

Denote by $\tilde{G}_2(\boldsymbol{\alpha}^{(0)})$ the group of transformations (of $\mathbf{z}$) of the form (3.74), so that a transformation (of $\mathbf{z}$) of the general form $\tilde{T}(\mathbf{z})$ is contained in $\tilde{G}_2(\boldsymbol{\alpha}^{(0)})$ if and only if, for some k (> 0), $\tilde{T}(\cdot) = \tilde{T}_2(\cdot\,; \boldsymbol{\alpha}^{(0)})$. Further, for $\boldsymbol{\tau}^{(0)} \in \mathcal{C}(\boldsymbol{\Lambda}')$, take $T_2(\mathbf{y}; \boldsymbol{\tau}^{(0)})$ to be the transformation of $\mathbf{y}$ determined from the transformation $\tilde{T}_2(\mathbf{z}; \mathbf{S}'\boldsymbol{\tau}^{(0)})$ in accordance with relationship (3.65); and take $G_2(\boldsymbol{\tau}^{(0)})$ to be the group of all such transformations, so that $T_2(\cdot\,; \boldsymbol{\tau}^{(0)}) \in G_2(\boldsymbol{\tau}^{(0)})$ if and only if, for some transformation $\tilde{T}_2(\cdot\,; \mathbf{S}'\boldsymbol{\tau}^{(0)}) \in \tilde{G}_2(\mathbf{S}'\boldsymbol{\tau}^{(0)})$, $T_2(\mathbf{y}; \boldsymbol{\tau}^{(0)}) = \mathbf{O}\tilde{T}_2(\mathbf{O}'\mathbf{y}; \mathbf{S}'\boldsymbol{\tau}^{(0)})$. The group $G_2(\boldsymbol{\tau}^{(0)})$ does not vary with the choice of the matrices $\mathbf{S}$, $\mathbf{U}$, and $\mathbf{L}$, as can be demonstrated via a relatively straightforward exercise. In the particularly simple special case where $\boldsymbol{\tau}^{(0)} = \mathbf{0}$, we have that

$$T_2(\mathbf{y}; \mathbf{0}) = k\mathbf{y}. \tag{3.76}$$

Clearly, the F test (of $\tilde{H}_0$ or H_0) is invariant with respect to the group $\tilde{G}_2(\boldsymbol{\alpha}^{(0)})$ or $G_2(\boldsymbol{\tau}^{(0)})$ of transformations of the form $\tilde{T}_2(\cdot\,; \boldsymbol{\alpha}^{(0)})$ or $T_2(\cdot\,; \boldsymbol{\tau}^{(0)})$. And the corresponding confidence sets $\tilde{A}_F(\mathbf{z})$ and $A_F(\mathbf{y})$ are equivariant with respect to $\tilde{G}_2(\boldsymbol{\alpha}^{(0)})$ or $G_2(\boldsymbol{\tau}^{(0)})$.

Invariance/equivariance with respect to groups of orthogonal transformations. The F test and the corresponding confidence sets are invariant with respect to various groups of orthogonal transformations (of the vector $\mathbf{z}$). For $\boldsymbol{\alpha}^{(0)} \in \mathcal{R}^{M_*}$, let $\tilde{T}_3(\mathbf{z}; \boldsymbol{\alpha}^{(0)})$ represent a one-to-one transformation from $\mathcal{R}^N$ onto $\mathcal{R}^N$ of the following form:

$$\tilde{T}_3(\mathbf{z}; \boldsymbol{\alpha}^{(0)}) = \begin{pmatrix} \boldsymbol{\alpha}^{(0)} + \mathbf{P}'(\mathbf{z}_1 - \boldsymbol{\alpha}^{(0)}) \\ \mathbf{z}_2 \\ \mathbf{z}_3 \end{pmatrix}, \tag{3.77}$$

where $\mathbf{P}$ is an $M_* \times M_*$ orthogonal matrix. In particular, $\tilde{T}_3(\mathbf{z}; \mathbf{0})$ represents a transformation of the form

$$\tilde{T}_3(\mathbf{z}; \mathbf{0}) = \begin{pmatrix} \mathbf{P}'\mathbf{z}_1 \\ \mathbf{z}_2 \\ \mathbf{z}_3 \end{pmatrix}. \tag{3.78}$$

Further, let $\tilde{T}_4(\mathbf{z})$ represent a one-to-one transformation from $\mathcal{R}^N$ onto $\mathcal{R}^N$ of the form

$$\tilde{T}_4(\mathbf{z}) = \begin{pmatrix} \mathbf{z}_1 \\ \mathbf{z}_2 \\ \mathbf{B}'\mathbf{z}_3 \end{pmatrix}, \tag{3.79}$$

where $\mathbf{B}$ is an $(N - P_*) \times (N - P_*)$ orthogonal matrix.

The transformations $\tilde{T}_3(\mathbf{z}; \boldsymbol{\alpha}^{(0)})$ and $\tilde{T}_4(\mathbf{z})$ are special cases of the transformation $\tilde{T}(\mathbf{z})$. In the special case $\tilde{T}_3(\mathbf{z}; \boldsymbol{\alpha}^{(0)})$,

$$\tilde{F}_1(\boldsymbol{\alpha}) = \boldsymbol{\alpha}^{(0)} + \mathbf{P}'(\boldsymbol{\alpha} - \boldsymbol{\alpha}^{(0)}), \quad \tilde{F}_2(\boldsymbol{\eta}) = \boldsymbol{\eta}, \quad \text{and} \quad \tilde{F}_3(\sigma) = \sigma.$$

And in the special case $\tilde{T}_4(\mathbf{z})$,

$$\tilde{F}_1(\boldsymbol{\alpha}) = \boldsymbol{\alpha}, \quad \tilde{F}_2(\boldsymbol{\eta}) = \boldsymbol{\eta}, \quad \text{and} \quad \tilde{F}_3(\sigma) = \sigma.$$

Denote by $\tilde{G}_3(\boldsymbol{\alpha}^{(0)})$ the group of transformations (of $\mathbf{z}$) of the form (3.77) and by $\tilde{G}_4$ the group of the form (3.79), so that a transformation (of $\mathbf{z}$) of the general form $\tilde{T}(\mathbf{z})$ is contained in $\tilde{G}_3(\boldsymbol{\alpha}^{(0)})$ if and only if, for some ($M_* \times M_*$ orthogonal matrix) $\mathbf{P}$, $\tilde{T}(\cdot) = \tilde{T}_3(\cdot\,; \boldsymbol{\alpha}^{(0)})$ and is contained in $\tilde{G}_4$ if and only if, for some $[(N-P_*)\times(N-P_*)$ orthogonal matrix$]$ $\mathbf{B}$, $\tilde{T}(\cdot) = \tilde{T}_4(\cdot)$. Further, take $T_4(\mathbf{y})$ to be the transformation of $\mathbf{y}$ determined from the transformation $\tilde{T}_4(\mathbf{z})$ in accordance with relationship (3.65), and [for $\boldsymbol{\tau}^{(0)} \in \mathcal{C}(\boldsymbol{\Lambda}')$] take $T_3(\mathbf{y}\,;\boldsymbol{\tau}^{(0)})$ to be that determined from the transformation $\tilde{T}_3(\mathbf{z}\,;\mathbf{S}'\boldsymbol{\tau}^{(0)})$. And take G_4 and $G_3(\boldsymbol{\tau}^{(0)})$ to be the respective groups of all such transformations, so that $T_4(\cdot) \in G_4$ if and only if, for some transformation $\tilde{T}_4(\cdot) \in \tilde{G}_4$, $T_4(\mathbf{y}) = \mathbf{O}\tilde{T}_4(\mathbf{O}'\mathbf{y})$ and $T_3(\cdot\,;\boldsymbol{\tau}^{(0)}) \in G_3(\boldsymbol{\tau}^{(0)})$ if and only if, for some transformation $\tilde{T}_3(\cdot\,;\mathbf{S}'\boldsymbol{\tau}^{(0)}) \in \tilde{G}_3(\mathbf{S}'\boldsymbol{\tau}^{(0)})$, $T_3(\mathbf{y}\,;\boldsymbol{\tau}^{(0)}) = \mathbf{O}\tilde{T}_3(\mathbf{O}'\mathbf{y}\,;\mathbf{S}'\boldsymbol{\tau}^{(0)})$. The groups $G_3(\boldsymbol{\tau}^{(0)})$ and G_4, like the groups G_0, G_1, G_{01}, and $G_2(\boldsymbol{\tau}^{(0)})$, do not vary with the choice of the matrices $\mathbf{S}$, $\mathbf{U}$, and $\mathbf{L}$.

Clearly, the F test (of $\tilde{H}_0$ or H_0) is invariant with respect to both the group $\tilde{G}_3(\boldsymbol{\alpha}^{(0)})$ or $G_3(\boldsymbol{\tau}^{(0)})$ of transformations of the form $\tilde{T}_3(\cdot\,;\boldsymbol{\alpha}^{(0)})$ or $T_3(\cdot\,;\boldsymbol{\tau}^{(0)})$ and the group $\tilde{G}_4$ or G_4 of transformations of the form $\tilde{T}_4(\cdot)$ or $T_4(\cdot)$. And the corresponding confidence sets $\tilde{A}_F(\mathbf{z})$ and $A_F(\mathbf{y})$ are equivariant with respect to $\tilde{G}_3(\boldsymbol{\alpha}^{(0)})$ or $G_3(\boldsymbol{\tau}^{(0)})$ and are invariant with respect to $\tilde{G}_4$ or G_4.

c. Simultaneous confidence intervals (or sets) and multiple comparisons: a generalized S method

Let us continue to take $\mathbf{y}$ to be an $N \times 1$ observable random vector that follows the G–M model. And let us continue to take $\tau_i = \boldsymbol{\lambda}_i'\boldsymbol{\beta}$ $(i = 1, 2, \ldots, M)$ to be estimable linear combinations of the elements of $\boldsymbol{\beta}$ (where at least one of the vectors $\boldsymbol{\lambda}_1, \boldsymbol{\lambda}_2, \ldots, \boldsymbol{\lambda}_M$ of coefficients is nonnull) and to take $\boldsymbol{\tau} = (\tau_1, \tau_2, \ldots, \tau_M)'$ and $\boldsymbol{\Lambda} = (\boldsymbol{\lambda}_1, \boldsymbol{\lambda}_2, \ldots, \boldsymbol{\lambda}_M)$ (in which case $\boldsymbol{\tau} = \boldsymbol{\Lambda}'\boldsymbol{\beta}$). Further, denote by D the distribution of the vector $\mathbf{e}$ of residual effects (which, by definition, is a distribution with mean vector $\mathbf{0}$ and variance-covariance matrix $\sigma^2\mathbf{I}$), and assume that $D \in \mathfrak{D}(\sigma)$, where $\mathfrak{D}(\sigma)$ is a specified subset of the set of all N-variate distributions with mean vector $\mathbf{0}$ and variance-covariance matrix $\sigma^2\mathbf{I}$—e.g., $\mathfrak{D}(\sigma)$ might consist of spherically symmetric distributions or only of the $N(\mathbf{0}, \sigma^2\mathbf{I})$ distribution.

Suppose that we wish to make inferences about each of the M linear combinations $\tau_1, \tau_2, \ldots, \tau_M$ or, more generally, about some or all linear combinations of $\tau_1, \tau_2, \ldots, \tau_M$. That is, suppose that we wish to make inferences about a linear combination $\tau = \boldsymbol{\delta}'\boldsymbol{\tau}$ and that we wish to do so for every $\boldsymbol{\delta} \in \Delta$, where $\Delta = \mathbb{R}^M$, where Δ is the M-dimensional set formed by the columns of $\mathbf{I}_M$, or, more generally, where Δ is a finite or infinite set of $M \times 1$ vectors—to avoid trivialities, it is assumed that Δ is such that $\boldsymbol{\Lambda\delta} \neq \mathbf{0}$ for some $\boldsymbol{\delta} \in \Delta$. Inference can take the form of a point estimate, of a confidence interval (or, more generally, a confidence set), and/or of a test that τ equals some hypothesized value (versus the alternative that it does not equal the hypothesized value).

One approach is to carry out the inferences for each of the linear combinations in isolation, that is, to ignore the fact that inferences are being carried out for other linear combinations. Such an approach is sometimes referred to as "one-at-a-time." In practice, there is a natural tendency to focus on those linear combinations (of $\tau_1, \tau_2, \ldots, \tau_M$) for which the results of the inferences are the "most extreme." The one-at-a-time approach does not account for any such tendency and, as a consequence, can result in unjustified and misleading conclusions.

In the case of confidence intervals or sets, one way to counter the deficiencies of the one-at-a-time approach is to require that the probability of simultaneous coverage equal $1-\dot{\gamma}$, where $\dot{\gamma}$ is a specified scalar between 0 and 1. That is, letting $A_{\boldsymbol{\delta}}(\mathbf{y})$ or simply $A_{\boldsymbol{\delta}}$ represent the confidence interval or set for the linear combination $\tau = \boldsymbol{\delta}'\boldsymbol{\tau}$, the requirement is that [for all $\boldsymbol{\beta} \in \mathbb{R}^P$, $\sigma > 0$, and $D \in \mathfrak{D}(\sigma)$]

$$\Pr[\boldsymbol{\delta}'\boldsymbol{\tau} \in A_{\boldsymbol{\delta}}(\mathbf{y}) \text{ for every } \boldsymbol{\delta} \in \Delta] = 1-\dot{\gamma}. \tag{3.80}$$

Similarly, in the case of hypothesis tests, an alternative to the one-at-a-time approach is to require that the probability of falsely rejecting one or more of the null hypotheses not exceed $\dot{\gamma}$. More specifically, letting $\tau_{\boldsymbol{\delta}}^{(0)}$ represent the hypothesized value of the linear combination $\tau = \boldsymbol{\delta}'\boldsymbol{\tau}$ and letting $C(\boldsymbol{\delta})$ represent the critical region for the test of the null hypothesis $H_0^{(\boldsymbol{\delta})} : \tau = \tau_{\boldsymbol{\delta}}^{(0)}$ versus the alternative hypothesis $H_1^{(\boldsymbol{\delta})} : \tau \neq \tau_{\boldsymbol{\delta}}^{(0)}$, the requirement is that

$$\max_{\boldsymbol{\beta}\in\mathbb{R}^P,\ \sigma>0,\ D\in\mathfrak{D}(\sigma)} \Pr\Bigl[\mathbf{y} \in \bigcup_{\{\boldsymbol{\delta}\in\Delta\,:\,\boldsymbol{\delta}'\boldsymbol{\tau}=\tau_{\boldsymbol{\delta}}^{(0)}\}} C(\boldsymbol{\delta})\Bigr] = \dot{\gamma}. \tag{3.81}$$

In regard to requirement (3.81) and in what follows, it is assumed that there exists an $M \times 1$ vector $\boldsymbol{\tau}^{(0)} \in \mathcal{C}(\boldsymbol{\Lambda}')$ such that $\tau_{\boldsymbol{\delta}}^{(0)} = \boldsymbol{\delta}'\boldsymbol{\tau}^{(0)}$, so that the collection $H_0^{(\boldsymbol{\delta})}$ $(\boldsymbol{\delta} \in \Delta)$ of null hypotheses is "internally consistent." Corresponding to a collection $A_{\boldsymbol{\delta}}$ $(\boldsymbol{\delta} \in \Delta)$ of confidence intervals or sets that satisfies requirement (3.80) (i.e., for which the probability of simultaneous coverage equals $1-\dot{\gamma}$) is a collection of tests of $H_0^{(\boldsymbol{\delta})}$ versus $H_1^{(\boldsymbol{\delta})}$ $(\boldsymbol{\delta} \in \Delta)$ with critical regions $C(\boldsymbol{\delta})$ $(\boldsymbol{\delta} \in \Delta)$ defined (implicitly) as follows:

$$\mathbf{y} \in C(\boldsymbol{\delta}) \quad \Leftrightarrow \quad \tau_{\boldsymbol{\delta}}^{(0)} \notin A_{\boldsymbol{\delta}}(\mathbf{y}). \tag{3.82}$$

As can be readily verified (via an argument that will subsequently be demonstrated), this collection of tests satisfies condition (3.81).

The null hypothesis $H_0^{(\boldsymbol{\delta})}$ can be thought of as representing a comparison between the "actual" or "true" value of some entity (e.g., a difference in effect between two "treatments") and a hypothesized value (e.g., 0). Accordingly, the null hypotheses forming the collection $H_0^{(\boldsymbol{\delta})}$ $(\boldsymbol{\delta} \in \Delta)$ may be referred to as *multiple comparisons*, and procedures for testing these null hypotheses in a way that accounts for the multiplicity may be referred to as *multiple-comparison procedures*.

A reformulation. The problem of making inferences about the linear combinations (of the elements $\tau_1, \tau_2, \ldots, \tau_M$ of $\boldsymbol{\tau}$) forming the collection $\boldsymbol{\delta}'\boldsymbol{\tau}$ $(\boldsymbol{\delta} \in \Delta)$ is reexpressible in terms associated with the canonical form of the G–M model. Adopting the notation and recalling the results of Subsection a, the linear combination $\tau = \boldsymbol{\delta}'\boldsymbol{\tau}$ (of $\tau_1, \tau_2, \ldots, \tau_M$) is reexpressible as a linear combination of the M_* elements of the vector $\boldsymbol{\alpha}$ $(= \mathbf{S}'\boldsymbol{\tau} = \mathbf{S}'\boldsymbol{\Lambda}'\boldsymbol{\beta})$. Making use of result (3.8), we find that

$$\tau = \boldsymbol{\delta}'\mathbf{W}'\boldsymbol{\alpha} = (\mathbf{W}\boldsymbol{\delta})'\boldsymbol{\alpha}. \tag{3.83}$$

Moreover, expression (3.83) is unique; that is, if $\tilde{\boldsymbol{\delta}}$ is an $M_* \times 1$ vector of constants such that (for every value of $\boldsymbol{\beta}$) $\tau = \tilde{\boldsymbol{\delta}}'\boldsymbol{\alpha}$, then

$$\tilde{\boldsymbol{\delta}} = \mathbf{W}\boldsymbol{\delta}. \tag{3.84}$$

To see this, observe that if $\tilde{\boldsymbol{\delta}}'\boldsymbol{\alpha} = \boldsymbol{\delta}'\boldsymbol{\tau}$ (for every value of $\boldsymbol{\beta}$), then

$$\tilde{\boldsymbol{\delta}}'\mathbf{S}'\boldsymbol{\Lambda}' = \boldsymbol{\delta}'\boldsymbol{\Lambda}' = \boldsymbol{\delta}'(\boldsymbol{\Lambda}\mathbf{S}\mathbf{W})' = \boldsymbol{\delta}'\mathbf{W}'\mathbf{S}'\boldsymbol{\Lambda}',$$

implying that

$$(\tilde{\boldsymbol{\delta}} - \mathbf{W}\boldsymbol{\delta})'(\boldsymbol{\Lambda}\mathbf{S})' = \mathbf{0}$$

and hence [since in light of result (3.3), the rows of $(\boldsymbol{\Lambda}\mathbf{S})'$ are linearly independent] that $\tilde{\boldsymbol{\delta}} - \mathbf{W}\boldsymbol{\delta} = \mathbf{0}$ or, equivalently, that $\tilde{\boldsymbol{\delta}} = \mathbf{W}\boldsymbol{\delta}$.

It is now clear that the problem of making inferences about $\tau = \boldsymbol{\delta}'\boldsymbol{\tau}$ for $\boldsymbol{\delta} \in \Delta$ can be recast as one of making inferences about $\tau = \tilde{\boldsymbol{\delta}}'\boldsymbol{\alpha}$ for $\tilde{\boldsymbol{\delta}} \in \tilde{\Delta}$, where

$$\tilde{\Delta} = \{\tilde{\boldsymbol{\delta}} \,:\, \tilde{\boldsymbol{\delta}} = \mathbf{W}\boldsymbol{\delta},\ \boldsymbol{\delta} \in \boldsymbol{\Delta}\}.$$

Note that if $\Delta = \mathcal{R}^M$, then

$$\tilde{\Delta} = \mathcal{R}^{M_*},$$

as is evident upon recalling that rank $\mathbf{W} = M_*$ and observing that $\mathcal{C}(\mathbf{W}) = \mathcal{R}^{M_*}$. Note also that corresponding to $\boldsymbol{\tau}^{(0)}$, there exists a unique $M_* \times 1$ vector $\boldsymbol{\alpha}^{(0)}$ such that

$$\boldsymbol{\tau}^{(0)} = \mathbf{W}'\boldsymbol{\alpha}^{(0)}. \tag{3.85}$$

Simultaneous confidence intervals: a general approach. Suppose that the distribution of the vector $\mathbf{e}$ of residual effects in the G–M model or, more generally, the distribution of the vector $\begin{pmatrix} \hat{\boldsymbol{\alpha}} - \boldsymbol{\alpha} \\ \mathbf{d} \end{pmatrix}$ is MVN or is some other absolutely continuous spherical distribution (with mean $\mathbf{0}$ and variance-covariance matrix $\sigma^2\mathbf{I}$). Then, making use of result (6.4.67) and letting $d_1, d_2, \ldots, d_{N-P_*}$ represent the elements of $\mathbf{d}$, we find that

$$\hat{\sigma}^{-1}(\hat{\boldsymbol{\alpha}} - \boldsymbol{\alpha}) = [(N-P_*)^{-1}\textstyle\sum_{i=1}^{N-P_*} d_i^2]^{-1/2}(\hat{\boldsymbol{\alpha}} - \boldsymbol{\alpha}) \sim MVt(N-P_*,\, \mathbf{I}_{M_*}). \tag{3.86}$$

And letting $\mathbf{t}$ represent an $M_* \times 1$ random vector that has an $MVt(N-P_*, \mathbf{I}_{M_*})$ distribution, it follows that

$$\max_{\{\tilde{\boldsymbol{\delta}}\in\tilde{\Delta}\,:\,\tilde{\boldsymbol{\delta}}\neq\mathbf{0}\}} \frac{|\tilde{\boldsymbol{\delta}}'(\hat{\boldsymbol{\alpha}}-\boldsymbol{\alpha})|}{(\tilde{\boldsymbol{\delta}}'\tilde{\boldsymbol{\delta}})^{1/2}\hat{\sigma}} \sim \max_{\{\tilde{\boldsymbol{\delta}}\in\tilde{\Delta}\,:\,\tilde{\boldsymbol{\delta}}\neq\mathbf{0}\}} \frac{|\tilde{\boldsymbol{\delta}}'\mathbf{t}|}{(\tilde{\boldsymbol{\delta}}'\tilde{\boldsymbol{\delta}})^{1/2}}. \tag{3.87}$$

Thus, letting (for any scalar $\dot{\gamma}$ such that $0 < \dot{\gamma} < 1$) $c_{\dot{\gamma}}$ represent the upper $100\dot{\gamma}\%$ point of the distribution of the random variable $\max_{\{\tilde{\boldsymbol{\delta}}\in\tilde{\Delta}\,:\,\tilde{\boldsymbol{\delta}}\neq\mathbf{0}\}} |\tilde{\boldsymbol{\delta}}'\mathbf{t}|/(\tilde{\boldsymbol{\delta}}'\tilde{\boldsymbol{\delta}})^{1/2}$, we find that

$$\Pr[|\tilde{\boldsymbol{\delta}}'(\hat{\boldsymbol{\alpha}}-\boldsymbol{\alpha})| \le (\tilde{\boldsymbol{\delta}}'\tilde{\boldsymbol{\delta}})^{1/2}\hat{\sigma}c_{\dot{\gamma}} \text{ for every } \tilde{\boldsymbol{\delta}}\in\tilde{\Delta}] = \Pr\Bigg[\max_{\{\tilde{\boldsymbol{\delta}}\in\tilde{\Delta}\,:\,\tilde{\boldsymbol{\delta}}\neq\mathbf{0}\}} \frac{|\tilde{\boldsymbol{\delta}}'(\hat{\boldsymbol{\alpha}}-\boldsymbol{\alpha})|}{(\tilde{\boldsymbol{\delta}}'\tilde{\boldsymbol{\delta}})^{1/2}\hat{\sigma}} \le c_{\dot{\gamma}}\Bigg] = 1-\dot{\gamma}. \tag{3.88}$$

For $\tilde{\boldsymbol{\delta}}\in\tilde{\Delta}$, denote by $\tilde{A}_{\tilde{\boldsymbol{\delta}}}(\mathbf{z})$ or simply by $\tilde{A}_{\tilde{\boldsymbol{\delta}}}$ a set of τ-values (i.e., a set of scalars), the contents of which may depend on the value of the vector $\mathbf{z}$ (defined in Subsection a). Further, suppose that

$$\tilde{A}_{\tilde{\boldsymbol{\delta}}} = \{\dot{\tau}\in\mathbb{R}^1 \,:\, \dot{\tau}=\tilde{\boldsymbol{\delta}}'\dot{\boldsymbol{\alpha}},\ |\tilde{\boldsymbol{\delta}}'(\hat{\boldsymbol{\alpha}}-\dot{\boldsymbol{\alpha}})| \le (\tilde{\boldsymbol{\delta}}'\tilde{\boldsymbol{\delta}})^{1/2}\hat{\sigma}c_{\dot{\gamma}}\}. \tag{3.89}$$

Clearly, the set (3.89) is reexpressible in the form

$$\tilde{A}_{\tilde{\boldsymbol{\delta}}} = \{\dot{\tau}\in\mathbb{R}^1 \,:\, \tilde{\boldsymbol{\delta}}'\hat{\boldsymbol{\alpha}} - (\tilde{\boldsymbol{\delta}}'\tilde{\boldsymbol{\delta}})^{1/2}\hat{\sigma}c_{\dot{\gamma}} \le \dot{\tau} \le \tilde{\boldsymbol{\delta}}'\hat{\boldsymbol{\alpha}} + (\tilde{\boldsymbol{\delta}}'\tilde{\boldsymbol{\delta}})^{1/2}\hat{\sigma}c_{\dot{\gamma}}\}, \tag{3.90}$$

that is, as an interval with upper and lower end points $\tilde{\boldsymbol{\delta}}'\hat{\boldsymbol{\alpha}} \pm (\tilde{\boldsymbol{\delta}}'\tilde{\boldsymbol{\delta}})^{1/2}\hat{\sigma}c_{\dot{\gamma}}$. And in light of result (3.88),

$$\Pr[\tilde{\boldsymbol{\delta}}'\boldsymbol{\alpha} \in \tilde{A}_{\tilde{\boldsymbol{\delta}}}(\mathbf{z}) \text{ for every } \tilde{\boldsymbol{\delta}}\in\tilde{\Delta}] = 1-\dot{\gamma}, \tag{3.91}$$

that is, the probability of simultaneous coverage of the intervals $\tilde{A}_{\tilde{\boldsymbol{\delta}}}(\mathbf{z})$ $(\tilde{\boldsymbol{\delta}}\in\tilde{\Delta})$ equals $1-\dot{\gamma}$.

For $\boldsymbol{\delta}\in\Delta$ [and for any choice of the sets $\tilde{A}_{\tilde{\boldsymbol{\delta}}}$ $(\tilde{\boldsymbol{\delta}}\in\tilde{\Delta})$], denote by $A_{\boldsymbol{\delta}}(\mathbf{y})$ or simply by $A_{\boldsymbol{\delta}}$ the set $\tilde{A}_{\mathbf{W}\boldsymbol{\delta}}(\mathbf{O}'\mathbf{y})$. Then, for any $M_* \times 1$ vector $\dot{\boldsymbol{\alpha}}$,

$$\tilde{\boldsymbol{\delta}}'\dot{\boldsymbol{\alpha}} \in \tilde{A}_{\tilde{\boldsymbol{\delta}}}(\mathbf{z}) \text{ for every } \tilde{\boldsymbol{\delta}}\in\tilde{\Delta} \ \Leftrightarrow\ \boldsymbol{\delta}'\mathbf{W}'\dot{\boldsymbol{\alpha}} \in A_{\boldsymbol{\delta}}(\mathbf{O}\mathbf{z}) \text{ for every } \boldsymbol{\delta}\in\Delta. \tag{3.92}$$

And upon observing that, for $\dot{\boldsymbol{\alpha}} = \boldsymbol{\alpha}$, $\boldsymbol{\delta}'\mathbf{W}'\dot{\boldsymbol{\alpha}} = \boldsymbol{\delta}'\boldsymbol{\tau}$ [as is evident from result (3.8)], it follows, in particular, that

$$\Pr[\boldsymbol{\delta}'\boldsymbol{\tau} \in A_{\boldsymbol{\delta}}(\mathbf{y}) \text{ for every } \boldsymbol{\delta}\in\Delta] = \Pr[\tilde{\boldsymbol{\delta}}'\boldsymbol{\alpha} \in \tilde{A}_{\tilde{\boldsymbol{\delta}}}(\mathbf{z}) \text{ for every } \tilde{\boldsymbol{\delta}}\in\tilde{\Delta}]. \tag{3.93}$$

Now, suppose that (for $\tilde{\boldsymbol{\delta}}\in\tilde{\Delta}$) $\tilde{A}_{\tilde{\boldsymbol{\delta}}}$ is interval (3.90). Then, we find [in light of results (3.7) and (3.9)] that (for $\boldsymbol{\delta}\in\Delta$)

$$A_{\boldsymbol{\delta}}(\mathbf{y}) = \tilde{A}_{\mathbf{W}\boldsymbol{\delta}}(\mathbf{O}'\mathbf{y}) = \{\dot{\tau}\in\mathbb{R}^1 \,:\, \boldsymbol{\delta}'\hat{\boldsymbol{\tau}} - (\boldsymbol{\delta}'\mathbf{C}\boldsymbol{\delta})^{1/2}\hat{\sigma}c_{\dot{\gamma}} \le \dot{\tau} \le \boldsymbol{\delta}'\hat{\boldsymbol{\tau}} + (\boldsymbol{\delta}'\mathbf{C}\boldsymbol{\delta})^{1/2}\hat{\sigma}c_{\dot{\gamma}}\} \tag{3.94}$$

and [in light of results (3.91) and (3.93)] that

$$\Pr[\boldsymbol{\delta}'\boldsymbol{\tau} \in A_{\boldsymbol{\delta}}(\mathbf{y}) \text{ for every } \boldsymbol{\delta}\in\Delta] = 1-\dot{\gamma}. \tag{3.95}$$

Thus, $A_{\boldsymbol{\delta}}(\mathbf{y})$ is an interval with upper and lower end points $\boldsymbol{\delta}'\hat{\boldsymbol{\tau}} \pm (\boldsymbol{\delta}'\mathbf{C}\boldsymbol{\delta})^{1/2}\hat{\sigma}c_{\dot{\gamma}}$, and the probability of simultaneous coverage of the intervals $A_{\boldsymbol{\delta}}(\mathbf{y})$ $(\boldsymbol{\delta}\in\Delta)$ equals $1-\dot{\gamma}$. In the special case where $M = M_* = 1$ and $\boldsymbol{\delta} = 1$, interval (3.94) simplifies to the t interval (3.55).

Note [in connection with the random variable $\max_{\{\tilde{\boldsymbol{\delta}}\in\tilde{\Delta}\,:\,\tilde{\boldsymbol{\delta}}\neq\mathbf{0}\}} |\tilde{\boldsymbol{\delta}}'\mathbf{t}|/(\tilde{\boldsymbol{\delta}}'\tilde{\boldsymbol{\delta}})^{1/2}$ and with the upper $100\dot{\gamma}\%$ point $c_{\dot{\gamma}}$ of its distribution] that for any $M_* \times 1$ vector $\dot{\mathbf{t}}$,

$$\max_{\{\tilde{\boldsymbol{\delta}}\in\tilde{\Delta}\,:\,\tilde{\boldsymbol{\delta}}\neq\mathbf{0}\}} \frac{|\tilde{\boldsymbol{\delta}}'\dot{\mathbf{t}}|}{(\tilde{\boldsymbol{\delta}}'\tilde{\boldsymbol{\delta}})^{1/2}} = \max\Bigg[-\min_{\{\tilde{\boldsymbol{\delta}}\in\tilde{\Delta}\,:\,\tilde{\boldsymbol{\delta}}\neq\mathbf{0}\}} \frac{\tilde{\boldsymbol{\delta}}'\dot{\mathbf{t}}}{(\tilde{\boldsymbol{\delta}}'\tilde{\boldsymbol{\delta}})^{1/2}},\ \max_{\{\tilde{\boldsymbol{\delta}}\in\tilde{\Delta}\,:\,\tilde{\boldsymbol{\delta}}\neq\mathbf{0}\}} \frac{\tilde{\boldsymbol{\delta}}'\dot{\mathbf{t}}}{(\tilde{\boldsymbol{\delta}}'\tilde{\boldsymbol{\delta}})^{1/2}}\Bigg], \tag{3.96}$$

as can be readily verified.

Multiple comparisons: some relationships. Let $\boldsymbol{\alpha}^{(0)}$ represent any particular value of $\boldsymbol{\alpha}$, and consider [for $\tilde{\boldsymbol{\delta}}\in\tilde{\Delta}$ and for an arbitrary choice of the set $\tilde{A}_{\tilde{\boldsymbol{\delta}}}(\mathbf{z})$] the test of the null hypothesis $\tilde{H}_0^{(\tilde{\boldsymbol{\delta}})}: \tilde{\boldsymbol{\delta}}'\boldsymbol{\alpha} = \tilde{\boldsymbol{\delta}}'\boldsymbol{\alpha}^{(0)}$ (versus the alternative hypothesis $\tilde{H}_1^{(\tilde{\boldsymbol{\delta}})}: \tilde{\boldsymbol{\delta}}'\boldsymbol{\alpha} \neq \tilde{\boldsymbol{\delta}}'\boldsymbol{\alpha}^{(0)}$) with critical region $\tilde{C}(\tilde{\boldsymbol{\delta}})$ defined as follows:

$$\tilde{C}(\tilde{\boldsymbol{\delta}}) = \{\mathbf{z} \,:\, \tilde{\boldsymbol{\delta}}'\boldsymbol{\alpha}^{(0)} \notin \tilde{A}_{\tilde{\boldsymbol{\delta}}}(\mathbf{z})\}. \tag{3.97}$$

When this test is used to test $\tilde{H}_0^{(\tilde{\boldsymbol{\delta}})}$ (versus $\tilde{H}_1^{(\tilde{\boldsymbol{\delta}})}$) for every $\tilde{\boldsymbol{\delta}} \in \tilde{\Delta}$, the probability of one or more false rejections is (by definition)

$$\Pr\Big[\mathbf{z} \in \bigcup_{\{\tilde{\boldsymbol{\delta}} \in \tilde{\Delta}\,:\, \tilde{\boldsymbol{\delta}}'\boldsymbol{\alpha} = \tilde{\boldsymbol{\delta}}'\boldsymbol{\alpha}^{(0)}\}} \tilde{C}(\tilde{\boldsymbol{\delta}})\Big].$$

Clearly,

$$\begin{aligned}
\Pr\Big[\mathbf{z} \in \bigcup_{\{\tilde{\boldsymbol{\delta}} \in \tilde{\Delta}\,:\, \tilde{\boldsymbol{\delta}}'\boldsymbol{\alpha} = \tilde{\boldsymbol{\delta}}'\boldsymbol{\alpha}^{(0)}\}} \tilde{C}(\tilde{\boldsymbol{\delta}})\Big] &= 1 - \Pr\Big[\mathbf{z} \notin \bigcup_{\{\tilde{\boldsymbol{\delta}} \in \tilde{\Delta}\,:\, \tilde{\boldsymbol{\delta}}'\boldsymbol{\alpha} = \tilde{\boldsymbol{\delta}}'\boldsymbol{\alpha}^{(0)}\}} \tilde{C}(\tilde{\boldsymbol{\delta}})\Big] \\
&= 1 - \Pr[\tilde{\boldsymbol{\delta}}'\boldsymbol{\alpha} \in \tilde{A}_{\tilde{\boldsymbol{\delta}}}(\mathbf{z}) \text{ for every } \tilde{\boldsymbol{\delta}} \in \tilde{\Delta} \text{ such that } \tilde{\boldsymbol{\delta}}'\boldsymbol{\alpha} = \tilde{\boldsymbol{\delta}}'\boldsymbol{\alpha}^{(0)}] \\
&\le 1 - \Pr[\tilde{\boldsymbol{\delta}}'\boldsymbol{\alpha} \in \tilde{A}_{\tilde{\boldsymbol{\delta}}}(\mathbf{z}) \text{ for every } \tilde{\boldsymbol{\delta}} \in \tilde{\Delta}],
\end{aligned}$$

with equality holding when $\tilde{\boldsymbol{\delta}}'\boldsymbol{\alpha} = \tilde{\boldsymbol{\delta}}'\boldsymbol{\alpha}^{(0)}$ for every $\tilde{\boldsymbol{\delta}} \in \tilde{\Delta}$. Thus, if the probability $\Pr[\tilde{\boldsymbol{\delta}}'\boldsymbol{\alpha} \in \tilde{A}_{\tilde{\boldsymbol{\delta}}}(\mathbf{z})$ for every $\tilde{\boldsymbol{\delta}} \in \tilde{\Delta}]$ of simultaneous coverage of the sets $\tilde{A}_{\tilde{\boldsymbol{\delta}}}(\mathbf{z})$ ($\tilde{\boldsymbol{\delta}} \in \tilde{\Delta}$) is greater than or equal to $1-\dot{\gamma}$, then

$$\Pr\Big[\mathbf{z} \in \bigcup_{\{\tilde{\boldsymbol{\delta}} \in \tilde{\Delta}\,:\, \tilde{\boldsymbol{\delta}}'\boldsymbol{\alpha} = \tilde{\boldsymbol{\delta}}'\boldsymbol{\alpha}^{(0)}\}} \tilde{C}(\tilde{\boldsymbol{\delta}})\Big] \le \dot{\gamma}. \tag{3.98}$$

Moreover, if the probability of simultaneous coverage of the sets $\tilde{A}_{\tilde{\boldsymbol{\delta}}}(\mathbf{z})$ ($\tilde{\boldsymbol{\delta}} \in \tilde{\Delta}$) equals $1-\dot{\gamma}$, then equality is attained in inequality (3.98) when $\tilde{\boldsymbol{\delta}}'\boldsymbol{\alpha} = \tilde{\boldsymbol{\delta}}'\boldsymbol{\alpha}^{(0)}$ for every $\tilde{\boldsymbol{\delta}} \in \tilde{\Delta}$.

Now, suppose that $\boldsymbol{\alpha}^{(0)}$ is the unique value of $\boldsymbol{\alpha}$ that satisfies condition (3.85) (i.e., the condition $\boldsymbol{\tau}^{(0)} = \mathbf{W}'\boldsymbol{\alpha}^{(0)}$). And observe that (by definition) $\tilde{\boldsymbol{\delta}} \in \tilde{\Delta}$ if and only if $\tilde{\boldsymbol{\delta}} = \mathbf{W}\boldsymbol{\delta}$ for some $\boldsymbol{\delta} \in \Delta$. Observe also [in light of result (3.8)] that for $\boldsymbol{\delta} \in \Delta$ and $\tilde{\boldsymbol{\delta}} = \mathbf{W}\boldsymbol{\delta}$,

$$\boldsymbol{\delta}'\boldsymbol{\tau} = \boldsymbol{\delta}'\mathbf{W}'\boldsymbol{\alpha} = \tilde{\boldsymbol{\delta}}'\boldsymbol{\alpha} \quad \text{and} \quad \boldsymbol{\delta}'\boldsymbol{\tau}^{(0)} = \boldsymbol{\delta}'\mathbf{W}'\boldsymbol{\alpha}^{(0)} = \tilde{\boldsymbol{\delta}}'\boldsymbol{\alpha}^{(0)},$$

in which case $H_0^{(\boldsymbol{\delta})}$ is equivalent to $\tilde{H}_0^{(\tilde{\boldsymbol{\delta}})}$ and $H_1^{(\boldsymbol{\delta})}$ to $\tilde{H}_1^{(\tilde{\boldsymbol{\delta}})}$.

The test of $H_0^{(\boldsymbol{\delta})}$ (versus $H_1^{(\boldsymbol{\delta})}$) with critical region $C(\boldsymbol{\delta})$ [defined (implicitly) by relationship (3.82)] is related to the test of $\tilde{H}_0^{(\mathbf{W}\boldsymbol{\delta})}$ (versus $\tilde{H}_1^{(\mathbf{W}\boldsymbol{\delta})}$) with critical region $\tilde{C}(\mathbf{W}\boldsymbol{\delta})$ [defined by expression (3.97)]. For $\boldsymbol{\delta} \in \Delta$,

$$C(\boldsymbol{\delta}) = \{\mathbf{y} \,:\, \boldsymbol{\delta}'\boldsymbol{\tau}^{(0)} \notin A_{\boldsymbol{\delta}}(\mathbf{y})\} = \{\mathbf{y} \,:\, \boldsymbol{\delta}'\mathbf{W}'\boldsymbol{\alpha}^{(0)} \notin \tilde{A}_{\mathbf{W}\boldsymbol{\delta}}(\mathbf{O}'\mathbf{y})\} = \{\mathbf{y} \,:\, \mathbf{O}'\mathbf{y} \in \tilde{C}(\mathbf{W}\boldsymbol{\delta})\}. \tag{3.99}$$

And the probability of one or more false rejections is expressible as

$$\Pr\Big[\mathbf{y} \in \bigcup_{\{\boldsymbol{\delta} \in \Delta\,:\, \boldsymbol{\delta}'\boldsymbol{\tau} = \boldsymbol{\delta}'\boldsymbol{\tau}^{(0)}\}} C(\boldsymbol{\delta})\Big] = \Pr\Big[\mathbf{z} \in \bigcup_{\{\tilde{\boldsymbol{\delta}} \in \tilde{\Delta}\,:\, \tilde{\boldsymbol{\delta}}'\boldsymbol{\alpha} = \tilde{\boldsymbol{\delta}}'\boldsymbol{\alpha}^{(0)}\}} \tilde{C}(\tilde{\boldsymbol{\delta}})\Big], \tag{3.100}$$

as is evident upon observing that

$$\{\tilde{\boldsymbol{\delta}} \in \tilde{\Delta} \,:\, \tilde{\boldsymbol{\delta}}'\boldsymbol{\alpha} = \tilde{\boldsymbol{\delta}}'\boldsymbol{\alpha}^{(0)}\} = \{\tilde{\boldsymbol{\delta}} \,:\, \tilde{\boldsymbol{\delta}} = \mathbf{W}\boldsymbol{\delta},\ \boldsymbol{\delta} \in \Delta,\ \boldsymbol{\delta}'\boldsymbol{\tau} = \boldsymbol{\delta}'\boldsymbol{\tau}^{(0)}\}.$$

Moreover, if the probability of simultaneous coverage of the sets $A_{\boldsymbol{\delta}}(\mathbf{y})$ ($\boldsymbol{\delta} \in \Delta$) equals $1-\dot{\gamma}$, then

$$\Pr\Big[\mathbf{y} \in \bigcup_{\{\boldsymbol{\delta} \in \Delta\,:\, \boldsymbol{\delta}'\boldsymbol{\tau} = \boldsymbol{\delta}'\boldsymbol{\tau}^{(0)}\}} C(\boldsymbol{\delta})\Big] \le \dot{\gamma}, \tag{3.101}$$

with equality holding when $\boldsymbol{\delta}'\boldsymbol{\tau} = \boldsymbol{\delta}'\boldsymbol{\tau}^{(0)}$ for every $\boldsymbol{\delta} \in \Delta$.

Multiple comparisons: a general method. When (for $\tilde{\boldsymbol{\delta}} \in \tilde{\Delta}$) the set $\tilde{A}_{\tilde{\boldsymbol{\delta}}}$ is taken to be the interval (3.90) [in which case the set $A_{\boldsymbol{\delta}}$ is the interval (3.94)], the critical region $\tilde{C}(\tilde{\boldsymbol{\delta}})$ [defined by equality (3.97)] and the critical region $C(\boldsymbol{\delta})$ [defined (implicitly) by relationship (3.82)] are expressible as follows:

$$\tilde{C}(\tilde{\boldsymbol{\delta}}) = \{\mathbf{z} \,:\, |\tilde{\boldsymbol{\delta}}'(\hat{\boldsymbol{\alpha}} - \boldsymbol{\alpha}^{(0)})| > (\tilde{\boldsymbol{\delta}}'\tilde{\boldsymbol{\delta}})^{1/2}\hat{\sigma}c_{\dot{\gamma}}\} \tag{3.102}$$

and

$$C(\boldsymbol{\delta}) = \{\mathbf{y} \,:\, |\boldsymbol{\delta}'(\hat{\boldsymbol{\tau}} - \boldsymbol{\tau}^{(0)})| > (\boldsymbol{\delta}'\mathbf{C}\boldsymbol{\delta})^{1/2}\hat{\sigma}c_{\dot{\gamma}}\}. \tag{3.103}$$

And when (in addition) the distribution of the vector $\mathbf{e}$ of residual effects in the G–M model or, more generally, the distribution of the vector $\begin{pmatrix}\hat{\boldsymbol{\alpha}}-\boldsymbol{\alpha}\\ \mathbf{d}\end{pmatrix}$ is MVN or is some other absolutely continuous spherical distribution [in which case the probability of simultaneous coverage of the intervals $\tilde{A}_{\tilde{\boldsymbol{\delta}}}(\mathbf{z})$ $(\tilde{\boldsymbol{\delta}} \in \tilde{\Delta})$ and of the intervals $A_{\boldsymbol{\delta}}(\mathbf{y})$ $(\boldsymbol{\delta} \in \Delta)$ is $1-\dot{\gamma}$],

$$\Pr\Big[\mathbf{z} \in \bigcup_{\{\tilde{\boldsymbol{\delta}}\in\tilde{\Delta}\,:\,\tilde{\boldsymbol{\delta}}'\boldsymbol{\alpha}=\tilde{\boldsymbol{\delta}}'\boldsymbol{\alpha}^{(0)}\}} \tilde{C}(\tilde{\boldsymbol{\delta}})\Big] \le \dot{\gamma}, \tag{3.104}$$

with equality holding when $\tilde{\boldsymbol{\delta}}'\boldsymbol{\alpha} = \tilde{\boldsymbol{\delta}}'\boldsymbol{\alpha}^{(0)}$ for every $\tilde{\boldsymbol{\delta}} \in \tilde{\Delta}$, and

$$\Pr\Big[\mathbf{y} \in \bigcup_{\{\boldsymbol{\delta}\in\Delta\,:\,\boldsymbol{\delta}'\boldsymbol{\tau}=\boldsymbol{\delta}'\boldsymbol{\tau}^{(0)}\}} C(\boldsymbol{\delta})\Big] \le \dot{\gamma}, \tag{3.105}$$

with equality holding when $\boldsymbol{\delta}'\boldsymbol{\tau} = \boldsymbol{\delta}'\boldsymbol{\tau}^{(0)}$ for every $\boldsymbol{\delta} \in \Delta$.

The S method. By definition, $c_{\dot{\gamma}}$ is the upper $100\,\dot{\gamma}\%$ point of the distribution of the random variable $\max_{\{\tilde{\boldsymbol{\delta}}\in\tilde{\Delta}\,:\,\tilde{\boldsymbol{\delta}}\neq\mathbf{0}\}} |\tilde{\boldsymbol{\delta}}'\mathbf{t}|/(\tilde{\boldsymbol{\delta}}'\tilde{\boldsymbol{\delta}})^{1/2}$ [where $\mathbf{t}$ is an $M_* \times 1$ random vector that has an $MVt(N-P_*,\, \mathbf{I}_{M_*})$ distribution]. The upper $100\,\dot{\gamma}\%$ point $c_{\dot{\gamma}}$ is such that

$$c_{\dot{\gamma}} \le [M_* \bar{F}_{\dot{\gamma}}(M_*,\, N-P_*)]^{1/2}, \tag{3.106}$$

with equality holding if $\Delta = \mathbb{R}^M$ or, more generally, if $\tilde{\Delta} = \mathbb{R}^{M_*}$ or, still more generally, if

$$\Pr(\mathbf{t} \in \dot{\Delta}) = 1, \tag{3.107}$$

where $\dot{\Delta}$ is the set $\{\dot{\boldsymbol{\delta}} \in \mathbb{R}^{M_*} : \exists$ a nonnull vector in $\tilde{\Delta}$ that is proportional to $\dot{\boldsymbol{\delta}}\}$.

For purposes of verification, observe (in light of Theorem 2.4.21, i.e., the Cauchy–Schwarz inequality) that for any $M_* \times 1$ vector $\dot{\mathbf{t}}$ and for any nonnull $M_* \times 1$ vector $\tilde{\boldsymbol{\delta}}$,

$$\frac{|\tilde{\boldsymbol{\delta}}'\dot{\mathbf{t}}|}{(\tilde{\boldsymbol{\delta}}'\tilde{\boldsymbol{\delta}})^{1/2}} \le (\dot{\mathbf{t}}'\dot{\mathbf{t}})^{1/2}, \tag{3.108}$$

with equality holding if and only if $\dot{\mathbf{t}} = \mathbf{0}$ or $\tilde{\boldsymbol{\delta}} = k\dot{\mathbf{t}}$ for some (nonzero) scalar k. And it follows that

$$\max_{\{\tilde{\boldsymbol{\delta}}\in\tilde{\Delta}\,:\,\tilde{\boldsymbol{\delta}}\neq\mathbf{0}\}} \frac{|\tilde{\boldsymbol{\delta}}'\dot{\mathbf{t}}|}{(\tilde{\boldsymbol{\delta}}'\tilde{\boldsymbol{\delta}})^{1/2}} \le (\dot{\mathbf{t}}'\dot{\mathbf{t}})^{1/2},$$

with equality holding for $\dot{\mathbf{t}} = \mathbf{0}$ and for every nonnull vector $\dot{\mathbf{t}}$ for which there exists a nonnull vector $\tilde{\boldsymbol{\delta}} \in \tilde{\Delta}$ such that $\tilde{\boldsymbol{\delta}} = k\dot{\mathbf{t}}$ for some scalar k. Observe also that $(\mathbf{t}'\mathbf{t})/M_* \sim SF(M_*,\, N-P_*)$ and hence that

$$\Pr\{(\mathbf{t}'\mathbf{t})^{1/2} > [M_* \bar{F}_{\dot{\gamma}}(M_*,\, N-P_*)]^{1/2}\} = \Pr[(\mathbf{t}'\mathbf{t})/M_* > \bar{F}_{\dot{\gamma}}(M_*,\, N-P_*)] = \dot{\gamma},$$

so that $[M_* \bar{F}_{\dot{\gamma}}(M_*,\, N-P_*)]^{1/2}$ is the upper $100\,\dot{\gamma}\%$ point of the distribution of $(\mathbf{t}'\mathbf{t})^{1/2}$. Thus, $c_{\dot{\gamma}}$ satisfies inequality (3.106), and equality holds in inequality (3.106) if $\tilde{\Delta}$ satisfies condition (3.107).

Now, suppose (as before) that the distribution of the vector $\mathbf{e}$ of residual effects in the G–M model or, more generally, the distribution of the vector $\begin{pmatrix}\hat{\boldsymbol{\alpha}}-\boldsymbol{\alpha}\\ \mathbf{d}\end{pmatrix}$ is MVN or is some other absolutely continuous spherical distribution (with mean $\mathbf{0}$ and variance-covariance matrix $\sigma^2\mathbf{I}$). Then, as a special case of result (3.95) (that where $\Delta = \mathbb{R}^M$), we have that

$$\Pr[\boldsymbol{\delta}'\boldsymbol{\tau} \in A_{\boldsymbol{\delta}}(\mathbf{y}) \text{ for every } \boldsymbol{\delta} \in \mathbb{R}^M] = 1-\dot{\gamma}, \tag{3.109}$$

where (for $\boldsymbol{\delta} \in \mathbb{R}^M$)

$$\begin{aligned} A_{\boldsymbol{\delta}}(\mathbf{y}) = \{\dot{\tau} \in \mathbb{R}^1 :\ & \boldsymbol{\delta}'\hat{\boldsymbol{\tau}} - (\boldsymbol{\delta}'\mathbf{C}\boldsymbol{\delta})^{1/2}\hat{\sigma}[M_* \bar{F}_{\dot{\gamma}}(M_*,\, N-P_*)]^{1/2} \\ & \le \dot{\tau} \le \boldsymbol{\delta}'\hat{\boldsymbol{\tau}} + (\boldsymbol{\delta}'\mathbf{C}\boldsymbol{\delta})^{1/2}\hat{\sigma}[M_* \bar{F}_{\dot{\gamma}}(M_*,\, N-P_*)]^{1/2}\}. \end{aligned} \tag{3.110}$$

And as a special case of result (3.105), we have that

TABLE 7.2. Value of $[M_* \bar{F}_{\dot{\gamma}}(M_*, N-P_*)]^{1/2}$ for selected values of M_*, $N-P_*$, and $\dot{\gamma}$.

	$N-P_*=10$			$N-P_*=25$			$N-P_*=\infty$		
M_*	$\dot{\gamma}$=.01	$\dot{\gamma}$=.10	$\dot{\gamma}$=.50	$\dot{\gamma}$=.01	$\dot{\gamma}$=.10	$\dot{\gamma}$=.50	$\dot{\gamma}$=.01	$\dot{\gamma}$=.10	$\dot{\gamma}$=.50
1	3.17	1.81	0.70	2.79	1.71	0.68	2.58	1.64	0.67
2	3.89	2.42	1.22	3.34	2.25	1.19	3.03	2.15	1.18
3	4.43	2.86	1.59	3.75	2.64	1.56	3.37	2.50	1.54
4	4.90	3.23	1.90	4.09	2.96	1.86	3.64	2.79	1.83
5	5.31	3.55	2.16	4.39	3.23	2.11	3.88	3.04	2.09
10	6.96	4.82	3.16	5.59	4.32	3.10	4.82	4.00	3.06
20	9.39	6.63	4.55	7.35	5.86	4.46	6.13	5.33	4.40
40	12.91	9.23	6.49	9.91	8.07	6.36	7.98	7.20	6.27

$$\Pr\left[\mathbf{y} \in \bigcup_{\{\boldsymbol{\delta} \in \mathbb{R}^M \,:\, \boldsymbol{\delta}'\boldsymbol{\tau} = \boldsymbol{\delta}'\boldsymbol{\tau}^{(0)}\}} C(\boldsymbol{\delta})\right] \le \dot{\gamma}, \tag{3.111}$$

where (for $\boldsymbol{\delta} \in \mathbb{R}^M$)

$$C(\boldsymbol{\delta}) = \{\mathbf{y} \,:\, |\boldsymbol{\delta}'(\hat{\boldsymbol{\tau}} - \boldsymbol{\tau}^{(0)})| > (\boldsymbol{\delta}'\mathbf{C}\boldsymbol{\delta})^{1/2}\hat{\sigma}[M_* \bar{F}_{\dot{\gamma}}(M_*, N-P_*)]^{1/2}\}, \tag{3.112}$$

and that equality holds in inequality (3.111) when $\boldsymbol{\delta}'\boldsymbol{\tau} = \boldsymbol{\delta}'\boldsymbol{\tau}^{(0)}$ for every $\boldsymbol{\delta} \in \mathbb{R}^M$.

The use of the interval (3.110) as a means for obtaining a confidence set for $\boldsymbol{\delta}'\boldsymbol{\tau}$ (for every $\boldsymbol{\delta} \in \mathbb{R}^M$) and the use of the critical region (3.112) as a means for obtaining a test of $H_0^{(\boldsymbol{\delta})}$ (versus $H_1^{(\boldsymbol{\delta})}$) (for every $\boldsymbol{\delta} \in \mathbb{R}^M$) are known as *Scheffé's method* or simply as the *S method*. The S method was proposed by Scheffé (1953, 1959).

The interval (3.110) (with end points of $\boldsymbol{\delta}'\hat{\boldsymbol{\tau}} \pm (\boldsymbol{\delta}'\mathbf{C}\boldsymbol{\delta})^{1/2}\hat{\sigma}[M_* \bar{F}_{\dot{\gamma}}(M_*, N-P_*)]^{1/2}$) is of length $2(\boldsymbol{\delta}'\mathbf{C}\boldsymbol{\delta})^{1/2}\hat{\sigma}[M_* \bar{F}_{\dot{\gamma}}(M_*, N-P_*)]^{1/2}$, which is proportional to $[M_* \bar{F}_{\dot{\gamma}}(M_*, N-P_*)]^{1/2}$ and depends on M_*, $N-P_*$, and $\dot{\gamma}$—more generally, the interval (3.94) (for a linear combination $\boldsymbol{\delta}'\boldsymbol{\tau}$ such that $\boldsymbol{\delta} \in \Delta$ and with end points of $\boldsymbol{\delta}'\hat{\boldsymbol{\tau}} \pm (\boldsymbol{\delta}'\mathbf{C}\boldsymbol{\delta})^{1/2}\hat{\sigma}c_{\dot{\gamma}}$) is of length $2(\boldsymbol{\delta}'\mathbf{C}\boldsymbol{\delta})^{1/2}\hat{\sigma}c_{\dot{\gamma}}$. Note that when $M_* = 1$, the interval (3.110) is identical to the $100(1-\dot{\gamma})\%$ confidence interval for $\boldsymbol{\delta}'\boldsymbol{\tau}$ $[= (\boldsymbol{\Lambda}\boldsymbol{\delta})'\boldsymbol{\beta}]$ obtained via a one-at-a-time approach by applying formula (3.55). Table 7.2 gives the value of the factor $[M_* \bar{F}_{\dot{\gamma}}(M_*, N-P_*)]^{1/2}$ for selected values of M_*, $N-P_*$, and $\dot{\gamma}$. As is evident from the tabulated values, this factor increases rapidly as M_* increases.

A connection. For $\boldsymbol{\delta} \in \mathbb{R}^M$, let $A_{\boldsymbol{\delta}}$ or $A_{\boldsymbol{\delta}}(\mathbf{y})$ represent the interval (3.110) (of $\boldsymbol{\delta}'\boldsymbol{\tau}$-values) associated with the S method for obtaining confidence intervals [for $\boldsymbol{\delta}'\boldsymbol{\tau}$ ($\boldsymbol{\delta} \in \mathbb{R}^M$)] having a probability of simultaneous coverage equal to $1-\dot{\gamma}$. And denote by A or $A(\mathbf{y})$ the set (3.38) (of $\boldsymbol{\tau}$-values), which as discussed earlier (in Part b) is a $100(1-\dot{\gamma})\%$ confidence set for the vector $\boldsymbol{\tau}$.

The sets $A_{\boldsymbol{\delta}}$ ($\boldsymbol{\delta} \in \mathbb{R}^M$) are related to the set A; their relationship is as follows: for every value of $\mathbf{y}$,

$$A = \{\dot{\boldsymbol{\tau}} \in \mathcal{C}(\boldsymbol{\Lambda}') \,:\, \boldsymbol{\delta}'\dot{\boldsymbol{\tau}} \in A_{\boldsymbol{\delta}} \text{ for every } \boldsymbol{\delta} \in \mathbb{R}^M\}, \tag{3.113}$$

so that [for $\dot{\boldsymbol{\tau}} \in \mathcal{C}(\boldsymbol{\Lambda}')$]

$$\Pr[\boldsymbol{\delta}'\dot{\boldsymbol{\tau}} \in A_{\boldsymbol{\delta}}(\mathbf{y}) \text{ for every } \boldsymbol{\delta} \in \mathbb{R}^M] = \Pr[\dot{\boldsymbol{\tau}} \in A(\mathbf{y})]. \tag{3.114}$$

Moreover, relationship (3.113) implies that the sets $C(\boldsymbol{\delta})$ ($\boldsymbol{\delta} \in \mathbb{R}^M$), where $C(\boldsymbol{\delta})$ is the critical region (3.112) (for testing the null hypothesis $H_0^{(\boldsymbol{\delta})}: \boldsymbol{\delta}'\boldsymbol{\tau} = \boldsymbol{\delta}'\boldsymbol{\tau}^{(0)}$ versus the alternative hypothesis $H_1^{(\boldsymbol{\delta})}: \boldsymbol{\delta}'\boldsymbol{\tau} \neq \boldsymbol{\delta}'\boldsymbol{\tau}^{(0)}$) associated with the S method of multiple comparisons, are related to the critical region C_F associated with the F test of the null hypothesis $H_0: \boldsymbol{\tau} = \boldsymbol{\tau}^{(0)}$ versus the alternative hypothesis $H_1: \boldsymbol{\tau} \neq \boldsymbol{\tau}^{(0)}$; their relationship is as follows:

$$C_F = \{\dot{\mathbf{y}} \in \mathbb{R}^N \,:\, \dot{\mathbf{y}} \in C(\boldsymbol{\delta}) \text{ for some } \boldsymbol{\delta} \in \mathbb{R}^M\}, \tag{3.115}$$

so that

$$\Pr[\mathbf{y} \in C(\boldsymbol{\delta}) \text{ for some } \boldsymbol{\delta} \in \mathbb{R}^M] = \Pr(\mathbf{y} \in C_F). \tag{3.116}$$

Let us verify relationship (3.113). Taking (for $\tilde{\boldsymbol{\delta}} \in \mathbb{R}^{M_*}$)

$$\tilde{A}_{\tilde{\boldsymbol{\delta}}} = \{\dot{\tau} \in \mathbb{R}^1 : \tilde{\boldsymbol{\delta}}'\hat{\boldsymbol{\alpha}} - (\tilde{\boldsymbol{\delta}}'\tilde{\boldsymbol{\delta}})^{1/2}\hat{\sigma}[M_* \bar{F}_{\dot{\gamma}}(M_*, N-P_*)]^{1/2} \le \dot{\tau} \le \tilde{\boldsymbol{\delta}}'\hat{\boldsymbol{\alpha}} + (\tilde{\boldsymbol{\delta}}'\tilde{\boldsymbol{\delta}})^{1/2}\hat{\sigma}[M_* \bar{F}_{\dot{\gamma}}(M_*, N-P_*)]^{1/2}\} \tag{3.117}$$

and denoting by $\tilde{A}$ the set (3.38), the relationship (3.113) is [in light of results (3.39) and (3.92)] equivalent to the following relationship: for every value of $\mathbf{z}$,

$$\tilde{A} = \{\dot{\boldsymbol{\alpha}} \in \mathbb{R}^{M_*} : \tilde{\boldsymbol{\delta}}'\dot{\boldsymbol{\alpha}} \in \tilde{A}_{\tilde{\boldsymbol{\delta}}} \text{ for every } \tilde{\boldsymbol{\delta}} \in \mathbb{R}^{M_*}\}. \tag{3.118}$$

Thus, it suffices to verify relationship (3.118).

Let $r = \hat{\sigma}[M_* \bar{F}_{\dot{\gamma}}(M_*, N-P_*)]^{1/2}$. And (letting $\dot{\boldsymbol{\alpha}}$ represent an arbitrary value of $\boldsymbol{\alpha}$) observe that

$$\dot{\boldsymbol{\alpha}} \in \tilde{A} \;\Leftrightarrow\; [(\dot{\boldsymbol{\alpha}}-\hat{\boldsymbol{\alpha}})'(\dot{\boldsymbol{\alpha}}-\hat{\boldsymbol{\alpha}})]^{1/2} \le r \tag{3.119}$$

and that (for $\tilde{\boldsymbol{\delta}} \in \mathbb{R}^{M_*}$)

$$\tilde{\boldsymbol{\delta}}'\dot{\boldsymbol{\alpha}} \in \tilde{A}_{\tilde{\boldsymbol{\delta}}} \;\Leftrightarrow\; |\tilde{\boldsymbol{\delta}}'(\dot{\boldsymbol{\alpha}}-\hat{\boldsymbol{\alpha}})| \le (\tilde{\boldsymbol{\delta}}'\tilde{\boldsymbol{\delta}})^{1/2} r. \tag{3.120}$$

Now, suppose that $\dot{\boldsymbol{\alpha}} \in \tilde{A}$. Then, upon applying result (2.4.10) (which is a special case of the Cauchy–Schwarz inequality) and result (3.119), we find (for every $M_* \times 1$ vector $\tilde{\boldsymbol{\delta}}$) that

$$|\tilde{\boldsymbol{\delta}}'(\dot{\boldsymbol{\alpha}}-\hat{\boldsymbol{\alpha}})| \le (\tilde{\boldsymbol{\delta}}'\tilde{\boldsymbol{\delta}})^{1/2}[(\dot{\boldsymbol{\alpha}}-\hat{\boldsymbol{\alpha}})'(\dot{\boldsymbol{\alpha}}-\hat{\boldsymbol{\alpha}})]^{1/2} \le (\tilde{\boldsymbol{\delta}}'\tilde{\boldsymbol{\delta}})^{1/2} r,$$

implying [in light of result (3.120)] that $\tilde{\boldsymbol{\delta}}'\dot{\boldsymbol{\alpha}} \in \tilde{A}_{\tilde{\boldsymbol{\delta}}}$.

Conversely, suppose that $\dot{\boldsymbol{\alpha}} \notin \tilde{A}$. Then, in light of result (3.119), $[(\dot{\boldsymbol{\alpha}}-\hat{\boldsymbol{\alpha}})'(\dot{\boldsymbol{\alpha}}-\hat{\boldsymbol{\alpha}})]^{1/2} > r$. And upon setting $\tilde{\boldsymbol{\delta}} = \dot{\boldsymbol{\alpha}}-\hat{\boldsymbol{\alpha}}$, we find that

$$|\tilde{\boldsymbol{\delta}}'(\dot{\boldsymbol{\alpha}}-\hat{\boldsymbol{\alpha}})| = (\dot{\boldsymbol{\alpha}}-\hat{\boldsymbol{\alpha}})'(\dot{\boldsymbol{\alpha}}-\hat{\boldsymbol{\alpha}}) = (\tilde{\boldsymbol{\delta}}'\tilde{\boldsymbol{\delta}})^{1/2}[(\dot{\boldsymbol{\alpha}}-\hat{\boldsymbol{\alpha}})'(\dot{\boldsymbol{\alpha}}-\hat{\boldsymbol{\alpha}})]^{1/2} > (\tilde{\boldsymbol{\delta}}'\tilde{\boldsymbol{\delta}})^{1/2} r,$$

thereby establishing [in light of result (3.120)] the existence of an $M_* \times 1$ vector $\tilde{\boldsymbol{\delta}}$ such that $\tilde{\boldsymbol{\delta}}'\dot{\boldsymbol{\alpha}} \notin \tilde{A}_{\tilde{\boldsymbol{\delta}}}$ and completing the verification of relationship (3.118).

In connection with relationship (3.113), it is worth noting that if the condition $\boldsymbol{\delta}'\dot{\boldsymbol{\tau}} \in A_{\boldsymbol{\delta}}$ is satisfied by any particular nonnull vector $\boldsymbol{\delta}$ in $\mathbb{R}^M$, then it is also satisfied by any vector $\dot{\boldsymbol{\delta}}$ in $\mathbb{R}^M$ such that $\boldsymbol{\Lambda}\dot{\boldsymbol{\delta}} = \boldsymbol{\Lambda}\boldsymbol{\delta}$ or such that $\dot{\boldsymbol{\delta}} \propto \boldsymbol{\delta}$.

An extension. Result (3.118) relates $\tilde{A}_{\tilde{\boldsymbol{\delta}}}$ ($\tilde{\boldsymbol{\delta}} \in \mathbb{R}^{M_*}$) to $\tilde{A}$, where $\tilde{A}_{\tilde{\boldsymbol{\delta}}}$ is the set (3.117) (of $\tilde{\boldsymbol{\delta}}'\boldsymbol{\alpha}$-values) and $\tilde{A}$ is the set (3.38) (of $\boldsymbol{\alpha}$-values). This relationship can be extended to a broader class of sets.

Suppose that the set Δ of $M \times 1$ vectors is such that the set $\tilde{\Delta} = \{\tilde{\boldsymbol{\delta}} : \tilde{\boldsymbol{\delta}} = \mathbf{W}\boldsymbol{\delta},\ \boldsymbol{\delta} \in \boldsymbol{\Delta}\}$ contains M_* linearly independent ($M_* \times 1$) vectors, say $\tilde{\boldsymbol{\delta}}_1, \tilde{\boldsymbol{\delta}}_2, \ldots, \tilde{\boldsymbol{\delta}}_{M_*}$. And for $\tilde{\boldsymbol{\delta}} \in \tilde{\Delta}$, take $\tilde{A}_{\tilde{\boldsymbol{\delta}}}$ to be interval (3.90) [which, when $\tilde{\Delta} = \mathbb{R}^{M_*}$, is identical to interval (3.117)], and take $\tilde{A}$ to be the set of $\boldsymbol{\alpha}$-values defined as follows:

$$\tilde{A} = \left\{\dot{\boldsymbol{\alpha}} : \max_{\{\tilde{\boldsymbol{\delta}} \in \tilde{\Delta} :\, \tilde{\boldsymbol{\delta}} \ne \mathbf{0}\}} \frac{|\tilde{\boldsymbol{\delta}}'(\hat{\boldsymbol{\alpha}}-\dot{\boldsymbol{\alpha}})|}{(\tilde{\boldsymbol{\delta}}'\tilde{\boldsymbol{\delta}})^{1/2}\hat{\sigma}} \le c_{\dot{\gamma}}\right\} \tag{3.121}$$

or, equivalently,

$$\tilde{A} = \{\dot{\boldsymbol{\alpha}} : \tilde{\boldsymbol{\delta}}'\hat{\boldsymbol{\alpha}} - (\tilde{\boldsymbol{\delta}}'\tilde{\boldsymbol{\delta}})^{1/2}\hat{\sigma} c_{\dot{\gamma}} \le \tilde{\boldsymbol{\delta}}'\dot{\boldsymbol{\alpha}} \le \tilde{\boldsymbol{\delta}}'\hat{\boldsymbol{\alpha}} + (\tilde{\boldsymbol{\delta}}'\tilde{\boldsymbol{\delta}})^{1/2}\hat{\sigma} c_{\dot{\gamma}} \text{ for every } \tilde{\boldsymbol{\delta}} \in \tilde{\Delta}\} \tag{3.122}$$

—in the special case where $\tilde{\Delta} = \mathbb{R}^{M_*}$, this set is identical to the set (3.38), as is evident from result (3.118). Further, for $\tilde{\boldsymbol{\delta}} \notin \tilde{\Delta}$, take

$$\tilde{A}_{\tilde{\boldsymbol{\delta}}} = \{\dot{\tau} \in \mathbb{R}^1 : \dot{\tau} = \tilde{\boldsymbol{\delta}}'\dot{\boldsymbol{\alpha}},\ \dot{\boldsymbol{\alpha}} \in \tilde{A}\}, \tag{3.123}$$

thereby extending the definition of $\tilde{A}_{\tilde{\boldsymbol{\delta}}}$ to every $\tilde{\boldsymbol{\delta}} \in \mathbb{R}^{M_*}$. Then, clearly,

$$\tilde{A} = \{\dot{\boldsymbol{\alpha}} \in \mathbb{R}^{M_*} : \tilde{\boldsymbol{\delta}}'\dot{\boldsymbol{\alpha}} \in \tilde{A}_{\tilde{\boldsymbol{\delta}}} \text{ for every } \tilde{\boldsymbol{\delta}} \in \mathbb{R}^{M_*}\}. \tag{3.124}$$

And [upon writing $\tilde{A}(\mathbf{z})$ for $\tilde{A}$ and $\tilde{A}_{\tilde{\boldsymbol{\delta}}}(\mathbf{z})$ for $\tilde{A}_{\tilde{\boldsymbol{\delta}}}$ and applying result (3.88)], we find that

$$\Pr[\tilde{\boldsymbol{\delta}}'\boldsymbol{\alpha} \in \tilde{A}_{\tilde{\boldsymbol{\delta}}}(\mathbf{z}) \text{ for every } \tilde{\boldsymbol{\delta}} \in \mathbb{R}^{M_*}] = \Pr[\boldsymbol{\alpha} \in \tilde{A}(\mathbf{z})] = 1-\dot{\gamma}, \tag{3.125}$$

so that $\tilde{A}$ is a $100(1-\dot{\gamma})\%$ confidence set for the vector $\boldsymbol{\alpha}$ and the probability of simultaneous coverage of the sets $\tilde{A}_{\tilde{\boldsymbol{\delta}}}$ $(\tilde{\boldsymbol{\delta}} \in \mathbb{R}^{M_*})$ equals $1-\dot{\gamma}$.

Results (3.124) and (3.125) can be reexpressed in terms of $A(\mathbf{y})$ and $A_{\boldsymbol{\delta}}(\mathbf{y})$ $(\boldsymbol{\delta} \in \mathbb{R}^M)$, where $A(\mathbf{y})$ is a set of $\boldsymbol{\tau}$-values and $A_{\boldsymbol{\delta}}(\mathbf{y})$ a set of $\boldsymbol{\delta}'\boldsymbol{\tau}$-values defined as follows:

$$A(\mathbf{y}) = \{\dot{\boldsymbol{\tau}} \,:\, \dot{\boldsymbol{\tau}} = \mathbf{W}'\dot{\boldsymbol{\alpha}},\ \dot{\boldsymbol{\alpha}} \in \tilde{A}(\mathbf{O}'\mathbf{y})\},$$

or equivalently

$$A = \{\dot{\boldsymbol{\tau}} \in \mathcal{C}(\boldsymbol{\Lambda}') :\, \boldsymbol{\delta}'\hat{\boldsymbol{\tau}} - (\boldsymbol{\delta}'\mathbf{C}\boldsymbol{\delta})^{1/2}\hat{\sigma}c_{\dot{\gamma}} \le \boldsymbol{\delta}'\dot{\boldsymbol{\tau}} \le \boldsymbol{\delta}'\hat{\boldsymbol{\tau}} + (\boldsymbol{\delta}'\mathbf{C}\boldsymbol{\delta})^{1/2}\hat{\sigma}c_{\dot{\gamma}} \text{ for every } \boldsymbol{\delta} \in \Delta\},$$

and

$$A_{\boldsymbol{\delta}}(\mathbf{y}) = \tilde{A}_{\mathbf{W}\boldsymbol{\delta}}(\mathbf{O}'\mathbf{y})$$

—for $\boldsymbol{\delta} \in \Delta$, $A_{\boldsymbol{\delta}}(\mathbf{y})$ is interval (3.94). Upon applying result (3.124), we find that

$$A(\mathbf{y}) = \{\dot{\boldsymbol{\tau}} \in \mathcal{C}(\boldsymbol{\Lambda}') \,:\, \boldsymbol{\delta}'\dot{\boldsymbol{\tau}} \in A_{\boldsymbol{\delta}}(\mathbf{y}) \text{ for every } \boldsymbol{\delta} \in \mathbb{R}^M\}. \tag{3.126}$$

Moreover,

$$\begin{aligned}\Pr[\boldsymbol{\delta}'\boldsymbol{\tau} \in A_{\boldsymbol{\delta}}(\mathbf{y}) \text{ for every } \boldsymbol{\delta} \in \mathbb{R}^M]\\ = \Pr[\boldsymbol{\tau} \in A(\mathbf{y})] = \Pr[\mathbf{W}'\boldsymbol{\alpha} \in A(\mathbf{y})] = \Pr[\boldsymbol{\alpha} \in \tilde{A}(\mathbf{z})] = 1-\dot{\gamma}.\end{aligned} \tag{3.127}$$

Let us now specialize to the case where (for M_* linearly independent $M_* \times 1$ vectors $\tilde{\boldsymbol{\delta}}_1, \tilde{\boldsymbol{\delta}}_2, \ldots, \tilde{\boldsymbol{\delta}}_{M_*}$) $\tilde{\Delta} = \{\tilde{\boldsymbol{\delta}}_1, \tilde{\boldsymbol{\delta}}_2, \ldots, \tilde{\boldsymbol{\delta}}_{M_*}\}$. For $i = 1, 2, \ldots, M_*$, let $\tau_i = \tilde{\boldsymbol{\delta}}_i'\boldsymbol{\alpha}$ and $\hat{\tau}_i = \tilde{\boldsymbol{\delta}}_i'\hat{\boldsymbol{\alpha}}$. And observe that corresponding to any $M_* \times 1$ vector $\tilde{\boldsymbol{\delta}}$, there exist (unique) scalars $k_1, k_2, \ldots, k_{M_*}$ such that $\tilde{\boldsymbol{\delta}} = \sum_{i=1}^{M_*} k_i \tilde{\boldsymbol{\delta}}_i$ and hence such that

$$\tilde{\boldsymbol{\delta}}'\boldsymbol{\alpha} = \textstyle\sum_{i=1}^{M_*} k_i \tau_i \quad \text{and} \quad \tilde{\boldsymbol{\delta}}'\hat{\boldsymbol{\alpha}} = \textstyle\sum_{i=1}^{M_*} k_i \hat{\tau}_i.$$

Observe also that the set $\tilde{A}$ [defined by expression (3.121) or (3.122)] is reexpressible as

$$\tilde{A} = \{\dot{\boldsymbol{\alpha}} \,:\, \hat{\tau}_i - (\tilde{\boldsymbol{\delta}}_i'\tilde{\boldsymbol{\delta}}_i)^{1/2}\hat{\sigma}c_{\dot{\gamma}} \le \tilde{\boldsymbol{\delta}}_i'\dot{\boldsymbol{\alpha}} \le \hat{\tau}_i + (\tilde{\boldsymbol{\delta}}_i'\tilde{\boldsymbol{\delta}}_i)^{1/2}\hat{\sigma}c_{\dot{\gamma}} \ (i = 1, 2, \ldots, M_*)\}. \tag{3.128}$$

Moreover, for $k_i \ne 0$,

$$\begin{aligned}&\hat{\tau}_i - (\tilde{\boldsymbol{\delta}}_i'\tilde{\boldsymbol{\delta}}_i)^{1/2}\hat{\sigma}c_{\dot{\gamma}} \le \tilde{\boldsymbol{\delta}}_i'\dot{\boldsymbol{\alpha}} \le \hat{\tau}_i + (\tilde{\boldsymbol{\delta}}_i'\tilde{\boldsymbol{\delta}}_i)^{1/2}\hat{\sigma}c_{\dot{\gamma}}\\ &\qquad\Leftrightarrow k_i\hat{\tau}_i - |k_i|(\tilde{\boldsymbol{\delta}}_i'\tilde{\boldsymbol{\delta}}_i)^{1/2}\hat{\sigma}c_{\dot{\gamma}} \le k_i\tilde{\boldsymbol{\delta}}_i'\dot{\boldsymbol{\alpha}} \le k_i\hat{\tau}_i + |k_i|(\tilde{\boldsymbol{\delta}}_i'\tilde{\boldsymbol{\delta}}_i)^{1/2}\hat{\sigma}c_{\dot{\gamma}}\end{aligned}$$

$(i = 1, 2, \ldots, M_*)$, so that the set $\tilde{A}_{\tilde{\boldsymbol{\delta}}}$ [defined for $\tilde{\boldsymbol{\delta}} \in \tilde{\Delta}$ by expression (3.90) and for $\tilde{\boldsymbol{\delta}} \notin \tilde{\Delta}$ by expression (3.123)] is expressible (for every $\tilde{\boldsymbol{\delta}} \in \mathbb{R}^{M_*}$) as

$$\begin{aligned}\tilde{A}_{\tilde{\boldsymbol{\delta}}} = \{\dot{\tau} \in \mathbb{R}^1 :\, &\textstyle\sum_{i=1}^{M_*} k_i\hat{\tau}_i - \sum_{i=1}^{M_*} |k_i|(\tilde{\boldsymbol{\delta}}_i'\tilde{\boldsymbol{\delta}}_i)^{1/2}\hat{\sigma}c_{\dot{\gamma}}\\ &\le \dot{\tau} \le \textstyle\sum_{i=1}^{M_*} k_i\hat{\tau}_i + \sum_{i=1}^{M_*} |k_i|(\tilde{\boldsymbol{\delta}}_i'\tilde{\boldsymbol{\delta}}_i)^{1/2}\hat{\sigma}c_{\dot{\gamma}}\}.\end{aligned} \tag{3.129}$$

Thus, it follows from result (3.125) that

$$\begin{aligned}\Pr\big[&\textstyle\sum_{i=1}^{M_*} k_i\hat{\tau}_i - \sum_{i=1}^{M_*} |k_i|(\tilde{\boldsymbol{\delta}}_i'\tilde{\boldsymbol{\delta}}_i)^{1/2}\hat{\sigma}c_{\dot{\gamma}} \le \sum_{i=1}^{M_*} k_i\tau_i \le \sum_{i=1}^{M_*} k_i\hat{\tau}_i\\ &+ \textstyle\sum_{i=1}^{M_*} |k_i|(\tilde{\boldsymbol{\delta}}_i'\tilde{\boldsymbol{\delta}}_i)^{1/2}\hat{\sigma}c_{\dot{\gamma}} \text{ for all scalars } k_1, k_2, \ldots, k_{M_*}\big] = 1-\dot{\gamma}.\end{aligned} \tag{3.130}$$

Suppose (for purposes of illustration) that $M_* = 2$ and that $\tilde{\boldsymbol{\delta}}_1 = (1,\ 0)'$ and $\tilde{\boldsymbol{\delta}}_2 = (0,\ 1)'$. Then, letting $\hat{\alpha}_1$ and $\hat{\alpha}_2$ represent the elements of $\hat{\boldsymbol{\alpha}}$, expression (3.128) for the set $\tilde{A}$ [defined by expression (3.121) or (3.122)] is reexpressible as

$$\tilde{A} = \{\dot{\boldsymbol{\alpha}} = (\dot{\alpha}_1,\ \dot{\alpha}_2)' \,:\, \hat{\alpha}_i - \hat{\sigma}c_{\dot{\gamma}} \le \dot{\alpha}_i \le \hat{\alpha}_i + \hat{\sigma}c_{\dot{\gamma}} \ (i = 1, 2)\}. \tag{3.131}$$

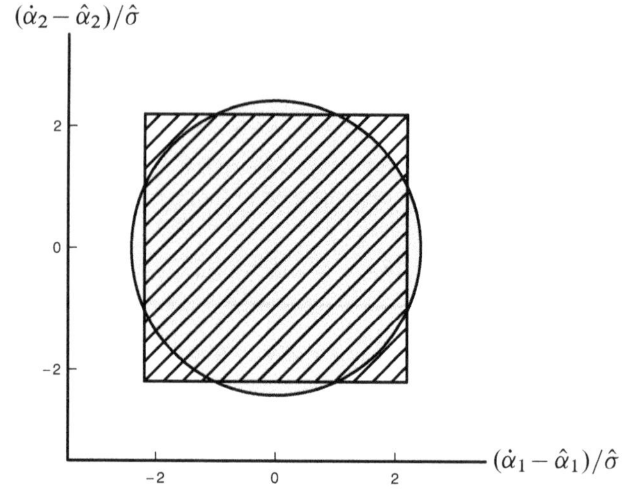

FIGURE 7.2. Display of the sets (3.131) and (3.133) [in terms of the transformed coordinates $(\dot{\alpha}_1 - \hat{\alpha}_1)/\hat{\sigma}$ and $(\dot{\alpha}_2 - \hat{\alpha}_2)/\hat{\sigma}$] for the case where $\dot{\gamma} = 0.10$ and $N - P_* = 10$; the set (3.131) is represented by the rectangular region and the set (3.133) by the circular region.

And expression (3.129) for the set $\tilde{A}_{\tilde{\boldsymbol{\delta}}}$ [defined for $\tilde{\boldsymbol{\delta}} \in \tilde{\Delta}$ by expression (3.90) and for $\tilde{\boldsymbol{\delta}} \notin \tilde{\Delta}$ by expression (3.123)] is reexpressible as

$$\tilde{A}_{\tilde{\boldsymbol{\delta}}} = \{\dot{\tau} \in \mathcal{R}^1 : \textstyle\sum_{i=1}^2 k_i \hat{\alpha}_i - \hat{\sigma} c_{\dot{\gamma}} \sum_{i=1}^2 |k_i| \le \dot{\tau} \le \sum_{i=1}^2 k_i \hat{\alpha}_i + \hat{\sigma} c_{\dot{\gamma}} \sum_{i=1}^2 |k_i|\} \tag{3.132}$$

[where $\tilde{\boldsymbol{\delta}} = (k_1, k_2)'$]. By way of comparison, consider the set $\tilde{A}_F$ (of $\boldsymbol{\alpha}$-values) given by expression (3.38), which (in the case under consideration) is reexpressible as

$$\{\dot{\boldsymbol{\alpha}} = (\dot{\alpha}_1, \dot{\alpha}_2)' : \textstyle\sum_{i=1}^2 (\dot{\alpha}_i - \hat{\alpha}_i)^2 \le 2\hat{\sigma}^2 \bar{F}_{\dot{\gamma}}(2, N - P_*)\}, \tag{3.133}$$

and the set (of $\tilde{\boldsymbol{\delta}}'\boldsymbol{\alpha}$-values) given by expression (3.117), which (in the present context) is reexpressible as

$$\tilde{A}_{\tilde{\boldsymbol{\delta}}} = \{\dot{\tau} \in \mathcal{R}^1 : \textstyle\sum_{i=1}^2 k_i \hat{\alpha}_i - \hat{\sigma}[2\bar{F}_{\dot{\gamma}}(2, N - P_*)]^{1/2}(k_1^2 + k_2^2)^{1/2} \le \dot{\tau} \le \sum_{i=1}^2 k_i \hat{\alpha}_i + \hat{\sigma}[2\bar{F}_{\dot{\gamma}}(2, N - P_*)]^{1/2}(k_1^2 + k_2^2)^{1/2}\}. \tag{3.134}$$

The two sets (3.131) and (3.133) of $\boldsymbol{\alpha}$-values are displayed in Figure 7.2 [in terms of the transformed coordinates $(\dot{\alpha}_1 - \hat{\alpha}_1)/\hat{\sigma}$ and $(\dot{\alpha}_2 - \hat{\alpha}_2)/\hat{\sigma}$] for the case where $\dot{\gamma} = 0.10$ and $N - P_* = 10$—in this case $c_{\dot{\gamma}} = 2.193$ and $\bar{F}_{\dot{\gamma}}(2, N - P_*) = 2.924$. The set (3.131) is represented by the rectangular region, and the set (3.133) by the circular region. For each of these two sets, the probability of coverage is $1 - \dot{\gamma} = 0.90$.

Interval (3.132) is of length $2\hat{\sigma} c_{\dot{\gamma}}(|k_1| + |k_2|)$, and interval (3.134) of length $2\hat{\sigma}[2\bar{F}_{\dot{\gamma}}(2, N - P_*)]^{1/2}(k_1^2 + k_2^2)^{1/2}$. Suppose that k_1 or k_2 is nonzero, in which case the length of both intervals is strictly positive—if both k_1 and k_2 were 0, both intervals would be of length 0. Further, let v represent the ratio of the length of interval (3.132) to the length of interval (3.134), and let $u = k_1^2/(k_1^2 + k_2^2)$. And observe that

$$v = \frac{c_{\dot{\gamma}}}{[2\bar{F}_{\dot{\gamma}}(2, N - P_*)]^{1/2}} \frac{|k_1| + |k_2|}{(k_1^2 + k_2^2)^{1/2}} = \frac{c_{\dot{\gamma}}}{[2\bar{F}_{\dot{\gamma}}(2, N - P_*)]^{1/2}}[u^{1/2} + (1 - u)^{1/2}],$$

so that v can be regarded as a function of u. Observe also that $0 \le u \le 1$, and that as u increases from 0 to $\frac{1}{2}$, $u^{1/2} + (1 - u)^{1/2}$ increases monotonically from 1 to $\sqrt{2}$ and that as u increases from $\frac{1}{2}$ to 1, $u^{1/2} + (1 - u)^{1/2}$ decreases monotonically from $\sqrt{2}$ to 1.

In Figure 7.3, v is plotted as a function of u for the case where $\dot{\gamma} = 0.10$ and $N - P_* = 10$. When $\dot{\gamma} = 0.10$ and $N - P_* = 10$, $v = 0.907[u^{1/2} + (1 - u)^{1/2}]$, and we find that $v > 1$ if

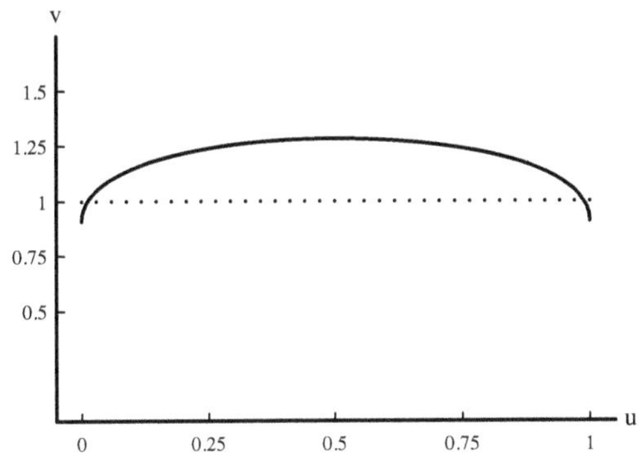

FIGURE 7.3. Plot (represented by the solid line) of $v = c_{\dot{\gamma}}[2\bar{F}_{\dot{\gamma}}(2,\, N-P_*)]^{-1/2}[u^{1/2} + (1-u)^{1/2}]$ as a function of $u = k_1^2/(k_1^2+k_2^2)$ for the case where $\dot{\gamma} = 0.10$ and $N-P_* = 10$.

$0.012 < u < 0.988$, that $v = 1$ if $u = 0.012$ or $u = 0.988$, and that $v < 1$ if $u < 0.012$ or $u > 0.988$—$u = 0.012$ when $|k_1/k_2| = 0.109$ and $u = 0.988$ when $|k_2/k_1| = 0.109$ (or, equivalently, when $|k_1/k_2| = 9.145$).

Computational issues. Consider the set $\tilde{A}_{\tilde{\boldsymbol{\delta}}}$ given for $\tilde{\boldsymbol{\delta}} \in \tilde{\Delta}$ by expression (3.90) and extended to $\tilde{\boldsymbol{\delta}} \notin \tilde{\Delta}$ by expression (3.123). These expressions involve (either explicitly or implicitly) the upper $100\,\dot{\gamma}\%$ point $c_{\dot{\gamma}}$ of the distribution of the random variable $\max_{\{\tilde{\boldsymbol{\delta}} \in \tilde{\Delta}\,:\, \tilde{\boldsymbol{\delta}} \neq \mathbf{0}\}} |\tilde{\boldsymbol{\delta}}'\mathbf{t}|/(\tilde{\boldsymbol{\delta}}'\tilde{\boldsymbol{\delta}})^{1/2}$. Let us consider the computation of $c_{\dot{\gamma}}$.

In certain special cases, the distribution of $\max_{\{\tilde{\boldsymbol{\delta}} \in \tilde{\Delta}\,:\, \tilde{\boldsymbol{\delta}} \neq \mathbf{0}\}} |\tilde{\boldsymbol{\delta}}'\mathbf{t}|/(\tilde{\boldsymbol{\delta}}'\tilde{\boldsymbol{\delta}})^{1/2}$ is sufficiently simple that the computation of $c_{\dot{\gamma}}$ is relatively tractable. One such special case has already been considered. If $\tilde{\Delta} = \mathbb{R}^{M_*}$ or, more generally, if $\tilde{\Delta}$ is such that condition (3.107) is satisfied, then

$$c_{\dot{\gamma}} = [M_* \bar{F}_{\dot{\gamma}}(M_*,\, N-P_*)]^{1/2}.$$

A second special case where the computation of $c_{\dot{\gamma}}$ is relatively tractable is that where for some nonnull $M_* \times 1$ orthogonal vectors $\tilde{\boldsymbol{\delta}}_1, \tilde{\boldsymbol{\delta}}_2, \ldots, \tilde{\boldsymbol{\delta}}_{M_*}$,

$$\tilde{\Delta} = \{\tilde{\boldsymbol{\delta}}_1, \tilde{\boldsymbol{\delta}}_2, \ldots, \tilde{\boldsymbol{\delta}}_{M_*}\}. \tag{3.135}$$

In that special case,

$$\max_{\{\tilde{\boldsymbol{\delta}} \in \tilde{\Delta}\,:\, \tilde{\boldsymbol{\delta}} \neq \mathbf{0}\}} \frac{|\tilde{\boldsymbol{\delta}}'\mathbf{t}|}{(\tilde{\boldsymbol{\delta}}'\tilde{\boldsymbol{\delta}})^{1/2}} \sim \max(|t_1|, |t_2|, \ldots, |t_{M_*}|),$$

where $t_1, t_2, \ldots, t_{M_*}$ are the elements of the M_*-dimensional random vector $\mathbf{t}$ [whose distribution is $MVt(N-P_*,\, \mathbf{I}_{M_*})$]. The distribution of $\max(|t_1|, |t_2|, \ldots, |t_{M_*}|)$ is referred to as a *Studentized maximum modulus distribution*.

In the special case where $\tilde{\Delta}$ is of the form (3.135) (and where $M_* \geq 2$)

$$c_{\dot{\gamma}} > [\bar{F}_{\dot{\gamma}}(M_*,\, N-P_*)]^{1/2}, \tag{3.136}$$

as is evident upon observing that

$$\begin{aligned}\max(|t_1|, |t_2|, \ldots, |t_{M_*}|) &= [\max(t_1^2, t_2^2, \ldots, t_{M_*}^2)]^{1/2} \\ &> \left[\textstyle\sum_{i=1}^{M_*} t_i^2/M_*\right]^{1/2} \quad \text{(with probability 1)}\end{aligned}$$

and that $\sum_{i=1}^{M_*} t_i^2/M_* \sim SF(M_*,\, N-P_*)$.

A third special case where the computation of $c_{\dot{\gamma}}$ is relatively tractable is that where for some nonzero scalar a and some $M_* \times 1$ orthonormal vectors $\dot{\boldsymbol{\delta}}_1, \dot{\boldsymbol{\delta}}_2, \ldots, \dot{\boldsymbol{\delta}}_{M_*}$,

$$\tilde{\Delta} = \{a(\dot{\boldsymbol{\delta}}_i - \dot{\boldsymbol{\delta}}_j) \;\; (j \neq i = 1, 2, \ldots, M_*)\}. \tag{3.137}$$

In that special case,

$$\max_{\{\tilde{\boldsymbol{\delta}}\in\tilde{\Delta}\,:\,\tilde{\boldsymbol{\delta}}\neq\mathbf{0}\}} \frac{|\tilde{\boldsymbol{\delta}}'\mathbf{t}|}{(\tilde{\boldsymbol{\delta}}'\tilde{\boldsymbol{\delta}})^{1/2}} \sim 2^{-1/2}[\max(t_1, t_2, \ldots, t_{M_*}) - \min(t_1, t_2, \ldots, t_{M_*})],$$

where (as in the second special case) $t_1, t_2, \ldots, t_{M_*}$ are the elements of the random vector $\mathbf{t}$ [whose distribution is $MVt(N - P_*, \mathbf{I}_{M_*})$]. The distribution of the random variable $\max_i t_i - \min_j t_j$ is known as the *distribution of the Studentized range*.

The special cases where $\tilde{\Delta} = \mathcal{R}^{M_*}$ or where $\tilde{\Delta}$ is of the form (3.135) or of the form (3.137) are among those considered by Scheffé (1959, chap. 3) in his discussion of multiple comparisons and simultaneous confidence intervals. In the special case where $\tilde{\Delta}$ is of the form (3.137), the use of interval (3.94) as a means for obtaining a confidence set for $\boldsymbol{\delta}'\boldsymbol{\tau}$ (for every $\boldsymbol{\delta} \in \Delta$) or of the critical region (3.103) as a means for obtaining a test of the null hypothesis $H_0^{(\boldsymbol{\delta})}$ (versus the alternative hypothesis $H_1^{(\boldsymbol{\delta})}$) (for every $\boldsymbol{\delta} \in \Delta$) is synonymous with what is known as *Tukey's method* or simply as the *T method*.

There are other special cases where the computation of $c_{\dot{\gamma}}$ is relatively tractable, including some that are encountered in making inferences about the points on a regression line or a response surface and that are discussed in some detail by Liu (2011). In less tractable cases, resort can be made to Monte Carlo methods. Repeated draws can be made from the distribution of the random variable $\max_{\{\tilde{\boldsymbol{\delta}}\in\tilde{\Delta}\,:\,\tilde{\boldsymbol{\delta}}\neq\mathbf{0}\}} |\tilde{\boldsymbol{\delta}}'\mathbf{t}|/(\tilde{\boldsymbol{\delta}}'\tilde{\boldsymbol{\delta}})^{1/2}$, and the results used to approximate $c_{\dot{\gamma}}$ to a high degree of accuracy; methods for using the draws in this way are discussed by, for example, Edwards and Berry (1987) and Liu (2011, app. A). To obtain a draw from the distribution of $\max_{\{\tilde{\boldsymbol{\delta}}\in\tilde{\Delta}\,:\,\tilde{\boldsymbol{\delta}}\neq\mathbf{0}\}} |\tilde{\boldsymbol{\delta}}'\mathbf{t}|/(\tilde{\boldsymbol{\delta}}'\tilde{\boldsymbol{\delta}})^{1/2}$, it suffices to obtain a draw, say $\dot{\mathbf{t}}$, from the distribution of $\mathbf{t}$ [i.e., from the $MVt(N - P_*, \mathbf{I}_{M_*})$ distribution], which is relatively easy, and to then find the maximum value of $|\tilde{\boldsymbol{\delta}}'\dot{\mathbf{t}}|/(\tilde{\boldsymbol{\delta}}'\tilde{\boldsymbol{\delta}})^{1/2}$ for $\tilde{\boldsymbol{\delta}} \in \tilde{\Delta}$, the difficulty of which depends on the characteristics of the set $\tilde{\Delta}$.

For $\tilde{\boldsymbol{\delta}} \notin \tilde{\Delta}$, the set $\tilde{A}_{\tilde{\boldsymbol{\delta}}}$ is defined by expression (3.123). What is the nature of this set, and how might it be constructed? These issues were considered in the preceding part of the present subsection for the special case where $\tilde{\Delta}$ consists of M_* linearly independent vectors. That special case is relatively simple; in that special case, $\tilde{A}_{\tilde{\boldsymbol{\delta}}}$ takes (for every $\tilde{\boldsymbol{\delta}} \in \mathcal{R}^{M_*}$) the form of interval (3.129), the end points of which are expressible in "closed" form.

Let us consider a more general case. Suppose that $\tilde{\Delta}$ consists of a finite number, say L ($\geq M_*$), vectors $\tilde{\boldsymbol{\delta}}_1, \tilde{\boldsymbol{\delta}}_2, \ldots, \tilde{\boldsymbol{\delta}}_L$. Suppose further that M_* of these vectors are linearly independent, and assume (for convenience) that the vectors have been numbered or renumbered so that $\tilde{\boldsymbol{\delta}}_1, \tilde{\boldsymbol{\delta}}_2, \ldots, \tilde{\boldsymbol{\delta}}_{M_*}$ are linearly independent. And consider a setting where $c_{\dot{\gamma}}$ has been computed and used to determine (for every $\tilde{\boldsymbol{\delta}} \in \tilde{\Delta}$) the end points of the interval $\tilde{A}_{\tilde{\boldsymbol{\delta}}}$ given by expression (3.90) and where we wish to construct (for $\tilde{\boldsymbol{\delta}} \notin \tilde{\Delta}$) the set $\tilde{A}_{\tilde{\boldsymbol{\delta}}}$ given by expression (3.123).

Let $\tilde{\boldsymbol{\delta}}$ represent any $M_* \times 1$ vector such that $\tilde{\boldsymbol{\delta}} \notin \tilde{\Delta}$. And let $\dot{\tau} = \tilde{\boldsymbol{\delta}}'\dot{\boldsymbol{\alpha}}$, where $\dot{\boldsymbol{\alpha}}$ is an $M_* \times 1$ vector, and regard $\dot{\tau}$ as a function (of $\dot{\boldsymbol{\alpha}}$) whose domain is restricted to the set $\tilde{A}$ defined by expression (3.122). The function $\dot{\tau}$ is linear, and its domain $\tilde{A}$ is a closed, bounded, convex set. Accordingly, $\dot{\tau}$ assumes a maximum value, say $\dot{\tau}_{\max}$, and a minimum value, say $\dot{\tau}_{\min}$, and the set $\tilde{A}_{\tilde{\boldsymbol{\delta}}}$ defined by expression (3.123) is the interval

$$\tilde{A}_{\tilde{\boldsymbol{\delta}}} = [\dot{\tau}_{\min}, \; \dot{\tau}_{\max}],$$

with upper and lower end points $\dot{\tau}_{\max}$ and $\dot{\tau}_{\min}$. In general, $\dot{\tau}_{\max}$ and $\dot{\tau}_{\min}$ must be determined by numerical methods.

Corresponding to $\tilde{\boldsymbol{\delta}}$ are scalars $k_1, k_2, \ldots, k_{M_*}$ such that $\tilde{\boldsymbol{\delta}} = \sum_{i=1}^{M_*} k_i \tilde{\boldsymbol{\delta}}_i$. Thus, letting (for $i = 1, 2, \ldots, M_*$) $x_i = (\tilde{\boldsymbol{\delta}}_i'\tilde{\boldsymbol{\delta}}_i)^{-1/2}\tilde{\boldsymbol{\delta}}_i'(\dot{\boldsymbol{\alpha}} - \hat{\boldsymbol{\alpha}})$, $\dot{\tau}$ is reexpressible as

$$\dot{\tau} = \tilde{\boldsymbol{\delta}}'\hat{\boldsymbol{\alpha}} + \tilde{\boldsymbol{\delta}}'(\dot{\boldsymbol{\alpha}} - \hat{\boldsymbol{\alpha}}) = \tilde{\boldsymbol{\delta}}'\hat{\boldsymbol{\alpha}} + \textstyle\sum_{i=1}^{M_*} k_i (\tilde{\boldsymbol{\delta}}_i'\tilde{\boldsymbol{\delta}}_i)^{1/2} x_i. \tag{3.138}$$

Moreover, $\dot{\boldsymbol{\alpha}} \in \tilde{A}$ [where $\tilde{A}$ is the set (3.121) or (3.122)] if and only if

$$-\hat{\sigma} c_{\dot{\gamma}} \leq (\tilde{\boldsymbol{\delta}}_i'\tilde{\boldsymbol{\delta}}_i)^{-1/2}\tilde{\boldsymbol{\delta}}_i'(\dot{\boldsymbol{\alpha}} - \hat{\boldsymbol{\alpha}}) \leq \hat{\sigma} c_{\dot{\gamma}} \qquad (i = 1, 2, \ldots, L). \tag{3.139}$$

Clearly, the first M_* of the L inequalities (3.139) are reexpressible as

$$-\hat{\sigma} c_{\dot{\gamma}} \le x_i \le \hat{\sigma} c_{\dot{\gamma}} \qquad (i = 1, 2, \ldots, M_*). \tag{3.140}$$

And (for $j = 1, 2, \ldots, L\text{–}M_*$) there exist scalars $a_{j1}, a_{j2}, \ldots, a_{jM_*}$ such that $\tilde{\boldsymbol{\delta}}_{M_*+j} = \sum_{i=1}^{M_*} a_{ji} \tilde{\boldsymbol{\delta}}_i$, so that the last $L - M_*$ of the L inequalities (3.139) are reexpressible as

$$-\hat{\sigma} c_{\dot{\gamma}} \le (\tilde{\boldsymbol{\delta}}'_{M_*+j} \tilde{\boldsymbol{\delta}}_{M_*+j})^{-1/2} \sum_{i=1}^{M_*} a_{ji} (\tilde{\boldsymbol{\delta}}'_i \tilde{\boldsymbol{\delta}}_i)^{1/2} x_i \le \hat{\sigma} c_{\dot{\gamma}} \qquad (j = 1, 2, \ldots, L - M_*). \tag{3.141}$$

We conclude that the problem of determining $\dot{\tau}_{\max}$ and $\dot{\tau}_{\min}$ is essentially that of determining the maximum and minimum values of the quantity (3.138) with respect to $x_1, x_2, \ldots, x_{M_*}$ subject to the constraints (3.140) and (3.141). The problem of maximizing or minimizing this quantity subject to these constraints can be formulated as a linear programming problem, and its solution can be effected by employing an algorithm for solving linear programming problems—refer, e.g., to Nocedal and Wright (2006, chaps. 13 & 14).

d. An illustration

Let us illustrate various of the results of Subsections a, b, and c by using them to add to the results obtained earlier (in Sections 7.1 and 7.2c) for the lettuce-yield data. Accordingly, let us take $\mathbf{y}$ to be the 20×1 random vector whose observed value is the vector of lettuce yields. Further, let us adopt the terminology and notation introduced in Section 7.1 along with those introduced in the present section. And let us restrict attention to the case where $\mathbf{y}$ is assumed to follow either the second-order or third-order model, that is, the G–M model obtained upon taking the function $\delta(\mathbf{u})$ (that defines the response surface) to be either the second-order polynomial (1.2) or the third-order polynomial (1.3) (and taking $\mathbf{u}$ to be the 3-dimensional column vector whose elements u_1, u_2, and u_3 represent the transformed amounts of Cu, Mo, and Fe). In what follows, the distribution of the vector $\mathbf{e}$ of residual effects (in the second- or third-order model) is taken to be $N(\mathbf{0}, \sigma^2 \mathbf{I})$.

Second-order model versus the third-order model. The second-order model has considerable appeal; it is relatively simple and relatively tractable. However, there may be a question as to whether the second-order polynomial (1.2) provides an "adequate" approximation to the response surface over the region of interest. A common way of addressing this question is to take the model to be the third-order model and to attempt to determine whether the data are consistent with the hypothesis that the coefficients of the third-order terms [i.e., the terms that appear in the third-order polynomial (1.3) but not the second-order polynomial (1.2)] equal 0.

Accordingly, suppose that $\mathbf{y}$ follows the third-order model (in which case $P = 20$, $P_* = 15$, and $N - P_* = 5$). There are 10 third-order terms, the coefficients of which are $\beta_{111}, \beta_{112}, \beta_{113}, \beta_{122}, \beta_{123}, \beta_{133}, \beta_{222}, \beta_{223}, \beta_{233}$, and β_{333}. Not all of these coefficients are estimable from the lettuce-yield data; only certain linear combinations are estimable. In fact, as discussed in Section 7.2c, a linear combination of the coefficients of the third-order terms is estimable (from these data) if and only if it is expressible as a linear combination of 5 linearly independent estimable linear combinations, and among the choices for the 5 linearly independent estimable linear combinations are the linear combinations $3.253\beta_{111} - 1.779\beta_{122} - 0.883\beta_{133}$, $1.779\beta_{112} - 3.253\beta_{222} + 0.883\beta_{233}$, $1.554\beta_{113} + 1.554\beta_{223} - 3.168\beta_{333}$, $2.116\beta_{123}$, and $0.471\beta_{333}$.

Let $\boldsymbol{\tau}$ represent the 5-dimensional column vector with elements $\tau_1 = 3.253\beta_{111} - 1.779\beta_{122} - 0.883\beta_{133}$, $\tau_2 = 1.779\beta_{112} - 3.253\beta_{222} + 0.883\beta_{233}$, $\tau_3 = 1.554\beta_{113} + 1.554\beta_{223} - 3.168\beta_{333}$, $\tau_4 = 2.116\beta_{123}$, and $\tau_5 = 0.471\beta_{333}$. And consider the null hypothesis $H_0 : \boldsymbol{\tau} = \boldsymbol{\tau}^{(0)}$, where $\boldsymbol{\tau}^{(0)} = \mathbf{0}$ (and where the alternative hypothesis is $H_1 : \boldsymbol{\tau} \ne \boldsymbol{\tau}^{(0)}$). Clearly, H_0 is testable.

In light of the results of Section 7.2c, the least squares estimator $\hat{\boldsymbol{\tau}}$ of $\boldsymbol{\tau}$ equals $(-1.97, 4.22, -5.11, -2.05, 2.07)'$, and $\text{var}(\hat{\boldsymbol{\tau}}) = \sigma^2 \mathbf{I}$—by construction, the linear combinations that form the elements of $\boldsymbol{\tau}$ are such that their least squares estimators are uncorrelated and have standard errors equal to σ. Further, $\hat{\sigma}^2 = 10.53$ (so that each element of $\hat{\boldsymbol{\tau}}$ has an estimated standard error of $\hat{\sigma} = 3.24$). And

$$F(\mathbf{0}) = \frac{(\hat{\boldsymbol{\tau}}-\mathbf{0})'\mathbf{C}^{-}(\hat{\boldsymbol{\tau}}-\mathbf{0})}{M_* \hat{\sigma}^2} = \frac{\hat{\boldsymbol{\tau}}'\hat{\boldsymbol{\tau}}}{5\hat{\sigma}^2} = 1.07.$$

The size-$\dot{\gamma}$ F test of $H_0: \boldsymbol{\tau} = \mathbf{0}$ (versus $H_1: \boldsymbol{\tau} \neq \mathbf{0}$) consists of rejecting H_0 if $F(\mathbf{0}) > \bar{F}_{\dot{\gamma}}(5, 5)$, and accepting H_0 otherwise. The p-value of the F test of H_0 (versus H_1), which (by definition) is the value of $\dot{\gamma}$ such that $F(\mathbf{0}) = \bar{F}_{\dot{\gamma}}(5, 5)$, equals 0.471. Thus, the size-$\dot{\gamma}$ F test rejects H_0 for values of $\dot{\gamma}$ larger than 0.471 and accepts H_0 for values less than or equal to 0.471. This result is more-or-less consistent with a hypothesis that the coefficients of the 10 third-order terms (of the third-order model) equal 0. However, there is a caveat: the power of the test depends on the values of the coefficients of the 10 third-order terms only through the values of the 5 linear combinations τ_1, τ_2, τ_3, τ_4, and τ_5. The distribution of $F(\mathbf{0})$ (under both H_0 and H_1), from which the power function of the size-$\dot{\gamma}$ F test of H_0 is determined, is $SF\big(5, 5, \sum_{i=1}^{5} \tau_i^2/\sigma^2\big)$.

Presence or absence of interactions. Among the stated objectives of the experimental study of lettuce yield was that of "determining the importance of interactions among Cu, Mo, and Fe." That is, to what extent (if any) does the change in yield effected by a change in the level of one of these three variables vary with the levels of the other two?

Suppose that $\mathbf{y}$ follows the second-order model, in which $\boldsymbol{\beta} = (\beta_1, \beta_2, \beta_3, \beta_4, \beta_{11}, \beta_{12}, \beta_{13}, \beta_{22}, \beta_{23}, \beta_{33})'$, $P_* = P = 10$, and $N - P_* = 10$. And take $\boldsymbol{\tau}$ to be the 3-dimensional column vector with elements $\tau_1 = \beta_{12}$, $\tau_2 = \beta_{13}$, and $\tau_3 = \beta_{23}$. Then, $M_* = M = 3$, and $\boldsymbol{\tau} = \boldsymbol{\Lambda}'\boldsymbol{\beta}$, where $\boldsymbol{\Lambda}$ is the 10×3 matrix whose columns are the 6th, 7th, and 9th columns of the 10×10 identity matrix.

Consider the problem of obtaining a $100(1-\dot{\gamma})\%$ confidence set for the vector $\boldsymbol{\tau}$ and that of obtaining confidence intervals for β_{12}, β_{13}, and β_{23} (and possibly for linear combinations of β_{12}, β_{13}, and β_{23}) for which the probability of simultaneous coverage is $1-\dot{\gamma}$. Consider also the problem of obtaining a size-$\dot{\gamma}$ test of the null hypothesis $H_0 : \boldsymbol{\tau} = \mathbf{0}$ (versus the alternative hypothesis $H_1: \boldsymbol{\tau} \neq \mathbf{0}$) and that of testing whether each of the quantities β_{12}, β_{13}, and β_{23} (and possibly each of their linear combinations) equals 0 (and of doing so in such a way that the probability of one or more false rejections equals $\dot{\gamma}$).

Letting $\hat{\beta}_{12}$, $\hat{\beta}_{13}$, and $\hat{\beta}_{23}$ represent the least squares estimators of β_{12}, β_{13}, and β_{23}, respectively, we find that $\hat{\beta}_{12} = -1.31$, $\hat{\beta}_{13} = -1.29$, and $\hat{\beta}_{23} = 0.66$, so that the least squares estimator $\hat{\boldsymbol{\tau}}$ of $\boldsymbol{\tau}$ equals $(-1.31, -1.29, 0.66)'$—refer to Table 7.1. Further, $\text{var}(\hat{\boldsymbol{\tau}}) = \sigma^2\mathbf{C}$, and

$$\mathbf{C} = \boldsymbol{\Lambda}'(\mathbf{X}'\mathbf{X})^{-}\boldsymbol{\Lambda} = \text{diag}(0.125, 0.213, 0.213).$$

And for the $M_* \times M$ matrix $\mathbf{T}$ of full row rank such that $\mathbf{C} = \mathbf{T}'\mathbf{T}$ and for the $M \times M_*$ matrix $\mathbf{S}$ such that $\mathbf{TS}$ is orthogonal, take

$$\mathbf{T} = \text{diag}(0.354, 0.462, 0.462) \quad \text{and} \quad \mathbf{S} = \mathbf{T}^{-1} = \text{diag}(2.83, 2.17, 2.17),$$

in which case

$$\boldsymbol{\alpha} = \mathbf{S}'\boldsymbol{\tau} = (2.83\beta_{12}, 2.17\beta_{13}, 2.17\beta_{23})' \text{ and } \hat{\boldsymbol{\alpha}} = \mathbf{S}'\hat{\boldsymbol{\tau}} = (-3.72, -2.80, 1.43)$$

(and the $M_* \times M$ matrix $\mathbf{W}$ such that $\boldsymbol{\Lambda}\mathbf{SW} = \boldsymbol{\Lambda}$ equals $\mathbf{T}$).

The "F statistic" $F(\mathbf{0})$ for testing the null hypothesis H_0: $\boldsymbol{\tau} = \mathbf{0}$ (versus the alternative hypothesis $H_1: \boldsymbol{\tau} \neq \mathbf{0}$) is expressible as

$$F(\mathbf{0}) = \frac{(\hat{\beta}_{12}, \hat{\beta}_{13}, \hat{\beta}_{23})'\mathbf{C}^{-1}(\hat{\beta}_{12}, \hat{\beta}_{13}, \hat{\beta}_{23})}{3\hat{\sigma}^2};$$

its value is 0.726, which is "quite small." Thus, if there is any "nonadditivity" in the effects of Cu, Mo, and Fe on lettuce yield, it is not detectable from the results obtained by carrying out an F test (on these data). Corresponding to the size-$\dot{\gamma}$ F test is the $100(1-\dot{\gamma})\%$ ellipsoidal confidence set A_F

(for the vector $\boldsymbol{\tau}$) given by expression (3.41); it consists of the values of $\boldsymbol{\tau} = (\beta_{12}, \beta_{13}, \beta_{23})'$ such that

$$0.244(\beta_{12}+1.31)^2 + 0.143(\beta_{13}+1.29)^2 + 0.143(\beta_{23}-0.66)^2 \leq \bar{F}_{\dot{\gamma}}(3, 10). \qquad (3.142)$$

For $\dot{\gamma}$ equal to .01, .10, and .50, the values of $\bar{F}_{\dot{\gamma}}(3, 10)$ are 6.552, 2.728, and 0.845, respectively.

Confidence intervals for which the probability of simultaneous coverage is $1-\dot{\gamma}$ can be obtained for β_{12}, β_{13}, and β_{23} and all linear combinations of β_{12}, β_{13}, and β_{23} by applying the S method. In the special case where $\dot{\gamma} = .10$, the intervals obtained for β_{12}, β_{13}, and β_{23} via the S method [using formula (3.110)] are:

$$-4.65 \leq \beta_{12} \leq 2.02, \quad -5.66 \leq \beta_{13} \leq 3.07, \quad \text{and} \quad -3.70 \leq \beta_{23} \leq 5.02.$$

By way of comparison, the 90% one-at-a-time confidence intervals obtained for β_{12}, β_{13}, and β_{23} [upon applying formula (3.55)] are:

$$-3.43 \leq \beta_{12} \leq 0.80, \quad -4.06 \leq \beta_{13} \leq 1.47, \quad \text{and} \quad -2.10 \leq \beta_{23} \leq 3.42.$$

Corresponding to the S method for obtaining (for all linear combinations of β_{12}, β_{13}, and β_{23} including β_{12}, β_{13}, and β_{23} themselves) confidence intervals for which the probability of simultaneous coverage is $1-\dot{\gamma}$ is the S method for obtaining for every linear combination of β_{12}, β_{13}, and β_{23} a test of the null hypothesis that the linear combination equals 0 versus the alternative hypothesis that it does not equal 0. The null hypothesis is either accepted or rejected depending on whether or not 0 is a member of the confidence interval for that linear combination. The tests are such that the probability of one or more false rejections is less than or equal to $\dot{\gamma}$.

The confidence intervals obtained for β_{12}, β_{13}, and β_{23} via the S method [using formula (3.110)] are "conservative." The probability of simultaneous coverage for these three intervals is greater than $1-\dot{\gamma}$, not equal to $1-\dot{\gamma}$—there are values of $\mathbf{y}$ for which coverage is achieved by these three intervals but for which coverage is not achieved by the intervals obtained [using formula (3.110)] for some linear combinations of β_{12}, β_{13}, and β_{23}.

Confidence intervals can be obtained for β_{12}, β_{13}, and β_{23} for which the probability of simultaneous coverage equals $1-\dot{\gamma}$. Letting $\boldsymbol{\delta}_1$, $\boldsymbol{\delta}_2$, and $\boldsymbol{\delta}_3$ represent the columns of the 3×3 identity matrix, take $\Delta = \{\boldsymbol{\delta}_1, \boldsymbol{\delta}_2, \boldsymbol{\delta}_3\}$, in which case $\tilde{\Delta} = \{\tilde{\boldsymbol{\delta}}_1, \tilde{\boldsymbol{\delta}}_2, \tilde{\boldsymbol{\delta}}_3\}$, where $\tilde{\boldsymbol{\delta}}_1$, $\tilde{\boldsymbol{\delta}}_2$, and $\tilde{\boldsymbol{\delta}}_3$ are the columns of the matrix $\mathbf{W}$ [which equals diag(0.354, 0.462, 0.462)]. Then, intervals can be obtained for β_{12}, β_{13}, and β_{23} for which the probability of simultaneous coverage equals $1-\dot{\gamma}$ by using formula (3.90) or (3.94). When $\tilde{\Delta} = \{\tilde{\boldsymbol{\delta}}_1, \tilde{\boldsymbol{\delta}}_2, \tilde{\boldsymbol{\delta}}_3\}$, $\tilde{\Delta}$ is of the form (3.135) and, consequently, $c_{\dot{\gamma}}$ is the upper $100\dot{\gamma}\%$ point of a Studentized maximum modulus distribution; specifically, it is the upper $100\dot{\gamma}\%$ point of the distribution of $\max(|t_1|, |t_2|, |t_3|)$, where t_1, t_2, and t_3 are the elements of a 3-dimensional random column vector whose distribution is $MVt(10, \mathbf{I}_3)$. The value of $c_{.10}$ is 2.410—refer, e.g., to Graybill (1976, p. 656). And [as obtained from formula (3.94)] confidence intervals for β_{12}, β_{13}, and β_{23} with a probability of simultaneous coverage equal to .90 are:

$$-4.13 \leq \beta_{12} \leq 1.50, \quad -4.97 \leq \beta_{13} \leq 2.38, \text{ and } -3.01 \leq \beta_{23} \leq 4.33. \qquad (3.143)$$

The values of $\boldsymbol{\tau}$ whose elements (β_{12}, β_{13}, and β_{23}) satisfy the three inequalities (3.143) form a 3-dimensional rectangular set A. The set A is a 90% confidence set for $\boldsymbol{\tau}$. It can be regarded as a "competitor" to the 90% ellipsoidal confidence set for $\boldsymbol{\tau}$ defined (upon setting $\dot{\gamma} = .10$) by inequality (3.142).

Starting with the confidence intervals (3.143) for β_{12}, β_{13}, and β_{23}, confidence intervals (with the same probability of simultaneous coverage) can be obtained for every linear combination of β_{12}, β_{13}, and β_{23}. Let $\boldsymbol{\delta}$ represent an arbitrary 3×1 vector and denote by k_1, k_2, and k_3 the elements of $\boldsymbol{\delta}$, so that $\boldsymbol{\delta}'\boldsymbol{\tau} = k_1\beta_{12} + k_2\beta_{13} + k_3\beta_{23}$. Further, take $A_{\boldsymbol{\delta}} = \tilde{A}_{\mathbf{W}\boldsymbol{\delta}}$ [where $\tilde{A}_{\tilde{\boldsymbol{\delta}}}$ is the set defined for $\tilde{\boldsymbol{\delta}} \in \tilde{\Delta}$ by expression (3.90) and for $\tilde{\boldsymbol{\delta}} \notin \tilde{\Delta}$ by expression (3.123)]. Then,

$$\Pr[\boldsymbol{\delta}'\boldsymbol{\tau} \in A_{\boldsymbol{\delta}}(\mathbf{y}) \text{ for every } \boldsymbol{\delta} \in \mathbb{R}^3] = 1-\dot{\gamma}.$$

And upon observing that $\mathbf{W}\boldsymbol{\delta} = k_1\tilde{\delta}_1 + k_2\tilde{\delta}_2 + k_3\tilde{\delta}_3$ and making use of formula (3.129), we find that $A_{\boldsymbol{\delta}}$ is the interval with end points

$$k_1\hat{\beta}_{12}+k_2\hat{\beta}_{13}+k_3\hat{\beta}_{23} \pm \textstyle\sum_{i=1}^{3} |k_i|(\boldsymbol{\delta}_i'\mathbf{C}\boldsymbol{\delta}_i)^{1/2}\hat{\sigma}c_{\dot{\gamma}}. \tag{3.144}$$

When $\dot{\gamma} = .10$,

$$\textstyle\sum_{i=1}^{3} |k_i|(\boldsymbol{\delta}_i'\mathbf{C}\boldsymbol{\delta}_i)^{1/2}\hat{\sigma}c_{\dot{\gamma}} = 7.954(0.354|k_1| + 0.462|k_2| + 0.462|k_3|).$$

The intervals obtained for all linear combinations of β_{12}, β_{13}, and β_{23} by taking the end points of each interval to be those given by expression (3.144) can be regarded as competitors to those obtained by applying the S method. When only one of the 3 coefficients k_1, k_2, and k_3 of the linear combination $k_1\beta_{12} + k_2\beta_{13} + k_3\beta_{23}$ is nonzero, the interval with end points (3.144) is shorter than the interval obtained by applying the S method.

Suppose however that k_1, k_2, and k_3 are such that for some nonzero scalar k, $k_i = k(\tilde{\boldsymbol{\delta}}_i'\tilde{\boldsymbol{\delta}}_i)^{-1/2}$ or, equivalently, $k_i = k(\boldsymbol{\delta}_i'\mathbf{C}\boldsymbol{\delta}_i)^{-1/2}$ $(i = 1, 2, 3)$. Then, the length of the interval with end points (3.144) is

$$2\textstyle\sum_{i=1}^{3} |k_i|(\boldsymbol{\delta}_i'\mathbf{C}\boldsymbol{\delta}_i)^{1/2}\hat{\sigma}c_{\dot{\gamma}} = 6|k|\hat{\sigma}c_{\dot{\gamma}} = 19.80\,|k|c_{\dot{\gamma}}$$

and that of the interval obtained by applying the S method is

$$\begin{aligned} 2(\boldsymbol{\delta}'\mathbf{C}\boldsymbol{\delta})^{1/2}\hat{\sigma}[M_*\bar{F}_{\dot{\gamma}}(M_*,\, N-P_*)]^{1/2} \\ &= 2\big(\textstyle\sum_{i=1}^{3} k_i^2\boldsymbol{\delta}_i'\mathbf{C}\boldsymbol{\delta}_i\big)^{1/2}\hat{\sigma}[3\bar{F}_{\dot{\gamma}}(3,\, 10)]^{1/2} \\ &= 2(3k^2)^{1/2}\hat{\sigma}[3\bar{F}_{\dot{\gamma}}(3, 10)]^{1/2} \\ &= 6|k|\hat{\sigma}[\bar{F}_{\dot{\gamma}}(3,\, 10)]^{1/2} = 19.80\,|k|[\bar{F}_{\dot{\gamma}}(3, 10)]^{1/2}. \end{aligned}$$

As a special case of result (3.136) (that where $M_* = 3$ and $N - P_* = 10$), we have that $c_{\dot{\gamma}}$ is greater than $[\bar{F}_{\dot{\gamma}}(3, 10)]^{1/2}$; in particular,

$$c_{.10} = 2.410 > 1.652 = [\bar{F}_{\dot{\gamma}}(3, 10)]^{1/2}.$$

Thus, when $k_i = k(\boldsymbol{\delta}_i'\mathbf{C}\boldsymbol{\delta}_i)^{-1/2}$ $(i = 1, 2, 3)$, the interval with end points (3.144) is lengthier than that obtained by applying the S method.

e. Some additional results on the generalized S method

Preliminary to a further discussion of the lettuce-yield data, it is convenient to introduce some additional results on the generalized S method (for obtaining simultaneous confidence intervals or sets and for making multiple comparisons). In what follows, let us adopt the notation employed in Subsection c.

Reformulation of a maximization problem. When in computing the upper $100\dot{\gamma}\%$ point $c_{\dot{\gamma}}$ [of the distribution of the random variable $\max_{\{\tilde{\boldsymbol{\delta}}\in\tilde{\Delta}\,:\,\tilde{\boldsymbol{\delta}}\neq\mathbf{0}\}} |\tilde{\boldsymbol{\delta}}'\mathbf{t}|/(\tilde{\boldsymbol{\delta}}'\tilde{\boldsymbol{\delta}})^{1/2}$] resort is made to Monte Carlo methods, the value of $\max_{\{\tilde{\boldsymbol{\delta}}\in\tilde{\Delta}\,:\,\tilde{\boldsymbol{\delta}}\neq\mathbf{0}\}} |\tilde{\boldsymbol{\delta}}'\mathbf{t}|/(\tilde{\boldsymbol{\delta}}'\tilde{\boldsymbol{\delta}})^{1/2}$ must be computed for each of a large number of values of $\mathbf{t}$. As is evident upon recalling result (3.96) and upon observing that

$$-\min_{\{\tilde{\boldsymbol{\delta}}\in\tilde{\Delta}\,:\,\tilde{\boldsymbol{\delta}}\neq\mathbf{0}\}} \tilde{\boldsymbol{\delta}}'\dot{\mathbf{t}}/(\tilde{\boldsymbol{\delta}}'\tilde{\boldsymbol{\delta}})^{1/2} = \max_{\{\tilde{\boldsymbol{\delta}}\in\tilde{\Delta}\,:\,\tilde{\boldsymbol{\delta}}\neq\mathbf{0}\}} \tilde{\boldsymbol{\delta}}'(-\dot{\mathbf{t}})/(\tilde{\boldsymbol{\delta}}'\tilde{\boldsymbol{\delta}})^{1/2}, \tag{3.145}$$

the value of $\max_{\{\tilde{\boldsymbol{\delta}}\in\tilde{\Delta}\,:\,\tilde{\boldsymbol{\delta}}\neq\mathbf{0}\}} |\tilde{\boldsymbol{\delta}}'\mathbf{t}|/(\tilde{\boldsymbol{\delta}}'\tilde{\boldsymbol{\delta}})^{1/2}$ can be computed for $\mathbf{t} = \dot{\mathbf{t}}$ by computing the value of $\max_{\{\tilde{\boldsymbol{\delta}}\in\tilde{\Delta}\,:\,\tilde{\boldsymbol{\delta}}\neq\mathbf{0}\}} \tilde{\boldsymbol{\delta}}'\mathbf{t}/(\tilde{\boldsymbol{\delta}}'\tilde{\boldsymbol{\delta}})^{1/2}$ for each of two values of $\mathbf{t}$ (namely, $\mathbf{t} = \dot{\mathbf{t}}$ and $\mathbf{t} = -\dot{\mathbf{t}}$) and by then selecting the larger of these two values.

The value of $\max_{\{\tilde{\boldsymbol{\delta}}\in\tilde{\Delta}\,:\,\tilde{\boldsymbol{\delta}}\neq\mathbf{0}\}} \tilde{\boldsymbol{\delta}}'\mathbf{t}/(\tilde{\boldsymbol{\delta}}'\tilde{\boldsymbol{\delta}})^{1/2}$ can be obtained for any particular value of $\mathbf{t}$ from the solution to a constrained nonlinear least squares problem. Letting $\dot{\mathbf{t}}$ represent an arbitrary value of $\mathbf{t}$ (i.e., an arbitrary $M_* \times 1$ vector) and letting λ represent an arbitrary scalar, consider the minimization (with respect to $\tilde{\boldsymbol{\delta}}$ and λ) of the sum of squares

$$(\dot{\mathbf{t}} - \lambda\tilde{\boldsymbol{\delta}})'(\dot{\mathbf{t}} - \lambda\tilde{\boldsymbol{\delta}}) \quad (= \dot{\mathbf{t}}'\dot{\mathbf{t}} - 2\lambda\tilde{\boldsymbol{\delta}}'\dot{\mathbf{t}} + \lambda^2\tilde{\boldsymbol{\delta}}'\tilde{\boldsymbol{\delta}}) \tag{3.146}$$

subject to the constraints $\tilde{\boldsymbol{\delta}} \in \tilde{\Delta}$ and $\lambda \geq 0$. Suppose that $\dot{\mathbf{t}}$ is such that for some vector $\ddot{\boldsymbol{\delta}} \in \tilde{\Delta}$, $\ddot{\boldsymbol{\delta}}'\dot{\mathbf{t}} > 0$—when no such vector exists, $\max_{\{\tilde{\boldsymbol{\delta}} \in \tilde{\Delta}\,:\, \tilde{\boldsymbol{\delta}} \neq \mathbf{0}\}} \tilde{\boldsymbol{\delta}}'\dot{\mathbf{t}}/(\tilde{\boldsymbol{\delta}}'\tilde{\boldsymbol{\delta}})^{1/2} \leq 0 \leq \max_{\{\tilde{\boldsymbol{\delta}} \in \tilde{\Delta}\,:\, \tilde{\boldsymbol{\delta}} \neq \mathbf{0}\}} \tilde{\boldsymbol{\delta}}'(-\dot{\mathbf{t}})/(\tilde{\boldsymbol{\delta}}'\tilde{\boldsymbol{\delta}})^{1/2}$, and (subject to the constraints $\tilde{\boldsymbol{\delta}} \in \tilde{\Delta}$ and $\lambda \geq 0$) the minimum value of the sum of squares (3.146) is the value $\dot{\mathbf{t}}'\dot{\mathbf{t}}$ attained at $\lambda = 0$. And let $\dot{\boldsymbol{\delta}}$ and $\dot{\lambda}$ represent any values of $\tilde{\boldsymbol{\delta}}$ and λ at which the sum of squares (3.146) attains its minimum value (for $\tilde{\boldsymbol{\delta}} \in \tilde{\Delta}$ and $\lambda \geq 0$). Then,

$$\dot{\lambda} = \frac{\dot{\boldsymbol{\delta}}'\dot{\mathbf{t}}}{\dot{\boldsymbol{\delta}}'\dot{\boldsymbol{\delta}}} > 0 \quad \text{and} \quad (\dot{\mathbf{t}} - \dot{\lambda}\dot{\boldsymbol{\delta}})'(\dot{\mathbf{t}} - \dot{\lambda}\dot{\boldsymbol{\delta}}) = \dot{\mathbf{t}}'\dot{\mathbf{t}} - \frac{(\dot{\boldsymbol{\delta}}'\dot{\mathbf{t}})^2}{\dot{\boldsymbol{\delta}}'\dot{\boldsymbol{\delta}}} < \dot{\mathbf{t}}'\dot{\mathbf{t}}. \tag{3.147}$$

Further,

$$\max_{\{\tilde{\boldsymbol{\delta}} \in \tilde{\Delta}\,:\, \tilde{\boldsymbol{\delta}} \neq \mathbf{0}\}} \frac{\tilde{\boldsymbol{\delta}}'\dot{\mathbf{t}}}{(\tilde{\boldsymbol{\delta}}'\tilde{\boldsymbol{\delta}})^{1/2}} = \frac{\dot{\boldsymbol{\delta}}'\dot{\mathbf{t}}}{(\dot{\boldsymbol{\delta}}'\dot{\boldsymbol{\delta}})^{1/2}} > 0. \tag{3.148}$$

Let us verify results (3.147) and (3.148). For any vector $\ddot{\boldsymbol{\delta}}$ in $\tilde{\Delta}$ such that $\ddot{\boldsymbol{\delta}}'\dot{\mathbf{t}} \neq 0$ and for $\ddot{\lambda} = \ddot{\boldsymbol{\delta}}'\dot{\mathbf{t}}/\ddot{\boldsymbol{\delta}}'\ddot{\boldsymbol{\delta}}$,

$$(\dot{\mathbf{t}} - \ddot{\lambda}\ddot{\boldsymbol{\delta}})'(\dot{\mathbf{t}} - \ddot{\lambda}\ddot{\boldsymbol{\delta}}) = \dot{\mathbf{t}}'\dot{\mathbf{t}} - \frac{(\ddot{\boldsymbol{\delta}}'\dot{\mathbf{t}})^2}{\ddot{\boldsymbol{\delta}}'\ddot{\boldsymbol{\delta}}} < \dot{\mathbf{t}}'\dot{\mathbf{t}}. \tag{3.149}$$

Thus, $\min_{\{\tilde{\boldsymbol{\delta}}, \lambda\,:\, \tilde{\boldsymbol{\delta}} \in \tilde{\Delta},\ \lambda \geq 0\}} (\dot{\mathbf{t}} - \lambda\tilde{\boldsymbol{\delta}})'(\dot{\mathbf{t}} - \lambda\tilde{\boldsymbol{\delta}})$ is less than $\dot{\mathbf{t}}'\dot{\mathbf{t}}$, which is the value of $(\dot{\mathbf{t}} - \lambda\tilde{\boldsymbol{\delta}})'(\dot{\mathbf{t}} - \lambda\tilde{\boldsymbol{\delta}})$ when $\lambda = 0$ or $\tilde{\boldsymbol{\delta}} = \mathbf{0}$. And it follows that $\dot{\lambda} > 0$ and $\dot{\boldsymbol{\delta}} \neq \mathbf{0}$.

To establish that $\max_{\{\tilde{\boldsymbol{\delta}} \in \tilde{\Delta}\,:\, \tilde{\boldsymbol{\delta}} \neq \mathbf{0}\}} \tilde{\boldsymbol{\delta}}'\dot{\mathbf{t}}/(\tilde{\boldsymbol{\delta}}'\tilde{\boldsymbol{\delta}})^{1/2} = \dot{\boldsymbol{\delta}}'\dot{\mathbf{t}}/(\dot{\boldsymbol{\delta}}'\dot{\boldsymbol{\delta}})^{1/2}$, assume (for purposes of establishing a contradiction) the contrary, that is, assume that there exists a nonnull vector $\ddot{\boldsymbol{\delta}} \in \tilde{\Delta}$ such that

$$\frac{\ddot{\boldsymbol{\delta}}'\dot{\mathbf{t}}}{(\ddot{\boldsymbol{\delta}}'\ddot{\boldsymbol{\delta}})^{1/2}} > \frac{\dot{\boldsymbol{\delta}}'\dot{\mathbf{t}}}{(\dot{\boldsymbol{\delta}}'\dot{\boldsymbol{\delta}})^{1/2}}.$$

Then, letting $\ddot{\lambda} = (\dot{\boldsymbol{\delta}}'\dot{\boldsymbol{\delta}}/\ddot{\boldsymbol{\delta}}'\ddot{\boldsymbol{\delta}})^{1/2}\dot{\lambda}$, we find that

$$\ddot{\lambda}\ddot{\boldsymbol{\delta}}'\dot{\mathbf{t}} > \dot{\lambda}\dot{\boldsymbol{\delta}}'\dot{\mathbf{t}} \quad \text{and} \quad \ddot{\lambda}^2\ddot{\boldsymbol{\delta}}'\ddot{\boldsymbol{\delta}} = \dot{\lambda}^2\dot{\boldsymbol{\delta}}'\dot{\boldsymbol{\delta}},$$

which implies that

$$(\dot{\mathbf{t}} - \ddot{\lambda}\ddot{\boldsymbol{\delta}})'(\dot{\mathbf{t}} - \ddot{\lambda}\ddot{\boldsymbol{\delta}}) < (\dot{\mathbf{t}} - \dot{\lambda}\dot{\boldsymbol{\delta}})'(\dot{\mathbf{t}} - \dot{\lambda}\dot{\boldsymbol{\delta}}),$$

thereby establishing the desired contradiction.

The verification of result (3.148) is complete upon observing that (since, by supposition, there exists a vector $\ddot{\boldsymbol{\delta}} \in \tilde{\Delta}$ such that $\ddot{\boldsymbol{\delta}}'\dot{\mathbf{t}} > 0$) $\max_{\{\tilde{\boldsymbol{\delta}} \in \tilde{\Delta}\,:\, \tilde{\boldsymbol{\delta}} \neq \mathbf{0}\}} \tilde{\boldsymbol{\delta}}'\dot{\mathbf{t}}/(\tilde{\boldsymbol{\delta}}'\tilde{\boldsymbol{\delta}})^{1/2} > 0$. Further, turning to result (3.147), $\dot{\lambda} = \dot{\boldsymbol{\delta}}'\dot{\mathbf{t}}/(\dot{\boldsymbol{\delta}}'\dot{\boldsymbol{\delta}})$, as is evident upon letting $\ddot{\lambda} = \dot{\boldsymbol{\delta}}'\dot{\mathbf{t}}/(\dot{\boldsymbol{\delta}}'\dot{\boldsymbol{\delta}})$ and upon observing that

$$(\dot{\mathbf{t}} - \lambda\dot{\boldsymbol{\delta}})'(\dot{\mathbf{t}} - \lambda\dot{\boldsymbol{\delta}}) = (\dot{\mathbf{t}} - \ddot{\lambda}\dot{\boldsymbol{\delta}})'(\dot{\mathbf{t}} - \ddot{\lambda}\dot{\boldsymbol{\delta}}) + (\lambda - \ddot{\lambda})^2\dot{\boldsymbol{\delta}}'\dot{\boldsymbol{\delta}}$$

and [in light of result (3.148)] that $\ddot{\lambda} > 0$. To complete the verification of result (3.147), it remains only to observe that $\dot{\boldsymbol{\delta}}'\dot{\mathbf{t}} \neq 0$ and to apply the special case of result (3.149) obtained upon setting $\ddot{\boldsymbol{\delta}} = \dot{\boldsymbol{\delta}}$ (and implicitly $\ddot{\lambda} = \dot{\lambda}$).

The constrained nonlinear least squares problem can be reformulated. Consider the minimization (with respect to $\boldsymbol{\delta}$ and λ) of the sum of squares

$$[\dot{\mathbf{t}} - \mathbf{W}(\lambda\boldsymbol{\delta})]'[\dot{\mathbf{t}} - \mathbf{W}(\lambda\boldsymbol{\delta})] \tag{3.150}$$

subject to the constaints $\boldsymbol{\delta} \in \Delta$ and $\lambda \geq 0$. And let $\ddot{\boldsymbol{\delta}}$ and $\ddot{\lambda}$ represent any solution to this constrained nonlinear least squares problem, that is, any values of $\boldsymbol{\delta}$ and λ at which the sum of squares (3.150) attains its minimum value (for $\boldsymbol{\delta} \in \Delta$ and $\lambda \geq 0$). Then, a solution $\dot{\boldsymbol{\delta}}$ and $\dot{\lambda}$ to the original constrained nonlinear least squares problem, that is, values of $\tilde{\boldsymbol{\delta}}$ and λ that minimize the sum of squares (3.146) subject to the constraints $\tilde{\boldsymbol{\delta}} \in \tilde{\Delta}$ and $\lambda \geq 0$, can be obtained by taking $\dot{\boldsymbol{\delta}} = \mathbf{W}\ddot{\boldsymbol{\delta}}$ and $\dot{\lambda} = \ddot{\lambda}$.

A variation. The computation of $\max_{\{\tilde{\boldsymbol{\delta}} \in \tilde{\Delta}\,:\, \tilde{\boldsymbol{\delta}} \neq \mathbf{0}\}} |\tilde{\boldsymbol{\delta}}'\dot{\mathbf{t}}|/(\tilde{\boldsymbol{\delta}}'\tilde{\boldsymbol{\delta}})^{1/2}$ can be approached in a way that differs somewhat from the preceding approach. Let $\dot{\lambda}$ and $\dot{\boldsymbol{\delta}}$ represent values of λ and $\tilde{\boldsymbol{\delta}}$ at which the sum of squares $(\dot{\mathbf{t}} - \lambda\tilde{\boldsymbol{\delta}})'(\dot{\mathbf{t}} - \lambda\tilde{\boldsymbol{\delta}})$ attains its minimum value subject to the constraint $\tilde{\boldsymbol{\delta}} \in \tilde{\Delta}$. And suppose that $\tilde{\Delta}$ contains one or more vectors that are not orthogonal (with respect to the usual inner product) to $\dot{\mathbf{t}}$—when every vector in $\tilde{\Delta}$ is orthogonal to $\dot{\mathbf{t}}$, $\max_{\{\tilde{\boldsymbol{\delta}}\in\tilde{\Delta}\,:\,\tilde{\boldsymbol{\delta}}\neq\mathbf{0}\}} |\tilde{\boldsymbol{\delta}}'\dot{\mathbf{t}}|/(\tilde{\boldsymbol{\delta}}'\tilde{\boldsymbol{\delta}})^{1/2} = 0$. Then, $\dot{\boldsymbol{\delta}} \neq \mathbf{0}$—refer to result (3.149)—and $\dot{\lambda} = \dot{\boldsymbol{\delta}}'\dot{\mathbf{t}}/(\dot{\boldsymbol{\delta}}'\dot{\boldsymbol{\delta}})$. Moreover, for any nonnull vector $\tilde{\boldsymbol{\delta}}$ in $\tilde{\Delta}$ and for $\lambda = \tilde{\boldsymbol{\delta}}'\dot{\mathbf{t}}/(\tilde{\boldsymbol{\delta}}'\tilde{\boldsymbol{\delta}})$,

$$\dot{\mathbf{t}}'\dot{\mathbf{t}} - (\dot{\mathbf{t}} - \lambda\tilde{\boldsymbol{\delta}})'(\dot{\mathbf{t}} - \lambda\tilde{\boldsymbol{\delta}}) = \dot{\mathbf{t}}'\dot{\mathbf{t}} - \left[\dot{\mathbf{t}}'\dot{\mathbf{t}} - \frac{(\tilde{\boldsymbol{\delta}}'\dot{\mathbf{t}})^2}{\tilde{\boldsymbol{\delta}}'\tilde{\boldsymbol{\delta}}}\right] = \frac{(\tilde{\boldsymbol{\delta}}'\dot{\mathbf{t}})^2}{\tilde{\boldsymbol{\delta}}'\tilde{\boldsymbol{\delta}}}. \tag{3.151}$$

Thus,

$$\max_{\{\tilde{\boldsymbol{\delta}}\in\tilde{\Delta}\,:\,\tilde{\boldsymbol{\delta}}\neq\mathbf{0}\}} \frac{(\tilde{\boldsymbol{\delta}}'\dot{\mathbf{t}})^2}{\tilde{\boldsymbol{\delta}}'\tilde{\boldsymbol{\delta}}} = \frac{(\dot{\boldsymbol{\delta}}'\dot{\mathbf{t}})^2}{\dot{\boldsymbol{\delta}}'\dot{\boldsymbol{\delta}}}, \tag{3.152}$$

since otherwise there would exist a nonnull value of $\tilde{\boldsymbol{\delta}}$, say $\ddot{\boldsymbol{\delta}}$, and a value of λ, namely, $\ddot{\lambda} = \ddot{\boldsymbol{\delta}}'\dot{\mathbf{t}}/(\ddot{\boldsymbol{\delta}}'\ddot{\boldsymbol{\delta}})$, such that $(\dot{\mathbf{t}} - \ddot{\lambda}\ddot{\boldsymbol{\delta}})'(\dot{\mathbf{t}} - \ddot{\lambda}\ddot{\boldsymbol{\delta}}) < (\dot{\mathbf{t}} - \dot{\lambda}\dot{\boldsymbol{\delta}})'(\dot{\mathbf{t}} - \dot{\lambda}\dot{\boldsymbol{\delta}})$.

Upon observing that

$$\max_{\{\tilde{\boldsymbol{\delta}}\in\tilde{\Delta}\,:\,\tilde{\boldsymbol{\delta}}\neq\mathbf{0}\}} \frac{|\tilde{\boldsymbol{\delta}}'\dot{\mathbf{t}}|}{(\tilde{\boldsymbol{\delta}}'\tilde{\boldsymbol{\delta}})^{1/2}} = \left[\max_{\{\tilde{\boldsymbol{\delta}}\in\tilde{\Delta}\,:\,\tilde{\boldsymbol{\delta}}\neq\mathbf{0}\}} \frac{(\tilde{\boldsymbol{\delta}}'\dot{\mathbf{t}})^2}{\tilde{\boldsymbol{\delta}}'\tilde{\boldsymbol{\delta}}}\right]^{1/2}, \tag{3.153}$$

upon applying result (3.152), and upon making use of the special case of result (3.151) obtained upon setting $\tilde{\boldsymbol{\delta}} = \dot{\boldsymbol{\delta}}$ and (implicitly) $\lambda = \dot{\lambda}$, we find that

$$\max_{\{\tilde{\boldsymbol{\delta}}\in\tilde{\Delta}\,:\,\tilde{\boldsymbol{\delta}}\neq\mathbf{0}\}} \frac{|\tilde{\boldsymbol{\delta}}'\dot{\mathbf{t}}|}{(\tilde{\boldsymbol{\delta}}'\tilde{\boldsymbol{\delta}})^{1/2}} = [\dot{\mathbf{t}}'\dot{\mathbf{t}} - (\dot{\mathbf{t}} - \dot{\lambda}\dot{\boldsymbol{\delta}})'(\dot{\mathbf{t}} - \dot{\lambda}\dot{\boldsymbol{\delta}})]^{1/2}. \tag{3.154}$$

Thus, $\max_{\{\tilde{\boldsymbol{\delta}}\in\tilde{\Delta}\,:\,\tilde{\boldsymbol{\delta}}\neq\mathbf{0}\}} |\tilde{\boldsymbol{\delta}}'\dot{\mathbf{t}}|/(\tilde{\boldsymbol{\delta}}'\tilde{\boldsymbol{\delta}})^{1/2}$ can be computed as the square root of the difference between the total sum of squares $\dot{\mathbf{t}}'\dot{\mathbf{t}}$ and the residual sum of squares $(\dot{\mathbf{t}} - \dot{\lambda}\dot{\boldsymbol{\delta}})'(\dot{\mathbf{t}} - \dot{\lambda}\dot{\boldsymbol{\delta}})$. Further,

$$\min_{\{\tilde{\boldsymbol{\delta}},\lambda\,:\,\tilde{\boldsymbol{\delta}}\in\tilde{\Delta}\}} (\dot{\mathbf{t}} - \lambda\tilde{\boldsymbol{\delta}})'(\dot{\mathbf{t}} - \lambda\tilde{\boldsymbol{\delta}}) = \min_{\{\boldsymbol{\delta},\lambda\,:\,\boldsymbol{\delta}\in\Delta\}} [\dot{\mathbf{t}} - \mathbf{W}(\lambda\boldsymbol{\delta})]'[\dot{\mathbf{t}} - \mathbf{W}(\lambda\boldsymbol{\delta})], \tag{3.155}$$

so that the residual sum of squares obtained by minimizing $(\dot{\mathbf{t}} - \lambda\tilde{\boldsymbol{\delta}})'(\dot{\mathbf{t}} - \lambda\tilde{\boldsymbol{\delta}})$ with respect to $\tilde{\boldsymbol{\delta}}$ and λ (subject to the constraint $\tilde{\boldsymbol{\delta}} \in \tilde{\Delta}$) is identical to that obtained by minimizing $[\dot{\mathbf{t}} - \mathbf{W}(\lambda\boldsymbol{\delta})]'[\dot{\mathbf{t}} - \mathbf{W}(\lambda\boldsymbol{\delta})]$ with respect to $\boldsymbol{\delta}$ and λ (subject to the constraint $\boldsymbol{\delta} \in \Delta$).

Constant-width simultaneous confidence intervals. For $\boldsymbol{\delta} \in \Delta$, $A_{\boldsymbol{\delta}}$ is (as defined in Part c) the confidence interval (for $\boldsymbol{\delta}'\boldsymbol{\tau}$) with end points $\boldsymbol{\delta}'\hat{\boldsymbol{\tau}} \pm (\boldsymbol{\delta}'\mathbf{C}\boldsymbol{\delta})^{1/2}\hat{\sigma}c_{\dot{\gamma}}$. The intervals $A_{\boldsymbol{\delta}}$ $(\boldsymbol{\delta} \in \Delta)$ were constructed in such a way that their probability of simultaneous coverage equals $1-\dot{\gamma}$. Even when the intervals corresponding to values of $\boldsymbol{\delta}$ for which $\boldsymbol{\Lambda\delta} = \mathbf{0}$ (i.e., values for which $\boldsymbol{\delta}'\boldsymbol{\tau} = 0$) are excluded, these intervals are (aside from various special cases) not all of the same width. Clearly, the width of the interval $A_{\boldsymbol{\delta}}$ is proportional to the standard error $(\boldsymbol{\delta}'\mathbf{C}\boldsymbol{\delta})^{1/2}\sigma$ or estimated standard error $(\boldsymbol{\delta}'\mathbf{C}\boldsymbol{\delta})^{1/2}\hat{\sigma}$ of the least squares estimator $\boldsymbol{\delta}'\hat{\boldsymbol{\tau}}$ of $\boldsymbol{\delta}'\boldsymbol{\tau}$.

Suppose that Δ is such that $\boldsymbol{\Lambda\delta} \neq \mathbf{0}$ for every $\boldsymbol{\delta} \in \Delta$, and suppose that $\tilde{\Delta}$ (which is the set $\{\tilde{\boldsymbol{\delta}} : \tilde{\boldsymbol{\delta}} = \mathbf{W}\boldsymbol{\delta},\ \boldsymbol{\delta} \in \boldsymbol{\Delta}\}$) is such that for every $M_* \times 1$ vector $\dot{\mathbf{t}}$, $\max_{\{\tilde{\boldsymbol{\delta}}\in\tilde{\Delta}\}} |\tilde{\boldsymbol{\delta}}'\dot{\mathbf{t}}|$ exists.. Then, confidence intervals can be obtained for the linear combinations $\boldsymbol{\delta}'\boldsymbol{\tau}$ $(\boldsymbol{\delta} \in \Delta)$ that have a probability of simultaneous coverage equal to $1-\dot{\gamma}$ and that are all of the same width.

Analogous to result (3.87), which underlies the generalized S method, we find that

$$\max_{\{\tilde{\boldsymbol{\delta}}\in\tilde{\Delta}\}} |\hat{\sigma}^{-1}\tilde{\boldsymbol{\delta}}'(\hat{\boldsymbol{\alpha}} - \boldsymbol{\alpha})| \sim \max_{\{\tilde{\boldsymbol{\delta}}\in\tilde{\Delta}\}} |\tilde{\boldsymbol{\delta}}'\mathbf{t}|, \tag{3.156}$$

where $\mathbf{t} \sim MVt(N - P_*, \mathbf{I}_{M_*})$. This result can be used to devise (for each $\tilde{\boldsymbol{\delta}} \in \tilde{\Delta}$) an interval, say $\tilde{A}^*_{\tilde{\boldsymbol{\delta}}}$, of $\tilde{\boldsymbol{\delta}}'\boldsymbol{\alpha}$-values such that the probability of simultaneous coverage of the intervals $\tilde{A}^*_{\tilde{\boldsymbol{\delta}}}$ $(\tilde{\boldsymbol{\delta}} \in \tilde{\Delta})$, like that of the intervals $\tilde{A}_{\tilde{\boldsymbol{\delta}}}$ $(\tilde{\boldsymbol{\delta}} \in \tilde{\Delta})$, equals $1-\dot{\gamma}$. Letting $c^*_{\dot{\gamma}}$ represent the upper $100\dot{\gamma}\%$ point of the distribution of $\max_{\{\tilde{\boldsymbol{\delta}}\in\tilde{\Delta}\}} |\tilde{\boldsymbol{\delta}}'\mathbf{t}|$, this interval is as follows:

$$\tilde{A}^*_{\tilde{\delta}} = \{\dot{\tau} \in \mathcal{R}^1 \;:\; \tilde{\delta}'\hat{\alpha} - \hat{\sigma}c^*_{\dot{\gamma}} \le \dot{\tau} \le \tilde{\delta}'\hat{\alpha} + \hat{\sigma}c^*_{\dot{\gamma}}\}, \tag{3.157}$$

Now, for $\boldsymbol{\delta} \in \Delta$, let $A^*_{\boldsymbol{\delta}} = \tilde{A}^*_{\mathbf{W}\boldsymbol{\delta}}$, so that $A^*_{\boldsymbol{\delta}}$ is an interval of $\boldsymbol{\delta}'\boldsymbol{\tau}$-values expressible as follows:

$$A^*_{\boldsymbol{\delta}} = \{\dot{\tau} \in \mathcal{R}^1 \;:\; \boldsymbol{\delta}'\hat{\boldsymbol{\tau}} - \hat{\sigma}c^*_{\dot{\gamma}} \le \dot{\tau} \le \boldsymbol{\delta}'\hat{\boldsymbol{\tau}} + \hat{\sigma}c^*_{\dot{\gamma}}\}. \tag{3.158}$$

Like the intervals $A_{\boldsymbol{\delta}}$ ($\boldsymbol{\delta} \in \Delta$) associated with the generalized S method, the probability of simultaneous coverage of the intervals $A^*_{\boldsymbol{\delta}}$ ($\boldsymbol{\delta} \in \Delta$) equals $1-\dot{\gamma}$. Unlike the intervals $A_{\boldsymbol{\delta}}$ ($\boldsymbol{\delta} \in \Delta$), the intervals $A^*_{\boldsymbol{\delta}}$ ($\boldsymbol{\delta} \in \Delta$) are all of the same width—each of them is of width $2\hat{\sigma}c^*_{\dot{\gamma}}$.

In practice, $c^*_{\dot{\gamma}}$ may have to be approximated by numerical means—refer, e.g., to Liu (2011, chaps. 3 & 7). If necessary, $c^*_{\dot{\gamma}}$ can be approximated by adopting a Monte Carlo approach in which repeated draws are made from the distribution of the random variable $\max_{\{\tilde{\delta}\in\tilde{\Delta}\}} |\tilde{\delta}'\mathbf{t}|$. Note that

$$\max_{\{\tilde{\delta}\in\tilde{\Delta}\}} |\tilde{\delta}'\mathbf{t}| = \max\big[\max_{\{\tilde{\delta}\in\tilde{\Delta}\}} \tilde{\delta}'\mathbf{t},\; \max_{\{\tilde{\delta}\in\tilde{\Delta}\}} \tilde{\delta}'(-\mathbf{t})\big], \tag{3.159}$$

so that to compute the value of $\max_{\{\tilde{\delta}\in\tilde{\Delta}\}} |\tilde{\delta}'\mathbf{t}|$ for any particular value of $\mathbf{t}$, say $\dot{\mathbf{t}}$, it suffices to compute the value of $\max_{\{\tilde{\delta}\in\tilde{\Delta}\}} \tilde{\delta}'\mathbf{t}$ for each of two values of $\mathbf{t}$ (namely, $\mathbf{t} = \dot{\mathbf{t}}$ and $\mathbf{t} = -\dot{\mathbf{t}}$) and to then select the larger of these two values.

Suppose that the set Δ is such that the set $\tilde{\Delta}$ contains M_* linearly independent ($M_* \times 1$) vectors, say $\tilde{\delta}_1, \tilde{\delta}_2, \ldots, \tilde{\delta}_{M_*}$. Then, in much the same way that the intervals $\tilde{A}_{\tilde{\delta}}$ ($\tilde{\delta} \in \tilde{\Delta}$) define a set $\tilde{A}$ of $\boldsymbol{\alpha}$-values—refer to expression (3.122)—and a set A of $\boldsymbol{\tau}$-values, the intervals $\tilde{A}^*_{\tilde{\delta}}$ ($\tilde{\delta} \in \tilde{\Delta}$) define a set

$$\tilde{A}^* = \{\dot{\boldsymbol{\alpha}} \;:\; \tilde{\delta}'\hat{\boldsymbol{\alpha}} - \hat{\sigma}c^*_{\dot{\gamma}} \le \tilde{\delta}'\dot{\boldsymbol{\alpha}} \le \tilde{\delta}'\hat{\boldsymbol{\alpha}} + \hat{\sigma}c^*_{\dot{\gamma}} \text{ for every } \tilde{\delta} \in \tilde{\Delta}\}$$

of $\boldsymbol{\alpha}$-values and a corresponding set $A^* = \{\dot{\boldsymbol{\tau}} \;:\; \dot{\boldsymbol{\tau}} = \mathbf{W}'\dot{\boldsymbol{\alpha}},\ \dot{\boldsymbol{\alpha}} \in \tilde{A}^*\}$ of $\boldsymbol{\tau}$-values. And analogous to the set $\tilde{A}_{\tilde{\delta}}$ defined for $\tilde{\delta} \notin \tilde{\Delta}$ in terms of $\tilde{A}$ by expression (3.123), we have the set $\tilde{A}^*_{\tilde{\delta}}$ defined in terms of the set $\tilde{A}^*$ as follows:

$$\tilde{A}^*_{\tilde{\delta}} = \{\dot{\tau} \in \mathcal{R}^1 \;:\; \dot{\tau} = \tilde{\delta}'\dot{\boldsymbol{\alpha}},\ \dot{\boldsymbol{\alpha}} \in \tilde{A}^*\}. \tag{3.160}$$

Further, analogous to the set $A_{\boldsymbol{\delta}}$ (of $\boldsymbol{\delta}'\boldsymbol{\tau}$-values) defined for every $\boldsymbol{\delta} \in \mathcal{R}^M$ by the relationship $A_{\boldsymbol{\delta}} = \tilde{A}_{\mathbf{W}\boldsymbol{\delta}}$, we have the set $A^*_{\boldsymbol{\delta}}$ defined (for every $\boldsymbol{\delta} \in \mathcal{R}^M$) by the relationship $A^*_{\boldsymbol{\delta}} = \tilde{A}^*_{\mathbf{W}\boldsymbol{\delta}}$.

Two lemmas (on order statistics). Subsequently, use is made of the following two lemmas.

Lemma 7.3.4. Let $X_1, X_2, \ldots, X_K$ and X represent $K+1$ statistically independent random variables, each of which has the same absolutely continuous distribution. Further, let α represent a scalar between 0 and 1, let $R = (K+1)(1-\alpha)$, and suppose that K and α are such that R is an integer. And denote by $X_{[1]}, X_{[2]}, \ldots, X_{[K]}$ the first through Kth order statistics of $X_1, X_2, \ldots, X_K$ (so that with probability one, $X_{[1]} < X_{[2]} < \cdots < X_{[K]}$). Then,

$$\Pr\big(X > X_{[R]}\big) = \alpha.$$

Proof. Denote by R' the unique (with probability 1) integer such that X ranks R'th (in magnitude) among the $K+1$ random variables $X_1, X_2, \ldots, X_K, X$ (so that $X < X_{[1]}$, $X_{[R'-1]} < X < X_{[R']}$, or $X > X_{[K]}$, depending on whether $R' = 1$, $2 \le R' \le K$, or $R' = K+1$). Then, upon observing that $\Pr(R'=k) = 1/(K+1)$ for $k = 1, 2, \ldots, K+1$, we find that

$$\Pr\big(X > X_{[R]}\big) = \Pr(R' > R) = \textstyle\sum_{k=R+1}^{K+1} \Pr(R'=k) = \dfrac{K-R+1}{K+1} = \alpha.$$

Q.E.D.

Lemma 7.3.5. Let $X_1, X_2, \ldots, X_K$ represent K statistically independent random variables. Further, suppose that $X_k \sim X$ ($k = 1, 2, \ldots, K$) for some random variable X whose distribution is absolutely continuous with a cdf $G(\cdot)$ that is strictly increasing over some finite or infinite interval I for which $\Pr(X \in I) = 1$. And denote by $X_{[1]}, X_{[2]}, \ldots, X_{[K]}$ the first through Kth order statistics of $X_1, X_2, \ldots, X_K$. Then, for any integer R betweeen 1 and K, inclusive,

$$G\big(X_{[R]}\big) \sim Be(R,\; K-R+1).$$

Proof. Let $U_k = G(X_k)$ $(k = 1, 2, \ldots, K)$. Then, $U_1, U_2, \ldots, U_K$ are statistically independent random variables, and each of them is distributed uniformly on the interval (0, 1)—refer, e.g., to Theorems 4.3.5 and 2.1.10 of Casella and Berger (2002).

Now, denote by $U_{[1]}, U_{[2]}, \ldots, U_{[K]}$ the first through Kth order statistics of $U_1, U_2, \ldots, U_K$. And observe that (with probability 1)

$$U_{[k]} = G\big(X_{[k]}\big) \quad (k = 1, 2, \ldots, K).$$

Thus, for any scalar u between 0 and 1 (and for $R = 1, 2, \ldots, K$),

$$\Pr\big[G\big(X_{[R]}\big) \le u\big] = \Pr\big(U_{[R]} \le u\big) = \textstyle\sum_{k=R}^{K} \binom{K}{k}\, u^k (1-u)^{K-k},$$

so that the distribution of $G\big(X_{[R]}\big)$ is an absolutely continuous distribution with a pdf $f(\cdot)$, where

$$f(u) = \begin{cases} \dfrac{d \sum_{k=R}^{K} \binom{K}{k}\, u^k (1-u)^{K-k}}{du}, & \text{if } 0 < u < 1, \\ 0, & \text{otherwise.} \end{cases} \tag{3.161}$$

Moreover, for $0 < u < 1$,

$$\frac{d \sum_{k=R}^{K} \binom{K}{k}\, u^k (1-u)^{K-k}}{du} = \frac{K!}{(R-1)!\,(K-R)!}\, u^{R-1} (1-u)^{K-R}, \tag{3.162}$$

as can be readily verified via a series of steps that are essentially the same as those taken by Casella and Berger (2002) in completing the proof of their Theorem 5.4.4. Based on result (3.162), we conclude that $f(\cdot)$ is the pdf of a $Be(R,\ K-R+1)$ distribution and hence that $G\big(X_{[R]}\big) \sim Be(R,\ K-R+1)$. Q.E.D.

Monte Carlo approximation. For purposes of devising a Monte Carlo approximation to $c_{\dot\gamma}$ or $c^*_{\dot\gamma}$, let $X = \max_{\{\tilde{\boldsymbol\delta} \in \tilde\Delta:\ \tilde{\boldsymbol\delta} \neq \mathbf{0}\}} |\tilde{\boldsymbol\delta}'\mathbf{t}| / (\tilde{\boldsymbol\delta}'\tilde{\boldsymbol\delta})^{1/2}$ (in the case of $c_{\dot\gamma}$) or $X = \max_{\{\tilde{\boldsymbol\delta} \in \tilde\Delta\}} |\tilde{\boldsymbol\delta}'\mathbf{t}|$ (in the case of $c^*_{\dot\gamma}$). Further, let $X_1, X_2, \ldots, X_K$ represent K statistically independent random variables, each of which has the same distribution as X, let $R = (K+1)(1-\dot\gamma)$, suppose that K and $\dot\gamma$ are such that R is an integer (which implies that $\dot\gamma$ is a rational number), and denote by $X_{[1]}, X_{[2]}, \ldots, X_{[K]}$ the first through Kth order statistics of $X_1, X_2, \ldots, X_K$. And observe that (as a consequence of Lemma 7.3.4)

$$\Pr\big(X > X_{[R]}\big) = \dot\gamma. \tag{3.163}$$

To obtain a Monte Carlo approximation to $c_{\dot\gamma}$ or $c^*_{\dot\gamma}$, make K draws, say $x_1, x_2, \ldots, x_K$, from the distribution of X. And letting $x_{[1]}, x_{[2]}, \ldots, x_{[K]}$ represent the values of X obtained upon rearranging $x_1, x_2, \ldots, x_K$ in increasing order from smallest to largest, take the Monte Carlo approximation to be as follows:

$$c_{\dot\gamma} \doteq x_{[R]} \quad \text{or} \quad c^*_{\dot\gamma} \doteq x_{[R]}. \tag{3.164}$$

Clearly, $x_1, x_2, \ldots, x_K$ can be regarded as realizations of the random variables $X_1, X_2, \ldots, X_K$ (and $x_{[1]}, x_{[2]}, \ldots, x_{[K]}$ as realizations of the random variables $X_{[1]}, X_{[2]}, \ldots, X_{[K]}$). Conceivably (and conceptually), the realizations $x_1, x_2, \ldots, x_K$ of $X_1, X_2, \ldots, X_K$ could be included (along with the realizations of the elements of $\mathbf{y}$) in what are regarded as the data. Then, in repeated sampling from the joint distribution of $X_1, X_2, \ldots, X_K$ and $\mathbf{y}$, the probability of simultaneous coverage of the intervals $A_{\boldsymbol\delta}$ $(\boldsymbol\delta \in \Delta)$ or $A^*_{\boldsymbol\delta}$ $(\boldsymbol\delta \in \Delta)$ when $X_{[R]}$ is substituted for $c_{\dot\gamma}$ or $c^*_{\dot\gamma}$ is (exactly) $1-\dot\gamma$, as is evident from result (3.163).

When $x_{[R]}$ is substituted for $c_{\dot\gamma}$ or $c^*_{\dot\gamma}$ and when the repeated sampling is restricted to the distribution of $\mathbf{y}$, the probability of simultaneous coverage of the intervals $A_{\boldsymbol\delta}$ $(\boldsymbol\delta \in \Delta)$ or $A^*_{\boldsymbol\delta}$ $(\boldsymbol\delta \in \Delta)$ is $G\big(x_{[R]}\big)$, where $G(\cdot)$ is the cdf of the random variable X. The difference between this probability and the specified probability is

$$G\big(x_{[R]}\big) - (1-\dot\gamma).$$

This difference can be regarded as the realization of the random variable

$$G(X_{[R]}) - (1-\dot{\gamma}).$$

The number K of draws from the distribution of X can be chosen so that for some specified "tolerance" $\epsilon > 0$ and some specified probability ω,

$$\Pr\big[\,\big|G(X_{[R]}) - (1-\dot{\gamma})\big| \le \epsilon\big] \ge \omega. \tag{3.165}$$

As an application of Lemma 7.3.5, we have that

$$G(X_{[R]}) \sim Be[(K+1)(1-\dot{\gamma}),\ (K+1)\dot{\gamma}].$$

Accordingly,

$$\mathrm{E}\big[G(X_{[R]})\big] = 1-\dot{\gamma} \quad \text{and} \quad \mathrm{var}\big[G(X_{[R]})\big] = \dot{\gamma}(1-\dot{\gamma})/(K+2),$$

as can be readily verified—refer to Exercise 6.4. Edwards and Berry (1987, p. 915) proposed (for purposes of deciding on a value for K and implicitly for R) an implementation of the criterion (3.165) in which the distribution of $G(X_{[R]})-(1-\dot{\gamma})$ is approximated by an $N[0,\ \dot{\gamma}(1-\dot{\gamma})/(K+2)]$ distribution—as $K \to \infty$, the pdf of the standardized random variable $[G(X_{[R]}) - (1-\dot{\gamma})]/\sqrt{\dot{\gamma}(1-\dot{\gamma})/(K+2)}$ tends to the pdf of the $N(0, 1)$ distribution (Johnson, Kotz, and Balakrishnan 1995, p. 240). When this implementation is adopted, the criterion (3.165) is replaced by the much simpler criterion

$$K \ge \dot{\gamma}(1-\dot{\gamma})(z_{(1-\omega)/2}/\epsilon)^2 - 2, \tag{3.166}$$

where (for any scalar α between 0 and 1) z_α is the upper $100\alpha\%$ point of the $N(0,\ 1)$ distribution. And the number K of draws is chosen to be such that inequality (3.166) is satisfied and such that $(K+1)(1-\dot{\gamma})$ is an integer.

Table 1 of Edwards and Berry gives choices for K that would be suitable if $\dot{\gamma}$ were taken to be .10, .05, or .01, ϵ were taken to be .01, .005, .002, or .001, and ω were taken to be .99—the entries in their Table 1 are for $K+1$. For example, if $\dot{\gamma}$ were taken to be .10, ϵ to be .001, and ω to be .99 [in which case $z_{(1-\omega)/2} = 2.5758$ and the right side of inequality (3.166) equals 597125], we could take $K = 599999$ [in which case $R = 600000 \times .90 = 540000$].

f. Confidence bands (as applied to the lettuce-yield data and in general)

Let us revisit the situation considered in Section 7.1 (and considered further in Subsection d of the present section), adopting the notation and terminology introduced therein. Accordingly, $\mathbf{u} = (u_1, u_2, \ldots, u_C)'$ is a column vector of C explanatory variables $u_1, u_2, \ldots, u_C$. And $\delta(\mathbf{u})$ is a function of $\mathbf{u}$ that defines a response surface and that is assumed to be of the form

$$\delta(\mathbf{u}) = \textstyle\sum_{j=1}^{P} \beta_j x_j(\mathbf{u}), \tag{3.167}$$

where $x_j(\mathbf{u})$ $(j = 1, 2, \ldots, P)$ are specified functions of $\mathbf{u}$ (and where $\beta_1, \beta_2, \ldots, \beta_P$ are unknown parameters). Further, the data are to be regarded as the realizations of the elements of an N-dimensional observable random column vector $\mathbf{y} = (y_1, y_2, \ldots, y_N)'$ that follows a G–M model with model matrix $\mathbf{X}$ whose ijth element is (for $i = 1, 2, \ldots, N$ and $j = 1, 2, \ldots, P$) $x_j(\mathbf{u}_i)$, where $\mathbf{u}_1, \mathbf{u}_2, \ldots, \mathbf{u}_N$ are the values of $\mathbf{u}$ corresponding to the first through Nth data points. Note that $\delta(\mathbf{u})$ is reexpressible in the form

$$\delta(\mathbf{u}) = [\mathbf{x}(\mathbf{u})]'\boldsymbol{\beta}, \tag{3.168}$$

where $\mathbf{x}(\mathbf{u}) = [x_1(\mathbf{u}), x_2(\mathbf{u}), \ldots, x_P(\mathbf{u})]'$ [and where $\boldsymbol{\beta} = (\beta_1, \beta_2, \ldots, \beta_P)'$]. In the case of the lettuce-yield data, $N = 20$, $C = 3$, u_1, u_2, and u_3 represent transformed amounts of Cu, Mo, and Fe, respectively, and among the choices for the function $\delta(\mathbf{u})$ are the first-, second-, and third-order polynomials (1.1), (1.2), and (1.3).

We may wish to make inferences about $\delta(\mathbf{u})$ for some or all values of $\mathbf{u}$. Assume that rank $\mathbf{X} = P$, in which case all of the elements $\beta_1, \beta_2, \ldots, \beta_P$ of $\boldsymbol{\beta}$ are estimable, and let $\hat{\boldsymbol{\beta}} = (\mathbf{X}'\mathbf{X})^{-1}\mathbf{X}'\mathbf{y}$, which is the least squares estimator of $\boldsymbol{\beta}$. Further, let $\hat{\delta}(\mathbf{u}) = [\mathbf{x}(\mathbf{u})]'\hat{\boldsymbol{\beta}}$. Then, for any particular value of $\mathbf{u}$,

$\delta(\mathbf{u})$ is estimable, $\hat{\delta}(\mathbf{u})$ is the least squares estimator of $\delta(\mathbf{u})$, $\text{var}[\hat{\delta}(\mathbf{u})] = \sigma^2[\mathbf{x}(\mathbf{u})]'(\mathbf{X}'\mathbf{X})^{-1}\mathbf{x}(\mathbf{u})$, and $\text{var}[\hat{\delta}(\mathbf{u})]$ is estimated unbiasedly by $\hat{\sigma}^2[\mathbf{x}(\mathbf{u})]'(\mathbf{X}'\mathbf{X})^{-1}\mathbf{x}(\mathbf{u})$ (where $\hat{\sigma}^2$ is the usual unbiased estimator of σ^2). In the case of the lettuce-yield data, the assumption that rank $\mathbf{X} = P$ is satisfied when $\delta(\mathbf{u})$ is taken to be the second-order polynomial (1.2) [though not when it is taken to be the third-order polynomial (1.3)], and [when $\delta(\mathbf{u})$ is taken to be the second-order polynomial (1.2)] $\hat{\delta}(\mathbf{u})$ is as depicted in Figure 7.1.

Suppose that inferences are to be made about $\delta(\mathbf{u})$ for every value of $\mathbf{u}$ in some subspace $\mathcal{U}$ of C-dimensional space [assuming that (at least for $\mathbf{u} \in \mathcal{U}$) $\delta(\mathbf{u})$ is of the form (3.167) or (3.168) and $\mathbf{x}(\mathbf{u})$ is nonnull]. In addition to obtaining a point estimate of $\delta(\mathbf{u})$ for every $\mathbf{u} \in \mathcal{U}$, we may wish to obtain a confidence interval for every such $\mathbf{u}$. With that in mind, assume that the distribution of the vector $\mathbf{e}$ of residual effects in the G–M model is MVN or is some other absolutely continuous spherical distribution (with mean $\mathbf{0}$ and variance-covariance matrix $\sigma^2\mathbf{I}$).

Some alternative procedures. As a one-at-a-time $100(1-\dot{\gamma})\%$ confidence interval for the value of $\delta(\mathbf{u})$ corresponding to any particular value of $\mathbf{u}$, we have the interval with end points

$$\hat{\delta}(\mathbf{u}) \pm \{[\mathbf{x}(\mathbf{u})]'(\mathbf{X}'\mathbf{X})^{-1}\mathbf{x}(\mathbf{u})\}^{1/2}\hat{\sigma}\,\bar{t}_{\dot{\gamma}/2}(N-P) \tag{3.169}$$

—refer to result (3.55). When an interval, say interval $I_{\mathbf{u}}^{(1)}(\mathbf{y})$, is obtained for every $\mathbf{u} \in \mathcal{U}$ by taking the end points of the interval to be those given by expression (3.169), the probability of simultaneous coverage $\Pr[\delta(\mathbf{u}) \in I_{\mathbf{u}}^{(1)}(\mathbf{y})$ for every $\mathbf{u} \in \mathcal{U}]$ is less than $1-\dot{\gamma}$—typically, it is much less than $1-\dot{\gamma}$.

At the opposite extreme from interval $I_{\mathbf{u}}^{(1)}(\mathbf{y})$ is the interval, say interval $I_{\mathbf{u}}^{(2)}(\mathbf{y})$, with end points

$$\hat{\delta}(\mathbf{u}) \pm \{[\mathbf{x}(\mathbf{u})]'(\mathbf{X}'\mathbf{X})^{-1}\mathbf{x}(\mathbf{u})\}^{1/2}\hat{\sigma}\,[P\,\bar{F}_{\dot{\gamma}}(P,\,N-P)]^{1/2}. \tag{3.170}$$

This interval is that obtained {for the linear combination $[\mathbf{x}(\mathbf{u})]'\boldsymbol{\beta}$} when the (ordinary) S method is used to obtain confidence intervals for every linear combination of $\beta_1, \beta_2, \ldots, \beta_P$ such that the probability of simultaneous coverage of the entire collection of intervals equals $1-\dot{\gamma}$. When attention is restricted to the confidence intervals for those of the linear combinations that are expressible in the form $[\mathbf{x}(\mathbf{u})]'\boldsymbol{\beta}$ for some $\mathbf{u} \in \mathcal{U}$, the probability of simultaneous coverage $\Pr[\delta(\mathbf{u}) \in I_{\mathbf{u}}^{(2)}(\mathbf{y})$ for every $\mathbf{u} \in \mathcal{U}]$ is greater than or equal to $1-\dot{\gamma}$. In fact, aside from special cases like those where every linear combination of $\beta_1, \beta_2, \ldots, \beta_P$ is expressible in the form $[\mathbf{x}(\mathbf{u})]'\boldsymbol{\beta}$ for some $\mathbf{u} \in \mathcal{U}$, the intervals $I_{\mathbf{u}}^{(2)}(\mathbf{y})$ ($\mathbf{u} \in \mathcal{U}$) are conservative, that is, the probability of simultaneous coverage of these intervals exceeds $1-\dot{\gamma}$. The extent to which these intervals are conservative depends on the space $\mathcal{U}$ as well as on the functional form of the elements of the vector $\mathbf{x}(\mathbf{u})$; they are less conservative when $\mathcal{U} = \mathcal{R}^C$ than when $\mathcal{U}$ is a proper subset of $\mathcal{R}^C$.

Intermediate to intervals $I_{\mathbf{u}}^{(1)}(\mathbf{y})$ and $I_{\mathbf{u}}^{(2)}(\mathbf{y})$ is the interval, say interval $I_{\mathbf{u}}^{(3)}(\mathbf{y})$, with end points

$$\hat{\delta}(\mathbf{u}) \pm \{[\mathbf{x}(\mathbf{u})]'(\mathbf{X}'\mathbf{X})^{-1}\mathbf{x}(\mathbf{u})\}^{1/2}\hat{\sigma}\,c_{\dot{\gamma}}, \tag{3.171}$$

where [letting $\mathbf{W}$ represent any $P \times P$ matrix such that $(\mathbf{X}'\mathbf{X})^{-1} = \mathbf{W}'\mathbf{W}$ and letting $\mathbf{t}$ represent a $P \times 1$ random vector that has an $MVt(N-P,\,\mathbf{I}_P)$ distribution] $c_{\dot{\gamma}}$ is the upper $100\,\dot{\gamma}\%$ point of the distribution of the random variable

$$\max_{\{\mathbf{u}\in\mathcal{U}\}} \frac{|[\mathbf{x}(\mathbf{u})]'\mathbf{W}'\mathbf{t}|}{\{[\mathbf{x}(\mathbf{u})]'(\mathbf{X}'\mathbf{X})^{-1}\mathbf{x}(\mathbf{u})\}^{1/2}}.$$

The collection of intervals $I_{\mathbf{u}}^{(3)}(\mathbf{y})$ ($\mathbf{u} \in \mathcal{U}$) is such that the probability of simultaneous coverage $\Pr[\delta(\mathbf{u}) \in I_{\mathbf{u}}^{(3)}(\mathbf{y})$ for every $\mathbf{u} \in \mathcal{U}]$ equals $1-\dot{\gamma}$.

As $\mathbf{u}$ ranges over the space $\mathcal{U}$, the lower and upper end points of interval $I_{\mathbf{u}}^{(3)}(\mathbf{y})$ form surfaces that define what is customarily referred to as a *confidence band*. The probability of this band covering the true response surface (in its entirety) equals $1-\dot{\gamma}$. The lower and upper end points of intervals $I_{\mathbf{u}}^{(1)}(\mathbf{y})$ and $I_{\mathbf{u}}^{(2)}(\mathbf{y})$ also form surfaces that define confidence bands; the first of these confidence bands has a probability of coverage that is typically much less than $1-\dot{\gamma}$, and the second is typically quite conservative (i.e., has a probability of coverage that is typically considerably greater than $1-\dot{\gamma}$).

The width of the confidence band defined by the surfaces formed by the end points of interval $I_{\mathbf{u}}^{(3)}(\mathbf{y})$ [or by the end points of interval $I_{\mathbf{u}}^{(1)}(\mathbf{y})$ or $I_{\mathbf{u}}^{(2)}(\mathbf{y})$] varies from one point in the space $\mathcal{U}$ to another. Suppose that $\mathcal{U}$ is such that for every $P \times 1$ vector $\dot{\mathbf{t}}$, $\max_{\{\mathbf{u}\in\mathcal{U}\}} |[\mathbf{x}(\mathbf{u})]'\mathbf{W}'\dot{\mathbf{t}}|$ exists [as would be the case if the functions $x_1(\mathbf{u}), x_2(\mathbf{u}), \ldots, x_P(\mathbf{u})$ are continuous and the set $\mathcal{U}$ is closed and bounded]. Then, as an alternative to interval $I_{\mathbf{u}}^{(3)}(\mathbf{y})$ with end points (3.171), we have the interval, say interval $I_{\mathbf{u}}^{(4)}(\mathbf{y})$, with end points

$$\hat{\delta}(\mathbf{u}) \pm \hat{\sigma} c_{\dot{\gamma}}^*, \tag{3.172}$$

where $c_{\dot{\gamma}}^*$ is the upper $100\dot{\gamma}\%$ point of the distribution of the random variable $\max_{\{\mathbf{u}\in\mathcal{U}\}} |[\mathbf{x}(\mathbf{u})]'\mathbf{W}'\mathbf{t}|$. Like the end points of interval $I_{\mathbf{u}}^{(3)}(\mathbf{y})$, the end points of interval $I_{\mathbf{u}}^{(4)}(\mathbf{y})$ form surfaces that define a confidence band having a probability of coverage equal to $1-\dot{\gamma}$. Unlike the confidence band formed by the end points of interval $I_{\mathbf{u}}^{(3)}(\mathbf{y})$, this confidence band is of uniform width (equal to $2\hat{\sigma}c_{\dot{\gamma}}^*$).

The end points (3.171) of interval $I_{\mathbf{u}}^{(3)}(\mathbf{y})$ depend on $c_{\dot{\gamma}}$, and the end points (3.172) of interval $I_{\mathbf{u}}^{(4)}(\mathbf{y})$ depend on $c_{\dot{\gamma}}^*$. Except for relatively simple special cases, $c_{\dot{\gamma}}$ and $c_{\dot{\gamma}}^*$ have to be replaced by approximations obtained via Monte Carlo methods. The computation of these approximations requires (in the case of $c_{\dot{\gamma}}$) the computation of $\max_{\{\mathbf{u}\in\mathcal{U}\}} |[\mathbf{x}(\mathbf{u})]'\mathbf{W}'\mathbf{t}|/\{[\mathbf{x}(\mathbf{u})]'(\mathbf{X}'\mathbf{X})^{-1}\mathbf{x}(\mathbf{u})\}^{1/2}$ and (in the case of $c_{\dot{\gamma}}^*$) the computation of $\max_{\{\mathbf{u}\in\mathcal{U}\}} |[\mathbf{x}(\mathbf{u})]'\mathbf{W}'\mathbf{t}|$ for each of a large number of values of $\mathbf{t}$.

In light of results (3.154) and (3.155), we find that for any value, say $\dot{\mathbf{t}}$, of $\mathbf{t}$, $\max_{\{\mathbf{u}\in\mathcal{U}\}} |[\mathbf{x}(\mathbf{u})]'\mathbf{W}'\dot{\mathbf{t}}|/\{[\mathbf{x}(\mathbf{u})]'(\mathbf{X}'\mathbf{X})^{-1}\mathbf{x}(\mathbf{u})\}^{1/2}$ can be determined by finding values, say $\dot{\lambda}$ and $\dot{\mathbf{u}}$, of the scalar λ and the vector $\mathbf{u}$ that minimize the sum of squares

$$[\dot{\mathbf{t}} - \lambda \mathbf{W}\mathbf{x}(\mathbf{u})]'[\dot{\mathbf{t}} - \lambda \mathbf{W}\mathbf{x}(\mathbf{u})],$$

subject to the constraint $\mathbf{u} \in \mathcal{U}$, and by then observing that

$$\max_{\{\mathbf{u}\in\mathcal{U}\}} \frac{|[\mathbf{x}(\mathbf{u})]'\mathbf{W}'\dot{\mathbf{t}}|}{\{[\mathbf{x}(\mathbf{u})]'(\mathbf{X}'\mathbf{X})^{-1}\mathbf{x}(\mathbf{u})\}^{1/2}} = \{\dot{\mathbf{t}}'\dot{\mathbf{t}} - [\dot{\mathbf{t}} - \dot{\lambda}\mathbf{W}\mathbf{x}(\dot{\mathbf{u}})]'[\dot{\mathbf{t}} - \dot{\lambda}\mathbf{W}\mathbf{x}(\dot{\mathbf{u}})]\}^{1/2}. \tag{3.173}$$

And $\max_{\{\mathbf{u}\in\mathcal{U}\}} |[\mathbf{x}(\mathbf{u})]'\mathbf{W}'\dot{\mathbf{t}}|$ can be determined by finding the maximum values of $[\mathbf{x}(\mathbf{u})]'\mathbf{W}'\dot{\mathbf{t}}$ and $[\mathbf{x}(\mathbf{u})]'\mathbf{W}'(-\dot{\mathbf{t}})$ with respect to $\mathbf{u}$ (subject to the constraint $\mathbf{u} \in \mathcal{U}$) and by then observing that $\max_{\{\mathbf{u}\in\mathcal{U}\}} |[\mathbf{x}(\mathbf{u})]'\mathbf{W}'\dot{\mathbf{t}}|$ equals the larger of these two values.

An illustration. Let us illustrate the four alternative procedures for constructing confidence bands by applying them to the lettuce-yield data. In this application (which adds to the results obtained for these data in Sections 7.1and 7.2c and in Subsection d of the present section), $N = 20$, $C = 3$, and u_1, u_2, and u_3 represent transformed amounts of Cu, Mo, and Fe, respectively. Further, let us take $\delta(\mathbf{u})$ to be the second-order polynomial (1.2), in which case rank $\mathbf{X} = P = 10$ (and N–rank $\mathbf{X} = N$–$P = 10$). And let us take $\dot{\gamma} = .10$ and take $\mathcal{U}$ to be the rectangular region defined by imposing on u_1, u_2, and u_3 upper and lower bounds as follows: $-1 \le u_i \le 1$ $(i = 1, 2, 3)$—the determination of the constants $c_{\dot{\gamma}}$ and $c_{\dot{\gamma}}^*$ needed to construct confidence bands $I_{\mathbf{u}}^{(3)}(\mathbf{y})$ and $I_{\mathbf{u}}^{(4)}(\mathbf{y})$ is considerably more straightforward when $\mathcal{U}$ is rectangular than when, e.g., it is spherical.

The values of the constants $\bar{t}_{.05}(10)$ and $[10\bar{F}_{.10}(10, 10)]^{1/2}$ needed to construct confidence bands $I_{\mathbf{u}}^{(1)}(\mathbf{y})$ and $I_{\mathbf{u}}^{(2)}(\mathbf{y})$ can be determined via well-known numerical methods and are readily available from multiple sources. They are as follows: $\bar{t}_{.05}(10) = 1.812461$ and $[10\bar{F}_{.10}(10, 10)]^{1/2} = 4.819340$.

In constructing confidence bands $I_{\mathbf{u}}^{(3)}(\mathbf{y})$ and $I_{\mathbf{u}}^{(4)}(\mathbf{y})$, resort was made to the approximate versions of those bands obtained upon replacing $c_{\dot{\gamma}}$ and $c_{\dot{\gamma}}^*$ with Monte Carlo approximations. The Monte Carlo approximations were determined from $K = 599999$ draws with the following results: $c_{.10} \doteq 3.448802$ and $c_{.10}^* \doteq 2.776452$—refer to the final 3 paragraphs of Subsection e for some discussion relevant to the "accuracy" of these approximations. Note that the approximation to $c_{.10}$ is considerably greater than $\bar{t}_{.05}(10)$ and significantly smaller than $[10\bar{F}_{.10}(10, 10)]^{1/2}$. If $\mathcal{U}$ had been taken to be $\mathcal{U} = \mathbb{R}^3$ rather than the rectangular region $\mathcal{U} = \{\mathbf{u} : |u_i| \le 1\ (i = 1, 2, 3)\}$, the Monte

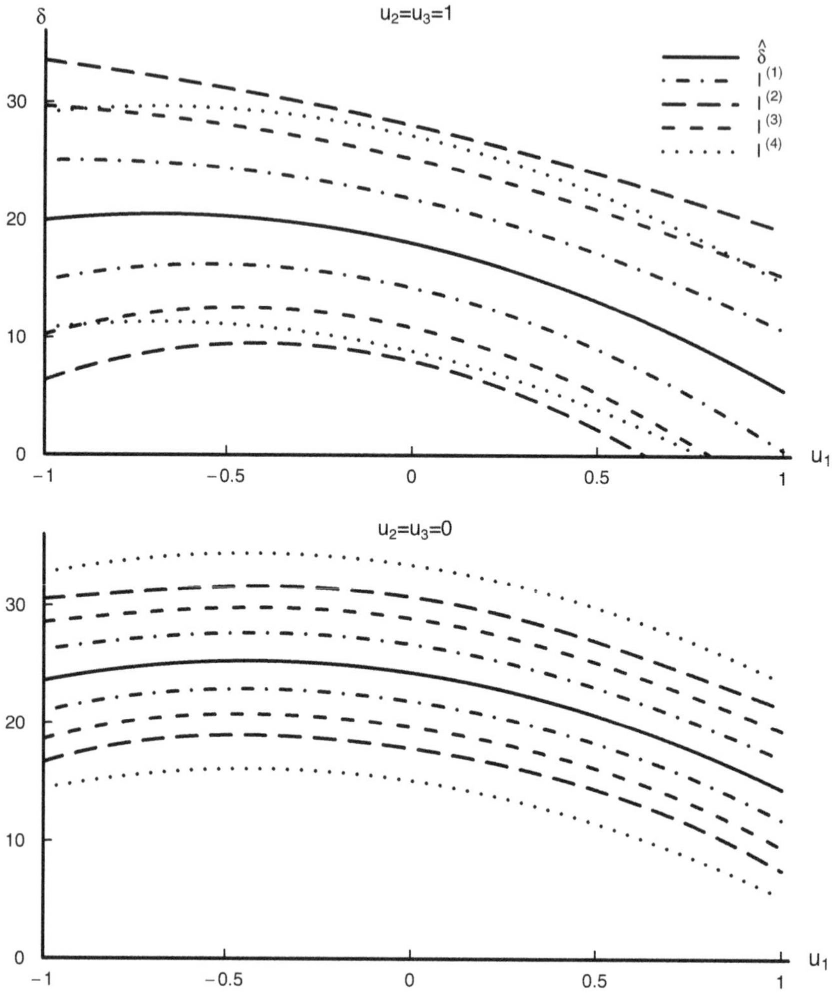

FIGURE 7.4. Two segments of the confidence bands $I_{\mathbf{u}}^{(1)}(\mathbf{y})$, $I_{\mathbf{u}}^{(2)}(\mathbf{y})$, $I_{\mathbf{u}}^{(3)}(\mathbf{y})$, and $I_{\mathbf{u}}^{(4)}(\mathbf{y})$ [and of the estimated response surface $\hat{\delta}(\mathbf{u})$] when the data are taken to be the lettuce-yield data, when $\delta(\mathbf{u})$ is taken to be the second-order polynomial (1.2) (where u_1, u_2, and u_3 are the transformed amounts of Cu, Mo, and Fe, respectively), when $\mathcal{U} = \{\mathbf{u} : |u_i| \le 1\ (i = 1, 2, 3)\}$, when $\dot{\gamma} = .10$, and when $c_{.10}$ and $c_{.10}^*$ are replaced by the Monte Carlo approximations $c_{.10} \doteq 3.448802$ and $c_{.10}^* \doteq 2.776452$.

Carlo approximation to $c_{.10}$ would have been $c_{.10} \doteq 3.520382$ (rather than $c_{.10} \doteq 3.448802$). The difference between the two Monte Carlo approximations to $c_{.10}$ is relatively small, suggesting that most of the difference between $c_{.10}$ and $[10\bar{F}_{.10}(10, 10)]^{1/2}$ is accounted for by the restriction of $\delta(\mathbf{u})$ to the form of the second-order polynomial (1.2) rather than the restriction of $\mathbf{u}$ to the region $\{\mathbf{u} : |u_i| \le 1\ (i = 1, 2, 3)\}$.

Segments of the various confidence bands [and of the estimated response surface $\hat{\delta}(\mathbf{u})$] are depicted in Figure 7.4; the segments depicted in the first plot are those for $\mathbf{u}$-values such that $u_2 = u_3 = 1$, and the segments depicted in the second plot are those for $\mathbf{u}$-values such that $u_2 = u_3 = 0$.

7.4 Some Optimality Properties

Suppose that $\mathbf{y}$ is an $N \times 1$ observable random vector that follows the G–M model. Suppose also that the distribution of the vector $\mathbf{e}$ of residual effects in the G–M model is $N(\mathbf{0}, \sigma^2\mathbf{I})$; or, more generally, suppose that the distribution of the vector $\mathbf{u} = \sigma^{-1}\mathbf{e}$ of standardized residual effects is an absolutely continuous distribution with a pdf $h(\cdot)$ of the form $h(\mathbf{u}) \propto g(\mathbf{u}'\mathbf{u})$, where $g(\cdot)$ is a known function such that $\int_0^\infty s^{N-1}g(s^2)\,ds < \infty$. Further, adopt the notation and terminology of Section 7.3. And partition $\mathbf{u}$ into subvectors $\mathbf{u}_1$, $\mathbf{u}_2$, and $\mathbf{u}_3$ (conformally with a partitioning of $\mathbf{z}$ into the subvectors $\mathbf{z}_1, \mathbf{z}_2$, and $\mathbf{z}_3$), so that

$$\mathbf{z} = \begin{pmatrix} \mathbf{z}_1 \\ \mathbf{z}_2 \\ \mathbf{z}_3 \end{pmatrix} \sim \begin{pmatrix} \boldsymbol{\alpha} + \sigma\mathbf{u}_1 \\ \boldsymbol{\eta} + \sigma\mathbf{u}_2 \\ \sigma\mathbf{u}_3 \end{pmatrix}.$$

Let us consider further the problem of testing the null hypothesis $H_0 : \boldsymbol{\tau} = \boldsymbol{\tau}^{(0)}$ or $\tilde{H}_0 : \boldsymbol{\alpha} = \boldsymbol{\alpha}^{(0)}$ versus the alternative hypothesis $H_1 : \boldsymbol{\tau} \neq \boldsymbol{\tau}^{(0)}$ or $\tilde{H}_1 : \boldsymbol{\alpha} \neq \boldsymbol{\alpha}^{(0)}$. Let us also consider further the problem of forming a confidence set for $\boldsymbol{\tau}$ or $\boldsymbol{\alpha}$.

Among the procedures for testing H_0 or $\tilde{H}_0$ versus H_1 or $\tilde{H}_1$ is the size-$\dot{\gamma}$ F test, with critical region C_F or $\tilde{C}_F$ and with critical (test) function $\phi_F(\mathbf{y})$ or $\tilde{\phi}_F(\mathbf{z})$. Corresponding to the size-$\dot{\gamma}$ F test are the $100(1-\dot{\gamma})\%$ confidence set A_F for $\boldsymbol{\tau}$ and the $100(1-\dot{\gamma})\%$ confidence set $\tilde{A}_F$ for $\boldsymbol{\alpha}$. In Section 7.3b, it was established that the size-$\dot{\gamma}$ F test and the corresponding $100(1-\dot{\gamma})\%$ confidence sets have various seemingly desirable properties. These properties serve to define certain classes of test procedures and certain classes of procedures for forming confidence sets. In what follows, the focus is on obtaining useful characterizations of these classes and on establishing the optimality of the F test and the corresponding confidence sets within these classes.

a. Some results on invariance

Denote by $\tilde{\phi}(\mathbf{z})$ or (when convenient) by $\tilde{\phi}(\mathbf{z}_1, \mathbf{z}_2, \mathbf{z}_3)$ an arbitrary function of the random vector $\mathbf{z}$ $(= \mathbf{O}'\mathbf{y})$. And let $\tilde{T}(\mathbf{z})$ represent a transformation (of $\mathbf{z}$) that satisfies the condition (3.57), and let $\tilde{G}$ represent a group of such transformations. Then, as discussed earlier [in the special case where $\tilde{\phi}(\cdot)$ is the critical function of a size-$\dot{\gamma}$ test of $\tilde{H}_0$ versus $\tilde{H}_1$], $\tilde{\phi}(\cdot)$ is said to be invariant with respect to $\tilde{G}$ if

$$\tilde{\phi}[\tilde{T}(\mathbf{z})] = \tilde{\phi}(\mathbf{z})$$

for every transformation $\tilde{T}(\cdot)$ in $\tilde{G}$ (and for every value of $\mathbf{z}$). What can be discerned about the characteristics and the distribution of $\tilde{\phi}(\mathbf{z})$ when the function $\tilde{\phi}(\cdot)$ is invariant with respect to various of the groups introduced in the final two parts of Section 7.2b?

In what follows, the primary results are presented as a series of propositions. Each of these propositions (after the first) builds on its predecessor. The verification of the propositions is deferred until the presentation of all of the propositions is complete.

The propositions are as follows:

(1) The function $\tilde{\phi}(\mathbf{z})$ is invariant with respect to the group $\tilde{G}_1$ of transformations of the form $\tilde{T}_1(\mathbf{z})$ if and only if $\tilde{\phi}(\mathbf{z}) = \tilde{\phi}_1(\mathbf{z}_1, \mathbf{z}_3)$ for some function $\tilde{\phi}_1(\cdot,\cdot)$—$\mathbf{z}_1$ and $\mathbf{z}_3$ form what is commonly referred to (e.g., Lehmann and Romano 2005b, chap. 6) as a maximal invariant. Moreover, the joint distribution of $\mathbf{z}_1$ and $\mathbf{z}_3$ depends on $\boldsymbol{\alpha}$, $\boldsymbol{\eta}$, and σ only through the values of $\boldsymbol{\alpha}$ and σ.

(2) The function $\tilde{\phi}(\mathbf{z})$ is invariant with respect to the group $\tilde{G}_3(\boldsymbol{\alpha}^{(0)})$ of transformations of the form $\tilde{T}_3(\mathbf{z};\boldsymbol{\alpha}^{(0)})$ as well as the group $\tilde{G}_1$ of transformations of the form $\tilde{T}_1(\mathbf{z})$ if and only if $\tilde{\phi}(\mathbf{z}) = \tilde{\phi}_{13}[(\mathbf{z}_1-\boldsymbol{\alpha}^{(0)})'(\mathbf{z}_1-\boldsymbol{\alpha}^{(0)})+\mathbf{z}_3'\mathbf{z}_3,\ \mathbf{z}_3]$ for some function $\tilde{\phi}_{13}(\cdot,\cdot)$. Moreover, the joint distribution of $(\mathbf{z}_1-\boldsymbol{\alpha}^{(0)})'(\mathbf{z}_1-\boldsymbol{\alpha}^{(0)})+\mathbf{z}_3'\mathbf{z}_3$ and $\mathbf{z}_3$ depends on $\boldsymbol{\alpha}$, $\boldsymbol{\eta}$, and σ only through the values of $(\boldsymbol{\alpha}-\boldsymbol{\alpha}^{(0)})'(\boldsymbol{\alpha}-\boldsymbol{\alpha}^{(0)})$ and σ.

(3) The function $\tilde{\phi}(\mathbf{z})$ is invariant with respect to the group $\tilde{G}_2(\boldsymbol{\alpha}^{(0)})$ of transformations of the form $\tilde{T}_2(\mathbf{z};\boldsymbol{\alpha}^{(0)})$ as well as the groups $\tilde{G}_1$ and $\tilde{G}_3(\boldsymbol{\alpha}^{(0)})$ of transformations of the form $\tilde{T}_1(\mathbf{z})$ and $\tilde{T}_3(\mathbf{z};\boldsymbol{\alpha}^{(0)})$ only if there exists a function $\tilde{\phi}_{132}(\cdot)$ [of an $(N-P_*)$-dimensional vector] such that

$$\tilde{\phi}(\mathbf{z}) = \tilde{\phi}_{132}\{[(\mathbf{z}_1-\boldsymbol{\alpha}^{(0)})'(\mathbf{z}_1-\boldsymbol{\alpha}^{(0)})+\mathbf{z}_3'\mathbf{z}_3]^{-1/2}\mathbf{z}_3\}$$

for those values of $\mathbf{z}$ for which $\mathbf{z}_3 \neq \mathbf{0}$. Moreover, $\Pr(\mathbf{z}_3 \neq \mathbf{0}) = 1$, and the distribution of $[(\mathbf{z}_1-\boldsymbol{\alpha}^{(0)})'(\mathbf{z}_1-\boldsymbol{\alpha}^{(0)})+\mathbf{z}_3'\mathbf{z}_3]^{-1/2}\mathbf{z}_3$ depends on $\boldsymbol{\alpha}$, $\boldsymbol{\eta}$, and σ only through the value of $(\boldsymbol{\alpha}-\boldsymbol{\alpha}^{(0)})'(\boldsymbol{\alpha}-\boldsymbol{\alpha}^{(0)})/\sigma^2$.

(4) The function $\tilde{\phi}(\mathbf{z})$ is invariant with respect to the group $\tilde{G}_4$ of transformations of the form $\tilde{T}_4(\mathbf{z})$ as well as the groups $\tilde{G}_1$, $\tilde{G}_3(\boldsymbol{\alpha}^{(0)})$, and $\tilde{G}_2(\boldsymbol{\alpha}^{(0)})$ of transformations of the form $\tilde{T}_1(\mathbf{z})$, $\tilde{T}_3(\mathbf{z};\boldsymbol{\alpha}^{(0)})$, and $\tilde{T}_2(\mathbf{z};\boldsymbol{\alpha}^{(0)})$ only if there exists a function $\tilde{\phi}_{1324}(\cdot)$ (of a single variable) such that

$$\tilde{\phi}(\mathbf{z}) = \tilde{\phi}_{1324}\{\mathbf{z}_3'\mathbf{z}_3/[(\mathbf{z}_1-\boldsymbol{\alpha}^{(0)})'(\mathbf{z}_1-\boldsymbol{\alpha}^{(0)})+\mathbf{z}_3'\mathbf{z}_3]\}$$

for those values of $\mathbf{z}$ for which $\mathbf{z}_3 \neq \mathbf{0}$. Moreover, $\Pr(\mathbf{z}_3 \neq \mathbf{0}) = 1$, and the distribution of $\mathbf{z}_3'\mathbf{z}_3/[(\mathbf{z}_1-\boldsymbol{\alpha}^{(0)})'(\mathbf{z}_1-\boldsymbol{\alpha}^{(0)})+\mathbf{z}_3'\mathbf{z}_3]$ depends on $\boldsymbol{\alpha}$, $\boldsymbol{\eta}$, and σ only through the value of $(\boldsymbol{\alpha}-\boldsymbol{\alpha}^{(0)})'(\boldsymbol{\alpha}-\boldsymbol{\alpha}^{(0)})/\sigma^2$.

Verification of Proposition (1). If $\tilde{\phi}(\mathbf{z}) = \tilde{\phi}_1(\mathbf{z}_1,\mathbf{z}_3)$ for some function $\tilde{\phi}_1(\cdot,\cdot)$, then it is clear that [for every transformation $\tilde{T}_1(\cdot)$ in the group $\tilde{G}_1$] $\tilde{\phi}[\tilde{T}_1(\mathbf{z})] = \tilde{\phi}_1(\mathbf{z}_1,\mathbf{z}_3) = \tilde{\phi}(\mathbf{z})$ and hence that the function $\tilde{\phi}(\mathbf{z})$ is invariant with respect to the group $\tilde{G}_1$. Conversely, suppose that $\tilde{\phi}(\mathbf{z})$ is invariant with respect to the group $\tilde{G}_1$. Then, for every choice of the vector $\mathbf{c}$,

$$\tilde{\phi}(\mathbf{z}_1,\mathbf{z}_2+\mathbf{c},\mathbf{z}_3) = \tilde{\phi}[\tilde{T}_1(\mathbf{z})] = \tilde{\phi}(\mathbf{z}_1,\mathbf{z}_2,\mathbf{z}_3).$$

And upon setting $\mathbf{c} = -\mathbf{z}_2$, we find that $\tilde{\phi}(\mathbf{z}) = \tilde{\phi}_1(\mathbf{z}_1,\mathbf{z}_3)$, where $\tilde{\phi}_1(\mathbf{z}_1,\mathbf{z}_3) = \tilde{\phi}(\mathbf{z}_1,\mathbf{0},\mathbf{z}_3)$. Moreover, the joint distribution of $\mathbf{z}_1$ and $\mathbf{z}_3$ does not depend on $\boldsymbol{\eta}$, as is evident upon observing that

$$\begin{pmatrix}\mathbf{z}_1\\ \mathbf{z}_3\end{pmatrix} \sim \begin{pmatrix}\boldsymbol{\alpha}+\sigma\mathbf{u}_1\\ \sigma\mathbf{u}_3\end{pmatrix}.$$

Verification of Proposition (2). Suppose that $\tilde{\phi}(\mathbf{z}) = \tilde{\phi}_{13}[(\mathbf{z}_1-\boldsymbol{\alpha}^{(0)})'(\mathbf{z}_1-\boldsymbol{\alpha}^{(0)})+\mathbf{z}_3'\mathbf{z}_3,\ \mathbf{z}_3]$ for some function $\tilde{\phi}_{13}(\cdot,\cdot)$. Then, $\tilde{\phi}(\mathbf{z})$ is invariant with respect to the group $\tilde{G}_1$ of transformations, as is evident from Proposition (1). And it is also invariant with respect to the group $\tilde{G}_3(\boldsymbol{\alpha}^{(0)})$, as is evident upon observing that

$$\begin{aligned}\tilde{\phi}[\tilde{T}_3(\mathbf{z};\boldsymbol{\alpha}^{(0)})] &= \tilde{\phi}_{13}[(\mathbf{z}_1-\boldsymbol{\alpha}^{(0)})'\mathbf{P}\mathbf{P}'(\mathbf{z}_1-\boldsymbol{\alpha}^{(0)})+\mathbf{z}_3'\mathbf{z}_3,\ \mathbf{z}_3]\\ &= \tilde{\phi}_{13}[(\mathbf{z}_1-\boldsymbol{\alpha}^{(0)})'(\mathbf{z}_1-\boldsymbol{\alpha}^{(0)})+\mathbf{z}_3'\mathbf{z}_3,\ \mathbf{z}_3] = \tilde{\phi}(\mathbf{z}).\end{aligned}$$

Conversely, suppose that $\tilde{\phi}(\mathbf{z})$ is invariant with respect to the group $\tilde{G}_3(\boldsymbol{\alpha}^{(0)})$ as well as the group $\tilde{G}_1$. Then, in light of Proposition (1), $\tilde{\phi}(\mathbf{z}) = \tilde{\phi}_1(\mathbf{z}_1,\mathbf{z}_3)$ for some function $\tilde{\phi}_1(\cdot,\cdot)$. And to establish that $\tilde{\phi}(\mathbf{z}) = \tilde{\phi}_{13}[(\mathbf{z}_1-\boldsymbol{\alpha}^{(0)})'(\mathbf{z}_1-\boldsymbol{\alpha}^{(0)})+\mathbf{z}_3'\mathbf{z}_3,\ \mathbf{z}_3]$ for some function $\tilde{\phi}_{13}(\cdot,\cdot)$, it suffices to observe that corresponding to any two values $\dot{\mathbf{z}}_1$ and $\ddot{\mathbf{z}}_1$ of $\mathbf{z}_1$ that satisfy the equality $(\ddot{\mathbf{z}}_1-\boldsymbol{\alpha}^{(0)})'(\ddot{\mathbf{z}}_1-\boldsymbol{\alpha}^{(0)}) = (\dot{\mathbf{z}}_1-\boldsymbol{\alpha}^{(0)})'(\dot{\mathbf{z}}_1-\boldsymbol{\alpha}^{(0)})$, there is an orthogonal matrix $\mathbf{P}$ such that $\ddot{\mathbf{z}}_1 = \boldsymbol{\alpha}^{(0)}+\mathbf{P}'(\dot{\mathbf{z}}_1-\boldsymbol{\alpha}^{(0)})$ (the existence of which follows from Lemma 5.9.9) and hence such that $\tilde{\phi}_1(\ddot{\mathbf{z}}_1,\mathbf{z}_3) = \tilde{\phi}_1[\boldsymbol{\alpha}^{(0)}+\mathbf{P}'(\dot{\mathbf{z}}_1-\boldsymbol{\alpha}^{(0)}),\ \mathbf{z}_3]$.

It remains to verify that the joint distribution of $(\mathbf{z}_1-\boldsymbol{\alpha}^{(0)})'(\mathbf{z}_1-\boldsymbol{\alpha}^{(0)})+\mathbf{z}_3'\mathbf{z}_3$ and $\mathbf{z}_3$ depends on $\boldsymbol{\alpha}$, $\boldsymbol{\eta}$, and σ only through the values of $(\boldsymbol{\alpha}-\boldsymbol{\alpha}^{(0)})'(\boldsymbol{\alpha}-\boldsymbol{\alpha}^{(0)})$ and σ. Denote by $\mathbf{O}$ an $M_*\times M_*$ orthogonal matrix defined as follows: if $\boldsymbol{\alpha}=\boldsymbol{\alpha}^{(0)}$, take $\mathbf{O}$ to be $\mathbf{I}_M$ or any other $M_*\times M_*$ orthogonal matrix; if $\boldsymbol{\alpha}\neq\boldsymbol{\alpha}^{(0)}$, take $\mathbf{O}$ to be the Helmert matrix whose first row is proportional to the vector $(\boldsymbol{\alpha}-\boldsymbol{\alpha}^{(0)})'$. Further, let $\tilde{\mathbf{u}}_1 = \mathbf{O}\mathbf{u}_1$, and take $\tilde{\mathbf{u}}$ to be the N-dimensional column vector whose transpose is $\tilde{\mathbf{u}}' = (\tilde{\mathbf{u}}_1', \mathbf{u}_2', \mathbf{u}_3')$. Then, upon observing that the joint distribution of $(\mathbf{z}_1-\boldsymbol{\alpha}^{(0)})'(\mathbf{z}_1-\boldsymbol{\alpha}^{(0)})+\mathbf{z}_3'\mathbf{z}_3$ and $\mathbf{z}_3$ is identical to that of the random variable $(\boldsymbol{\alpha}-\boldsymbol{\alpha}^{(0)}+\sigma\mathbf{u}_1)'(\boldsymbol{\alpha}-\boldsymbol{\alpha}^{(0)}+\sigma\mathbf{u}_1)+\sigma^2\mathbf{u}_3'\mathbf{u}_3$ and the random vector $\sigma\mathbf{u}_3$, observing that

$$
\begin{aligned}
&(\boldsymbol{\alpha}-\boldsymbol{\alpha}^{(0)}+\sigma\mathbf{u}_1)'(\boldsymbol{\alpha}-\boldsymbol{\alpha}^{(0)}+\sigma\mathbf{u}_1)+\sigma^2\mathbf{u}_3'\mathbf{u}_3\\
&\quad=(\boldsymbol{\alpha}-\boldsymbol{\alpha}^{(0)})'(\boldsymbol{\alpha}-\boldsymbol{\alpha}^{(0)})+2\sigma(\boldsymbol{\alpha}-\boldsymbol{\alpha}^{(0)})'\mathbf{u}_1+\sigma^2\mathbf{u}_1'\mathbf{u}_1+\sigma^2\mathbf{u}_3'\mathbf{u}_3\\
&\quad=(\boldsymbol{\alpha}-\boldsymbol{\alpha}^{(0)})'(\boldsymbol{\alpha}-\boldsymbol{\alpha}^{(0)})+2\sigma(\boldsymbol{\alpha}-\boldsymbol{\alpha}^{(0)})'\mathbf{O}'\mathbf{O}\mathbf{u}_1+\sigma^2\mathbf{u}_1'\mathbf{O}'\mathbf{O}\mathbf{u}_1+\sigma^2\mathbf{u}_3'\mathbf{u}_3\\
&\quad=(\boldsymbol{\alpha}-\boldsymbol{\alpha}^{(0)})'(\boldsymbol{\alpha}-\boldsymbol{\alpha}^{(0)})+\sigma^2(\tilde{\mathbf{u}}_1'\tilde{\mathbf{u}}_1+\mathbf{u}_3'\mathbf{u}_3)+2\sigma\{[(\boldsymbol{\alpha}-\boldsymbol{\alpha}^{(0)})'(\boldsymbol{\alpha}-\boldsymbol{\alpha}^{(0)})]^{1/2},0,0,\dots,0\}\tilde{\mathbf{u}}_1,
\end{aligned}
$$

and observing that $\tilde{\mathbf{u}}=\operatorname{diag}(\mathbf{O},\mathbf{I},\mathbf{I})\mathbf{u}\sim\mathbf{u}$, it is evident that the joint distribution of $(\mathbf{z}_1-\boldsymbol{\alpha}^{(0)})'(\mathbf{z}_1-\boldsymbol{\alpha}^{(0)})+\mathbf{z}_3'\mathbf{z}_3$ and $\mathbf{z}_3$ depends on $\boldsymbol{\alpha}$, $\boldsymbol{\eta}$, and σ only through the values of $(\boldsymbol{\alpha}-\boldsymbol{\alpha}^{(0)})'(\boldsymbol{\alpha}-\boldsymbol{\alpha}^{(0)})$ and σ.

Verification of Proposition (3). Suppose that $\tilde{\phi}(\mathbf{z})$ is invariant with respect to the group $\tilde{G}_2(\boldsymbol{\alpha}^{(0)})$ as well as the groups $\tilde{G}_1$ and $\tilde{G}_3(\boldsymbol{\alpha}^{(0)})$. Then, in light of Proposition (2), $\tilde{\phi}(\mathbf{z})=\tilde{\phi}_{13}[(\mathbf{z}_1-\boldsymbol{\alpha}^{(0)})'(\mathbf{z}_1-\boldsymbol{\alpha}^{(0)})+\mathbf{z}_3'\mathbf{z}_3,\ \mathbf{z}_3]$ for some function $\tilde{\phi}_{13}(\cdot,\cdot)$, in which case $\tilde{\phi}[\tilde{T}_2(\mathbf{z};\boldsymbol{\alpha}^{(0)})]=\tilde{\phi}_{13}\{k^2[(\mathbf{z}_1-\boldsymbol{\alpha}^{(0)})'(\mathbf{z}_1-\boldsymbol{\alpha}^{(0)})+\mathbf{z}_3'\mathbf{z}_3],\ k\mathbf{z}_3\}$. Thus, to establish the existence of a function $\tilde{\phi}_{132}(\cdot)$ such that $\tilde{\phi}(\mathbf{z})=\tilde{\phi}_{132}\{[(\mathbf{z}_1-\boldsymbol{\alpha}^{(0)})'(\mathbf{z}_1-\boldsymbol{\alpha}^{(0)})+\mathbf{z}_3'\mathbf{z}_3]^{-1/2}\mathbf{z}_3\}$ for those values of $\mathbf{z}$ for which $\mathbf{z}_3\neq\mathbf{0}$, it suffices to take $\dot{\mathbf{z}}_1$ and $\ddot{\mathbf{z}}_1$ to be values of $\mathbf{z}_1$ and $\dot{\mathbf{z}}_3$ and $\ddot{\mathbf{z}}_3$ nonnull values of $\mathbf{z}_3$ such that

$$[(\ddot{\mathbf{z}}_1-\boldsymbol{\alpha}^{(0)})'(\ddot{\mathbf{z}}_1-\boldsymbol{\alpha}^{(0)})+\ddot{\mathbf{z}}_3'\ddot{\mathbf{z}}_3]^{-1/2}\ddot{\mathbf{z}}_3=[(\dot{\mathbf{z}}_1-\boldsymbol{\alpha}^{(0)})'(\dot{\mathbf{z}}_1-\boldsymbol{\alpha}^{(0)})+\dot{\mathbf{z}}_3'\dot{\mathbf{z}}_3]^{-1/2}\dot{\mathbf{z}}_3$$

and to observe that

$$\ddot{\mathbf{z}}_3=k\dot{\mathbf{z}}_3\quad\text{and}\quad(\ddot{\mathbf{z}}_1-\boldsymbol{\alpha}^{(0)})'(\ddot{\mathbf{z}}_1-\boldsymbol{\alpha}^{(0)})+\ddot{\mathbf{z}}_3'\ddot{\mathbf{z}}_3=k^2[(\dot{\mathbf{z}}_1-\boldsymbol{\alpha}^{(0)})'(\dot{\mathbf{z}}_1-\boldsymbol{\alpha}^{(0)})+\dot{\mathbf{z}}_3'\dot{\mathbf{z}}_3],$$

where $k=[(\ddot{\mathbf{z}}_1-\boldsymbol{\alpha}^{(0)})'(\ddot{\mathbf{z}}_1-\boldsymbol{\alpha}^{(0)})+\ddot{\mathbf{z}}_3'\ddot{\mathbf{z}}_3]^{1/2}/[(\dot{\mathbf{z}}_1-\boldsymbol{\alpha}^{(0)})'(\dot{\mathbf{z}}_1-\boldsymbol{\alpha}^{(0)})+\dot{\mathbf{z}}_3'\dot{\mathbf{z}}_3]^{1/2}$. It remains only to observe [as in the verification of Proposition (2)] that the joint distribution of $(\mathbf{z}_1-\boldsymbol{\alpha}^{(0)})'(\mathbf{z}_1-\boldsymbol{\alpha}^{(0)})+\mathbf{z}_3'\mathbf{z}_3$ and $\mathbf{z}_3$ is identical to that of the random variable

$$(\boldsymbol{\alpha}-\boldsymbol{\alpha}^{(0)})'(\boldsymbol{\alpha}-\boldsymbol{\alpha}^{(0)})+\sigma^2(\mathbf{u}_1'\mathbf{u}_1+\mathbf{u}_3'\mathbf{u}_3)+2\sigma\{[(\boldsymbol{\alpha}-\boldsymbol{\alpha}^{(0)})'(\boldsymbol{\alpha}-\boldsymbol{\alpha}^{(0)})]^{1/2},0,0,\dots,0\}\mathbf{u}_1$$

and the random vector $\sigma\mathbf{u}_3$ and hence that

$$
\begin{aligned}
&[(\mathbf{z}_1-\boldsymbol{\alpha}^{(0)})'(\mathbf{z}_1-\boldsymbol{\alpha}^{(0)})+\mathbf{z}_3'\mathbf{z}_3]^{-1/2}\mathbf{z}_3\\
&\qquad\sim\Bigg\{\frac{(\boldsymbol{\alpha}-\boldsymbol{\alpha}^{(0)})'(\boldsymbol{\alpha}-\boldsymbol{\alpha}^{(0)})}{\sigma^2}+\mathbf{u}_1'\mathbf{u}_1+\mathbf{u}_3'\mathbf{u}_3\\
&\qquad\qquad+2\left[\frac{[(\boldsymbol{\alpha}-\boldsymbol{\alpha}^{(0)})'(\boldsymbol{\alpha}-\boldsymbol{\alpha}^{(0)})]^{1/2}}{\sigma},0,0,\dots,0\right]\mathbf{u}_1\Bigg\}^{-1/2}\mathbf{u}_3.
\end{aligned}
$$

Verification of Proposition (4). Suppose that $\tilde{\phi}(\mathbf{z})$ is invariant with respect to the group $\tilde{G}_4$ as well as the groups $\tilde{G}_1$, $\tilde{G}_3(\boldsymbol{\alpha}^{(0)})$, and $\tilde{G}_2(\boldsymbol{\alpha}^{(0)})$. Then, in light of Proposition (3), there exists a function $\tilde{\phi}_{132}(\cdot)$ such that $\tilde{\phi}(\mathbf{z})=\tilde{\phi}_{132}\{[(\mathbf{z}_1-\boldsymbol{\alpha}^{(0)})'(\mathbf{z}_1-\boldsymbol{\alpha}^{(0)})+\mathbf{z}_3'\mathbf{z}_3]^{-1/2}\mathbf{z}_3\}$ for those values of $\mathbf{z}$ for which $\mathbf{z}_3\neq\mathbf{0}$. Moreover, there exists a function $\tilde{\phi}_{1324}(\cdot)$ such that for every value of $\mathbf{z}_1$ and every nonnull value of $\mathbf{z}_3$,

$$\tilde{\phi}_{132}\{[(\mathbf{z}_1-\boldsymbol{\alpha}^{(0)})'(\mathbf{z}_1-\boldsymbol{\alpha}^{(0)})+\mathbf{z}_3'\mathbf{z}_3]^{-1/2}\mathbf{z}_3\}=\tilde{\phi}_{1324}\{\mathbf{z}_3'\mathbf{z}_3/[(\mathbf{z}_1-\boldsymbol{\alpha}^{(0)})'(\mathbf{z}_1-\boldsymbol{\alpha}^{(0)})+\mathbf{z}_3'\mathbf{z}_3]\}.$$

To confirm this, it suffices to take $\dot{\mathbf{z}}_1$ and $\ddot{\mathbf{z}}_1$ to be values of $\mathbf{z}_1$ and $\dot{\mathbf{z}}_3$ and $\ddot{\mathbf{z}}_3$ nonnull values of $\mathbf{z}_3$ such that

$$\frac{\ddot{\mathbf{z}}_3'\ddot{\mathbf{z}}_3}{(\ddot{\mathbf{z}}_1-\boldsymbol{\alpha}^{(0)})'(\ddot{\mathbf{z}}_1-\boldsymbol{\alpha}^{(0)})+\ddot{\mathbf{z}}_3'\ddot{\mathbf{z}}_3}=\frac{\dot{\mathbf{z}}_3'\dot{\mathbf{z}}_3}{(\dot{\mathbf{z}}_1-\boldsymbol{\alpha}^{(0)})'(\dot{\mathbf{z}}_1-\boldsymbol{\alpha}^{(0)})+\dot{\mathbf{z}}_3'\dot{\mathbf{z}}_3}\tag{4.1}$$

and to observe that equality (4.1) is reexpressible as the equality

$$
\begin{aligned}
&\{[(\ddot{\mathbf{z}}_1-\boldsymbol{\alpha}^{(0)})'(\ddot{\mathbf{z}}_1-\boldsymbol{\alpha}^{(0)})+\ddot{\mathbf{z}}_3'\ddot{\mathbf{z}}_3]^{-1/2}\ddot{\mathbf{z}}_3\}'\{[(\ddot{\mathbf{z}}_1-\boldsymbol{\alpha}^{(0)})'(\ddot{\mathbf{z}}_1-\boldsymbol{\alpha}^{(0)})+\ddot{\mathbf{z}}_3'\ddot{\mathbf{z}}_3]^{-1/2}\ddot{\mathbf{z}}_3\}\\
&\quad=\{[(\dot{\mathbf{z}}_1-\boldsymbol{\alpha}^{(0)})'(\dot{\mathbf{z}}_1-\boldsymbol{\alpha}^{(0)})+\dot{\mathbf{z}}_3'\dot{\mathbf{z}}_3]^{-1/2}\dot{\mathbf{z}}_3\}'\{[(\dot{\mathbf{z}}_1-\boldsymbol{\alpha}^{(0)})'(\dot{\mathbf{z}}_1-\boldsymbol{\alpha}^{(0)})+\dot{\mathbf{z}}_3'\dot{\mathbf{z}}_3]^{-1/2}\dot{\mathbf{z}}_3\}
\end{aligned}
$$

and hence that equality (4.1) implies (in light of Lemma 5.9.9) the existence of an orthogonal matrix $\mathbf{B}$ for which

$$[(\ddot{\mathbf{z}}_1-\boldsymbol{\alpha}^{(0)})'(\ddot{\mathbf{z}}_1-\boldsymbol{\alpha}^{(0)})+\ddot{\mathbf{z}}_3'\ddot{\mathbf{z}}_3]^{-1/2}\ddot{\mathbf{z}}_3 = \mathbf{B}'\{[(\dot{\mathbf{z}}_1-\boldsymbol{\alpha}^{(0)})'(\dot{\mathbf{z}}_1-\boldsymbol{\alpha}^{(0)})+\dot{\mathbf{z}}_3'\dot{\mathbf{z}}_3]^{-1/2}\dot{\mathbf{z}}_3\}$$
$$= [(\dot{\mathbf{z}}_1-\boldsymbol{\alpha}^{(0)})'(\dot{\mathbf{z}}_1-\boldsymbol{\alpha}^{(0)})+(\mathbf{B}'\dot{\mathbf{z}}_3)'\mathbf{B}'\dot{\mathbf{z}}_3]^{-1/2}\mathbf{B}'\dot{\mathbf{z}}_3.$$

That the distribution of $\mathbf{z}_3'\mathbf{z}_3/[(\mathbf{z}_1-\boldsymbol{\alpha}^{(0)})'(\mathbf{z}_1-\boldsymbol{\alpha}^{(0)})+\mathbf{z}_3'\mathbf{z}_3]$ depends on $\boldsymbol{\alpha}$, $\boldsymbol{\eta}$, and σ only through the value of $(\boldsymbol{\alpha}-\boldsymbol{\alpha}^{(0)})'(\boldsymbol{\alpha}-\boldsymbol{\alpha}^{(0)})/\sigma^2$follows from Proposition (3) upon observing that

$$\mathbf{z}_3'\mathbf{z}_3/[(\mathbf{z}_1-\boldsymbol{\alpha}^{(0)})'(\mathbf{z}_1-\boldsymbol{\alpha}^{(0)})+\mathbf{z}_3'\mathbf{z}_3]$$
$$= \{[(\mathbf{z}_1-\boldsymbol{\alpha}^{(0)})'(\mathbf{z}_1-\boldsymbol{\alpha}^{(0)})+\mathbf{z}_3'\mathbf{z}_3]^{-1/2}\mathbf{z}_3\}'\{[(\mathbf{z}_1-\boldsymbol{\alpha}^{(0)})'(\mathbf{z}_1-\boldsymbol{\alpha}^{(0)})+\mathbf{z}_3'\mathbf{z}_3]^{-1/2}\mathbf{z}_3\}.$$

b. A relationship between sufficiency and invariance

As is evident from the final two parts of Section 7.3a, $\mathbf{z}_1$, $\mathbf{z}_2$, and $\mathbf{z}_3'\mathbf{z}_3$ form a sufficient statistic. The following proposition has the effect of establishing a relationship between sufficiency and invariance.

Proposition. The function $\tilde{\phi}(\mathbf{z})$ is invariant with respect to the group $\tilde{G}_4$ of transformations of the form $\tilde{T}_4(\mathbf{z})$ if and only if $\tilde{\phi}(\mathbf{z}) = \tilde{\phi}_4(\mathbf{z}_1, \mathbf{z}_2, \mathbf{z}_3'\mathbf{z}_3)$ for some function $\tilde{\phi}_4(\cdot,\cdot,\cdot)$.

Verification of the proposition. Suppose that $\tilde{\phi}(\mathbf{z}) = \tilde{\phi}_4(\mathbf{z}_1, \mathbf{z}_2, \mathbf{z}_3'\mathbf{z}_3)$ for some function $\tilde{\phi}_4(\cdot,\cdot,\cdot)$. Then,

$$\tilde{\phi}[\tilde{T}_4(\mathbf{z})] = \tilde{\phi}_4[\mathbf{z}_1, \mathbf{z}_2, (\mathbf{B}'\mathbf{z}_3)'\mathbf{B}'\mathbf{z}_3] = \tilde{\phi}_4(\mathbf{z}_1, \mathbf{z}_2, \mathbf{z}_3'\mathbf{z}_3) = \tilde{\phi}(\mathbf{z}).$$

Conversely, suppose that $\tilde{\phi}(\mathbf{z})$ is invariant with respect to the group $\tilde{G}_4$. Then, to establish that $\tilde{\phi}(\mathbf{z}) = \tilde{\phi}_4(\mathbf{z}_1, \mathbf{z}_2, \mathbf{z}_3'\mathbf{z}_3)$ for some function $\tilde{\phi}_4(\cdot,\cdot,\cdot)$, it suffices to observe that corresponding to any two values $\dot{\mathbf{z}}_3$ and $\ddot{\mathbf{z}}_3$ of $\mathbf{z}_3$ that satisfy the equality $\ddot{\mathbf{z}}_3'\ddot{\mathbf{z}}_3 = \dot{\mathbf{z}}_3'\dot{\mathbf{z}}_3$, there is an orthogonal matrix $\mathbf{B}$ such that $\ddot{\mathbf{z}}_3 = \mathbf{B}'\dot{\mathbf{z}}_3$ (the existence of which follows from Lemma 5.9.9).

c. The Neyman–Pearson fundamental lemma and its implications

Preliminary to comparing the power of the size-$\dot{\gamma}$ F test of $\tilde{H}_0$ or H_0 (versus $\tilde{H}_1$ or H_1) with that of other invariant tests of $\tilde{H}_0$ or H_0, it is convenient to introduce the following theorem.

Theorem 7.4.1. Let X represent an observable random variable with an absolutely continuous distribution that depends on a parameter θ (of unknown value). Further, let Θ represent the parameter space, let $\theta^{(0)}$ represent an hypothesized value of θ, let C represent an arbitrary critical (rejection) region for testing the null hypothesis that $\theta = \theta^{(0)}$ versus the alternative hypothesis that $\theta \neq \theta^{(0)}$, define $\gamma(\theta;\, C) = \Pr(X \in C)$ [so that $\gamma(\theta;\, C)$ is the power function of the test with critical region C], and (for $\theta \in \Theta$) denote by $f(\cdot;\, \theta)$ the pdf of the distribution of X. And let $\dot{\gamma}$ represent a scalar in the interval $0 < \dot{\gamma} < 1$. Then, subject to the constraint

$$\gamma(\theta^{(0)};\, C) \leq \dot{\gamma}$$

(i.e., the constraint that the size of the test not exceed $\dot{\gamma}$), $\gamma(\theta^*;\, C)$ attains its maximum value for any specified value θ^* of θ (other than $\theta^{(0)}$) when C is taken to be a critical region C^* that satisfies the following conditions:

$$x \in C^* \text{ if } f(x;\, \theta^*) > k f(x;\, \theta^{(0)}) \quad \text{and} \quad x \notin C^* \text{ if } f(x;\, \theta^*) < k f(x;\, \theta^{(0)}) \tag{4.2}$$

(for some nonnegative constant k) and

$$\gamma(\theta^{(0)};\, C^*) = \dot{\gamma}. \tag{4.3}$$

The result of Theorem 7.4.1 constitutes part of a version of what is known as the Neyman–Pearson fundamental lemma or simply as the Neyman–Pearson lemma. For a proof of this result, refer, for example, to Casella and Berger (2002, sec. 8.3).

In regard to Theorem 7.4.1, the test with critical region (set) C can be identified by the indicator function of C rather than by C itself; this function is the so-called critical (test) function. The definition of a test can be extended to include "randomized" tests; this can be done by extending the definition of a critical function to include any function $\phi(\cdot)$ such that $0 \le \phi(x) \le 1$ for every scalar x—when $\phi(\cdot)$ is an indicator function of a set C, $\phi(x)$ equals either 0 or 1. Under the extended definition, the test with critical function $\phi(\cdot)$ consists of rejecting the null hypothesis $\theta = \theta^{(0)}$ with probability $\phi(x)$, where x is the observed value of X. This test has a power function, say $\gamma(\theta; \phi)$, that is expressible as $\gamma(\theta; \phi) = \mathrm{E}[\phi(X)]$. The coverage of Theorem 7.4.1 can be extended: it can be shown that among tests (randomized as well as nonrandomized) whose size does not exceed $\dot{\gamma}$, the power function attains its maximum value for any specified value θ^* of θ (other than $\theta^{(0)}$) when the test is taken to be a nonrandomized test with critical region C^* that satisfies conditions (4.2) and (4.3).

In the context of Theorem 7.4.1, the (nonrandomized) test with critical region C^* that satisfies conditions (4.2) and (4.3) is optimal (in the sense that the value attained by its power function at the specified value θ^* of θ is a maximum among all tests whose size does not exceed $\dot{\gamma}$). In general, this test varies with θ^*. Suppose, however, that the set $\mathcal{X} = \{x : f(x; \theta) > 0\}$ does not vary with the value of θ and that for every value of θ in Θ and for $x \in \mathcal{X}$, the "likelihood ratio" $f(x; \theta)/f(x; \theta^{(0)})$ is a nondecreasing function of x or, alternatively, is a nonincreasing function of x. Then, there is a critical region C^* that satisfies conditions (4.2) and (4.3) and that does not vary with θ^*: depending on whether the ratio $f(x; \theta)/f(x; \theta^{(0)})$ is a nondecreasing function of x or a nonincreasing function of x, we can take $C^* = \{x : x > k'\}$, where k' is the upper $100\dot{\gamma}\%$ point of the distribution with pdf $f(\cdot; \theta^{(0)})$, or take $C^* = \{x : x < k'\}$, where k' is the lower $100\dot{\gamma}\%$ point of the distribution with pdf $f(\cdot; \theta^{(0)})$. In either case, the test with critical region C^* constitutes what is referred to as a UMP (uniformly most powerful) test.

d. A UMP invariant test and a UMA invariant confidence set

Let us now resume the discussion begun in Subsections a and b (pertaining to the problem of testing the null hypothesis $H_0 : \boldsymbol{\tau} = \boldsymbol{\tau}^{(0)}$ or $\tilde{H}_0 : \boldsymbol{\alpha} = \boldsymbol{\alpha}^{(0)}$ versus the alternative hypothesis $H_1 : \boldsymbol{\tau} \neq \boldsymbol{\tau}^{(0)}$ or $\tilde{H}_1 : \boldsymbol{\alpha} \neq \boldsymbol{\alpha}^{(0)}$). And let us continue to denote by $\tilde{\phi}(\mathbf{z})$ or (when convenient) by $\tilde{\phi}(\mathbf{z}_1, \mathbf{z}_2, \mathbf{z}_3)$ an arbitrary function of the random vector $\mathbf{z}$ $(= \mathbf{O}'\mathbf{y})$.

Invariant functions: an alternative representation. If $\tilde{\phi}(\mathbf{z})$ is invariant with respect to the four groups $\tilde{G}_1$, $\tilde{G}_3(\boldsymbol{\alpha}^{(0)})$, $\tilde{G}_2(\boldsymbol{\alpha}^{(0)})$, and $\tilde{G}_4$ of transformations, then [according to Proposition (4) of Subsection a] there exists a function $\tilde{\phi}_{1324}(\cdot)$ such that $\tilde{\phi}(\mathbf{z}) = \tilde{\phi}_{1324}\{\mathbf{z}_3'\mathbf{z}_3/[(\mathbf{z}_1 - \boldsymbol{\alpha}^{(0)})'(\mathbf{z}_1 - \boldsymbol{\alpha}^{(0)}) + \mathbf{z}_3'\mathbf{z}_3]\}$ for those values of $\mathbf{z}$ for which $\mathbf{z}_3 \neq \mathbf{0}$. And the distribution of $\mathbf{z}_3'\mathbf{z}_3/[(\mathbf{z}_1 - \boldsymbol{\alpha}^{(0)})'(\mathbf{z}_1 - \boldsymbol{\alpha}^{(0)}) + \mathbf{z}_3'\mathbf{z}_3]$ depends on $\boldsymbol{\alpha}$, $\boldsymbol{\eta}$, and σ only through the value of $(\boldsymbol{\alpha} - \boldsymbol{\alpha}^{(0)})'(\boldsymbol{\alpha} - \boldsymbol{\alpha}^{(0)})/\sigma^2$. Moreover, for those values of $\mathbf{z}$ for which $\mathbf{z}_1 \neq \boldsymbol{\alpha}^{(0)}$ or $\mathbf{z}_3 \neq \mathbf{0}$,

$$\frac{\mathbf{z}_3'\mathbf{z}_3}{(\mathbf{z}_1 - \boldsymbol{\alpha}^{(0)})'(\mathbf{z}_1 - \boldsymbol{\alpha}^{(0)}) + \mathbf{z}_3'\mathbf{z}_3} = 1 - \frac{(\mathbf{z}_1 - \boldsymbol{\alpha}^{(0)})'(\mathbf{z}_1 - \boldsymbol{\alpha}^{(0)})}{(\mathbf{z}_1 - \boldsymbol{\alpha}^{(0)})'(\mathbf{z}_1 - \boldsymbol{\alpha}^{(0)}) + \mathbf{z}_3'\mathbf{z}_3}. \tag{4.4}$$

Now, take X to be an observable random variable defined as follows:

$$X = \begin{cases} \dfrac{(\mathbf{z}_1 - \boldsymbol{\alpha}^{(0)})'(\mathbf{z}_1 - \boldsymbol{\alpha}^{(0)})}{(\mathbf{z}_1 - \boldsymbol{\alpha}^{(0)})'(\mathbf{z}_1 - \boldsymbol{\alpha}^{(0)}) + \mathbf{z}_3'\mathbf{z}_3}, & \text{if } \mathbf{z}_1 \neq \boldsymbol{\alpha}^{(0)} \text{ or } \mathbf{z}_3 \neq \mathbf{0}, \\ 0, & \text{if } \mathbf{z}_1 = \boldsymbol{\alpha}^{(0)} \text{ and } \mathbf{z}_3 = \mathbf{0}. \end{cases} \tag{4.5}$$

And let $\lambda = (\boldsymbol{\alpha} - \boldsymbol{\alpha}^{(0)})'(\boldsymbol{\alpha} - \boldsymbol{\alpha}^{(0)})/\sigma^2$. When regarded as a function of $\mathbf{z}$, X is invariant with respect to the groups $\tilde{G}_1$, $\tilde{G}_3(\boldsymbol{\alpha}^{(0)})$, $\tilde{G}_2(\boldsymbol{\alpha}^{(0)})$, and $\tilde{G}_4$ (as can be readily verified). Thus, if $\tilde{\phi}(\mathbf{z})$ depends on $\mathbf{z}$ only through the value of X, then it is invariant with respect to the groups $\tilde{G}_1$, $\tilde{G}_3(\boldsymbol{\alpha}^{(0)})$, $\tilde{G}_2(\boldsymbol{\alpha}^{(0)})$, and $\tilde{G}_4$. Conversely, if $\tilde{\phi}(\mathbf{z})$ is invariant with respect to the groups $\tilde{G}_1$, $\tilde{G}_3(\boldsymbol{\alpha}^{(0)})$, $\tilde{G}_2(\boldsymbol{\alpha}^{(0)})$, and $\tilde{G}_4$, then [in light of equality (4.4)] there exists a function $\tilde{\phi}^*_{1324}(\cdot)$ such that

$$\tilde{\phi}(\mathbf{z}) = \tilde{\phi}^*_{1324}(X) \tag{4.6}$$

for those values of $\mathbf{z}$ for which $\mathbf{z}_3 \neq \mathbf{0}$, in which case [since $\Pr(\mathbf{z}_3 \neq \mathbf{0}) = 1$] $\tilde{\phi}(\mathbf{z}) \sim \tilde{\phi}^*_{1324}(X)$. Moreover, the distribution of X or any function of X depends on $\boldsymbol{\alpha}$, $\boldsymbol{\eta}$, and σ only through the value of the nonnegative parametric function λ. In fact,

$$X \sim \frac{\lambda + 2(\sqrt{\lambda}, 0, 0, \ldots, 0)\mathbf{u}_1 + \mathbf{u}_1'\mathbf{u}_1}{\lambda + 2(\sqrt{\lambda}, 0, 0, \ldots, 0)\mathbf{u}_1 + \mathbf{u}_1'\mathbf{u}_1 + \mathbf{u}_3'\mathbf{u}_3}, \tag{4.7}$$

as can be readily verified via a development similar to that employed in Subsection a in the verification of Proposition (2).

Applicability of the Neyman–Pearson lemma. As is evident from the preceding part of the present subsection, a test of $\tilde{H}_0$ versus $\tilde{H}_1$ with critical function $\tilde{\phi}(\mathbf{z})$ is invariant with respect to the groups $\tilde{G}_1$, $\tilde{G}_3(\boldsymbol{\alpha}^{(0)})$, $\tilde{G}_2(\boldsymbol{\alpha}^{(0)})$, and $\tilde{G}_4$ if and only if there exists a test with critical function $\tilde{\phi}^*_{1324}(X)$ such that $\tilde{\phi}(\mathbf{z}) = \tilde{\phi}^*_{1324}(X)$ with probability 1, in which case

$$\mathrm{E}[\tilde{\phi}(\mathbf{z})] = \mathrm{E}[\tilde{\phi}^*_{1324}(X)], \tag{4.8}$$

that is, the two tests have the same power function. The upshot of this remark is that the Neyman–Pearson lemma can be used to address the problem of finding a test of $\tilde{H}_0$ versus $\tilde{H}_1$ that is "optimal" among tests that are invariant with respect to the groups $\tilde{G}_1$, $\tilde{G}_3(\boldsymbol{\alpha}^{(0)})$, $\tilde{G}_2(\boldsymbol{\alpha}^{(0)})$, and $\tilde{G}_4$ and whose size does not exceed $\dot{\gamma}$. Among such tests, the power function of the test attains its maximum value for values of $\boldsymbol{\alpha}$, $\boldsymbol{\eta}$, and σ such that λ equals some specified value λ^* when the critical region of the test is a critical region obtained upon applying the Neyman–Pearson lemma, taking X to be the observable random variable defined by expression (4.5) and taking $\theta = \lambda$, $\Theta = [0, \infty)$, $\theta^{(0)} = 0$, and $\theta^* = \lambda^*$.

Special case: $\sigma^{-1}\mathbf{e} \sim N(\mathbf{0}, \mathbf{I})$. Suppose that the distribution of the vector $\mathbf{u} = \sigma^{-1}\mathbf{e}$ is $N(\mathbf{0}, \mathbf{I})$, in which case $\mathbf{z} \sim N\left[\begin{pmatrix}\boldsymbol{\alpha}\\ \boldsymbol{\eta}\\ \mathbf{0}\end{pmatrix}, \sigma^2\mathbf{I}\right]$. Then,

$$X \sim \frac{(\mathbf{z}_1 - \boldsymbol{\alpha}^{(0)})'(\mathbf{z}_1 - \boldsymbol{\alpha}^{(0)})}{(\mathbf{z}_1 - \boldsymbol{\alpha}^{(0)})'(\mathbf{z}_1 - \boldsymbol{\alpha}^{(0)}) + \mathbf{z}_3'\mathbf{z}_3} \sim Be\left(\frac{M_*}{2}, \frac{N-P_*}{2}, \frac{\lambda}{2}\right), \tag{4.9}$$

as is evident upon observing that $\mathbf{z}_3'\mathbf{z}_3/\sigma^2 \sim \chi^2(N-P_*)$ and that $(\mathbf{z}_1 - \boldsymbol{\alpha}^{(0)})'(\mathbf{z}_1 - \boldsymbol{\alpha}^{(0)})/\sigma^2$ is statistically independent of $\mathbf{z}_3'\mathbf{z}_3/\sigma^2$ and has a $\chi^2(M_*, \lambda)$ distribution. Further, let (for $\lambda \geq 0$) $f(\cdot; \lambda)$ represent the pdf of the $Be[M_*/2, (N-P_*)/2, \lambda/2]$ distribution, and observe [in light of result (6.3.24)] that (for $\lambda > 0$) the ratio $f(x; \lambda)/f(x; 0)$ is a strictly increasing function of x (over the interval $0 < x < 1$). And consider the test (of the null hypothesis $\tilde{H}_0 : \boldsymbol{\alpha} = \boldsymbol{\alpha}^{(0)}$ versus the alternative hypothesis $\tilde{H}_1 : \boldsymbol{\alpha} \neq \boldsymbol{\alpha}^{(0)}$) that rejects $\tilde{H}_0$ if and only if

$$X > \bar{B}_{\dot{\gamma}}[M_*/2, (N-P_*)/2], \tag{4.10}$$

where $\bar{B}_{\dot{\gamma}}[M_*/2, (N-P_*)/2]$ is the upper $100\dot{\gamma}\%$ point of the $Be[M_*/2, (N-P_*)/2]$ distribution. This test is of size $\dot{\gamma}$, is invariant with respect to the groups $\tilde{G}_1$, $\tilde{G}_3(\boldsymbol{\alpha}^{(0)})$, $\tilde{G}_2(\boldsymbol{\alpha}^{(0)})$, and $\tilde{G}_4$, and among tests whose size does not exceed $\dot{\gamma}$ and that are invariant with respect to the groups $\tilde{G}_1$, $\tilde{G}_3(\boldsymbol{\alpha}^{(0)})$, $\tilde{G}_2(\boldsymbol{\alpha}^{(0)})$, and $\tilde{G}_4$, is UMP (as is evident from the discussion in Subsection c and from the discussion in the preceding part of the present subsection).

The F ***test***. Note that (for $\mathbf{z}_3 \neq \mathbf{0}$)

$$\frac{(\mathbf{z}_1 - \boldsymbol{\alpha}^{(0)})'(\mathbf{z}_1 - \boldsymbol{\alpha}^{(0)})}{(\mathbf{z}_1 - \boldsymbol{\alpha}^{(0)})'(\mathbf{z}_1 - \boldsymbol{\alpha}^{(0)}) + \mathbf{z}_3'\mathbf{z}_3} = \frac{[M_*/(N-P_*)]\tilde{F}(\boldsymbol{\alpha}^{(0)})}{1 + [M_*/(N-P_*)]\tilde{F}(\boldsymbol{\alpha}^{(0)})}, \tag{4.11}$$

where (for an arbitrary $M_* \times 1$ vector $\dot{\boldsymbol{\alpha}}$)

$$\tilde{F}(\dot{\alpha}) = \frac{(\mathbf{z}_1 - \dot{\alpha})'(\mathbf{z}_1 - \dot{\alpha})/M_*}{\mathbf{z}_3'\mathbf{z}_3/(N - P_*)}.$$

Note also that expression (4.11) is [for $\tilde{F}(\boldsymbol{\alpha}^{(0)}) \geq 0$] a strictly increasing function of $\tilde{F}(\boldsymbol{\alpha}^{(0)})$. Thus, the set of $\mathbf{z}$-values that satisfy inequality (4.10) is essentially (wp1) identical to the critical region $\tilde{C}_F$ of the size-$\dot{\gamma}$ F test. Accordingly, the size-$\dot{\gamma}$ F test (of $\tilde{H}_0$ versus $\tilde{H}_1$) is equivalent to a size-$\dot{\gamma}$ test that is invariant with respect to the groups $\tilde{G}_1$, $\tilde{G}_3(\boldsymbol{\alpha}^{(0)})$, $\tilde{G}_2(\boldsymbol{\alpha}^{(0)})$, and $\tilde{G}_4$ and that [when $\sigma^{-1}\mathbf{e} \sim N(\mathbf{0}, \mathbf{I})$] is UMP among all tests (of $\tilde{H}_0$ versus $\tilde{H}_1$) whose size does not exceed $\dot{\gamma}$ and that are invariant with respect to the groups $\tilde{G}_1$, $\tilde{G}_3(\boldsymbol{\alpha}^{(0)})$, $\tilde{G}_2(\boldsymbol{\alpha}^{(0)})$, and $\tilde{G}_4$; the equivalence of the two tests is in the sense that their critical functions are equal wp1—earlier (in Section 7.3b), it was noted that the F test is invariant [with respect to the groups $\tilde{G}_1$, $\tilde{G}_3(\boldsymbol{\alpha}^{(0)})$, $\tilde{G}_2(\boldsymbol{\alpha}^{(0)})$, and $\tilde{G}_4$] for those values of $\mathbf{z}$ in the set $\{\mathbf{z} : \mathbf{z}_3 \neq \mathbf{0}\}$ (which are the values for which the F test is defined).

A stronger result. Suppose (as in the preceding 2 parts of the present subsection) that $\sigma^{-1}\mathbf{e} \sim N(\mathbf{0}, \mathbf{I})$. Then, as is evident from the results in the next-to-last part of Section 7.3a, $\mathbf{z}_1$, $\mathbf{z}_2$, and $\mathbf{z}_3'\mathbf{z}_3$ form a (vector-valued) sufficient statistic. And in light of the proposition of Subsection b, a critical function $\tilde{\phi}(\mathbf{z})$ is reexpressible as a function of this statistic if and only if it is invariant with respect to the group $\tilde{G}_4$ of transformations of the form $\tilde{T}_4(\mathbf{z})$. Thus, a test (of $\tilde{H}_0$ versus $\tilde{H}_1$) is invariant with respect to the groups $\tilde{G}_1$, $\tilde{G}_3(\boldsymbol{\alpha}^{(0)})$, $\tilde{G}_2(\boldsymbol{\alpha}^{(0)})$, and $\tilde{G}_4$ of transformations if and only if it is invariant with respect to the groups $\tilde{G}_1$, $\tilde{G}_3(\boldsymbol{\alpha}^{(0)})$, and $\tilde{G}_2(\boldsymbol{\alpha}^{(0)})$ and, in addition, its critical function depends on $\mathbf{z}$ only through the value of the sufficient statistic formed by $\mathbf{z}_1$, $\mathbf{z}_2$, and $\mathbf{z}_3'\mathbf{z}_3$.

Let $\tilde{\phi}(\mathbf{z})$ represent the critical function of any (possibly randomized) test (of $\tilde{H}_0$ versus $\tilde{H}_1$). Corresponding to $\tilde{\phi}(\mathbf{z})$ is a critical function $\bar{\phi}(\mathbf{z}) = \mathrm{E}[\tilde{\phi}(\mathbf{z}) \,|\, \mathbf{z}_1, \mathbf{z}_2, \mathbf{z}_3'\mathbf{z}_3]$ that depends on $\mathbf{z}$ only through the value of the sufficient statistic—here, the conditional expectation of any function of $\mathbf{z}$ is taken to be that determined from any particular version of the conditional distribution of $\mathbf{z}$. Moreover,

$$\mathrm{E}[\bar{\phi}(\mathbf{z})] = \mathrm{E}[\tilde{\phi}(\mathbf{z})],$$

so that the power function of the test with critical function $\bar{\phi}(\mathbf{z})$ is identical to that of the test with critical function $\tilde{\phi}(\mathbf{z})$. And for any transformation $\tilde{T}(\mathbf{z})$ such that $\tilde{\phi}[\tilde{T}(\mathbf{z})] = \tilde{\phi}(\mathbf{z})$,

$$\bar{\phi}[\tilde{T}(\mathbf{z})] = \mathrm{E}\{\tilde{\phi}[\tilde{T}(\mathbf{z})] \,|\, \mathbf{z}_1, \mathbf{z}_2, \mathbf{z}_3'\mathbf{z}_3\} = \bar{\phi}(\mathbf{z}).$$

Thus, if the test with critical function $\tilde{\phi}(\mathbf{z})$ is invariant with respect to the groups $\tilde{G}_1$, $\tilde{G}_3(\boldsymbol{\alpha}^{(0)})$, and $\tilde{G}_2(\boldsymbol{\alpha}^{(0)})$ of transformations, then so is the test with critical function $\bar{\phi}(\mathbf{z})$.

There is an implication that the size-$\dot{\gamma}$ test (of $\tilde{H}_0$ versus $\tilde{H}_1$) that is invariant with respect to the groups $\tilde{G}_1$, $\tilde{G}_3(\boldsymbol{\alpha}^{(0)})$, and $\tilde{G}_2(\boldsymbol{\alpha}^{(0)})$ and whose critical region is determined by inequality (4.10) or (for values of $\mathbf{z}$ such that $\mathbf{z}_3 \neq \mathbf{0}$) by the inequality $\tilde{F}(\boldsymbol{\alpha}^{(0)}) > \bar{F}_{\dot{\gamma}}(M_*, N - P_*)$ is UMP among all tests whose size does not exceed $\dot{\gamma}$ and that are invariant with respect to the groups $\tilde{G}_1$, $\tilde{G}_3(\boldsymbol{\alpha}^{(0)})$, and $\tilde{G}_2(\boldsymbol{\alpha}^{(0)})$. The restriction to tests that are invariant with respect to the group $\tilde{G}_4$ is unnecessary.

General case. The results of the preceding three parts of the present subsection were obtained under a supposition that the distribution of the vector $\mathbf{u} = \sigma^{-1}\mathbf{e}$ is $N(\mathbf{0}, \mathbf{I})$. Let us now consider the general case where the distribution of $\mathbf{u}$ is an absolutely continuous spherically symmetric distribution that may differ from the $N(\mathbf{0}, \mathbf{I})$ distribution. Specifically, let us consider the extent to which the results obtained in the special case where $\mathbf{u} \sim N(\mathbf{0}, \mathbf{I})$ extend to the general case.

In the general case as well as the special case, the set of $\mathbf{z}$-values that satisfy inequality (4.10) form the critical region of a test of $\tilde{H}_0$ versus $\tilde{H}_1$ that is invariant with respect to the groups $\tilde{G}_1$, $\tilde{G}_3(\boldsymbol{\alpha}^{(0)})$, $\tilde{G}_2(\boldsymbol{\alpha}^{(0)})$, and $\tilde{G}_4$ of transformations and that is of size $\dot{\gamma}$. This observaion (as it pertains to the size of the test) is consistent with one made earlier (in Section 7.3b) in a discussion of the F test—the size of this test is the same as that of the F test.

Is this test UMP among all tests that are invariant with respect to the groups $\tilde{G}_1$, $\tilde{G}_3(\boldsymbol{\alpha}^{(0)})$, and $\tilde{G}_2(\boldsymbol{\alpha}^{(0)})$ and whose size does not exceed $\dot{\gamma}$ (as it is in the special case of normality)? As in the special case, this question can be addressed by applying the Neyman–Pearson lemma.

Let us continue to take X to be the observable random variable defined by equality (4.5), and let us take $f(\cdot\,;\lambda)$ to be the pdf of the distribution of X [as determined from the distribution of $\mathbf{u}$ on the basis of expression (4.7)]. When $\lambda = 0$, $X \sim Be[M_*/2, (N-P_*)/2]$, so that $f(\cdot\,;0)$ is the same in the general case as in the special case where $\mathbf{u} \sim N(\mathbf{0}, \mathbf{I})$—refer, e.g., to Part (1) of Theorem 6.3.1. However, when $\lambda > 0$, the distribution of X varies with the distribution of $\mathbf{u}$—in the special case where $\mathbf{u} \sim N(\mathbf{0}, \mathbf{I})$, this distribution is $Be[M_*/2, (N-P_*)/2, \lambda/2]$ (a noncentral beta distribution). Further, let C represent an arbitrary critical region for testing (on the basis of X) the null hypothesis that $\lambda = 0$ versus the alternative hypothesis that $\lambda > 0$. And denote by $\gamma(\cdot\,; C)$ the power function of the test with critical region C, so that (for $\lambda \geq 0$) $\gamma(\lambda; C) = \int_C f(x;\lambda)\,dx$.

Now, consider a critical region C^* such that subject to the constraint $\gamma(0; C) \leq \dot{\gamma}$, $\gamma(\lambda; C)$ attains its maximum value for any particular (strictly positive) value λ^* of λ when $C = C^*$. At least in principle, such a critical region can be determined by applying the Neyman–Pearson lemma. The critical region C^* defines a set of $\mathbf{z}$-values that form the critical region of a size-$\dot{\gamma}$ test of $\tilde{H}_0$ versus $\tilde{H}_1$ that is invariant with respect to the groups $\tilde{G}_1$, $\tilde{G}_3(\boldsymbol{\alpha}^{(0)})$, $\tilde{G}_2(\boldsymbol{\alpha}^{(0)})$, and $\tilde{G}_4$ of transformations. Moreover, upon recalling (from the final part of Section 7.3a) that [as in the special case where $\mathbf{u} \sim N(\mathbf{0}, \mathbf{I})$] $\mathbf{z}_1$, $\mathbf{z}_2$, and $\mathbf{z}_3'\mathbf{z}_3$ form a (vector-valued) sufficient statistic and upon employing essentially the same reasoning as in the preceding part of the present subsection, we find that the value attained for $\lambda = \lambda^*$ by the power function of that test is greater than or equal to that attained for $\lambda = \lambda^*$ by the power function of any test of $\tilde{H}_0$ versus $\tilde{H}_1$ that is invariant with respect to the groups $\tilde{G}_1$, $\tilde{G}_3(\boldsymbol{\alpha}^{(0)})$, and $\tilde{G}_2(\boldsymbol{\alpha}^{(0)})$ and whose size does not exceed $\dot{\gamma}$.

If [as in the special case where $\mathbf{u} \sim N(\mathbf{0}, \mathbf{I})$] the ratio $f(x;\lambda^*)/f(x;0)$ is a nondecreasing function of x, then C^* can be taken to be the critical region defined by inequality (4.10) or (aside from a set of $\mathbf{z}$-values of probability 0) by the inequality $\tilde{F}(\boldsymbol{\alpha}^{(0)}) > \bar{F}_{\dot{\gamma}}(M_*, N-P_*)$. And if [as in the special case where $\mathbf{u} \sim N(\mathbf{0}, \mathbf{I})$] the ratio $f(x;\lambda^*)/f(x;0)$ is a nondecreasing function of x for every choice of λ^*, then the size-$\dot{\gamma}$ test (of $\tilde{H}_0$ versus $\tilde{H}_1$) that is invariant with respect to the groups $\tilde{G}_1$, $\tilde{G}_3(\boldsymbol{\alpha}^{(0)})$, $\tilde{G}_2(\boldsymbol{\alpha}^{(0)})$, and $\tilde{G}_4$ and whose critical region is defined by inequality (4.10) or (for values of $\mathbf{z}$ such that $\mathbf{z}_3 \neq \mathbf{0}$) by the inequality $\tilde{F}(\boldsymbol{\alpha}^{(0)}) > \bar{F}_{\dot{\gamma}}(M_*, N-P_*)$ is UMP among all tests that are invariant with respect to the groups $\tilde{G}_1$, $\tilde{G}_3(\boldsymbol{\alpha}^{(0)})$, and $\tilde{G}_2(\boldsymbol{\alpha}^{(0)})$ and whose size does not exceed $\dot{\gamma}$. However, in general, C^* may differ nontrivially (i.e., for a set of $\mathbf{z}$-values having nonzero probability) from the critical region defined by inequality (4.10) or by the inequality $\tilde{F}(\boldsymbol{\alpha}^{(0)}) > \bar{F}_{\dot{\gamma}}(M_*, N-P_*)$ and from one choice of λ^* to another, and there may not be any test that is UMP among all tests that are invariant with respect to the groups $\tilde{G}_1$, $\tilde{G}_3(\boldsymbol{\alpha}^{(0)})$, and $\tilde{G}_2(\boldsymbol{\alpha}^{(0)})$ and whose size does not exceed $\dot{\gamma}$.

Restatement of results. The results of the preceding parts of the present subsection are stated in terms of the transformed observable random vector $\mathbf{z}$ $(= \mathbf{O}'\mathbf{y})$ and in terms of the transformed parametric vector $\boldsymbol{\alpha}$ $(= \mathbf{S}'\boldsymbol{\tau})$ associated with the canonical form of the G–M model. These results can be restated in terms of $\mathbf{y}$ and $\boldsymbol{\tau}$.

The null hypothesis $\tilde{H}_0 : \boldsymbol{\alpha} = \boldsymbol{\alpha}^{(0)}$ is equivalent to the null hypothesis $H_0 : \boldsymbol{\tau} = \boldsymbol{\tau}^{(0)}$—the parameter vector $\boldsymbol{\beta}$ satisfies the condition $\boldsymbol{\tau} = \boldsymbol{\tau}^{(0)}$ if and only if it satisfies the condition $\boldsymbol{\alpha} = \boldsymbol{\alpha}^{(0)}$—and the alternative hypothesis $\tilde{H}_1 : \boldsymbol{\alpha} \neq \boldsymbol{\alpha}^{(0)}$ is equivalent to the alternative hypothesis $H_1 : \boldsymbol{\tau} \neq \boldsymbol{\tau}^{(0)}$. Moreover, $\lambda = (\boldsymbol{\alpha}-\boldsymbol{\alpha}^{(0)})'(\boldsymbol{\alpha}-\boldsymbol{\alpha}^{(0)})/\sigma^2$ is reexpressible as

$$\lambda = (\boldsymbol{\tau}-\boldsymbol{\tau}^{(0)})'\mathbf{C}^-(\boldsymbol{\tau}-\boldsymbol{\tau}^{(0)})/\sigma^2 \tag{4.12}$$

—refer to result (3.34)—and $\mathbf{z}_3'\mathbf{z}_3$ and $(\mathbf{z}_1-\boldsymbol{\alpha}^{(0)})'(\mathbf{z}_1-\boldsymbol{\alpha}^{(0)})$ are reexpressible as

$$\mathbf{z}_3'\mathbf{z}_3 = \mathbf{d}'\mathbf{d} = \mathbf{y}'(\mathbf{I}-\mathbf{P_X})\mathbf{y} \tag{4.13}$$

and

$$(\mathbf{z}_1-\boldsymbol{\alpha}^{(0)})'(\mathbf{z}_1-\boldsymbol{\alpha}^{(0)}) = (\hat{\boldsymbol{\alpha}}-\boldsymbol{\alpha}^{(0)})'(\hat{\boldsymbol{\alpha}}-\boldsymbol{\alpha}^{(0)}) = (\hat{\boldsymbol{\tau}}-\boldsymbol{\tau}^{(0)})'\mathbf{C}^-(\hat{\boldsymbol{\tau}}-\boldsymbol{\tau}^{(0)}) \tag{4.14}$$

—refer to result (3.21) or (3.26) and to result (3.28). Results (4.13) and (4.14) can be used to reexpress (in terms of $\mathbf{y}$) the observable random variable X defined by expression (4.5) and as noted earlier (in Section 7.3b) to reexpress $\tilde{F}(\boldsymbol{\alpha}^{(0)})$.

A function $\tilde{\phi}(\mathbf{z})$ of $\mathbf{z}$ that is the critical function of a test of $\tilde{H}_0$ or H_0 versus $\tilde{H}_1$ or H_1 is reexpressible as a function $\phi(\mathbf{y})$ [$= \tilde{\phi}(\mathbf{O}'\mathbf{y})$] of $\mathbf{y}$. And corresponding to any one-to-one transformation $\tilde{T}(\mathbf{z})$ of $\mathbf{z}$ (from $\mathbb{R}^N$ onto $\mathbb{R}^N$) is a one-to-one transformation $T(\mathbf{y})$ [$= \mathbf{O}\tilde{T}(\mathbf{O}'\mathbf{y})$] of $\mathbf{y}$ (from $\mathbb{R}^N$ onto $\mathbb{R}^N$), and corresponding to any group $\tilde{G}$ of such transformations of $\mathbf{z}$ is a group G consisting of the corresponding transformations of $\mathbf{y}$. Further, a test (of $\tilde{H}_0$ or H_0 versus $\tilde{H}_1$ or H_1) is invariant with respect to the group $\tilde{G}$ if and only if it is invariant with respect to the group G—if $\tilde{\phi}(\mathbf{z})$ is the critical function expressed as a function of $\mathbf{z}$ and $\phi(\mathbf{y})$ the critical function expressed as a function of $\mathbf{y}$, then $\tilde{\phi}[\tilde{T}(\mathbf{z})] = \tilde{\phi}(\mathbf{z})$ for every transformation $\tilde{T}(\mathbf{z})$ in $\tilde{G}$ if and only if $\phi[T(\mathbf{y})] = \phi(\mathbf{y})$ for every transformation $T(\mathbf{y})$ in G. Refer to the final two parts of Section 7.3b for some specifics as they pertain to the groups $\tilde{G}_1$, $\tilde{G}_2(\boldsymbol{\alpha}^{(0)})$, $\tilde{G}_3(\boldsymbol{\alpha}^{(0)})$, and $\tilde{G}_4$.

Confidence sets. Take (for each value $\dot{\boldsymbol{\alpha}}$ of $\boldsymbol{\alpha}$) $\tilde{X}(\dot{\boldsymbol{\alpha}})$ to be the random variable defined as follows:

$$\tilde{X}(\dot{\boldsymbol{\alpha}}) = \begin{cases} \dfrac{(\mathbf{z}_1-\dot{\boldsymbol{\alpha}})'(\mathbf{z}_1-\dot{\boldsymbol{\alpha}})}{(\mathbf{z}_1-\dot{\boldsymbol{\alpha}})'(\mathbf{z}_1-\dot{\boldsymbol{\alpha}})+\mathbf{z}_3'\mathbf{z}_3}, & \text{if } \mathbf{z}_1 \neq \dot{\boldsymbol{\alpha}} \text{ or } \mathbf{z}_3 \neq \mathbf{0}, \\ 0, & \text{if } \mathbf{z}_1 = \dot{\boldsymbol{\alpha}} \text{ and } \mathbf{z}_3 = \mathbf{0}. \end{cases}$$

And take $\tilde{A}_F(\mathbf{z})$ to be the set (of $\boldsymbol{\alpha}$-values)

$$\tilde{A}_F(\mathbf{z}) = \{\dot{\boldsymbol{\alpha}} : \tilde{X}(\dot{\boldsymbol{\alpha}}) \le \bar{B}_{\dot{\gamma}}[M_*/2,\ (N-P_*)/2]\}$$

or, equivalently (for those values of $\mathbf{z}$ such that $\mathbf{z}_3 \neq \mathbf{0}$), the set

$$\tilde{A}_F(\mathbf{z}) = \{\dot{\boldsymbol{\alpha}} : \tilde{F}(\dot{\boldsymbol{\alpha}}) \le \bar{F}_{\dot{\gamma}}(M_*,\ N-P_*)\}.$$

As discussed in Section 7.3b, $\tilde{A}_F(\mathbf{z})$ is a $100(1-\dot{\gamma})\%$ confidence set for $\boldsymbol{\alpha}$.

Let $\tilde{G}_2$ represent the group of transformations consisting of the totality of the groups $\tilde{G}_2(\boldsymbol{\alpha}^{(0)})$ ($\boldsymbol{\alpha}^{(0)} \in \mathbb{R}^{M_*}$), and, similarly, let $\tilde{G}_3$ represent the group of transformations consisting of the totality of the groups $\tilde{G}_3(\boldsymbol{\alpha}^{(0)})$ ($\boldsymbol{\alpha}^{(0)} \in \mathbb{R}^{M_*}$). Then, the $100(1-\dot{\gamma})\%$ confidence set $\tilde{A}_F(\mathbf{z})$ is invariant with respect to the groups $\tilde{G}_1$ and $\tilde{G}_4$ and is equivariant with respect to the groups $\tilde{G}_2$ and $\tilde{G}_3$—refer to the discussion in the final 3 parts of Section 7.3b. And for purposes of comparison, let $\tilde{A}(\mathbf{z})$ represent any confidence set for $\boldsymbol{\alpha}$ whose probability of coverage equals or exceeds $1-\dot{\gamma}$ and that is invariant with respect to the group $\tilde{G}_1$ and equivariant with respect to the groups $\tilde{G}_2$ and $\tilde{G}_3$. Further, for $\boldsymbol{\alpha}^{(0)} \in \mathbb{R}^{M_*}$, take $\delta(\boldsymbol{\alpha}^{(0)}; \boldsymbol{\alpha})$ to be the probability $\Pr[\boldsymbol{\alpha}^{(0)} \in \tilde{A}(\mathbf{z})]$ of $\boldsymbol{\alpha}^{(0)}$ being "covered" by the confidence set $\tilde{A}(\mathbf{z})$—$\delta(\boldsymbol{\alpha}; \boldsymbol{\alpha}) \ge 1-\dot{\gamma}$, and for $\boldsymbol{\alpha}^{(0)} \neq \boldsymbol{\alpha}$, $\delta(\boldsymbol{\alpha}^{(0)}; \boldsymbol{\alpha})$ is referred to as the probability of false coverage.

Corresponding to $\tilde{A}(\mathbf{z})$ is the test of the null hypothesis $\tilde{H}_0 : \boldsymbol{\alpha} = \boldsymbol{\alpha}^{(0)}$ versus the alternative hypothesis $\tilde{H}_1 : \boldsymbol{\alpha} \neq \boldsymbol{\alpha}^{(0)}$ with critical function $\tilde{\phi}(\cdot\,; \boldsymbol{\alpha}^0)$ defined as follows:

$$\tilde{\phi}(\mathbf{z}; \boldsymbol{\alpha}^0) = \begin{cases} 1, & \text{if } \boldsymbol{\alpha}^{(0)} \notin \tilde{A}(\mathbf{z}), \\ 0, & \text{if } \boldsymbol{\alpha}^{(0)} \in \tilde{A}(\mathbf{z}). \end{cases}$$

And in light of the invariance of $\tilde{A}(\mathbf{z})$ with respect to the group $\tilde{G}_1$ and the equivariance of $\tilde{A}(\mathbf{z})$ with respect to the groups $\tilde{G}_2$ and $\tilde{G}_3$,

$$\tilde{\phi}[\tilde{T}_1(\mathbf{z}); \boldsymbol{\alpha}^{(0)}] = \tilde{\phi}[\tilde{T}_2(\mathbf{z}); \boldsymbol{\alpha}^{(0)}] = \tilde{\phi}[\tilde{T}_3(\mathbf{z}); \boldsymbol{\alpha}^{(0)}] = \tilde{\phi}(\mathbf{z}; \boldsymbol{\alpha}^0), \tag{4.15}$$

as can be readily verified. Further, let $\gamma(\boldsymbol{\alpha}; \boldsymbol{\alpha}^{(0)}) = \mathrm{E}[\tilde{\phi}(\mathbf{z}; \boldsymbol{\alpha}^{(0)})]$, and observe that

$$\gamma(\boldsymbol{\alpha}; \boldsymbol{\alpha}^{(0)}) = 1 - \delta(\boldsymbol{\alpha}^{(0)}; \boldsymbol{\alpha}), \tag{4.16}$$

implying in particular that

$$\gamma(\boldsymbol{\alpha}; \boldsymbol{\alpha}) = 1 - \delta(\boldsymbol{\alpha}; \boldsymbol{\alpha}) \le \dot{\gamma}. \tag{4.17}$$

Now, suppose that the distribution of the random vector $\mathbf{u} = \sigma^{-1}\mathbf{e}$ is $N(\mathbf{0}, \mathbf{I})$ or, more generally, that the pdf $f(\cdot\,; \lambda)$ of the distribution of the random variable (4.7) is such that for every $\lambda > 0$ the ratio $f(x; \lambda)/f(x; 0)$ is (for $0 < x < 1$) a nondecreasing function of x. Then, in light of equalities (4.15) and (4.17), it follows from our previous results (on the optimality of the F test) that (for every value of $\boldsymbol{\alpha}$ and regardless of the choice for $\boldsymbol{\alpha}^{(0)}$) $\gamma(\boldsymbol{\alpha};\, \boldsymbol{\alpha}^{(0)})$ attains its maximum value when $\tilde{A}(\mathbf{z})$ is taken to be the set $\tilde{A}_F(\mathbf{z})$ [in which case $\tilde{\phi}(\mathbf{z};\, \boldsymbol{\alpha}^{(0)})$ is the critical function of the test that rejects $\tilde{H}_0$ if $\tilde{X}(\boldsymbol{\alpha}^{(0)}) > \bar{B}_{\dot{\gamma}}[M_*/2,\, (N-P_*)/2]$ or (for values of $\mathbf{z}$ such that $\mathbf{z}_3 \neq \mathbf{0}$) the critical function of the F test (of $\tilde{H}_0$ versus $\tilde{H}_1$)]. And in light of equality (4.16), we conclude that among confidence sets for $\boldsymbol{\alpha}$ whose probability of coverage equals or exceeds $1-\dot{\gamma}$ and that are invariant with respect to the group $\tilde{G}_1$ and equivariant with respect to the groups $\tilde{G}_2$ and $\tilde{G}_3$, the confidence set $\tilde{A}_F(\mathbf{z})$ is UMA (uniformly most accurate) in the sense that (for every false value of $\boldsymbol{\alpha}$) its probability of false coverage is a minimum.

This result on the optimality of the $100(1-\dot{\gamma})\%$ confidence set $\tilde{A}_F(\mathbf{z})$ (for the vector $\boldsymbol{\alpha}$) can be translated into a result on the optimality of the following $100(1-\dot{\gamma})\%$ confidence set for the vector $\boldsymbol{\tau}$:

$$A_F(\mathbf{y}) = \{\dot{\boldsymbol{\tau}} \,:\, \dot{\boldsymbol{\tau}} = \mathbf{W}'\dot{\boldsymbol{\alpha}},\ \dot{\boldsymbol{\alpha}} \in \tilde{A}_F(\mathbf{O}'\mathbf{y})\} = \{\dot{\boldsymbol{\tau}} \in \boldsymbol{\Lambda}' \,:\, X(\dot{\boldsymbol{\tau}}) \le \bar{B}_{\dot{\gamma}}[M_*/2,\, (N-P_*)/2]\}$$

where (for $\dot{\boldsymbol{\tau}} \in \boldsymbol{\Lambda}'$)

$$X(\dot{\boldsymbol{\tau}}) = \begin{cases} \dfrac{(\hat{\boldsymbol{\tau}}-\dot{\boldsymbol{\tau}})'\mathbf{C}^-(\hat{\boldsymbol{\tau}}-\dot{\boldsymbol{\tau}})}{(\hat{\boldsymbol{\tau}}-\dot{\boldsymbol{\tau}})'\mathbf{C}^-(\hat{\boldsymbol{\tau}}-\dot{\boldsymbol{\tau}})+\mathbf{y}'(\mathbf{I}-\mathbf{P_X})\mathbf{y}}, & \text{if } \hat{\boldsymbol{\tau}} \neq \dot{\boldsymbol{\tau}} \text{ or } \mathbf{y}'(\mathbf{I}-\mathbf{P_X})\mathbf{y} > 0, \\ 0, & \text{if } \hat{\boldsymbol{\tau}} = \dot{\boldsymbol{\tau}} \text{ and } \mathbf{y}'(\mathbf{I}-\mathbf{P_X})\mathbf{y} = 0. \end{cases}$$

The set $A_F(\mathbf{y})$ is invariant with respect to the group G_1 and equivariant with respect to the groups G_2 and G_3 consisting respectively of the totality of the groups $G_2(\tau^{(0)})$ $[\boldsymbol{\tau}^{(0)} \in \mathcal{C}(\boldsymbol{\Lambda}')]$ and the totality of the groups $G_3(\tau^{(0)})$ $[\boldsymbol{\tau}^{(0)} \in \mathcal{C}(\boldsymbol{\Lambda}')]$. And the probability $\Pr[\boldsymbol{\tau}^{(0)} \in A_F(\mathbf{y})]$ of the set $A_F(\mathbf{y})$ covering any particular vector $\boldsymbol{\tau}^{(0)}$ in $\mathcal{C}(\boldsymbol{\Lambda}')$ equals the probability $\Pr[\mathbf{S}'\boldsymbol{\tau}^{(0)} \in \tilde{A}_F(\mathbf{z})]$ of the set $\tilde{A}_F(\mathbf{z})$ covering the vector $\mathbf{S}'\boldsymbol{\tau}^{(0)}$.

Corresponding to any confidence set $A(\mathbf{y})$ (for $\boldsymbol{\tau}$) whose probability of coverage equals or exceeds $1-\dot{\gamma}$ and that is invariant with respect to the group G_1 and equivariant with respect to the groups G_2 and G_3 is a confidence set $\tilde{A}(\mathbf{z})$ (for $\boldsymbol{\alpha}$) defined as follows:

$$\tilde{A}(\mathbf{z}) = \{\dot{\boldsymbol{\alpha}} \,:\, \dot{\boldsymbol{\alpha}} = \mathbf{S}'\dot{\boldsymbol{\tau}},\ \dot{\boldsymbol{\tau}} \in A(\mathbf{O}\mathbf{z})\}.$$

This set is such that

$$A(\mathbf{y}) = \{\dot{\boldsymbol{\tau}} \,:\, \dot{\boldsymbol{\tau}} = \mathbf{W}'\dot{\boldsymbol{\alpha}},\ \dot{\boldsymbol{\alpha}} \in \tilde{A}(\mathbf{O}'\mathbf{y})\},$$

and it is invariant with respect to the group $\tilde{G}_1$ and equivariant with respect to the groups $\tilde{G}_2$ and $\tilde{G}_3$. Moreover, for any vector $\boldsymbol{\tau}^{(0)} \in \mathcal{C}(\boldsymbol{\Lambda}')$,

$$\Pr[\boldsymbol{\tau}^{(0)} \in A(\mathbf{y})] = \Pr[\mathbf{S}'\boldsymbol{\tau}^{(0)} \in \tilde{A}(\mathbf{z})]$$

—in particular, $\Pr[\boldsymbol{\tau} \in A(\mathbf{y})] = \Pr[\boldsymbol{\alpha} \in \tilde{A}(\mathbf{z})]$. Since the confidence set $\tilde{A}_F(\mathbf{z})$ is UMA among confidence sets for $\boldsymbol{\alpha}$ whose probability of coverage equals or exceeds $1-\dot{\gamma}$ and that are invariant with respect to the group $\tilde{G}_1$ and equivariant with respect to the groups $\tilde{G}_2$ and $\tilde{G}_3$, we conclude that the confidence set $A_F(\mathbf{y})$ is UMA among confidence sets for $\boldsymbol{\tau}$ whose probability of coverage equals or exceeds $1-\dot{\gamma}$ and that are invariant with respect to the group G_1 and equivariant with respect to the groups G_2 and G_3.

e. Average power and average probability of false coverage

Let us consider further the problem of testing the null hypothesis $H_0 : \boldsymbol{\tau} = \boldsymbol{\tau}^{(0)}$ or $\tilde{H}_0 : \boldsymbol{\alpha} = \boldsymbol{\alpha}^{(0)}$ versus the alternative hypothesis $H_1 : \boldsymbol{\tau} \neq \boldsymbol{\tau}^{(0)}$ or $\tilde{H}_1 : \boldsymbol{\alpha} \neq \boldsymbol{\alpha}^{(0)}$. In Subsection d, it was shown (under an assumption of normality) that the size-$\dot{\gamma}$ F test is optimal in the sense that it is UMP among tests whose size does not exceed $\dot{\gamma}$ and that are invariant with respect to certain groups of transformations. In what follows, it is shown that the size-$\dot{\gamma}$ F test is also optimal in another sense. To proceed, we require certain definitions/results involving the integration of a function over a hypersphere.

Integration of a function over a hypersphere. Let N represent a positive integer greater than or equal to 2. And consider the integration of a function $g(\mathbf{s})$ of an $N \times 1$ vector $\mathbf{s}$ over a set $S_N(\mathbf{s}^{(0)}, \rho)$ defined for $\mathbf{s}^{(0)} \in \mathcal{R}^N$ and $\rho > 0$ as follows:

$$S_N(\mathbf{s}^{(0)}, \rho) = \{\mathbf{s} \in \mathcal{R}^N : (\mathbf{s}-\mathbf{s}^{(0)})'(\mathbf{s}-\mathbf{s}^{(0)}) = \rho^2\}.$$

When $N = 2$, $S_N(\mathbf{s}^{(0)}, \rho)$ is a circle, and when $N = 3$, it is a sphere. More generally, $S_N(\mathbf{s}^{(0)}, \rho)$ is referred to as a hypersphere (of dimension $N-1$). This circle, sphere, or hypersphere is centered at the point $\mathbf{s}^{(0)}$ and is of radius ρ.

Integration over the set $S_N(\mathbf{s}^{(0)}, \rho)$ is related to integration over a set B_N defined as follows:

$$B_N = \{\mathbf{x} \in \mathcal{R}^N : \mathbf{x}'\mathbf{x} \leq 1\}.$$

This set is a closed ball; it is centered at the origin $\mathbf{0}$ and is of radius 1.

Let us write $\int_{S_N(\mathbf{s}^{(0)}, \rho)} g(\mathbf{s})\, d\mathbf{s}$ for the integral of the function $g(\cdot)$ over the hypersphere $S_N(\mathbf{s}^{(0)}, \rho)$ centered at $\mathbf{s}^{(0)}$ and of radius ρ. In the special case of a hypersphere centered at the origin and of radius 1,

$$\int_{S_N(\mathbf{0}, 1)} g(\mathbf{s})\, d\mathbf{s} = N \int_{B_N} g[(\mathbf{x}'\mathbf{x})^{-1/2}\mathbf{x}]\, d\mathbf{x} \tag{4.18}$$

—for $\mathbf{x} = \mathbf{0}$, define $(\mathbf{x}'\mathbf{x})^{-1/2}\mathbf{x} = \mathbf{0}$. More generally (in the special case of a hypersphere centered at the origin and of radius ρ),

$$\int_{S_N(\mathbf{0}, \rho)} g(\mathbf{s})\, d\mathbf{s} = N\rho^{N-1} \int_{B_N} g[\rho\,(\mathbf{x}'\mathbf{x})^{-1/2}\mathbf{x}]\, d\mathbf{x}. \tag{4.19}$$

And still more generally,

$$\int_{S_N(\mathbf{s}^{(0)}, \rho)} g(\mathbf{s})\, d\mathbf{s} = \int_{S_N(\mathbf{0}, \rho)} g(\mathbf{s}^{(0)} + \tilde{\mathbf{s}})\, d\tilde{\mathbf{s}} \tag{4.20}$$

$$= N\rho^{N-1} \int_{B_N} g[\mathbf{s}^{(0)} + \rho\,(\mathbf{x}'\mathbf{x})^{-1/2}\mathbf{x}]\, d\mathbf{x}. \tag{4.21}$$

Following Baker (1997), let us regard equality (4.18) or, more generally, equality (4.19) or (4.21) as a definition—Baker indicated that he suspects equality (4.18) "is folkloric, and likely has appeared as a theorem rather than a definition." Various basic results on integration over a hypersphere follow readily from equality (4.18), (4.19), or (4.21). In particular, for any two constants a and b and for "any" two functions $g_1(\mathbf{s})$ and $g_2(\mathbf{s})$ of an $N \times 1$ vector $\mathbf{s}$,

$$\int_{S_N(\mathbf{s}^{(0)}, \rho)} [a g_1(\mathbf{s}) + b g_2(\mathbf{s})]\, d\mathbf{s} = a \int_{S_N(\mathbf{s}^{(0)}, \rho)} g_1(\mathbf{s})\, d\mathbf{s} + b \int_{S_N(\mathbf{s}^{(0)}, \rho)} g_2(\mathbf{s})\, d\mathbf{s}, \tag{4.22}$$

and for any $N \times N$ orthogonal matrix $\mathbf{O}$ [and any function $g(\mathbf{s})$ of an $N \times 1$ vector $\mathbf{s}$],

$$\int_{S_N(\mathbf{0}, \rho)} g(\mathbf{O}\mathbf{s})\, d\mathbf{s} = \int_{S_N(\mathbf{0}, \rho)} g(\mathbf{s})\, d\mathbf{s}. \tag{4.23}$$

In the special case where $g(\mathbf{s}) = 1$ (for every $N \times 1$ vector $\mathbf{s}$), $\int_{S_N(\mathbf{s}^{(0)}, \rho)} g(\mathbf{s})\, d\mathbf{s}$ represents the "surface area" of the $(N-1)$-dimensional hypersphere $S_N(\mathbf{s}^{(0)}, \rho)$. As demonstrated by Baker (1997),

$$\int_{S_N(\mathbf{s}^{(0)}, \rho)} d\mathbf{s} = \frac{2\pi^{N/2}}{\Gamma(N/2)} \rho^{N-1}. \tag{4.24}$$

The integration of a function over a hypersphere is defined for hyperspheres of dimension 1 or more by equality (4.21) or, in special cases, by equality (4.18) or (4.19). It is convenient to also define the integration of a function over a hypersphere for a "hypersphere" $S_1(s^{(0)}, \rho)$ of dimension 0 (centered at $s^{(0)}$ and of radius ρ), that is, for the "hypersphere"

$$S_1(s^{(0)}, \rho) = \{s \in \mathcal{R}^1 : (s-s^{(0)})^2 = \rho^2\} = \{s \in \mathcal{R}^1 : s = s^{(0)} \pm \rho\}. \tag{4.25}$$

Let us write $\int_{S_1(s^{(0)}, \rho)} g(s)\, ds$ for the integral of a function $g(\cdot)$ over the set $S_1(s^{(0)}, \rho)$, and define

$$\int_{S_1(s^{(0)}, \rho)} g(s)\, ds = g(s^{(0)} + \rho) + g(s^{(0)} - \rho). \tag{4.26}$$

It is also convenient to extend the definition of the integral $\int_{S_N(\mathbf{s}^{(0)}, \rho)} g(\mathbf{s})\, d\mathbf{s}$ to "hyperspheres" of radius 0, that is, to $\rho = 0$. For $N \geq 1$, let us take

$$\int_{S_N(\mathbf{s}^{(0)}, 0)} g(\mathbf{s})\, d\mathbf{s} = g(\mathbf{s}^{(0)}). \tag{4.27}$$

Note that when the definition of the integral of a function $g(\cdot)$ over the set $S_N(\mathbf{s}^{(0)}, \rho)$ is extended to $N = 1$ and/or $\rho = 0$ via equalitiies (4.26) and/or (4.27), properties (4.22) and (4.23) continue to apply. Note also that

$$\int_{S_N(\mathbf{s}^{(0)},\,0)} d\mathbf{s} = 1 \tag{4.28}$$

and that (for $\rho > 0$)

$$\int_{S_1(s^{(0)},\,\rho)} ds = 2. \tag{4.29}$$

An optimality criterion: average power. Let us now resume discussion of the problem of testing the null hypothesis $H_0 : \boldsymbol{\tau} = \boldsymbol{\tau}^{(0)}$ or $\tilde{H}_0 : \boldsymbol{\alpha} = \boldsymbol{\alpha}^{(0)}$ versus the alternative hypothesis $H_1 : \boldsymbol{\tau} \neq \boldsymbol{\tau}^{(0)}$ or $\tilde{H}_1 : \boldsymbol{\alpha} \neq \boldsymbol{\alpha}^{(0)}$. In doing so, let us take the context to be that of Section 7.3, and let us adopt the notation and terminology introduced therein. Thus, $\mathbf{y}$ is taken to be an $N\times 1$ observable random vector that follows the G–M model, and $\mathbf{z} = \mathbf{O}'\mathbf{y} = (\mathbf{z}_1', \mathbf{z}_2', \mathbf{z}_3')' = (\hat{\boldsymbol{\alpha}}', \hat{\boldsymbol{\eta}}', \mathbf{d}')'$. Moreover, in what follows, it is assumed that the distribution of the vector $\mathbf{e}$ of residual effects in the G–M model is $N(\mathbf{0}, \sigma^2\mathbf{I})$, in which case $\mathbf{y} \sim N(\mathbf{X}\boldsymbol{\beta}, \sigma^2\mathbf{I})$ and $\mathbf{z} \sim N[(\boldsymbol{\alpha}', \boldsymbol{\eta}', \mathbf{0}')', \sigma^2\mathbf{I}]$.

Let $\tilde{\phi}(\mathbf{z})$ represent (in terms of the transformed vector $\mathbf{z}$) the critical function of an arbitrary (possibly randomized) test of the null hypothesis $\tilde{H}_0 : \boldsymbol{\alpha} = \boldsymbol{\alpha}^{(0)}$ [so that $0 \le \tilde{\phi}(\mathbf{z}) \le 1$ for every value of $\mathbf{z}$]. And let $\tilde{\gamma}(\boldsymbol{\alpha}, \boldsymbol{\eta}, \sigma)$ represent the power function of this test. By definition,

$$\tilde{\gamma}(\boldsymbol{\alpha}, \boldsymbol{\eta}, \sigma) = \mathrm{E}[\tilde{\phi}(\mathbf{z})].$$

Further, let $\tilde{\boldsymbol{\alpha}} = \boldsymbol{\alpha} - \boldsymbol{\alpha}^{(0)}$, and take $\bar{\gamma}(\rho, \boldsymbol{\eta}, \sigma)$ to be the function of ρ, $\boldsymbol{\eta}$, and σ defined (for $\rho \ge 0$) as

$$\bar{\gamma}(\rho, \boldsymbol{\eta}, \sigma) = \frac{\int_{S(\boldsymbol{\alpha}^{(0)},\,\rho)} \tilde{\gamma}(\boldsymbol{\alpha}, \boldsymbol{\eta}, \sigma)\, d\boldsymbol{\alpha}}{\int_{S(\boldsymbol{\alpha}^{(0)},\,\rho)} d\boldsymbol{\alpha}},$$

where $S(\boldsymbol{\alpha}^{(0)}, \rho) = \{\boldsymbol{\alpha} : (\boldsymbol{\alpha}-\boldsymbol{\alpha}^{(0)})'(\boldsymbol{\alpha}-\boldsymbol{\alpha}^{(0)}) = \rho^2\}$, or equivalently as

$$\bar{\gamma}(\rho, \boldsymbol{\eta}, \sigma) = \frac{\int_{S(\mathbf{0},\,\rho)} \tilde{\gamma}(\alpha^{(0)}+\tilde{\boldsymbol{\alpha}}, \boldsymbol{\eta}, \sigma)\, d\tilde{\boldsymbol{\alpha}}}{\int_{S(\mathbf{0},\,\rho)} d\tilde{\boldsymbol{\alpha}}},$$

where $S(\mathbf{0}, \rho) = \{\tilde{\boldsymbol{\alpha}} : \tilde{\boldsymbol{\alpha}}'\tilde{\boldsymbol{\alpha}} = \rho^2\}$—for convenience, the same symbol (e.g., $\boldsymbol{\alpha}$) is sometimes used for more than one purpose.

The function $\bar{\gamma}(\rho, \boldsymbol{\eta}, \sigma)$ provides in whole or in part a possible basis for the evaluation of the test. For $\rho > 0$, the value of $\bar{\gamma}(\rho, \boldsymbol{\eta}, \sigma)$ represents the average (with respect to the value of $\boldsymbol{\alpha}$) power of the test over a hypersphere centered at (the hypothesized value) $\boldsymbol{\alpha}^{(0)}$ and of radius ρ. When $\rho = 0$, this value equals the Type-I error of the test, that is, the probability of falsely rejecting $\tilde{H}_0$—in general, this value can vary with $\boldsymbol{\eta}$ and/or σ.

Upon writing $f(\cdot\,; \boldsymbol{\alpha}, \boldsymbol{\eta}, \sigma)$ for the pdf of the distribution of $\mathbf{z}$, we find that

$$\begin{aligned}\bar{\gamma}(\rho, \boldsymbol{\eta}, \sigma) &= \frac{\int_{S(\mathbf{0},\,\rho)} \bigl[\int_{\mathbb{R}^{M_*}} \tilde{\phi}(\mathbf{z}) f(\mathbf{z}\,; \alpha^{(0)}+\tilde{\boldsymbol{\alpha}}, \boldsymbol{\eta}, \sigma)\, d\mathbf{z}\bigr]\, d\tilde{\boldsymbol{\alpha}}}{\int_{S(\mathbf{0},\,\rho)} d\tilde{\boldsymbol{\alpha}}} \\ &= \frac{\int_{\mathbb{R}^{M_*}} \tilde{\phi}(\mathbf{z}) \bigl[\int_{S(\mathbf{0},\,\rho)} f(\mathbf{z}\,; \alpha^{(0)}+\tilde{\boldsymbol{\alpha}}, \boldsymbol{\eta}, \sigma)\, d\tilde{\boldsymbol{\alpha}}\bigr]\, d\mathbf{z}}{\int_{S(\mathbf{0},\,\rho)} d\tilde{\boldsymbol{\alpha}}}.\end{aligned} \tag{4.30}$$

And upon writing $f_1(\cdot\,; \boldsymbol{\alpha}, \sigma)$, $f_2(\cdot\,; \boldsymbol{\eta}, \sigma)$, and $f_3(\cdot\,; \sigma)$ for the pdfs of the distributions of $\hat{\boldsymbol{\alpha}}$, $\hat{\boldsymbol{\eta}}$, and $\mathbf{d}$, we find that

$$\frac{\int_{S(\mathbf{0},\,\rho)} f(\mathbf{z}\,; \alpha^{(0)}+\tilde{\boldsymbol{\alpha}}, \boldsymbol{\eta}, \sigma)\, d\tilde{\boldsymbol{\alpha}}}{\int_{S(\mathbf{0},\,\rho)} d\tilde{\boldsymbol{\alpha}}} = f_1^*(\hat{\boldsymbol{\alpha}}\,; \rho, \sigma)\, f_2(\hat{\boldsymbol{\eta}}\,; \boldsymbol{\eta}, \sigma)\, f_3(\mathbf{d}\,; \sigma), \tag{4.31}$$

where

$$\begin{aligned}f_1^*(\hat{\boldsymbol{\alpha}}\,; \rho, \sigma) &= \frac{\int_{S(\mathbf{0},\,\rho)} f_1(\hat{\boldsymbol{\alpha}}\,; \alpha^{(0)}+\tilde{\boldsymbol{\alpha}}, \sigma)\, d\tilde{\boldsymbol{\alpha}}}{\int_{S(\mathbf{0},\,\rho)} d\tilde{\boldsymbol{\alpha}}} \\ &= (2\pi\sigma^2)^{-M_*/2} e^{-(\hat{\boldsymbol{\alpha}}-\boldsymbol{\alpha}^{(0)})'(\hat{\boldsymbol{\alpha}}-\boldsymbol{\alpha}^{(0)})/(2\sigma^2)} e^{-\rho^2/(2\sigma^2)}\, \psi(\hat{\boldsymbol{\alpha}}-\boldsymbol{\alpha}^{(0)}\,; \rho, \sigma)\end{aligned}$$

and where $\psi(\cdot\,;\rho,\sigma)$ is a function whose value is defined for every $M_* \times 1$ vector $\mathbf{x}$ as follows:

$$\psi(\mathbf{x}\,;\rho,\sigma) = \frac{\int_{S(\mathbf{0},\rho)} e^{\mathbf{x}'\tilde{\boldsymbol{\alpha}}/\sigma^2}\, d\tilde{\boldsymbol{\alpha}}}{\int_{S(\mathbf{0},\rho)} d\tilde{\boldsymbol{\alpha}}}.$$

Note that

$$f_1^*(\hat{\boldsymbol{\alpha}}\,;0,\sigma) = (2\pi\sigma^2)^{-M_*/2} e^{-(\hat{\boldsymbol{\alpha}}-\boldsymbol{\alpha}^{(0)})'(\hat{\boldsymbol{\alpha}}-\boldsymbol{\alpha}^{(0)})/(2\sigma^2)} = f_1(\hat{\boldsymbol{\alpha}}\,;\boldsymbol{\alpha}^{(0)},\sigma). \tag{4.32}$$

The function $\psi(\cdot\,;\rho,\sigma)$ is such that for every $M_* \times M_*$ orthogonal matrix $\mathbf{O}$ and every $M_* \times 1$ vector $\mathbf{x}$,

$$\psi(\mathbf{O}\mathbf{x}\,;\rho,\sigma) = \psi(\mathbf{x}\,;\rho,\sigma),$$

as is evident from result (4.23). Moreover, corresponding to any $M_* \times 1$ vectors $\mathbf{x}_1$ and $\mathbf{x}_2$ such that $\mathbf{x}_2'\mathbf{x}_2 = \mathbf{x}_1'\mathbf{x}_1$, there exists an orthogonal matrix $\mathbf{O}$ such that $\mathbf{x}_2 = \mathbf{O}\mathbf{x}_1$, as is evident from Lemma 5.9.9. Thus, $\psi(\mathbf{x}\,;\rho,\sigma)$ depends on $\mathbf{x}$ only through the value of $\mathbf{x}'\mathbf{x}$. That is,

$$\psi(\mathbf{x}\,;\rho,\sigma) = \tilde{\psi}(\mathbf{x}'\mathbf{x}\,;\rho,\sigma)$$

for some function $\tilde{\psi}(\cdot\,;\rho,\sigma)$ of a single (nonnegative) variable. And $f_1^*(\hat{\boldsymbol{\alpha}}\,;\rho,\sigma)$ is reexpressible as

$$f_1^*(\hat{\boldsymbol{\alpha}}\,;\rho,\sigma) = (2\pi\sigma^2)^{-M_*/2} e^{-(\hat{\boldsymbol{\alpha}}-\boldsymbol{\alpha}^{(0)})'(\hat{\boldsymbol{\alpha}}-\boldsymbol{\alpha}^{(0)})/(2\sigma^2)} e^{-\rho^2/(2\sigma^2)} \tilde{\psi}[(\hat{\boldsymbol{\alpha}}-\boldsymbol{\alpha}^{(0)})'(\hat{\boldsymbol{\alpha}}-\boldsymbol{\alpha}^{(0)})\,;\rho,\sigma]. \tag{4.33}$$

Letting $\mathbf{x}_*$ represent any $M_* \times 1$ vector such that $\mathbf{x}_*'\mathbf{x}_* = 1$, we find that for any nonnegative scalar t,

$$\tilde{\psi}(t\,;\rho,\sigma) = \tilde{\psi}(t\mathbf{x}_*'\mathbf{x}_*\,;\rho,\sigma) = \tfrac{1}{2}\big[\psi(\sqrt{t}\mathbf{x}_*\,;\rho,\sigma) + \psi(-\sqrt{t}\mathbf{x}_*\,;\rho,\sigma)\big].$$

Thus,

$$\tilde{\psi}(0\,;\rho,\sigma) = 1. \tag{4.34}$$

Moreover, for $t > 0$,

$$\frac{d\tilde{\psi}(t\,;\rho,\sigma)}{dt} = \frac{\frac{1}{2}\int_{S(\mathbf{0},\rho)} d\big[e^{\sqrt{t}\mathbf{x}_*'\tilde{\boldsymbol{\alpha}}/\sigma^2} + e^{-\sqrt{t}\mathbf{x}_*'\tilde{\boldsymbol{\alpha}}/\sigma^2}\big]/dt\; d\tilde{\boldsymbol{\alpha}}}{\int_{S(\mathbf{0},\rho)} d\tilde{\boldsymbol{\alpha}}};$$

and upon observing that (for $t > 0$)

$$\begin{aligned} d\big[e^{\sqrt{t}\mathbf{x}_*'\tilde{\boldsymbol{\alpha}}/\sigma^2} + e^{-\sqrt{t}\mathbf{x}_*'\tilde{\boldsymbol{\alpha}}/\sigma^2}\big]/dt &= \tfrac{1}{2}\big(t^{-1/2}\mathbf{x}_*'\tilde{\boldsymbol{\alpha}}/\sigma^2\big)\big(e^{\sqrt{t}\mathbf{x}_*'\tilde{\boldsymbol{\alpha}}/\sigma^2} + e^{-\sqrt{t}\mathbf{x}_*'\tilde{\boldsymbol{\alpha}}/\sigma^2}\big) \\ &> 0 \text{ (unless } \mathbf{x}_*'\tilde{\boldsymbol{\alpha}} = 0), \end{aligned}$$

it follows that (for $t > 0$ and $\rho > 0$)

$$\frac{d\tilde{\psi}(t\,;\rho,\sigma)}{dt} > 0 \tag{4.35}$$

and hence that (unless $\rho = 0$) $\tilde{\psi}(\cdot\,;\rho,\sigma)$ is a strictly increasing function.

An equivalence. Let $\mathcal{T}$ represent a collection of (possibly randomized) tests of H_0 or $\tilde{H}_0$ (versus H_1 or $\tilde{H}_1$). And consider the problem of identifying a test in $\mathcal{T}$ for which the value of $\bar{\gamma}(\rho,\boldsymbol{\eta},\sigma)$ (at any particular values of ρ, $\boldsymbol{\eta}$, and σ) is greater than or equal to its value for every other test in $\mathcal{T}$. This problem can be transformed into a problem that is equivalent to the original but that is more directly amenable to solution by conventional means.

Together, results (4.30) and (4.31) imply that

$$\bar{\gamma}(\rho,\boldsymbol{\eta},\sigma) = \int_{\mathbb{R}^N} \tilde{\phi}(\mathbf{z}) f^*(\mathbf{z}\,;\rho,\boldsymbol{\eta},\sigma)\, d\mathbf{z}, \tag{4.36}$$

where $f^*(\mathbf{z}\,;\rho,\boldsymbol{\eta},\sigma) = f_1^*(\hat{\boldsymbol{\alpha}}\,;\rho,\sigma) f_2(\hat{\boldsymbol{\eta}}\,;\boldsymbol{\eta},\sigma) f_3(\mathbf{d}\,;\sigma)$. Moreover, $f_1^*(\hat{\boldsymbol{\alpha}}\,;\rho,\sigma) \geq 0$ (for every value of $\hat{\boldsymbol{\alpha}}$) and $\int_{\mathbb{R}^{M_*}} f_1^*(\hat{\boldsymbol{\alpha}}\,;\rho,\sigma)\, d\hat{\boldsymbol{\alpha}} = 1$, so that $f_1^*(\cdot\,;\rho,\sigma)$ can serve as the pdf of an absolutely continuous distribution. And it follows that

$$\bar{\gamma}(\rho,\boldsymbol{\eta},\sigma) = \mathrm{E}^*[\tilde{\phi}(\mathbf{z})], \tag{4.37}$$

where the symbol E^* denotes an expected value obtained when the underlying distribution of $\mathbf{z}$ is taken to be that with pdf $f^*(\cdot\,;\rho,\boldsymbol{\eta},\sigma)$ rather than that with pdf $f(\cdot\,;\boldsymbol{\alpha},\boldsymbol{\eta},\sigma)$.

In the transformed version of the problem of identifying a test of H_0 or $\tilde{H}_0$ that maximizes the value of $\bar{\gamma}(\rho, \boldsymbol{\eta}, \sigma)$, the distribution of $\hat{\boldsymbol{\alpha}}$ is taken to be that with pdf $f_1^*(\cdot\,; \rho, \sigma)$ and the distribution of the observable random vector $\mathbf{z} = (\hat{\boldsymbol{\alpha}}', \hat{\boldsymbol{\eta}}', \mathbf{d}')'$ is taken to be that with pdf $f^*(\cdot\,; \rho, \boldsymbol{\eta}, \sigma)$. Further, the nonnegative scalar ρ (as well as σ and each element of $\boldsymbol{\eta}$) is regarded as an unknown parameter. And the test with critical function $\tilde{\phi}(\cdot)$ is regarded as a test of the null hypothesis $H_0^* : \rho = 0$ versus the alternative hypothesis $H_1^* : \rho > 0$ (and $\mathcal{T}$ is regarded as a collection of tests of H_0^*). Note [in light of result (4.37)] that in this context, $\bar{\gamma}(\rho, \boldsymbol{\eta}, \sigma)$ is interpretable as the power function of the test with critical function $\tilde{\phi}(\cdot)$.

The transformed version of the problem consists of identifying a test in $\mathcal{T}$ for which the value of the power function (at the particular values of ρ, $\boldsymbol{\eta}$, and σ) is greater than or equal to the value of the power function for every other test in $\mathcal{T}$. The transformed version is equivalent to the original version in that a test of H_0 or $\tilde{H}_0$ is a solution to the original version if and only if its critical function is the critical function of a test of H_0^* that is a solution to the transformed version.

In what follows, $\mathcal{T}$ is taken to be the collection of all size-$\dot{\gamma}$ similar tests of H_0 or $\tilde{H}_0$. And a solution to the problem of identifying a test in $\mathcal{T}$ for which the value of $\bar{\gamma}(\rho, \boldsymbol{\eta}, \sigma)$ is greater than or equal to its value for every other test in $\mathcal{T}$ is effected by solving the transformed version of this problem. Note that in the context of the transformed version, $\mathcal{T}$ represents the collection of all size-$\dot{\gamma}$ similar tests of H_0^*, as is evident upon observing [in light of result (4.32)] that a test of H_0^* and a test of H_0 or $\tilde{H}_0$ that have the same critical function have the same Type-I error.

A sufficient statistic. Suppose that the distribution of $\hat{\boldsymbol{\alpha}}$ is taken to be that with pdf $f_1^*(\cdot\,; \rho, \sigma)$ and the distribution of $\mathbf{z} = (\hat{\boldsymbol{\alpha}}', \hat{\boldsymbol{\eta}}', \mathbf{d}')'$ to be that with pdf $f^*(\cdot\,; \rho, \boldsymbol{\eta}, \sigma)$. Further, let

$$u = (\hat{\boldsymbol{\alpha}} - \boldsymbol{\alpha}^{(0)})'(\hat{\boldsymbol{\alpha}} - \boldsymbol{\alpha}^{(0)}) \quad \text{and} \quad v = \mathbf{d}'\mathbf{d}.$$

And define

$$s = \frac{u}{u+v} \quad \text{and} \quad w = u + v.$$

Then, clearly, u, v, and $\hat{\boldsymbol{\eta}}$ form a sufficient statistic—recall result (4.33)—and s, w, and $\hat{\boldsymbol{\eta}}$ also form a sufficient statistic.

The random vectors $\hat{\boldsymbol{\alpha}}$, $\hat{\boldsymbol{\eta}}$, and $\mathbf{d}$ are distributed independently and, consequently, u, v, and $\hat{\boldsymbol{\eta}}$ are distributed independently. And the random variable u has an absolutely continuous distribution with a pdf $g_1^*(\cdot\,; \rho, \sigma)$ that is derivable from the pdf $f_1^*(\cdot\,; \rho, \sigma)$ of $\hat{\boldsymbol{\alpha}}$ by introducing successive changes in variables as follows: from $\hat{\boldsymbol{\alpha}}$ to the vector $\mathbf{x} = (x_1, x_2, \ldots, x_{M_*})' = \hat{\boldsymbol{\alpha}} - \boldsymbol{\alpha}^{(0)}$, from $\mathbf{x}$ to the $M_* \times 1$ vector $\mathbf{t}$ whose ith element is $t_i = x_i^2$, from $\mathbf{t}$ to the $M_* \times 1$ vector $\mathbf{y}$ whose first $M_* - 1$ elements are $y_i = t_i$ $(i = 1, 2, \ldots, M_* - 1)$ and whose M_*th element is $y_{M_*} = \sum_{i=1}^{M_*} t_i$, and finally from $\mathbf{y}$ to the vector whose first $M_* - 1$ elements are y_i / y_{M_*} $(i = 1, 2, \ldots, M_* - 1)$ and whose M_*th element is y_{M_*}—refer to Section 6.1g for the details. Upon introducing these changes of variables and upon observing that $u = y_{M_*}$, we find that (for $u > 0$)

$$g_1^*(u\,; \rho, \sigma) \propto u^{(M_*/2)-1} e^{-u/(2\sigma^2)} \tilde{\psi}(u\,; \rho, \sigma). \tag{4.38}$$

Moreover, the random variable v has an absolutely continuous distribution with a pdf $g_2(\cdot\,; \sigma)$ such that (for $v > 0$)

$$g_2(v\,; \sigma) = \frac{1}{\Gamma[(N - P_*)/2](2\sigma^2)^{(N-P_*)/2}} v^{[(N-P_*)/2]-1} e^{-v/(2\sigma^2)}, \tag{4.39}$$

as is evident upon observing that v/σ^2 has a chi-square distribution with $N - P_*$ degrees of freedom.

The random variables s and w have a joint distribution that is absolutely continuous with a pdf $h^*(\cdot\,, \cdot\,; \rho, \sigma)$ that is determinable (via a change of variables) from the pdfs of the distributions of u and v. For $0 < s < 1$ and $w > 0$, we find that

$$\begin{aligned} h^*(s\,, w\,; \rho, \sigma) &= g_1^*(sw\,; \rho, \sigma)\, g_2[(1-s)w\,; \sigma]\, w \\ &\propto s^{(M_*/2)-1}(1-s)^{[(N-P_*)/2]-1} w^{[(N-P_*+M_*)/2]-1} e^{-w/(2\sigma^2)} \tilde{\psi}(sw\,; \rho, \sigma). \end{aligned} \tag{4.40}$$

Corresponding to a test of H_0^* versus H_1^* with critical function $\tilde{\phi}(\mathbf{z})$ is the test with the critical function $\mathrm{E}^*[\tilde{\phi}(\mathbf{z}) \mid s, w, \hat{\boldsymbol{\eta}}]$ obtained upon taking the expected value of $\tilde{\phi}(\mathbf{z})$ conditional on s, w, and $\hat{\boldsymbol{\eta}}$. Moreover,

$$\mathrm{E}^*\{\mathrm{E}^*[\tilde{\phi}(\mathbf{z}) \mid s, w, \hat{\boldsymbol{\eta}}]\} = \mathrm{E}^*[\tilde{\phi}(\mathbf{z})], \tag{4.41}$$

so that the test with critical function $\mathrm{E}^*[\tilde{\phi}(\mathbf{z}) \mid s, w, \hat{\boldsymbol{\eta}}]$ has the same power function as the test with critical function $\tilde{\phi}(\mathbf{z})$. Thus, for purposes of identifying a size-$\dot{\gamma}$ similar test of H_0^* for which the value of the power function (at particular values of ρ, $\boldsymbol{\eta}$, and σ) is greater than or equal to the value of the power function for every other size-$\dot{\gamma}$ similar test of H_0^*, it suffices to restrict attention to tests that depend on $\mathbf{z} = (\hat{\boldsymbol{\alpha}}', \hat{\boldsymbol{\eta}}', \mathbf{d}')'$ only through the values of s, w, and $\hat{\boldsymbol{\eta}}$.

Conditional Type-I error and conditional power. Let $\tilde{\phi}(s, w, \hat{\boldsymbol{\eta}})$ represent the critical function of a (possibly randomized) test (of H_0^* versus H_1^*) that depends on $\mathbf{z}$ only through the value of the sufficient statistic formed by s, w, and $\hat{\boldsymbol{\eta}}$. And let $\gamma^*(\rho, \boldsymbol{\eta}, \sigma)$ represent the power function of the test. Then,

$$\gamma^*(\rho, \boldsymbol{\eta}, \sigma) = \mathrm{E}^*[\tilde{\phi}(s, w, \hat{\boldsymbol{\eta}})]. \tag{4.42}$$

And the test is a size-$\dot{\gamma}$ similar test if and only if

$$\gamma^*(0, \boldsymbol{\eta}, \sigma) = \dot{\gamma} \tag{4.43}$$

for all values of $\boldsymbol{\eta}$ and σ.

The conditional expected value $\mathrm{E}^*[\tilde{\phi}(s, w, \hat{\boldsymbol{\eta}}) \mid w, \hat{\boldsymbol{\eta}}]$ represents the conditional (on w and $\hat{\boldsymbol{\eta}}$) probability of the test with critical function $\tilde{\phi}(s, w, \hat{\boldsymbol{\eta}})$ rejecting the null hypothesis H_0^*. The power function $\gamma^*(\rho, \boldsymbol{\eta}, \sigma)$ of this test can be expressed in terms of $\mathrm{E}^*[\tilde{\phi}(s, w, \hat{\boldsymbol{\eta}}) \mid w, \hat{\boldsymbol{\eta}}]$. Upon reexpressing the right side of equality (4.42) in terms of this conditional expected value, we find that

$$\gamma^*(\rho, \boldsymbol{\eta}, \sigma) = \mathrm{E}^*\{\mathrm{E}^*[\tilde{\phi}(s, w, \hat{\boldsymbol{\eta}}) \mid w, \hat{\boldsymbol{\eta}}]\}. \tag{4.44}$$

Let us write $\mathrm{E}_0^*[\tilde{\phi}(s, w, \hat{\boldsymbol{\eta}}) \mid w, \hat{\boldsymbol{\eta}}]$ for $\mathrm{E}^*[\tilde{\phi}(s, w, \hat{\boldsymbol{\eta}}) \mid w, \hat{\boldsymbol{\eta}}]$ in the special case where $\rho = 0$, so that $\mathrm{E}_0^*[\tilde{\phi}(s, w, \hat{\boldsymbol{\eta}}) \mid w, \hat{\boldsymbol{\eta}}]$ represents the conditional (on w and $\hat{\boldsymbol{\eta}}$) probability of a Type-I error. When $\rho = 0$, s is distributed independently of w and $\hat{\boldsymbol{\eta}}$ as a $Be[M_*/2,\ (N{-}P_*)/2]$ random variable, so that (under H_0^*) the conditional distribution of s given w and $\hat{\boldsymbol{\eta}}$ is an absolutely continuous distribution with a pdf $h_0^*(s)$ that is expressible (for $0 < s < 1$) as

$$h_0^*(s) = \frac{1}{B[M_*/2,\ (N{-}P_*)/2]}\, s^{(M_*/2)-1}(1-s)^{[(N-P_*)/2]-1}$$

—refer to result (4.40) and note that $\tilde{\psi}(sw\,; 0, \sigma) = 1$. And (for every value of w and every value of $\hat{\boldsymbol{\eta}}$)

$$\mathrm{E}_0^*[\tilde{\phi}(s, w, \hat{\boldsymbol{\eta}}) \mid w, \hat{\boldsymbol{\eta}}] = \int_0^1 \tilde{\phi}(s, w, \hat{\boldsymbol{\eta}})h_0^*(s)\, ds.$$

Under H_0^* (i.e., when $\rho = 0$), w and $\hat{\boldsymbol{\eta}}$ form a complete sufficient statistic [for distributions of $\mathbf{z}$ with a pdf of the form $f^*(\mathbf{z}\,; \rho, \boldsymbol{\eta}, \sigma)$]—refer to Section 5.8. And in light of result (4.44), it follows that the test with critical function $\tilde{\phi}(s, w, \hat{\boldsymbol{\eta}})$ is a size-$\dot{\gamma}$ similar test if and only if

$$\mathrm{E}_0^*[\tilde{\phi}(s, w, \hat{\boldsymbol{\eta}}) \mid w, \hat{\boldsymbol{\eta}}] = \dot{\gamma} \qquad \text{(wp1)}. \tag{4.45}$$

Thus, a size-$\dot{\gamma}$ similar test of H_0^* for which the value of $\gamma^*(\rho, \boldsymbol{\eta}, \sigma)$ (at any particular values of ρ, $\boldsymbol{\eta}$, and σ) is greater than or equal to its value for any other size-$\dot{\gamma}$ similar test can be obtained by taking the critical function $\tilde{\phi}(s, w, \hat{\boldsymbol{\eta}})$ of the test to be that derived by regarding (for each value of w and each value of $\hat{\boldsymbol{\eta}}$) $\tilde{\phi}(s, w, \hat{\boldsymbol{\eta}})$ as a function of s alone and by maximizing the value of $\mathrm{E}^*[\tilde{\phi}(s, w, \hat{\boldsymbol{\eta}}) \mid w, \hat{\boldsymbol{\eta}}]$ (with respect to the choice of that function) subject to the constraint $\mathrm{E}_0^*[\tilde{\phi}(s, w, \hat{\boldsymbol{\eta}}) \mid w, \hat{\boldsymbol{\eta}}] = \dot{\gamma}$.

When $\rho > 0$ (as when $\rho = 0$), $\hat{\boldsymbol{\eta}}$ is distributed independently of s and w. However, when $\rho > 0$ (unlike when $\rho = 0$), s and w are statistically dependent. And in general (i.e., for $\rho \geq 0$), the conditional distribution of s given w and $\hat{\boldsymbol{\eta}}$ is an absolutely continuous distribution with a pdf $h_+^*(s \mid w)$ such that

$$h_+^*(s \mid w) \propto h^*(s\,, w\,; \rho, \sigma) \tag{4.46}$$

—this distribution varies with the values of ρ and σ as well as the value of w.

Application of the Neyman–Pearson lemma. Observe [in light of results (4.46) and (4.40)] that (for $0 < s < 1$)

$$\frac{h_+^*(s \mid w)}{h_0^*(s)} \propto \tilde{\psi}(sw\,; \rho, \sigma).$$

Observe also that (when $w > 0$ and $\rho > 0$) $\tilde{\psi}(sw\,; \rho, \sigma)$ is strictly increasing in s [as is evident upon recalling that (when $\rho > 0$) $\tilde{\psi}(\cdot\,; \rho, \sigma)$ is a strictly increasing function]. Further, let $\tilde{\phi}^*(s)$ represent the critical function of a (size-$\dot{\gamma}$) test (of H_0^*) that depends on $\mathbf{z}$ only through the value of s and that is defined as follows:

$$\tilde{\phi}^*(s) = \begin{cases} 1, & \text{if } s > k, \\ 0, & \text{if } s \le k, \end{cases}$$

where $\int_k^1 h_0^*(s)\,ds = \dot{\gamma}$. Then, it follows from an extended (to cover randomized tests) version of Theorem 7.4.1 (the Neyman–Pearson lemma) that (for every strictly positive value of ρ and for all values of $\boldsymbol{\eta}$ and σ) the "conditional power" $\mathrm{E}^*[\tilde{\phi}(s, w, \hat{\boldsymbol{\eta}}) \mid w, \hat{\boldsymbol{\eta}}]$ of the test with critical function $\tilde{\phi}(s, w, \hat{\boldsymbol{\eta}})$ attains its maximum value, subject to the constraint $\mathrm{E}_0^*[\tilde{\phi}(s, w, \hat{\boldsymbol{\eta}}) \mid w, \hat{\boldsymbol{\eta}}] = \dot{\gamma}$, when $\tilde{\phi}(s, w, \hat{\boldsymbol{\eta}})$ is taken to be the function $\tilde{\phi}^*(s)$. And (in light of the results of the preceding two parts of the present subsection) we conclude that the test of H_0^* with critical function $\tilde{\phi}^*(s)$ is UMP among all size-$\dot{\gamma}$ similar tests.

Main result. Based on relationship (4.37) and equality (4.32), the result on the optimality of the test of H_0^* with critical function $\tilde{\phi}^*(s)$ can be reexpressed as a result on the optimality of the test of H_0 or $\tilde{H}_0$ with the same critical function. Moreover, the test of H_0 or $\tilde{H}_0$ with critical function $\tilde{\phi}^*(s)$ is equivalent to the size-$\dot{\gamma}$ F test of H_0 or $\tilde{H}_0$. Thus, among size-$\dot{\gamma}$ similar tests of $H_0: \boldsymbol{\tau} = \boldsymbol{\tau}^{(0)}$ or $\tilde{H}_0: \boldsymbol{\alpha} = \boldsymbol{\alpha}^{(0)}$, the average value $\bar{\gamma}(\rho, \boldsymbol{\eta}, \sigma)$ of the power function $\tilde{\gamma}(\boldsymbol{\alpha}, \boldsymbol{\eta}, \sigma)$ over those $\boldsymbol{\alpha}$-values located on the sphere $S(\boldsymbol{\alpha}^{(0)}, \rho)$, centered at $\boldsymbol{\alpha}^{(0)}$ and of radius ρ, is maximized (for every $\rho > 0$ and for all $\boldsymbol{\eta}$ and σ) by taking the test of H_0 or $\tilde{H}_0$ to be the size-$\dot{\gamma}$ F test.

A corollary. Suppose that a test of $H_0: \boldsymbol{\tau} = \boldsymbol{\tau}^{(0)}$ or $\tilde{H}_0: \boldsymbol{\alpha} = \boldsymbol{\alpha}^{(0)}$ versus $H_1: \boldsymbol{\tau} \neq \boldsymbol{\tau}^{(0)}$ or $\tilde{H}_1: \boldsymbol{\alpha} \neq \boldsymbol{\alpha}^{(0)}$ is such that the power function $\tilde{\gamma}(\boldsymbol{\alpha}, \boldsymbol{\eta}, \sigma)$ depends on $\boldsymbol{\alpha}$, $\boldsymbol{\eta}$, and σ only through the value of $(\boldsymbol{\alpha}-\boldsymbol{\alpha}^0)'(\boldsymbol{\alpha}-\boldsymbol{\alpha}^0)/\sigma^2$, so that

$$\tilde{\gamma}(\boldsymbol{\alpha}, \boldsymbol{\eta}, \sigma) = \ddot{\gamma}[(\boldsymbol{\alpha}-\boldsymbol{\alpha}^0)'(\boldsymbol{\alpha}-\boldsymbol{\alpha}^0)/\sigma^2] \tag{4.47}$$

for some function $\ddot{\gamma}(\cdot)$. Then, for all $\boldsymbol{\eta}$ and σ,

$$\tilde{\gamma}(\boldsymbol{\alpha}^{(0)}, \boldsymbol{\eta}, \sigma) = \ddot{\gamma}(0). \tag{4.48}$$

And for $\rho \ge 0$ and for all $\boldsymbol{\eta}$ and σ,

$$\bar{\gamma}(\rho, \boldsymbol{\eta}, \sigma) = \ddot{\gamma}(\rho^2/\sigma^2). \tag{4.49}$$

As noted in Section 7.3b, the size-$\dot{\gamma}$ F test is among those tests of H_0 or $\tilde{H}_0$ for which the power function $\tilde{\gamma}(\boldsymbol{\alpha}, \boldsymbol{\eta}, \sigma)$ is of the form (4.47). Moreover, any size-$\dot{\gamma}$ test of H_0 or $\tilde{H}_0$ for which $\tilde{\gamma}(\boldsymbol{\alpha}, \boldsymbol{\eta}, \sigma)$ is of the form (4.47) is a size-$\dot{\gamma}$ similar test, as is evident from result (4.48). And, together, results (4.47) and (4.49) imply that (for all $\boldsymbol{\alpha}$, $\boldsymbol{\eta}$, and σ)

$$\tilde{\gamma}(\boldsymbol{\alpha}, \boldsymbol{\eta}, \sigma) = \bar{\gamma}\{[(\boldsymbol{\alpha}-\boldsymbol{\alpha}^0)'(\boldsymbol{\alpha}-\boldsymbol{\alpha}^0)]^{1/2}, \boldsymbol{\eta}, \sigma\}.$$

Thus, it follows from what has already been proven that among size-$\dot{\gamma}$ tests of H_0 or $\tilde{H}_0$ (versus H_1 or $\tilde{H}_1$) with a power function of the form (4.47), the size-$\dot{\gamma}$ F test is a UMP test.

The power function of a test of H_0 or $\tilde{H}_0$ versus H_1 or $\tilde{H}_1$ can be expressed as either a function $\tilde{\gamma}(\boldsymbol{\alpha}, \boldsymbol{\eta}, \sigma)$ of $\boldsymbol{\alpha}$, $\boldsymbol{\eta}$, and σ or as a function $\gamma(\boldsymbol{\tau}, \boldsymbol{\eta}, \sigma) = \tilde{\gamma}(\mathbf{S}'\boldsymbol{\tau}, \boldsymbol{\eta}, \sigma)$ of $\boldsymbol{\tau}$, $\boldsymbol{\eta}$, and σ. Note [in light of result (3.48)] that for $\tilde{\gamma}(\boldsymbol{\alpha}, \boldsymbol{\eta}, \sigma)$ to depend on $\boldsymbol{\alpha}$, $\boldsymbol{\eta}$, and σ only through the value of $(\boldsymbol{\alpha}-\boldsymbol{\alpha}^0)'(\boldsymbol{\alpha}-\boldsymbol{\alpha}^0)/\sigma^2$, it is necessary and sufficient that $\gamma(\boldsymbol{\tau}, \boldsymbol{\eta}, \sigma)$ depend on $\boldsymbol{\tau}$, $\boldsymbol{\eta}$, and σ only through the value of $(\boldsymbol{\tau}-\boldsymbol{\tau}^{(0)})'\mathbf{C}^-(\boldsymbol{\tau}-\boldsymbol{\tau}^{(0)})/\sigma^2$.

Special case: $M = M_* = 1$. Suppose that $M = M_* = 1$. And recall (from Section 7.3b) that in this special case, the size-$\dot{\gamma}$ F test of $H_0 : \boldsymbol{\tau} = \boldsymbol{\tau}^{(0)}$ or $\tilde{H}_0 : \boldsymbol{\alpha} = \boldsymbol{\alpha}^{(0)}$ versus $H_1 : \boldsymbol{\tau} \neq \boldsymbol{\tau}^{(0)}$ or $\tilde{H}_1 : \boldsymbol{\alpha} \neq \boldsymbol{\alpha}^{(0)}$ simplifies to the (size-$\dot{\gamma}$) two-sided t test. Further, let us write α and τ for $\boldsymbol{\alpha}$ and $\boldsymbol{\tau}$ and $\alpha^{(0)}$ and $\tau^{(0)}$ for $\boldsymbol{\alpha}^{(0)}$ and $\boldsymbol{\tau}^{(0)}$.

For $\rho > 0$, $S(\alpha^{(0)}, \rho) = \{\alpha : |\alpha - \alpha^{(0)}| = \rho\}$, so that $S(\alpha^{(0)}, \rho)$ consists of the two points $\alpha^{(0)} \pm \rho$. And we find that among size-$\dot{\gamma}$ similar tests of H_0 or $\tilde{H}_0$ versus H_1 or $\tilde{H}_1$, the average power at any two points that are equidistant from $\alpha^{(0)}$ or $\tau^{(0)}$ is maximized by taking the test to be the (size-$\dot{\gamma}$) two-sided t test. Moreover, among all size-$\dot{\gamma}$ tests of H_0 or $\tilde{H}_0$ whose power functions depend on α, $\boldsymbol{\eta}$, and σ or τ, $\boldsymbol{\eta}$, and σ only through the value of $|\alpha - \alpha^{(0)}|/\sigma$ or $|\tau - \tau^{(0)}|/\sigma$, the (size-$\dot{\gamma}$) two-sided t test is a UMP test.

Since the size-$\dot{\gamma}$ two-sided t test is equivalent to the size-$\dot{\gamma}$ F test and since (as discussed in Section 7.3b) the size-$\dot{\gamma}$ F test is strictly unbiased, the size-$\dot{\gamma}$ two-sided t test is a strictly unbiased test. Moreover, it can be shown that among all level-$\dot{\gamma}$ unbiased tests of H_0 or $\tilde{H}_0$ versus H_1 or $\tilde{H}_1$, the size-$\dot{\gamma}$ two-sided t test is a UMP test.

Confidence sets. Let $\tilde{A}(\hat{\boldsymbol{\alpha}}, \hat{\boldsymbol{\eta}}, \mathbf{d})$ or simply $\tilde{A}$ represent an arbitrary confidence set for $\boldsymbol{\alpha}$ with confidence coefficient $1 - \dot{\gamma}$. And let $\tilde{A}_F(\hat{\boldsymbol{\alpha}}, \hat{\boldsymbol{\eta}}, \mathbf{d})$ or simply $\tilde{A}_F$ represent the $100(1-\dot{\gamma})\%$ confidence set

$$\{\dot{\boldsymbol{\alpha}} : (\dot{\boldsymbol{\alpha}} - \hat{\boldsymbol{\alpha}})'(\dot{\boldsymbol{\alpha}} - \hat{\boldsymbol{\alpha}}) \leq M_* \hat{\sigma}^2 \bar{F}_{\dot{\gamma}}(M_*, N - P_*)\}.$$

Our results on the optimality of the size-$\dot{\gamma}$ F test of $\tilde{H}_0 : \boldsymbol{\alpha} = \boldsymbol{\alpha}^{(0)}$ can be reexpressed as results on the optimality of $\tilde{A}_F$. When $\tilde{A}$ is required to be such that $\Pr(\boldsymbol{\alpha} \in \tilde{A}) = 1 - \dot{\gamma}$ for all $\boldsymbol{\alpha}$, $\boldsymbol{\eta}$, and σ, the optimal choice for $\tilde{A}$ is $\tilde{A}_F$; this choice is optimal in the sense that (for $\rho > 0$) it minimizes the average value (with respect to $\boldsymbol{\alpha}^{(0)}$) of $\Pr(\boldsymbol{\alpha}^{(0)} \in \tilde{A})$ (the probability of false coverage) over the sphere $S(\boldsymbol{\alpha}, \rho)$ centered at $\boldsymbol{\alpha}$ with radius ρ. And $\tilde{A}_F$ is UMA (uniformly most accurate) when the choice of the set $\tilde{A}$ is restricted to a set for which the probability $\Pr(\boldsymbol{\alpha}^{(0)} \in \tilde{A})$ of $\tilde{A}$ covering a vector $\boldsymbol{\alpha}^{(0)}$ depends on $\boldsymbol{\alpha}$, $\boldsymbol{\eta}$, and σ only through the value $(\boldsymbol{\alpha}^{(0)} - \boldsymbol{\alpha})'(\boldsymbol{\alpha}^{(0)} - \boldsymbol{\alpha})/\sigma^2$; it is UMA in the sense that the probability of false coverage is minimized for every $\boldsymbol{\alpha}^{(0)}$ and for all $\boldsymbol{\alpha}$, $\boldsymbol{\eta}$, and σ. Moreover, among those $100(1-\dot{\gamma})$ confidence sets for which the probability of covering a vector $\boldsymbol{\tau}^{(0)}$ [in $\mathcal{C}(\boldsymbol{\Lambda}')$] depends on $\boldsymbol{\tau}$, $\boldsymbol{\eta}$, and σ only through the value of $(\boldsymbol{\tau}^{(0)} - \boldsymbol{\tau})'\mathbf{C}^-(\boldsymbol{\tau}^{(0)} - \boldsymbol{\tau})/\sigma^2$, the UMA set is the set

$$A_F = \{\dot{\boldsymbol{\tau}} \in \mathcal{C}(\boldsymbol{\Lambda}') : (\dot{\boldsymbol{\tau}} - \hat{\boldsymbol{\tau}})'\mathbf{C}^-(\dot{\boldsymbol{\tau}} - \hat{\boldsymbol{\tau}}) \leq M_* \hat{\sigma}^2 \bar{F}_{\dot{\gamma}}(M_*, N - P_*)\}.$$

7.5 One-Sided *t* Tests and the Corresponding Confidence Bounds

Suppose (as in Sections 7.3 and 7.4) that $\mathbf{y}$ is an $N \times 1$ observable random vector that follows the G–M model. And suppose that the distribution of the vector $\mathbf{e}$ of residual effects in the G–M model is $N(\mathbf{0}, \sigma^2\mathbf{I})$, so that $\mathbf{y} \sim N(\mathbf{X}\boldsymbol{\beta}, \sigma^2\mathbf{I})$. Let us consider further the problem of making inferences for estimable linear combinations of the elements of $\boldsymbol{\beta}$.

Procedures were presented and discussed in Sections 7.3 and 7.4 for constructing a confidence interval for any particular estimable linear combination or for each of a number of estimable linear combinations in such a way that the probability of coverage or probability of simultaneous coverage equals $1 - \dot{\gamma}$. The end points of the interval are interpretable as upper and lower bounds; these bounds are equidistant from the least squares estimate. In some cases, there may be no interest in bounding the linear combination both from above and from below; rather, the objective may be to obtain as tight an upper bound as possible or, alternatively, as tight a lower bound as possible.

a. Some basic results

Let us consider inference about a single estimable linear combination $\tau = \boldsymbol{\lambda}'\boldsymbol{\beta}$ of the elements of $\boldsymbol{\beta}$ (where $\boldsymbol{\lambda} \neq \mathbf{0}$). Further, let us take the context to be that of Section 7.3, and let us adopt the notation and terminology employed therein. Accordingly, suppose that $M = M_* = 1$, regard τ as the lone element of the vector $\boldsymbol{\tau}$, and write $\hat{\tau}$ for the lone element of the vector $\hat{\boldsymbol{\tau}}$ (of least squares estimators) and c for the lone element of the matrix $\mathbf{C}$ $[= \boldsymbol{\Lambda}'(\mathbf{X}'\mathbf{X})^{-}\boldsymbol{\Lambda}]$. And observe that $\mathbf{S}$ is the matrix with lone element $c^{-1/2}$, $\boldsymbol{\alpha}$ is the vector with lone element $\alpha = c^{-1/2}\tau$, and $\hat{\boldsymbol{\alpha}}$ is the vector with lone element $\hat{\alpha} = c^{-1/2}\hat{\tau}$.

Confidence bounds. Letting $\hat{\sigma}^2 = \mathbf{y}'(\mathbf{I}-\mathbf{P}_{\mathbf{X}})\mathbf{y}/(N-P_*)$, letting $\dot{\tau}$ represent an arbitrary value of τ and defining $\dot{\alpha} = c^{-1/2}\dot{\tau}$, and recalling results (3.21) and (3.51), we find that

$$\frac{\hat{\tau}-\dot{\tau}}{\hat{\sigma}\sqrt{c}} = \frac{(1/\sigma)(\hat{\alpha}-\dot{\alpha})}{\sqrt{(1/\sigma^2)\mathbf{d}'\mathbf{d}/(N-P_*)}} \sim St[N-P_*,\ (\alpha-\dot{\alpha})/\sigma]. \tag{5.1}$$

And upon applying this result in the special case where $\dot{\tau} = \tau$ or equivalently where $\dot{\alpha} = \alpha$, we find that the interval

$$[\hat{\tau} - \sqrt{c}\,\hat{\sigma}\,\bar{t}_{\dot{\gamma}}(N-P_*),\ \infty) \tag{5.2}$$

is a $100(1-\dot{\gamma})\%$ confidence interval for τ. Similarly, the interval

$$(-\infty,\ \hat{\tau} + \sqrt{c}\,\hat{\sigma}\,\bar{t}_{\dot{\gamma}}(N-P_*)] \tag{5.3}$$

is a $100(1-\dot{\gamma})\%$ confidence interval for τ. Thus, $\hat{\tau}-\sqrt{c}\,\hat{\sigma}\,\bar{t}_{\dot{\gamma}}(N-P_*)$ is a $100(1-\dot{\gamma})\%$ lower "confidence bound" for τ, and $\hat{\tau} + \sqrt{c}\,\hat{\sigma}\,\bar{t}_{\dot{\gamma}}(N-P_*)$ is a $100(1-\dot{\gamma})\%$ upper "confidence bound" for τ.

One-sided t tests. Let $\tau^{(0)}$ represent any hypothesized value of τ. Then, corresponding to the confidence interval (5.2) is a "one-sided t test" of the null hypothesis $H_0^+ : \tau \leq \tau^{(0)}$ versus the alternative hypothesis $H_1^+ : \tau > \tau^{(0)}$ with critical region C^+ defined as follows: $C^+ = \{\mathbf{y} : \tau^{(0)} < \hat{\tau} - \sqrt{c}\,\hat{\sigma}\,\bar{t}_{\dot{\gamma}}(N-P_*)\}$ or, equivalently,

$$C^+ = \left\{\mathbf{y} : \frac{\hat{\tau}-\tau^{(0)}}{\hat{\sigma}\sqrt{c}} > \bar{t}_{\dot{\gamma}}(N-P_*)\right\}$$

—when $\hat{\sigma} = 0$, interpret the value of $(\hat{\tau}-\tau^{(0)})/(\hat{\sigma}\sqrt{c})$ as $-\infty$, 0, or ∞ depending on whether $\hat{\tau} < \tau^{(0)}$, $\hat{\tau} = \tau^{(0)}$, or $\hat{\tau} > \tau^{(0)}$. This test is such that the probability $\Pr(\mathbf{y} \in C^+)$ of rejecting H_0^+ depends on $\boldsymbol{\beta}$ and σ only through the value of $(\tau-\tau^{(0)})/\sigma$ and in fact is a strictly increasing function of $(\tau-\tau^{(0)})/\sigma$, as is evident from result (5.1) and from the very definition of the noncentral t distribution. Thus, the test of H_0^+ versus H_1^+ with critical region C^+ is of size $\dot{\gamma}$ and is (strictly) unbiased.

There is a second one-sided t test. Corresponding to the confidence interval (5.3) is the one-sided t test of the null hypothesis $H_0^- : \tau \geq \tau^{(0)}$ versus the alternative hypothesis $H_1^- : \tau < \tau^{(0)}$ with critical region C^- defined as follows: $C^- = \{\mathbf{y} : \tau^{(0)} > \hat{\tau} + \sqrt{c}\,\hat{\sigma}\,\bar{t}_{\dot{\gamma}}(N-P_*)\}$ or, equivalently,

$$C^- = \left\{\mathbf{y} : \frac{\hat{\tau}-\tau^{(0)}}{\hat{\sigma}\sqrt{c}} < -\bar{t}_{\dot{\gamma}}(N-P_*)\right\}.$$

This test is such that the probability $\Pr(\mathbf{y} \in C^-)$ of rejecting H_0^- depends on $\boldsymbol{\beta}$ and σ only through the value of $(\tau-\tau^{(0)})/\sigma$ and in fact is a strictly decreasing function of $(\tau-\tau^{(0)})/\sigma$. And it is of size $\dot{\gamma}$ and is (strictly) unbiased.

Recharacterization of the one-sided t tests. Upon letting $\alpha^{(0)} = c^{-1/2}\tau^{(0)}$, the size-$\dot{\gamma}$ one-sided t test of H_0^+ can be recharacterized as a test of the null hypothesis $\tilde{H}_0^+ : \alpha \leq \alpha^{(0)}$ versus the alternative hypothesis $\tilde{H}_1^+ : \alpha > \alpha^{(0)}$, and its critical region C^+ can be reexpressed as a set $\tilde{C}^+ = \{\mathbf{z} : (\hat{\alpha}-\alpha^{(0)})/\hat{\sigma} > \bar{t}_{\dot{\gamma}}(N-P_*)\}$ of values of the transformed vector $\mathbf{z} = (\hat{\boldsymbol{\alpha}}', \hat{\boldsymbol{\eta}}', \mathbf{d}')'$—clearly, $\mathbf{z} \in \tilde{C}^+ \Leftrightarrow \mathbf{y} \in C^+$. Similarly, the size-$\dot{\gamma}$ one-sided t test of H_0^- can be recharacterized as a test of the null hypothesis $\tilde{H}_0^-: \alpha \geq \alpha^{(0)}$ versus the alternative hypothesis $\tilde{H}_1^-: \alpha < \alpha^{(0)}$, and its critical region C^- can be reexpressed in the form of the set $\tilde{C}^- = \{\mathbf{z} : (\hat{\alpha}-\alpha^{(0)})/\hat{\sigma} < \bar{t}_{\dot{\gamma}}(N-P_*)\}$.

Invariance. Both of the two size-$\dot{\gamma}$ one-sided t tests are invariant with respect to groups of transformations (of $\mathbf{z}$) of the form $\tilde{T}_1(\mathbf{z})$, of the form $\tilde{T}_2(\mathbf{z}; \boldsymbol{\alpha}^{(0)})$, and of the form $\tilde{T}_4(\mathbf{z})$—transformations of these forms are among those discussed in Section 7.3b. And the $100(1-\dot{\gamma})\%$ confidence intervals (5.2) and (5.3) are invariant with respect to groups of transformations of the form $\tilde{T}_1(\mathbf{z})$ and of the form $\tilde{T}_4(\mathbf{z})$ and are equivariant with respect to groups of transformations of the form $\tilde{T}_2(\mathbf{z}; \boldsymbol{\alpha}^{(0)})$.

b. Confidence intervals of a more general form

The $100(1-\dot{\gamma})\%$ confidence intervals (5.2) and (5.3) and the $100(1-\dot{\gamma})\%$ confidence interval (3.55) can be regarded as special cases of a $100(1-\dot{\gamma})\%$ confidence interval (for τ) of a more general form. Let $\dot{\gamma}_\ell$ and $\dot{\gamma}_u$ represent any two nonnegative scalars such that $\dot{\gamma}_\ell + \dot{\gamma}_u = \dot{\gamma}$ (and define $\bar{t}_0 = \infty$). Then, clearly, the interval

$$\{\dot{\tau} \in \mathcal{R}^1 : \hat{\tau} - \sqrt{c}\,\hat{\sigma}\bar{t}_{\dot{\gamma}_\ell}(N-P_*) \leq \dot{\tau} \leq \hat{\tau} + \sqrt{c}\,\hat{\sigma}\bar{t}_{\dot{\gamma}_u}(N-P_*)\} \tag{5.4}$$

is a $100(1-\dot{\gamma})\%$ confidence interval for τ. And interval (3.55) is the special case of interval (5.4) where $\dot{\gamma}_\ell = \dot{\gamma}_u = \dot{\gamma}/2$, interval (5.2) is the special case where $\dot{\gamma}_\ell = \dot{\gamma}$ and $\dot{\gamma}_u = 0$, and interval (5.3) is the special case where $\dot{\gamma}_\ell = 0$ and $\dot{\gamma}_u = \dot{\gamma}$.

c. Invariant tests

Let us add to the brief "discussion" (in the final part of Subsection a) of the invariance of one-sided t tests by establishing some properties possessed by the critical functions of tests that are invariant with respect to various groups of transformations but not by the critical functions of other tests.

Two propositions. The four propositions introduced previously (in Section 7.4a) serve to characterize functions of $\mathbf{z}$ that are invariant with respect to the four groups $\tilde{G}_1$, $\tilde{G}_3(\alpha^{(0)})$, $\tilde{G}_2(\alpha^{(0)})$, and $\tilde{G}_4$. In the present context, the group $\tilde{G}_3(\alpha^{(0)})$ is irrelevant. The following two propositions [in which $\tilde{\phi}(\mathbf{z})$ represents an arbitrary function of $\mathbf{z}$ and in which z_1 represents the lone element of the vector $\mathbf{z}_1$] take the place of Propositions (2), (3), and (4) of Section 7.4a and provide the characterization needed for present purposes:

(2′) The function $\tilde{\phi}(\mathbf{z})$ is invariant with respect to the group $\tilde{G}_2(\boldsymbol{\alpha}^{(0)})$ of transformations of the form $\tilde{T}_2(\mathbf{z}; \boldsymbol{\alpha}^{(0)})$ as well as the group $\tilde{G}_1$ of transformations of the form $\tilde{T}_1(\mathbf{z})$ only if there exists a function $\tilde{\phi}_{12}(\cdot)$ [of an $(N-P_*+1)$-dimensional vector] such that

$$\tilde{\phi}(\mathbf{z}) = \tilde{\phi}_{12}\left\{[(z_1-\alpha^{(0)})^2 + \mathbf{z}_3'\mathbf{z}_3]^{-1/2}\begin{pmatrix} z_1-\alpha^{(0)} \\ \mathbf{z}_3 \end{pmatrix}\right\}$$

for those values of $\mathbf{z}$ for which $\mathbf{z}_3 \neq \mathbf{0}$. Moreover, $\Pr(\mathbf{z}_3 \neq \mathbf{0}) = 1$, and the distribution of $[(z_1-\alpha^{(0)})^2 + \mathbf{z}_3'\mathbf{z}_3]^{-1/2}\begin{pmatrix} z_1-\alpha^{(0)} \\ \mathbf{z}_3 \end{pmatrix}$ depends on α, $\boldsymbol{\eta}$, and σ only through the value of $(\alpha-\alpha^{(0)})/\sigma$.

(3′) The function $\tilde{\phi}(\mathbf{z})$ is invariant with respect to the group $\tilde{G}_4$ of transformations of the form $\tilde{T}_4(\mathbf{z})$ as well as the groups $\tilde{G}_1$ and $\tilde{G}_2(\boldsymbol{\alpha}^{(0)})$ of transformations of the form $\tilde{T}_1(\mathbf{z})$ and $\tilde{T}_2(\mathbf{z}; \boldsymbol{\alpha}^{(0)})$ only if there exists a function $\tilde{\phi}_{124}(\cdot)$ (of a single variable) such that

$$\tilde{\phi}(\mathbf{z}) = \tilde{\phi}_{124}\{[(z_1-\alpha^{(0)})^2 + \mathbf{z}_3'\mathbf{z}_3]^{-1/2}(z_1-\alpha^{(0)})\}$$

for those values of $\mathbf{z}$ for which $\mathbf{z}_3 \neq \mathbf{0}$. Moreover, $\Pr(\mathbf{z}_3 \neq \mathbf{0}) = 1$, and the distribution of $[(z_1-\alpha^{(0)})^2 + \mathbf{z}_3'\mathbf{z}_3]^{-1/2}(z_1-\alpha^{(0)})$ depends on α, $\boldsymbol{\eta}$, and σ only through the value of $(\alpha-\alpha^{(0)})/\sigma$.

Verification of Proposition (2′). Suppose that $\tilde{\phi}(\mathbf{z})$ is invariant with respect to the group $\tilde{G}_2(\boldsymbol{\alpha}^{(0)})$ as well as the group $\tilde{G}_1$. Then, in light of Proposition (1) (of Section 7.4a), $\tilde{\phi}(\mathbf{z}) = \tilde{\phi}_1(\mathbf{z}_1, \mathbf{z}_3)$ for some function $\tilde{\phi}_1(\cdot, \cdot)$. Thus, to establish the existence of a function $\tilde{\phi}_{12}(\cdot)$ such that $\tilde{\phi}(\mathbf{z}) = \tilde{\phi}_{12}\left\{[(z_1-\alpha^{(0)})^2 + \mathbf{z}_3'\mathbf{z}_3]^{-1/2}\begin{pmatrix} z_1-\alpha^{(0)} \\ \mathbf{z}_3 \end{pmatrix}\right\}$ for those values of $\mathbf{z}$ for which $\mathbf{z}_3 \neq \mathbf{0}$, it suffices to take $\dot{z}_1$ and $\ddot{z}_1$ to be values of z_1 and $\dot{\mathbf{z}}_3$ and $\ddot{\mathbf{z}}_3$ nonnull values of $\mathbf{z}_3$ such that

$$[(\ddot{z}_1-\alpha^{(0)})^2 + \ddot{\mathbf{z}}_3'\ddot{\mathbf{z}}_3]^{-1/2}\begin{pmatrix} \ddot{z}_1-\alpha^{(0)} \\ \ddot{\mathbf{z}}_3 \end{pmatrix} = [(\dot{z}_1-\alpha^{(0)})^2 + \dot{\mathbf{z}}_3'\dot{\mathbf{z}}_3]^{-1/2}\begin{pmatrix} \dot{z}_1-\alpha^{(0)} \\ \dot{\mathbf{z}}_3 \end{pmatrix}$$

and to observe that

$$\ddot{\mathbf{z}}_3 = k\dot{\mathbf{z}}_3 \quad \text{and} \quad \ddot{z}_1 = \alpha^{(0)} + k(\dot{z}_1-\alpha^{(0)}),$$

where $k = [(\ddot{z}_1-\alpha^{(0)})^2 + \ddot{\mathbf{z}}_3'\ddot{\mathbf{z}}_3]^{1/2}/[(\dot{z}_1-\alpha^{(0)})^2 + \dot{\mathbf{z}}_3'\dot{\mathbf{z}}_3]^{1/2}$. And upon letting u_1 represent a random variable and $\mathbf{u}_3$ an $(N-P_*)\times 1$ random vector such that $\begin{pmatrix} u_1 \\ \mathbf{u}_3 \end{pmatrix} \sim N(\mathbf{0}, \mathbf{I})$, it remains only to observe that

$$\begin{aligned} &[(z_1-\alpha^{(0)})^2 + \mathbf{z}_3'\mathbf{z}_3]^{-1/2}\begin{pmatrix} z_1-\alpha^{(0)} \\ \mathbf{z}_3 \end{pmatrix} \\ &\qquad \sim [(\alpha-\alpha^{(0)}+\sigma u_1)^2 + \sigma^2\mathbf{u}_3'\mathbf{u}_3]^{-1/2}\begin{pmatrix} \alpha-\alpha^{(0)}+\sigma u_1 \\ \sigma\mathbf{u}_3 \end{pmatrix} \\ &\qquad = \{[\sigma^{-1}(\alpha-\alpha^{(0)})+u_1]^2 + \mathbf{u}_3'\mathbf{u}_3\}^{-1/2}\begin{pmatrix} \sigma^{-1}(\alpha-\alpha^{(0)})+u_1 \\ \mathbf{u}_3 \end{pmatrix}. \end{aligned}$$

Verification of Proposition (3′). Suppose that $\tilde{\phi}(\mathbf{z})$ is invariant with respect to the group $\tilde{G}_4$ as well as the groups $\tilde{G}_1$ and $\tilde{G}_2(\boldsymbol{\alpha}^{(0)})$. Then, in light of Proposition (2′), there exists a function $\tilde{\phi}_{12}(\cdot)$ such that $\tilde{\phi}(\mathbf{z}) = \tilde{\phi}_{12}\left\{[(z_1-\alpha^{(0)})^2 + \mathbf{z}_3'\mathbf{z}_3]^{-1/2}\begin{pmatrix} z_1-\alpha^{(0)} \\ \mathbf{z}_3 \end{pmatrix}\right\}$ for those values of $\mathbf{z}$ for which $\mathbf{z}_3 \neq \mathbf{0}$. Moreover, there also exists a function $\tilde{\phi}_{124}(\cdot)$ such that

$$\tilde{\phi}_{12}\left\{[(z_1-\alpha^{(0)})^2 + \mathbf{z}_3'\mathbf{z}_3]^{-1/2}\begin{pmatrix} z_1-\alpha^{(0)} \\ \mathbf{z}_3 \end{pmatrix}\right\} = \tilde{\phi}_{124}\{[(z_1-\alpha^{(0)})^2 + \mathbf{z}_3'\mathbf{z}_3]^{-1/2}(z_1-\alpha^{(0)})\}$$

(for those values of $\mathbf{z}$ for which $\mathbf{z}_3 \neq \mathbf{0}$). To confirm this, it suffices to take $\dot{z}_1$ and $\ddot{z}_1$ to be values of z_1 and $\dot{\mathbf{z}}_3$ and $\ddot{\mathbf{z}}_3$ nonnull values of $\mathbf{z}_3$ such that

$$[(\ddot{z}_1-\alpha^{(0)})^2 + \ddot{\mathbf{z}}_3'\ddot{\mathbf{z}}_3]^{-1/2}(\ddot{z}_1-\alpha^{(0)}) = [(\dot{z}_1-\alpha^{(0)})^2 + \dot{\mathbf{z}}_3'\dot{\mathbf{z}}_3]^{-1/2}(\dot{z}_1-\alpha^{(0)}) \tag{5.5}$$

and to observe that equality (5.5) implies that

$$\begin{aligned} &\{[(\ddot{z}_1-\alpha^{(0)})^2 + \ddot{\mathbf{z}}_3'\ddot{\mathbf{z}}_3]^{-1/2}\ddot{\mathbf{z}}_3\}'\{[(\ddot{z}_1-\alpha^{(0)})^2 + \ddot{\mathbf{z}}_3'\ddot{\mathbf{z}}_3]^{-1/2}\ddot{\mathbf{z}}_3\} \\ &\qquad = \{[(\dot{z}_1-\alpha^{(0)})^2 + \dot{\mathbf{z}}_3'\dot{\mathbf{z}}_3]^{-1/2}\dot{\mathbf{z}}_3\}'\{[(\dot{z}_1-\alpha^{(0)})^2 + \dot{\mathbf{z}}_3'\dot{\mathbf{z}}_3]^{-1/2}\dot{\mathbf{z}}_3\} \end{aligned}$$

and hence implies (in light of Lemma 5.9.9) the existence of an orthogonal matrix $\mathbf{B}$ such that

$$\begin{aligned} [(\ddot{z}_1-\alpha^{(0)})^2 + \ddot{\mathbf{z}}_3'\ddot{\mathbf{z}}_3]^{-1/2}\ddot{\mathbf{z}}_3 &= \mathbf{B}'\{[(\dot{z}_1-\alpha^{(0)})^2 + \dot{\mathbf{z}}_3'\dot{\mathbf{z}}_3]^{-1/2}\dot{\mathbf{z}}_3\} \\ &= [(\dot{z}_1-\alpha^{(0)})^2 + (\mathbf{B}'\dot{\mathbf{z}}_3)'\mathbf{B}'\dot{\mathbf{z}}_3]^{-1/2}\mathbf{B}'\dot{\mathbf{z}}_3 \end{aligned}$$

and

$$[(\ddot{z}_1-\alpha^{(0)})^2 + \ddot{\mathbf{z}}_3'\ddot{\mathbf{z}}_3]^{-1/2}(\ddot{z}_1-\alpha^{(0)}) = [(\dot{z}_1-\alpha^{(0)})^2 + (\mathbf{B}'\dot{\mathbf{z}}_3)'\mathbf{B}'\dot{\mathbf{z}}_3]^{-1/2}(\dot{z}_1-\alpha^{(0)}).$$

That the distribution of $[(z_1-\alpha^{(0)})^2 + \mathbf{z}_3'\mathbf{z}_3]^{-1/2}(z_1-\alpha^{(0)})$ depends on α, $\boldsymbol{\eta}$, and σ only through the value of $(\alpha-\alpha^{(0)})/\sigma$ follows from Proposition (2′).

d. Optimality of a one-sided t test

Let us consider the extent to which the one-sided t test of the null hypothesis H_0^+ or $\tilde{H}_0^+$ (versus H_1^+ or $\tilde{H}_1^+$) and the one-sided t test of the null hypothesis H_0^- or $\tilde{H}_0^-$ (versus H_1^- or $\tilde{H}_1^-$) compare favorably with various other tests of these null hypotheses.

Best invariant test of H_0^+ or $\tilde{H}_0^+$ (versus H_1^+ or $\tilde{H}_1^+$). The size-$\dot{\gamma}$ one-sided t test of H_0^+ can be recharacterized as a test of $\tilde{H}_0^+$, and its critical region and critical function can be reexpressed as a set of $\mathbf{z}$-values or as a function of $\mathbf{z}$ rather than as a set of $\mathbf{y}$-values or a function of $\mathbf{y}$—refer to Subsection a. As noted earlier (in Subsection a), the size-$\dot{\gamma}$ one-sided t test of H_0^+ or $\tilde{H}_0^+$ is invariant to the groups $\tilde{G}_1$, $\tilde{G}_2(\alpha^{(0)})$, and $\tilde{G}_4$ of transformations (of $\mathbf{z}$) of the form $\tilde{T}_1(\mathbf{z})$, $\tilde{T}_2(\mathbf{z};\boldsymbol{\alpha}^{(0)})$, and $\tilde{T}_4(\mathbf{z})$. In fact, as we now proceed to show, the size-$\dot{\gamma}$ one-sided t test is UMP among all level-$\dot{\gamma}$ tests of H_0^+ or $\tilde{H}_0^+$ (versus H_1^+ or $\tilde{H}_1^+$) that are invariant with respect to these groups of transformations; under H_1^+ or $\tilde{H}_1^+$, its power is uniformly (i.e., for all $\alpha > \alpha^{(0)}$ and for all $\boldsymbol{\eta}$ and σ) greater than or equal to that of every test of H_0^+ or $\tilde{H}_0^+$ whose size does not exceed $\dot{\gamma}$ and that is invariant with respect to the groups $\tilde{G}_1$, $\tilde{G}_2(\alpha^{(0)})$, and $\tilde{G}_4$.

Let $\tilde{\phi}(\mathbf{z})$ represent [in terms of the transformed observable random vector $\mathbf{z} = (z_1, \mathbf{z}_2', \mathbf{z}_3')'$] the critical function of an arbitrary (possibly randomized) test of the null hypothesis H_0^+ or $\tilde{H}_0^+$ (versus the alternative hypothesis H_1^+ or $\tilde{H}_1^+$). Further, let Y represent an observable random variable defined as follows:

$$Y = \begin{cases} [(z_1-\alpha^{(0)})^2 + \mathbf{z}_3'\mathbf{z}_3]^{-1/2}(z_1-\alpha^{(0)}), & \text{if } z_1 \neq \alpha^{(0)} \text{ or } \mathbf{z}_3 \neq \mathbf{0}, \\ 0, & \text{if } z_1 = \alpha^{(0)} \text{ and } \mathbf{z}_3 = \mathbf{0}. \end{cases}$$

And let $\lambda = (\alpha-\alpha^{(0)})/\sigma$.

When regarded as a function of $\mathbf{z}$, Y is invariant with respect to the groups $\tilde{G}_1$, $\tilde{G}_2(\alpha^{(0)})$, and $\tilde{G}_4$ (as can be readily verified). Thus, if $\tilde{\phi}(\mathbf{z})$ depends on $\mathbf{z}$ only through the value of Y, then $\tilde{\phi}(\mathbf{z})$ is invariant with respect to the groups $\tilde{G}_1$, $\tilde{G}_2(\alpha^{(0)})$, and $\tilde{G}_4$. Conversely, if $\tilde{\phi}(\mathbf{z})$ is invariant with respect to the groups $\tilde{G}_1$, $\tilde{G}_2(\alpha^{(0)})$, and $\tilde{G}_4$, then [in light of Proposition (3′) of Subsection c] there exists a test with a critical function $\tilde{\phi}_{124}(Y)$ for which

$$\tilde{\phi}(\mathbf{z}) = \tilde{\phi}_{124}(Y) \quad \text{(wp1)}$$

and hence for which

$$\mathrm{E}[\tilde{\phi}(\mathbf{z})] = \mathrm{E}[\tilde{\phi}_{124}(Y)],$$

so that the test with critical function $\tilde{\phi}_{124}(Y)$ has the same power function as the test with critical function $\tilde{\phi}(\mathbf{z})$. And the distribution of Y or "any" function of Y depends on α, $\boldsymbol{\eta}$, and σ only through the value of λ {as is evident from Proposition (3′) of Subsection c upon observing that $Y \sim [(z_1-\alpha^{(0)})^2 + \mathbf{z}_3'\mathbf{z}_3]^{-1/2}(z_1-\alpha^{(0)})$}. Moreover, Y is a strictly increasing function of the quantity $(\hat{\alpha}-\alpha^{(0)})/\hat{\sigma}$ [as can be readily verified by making use of relationship (6.4.4)], so that

$$\mathbf{z} \in \tilde{C}^+ \quad \Leftrightarrow \quad Y \in \left\{ Y : Y \le \frac{\bar{t}_{\dot{\gamma}}(N-P_*)}{\sqrt{N-P_*+[\bar{t}_{\dot{\gamma}}(N-P_*)]^2}} \right\}. \tag{5.6}$$

The upshot of these remarks is that Theorem 7.4.1 (the Neyman–Pearson lemma) can be used to show that the size-$\dot{\gamma}$ one-sided t test is UMP among all level-$\dot{\gamma}$ tests of H_0^+ or $\tilde{H}_0^+$ (versus H_1^+ or $\tilde{H}_1^+$) that are invariant with respect to the groups $\tilde{G}_1$, $\tilde{G}_2(\alpha^{(0)})$, and $\tilde{G}_4$. Accordingly, let λ^* represent any strictly positive scalar, and let $f(\cdot\,;\lambda)$ represent the pdf of the distribution of Y—clearly, this distribution is absolutely continuous. Further, upon letting $U = [(z_1-\alpha^{(0)})^2 + \mathbf{z}_3'\mathbf{z}_3]/\sigma^2$ and letting $q(\cdot\,,\cdot\,;\lambda)$ represent the pdf of the joint distribution of U and Y (which is absolutely continuous and depends on α, $\boldsymbol{\eta}$, and σ only through the value of λ) and upon observing that (for $-1 < y < 1$)

$$\frac{f(y\,;\lambda)}{f(y\,;0)} = \int_0^\infty \frac{q(u\,,y\,;\lambda)}{f(y\,;0)}\,du$$

and [in light of results (6.4.32) and (6.4.7)] that (for some strictly positive scalar c that does not depend on u, y, or λ and for $0 < u < \infty$ and $-1 < y < 1$)

$$\frac{q(u, y; \lambda)}{f(y; 0)} = c u^{(N-P_*-1)/2} e^{-u/2} e^{u^{1/2} y\lambda} e^{-\lambda^2/2},$$

we find that (for $-1 < y < 1$)

$$\begin{aligned} \frac{d[f(y;\lambda)/f(y;0)]}{dy} &= \int_0^\infty \frac{\partial[q(u, y; \lambda)/f(y;0)]}{\partial y}\, du \\ &= c\lambda e^{-\lambda^2/2} \int_0^\infty u^{(N-P_*)/2} e^{-u/2} e^{u^{1/2} y\lambda}\, du \\ &> 0 \quad \text{if } \lambda > 0, \end{aligned}$$

so that the ratio $f(y; \lambda^*)/f(y; 0)$ is a strictly increasing function of y.

Thus, upon applying Theorem 7.4.1 (with $X = Y$, $\theta = \lambda$, $\Theta = (-\infty, \infty)$, $\theta^{(0)} = 0$, and $\theta^* = \lambda^*$, we find [in light of the equivalence (5.6)] that among tests of the null hypothesis $\lambda = 0$ (versus the alternative hypothesis $\lambda \neq 0$) that are of level $\dot{\gamma}$ and that are invariant with respect to the groups $\tilde{G}_1$, $\tilde{G}_2(\alpha^{(0)})$, and $\tilde{G}_4$, the power of the test at the point $\lambda = \lambda^*$ attains its maximum value (for every choice of the strictly positive scalar λ^*) when the critical region of the test is taken to be the region $\tilde{C}^+$. Moreover, tests of the null hypothesis $\lambda = 0$ that are of level $\dot{\gamma}$ and that are invariant with respect to the groups $\tilde{G}_1$, $\tilde{G}_2(\alpha^{(0)})$, and $\tilde{G}_4$ include as a subset tests of the null hypothesis $\lambda \leq 0$ that are of level $\dot{\gamma}$ and that are invariant with respect to those groups. And we conclude that among all tests of the null hypothesis $H_0^+ : \boldsymbol{\tau} \leq \boldsymbol{\tau}^{(0)}$ or $\tilde{H}_0^+ : \boldsymbol{\alpha} \leq \boldsymbol{\alpha}^{(0)}$ (versus the alternative hypothesis $H_1^+ : \boldsymbol{\tau} > \boldsymbol{\tau}^{(0)}$ or $\tilde{H}_1^+ : \boldsymbol{\alpha} > \boldsymbol{\alpha}^{(0)}$) that are of level $\dot{\gamma}$ and that are invariant with respect to the groups $\tilde{G}_1$, $\tilde{G}_2(\alpha^{(0)})$, and $\tilde{G}_4$, the size-$\dot{\gamma}$ one-sided t test of H_0^+ is a UMP test.

A stronger result. As is evident from the next-to-last part of Section 7.3a, z_1, $\mathbf{z}_2$, and $\mathbf{z}_3'\mathbf{z}_3$ form a (vector-valued) sufficient statistic. And in light of the proposition of Section 7.4b, the critical function of a test or H_0^+ or $\tilde{H}_0^+$ is expressible as a function of this statistic if and only if the test is invariant with respect to the group $\tilde{G}_4$ of transformations (of $\mathbf{z}$) of the form $\tilde{T}_4(\mathbf{z})$. Thus, a test of H_0^+ or $\tilde{H}_0^+$ is invariant with respect to the groups $\tilde{G}_1$, $\tilde{G}_2(\alpha^{(0)})$, and $\tilde{G}_4$ of transformations if and only if it is invariant with respect to the groups $\tilde{G}_1$ and $\tilde{G}_2(\alpha^{(0)})$ and, in addition, its critical function is expressible as a function of z_1, $\mathbf{z}_2$, and $\mathbf{z}_3'\mathbf{z}_3$.

Let $\tilde{\phi}(\mathbf{z})$ represent a function of $\mathbf{z}$ that is the critical function of a (possibly randomized) test of H_0^+ or $\tilde{H}_0^+$. Then, corresponding to $\tilde{\phi}(\mathbf{z})$ is a critical function $\bar{\phi}(\mathbf{z}) = \mathrm{E}[\tilde{\phi}(\mathbf{z}) \mid z_1, \mathbf{z}_2, \mathbf{z}_3'\mathbf{z}_3]$ that depends on $\mathbf{z}$ only through the value of the sufficient statistic. Moreover, the test with critical function $\bar{\phi}(\mathbf{z})$ has the same power function as that with critical function $\tilde{\phi}(\mathbf{z})$; and if the test with critical function $\tilde{\phi}(\mathbf{z})$ is invariant with respect to the groups $\tilde{G}_1$ and $\tilde{G}_2(\alpha^{(0)})$ of transformations, then so is the test with critical function $\bar{\phi}(\mathbf{z})$—refer to the fifth part of Section 7.4d.

Thus, the results (obtained in the preceding part of the present subsection) on the optimality of the size-$\dot{\gamma}$ one-sided t test can be strengthened. We find that among all tests of the null hypothesis $H_0^+ : \boldsymbol{\tau} \leq \boldsymbol{\tau}^{(0)}$ or $\tilde{H}_0^+ : \boldsymbol{\alpha} \leq \boldsymbol{\alpha}^{(0)}$ (versus the alternative hypothesis $H_1^+ : \boldsymbol{\tau} > \boldsymbol{\tau}^{(0)}$ or $\tilde{H}_1^+ : \boldsymbol{\alpha} > \boldsymbol{\alpha}^{(0)}$) that are of level $\dot{\gamma}$ and that are invariant with respect to the groups $\tilde{G}_1$ and $\tilde{G}_2(\alpha^{(0)})$, the size-$\dot{\gamma}$ one-sided t test of H_0^+ is a UMP test. The restriction to tests that are invariant with respect to the group $\tilde{G}_4$ is unnecessary.

Best invariant test of H_0^- or $\tilde{H}_0^-$ (versus H_1^- or $\tilde{H}_1^-$): an analogous result. By proceeding in much the same fashion as in arriving at the result (on the optimality of the size-$\dot{\gamma}$ one-sided t test of the null hypothesis H_0^+) presented in the preceding part of the present subsection, one can establish the following result on the optimality of the size-$\dot{\gamma}$ one-sided t test of the null hypothesis H_0^-: Among all tests of the null hypothesis $H_0^- : \boldsymbol{\tau} \geq \boldsymbol{\tau}^{(0)}$ or $\tilde{H}_0^- : \boldsymbol{\alpha} \geq \boldsymbol{\alpha}^{(0)}$ (versus the alternative hypothesis

$H_1^-: \boldsymbol{\tau} < \boldsymbol{\tau}^{(0)}$ or $\tilde{H}_1^-: \boldsymbol{\alpha} < \boldsymbol{\alpha}^{(0)}$) that are of level $\dot{\gamma}$ and that are invariant with respect to the groups $\tilde{G}_1$ and $\tilde{G}_2(\alpha^{(0)})$, the size-$\dot{\gamma}$ one-sided t test of H_0^- is a UMP test.

Confidence intervals. By proceeding in much the same fashion as in the final part of Section 7.4d (in translating results on the optimality of the F test into results on the optimality of the corresponding confidence set), the results of the preceding parts of the present subsection (of Section 7.5) on the optimality of the one-sided t tests can be translated into results on the optimality of the corresponding confidence intervals.

Clearly, the $100(1-\dot{\gamma})\%$ confidence intervals (5.2) and (5.3) (for τ) and the corresponding intervals $[\hat{\tau} - \hat{\sigma}\bar{t}_{\dot{\gamma}}(N-P_*),\ \infty)$ and $(-\infty,\ \hat{\tau} + \hat{\sigma}\bar{t}_{\dot{\gamma}}(N-P_*)]$ for α are equivariant with respect to the group $\tilde{G}_2$ of transformations [the group formed by the totality of the groups $\tilde{G}_2(\alpha^{(0)})$ $(\alpha^{(0)} \in \mathcal{R}^1)$], and they are invariant with respect to the group $\tilde{G}_1$ (and also with respect to the group $\tilde{G}_4$). Now, let $\tilde{A}(\mathbf{z})$ represent any confidence set for α whose probability of coverage equals or exceeds $1-\dot{\gamma}$ and that is invariant with respect to the group $\tilde{G}_1$ and equivariant with respect to the group $\tilde{G}_2$. Further, for $\alpha^{(0)} \in \mathcal{R}^1$, denote by $\delta(\alpha^{(0)};\, \alpha)$ the probability $\Pr[\alpha^{(0)} \in \tilde{A}(\mathbf{z})]$ of $\alpha^{(0)}$ being covered by the confidence set $\tilde{A}(\mathbf{z})$ [and note that $\delta(\alpha;\, \alpha) \geq 1-\dot{\gamma}$]. Then, the interval $[\hat{\tau} - \hat{\sigma}\bar{t}_{\dot{\gamma}}(N-P_*),\ \infty)$ is the optimal choice for $\tilde{A}(\mathbf{z})$ in the sense that for every scalar $\alpha^{(0)}$ such that $\alpha^{(0)} < \alpha$, it is the choice that minimizes $\delta(\alpha^{(0)};\, \alpha)$, that is, the choice that minimizes the probability of the confidence set covering any value of α smaller than the true value.

This result on the optimality of the $100(1-\dot{\gamma})\%$ confidence interval $[\hat{\tau} - \hat{\sigma}\bar{t}_{\dot{\gamma}}(N-P_*),\ \infty)$ (for α) can be reexpressed as a result on the optimality of the $100(1-\dot{\gamma})\%$ confidence interval (5.2) (for τ), and/or (as discussed in Section 7.3b) conditions (pertaining to the invariance or equivariance of confidence sets) that are expressed in terms of groups of transformations of $\mathbf{z}$ can be reexpressed in terms of the corresponding groups of transformations of $\mathbf{y}$. Corresponding to the groups $\tilde{G}_1$ and $\tilde{G}_4$ and the group $\tilde{G}_2$ of transformations of $\mathbf{z}$ are the groups G_1 and G_4 and the group G_2 [consisting of the totality of the groups $G_2(\tau^{(0)})$ $(\tau^{(0)} \in \mathcal{R}^1)$] of transformations of $\mathbf{y}$.

Let $A(\mathbf{y})$ represent any confidence set for τ for which $\Pr[\tau \in A(\mathbf{y})] \geq 1-\dot{\gamma}$ and that is invariant with respect to the group G_1 and equivariant with respect to the group G_2. Then, the interval (5.2) is the optimal choice for $A(\mathbf{y})$ in the sense that for every scalar $\tau^{(0)} < \tau$, it is the choice that minimizes $\Pr[\tau^{(0)} \in A(\mathbf{y})]$; that is, the choice that minimizes the probability of the confidence set covering any value of τ smaller than the true value. Moreover, the probability of the interval (5.2) covering a scalar $\tau^{(0)}$ is less than, equal to, or greater than $1-\dot{\gamma}$ depending on whether $\tau^{(0)} < \tau$, $\tau^{(0)} = \tau$, or $\tau^{(0)} > \tau$. While the interval (5.2) is among those choices for the confidence set $A(\mathbf{y})$ that are invariant with respect to the group G_4 of transformations (of $\mathbf{y}$) and is also among those choices for which $\Pr[\tau^{(0)} \in A(\mathbf{y})] > 1-\dot{\gamma}$ for every scalar $\tau^{(0)} > \tau$, the optimality of the interval (5.2) is not limited to choices for $A(\mathbf{y})$ that have either or both of those properties.

If the objective is to minimize the probability of $A(\mathbf{y})$ covering values of τ that are larger than the true value rather than smaller, the optimal choice for $A(\mathbf{y})$ is interval (5.3) rather than interval (5.2). For every scalar $\tau^{(0)} > \tau$, interval (5.3) is the choice that minimizes $\Pr[\tau^{(0)} \in A(\mathbf{y})]$. And the probability of the interval (5.3) covering a scalar $\tau^{(0)}$ is greater than, equal to, or less than $1-\dot{\gamma}$ depending on whether $\tau^{(0)} < \tau$, $\tau^{(0)} = \tau$, or $\tau^{(0)} > \tau$.

Best unbiased tests. It can be shown that among all tests of the null hypothesis $H_0^+: \boldsymbol{\tau} \leq \boldsymbol{\tau}^{(0)}$ or $\tilde{H}_0^+: \boldsymbol{\alpha} \leq \boldsymbol{\alpha}^{(0)}$ (versus the alternative hypothesis $H_1^+: \boldsymbol{\tau} > \boldsymbol{\tau}^{(0)}$ or $\tilde{H}_1^+: \boldsymbol{\alpha} > \boldsymbol{\alpha}^{(0)}$) that are of level $\dot{\gamma}$ and that are unbiased, the size-$\dot{\gamma}$ one-sided t test of H_0^+ is a UMP test. Similarly, among all tests of the null hypothesis $H_0^-: \boldsymbol{\tau} \geq \boldsymbol{\tau}^{(0)}$ or $\tilde{H}_0^-: \boldsymbol{\alpha} \geq \boldsymbol{\alpha}^{(0)}$ (versus the alternative hypothesis $H_1^-: \boldsymbol{\tau} < \boldsymbol{\tau}^{(0)}$ or $\tilde{H}_1^-: \boldsymbol{\alpha} < \boldsymbol{\alpha}^{(0)}$) that are of level $\dot{\gamma}$ and that are unbiased, the size-$\dot{\gamma}$ one-sided t test of H_0^- is a UMP test.

e. Simultaneous inference

Suppose that we wish to obtain either a lower confidence bound or an upper confidence bound for each of a number of estimable linear combinations of the elements of $\boldsymbol{\beta}$. And suppose that we wish for these confidence bounds to be such that the probability of simultaneous coverage equals $1-\dot{\gamma}$—simultaneous coverage occurs when for every one of the linear combinations, the true value of the linear combination is covered by the interval formed (in the case of a lower confidence bound) by the scalars greater than or equal to the confidence bound or alternatively (in the case of an upper confidence bound) by the scalars less than or equal to the confidence bound. Or suppose that for each of a number of estimable linear combinations, we wish to test the null hypothesis that the true value of the linear combination is less than or equal to some hypothesized value (versus the alternative that it exceeds the hypothesized value) or the null hypothesis that the true value is greater than or equal to some hypothesized value (versus the alternative that it is less than the hypothesized value), and suppose that we wish to do so in such a way that the probability of falsely rejecting one or more of the null hypotheses is less than or equal to $\dot{\gamma}$.

Let $\tau_1, \tau_2, \ldots, \tau_M$ represent estimable linear combinations of the elements of $\boldsymbol{\beta}$, and let $\boldsymbol{\tau} = (\tau_1, \tau_2, \ldots, \tau_M)'$. And suppose that the linear combinations (of the elements of $\boldsymbol{\beta}$) for which we wish to obtain confidence bounds or to subject to hypothesis tests are expressible in the form $\tau = \boldsymbol{\delta}'\boldsymbol{\tau}$, where $\boldsymbol{\delta}$ is an arbitrary member of a specified collection Δ of $M \times 1$ vectors—it is assumed that $\boldsymbol{\Lambda\delta} \neq \mathbf{0}$ for some $\boldsymbol{\delta} \in \Delta$. Further, in connection with the hypothesis tests, denote by $\tau_{\boldsymbol{\delta}}^{(0)}$ the hypothesized value of $\tau = \boldsymbol{\delta}'\boldsymbol{\tau}$ and by $H_0^{(\boldsymbol{\delta})}$ and $H_1^{(\boldsymbol{\delta})}$ the null and alternative hypotheses, so that either $H_0^{(\boldsymbol{\delta})} : \tau \leq \tau_{\boldsymbol{\delta}}^{(0)}$ and $H_1^{(\boldsymbol{\delta})} : \tau > \tau_{\boldsymbol{\delta}}^{(0)}$ or $H_0^{(\boldsymbol{\delta})} : \tau \geq \tau_{\boldsymbol{\delta}}^{(0)}$ and $H_1^{(\boldsymbol{\delta})} : \tau < \tau_{\boldsymbol{\delta}}^{(0)}$. It is assumed that the various hypothesized values $\tau_{\boldsymbol{\delta}}^{(0)}$ $(\boldsymbol{\delta} \in \Delta)$ are simultaneously achievable in the sense that there exists an $M \times 1$ vector $\boldsymbol{\tau}^{(0)} \in \mathcal{C}(\boldsymbol{\Lambda}')$ such that (for every $\boldsymbol{\delta} \in \Delta$) $\tau_{\boldsymbol{\delta}}^{(0)} = \boldsymbol{\delta}'\boldsymbol{\tau}^{(0)}$.

Note that for any scalar $\dot{\tau}$,

$$\boldsymbol{\delta}'\boldsymbol{\tau} \leq \dot{\tau} \;\Leftrightarrow\; (-\boldsymbol{\delta})'\boldsymbol{\tau} \geq -\dot{\tau} \qquad \text{and} \qquad \boldsymbol{\delta}'\boldsymbol{\tau} > \dot{\tau} \;\Leftrightarrow\; (-\boldsymbol{\delta})'\boldsymbol{\tau} < -\dot{\tau}.$$

Thus, for purposes of obtaining for every one of the linear combinations $\boldsymbol{\delta}'\boldsymbol{\tau}$ $(\boldsymbol{\delta} \in \Delta)$ either a lower or an upper confidence bound (and for doing so in such a way that the probability of simultaneous coverage equals $1-\dot{\gamma}$), there is no real loss of generality in restricting attention to the case where all of the confidence bounds are lower bounds. Similarly, for purposes of obtaining for every one of the linear combinations $\boldsymbol{\delta}'\boldsymbol{\tau}$ $(\boldsymbol{\delta} \in \Delta)$ a test of the null hypothesis $H_0^{(\boldsymbol{\delta})}$ versus the alternative hypothesis $H_1^{(\boldsymbol{\delta})}$ (and for doing so in such a way that the probability of falsely rejecting one or more of the null hypotheses is less than or equal to $\dot{\gamma}$), there is no real loss of generality in restricting attention to the case where for every $\boldsymbol{\delta} \in \Delta$, $H_0^{(\boldsymbol{\delta})}$ and $H_1^{(\boldsymbol{\delta})}$ are $H_0^{(\boldsymbol{\delta})} : \tau \leq \tau_{\boldsymbol{\delta}}^{(0)}$ and $H_1^{(\boldsymbol{\delta})} : \tau > \tau_{\boldsymbol{\delta}}^{(0)}$.

Simultaneous confidence bounds. Confidence bounds with a specified probability of simultaneous coverage can be obtained by adopting an approach similar to that employed in Sections 7.3c and 7.3e in obtaining confidence intervals (each of which has end points that are equidistant from the least squares estimator) with a specified probability of simultaneous coverage. As before, let $\tilde{\Delta}$ represent the set of $M_* \times 1$ vectors defined as follows:

$$\tilde{\Delta} = \{\tilde{\boldsymbol{\delta}} \,:\, \tilde{\boldsymbol{\delta}} = \mathbf{W}\boldsymbol{\delta},\ \boldsymbol{\delta} \in \boldsymbol{\Delta}\}.$$

Further, letting $\mathbf{t}$ represent an $M_* \times 1$ random vector that has an $MVt(N-P_*, \mathbf{I}_{M_*})$ distribution, denote by $a_{\dot{\gamma}}$ the upper $100\,\dot{\gamma}\%$ point of the distribution of the random variable

$$\max_{\{\tilde{\boldsymbol{\delta}} \in \tilde{\Delta}\,:\, \tilde{\boldsymbol{\delta}} \neq \mathbf{0}\}} \frac{\tilde{\boldsymbol{\delta}}'\mathbf{t}}{(\tilde{\boldsymbol{\delta}}'\tilde{\boldsymbol{\delta}})^{1/2}}. \tag{5.7}$$

And (for $\boldsymbol{\delta} \in \Delta$) take $A_{\boldsymbol{\delta}}(\mathbf{y})$ to be the set

$$\{\dot{\tau} \in \mathbb{R}^1 \,:\, \boldsymbol{\delta}'\hat{\boldsymbol{\tau}} - (\boldsymbol{\delta}'\mathbf{C}\boldsymbol{\delta})^{1/2}\hat{\sigma} a_{\dot{\gamma}} \leq \dot{\tau} < \infty\}$$

of $\boldsymbol{\delta}'\boldsymbol{\tau}$-values obtained upon regarding $\boldsymbol{\delta}'\hat{\boldsymbol{\tau}} - (\boldsymbol{\delta}'\mathbf{C}\boldsymbol{\delta})^{1/2}\hat{\sigma}a_{\dot{\gamma}}$ as a lower bound. Then,

$$\Pr[\boldsymbol{\delta}'\boldsymbol{\tau} \in A_{\boldsymbol{\delta}}(\mathbf{y}) \text{ for every } \boldsymbol{\delta} \in \Delta] = 1-\dot{\gamma},$$

so that the probability of simultaneous coverage of the intervals $A_{\boldsymbol{\delta}}(\mathbf{y})$ $(\boldsymbol{\delta} \in \Delta)$ defined by the lower bounds $\boldsymbol{\delta}'\hat{\boldsymbol{\tau}} - (\boldsymbol{\delta}'\mathbf{C}\boldsymbol{\delta})^{1/2}\hat{\sigma}a_{\dot{\gamma}}$ $(\boldsymbol{\delta} \in \Delta)$ is equal to $1-\dot{\gamma}$.

To make use of the lower bounds $\boldsymbol{\delta}'\hat{\boldsymbol{\tau}} - (\boldsymbol{\delta}'\mathbf{C}\boldsymbol{\delta})^{1/2}\hat{\sigma}a_{\dot{\gamma}}$ $(\boldsymbol{\delta} \in \Delta)$, we need the upper $100\,\dot{\gamma}\%$ point $a_{\dot{\gamma}}$ of the distribution of the random variable (5.7). Aside from relatively simple special cases, the distribution of this random variable is sufficiently complex that the computation of $a_{\dot{\gamma}}$ requires the use of Monte Carlo methods. When resort is made to Monte Carlo methods, the maximum value of the quantity $\tilde{\boldsymbol{\delta}}'\mathbf{t}/(\tilde{\boldsymbol{\delta}}'\tilde{\boldsymbol{\delta}})^{1/2}$ (where the maximization is with respect to $\tilde{\boldsymbol{\delta}}$ and is subject to the constraint $\tilde{\boldsymbol{\delta}} \in \tilde{\Delta}$) must be determined for each of a very large number of values of $\mathbf{t}$. As discussed in Section 7.3e, the maximum value of this quantity can be determined from the solution to a constrained nonlinear least squares problem.

A variation. Suppose that Δ is such that $\boldsymbol{\Lambda}\boldsymbol{\delta} \neq \mathbf{0}$ for every $\boldsymbol{\delta} \in \Delta$ (in which case, $\mathbf{0} \notin \tilde{\Delta}$, as is evident upon observing that $\boldsymbol{\Lambda}\boldsymbol{\delta} = \boldsymbol{\Lambda}\mathbf{S}\mathbf{W}\boldsymbol{\delta}$ and hence that $\boldsymbol{\Lambda}\boldsymbol{\delta} \neq \mathbf{0} \Rightarrow \mathbf{W}\boldsymbol{\delta} \neq \mathbf{0}$). Aside from special cases, the interval $A_{\boldsymbol{\delta}}(\mathbf{y})$ is such that the distance $(\boldsymbol{\delta}'\mathbf{C}\boldsymbol{\delta})^{1/2}\hat{\sigma}a_{\dot{\gamma}}$ between the least squares estimator $\boldsymbol{\delta}'\hat{\boldsymbol{\tau}}$ (of $\boldsymbol{\delta}'\boldsymbol{\tau}$) and the lower confidence bound $\boldsymbol{\delta}'\hat{\boldsymbol{\tau}} - (\boldsymbol{\delta}'\mathbf{C}\boldsymbol{\delta})^{1/2}\hat{\sigma}a_{\dot{\gamma}}$ varies with $\boldsymbol{\delta}$.

Now, suppose that $\tilde{\Delta}$ is such that for every $M_* \times 1$ vector $\dot{\mathbf{t}}$, $\max_{\{\tilde{\boldsymbol{\delta}} \in \tilde{\Delta}\}} \tilde{\boldsymbol{\delta}}'\dot{\mathbf{t}}$ exists, and denote by $a^*_{\dot{\gamma}}$ the upper $100\dot{\gamma}\%$ point of the distribution of the random variable $\max_{\{\tilde{\boldsymbol{\delta}} \in \tilde{\Delta}\}} \tilde{\boldsymbol{\delta}}'\mathbf{t}$. Then, as an alternative to the interval $A_{\boldsymbol{\delta}}(\mathbf{y})$, we have the interval $A^*_{\boldsymbol{\delta}}(\mathbf{y})$ defined as follows:

$$A^*_{\boldsymbol{\delta}}(\mathbf{y}) = \{\dot{\tau} \in \mathcal{R}^1 \;:\; \boldsymbol{\delta}'\hat{\boldsymbol{\tau}} - \hat{\sigma}a^*_{\dot{\gamma}} \leq \dot{\tau} < \infty\}.$$

The probability of simultaneous coverage of the intervals $A^*_{\boldsymbol{\delta}}(\mathbf{y})$ $(\boldsymbol{\delta} \in \Delta)$, like that of the intervals $A_{\boldsymbol{\delta}}(\mathbf{y})$ $(\boldsymbol{\delta} \in \Delta)$, is equal to $1-\dot{\gamma}$. However, in the case of the interval $A^*_{\boldsymbol{\delta}}(\mathbf{y})$, the lower confidence bound $\boldsymbol{\delta}'\hat{\boldsymbol{\tau}} - \hat{\sigma}a^*_{\dot{\gamma}}$ is such that the distance $\hat{\sigma}a^*_{\dot{\gamma}}$ between it and the least squares estimator $\boldsymbol{\delta}'\hat{\boldsymbol{\tau}}$ does not vary with $\boldsymbol{\delta}$.

Application to the points on a regression line or on a response surface. Refer to Liu (2011) for a detailed and well-illustrated discussion of the use of the upper or lower confidence bounds $\boldsymbol{\delta}'\hat{\boldsymbol{\tau}} \pm (\boldsymbol{\delta}'\mathbf{C}\boldsymbol{\delta})^{1/2}\hat{\sigma}a_{\dot{\gamma}}$ and $\boldsymbol{\delta}'\hat{\boldsymbol{\tau}} \pm \hat{\sigma}a^*_{\dot{\gamma}}$ in making inferences about the points on a regression line or on a response surface.

One-sided tests. Corresponding to the confidence intervals $A_{\boldsymbol{\delta}}(\mathbf{y})$ $(\boldsymbol{\delta} \in \Delta)$, for which the probability of simultaneous coverage is equal to $1-\dot{\gamma}$, are the tests of the null hypotheses $H_0^{(\boldsymbol{\delta})} : \tau \leq \tau_{\boldsymbol{\delta}}^{(0)}$ $(\boldsymbol{\delta} \in \Delta)$ with critical regions $C(\boldsymbol{\delta})$ $(\boldsymbol{\delta} \in \Delta)$, respectively, defined (for each $\boldsymbol{\delta} \in \Delta$) as follows:

$$C(\boldsymbol{\delta}) = \{\mathbf{y} \;:\; \tau_{\boldsymbol{\delta}}^{(0)} \notin A_{\boldsymbol{\delta}}(\mathbf{y})\}.$$

Similarly, if Δ is such that $\boldsymbol{\Lambda}\boldsymbol{\delta} \neq \mathbf{0}$ for every $\boldsymbol{\delta} \in \Delta$ and if $\tilde{\Delta} = \{\tilde{\boldsymbol{\delta}} \;:\; \tilde{\boldsymbol{\delta}} = \mathbf{W}\boldsymbol{\delta},\, \boldsymbol{\delta} \in \boldsymbol{\Delta}\}$ is such that (for every $M_* \times 1$ vector $\dot{\mathbf{t}}$) $\max_{\{\tilde{\boldsymbol{\delta}} \in \tilde{\Delta}\}} \tilde{\boldsymbol{\delta}}'\dot{\mathbf{t}}$ exists, then corresponding to the confidence intervals $A^*_{\boldsymbol{\delta}}(\mathbf{y})$ $(\boldsymbol{\delta} \in \Delta)$ are the tests of the null hypotheses $H_0^{(\boldsymbol{\delta})} : \tau \leq \tau_{\boldsymbol{\delta}}^{(0)}$ $(\boldsymbol{\delta} \in \Delta)$ with critical regions $C^*(\boldsymbol{\delta})$ $(\boldsymbol{\delta} \in \Delta)$, respectively, defined (for each $\boldsymbol{\delta} \in \Delta$) as

$$C^*(\boldsymbol{\delta}) = \{\mathbf{y} \;:\; \tau_{\boldsymbol{\delta}}^{(0)} \notin A^*_{\boldsymbol{\delta}}(\mathbf{y})\}.$$

The tests of the null hypotheses $H_0^{(\boldsymbol{\delta})}$ $(\boldsymbol{\delta} \in \Delta)$ with critical regions $C(\boldsymbol{\delta})$ $(\boldsymbol{\delta} \in \Delta)$ are such that the probability of one or more false rejections is less than or equal to $\dot{\gamma}$, with equality holding when $\boldsymbol{\delta}'\boldsymbol{\tau} = \tau_{\boldsymbol{\delta}}^{(0)}$ for every $\boldsymbol{\delta} \in \Delta$. To see this, let (for $\boldsymbol{\delta} \in \Delta$)

$$C'(\boldsymbol{\delta}) = \{\mathbf{y} \;:\; \boldsymbol{\delta}'\boldsymbol{\tau} \notin A_{\boldsymbol{\delta}}(\mathbf{y})\}.$$

Then, for $\boldsymbol{\delta} \in \Delta$ such that $\boldsymbol{\delta}'\boldsymbol{\tau} \leq \tau_{\boldsymbol{\delta}}^{(0)}$, $C(\boldsymbol{\delta}) \subset C'(\boldsymbol{\delta})$. Thus,

$$\Pr\Bigl[\mathbf{y} \in \bigcup_{\{\boldsymbol{\delta}\in\Delta:\, \boldsymbol{\delta}'\boldsymbol{\tau}\le\tau_{\boldsymbol{\delta}}^{(0)}\}} C(\boldsymbol{\delta})\Bigr] \le \Pr\Bigl[\mathbf{y} \in \bigcup_{\{\boldsymbol{\delta}\in\Delta:\, \boldsymbol{\delta}'\boldsymbol{\tau}\le\tau_{\boldsymbol{\delta}}^{(0)}\}} C'(\boldsymbol{\delta})\Bigr] \le \Pr\Bigl[\mathbf{y} \in \bigcup_{\{\boldsymbol{\delta}\in\Delta\}} C'(\boldsymbol{\delta})\Bigr] = \dot{\gamma},$$

so that

$$\Pr\Bigl[\mathbf{y} \in \bigcup_{\{\boldsymbol{\delta}\in\Delta:\, \boldsymbol{\delta}'\boldsymbol{\tau}\le\tau_{\boldsymbol{\delta}}^{(0)}\}} C(\boldsymbol{\delta})\Bigr] \le \dot{\gamma},$$

with equality holding when $\boldsymbol{\delta}'\boldsymbol{\tau} = \tau_{\boldsymbol{\delta}}^{(0)}$ for every $\boldsymbol{\delta} \in \Delta$.

Now, suppose that Δ is such that $\boldsymbol{\Lambda\delta} \neq \mathbf{0}$ for every $\boldsymbol{\delta} \in \Delta$ and is such that (for every $M_*\times 1$ vector $\dot{\mathbf{t}}$) $\max_{\{\tilde{\boldsymbol{\delta}}\in\tilde{\Delta}\}} \tilde{\boldsymbol{\delta}}'\dot{\mathbf{t}}$ exists. Then, by employing essentially the same argument as in the case of the tests with critical regions $C(\boldsymbol{\delta})$ $(\boldsymbol{\delta} \in \Delta)$, it can be shown that the tests of the null hypotheses $H_0^{(\boldsymbol{\delta})}$ $(\boldsymbol{\delta} \in \Delta)$ with critical regions $C^*(\boldsymbol{\delta})$ $(\boldsymbol{\delta} \in \Delta)$ are such that the probability of one or more false rejections is less than or equal to $\dot{\gamma}$, with equality holding when $\boldsymbol{\delta}'\boldsymbol{\tau} = \tau_{\boldsymbol{\delta}}^{(0)}$ for every $\boldsymbol{\delta} \in \Delta$.

f. Nonnormality

In Subsections a and b, it was established that each of the confidence sets (5.2), (5.3), and more generally (5.4) has a probability of coverage equal to $1-\dot{\gamma}$. It was also established that the test of the null hypothesis H_0^+ or alternatively H_0^- with critical region C^+ or C^-, respectively, is such that the probability of falsely rejecting H_0^+ or H_0^- is equal to $\dot{\gamma}$. And in Subsection e, it was established that the probability of simultaneous coverage of the confidence intervals $A_{\boldsymbol{\delta}}(\mathbf{y})$ $(\boldsymbol{\delta} \in \Delta)$ and the probability of simultaneous coverage of the confidence intervals $A_{\boldsymbol{\delta}}^*(\mathbf{y})$ $(\boldsymbol{\delta} \in \Delta)$ are both equal to $1-\dot{\gamma}$. In addition, it was established that the tests of the null hypotheses $H_0^{(\boldsymbol{\delta})}$ $(\boldsymbol{\delta} \in \Delta)$ with critical regions $C(\boldsymbol{\delta})$ $(\boldsymbol{\delta} \in \Delta)$ and the tests of $H_0^{(\boldsymbol{\delta})}$ $(\boldsymbol{\delta} \in \Delta)$ with critical regions $C^*(\boldsymbol{\delta})$ $(\boldsymbol{\delta} \in \Delta)$ are both such that the probability of falsely rejecting one or more of the null hypotheses is less than or equal to $\dot{\gamma}$.

A supposition of normality (made at the beginning of Section 7.5) underlies those results. However, this supposition is stronger than necessary. It suffices to take the distribution of the vector $\mathbf{e}$ of residual effects in the G–M model to be an absolutely continuous spherical distribution. In fact, as is evident upon observing that

$$\hat{\sigma}^{-1}(\hat{\boldsymbol{\alpha}}-\boldsymbol{\alpha}) = [\mathbf{d}'\mathbf{d}/(N-P_*)]^{-1/2}(\hat{\boldsymbol{\alpha}}-\boldsymbol{\alpha})$$

and recalling result (6.4.67), it suffices to take the distribution of the vector $\begin{pmatrix}\hat{\boldsymbol{\alpha}}-\boldsymbol{\alpha}\\ \mathbf{d}\end{pmatrix}$ to be an absolutely continuous spherical distribution. A supposition of normality is also stronger than what is needed (in Subsection c) in establishing the results of Propositions (2′) and (3′).

7.6 The Residual Variance σ^2: Confidence Intervals and Tests of Hypotheses

Suppose (as in Sections 7.3, 7.4, and 7.5) that $\mathbf{y} = (y_1, y_2, \ldots, y_N)'$ is an observable random vector that follows the G–M model. Suppose further that the distribution of the vector $\mathbf{e} = (e_1, e_2, \ldots, e_N)'$ of residual effects in the G–M model is $N(\mathbf{0}, \sigma^2\mathbf{I})$, in which case $\mathbf{y} \sim N(\mathbf{X}\boldsymbol{\beta}, \sigma^2\mathbf{I})$.

Let us consider the problem of constructing a confidence interval for the variance σ^2 (of the residual effects $e_1, e_2, \ldots, e_N$ and of the observable random variables $y_1, y_2, \ldots, y_N$) or its (positive) square root σ (the standard deviation). Let us also consider the closely related problem of testing hypotheses about σ^2 or σ. In doing so, let us adopt the notation and terminology employed in Section 7.3 (and subsequently in Sections 7.4 and 7.5).

a. The basics

Let S represent the sum of squares $\tilde{\mathbf{e}}'\tilde{\mathbf{e}} = \mathbf{y}'(\mathbf{I} - \mathbf{P_X})\mathbf{y}$ of the elements of the vector $\tilde{\mathbf{e}} = (\mathbf{I} - \mathbf{P_X})\mathbf{y}$ of least squares residuals. And letting (as in Section 7.3a) $\mathbf{d} = \mathbf{L}'\mathbf{y}$ [where $\mathbf{L}$ is an $N \times (N - P_*)$ matrix of rank $N - P_*$ such that $\mathbf{X}'\mathbf{L} = \mathbf{0}$ and $\mathbf{L}'\mathbf{L} = \mathbf{I}$] and recalling result (3.21), observe that

$$S = \mathbf{d}'\mathbf{d}.$$

Moreover,

$$\mathbf{d} \sim N(\mathbf{0}, \sigma^2\mathbf{I})$$

and hence $(1/\sigma)\mathbf{d} \sim N(\mathbf{0}, \mathbf{I})$, so that

$$S/(\sigma^2) = [(1/\sigma)\mathbf{d}]'(1/\sigma)\mathbf{d} \sim \chi^2(N - P_*). \tag{6.1}$$

Confidence intervals/bounds. For $0 < \alpha < 1$, denote by $\bar{\chi}^2_\alpha(N - P_*)$ or simply by $\bar{\chi}^2_\alpha$ the upper $100\alpha\%$ point of the $\chi^2(N - P_*)$ distribution. Then, upon applying result (6.1), we obtain as a $100(1 - \dot{\gamma})\%$ "one-sided" confidence interval for σ, the interval

$$\sqrt{\frac{S}{\bar{\chi}^2_{\dot{\gamma}}}} \leq \sigma < \infty, \tag{6.2}$$

so that $\sqrt{S/\bar{\chi}^2_{\dot{\gamma}}}$ is a $100(1 - \dot{\gamma})\%$ lower "confidence bound" for σ. Similarly, the one-sided interval

$$0 < \sigma \leq \sqrt{\frac{S}{\bar{\chi}^2_{1-\dot{\gamma}}}} \tag{6.3}$$

constitutes a $100(1 - \dot{\gamma})\%$ confidence interval for σ, and $\sqrt{S/\bar{\chi}^2_{1-\dot{\gamma}}}$ is interpretable as a $100(1 - \dot{\gamma})\%$ upper confidence bound. And upon letting $\dot{\gamma}_1$ and $\dot{\gamma}_2$ represent any two strictly positive scalars such that $\dot{\gamma}_1 + \dot{\gamma}_2 = \dot{\gamma}$, we obtain as a $100(1 - \dot{\gamma})\%$ "two-sided" confidence interval for σ, the interval

$$\sqrt{\frac{S}{\bar{\chi}^2_{\dot{\gamma}_1}}} \leq \sigma \leq \sqrt{\frac{S}{\bar{\chi}^2_{1-\dot{\gamma}_2}}}. \tag{6.4}$$

Tests of hypotheses. Let σ_0 represent a hypothesized value of σ (where $0 < \sigma_0 < \infty$). Further, let $T = S/(\sigma_0^2)$. Then, corresponding to the confidence interval (6.4) is a size-$\dot{\gamma}$ (nonrandomized) test of the null hypothesis $H_0 : \sigma = \sigma_0$ versus the alternative hypothesis $H_1 : \sigma \neq \sigma_0$ with critical region

$$C = \{\mathbf{y} : T < \bar{\chi}^2_{1-\dot{\gamma}_2} \text{ or } T > \bar{\chi}^2_{\dot{\gamma}_1}\} \tag{6.5}$$

consisting of all values of $\mathbf{y}$ for which σ_0 is larger than $\sqrt{S/\bar{\chi}^2_{1-\dot{\gamma}_2}}$ or smaller than $\sqrt{S/\bar{\chi}^2_{\dot{\gamma}_1}}$. And corresponding to the confidence interval (6.2) is a (nonrandomized) test of the null hypothesis $H_0^+ : \sigma \leq \sigma_0$ versus the alternative hypothesis $H_1^+ : \sigma > \sigma_0$ with critical region

$$C^+ = \{\mathbf{y} : T > \bar{\chi}^2_{\dot{\gamma}}\} \tag{6.6}$$

consisting of all values of $\mathbf{y}$ for which the lower confidence bound $\sqrt{S/\bar{\chi}^2_{\dot{\gamma}}}$ exceeds σ_0. Similarly, corresponding to the confidence interval (6.3) is a (nonrandomized) test of the null hypothesis $H_0^- : \sigma \geq \sigma_0$ versus the alternative hypothesis $H_1^- : \sigma < \sigma_0$ with critical region

$$C^- = \{\mathbf{y} : T < \bar{\chi}^2_{1-\dot{\gamma}}\} \tag{6.7}$$

consisting of all values of $\mathbf{y}$ for which σ_0 exceeds the upper confidence bound $\sqrt{S/\bar{\chi}^2_{1-\dot{\gamma}}}$.

The tests of H_0^+ and H_0^- with critical regions C^+ and C^-, respectively, like the test of H_0 with critical region C, are of size $\dot{\gamma}$. To see this, it suffices to observe that

$$\Pr(T > \bar{\chi}^2_{\dot{\gamma}}) = \Pr[S/(\sigma^2) > (\sigma_0/\sigma)^2\bar{\chi}^2_{\dot{\gamma}}] \tag{6.8}$$

and

$$\Pr(T < \bar{\chi}^2_{1-\dot{\gamma}}) = \Pr[S/(\sigma^2) < (\sigma_0/\sigma)^2 \bar{\chi}^2_{1-\dot{\gamma}}], \tag{6.9}$$

which implies that $\Pr(T > \bar{\chi}^2_{\dot{\gamma}})$ is greater than, equal to, or less than $\dot{\gamma}$ and $\Pr(T < \bar{\chi}^2_{1-\dot{\gamma}})$ less than, equal to, or greater than $\dot{\gamma}$ depending on whether $\sigma > \sigma_0$, $\sigma = \sigma_0$, or $\sigma < \sigma_0$.

Translation invariance. The confidence intervals (6.2), (6.3), and (6.4) and the tests of H_0, H_0^+, and H_0^- with critical regions C, C^+, and C^- are translation invariant. That is, the results produced by these procedures are unaffected when for any $P \times 1$ vector $\mathbf{k}$, the value of the vector $\mathbf{y}$ is replaced by the value of the vector $\mathbf{y} + \mathbf{X}\mathbf{k}$. To see this, it suffices to observe that the procedures depend on $\mathbf{y}$ only through the value of the vector $\tilde{\mathbf{e}} = (\mathbf{I} - \mathbf{P_X})\mathbf{y}$ of least squares residuals and that

$$(\mathbf{I} - \mathbf{P_X})(\mathbf{y} + \mathbf{X}\mathbf{k}) = \mathbf{y}.$$

Unbiasedness. The test of H_0^+ versus H_1^+ with critical region C^+ and the test of H_0^- versus H_1^- with critical region C^- are both unbiased, as is evident from results (6.8) and (6.9). In fact, they are both strictly unbiased. In contrast, the test of H_0 versus H_1 with critical region C is unbiased only if $\dot{\gamma}_1$ (and $\dot{\gamma}_2 = \dot{\gamma} - \dot{\gamma}_1$) are chosen judiciously. Let us consider how to choose $\dot{\gamma}_1$ so as to achieve unbiasedness—it is not a simple matter of setting $\dot{\gamma}_1 = \dot{\gamma}/2$.

Let $\gamma(\sigma)$ represent the power function of a (possibly randomized) size-$\dot{\gamma}$ test of H_0 versus H_1. And suppose that (as in the case of the test with critical region C), the test is among those size-$\dot{\gamma}$ tests with a critical function that is expressible as a function, say $\phi(T)$, of T. Then, $\gamma(\sigma) = \mathrm{E}[\phi(T)]$, and

$$1 - \gamma(\sigma) = \mathrm{E}[1 - \phi(T)] = \int_0^\infty [1 - \phi(t)]\, h(t)\, dt, \tag{6.10}$$

where $h(\cdot)$ is the pdf of the distribution of T.

The pdf of the distribution of T is derivable from the pdf of the $\chi^2(N - P_*)$ distribution. Let $U = S/(\sigma^2)$, so that $U \sim \chi^2(N - P_*)$. And observe that $U = (\sigma_0/\sigma)^2 T$. Further, let $g(\cdot)$ represent the pdf of the $\chi^2(N - P_*)$ distribution. Then, upon recalling result (6.1.16), we find that for $t > 0$,

$$\begin{aligned} h(t) &= (\sigma_0/\sigma)^2 g[(\sigma_0/\sigma)^2 t] \\ &= \frac{1}{\Gamma[(N - P_*)/2]\, 2^{(N-P_*)/2}} (\sigma_0/\sigma)^{N-P_*}\, t^{[(N-P_*)/2]-1} e^{-(\sigma_0/\sigma)^2 t/2} \end{aligned} \tag{6.11}$$

—for $t \le 0$, $h(t) = 0$.

A necessary and sufficient condition for the test to be unbiased is that $\gamma(\sigma)$ attain its minimum value (with respect to σ) at $\sigma = \sigma_0$ or, equivalently, that $1 - \gamma(\sigma)$ attain its maximum value at $\sigma = \sigma_0$—for the test to be strictly unbiased, it is necessary and sufficient that $1 - \gamma(\sigma)$ attain its maximum value at $\sigma = \sigma_0$ and at no other value of σ. And if $1 - \gamma(\sigma)$ attains its maximum value at $\sigma = \sigma_0$, then

$$\left.\frac{d[1 - \gamma(\sigma)]}{d\sigma}\right|_{\sigma = \sigma_0} = 0. \tag{6.12}$$

Now, suppose that the test is such that $\lim_{\sigma \to 0} 1 - \gamma(\sigma)$ and $\lim_{\sigma \to \infty} 1 - \gamma(\sigma)$ are both equal to 0 (as in the case of the test with critical region C) or, more generally, that both of these limits are smaller than $1 - \dot{\gamma}$. Suppose also that the test is such that condition (6.12) is satisfied and such that $\dfrac{d[1 - \gamma(\sigma)]}{d\sigma} \neq 0$ for $\sigma \neq \sigma_0$. Then, $1 - \gamma(\sigma)$ attains its maximum value at $\sigma = \sigma_0$ and at no other value of σ, and hence the test is unbiased and, in fact, is strictly unbiased.

The derivative of $1 - \gamma(\sigma)$ is expressible as follows:

$$\begin{aligned} \frac{d[1 - \gamma(\sigma)]}{d\sigma} &= \int_0^\infty [1 - \phi(t)]\, \frac{dh(t)}{d\sigma}\, dt \\ &= \sigma^{-1}(N - P_*) \int_0^\infty [1 - \phi(t)] \left[\left(\frac{\sigma_0}{\sigma}\right)^2 \frac{t}{N - P_*} - 1\right] h(t)\, dt. \end{aligned} \tag{6.13}$$

Further, upon recalling the relationship $U = (\sigma_0/\sigma)^2 T$ between the random variable U [$= S/(\sigma^2)$] that has a $\chi^2(N-P_*)$ distribution with pdf $g(\cdot)$ and the random variable T [$= S/(\sigma_0^2)$] and upon introducing a change of variable, we find that $1-\gamma(\sigma)$ and its derivative are reexpressible in the form

$$1-\gamma(\sigma) = \int_0^\infty \{1-\phi[(\sigma/\sigma_0)^2 u]\}\, g(u)\, du, \tag{6.14}$$

and

$$\frac{d[1-\gamma(\sigma)]}{d\sigma} = \sigma^{-1}(N-P_*) \int_0^\infty \{1-\phi[(\sigma/\sigma_0)^2 u]\} \left(\frac{u}{N-P_*} - 1\right) g(u)\, du. \tag{6.15}$$

And in the special case of the test with critical region C,

$$1-\gamma(\sigma) = \int_{\bar{\chi}^2_{1-\dot{\gamma}+\dot{\gamma}_1}(\sigma_0/\sigma)^2}^{\bar{\chi}^2_{\dot{\gamma}_1}(\sigma_0/\sigma)^2} g(u)\, du, \tag{6.16}$$

and

$$\frac{d[1-\gamma(\sigma)]}{d\sigma} = \sigma^{-1}(N-P_*) \int_{\bar{\chi}^2_{1-\dot{\gamma}+\dot{\gamma}_1}(\sigma_0/\sigma)^2}^{\bar{\chi}^2_{\dot{\gamma}_1}(\sigma_0/\sigma)^2} \left(\frac{u}{N-P_*} - 1\right) g(u)\, du. \tag{6.17}$$

In the further special case where $\sigma = \sigma_0$, result (6.17) is reexpressible as

$$\left.\frac{d[1-\gamma(\sigma)]}{d\sigma}\right|_{\sigma=\sigma_0} = \sigma_0^{-1}(N-P_*) \int_{\bar{\chi}^2_{1-\dot{\gamma}+\dot{\gamma}_1}}^{\bar{\chi}^2_{\dot{\gamma}_1}} \left(\frac{u}{N-P_*} - 1\right) g(u)\, du. \tag{6.18}$$

For what value or values of $\dot{\gamma}_1$ (in the interval $0 < \dot{\gamma}_1 < \dot{\gamma}$) is expression (6.18) equal to 0? Note that the integrand of the integral in expression (6.18) is greater than 0 for $u > N-P_*$ and less than 0 for $u < N-P_*$ [and that $N-P_* = \mathrm{E}(U)$]. And assume that $\dot{\gamma}$ is sufficiently small that $\bar{\chi}^2_{\dot{\gamma}} > N-P_*$—the median of a chi-square distribution is smaller than the mean (e.g., Sen 1989), so that $\bar{\chi}^2_{\dot{\gamma}} > N-P_* \Rightarrow \bar{\chi}^2_{1-\dot{\gamma}} < N-P_*$. Then, the value or values of $\dot{\gamma}_1$ for which expression (6.18) equals 0 are those for which

$$\int_{N-P_*}^{\bar{\chi}^2_{\dot{\gamma}_1}} \left(\frac{u}{N-P_*} - 1\right) g(u)\, du = \int_{\bar{\chi}^2_{1-\dot{\gamma}+\dot{\gamma}_1}}^{N-P_*} \left(1 - \frac{u}{N-P_*}\right) g(u)\, du. \tag{6.19}$$

Both the left and right sides of equation (6.19) are strictly positive. As $\dot{\gamma}_1$ increases from 0 to $\dot{\gamma}$ (its upper limit), the left side of equation (6.19) decreases from $\int_{N-P_*}^{\infty} \left(\frac{u}{N-P_*} - 1\right) g(u)\, du$ to $\int_{N-P_*}^{\bar{\chi}^2_{\dot{\gamma}}} \left(\frac{u}{N-P_*} - 1\right) g(u)\, du$ and the right side increases from $\int_{\bar{\chi}^2_{1-\dot{\gamma}}}^{N-P_*} \left(1 - \frac{u}{N-P_*}\right) g(u)\, du$ to $\int_0^{N-P_*} \left(1 - \frac{u}{N-P_*}\right) g(u)\, du$. Assume that $\dot{\gamma}$ is such that

$$\int_{N-P_*}^{\infty} \left(\frac{u}{N-P_*} - 1\right) g(u)\, du \ge \int_{\bar{\chi}^2_{1-\dot{\gamma}}}^{N-P_*} \left(1 - \frac{u}{N-P_*}\right) g(u)\, du$$

and is also such that

$$\int_0^{N-P_*} \left(1 - \frac{u}{N-P_*}\right) g(u)\, du \ge \int_{N-P_*}^{\bar{\chi}^2_{\dot{\gamma}}} \left(\frac{u}{N-P_*} - 1\right) g(u)\, du$$

—otherwise, there would not exist any solution (for $\dot{\gamma}_1$) to equation (6.19). Then, there exists a unique value, say $\dot{\gamma}_1^*$, of $\dot{\gamma}_1$ that is a solution to equation (6.19).

Suppose the test (of H_0 versus H_1) is that with critical region C and that $\dot{\gamma}_1 = \dot{\gamma}_1^*$. That is, suppose the test is that with critical region

$$C^* = \{\mathbf{y} : T < \bar{\chi}^2_{1-\dot{\gamma}_2^*} \text{ or } T > \bar{\chi}^2_{\dot{\gamma}_1^*}\}, \tag{6.20}$$

where $\dot{\gamma}_2^* = \dot{\gamma} - \dot{\gamma}_1^*$. Then,

$$\frac{d[1-\gamma(\sigma)]}{d\sigma} = \sigma^{-1}(N-P_*)\int_{\bar{\chi}^2_{1-\dot{\gamma}+\dot{\gamma}_1^*}(\sigma_0/\sigma)^2}^{\bar{\chi}^2_{\dot{\gamma}_1^*}(\sigma_0/\sigma)^2} \left(\frac{u}{N-P_*} - 1\right) g(u)\, du.$$

and

$$\left.\frac{d[1-\gamma(\sigma)]}{d\sigma}\right|_{\sigma=\sigma_0} = 0.$$

To conclude that the test is unbiased (and, in fact, strictly unbiased), it remains only to show that $\dfrac{d[1-\gamma(\sigma)]}{d\sigma} \neq 0$ for $\sigma \neq \sigma_0$ or, equivalently, that

$$\int_{\bar{\chi}^2_{1-\dot{\gamma}+\dot{\gamma}_1^*}(\sigma_0/\sigma)^2}^{\bar{\chi}^2_{\dot{\gamma}_1^*}(\sigma_0/\sigma)^2} \left(\frac{u}{N-P_*} - 1\right) g(u)\, du = 0 \tag{6.21}$$

implies that $\sigma = \sigma_0$.

Suppose that σ satisfies condition (6.21). Then,

$$\bar{\chi}^2_{\dot{\gamma}_1^*}(\sigma_0/\sigma)^2 > N-P_* > \bar{\chi}^2_{1-\dot{\gamma}+\dot{\gamma}_1^*}(\sigma_0/\sigma)^2,$$

since otherwise $\dfrac{u}{N-P_*} - 1$ would either be less than 0 for all values of u between $\bar{\chi}^2_{1-\dot{\gamma}+\dot{\gamma}_1^*}(\sigma_0/\sigma)^2$ and $\bar{\chi}^2_{\dot{\gamma}_1^*}(\sigma_0/\sigma)^2$ or greater than 0 for all such values. Thus, σ is such that

$$\int_{N-P_*}^{\bar{\chi}^2_{\dot{\gamma}_1^*}(\sigma_0/\sigma)^2} \left(\frac{u}{N-P_*} - 1\right) g(u)\, du = \int_{\bar{\chi}^2_{1-\dot{\gamma}+\dot{\gamma}_1^*}(\sigma_0/\sigma)^2}^{N-P_*} \left(1 - \frac{u}{N-P_*}\right) g(u)\, du.$$

Moreover, if $\sigma < \sigma_0$ (in which case $\sigma_0/\sigma > 1$), then

$$\begin{aligned}\int_{N-P_*}^{\bar{\chi}^2_{\dot{\gamma}_1^*}(\sigma_0/\sigma)^2} \left(\frac{u}{N-P_*} - 1\right) g(u)\, du &> \int_{N-P_*}^{\bar{\chi}^2_{\dot{\gamma}_1^*}} \left(\frac{u}{N-P_*} - 1\right) g(u)\, du \\ &= \int_{\bar{\chi}^2_{1-\dot{\gamma}+\dot{\gamma}_1^*}}^{N-P_*} \left(1 - \frac{u}{N-P_*}\right) g(u)\, du \\ &> \int_{\bar{\chi}^2_{1-\dot{\gamma}+\dot{\gamma}_1^*}(\sigma_0/\sigma)^2}^{N-P_*} \left(1 - \frac{u}{N-P_*}\right) g(u)\, du.\end{aligned}$$

Similarly, if $\sigma > \sigma_0$ (in which case $\sigma_0/\sigma < 1$), then

$$\begin{aligned}\int_{N-P_*}^{\bar{\chi}^2_{\dot{\gamma}_1^*}(\sigma_0/\sigma)^2} \left(\frac{u}{N-P_*} - 1\right) g(u)\, du &< \int_{N-P_*}^{\bar{\chi}^2_{\dot{\gamma}_1^*}} \left(\frac{u}{N-P_*} - 1\right) g(u)\, du \\ &= \int_{\bar{\chi}^2_{1-\dot{\gamma}+\dot{\gamma}_1^*}}^{N-P_*} \left(1 - \frac{u}{N-P_*}\right) g(u)\, du \\ &< \int_{\bar{\chi}^2_{1-\dot{\gamma}+\dot{\gamma}_1^*}(\sigma_0/\sigma)^2}^{N-P_*} \left(1 - \frac{u}{N-P_*}\right) g(u)\, du.\end{aligned}$$

Thus, $\sigma \geq \sigma_0$ and $\sigma \leq \sigma_0$ (since if $\sigma < \sigma_0$, we arrive at a contradiction, and if $\sigma > \sigma_0$, we also arrive at a contradiction), and hence $\sigma = \sigma_0$.

We have established that the size-$\dot{\gamma}$ test of H_0 versus H_1 with critical region C^* is a strictly unbiased test. The value $\dot{\gamma}_1^*$ of $\dot{\gamma}_1$ that is a solution to equation (6.19) and that is needed to implement this test can be determined by, for example, employing the method of bisection. In that regard, it can be shown (and is worth noting) that for any constants c_0 and c_1 such that $\infty \geq c_1 > c_0 \geq 0$,

$$\begin{aligned}\int_{c_0}^{c_1} \left(\frac{u}{N-P_*} - 1\right) g(u)\, du &= -\int_{c_0}^{c_1} \left(1 - \frac{u}{N-P_*}\right) g(u)\, du \\ &= G^*(c_1) - G(c_1) - [G^*(c_0) - G(c_0)],\end{aligned} \tag{6.22}$$

where $G(\cdot)$ is the cdf of the $\chi^2(N-P_*)$ distribution and $G^*(\cdot)$ is the cdf of the $\chi^2(N-P_*+2)$ distribution. It is also worth noting that equation (6.19) does not involve σ_0 and hence that $\dot{\gamma}_1^*$ does not vary with the choice of σ_0.

Corresponding to the size-$\dot{\gamma}$ strictly unbiased test of H_0 versus H_1 with critical region C^* is the following $100(1-\dot{\gamma})\%$ confidence interval for σ:

$$\sqrt{\frac{S}{\bar{\chi}^2_{\dot{\gamma}_1^*}}} \le \sigma \le \sqrt{\frac{S}{\bar{\chi}^2_{1-\dot{\gamma}_2^*}}}. \tag{6.23}$$

This interval is the special case of the $100(1-\dot{\gamma})\%$ confidence interval (6.4) obtained upon setting $\dot{\gamma}_1 = \dot{\gamma}_1^*$ and $\dot{\gamma}_2 = \dot{\gamma}_2^*$ $(= \dot{\gamma} - \dot{\gamma}_1^*)$. As is evident from the (strict) unbiasedness of the corresponding test, the $100(1-\dot{\gamma})\%$ confidence interval (6.23) is strictly unbiased in the sense that the probability $1-\dot{\gamma}$ of its covering the true value of σ is greater than the probability of its covering any value other than the true value.

Like the $100(1-\dot{\gamma})\%$ confidence interval (6.23), the $100(1-\dot{\gamma})\%$ lower confidence bound $\sqrt{S/\bar{\chi}^2_{\dot{\gamma}}}$ and the $100(1-\dot{\gamma})\%$ upper confidence bound $\sqrt{S/\bar{\chi}^2_{1-\dot{\gamma}}}$ (for σ) are strictly unbiased. The strict unbiasedness of the lower confidence bound follows from the strict unbiasedness of the size-$\dot{\gamma}$ test of H_0^+ versus H_1^+ with critical region C^+ and is in the sense that $\Pr\left(\sqrt{S/\bar{\chi}^2_{\dot{\gamma}}} \le \sigma_0\right) < 1-\dot{\gamma}$ for any positive scalar σ_0 such that $\sigma_0 < \sigma$. Similarly, the strict unbiasedness of the upper confidence bound follows from the strict unbiasedness of the size-$\dot{\gamma}$ test of H_0^- versus H_1^- with critical region C^- and is in the sense that $\Pr\left(\sqrt{S/\bar{\chi}^2_{1-\dot{\gamma}}} \ge \sigma_0\right) < 1-\dot{\gamma}$ for any scalar σ_0 such that $\sigma_0 > \sigma$.

b. An illustration

Let us illustrate various of the results of Subsection a by using them to add to the results obtained earlier (in Sections 7.1, 7.2c, and 7.3d and in the final part of Section 7.3f) for the lettuce-yield data. Accordingly, let us take $\mathbf{y}$ to be the 20×1 random vector whose observed value is the vector of lettuce yields. And suppose that $\mathbf{y}$ follows the G–M model obtained upon taking the function $\delta(\mathbf{u})$ (that defines the response surface) to be the second-order polynomial (1.2) (where $\mathbf{u}$ is the 3-dimensional column vector whose elements represent transformed amounts of Cu, Mo, and Fe). Suppose further that the distribution of the vector $\mathbf{e}$ of residual effects is $N(\mathbf{0}, \sigma^2\mathbf{I})$.

Then, as is evident from the results of Section 7.1, S (the residual sum of squares) equals 108.9407. And $N-P_* = N-P = 10$. Further, the usual (unbiased) point estimator $\hat{\sigma}^2 = S/(N-P_*)$ of σ^2 equals 10.89, and upon taking the square root of this value, we obtain 3.30 as an estimate of σ.

When $\dot{\gamma} = 0.10$, the value $\dot{\gamma}_1^*$ of $\dot{\gamma}_1$ that is a solution to equation (6.19) is found to be 0.03495, and the corresponding value $\dot{\gamma}_2^*$ of $\dot{\gamma}_2$ is $\dot{\gamma}_2^* = \dot{\gamma} - \dot{\gamma}_1^* = 0.06505$. And $\bar{\chi}^2_{.03495} = 19.446$, and $\bar{\chi}^2_{1-.06505} = \bar{\chi}^2_{.93495} = 4.258$. Thus, upon setting $S = 108.9407$, $\dot{\gamma}_1^* = 0.03495$, and $\dot{\gamma}_2^* = 0.06505$ in the interval (6.23), we obtain as a 90% strictly unbiased confidence interval for σ the interval

$$2.37 \le \sigma \le 5.06.$$

By way of comparison, the 90% confidence interval for σ obtained upon setting $S = 108.9407$ and $\dot{\gamma}_1 = \dot{\gamma}_2 = 0.5$ in the interval (6.4) is

$$2.44 \le \sigma \le 5.26$$

—this interval is not unbiased.

If (instead of obtaining a two-sided confidence interval for σ) we had chosen to obtain [as an application of interval (6.2)] a 90% "lower confidence bound," we would have obtained

$$2.61 \le \sigma < \infty.$$

Similarly, if we had chosen to obtain [as an application of interval (6.3)] a 90% "upper confidence bound," we would have obtained

$$0 < \sigma \le 4.73.$$

c. Optimality

Are the tests of H_0^+, H_0^-, and H_0 (versus H_1^+, H_1^-, and H_1) with critical regions C^+, C^-, and C^* and the corresponding confidence intervals (6.2), (6.3), and (6.23) optimal and if so, in what sense? These questions are addressed in what follows. In the initial treatment, attention is restricted to translation-invariant procedures. Then, the results obtained in that context are extended to a broader class of procedures.

Translation-invariant procedures. As noted earlier (in Section 7.3a) the vector $\mathbf{d} = \mathbf{L}'\mathbf{y}$ [where $\mathbf{L}$ is an $N \times (N - P_*)$ matrix of full column rank such that $\mathbf{X}'\mathbf{L} = \mathbf{0}$ and $\mathbf{L}'\mathbf{L} = \mathbf{I}$] is an $(N - P_*)$-dimensional vector of linearly independent error contrasts—an error contrast is a linear combination (of the elements of $\mathbf{y}$) with an expected value equal to 0. And as is evident from the discussion of error contrasts in Section 5.9b, a (possibly randomized) test of H_0^+, H_0^-, or H_0 (versus H_1^+, H_1^-, or H_1) is translation invariant if and only if its critical function is expressible as a function, say $\phi(\mathbf{d})$, of $\mathbf{d}$. Moreover, when the observed value of $\mathbf{d}$ (rather than that of $\mathbf{y}$) is regarded as the data vector, $S = \mathbf{d}'\mathbf{d}$ or, alternatively, $T = S/\sigma_0^2$ is a sufficient statistic—refer to the next-to-last part of Section 7.3a for some relevant discussion. Thus, corresponding to the test with critical function $\phi(\mathbf{d})$ is a (possibly randomized) test with critical function $\mathrm{E}[\phi(\mathbf{d}) \mid S]$ or $\mathrm{E}[\phi(\mathbf{d}) \mid T]$ that depends on $\mathbf{d}$ only through the value of S or T and that has the same power function.

Now, consider the size-$\dot{\gamma}$ translation-invariant test of the null hypothesis $H_0^+ : \sigma \leq \sigma_0$ (versus the alternative hypothesis $H_1^+ : \sigma > \sigma_0$) with critical region $C^+ = \{\mathbf{y} \,:\, T > \bar{\chi}^2_{\dot{\gamma}}\}$. Further, let σ_* represent any particular value of σ greater than σ_0, let $h_0(\cdot)$ represent the pdf of the $\chi^2(N - P_*)$ distribution (which is the distribution of T when $\sigma = \sigma_0$), let $h_*(\cdot)$ represent the pdf of the distribution of T when $\sigma = \sigma_*$, and observe [in light of result (6.11)] that (for $t > 0$)

$$\frac{h_*(t)}{h_0(t)} = \left(\frac{\sigma_0}{\sigma_*}\right)^{N-P_*} e^{[1-(\sigma_0/\sigma_*)^2]t/2}. \tag{6.24}$$

Then, upon applying Theorem 7.4.1 (the Neyman–Pearson lemma) with $X = T$, $\theta = \sigma$, $\Theta = [\sigma_0, \infty)$, $\theta^{(0)} = \sigma_0$, and $\theta^* = \sigma_*$, we find that conditions (4.2) and (4.3) (of Theorem 7.4.1) are satisfied when the critical region is taken to be the set consisting of all values of T for which $T > \bar{\chi}^2_{\dot{\gamma}}$. And upon observing that the test of H_0^+ with critical region C^+ is such that $\Pr(\mathbf{y} \in C^+) \leq \dot{\gamma}$ for $\sigma < \sigma_0$ as well as for $\sigma = \sigma_0$ and upon recalling the discussion following Theorem 7.4.1, we find that the test of H_0^+ with critical region C^+ is UMP among all (possibly randomized) level-$\dot{\gamma}$ translation-invariant tests—note that the set consisting of all such tests is a subset of the set consisting of all (possibly randomized) translation-invariant tests for which the probability of rejecting H_0^+ is less than or equal to $\dot{\gamma}$ when $\sigma = \sigma_0$.

By employing a similar argument, it can be shown that the size-$\dot{\gamma}$ translation-invariant test of the null hypothesis $H_0^- : \sigma \geq \sigma_0$ (versus the alternative hypothesis $H_1^- : \sigma < \sigma_0$) with critical region $C^- = \{\mathbf{y} \,:\, T < \bar{\chi}^2_{1-\dot{\gamma}}\}$ is UMP among all (possibly randomized) level-$\dot{\gamma}$ translation-invariant tests (of H_0^- versus H_1^-).

It remains to consider the size-$\dot{\gamma}$ translation-invariant test of the null hypothesis $H_0 : \sigma = \sigma_0$ (versus the alternative hypothesis $H_1 : \sigma \neq \sigma_0$) with critical region C. In the special case where $C = C^*$ (i.e., the special case where $\dot{\gamma}_1 = \dot{\gamma}_1^*$ and $\dot{\gamma}_2 = \dot{\gamma}_2^* = \dot{\gamma} - \dot{\gamma}_1^*$), this test is (strictly) unbiased. In fact, the test of H_0 with critical region C^* is optimal in the sense that it is UMP among all (possibly randomized) level-$\dot{\gamma}$ translation-invariant unbiased tests of H_0 (versus H_1).

Let us verify the optimality of this test. Accordingly, take $\phi(T)$ to be a function of T that represents the critical function of a (translation-invariant possibly randomized) test of H_0 (versus H_1). Further, denote by $\gamma(\sigma)$ the power function of the test with critical function $\phi(T)$. And observe that (by definition) this test is of level $\dot{\gamma}$ if $\gamma(\sigma_0) \leq \dot{\gamma}$ or, equivalently, if

$$\int_0^\infty \phi(t)\, h_0(t)\, dt \leq \dot{\gamma} \tag{6.25}$$

[where, as before, $h_0(\cdot)$ represents the pdf of the $\chi^2(N-P_*)$ distribution].

If the test with critical function $\phi(T)$ is of level $\dot{\gamma}$ and is unbiased, then $\gamma(\sigma_0) = \dot{\gamma}$ (i.e., the test is of size $\dot{\gamma}$) or, equivalently,

$$\int_0^\infty \phi(t)\, h_0(t)\; dt = \dot{\gamma}, \tag{6.26}$$

as becomes evident upon observing that $\gamma(\cdot)$ is a continuous function—refer, e.g., to Lehmann and Romano (2005b, sec. 3.1). And if the test is of size $\dot{\gamma}$ and is unbiased, then $\left.\dfrac{d\gamma(\sigma)}{d\sigma}\right|_{\sigma=\sigma_0} = 0$ or, equivalently,

$$\int_0^\infty \phi(t) \left(\frac{t}{N-P_*} - 1\right) h_0(t)\; dt = 0 \tag{6.27}$$

—refer to Subsection a.

Denote by σ_* an arbitrary value of σ other than σ_0. And consider the problem of determining the choice of the critical function $\phi(\cdot)$ that maximizes $\gamma(\sigma_*)$ subject to the constraint that $\phi(\cdot)$ satisfy conditions (6.26) and (6.27). Note [in regard to the maximization of $\gamma(\sigma_*)$] that

$$\gamma(\sigma_*) = \int_0^\infty \phi(t)\, h_*(t)\; dt, \tag{6.28}$$

where $h_*(\cdot)$ represents the pdf of the distribution of T when $\sigma = \sigma_*$.

To show that any particular choice for $\phi(\cdot)$, say $\phi_*(\cdot)$, maximizes $\gamma(\sigma_*)$ subject to the constraints imposed by conditions (6.26) and (6.27), it suffices [according to a generalized version of the Neyman–Pearson lemma stated by Lehmann and Romano (2005b) in the form of their Theorem 3.6.1] to demonstrate that $\phi_*(\cdot)$ satisfies conditions (6.26) and (6.27) and to establish the existence of constants k_1 and k_2 such that

$$\phi_*(t) = \begin{cases} 1, & \text{when } h_*(t) > k_1 h_0(t) + k_2 h_0(t)\{[t/(N-P_*)] - 1\}, \\ 0, & \text{when } h_*(t) < k_1 h_0(t) + k_2 h_0(t)\{[t/(N-P_*)] - 1\}. \end{cases} \tag{6.29}$$

And condition (6.29) is reexpressible in the form

$$\phi_*(t) = \begin{cases} 1, & \text{when } k_1^* + k_2^* t < e^{bt}, \\ 0, & \text{when } k_1^* + k_2^* t > e^{bt}, \end{cases} \tag{6.30}$$

where $k_1^* = (\sigma_0/\sigma_*)^{-(N-P_*)}(k_1 - k_2)$, $k_2^* = (\sigma_0/\sigma_*)^{-(N-P_*)} k_2/(N-P_*)$, and $b = (1/2)[1 - (\sigma_0/\sigma_*)^2]$. Moreover, among the choices for the critical function $\phi(\cdot)$ are choices that satisfy conditions (6.26) and (6.27) and for which $\gamma(\sigma_*) \geq \dot{\gamma}$, as is evident upon observing that one such choice is that obtained upon setting $\phi(t) = \dot{\gamma}$ (for all t). Thus, if the critical function $\phi_*(\cdot)$ satisfies conditions (6.26) and (6.27) and is such that corresponding to every choice of σ_* there exist constants k_1 and k_2 that satisfy condition (6.29) or, equivalently, condition (6.30), then the test with critical function $\phi_*(\cdot)$ is a size-$\dot{\gamma}$ translation-invariant unbiased test and [since the tests for which the critical function $\phi(\cdot)$ is such that the test is of level $\dot{\gamma}$ and is unbiased constitute a subset of those tests for which $\phi(\cdot)$ satisfies conditions (6.26) and (6.27)] is UMP among all level-$\dot{\gamma}$ translation-invariant unbiased tests.

Suppose that the choice $\phi_*(\cdot)$ for $\phi(\cdot)$ is as follows:

$$\phi_*(t) = \begin{cases} 1, & \text{when } t < \bar{\chi}^2_{1-\dot{\gamma}_2^*} \text{ or } t > \bar{\chi}^2_{\dot{\gamma}_1^*}, \\ 0, & \text{when } \bar{\chi}^2_{1-\dot{\gamma}_2^*} \leq t \leq \bar{\chi}^2_{\dot{\gamma}_1^*}. \end{cases} \tag{6.31}$$

And observe that for this choice of $\phi(\cdot)$, the test with critical function $\phi(\cdot)$ is identical to the (nonrandomized) test of H_0 with critical region C^*. By construction, this test is such that $\phi_*(\cdot)$ satisfies conditions (6.26) and (6.27). Thus, to verify that the test of H_0 with critical region C^* is UMP among all (possibly randomized) level-$\dot{\gamma}$ translation-invariant unbiased tests, it suffices to show that (corresponding to every choice of σ_*) there exist constants k_1 and k_2 such that $\phi_*(\cdot)$ is expressible in the form (6.30).

Accordingly, suppose that k_1 and k_2 are the constants defined (implicitly) by taking k_1^* and k_2^* to be such that

$$k_1^* + k_2^* c_0 = e^{bc_0} \quad \text{and} \quad k_1^* + k_2^* c_1 = e^{bc_1},$$

where $c_0 = \bar{\chi}^2_{1-\dot{\gamma}_2^*}$ and $c_1 = \bar{\chi}^2_{\dot{\gamma}_1^*}$. Further, let $u(t) = e^{bt} - (k_1^* + k_2^* t)$ (a function of t with domain $0 < t < \infty$), and observe that $u(c_1) = u(c_0) = 0$. Observe also that

$$\frac{du(t)}{dt} = be^{bt} - k_2^* \quad \text{and} \quad \frac{d^2u(t)}{dt^2} = b^2 e^{bt} > 0,$$

so that $u(\cdot)$ is a strictly convex function and its derivative is a strictly increasing function. Then, clearly,

$$u(t) < 0 \quad \text{for } c_0 < t < c_1. \tag{6.32}$$

And $\frac{du(t)}{dt} < 0$ for $t \le c_0$ and $\frac{du(t)}{dt} > 0$ for $t \ge c_1$, which implies that

$$u(t) > 0 \quad \text{for } t < c_0 \text{ and } t > c_1$$

and hence in combination with result (6.32) implies that $\phi_*(t)$ is expressible in the form (6.30) and which in doing so completes the verification that the test of H_0 (versus H_1) with critical region C^* is UMP among all level-$\dot{\gamma}$ translation-invariant unbiased tests.

Corresponding to the test of H_0 with critical region C^* is the $100(1-\dot{\gamma})\%$ translation-invariant strictly unbiased confidence interval (6.23). Among translation-invariant confidence sets (for σ) that have a probability of coverage greater than or equal to $1-\dot{\gamma}$ and that are unbiased (in the sense that the probability of covering the true value of σ is greater than or equal to the probability of covering any value other than the true value), the confidence interval (6.23) is optimal; it is optimal in the sense that the probability of covering any value of σ other than the true value is minimized.

The $100(1-\dot{\gamma})\%$ translation-invariant confidence interval (6.2) is optimal in a different sense; among all translation-invariant confidence sets (for σ) that have a probability of coverage greater than or equal to $1-\dot{\gamma}$, it is optimal in the sense that it is the confidence set that minimizes the probability of covering values of σ smaller than the true value. Analogously, among all translation-invariant confidence sets that have a probability of coverage greater than or equal to $1-\dot{\gamma}$, the $100(1-\dot{\gamma})\%$ translation-invariant confidence interval (6.3) is optimal in the sense that it is the confidence set that minimizes the probability of covering values of σ larger than the true value.

Optimality in the absence of a restriction to translation-invariant procedures. Let $\mathbf{z} = \mathbf{O}'\mathbf{y}$ represent an observable N-dimensional random column vector that follows the canonical form of the G–M model (as defined in Section 7.3a) in the special case where $M_* = P_*$. Then, $\boldsymbol{\alpha}$ and its least squares estimator $\hat{\boldsymbol{\alpha}}$ are P_*-dimensional, and $\mathbf{z} = (\hat{\boldsymbol{\alpha}}', \mathbf{d}')'$, where (as before) $\mathbf{d} = \mathbf{L}'\mathbf{y}$, so that the critical function of any (possibly randomized) test of H_0^+, H_0^-, or H_0 is expressible as a function of $\hat{\boldsymbol{\alpha}}$ and $\mathbf{d}$. Moreover, $\hat{\boldsymbol{\alpha}}$ and $T = S/\sigma_0^2 = \mathbf{d}'\mathbf{d}/\sigma_0^2 = \mathbf{y}'(\mathbf{I}-\mathbf{P_X})\mathbf{y}/\sigma_0^2$ form a sufficient statistic, as is evident upon recalling the results of the next-to-last part of Section 7.3a. And corresponding to any (possibly randomized) test of H_0^+, H_0^-, or H_0, say one with critical function $\tilde{\phi}(\hat{\boldsymbol{\alpha}}, \mathbf{d})$, there is a test with a critical function, say $\phi(T, \hat{\boldsymbol{\alpha}})$, that depends on $\mathbf{d}$ only through the value of T and that has the same power function—take $\phi(T, \hat{\boldsymbol{\alpha}}) = \mathrm{E}[\tilde{\phi}(\hat{\boldsymbol{\alpha}}, \mathbf{d}) \mid T, \hat{\boldsymbol{\alpha}}]$. Thus, for present purposes, it suffices to restrict attention to tests with critical functions that are expressible in the form $\phi(T, \hat{\boldsymbol{\alpha}})$.

Suppose that $\phi(T, \hat{\boldsymbol{\alpha}})$ is the critical function of a level-$\dot{\gamma}$ test of the null hypothesis $H_0^+ : \sigma \le \sigma_0$ versus the alternative hypothesis $H_1^+ : \sigma > \sigma_0$, and consider the choice of the function $\phi(T, \hat{\boldsymbol{\alpha}})$. Further, let $\gamma(\sigma, \boldsymbol{\alpha})$ represent the power function of the test. Then, by definition, $\gamma(\sigma, \boldsymbol{\alpha}) = \mathrm{E}[\phi(T, \hat{\boldsymbol{\alpha}})]$, and, in particular, $\gamma(\sigma_0, \boldsymbol{\alpha}) = \mathrm{E}_0[\phi(T, \hat{\boldsymbol{\alpha}})]$, where E_0 represents the expectation operator in the special case where $\sigma = \sigma_0$. Since the test is of level $\dot{\gamma}$, $\gamma(\sigma_0, \boldsymbol{\alpha}) \le \dot{\gamma}$ (for all $\boldsymbol{\alpha}$).

Now, suppose that the level-$\dot{\gamma}$ test with critical function $\phi(T, \hat{\boldsymbol{\alpha}})$ and power function $\gamma(\sigma, \boldsymbol{\alpha})$ is unbiased. Then, upon observing that $\gamma(\cdot, \cdot)$ is a continuous function, it follows—refer, e.g., to Lehmann and Romano (2005b, sec. 4.1)—that

$$\gamma(\sigma_0, \boldsymbol{\alpha}) = \dot{\gamma} \quad \text{(for all } \boldsymbol{\alpha}\text{)}. \tag{6.33}$$

Clearly,

$$\gamma(\sigma, \boldsymbol{\alpha}) = \mathrm{E}\{\mathrm{E}[\phi(T, \hat{\boldsymbol{\alpha}}) \,|\, \hat{\boldsymbol{\alpha}}]\}. \tag{6.34}$$

In particular, $\gamma(\sigma_0, \boldsymbol{\alpha}) = \mathrm{E}_0\{\mathrm{E}_0[\phi(T, \hat{\boldsymbol{\alpha}}) \,|\, \hat{\boldsymbol{\alpha}}]\}$, so that result (6.33) can be restated as

$$\mathrm{E}_0\{\mathrm{E}_0[\phi(T, \hat{\boldsymbol{\alpha}}) \,|\, \hat{\boldsymbol{\alpha}}]\} = \dot{\gamma} \quad \text{(for all } \boldsymbol{\alpha}). \tag{6.35}$$

Moreover, with σ fixed (at σ_0), $\hat{\boldsymbol{\alpha}}$ is a complete sufficient statistic—refer to the next-to-last part of Section 7.3a. Thus, $\mathrm{E}_0[\phi(T, \hat{\boldsymbol{\alpha}}) \,|\, \hat{\boldsymbol{\alpha}}]$ does not depend on $\boldsymbol{\alpha}$, and condition (6.35) is equivalent to the condition

$$\mathrm{E}_0[\phi(T, \hat{\boldsymbol{\alpha}}) \,|\, \hat{\boldsymbol{\alpha}}] = \dot{\gamma} \quad \text{(wp1)}. \tag{6.36}$$

Let σ_* represent any particular value of σ greater than σ_0, let $\boldsymbol{\alpha}_*$ represent any particular value of $\boldsymbol{\alpha}$, and let E_* represent the expectation operator in the special case where $\sigma = \sigma_*$ and $\boldsymbol{\alpha} = \boldsymbol{\alpha}_*$. Then, in light of result (6.34), the choice of the critical function $\phi(T, \hat{\boldsymbol{\alpha}})$ that maximizes $\gamma(\sigma_*, \boldsymbol{\alpha}_*)$ subject to the constraint (6.36), and hence subject to the constraint (6.33), is that obtained by choosing (for each value of $\hat{\boldsymbol{\alpha}}$) $\phi(\cdot, \hat{\boldsymbol{\alpha}})$ so as to maximize $\mathrm{E}_*[\phi(T, \hat{\boldsymbol{\alpha}}) \,|\, \hat{\boldsymbol{\alpha}}]$ subject to the constraint $\mathrm{E}_0[\phi(T, \hat{\boldsymbol{\alpha}}) \,|\, \hat{\boldsymbol{\alpha}}] = \dot{\gamma}$. Moreover, upon observing that T is distributed independently of $\hat{\boldsymbol{\alpha}}$ and hence that the distribution of T conditional on $\hat{\boldsymbol{\alpha}}$ is the same as the unconditional distribution of T and upon proceeding as in the preceding part of the present subsection (in determining the optimal translation-invariant test), we find that (for every choice of σ_* and $\boldsymbol{\alpha}_*$ and for every value of $\hat{\boldsymbol{\alpha}}$) $\mathrm{E}_*[\phi(T, \hat{\boldsymbol{\alpha}}) \,|\, \hat{\boldsymbol{\alpha}}]$ can be maximized subject to the constraint $\mathrm{E}_0[\phi(T, \hat{\boldsymbol{\alpha}}) \,|\, \hat{\boldsymbol{\alpha}}] = \dot{\gamma}$ by taking

$$\phi(t, \hat{\boldsymbol{\alpha}}) = \begin{cases} 1, & \text{when } t > \bar{\chi}^2_{\dot{\gamma}}, \\ 0, & \text{when } t \le \bar{\chi}^2_{\dot{\gamma}}. \end{cases} \tag{6.37}$$

And it follows that the test with critical function (6.37) is UMP among all tests of H_0^+ (versus H_1^+) with a critical function that satisfies condition (6.33).

Clearly, the test with critical function (6.37) is identical to the test with critical region C^+, which is the UMP level-$\dot{\gamma}$ translation-invariant test. And upon recalling (from Subsection a) that the test with critical region C^+ is unbiased and upon observing that those tests with a critical function for which the test is of level-$\dot{\gamma}$ and is unbiased is a subset of those tests with a critical function that satisfies condition (6.33), we conclude that the test with critical region C^+ is UMP among all level-$\dot{\gamma}$ unbiased tests of H_0^+ (versus H_1^+).

It can be shown in similar fashion that the size-$\dot{\gamma}$ translation-invariant unbiased test of H_0^- versus H_1^- with critical region C^- is UMP among all level-$\dot{\gamma}$ unbiased tests of H_0^- versus H_1^-. However, as pointed out by Lehmann and Romano (2005b, sec. 3.9.1), the result on the optimality of the test of H_0^+ with critical region C^+ can be strengthened in a way that does not extend to the test of H_0^- with critical region C^-. In the case of the test of H_0^+, the restriction to unbiased tests is unnecessary. It can be shown that the test of H_0^+ versus H_1^+ with critical region C^+ is UMP among all level-$\dot{\gamma}$ tests, not just among those level-$\dot{\gamma}$ tests that are unbiased.

It remains to consider the optimality of the test of the null hypothesis $H_0: \sigma = \sigma_0$ (versus the alternative hypothesis $H_1: \sigma \neq \sigma_0$) with critical region C^*. Accordingly, suppose that $\phi(T, \hat{\boldsymbol{\alpha}})$ is the critical function of a (possibly randomized) test of H_0 (versus H_1) with power function $\gamma(\sigma, \boldsymbol{\alpha})$.

If the test is of level $\dot{\gamma}$ and is unbiased, then [in light of the continuity of the function $\gamma(\cdot, \cdot)$] $\gamma(\sigma_0, \boldsymbol{\alpha}) = \dot{\gamma}$ (for all $\boldsymbol{\alpha}$) or, equivalently,

$$\int_{\mathbb{R}^{P_*}} \int_0^\infty \phi(t, \hat{\boldsymbol{\alpha}}) h_0(t) f_0(\hat{\boldsymbol{\alpha}}; \boldsymbol{\alpha}) \, dt \, d\hat{\boldsymbol{\alpha}} = \dot{\gamma} \quad \text{(for all } \boldsymbol{\alpha}), \tag{6.38}$$

where $f_0(\cdot; \boldsymbol{\alpha})$ represents the pdf of the $N(\boldsymbol{\alpha}, \sigma_0^2\mathbf{I})$ distribution (which is the distribution of $\hat{\boldsymbol{\alpha}}$ when $\sigma = \sigma_0$) and where (as before) $h_0(\cdot)$ represents the pdf of the $\chi^2(N - P_*)$ distribution (which is the distribution of T when $\sigma = \sigma_0$)—condition (6.38) is analogous to condition (6.26). Moreover, if the test is such that condition (6.38) is satisfied and if the test is unbiased, then $\left.\frac{d\gamma(\sigma, \boldsymbol{\alpha})}{d\sigma}\right|_{\sigma=\sigma_0} = 0$ (for all $\boldsymbol{\alpha}$) or, equivalently,

$$\int_{\mathbb{R}^{P_*}} \int_0^{\infty} \phi(t, \hat{\boldsymbol{\alpha}}) \left(\frac{t}{N-P_*} - 1\right) h_0(t) f_0(\hat{\boldsymbol{\alpha}}; \boldsymbol{\alpha})\, dt\, d\hat{\boldsymbol{\alpha}} = 0 \quad (\text{for all } \boldsymbol{\alpha}), \tag{6.39}$$

analogous to condition (6.27)—the equivalence of condition (6.39) can be verified via a relatively straightforward exercise.

As in the case of testing the null hypothesis H_0^+,

$$\gamma(\sigma, \boldsymbol{\alpha}) = \mathrm{E}\{\mathrm{E}[\phi(T, \hat{\boldsymbol{\alpha}}) \mid \hat{\boldsymbol{\alpha}}]\}. \tag{6.40}$$

Moreover, condition (6.38) is equivalent to the condition

$$\int_0^{\infty} \phi(t, \hat{\boldsymbol{\alpha}})\, h_0(t)\, dt = \dot{\gamma} \quad (\text{wp1}), \tag{6.41}$$

and condition (6.39) is equivalent to the condition

$$\int_0^{\infty} \phi(t, \hat{\boldsymbol{\alpha}}) \left(\frac{t}{N-P_*} - 1\right) h_0(t)\, dt = 0 \quad (\text{wp1}), \tag{6.42}$$

as is evident upon recalling that with σ fixed (at σ_0), $\hat{\boldsymbol{\alpha}}$ is a complete sufficient statistic.

Denote by $\boldsymbol{\alpha}_*$ any particular value of $\boldsymbol{\alpha}$ and by σ_* any particular value of σ other than σ_0, and (as before) let $h_*(\cdot)$ represent the pdf of the distribution of T in the special case where $\sigma = \sigma_*$. And observe (in light of the statistical independence of T and $\hat{\boldsymbol{\alpha}}$) that when $\sigma = \sigma_*$,

$$\mathrm{E}[\phi(T, \hat{\boldsymbol{\alpha}}) \mid \hat{\boldsymbol{\alpha}}] = \int_0^{\infty} \phi(t, \hat{\boldsymbol{\alpha}})\, h_*(t)\, dt. \tag{6.43}$$

Observe also [in light of result (6.43) along with result (6.40) and in light of the equivalence of conditions (6.41) and (6.42) to conditions (6.38) and (6.39)] that to maximize $\gamma(\sigma_*, \boldsymbol{\alpha}_*)$ [with respect to the choice of the critical function $\phi(\cdot, \cdot)$] subject to the constraint that $\phi(\cdot, \cdot)$ satisfy conditions (6.38) and (6.39), it suffices to take (for each value of $\hat{\boldsymbol{\alpha}}$) $\phi(\cdot, \hat{\boldsymbol{\alpha}})$ to be the critical function that maximizes

$$\int_0^{\infty} \phi(t, \hat{\boldsymbol{\alpha}})\, h_*(t)\, dt \tag{6.44}$$

subject to the constraints imposed by the conditions

$$\int_0^{\infty} \phi(t, \hat{\boldsymbol{\alpha}})\, h_0(t)\, dt = \dot{\gamma} \quad \text{and} \quad \int_0^{\infty} \phi(t, \hat{\boldsymbol{\alpha}}) \left(\frac{t}{N-P_*} - 1\right) h_0(t)\, dt = 0. \tag{6.45}$$

A solution for $\phi(\cdot, \hat{\boldsymbol{\alpha}})$ to the latter constrained maximization problem can be obtained by applying the results obtained earlier (in the first part of the present subsection) in choosing a translation-invariant critical function $\phi(\cdot)$ so as to maximize the quantity (6.28) subject to the constraints imposed by conditions (6.26) and (6.27). Upon doing so, we find that among those choices for the critical function $\phi(T, \hat{\boldsymbol{\alpha}})$ that satisfy conditions (6.38) and (6.39), $\gamma(\sigma_*, \boldsymbol{\alpha}_*)$ can be maximized (for every choice of σ_* and $\boldsymbol{\alpha}_*$) by taking

$$\phi(t, \hat{\boldsymbol{\alpha}}) = \begin{cases} 1, & \text{when } t < \bar{\chi}^2_{1-\dot{\gamma}_2^*} \text{ or } t > \bar{\chi}^2_{\dot{\gamma}_1^*}, \\ 0, & \text{when } \bar{\chi}^2_{1-\dot{\gamma}_2^*} \le t \le \bar{\chi}^2_{\dot{\gamma}_1^*}, \end{cases} \tag{6.46}$$

which is the critical function of the size-$\dot{\gamma}$ translation-invariant unbiased test of H_0 (versus H_1) with critical region C^*. Since the set consisting of all level-$\dot{\gamma}$ unbiased tests of H_0 versus H_1 is a subset of the set consisting of all tests with a critical function that satisfies conditions (6.38) and (6.39), it follows that the size-$\dot{\gamma}$ translation-invariant unbiased test with critical region C^* is UMP among all level-$\dot{\gamma}$ unbiased tests.

The optimality properties of the various tests can be reexpressed as optimality properties of the corresponding confidence intervals. Each of the confidence intervals (6.2) and (6.23) is optimal in essentially the same sense as when attention is restricted to translation-invariant procedures. The confidence interval (6.3) is optimal in the sense that among all confidence sets for σ with a probability of coverage greater than or equal to $1-\dot{\gamma}$ and that are unbiased (in the sense that the probability of covering the true value of σ is greater than or equal to the probability of covering any value larger than the true value), the probability of covering any value larger than the true value is minimized.

7.7 Multiple Comparisons and Simultaneous Confidence Intervals: Some Enhancements

Let us revisit the topic of multiple comparisons and simultaneous confidence intervals, which was considered earlier in Section 7.3c. As before, let us take $\mathbf{y}$ to be an $N \times 1$ observable random vector that follows the G–M model, take $\tau_i = \boldsymbol{\lambda}_i'\boldsymbol{\beta}$ $(i = 1, 2, \ldots, M)$ to be estimable linear combinations of the elements of $\boldsymbol{\beta}$, and take $\boldsymbol{\tau} = (\tau_1, \tau_2, \ldots, \tau_M)'$ and $\boldsymbol{\Lambda} = (\boldsymbol{\lambda}_1, \boldsymbol{\lambda}_2, \ldots, \boldsymbol{\lambda}_M)$ (in which case $\boldsymbol{\tau} = \boldsymbol{\Lambda}'\boldsymbol{\beta}$). Further, let us assume that none of the columns $\boldsymbol{\lambda}_1, \boldsymbol{\lambda}_2, \ldots, \boldsymbol{\lambda}_M$ of $\boldsymbol{\Lambda}$ is null or is a scalar multiple of another column of $\boldsymbol{\Lambda}$. And let us assume (at least initially) that the distribution of the vector $\mathbf{e}$ of residual effects in the G–M model is MVN, in which case $\mathbf{y} \sim N(\mathbf{X}\boldsymbol{\beta}, \sigma^2\mathbf{I})$.

Suppose that we wish to make inferences about each of the linear combinations $\tau_1, \tau_2, \ldots, \tau_M$. Among the forms the inferences may take is that of multiple comparisons. For $i = 1, 2, \ldots, M$, let $\tau_i^{(0)}$ represent a hypothesized value of τ_i. And suppose that each of M null hypotheses $H_i^{(0)} : \tau_i = \tau_i^{(0)}$ $(i = 1, 2, \ldots, M)$ is to be tested against the corresponding one of the M alternative hypotheses $H_i^{(1)} : \tau_i \neq \tau_i^{(0)}$ $(i = 1, 2, \ldots, M)$.

Let $\boldsymbol{\tau}^{(0)} = (\tau_1^{(0)}, \tau_2^{(0)}, \ldots, \tau_M^{(0)})'$, and assume that $\boldsymbol{\tau}^{(0)} = \boldsymbol{\Lambda}'\boldsymbol{\beta}^{(0)}$ for some $P \times 1$ vector $\boldsymbol{\beta}^{(0)}$ or, equivalently, that $\boldsymbol{\tau}^{(0)} \in \mathcal{C}(\boldsymbol{\Lambda}')$ (which insures that the collection of null hypotheses $H_1^{(0)}, H_2^{(0)}, \ldots, H_M^{(0)}$ is "internally consistent"). And (as is customary) let us refer to the probability of falsely rejecting one or more of the M null hypotheses as the familywise error rate (FWER). In devising multiple-comparison procedures, the traditional approach has been to focus on those alternatives (to the so-called one-at-a-time test procedures) that control the FWER (in the sense that FWER $\leq \dot{\gamma}$ for some specified scalar $\dot{\gamma}$ such as 0.01 or 0.05), that are relatively simple in form, and that are computationally tractable. The multiple-comparison procedures that form what in Section 7.3c is referred to as the generalized S method are obtainable via such an approach. Corresponding to those multiple-comparison procedures are the procedures (discussed in Section 7.3c) for obtaining confidence intervals for the linear combinations $\tau_1, \tau_2, \ldots, \tau_M$ that have a probability of simultaneous coverage equal to $1-\dot{\gamma}$.

While tests of $H_1^{(0)}, H_2^{(0)}, \ldots, H_M^{(0)}$ with an FWER equal to $\dot{\gamma}$ and confidence intervals for $\tau_1, \tau_2, \ldots, \tau_M$ with a probability of simultaneous coverage equal to $1 - \dot{\gamma}$ can be achieved via the methods discussed in Section 7.3c, there is a downside to the adoption of such methods. For even relatively small values of M_* $(= \text{rank}\,\boldsymbol{\Lambda})$ and for the customary values of $\dot{\gamma}$ (such as 0.01, 0.05, and 0.10), the probability of rejecting any particular one (say the ith) of the null hypotheses $H_1^{(0)}, H_2^{(0)}, \ldots, H_M^{(0)}$ can be quite small, even when τ_i differs substantially from $\tau_i^{(0)}$. And the confidence intervals for $\tau_1, \tau_2, \ldots, \tau_M$ are likely to be very wide. As is discussed in Subsection a, test procedures that are much more likely to reject any of the M null hypotheses $H_1^{(0)}, H_2^{(0)}, \ldots, H_M^{(0)}$ can be obtained by imposing a restriction on the false rejection of multiple null hypotheses less severe than that imposed by a requirement that the FWER be less than or equal to $\dot{\gamma}$. And much shorter confidence intervals for $\tau_1, \tau_2, \ldots, \tau_M$ can be obtained by adopting a criterion less stringent than that inherent in a requirement that the probability of simultaneous coverage be greater than or equal to $1-\dot{\gamma}$. Moreover, as is discussed in Subsection b, improvements can be effected in the various test procedures (at the expense of additional complexity and additional computational demands) by employing "step-down" methods.

In some applications of the testing of the M null hypotheses, the linear combinations $\tau_1, \tau_2, \ldots, \tau_M$ may represent the "effects" of "genes" or other such entities, and the object may be to "discover" or "detect" those entities whose effects are nonnegligible and that should be subjected to further evaluation and/or future investigation. In such applications, M can be very large. And limiting the number of rejections of true null hypotheses may be less of a point of emphasis than rejecting a high proportion of the false null hypotheses. In Subsection c, an example is presented of an application where M is in the thousands. Methods that are well suited for such applications are discussed in Subsections d and e.

a. A further generalization of the S method

Some preliminaries. Let us incorporate the notation introduced in Section 7.3a (in connection with the canonical form of the G–M model) and take advantage of the results introduced therein. Accordingly, $\hat{\boldsymbol{\tau}} = \boldsymbol{\Lambda}'(\mathbf{X}'\mathbf{X})^{-}\mathbf{X}'\mathbf{y}$ is the least squares estimator of the vector $\boldsymbol{\tau}$. And $\text{var}(\hat{\boldsymbol{\tau}}) = \sigma^2\mathbf{C}$, where $\mathbf{C} = \boldsymbol{\Lambda}'(\mathbf{X}'\mathbf{X})^{-}\boldsymbol{\Lambda}$.

Corresponding to $\boldsymbol{\tau}$ is the transformed vector $\boldsymbol{\alpha} = \mathbf{S}'\boldsymbol{\tau}$, where $\mathbf{S}$ is an $M \times M_*$ matrix such that $\mathbf{S}'\mathbf{C}\mathbf{S} = \mathbf{I}$. The least squares estimator of $\boldsymbol{\alpha}$ is $\hat{\boldsymbol{\alpha}} = \mathbf{S}'\hat{\boldsymbol{\tau}}$. And $\hat{\boldsymbol{\alpha}} \sim N(\boldsymbol{\alpha}, \sigma^2\mathbf{I})$. Further, $\boldsymbol{\tau} = \mathbf{W}'\boldsymbol{\alpha}$, $\hat{\boldsymbol{\tau}} = \mathbf{W}'\hat{\boldsymbol{\alpha}}$, and $\mathbf{C} = \mathbf{W}'\mathbf{W}$. where $\mathbf{W}$ is the unique $M_* \times M$ matrix that satisfies the equality $\boldsymbol{\Lambda}\mathbf{S}\mathbf{W} = \boldsymbol{\Lambda}$; and as an estimator of σ, we have the (positive) square root $\hat{\sigma}$ of $\hat{\sigma}^2 = \mathbf{d}'\mathbf{d}/(N-P_*) = \mathbf{y}'(\mathbf{I}-\mathbf{P}_{\mathbf{X}})\mathbf{y}/(N-P_*)$.

Let $\hat{\tau}_1, \hat{\tau}_2, \ldots, \hat{\tau}_M$ represent the elements of $\hat{\boldsymbol{\tau}}$, so that (for $i = 1, 2, \ldots, M$) $\hat{\tau}_i = \boldsymbol{\lambda}_i'(\mathbf{X}'\mathbf{X})^{-}\mathbf{X}'\mathbf{y}$ is the least squares estimator of τ_i. And observe that τ_i and its least squares estimator are reexpressible as $\tau_i = \mathbf{w}_i'\boldsymbol{\alpha}$ and $\hat{\tau}_i = \mathbf{w}_i'\hat{\boldsymbol{\alpha}}$, where $\mathbf{w}_i$ represents the ith column of $\mathbf{W}$. Moreover, (in light of the assumption that no column of $\boldsymbol{\Lambda}$ is null or is a scalar multiple of another column of $\boldsymbol{\Lambda}$) no column of $\mathbf{W}$ is null or is a scalar multiple of another column of $\mathbf{W}$; and upon observing (in light of Theorem 2.4.21) that (for $i \neq j = 1, 2, \ldots, M$)

$$|\text{corr}(\hat{\tau}_i, \hat{\tau}_j)| = \frac{|\mathbf{w}_i'\mathbf{w}_j|}{(\mathbf{w}_i'\mathbf{w}_i)^{1/2}(\mathbf{w}_j'\mathbf{w}_j)^{1/2}} < 1, \tag{7.1}$$

it follows that (for $i \neq j = 1, 2, \ldots, M$ and for any constants $\dot{\tau}_i$ and $\dot{\tau}_j$ and any nonzero constants a_i and a_j)

$$a_i(\hat{\tau}_i - \dot{\tau}_i) \neq a_j(\hat{\tau}_j - \dot{\tau}_j) \quad (\text{wp1}). \tag{7.2}$$

For $i = 1, 2, \ldots, M$, define

$$t_i = \frac{\hat{\tau}_i - \tau_i}{[\boldsymbol{\lambda}_i'(\mathbf{X}'\mathbf{X})^{-}\boldsymbol{\lambda}_i]^{1/2}\hat{\sigma}} \quad \text{and} \quad t_i^{(0)} = \frac{\hat{\tau}_i - \tau_i^{(0)}}{[\boldsymbol{\lambda}_i'(\mathbf{X}'\mathbf{X})^{-}\boldsymbol{\lambda}_i]^{1/2}\hat{\sigma}}. \tag{7.3}$$

And observe that t_i and $t_i^{(0)}$ are reexpressible as

$$t_i = \frac{\mathbf{w}_i'(\hat{\boldsymbol{\alpha}}-\boldsymbol{\alpha})}{(\mathbf{w}_i'\mathbf{w}_i)^{1/2}\hat{\sigma}} \quad \text{and} \quad t_i^{(0)} = \frac{\mathbf{w}_i'(\hat{\boldsymbol{\alpha}}-\boldsymbol{\alpha}^{(0)})}{(\mathbf{w}_i'\mathbf{w}_i)^{1/2}\hat{\sigma}}, \tag{7.4}$$

where $\boldsymbol{\alpha}^{(0)} = \mathbf{S}'\boldsymbol{\tau}^{(0)} = (\boldsymbol{\Lambda}\mathbf{S})'\boldsymbol{\beta}^{(0)}$. Further, let $\mathbf{t} = (t_1, t_2, \ldots, t_M)'$, and observe that

$$\mathbf{t} = \mathbf{D}^{-1}\mathbf{W}'[\hat{\sigma}^{-1}(\hat{\boldsymbol{\alpha}}-\boldsymbol{\alpha})], \tag{7.5}$$

where $\mathbf{D} = \text{diag}[(\mathbf{w}_1'\mathbf{w}_1)^{1/2}, (\mathbf{w}_2'\mathbf{w}_2)^{1/2}, \ldots, (\mathbf{w}_M'\mathbf{w}_M)^{1/2}]$, and that

$$\hat{\sigma}^{-1}(\hat{\boldsymbol{\alpha}}-\boldsymbol{\alpha}) \sim MVt(N-P_*, \mathbf{I}). \tag{7.6}$$

Multiple comparisons. Among the procedures for testing each of the M null hypotheses $H_1^{(0)}, H_2^{(0)}, \ldots, H_M^{(0)}$ (and of doing so in a way that accounts for the multiplicity of tests) is that provided by the generalized S method—refer to Section 7.3. The generalized S method controls the FWER. That control comes at the expense of the power of the tests, which for even moderately large values of M can be quite low.

A less conservative approach (i.e., one that strikes a better balance between the probability of false rejections and the power of the tests) can be achieved by adopting a criterion that is based on controlling what has been referred to by Lehmann and Romano (2005a) as the k-FWER (where k

is a positive integer). In the present context, the k-FWER is the probability of falsely rejecting k or more of the null hypotheses $H_1^{(0)}, H_2^{(0)}, \ldots, H_M^{(0)}$, that is, the probability of rejecting $H_i^{(0)}$ for k or more of the values of i (between 1 and M, inclusive) for which $\tau_i = \tau_i^{(0)}$. A procedure for testing $H_1^{(0)}, H_2^{(0)}, \ldots, H_M^{(0)}$ is said to control the k-FWER at level $\dot{\gamma}$ (where $0 < \dot{\gamma} < 1$) if k-FWER $\leq \dot{\gamma}$. Clearly, the FWER is a special case of the k-FWER; it is the special case where $k = 1$. And (for any k) FWER $\leq \dot{\gamma} \Rightarrow k$-FWER $\leq \dot{\gamma}$ (so that k-FWER $\leq \dot{\gamma}$ is a less stringent criterion than FWER $\leq \dot{\gamma}$); more generally, for any $k' < k$, k'-FWER $\leq \dot{\gamma} \Rightarrow k$-FWER $\leq \dot{\gamma}$ (so that k-FWER $\leq \dot{\gamma}$ is a less stringent criterion than k'-FWER $\leq \dot{\gamma}$).

For purposes of devising a procedure for testing $H_1^{(0)}, H_2^{(0)}, \ldots, H_M^{(0)}$ that controls the k-FWER at level $\dot{\gamma}$, let $i_1, i_2, \ldots, i_M$ represent a permutation of the first M positive integers $1, 2, \ldots, M$ such that

$$|t_{i_1}| \geq |t_{i_2}| \geq \cdots \geq |t_{i_M}| \tag{7.7}$$

—as is evident from result (7.2), this permutation is unique (wp1). And (for $j = 1, 2, \ldots, M$) define $t_{(j)} = t_{i_j}$. Further, denote by $c_{\dot{\gamma}}(j)$ the upper $100\dot{\gamma}\%$ point of the distribution of $|t_{(j)}|$.

Now, consider the procedure for testing $H_1^{(0)}, H_2^{(0)}, \ldots, H_M^{(0)}$ that (for $i = 1, 2, \ldots, M$) rejects $H_i^{(0)}$ if and only if $\mathbf{y} \in C_i$, where the critical region C_i is defined as follows:

$$C_i = \{\mathbf{y} : |t_i^{(0)}| > c_{\dot{\gamma}}(k)\}. \tag{7.8}$$

Clearly,

$$\begin{aligned} \Pr\big(\mathbf{y} \in C_i &\text{ for } k \text{ or more values of } i \text{ with } \tau_i = \tau_i^{(0)}\big) \\ &= \Pr\big(|t_i| > c_{\dot{\gamma}}(k) \text{ for } k \text{ or more values of } i \text{ with } \tau_i = \tau_i^{(0)}\big) \\ &\leq \Pr[|t_i| > c_{\dot{\gamma}}(k) \text{ for } k \text{ or more values of } i] \\ &= \Pr[|t_{(k)}| > c_{\dot{\gamma}}(k)] = \dot{\gamma}. \end{aligned} \tag{7.9}$$

Thus, the procedure that tests each of the null hypotheses $H_1^{(0)}, H_2^{(0)}, \ldots, H_M^{(0)}$ on the basis of the corresponding one of the critical regions $C_1, C_2, \ldots, C_M$ controls the k-FWER at level $\dot{\gamma}$; its k-FWER is less than or equal to $\dot{\gamma}$. In the special case where $k = 1$, this procedure is identical to that obtained via the generalized S method, which was discussed earlier (in Section 7.3c).

Simultaneous confidence intervals. Corresponding to the test of $H_i^{(0)}$ with critical region C_i is the confidence interval, say $A_i(\mathbf{y})$, with end points

$$\hat{\tau}_i \pm [\boldsymbol{\lambda}_i'(\mathbf{X}'\mathbf{X})^{-}\boldsymbol{\lambda}_i]^{1/2}\hat{\sigma} c_{\dot{\gamma}}(k) \tag{7.10}$$

($i = 1, 2, \ldots, M$). The correspondence is that implicit in the following relationship:

$$\tau_i^{(0)} \in A_i(\mathbf{y}) \Leftrightarrow \mathbf{y} \notin C_i. \tag{7.11}$$

The confidence intervals $A_1(\mathbf{y}), A_2(\mathbf{y}), \ldots, A_M(\mathbf{y})$ are [in light of result (7.9)] such that

$$\begin{aligned} \Pr[\tau_i \in A_i(\mathbf{y}) &\text{ for at least } M{-}k{+}1 \text{ values of } i] \\ &= \Pr[|t_i| \leq c_{\dot{\gamma}}(k) \text{ for at least } M{-}k{+}1 \text{ values of } i] \\ &= \Pr[|t_i| > c_{\dot{\gamma}}(k) \text{ for no more than } k{-}1 \text{ values of } i] \\ &= 1 - \Pr[|t_i| > c_{\dot{\gamma}}(k) \text{ for } k \text{ or more values of } i] \\ &= 1 - \dot{\gamma}. \end{aligned}$$

In the special case where $k = 1$, the confidence intervals $A_1(\mathbf{y}), A_2(\mathbf{y}), \ldots, A_M(\mathbf{y})$ are identical (when Δ is taken to be the set whose members are the columns of $\mathbf{I}_M$) to the confidence intervals $A_{\boldsymbol{\delta}}(\mathbf{y})$ ($\boldsymbol{\delta} \in \Delta$) of Section 7.3c—refer to the representation (3.94)—and (in that special case) the probability of simultaneous coverage by all M of the intervals is equal to $1-\dot{\gamma}$.

Computations/approximations. To implement the test and/or interval procedures, we require the upper $100\dot{\gamma}\%$ point $c_{\dot{\gamma}}(k)$ of the distribution of $|t_{(k)}|$. As discussed in Section 7.3c in the special case of the computation of $c_{\dot{\gamma}}$—when $k=1$, $c_{\dot{\gamma}}(k) = c_{\dot{\gamma}}$—Monte Carlo methods can (at least in principle) be used to compute $c_{\dot{\gamma}}(k)$. Whether the use of Monte Carlo methods is feasible depends on the feasibility of making a large number of draws from the distribution of $|t_{(k)}|$. The process of making a large number of such draws can be facilitated by taking advantage of results (7.5) and (7.6). And by employing methods like those discussed by Edwards and Berry (1987), the resultant draws can be used to approximate $c_{\dot{\gamma}}(k)$ to a high degree of accuracy.

If the use of Monte Carlo methods is judged to be infeasible, overly burdensome, or aesthetically unacceptable, there remains the option of replacing $c_{\dot{\gamma}}(k)$ with an upper bound. In that regard, it can be shown that for any M random variables $x_1, x_2, \ldots, x_M$ and any constant c,

$$\Pr(x_i > c \text{ for } k \text{ or more values of } i) \le (1/k)\textstyle\sum_{i=1}^{M} \Pr(x_i > c). \tag{7.12}$$

And upon applying inequality (7.12) in the special case where $x_i = |t_i|$ $(i = 1, 2, \ldots, M)$ and where $c = \bar{t}_{k\dot{\gamma}/(2M)}(N-P_*)$ and upon observing that (for $i = 1, 2, \ldots, M$) $t_i \sim St(N-P_*)$ and hence that

$$\begin{aligned}(1/k)\textstyle\sum_{i=1}^{M} &\Pr[|t_i| > \bar{t}_{k\dot{\gamma}/(2M)}(N-P_*)] \\ &= (2/k)\textstyle\sum_{i=1}^{M} \Pr[t_i > \bar{t}_{k\dot{\gamma}/(2M)}(N-P_*)] = \frac{2M}{k}\frac{k\dot{\gamma}}{2M} = \dot{\gamma},\end{aligned}$$

we find that

$$\Pr(|t_i| > \bar{t}_{k\dot{\gamma}/(2M)}(N-P_*) \text{ for } k \text{ or more values of } i) \le \dot{\gamma}. \tag{7.13}$$

Together, results (7.13) and (7.9) imply that

$$\bar{t}_{k\dot{\gamma}/(2M)}(N-P_*) \ge c_{\dot{\gamma}}(k). \tag{7.14}$$

Thus, $\bar{t}_{k\dot{\gamma}/(2M)}(N-P_*)$ is an upper bound for $c_{\dot{\gamma}}(k)$, and upon replacing $c_{\dot{\gamma}}(k)$ with $\bar{t}_{k\dot{\gamma}/(2M)}(N-P_*)$ in the definitions of the critical regions of the tests of $H_1^{(0)}, H_2^{(0)}, \ldots, H_M^{(0)}$ and in the definitions of the confidence intervals for $\tau_1, \tau_2, \ldots, \tau_M$, we obtain tests and confidence intervals that are conservative; they are conservative in the sense that typically the tests are less sensitive and the confidence intervals wider than before.

Some numerical results For purposes of comparison, some results were obtained on the values of $c_{\dot{\gamma}}(k)$ and of (the upper bound) $\bar{t}_{k\dot{\gamma}/(2M)}(N-P_*)$ for selected values of M, k, and $\dot{\gamma}$—the value of $N-P_*$ was taken to be 25. These results are presented in Table 7.3. The values of $c_{\dot{\gamma}}(k)$ recorded in the table are approximations that were determined by Monte Carlo methods from 599999 draws—refer to Section 7.3e for some discussion pertaining to the nature and the accuracy of the Monte Carlo approximations. These values are those for the special case where $M_* = M$ and where $\hat{\tau}_1, \hat{\tau}_2, \ldots, \hat{\tau}_M$ are uncorrelated.

Extensions. The tests with critical regions $C_1, C_2, \ldots, C_M$ can be modified for use when the null and alternative hypotheses are either $H_i^{(0)} : \tau_i \le \tau_i^{(0)}$ and $H_i^{(1)} : \tau_i > \tau_i^{(0)}$ $(i = 1, 2, \ldots, M)$ or $H_i^{(0)} : \tau_i \ge \tau_i^{(0)}$ and $H_i^{(1)} : \tau_i < \tau_i^{(0)}$ $(i = 1, 2, \ldots, M)$ rather than $H_i^{(0)} : \tau_i = \tau_i^{(0)}$ and $H_i^{(1)} : \tau_i \ne \tau_i^{(0)}$ $(i = 1, 2, \ldots, M)$. And the confidence intervals $A_1(\mathbf{y}), A_2(\mathbf{y}), \ldots, A_M(\mathbf{y})$ can be modified for use in obtaining upper or lower confidence bounds for $\tau_1, \tau_2, \ldots, \tau_M$.

In defining (for $j = 1, 2, \ldots, M$) $t_{(j)} = t_{i_j}$, take $i_1, i_2, \ldots, i_M$ to be a permutation of the first M positive integers $1, 2, \ldots, M$ such that $t_{i_1} \ge t_{i_2} \ge \cdots \ge t_{i_M}$ rather than (as before) a permutation such that $|t_{i_1}| \ge |t_{i_2}| \ge \cdots \ge |t_{i_M}|$. Further, take $a_{\dot{\gamma}}(j)$ to be the upper $100\dot{\gamma}\%$ point of the distribution of the redefined random variable $t_{(j)}$. Then, the modifications needed to obtain the critical regions for testing the null hypotheses $H_i^{(0)} : \tau_i \le \tau_i^{(0)}$ $(i = 1, 2, \ldots, M)$ and to obtain the lower confidence bounds for $\tau_1, \tau_2, \ldots, \tau_M$ are those that result in procedures identical to the procedures obtained by proceeding as in Section 7.5e (in the special case where $k = 1$) and by inserting $a_{\dot{\gamma}}(k)$ in place of $a_{\dot{\gamma}}$. Moreover, by introducing modifications analogous to those described

TABLE 7.3. Values of $\bar{t}_{k\dot{\gamma}/(2M)}(N{-}P_*)$ and $c_{\dot{\gamma}}(k)$ for selected values of M, k, and $\dot{\gamma}$ (and for $N{-}P_* = 25$).

		$\bar{t}_{k\dot{\gamma}/(2M)}(N{-}P_*)$			$c_{\dot{\gamma}}(k)$		
M	k	$\dot{\gamma}$=.10	$\dot{\gamma}$=.20	$\dot{\gamma}$=.50	$\dot{\gamma}$=.10	$\dot{\gamma}$=.20	$\dot{\gamma}$=.50
10	1	2.79	2.49	2.06	2.74	2.41	1.86
	2	2.49	2.17	1.71	2.06	1.82	1.42
100	1	3.73	3.45	3.08	3.61	3.28	2.76
	2	3.45	3.17	2.79	3.11	2.85	2.43
	5	3.08	2.79	2.38	2.54	2.34	2.02
1,000	1	4.62	4.35	4.00	4.37	4.02	3.46
	5	4.00	3.73	3.36	3.53	3.28	2.87
	10	3.73	3.45	3.08	3.21	2.99	2.62
10,000	1	5.51	5.24	4.89	5.05	4.67	4.06
	10	4.62	4.35	4.00	4.08	3.80	3.35
	50	4.00	3.73	3.36	3.46	3.23	2.85
	100	3.73	3.45	3.08	3.18	2.96	2.61

in the third part of Section 7.3e and the second part of Section 7.5e, versions of the confidence intervals and confidence bounds can be obtained such that all M intervals are of equal length and such that all M bounds are equidistant from the least squares estimates.

Nonnormality. The assumption that the vector $\mathbf{e}$ of residual effects in the G–M model has an MVN distribution is stronger than necessary. To insure that the probability of the tests falsely rejecting k or more of the M null hypotheses does not exceed $\dot{\gamma}$ and to insure that the probability of the confidence intervals or bounds covering at least $M{-}k{+}1$ of the M linear combinations $\tau_1, \tau_2, \ldots, \tau_M$ is equal to $1{-}\dot{\gamma}$, it is sufficient that $\mathbf{e}$ have an absolutely continuous spherical distribution. In fact, it is sufficient that the vector $\begin{pmatrix} \hat{\boldsymbol{\alpha}} - \boldsymbol{\alpha} \\ \mathbf{d} \end{pmatrix}$ have an absolutely continuous spherical distribution.

b. Multiple comparisons: use of step-down methods to control the FWER or k-FWER

Let us consider further the multiple-comparison problem considered in Subsection a; that is, the problem of testing the M null hypotheses $H_1^{(0)}, H_2^{(0)}, \ldots, H_M^{(0)}$ versus the M alternative hypotheses $H_1^{(1)}, H_2^{(1)}, \ldots, H_M^{(1)}$ (and of doing so in a way that accounts for the multiplicity of tests). The tests considered in Subsection a are those with the critical regions

$$C_i = \{\mathbf{y} : |t_i^{(0)}| > c_{\dot{\gamma}}(k)\} \quad (i = 1, 2, \ldots, M) \tag{7.15}$$

and those with the critical regions

$$\{\mathbf{y} : |t_i^{(0)}| > \bar{t}_{k\dot{\gamma}/(2M)}(N - P_*)\} \quad (i = 1, 2, \ldots, M) \tag{7.16}$$

obtained upon replacing $c_{\dot{\gamma}}(k)$ with the upper bound $\bar{t}_{k\dot{\gamma}/(2M)}(N - P_*)$. The critical regions (7.15) and (7.16) of these tests are relatively simple in form. In what follows, some alternative procedures for testing $H_1^{(0)}, H_2^{(0)}, \ldots, H_M^{(0)}$ are considered. The critical regions of the alternative tests are of a more complex form; however like the tests with critical regions (7.15) or (7.16), they control the FWER or, more generally, the k-FWER and, at the same time, they are more powerful than the tests with critical regions (7.15) or (7.16)—their critical regions are larger than the critical regions (7.15) or (7.16).

Some additional notation and some preliminaries. Let us continue to employ the notation introduced in Subsection a. In particular, let us continue (for $i = 1, 2, \ldots, M$) to take t_i and $t_i^{(0)}$ to be the random variables defined by expressions (7.3). And let us continue to take $\mathbf{t} = (t_1, t_2, \ldots, t_M)'$, to take $i_1, i_2, \ldots, i_M$ to be as defined by inequalities (7.7), and (for $j = 1, 2, \ldots, M$) to define $t_{(j)} = t_{i_j}$. Further, let us extend these definitions to the $t_i^{(0)}$'s by taking $\mathbf{t}^{(0)} = (t_1^{(0)}, t_2^{(0)}, \ldots, t_M^{(0)})'$, taking $\tilde{i}_1, \tilde{i}_2, \ldots, \tilde{i}_M$ to be a permutation of the integers $1, 2, \ldots, M$ such that

$$|t_{\tilde{i}_1}^{(0)}| \geq |t_{\tilde{i}_2}^{(0)}| \geq \cdots \geq |t_{\tilde{i}_M}^{(0)}|, \tag{7.17}$$

and (for $j = 1, 2, \ldots, M$) letting $t_{(j)}^{(0)} = t_{\tilde{i}_j}^{(0)}$.

Let us use the symbol I to represent the set $\{1, 2, \ldots, M\}$. Also, let us denote by T the subset of I consisting of those values of the integer i ($1 \leq i \leq M$) for which $H_i^{(0)}$ is true, that is, those values for which $\tau_i = \tau_i^{(0)}$—the number of elements in this subset and the identity of the elements are of course unknown. Further, for any subset S of I, let M_S represent the size of S, that is, the number of elements in S.

Now, for an arbitrary (nonempty) subset $S = \{j_1, j_2, \ldots, j_{M_S}\}$ of I, let $\mathbf{t}_S$ represent the M_S-dimensional subvector of $\mathbf{t}$ whose elements are $t_{j_1}, t_{j_2}, \ldots, t_{j_{M_S}}$, and let $i_1^*(S), i_2^*(S), \ldots, i_{M_S}^*(S)$ represent a permutation of the elements of S such that

$$|t_{i_1^*(S)}| \geq |t_{i_2^*(S)}| \geq \cdots \geq |t_{i_{M_S}^*(S)}|.$$

Similarly, let $\mathbf{t}_S^{(0)}$ represent the M_S-dimensional subvector of $\mathbf{t}^{(0)}$ whose elements are $t_{j_1}^{(0)}, t_{j_2}^{(0)}, \ldots, t_{j_{M_S}}^{(0)}$, and let $\tilde{i}_1^*(S), \tilde{i}_2^*(S), \ldots, \tilde{i}_{M_S}^*(S)$ represent a permutation of the elements of S such that

$$|t_{\tilde{i}_1^*(S)}^{(0)}| \geq |t_{\tilde{i}_2^*(S)}^{(0)}| \geq \cdots \geq |t_{\tilde{i}_{M_S}^*(S)}^{(0)}|.$$

And note that when $S = T$ or more generally when $S \subset T$, $\mathbf{t}_S^{(0)} = \mathbf{t}_S$. Note also that the (marginal) distribution of $\mathbf{t}_S$ is a multivariate t distribution with $N - P_*$ degrees of freedom and with a correlation matrix that is the submatrix of the correlation matrix of $\mathbf{t}$ formed by its $j_1, j_2, \ldots, j_{M_S}$th rows and columns—refer to result (6.4.48). Moreover, $i_1^*(S), i_2^*(S), \ldots, i_{M_S}^*(S)$ is a subsequence of the sequence $i_1, i_2, \ldots, i_M$ and $t_{i_1^*(S)}, t_{i_2^*(S)}, \ldots, t_{i_{M_S}^*(S)}$ a subsequence of the sequence $t_{(1)}, t_{(2)}, \ldots, t_{(M)}$ (or, equivalently, of $t_{i_1}, t_{i_2}, \ldots, t_{i_M}$)—specifically, they are the subsequences obtained upon striking out the jth member of the sequence for every $j \in I$ for which $i_j \notin S$. Similarly, $\tilde{i}_1^*(S), \tilde{i}_2^*(S), \ldots, \tilde{i}_{M_S}^*(S)$ is a subsequence of the sequence $\tilde{i}_1, \tilde{i}_2, \ldots, \tilde{i}_M$ and $t_{\tilde{i}_1^*(S)}^{(0)}, t_{\tilde{i}_2^*(S)}^{(0)}, \ldots, t_{\tilde{i}_{M_S}^*(S)}^{(0)}$ a subsequence of the sequence $t_{(1)}^{(0)}, t_{(2)}^{(0)}, \ldots, t_{(M)}^{(0)}$.

Additionally, for an arbitrary subset S of I (of size $M_S \geq k$), let $t_{k;S} = t_{i_k^*(S)}$ and $t_{k;S}^{(0)} = t_{\tilde{i}_k^*(S)}^{(0)}$. Further, let $c_{\dot{\gamma}}(k; S)$ represent the upper $100\dot{\gamma}\%$ point of the distribution of $|t_{k;S}|$. And observe that

$$t_{k;S}^{(0)} = t_{k;S} \quad \text{when } S \subset T \tag{7.18}$$

and that

$$c_{\dot{\gamma}}(k; S_*) < c_{\dot{\gamma}}(k; S) \quad \text{for any (proper) subset } S_* \text{ of } S \text{ (of size } M_{S_*} \geq k). \tag{7.19}$$

An alternative procedure for testing the null hypotheses $H_1^{(0)}, H_2^{(0)}, \ldots, H_M^{(0)}$ in a way that controls the FWER or, more generally, the k-FWER: definition, characteristics, terminology, and properties. The null hypotheses $H_1^{(0)}, H_2^{(0)}, \ldots, H_M^{(0)}$ can be tested in a way that controls the FWER or, more generally, the k-FWER by adopting the procedure with critical regions $C_1, C_2, \ldots, C_M$. For purposes of defining an alternative to this procedure, let (for $j = k, k+1, \ldots, M$) $\tilde{\Omega}_{k;j}$ represent a collection of subsets of $I = \{1, 2, \ldots, M\}$ consisting of every $S \subset I$ for which $M_S \geq k$ and for which $\tilde{i}_k^*(S) = \tilde{i}_j$. And let $\tilde{\Omega}_{k;j}^+$ represent a collection (of subsets of I) consisting of those subsets in $\tilde{\Omega}_{k;j}$ whose elements include all $M - j$ of the integers $\tilde{i}_{j+1}, \tilde{i}_{j+2}, \ldots, \tilde{i}_M$, that is, $\tilde{\Omega}_{k;j}^+$ is the collection of those subsets of I whose elements consist of $k-1$ of the integers $\tilde{i}_1, \tilde{i}_2, \ldots, \tilde{i}_{j-1}$ and all $M - j + 1$ of the integers $\tilde{i}_j, \tilde{i}_{j+1}, \ldots, \tilde{i}_M$. By definition,

$$\tilde{\Omega}^+_{k;j} \subset \tilde{\Omega}_{k;j} \quad (j = k, k+1, \ldots, M) \quad \text{and} \quad \tilde{\Omega}^+_{k;k} = \{I\}. \tag{7.20}$$

Moreover,

$$S \in \tilde{\Omega}_{k;j} \;\Rightarrow\; S \subset S^+ \text{ for some } S^+ \in \tilde{\Omega}^+_{k;j}, \tag{7.21}$$

and (for $j > k$)

$$S \in \tilde{\Omega}_{k;j} \;\Rightarrow\; S \subset S^+ \text{ for some } S^+ \in \tilde{\Omega}_{k;j-1} \tag{7.22}$$

[where in result (7.22), S^+ is such that S is a proper subset of S^+].

Now, for $j = k, k+1, \ldots, M$, define

$$\alpha_j = \max_{S \in \tilde{\Omega}_{k;j}} c_{\dot{\gamma}}(k; S). \tag{7.23}$$

And [recalling result (7.19)] note [in light of results (7.20) and (7.21)] that α_j is reexpressible as

$$\alpha_j = \max_{S \in \tilde{\Omega}^+_{k;j}} c_{\dot{\gamma}}(k; S) \tag{7.24}$$

—in the special case where $k = 1$, $\tilde{\Omega}^+_{k;j}$ contains a single set, say $\tilde{S}_j$ (the elements of which are $\tilde{i}_j, \tilde{i}_{j+1}, \ldots, \tilde{i}_M$), and in that special case, equality (7.24) simplifies to $\alpha_j = c_{\dot{\gamma}}(1; \tilde{S}_j)$. Note also that

$$\alpha_k = c_{\dot{\gamma}}(k; I) = c_{\dot{\gamma}}(k). \tag{7.25}$$

Moreover, for $k \le j' < j \le M$,

$$\alpha_{j'} > \alpha_j, \tag{7.26}$$

as is evident from result (7.22) [upon once again recalling result (7.19)].

Let us extend the definition (7.23) of the α_j's by taking

$$\alpha_1 = \alpha_2 = \cdots = \alpha_{k-1} = \alpha_k, \tag{7.27}$$

so that [in light of inequality (7.26)]

$$\alpha_1 = \cdots = \alpha_{k-1} = \alpha_k > \alpha_{k+1} > \alpha_{k+2} > \cdots > \alpha_M. \tag{7.28}$$

Further, take J to be an integer (between 0 and M, inclusive) such that $|t^{(0)}_{(j)}| > \alpha_j$ for $j = 1, 2, \ldots, J$ and $|t^{(0)}_{(J+1)}| \le \alpha_{J+1}$—if $|t^{(0)}_{(j)}| > \alpha_j$ for $j = 1, 2, \ldots, M$, take $J = M$; and if $|t^{(0)}_{(1)}| \le \alpha_1$, take $J = 0$. And consider the following multiple-comparison procedure for testing the null hypotheses $H^{(0)}_1, H^{(0)}_2, \ldots, H^{(0)}_M$: when $J \ge 1$, $H^{(0)}_{\tilde{i}_1}, H^{(0)}_{\tilde{i}_2}, \ldots, H^{(0)}_{\tilde{i}_J}$ are rejected; when $J = 0$, none of the null hypotheses are rejected. This procedure can be regarded as a stepwise procedure, one of a kind known as a step-down procedure. Specifically, the procedure can be regarded as one in which (starting with $H^{(0)}_{\tilde{i}_1}$) the null hypotheses are tested sequentially in the order $H^{(0)}_{\tilde{i}_1}, H^{(0)}_{\tilde{i}_2}, \ldots, H^{(0)}_{\tilde{i}_M}$ by comparing the $|t^{(0)}_{(j)}|$'s with the α_j's; the testing ceases upon encountering a null hypothesis $H^{(0)}_{\tilde{i}_j}$ for which $|t^{(0)}_{(j)}|$ does not exceed α_j.

The step-down procedure for testing the null hypotheses $H^{(0)}_1, H^{(0)}_2, \ldots, H^{(0)}_M$ can be characterized in terms of its critical regions. The critical regions of this procedure, say $C^*_1, C^*_2, \ldots, C^*_M$, are expressible as follows:

$$C^*_i = \{\mathbf{y} : J \ge 1 \text{ and } i = \tilde{i}_{j'} \text{ for some integer } j' \text{ between 1 and } J \text{, inclusive}\} \tag{7.29}$$

$(i = 1, 2, \ldots, M)$. Alternatively, C^*_i is expressible in the form

$$C^*_i = \bigcup_{j'=1}^{M} \left\{\mathbf{y} : i = \tilde{i}_{j'} \text{ and for every } j \le j',\; t^{(0)}_{(j)} > \alpha_j\right\} \tag{7.30}$$

—for any particular value of $\mathbf{y}$, $t^{(0)}_{(j)} > \alpha_j$ for every $j \le j'$ if and only if $j' \le J$.

It can be shown (and subsequently will be shown) that the step-down procedure for testing the null hypotheses $H^{(0)}_1, H^{(0)}_2, \ldots, H^{(0)}_M$, like the test procedure considered in Subsection a, controls the k-FWER at level $\dot{\gamma}$ (and in the special case where $k = 1$, controls the FWER at level $\dot{\gamma}$). Moreover, in light of result (7.25) and definition (7.27), the critical regions $C_1, C_2, \ldots, C_M$ of the test procedure considered in Subsection a are reexpressible in the form

$$C_i = \bigcup_{j'=1}^{M} \{\mathbf{y} : i = \tilde{i}_{j'} \text{ and } t_{(j')}^{(0)} > \alpha_1\} \tag{7.31}$$

($i = 1, 2, \ldots, M$). And upon comparing expression (7.31) with expression (7.30) and upon observing [in light of the relationships (7.28)] that $t_{(j')}^{(0)} > \alpha_1$ implies that $t_{(j)}^{(0)} > \alpha_j$ for every $j \le j'$, we find that

$$C_i \subset C_i^* \quad (i = 1, 2, \ldots, M). \tag{7.32}$$

That is, the critical regions $C_1, C_2, \ldots, C_M$ of the test procedure considered in Subsection a are subsets of the corresponding critical regions $C_1^*, C_2^*, \ldots, C_M^*$ of the step-down procedure. In fact, C_i is a proper subset of C_i^* ($i = 1, 2, \ldots, M$).

We conclude that while both the step-down procedure and the procedure with critical regions $C_1, C_2, \ldots, C_M$ control the k-FWER (or when $k = 1$, the FWER), the step-down procedure is more powerful in that its adoption can result in additional rejections. However, at the same time, it is worth noting that the increased power comes at the expense of some increase in complexity and computational intensity.

Verification that the step-down procedure for testing the null hypotheses $H_1^{(0)}, H_2^{(0)}, \ldots, H_M^{(0)}$ controls the k-FWER (at level $\dot{\gamma}$). Suppose (for purposes of verifying that the step-down procedure controls the k-FWER) that $M_T \ge k$—if $M_T < k$, then fewer than k of the null hypotheses are true and hence at most there can be $k-1$ false rejections. Then, there exists an integer j' (where $k \le j' \le M - M_T + k$) such that $\tilde{i}_{j'} = \tilde{i}_k^*(T)$—$j' = k$ when T is the k-dimensional set whose elements are $\tilde{i}_1, \tilde{i}_2, \ldots, \tilde{i}_k$, and $j' = M - M_T + k$ when T is the set whose M_T elements are $\tilde{i}_{M-M_T+1}, \tilde{i}_{M-M_T+2}, \ldots, \tilde{i}_M$.

The step-down procedure results in k or more false rejections if and only if

$$|t_{(1)}^{(0)}| > \alpha_1, \quad |t_{(2)}^{(0)}| > \alpha_2, \quad \ldots, \quad |t_{(j'-1)}^{(0)}| > \alpha_{j'-1}, \quad \text{and} \quad |t_{(j')}^{(0)}| > \alpha_{j'}. \tag{7.33}$$

Thus, the step-down procedure is such that

$$\Pr(k \text{ or more false rejections}) \le \Pr\left(|t_{(j')}^{(0)}| > \alpha_{j'}\right). \tag{7.34}$$

Moreover,

$$t_{(j')}^{(0)} = t_{\tilde{i}_{j'}}^{(0)} = t_{\tilde{i}_k^*(T)}^{(0)} = t_{k;T}^{(0)} = t_{k;T}, \tag{7.35}$$

as is evident upon recalling result (7.18), and

$$\alpha_{j'} = \max_{S \in \tilde{\Omega}_{k;j'}} c_{\dot{\gamma}}(k; S) \ge c_{\dot{\gamma}}(k; T). \tag{7.36}$$

Together, results (7.35) and (7.36) imply that

$$\Pr\left(|t_{(j')}^{(0)}| > \alpha_{j'}\right) = \Pr\left(|t_{k;T}| > \alpha_{j'}\right) \le \Pr\left[|t_{k;T}| > c_{\dot{\gamma}}(k; T)\right] = \dot{\gamma}. \tag{7.37}$$

And upon combining result (7.37) with result (7.34), we conclude that the step-down procedure is such that

$$\Pr(k \text{ or more false rejections}) \le \dot{\gamma}, \tag{7.38}$$

thereby completing the verification that the step-down procedure controls the k-FWER (at level $\dot{\gamma}$).

A caveat. In the specification of the step-down procedure, $\alpha_1, \alpha_2, \ldots, \alpha_{k-1}$ were set equal to α_k. As is evident from the verification (in the preceding part of the present subsection) that the step-down procedure controls the k-FWER (at level $\dot{\gamma}$), its ability to do so is not affected by the choice of $\alpha_1, \alpha_2, \ldots, \alpha_{k-1}$. In fact, if the procedure were modified in such a way that $k-1$ of the null hypotheses [specifically, the null hypotheses $H_i^{(0)}$ ($i = \tilde{i}_1, \tilde{i}_2, \ldots, \tilde{i}_{k-1}$)] were always rejected, the ability of the procedure to control the k-FWER would be unaffected—this modification corresponds to setting $\alpha_1, \alpha_2, \ldots, \alpha_{k-1}$ equal to $-\infty$ rather than to α_k. However, convention (in the case of step-down procedures) suggests that $\alpha_1, \alpha_2, \ldots, \alpha_{k-1}$ be chosen in such a way that the sequence $\alpha_1, \alpha_2, \ldots, \alpha_{k-1}, \alpha_k, \alpha_{k+1}, \ldots, \alpha_M$ is nonincreasing—refer, e.g., to Lehmann and Romano (2005a, p. 1143) for some related remarks. Within the confines of that convention, the choice $\alpha_1 = \alpha_2 = \cdots = \alpha_{k-1} = \alpha_k$ is the best choice; it is best in the sense that it maximizes the size of the critical regions $C_1^*, C_2^*, \ldots, C_M^*$.

A potential improvement. Let us continue to denote by I the set $\{1, 2, \ldots, M\}$ and to denote by T the subset of I consisting of those values of the integer i $(1 \le i \le M)$ for which $H_i^{(0)}$ is true, that is, those values for which $\tau_i = \tau_i^{(0)}$. Further, let us denote by F the subset of I consisting of those values of i for which $H_i^{(0)}$ is false, that is, those values for which $\tau_i \neq \tau_i^{(0)}$. And denote by Ω the collection of all 2^M subsets of I (including the empty set). Then, by definition, both T and F belong to the collection Ω, T and F are disjoint (i.e., have no members in common), and $T \cup F = I$.

When the coefficient vectors $\boldsymbol{\lambda}_1, \boldsymbol{\lambda}_2, \ldots, \boldsymbol{\lambda}_M$ of $\tau_1, \tau_2, \ldots, \tau_M$ are linearly independent, T could potentially be any one of the 2^M subsets (of I) that form the collection Ω. When $\boldsymbol{\lambda}_1, \boldsymbol{\lambda}_2, \ldots, \boldsymbol{\lambda}_M$ are linearly dependent, that is no longer the case.

Suppose that $\boldsymbol{\lambda}_1, \boldsymbol{\lambda}_2, \ldots, \boldsymbol{\lambda}_M$ are linearly dependent. Suppose further that the i^*th of the vectors $\boldsymbol{\lambda}_1, \boldsymbol{\lambda}_2, \ldots, \boldsymbol{\lambda}_M$ is expressible in the form

$$\boldsymbol{\lambda}_{i^*} = \textstyle\sum_{i \in S} a_i \boldsymbol{\lambda}_i,$$

where $i^* \notin S \subset I$ and where the a_i's are scalars—when the vectors $\boldsymbol{\lambda}_1, \boldsymbol{\lambda}_2, \ldots, \boldsymbol{\lambda}_M$ are linearly dependent, at least one of them is expressible in terms of the others. Then, we find [upon recalling that for some $P \times 1$ vector $\boldsymbol{\beta}^{(0)}$, $\tau_i^{(0)} = \boldsymbol{\lambda}_i' \boldsymbol{\beta}^{(0)}$ $(i = 1, 2, \ldots, M)$] that if T were equal to S, it would be the case that

$$\begin{aligned} \tau_{i^*} = \boldsymbol{\lambda}_{i^*}' \boldsymbol{\beta} = \textstyle\sum_{i \in T} a_i \boldsymbol{\lambda}_i' \boldsymbol{\beta} = \textstyle\sum_{i \in T} a_i \tau_i \\ = \textstyle\sum_{i \in T} a_i \tau_i^{(0)} = \textstyle\sum_{i \in T} a_i \boldsymbol{\lambda}_i' \boldsymbol{\beta}^{(0)} = \boldsymbol{\lambda}_{i^*}' \boldsymbol{\beta}^{(0)} = \tau_{i^*}^{(0)}. \end{aligned} \tag{7.39}$$

Thus, T cannot equal S [since $T = S \Rightarrow i^* \in F$, contrary to what is implied by result (7.39)]. In effect, the linear dependence of $\boldsymbol{\lambda}_1, \boldsymbol{\lambda}_2, \ldots, \boldsymbol{\lambda}_M$ and the resultant existence of a relationship of the form $\boldsymbol{\lambda}_{i^*} = \sum_{i \in S} a_i \boldsymbol{\lambda}_i$ imposes a constraint on T.

Based on the immediately preceding development, we conclude that $T \in \Omega^*$, where Ω^* is a collection of subsets (of I) defined as follows: a subset S is a member of Ω^* if for every integer $i^* \in I$ such that $i^* \notin S$, $\boldsymbol{\lambda}_{i^*}$ is linearly independent of the M_S vectors $\boldsymbol{\lambda}_i$ $(i \in S)$ in the sense that $\boldsymbol{\lambda}_{i^*}$ is not expressible as a linear combination of $\boldsymbol{\lambda}_i$ $(i \in S)$ or, equivalently, the rank of the $P \times (M_S + 1)$ matrix with columns $\boldsymbol{\lambda}_i$ $(i \in S)$ and $\boldsymbol{\lambda}_{i^*}$ is greater (by 1) than that of the $P \times M_S$ matrix with columns $\boldsymbol{\lambda}_i$ $(i \in S)$—refer to the discussion in Section 2.9a (on the consistency of linear systems) for some relevant results. Note that the definition of Ω^* is such that $\Omega^* \subset \Omega$ and is such that Ω^* includes I and also includes the empty set. When $\boldsymbol{\lambda}_1, \boldsymbol{\lambda}_2, \ldots, \boldsymbol{\lambda}_M$ are linearly independent, $\Omega^* = \Omega$; when $\boldsymbol{\lambda}_1, \boldsymbol{\lambda}_2, \ldots, \boldsymbol{\lambda}_M$ are linearly dependent, there are some subsets of I (and hence some members of Ω) that are not members of Ω^*.

Suppose that T is subject to the constraint $T \in \Omega^*$ or, more generally, to the constraint $T \in \Omega^-$, where either $\Omega^- = \Omega^*$ or Ω^- is some collection (of known identity) of subsets (of I) other than Ω^*. A very simple special case (of mostly hypothetical interest) is that where the collection Ω^- (to which T is constrained) is the collection whose only members are I and the empty set. Let us consider how (when Ω^- does not contain every member of Ω) the information inherent in the constraint $T \in \Omega^-$ might be used to effect improvements in the step-down procedure (for testing the null hypotheses $H_1^{(0)}, H_2^{(0)}, \ldots, H_M^{(0)}$).

Consider the generalization of the step-down procedure obtained upon replacing (for $j = k, k+1, \ldots, M$) the definition of α_j given by expression (7.23) with the definition

$$\alpha_j = \max_{S \in (\tilde{\Omega}_{k;j} \cap \Omega^-)} c_{\dot{\gamma}}(k; S) \tag{7.40}$$

—when $\tilde{\Omega}_{k;j} \cap \Omega^- = \varnothing$ (the empty set), set $\alpha_j = -\infty$. And in the generalization, continue [as in definition (7.27)] to set

$$\alpha_1 = \alpha_2 = \cdots = \alpha_{k-1} = \alpha_k.$$

Under the extended definition (7.40), it is no longer necessarily the case that $\alpha_{j'} > \alpha_j$ for every j and j' for which $k \le j' < j \le M$ and hence no longer necessarily the case that the sequence $\alpha_1, \alpha_2, \ldots, \alpha_M$ is nonincreasing. Nor is it necessarily the case that α_j is reexpressible (for $k \le j \le M$) as $\alpha_j = \max_{S \in (\tilde{\Omega}^+_{k;j} \cap \Omega^-)} c_{\dot{\gamma}}(k; S)$, contrary to what might have been conjectured on the basis of result (7.24).

Like the original version of the step-down procedure, the generalized (to account for the constraint $T \in \Omega^-$) version controls the k-FWER or (in the special case where $k = 1$) the FWER (at level $\dot{\gamma}$). That the generalized version has that property can be verified by proceeding in essentially the same way as in the verification (in a preceding part of the present subsection) that the original version has that property. In that regard, it is worth noting that for $T \in \Omega^-$, $\tilde{\Omega}_{k;j'} \cap \Omega^-$ is nonempty and hence $\alpha_{j'}$ is finite. And in the extension of the verification to the generalized version, the maximization (with respect to S) in result (7.36) is over the intersection of $\tilde{\Omega}_{k;j'}$ with Ω^- rather than over $\tilde{\Omega}_{k;j'}$ itself.

When $\Omega^- = \Omega$, the generalized version of the step-down procedure is identical to the original version. When Ω^- is "smaller" than Ω (as when $\boldsymbol{\lambda}_1, \boldsymbol{\lambda}_2, \ldots, \boldsymbol{\lambda}_M$ are linearly dependent and $\Omega^- = \Omega^*$), some of the α_j's employed in the generalized version are smaller than (and the rest equal to) those employed in the original version. Thus, when Ω^- is smaller than Ω,the generalized version is more powerful than the original (in that its use can result in additional rejections).

It is informative to consider the generalized version of the step-down procedure in the aforementioned simple special case where the only members of Ω^- are I and the empty set. In that special case, the generalized version is such that $\alpha_k = c_{\dot{\gamma}}(k; I) = c_{\dot{\gamma}}(k)$—refer to result (7.25)—and $\alpha_{k+1} = \alpha_{k+2} = \cdots = \alpha_M = -\infty$. Thus, in that special case, the generalized version of the step-down procedure rejects none of the null hypotheses $H_1^{(0)}, H_2^{(0)}, \ldots, H_M^{(0)}$ if $|t^{(0)}_{(1)}| \le c_{\dot{\gamma}}(k)$, rejects $H^{(0)}_{\tilde{i}_1}, H^{(0)}_{\tilde{i}_2}, \ldots, H^{(0)}_{\tilde{i}_{j'}}$ if $|t^{(0)}_{(j)}| > c_{\dot{\gamma}}(k)$ $(j = 1, 2, \ldots, j')$ and $|t^{(0)}_{(j'+1)}| \le c_{\dot{\gamma}}(k)$ for some integer j' between 1 and $k-1$, inclusive, and rejects all M of the null hypotheses if $|t^{(0)}_{(j)}| > c_{\dot{\gamma}}(k)$ $(j = 1, 2, \ldots, k)$.

An illustrative example. For purposes of illustration, consider a setting where $M = P = 3$, with $\boldsymbol{\lambda}_1' = (1, -1, 0)$, $\boldsymbol{\lambda}_2' = (1, 0, -1)$, and $\boldsymbol{\lambda}_3' = (0, 1, -1)$. This setting is of a kind that is encountered in applications where pairwise comparisons are to be made among some number of "treatments" (3 in this case).

Clearly, any two of the three vectors $\boldsymbol{\lambda}_1$, $\boldsymbol{\lambda}_2$, and $\boldsymbol{\lambda}_3$ are linearly independent. Moreover, each of these three vectors is expressible as a difference between the other two (e.g., $\boldsymbol{\lambda}_3 = \boldsymbol{\lambda}_2 - \boldsymbol{\lambda}_1$), implying in particular that $M_* = 2$ and that $\boldsymbol{\lambda}_1$, $\boldsymbol{\lambda}_2$, and $\boldsymbol{\lambda}_3$ are linearly dependent. And the $2^M = 8$ members of Ω are $\varnothing$ (the empty set), $\{1\}$, $\{2\}$, $\{3\}$, $\{1, 2\}$, $\{1, 3\}$, $\{2, 3\}$, and $I = \{1, 2, 3\}$; and the members of Ω^* are $\varnothing$, $\{1\}$, $\{2\}$, $\{3\}$, and $\{1, 2, 3\}$.

Now, suppose that $k = 1$. Then, we find that (for $j = 1, 2, 3$) the members of the collections $\tilde{\Omega}_{k;j}$, $\tilde{\Omega}^+_{k;j}$, and $\tilde{\Omega}_{k;j} \cap \Omega^*$ are

$$\begin{aligned}
\tilde{\Omega}_{1;1}&: \{\tilde{i}_1\}, \{\tilde{i}_1, \tilde{i}_2\}, \{\tilde{i}_1, \tilde{i}_3\}, \text{ and } \{\tilde{i}_1, \tilde{i}_2, \tilde{i}_3\},\\
\tilde{\Omega}^+_{1;1}&: \{\tilde{i}_1, \tilde{i}_2, \tilde{i}_3\},\\
\tilde{\Omega}_{1;1} \cap \Omega^*&: \{\tilde{i}_1\} \text{ and } \{\tilde{i}_1, \tilde{i}_2, \tilde{i}_3\},\\
\tilde{\Omega}_{1;2}&: \{\tilde{i}_2\} \text{ and } \{\tilde{i}_2, \tilde{i}_3\},\\
\tilde{\Omega}^+_{1;2}&: \{\tilde{i}_2, \tilde{i}_3\},\\
\tilde{\Omega}_{1;2} \cap \Omega^*&: \{\tilde{i}_2\},\\
\tilde{\Omega}_{1;3}&: \{\tilde{i}_3\},
\end{aligned}$$

$$\tilde{\Omega}^+_{1;3}\colon \{\tilde{i}_3\},$$
$$\tilde{\Omega}_{1;3} \cap \Omega^*\colon \{\tilde{i}_3\}.$$

Alternatively, suppose that $k = 2$. Then, we find that (for $j = 2, 3$) the members of the collections $\tilde{\Omega}_{k;j}$, $\tilde{\Omega}^+_{k;j}$, and $\tilde{\Omega}_{k;j} \cap \Omega^*$ are

$$\begin{aligned}
\tilde{\Omega}_{2:2}&\colon \{\tilde{i}_1, \tilde{i}_2\} \text{ and } \{\tilde{i}_1, \tilde{i}_2, \tilde{i}_3\},\\
\tilde{\Omega}^+_{2:2}&\colon \{\tilde{i}_1, \tilde{i}_2, \tilde{i}_3\},\\
\tilde{\Omega}_{2:2} \cap \Omega^*&\colon \{\tilde{i}_1, \tilde{i}_2, \tilde{i}_3\},\\
\tilde{\Omega}_{2:3}&\colon \{\tilde{i}_1, \tilde{i}_3\} \text{ and } \{\tilde{i}_2, \tilde{i}_3\},\\
\tilde{\Omega}^+_{2:3}&\colon \{\tilde{i}_1, \tilde{i}_3\} \text{ and } \{\tilde{i}_2, \tilde{i}_3\},\\
\tilde{\Omega}_{2:3} \cap \Omega^*&\colon \varnothing \text{ (the empty set).}
\end{aligned}$$

Thus, when $\Omega^- = \Omega^*$ and $k = 1$, $\alpha_1 = c_{\dot{\gamma}}(1; I) = c_{\dot{\gamma}}(1)$, $\alpha_2 = c_{\dot{\gamma}}(1; \{\tilde{i}_2\})$ or $\alpha_2 = c_{\dot{\gamma}}(1; \{\tilde{i}_2, \tilde{i}_3\})$ depending on whether T is subject to the constraint $T \in \Omega^*$ or T is unconstrained, and $\alpha_3 = c_{\dot{\gamma}}(1; \{\tilde{i}_3\})$. And when $\Omega^- = \Omega^*$ and $k = 2$, $\alpha_1 = \alpha_2 = c_{\dot{\gamma}}(2; I) = c_{\dot{\gamma}}(2)$ and $\alpha_3 = -\infty$ or $\alpha_3 = \max[c_{\dot{\gamma}}(2; \{\tilde{i}_1, \tilde{i}_3\}),\ c_{\dot{\gamma}}(2; \{\tilde{i}_2, \tilde{i}_3\})]$ depending on whether T is subject to the constraint $T \in \Omega^*$ or T is unconstrained.

Computations/approximations. To implement the step-down procedure for testing the null hypotheses $H_1^{(0)}, H_2^{(0)}, \ldots, H_M^{(0)}$, we require some or all of the constants $\alpha_k, \alpha_{k+1}, \ldots, \alpha_M$ defined by expression (7.23) or, more generally, by expression (7.40). And to obtain the requisite α_j's, we require the values of $c_{\dot{\gamma}}(k; S)$ corresponding to various choices for S. As in the case of the constant $c_{\dot{\gamma}}(k)$ required to implement the test procedure with critical regions (7.15), the requisite values of $c_{\dot{\gamma}}(k; S)$ can be computed to a high degree of accuracy by employing Monte Carlo methods of a kind discussed by Edwards and Berry (1987). In that regard, note that (for any subset S of I) a draw from the distribution of $t_{i_k^*(S)}$ can be obtained from a draw from the distribution of the vector $\mathbf{t}$, and recall (from the discussion in Subsection a) that the feasibility of making a large number of draws from the distribution of $\mathbf{t}$ can be enhanced by making use of results (7.5) and (7.6).

As in the case of the test procedure with critical regions (7.15), a computationally less demanding (but less powerful) version of the step-down procedure can be devised. In fact, such a version can be devised by making use of the inequality

$$\bar{t}_{k\dot{\gamma}/(2M_S)}(N - P_*) \geq c_{\dot{\gamma}}(k; S) \qquad (\text{where } S \subset I), \tag{7.41}$$

which is a generalization of inequality (7.14)—when $S = I$, inequality (7.41) is equivalent to inequality (7.14)—and which can be devised from inequality (7.12) by proceeding in essentially the same way as in the derivation (in Subsection a) of inequality (7.14).

In the case of the original (unconstrained) version of the step-down procedure [where (for $j = k, k{+}1, \ldots, M$] α_j is defined by expression (7.23) and is reexpressible in the form (7.24)], it follows from inequality (7.41) that (for $j = k, k+1, \ldots, M$)

$$\alpha_j = \max_{S \in \tilde{\Omega}^+_{k;j}} c_{\dot{\gamma}}(k; S) \leq \max_{S \in \tilde{\Omega}^+_{k;j}} \bar{t}_{k\dot{\gamma}/(2M_S)}(N - P_*) \tag{7.42}$$

and hence (since for $S \in \tilde{\Omega}^+_{k;j}$, $M_S = k - 1 + M - j + 1 = M + k - j$) that

$$\alpha_j \leq \bar{t}_{k\dot{\gamma}/[2(M+k-j)]}(N - P_*). \tag{7.43}$$

In the case of the generalized version of the step-down procedure [where T is subject to the constraint $T \in \Omega^-$ and where (for $j = k, k+1, \ldots, M$) α_j is defined by expression (7.40)], it follows from inequality (7.41) that (for $j = k, k+1, \ldots, M$)

$$\alpha_j = \max_{S \in (\tilde{\Omega}_{k;j} \cap \Omega^-)} c_{\dot{\gamma}}(k; S) \leq \max_{S \in (\tilde{\Omega}_{k;j} \cap \Omega^-)} \bar{t}_{k\dot{\gamma}/(2M_S)}(N - P_*) = \bar{t}_{k\dot{\gamma}/[2M_*(k;j)]}(N - P_*), \tag{7.44}$$

where $M_*(k; j) = \max_{S \in (\tilde{\Omega}_{k;j} \cap \Omega^-)} M_S$.

Now, consider a modification of the step-down procedure in which (for $j = k, k+1, \ldots, M$) α_j is replaced by $\bar{t}_{k\dot{\gamma}/[2(M+k-j)]}(N-P_*)$ or, more generally, by $\bar{t}_{k\dot{\gamma}/[2M_*(k;j)]}(N-P_*)$ (and in which $\alpha_1, \alpha_2, \ldots, \alpha_{k-1}$ are replaced by the replacement for α_k). This modification results in a version of the step-down procedure for controlling the k-FWER that can be much less demanding from a computational standpoint but that is less powerful. When the replacement for α_j is $\bar{t}_{k\dot{\gamma}/[2(M+k-j)]}(N-P_*)$ ($j = k, k+1, \ldots, M$), the modified version of the step-down procedure is equivalent to the step-down procedure for controlling the k-FWER proposed by Lehmann and Romano (2005a, sec. 2) and in the special case where $k = 1$, it is equivalent to the step-down procedure for controlling the FWER proposed by Holm (1979). The modified version of the step-down procedure is more powerful than the test procedure with critical regions (7.16) in that its adoption can result in more rejections than the adoption of the latter procedure.

Extensions. By introducing some relatively simple and transparent modifications, the coverage of the present subsection (on the use of step-down methods to control the FWER or k-FWER) can be extended to the case where the null and alternative hypotheses are either $H_i^{(0)} : \tau_i \le \tau_i^{(0)}$ and $H_i^{(1)} : \tau_i > \tau_i^{(0)}$ ($i = 1, 2, \ldots, M$) or $H_i^{(0)} : \tau_i \ge \tau_i^{(0)}$ and $H_i^{(1)} : \tau_i < \tau_i^{(0)}$ ($i = 1, 2, \ldots, M$) rather than $H_i^{(0)} : \tau_i = \tau_i^{(0)}$ and $H_i^{(1)} : \tau_i \ne \tau_i^{(0)}$ ($i = 1, 2, \ldots, M$). Suppose, in particular, that the null and alternative hypotheses are $H_i^{(0)} : \tau_i \le \tau_i^{(0)}$ and $H_i^{(1)} : \tau_i > \tau_i^{(0)}$ ($i = 1, 2, \ldots, M$). Then, instead of taking the permutation $i_1, i_2, \ldots, i_M$ and the permutation $\tilde{i}_1, \tilde{i}_2, \ldots, \tilde{i}_M$ to be as defined by inequalities (7.7) and (7.17), we need to take them to be as defined by the inequalities

$$t_{i_1} \ge t_{i_2} \ge \cdots \ge t_{i_M} \quad \text{and} \quad t_{\tilde{i}_1}^{(0)} \ge t_{\tilde{i}_2}^{(0)} \ge \cdots \ge t_{\tilde{i}_M}^{(0)}.$$

Similarly, we need to redefine $i_1^*(S), i_2^*(S), \ldots, i_{M_S}^*(S)$ and $\tilde{i}_1^*(S), \tilde{i}_2^*(S), \ldots, \tilde{i}_{M_S}^*(S)$ to be permutations of the elements of the subset S ($\subset I$) that satisfy the inequalities

$$t_{i_1^*(S)} \ge t_{i_2^*(S)} \ge \cdots \ge t_{i_{M_S}^*(S)} \quad \text{and} \quad t_{\tilde{i}_1^*(S)}^{(0)} \ge t_{\tilde{i}_2^*(S)}^{(0)} \ge \cdots \ge t_{\tilde{i}_{M_S}^*(S)}^{(0)}.$$

And $t_{(j)}$ and $t_{(j)}^{(0)}$ ($j = 1, 2, \ldots, M$) and $t_{k;S}$ and $t_{k:S}^{(0)}$ ($S \subset I$; $M_S \ge k$) need to be redefined accordingly. Further, redefine $c_{\dot{\gamma}}(k; S)$ to be the upper $100\dot{\gamma}\%$ point of the distribution of $t_{k;S}$, and redefine the α_j's in terms of the redefined $c_{\dot{\gamma}}(k; S)$'s.

In the form taken following its redefinition in terms of the redefined $\tilde{i}_j$'s, $t_j^{(0)}$'s, and α_j's, the step-down procedure rejects the null hypotheses $H_{\tilde{i}_1}^{(0)}, H_{\tilde{i}_2}^{(0)}, \ldots, H_{\tilde{i}_J}^{(0)}$, where J is an integer (between 0 and M, inclusive) such that $t_{(j)}^{(0)} > \alpha_j$ for $j = 1, 2, \ldots, J$ and $t_{(J+1)}^{(0)} \le \alpha_{J+1}$—when $t_{(j)}^{(0)} > \alpha_j$ for $j = 1, 2, \ldots, M$, $J = M$ and all M of the null hypotheses are rejected; when $t_{(1)}^{(0)} \le \alpha_1$, $J = 0$ and none of the null hypotheses are rejected. In effect, the redefined step-down procedure tests the M null hypotheses sequentially in the order $H_{\tilde{i}_1}^{(0)}, H_{\tilde{i}_2}^{(0)}, \ldots, H_{\tilde{i}_M}^{(0)}$ by comparing the $t_{(j)}^{(0)}$'s with the α_j's; the testing ceases upon encountering a null hypothesis $H_{\tilde{i}_j}^{(0)}$ for which $t_{(j)}^{(0)}$ does not exceed α_j.

The redefined $c_{\dot{\gamma}}(k; S)$'s satisfy the inequality

$$\bar{t}_{k\dot{\gamma}/M_S}(N-P_*) \ge c_{\dot{\gamma}}(k; S).$$

This inequality takes the place of inequality (7.41). And to obtain upper bounds for the redefined α_j's, results (7.42), (7.43), and (7.44) need to be revised accordingly. Upon replacing the redefined α_j's with the redefined upper bounds, we obtain versions of the redefined step-down procedure that are computationally less demanding but that are less powerful.

The various versions of the redefined step-down procedure control the k-FWER (at level $\dot{\gamma}$). By way of comparison, we have the tests of the null hypotheses with critical regions

$$\{\mathbf{y} : t_i^{(0)} > c_{\dot{\gamma}}(k)\} \quad (i = 1, 2, \ldots, M),$$

where $c_{\dot{\gamma}}(k)$ is redefined to be the upper $100\dot{\gamma}\%$ point of the distribution of the redefined random variable $t_{(k)}$, and the tests with critical regions

$$\{\mathbf{y} : t_i^{(0)} > \bar{t}_{k\dot{\gamma}/M}(N - P_*)\} \quad (i = 1, 2, \ldots, M).$$

These tests are the redefined versions of the tests with critical regions (7.15) and (7.16). Like the redefined step-down procedures, they control the k-FWER; however, the redefined step-down procedures are more powerful.

Nonnormality. The assumption that the vector $\mathbf{e}$ of residual effects in the G–M model has an MVN distribution is stronger than necessary. To insure the validity of the various results on step-down methods, it is sufficient that $\mathbf{e}$ have an absolutely continuous spherical distribution. In fact, it is sufficient that the vector $\begin{pmatrix} \hat{\alpha} - \alpha \\ \mathbf{d} \end{pmatrix}$ have an absolutely continuous spherical distribution.

c. Alternative criteria for devising and evaluating multiple-comparison procedures: the false discovery proportion and the false discovery rate

Let us consider further alternatives to the control of the FWER as a basis for testing the M null hypotheses $H_1^{(0)}, H_2^{(0)}, \ldots, H_M^{(0)}$ in a way that accounts for the multiplicity of tests. And in doing so, let us continue to make use of the notation and terminology employed in the preceding parts of the present section.

Multiple-comparison procedures are sometimes used as a "screening device." That is, the objective may be to identify or "discover" which of the M linear combinations $\tau_1, \tau_2, \ldots, \tau_M$ are worthy of further investigation. In such a case, the rejection of a null hypothesis represents a "(true) discovery" or a "false discovery" depending on whether the null hypothesis is false or true. And the success of an application of a multiple-comparison procedure may be judged (at least in part) on the basis of whether or not a large proportion of the false null hypotheses are rejected (discovered) and on whether or not the proportion of rejected null hypotheses that are false rejections (false discoveries) is small.

An example: microarray data. Data from so-called microarrays are sometimes used to identify or discover which among a large number of "genes" are associated with some disease and are worthy of further study. Among the examples of such data are the microarray data from a study of prostate cancer that were obtained, analyzed, and discussed by Efron (2010). In that example, the data consist of the "expression levels" obtained for 6033 genes on 102 men; 50 of the men were "normal control subjects" and 52 were "prostate cancer patients." According to Efron, "the principal goal of the study was to discover a small number of 'interesting' genes, that is, genes whose expression levels differ between the prostate and normal subjects." He went on to say that "such genes, once identified, might be further investigated for a causal link to prostate cancer development."

The prostate data can be regarded as multivariate in nature, and as such could be formulated and modeled in the way described and discussed in Section 4.5. Specifically, the subjects could be regarded as $R = 102$ observational units and the genes (or, more precisely, their expression levels) as $S = 6033$ response variables. Further, the data on the sth gene could be regarded as the observed values of the elements of a 102-dimensional random vector $\mathbf{y}_s$ ($s = 1, 2, \ldots, 6033$). And the ($N = 6033 \times 102 = 615{,}366$)-dimensional vector $\mathbf{y} = (\mathbf{y}_1', \mathbf{y}_2', \ldots, \mathbf{y}_S')'$ could be regarded as following a general linear model, where the model equation is as defined by expression (4.5.2) or (4.5.3) and where (assuming that observational units 1 through 50 correspond to the normal subjects)

$$\mathbf{X}_1 = \mathbf{X}_2 = \cdots = \mathbf{X}_S = \mathrm{diag}(\mathbf{1}_{50}, \mathbf{1}_{52}).$$

This model is such that each of the $S = 6033$ subvectors $\boldsymbol{\beta}_1, \boldsymbol{\beta}_2, \ldots, \boldsymbol{\beta}_S$ of the parameter vector $\boldsymbol{\beta}$ has two elements. Clearly, the first element of $\boldsymbol{\beta}_s$ represents the expected value of the response variable (expression level) for the sth gene when the subject is a normal subject, and the second element represents the expected value of that variable when the subject is a cancer patient; let us write μ_{s1} for the first element of β_s and μ_{s2} for the second element. The quantities of interest are represented by the $M = 6033$ linear combinations $\tau_1, \tau_2, \ldots, \tau_M$ defined as follows:

$$\tau_s = \mu_{s2} - \mu_{s1} \quad (s = 1, 2, \ldots, 6033).$$

And, conceivably, the problem of discovering which genes are worthy of further study (i.e., might be associated with prostate cancer development) could be formulated as one of testing the 6033 null hypotheses $H_s^{(0)} : \tau_s = 0$ $(s = 1, 2, \ldots, 6033)$ versus the alternative hypotheses $H_s^{(1)} : \tau_s \neq 0$ $(s = 1, 2, \ldots, 6033)$. In such an approach, the genes corresponding to whichever null hypotheses are rejected would be deemed to be the ones of interest.

Note that the model for the prostate data is such that the variance-covariance matrix of the residual effects is of the form (4.5.4). And recall that the variance-covariance matrix of the residual effects in the G–M model is of the form (4.1.17). Thus, to obtain a model for the prostate data that is a G–M model, we would need to introduce the simplifying assumptions that $\sigma_{ss'} = 0$ for $s' \neq s = 1, 2, \ldots, S$ and that $\sigma_{11} = \sigma_{22} = \cdots = \sigma_{SS}$.

Alternative (to the FWER or k-FWER) criteria for devising and evaluating multiple-comparison procedures: the false discovery proportion and false discovery rate. In applications of multiple-comparison procedures where the procedure is to be used as a screening device, the number M of null hypotheses is generally large and sometimes (as in the example) very large. Multiple-comparison procedures (like those described and discussed in Section 7.3c) that restrict the FWER (familywise error rate) to the customary levels (such as 0.01 or 0.05) are not well suited for such applications. Those procedures are such that among the linear combinations τ_i $(i \in F)$, only those for which the true value differs from the hypothesized value by a very large margin have a reasonable chance of being rejected (discovered). It would seem that the situation could be improved to at least some extent by taking the level to which the FWER is restricted to be much higher than what is customary. Or one could turn to procedures [like those considered in Subsections a and b (of the present section)] that restrict the k-FWER to a specified level, taking the value of k to be larger (and perhaps much larger) than 1 and perhaps taking the level to which the k-FWER is restricted to be higher or much higher than the customary levels. An alternative (to be considered in what follows) is to adopt a procedure like that devised by Benjamini and Hochberg (1995) on the basis of a criterion that more directly reflects the objectives underlying the use of the procedure as a screening device.

The criterion employed by Benjamini and Hochberg is defined in terms of the false discovery proportion as is a related, but somewhat different, criterion considered by Lehmann and Romano (2005a). By definition, the false discovery proportion is the number of rejected null hypotheses that are true (number of false discoveries) divided by the total number of rejections—when the total number of rejections equals 0, the value of the false discovery proportion is by convention (e.g., Benjamini and Hochberg 1995; Lehmann and Romano 2005a) taken to be 0. Let us write FDP for the false discovery proportion. And let us consider the problem of devising multiple-comparison procedures for which that proportion is likely to be small.

In what follows (in Subsections e and d), two approaches to this problem are described, one of which is that of Benjamini and Hochberg (1995) (and of Benjamini and Yekutieli 2001) and the other of which is that of Lehmann and Romano (2005a). The difference between the two approaches is attributable to a difference in the criterion adopted as a basis for exercising control over the FDP. In the Benjamini and Hochberg approach, that control takes the form of a requirement that for some specified constant δ $(0 < \delta < 1)$,

$$\mathrm{E(FDP)} \leq \delta. \tag{7.45}$$

As has become customary, let us refer to the expected value E(FDP) of the false discovery proportion as the false discovery rate and denote it by the symbol FDR. In the Lehmann and Romano approach, control over the FDP takes a different form; it takes the form of a requirement that for some constants κ and ϵ [in the interval (0, 1)],

$$\Pr(\mathrm{FDP} > \kappa) \leq \epsilon \tag{7.46}$$

—typically, κ and ϵ are chosen to be much closer to 0 than to 1, so that the effect of the requirement

(7.46) is to impose on FDP a rather stringent upper bound κ that is violated only infrequently. As is to be demonstrated herein, multiple-comparison procedures that satisfy requirement (7.45) or requirement (7.46) can be devised; these procedures are stepwise in nature.

d. Step-down multiple-comparison procedures that bound the FDP (from above) with a high probability

Let us tackle the problem of devising a multiple-comparison procedure that controls the FDP in the sense defined by inequality (7.46). And in doing so, let us make further use of the notation and terminology employed in the preceding parts of the present section.

A step-down procedure: general form. Corresponding to any nonincreasing sequence of M strictly positive scalars $\alpha_1 \geq \alpha_2 \geq \cdots \geq \alpha_M$ (> 0), there is a step-down procedure for testing the M null hypotheses $H_i^{(0)}: \tau_i = \tau_i^{(0)}$ $(i = 1, 2, \ldots, M)$ [versus the alternative hypotheses $H_i^{(1)}: \tau_i \neq \tau_i^{(0)}$ $(i = 1, 2, \ldots, M)$]. This procedure can be regarded as one in which the null hypotheses are tested sequentially in the order $H_{\tilde{i}_1}^{(0)}, H_{\tilde{i}_2}^{(0)}, \ldots, H_{\tilde{i}_M}^{(0)}$ [where $\tilde{i}_1, \tilde{i}_2, \ldots, \tilde{i}_M$ is a permutation of the integers $1, 2, \ldots, M$ defined implicitly (wp1) by the inequalities $|t_{\tilde{i}_1}^{(0)}| \geq |t_{\tilde{i}_2}^{(0)}| \geq \cdots \geq |t_{\tilde{i}_M}^{(0)}|$]. Specifically, this procedure consists of rejecting the first J of these null hypotheses (i.e., the null hypotheses $H_{\tilde{i}_1}^{(0)}, H_{\tilde{i}_2}^{(0)}, \ldots, H_{\tilde{i}_J}^{(0)}$), where J is an integer (between 0 and M, inclusive) such that $|t_{\tilde{i}_j}^{(0)}| > \alpha_j$ for $j = 1, 2, \ldots, J$ and $|t_{\tilde{i}_{J+1}}^{(0)}| \leq \alpha_{J+1}$—if $|t_{\tilde{i}_1}^{(0)}| \leq \alpha_1$, take $J = 0$ (and reject none of the null hypotheses), and if $|t_{\tilde{i}_j}^{(0)}| > \alpha_j$ for $j = 1, 2, \ldots, M$, take $J = M$ (and reject all M of the null hypotheses).

Some general results pertaining to the dependence of the FDP on the choice of the α_j's. Let us consider how the FDP is affected by the choice of $\alpha_1, \alpha_2, \ldots, \alpha_M$ (where $\alpha_1, \alpha_2, \ldots, \alpha_M$ are constants). For purposes of doing so, suppose that $M_T > 0$ (i.e., at least one of the M null hypotheses $H_1^{(0)}, H_2^{(0)}, \ldots, H_M^{(0)}$ is true)—if $M_T = 0$, then the FDP $= 0$ regardless of how the α_j's are chosen. And let R_1 represent the number of values of $s \in F$ for which $|t_s^{(0)}| > \alpha_1$, and for $j = 2, 3, \ldots, M$, let R_j represent the number of values of $s \in F$ for which $\alpha_{j-1} \geq |t_s^{(0)}| > \alpha_j$—recall that, by definition, $s \in F$ if $H_s^{(0)}$ is false. Further, for $j = 1, 2, \ldots, M$, let $R_j^+ = \sum_{i=1}^{j} R_i$, so that R_j^+ represents the number of values of $s \in F$ for which $|t_s^{(0)}| > \alpha_j$. Clearly, $R_M^+ \leq M_F$; in fact, $M_F - R_M^+$ equals the number of values of $s \in F$ for which $|t_s^{(0)}| \leq \alpha_M$.

The ith of the M null hypotheses $H_1^{(0)}, H_2^{(0)}, \ldots, H_M^{(0)}$ is among the J null hypotheses rejected by the step-down procedure if and only if $|t_i^{(0)}| > \alpha_J$. Thus, of the J rejected null hypotheses, R_J^+ of them are false, and the other $J - R_J^+$ of them are true. And it follows that the false discovery proportion is expressible as

$$\text{FDP} = \begin{cases} (J - R_J^+)/J, & \text{if } J > 0, \\ 0, & \text{if } J = 0. \end{cases} \tag{7.47}$$

Now, let j' represent an integer between 0 and M, inclusive, defined as follows: if there exists an integer $j \in I$ for which $(j - R_j^+)/j > \kappa$, take j' to be the smallest such integer, that is, take

$$j' = \min\{j \in I : (j - R_j^+)/j > \kappa\}$$

or, equivalently, take

$$j' = \min\{j \in I : j - R_j^+ > j\kappa\};$$

otherwise [i.e., if there exists no $j \in I$ for which $(j - R_j^+)/j > \kappa$] take $j' = 0$. And observe [in light of expression (7.47)] that FDP $\leq \kappa$ if $j' = 0$ or, alternatively, if $j' \geq 1$ and $J < j'$ and hence that

$$\text{FDP} > \kappa \;\Rightarrow\; J \geq j' \geq 1 \;\Rightarrow\; j' \geq 1 \text{ and } |t_{\tilde{i}_j}^{(0)}| > \alpha_j \;\; (j = 1, 2, \ldots, j').$$

Observe also that j' is a random variable—its value depends on the values of $R_1, R_2, \ldots, R_M$. These observations lead to the inequality

$$\Pr(\text{FDP} > \kappa) \le \Pr[j' \ge 1 \text{ and } |t_{\tilde{i}_j}^{(0)}| > \alpha_j \; (j = 1, 2, \ldots, j')] \tag{7.48}$$

and ultimately to the conclusion that

$$\Pr[j' \ge 1 \text{ and } |t_{\tilde{i}_j}^{(0)}| > \alpha_j \; (j = 1, 2, \ldots, j')] \le \epsilon \;\Rightarrow\; \Pr(\text{FDP} > \kappa) \le \epsilon. \tag{7.49}$$

The significance of result (7.49) is that it can be exploited for purposes of obtaining relatively tractable conditions that when satisfied by the step-down procedure insure that $\Pr(\text{FDP} > \kappa) \le \epsilon$. Suppose [for purposes of exploiting result (7.49) for such purposes] that $j' \ge 1$. Then, as can with some effort be verified (and as is to be verified in a subsequent part of the present subsection),

$$R_{j'} = 0 \tag{7.50}$$

and

$$j' - R_{j'}^{+} = [j'\kappa] + 1, \tag{7.51}$$

where (for any real number x) $[x]$ denotes the largest integer that is less than or equal to x.

Let us denote by k' the random variable defined (in terms of j') as follows: $k' = [j'\kappa] + 1$ for values of $j' \ge 1$ and $k' = 0$ for $j' = 0$. And observe [in light of result (7.51)] that (for $j' \ge 1$)

$$M_F \ge R_{j'}^{+} = j' - k'. \tag{7.52}$$

Observe also (in light of the equality $M_T = M - M_F$) that

$$M_T \le M - (j' - k') = M + k' - j'. \tag{7.53}$$

Further, let us (for $j' \ge 1$) denote by k the number of members in the set $\{\tilde{i}_1, \tilde{i}_2, \ldots, \tilde{i}_{j'}\}$ that are members of T, so that k of the null hypotheses $H_{\tilde{i}_1}^{(0)}, H_{\tilde{i}_2}^{(0)}, \ldots, H_{\tilde{i}_{j'}}^{(0)}$ are true (and the other $j' - k$ are false). And suppose (in addition to $j' \ge 1$) that $|t_{\tilde{i}_j}^{(0)}| > \alpha_j$ $(j = 1, 2, \ldots, j')$. Then, clearly, $R_{j'}^{+} \ge j' - k$ (or, equivalently, $k \ge j' - R_{j'}^{+}$), so that [in light of result (7.51)]

$$k' \le k \le j'. \tag{7.54}$$

Thus, for some strictly positive integer $s \le j'$,

$$\tilde{i}_{k'}^{*}(T) = \tilde{i}_s.$$

And [upon observing that $\tilde{i}_{k'}^{*}(T) = i_{k'}^{*}(T)$ and that $\tilde{i}_s \in T$] it follows that

$$|t_{k';T}| = |t_{i_{k'}^{*}(T)}| = |t_{\tilde{i}_s}| = |t_{\tilde{i}_s}^{(0)}| > \alpha_s \ge \alpha_{j'}. \tag{7.55}$$

Upon applying result (7.55) [and realizing that result (7.54) implies that $k' \le M_T$], we obtain the inequality

$$\Pr[j' \ge 1 \text{ and } |t_{\tilde{i}_j}^{(0)}| > \alpha_j \; (j = 1, 2, \ldots, j')] \le \Pr(j' \ge 1,\; k' \le M_T, \text{ and } |t_{k';T}| > \alpha_{j'}). \tag{7.56}$$

The relevance of inequality (7.56) is the implication that to obtain a step-down procedure for which $\Pr[j' \ge 1 \text{ and } |t_{\tilde{i}_j}^{(0)}| > \alpha_j \; (j = 1, 2, \ldots, j')] \le \epsilon$ and ultimately [in light of relationship (7.49)] one for which $\Pr(\text{FDP} > \kappa) \le \epsilon$, it suffices to obtain a procedure for which

$$\Pr(j' \ge 1,\; k' \le M_T, \text{ and } |t_{k';T}| > \alpha_{j'}) \le \epsilon. \tag{7.57}$$

For purposes of obtaining a more tractable sufficient condition than condition (7.57), observe that (for $j' \ge 1$)

$$k' - 1 \le j'\kappa < k' \tag{7.58}$$

and that

$$1 \le k' \le [M\kappa] + 1. \tag{7.59}$$

Accordingly, the nonzero values of the random variable j' can be partitioned into mutually exclusive categories based on the value of k': for $u = 1, 2, \ldots, [M\kappa] + 1$, let

$$I_u = \{j \in I \;:\; (u-1)/\kappa \le j < u/\kappa\},$$

and observe that (for $j' \ge 1$)

$$k' = u \;\Leftrightarrow\; j' \in I_u. \tag{7.60}$$

Further, for $u = 1, 2, \ldots, [M\kappa]+1$, let j_u^* represent the largest member of the set I_u, so that for $u = 1, 2, \ldots, [M\kappa]$, $j_u^* = [u/\kappa]-1$ or $j_u^* = [u/\kappa]$ depending on whether or not u/κ is an integer (and for $u = [M\kappa]+1$, $j_u^* = M$). And define $\alpha_u^* = \alpha_{j_u^*}$ or, equivalently,

$$\alpha_u^* = \min_{j \in I \,:\, j < u/\kappa} \alpha_j.$$

Let $K = \min([M\kappa]+1,\; M_T)$. Then, based on the partitioning of the nonzero values of j' into the mutually exclusive categories defined by relationship (7.60), we find that

$$\begin{aligned}
\Pr(j' \ge 1,\; k' \le M_T,\; &\text{and } |t_{k';T}| > \alpha_{j'}) \\
&= \textstyle\sum_{u=1}^{K} \Pr(|t_{u;T}| > \alpha_{j'} \text{ and } j' \in I_u) \\
&\le \textstyle\sum_{u=1}^{K} \Pr(|t_{u;T}| > \alpha_u^* \text{ and } j' \in I_u) \\
&\le \textstyle\sum_{u=1}^{K} \Pr(|t_{r;T}| > \alpha_r^* \text{ for 1 or more values of } r \in \{1, 2, \ldots, K\} \text{ and } j' \in I_u) \\
&\le \Pr(|t_{r;T}| > \alpha_r^* \text{ for 1 or more values of } r \in \{1, 2, \ldots, K\}).
\end{aligned} \tag{7.61}$$

Thus, to obtain a step-down procedure for which $\Pr(\text{FDP} > \kappa) \le \epsilon$, it suffices [in light of the sufficiency of condition (7.57)] to obtain a procedure for which

$$\Pr(|t_{r;T}| > \alpha_r^* \text{ for 1 or more values of } r \in \{1, 2, \ldots, K\}) \le \epsilon. \tag{7.62}$$

Verification of results (7.50) and (7.51). Let us verify result (7.50). Suppose that $j' \ge 2$—if $j' = 1$, then $j' - R_{j'}^+ = 1 - R_{j'}$ and $j'\kappa = \kappa$, so that $1 - R_{j'} > \kappa$, which implies that $R_{j'} < 1-\kappa$ and hence (since $1-\kappa < 1$ and $R_{j'}$ is a nonnegative integer) that $R_{j'} = 0$. And observe that (since $j' - R_{j'}^+ > j'\kappa$ and since $R_{j'}^+ = R_{j'-1}^+ + R_{j'}$)

$$j' - 1 - R_{j'-1}^+ - (R_{j'} - 1) > (j'-1)\kappa + \kappa$$

and hence that

$$j' - 1 - R_{j'-1}^+ > (j'-1)\kappa + R_{j'} - 1 + \kappa.$$

Thus, $R_{j'} = 0$, since otherwise (i.e., if $R_{j'} \ge 1$), it would be the case that $R_{j'} - 1 + \kappa \ge \kappa > 0$ and hence that

$$j' - 1 - R_{j'-1}^+ > (j'-1)\kappa,$$

contrary to the definition of j' as the smallest integer $j \in I$ for which $j - R_j^+ > j\kappa$.

Turning now to the verification of result (7.51), we find that

$$j - R_j^+ \ge [j\kappa]+1$$

for every integer $j \in I$ for which $j - R_j^+ > j\kappa$, as is evident upon observing that (for $j \in I$) $j - R_j^+$ is an integer and hence that $j - R_j^+ > j\kappa$ implies that either $j - R_j^+ = [j\kappa]+1$ or $j - R_j^+ > [j\kappa]+1$ depending on whether $j\kappa \ge j - R_j^+ - 1$ or $j\kappa < j - R_j^+ - 1$. Thus,

$$j' - R_{j'}^+ \ge [j'\kappa]+1. \tag{7.63}$$

Moreover, if $j' - R_{j'}^+ > [j'\kappa]+1$, it would (since $R_{j'} = 0$) be the case that

$$j' - 1 - R_{j'-1}^+ > [j'\kappa] \tag{7.64}$$

and hence [since both sides of inequality (7.64) are integers] that

$$j' - 1 - R_{j'-1}^+ \ge [j'\kappa]+1 > [j'\kappa]+1-\kappa > j'\kappa - \kappa = (j'-1)\kappa,$$

contrary to the definition of j' as the smallest integer $j \in I$ for which $j - R_j^+ > j\kappa$. And it follows that inequality (7.63) holds as an equality, that is,

$$j' - R_{j'}^{+} = [j'\kappa] + 1.$$

Step-down procedures of a particular kind. If and when the first $j-1$ of the (ordered) null hypotheses $H_{\tilde{i}_1}^{(0)}, H_{\tilde{i}_2}^{(0)}, \ldots, H_{\tilde{i}_M}^{(0)}$ are rejected by the step-down procedure, the decision as to whether or not $H_{\tilde{i}_j}^{(0)}$ is rejected is determined by whether or not $|t_{\tilde{i}_j}^{(0)}| > \alpha_j$ and hence depends on the choice of α_j. If (following the rejection of $H_{\tilde{i}_1}^{(0)}, H_{\tilde{i}_2}^{(0)}, \ldots, H_{\tilde{i}_{j-1}}^{(0)}$) $H_{\tilde{i}_j}^{(0)}$ is rejected, then (as of the completion of the jth step of the step-down procedure) there are j total rejections (total discoveries) and the proportion of those that are false rejections (false discoveries) equals $k(j)/j$, where $k(j)$ represents the number of null hypotheses among the j rejected null hypotheses that are true. Whether $k(j)/j > \kappa$ or $k(j)/j \le \kappa$ [i.e., whether or not $k(j)/j$ exceeds the prescribed upper bound] depends on whether $k(j) \ge [j\kappa]+1$ or $k(j) \le [j\kappa]$. As discussed by Lehmann and Romano (2005a, p. 1147), that suggests taking the step-down procedure to be of the form of a step-down procedure for controlling the k-FWER at some specified level $\dot{\gamma}$ and of taking $k = [j\kappa]+1$ (in which case k varies with j). That line of reasoning leads (upon recalling the results of Subsection b) to taking α_j to be of the form

$$\alpha_j = \bar{t}_{([j\kappa]+1)\dot{\gamma}/\{2(M+[j\kappa]+1-j)\}}(N-P_*). \tag{7.65}$$

Taking (for $j = 1, 2, \ldots, M$) α_j to be of the form (7.65) reduces the task of choosing the α_j's to one of choosing $\dot{\gamma}$.

Note that if (for $j = 1, 2, \ldots, M$) α_j is taken to be of the form (7.65), then (for $j' \ge 1$)

$$\alpha_{j'} = \bar{t}_{k'\dot{\gamma}/[2(M+k'-j')]}(N-P_*) \ge \bar{t}_{k'\dot{\gamma}/(2M_T)}(N-P_*), \tag{7.66}$$

as is evident upon recalling result (7.53).

Control of the probability of the FDP exceeding κ: a special case. Let $\mathbf{\Sigma}$ represent the correlation matrix of the least squares estimators $\hat{\tau}_i$ $(i \in T)$ of the M_T estimable linear combinations τ_i $(i \in T)$ of the elements of $\boldsymbol{\beta}$. And suppose that τ_i $(i \in T)$ are linearly independent (in which case $\mathbf{\Sigma}$ is nonsingular). Suppose further that $\mathbf{\Sigma} = \mathbf{I}$ [in which case $\hat{\tau}_i$ $(i \in T)$ are (in light of the normality assumption) statistically independent] or, more generally, that there exists a diagonal matrix $\mathbf{D}$ with diagonal elements of ± 1 for which all of the off-diagonal elements of the matrix $-\mathbf{D\Sigma}^{-1}\mathbf{D}$ are nonnegative. Then, the version of the step-down procedure obtained upon taking (for $j = 1, 2, \ldots, M$) α_j to be of the form (7.65) and upon setting $\dot{\gamma} = \epsilon$ is such that $\Pr(\text{FDP} > \kappa) \le \epsilon$.

Let us verify that this is the case. The verification makes use of an inequality known as the Simes inequality. There are multiple versions of this inequality. The version best suited for present purposes is expressible in the following form:

$$\Pr(X_{(r)} > a_r \text{ for 1 or more values of } r \in \{1, 2, \ldots, n\}) \le (1/n)\textstyle\sum_{r=1}^{n} \Pr(X_r > a_n), \tag{7.67}$$

where $X_1, X_2, \ldots, X_n$ are absolutely continuous random variables whose joint distribution satisfies the so-called PDS condition, where $X_{(1)}, X_{(2)}, \ldots, X_{(n)}$ are the random variables whose values are those obtained by ordering the values of $X_1, X_2, \ldots, X_n$ from largest to smallest, and where $a_1, a_2, \ldots, a_n$ are any constants for which $a_1 \ge a_2 \ge \cdots \ge a_n \ge 0$ and for which $(1/r)\Pr(X_j > a_r)$ is nondecreasing in r (for $r = 1, 2, \ldots, n$ and for every $j \in \{1, 2, \ldots, n\}$)—refer, e.g., to Sarkar (2008, sec. 1).

Now, suppose that (for $j = 1, 2, \ldots, M$) α_j is of the form (7.65). To verify that the version of the step-down procedure with α_j's of this form is such that $\Pr(\text{FDP} > \kappa) \le \epsilon$ when $\dot{\gamma} = \epsilon$, it suffices to verify that (for α_j's of this form) condition (7.62) can be satisfied by taking $\dot{\gamma} = \epsilon$. For purposes of doing so, observe [in light of inequality (7.66)] that

$$\begin{aligned}
&\Pr(|t_{r;T}| > \alpha_r^* \text{ for 1 or more values of } r \in \{1, 2, \ldots, K\}) \\
&\qquad \le \Pr[|t_{r;T}| > \bar{t}_{r\dot{\gamma}/(2M_T)}(N-P_*) \text{ for 1 or more values of } r \in \{1, 2, \ldots, K\}] \\
&\qquad \le \Pr[|t_{r;T}| > \bar{t}_{r\dot{\gamma}/(2M_T)}(N-P_*) \text{ for 1 or more values of } r \in \{1, 2, \ldots, M_T\}].
\end{aligned} \tag{7.68}$$

Next, apply the Simes inequality (7.67) to the "rightmost" side of inequality (7.68), taking $n = M_T$ and (for $r = 1, 2, \ldots, M_T$) taking $a_r = \bar{t}_{r\dot{\gamma}/(2M_T)}(N - P_*)$ and taking X_r to be the rth of the M_T random variables $|t_i|$ ($i \in T$). That the distributional assumptions underlying the Simes inequality are satisfied in this application follows from Theorem 3.1 of Sarkar (2008). Moreover, the assumption that $(1/r)\Pr(X_j > a_r)$ is nondecreasing in r is also satisfied, as is evident upon observing that $\Pr[|t_j| > \bar{t}_{r\dot{\gamma}/(2M_T)}(N-P_*)] = r\dot{\gamma}/M_T$. Thus, having justified the application of the Simes inequality, we find that

$$\Pr[|t_{r;T}| > \bar{t}_{r\dot{\gamma}/(2M_T)}(N-P_*) \text{ for 1 or more values of } r \in \{1, 2, \ldots, M_T\}] \\ \leq (1/M_T)\textstyle\sum_{i\in T} \Pr[|t_i| > \bar{t}_{M_T\dot{\gamma}/(2M_T)}(N-P_*)] = M_T\dot{\gamma}/M_T = \dot{\gamma}. \quad (7.69)$$

In combination with result (7.68), result (7.69) implies that

$$\Pr(|t_{r;T}| > \alpha_r^* \text{ for 1 or more values of } r \in \{1, 2, \ldots, K\}) \leq \dot{\gamma}.$$

And it follows that condition (7.62) can be satisfied by taking $\dot{\gamma} = \epsilon$.

Control of the probability of the FDP exceeding κ***: the general case.*** To insure that the step-down procedure is such that $\Pr(\text{FDP} > \kappa) \leq \epsilon$, it suffices to restrict the choice of the α_j's to a choice that satisfies condition (7.62), that is, to a choice for which $\Pr(|t_{r;T}| > \alpha_r^*$ for 1 or more values of $r \in \{1, 2, \ldots, K\}) \leq \epsilon$. The probability of $|t_{r;T}|$ exceeding α_r^* for 1 or more values of $r \in \{1, 2, \ldots, K\}$ depends on T, which is a set of unknown identity (and unknown size); it depends on T through the value of K as well as through the (absolute) values of the $t_{r;T}$'s. However, $|t_{r;T}| \leq |t_{i_r}| = |t_{(r)}|$ and $K \leq [M\kappa] + 1$, so that by substituting $t_{(r)}$ for $t_{r;T}$ and $[M\kappa] + 1$ for K, we can obtain an upper bound for this probability that does not depend on T—clearly, the value of this probability is at least as great following these substitutions as before. And by restricting the choice of the α_j's to a choice for which the resultant upper bound does not exceed ϵ, that is, to one that satisfies the condition

$$\Pr(|t_{(r)}| > \alpha_r^* \text{ for 1 or more values of } r \in \{1, 2, \ldots, [M\kappa]+1\}) \leq \epsilon \quad (7.70)$$

[and hence also satisfies condition (7.62)], we can insure that the step-down procedure is such that $\Pr(\text{FDP} > \kappa) \leq \epsilon$.

Let us consider the choice of the α_j's when (for $j = 1, 2, \ldots, M$) α_j is taken to be of the form (7.65). When α_j is taken to be of that form, the value of α_j corresponding to any particular value of $\dot{\gamma}$ can be regarded as the value of a function, say $\dot{\alpha}_j(\dot{\gamma})$, of $\dot{\gamma}$. And corresponding to the functions $\dot{\alpha}_1(\cdot), \dot{\alpha}_2(\cdot), \ldots, \dot{\alpha}_M(\cdot)$ is the function, say $\dot{\ell}(\dot{\gamma})$, of $\dot{\gamma}$ defined by the left side of inequality (7.70) when $\alpha_1 = \dot{\alpha}_1(\dot{\gamma})$, $\alpha_2 = \dot{\alpha}_2(\dot{\gamma})$, $\ldots$, $\alpha_M = \dot{\alpha}_M(\dot{\gamma})$. Clearly, $\dot{\alpha}_j(\dot{\gamma})$ is a decreasing function of $\dot{\gamma}$, and $\dot{\ell}(\dot{\gamma})$ is an increasing function.

Now, let $\dot{\gamma}_\epsilon$ represent the solution (for $\dot{\gamma}$) to the equation $\dot{\ell}(\dot{\gamma}) = \epsilon$, so that $\dot{\gamma}_\epsilon$ is the largest value of $\dot{\gamma}$ that satisfies the inequality $\dot{\ell}(\dot{\gamma}) \leq \epsilon$. Then, upon setting $\alpha_j = \dot{\alpha}_j(\dot{\gamma}_\epsilon)$ ($j = 1, 2, \ldots, M$), we obtain a choice for the α_j's that satisfies condition (7.70) and hence one for which the step-down procedure is such that $\Pr(\text{FDP} > \kappa) \leq \epsilon$.

To implement the step-down procedure when $\alpha_j = \dot{\alpha}_j(\dot{\gamma}_\epsilon)$ ($j = 1, 2, \ldots, M$), we must be able to carry out the requisite computations for obtaining the solution $\dot{\gamma}_\epsilon$ to the equation $\dot{\ell}(\dot{\gamma}) = \epsilon$. As in the case of the computation of $c_{\dot{\gamma}}(k; S)$ (for any particular $\dot{\gamma}$, k, and S) resort can be made to Monte Carlo methods of the kind discussed by Edwards and Berry (1987).

Suppose that a large number of draws are made from the distribution of the vector $\mathbf{t} = (t_1, t_2, \ldots, t_M)'$—the feasibility of making a large number of draws from that distribution can be enhanced by taking advantage of results (7.5) and (7.6). And observe that (for any choice of the α_j's)

$$|t_{(r)}| > \alpha_r^* \text{ for 1 or more values of } r \in \{1, 2, \ldots, [M\kappa]+1\} \\ \Leftrightarrow \max_{r\in\{1,2,\ldots,[M\kappa]+1\}} (|t_{(r)}| - \alpha_r^*) > 0, \quad (7.71)$$

so that

$$\dot{\ell}(\dot{\gamma}) = \epsilon \Leftrightarrow \dot{c}_\epsilon(\dot{\gamma}) = 0, \tag{7.72}$$

where $\dot{c}_\epsilon(\dot{\gamma})$ is the upper $100\epsilon\%$ point of the distribution of the random variable $\max_{r\in\{1,2,\ldots,[M\kappa]+1\}}(|t_{(r)}| - \alpha_r^*)$ when $\alpha_1 = \dot{\alpha}_1(\dot{\gamma}), \alpha_2 = \dot{\alpha}_2(\dot{\gamma}), \ldots, \alpha_M = \dot{\alpha}_M(\dot{\gamma})$. Clearly, $\dot{c}_\epsilon(\dot{\gamma})$ is an increasing function (of $\dot{\gamma}$). Moreover, by making use of Monte Carlo methods of the kind discussed by Edwards and Berry (1987), the draws from the distribution of $\mathbf{t}$ can be used to approximate the values of $\dot{c}_\epsilon(\dot{\gamma})$ corresponding to the various values of $\dot{\gamma}$. Thus, by solving the equation obtained from the equation $\dot{c}_\epsilon(\dot{\gamma}) = 0$ upon replacing the values of $\dot{c}_\epsilon(\dot{\gamma})$ with their approximations, we can [in light of result (7.72)] obtain an approximation to the solution $\dot{\gamma}_\epsilon$ to the equation $\dot{\ell}(\dot{\gamma}) = \epsilon$.

A potential improvement. By definition (specifically, the definition of $j_1^*, j_2^*, \ldots, j_{[M\kappa]+1}^*$),

$$\alpha_j \geq \alpha_{j_u^*} \qquad (j \in I_u;\ u = 1,2,\ldots,[M\kappa]+1). \tag{7.73}$$

If (for 1 or more values of u) there exist values of $j \in I_u$ for which $\alpha_j > \alpha_{j_u^*}$, then the step-down procedure can be "improved" by setting $\alpha_j = \alpha_{j_u^*}$ for every such j, in which case

$$\alpha_j = \alpha_{j_u^*} \qquad (j \in I_u;\ u = 1,2,\ldots,[M\kappa]+1). \tag{7.74}$$

This change in the α_j's can be expected to result in additional rejections (discoveries). Moreover, the change in the α_j's does not affect inequality (7.62) or inequality (7.70); so that if inequality (7.62) or inequality (7.70) and (as a consequence) the condition $\Pr(\text{FDP} > \kappa) \leq \epsilon$ are satisfied prior to the change, they are also satisfied following the change. Thus, when the objective is to produce a "large" number of rejections (discoveries) while simultaneously controlling $\Pr(\text{FDP} > \kappa)$ at level ϵ, the step-down procedure with

$$\alpha_j = \dot{\alpha}_{j_u^*}(\dot{\gamma}_\epsilon) \qquad (j \in I_u;\ u = 1,2,\ldots,[M\kappa]+1) \tag{7.75}$$

[where $\dot{\alpha}_j(\cdot)$ $(j = 1,2,\ldots,M)$ and $\dot{\gamma}_\epsilon$ are as defined in the preceding part of the present subsection] may be preferable to that with $\alpha_j = \dot{\alpha}_j(\dot{\gamma}_\epsilon)$ $(j = 1,2,\ldots,M)$.

Extensions. By introducing some relatively simple and (for the most part) transparent modifications, the coverage of the present subsection [pertaining to step-down methods for controlling $\Pr(\text{FDP} > \kappa)$ at a specified level ϵ] can be extended to the case where the null and alternative hypotheses are either $H_i^{(0)} : \tau_i \leq \tau_i^{(0)}$ and $H_i^{(1)} : \tau_i > \tau_i^{(0)}$ $(i = 1,2,\ldots,M)$ or $H_i^{(0)} : \tau_i \geq \tau_i^{(0)}$ and $H_i^{(1)} : \tau_i < \tau_i^{(0)}$ $(i = 1,2,\ldots,M)$ rather than $H_i^{(0)} : \tau_i = \tau_i^{(0)}$ and $H_i^{(1)} : \tau_i \neq \tau_i^{(0)}$ $(i = 1,2,\ldots,M)$.

Suppose, in particular, that the null and alternative hypotheses are $H_i^{(0)} : \tau_i \leq \tau_i^{(0)}$ and $H_i^{(1)} : \tau_i > \tau_i^{(0)}$ $(i = 1,2,\ldots,M)$. Then, the requisite modifications are similar in nature to those described in some detail (in Part 8 of Subsection b) for extending the coverage (provided in the first 7 parts of Subsection b) of step-down methods for controlling the FWER or k-FWER. In particular, the permutations $i_1, i_2, \ldots, i_M$ and $\tilde{i}_1, \tilde{i}_2, \ldots, \tilde{i}_M$ and the permutations $i_1^*(S), i_2^*(S), \ldots, i_{M_S}^*(S)$ and $\tilde{i}_1^*(S), \tilde{i}_2^*(S), \ldots, \tilde{i}_{M_S}^*(S)$ are redefined and the step-down procedure subjected to the same kind of modifications as in Part 8 of Subsection b.

The definitions of the R_j's are among the other items that require modification. The quantity R_1 is redefined to be the number of values of $s \in F$ for which $t_s^{(0)} > \alpha_1$; and for $j = 2,3,\ldots,M$, R_j is redefined to be the number of values of $s \in F$ for which $\alpha_{j-1} \geq t_s^{(0)} > \alpha_j$. Then, for $j = 1,2, \ldots, M$, $R_j^+ = \sum_{i=1}^{j} R_i$ represents the number of values of $s \in F$ for which $t_s^{(0)} > \alpha_j$. Clearly, various other quantities (such as k' and j') are affected implicitly or explicitly by the redefinition of the R_j's.

In lieu of condition (7.62), we have the condition

$$\Pr(t_{r;T} > \alpha_r^* \text{ for 1 or more values of } r \in \{1,2,\ldots,K\}) \leq \epsilon, \tag{7.76}$$

where $t_{r;T} = t_{i_r^*(T)}$. And in lieu of condition (7.70), we have the condition

$$\Pr(t_{(r)} > \alpha_r^* \text{ for 1 or more values of } r \in \{1,2,\ldots,[M\kappa]+1\}) \leq \epsilon, \tag{7.77}$$

where $t_{(r)} = t_{i_r}$—unlike condition (7.76), this condition does not involve T. When the α_j's satisfy condition (7.76) or (7.77), the step-down procedure is such that $\Pr(\text{FDP} > \kappa) \leq \epsilon$.

The analogue of taking α_j to be of the form (7.65) is to take α_j to be of the form

$$\alpha_j = \bar{t}_{([j\kappa]+1)\dot{\gamma}/(M+[j\kappa]+1-j)}(N-P_*), \tag{7.78}$$

where $0 < \dot{\gamma} < 1/2$. When (for $j = 1, 2, \ldots, M$) α_j is taken to be of the form (7.78), it is the case that (for $j' \geq 1$)

$$\alpha_{j'} = \bar{t}_{k'\dot{\gamma}/(M+k'-j')}(N-P_*) \geq \bar{t}_{k'\dot{\gamma}/M_T}(N-P_*) \tag{7.79}$$

—this result is the analogue of result (7.66). And as before, results on controlling $\Pr(\text{FDP} > \kappa)$ at a specified level ϵ are obtainable (under certain conditions) by making use of the Simes inequality.

Let $\boldsymbol{\Sigma}$ represent the correlation matrix of the least squares estimators $\hat{\tau}_i$ $(i \in T)$ of the M_T estimable linear combinations τ_i $(i \in T)$ of the elements of $\boldsymbol{\beta}$. And suppose that τ_i $(i \in T)$ are linearly independent (in which case $\boldsymbol{\Sigma}$ is nonsingular). Then, in the special case where (for $j = 1, 2, \ldots, M$) α_j is taken to be of the form (7.78) and where $\dot{\gamma} = \epsilon < 1/2$, the step-down procedure is such that $\Pr(\text{FDP} > \kappa) \leq \epsilon$ provided that the off-diagonal elements of the correlation matrix $\boldsymbol{\Sigma}$ are nonnegative—it follows from Theorem 3.1 of Sarkar (2008) that the distributional assumptions needed to justify the application of the Simes inequality are satisfied when the off-diagonal elements of $\boldsymbol{\Sigma}$ are nonnegative.

More generally, when (for $j = 1, 2, \ldots, M$) α_j is taken to be of the form (7.78), the value of $\dot{\gamma}$ needed to achieve control of $\Pr(\text{FDP} > \kappa)$ at a specified level ϵ can be determined "numerically" via an approach similar to that described in a preceding part of the present subsection. Instead of taking (for $j = 1, 2, \ldots, M$) $\dot{\alpha}_j(\dot{\gamma})$ to be the function of $\dot{\gamma}$ whose values are those of expression (7.65), take it to be the function whose values are those of expression (7.78). And take $\dot{\ell}(\dot{\gamma})$ to be the function of $\dot{\gamma}$ whose values are those of the left side of inequality (7.77) when (for $j = 1, 2, \ldots, M$) $\alpha_j = \dot{\alpha}_j(\dot{\gamma})$. Further, take $\dot{\gamma}_\epsilon$ to be the solution (for $\dot{\gamma}$) to the equation $\dot{\ell}(\dot{\gamma}) = \epsilon$ (if ϵ is sufficiently small that a solution exists) or, more generally, take $\dot{\gamma}_\epsilon$ to be a value of $\dot{\gamma}$ small enough that $\dot{\ell}(\dot{\gamma}) \leq \epsilon$. Then, upon taking (for $j = 1, 2, \ldots, M$) $\alpha_j = \dot{\alpha}_j(\dot{\gamma}_\epsilon)$, we obtain a version of the step-down procedure for which $\Pr(\text{FDP} > \kappa) \leq \epsilon$. It is worth noting that an "improvement" in the resultant procedure can be achieved by introducing a modification of the α_j's analogous to modification (7.75).

Nonnormality. The approach taken herein [in insuring that the step-down procedure is such that $\Pr(\text{FDP} > \kappa) \leq \epsilon$] is based on insuring that the α_j's satisfy condition (7.62) or condition (7.70). The assumption (introduced at the beginning of Section 7.7) that the distribution of the vector $\mathbf{e}$ (of the residual effects in the G–M model) is MVN can be relaxed without invalidating that approach.

Whether or not condition (7.62) is satisfied by any particular choice for the α_j's is completely determined by the distribution of the M_T random variables t_i $(i \in T)$, and whether or not condition (7.70) is satisfied is completely determined by the distribution of the M random variables t_i $(i = 1, 2, \ldots, M)$. And each of the t_i's is expressible as a linear combination of the elements of the vector $\hat{\sigma}^{-1}(\hat{\boldsymbol{\alpha}} - \boldsymbol{\alpha})$, as is evident from result (7.5). Moreover, $\hat{\sigma}^{-1}(\hat{\boldsymbol{\alpha}} - \boldsymbol{\alpha}) \sim MVt(N-P_*, \mathbf{I})$—refer to result (7.6)—not only when the distribution of the vector $\begin{pmatrix} \hat{\boldsymbol{\alpha}} - \boldsymbol{\alpha} \\ \mathbf{d} \end{pmatrix}$ is MVN (as is the case when the distribution of $\mathbf{e}$ is MVN) but, more generally, when $\begin{pmatrix} \hat{\boldsymbol{\alpha}} - \boldsymbol{\alpha} \\ \mathbf{d} \end{pmatrix}$ has an absolutely continuous spherical distribution (as is the case when $\mathbf{e}$ has an absolutely continuous spherical distribution)—refer to result (6.4.67). Thus, if the α_j's satisfy condition (7.62) or condition (7.70) when the distribution of $\mathbf{e}$ is MVN, then $\Pr(\text{FDP} > \kappa) \leq \epsilon$ not only in the case where the distribution of $\mathbf{e}$ is MVN but also in various other cases.

e. Step-up multiple-comparison procedures for controlling the FDR

Having considered (in Subsection d) the problem of devising a multiple-comparison procedure that controls $\Pr(\text{FDP} > \kappa)$ at a specified level, let us consider the problem of devising a multiple-comparison procedure that controls the FDR [i.e., controls E(FDP)] at a specified level. And in doing so, let us make further use of various of the notation and terminology employed in the preceding parts of the present section.

A step-up procedure: general form. Let $\alpha_1, \alpha_2, \ldots, \alpha_M$ represent a nonincreasing sequence of M strictly positive scalars (so that $\alpha_1 \geq \alpha_2 \geq \cdots \geq \alpha_M > 0$). Then, corresponding to this sequence, there is (as discussed in Subsection d) a step-down procedure for testing the M null hypotheses $H_i^{(0)}: \tau_i = \tau_i^{(0)}$ $(i = 1, 2, \ldots, M)$ [versus the alternative hypotheses $H_i^{(1)}: \tau_i \neq \tau_i^{(0)}$ $(i = 1, 2, \ldots, M)$]. There is also a step-up procedure for testing these M null hypotheses. Like the step-down procedure, the step-up procedure rejects (for some integer J between 0 and M, inclusive) the first J of the null hypotheses in the sequence $H_{\tilde{i}_1}^{(0)}, H_{\tilde{i}_2}^{(0)}, \ldots, H_{\tilde{i}_M}^{(0)}$ [where $\tilde{i}_1, \tilde{i}_2, \ldots, \tilde{i}_M$ is a permutation of the integers $1, 2, \ldots, M$ defined implicitly (wp1) by the inequalities $|t_{\tilde{i}_1}^{(0)}| \geq |t_{\tilde{i}_2}^{(0)}| \geq \cdots \geq |t_{\tilde{i}_M}^{(0)}|$]. Where it differs from the step-down procedure is in the choice of J. In the step-up procedure, J is taken to be the largest value of j for which $|t_{\tilde{i}_j}^{(0)}| > \alpha_j$—if $|t_{\tilde{i}_j}^{(0)}| \leq \alpha_j$ for $j = 1, 2, \ldots, M$, set $J = 0$. The step-up procedure can be visualized as a procedure in which the null hypotheses are tested sequentially in the order $H_{\tilde{i}_M}^{(0)}, H_{\tilde{i}_{M-1}}^{(0)}, \ldots, H_{\tilde{i}_1}^{(0)}$; by definition, $|t_{\tilde{i}_M}^{(0)}| \leq \alpha_M$, $|t_{\tilde{i}_{M-1}}^{(0)}| \leq \alpha_{M-1}, \ldots, |t_{\tilde{i}_{J+1}}^{(0)}| \leq \alpha_{J+1}$, and $|t_{\tilde{i}_J}^{(0)}| > \alpha_J$.

Note that in contrast to the step-down procedure (where the choice of J is such that $|t_{\tilde{i}_j}^{(0)}| > \alpha_j$ for $j = 1, 2, \ldots, J$), the choice of J in the step-up procedure is such that $|t_{\tilde{i}_j}^{(0)}|$ does not necessarily exceed α_j for every value of $j \leq J$ (though $|t_{\tilde{i}_j}^{(0)}|$ does necessarily exceed α_J for every value of $j \leq J$). Note also that (when the α_j's are the same in both cases) the number of rejections produced by the step-up procedure is at least as great as the number produced by the step-down procedure.

For $j = 1, 2, \ldots, M$, let

$$\alpha'_j = \Pr(|t| > \alpha_j), \tag{7.80}$$

where $t \sim St(N - P_*)$. In their investigation of the FDR as a criterion for evaluating and devising multiple-comparison procedures, Benjamini and Hochberg (1995) proposed a step-up procedure in which (as applied to the present setting) $\alpha_1, \alpha_2, \ldots, \alpha_M$ are of the form defined implicitly by equality (7.80) upon taking (for $j = 1, 2, \ldots, M$) α'_j to be of the form

$$\alpha'_j = j\dot{\gamma}/M, \tag{7.81}$$

where $0 < \dot{\gamma} < 1$. Clearly, taking the α_j's to be of that form is equivalent to taking them to be of the form

$$\alpha_j = \bar{t}_{j\dot{\gamma}/(2M)}(N - P_*) \tag{7.82}$$

$(j = 1, 2, \ldots, M)$.

The FDR of a step-up procedure. Let B_i represent the event consisting of those values of the vector $\mathbf{t}^{(0)}$ (with elements $t_1^{(0)}, t_2^{(0)}, \ldots, t_M^{(0)}$) for which $H_i^{(0)}$ is rejected by the step-up procedure $(i = 1, 2, \ldots, M)$. Further, let X_i represent a random variable defined as follows:

$$X_i = \begin{cases} 1, & \text{if } \mathbf{t}^{(0)} \in B_i, \\ 0, & \text{if } \mathbf{t}^{(0)} \notin B_i. \end{cases}$$

Then, $\sum_{i \in T} X_i$ equals the number of falsely rejected null hypotheses, and $\sum_{j \in I} X_j$ equals the total number of rejections—recall that $I = \{1, 2, \ldots, M\}$ and that T is the subset of I consisting of those values of $i \in I$ for which $H_i^{(0)}$ is true. And FDP > 0 only if $\sum_{j \in I} X_j > 0$, in which case

$$\text{FDP} = \frac{\sum_{i \in T} X_i}{\sum_{j \in I} X_j}. \tag{7.83}$$

Observe [in light of expression (7.83)] that the false discovery rate is expressible as

$$\mathrm{FDR} = \mathrm{E(FDP)} = \textstyle\sum_{i\in T} \mathrm{E}(\mathrm{FDP}_i), \tag{7.84}$$

where $\mathrm{FDP}_i = \dfrac{X_i}{\sum_{j\in I} X_j}$ if $\sum_{j\in I} X_j > 0$ and $\mathrm{FDP}_i = 0$ if $\sum_{j\in I} X_j = 0$. Observe also that $\mathrm{FDP}_i > 0$ only if $\mathbf{t}^{(0)} \in B_i$, in which case $\mathrm{FDP}_i = 1/\sum_{j\in I} X_j$, and that (for $k = 1, 2, \ldots, M$)

$$\textstyle\sum_{j\in I} X_j = k \;\Leftrightarrow\; \mathbf{t}^{(0)} \in A_k,$$

where A_k is the event consisting of those values of $\mathbf{t}^{(0)}$ for which exactly k of the M null hypotheses $H_1^{(0)}, H_2^{(0)}, \ldots, H_M^{(0)}$ are rejected by the step-up procedure. Accordingly,

$$\mathrm{FDR} = \textstyle\sum_{i\in T} \sum_{k=1}^{M} (1/k) \Pr\big(\mathbf{t}^{(0)} \in B_i \cap A_k\big). \tag{7.85}$$

Clearly, $A_M = \{\mathbf{t}^{(0)} : |t_{\tilde{i}_M}^{(0)}| > \alpha_M\}$; and for $k = M-1, M-2, \ldots, 1$,

$$A_k = \{\mathbf{t}^{(0)} : |t_{\tilde{i}_j}^{(0)}| \le \alpha_j \; (j = M, M-1, \ldots, k+1);\; |t_{\tilde{i}_k}^{(0)}| > \alpha_k\}. \tag{7.86}$$

Or, equivalently,

$$A_k = \{\mathbf{t}^{(0)} : \max_{j\in I\,:\,|t_{\tilde{i}_j}^{(0)}| > \alpha_j} j = k\} \tag{7.87}$$

$(k = M, M-1, \ldots, 1)$.

Now, for purposes of obtaining an expression for the FDR that is "more useful" than expression (7.85), let S_i represent the $(M-1)$-dimensional subset of the set $I = \{1, 2, \ldots, M\}$ obtained upon deleting the integer i, and denote by $\mathbf{t}^{(-i)}$ the $(M-1)$-dimensional subvector of $\mathbf{t}^{(0)}$ obtained upon striking out the ith element $t_i^{(0)}$. And recall that for any (nonempty) subset S of I, $\tilde{i}_1^*(S), \tilde{i}_2^*(S), \ldots, \tilde{i}_{M_S}^*(S)$ is a permutation of the elements of S such that $|t_{\tilde{i}_1^*(S)}^{(0)}| \ge |t_{\tilde{i}_2^*(S)}^{(0)}| \ge \cdots \ge |t_{\tilde{i}_{M_S}^*(S)}^{(0)}|$; and observe that for $i \in I$ and for j' such that $\tilde{i}_{j'} = i$,

$$\tilde{i}_j = \begin{cases} \tilde{i}_j^*(S_i), & \text{for } j = 1, 2, \ldots, j'-1, \\ i, & \text{for } j = j', \\ \tilde{i}_{j-1}^*(S_i), & \text{for } j = j'+1, j'+2, \ldots, M, \end{cases} \tag{7.88}$$

and, conversely,

$$\tilde{i}_{j-1}^*(S_i) = \begin{cases} \tilde{i}_{j-1}, & \text{for } j = 2, 3, \ldots, j', \\ \tilde{i}_j, & \text{for } j = j'+1, j'+2, \ldots, M. \end{cases} \tag{7.89}$$

Further, for $i = 1, 2, \ldots, M$, define $A_{M;i}^* = \{\mathbf{t}^{(-i)} : |t_{\tilde{i}_{M-1}^*(S_i)}^{(0)}| > \alpha_M\}$,

$$A_{k;i}^* = \{\mathbf{t}^{(-i)} : |t_{\tilde{i}_{j-1}^*(S_i)}^{(0)}| \le \alpha_j \text{ for } j = M, M-1, \ldots, k+1;\; |t_{\tilde{i}_{k-1}^*(S_i)}^{(0)}| > \alpha_k\}$$

$(k = M-1, M-2, \ldots, 2)$, and

$$A_{1;i}^* = \{\mathbf{t}^{(-i)} : |t_{\tilde{i}_{j-1}^*(S_i)}^{(0)}| \le \alpha_j \text{ for } j = M, M-1, \ldots, 2\}.$$

For $i, k = 1, 2, \ldots, M$, $A_{k;i}^*$ is interpretable in terms of the results that would be obtained if the step-up procedure were applied to the $M-1$ null hypotheses $H_1^{(0)}, H_2^{(0)}, \ldots, H_{i-1}^{(0)}, H_{i+1}^{(0)}, \ldots, H_M^{(0)}$ (rather than to all M of the null hypotheses) with the role of $\alpha_1, \alpha_2, \ldots, \alpha_M$ being assumed by $\alpha_2, \alpha_3, \ldots, \alpha_M$; if $\mathbf{t}^{(-i)} \in A_{k;i}^*$, then exactly $k-1$ of these $M-1$ null hypotheses would be rejected.

It can be shown and (in the next part of the present subsection) will be shown that for $i, k = 1, 2, \ldots, M$,

$$\mathbf{t}^{(0)} \in B_i \cap A_k \;\Leftrightarrow\; t_i^{(0)} \in B_{k;i}^* \text{ and } \mathbf{t}^{(-i)} \in A_{k;i}^*, \tag{7.90}$$

where $B_{k;i}^* = \{t_i^{(0)} : |t_i^{(0)}| > \alpha_k\}$. And upon recalling result (7.85) and making use of relationship

(7.90), we find that

$$\begin{aligned}\text{FDR} &= \textstyle\sum_{i\in T}\sum_{k=1}^{M}(1/k)\Pr\big(t_i \in B_{k;i}^* \text{ and } \mathbf{t}^{(-i)} \in A_{k;i}^*\big)\\ &= \textstyle\sum_{i\in T}\sum_{k=1}^{M}(\alpha_k'/k)\Pr\big(\mathbf{t}^{(-i)} \in A_{k;i}^* \mid |t_i| > \alpha_k\big)\end{aligned} \tag{7.91}$$

[where α_k' is as defined by expression (7.80) and t_i as defined by expression (7.3)].

Note that (for $k = M, M-1, \ldots, 2$)

$$A_{k;i}^* = \{\mathbf{t}^{(-i)} : \max_{j\in\{M,M-1,\ldots,2\}\,:\,|t^{(0)}_{\tilde{i}^*_{j-1}(S_i)}|>\alpha_j} j = k\}, \tag{7.92}$$

analogous to result (7.87). Note also that the sets $A_{1;i}^*, A_{2;i}^*, \ldots, A_{M;i}^*$ are mutually disjoint and that

$$\textstyle\bigcup_{k=1}^{M} A_{k;i}^* = \mathcal{R}^{M-1} \tag{7.93}$$

$(i = 1, 2, \ldots, M)$.

Verification of result (7.90). Suppose that $\mathbf{t}^{(0)} \in B_i \cap A_k$. Then,

$$|t^{(0)}_{\tilde{i}_j}| \le \alpha_j \text{ for } j > k \quad\text{and}\quad |t^{(0)}_{\tilde{i}_k}| > \alpha_k. \tag{7.94}$$

Further,

$$i = \tilde{i}_{j'} \text{ for some } j' \le k, \tag{7.95}$$

and [in light of result (7.89)]

$$\tilde{i}^*_{j-1}(S_i) = \tilde{i}_j \quad\text{for } j > j' \text{ and hence for } j > k. \tag{7.96}$$

Results (7.95) and (7.94) imply that

$$|t_i^{(0)}| = |t^{(0)}_{\tilde{i}_j}| \ge |t^{(0)}_{\tilde{i}_k}| > \alpha_k$$

and hence that $t_i^{(0)} \in B_{k;i}^*$. Moreover, it follows from results (7.96) and (7.94) that for $j > k$,

$$|t^{(0)}_{\tilde{i}^*_{j-1}(S_i)}| = |t^{(0)}_{\tilde{i}_j}| \le \alpha_j. \tag{7.97}$$

And for $k > 1$,

$$\tilde{i}^*_{k-1}(S_i) = \begin{cases}\tilde{i}_k, & \text{if } k > j',\\ \tilde{i}_{k-1}, & \text{if } k = j',\end{cases}$$

as is evident from result (7.89), so that [observing that $|t^{(0)}_{\tilde{i}_{k-1}}| \ge |t^{(0)}_{\tilde{i}_k}|$ and making use of result (7.94)]

$$|t^{(0)}_{\tilde{i}^*_{k-1}(S_i)}| \ge |t^{(0)}_{\tilde{i}_k}| > \alpha_k \qquad (k > 1). \tag{7.98}$$

Together, results (7.97) and (7.98) imply that $\mathbf{t}^{(-i)} \in A_{k;i}^*$.

Conversely, suppose that $t_i^{(0)} \in B_{k;i}^*$ and $\mathbf{t}^{(-i)} \in A_{k;i}^*$. Further, denote by j' the integer defined (implicitly and uniquely) by the equality $i = \tilde{i}_{j'}$. And observe that $j' \le k$, since otherwise (i.e., if $j' > k$) it would [in light of result (7.88)] be the case that

$$|t_i^{(0)}| = |t^{(0)}_{\tilde{i}_{j'}}| \le |t^{(0)}_{\tilde{i}_{j'-1}}| = |t^{(0)}_{\tilde{i}^*_{j'-1}(S_i)}| \le \alpha_{j'} \le \alpha_k,$$

contrary to the supposition (which implies that $|t_i^{(0)}| > \alpha_k$). Then, making use of result (7.88), we find that $t^{(0)}_{\tilde{i}_k} = t_i^{(0)}$ or $t^{(0)}_{\tilde{i}_k} = t^{(0)}_{\tilde{i}^*_{k-1}(S_i)}$ (depending on whether $j' = k$ or $j' < k$) and that in either case

$$|t^{(0)}_{\tilde{i}_k}| > \alpha_k,$$

and we also find that for $j > k\ \ (\ge j')$

$$|t^{(0)}_{\tilde{i}_j}| = |t^{(0)}_{\tilde{i}^*_{j-1}(S_i)}| \le \alpha_j,$$

leading to the conclusion that $\mathbf{t}^{(0)} \in A_k$ and (since $i = \tilde{i}_{j'}$ and $j' \leq k$) to the further conclusion that $\mathbf{t}^{(0)} \in B_i$ and ultimately to the conclusion that $\mathbf{t}^{(0)} \in B_i \cap A_k$.

Control of the FDR in a special case. For $i = 1, 2, \ldots, M$, let

$$z_i^{(0)} = \frac{\hat{\tau}_i - \tau_i^{(0)}}{[\boldsymbol{\lambda}_i'(\mathbf{X}'\mathbf{X})^- \boldsymbol{\lambda}_i]^{1/2}\sigma} \quad \text{and} \quad z_i = \frac{\hat{\tau}_i - \tau_i}{[\boldsymbol{\lambda}_i'(\mathbf{X}'\mathbf{X})^- \boldsymbol{\lambda}_i]^{1/2}\sigma}.$$

And define $\mathbf{z}^{(0)} = (z_1^{(0)}, z_2^{(0)}, \ldots, z_M^{(0)})'$, and observe that $z_i^{(0)} = z_i$ for $i \in T$. Further, for $i = 1, 2, \ldots, M$, denote by $\mathbf{z}^{(-i)}$ the $(M-1)$-dimensional subvector of $\mathbf{z}^{(0)}$ obtained upon deleting the ith element; and let $u = \hat{\sigma}/\sigma$. Then, $t_i^{(0)} = z_i^{(0)}/u$, $t_i = z_i/u$, $\mathbf{t}^{(0)} = u^{-1}\mathbf{z}^{(0)}$, and $\mathbf{t}^{(-i)} = u^{-1}\mathbf{z}^{(-i)}$; u and $\mathbf{z}^{(0)}$ are statistically independent; and $(N-P_*)u^2 \sim \chi^2(N-P_*)$ and $\mathbf{z}^{(0)} \sim N(\boldsymbol{\mu}, \boldsymbol{\Sigma})$, where $\boldsymbol{\mu}$ is the $M \times 1$ vector with ith element $\mu_i = [\sigma^2 \boldsymbol{\lambda}_i'(\mathbf{X}'\mathbf{X})^- \boldsymbol{\lambda}_i]^{-1/2}(\tau_i - \tau_i^{(0)})$ and $\boldsymbol{\Sigma}$ is the $M \times M$ matrix with ijth element

$$\rho_{ij} = \frac{\boldsymbol{\lambda}_i'(\mathbf{X}'\mathbf{X})^- \boldsymbol{\lambda}_j}{[\boldsymbol{\lambda}_i'(\mathbf{X}'\mathbf{X})^- \boldsymbol{\lambda}_i]^{1/2}[\boldsymbol{\lambda}_j'(\mathbf{X}'\mathbf{X})^- \boldsymbol{\lambda}_j]^{1/2}}$$

—recall the results of Section 7.3a, including results (3.9) and (3.21).

Let $f(\cdot)$ represent the pdf of the $N(0, 1)$ distribution, and denote by $h(\cdot)$ the pdf of the distribution of the random variable u. Then, the conditional probability $\Pr(\mathbf{t}^{(-i)} \in A_{k;i}^* \mid |t_i| > \alpha_k)$, which appears in expression (7.91) (for the FDR), is expressible as follows:

$$\begin{aligned}
\Pr(\mathbf{t}^{(-i)} \in A_{k;i}^* \mid |t_i| > \alpha_k) &= \Pr(u^{-1}\mathbf{z}^{(-i)} \in A_{k;i}^* \mid |z_i| > u\alpha_k) \\
&= \int_0^\infty \int_{\underline{z}_i \,:\, |\underline{z}_i| > \underline{u}\alpha_k} \Pr(u^{-1}\mathbf{z}^{(-i)} \in A_{k;i}^* \mid z_i = \underline{z}_i, u = \underline{u}) \\
&\qquad\qquad \frac{f(\underline{z}_i)}{\Pr(|z_i| > \underline{u}\alpha_k)}\, d\underline{z}_i\, h(\underline{u})\, d\underline{u}. \qquad (7.99)
\end{aligned}$$

Moreover, the distribution of $\mathbf{z}^{(0)}$ conditional on $u = \underline{u}$ does not depend on $\underline{u}$; and if $\rho_{ij} = 0$ for all $j \neq i$, then the distribution of $\mathbf{z}^{(-i)}$ conditional on $z_i = \underline{z}_i$ does not depend on $\underline{z}_i$. Thus, if $\rho_{ij} = 0$ for all $j \neq i$, then

$$\Pr(u^{-1}\mathbf{z}^{(-i)} \in A_{k;i}^* \mid z_i = \underline{z}_i, u = \underline{u}) = \Pr(u^{-1}\mathbf{z}^{(-i)} \in A_{k;i}^* \mid u = \underline{u}). \qquad (7.100)$$

And upon substituting expression (7.100) in formula (7.99), we find that (in the special case where $\rho_{ij} = 0$ for all $j \neq i$)

$$\Pr(\mathbf{t}^{(-i)} \in A_{k;i}^* \mid |t_i| > \alpha_k) = \mathrm{E}[\Pr(u^{-1}\mathbf{z}^{(-i)} \in A_{k;i}^* \mid u)] = \Pr(\mathbf{t}^{(-i)} \in A_{k;i}^*). \qquad (7.101)$$

Now, suppose that $\rho_{ij} = 0$ for all $i \in T$ and $j \in I$ such that $j \neq i$. Then, in light of result (7.101), formula (7.91) for the FDR can be reexpressed as follows:

$$\mathrm{FDR} = \textstyle\sum_{i \in T} \sum_{k=1}^M (\alpha_k'/k) \Pr(\mathbf{t}^{(-i)} \in A_{k;i}^*). \qquad (7.102)$$

Moreover, in the special case where (for $j = 1, 2, \ldots, M$) α_j is of the form (7.82) [and hence where α_j' is of the form (7.81)], expression (7.102) can be simplified. In that special case, we find (upon recalling that the sets $A_{1;i}^*, A_{2;i}^*, \ldots, A_{M;i}^*$ are mutually disjoint and that their union is $\mathcal{R}^{M-1}$) that

$$\begin{aligned}
\mathrm{FDR} &= \textstyle\sum_{i \in T} (\dot{\gamma}/M) \sum_{k=1}^M \Pr(\mathbf{t}^{(-i)} \in A_{k;i}^*) \\
&= (M_T/M)\, \dot{\gamma} \Pr(\mathbf{t}^{(-i)} \in \mathcal{R}^{M-1}) \\
&= (M_T/M)\, \dot{\gamma} \leq \dot{\gamma}. \qquad (7.103)
\end{aligned}$$

Based on inequality (7.103), we conclude that if $\rho_{ij} = 0$ for $j \neq i = 1, 2, \ldots, M$ (so that $\rho_{ij} = 0$ for all $i \in T$ and $j \in I$ such that $j \neq i$ regardless of the unknown identity of the set T), then the FDR can be controlled at level δ (in the sense that $\mathrm{FDR} \leq \delta$) by taking (for $j = 1, 2, \ldots, M$) α_j to be of the form (7.82) and by setting $\dot{\gamma} = \delta$. The resultant step-up procedure is that proposed by Benjamini and Hochberg (1995).

Note that when (for $j = 1, 2, \ldots, M$) α_j is taken to be of the form (7.82), the FDR could be reduced by decreasing the value of $\dot{\gamma}$, but that the reduction in FDR would come at the expense of a potential reduction in the number of rejections (discoveries)—there would be a potential reduction in the number of rejections of false null hypotheses (number of true discoveries) as well as in the number of true null hypotheses (number of false discoveries). Note also that the validity of results (7.102) and (7.103) depends only on various characteristics of the distribution of the random variables z_j ($j \in T$) and u and on those random variables being distributed independently of the random variables z_j ($j \in F$); the (marginal) distribution of the random variables z_j ($j \in F$) is "irrelevant."

An extension. The step-up procedure and the various results on its FDR can be readily extended to the case where the null and alternative hypotheses are either $H_i^{(0)} : \tau_i \le \tau_i^{(0)}$ and $H_i^{(1)} : \tau_i > \tau_i^{(0)}$ ($i = 1, 2, \ldots, M$) or $H_i^{(0)} : \tau_i \ge \tau_i^{(0)}$ and $H_i^{(1)} : \tau_i < \tau_i^{(0)}$ ($i = 1, 2, \ldots, M$) rather than $H_i^{(0)} : \tau_i = \tau_i^{(0)}$ and $H_i^{(1)} : \tau_i \ne \tau_i^{(0)}$ ($i = 1, 2, \ldots, M$). Suppose, in particular, that the null and alternative hypotheses are $H_i^{(0)} : \tau_i \le \tau_i^{(0)}$ and $H_i^{(1)} : \tau_i > \tau_i^{(0)}$ ($i = 1, 2, \ldots, M$), so that the set T of values of $i \in I$ for which $H_i^{(0)}$ is true and the set F for which it is false are $T = \{i \in I : \tau_i \le \tau_i^{(0)}\}$ and $F = \{i \in I : \tau_i > \tau_i^{(0)}\}$. And consider the extension of the step-up procedure and of the various results on its FDR to this case.

In regard to the procedure itself, it suffices to redefine the permutation $\tilde{i}_1, \tilde{i}_2, \ldots, \tilde{i}_M$ and the integer J: take $\tilde{i}_1, \tilde{i}_2, \ldots, \tilde{i}_M$ to be the permutation (of the integers $1, 2, \ldots, M$) defined implicitly (wp1) by the inequalities $t_{\tilde{i}_1}^{(0)} \ge t_{\tilde{i}_2}^{(0)} \ge \cdots \ge t_{\tilde{i}_M}^{(0)}$, and take J to be the largest value of j for which $t_{\tilde{i}_j}^{(0)} > \alpha_j$—if $t_{\tilde{i}_j}^{(0)} \le \alpha_j$ for $j = 1, 2, \ldots, M$, set $J = 0$. As before, $\alpha_1, \alpha_2, \ldots, \alpha_M$ represents a nonincreasing sequence of scalars (so that $\alpha_1 \ge \alpha_2 \ge \cdots \ge \alpha_M$); however, unlike before, some or all of the α_j's can be negative.

In regard to the FDR of the step-up procedure, we find by proceeding in the same fashion as in arriving at expression (7.91) and by redefining (for an arbitrary nonempty subset S of I) $\tilde{i}_1^*(S), \tilde{i}_2^*(S), \ldots, \tilde{i}_{M_S}^*(S)$ to be a permutation of the elements of S for which

$$t_{\tilde{i}_1^*(S)}^{(0)} \ge t_{\tilde{i}_2^*(S)}^{(0)} \ge \cdots \ge t_{\tilde{i}_{M_S}^*(S)}^{(0)}, \tag{7.104}$$

by redefining $A_{k;i}^*$ for $k = M$ as

$$A_{M;i}^* = \{\mathbf{t}^{(-i)} : t_{\tilde{i}_{M-1}^*(S_i)}^{(0)} > \alpha_M\}, \tag{7.105}$$

for $k = M-1, M-2, \ldots, 2$ as

$$A_{k;i}^* = \{\mathbf{t}^{(-i)} : t_{\tilde{i}_{j-1}^*(S_i)}^{(0)} \le \alpha_j \text{ for } j = M, M-1, \ldots, k+1;\ t_{\tilde{i}_{k-1}^*(S_i)}^{(0)} > \alpha_k\}, \tag{7.106}$$

and for $k = 1$ as

$$A_{1;i}^* = \{\mathbf{t}^{(-i)} : t_{\tilde{i}_{j-1}^*(S_i)}^{(0)} \le \alpha_j \text{ for } j = M, M-1, \ldots, 2\}, \tag{7.107}$$

by redefining $B_{k;i}^*$ (for $k = 1, 2, \ldots, M$) as

$$B_{k;i}^* = \{t_i^{(0)} : t_i^{(0)} > \alpha_k\}, \tag{7.108}$$

and by redefining α_j' (for $j = 1, 2, \ldots, M$) as

$$\alpha_j' = \Pr(t > \alpha_j) \tag{7.109}$$

[where $t \sim St(N - P_*)$], that

$$\begin{aligned}
\text{FDR} &= \textstyle\sum_{i\in T}\sum_{k=1}^{M}(1/k)\Pr\big(t_i^{(0)}\in B_{k;i}^{*}\ \text{and}\ \mathbf{t}^{(-i)}\in A_{k;i}^{*}\big)\\
&= \textstyle\sum_{i\in T}\sum_{k=1}^{M}(1/k)\Pr\big(t_i^{(0)}>\alpha_k\big)\Pr\big(\mathbf{t}^{(-i)}\in A_{k;i}^{*}\mid t_i^{(0)}>\alpha_k\big)\\
&\le \textstyle\sum_{i\in T}\sum_{k=1}^{M}(\alpha_k'/k)\Pr\big(\mathbf{t}^{(-i)}\in A_{k;i}^{*}\mid t_i^{(0)}>\alpha_k\big).
\end{aligned} \tag{7.110}$$

Note that (for $k = M, M-1, \ldots, 2$) $A_{k;i}^{*}$ [as redefined by expression (7.105) or (7.106)] is reexpressible as

$$A_{k;i}^{*} = \{\mathbf{t}^{(-i)} : \max_{j\in\{M,M-1,\ldots,2\}\,:\, t^{(0)}_{\tilde{i}^{*}_{j-1}(S_i)}>\alpha_j} j = k\},$$

analogous to expression (7.92). Note also that subsequent to the redefinition of the sets $A_{1;i}^{*}, A_{2;i}^{*}, \ldots, A_{M;i}^{*}$, it is still the case that they are mutually disjoint and that $\bigcup_{k=1}^{M} A_{k;i}^{*} = \mathbb{R}^{M-1}$.

Take u, z_i, $\mathbf{z}^{(-i)}$, $f(\cdot)$, and $h(\cdot)$ to be as defined in the preceding part of the present subsection, and take $A_{k;i}^{*}$ to be as redefined by expression (7.105), (7.106), or (7.107). Then, analogous to result (7.99), we find that

$$\begin{aligned}
&\Pr\big(\mathbf{t}^{(-i)}\in A_{k;i}^{*}\mid t_i^{(0)}>\alpha_k\big)\\
&\quad= \Pr\big(u^{-1}\mathbf{z}^{(-i)}\in A_{k;i}^{*}\mid z_i>u\,\alpha_k-\mu_i\big)\\
&\quad= \int_0^\infty\int_{\underline{u}\alpha_k-\mu_i}^{\infty}\Pr\big(u^{-1}\mathbf{z}^{(-i)}\in A_{k;i}^{*}\mid z_i=\underline{z}_i, u=\underline{u}\big)\\
&\qquad\qquad\qquad\frac{f(\underline{z}_i)}{\Pr(z_i>\underline{u}\,\alpha_k-\mu_i)}\,d\underline{z}_i\, h(\underline{u})\,d\underline{u}.
\end{aligned} \tag{7.111}$$

Now, suppose that $\rho_{ij} = 0$ for all $i \in T$ and $j \in I$ such that $j \neq i$. Then, by proceeding in much the same fashion as in arriving at result (7.103), we find that in the special case where (for $j = 1, 2, \ldots, M$) α_j' [as redefined by expression (7.109)] is taken to be of the form (7.81) and hence where (for $j = 1, 2, \ldots, M$) $\alpha_j = \bar{t}_{j\dot{\gamma}/M}(N-P_*)$,

$$\text{FDR} \le (M_T/M)\,\dot{\gamma} \le \dot{\gamma}.$$

Thus, as in the case of the step-up procedure for testing the null hypotheses $H_i^{(0)} : \tau_i = \tau_i^{(0)}$ $(i = 1, 2, \ldots, M)$ when (for $i = 1, 2, \ldots, M$) $\alpha_j = \bar{t}_{j\delta/(2M)}(N-P_*)$, we find that if $\rho_{ij} = 0$ for $j \neq i = 1, 2, \ldots, M$, then in the special case where (for $j = 1, 2, \ldots, M$) $\alpha_j = \bar{t}_{j\delta/M}(N-P_*)$, the step-up procedure for testing the null hypotheses $H_i^{(0)} : \tau_i \le \tau_i^{(0)}$ $(i = 1, 2, \ldots, M)$ controls the FDR at level δ (in the sense that FDR $\le \delta$) and does so regardless of the unknown identity of the set T.

Nonindependence. Let us consider further the step-up procedures for testing the null hypotheses $H_i^{(0)} : \tau_i = \tau_i^{(0)}$ $(i = 1, 2, \ldots, M)$ and for testing the null hypotheses $H_i^{(0)} : \tau_i \le \tau_i^{(0)}$ $(i = 1, 2, \ldots, M)$. Suppose that the α_j's are those that result from taking (for $j = 1, 2, \ldots, M$) α_j' to be of the form $\alpha_j' = j\dot{\gamma}/M$ and from setting $\dot{\gamma} = \delta$. If $\rho_{ij} = 0$ for $j \neq i = 1, 2, \ldots, M$, then (as indicated in the preceding 2 parts of the present subsection) the step-up procedures control the FDR at level δ. To what extent does this property (i.e., control of the FDR at level δ) extend to cases where $\rho_{ij} \neq 0$ for some or all $j \neq i = 1, 2, \ldots, M$?

Suppose that $\tau_1, \tau_2, \ldots, \tau_M$ are linearly independent and that $\mathbf{\Sigma}$ is nonsingular (as would necessarily be the case if $\rho_{ij} = 0$ for $j \neq i = 1, 2, \ldots, M$). Then, it can be shown (and subsequently will be shown) that in the case of the step-up procedure for testing the null hypotheses $H_i^{(0)} : \tau_i \le \tau_i^{(0)}$ $(i = 1, 2, \ldots, M)$ with $\alpha_j' = j\dot{\gamma}/M$ or, equivalently, $\alpha_j = \bar{t}_{j\dot{\gamma}/M}(N-P_*)$ (for $j = 1, 2, \ldots, M$),

$$\rho_{ij} \ge 0 \text{ for all } i \in T \text{ and } j \in I \text{ such that } j \neq i \;\Rightarrow\; \text{FDR} \le (M_T/M)\,\dot{\gamma}. \tag{7.112}$$

Thus, when $\alpha_j = \bar{t}_{j\dot{\gamma}/M}(N-P_*)$ for $j = 1, 2, \ldots, M$ and when $\dot{\gamma} = \delta$, the step-up procedure for testing the null hypotheses $H_i^{(0)} : \tau_i \le \tau_i^{(0)}$ $(i = 1, 2, \ldots, M)$ controls the FDR at level δ (regardless of the unknown identity of T) provided that $\rho_{ij} \ge 0$ for $j \neq i = 1, 2, \ldots, M$.

Turning now to the case of the step-up procedure for testing the null hypotheses $H_i^{(0)}: \tau_i = \tau_i^{(0)}$ $(i = 1, 2, \ldots, M)$, let $\boldsymbol{\Sigma}_T$ represent the $M_T \times M_T$ submatrix of $\boldsymbol{\Sigma}$ obtained upon striking out the ith row and ith column for every $i \in F$, and suppose that (for $j = 1, 2, \ldots, M$) $\alpha_j' = j\dot{\gamma}/M$ or, equivalently, $\alpha_j = \bar{t}_{j\dot{\gamma}/(2M)}(N - P_*)$. Then, it can be shown (and will be shown) that in this case,

> the existence of an $M_T \times M_T$ diagonal matrix $\mathbf{D}_T$ with diagonal elements of ± 1 for which all of the off-diagonal elements of the matrix $-\mathbf{D}_T \boldsymbol{\Sigma}_T^{-1} \mathbf{D}_T$ are nonnegative, together with the condition $\rho_{ij} = 0$ for all $i \in T$ and $j \in F$

$$\Rightarrow FDR \le (M_T/M)\,\dot{\gamma}. \quad (7.113)$$

Relationship (7.113) serves to define (for every T) a collection of values of $\boldsymbol{\Sigma}$ for which FDR $\le (M_T/M)\,\dot{\gamma}$; this collection may include values of $\boldsymbol{\Sigma}$ in addition to those for which $\rho_{ij} = 0$ for all $i \in T$ and $j \in I$ such that $j \ne i$. Note, however, that when T contains only a single member, say the ith member, of the set $\{1, 2, \ldots, M\}$, this collection consists of those values of $\boldsymbol{\Sigma}$ for which $\rho_{ij} = 0$ for every $j \ne i$. Thus, relationship (7.113) does not provide a basis for adding to the collection of values of $\boldsymbol{\Sigma}$ for which the step-up procedure [for testing the null hypotheses $H_i^{(0)}: \tau_i = \tau_i^{(0)}$ $(i = 1, 2, \ldots, M)$ with (for $j = 1, 2, \ldots, M$) $\alpha_j = \bar{t}_{j\dot{\gamma}/(2M)}(N - P_*)$ and with $\dot{\gamma} = \delta$] controls the FDR at level δ (regardless of the unknown value of T).

In the case of the step-up procedure for testing the null hypotheses $H_i^{(0)}: \tau_i = \tau_i^{(0)}$ $(i = 1, 2, \ldots, M)$ or the null hypotheses $H_i^{(0)}: \tau_i \le \tau_i^{(0)}$ $(i = 1, 2, \ldots, M)$ with the α_j's chosen so that (for $j = 1, 2, \ldots, M$) α_j' is of the form $\alpha_j' = j\dot{\gamma}/M$, control of the FDR at level δ can be achieved by setting $\dot{\gamma} = \delta$ only when $\boldsymbol{\Sigma}$ satisfies certain relatively restrictive conditions. However, such control can be achieved regardless of the value of $\boldsymbol{\Sigma}$ by setting $\dot{\gamma}$ equal to a value that is sufficiently smaller than δ. Refer to Exercise 31 for some specifics.

Verification of results (7.112) and (7.113). Suppose [for purposes of verifying result (7.112)] that $\rho_{ij} \ge 0$ for all $i \in T$ and $j \in I$ such that $j \ne i$. And consider the function $g(\mathbf{z}^{(-i)}; k', \underline{u})$ of $\mathbf{z}^{(-i)}$ defined for each strictly positive scalar $\underline{u}$ and for each integer k' between 1 and M, inclusive, as follows:

$$g(\mathbf{z}^{(-i)}; k', \underline{u}) = \begin{cases} 1, & \text{if } \underline{u}\mathbf{z}^{(-i)} \in \bigcup_{k \le k'} A_{k;i}^*, \\ 0, & \text{otherwise,} \end{cases}$$

where $A_{k;i}^*$ is the set of $\mathbf{t}^{(-i)}$-values defined by expression (7.105), (7.106), or (7.107).

Clearly, the function $g(\cdot\,; k', \underline{u})$ is a nonincreasing function [in the sense that $g(\mathbf{z}_2^{(-i)}; k', \underline{u}) \le g(\mathbf{z}_1^{(-i)}; k', \underline{u})$ for any 2 values $\mathbf{z}_1^{(-i)}$ and $\mathbf{z}_2^{(-i)}$ of $\mathbf{z}^{(-i)}$ for which the elements of $\mathbf{z}_2^{(-i)}$ are greater than or equal to the corresponding elements of $\mathbf{z}_1^{(-i)}$]. And the distribution of $\mathbf{z}^{(-i)}$ conditional on $z_i = \underline{z}_i$ is MVN with

$$\mathrm{E}\big(\mathbf{z}^{(-i)} \mid z_i = \underline{z}_i\big) = \boldsymbol{\mu}^{(-i)} + \boldsymbol{\rho}^{(-i)}(\underline{z}_i - \mu_i) \quad \text{and} \quad \mathrm{var}\big(\mathbf{z}^{(-i)} \mid z_i = \underline{z}_i\big) = \boldsymbol{\Sigma}^{(-i)} - \boldsymbol{\rho}^{(-i)}\boldsymbol{\rho}^{(-i)\prime},$$

where $\boldsymbol{\mu}^{(-i)}$ is the $(M-1)$-dimensional subvector of the vector $\boldsymbol{\mu}$ and $\boldsymbol{\rho}^{(-i)}$ the $(M-1)$-dimensional subvector of the vector $(\rho_{1i}, \rho_{2i}, \ldots, \rho_{Mi})'$ obtained upon excluding the ith element and where $\boldsymbol{\Sigma}^{(-i)}$ is the $(M-1) \times (M-1)$ submatrix of $\boldsymbol{\Sigma}$ obtained upon excluding the ith row and ith column.

Now, let

$$q(\underline{z}_i; k', \underline{u}) = \Pr\big(u^{-1}\mathbf{z}^{(-i)} \in \textstyle\bigcup_{k \le k'} A_{k;i}^* \mid z_i = \underline{z}_i, u = \underline{u}\big).$$

And regard $q(\underline{z}_i; k', \underline{u})$ as a function of $\underline{z}_i$, and observe (in light of the statistical independence of u and $\mathbf{z}^{(0)}$) that

$$q(\underline{z}_i; k', \underline{u}) = \Pr\big(\underline{u}^{-1}\mathbf{z}^{(-i)} \in \textstyle\bigcup_{k \le k'} A_{k;i}^* \mid z_i = \underline{z}_i\big) = \mathrm{E}\big[g(\mathbf{z}^{(-i)}; k', \underline{u}) \mid z_i = \underline{z}_i\big].$$

Then, based on a property of the MVN distribution that is embodied in Theorem 5 of Müller (2001), it can be deduced that $q(\cdot\,; k', \underline{u})$ is a nonincreasing function. Moreover, for "any" nonincreasing function, say $q(\underline{z}_i)$, of $\underline{z}_i$ (and for $k' = 1, 2, \ldots, M-1$),

$$\int_{\underline{u}\alpha_{k'+1}-\mu_i}^{\infty} q(\underline{z}_i)\frac{f(\underline{z}_i)}{\Pr(z_i > \underline{u}\alpha_{k'+1}-\mu_i)}\,d\underline{z}_i \ge \int_{\underline{u}\alpha_{k'}-\mu_i}^{\infty} q(\underline{z}_i)\frac{f(\underline{z}_i)}{\Pr(z_i > \underline{u}\alpha_{k'}-\mu_i)}\,d\underline{z}_i, \tag{7.114}$$

as can be readily verified. Thus,

$$\begin{aligned}
&\int_{\underline{u}\alpha_{k'+1}-\mu_i}^{\infty} \Pr\big(u^{-1}\mathbf{z}^{(-i)} \in A^*_{k'+1;i} \mid z_i=\underline{z}_i, u=\underline{u}\big)\frac{f(\underline{z}_i)}{\Pr(z_i > \underline{u}\alpha_{k'+1}-\mu_i)}\,d\underline{z}_i \\
&\quad= \int_{\underline{u}\alpha_{k'+1}-\mu_i}^{\infty} \Pr\Big(u^{-1}\mathbf{z}^{(-i)} \in \bigcup_{k\le k'+1} A^*_{k;i} \mid z_i=\underline{z}_i, u=\underline{u}\Big)\frac{f(\underline{z}_i)}{\Pr(z_i > \underline{u}\alpha_{k'+1}-\mu_i)}\,d\underline{z}_i \\
&\qquad - \int_{\underline{u}\alpha_{k'+1}-\mu_i}^{\infty} \Pr\Big(u^{-1}\mathbf{z}^{(-i)} \in \bigcup_{k\le k'} A^*_{k;i} \mid z_i=\underline{z}_i, u=\underline{u}\Big)\frac{f(\underline{z}_i)}{\Pr(z_i > \underline{u}\alpha_{k'+1}-\mu_i)}\,d\underline{z}_i \\
&\quad\le \int_{\underline{u}\alpha_{k'+1}-\mu_i}^{\infty} \Pr\Big(u^{-1}\mathbf{z}^{(-i)} \in \bigcup_{k\le k'+1} A^*_{k;i} \mid z_i=\underline{z}_i, u=\underline{u}\Big)\frac{f(\underline{z}_i)}{\Pr(z_i > \underline{u}\alpha_{k'+1}-\mu_i)}\,d\underline{z}_i \\
&\qquad - \int_{\underline{u}\alpha_{k'}-\mu_i}^{\infty} \Pr\Big(u^{-1}\mathbf{z}^{(-i)} \in \bigcup_{k\le k'} A^*_{k;i} \mid z_i=\underline{z}_i, u=\underline{u}\Big)\frac{f(\underline{z}_i)}{\Pr(z_i > \underline{u}\alpha_{k'}-\mu_i)}\,d\underline{z}_i.
\end{aligned} \tag{7.115}$$

Finally, upon summing both "sides" of inequality (7.115) over k' (from 1 to $M-1$), we find that

$$\begin{aligned}
&\sum_{k=1}^{M} \int_{\underline{u}\alpha_k-\mu_i}^{\infty} \Pr\big(u^{-1}\mathbf{z}^{(-i)} \in A^*_{k;i} \mid z_i=\underline{z}_i, u=\underline{u}\big)\frac{f(\underline{z}_i)}{\Pr(z_i > \underline{u}\alpha_k-\mu_i)}\,d\underline{z}_i \\
&\quad= \int_{\underline{u}\alpha_1-\mu_i}^{\infty} \Pr\big(u^{-1}\mathbf{z}^{(-i)} \in A^*_{1;i} \mid z_i=\underline{z}_i, u=\underline{u}\big)\frac{f(\underline{z}_i)}{\Pr(z_i > \underline{u}\alpha_1-\mu_i)}\,d\underline{z}_i \\
&\qquad + \sum_{k'=1}^{M-1} \int_{\underline{u}\alpha_{k'+1}-\mu_i}^{\infty} \Pr\big(u^{-1}\mathbf{z}^{(-i)} \in A^*_{k'+1;i} \mid z_i=\underline{z}_i, u=\underline{u}\big)\frac{f(\underline{z}_i)}{\Pr(z_i > \underline{u}\alpha_{k'+1}-\mu_i)}\,d\underline{z}_i \\
&\quad\le \int_{\underline{u}\alpha_M-\mu_i}^{\infty} \Pr\Big(u^{-1}\mathbf{z}^{(-i)} \in \bigcup_{k\le M} A^*_{k;i} \mid z_i=\underline{z}_i, u=\underline{u}\Big)\frac{f(\underline{z}_i)}{\Pr(z_i > \underline{u}\alpha_M-\mu_i)}\,d\underline{z}_i \\
&\quad= \int_{\underline{u}\alpha_M-\mu_i}^{\infty} 1\,\frac{f(\underline{z}_i)}{\Pr(z_i > \underline{u}\alpha_M-\mu_i)}\,d\underline{z}_i = 1;
\end{aligned}$$

so that to complete the verification of result (7.112), it remains only to observe [in light of expressions (7.110) and (7.111)] that

$$\begin{aligned}
\text{FDR} &\le \textstyle\sum_{i\in T}\sum_{k=1}^{M} \dfrac{k\dot{\gamma}/M}{k}\Pr\big(\mathbf{t}^{(-i)} \in A^*_{k;i} \mid t_i^{(0)} > \alpha_k\big) \\
&= (\dot{\gamma}/M)\textstyle\sum_{i\in T}\sum_{k=1}^{M} \Pr\big(\mathbf{t}^{(-i)} \in A^*_{k;i} \mid t_i^{(0)} > \alpha_k\big) \\
&\le (\dot{\gamma}/M)\textstyle\sum_{i\in T}\int_0^\infty 1\,h(\underline{u})\,d\underline{u} \\
&= (M_T/M)\,\dot{\gamma}.
\end{aligned}$$

Turning to the verification of result (7.113), suppose that there exists an $M_T \times M_T$ diagonal matrix $\mathbf{D}_T$ with diagonal elements of ± 1 for which all of the off-diagonal elements of the matrix $-\mathbf{D}_T\boldsymbol{\Sigma}_T^{-1}\mathbf{D}_T$ are nonnegative, and suppose in addition that $\rho_{ij}=0$ for all $i \in T$ and $j \in F$. Then,

$$\begin{aligned}
&\Pr\big(\mathbf{t}^{(-i)} \in A^*_{k;i} \mid |t_i| > \alpha_k\big) \\
&\quad= \Pr\big(u^{-1}\mathbf{z}^{(-i)} \in A^*_{k;i} \mid |z_i| > u\alpha_k\big) \\
&\quad= \int_0^\infty \int_{\mathcal{R}^{M_F}} \int_{\underline{u}\alpha_k}^{\infty} \Pr\big(u^{-1}\mathbf{z}^{(-i)} \in A^*_{k;i} \mid |z_i|=\underline{z}_i, \mathbf{z}_F^{(0)}=\underline{\mathbf{z}}_F, u=\underline{u}\big) \\
&\qquad\qquad \frac{f^*(\underline{z}_i)}{\Pr(|z_i| > \underline{u}\alpha_k)}\,d\underline{z}_i\; p(\underline{\mathbf{z}}_F)\,d\underline{\mathbf{z}}_F\; h(\underline{u})\,d\underline{u},
\end{aligned}$$

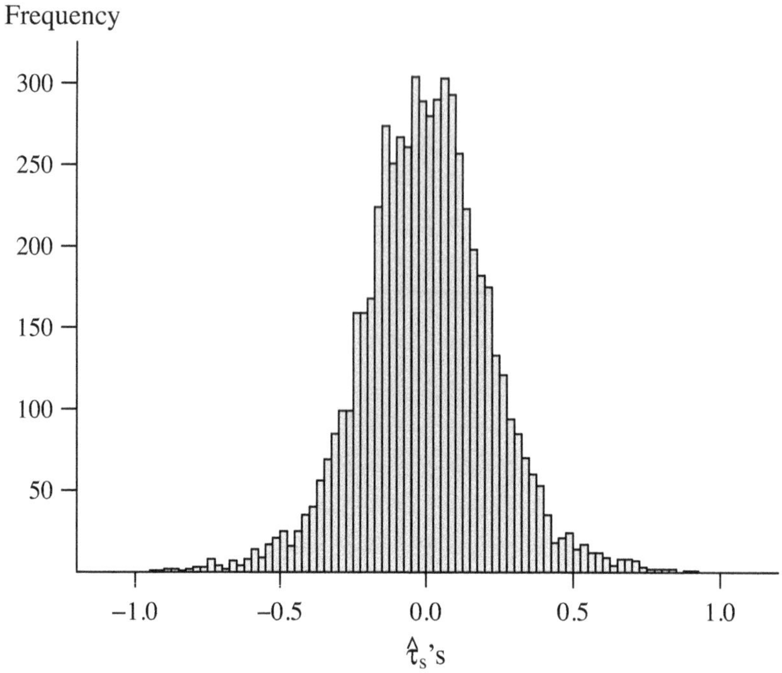

FIGURE 7.5. A display (in the form of a histogram with intervals of width 0.025) of the frequencies (among the 6033 genes) of the various values of the $\hat{\tau}_s$'s.

where $\mathbf{z}_F^{(0)}$ is the $M_F \times 1$ random vector whose elements are $z_j^{(0)}$ $(j \in F)$, where $p(\cdot)$ is the pdf of the distribution of $\mathbf{z}_F^{(0)}$, and where $f^*(\cdot) = 2f(\cdot)$ is the pdf of the distribution of the absolute value of a random variable that has an $N(0, 1)$ distribution. And whether or not $u^{-1}\mathbf{z}^{(-i)} \in A^*_{k;i}$ when $\mathbf{z}_F^{(0)} = \underline{\mathbf{z}}_F$ and $u = \underline{u}$ is determined by the absolute values $|z_j|$ $(j \in T, j \neq i)$ of the $M_T - 1$ random variables z_j $(j \in T, j \neq i)$. Moreover, it follows from the results of Karlin and Rinott (1980, theorem 4.1; 1981, theorem 3.1) that for "any" nonincreasing function $g[|z_j|\ (j \in T,\ j \neq i)]$ of the absolute values of z_j $(j \in T,\ j \neq i)$, the conditional expected value of $g[|z_j|\ (j \in T,\ j \neq i)]$ given that $|z_i| = \underline{z}_i$ is a nonincreasing function of $\underline{z}_i$. Accordingly, result (7.113) can be verified by proceeding in much the same way as in the verification of result (7.112).

f. An illustration

Let us use the example from Part 1 of Subsection c to illustrate various of the alternative multiple-comparison procedures. In that example, the data consist of the expression levels obtained for 6033 genes on 102 men, 50 of whom were normal (control) subjects and 52 of whom were prostate cancer patients. And the objective was presumed to be that of testing each of the 6033 null hypotheses $H_s^{(0)} : \tau_s = 0$ $(s = 1, 2, \ldots, 6033)$ versus the corresponding one of the alternative hypotheses $H_s^{(1)} : \tau_s \neq 0$ $(s = 1, 2, \ldots, 6033)$, where $\tau_s = \mu_{s2} - \mu_{s1}$ represents the expected difference (between the cancer patients and the normal subjects) in the expression level of the sth gene.

Assume that the subjects have been numbered in such a way that the first through 50th subjects are the normal (control) subjects and the 51st through 102nd subjects are the cancer patients. And for $s = 1, 2, \ldots, 6033$ and $j = 1, 2, \ldots, 102$, denote by y_{sj} the random variable whose value is the value obtained for the expression level of the sth gene on the jth subject. Then, the least squares estimator of τ_s is $\hat{\tau}_s = \hat{\mu}_{s2} - \hat{\mu}_{s1}$, where $\hat{\mu}_{s1} = (1/50)\sum_{j=1}^{50} y_{sj}$ and $\hat{\mu}_{s2} = (1/52)\sum_{j=51}^{102} y_{sj}$. The values of $\hat{\tau}_1, \hat{\tau}_2, \ldots, \hat{\tau}_{6033}$ are displayed (in the form of a histogram) in Figure 7.5.

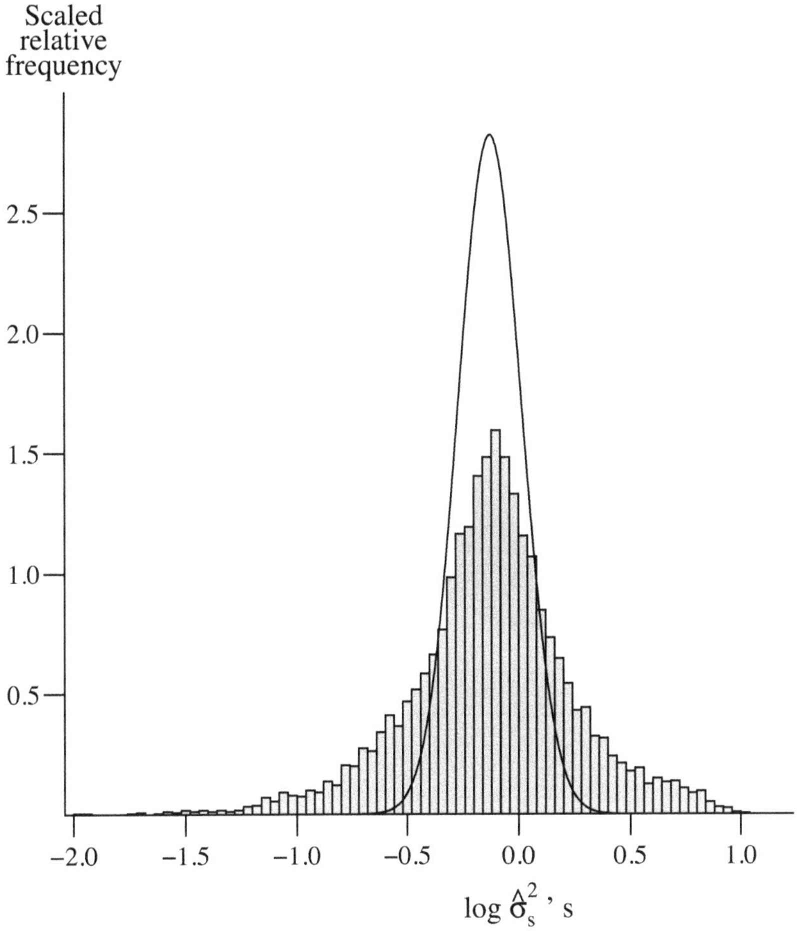

FIGURE 7.6. A display (in the form of a histogram with intervals of width 0.04 that has been rescaled so that it encloses an area equal to 1 and that has been overlaid with a plot of the pdf of the $N\{-0.127,\ 1/50\}$ distribution) of the relative frequencies (among the 6033 genes) of the various values of the $\log \hat{\sigma}_s^2$'s.

Assume that $\sigma_{ss'} = 0$ for $s' \neq s = 1, 2, \ldots, S$, as would be the case if the results obtained for each of the S genes are "unrelated" to those obtained for each of the others—this assumption is consistent with Efron's (2010) assumptions about these data. Further, for $s = 1, 2, \ldots, S$, write σ_s^2 for σ_{ss}; and take $\hat{\sigma}_s^2$ to be the unbiased estimator of σ_s^2 defined as

$$\hat{\sigma}_s^2 = \left[\textstyle\sum_{j=1}^{N_1} (y_{sj} - \hat{\mu}_{s1})^2 + \sum_{j=N_1+1}^{N_1+N_2} (y_{sj} - \hat{\mu}_{s2})^2\right] / (N_1 + N_2 - 2),$$

where $N_1 = 50$ and $N_2 = 52$.

The results of Lehmann (1986, sec. 7.3), which are asymptotic in nature, suggest that (at least in the case where the joint distribution of $y_{s1}, y_{s2}, \ldots, y_{s,102}$ is MVN) the distribution of $\log \hat{\sigma}_s^2$ can be approximated by the $N\{\log \sigma_s^2,\ 1/50\ [= 2/(N_1 + N_2 - 2)]\}$ distribution. The values of $\log \hat{\sigma}_1^2, \log \hat{\sigma}_2^2, \ldots, \log \hat{\sigma}_{6033}^2$ are displayed in Figure 7.6 in the form of a histogram that has been rescaled so that it encloses an area equal to 1 and that has been overlaid with a plot of the pdf of the $N\{-0.127\ [= (1/6033) \sum_{s=1}^{6033} \log \hat{\sigma}_s^2],\ 1/50\}$ distribution. As is readily apparent from Figure 7.6, it would be highly unrealistic to assume that $\sigma_1^2 = \sigma_2^2 = \cdots \sigma_S^2$, that is, to assume that the variability of the expression levels (from one normal subject to another or one cancer patient to another) is the same for all S genes. The inappropriateness of any such assumption is reflected in the results obtained upon applying various of the many procedures proposed for testing for the homogeneity of

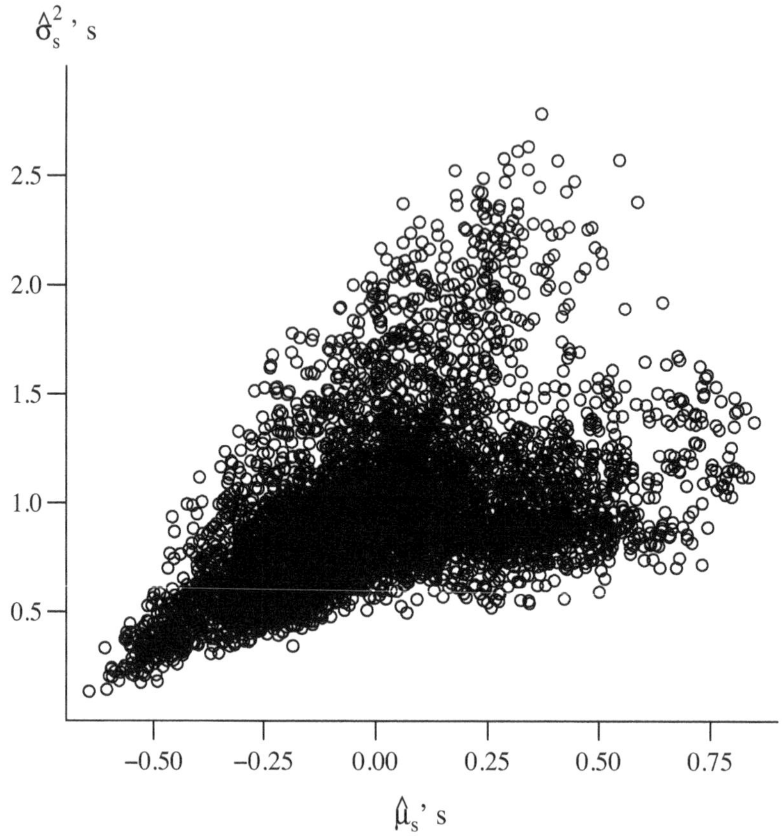

FIGURE 7.7. Plot of the values of the $\hat{\sigma}_s^2$'s against the values of the corresponding $\hat{\mu}_s$'s.

variances, including that proposed by Hartley (1950) as well as that proposed by Lehmann (1986, sec. 7.3).

Not only does the variance σ_s^2 (among the normal subjects or the cancer patients) of the expression levels of the sth gene appear to depend on s, but there appears to be a strong tendency for σ_s^2 to increase with the mean (μ_{s1} in the case of the normal subjects and μ_{s2} in the case of the cancer patients). Let $\hat{\mu}_s = (\hat{\mu}_{s1} + \hat{\mu}_{s2})/2$. Then, the tendency for σ_s^2 to increase with μ_{s1} or μ_{s2} is clearly evident in Figure 7.7, in which the values of the $\hat{\sigma}_s^2$'s are plotted against the values of the corresponding $\hat{\mu}_s$'s.

In Figure 7.8, the values of the $\hat{\sigma}_s^2$'s are plotted against the values of the corresponding $\hat{\tau}_s$'s. This figure suggests that while σ_s^2 may vary to a rather considerable extent with μ_{s1} and μ_{s2} individually and with their average (and while small values of σ_s^2 may be somewhat more likely when $|\tau_s|$ is small), any tendency for σ_s^2 to vary with τ_s is relatively inconsequential.

Consider (in the context of the present application) the quantities $t_1, t_2, \ldots, t_M$ and $t_1^{(0)}, t_2^{(0)}, \ldots, t_M^{(0)}$ defined by equalities (7.3)—in the present application, $M = S$. The various multiple-comparison procedures described and discussed in Subsections a, b, d, and e depend on the data only through the (absolute) values of $t_1^{(0)}, t_2^{(0)}, \ldots, t_M^{(0)}$. And the justification for those procedures is based on their ability to satisfy criteria defined in terms of the distribution of the random vector $\mathbf{t} = (t_1, t_2, \ldots, t_M)'$. Moreover, the vector $\mathbf{t}$ is expressible in the form (7.5), so that its distribution is determined by the distribution of the random vector $\hat{\sigma}^{-1}(\hat{\alpha} - \alpha)$. When the observable random vector $\mathbf{y}$ follows the G–M model and when in addition the distribution of the vector $\mathbf{e}$ of residual effects is MVN (or, more generally, is any absolutely continuous spherical distribution), the distribution of $\hat{\sigma}^{-1}(\hat{\alpha} - \alpha)$ is $MVt(N - P_*, \mathbf{I})$—refer to result (7.6).

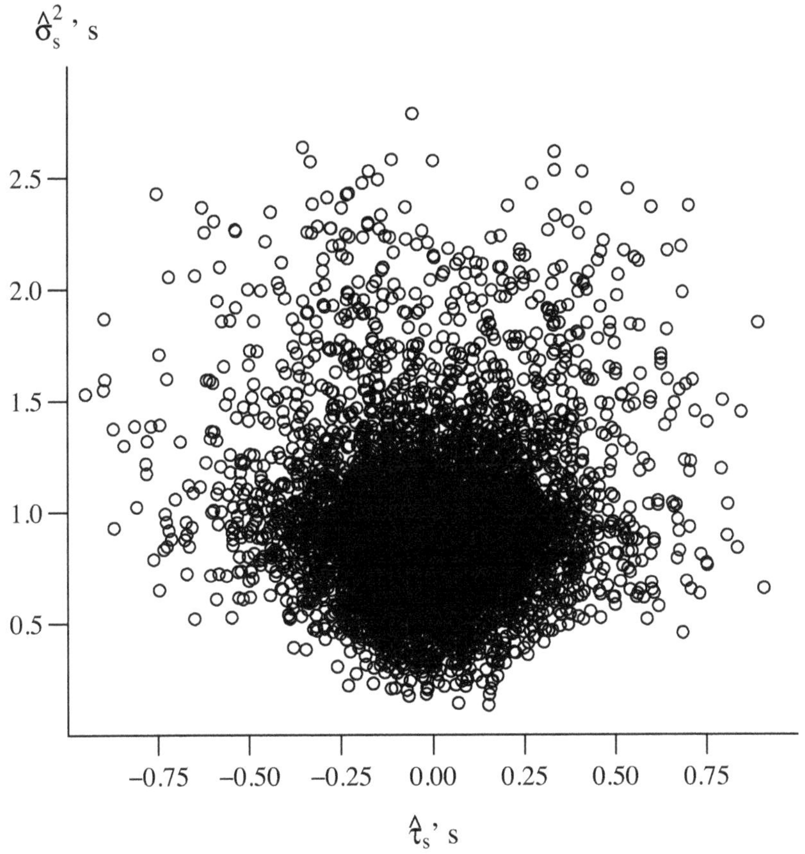

FIGURE 7.8. Plot of the values of the $\hat{\sigma}_s^2$'s against the values of the corresponding $\hat{\tau}_s$'s.

In the present application, the assumption (inherent in the G–M model) of the homogeneity of the variances of the residual effects appears to be highly unrealistic, and consequently the distribution of the vector $\hat{\sigma}^{-1}(\hat{\alpha}-\alpha)$ may differ appreciably from the $MVt(N-P_*,\, \mathbf{I})$ distribution. Allowance can be made for the heterogeneity of the variances of the residual effects by redefining the t_i's and the $t_1^{(0)}$'s (and by modifying the various multiple-comparison procedures accordingly).

For $s = 1, 2, \ldots, S$, redefine t_s and $t_s^{(0)}$ as

$$t_s = \frac{\hat{\tau}_s - \tau_s}{[(1/50) + (1/52)]^{1/2}\,\hat{\sigma}_s} \quad \text{and} \quad t_s^{(0)} = \frac{\hat{\tau}_s}{[(1/50) + (1/52)]^{1/2}\,\hat{\sigma}_s} \tag{7.116}$$

(where $\hat{\sigma}_s$ represents the positive square root of $\hat{\sigma}_s^2$). Further, let $\mathbf{y}_{s1} = (y_{s1}, y_{s2}, \ldots, y_{s,50})'$ and $\mathbf{y}_{s2} = (y_{s,51}, y_{s,52}, \ldots, y_{s,102})'$, take $\mathbf{L}_1$ to be any 50×49 matrix whose columns form an orthonormal basis for $\mathfrak{N}(\mathbf{1}_{50}')$ and $\mathbf{L}_2$ to be any 52×51 matrix whose columns form an orthonormal basis for $\mathfrak{N}(\mathbf{1}_{52}')$, and assume that the 6033 101-dimensional vectors

$$\begin{pmatrix} (\hat{\tau}_s - \tau_s)/[(1/50)+(1/52)]^{1/2} \\ \mathbf{L}_1'\mathbf{y}_{s1} \\ \mathbf{L}_2'\mathbf{y}_{s2} \end{pmatrix} \qquad (s = 1, 2, \ldots, 6033)$$

are distributed independently and that each of them has an $N(\mathbf{0},\, \sigma_s^2 \mathbf{I}_{101})$ distribution or, more generally, has an absolutely continuous spherical distribution with variance-covariance matrix $\sigma_s^2 \mathbf{I}_{101}$. And observe that under that assumption, the random variables $t_1, t_2, \ldots, t_{6033}$ are statistically independent and each of them has an $St[100\,(= N_1 + N_2 - 2)]$ distribution—refer to the final part of Section 6.6.4d.

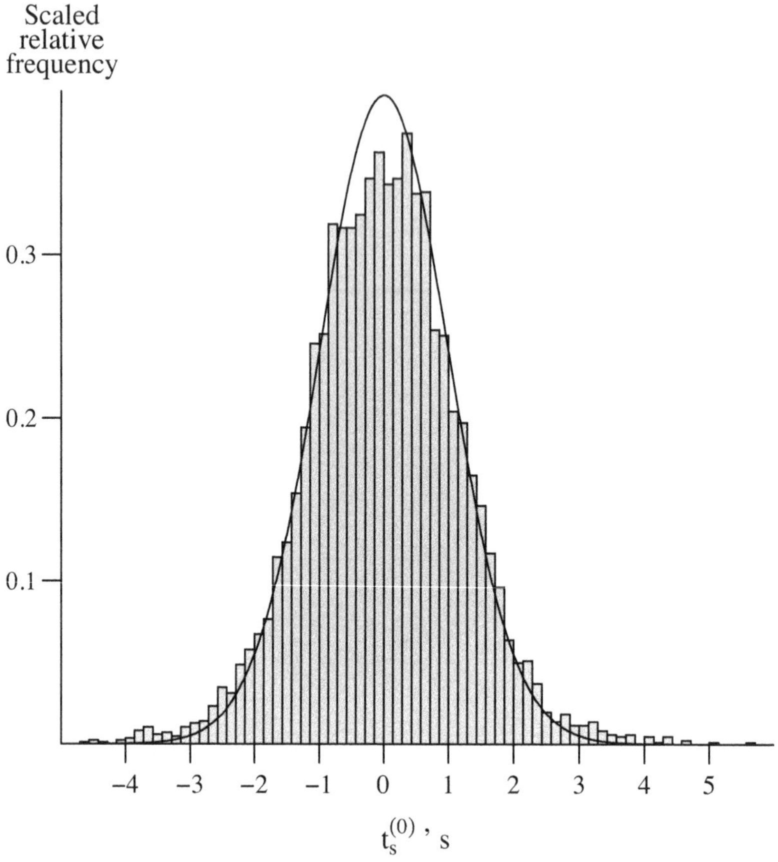

FIGURE 7.9. A display (in the form of a histogram with intervals of width 0.143 that has been rescaled so that it encloses an area equal to 1 and that has been overlaid with a plot of the pdf of the $St(100)$ distribution) of the relative frequencies (among the 6033 genes) of the various values of the $t_s^{(0)}$'s.

In what follows, modified versions of the various multiple-comparison procedures [in which t_s and $t_s^{(0)}$ have (for $s = 1, 2, \ldots, 6033$) been redefined by equalities (7.116)] are described and applied. It is worth noting that the resultant procedures do not take advantage of any "relationships" among the μ_{s1}'s, μ_{s2}'s and σ_s^2's of the kind reflected in Figures 7.5, 7.6, 7.7, and (to a lesser extent) 7.8. At least in principle, a more sophisticated model that reflects those relationships could be devised and could serve as a basis for constructing improved procedures—refer, e.g., to Efron (2010). And/or one could seek to transform the data in such a way that the assumptions underlying the various unmodified multiple-comparison procedures (including that of the homogeneity of the residual variances) are applicable—refer, e.g., to Durbin et al. (2002).

The values of the redefined $t_s^{(0)}$'s are displayed in Figure 7.9 in the form of a histogram that has been rescaled so that it encloses an area equal to 1 and that has been overlaid with a plot of the pdf of the $St(100)$ distribution. The genes with the most extreme $t^{(0)}$-values are listed in Table 7.4.

FWER and k-FWER. Let us consider the modification of the multiple-comparison procedures described in Subsections a and b for controlling the FWER or k-FWER. Among the quantities affected by the redefinition of the t_i's (either directly or indirectly) are the following: the permutations $i_1, i_2, \ldots, i_M$ and [for any (nonempty) subset S of I] $i_1^*(S), i_2^*(S), \ldots, i_{M_S}^*(S)$; the quantities $t_{(j)} = t_{i_j}$ $(j = 1, 2, \ldots, M)$ and $t_{j;S} = t_{i_j^*(S)}$ $(j = 1, 2, \ldots, M_S)$; the upper $100\dot{\gamma}\%$ point $c_{\dot{\gamma}}(j)$ of the distribution of $|t_{(j)}|$ and the upper $100\dot{\gamma}\%$ point $c_{\dot{\gamma}}(j; S)$ of the distribution of $|t_{j;S}|$; and (for

TABLE 7.4. The 100 most extreme among the values of the $t_s^{(0)}$'s obtained for the 6033 genes represented in the prostate data.

s	$t_s^{(0)}$	s	$t_s^{(0)}$	s	$t_s^{(0)}$	s	$t_s^{(0)}$	s	$t_s^{(0)}$
610	5.65	4073	−3.98	2945	−3.65	4515	−3.35	2811	3.19
1720	5.11	735	−3.87	2856	−3.64	637	3.34	478	−3.19
364	−4.67	3665	−3.84	3017	−3.61	4496	−3.34	1507	3.18
332	4.64	1130	3.83	698	−3.59	298	−3.32	3313	−3.18
914	4.61	1346	−3.82	3292	−3.59	292	3.31	3585	−3.16
3940	−4.57	1589	−3.81	905	3.59	1659	3.31	2852	3.16
4546	−4.54	921	−3.80	4396	−3.57	1491	3.30	3242	3.14
1068	4.40	4549	3.79	4552	3.56	718	−3.30	5159	−3.13
579	4.35	739	−3.77	3930	3.56	3208	3.29	4378	3.12
4331	−4.34	4104	−3.75	1588	3.55	1966	3.29	1329	3.12
1089	4.31	4981	3.75	721	3.54	452	3.28	4500	−3.07
3647	4.31	1314	3.75	3260	−3.49	3879	3.27	4671	3.06
1113	4.22	702	−3.74	4154	−3.48	3200	3.27	354	−3.06
1077	4.13	2897	3.72	805	−3.46	1647	3.26	341	3.05
4518	4.10	4000	−3.71	4040	−3.44	2968	3.25	995	−3.04
1557	4.10	2	3.68	11	3.44	4013	3.22	3961	−3.04
4088	−4.06	3282	3.66	3505	−3.44	2912	3.20	3696	−3.03
3991	−4.05	694	−3.66	3269	−3.42	684	3.20	1097	3.03
3375	4.01	2370	3.66	4492	−3.40	1572	3.19	3343	−3.03
4316	−3.99	3600	−3.65	377	3.38	913	−3.19	641	3.03

$j = k, k+1, \ldots, M$) the quantity

$$\alpha_j = \max_{S \in \tilde{\Omega}^+_{k;j}} c_{\dot{\gamma}}(k; S) \tag{7.117}$$

(where $\tilde{\Omega}^+_{k;j}$ is the collection of subsets of I whose elements include $k-1$ of the integers $\tilde{i}_1, \tilde{i}_2, \ldots, \tilde{i}_{j-1}$ and all $M-j+1$ of the integers $\tilde{i}_j, \tilde{i}_{j+1}, \ldots, \tilde{i}_M$ and where $\tilde{i}_1, \tilde{i}_2, \ldots, \tilde{i}_M$ is a permutation of the integers $1, 2, \ldots, M$ such that $|t^{(0)}_{\tilde{i}_1}| \geq |t^{(0)}_{\tilde{i}_2}| \geq \cdots \geq |t^{(0)}_{\tilde{i}_M}|$).

In regard to expression (7.117), note that subsequent to the redefinition of the t_i's, $c_{\dot{\gamma}}(k; S)$ depends on the subset S only through the size M_S of S. Moreover, for $S \in \tilde{\Omega}^+_{k;j}$,

$$M_S = M - j + 1 + k - 1 = M + k - j.$$

Thus, expression (7.117) simplifies to the following expression:

$$\alpha_j = c_{\dot{\gamma}}(k; S) \text{ for any } S \subset I \text{ such that } M_S = M + k - j. \tag{7.118}$$

And in lieu of inequalities (7.14) and (7.43), we have (subsequent to the redefinition of the t_i's) the inequalities

$$c_{\dot{\gamma}}(k) \leq \bar{t}_{k\dot{\gamma}/(2M)}(100) \tag{7.119}$$

and (for $j \geq k$)

$$\alpha_j \leq \bar{t}_{k\dot{\gamma}/[2(M+k-j)]}(100). \tag{7.120}$$

Based on these results, we can obtain suitably modified versions of the multiple-comparison procedures described in Subsections a and b for controlling the FWER or k-FWER. As a multiple-comparison procedure (for testing $H_1^{(0)}, H_2^{(0)}, \ldots, H_M^{(0)}$) that controls the k-FWER at level $\dot{\gamma}$, we have the procedure that rejects $H_i^{(0)}$ if and only if $\mathbf{y} \in C_i$, where

$$C_i = \{\mathbf{y} : |t_i^{(0)}| > c_{\dot{\gamma}}(k)\} \tag{7.121}$$

TABLE 7.5. The number (No.) of rejected null hypotheses (discoveries) obtained (for $k = 1, 2, \ldots, 20$ and for 3 values of $\dot{\gamma}$) upon applying the multiple-comparison procedure for controlling the k-FWER and the number obtained upon applying its more conservative counterpart, along with the values of $c_{\dot{\gamma}}(k)$ and $\bar{t} = \bar{t}_{k\dot{\gamma}/(2M)}(100)$ and along with the number (if any) of additional rejections (discoveries) obtained upon applying the step-down versions of those procedures.

	$\dot{\gamma} = .05$				$\dot{\gamma} = .10$				$\dot{\gamma} = .20$			
k	$c_{\dot{\gamma}}(k)$	No.	$\bar{t}$	No.	$c_{\dot{\gamma}}(k)$	No.	$\bar{t}$	No.	$c_{\dot{\gamma}}(k)$	No.	$\bar{t}$	No.
1	4.70	2	4.70	2	4.52	7	4.53	7	4.32	10	4.35	9
2	4.20	13	4.53	7	4.09	16	4.35	9	3.97	21	4.16	13
3	3.97	21	4.42	7	3.89	21	4.24	12	3.79	28	4.05	17
4	3.83	23+1	4.35	9	3.76	29	4.16	13	3.68	35+1	3.98	21
5	3.72	33	4.29	12	3.66	36	4.10	14	3.59	43+1	3.91	21
6	3.64	42	4.24	12	3.59	46	4.05	17	3.52	51	3.86	22
7	3.57	46+1	4.20	13	3.52	51	4.01	19	3.46	53	3.82	24+1
8	3.52	51	4.16	13	3.47	53	3.98	21	3.41	58	3.78	28
9	3.47	53	4.13	13	3.42	58	3.94	21	3.37	60	3.75	31
10	3.42	58	4.10	14	3.38	60	3.91	21	3.33	63	3.72	33+1
11	3.38	59+1	4.08	16	3.34	61+1	3.89	21	3.30	67+1	3.69	35
12	3.35	61	4.05	17	3.31	66	3.86	22	3.27	73+1	3.67	36
13	3.32	64	4.03	18	3.28	70+1	3.84	23	3.24	75	3.64	42
14	3.29	70	4.01	19	3.25	74+1	3.82	24+1	3.21	76	3.62	42
15	3.26	74	3.99	19	3.22	75+1	3.80	26	3.18	83+1	3.60	43
16	3.23	75	3.98	21	3.20	78	3.78	28	3.16	84+2	3.58	46
17	3.21	76+1	3.96	21	3.18	84	3.76	29	3.14	86+1	3.56	48
18	3.18	82+2	3.94	21	3.15	86	3.75	31	3.12	89+1	3.55	50
19	3.16	84	3.93	21	3.13	87+1	3.73	33	3.10	90	3.53	51
20	3.14	86	3.91	21	3.11	90	3.72	33+1	3.08	90	3.51	51

(and where the definition of $c_{\dot{\gamma}}(k)$ is in terms of the distribution of the redefined t_i's). And as a less computationally intensive but more "conservative" variation on this procedure, we have the procedure obtained by replacing $c_{\dot{\gamma}}(k)$ with the upper bound $\bar{t}_{k\dot{\gamma}/(2M)}(100)$, that is, the procedure obtained by taking

$$C_i = \{\mathbf{y} : |t_i^{(0)}| > \bar{t}_{k\dot{\gamma}/(2M)}(100)\} \tag{7.122}$$

rather than taking C_i to be the set (7.121). Further, suitably modified versions of the step-down procedures (described in Subsection b) for controlling the k-FWER at level $\dot{\gamma}$ are obtained by setting $\alpha_1 = \alpha_2 = \cdots = \alpha_k$ and (for $j = k, k+1, \ldots, M$) taking α_j to be as in equality (7.118) [where the definition of $c_{\dot{\gamma}}(k; S)$ is in terms of the redefined t_i's] and taking the replacement for α_j (in the more conservative of the step-down procedures) to be $\bar{t}_{k\dot{\gamma}/[2(M+k-j)]}(100)$. As before, the computation of $c_{\dot{\gamma}}(k)$ and $c_{\dot{\gamma}}(k; S)$ is amenable to the use of Monte Carlo methods.

The modified versions of the various procedures for controlling the k-FWER were applied to the prostate data. Results were obtained for $k = 1, 2, \ldots, 20$ and for three different choices for $\dot{\gamma}$ ($\dot{\gamma} = .05, .10$, and $.20$). The requisite values of $c_{\dot{\gamma}}(k)$ and $c_{\dot{\gamma}}(k; S)$ were determined by Monte Carlo methods from 149999 draws (from the joint distribution of the redefined t_i's). The number of rejected null hypotheses (discoveries), along with the values of $c_{\dot{\gamma}}(k)$ and $\bar{t}_{k\dot{\gamma}/(2M)}(100)$, is listed (for each of the various combinations of k- and $\dot{\gamma}$-values) in Table 7.5.

The results clearly indicate that (at least in this kind of application and at least for $k > 1$) the adoption of the more conservative versions of the procedures for controlling the k-FWER can result in a drastic reduction in the total number of rejections (discoveries). Also, while the total number of rejections (discoveries) increases with the value of k, it appears to do so at a mostly decreasing rate.

Controlling the probability of the FDP exceeding a specified level. Consider the modification of the step-down procedure described in Subsection d for controlling Pr(FDP $> \kappa$) (where κ is a

TABLE 7.6. The number of rejected null hypotheses (discoveries) obtained (for each of 4 values of κ and each of 3 values of ϵ) upon applying the step-down multiple-comparison procedure for controlling $\Pr(\text{FDP} > \kappa)$ at level ϵ (along with the value of $\dot{\gamma}_\epsilon$) when (for $j = 1, 2, \ldots, M$) α_j is taken to be of the form (7.123) and when $\dot{\gamma}$ is set equal to $\dot{\gamma}_\epsilon$ and the value of $\dot{\gamma}_\epsilon$ taken to be an approximate value determined by Monte Carlo methods.

	$\epsilon = .05$		$\epsilon = .10$		$\epsilon = .20$	
κ	$\dot{\gamma}_\epsilon$	Number of rejections	$\dot{\gamma}_\epsilon$	Number of rejections	$\dot{\gamma}_\epsilon$	Number of rejections
.05	.04	2	.05	2	.05	2
.10	.05	2	.09	6	.10	7
.20	.05	2	.10	8	.19	21
.50	.05	13	.10	22	.20	60

specified constant) at a specified level ϵ. And for purposes of doing so, continue to take t_s and $t_s^{(0)}$ (for $s = 1, 2, \ldots, S$) to be as redefined by equalities (7.116). By employing essentially the same reasoning as before (i.e., in Subsection d), it can be shown that the step-down procedure controls $\Pr(\text{FDP} > \kappa)$ at level ϵ if the α_j's satisfy condition (7.62), where now the random variable $t_{r;S}$ is (for any $S \subset I$ including $S = T$) as follows its modification to reflect the redefinition of the t_s's. Further, the same line of reasoning that led before to taking α_j to be of the form (7.65) leads now to taking α_j to be of the form

$$\alpha_j = \bar{t}_{([j\kappa]+1)\dot{\gamma}/\{2(M+[j\kappa]+1-j)\}}(100) \tag{7.123}$$

($j = 1, 2, \ldots, M$), following which the problem of choosing the α_j's so as to satisfy condition (7.62) is reduced to that of choosing $\dot{\gamma}$. And to insure that α_j's of the form (7.123) satisfy condition (7.62) and hence that the step-down procedure controls $\Pr(\text{FDP} > \kappa)$ at level ϵ, it suffices to take $\dot{\gamma} = \dot{\gamma}_\epsilon$, where $\dot{\gamma}_\epsilon$ is the solution to the equation $\dot{c}_\epsilon(\dot{\gamma}) = 0$ and where $\dot{c}_\epsilon(\cdot)$ is a modified (to account for the redefinition of the t_s's) version of the function $\dot{c}_\epsilon(\cdot)$ defined in Subsection d.

The step-down procedure [in which (for $j = 1, 2, \ldots, M$) α_j was taken to be of the form (7.123) and $\dot{\gamma}$ was set equal to $\dot{\gamma}_\epsilon$] was applied to the prostate data. Results were obtained for four different values of κ ($\kappa = .05, .10, .20,$ and $.50$) and three different values of ϵ ($\epsilon = .05, .10,$ and $.20$). The value of $\dot{\gamma}_\epsilon$ was taken to be that of the approximation provided by the solution to the equation $\ddot{c}_\epsilon(\dot{\gamma}) = 0$, where $\ddot{c}_\epsilon(\cdot)$ is a function whose values are approximations to the values of $\dot{c}_\epsilon(\cdot)$ determined by Monte Carlo methods from 149999 draws. For each of the various combinations of κ- and ϵ-values, the number of rejected null hypotheses (discoveries) is reported in Table 7.6 along with the value of $\dot{\gamma}_\epsilon$. It is worth noting that the potential improvement that comes from resetting various of the α_j's [so as to achieve conformance with condition (7.74)] did not result in any additional rejections in any of these cases.

An improvement. In the present setting [where (for $s = 1, 2, \ldots, S$) t_s and $t_s^{(0)}$ have been redefined as in equalities (7.116)], the t_s's are statistically independent and the $t_s^{(0)}$'s are statistically independent. By taking advantage of the statistical independence, we can attempt to devise improved versions of the step-down procedure for controlling $\Pr(\text{FDP} > \kappa)$. In that regard, recall (from Subsection d) that

$$\Pr(j' \geq 1,\ k' \leq M_T,\ \text{and}\ |t_{k';T}| > \alpha_{j'}) \leq \epsilon \quad \Rightarrow \quad \Pr(\text{FDP} > \kappa) \leq \epsilon \tag{7.124}$$

and that $k' = [j'\kappa] + 1$ for values of $j' \geq 1$. And observe that

$$\begin{aligned} &\Pr(j' \geq 1,\ k' \leq M_T,\ \text{and}\ |t_{k';T}| > \alpha_{j'}) \\ &\quad = \Pr(j' \geq 1 \text{ and } k' \leq M_T)\Pr(|t_{k';T}| > \alpha_{j'} \mid j' \geq 1 \text{ and } k' \leq M_T) \\ &\quad \leq \Pr(|t_{k';T}| > \alpha_{j'} \mid j' \geq 1 \text{ and } k' \leq M_T) \\ &\quad = \sum \Pr(|t_{k';T}| > \alpha_{j'} \mid j' = \underline{j}')\Pr(j' = \underline{j}' \mid j' \geq 1 \text{ and } k' \leq M_T), \end{aligned} \tag{7.125}$$

TABLE 7.7. The number of rejected null hypotheses (discoveries) obtained (for each of 4 values of κ and each of 3 values of ϵ) upon applying the step-down multiple-comparison procedure for controlling $\Pr(\text{FDP} > \kappa)$ at level ϵ when (for $j = 1, 2, \ldots, M$) α_j is as specified by equality (7.129) or, alternatively, as specified by equality (7.130).

	$\alpha_j = c_\epsilon(k_j, S)$ where $M_S = M + k_j - j$			$\alpha_j = \bar{t}_{k_j\epsilon/[2(M+k_j-j)]}(100)$		
κ	$\epsilon = .05$	$\epsilon = .10$	$\epsilon = .20$	$\epsilon = .05$	$\epsilon = .10$	$\epsilon = .20$
.05	2	7	10	2	7	9
.10	2	7	28	2	7	13
.20	2	78	90	2	12	21
.50	230	248	262	13	22	60

where the summation is with respect to $\underline{j}'$ and is over the set $\{\underline{j}' : \underline{j}' \geq 1 \text{ and } [\underline{j}'\kappa] + 1 \leq M_T\}$.

For $s \in T$, $t_s = t_s^{(0)}$; and (by definition) the values of the random variables j' and k' are completely determined by the values of the M_F random variables $t_s^{(0)}$ $(s \in F)$. Thus, in the present setting (where the $t_s^{(0)}$'s are statistically independent) we find that (for $\underline{j}'$ such that $\underline{j}' \geq 1$ and $[\underline{j}'\kappa] + 1 \leq M_T$)

$$\Pr(|t_{k';T}| > \alpha_{j'} \mid j' = \underline{j}') = \Pr(|t_{\underline{k}';T}| > \alpha_{\underline{j}'}), \tag{7.126}$$

where $\underline{k}' = [\underline{j}'\kappa] + 1$. Moreover,

$$\Pr(|t_{\underline{k}';T}| > c_\epsilon(\underline{k}'; S) \leq \epsilon \tag{7.127}$$

for any $S \subset I$ such that $T \subset S$ and hence (when as in the present setting, the t_s's are statistically independent) for any $S \subset I$ such that $M_S \geq M_T$.

Now, recall [from result (7.53)] that

$$M + k' - j' \geq M_T. \tag{7.128}$$

And observe that, together, results (7.125), (7.126), (7.127), and (7.128) imply that the condition

$$\Pr(j' \geq 1,\ k' \leq M_T, \text{ and } |t_{k';T}| > \alpha_{j'}) \leq \epsilon$$

and hence [in light of result (7.124)] the condition $\Pr(\text{FDP} > \kappa) \leq \epsilon$ can be satisfied by taking (for $j = 1, 2, \ldots, M$)

$$\alpha_j = c_\epsilon(k_j; S) \text{ for any } S \subset I \text{ such that } M_S = M + k_j - j, \tag{7.129}$$

where $k_j = [j\kappa] + 1$.

Resort to Monte Carlo methods may be needed to effect the numerical evaluation of expression (7.129). A more conservative but less computationally intensive version of the step-down procedure for controlling $\Pr(\text{FDP} > \kappa)$ at level ϵ can be effected [on the basis of inequality (7.120)] by taking (for $j = 1, 2, \ldots, M$)

$$\alpha_j = \bar{t}_{k_j\epsilon/[2(M+k_j-j)]}(100). \tag{7.130}$$

The number of rejections (discoveries) resulting from the application of the step-down procedure (to the prostate data) was determined for the case where (for $j = 1, 2, \ldots, M$) α_j is as specified by equality (7.129) and also for the case where α_j is as specified by equality (7.130). The values of the $c_\epsilon(k_j; S)$'s (required in the first of the 2 cases) were taken to be approximate values determined by Monte Carlo methods from 149999 draws. The number of rejections was determined for each of four values of κ ($\kappa = .05$, .10, .20, and .50) and for each of three values of ϵ ($\epsilon = .05$, .10, and .20). The results are presented in Table 7.7.

Both the results presented in Table 7.6 and those presented in the right half of Table 7.7 are for cases where (for $j = 1, 2, \ldots, M$) α_j is of the form (7.123). The latter results are those obtained when $\dot{\gamma} = \epsilon$, and the former those obtained when $\dot{\gamma} = \dot{\gamma}_\epsilon$—$\dot{\gamma}_\epsilon$ depends on κ as well as on ϵ. For $\dot{\gamma}_\epsilon < \epsilon$, the number of rejections (discoveries) obtained when $\dot{\gamma} = \epsilon$ is at least as great as the number obtained when $\dot{\gamma} = \dot{\gamma}_\epsilon$. In the application to the prostate data, there are three combinations of κ- and

ϵ-values for which ϵ exceeds $\dot{\gamma}_\epsilon$ by a substantial amount (that where $\kappa = .05$ and $\epsilon = .10$, that where $\kappa = .05$ and $\epsilon = .20$, and that where $\kappa = .10$ and $\epsilon = .20$). When $\kappa = .05$ and $\epsilon = .20$, $\dot{\gamma}_\epsilon = .05$ and setting $\dot{\gamma} = \epsilon$ rather than $\dot{\gamma} = \dot{\gamma}_\epsilon$ results in an additional 7 rejections (discoveries); similarly, when $\kappa = .10$ and $\epsilon = .20$, $\dot{\gamma}_\epsilon = .10$ and setting $\dot{\gamma} = \epsilon$ rather than $\dot{\gamma} = \dot{\gamma}_\epsilon$ results in an additional 6 rejections.

Based on the results presented in Table 7.7, it appears that the difference in the number of rejections produced by the step-down procedure in the case where (for $j = 1, 2, \ldots, M$) α_j is as specified by equality (7.129) and the number produced in the case where α_j is as specified by equality (7.130) can be either negligible or extremely large. This difference tends to be larger for the larger values of κ and also tends to be larger for the larger values of ϵ.

Controlling the FDR. Consider the multiple-comparison procedure obtained upon modifying the step-up procedure described in Subsection e (for controlling the FDR) so as to reflect the redefinition of the t_j's and $t_j^{(0)}$'s. The requisite modifications include that of taking the distribution of the random variable t in definition (7.80) (of α_j') to be $St(100)$ rather than $St(N - P_*)$. As a consequence of this modification, taking α_j' to be of the form (7.81) leads to taking α_j to be of the form

$$\alpha_j = \bar{t}_{j\dot{\gamma}/(2M)}(100), \tag{7.131}$$

rather than of the form (7.82).

The FDR of the modified step-up procedure is given by expression (7.91) (suitably reinterpreted to reflect the redefinition of the t_j's and $t_j^{(0)}$'s). Moreover, subsequent to the redefinition of the t_j's and $t_j^{(0)}$'s, the $t_j^{(0)}$'s's are statistically independent and hence (since $t_j^{(0)} = t_j$ for $j \in T$)

$$\Pr\big(\mathbf{t}^{(-i)} \in A_{k;i}^* \mid |t_i| > \alpha_k\big) = \Pr\big(\mathbf{t}^{(-i)} \in A_{k;i}^*\big).$$

And upon recalling that the sets $A_{1;i}^*, A_{2;i}^*, \ldots, A_{M;i}^*$ are mutually disjoint and that their union equals $\mathcal{R}^{M-1}$ and upon proceeding in the same way as in the derivation of result (7.103), we find that when (for $j = 1, 2, \ldots, M$) α_j' is taken to be of the form (7.81) [in which case α_j is of the form (7.131)], the revised step-up procedure is such that

$$\text{FDR} = (M_T/M)\dot{\gamma} \leq \dot{\gamma}.$$

Thus, when α_j' is taken to be of the form (7.81) and when $\dot{\gamma}$ is set equal to δ, the revised step-up procedure controls the FDR at level δ.

When α_j' is taken to be of the form (7.81), the number of rejections (discoveries) obtained upon applying the modified step-up procedure to the prostate data is (for each of 3 values of $\dot{\gamma}$) as follows:

$\dot{\gamma} = .05$	$\dot{\gamma} = .10$	$\dot{\gamma} = .20$
21	59	105

It is of interest to compare these results with the results presented in the preceding part of the present subsection, which are those obtained from the application of the step-down procedure for controlling Pr(FDP $> \kappa$) at level ϵ. In that regard, it can be shown that (for $j = 1, 2, \ldots, M$ and $0 < \kappa < 1$)

$$\bar{t}_{j\dot{\gamma}/(2M)}(100) \leq \bar{t}_{k_j\dot{\gamma}/[2(M+k_j-j)]}(100), \tag{7.132}$$

where $k_j = [j\kappa] + 1$—refer to Exercise 32. Thus, the application of a step-up procedure in which (for $j = 1, 2, \ldots, M$) α_j is of the form (7.131) and $\dot{\gamma} = \epsilon$ produces as least as many rejections (discoveries) as the application of a step-down procedure in which α_j is as specified by equality (7.130)—even if the α_j's were the same in both cases, the application of the step-up procedure would result in at least as many rejections as the application of the step-down procedure.

In the application of these two stepwise procedures to the prostate data, the step-up procedure produced substantially more rejections than the step-down procedure and did so even for κ as large as .50—refer to the entries in the right half of Table 7.7. It is worth noting that (in the case of the prostate data) the step-up procedure produced no more rejections than would have been produced by a step-down procedure with the same α_j's, so that (in this case) the difference in the number of rejections produced by the two stepwise procedures is due entirely to the differences between the α_j's.

The number of rejections (discoveries) produced by the step-up procedure [that in which α_j is of the form (7.131) and $\dot{\gamma} = \epsilon$] may or may not exceed the number produced by the step-down procedure in which α_j is as specified by equality (7.129) rather than by equality (7.130). When (in the application to the prostate data) $\kappa = .50$ or when $\epsilon = .10$ and $\kappa = .20$, this step-down procedure [which, like that in which α_j is specified by equality (7.130), controls Pr(FDP $> \kappa$) at level ϵ] produced substantially more rejections than the step-up procedure (which controls the FDR at level ϵ)—refer to the entries in the left half of Table 7.7.

7.8 Prediction

The prediction of the realization of an unobservable random variable or vector by a single observable "point" was discussed in Section 5.10, both in general terms and in the special case where the relevant information is the information provided by an observable random vector that follows a G–M, Aitken or general linear model. Let us extend the discussion of prediction to include predictions that take the form of intervals or sets and to include tests of hypotheses about the realizations of unobservable random variables.

a. Some general results

Let $\mathbf{y}$ represent an $N \times 1$ observable random vector. And as in Section 5.10a, consider the use of $\mathbf{y}$ in making inferences about the realization of an unobservable random variable or, more generally, the realization of an unobservable random vector, say the realization of an $M \times 1$ unobservable random vector $\mathbf{w} = (w_1, w_2, \ldots, w_M)'$. Further, assume the existence of the first- and second-order moments of the joint distribution of $\mathbf{w}$ and $\mathbf{y}$; define $\boldsymbol{\mu}_y = \mathrm{E}(\mathbf{y})$, $\boldsymbol{\mu}_w = \mathrm{E}(\mathbf{w})$, $\mathbf{V}_y = \mathrm{var}(\mathbf{y})$, $\mathbf{V}_{yw} = \mathrm{cov}(\mathbf{y}, \mathbf{w})$, and $\mathbf{V}_w = \mathrm{var}(\mathbf{w})$; and assume that $\mathrm{rank}\begin{pmatrix} \mathbf{V}_y & \mathbf{V}_{yw} \\ \mathbf{V}_{yw}' & \mathbf{V}_w \end{pmatrix} = N+M$. In considering the special case $M = 1$, let us write w, μ_w, $\mathbf{v}_{yw}$, and v_w for $\mathbf{w}$, $\boldsymbol{\mu}_w$, $\mathbf{V}_{yw}$, and $\mathbf{V}_w$, respectively.

Prediction intervals and sets. Let $S(\mathbf{y})$ represent a set of $\mathbf{w}$-values that varies with the value of $\mathbf{y}$ and that could potentially be used to predict the realization of $\mathbf{w}$. And consider each of the following two conditions:

$$\Pr[\mathbf{w} \in S(\mathbf{y}) \mid \mathbf{y}] = 1 - \dot{\gamma} \quad \text{(for "every" value of } \mathbf{y}) \tag{8.1}$$

and

$$\Pr[\mathbf{w} \in S(\mathbf{y})] = 1 - \dot{\gamma}. \tag{8.2}$$

Clearly, condition (8.1) implies condition (8.2), but (in general) condition (8.2) does not imply condition (8.1).

Now, suppose that the joint distribution of $\mathbf{w}$ and $\mathbf{y}$ is known or, more generally, that (for "every" value of $\mathbf{y}$) the conditional distribution of $\mathbf{w}$ given $\mathbf{y}$ is known. Then, among the choices for the prediction set $S(\mathbf{y})$ are sets that reflect the characteristics of the conditional distribution of $\mathbf{w}$ given $\mathbf{y}$ and that satisfy condition (8.1).

In a Bayesian framework, the probability $\Pr[\mathbf{w} \in S(\mathbf{y}) \mid \mathbf{y}]$ would be referred to as a posterior probability; and in the case of a choice for $S(\mathbf{y})$ that satisfies condition (8.1), $S(\mathbf{y})$ would be referred to as a $100(1-\dot{\gamma})$ percent credible set. Any choice for $S(\mathbf{y})$ that satisfies condition (8.1) is such that in repeated sampling from the conditional distribution of $\mathbf{w}$ given $\mathbf{y}$ or from the joint distribution of $\mathbf{w}$ and $\mathbf{y}$, the value of $\mathbf{w}$ would be included in $S(\mathbf{y})$ $100(1-\dot{\gamma})$ percent of the time.

Alternatively, suppose that what is known about the joint distribution of $\mathbf{w}$ and $\mathbf{y}$ is more limited.

Specifically, suppose that (while insufficient to determine the conditional distribution of $\mathbf{w}$ given $\mathbf{y}$) knowledge about the joint distribution is sufficient to form an unbiased (point) predictor $\tilde{\mathbf{w}}(\mathbf{y})$ (of the realization of $\mathbf{w}$) with a prediction error, say $\mathbf{e} = \tilde{\mathbf{w}}(\mathbf{y}) - \mathbf{w}$, whose (unconditional) distribution is known—clearly, $\mathrm{E}(\mathbf{e}) = \mathbf{0}$. For example, depending on the state of knowledge, $\tilde{\mathbf{w}}(\mathbf{y})$ might be

$$\tilde{\mathbf{w}}(\mathbf{y}) = \mathrm{E}(\mathbf{w} \mid \mathbf{y}), \tag{8.3}$$

$$\tilde{\mathbf{w}}(\mathbf{y}) = \boldsymbol{\tau} + \mathbf{V}_{yw}' \mathbf{V}_y^{-1} \mathbf{y}, \tag{8.4}$$

where $\boldsymbol{\tau} = \boldsymbol{\mu}_w - \mathbf{V}_{yw}' \mathbf{V}_y^{-1} \boldsymbol{\mu}_y$, or

$$\tilde{\mathbf{w}}(\mathbf{y}) = \tilde{\boldsymbol{\tau}}(\mathbf{y}) + \mathbf{V}_{yw}' \mathbf{V}_y^{-1} \mathbf{y}, \tag{8.5}$$

where $\tilde{\boldsymbol{\tau}}(\mathbf{y})$ is a function of $\mathbf{y}$ with an expected value of $\boldsymbol{\tau}$—refer to Section 5.10. Then, the knowledge about the joint distribution of $\mathbf{w}$ and $\mathbf{y}$ may be insufficient to obtain a prediction set that satisfies condition (8.1). However, a prediction set $S(\mathbf{y})$ that satisfies condition (8.2) can be obtained by taking

$$S(\mathbf{y}) = \{\, \underline{\mathbf{w}} : \underline{\mathbf{w}} = \tilde{\mathbf{w}}(\mathbf{y}) - \underline{\mathbf{e}},\ \underline{\mathbf{e}} \in S^* \,\}, \tag{8.6}$$

where S^* is any set of $M \times 1$ vectors for which

$$\Pr[\tilde{\mathbf{w}}(\mathbf{y}) - \mathbf{w} \in S^*] = 1 - \dot{\gamma}. \tag{8.7}$$

Let us refer to any prediction set that satisfies condition (8.2) as a $100(1-\dot{\gamma})\%$ prediction set. In repeated sampling from the joint distribution of $\mathbf{w}$ and $\mathbf{y}$, the value of $\mathbf{w}$ would be included in a $100(1-\dot{\gamma})\%$ prediction set $100(1-\dot{\gamma})$ percent of the time.

HPD prediction sets. Suppose that (for "every" value of $\mathbf{y}$) the conditional distribution of $\mathbf{w}$ given $\mathbf{y}$ is known and is absolutely continuous with pdf $f(\cdot \mid \mathbf{y})$. Among the various choices for a set $S(\mathbf{y})$ that satisfies condition (8.1) is a set of the form

$$S(\mathbf{y}) = \{\mathbf{w} : f(\mathbf{w} \mid \mathbf{y}) \ge k\}, \tag{8.8}$$

where k is a (strictly) positive constant. In a Bayesian framework, a set of the form (8.8) that satisfies condition (8.1) would be referred to as a $100(1-\dot{\gamma})\%$ HPD (highest posterior density) credible set. In the present setting, let us refer to such a set as a $100(1-\dot{\gamma})\%$ HPD prediction set.

A $100(1-\dot{\gamma})\%$ HPD prediction set has the following property: Among all choices for the set $S(\mathbf{y})$ that satisfy the condition

$$\Pr[\mathbf{w} \in S(\mathbf{y}) \mid \mathbf{y}] \ge 1 - \dot{\gamma}, \tag{8.9}$$

it is the smallest, that is, it minimizes the quantity

$$\int_{S(\mathbf{y})} d\underline{\mathbf{w}}. \tag{8.10}$$

Let us verify that a $100(1-\dot{\gamma})\%$ HPD prediction set has this property. For purposes of doing so, take $\phi(\cdot)$ to be a function defined (for $M \times 1$ vectors) as follows:

$$\phi(\underline{\mathbf{w}}) = \begin{cases} 1, & \text{if } \underline{\mathbf{w}} \in S(\mathbf{y}), \\ 0, & \text{if } \underline{\mathbf{w}} \notin S(\mathbf{y}). \end{cases}$$

And observe that the quantity (8.10) is reexpressible as

$$\int_{S(\mathbf{y})} d\underline{\mathbf{w}} = \int_{\mathbb{R}^M} \phi(\underline{\mathbf{w}})\, d\underline{\mathbf{w}}$$

and that condition (8.9) is reexpressible in the form

$$\int_{\mathbb{R}^M} \phi(\underline{\mathbf{w}}) f(\underline{\mathbf{w}} \mid \mathbf{y})\, d\underline{\mathbf{w}} \ge 1 - \dot{\gamma}.$$

Further, let $S^*(\mathbf{y})$ represent a $100(1-\dot{\gamma})\%$ HPD prediction set, and define

$$\phi^*(\underline{\mathbf{w}}) = \begin{cases} 1, & \text{if } \underline{\mathbf{w}} \in S^*(\mathbf{y}), \\ 0, & \text{if } \underline{\mathbf{w}} \notin S^*(\mathbf{y}). \end{cases}$$

Now, define sets $S^+(\mathbf{y})$ and $S^-(\mathbf{y})$ as follows:

$$S^+(\mathbf{y}) = \{\mathbf{w} : \phi^*(\mathbf{w}) - \phi(\mathbf{w}) > 0\}$$

and

$$S^-(\mathbf{y}) = \{\mathbf{w} : \phi^*(\mathbf{w}) - \phi(\mathbf{w}) < 0\}.$$

And observe that if $\mathbf{w} \in S^+(\mathbf{y})$, then $\phi^*(\mathbf{w}) = 1$ and hence $f(\mathbf{w}\,|\,\mathbf{y}) - k \geq 0$; and similarly if $\mathbf{w} \in S^-(\mathbf{y})$, then $\phi^*(\mathbf{w}) = 0$ and hence $f(\mathbf{w}\,|\,\mathbf{y}) - k < 0$. Thus,

$$\int_{\mathcal{R}^M} [\phi^*(\underline{\mathbf{w}}) - \phi(\underline{\mathbf{w}})][f(\underline{\mathbf{w}}\,|\,\mathbf{y}) - k]\, d\underline{\mathbf{w}} = \int_{S^+(\mathbf{y}) \cup S^-(\mathbf{y})} [\phi^*(\underline{\mathbf{w}}) - \phi(\underline{\mathbf{w}})][f(\underline{\mathbf{w}}\,|\,\mathbf{y}) - k]\, d\underline{\mathbf{w}} \geq 0,$$

so that

$$k \int_{\mathcal{R}^M} [\phi^*(\underline{\mathbf{w}}) - \phi(\underline{\mathbf{w}})]\, d\underline{\mathbf{w}} \leq \int_{\mathcal{R}^M} [\phi^*(\underline{\mathbf{w}}) - \phi(\underline{\mathbf{w}})]\, f(\underline{\mathbf{w}}\,|\,\mathbf{y})\, d\underline{\mathbf{w}} \leq 1 - \dot{\gamma} - (1 - \dot{\gamma}) = 0,$$

in which case

$$\int_{\mathcal{R}^M} \phi^*(\underline{\mathbf{w}})\, d\underline{\mathbf{w}} \leq \int_{\mathcal{R}^M} \phi(\underline{\mathbf{w}})\, d\underline{\mathbf{w}}$$

or, equivalently,

$$\int_{S^*(\mathbf{y})} d\underline{\mathbf{w}} \leq \int_{S(\mathbf{y})} d\underline{\mathbf{w}}$$

(as was to be verified).

$100(1-\dot{\gamma})\%$ ***prediction sets of the form (8.6): minimum size***. Suppose that knowledge about the joint distribution of $\mathbf{w}$ and $\mathbf{y}$ is sufficient to form an unbiased (point) predictor $\tilde{\mathbf{w}}(\mathbf{y})$ (of the realization of $\mathbf{w}$) with a prediction error $\mathbf{e} = \tilde{\mathbf{w}}(\mathbf{y}) - \mathbf{w}$ whose distribution is known. Then, as discussed earlier (in Part 1 of the present subsection), any set $S(\mathbf{y})$ of the form (8.6) is a $100(1-\dot{\gamma})\%$ prediction set.

How do the various $100(1-\dot{\gamma})\%$ prediction sets of the form (8.6) compare with each other? And how do they compare with the additional prediction sets that would result if condition (8.7) were replaced by the condition

$$\Pr[\tilde{\mathbf{w}}(\mathbf{y}) - \mathbf{w} \in S^*] \geq 1 - \dot{\gamma} \tag{8.11}$$

—the additional prediction sets are ones whose (unconditional) probability of coverage exceeds $1-\dot{\gamma}$. Specifically, how do they compare with respect to size.

Clearly, for any of the prediction sets of the form (8.6) and for any of the additional prediction sets that would result if condition (8.7) were replaced by condition (8.11), the size of the prediction set is the same as the size of the set S^*. Thus, for purposes of comparing any of these prediction sets with respect to size, it suffices to compare (with respect to size) the choices for the set S^* that gave rise to the prediction sets.

Now, suppose that the (unconditional) distribution of the vector $\mathbf{e}$ is absolutely continuous, say absolutely continuous with pdf $g(\cdot)$. Then, among the choices for the set S^* that satisfy condition (8.7) is a choice of the form

$$S^* = \{\mathbf{e} : g(\mathbf{e}) > k\} \tag{8.12}$$

(where $0 < k < 1$). Among all choices for S^* that satisfy condition (8.11), this choice is of minimum size—by definition, the size of a set S of $\mathbf{e}$-values is $\int_S d\underline{\mathbf{e}}$. That this choice is of minimum size can be verified by proceeding in much the same fashion as in the preceding part of the present subsection (in verifying that HPD prediction sets are of minimum size).

A special case: multivariate normality. If the joint distribution of $\mathbf{w}$ and $\mathbf{y}$ were MVN, the conditional distribution of $\mathbf{w}$ given $\mathbf{y}$ would be the $N(\boldsymbol{\tau} + \mathbf{V}_{yw}'\mathbf{V}_y^{-1}\mathbf{y},\, \mathbf{V}_w - \mathbf{V}_{yw}'\mathbf{V}_y^{-1}\mathbf{V}_{yw})$ distribution (where $\boldsymbol{\tau} = \boldsymbol{\mu}_w - \mathbf{V}_{yw}'\mathbf{V}_y^{-1}\boldsymbol{\mu}_y$). Accordingly, suppose that the joint distribution of $\mathbf{w}$ and $\mathbf{y}$ is MVN or of some other form for which the conditional distribution of $\mathbf{w}$ given $\mathbf{y}$ is the $N(\boldsymbol{\tau} + \mathbf{V}_{yw}'\mathbf{V}_y^{-1}\mathbf{y},\, \mathbf{V}_w - \mathbf{V}_{yw}'\mathbf{V}_y^{-1}\mathbf{V}_{yw})$ distribution. Suppose further that the values of the vector $\boldsymbol{\tau}$ and of the matrices $\mathbf{V}_{yw}'\mathbf{V}_y^{-1}$ and $\mathbf{V}_w - \mathbf{V}_{yw}'\mathbf{V}_y^{-1}\mathbf{V}_{yw}$ are known, in which case the conditional distribution of $\mathbf{w}$ given $\mathbf{y}$ would be known. Then, a $100(1-\dot{\gamma})\%$ HPD prediction set (for the realization of $\mathbf{w}$) would be

$$\{\underline{\mathbf{w}} : [\underline{\mathbf{w}} - \eta(\mathbf{y})]'(\mathbf{V}_w - \mathbf{V}_{yw}'\mathbf{V}_y^{-1}\mathbf{V}_{yw})^{-1}[\underline{\mathbf{w}} - \eta(\mathbf{y})] \leq \bar{\chi}^2_{\dot{\gamma}}(M)\}, \tag{8.13}$$

where $\eta(\mathbf{y}) = \boldsymbol{\tau} + \mathbf{V}_{yw}'\mathbf{V}_y^{-1}\mathbf{y}$ and where $\bar{\chi}^2_{\dot{\gamma}}(M)$ is the upper $100\dot{\gamma}\%$ point of the $\chi^2(M)$ distribution [as can be readily verified by making use of formula (3.5.32) for the pdf of an MVN distribution and of result (6.6.14) on the distribution of quadratic forms].

In the special case where $M = 1$, the $100(1-\dot{\gamma})\%$ HPD prediction set is reexpressible as the interval

$$\{\underline{w} : \eta(\mathbf{y}) - \bar{z}_{\dot{\gamma}/2}(v_w - \mathbf{v}'_{yw}\mathbf{V}_y^{-1}\mathbf{v}_{yw})^{1/2} \le \underline{w} \le \eta(\mathbf{y}) + \bar{z}_{\dot{\gamma}/2}(v_w - \mathbf{v}'_{yw}\mathbf{V}_y^{-1}\mathbf{v}_{yw})^{1/2}\}, \tag{8.14}$$

where (in this special case) $\eta(\mathbf{y}) = \tau + \mathbf{v}'_{yw}\mathbf{V}_y^{-1}\mathbf{y}$ (with $\tau = \mu_w - \mathbf{v}'_{yw}\mathbf{V}_y^{-1}\boldsymbol{\mu}_y$) and where (for $0 < \alpha < 1$) z_α represents the upper $100\alpha\%$ point of the $N(0, 1)$ distribution. The prediction interval (8.14) satisfies condition (8.1); among all prediction intervals that satisfy condition (8.1), it is the shortest. Among the other prediction intervals (in the special case where $M = 1$) that satisfy condition (8.1) are

$$\{\underline{w} : -\infty < \underline{w} \le \eta(\mathbf{y}) + \bar{z}_{\dot{\gamma}}(v_w - \mathbf{v}'_{yw}\mathbf{V}_y^{-1}\mathbf{v}_{yw})^{1/2}\} \tag{8.15}$$

and

$$\{\underline{w} : \eta(\mathbf{y}) - \bar{z}_{\dot{\gamma}}(v_w - \mathbf{v}'_{yw}\mathbf{V}_y^{-1}\mathbf{v}_{yw})^{1/2} \le \underline{w} < \infty\}. \tag{8.16}$$

Alternatively, suppose that knowledge about the joint distribution of $\mathbf{w}$ and $\mathbf{y}$ is limited to that obtained from knowing the values of the vector $\boldsymbol{\tau} = \boldsymbol{\mu}_w - \mathbf{V}'_{yw}\mathbf{V}_y^{-1}\boldsymbol{\mu}_y$ and the matrix $\mathbf{V}'_{yw}\mathbf{V}_y^{-1}$ and from knowing the distribution of the random vector $\mathbf{e} = \tilde{\mathbf{w}}(\mathbf{y}) - \mathbf{w}$, where $\tilde{\mathbf{w}}(\mathbf{y}) = \boldsymbol{\tau} + \mathbf{V}'_{yw}\mathbf{V}_y^{-1}\mathbf{y}$. And observe that regardless of the form of the distribution of $\mathbf{e}$, it is the case that

$$\operatorname{cov}(\mathbf{y}, \mathbf{e}) = \mathbf{0} \quad \text{and} \quad \operatorname{var}(\mathbf{e}) = \mathbf{V}_w - \mathbf{V}'_{yw}\mathbf{V}_y^{-1}\mathbf{V}_{yw}$$

[and that $\mathrm{E}(\mathbf{e}) = \mathbf{0}$]. Further, suppose that the (unconditional) distribution of $\mathbf{e}$ is MVN, which (in light of the supposition that the distribution of $\mathbf{e}$ is known) implies that the matrix $\mathbf{V}_w - \mathbf{V}'_{yw}\mathbf{V}_y^{-1}\mathbf{V}_{yw}$ is known.

If the joint distribution of $\mathbf{w}$ and $\mathbf{y}$ were MVN, then $\mathbf{e}$ would be distributed independently of $\mathbf{y}$ [and hence independently of $\tilde{\mathbf{w}}(\mathbf{y})$], in which case the conditional distribution of $\tilde{\mathbf{w}}(\mathbf{y}) - \mathbf{e}$ given $\mathbf{y}$ [or given $\tilde{\mathbf{w}}(\mathbf{y})$] would be identical to the conditional distribution of $\mathbf{w}$ given $\mathbf{y}$—both conditional distributions would be $N[\tilde{\mathbf{w}}(\mathbf{y}), \mathbf{V}_w - \mathbf{V}'_{yw}\mathbf{V}_y^{-1}\mathbf{V}_{yw}]$—and any prediction set of the form (8.6) would satisfy condition (8.1) as well as condition (8.2). More generally, a prediction set of the form (8.6) would satisfy condition (8.2) [and hence would qualify as a $100(1-\dot{\gamma})\%$ prediction set], but would not necessarily satisfy condition (8.1). In particular, the prediction set (8.13) and (in the special case where $M = 1$) prediction intervals (8.14), (8.15), and (8.16) would satisfy condition (8.2), however aside from special cases (like that where the joint distribution of $\mathbf{w}$ and $\mathbf{y}$ is MVN) they would not necessarily satisfy condition (8.1).

Simultaneous prediction intervals. Let us consider further the case where conditionally on $\mathbf{y}$, the distribution of the vector $\mathbf{w} = (w_1, w_2, \ldots, w_M)'$ is $N[\eta(\mathbf{y}), \mathbf{V}_w - \mathbf{V}'_{yw}\mathbf{V}_y^{-1}\mathbf{V}_{yw}]$, with $\eta(\mathbf{y}) = \boldsymbol{\tau} + \mathbf{V}'_{yw}\mathbf{V}_y^{-1}\mathbf{y}$ and with $\boldsymbol{\tau} = \boldsymbol{\mu}_w - \mathbf{V}'_{yw}\mathbf{V}_y^{-1}\boldsymbol{\mu}_y$, and where the value of $\boldsymbol{\tau}$ and of the matrices $\mathbf{V}'_{yw}\mathbf{V}_y^{-1}$ and $\mathbf{V}_w - \mathbf{V}'_{yw}\mathbf{V}_y^{-1}\mathbf{V}_{yw}$ are known. Interest in this case might include the prediction of the realizations of some or all of the random variables $w_1, w_2, \ldots, w_M$ or, more generally, the realizations of some or all linear combinations of these random variables.

Predictive inference for the realization of any particular linear combination of the random variables $w_1, w_2, \ldots, w_M$, say for that of the linear combination $\boldsymbol{\delta}'\mathbf{w}$ (where $\boldsymbol{\delta} \ne \mathbf{0}$), might take the form of one of the following intervals:

$$\begin{aligned}\{\underline{w} : \boldsymbol{\delta}'\eta(\mathbf{y}) - \bar{z}_{\dot{\gamma}/2}[\boldsymbol{\delta}'(\mathbf{V}_w - \mathbf{V}'_{yw}\mathbf{V}_y^{-1}\mathbf{V}_{yw})\boldsymbol{\delta}]^{1/2} \\ \le \underline{w} \le \boldsymbol{\delta}'\eta(\mathbf{y}) + \bar{z}_{\dot{\gamma}/2}[\boldsymbol{\delta}'(\mathbf{V}_w - \mathbf{V}'_{yw}\mathbf{V}_y^{-1}\mathbf{V}_{yw})\boldsymbol{\delta}]^{1/2}\},\end{aligned} \tag{8.17}$$

$$\{\underline{w} : -\infty < \underline{w} \le \boldsymbol{\delta}'\eta(\mathbf{y}) + \bar{z}_{\dot{\gamma}}[\boldsymbol{\delta}'(\mathbf{V}_w - \mathbf{V}'_{yw}\mathbf{V}_y^{-1}\mathbf{V}_{yw})\boldsymbol{\delta}]^{1/2}\}, \tag{8.18}$$

or

$$\{\underline{w} : \boldsymbol{\delta}'\eta(\mathbf{y}) - \bar{z}_{\dot{\gamma}}[\boldsymbol{\delta}'(\mathbf{V}_w - \mathbf{V}'_{yw}\mathbf{V}_y^{-1}\mathbf{V}_{yw})\boldsymbol{\delta}]^{1/2} \le \underline{w} < \infty\}. \tag{8.19}$$

In an application to a single linear combination or in an application to any one of a number of linear combinations that is assessed independently of the application to any of the others, the probability of coverage (both conditionally on $\mathbf{y}$ and unconditionally) of interval (8.17), (8.18), or (8.19) equals

$1-\dot{\gamma}$. However, when prediction intervals are obtained for each of a number of linear combinations, it is often the case that those linear combinations identified with the "most extreme" intervals receive the most attention. "One-at-a-time" prediction intervals like intervals (8.17), (8.18), and (8.19) do not account for any such identification and, as a consequence, their application can sometimes lead to erroneous conclusions. In the case of intervals (8.17), (8.18), and (8.19), this potential pitfall can be avoided by introducing modifications that convert these one-at-a-time intervals into prediction intervals that provide for control of the probability of simultaneous coverage.

Accordingly, let Δ represent a finite or infinite collection of (nonnull) M-dimensional column vectors. And suppose that we wish to obtain a prediction interval for the realization of each of the linear combinations $\boldsymbol{\delta}'\mathbf{w}$ ($\boldsymbol{\delta} \in \Delta$). Suppose further that we wish for the intervals to be such that the probability of simultaneous coverage equals $1-\dot{\gamma}$. Such intervals can be obtained by taking (for each $\boldsymbol{\delta} \in \Delta$) the interval for the realization of $\boldsymbol{\delta}'\mathbf{w}$ to be a modified version of interval (8.17), (8.18), or (8.19); the requisite modification consists of introducing a suitable replacement for $\bar{z}_{\dot{\gamma}/2}$ or $\bar{z}_{\dot{\gamma}}$.

Let $\mathbf{R}$ represent an $M \times M$ nonsingular matrix such that $\mathbf{V}_w - \mathbf{V}_{yw}'\mathbf{V}_y^{-1}\mathbf{V}_{yw} = \mathbf{R}'\mathbf{R}$—upon observing [in light of Corollary 2.13.33 and result (2.5.5)] that $\mathbf{V}_w - \mathbf{V}_{yw}'\mathbf{V}_y^{-1}\mathbf{V}_{yw}$ is a symmetric positive definite matrix, the existence of the matrix $\mathbf{R}$ follows from Corollary 2.13.29. Further, let $\mathbf{z} = (\mathbf{R}^{-1})'[\eta(\mathbf{y})-\mathbf{w}]$, so that $\mathbf{z} \sim N(\mathbf{0}, \mathbf{I})$ (both conditionally on $\mathbf{y}$ and unconditionally) and (for every nonnull $M \times 1$ vector $\boldsymbol{\delta}$)

$$\frac{\boldsymbol{\delta}'\eta(\mathbf{y}) - \boldsymbol{\delta}'\mathbf{w}}{[\boldsymbol{\delta}'(\mathbf{V}_w - \mathbf{V}_{yw}'\mathbf{V}_y^{-1}\mathbf{V}_{yw})\boldsymbol{\delta}]^{1/2}} = \frac{(\mathbf{R}\boldsymbol{\delta})'\mathbf{z}}{[(\mathbf{R}\boldsymbol{\delta})'\mathbf{R}\boldsymbol{\delta}]^{1/2}} \sim N(0, 1). \tag{8.20}$$

And take the replacement for $z_{\dot{\gamma}/2}$ in interval (8.17) to be the upper $100\dot{\gamma}\%$ point of the distribution of the random variable

$$\max_{\boldsymbol{\delta} \in \Delta} \frac{|(\mathbf{R}\boldsymbol{\delta})'\mathbf{z}|}{[(\mathbf{R}\boldsymbol{\delta})'\mathbf{R}\boldsymbol{\delta}]^{1/2}}. \tag{8.21}$$

Similarly, take the replacement for $z_{\dot{\gamma}}$ in intervals (8.18) and (8.19) to be the upper $100\dot{\gamma}\%$ point of the distribution of the random variable

$$\max_{\boldsymbol{\delta} \in \Delta} \frac{(\mathbf{R}\boldsymbol{\delta})'\mathbf{z}}{[(\mathbf{R}\boldsymbol{\delta})'\mathbf{R}\boldsymbol{\delta}]^{1/2}}. \tag{8.22}$$

Then, as is evident from result (8.20), the prediction intervals obtained for the realizations of the linear combinations $\boldsymbol{\delta}'\mathbf{w}$ ($\boldsymbol{\delta} \in \Delta$) upon the application of the modified version of interval (8.17), (8.18), or (8.19), are such that the probability of simultaneous coverage equals $1-\dot{\gamma}$ (both conditionally on $\mathbf{y}$ and unconditionally). In fact, when the unconditional distribution of $\eta(\mathbf{y})-\mathbf{w}$ is MVN, the unconditional probability of simultaneous coverage of the prediction intervals obtained upon the application of the modified version of interval (8.17), (8.18), or (8.19) would equal $1-\dot{\gamma}$ even if the conditional distribution of $\mathbf{w}$ given $\mathbf{y}$ differed from the $N[\eta(\mathbf{y}),\ \mathbf{V}_w - \mathbf{V}_{yw}'\mathbf{V}_y^{-1}\mathbf{V}_{yw}]$ distribution.

When $\Delta = \{\boldsymbol{\delta} \in \mathbb{R}^M : \boldsymbol{\delta} \neq \mathbf{0}\}$, the upper $100\dot{\gamma}\%$ point of the distribution of the random variable (8.21) equals $\sqrt{\bar{\chi}^2_{\dot{\gamma}}(M)}$, and $\boldsymbol{\delta}'\mathbf{w}$ is contained in the modified version of interval (8.17) for every $\boldsymbol{\delta} \in \Delta$ if and only if $\mathbf{w}$ is contained in the set (8.13), as can be verified by proceeding in much the same way as in Section 7.3c in the verification of some similar results. When the members of Δ consist of the columns of the $M \times M$ identity matrix, the linear combinations $\boldsymbol{\delta}'\mathbf{w}$ ($\boldsymbol{\delta} \in \Delta$) consist of the M random variables $w_1, w_2, \ldots, w_M$.

For even moderately large values of M, a requirement that the prediction intervals achieve simultaneous coverage with a high probability can be quite severe. A less stringent alternative would be to require (for some integer k greater than 1) that with a high probability, no more than k of the intervals fail to cover. Thus, in modifying interval (8.17) for use in obtaining prediction intervals for all of the linear combinations $\boldsymbol{\delta}'\mathbf{w}$ ($\boldsymbol{\delta} \in \Delta$), we could replace $z_{\dot{\gamma}/2}$ with the upper $100\dot{\gamma}\%$ point of the distribution of the kth largest of the random variables $|(\mathbf{R}\boldsymbol{\delta})'\mathbf{z}|/[(\mathbf{R}\boldsymbol{\delta})'\mathbf{R}\boldsymbol{\delta}]^{1/2}$ ($\boldsymbol{\delta} \in \Delta$), rather than with the upper $100\dot{\gamma}\%$ point of the distribution of the largest of these random variables. Similarly, in modifying interval (8.18) or (8.19), we could replace $z_{\dot{\gamma}}$ with the upper $100\dot{\gamma}\%$ point of the distribution of the kth largest (rather than the largest) of the random variables $(\mathbf{R}\boldsymbol{\delta})'\mathbf{z}/[(\mathbf{R}\boldsymbol{\delta})'\mathbf{R}\boldsymbol{\delta}]^{1/2}$ $(\boldsymbol{\delta} \in \Delta)$. In either case [and in the case of the distribution of the random variable (8.21) or (8.22), the upper $100\dot{\gamma}\%$ point of the relevant distribution could be determined numerically via Monte Carlo methods.

Hypothesis tests. Let S_0 represent a (nonempty but proper) subset of the set $\mathcal{R}^M$ of all M-dimensional column vectors, and let S_1 represent the complement of S_0, that is, the subset of $\mathcal{R}^M$ consisting of all M-dimensional column vectors other than those in S_0 (so that $S_0 \cap S_1$ is the empty set, and $S_0 \cup S_1 = \mathcal{R}^M$). And consider the problem of testing the null hypothesis $H_0 : \mathbf{w} \in S_0$ versus the alternative hypothesis $H_1 : \mathbf{w} \in S_1$—note that $\mathbf{w} \in S_1 \Leftrightarrow \mathbf{w} \notin S_0$.

Let $\phi(\cdot)$ represent the critical function of a (nonrandomized) test of H_0 versus H_1, in which case $\phi(\cdot)$ is of the form

$$\phi(\underset{\sim}{y}) = \begin{cases} 1, & \text{if } \underset{\sim}{y} \in A, \\ 0, & \text{if } \underset{\sim}{y} \notin A, \end{cases}$$

where A is a (nonempty but proper) subset of $\mathcal{R}^N$ known as the critical region, and H_0 is rejected if and only if $\phi(\mathbf{y}) = 1$. Or, more generally, let $\phi(\cdot)$ represent the critical function of a possibly randomized test, in which case $0 \le \phi(\underset{\sim}{y}) \le 1$ (for $\underset{\sim}{y} \in \mathcal{R}^N$) and H_0 is rejected with probability $\phi(\underset{\sim}{y})$ when $\mathbf{y} = \underset{\sim}{y}$. Further, let

$$\pi_0 = \Pr(\mathbf{w} \in S_0) \quad \text{and} \quad \pi_1 = \Pr(\mathbf{w} \in S_1) \ \ (= 1 - \pi_0),$$

and take $p_0(\cdot)$ and $p_1(\cdot)$ to be functions defined (on $\mathcal{R}^N$) as follows:

$$p_0(\mathbf{y}) = \Pr(\mathbf{w} \in S_0 \mid \mathbf{y}) \quad \text{and} \quad p_1(\mathbf{y}) = \Pr(\mathbf{w} \in S_1 \mid \mathbf{y}) \ \ [= 1 - p_0(\mathbf{y})].$$

Now, suppose that the joint distribution of $\mathbf{w}$ and $\mathbf{y}$ is such that the (marginal) distribution of $\mathbf{y}$ is known and is absolutely continuous with pdf $f(\cdot)$ and is such that $p_0(\cdot)$ is known (in which case π_0, π_1, and $p_1(\cdot)$ would also be known and π_0 and π_1 would constitute what in a Bayesian framework would be referred to as prior probabilities and $p_0(\mathbf{y})$ and $p_1(\mathbf{y})$ what would be referred to as posterior probabilities). Suppose further that $0 < \pi_0 < 1$ (in which case $0 < \pi_1 < 1$)—if $\pi_0 = 0$ or $\pi_0 = 1$, then a test for which the probability of an error of the first kind (i.e., of falsely rejecting H_0) and the probability of an error of the second kind (i.e., of falsely accepting H_0) are both 0 could be achieved by taking $\phi(\mathbf{y}) = 1$ for every value of $\mathbf{y}$ or by taking $\phi(\mathbf{y}) = 0$ for every value of $\mathbf{y}$.

Note that when the distribution of $\mathbf{w}$ is absolutely continuous, the supposition that $\pi_0 > 0$ rules out the case where S_0 consists of a single point—refer, e.g., to Berger (2002, sec. 4.3.3) for some related discussion. Further, take $f_0(\cdot)$ and $f_1(\cdot)$ to be functions defined (on $\mathcal{R}^N$) as follows: for $\underset{\sim}{y} \in \mathcal{R}^N$,

$$f_0(\underset{\sim}{y}) = p_0(\underset{\sim}{y}) f(\underset{\sim}{y}) / \pi_0 \quad \text{and} \quad f_1(\underset{\sim}{y}) = p_1(\underset{\sim}{y}) f(\underset{\sim}{y}) / \pi_1$$

—when (as is being implicitly assumed herein) the function $p_0(\cdot)$ is "sufficiently well-behaved," the conditional distribution of $\mathbf{y}$ given that $\mathbf{w} \in S_9$ is absolutely continuous with pdf $f_0(\cdot)$ and the conditional distribution of $\mathbf{y}$ given that $\mathbf{w} \in S_1$ is absolutely continuous with pdf $f_1(\cdot)$. And observe that the probability of rejecting the null hypothesis H_0 when H_0 is true is expressible as

$$\text{E}[\phi(\mathbf{y}) \mid \mathbf{w} \in S_0] = \int_{\mathcal{R}^N} \phi(\underset{\sim}{y}) f_0(\underset{\sim}{y}) \, d\underset{\sim}{y}, \tag{8.23}$$

and the probability of rejecting H_0 when H_0 is false is expressible as

$$\text{E}[\phi(\mathbf{y}) \mid \mathbf{w} \in S_1] = \int_{\mathcal{R}^N} \phi(\underset{\sim}{y}) f_1(\underset{\sim}{y}) \, d\underset{\sim}{y}. \tag{8.24}$$

Upon applying a version of the Neyman–Pearson lemma stated by Lehmann and Romano (2005b, sec. 3.2) in the form of their theorem 3.2.1, we find that there exists a critical function $\phi^*(\cdot)$ defined by taking

$$\phi^*(\underset{\sim}{y}) = \begin{cases} 1, & \text{when } f_1(\underset{\sim}{y}) > k f_0(\underset{\sim}{y}), \\ c, & \text{when } f_1(\underset{\sim}{y}) = k f_0(\underset{\sim}{y}), \\ 0, & \text{when } f_1(\underset{\sim}{y}) < k f_0(\underset{\sim}{y}), \end{cases} \tag{8.25}$$

and by taking c and k $(0 \le c \le 1,\ 0 \le k < \infty)$ to be constants for which

$$\mathrm{E}[\phi^*(\mathbf{y}) \mid \mathbf{w} \in S_0] = \dot{\gamma}; \tag{8.26}$$

and we find that among all choices for the critical function $\phi(\cdot)$ for which

$$\mathrm{E}[\phi(\mathbf{y}) \mid \mathbf{w} \in S_0] \leq \dot{\gamma},$$

the power $\mathrm{E}[\phi(\mathbf{y}) \mid \mathbf{w} \in S_1]$ of the test attains its maximum value when $\phi(\cdot)$ is taken to be $\phi^*(\cdot)$.

In regard to expression (8.25), note that for every $N \times 1$ vector $\underline{\mathbf{y}}$ for which $f_0(\underline{\mathbf{y}}) > 0$,

$$f_1(\underline{\mathbf{y}}) \begin{Bmatrix} > \\ = \\ < \end{Bmatrix} k f_0(\underline{\mathbf{y}}) \;\Leftrightarrow\; B \begin{Bmatrix} > \\ = \\ < \end{Bmatrix} k \;\Leftrightarrow\; p_1(\underline{\mathbf{y}})/p_0(\underline{\mathbf{y}}) \begin{Bmatrix} > \\ = \\ < \end{Bmatrix} k' \;\Leftrightarrow\; p_0(\underline{\mathbf{y}}) \begin{Bmatrix} < \\ = \\ > \end{Bmatrix} k'', \tag{8.27}$$

where

$$B = \frac{f_1(\underline{\mathbf{y}})}{f_0(\underline{\mathbf{y}})} = \frac{p_1(\underline{\mathbf{y}})/\pi_1}{p_0(\underline{\mathbf{y}})/\pi_0} = \frac{p_1(\underline{\mathbf{y}})/p_0(\underline{\mathbf{y}})}{\pi_1/\pi_0}, \quad k' = (\pi_1/\pi_0)k, \text{ and } k'' = (1+k')^{-1}.$$

In Bayesian terms, the ratio π_1/π_0 represents the prior odds in favor of H_1 and against H_0 and (when $\underline{\mathbf{y}}$ is regarded as the observed value of $\mathbf{y}$) $p_1(\underline{\mathbf{y}})/p_0(\underline{\mathbf{y}})$ represents the posterior odds in favor of H_1 and against H_0 and B represents the Bayes factor in favor of H_1 (e.g., Berger 1985, sec 4.3.3).

The result on the existence of the critical function $\phi^*(\cdot)$ and on the optimality of the test with critical function $\phi^*(\cdot)$ can be generalized. Let $\mathbf{z} = \mathbf{z}(\mathbf{y})$ represent an N^*-dimensional column vector whose elements are (known) functions of the N-dimensional observable random column vector $\mathbf{y}$ (where $N^* \leq N$). And suppose that the (marginal) distribution of $\mathbf{z}$ is known and is absolutely continuous with pdf $f^*(\cdot)$ and also that $\Pr(\mathbf{w} \in S_0 \mid \mathbf{z})$ is a known function of the value of $\mathbf{z}$. Suppose further that the choice of a test procedure is restricted to procedures that depend on $\mathbf{y}$ only through the value of $\mathbf{z}$.

The result on existence and optimality can be generalized so as to achieve coverage of this situation. It is a simple matter of replacing [in the definitions of $p_0(\cdot)$, $p_1(\cdot)$, $f_0(\cdot)$, and $f_1(\cdot)$ and in the statement of the result] the observable random vector $\mathbf{y}$ and the pdf $f(\cdot)$ with the observable random vector $\mathbf{z}$ and the pdf $f^*(\cdot)$. In the generalized version of the result, the optimality is in regard to $\dot{\gamma}$-level test procedures that depend on $\mathbf{y}$ only through the value of $\mathbf{z}$ rather than (as in the original version) in regard to all $\dot{\gamma}$-level test procedures.

Suppose, in particular, that enough is known about the joint distribution of $\mathbf{w}$ and $\mathbf{y}$ that there exists an unbiased predictor $\tilde{\mathbf{w}}(\mathbf{y})$ of the realization of $\mathbf{w}$. And writing $\tilde{\mathbf{w}}$ for $\tilde{\mathbf{w}}(\mathbf{y})$, suppose further that the distribution of $\tilde{\mathbf{w}}$ is known and is absolutely continuous and that $\Pr(\mathbf{w} \in S_0 \mid \tilde{\mathbf{w}})$ is a known function of the value of $\tilde{\mathbf{w}}$. Then, as a special case of the generalized version of the result on the existence of the critical function $\phi^*(\cdot)$ and on the optimality of the test with critical function $\phi^*(\cdot)$, we have the case where $\mathbf{z} = \tilde{\mathbf{w}}$. In that special case, the functions $p_0(\cdot)$ and $p_1(\cdot)$ can be expressed in terms of the conditional (on $\tilde{\mathbf{w}}$) distribution of $\mathbf{w}$ and reexpressed in terms of the conditional distribution of the prediction error $\mathbf{e} = \tilde{\mathbf{w}} - \mathbf{w}$ as follows:

$$p_0(\tilde{\mathbf{w}}) = \Pr(\mathbf{w} \in S_0 \mid \tilde{\mathbf{w}}) = \Pr[\mathbf{e} \in S_0^*(\tilde{\mathbf{w}}) \mid \tilde{\mathbf{w}}], \tag{8.28}$$

where (for $\underline{\tilde{\mathbf{w}}} \in \mathcal{R}^M$) $S_0^*(\underline{\tilde{\mathbf{w}}}) = \{\underline{\mathbf{e}} \in \mathcal{R}^M : \underline{\mathbf{e}} = \underline{\tilde{\mathbf{w}}} - \underline{\mathbf{w}},\; \underline{\mathbf{w}} \in S_0\}$, and, similarly,

$$p_1(\tilde{\mathbf{w}}) = \Pr(\mathbf{w} \in S_1 \mid \tilde{\mathbf{w}}) = \Pr[\mathbf{e} \in S_1^*(\tilde{\mathbf{w}}) \mid \tilde{\mathbf{w}}], \tag{8.29}$$

where $S_1^*(\underline{\tilde{\mathbf{w}}}) = \{\underline{\mathbf{e}} \in \mathcal{R}^M : \underline{\mathbf{e}} = \underline{\tilde{\mathbf{w}}} - \underline{\mathbf{w}},\; \underline{\mathbf{w}} \in S_1\}$.

Multiple comparisons. Suppose that we wish to make inferences about the realization of each of the M random variables $w_1, w_2, \ldots, w_M$. Suppose, in particular, that (for $j = 1, 2, \ldots, M$) we wish to test the null hypothesis $H_j^{(0)}: w_j \in S_j^{(0)}$ versus the alternative hypothesis $H_j^{(1)}: w_j \notin S_j^{(0)}$, where $S_j^{(0)}$ is "any" particular (nonempty) subset of $\mathcal{R}^1$—the M subsets $S_1^{(0)}, S_2^{(0)}, \ldots, S_M^{(0)}$ may or may not be identical.

In a "one-at-a-time" approach, each of the M null hypotheses $H_1^{(0)}, H_2^{(0)}, \ldots, H_M^{(0)}$ would be tested individually at a specified level, in which case the probability of falsely rejecting one or more of the M null hypotheses would be larger (and for even moderately large values of M, could be much larger) than the specified level. Let us consider an alternative approach in which the probability of falsely rejecting one or more of the null hypotheses (the FWER) or, more generally, the probability of falsely rejecting k or more of the null hypotheses (the k-FWER) is controlled at a specified level, say $\dot{\gamma}$. Such an approach can be devised by making use of the results of the preceding part of the present subsection and by invoking the so-called closure principle (e.g., Bretz, Hothorn, and Westfall 2011, sec. 2.2.3; Efron 2010, p. 38), as is to be demonstrated in what follows.

Let $\mathbf{z} = \mathbf{z}(\mathbf{y})$ represent an N^*-dimensional column vector whose elements are (known) functions of $\mathbf{y}$—e.g., $\mathbf{z}(\mathbf{y}) = \mathbf{y}$ or (at the other extreme) $\mathbf{z}(\mathbf{y}) = \tilde{\mathbf{w}}(\mathbf{y})$, where $\tilde{\mathbf{w}}(\mathbf{y})$ is an unbiased (point) predictor of the realization of the vector $\mathbf{w} = (w_1, w_2, \ldots, w_M)'$. And suppose that the (marginal) distribution of $\mathbf{z}$ is known and is absolutely continuous with pdf $f^*(\cdot)$.

Now, let Ω_k represent the collection of all subsets of $I = \{1, 2, \ldots, M\}$ of size k or greater, and let Ω_{kj} represent the collection consisting of those subsets of I of size k or greater that include the integer j. Further, for $I_* \in \Omega_k$; let

$$S(I_*) = \{\underline{\mathbf{w}} = (\underline{w}_1, \underline{w}_2, \ldots, \underline{w}_M)' \in \mathbb{R}^M : \underline{w}_j \in S_j^{(0)} \text{ for } j \in I_*\};$$

and suppose that (for $I_* \in \Omega_k$) the joint distribution of $\mathbf{w}$ and $\mathbf{z}$ is such that $\Pr[\mathbf{w} \in S(I_*) \mid \mathbf{z}]$ is a known function of the value of $\mathbf{z}$—in which case $\Pr[\mathbf{w} \in S(I_*)]$ would be known—and is such that $0 < \Pr[\mathbf{w} \in S(I_*)] < 1$. And (for $I_* \in \Omega_k$) take $\phi^*(\cdot\,; I_*)$ to be the critical function (defined on $\mathbb{R}^{N^*}$) of the most-powerful $\dot{\gamma}$-level procedure for testing the null hypothesis $H_0(I_*) : \mathbf{w} \in S(I_*)$ [versus the alternative hypothesis $H_1(I_*) : \mathbf{w} \notin S(I_*)$] on the basis of $\mathbf{z}$; this procedure is that obtained upon applying [with $S_0 = S(I_*)$] the generalized version of the result of the preceding part of the present subsection (i.e., the version of the result in which the choice of a test procedure is restricted to procedures that depend on $\mathbf{y}$ only through the value of $\mathbf{z}$). Then, the k-FWER (and in the special case where $k = 1$, the FWER) can be controlled at level $\dot{\gamma}$ by employing a (multiple-comparison) procedure in which (for $j = 1, 2, \ldots, M$) $H_j^{(0)}: w_j \in S_j^{(0)}$ is rejected (in favor of $H_j^{(1)}: w_j \notin S_j^{(0)}$) if and only if for every $I_* \in \Omega_{kj}$ the null hypothesis $H_0(I_*) : \mathbf{w} \in S(I_*)$ is rejected by the $\dot{\gamma}$-level test with critical function $\phi^*(\cdot\,; I_*)$.

Let us verify that this procedure controls the k-FWER at level $\dot{\gamma}$. Define

$$T = \{j \in I : H_j^{(0)} \text{ is true}\} = \{j \in I : w_j \in S_j^{(0)}\}.$$

Further, denote by RT the subset of T defined as follows: for $j \in T$, $j \in RT$ if $H_j^{(0)}$ is among the null hypotheses rejected by the multiple-comparison procedure. And (denoting by M_S the size of an arbitrary set S) suppose that $M_{RT} \geq k$, in which case $M_T \geq k$ and it follows from the very definition of the multiple-comparison procedure that the null hypothesis $H_0(T)$ is rejected by the $\dot{\gamma}$-level test with critical function $\phi^*(\cdot\,; T)$. Thus,

$$\Pr(M_{RT} \geq k) \leq \dot{\gamma}$$

(so that the multiple-comparison procedure controls the k-FWER at level $\dot{\gamma}$ as was to be verified).

For some discussion (in the context of the special case where $k = 1$) of shortcut procedures for achieving an efficient implementation of this kind of multiple-comparison procedure, refer, for example, to Bretz, Hothorn, and Westfall (2011).

False discovery proportion: control of the probability of its exceeding a specified constant. Let us consider further the multiple-comparison procedure (described in the preceding part of the present subsection) for testing the null hypotheses $H_j^{(0)}$ ($j = 1, 2, \ldots, M$); that procedure controls the k-FWER at level $\dot{\gamma}$. Denote by R_k the total number of null hypotheses rejected by the multiple-comparison procedure—previously (in Section 7.7d) that symbol was used to denote something else—and denote by RT_k (rather than simply by RT, as in the preceding part of the present subsection) the number of true null hypotheses rejected by the procedure. Further, define

$$\text{FDP}_k = \begin{cases} RT_k / R_k, & \text{if } R_k > 0, \\ 0, & \text{if } R_k = 0; \end{cases}$$

this quantity represents what (in the present context) constitutes the false discovery proportion—refer to Section 7.7c.

For any scalar κ in the interval (0, 1), the multiple-comparison procedure can be used to control the probability $\Pr(\text{FDP}_k > \kappa)$ at level $\dot{\gamma}$. As is to be shown in what follows, such control can be achieved by making a judicious choice for k (a choice based on the value of the observable random vector $\mathbf{z}$).

Upon observing that $R_k \geq RT_k$, we find that

$$\text{FDP}_k > \kappa \;\Leftrightarrow\; RT_k > \kappa R_k \;\Leftrightarrow\; RT_k \geq [\kappa R_k]+1 \tag{8.30}$$

(where for any real number x, $[x]$ denotes the largest integer that is less than or equal to x). Moreover, for any k for which $k \leq [\kappa R_k]+1$,

$$RT_k \geq [\kappa R_k]+1 \;\Rightarrow\; \begin{array}{l}\text{the null hypothesis } H_0(T) \text{ is rejected by the}\\ \dot{\gamma}\text{-level test with critical function } \phi^*(\,\cdot\,; T).\end{array} \tag{8.31}$$

The quantity $[\kappa R_k]+1$ is nondecreasing in k and is bounded from below by 1 and from above by $[\kappa M]+1$. Let $K = K(\mathbf{z})$ represent the largest value of k for which $k \leq [\kappa R_k]+1$ (or, equivalently, the largest value for which $k = [\kappa R_k]+1$). Then, it follows from results (8.30) and (8.31) that

$$\Pr[\text{FDP}_{K(\mathbf{z})} > \kappa] \leq \dot{\gamma}. \tag{8.32}$$

Thus, by taking $k = K$, the multiple-comparison procedure for controlling the k-FWER at level $\dot{\gamma}$ can be configured to control (at the same level $\dot{\gamma}$) the probability of the false discovery proportion exceeding κ. Moreover, it is also the case—refer to Exercise 28—that when $k = K$, the false discovery rate $\text{E}(\text{FDP}_{K(\mathbf{z})})$ of this procedure is controlled at level $\delta = \dot{\gamma} + \kappa(1-\dot{\gamma})$.

b. Prediction on the basis of a G–M, Aitken, or general linear model

Suppose that $\mathbf{y}$ is an $N\times 1$ observable random vector that follows a general linear model. And consider the use of the observed value of $\mathbf{y}$ in making predictive inferences about the realization of an $M\times 1$ vector $\mathbf{w} = (w_1, w_2, \ldots, w_M)'$ of unobservable random variables that is expressible in the form

$$\mathbf{w} = \boldsymbol{\Lambda}'\boldsymbol{\beta} + \mathbf{d}, \tag{8.33}$$

where $\boldsymbol{\Lambda}$ is a $P\times M$ matrix of (known) constants (such that $\boldsymbol{\Lambda}'\boldsymbol{\beta}$ is estimable) and where $\mathbf{d}$ is an $M\times 1$ (unobservable) random vector with $\text{E}(\mathbf{d}) = \mathbf{0}$, $\text{var}(\mathbf{d}) = \mathbf{V}_w(\boldsymbol{\theta})$, and (denoting by $\mathbf{e}$ the vector of residual effects in the general linear model) $\text{cov}(\mathbf{e}, \mathbf{d}) = \mathbf{V}_{yw}(\boldsymbol{\theta})$ for some matrices $\mathbf{V}_w(\boldsymbol{\theta})$ and $\mathbf{V}_{yw}(\boldsymbol{\theta})$ whose elements [like those of the matrix $\text{var}(\mathbf{e}) = \mathbf{V}(\boldsymbol{\theta})$] are known functions of the parametric vector $\boldsymbol{\theta}$. Further, write $\mathbf{V}_y(\boldsymbol{\theta})$ for $\mathbf{V}(\boldsymbol{\theta})$, and observe that

$$\text{E}\begin{pmatrix}\mathbf{y}\\ \mathbf{w}\end{pmatrix} = \begin{pmatrix}\mathbf{X}\\ \boldsymbol{\Lambda}'\end{pmatrix}\boldsymbol{\beta} \quad\text{and}\quad \text{var}\begin{pmatrix}\mathbf{y}\\ \mathbf{w}\end{pmatrix} = \begin{bmatrix}\mathbf{V}_y(\boldsymbol{\theta}) & \mathbf{V}_{yw}(\boldsymbol{\theta})\\ [\mathbf{V}_{yw}(\boldsymbol{\theta})]' & \mathbf{V}_w(\boldsymbol{\theta})\end{bmatrix}.$$

The Aitken model can be regarded as the special case of the general linear model where $\boldsymbol{\theta} = (\sigma)$ and where $\mathbf{V}(\boldsymbol{\theta}) = \sigma^2\mathbf{H}$, and the G–M model can be regarded as the further special case where $\mathbf{H} = \mathbf{I}$. Let us suppose that when (in making predictive inferences about the realization of $\mathbf{w}$) the model is taken to be the Aitken model or the G–M model, $\mathbf{V}_w(\boldsymbol{\theta})$ and $\mathbf{V}_{yw}(\boldsymbol{\theta})$ are taken to be of the form

$$\mathbf{V}_w(\boldsymbol{\theta}) = \sigma^2\mathbf{H}_w \quad\text{and}\quad \mathbf{V}_{yw}(\boldsymbol{\theta}) = \sigma^2\mathbf{H}_{yw},$$

where $\mathbf{H}_w$ and $\mathbf{H}_{yw}$ are (known) matrices of constants. Thus, when the model is taken to be the Aitken model or the G–M model,

$$\text{E}\begin{pmatrix}\mathbf{y}\\ \mathbf{w}\end{pmatrix} = \begin{pmatrix}\mathbf{X}\\ \boldsymbol{\Lambda}'\end{pmatrix}\boldsymbol{\beta} \quad\text{and}\quad \text{var}\begin{pmatrix}\mathbf{y}\\ \mathbf{w}\end{pmatrix} = \sigma^2\begin{pmatrix}\mathbf{H}_y & \mathbf{H}_{yw}\\ \mathbf{H}'_{yw} & \mathbf{H}_w\end{pmatrix}, \tag{8.34}$$

where (in the case of the Aitken model) $\mathbf{H}_y = \mathbf{H}$ and (in the case of the G–M model) $\mathbf{H}_y = \mathbf{I}$.

The joint distribution of $\mathbf{y}$ and $\mathbf{w}$ is such that (regardless of the specific form of the distribution) the variance-covariance matrix of the vector $\begin{pmatrix} \mathbf{y} \\ \mathbf{w} \end{pmatrix}$ depends on unknown parameters (σ or, more generally, the elements of $\boldsymbol{\theta}$). Thus, aside from trivial special cases and in the absence of "additional information" about the value of σ or $\boldsymbol{\theta}$, the information available about the joint distribution of $\mathbf{y}$ and $\mathbf{w}$ falls short of what would be needed to apply the prediction procedures described in Subsection a. Among the kinds of additional information that would allow the application of some or all of those procedures, the simplest and most extreme is the kind that imparts "exact" knowledge of the value of σ or $\boldsymbol{\theta}$. In some cases, the information about the value of σ or $\boldsymbol{\theta}$ provided by the value of $\mathbf{y}$ (perhaps in combination with information available from "external sources") may be adequate to provide justification for proceeding as though the value of σ or $\boldsymbol{\theta}$ is known.

A special case: G–M model. Let us focus on the special case where $\mathbf{y}$ follows the G–M model and hence where the expected value and the variance-covariance matrix of $\begin{pmatrix} \mathbf{y} \\ \mathbf{w} \end{pmatrix}$ are of the form (8.34) with $\mathbf{H}_y = \mathbf{I}$. In that special case, an unbiased (point) predictor of the realization of $\mathbf{w}$ is

$$\tilde{\mathbf{w}}(\mathbf{y}) = \boldsymbol{\Lambda}'(\mathbf{X}'\mathbf{X})^{-}\mathbf{X}'\mathbf{y} + \mathbf{H}_{yw}'(\mathbf{I} - \mathbf{P}_{\mathbf{X}})\mathbf{y}. \tag{8.35}$$

In fact, as discussed in Section 5.10b, this predictor is the BLUP (best linear unbiased predictor). And the variance-covariance matrix of its prediction error is

$$\text{var}[\tilde{\mathbf{w}}(\mathbf{y}) - \mathbf{w}] = \sigma^2 \mathbf{H}_*, \tag{8.36}$$

where $\mathbf{H}_* = (\boldsymbol{\Lambda} - \mathbf{X}'\mathbf{H}_{yw})'(\mathbf{X}'\mathbf{X})^{-}(\boldsymbol{\Lambda} - \mathbf{X}'\mathbf{H}_{yw}) + \mathbf{H}_w - \mathbf{H}_{yw}'\mathbf{H}_{yw}$—refer to result (10.38). Then, clearly, the distribution of the prediction error is not known; regardless of its form, it depends on σ, the value of which is unknown. Thus, formula (8.6) for obtaining $100(1-\dot{\gamma})\%$ prediction sets is not applicable.

In effect, formula (8.6) is based on regarding the prediction error $\tilde{\mathbf{w}}(\mathbf{y})$–$\mathbf{w}$ as a "pivotal quantity." In the present setting, the prediction error cannot be so regarded. Suppose, however, that we divide the prediction error by $\hat{\sigma}$ [where $\hat{\sigma}^2 = \mathbf{y}'(\mathbf{I} - \mathbf{P}_{\mathbf{X}})\mathbf{y}/(N - P_*)$ is the usual unbiased estimator of σ^2], thereby obtaining a vector that is expressible as

$$(1/\hat{\sigma})[\tilde{\mathbf{w}}(\mathbf{y}) - \mathbf{w}] = [(1/\sigma^2)\mathbf{y}'(\mathbf{I} - \mathbf{P}_{\mathbf{X}})\mathbf{y}/(N - P_*)]^{-1/2}(1/\sigma)[\tilde{\mathbf{w}}(\mathbf{y}) - \mathbf{w}]. \tag{8.37}$$

Moreover,

$$\mathbf{y}'(\mathbf{I} - \mathbf{P}_{\mathbf{X}})\mathbf{y} = [(\mathbf{I} - \mathbf{P}_{\mathbf{X}})\mathbf{y}]'(\mathbf{I} - \mathbf{P}_{\mathbf{X}})\mathbf{y}, \tag{8.38}$$

so that $\mathbf{y}'(\mathbf{I} - \mathbf{P}_{\mathbf{X}})\mathbf{y}$ depends on $\mathbf{y}$ only through the value of the vector $(\mathbf{I} - \mathbf{P}_{\mathbf{X}})\mathbf{y}$, and

$$\begin{aligned} &\text{cov}[\tilde{\mathbf{w}}(\mathbf{y}) - \mathbf{w}, (\mathbf{I} - \mathbf{P}_{\mathbf{X}})\mathbf{y}] \\ &\quad = \sigma^2[\boldsymbol{\Lambda}'(\mathbf{X}'\mathbf{X})^{-}\mathbf{X}'(\mathbf{I} - \mathbf{P}_{\mathbf{X}}) + \mathbf{H}_{yw}'(\mathbf{I} - \mathbf{P}_{\mathbf{X}})(\mathbf{I} - \mathbf{P}_{\mathbf{X}}) - \mathbf{H}_{yw}'(\mathbf{I} - \mathbf{P}_{\mathbf{X}})] = \mathbf{0}. \end{aligned} \tag{8.39}$$

Now, suppose that the joint distribution of $\mathbf{y}$ and $\mathbf{w}$ is MVN. Then, $(1/\sigma^2)\mathbf{y}'(\mathbf{I} - \mathbf{P}_{\mathbf{X}})\mathbf{y}$ and $(1/\sigma)[\tilde{\mathbf{w}}(\mathbf{y}) - \mathbf{w}]$ are statistically independent [as is evident from results (8.38) and (8.39)]. Further,

$$(1/\sigma)[\tilde{\mathbf{w}}(\mathbf{y}) - \mathbf{w}] \sim N(\mathbf{0}, \mathbf{H}_*). \tag{8.40}$$

And

$$(1/\sigma^2)\mathbf{y}'(\mathbf{I} - \mathbf{P}_{\mathbf{X}})\mathbf{y} \sim \chi^2(N - P_*) \tag{8.41}$$

(as is evident upon recalling the results of Section 7.3). Thus, the distribution of $(1/\hat{\sigma})[\tilde{\mathbf{w}}(\mathbf{y})$–$\mathbf{w}]$ does not depend on any unknown parameters, and hence $(1/\hat{\sigma})[\tilde{\mathbf{w}}(\mathbf{y}) - \mathbf{w}]$ can serve [in lieu of $\tilde{\mathbf{w}}(\mathbf{y}) - \mathbf{w}$] as a "pivotal quantity."

It is now clear that a $100(1-\dot{\gamma})\%$ prediction set $S(\mathbf{y})$ (for the realization of $\mathbf{w}$) can be obtained by taking

$$S(\mathbf{y}) = \{\underline{\mathbf{w}} : \underline{\mathbf{w}} = \tilde{\mathbf{w}}(\mathbf{y}) - \hat{\sigma}\underline{\mathbf{u}},\ \underline{\mathbf{u}} \in S^*\}, \tag{8.42}$$

where S^* is any set of $M \times 1$ vectors for which

$$\Pr[(1/\hat{\sigma})[\tilde{\mathbf{w}}(\mathbf{y}) - \mathbf{w}] \in S^*] = 1 - \dot{\gamma}. \tag{8.43}$$

A prediction set of a particular kind. Suppose that the matrix $\mathbf{H}_*$ is nonsingular, as would be the case if $\mathbf{H}_w - \mathbf{H}'_{yw}\mathbf{H}_{yw}$ were nonsingular or equivalently if the matrix $\begin{pmatrix} \mathbf{I} & \mathbf{H}_{yw} \\ \mathbf{H}'_{yw} & \mathbf{H}_w \end{pmatrix}$ were nonsingular. Further, define $\mathbf{u} = (1/\hat{\sigma})[\tilde{\mathbf{w}}(\mathbf{y})-\mathbf{w}]$, and let

$$R = (1/M)\mathbf{u}'\mathbf{H}_*^{-1}\mathbf{u} = (1/M)[\tilde{\mathbf{w}}(\mathbf{y})-\mathbf{w}]'\mathbf{H}_*^{-1}[\tilde{\mathbf{w}}(\mathbf{y})-\mathbf{w}]/\hat{\sigma}^2. \tag{8.44}$$

And observe [in light of result (6.6.15) and the results of the preceding part of the present subsection]

$$R \sim SF(M,\, N-P_*), \tag{8.45}$$

and write $\bar{F}_{\dot{\gamma}}$ for $\bar{F}_{\dot{\gamma}}(M,\, N-P_*)$ [the upper $100\dot{\gamma}\%$ point of the $SF(M,\, N-P_*)$ distribution]. Then, among the choices for the set S^* defined by equality (8.43) is the set

$$S^* = \{\underline{\mathbf{u}} : (1/M)\underline{\mathbf{u}}'\mathbf{H}_*^{-1}\underline{\mathbf{u}} \le \bar{F}_{\dot{\gamma}}\}. \tag{8.46}$$

Corresponding to this choice for S^* is the $100(1-\dot{\gamma})\%$ prediction set

$$S(\mathbf{y}) = \{\underline{\mathbf{w}} : (1/M)[\underline{\mathbf{w}}-\tilde{\mathbf{w}}(\mathbf{y})]'\mathbf{H}_*^{-1}[\underline{\mathbf{w}}-\tilde{\mathbf{w}}(\mathbf{y})]/\hat{\sigma}^2 \le \bar{F}_{\dot{\gamma}}\} \tag{8.47}$$

defined by equality (8.42). This set is similar in form to the set (8.13).

Special case: $M = 1$. Let us continue the discussion of the preceding two parts of the present subsection (pertaining to the prediction of the realization of $\mathbf{w}$ in the special case where $\mathbf{y}$ follows a G–M model) by further specializing to the case where $M = 1$. In this further special case, let us write h_w, $\mathbf{h}_{yw}$, h_*, w, $\tilde{w}(\mathbf{y})$, and u for $\mathbf{H}_w$, $\mathbf{H}_{yw}$, $\mathbf{H}_*$, $\mathbf{w}$, $\tilde{\mathbf{w}}(\mathbf{y})$, and $\mathbf{u}$, respectively. Then, $u = \hat{\sigma}^{-1}[\tilde{w}(\mathbf{y})-w]$ and $u/h_*^{1/2} \sim St(N-P_*)$. And R is reexpressible as

$$R = u^2/h_* = (u/h_*^{1/2})^2. \tag{8.48}$$

For $0 < \alpha < 1$, let us write $\bar{t}_\alpha$ for $\bar{t}_\alpha(N-P_*)$ [i.e., for the upper $100\alpha\%$ point of the $St(N-P_*)$ distribution]. And observe [in light of result (6.4.26)] that (when $M=1$)

$$\bar{F}_{\dot{\gamma}}^{1/2} = \bar{t}_{\dot{\gamma}/2}.$$

Thus, when $M=1$, the $100(1-\dot{\gamma})\%$ prediction set (8.47) takes the form of the interval

$$\{\underline{w} : \tilde{w}(\mathbf{y}) - \bar{t}_{\dot{\gamma}/2}h_*^{1/2}\hat{\sigma} \le \underline{w} \le \tilde{w}(\mathbf{y}) + \bar{t}_{\dot{\gamma}/2}h_*^{1/2}\hat{\sigma}\}. \tag{8.49}$$

This interval is analogous to interval (8.14).

Two other $100(1-\dot{\gamma})\%$ prediction intervals are

$$\{\underline{w} : -\infty < \underline{w} \le \tilde{w}(\mathbf{y}) + \bar{t}_{\dot{\gamma}}h_*^{1/2}\hat{\sigma}\} \tag{8.50}$$

and

$$\{\underline{w} : \tilde{w}(\mathbf{y}) - \bar{t}_{\dot{\gamma}}h_*^{1/2}\hat{\sigma} \le \underline{w} < \infty\}. \tag{8.51}$$

These intervals are analogous to intervals (8.15) and (8.16). They represent the special cases of the $100(1-\dot{\gamma})\%$ prediction set (8.42) obtained when $M = 1$ and when [in the case of interval (8.50)] S^* is taken to be the set $\{\underline{u} : -\underline{u}/h_*^{1/2} \le \bar{t}_{\dot{\gamma}}\}$ or [in the case of interval (8.51)] to be the set $\{\underline{u} : \underline{u}/h_*^{1/2} \le \bar{t}_{\dot{\gamma}}\}$.

Simultaneous prediction intervals. Let us continue the discussion of the preceding three parts of the present subsection [pertaining to predictive inference about the realization of $\mathbf{w} = (w_1, w_2, \ldots, w_M)'$ when $\mathbf{y}$ follows a G–M model]. And in doing so, let us continue to suppose that the matrix $\mathbf{H}_* = \mathbf{H}_w - \mathbf{H}'_{yw}\mathbf{H}_{yw}$ is nonsingular.

Suppose that we wish to undertake predictive inference for the realizations of some or all of the random variables $w_1, w_2, \ldots, w_M$ or, more generally, for the realizations of some or all linear combinations of these random variables. Predictive inference for the realization of any particular linear combination, say for that of the linear combination $\boldsymbol{\delta}'\mathbf{w}$ (where $\boldsymbol{\delta} \ne \mathbf{0}$), might take the form of the interval

$$\{\underline{w} : \boldsymbol{\delta}'\tilde{\mathbf{w}}(\mathbf{y}) - \bar{t}_{\dot{\gamma}/2}(\boldsymbol{\delta}'\mathbf{H}_*\boldsymbol{\delta})^{1/2}\hat{\sigma} \le \underline{w} \le \boldsymbol{\delta}'\tilde{\mathbf{w}}(\mathbf{y}) + \bar{t}_{\dot{\gamma}/2}(\boldsymbol{\delta}'\mathbf{H}_*\boldsymbol{\delta})^{1/2}\hat{\sigma}\}, \tag{8.52}$$

$$\{\underline{w} : -\infty < \underline{w} \le \boldsymbol{\delta}'\tilde{\mathbf{w}}(\mathbf{y}) + \bar{t}_{\dot{\gamma}}(\boldsymbol{\delta}'\mathbf{H}_*\boldsymbol{\delta})^{1/2}\hat{\sigma}\}, \tag{8.53}$$

or

$$\{\underline{w} : \boldsymbol{\delta}'\tilde{\mathbf{w}}(\mathbf{y}) - \bar{t}_{\dot{\gamma}}(\boldsymbol{\delta}'\mathbf{H}_*\boldsymbol{\delta})^{1/2}\hat{\sigma} \le \underline{w} < \infty\}. \tag{8.54}$$

It follows from the results of the preceding part of the present subsection that when considered in "isolation," interval (8.52), (8.53), or (8.54) has a probability of coverage equal to $1-\dot{\gamma}$. However, when such an interval is obtained for the realization of each of a number of linear combinations, the probability of all of the intervals covering (or even that of all but a "small number" of the intervals covering) can be much less than $1-\dot{\gamma}$.

Accordingly, let Δ represent a finite or infinite collection of (nonnull) M-dimensional column vectors. And suppose that we wish to obtain a prediction interval for the realization of each of the linear combinations $\boldsymbol{\delta}'\mathbf{w}$ ($\boldsymbol{\delta} \in \Delta$) and that we wish to do so in such a way that the probability of simultaneous coverage equals $1-\dot{\gamma}$ or, more generally, that the probability of coverage by all but at most some number k of the intervals equals $1-\dot{\gamma}$. Such intervals can be obtained by taking (for each $\boldsymbol{\delta} \in \Delta$) the interval for the realization of $\boldsymbol{\delta}'\mathbf{w}$ to be a modified version of interval (8.52), (8.53), or (8.54) in which the constant $\bar{t}_{\dot{\gamma}/2}$ or $\bar{t}_{\dot{\gamma}}$ is replaced by a larger constant.

Let $\mathbf{R}_*$ represent an $M \times M$ nonsingular matrix such that $\mathbf{H}_* = \mathbf{R}_*'\mathbf{R}_*$. Further, let

$$\mathbf{z} = (1/\sigma)(\mathbf{R}_*^{-1})'[\tilde{\mathbf{w}}(\mathbf{y}) - \mathbf{w}] \quad \text{and} \quad v = (1/\sigma^2)\mathbf{y}'(\mathbf{I}-\mathbf{P}_{\mathbf{X}})\mathbf{y},$$

so that $\mathbf{z} \sim N(\mathbf{0}, \mathbf{I})$, $v \sim \chi^2(N-P_*)$, $\mathbf{z}$ and v are statistically independent, and (for every $M \times 1$ vector $\boldsymbol{\delta}$)

$$\frac{\boldsymbol{\delta}'\tilde{\mathbf{w}}(\mathbf{y}) - \boldsymbol{\delta}'\mathbf{w}}{(\boldsymbol{\delta}'\mathbf{H}_*\boldsymbol{\delta})^{1/2}\hat{\sigma}} = \frac{(\mathbf{R}_*\boldsymbol{\delta})'\mathbf{z}}{[(\mathbf{R}_*\boldsymbol{\delta})'\mathbf{R}_*\boldsymbol{\delta}]^{1/2}\sqrt{v/(N-P_*)}} \sim St(N-P_*) \tag{8.55}$$

—result (8.55) is analogous to result (8.20). Then, prediction intervals for the realizations of the random variables $\boldsymbol{\delta}'\mathbf{w}$ ($\boldsymbol{\delta} \in \Delta$) having a probability of simultaneous coverage equal to $1-\dot{\gamma}$ can be obtained from the application of a modified version of interval (8.52) in which $\bar{t}_{\dot{\gamma}/2}$ is replaced by the upper $100\dot{\gamma}\%$ point of the distribution of the random variable

$$\max_{\boldsymbol{\delta}\in\Delta} \frac{|(\mathbf{R}_*\boldsymbol{\delta})'\mathbf{z}|}{[(\mathbf{R}_*\boldsymbol{\delta})'\mathbf{R}_*\boldsymbol{\delta}]^{1/2}\sqrt{v/(N-P_*)}} \tag{8.56}$$

or from the application of a modified version of interval (8.53) or (8.54) in which $\bar{t}_{\dot{\gamma}}$ is replaced by the upper $100\dot{\gamma}\%$ point of the distribution of the random variable

$$\max_{\boldsymbol{\delta}\in\Delta} \frac{(\mathbf{R}_*\boldsymbol{\delta})'\mathbf{z}}{[(\mathbf{R}_*\boldsymbol{\delta})'\mathbf{R}_*\boldsymbol{\delta}]^{1/2}\sqrt{v/(N-P_*)}}. \tag{8.57}$$

More generally, for $k \ge 1$, prediction intervals for which the probability of coverage by all but at most k of the intervals can be obtained from the application of a modified version of interval (8.52) in which $\bar{t}_{\dot{\gamma}/2}$ is replaced by the upper $100\dot{\gamma}\%$ point of the distribution of the kth largest of the random variables $\dfrac{|(\mathbf{R}_*\boldsymbol{\delta})'\mathbf{z}|}{[(\mathbf{R}_*\boldsymbol{\delta})'\mathbf{R}_*\boldsymbol{\delta}]^{1/2}\sqrt{v/(N-P_*)}}$ ($\boldsymbol{\delta} \in \Delta$) or from the application of a modified version of interval (8.53) or (8.54) in which $\bar{t}_{\dot{\gamma}}$ is replaced by the upper $100\dot{\gamma}\%$ point of the distribution of the kth largest of the random variables $\dfrac{(\mathbf{R}_*\boldsymbol{\delta})'\mathbf{z}}{[(\mathbf{R}_*\boldsymbol{\delta})'\mathbf{R}_*\boldsymbol{\delta}]^{1/2}\sqrt{v/(N-P_*)}}$ ($\boldsymbol{\delta} \in \Delta$). The resultant intervals are analogous to those devised in Part 5 of Subsection a by employing a similar approach. And as in the case of the latter intervals, the requisite percentage points could be determined via Monte Carlo methods.

When $\Delta = \{\boldsymbol{\delta} \in \mathbb{R}^M : \boldsymbol{\delta} \neq \mathbf{0}\}$, the upper $100\dot{\gamma}\%$ point of the distribution of the random variable (8.56) equals $[M\bar{F}_{\dot{\gamma}}(M, N-P_*)]^{1/2}$, and $\boldsymbol{\delta}'\mathbf{w}$ is contained in the modified version of interval (8.52) [in which $\bar{t}_{\dot{\gamma}/2}$ has been replaced by the upper $100\dot{\gamma}\%$ point of the distribution of the random variable (8.56)] if and only if $\mathbf{w}$ is contained in the set (8.47)—refer to results (3.106) and (3.113) and to the ensuing discussion. When the members of Δ consist of the columns of $\mathbf{I}_M$, the linear combinations $\boldsymbol{\delta}'\mathbf{w}$ ($\boldsymbol{\delta} \in \Delta$) consist of the M random variables $w_1, w_2, \ldots, w_M$.

Extensions and limitations. Underlying the results of the preceding four parts of the present subsection is a supposition that the joint distribution of $\mathbf{y}$ and $\mathbf{w}$ is MVN. This supposition is stronger than necessary.

As in Section 7.3, let $\mathbf{L}$ represent an $N \times (N-P_*)$ matrix whose columns form an orthonormal basis for $\mathcal{N}(\mathbf{X}')$. And let $\mathbf{x} = (1/\sigma)\mathbf{L}'\mathbf{y}$, and continue to define $\mathbf{z} = (1/\sigma)(\mathbf{R}_*^{-1})'[\tilde{\mathbf{w}}(\mathbf{y})-\mathbf{w}]$. Further, let $\mathbf{t} = (1/\hat{\sigma})(\mathbf{R}_*^{-1})'[\tilde{\mathbf{w}}(\mathbf{y})-\mathbf{w}]$; and recalling result (3.21), observe that $\mathbf{x}'\mathbf{x} = (1/\sigma^2)\mathbf{y}'(\mathbf{I}-\mathbf{P_X})\mathbf{y}$ and hence that

$$\mathbf{t} = [\mathbf{x}'\mathbf{x}/(N-P_*)]^{-1/2}\mathbf{z}. \tag{8.58}$$

Observe also that

$$(1/\hat{\sigma})[\tilde{\mathbf{w}}(\mathbf{y})-\mathbf{w}] = \mathbf{R}_*'\mathbf{t}, \tag{8.59}$$

so that the distribution of the vector $(1/\hat{\sigma})[\tilde{\mathbf{w}}(\mathbf{y})-\mathbf{w}]$ is determined by the distribution of the vector $\mathbf{t}$.

When the joint distribution of $\mathbf{y}$ and $\mathbf{w}$ is MVN, then

$$\begin{pmatrix}\mathbf{z}\\ \mathbf{x}\end{pmatrix} \sim N(\mathbf{0}, \mathbf{I}), \tag{8.60}$$

as is evident upon observing the $\mathbf{L}'\mathbf{L} = \mathbf{I}$ and upon recalling expression (8.35) and recalling [from result (3.10)] that $\mathbf{L}'\mathbf{X} = \mathbf{0}$. And when $\begin{pmatrix}\mathbf{z}\\ \mathbf{x}\end{pmatrix} \sim N(\mathbf{0}, \mathbf{I})$,

$$\mathbf{t} \sim MVt(N-P_*, \mathbf{I}), \tag{8.61}$$

as is evident from expression (8.58). More generally, $\mathbf{t} \sim MVt(N-P_*, \mathbf{I})$ when the vector $\begin{pmatrix}(\mathbf{R}_*^{-1})'[\tilde{\mathbf{w}}(\mathbf{y})-\mathbf{w}]\\ \mathbf{L}'\mathbf{y}\end{pmatrix}$ has an absolutely continuous spherical distribution—refer to result (6.4.67).

Thus, the supposition (underlying the results of the preceding four parts of the present subsection) that the joint distribution of $\mathbf{y}$ and $\mathbf{w}$ is MVN could be replaced by the weaker supposition that the vector $\begin{pmatrix}(\mathbf{R}_*^{-1})'[\tilde{\mathbf{w}}(\mathbf{y})-\mathbf{w}]\\ \mathbf{L}'\mathbf{y}\end{pmatrix}$ has an MVN distribution or, more generally, that it has an absolutely continuous spherical distribution. In fact, ultimately, it could be replaced by a supposition that the vector $\mathbf{t} = (1/\hat{\sigma})(\mathbf{R}_*^{-1})'[\tilde{\mathbf{w}}(\mathbf{y})-\mathbf{w}]$ has an $MVt(N-P_*, \mathbf{I})$ distribution.

The results of the preceding four parts of the present subsection (which are for the case where $\mathbf{y}$ follows a G–M model) can be readily extended to the more general case where $\mathbf{y}$ follows an Aitken model. Let $N_* = \operatorname{rank}\mathbf{H}$. And let $\mathbf{Q}$ represent an $N \times N_*$ matrix such that $\mathbf{Q}'\mathbf{H}\mathbf{Q} = \mathbf{I}_{N_*}$—the existence of such a matrix follows, e.g., from Corollary 2.13.23, which implies the existence of an $N_* \times N$ matrix $\mathbf{P}$ (of full row rank) for which $\mathbf{H} = \mathbf{P}'\mathbf{P}$, and from Lemma 2.5.1, which implies the existence of a right inverse of $\mathbf{P}$. Then,

$$\operatorname{var}\begin{pmatrix}\mathbf{Q}'\mathbf{y}\\ \mathbf{w}\end{pmatrix} = \sigma^2\begin{bmatrix}\mathbf{I}_{N_*} & \mathbf{Q}'\mathbf{H}_{yw}\\ (\mathbf{Q}'\mathbf{H}_{yw})' & \mathbf{H}_w\end{bmatrix}.$$

Thus, by applying the results of the preceding four parts of the present subsection with $\mathbf{Q}'\mathbf{y}$ in place of $\mathbf{y}$ and with $\mathbf{Q}'\mathbf{H}_{yw}$ in place of $\mathbf{H}_{yw}$ (and with N_* in place of N), those results can be extended to the case where $\mathbf{y}$ follows an Aitken model.

In the case of the general linear model, the existence of a pivotal quantity for the realization of $\mathbf{w}$ is restricted to special cases. These special cases are ones where (as in the special case of the G–M or Aitken model) the dependence of the values of the matrices $\mathbf{V}_y(\boldsymbol{\theta})$, $\mathbf{V}_{yw}(\boldsymbol{\theta})$, and $\mathbf{V}_w(\boldsymbol{\theta})$ on the elements of the vector $\boldsymbol{\theta}$ is of a relatively simple form.

Exercises

Exercise 1. Take the context to be that of Section 7.1 (where a second-order G–M model is applied to the results of an experimental study of the yield of lettuce plants for purposes of making inferences about the response surface and various of its characteristics). Assume that the second-order regression coefficients $\beta_{11}, \beta_{12}, \beta_{13}, \beta_{22}, \beta_{23}$, and β_{33} are such that the matrix $\mathbf{A}$ is nonsingular. Assume also that the distribution of the vector $\mathbf{e}$ of residual effects in the G–M model is MVN (in which case the matrix $\hat{\mathbf{A}}$ is nonsingular with probability 1). Show that the large-sample distribution of the estimator $\hat{\mathbf{u}}_0$ of the stationary point $\mathbf{u}_0$ of the response surface is MVN with mean vector $\mathbf{u}_0$ and variance-covariance matrix

$$(1/4)\,\mathbf{A}^{-1}\mathrm{var}(\hat{\mathbf{a}} + 2\hat{\mathbf{A}}\mathbf{u}_0)\,\mathbf{A}^{-1}. \tag{E.1}$$

Do so by applying standard results on multi-parameter maximum likelihood estimation—refer, e.g., to McCulloch, Searle, and Neuhaus (2008, sec. S.4) and Zacks (1971, chap. 5).

Exercise 2. Taking the context to be that of Section 7.2a (and adopting the same notation and terminology as in Section 7.2a), show that for $\tilde{\mathbf{R}}'\mathbf{X}'\mathbf{y}$ to have the same expected value under the augmented G–M model as under the original G–M model, it is necessay (as well as sufficient) that $\mathbf{X}'\mathbf{Z} = \mathbf{0}$.

Exercise 3. Adopting the same notation and terminology as in Section 7.2, consider the expected value of the usual estimator $\mathbf{y}'(\mathbf{I}-\mathbf{P}_{\mathbf{X}})\mathbf{y}/[N-\mathrm{rank}(\mathbf{X})]$ of the variance of the residual effects of the (original) G–M model $\mathbf{y} = \mathbf{X}\boldsymbol{\beta} + \mathbf{e}$. How is the expected value of this estimator affected when the model equation is augmented via the inclusion of the additional "term" $\mathbf{Z}\boldsymbol{\tau}$? That is, what is the expected value of this estimator when its expected value is determined under the augmented G–M model (rather than under the original G–M model)?

Exercise 4. Adopting the same notation and terminology as in Sections 7.1 and 7.2, regard the lettuce yields as the observed values of the N $(= 20)$ elements of the random column vector $\mathbf{y}$, and take the model to be the "reduced" model derived from the second-order G–M model (in the 3 variables Cu, Mo, and Fe) by deleting the four terms involving the variable Mo—such a model would be consistent with an assumption that Mo is "more-or-less inert," i.e., has no discernible effect on the yield of lettuce.

(a) Compute the values of the least squares estimators of the regression coefficients ($\beta_1, \beta_2, \beta_4, \beta_{11}, \beta_{13}$, and β_{33}) of the reduced model, and determine the standard errors, estimated standard errors, and correlation matrix of these estimators.

(b) Determine the expected values of the least squares estimators (of $\beta_1, \beta_2, \beta_4, \beta_{11}, \beta_{13}$, and β_{33}) from Part (a) under the complete second-order G–M model (i.e., the model that includes the 4 terms involving the variable Mo), and determine (on the basis of the complete model) the estimated standard errors of these estimators.

(c) Find four linearly independent linear combinations of the four deleted regression coefficients ($\beta_3, \beta_{12}, \beta_{22}$, and β_{23}) that, under the complete second-order G–M model, would be estimable and whose least squares estimators would be uncorrelated, each with a standard error of σ; and compute the values of the least squares estimators of these linearly independent linear combinations, and determine the estimated standard errors of the least squares estimators.

Exercise 5. Suppose that $\mathbf{y}$ is an $N \times 1$ observable random vector that follows the G–M model. Show that

$$E(\mathbf{y}) = \mathbf{X}\tilde{\mathbf{R}}\mathbf{S}\boldsymbol{\alpha} + \mathbf{U}\boldsymbol{\eta},$$

where $\tilde{\mathbf{R}}, \mathbf{S}, \mathbf{U}, \boldsymbol{\alpha}$, and $\boldsymbol{\eta}$ are as defined in Section 7.3a.

Exercise 6. Taking the context to be that of Section 7.3, adopting the notation employed therein, supposing that the distribution of the vector $\mathbf{e}$ of residual effects (in the G–M model) is MVN, and assuming that $N > P_* + 2$, show that

$$\mathrm{E}[\tilde{F}(\boldsymbol{\alpha}^{(0)})] = \frac{N-P_*}{N-P_*-2}\left[1+\frac{(\boldsymbol{\alpha}-\boldsymbol{\alpha}^{(0)})'(\boldsymbol{\alpha}-\boldsymbol{\alpha}^{(0)})}{M_*\sigma^2}\right]$$
$$= \frac{N-P_*}{N-P_*-2}\left[1+\frac{(\boldsymbol{\tau}-\boldsymbol{\tau}^{(0)})'\mathbf{C}^-(\boldsymbol{\tau}-\boldsymbol{\tau}^{(0)})}{M_*\sigma^2}\right].$$

Exercise 7. Take the context to be that of Section 7.3, adopt the notation employed therein, and suppose that the distribution of the vector $\mathbf{e}$ of residual effects (in the G–M model) is MVN. For $\dot{\boldsymbol{\tau}} \in \mathcal{C}(\boldsymbol{\Lambda}')$, the distribution of $F(\dot{\boldsymbol{\tau}})$ is obtainable (upon setting $\dot{\boldsymbol{\alpha}} = \mathbf{S}'\dot{\boldsymbol{\tau}}$) from that of $\tilde{F}(\dot{\boldsymbol{\alpha}})$: in light of the relationship (3.32) and results (3.24) and (3.34),

$$F(\dot{\boldsymbol{\tau}}) \sim SF[M_*,\ N-P_*,\ (1/\sigma^2)(\dot{\boldsymbol{\tau}}-\boldsymbol{\tau})'\mathbf{C}^-(\dot{\boldsymbol{\tau}}-\boldsymbol{\tau})].$$

Provide an alternative derivation of the distribution of $F(\dot{\boldsymbol{\tau}})$ by (1) taking $\mathbf{b}$ to be a $P \times 1$ vector such that $\dot{\boldsymbol{\tau}} = \boldsymbol{\Lambda}'\mathbf{b}$ and establishing that $F(\dot{\boldsymbol{\tau}})$ is expressible in the form

$$F(\dot{\boldsymbol{\tau}}) = \frac{(1/\sigma^2)(\mathbf{y}-\mathbf{Xb})'\mathbf{P}_{\mathbf{X}\tilde{\mathbf{R}}}(\mathbf{y}-\mathbf{Xb})/M_*}{(1/\sigma^2)(\mathbf{y}-\mathbf{Xb})'(\mathbf{I}-\mathbf{P_X})(\mathbf{y}-\mathbf{Xb})/(N-P_*)}$$

and by (2) regarding $(1/\sigma^2)(\mathbf{y}-\mathbf{Xb})'\mathbf{P}_{\mathbf{X}\tilde{\mathbf{R}}}(\mathbf{y}-\mathbf{Xb})$ and $(1/\sigma^2)(\mathbf{y}-\mathbf{Xb})'(\mathbf{I}-\mathbf{P_X})(\mathbf{y}-\mathbf{Xb})$ as quadratic forms (in $\mathbf{y}-\mathbf{Xb}$) and making use of Corollaries 6.6.4 and 6.8.2.

Exercise 8. Take the context to be that of Section 7.3, and adopt the notation employed therein. Taking the model to be the canonical form of the G–M model and taking the distribution of the vector of residual effects to be $N(\mathbf{0},\ \sigma^2\mathbf{I})$, derive (in terms of the transformed vector $\mathbf{z}$) the size-$\dot{\gamma}$ likelihood ratio test of the null hypothesis $\tilde{H}_0 : \boldsymbol{\alpha} = \boldsymbol{\alpha}^{(0)}$ (versus the alternative hypothesis $\tilde{H}_1 : \boldsymbol{\alpha} \neq \boldsymbol{\alpha}^{(0)}$)—refer, e.g., to Casella and Berger (2002, sec. 8.2) for a discussion of likelihood ratio tests. Show that the size-$\dot{\gamma}$ likelihood ratio test is identical to the size-$\dot{\gamma}$ F test.

Exercise 9. Verify result (3.67).

Exercise 10. Verify the equivalence of conditions (3.59) and (3.68) and the equivalence of conditions (3.61) and (3.69).

Exercise 11. Taking the context to be that of Section 7.3 and adopting the notation employed therein, show that, corresponding to any two choices $\mathbf{S}_1$ and $\mathbf{S}_2$ for the matrix $\mathbf{S}$ (i.e., any two $M \times M_*$ matrices $\mathbf{S}_1$ and $\mathbf{S}_2$ such that $\mathbf{T}\mathbf{S}_1$ and $\mathbf{T}\mathbf{S}_2$ are orthogonal), there exists a unique $M_* \times M_*$ matrix $\mathbf{Q}$ such that

$$\mathbf{X}\tilde{\mathbf{R}}\mathbf{S}_2 = \mathbf{X}\tilde{\mathbf{R}}\mathbf{S}_1\mathbf{Q},$$

and show that this matrix is orthogonal.

Exercise 12. Taking the context to be that of Section 7.3, adopting the notation employed therein, and making use of the results of Exercise 11 (or otherwise), verify that none of the groups G_0, G_1, G_{01}, $G_2(\boldsymbol{\tau}^{(0)})$, $G_3(\boldsymbol{\tau}^{(0)})$, and G_4 (of transformations of $\mathbf{y}$) introduced in the final three parts of Subsection b of Section 7.3 vary with the choice of the matrices $\mathbf{S}$, $\mathbf{U}$, and $\mathbf{L}$.

Exercise 13. Consider the set $\tilde{A}_{\tilde{\boldsymbol{\delta}}}$ (of $\tilde{\boldsymbol{\delta}}'\boldsymbol{\alpha}$-values) defined (for $\tilde{\boldsymbol{\delta}} \in \tilde{\Delta}$) by expression (3.89) or (3.90). Underlying this definition is an implicit assumption that (for any $M_*{\times}1$ vector $\dot{\mathbf{t}}$) the function $f(\tilde{\boldsymbol{\delta}}) = |\tilde{\boldsymbol{\delta}}'\dot{\mathbf{t}}|/(\tilde{\boldsymbol{\delta}}'\tilde{\boldsymbol{\delta}})^{1/2}$, with domain $\{\tilde{\boldsymbol{\delta}} \in \tilde{\Delta} : \tilde{\boldsymbol{\delta}} \neq \mathbf{0}\}$, attains a maximum value. Show (1) that this function has a supremum and (2) that if the set

$$\ddot{\Delta} = \{\ddot{\boldsymbol{\delta}} \in \mathcal{R}^{M_*} : \exists \text{ a nonnull vector } \tilde{\boldsymbol{\delta}} \text{ in } \tilde{\Delta} \text{ such that } \ddot{\boldsymbol{\delta}} = (\tilde{\boldsymbol{\delta}}'\tilde{\boldsymbol{\delta}})^{-1/2}\tilde{\boldsymbol{\delta}}\}$$

is closed, then there exists a nonnull vector in $\tilde{\Delta}$ at which this function attains a maximum value.

Exercise 14. Take the context to be that of Section 7.3, and adopt the notation employed therein. Further, let $r = \hat{\sigma}c_{\dot{\gamma}}$, and for $\tilde{\boldsymbol{\delta}} \in \mathcal{R}^{M_*}$, let

$$\dot{A}_{\tilde{\delta}} = \{\dot{\tau} \in \mathcal{R}^1 : \dot{\tau} = \tilde{\delta}'\dot{\alpha},\ \dot{\alpha} \in \tilde{A}\},$$

where $\tilde{A}$ is the set (3.121) or (3.122) and is expressible in the form

$$\tilde{A} = \left\{\dot{\alpha} \in \mathcal{R}^{M_*} : \frac{|\dot{\delta}'(\dot{\alpha}-\hat{\alpha})|}{(\dot{\delta}'\dot{\delta})^{1/2}} \le r \text{ for every nonnull } \dot{\delta} \in \tilde{\Delta}\right\}.$$

For $\tilde{\delta} \notin \tilde{\Delta}$, $\dot{A}_{\tilde{\delta}}$ is identical to the set $\tilde{A}_{\tilde{\delta}}$ defined by expression (3.123). Show that for $\tilde{\delta} \in \tilde{\Delta}$, $\dot{A}_{\tilde{\delta}}$ is identical to the set $\tilde{A}_{\tilde{\delta}}$ defined by expression (3.89) or (3.90) or, equivalently, by the expression

$$\tilde{A}_{\tilde{\delta}} = \{\dot{\tau} \in \mathcal{R}^1 : |\dot{\tau} - \tilde{\delta}'\hat{\alpha})| \le (\tilde{\delta}'\tilde{\delta})^{1/2} r\}. \tag{E.2}$$

Exercise 15. Taking the sets Δ and $\tilde{\Delta}$, the matrix $\mathbf{C}$, and the random vector $\mathbf{t}$ to be as defined in Section 7.3, supposing that the set $\{\delta \in \Delta : \delta'\mathbf{C}\delta \ne 0\}$ consists of a finite number of vectors $\delta_1, \delta_2, \ldots, \delta_Q$, and letting $\mathbf{K}$ represent a $Q \times Q$ (correlation) matrix with ijth element $(\delta_i'\mathbf{C}\delta_i)^{-1/2}(\delta_j'\mathbf{C}\delta_j)^{-1/2}\delta_i'\mathbf{C}\delta_j$, show that

$$\max_{\{\tilde{\delta} \in \tilde{\Delta} :\, \tilde{\delta} \ne \mathbf{0}\}} \frac{|\tilde{\delta}'\mathbf{t}|}{(\tilde{\delta}'\tilde{\delta})^{1/2}} = \max(|u_1|, |u_2|, \ldots, |u_Q|),$$

where $u_1, u_2, \ldots, u_Q$ are the elements of a random vector $\mathbf{u}$ that has an $MVt(N{-}P_*,\ \mathbf{K})$ distribution.

Exercise 16. Define $\tilde{\Delta}$, $c_{\dot{\gamma}}$, and $\mathbf{t}$ as in Section 7.3c [so that $\mathbf{t}$ is an $M_* \times 1$ random vector that has an $MVt(N-P_*,\ \mathbf{I}_{M_*})$ distribution]. Show that

$$\Pr\left[\max_{\{\tilde{\delta} \in \tilde{\Delta} :\, \tilde{\delta} \ne \mathbf{0}\}} \frac{\tilde{\delta}'\mathbf{t}}{(\tilde{\delta}'\tilde{\delta})^{1/2}} > c_{\dot{\gamma}}\right] \ge \dot{\gamma}/2,$$

with equality holding if and only if there exists a nonnull $M_* \times 1$ vector $\ddot{\delta}$ (of norm 1) such that $(\tilde{\delta}'\tilde{\delta})^{-1/2}\tilde{\delta} = \ddot{\delta}$ for every nonnull vector $\tilde{\delta}$ in $\tilde{\Delta}$.

Exercise 17.

(a) Letting $E_1, E_2, \ldots, E_L$ represent any events in a probability space and (for any event E) denoting by $\overline{E}$ the complement of E, verify the following (Bonferroni) inequality:

$$\Pr(E_1 \cap E_2 \cap \cdots \cap E_L) \ge 1 - \textstyle\sum_{i=1}^{L} \Pr(\overline{E_i}).$$

(b) Take the context to be that of Section 7.3c, where $\mathbf{y}$ is an $N \times 1$ observable random vector that follows a G–M model with $N \times P$ model matrix $\mathbf{X}$ of rank P_* and where $\boldsymbol{\tau} = \boldsymbol{\Lambda}'\boldsymbol{\beta}$ is an $M \times 1$ vector of estimable linear combinations of the elements of $\boldsymbol{\beta}$ (such that $\boldsymbol{\Lambda} \ne \mathbf{0}$). Further, suppose that the distribution of the vector $\mathbf{e}$ of residual effects is $N(\mathbf{0}, \sigma^2\mathbf{I})$ (or is some other spherically symmetric distribution with mean vector $\mathbf{0}$ and variance-covariance matrix $\sigma^2\mathbf{I}$), let $\hat{\boldsymbol{\tau}}$ represent the least squares estimator of $\boldsymbol{\tau}$, let $\mathbf{C} = \boldsymbol{\Lambda}'(\mathbf{X}'\mathbf{X})^-\boldsymbol{\Lambda}$, let $\delta_1, \delta_2, \ldots, \delta_L$ represent $M \times 1$ vectors of constants such that (for $i = 1, 2, \ldots, L$) $\delta_i'\mathbf{C}\delta_i > 0$, let $\hat{\sigma}$ represent the positive square root of the usual estimator of σ^2 (i.e., the estimator obtained upon dividing the residual sum of squares by $N - P_*$), and let $\dot{\gamma}_1, \dot{\gamma}_2, \ldots, \dot{\gamma}_L$ represent positive scalars such that $\sum_{i=1}^{L} \dot{\gamma}_i = \dot{\gamma}$. And (for $i = 1, 2, \ldots, L$) denote by $A_i(\mathbf{y})$ a confidence interval for $\delta_i'\boldsymbol{\tau}$ with end points

$$\delta_i'\hat{\boldsymbol{\tau}} \pm (\delta_i'\mathbf{C}\delta_i)^{1/2}\hat{\sigma}\,\bar{t}_{\dot{\gamma}_i/2}(N - P_*).$$

Use the result of Part (a) to show that

$$\Pr[\delta_i'\boldsymbol{\tau} \in A_i(\mathbf{y})\ \ (i = 1, 2, \ldots, L)] \ge 1 - \dot{\gamma}$$

and hence that the intervals $A_1(\mathbf{y}), A_2(\mathbf{y}), \ldots, A_L(\mathbf{y})$ are conservative in the sense that their probability of simultaneous coverage is greater than or equal to $1{-}\dot{\gamma}$—when $\dot{\gamma}_1 = \dot{\gamma}_2 = \cdots = \dot{\gamma}_L$, the end points of interval $A_i(\mathbf{y})$ become

$$\delta_i'\hat{\boldsymbol{\tau}} \pm (\delta_i'\mathbf{C}\delta_i)^{1/2}\hat{\sigma}\,\bar{t}_{\dot{\gamma}/(2L)}(N - P_*),$$

and the intervals $A_1(\mathbf{y}), A_2(\mathbf{y}), \ldots, A_L(\mathbf{y})$ are referred to as Bonferroni t-intervals.

Exercise 18. Suppose that the data (of Section 4.2b) on the lethal dose of ouabain in cats are regarded as the observed values of the elements $y_1, y_2, \ldots, y_N$ of an $N(=41)$-dimensional observable random vector $\mathbf{y}$ that follows a G–M model. Suppose further that (for $i = 1, 2, \ldots, 41$) $\mathrm{E}(y_i) = \delta(u_i)$, where $u_1, u_2, \ldots, u_{41}$ are the values of the rate u of injection and where $\delta(u)$ is the third-degree polynomial

$$\delta(u) = \beta_1 + \beta_2 u + \beta_3 u^2 + \beta_4 u^3.$$

And suppose that the distribution of the vector $\mathbf{e}$ of residual effects is $N(\mathbf{0}, \sigma^2\mathbf{I})$ (or is some other spherically symmetric distribution with mean vector $\mathbf{0}$ and variance-covariance matrix $\sigma^2\mathbf{I}$).

(a) Compute the values of the least squares estimators $\hat{\beta}_1, \hat{\beta}_2, \hat{\beta}_3$, and $\hat{\beta}_4$ of $\beta_1, \beta_2, \beta_3$, and β_4, respectively, and the value of the positive square root $\hat{\sigma}$ of the usual unbiased estimator of σ^2—it follows from the results of Section 5.3d that P_* $(= \text{rank}\,\mathbf{X}) = P = 4$, in which case $N - P_* = N - P = 37$, and that $\beta_1, \beta_2, \beta_3$, and β_4 are estimable.

(b) Find the values of $\bar{t}_{.05}(37)$ and $[4\bar{F}_{.10}(4, 37)]^{1/2}$, which would be needed if interval $I_{\mathbf{u}}^{(1)}(\mathbf{y})$ with end points (3.169) and interval $I_{\mathbf{u}}^{(2)}(\mathbf{y})$ with end points (3.170) (where in both cases $\dot{\gamma}$ is taken to be .10) were used to construct confidence bands for the response surface $\delta(u)$.

(c) By (for example) making use of the results in Liu's (2011) Appendix E, compute Monte Carlo approximations to the constants $c_{.10}$ and $c_{.10}^*$ that would be needed if interval $I_{\mathbf{u}}^{(3)}(\mathbf{y})$ with end points (3.171) and interval $I_{\mathbf{u}}^{(4)}(\mathbf{y})$ with end points (3.172) were used to construct confidence bands for $\delta(u)$; compute the approximations for the case where u is restricted to the interval $1 \leq u \leq 8$, and (for purposes of comparison) also compute $c_{.10}$ for the case where u is unrestricted.

(d) Plot (as a function of u) the value of the least squares estimator $\hat{\delta}(u) = \hat{\beta}_1 + \hat{\beta}_2 u + \hat{\beta}_3 u^2 + \hat{\beta}_4 u^3$ and (taking $\dot{\gamma} = .10$) the values of the end points (3.169) and (3.170) of intervals $I_{\mathbf{u}}^{(1)}(\mathbf{y})$ and $I_{\mathbf{u}}^{(2)}(\mathbf{y})$ and the values of the approximations to the end points (3.171) and (3.172) of intervals $I_{\mathbf{u}}^{(3)}(\mathbf{y})$ and $I_{\mathbf{u}}^{(4)}(\mathbf{y})$ obtained upon replacing $c_{.10}$ and $c_{.10}^*$ with their Monte Carlo approximations—assume (for purposes of creating the plot and for approximating $c_{.10}$ and $c_{.10}^*$) that u is restricted to the interval $1 \leq u \leq 8$.

Exercise 19. Taking the setting to be that of the final four parts of Section 7.3b (and adopting the notation and terminology employed therein) and taking $\tilde{G}_2$ to be the group of transformations consisting of the totality of the groups $\tilde{G}_2(\boldsymbol{\alpha}^{(0)})$ $(\boldsymbol{\alpha}^{(0)} \in \mathbb{R}^{M_*})$ and $\tilde{G}_3$ the group consisting of the totality of the groups $\tilde{G}_3(\boldsymbol{\alpha}^{(0)})$ $(\boldsymbol{\alpha}^{(0)} \in \mathbb{R}^{M_*})$, show that (1) if a confidence set $\tilde{A}(\mathbf{z})$ for $\boldsymbol{\alpha}$ is equivariant with respect to the groups $\tilde{G}_0$ and $\tilde{G}_2(\mathbf{0})$, then it is equivariant with respect to the group $\tilde{G}_2$ and (2) if a confidence set $\tilde{A}(\mathbf{z})$ for $\boldsymbol{\alpha}$ is equivariant with respect to the groups $\tilde{G}_0$ and $\tilde{G}_3(\mathbf{0})$, then it is equivariant with respect to the group $\tilde{G}_3$.

Exercise 20. Taking the setting to be that of Section 7.4e (and adopting the assumption of normality and the notation and terminology employed therein), suppose that $M_* = 1$, and write $\hat{\alpha}$ for $\hat{\boldsymbol{\alpha}}$, α for $\boldsymbol{\alpha}$, and $\alpha^{(0)}$ for $\boldsymbol{\alpha}^{(0)}$. Further, let $\tilde{\phi}(\hat{\alpha}, \hat{\boldsymbol{\eta}}, \mathbf{d})$ represent the critical function of an arbitrary (possibly randomized) level-$\dot{\gamma}$ test of the null hypothesis $\tilde{H}_0 : \alpha = \alpha^{(0)}$ versus the alternative hypothesis $\tilde{H}_1 : \alpha \neq \alpha^{(0)}$, and let $\tilde{\gamma}(\alpha, \boldsymbol{\eta}, \sigma)$ represent its power function $\{$so that $\tilde{\gamma}(\alpha, \boldsymbol{\eta}, \sigma) = \mathrm{E}[\tilde{\phi}(\hat{\alpha}, \hat{\boldsymbol{\eta}}, \mathbf{d})]\}$. And define $s = (\hat{\alpha} - \alpha^{(0)})/[(\hat{\alpha} - \alpha^{(0)})^2 + \mathbf{d}'\mathbf{d}]^{1/2}$ and $w = (\hat{\alpha} - \alpha^{(0)})^2 + \mathbf{d}'\mathbf{d}$, denote by $\ddot{\phi}(s, w, \hat{\boldsymbol{\eta}})$ the critical function of a level-$\dot{\gamma}$ test (of $\tilde{H}_0$ versus $\tilde{H}_1$) that depends on $\hat{\alpha}$, $\hat{\boldsymbol{\eta}}$, and $\mathbf{d}$ only through the values of s, w, and $\hat{\boldsymbol{\eta}}$, and write E_0 for the expectation operator E in the special case where $\alpha = \alpha^{(0)}$.

(a) Show that if the level-$\dot{\gamma}$ test with critical function $\tilde{\phi}(\hat{\alpha}, \hat{\boldsymbol{\eta}}, \mathbf{d})$ is an unbiased test, then

$$\tilde{\gamma}(\alpha^{(0)}, \boldsymbol{\eta}, \sigma) = \dot{\gamma} \quad \text{for all } \boldsymbol{\eta} \text{ and } \sigma \tag{E.3}$$

and

$$\left.\frac{\partial \tilde{\gamma}(\alpha, \boldsymbol{\eta}, \sigma)}{\partial \alpha}\right|_{\alpha = \alpha^{(0)}} = 0 \quad \text{for all } \boldsymbol{\eta} \text{ and } \sigma. \tag{E.4}$$

(b) Show that

$$\frac{\partial \dot{\gamma}(\alpha, \boldsymbol{\eta}, \sigma)}{\partial \alpha} = \mathrm{E}\left[\frac{\hat{\alpha} - \alpha}{\sigma^2} \tilde{\phi}(\hat{\alpha}, \hat{\boldsymbol{\eta}}, \mathbf{d})\right]. \tag{E.5}$$

(c) Show that corresponding to the level-$\dot{\gamma}$ test (of $\tilde{H}_0$) with critical function $\tilde{\phi}(\hat{\alpha}, \hat{\boldsymbol{\eta}}, \mathbf{d})$, there is a (level-$\dot{\gamma}$) test that depends on $\hat{\alpha}$, $\hat{\boldsymbol{\eta}}$, and $\mathbf{d}$ only through the values of s, w, and $\hat{\boldsymbol{\eta}}$ and that has the same power function as the test with critical function $\tilde{\phi}(\hat{\alpha}, \hat{\boldsymbol{\eta}}, \mathbf{d})$.

(d) Show that when $\alpha = \alpha^{(0)}$, (1) w and $\hat{\boldsymbol{\eta}}$ form a complete sufficient statistic and (2) s is statistically independent of w and $\hat{\boldsymbol{\eta}}$ and has an absolutely continuous distribution, the pdf of which is the pdf $h^*(\cdot)$ given by result (6.4.7).

(e) Show that when the critical function $\tilde{\phi}(\hat{\alpha}, \hat{\boldsymbol{\eta}}, \mathbf{d})$ (of the level-$\dot{\gamma}$ test) is of the form $\ddot{\phi}(s, w, \hat{\boldsymbol{\eta}})$, condition (E.3) is equivalent to the condition

$$\mathrm{E}_0[\ddot{\phi}(s, w, \hat{\boldsymbol{\eta}}) \mid w, \hat{\boldsymbol{\eta}}] = \dot{\gamma} \quad (\text{wp1}) \tag{E.6}$$

and condition (E.4) is equivalent to the condition

$$\mathrm{E}_0[s w^{1/2} \ddot{\phi}(s, w, \hat{\boldsymbol{\eta}}) \mid w, \hat{\boldsymbol{\eta}}] = 0 \quad (\text{wp1}). \tag{E.7}$$

(f) Using the generalized Neyman–Pearson lemma (Lehmann and Romano 2005b, sec. 3.6; Shao 2010, sec. 6.1.1), show that among critical functions of the form $\ddot{\phi}(s, w, \hat{\boldsymbol{\eta}})$ that satisfy (for any particular values of w and $\hat{\boldsymbol{\eta}}$) the conditions

$$\mathrm{E}_0[\ddot{\phi}(s, w, \hat{\boldsymbol{\eta}}) \mid w, \hat{\boldsymbol{\eta}}] = \dot{\gamma} \quad \text{and} \quad \mathrm{E}_0[s w^{1/2} \ddot{\phi}(s, w, \hat{\boldsymbol{\eta}}) \mid w, \hat{\boldsymbol{\eta}}] = 0, \tag{E.8}$$

the value of $\mathrm{E}[\ddot{\phi}(s, w, \hat{\boldsymbol{\eta}}) \mid w, \hat{\boldsymbol{\eta}}]$ (at those particular values of w and $\hat{\boldsymbol{\eta}}$) is maximized [for any particular value of α ($\neq \alpha^{(0)}$) and any particular values of $\boldsymbol{\eta}$ and σ] when the critical function is taken to be the critical function $\ddot{\phi}^*(s, w, \hat{\boldsymbol{\eta}})$ defined (for all s, w, and $\hat{\boldsymbol{\eta}}$) as follows:

$$\ddot{\phi}^*(s, w, \hat{\boldsymbol{\eta}}) = \begin{cases} 1, & \text{if } s < -c \text{ or } s > c, \\ 0, & \text{if } -c \leq s \leq c, \end{cases}$$

where c is the upper $100(\dot{\gamma}/2)\%$ point of the distribution with pdf $h^*(\cdot)$ [given by result (6.4.7)].

(g) Use the results of the preceding parts to conclude that among all level-$\dot{\gamma}$ tests of $\tilde{H}_0$ versus $\tilde{H}_1$ that are unbiased, the size-$\dot{\gamma}$ two-sided t test is a UMP test.

Exercise 21. Taking the setting to be that of Section 7.4e and adopting the assumption of normality and the notation and terminology employed therein, let $\tilde{\gamma}(\boldsymbol{\alpha}, \boldsymbol{\eta}, \sigma)$ represent the power function of a size-$\dot{\gamma}$ similar test of $H_0: \boldsymbol{\tau} = \boldsymbol{\tau}^{(0)}$ or $\tilde{H}_0: \boldsymbol{\alpha} = \boldsymbol{\alpha}^{(0)}$ versus $H_1: \boldsymbol{\tau} \neq \boldsymbol{\tau}^{(0)}$ or $\tilde{H}_1: \boldsymbol{\alpha} \neq \boldsymbol{\alpha}^{(0)}$. Show that $\min_{\boldsymbol{\alpha} \in S(\boldsymbol{\alpha}^{(0)}, \rho)} \tilde{\gamma}(\boldsymbol{\alpha}, \boldsymbol{\eta}, \sigma)$ attains its maximum value when the size-$\dot{\gamma}$ similar test is taken to be the size-$\dot{\gamma}$ F test.

Exercise 22. Taking the setting to be that of Section 7.4e and adopting the assumption of normality and the notation and terminology employed therein, let $\tilde{\phi}(\hat{\boldsymbol{\alpha}}, \hat{\boldsymbol{\eta}}, \mathbf{d})$ represent the critical function of an arbitrary size-$\dot{\gamma}$ test of the null hypothesis $\tilde{H}_0 : \boldsymbol{\alpha} = \boldsymbol{\alpha}^{(0)}$ versus the alternative hypothesis $\tilde{H}_1: \boldsymbol{\alpha} \neq \boldsymbol{\alpha}^{(0)}$. Further, let $\tilde{\gamma}(\cdot, \cdot, \cdot; \tilde{\phi})$ represent the power function of the test with critical function $\tilde{\phi}(\cdot, \cdot, \cdot)$, so that $\tilde{\gamma}(\boldsymbol{\alpha}, \boldsymbol{\eta}, \sigma; \tilde{\phi}) = \mathrm{E}[\tilde{\phi}(\hat{\boldsymbol{\alpha}}, \hat{\boldsymbol{\eta}}, \mathbf{d})]$. And take $\tilde{\gamma}^*(\cdot, \cdot, \cdot)$ to be the function defined as follows:

$$\tilde{\gamma}^*(\boldsymbol{\alpha}, \boldsymbol{\eta}, \sigma) = \sup_{\tilde{\phi}} \tilde{\gamma}(\boldsymbol{\alpha}, \boldsymbol{\eta}, \sigma; \tilde{\phi}).$$

This function is called the *envelope power function*.

(a) Show that $\tilde{\gamma}^*(\boldsymbol{\alpha}, \boldsymbol{\eta}, \sigma)$ depends on $\boldsymbol{\alpha}$ only through the value of $(\boldsymbol{\alpha} - \boldsymbol{\alpha}^{(0)})'(\boldsymbol{\alpha} - \boldsymbol{\alpha}^{(0)})$.

(b) Let $\tilde{\phi}_F(\hat{\boldsymbol{\alpha}}, \hat{\boldsymbol{\eta}}, \mathbf{d})$ represent the critical function of the size-$\dot{\gamma}$ F test of $\tilde{H}_0$ versus $\tilde{H}_1$. And as a basis for evaluating the test with critical function $\tilde{\phi}(\cdot, \cdot, \cdot)$, consider the use of the criterion

$$\max_{\boldsymbol{\alpha} \in S(\boldsymbol{\alpha}^{(0)}, \rho)} [\tilde{\gamma}^*(\boldsymbol{\alpha}, \boldsymbol{\eta}, \sigma) - \tilde{\gamma}(\boldsymbol{\alpha}, \boldsymbol{\eta}, \sigma; \tilde{\phi})], \tag{E.9}$$

which reflects [for $\boldsymbol{\alpha} \in S(\boldsymbol{\alpha}^{(0)}, \rho)$] the extent to which the power function of the test deviates from the envelope power function. Using the result of Exercise 21 (or otherwise), show that the size-$\dot{\gamma}$ F test is the "most stringent" size-$\dot{\gamma}$ similar test in the sense that (for "every" value of ρ) the value attained by the quantity (E.9) when $\tilde{\phi} = \tilde{\phi}_F$ is a minimum among those attained when $\tilde{\phi}$ is the critical function of some (size-$\dot{\gamma}$) similar test.

Exercise 23. Take the setting to be that of Section 7.5a (and adopt the assumption of normality and the notation and terminology employed therein). Show that among all tests of the null hypothesis $H_0^+ : \tau \le \tau^{(0)}$ or $\tilde{H}_0^+ : \alpha \le \alpha^{(0)}$ (versus the alternative hypothesis $H_1^+ : \tau > \tau^{(0)}$ or $\tilde{H}_1^+ : \alpha > \alpha^{(0)}$) that are of level $\dot{\gamma}$ and that are unbiased, the size-$\dot{\gamma}$ one-sided t test is a UMP test. (*Hint*. Proceed stepwise as in Exercise 20.)

Exercise 24.

(a) Let (for an arbitrary positive integer M) $f_M(\cdot)$ represent the pdf of a $\chi^2(M)$ distribution. Show that (for $0 < x < \infty$)
$$x f_M(x) = M f_{M+2}(x).$$

(b) Verify [by using Part (a) or by other means] result (6.22).

Exercise 25. This exercise is to be regarded as a continuation of Exercise 18. Suppose (as in Exercise 18) that the data (of Section 4.2b) on the lethal dose of ouabain in cats are regarded as the observed values of the elements $y_1, y_2, \ldots, y_N$ of an $N(= 41)$-dimensional observable random vector $\mathbf{y}$ that follows a G–M model. Suppose further that (for $i = 1, 2, \ldots, 41$) $E(y_i) = \delta(u_i)$, where $u_1, u_2, \ldots, u_{41}$ are the values of the rate u of injection and where $\delta(u)$ is the third-degree polynomial
$$\delta(u) = \beta_1 + \beta_2 u + \beta_3 u^2 + \beta_4 u^3.$$
And suppose that the distribution of the vector $\mathbf{e}$ of residual effects is $N(\mathbf{0}, \sigma^2\mathbf{I})$.

(a) Determine for $\dot{\gamma} = 0.10$ and also for $\dot{\gamma} = 0.05$ (1) the value of the $100(1-\dot{\gamma})\%$ lower confidence bound for σ provided by the left end point of interval (6.2) and (2) the value of the $100(1-\dot{\gamma})\%$ upper confidence bound for σ provided by the right end point of interval (6.3).

(b) Obtain [via an implementation of interval (6.23)] a 90% two-sided strictly unbiased confidence interval for σ.

Exercise 26. Take the setting to be that of the final part of Section 7.6c, and adopt the notation and terminology employed therein. In particular, take the canonical form of the G–M model to be that identified with the special case where $M_* = P_*$, so that $\boldsymbol{\alpha}$ and $\hat{\boldsymbol{\alpha}}$ are P_*-dimensional. Show that the (size-$\dot{\gamma}$) test of the null hypothesis $H_0^+ : \sigma \le \sigma_0$ (versus the alternative hypothesis $H_1^+ : \sigma > \sigma_0$) with critical region C^+ is UMP among all level-$\dot{\gamma}$ tests. Do so by carrying out the following steps.

(a) Let $\phi(T, \hat{\boldsymbol{\alpha}})$ represent the critical function of a level-$\dot{\gamma}$ test of H_0^+ versus H_1^+ [that depends on the vector $\mathbf{d}$ only through the value of T $(= \mathbf{d}'\mathbf{d}/\sigma_0^2)$]. And let $\gamma(\sigma, \boldsymbol{\alpha})$ represent the power function of the test with critical function $\phi(T, \hat{\boldsymbol{\alpha}})$. Further, let σ_* represent any particular value of σ greater than σ_0, let $\boldsymbol{\alpha}_*$ represent any particular value of $\boldsymbol{\alpha}$, and denote by $h(\cdot\,; \sigma)$ the pdf of the distribution of T, by $f(\cdot\,; \boldsymbol{\alpha}, \sigma)$ the pdf of the distribution of $\hat{\boldsymbol{\alpha}}$, and by $s(\cdot)$ the pdf of the $N(\boldsymbol{\alpha}_*, \sigma_*^2 - \sigma_0^2)$ distribution. Show (1) that
$$\int_{\mathcal{R}^{P_*}} \gamma(\sigma_0, \boldsymbol{\alpha})\, s(\boldsymbol{\alpha})\, d\boldsymbol{\alpha} \le \dot{\gamma} \tag{E.10}$$
and (2) that
$$\int_{\mathcal{R}^{P_*}} \gamma(\sigma_0, \boldsymbol{\alpha})\, s(\boldsymbol{\alpha})\, d\boldsymbol{\alpha} = \int_{\mathcal{R}^{P_*}} \int_0^\infty \phi(t, \hat{\boldsymbol{\alpha}})\, h(t; \sigma_0)\, f(\hat{\boldsymbol{\alpha}}; \boldsymbol{\alpha}_*, \sigma_*)\, dt\, d\hat{\boldsymbol{\alpha}}. \tag{E.11}$$

(b) By, for example, using a version of the Neyman–Pearson lemma like that stated by Casella and Berger (2002) in the form of their Theorem 8.3.12, show that among those choices for the critical function $\phi(T, \hat{\boldsymbol{\alpha}})$ for which the power function $\gamma(\cdot, \cdot)$ satisfies condition (E.10), $\gamma(\sigma_*, \boldsymbol{\alpha}_*)$ can be maximized by taking $\phi(T, \hat{\boldsymbol{\alpha}})$ to be the critical function $\phi_*(T, \hat{\boldsymbol{\alpha}})$ defined as follows:

$$\phi_*(t, \hat{\boldsymbol{\alpha}}) = \begin{cases} 1, & \text{when } t > \bar{\chi}^2_{\dot{\gamma}}, \\ 0, & \text{when } t \leq \bar{\chi}^2_{\dot{\gamma}}. \end{cases}$$

(c) Use the results of Parts (a) and (b) to reach the desired conclusion, that is, to show that the test of H_0^+ (versus H_1^+) with critical region C^+ is UMP among all level-$\dot{\gamma}$ tests.

Exercise 27. Take the context to be that of Section 7.7a, and adopt the notation employed therein. Using Markov's inequality (e.g., Casella and Berger 2002, lemma 3.8.3; Bickel and Doksum 2001, sec. A.15) or otherwise, verify inequality (7.12), that is, the inequality

$$\Pr(x_i > c \text{ for } k \text{ or more values of } i) \leq (1/k)\textstyle\sum_{i=1}^{M} \Pr(x_i > c).$$

Exercise 28.

(a) Letting X represent any random variable whose values are confined to the interval $[0, 1]$ and letting κ $(0 < \kappa < 1)$ represent a constant, show (1) that

$$\mathrm{E}(X) \leq \kappa \Pr(X \leq \kappa) + \Pr(X > \kappa) \tag{E.12}$$

and then use inequality (E.12) along with Markov's inequality (e.g., Casella and Berger 2002, sec. 3.8) to (2) show that

$$\frac{\mathrm{E}(X) - \kappa}{1 - \kappa} \leq \Pr(X > \kappa) \leq \frac{\mathrm{E}(X)}{\kappa}. \tag{E.13}$$

(b) Show that the requirement that the false discovery rate (FDR) satisfy condition (7.45) and the requirement that the false discovery proportion (FDP) satisfy condition (7.46) are related as follows:
(1) if FDR $\leq \delta$, then $\Pr(\text{FDP} > \kappa) \leq \delta/\kappa$; and
(2) if $\Pr(\text{FDP} > \kappa) \leq \epsilon$, then FDR $\leq \epsilon + \kappa(1 - \epsilon)$.

Exercise 29. Taking the setting to be that of Section 7.7 and adopting the terminology and notation employed therein, consider the use of a multiple-comparison procedure in testing (for every $i \in I = \{1, 2, \ldots, M\}$) the null hypothesis $H_i^{(0)} : \tau_i = \tau_i^{(0)}$ versus the alternative hypothesis $H_i^{(1)} : \tau_i \neq \tau_i^{(0)}$ (or $H_i^{(0)} : \tau_i \leq \tau_i^{(0)}$ versus $H_i^{(1)} : \tau_i > \tau_i^{(0)}$). And denote by T the set of values of $i \in I$ for which $H_i^{(0)}$ is true and by F the set for which $H_i^{(0)}$ is false. Further, denote by M_T the size of the set T and by X_T the number of values of $i \in T$ for which $H_i^{(0)}$ is rejected. Similarly, denote by M_F the size of the set F and by X_F the number of values of $i \in F$ for which $H_i^{(0)}$ is rejected. Show that

(a) in the special case where $M_T = 0$, FWER = FDR = 0;

(b) in the special case where $M_T = M$, FWER = FDR; and

(c) in the special case where $0 < M_T < M$, FWER $\geq$ FDR, with equality holding if and only if $\Pr(X_T > 0 \text{ and } X_F > 0) = 0$.

Exercise 30.

(a) Let $\hat{p}_1, \hat{p}_2, \ldots, \hat{p}_t$ represent p-values [so that $\Pr(\hat{p}_i \leq u) \leq u$ for $i = 1, 2, \ldots, t$ and for every $u \in (0, 1)$]. Further, let $\hat{p}_{(j)} = \hat{p}_{i_j}$ $(j = 1, 2, \ldots, t)$, where $i_1, i_2, \ldots, i_t$ is a permutation of the first t positive integers $1, 2, \ldots, t$ such that $\hat{p}_{i_1} \leq \hat{p}_{i_2} \leq \cdots \leq \hat{p}_{i_t}$. And let s represent a positive integer such that $s \le t$, and let $c_0, c_1, \ldots, c_s$ represent constants such that $0 = c_0 \le c_1 \le \cdots \le c_s \le 1$. Show that

$$\Pr(\hat{p}_{(j)} \le c_j \text{ for 1 or more values of } j \in \{1, 2, \ldots, s\}) \le t \textstyle\sum_{j=1}^{s} (c_j - c_{j-1})/j. \qquad \text{(E.14)}$$

(b) Take the setting to be that of Section 7.7d, and adopt the notation and terminology employed therein. And suppose that the α_j's of the step-down multiple-comparison procedure for testing the null hypotheses $H_1^{(0)}, H_2^{(0)}, \ldots, H_M^{(0)}$ are of the form

$$\alpha_j = \bar{t}_{([j\kappa]+1)\dot{\gamma}/\{2(M+[j\kappa]+1-j)\}}(N - P_*) \quad (j = 1, 2, \ldots, M) \qquad \text{(E.15)}$$

[where $\dot{\gamma} \in (0, 1)$].

(1) Show that if

$$\Pr(|t_{u;T}| > \bar{t}_{u\dot{\gamma}/(2M_T)}(N - P_*) \text{ for 1 or more values of } u \in \{1, 2, \ldots, K\}) \le \epsilon, \qquad \text{(E.16)}$$

then the step-down procedure [with α_j's of the form (E.15)] is such that $\Pr(\text{FDP} > \kappa) \le \epsilon$.

(2) Reexpress the left side of inequality (E.16) in terms of the left side of inequality (E.14).

(3) Use Part (a) to show that

$$\Pr(|t_{u;T}| > \bar{t}_{u\dot{\gamma}/(2M_T)}(N - P_*) \text{ for 1 or more values of } u \in \{1, 2, \ldots, K\}) \le \dot{\gamma} \textstyle\sum_{u=1}^{[M\kappa]+1} 1/u. \qquad \text{(E.17)}$$

(4) Show that the version of the step-down procedure [with α_j's of the form (E.15)] obtained upon setting $\dot{\gamma} = \epsilon / \sum_{u=1}^{[M\kappa]+1} 1/u$ is such that $\Pr(\text{FDP} > \kappa) \le \epsilon$.

Exercise 31. Take the setting to be that of Section 7.7e, and adopt the notation and terminology employed therein. And take $\alpha_1^*, \alpha_2^*, \ldots, \alpha_M^*$ to be scalars defined implicitly (in terms of $\alpha_1, \alpha_2, \ldots, \alpha_M$) by the equalities

$$\alpha'_k = \textstyle\sum_{j=1}^{k} \alpha_j^* \qquad (k = 1, 2, \ldots, M) \qquad \text{(E.18)}$$

or explicitly as

$$\alpha_j^* = \begin{cases} \Pr(\alpha_{j-1} \ge |t| > \alpha_j), & \text{for } j = 2, 3, \ldots, M, \\ \Pr(|t| > \alpha_j), & \text{for } j = 1, \end{cases}$$

where $t \sim St(N - P_*)$.

(a) Show that the step-up procedure for testing the null hypotheses $H_i^{(0)} : \tau_i = \tau_i^{(0)}$ $(i = 1, 2, \ldots, M)$ is such that (1) the FDR is less than or equal to $M_T \sum_{j=1}^{M} \alpha_j^*/j$; (2) when $M \sum_{j=1}^{M} \alpha_j^*/j < 1$, the FDR is controlled at level $M \sum_{j=1}^{M} \alpha_j^*/j$ (regardless of the identity of the set T); and (3) in the special case where (for $j = 1, 2, \ldots, M$) α'_j is of the form $\alpha'_j = j\dot{\gamma}/M$, the FDR is less than or equal to $\dot{\gamma}\,(M_T/M) \sum_{j=1}^{M} 1/j$ and can be controlled at level δ by taking $\dot{\gamma} = \delta \left(\sum_{j=1}^{M} 1/j\right)^{-1}$.

(b) The sum $\sum_{j=1}^{M} 1/j$ is "tightly" bounded from above by the quantity

$$\gamma + \log(M + 0.5) + [24(M + 0,5)^2]^{-1}, \qquad \text{(E.19)}$$

where γ is the Euler–Mascheroni constant (e.g., Chen 2010)—to 10 significant digits, $\gamma = 0.5772156649$. Determine the value of $\sum_{j=1}^{M} 1/j$ and the amount by which this value is exceeded by the value of expression (E.19). Do so for each of the following values of M: 5, 10, 50, 100, 500, 1,000, 5,000, 10,000, 20,000, and 50,000.

(c) What modifications are needed to extend the results encapsulated in Part (a) to the step-up procedure for testing the null hypotheses $H_i^{(0)} : \tau_i \le \tau_i^{(0)}$ $(i = 1, 2, \ldots, M)$.

Exercise 32. Take the setting to be that of Section 7.7, and adopt the terminology and notation employed therein. Further, for $j = 1, 2, \ldots, M$, let

$$\dot{\alpha}_j = \bar{t}_{k_j \dot{\gamma}/[2(M + k_j - j)]}(N - P_*),$$

where [for some scalar κ $(0<\kappa<1)$] $k_j = [j\kappa]+1$, and let

$$\ddot{\alpha}_j = \bar{t}_{j\dot{\gamma}/(2M)}(N-P_*).$$

And consider two stepwise multiple-comparison procedures for testing the null hypotheses $H_1^{(0)}, H_2^{(0)}, \ldots, H_M^{(0)}$: a stepwise procedure for which α_j is taken to be of the form $\alpha_j = \dot{\alpha}_j$ [as in Section 7.7d in devising a step-down procedure for controlling Pr(FDP $>\kappa$)] and a stepwise procedure for which α_j is taken to be of the form $\alpha_j = \ddot{\alpha}_j$ (as in Section 7.7e in devising a step-up procedure for controlling the FDR). Show that (for $j = 1,2,\ldots,M$) $\dot{\alpha}_j \geq \ddot{\alpha}_j$, with equality holding if and only if $j \leq 1/(1-\kappa)$ or $j = M$.

Exercise 33. Take the setting to be that of Section 7.7f, and adopt the terminology and notation employed therein. And consider a multiple-comparison procedure in which (for $i = 1,2,\ldots,M$) the ith of the M null hypotheses $H_1^{(0)}, H_2^{(0)}, \ldots, H_M^{(0)}$ is rejected if $|t_i^{(0)}|>c$, where c is a strictly positive constant. Further, recall that T is the subset of the set $I=\{1,2,\ldots,M\}$ such that $i \in T$ if $H_i^{(0)}$ is true, denote by R the subset of I such that $i \in R$ if $H_i^{(0)}$ is rejected, and (for $i = 1,2,\ldots,M$) take X_i to be a random variable defined as follows:

$$X_i = \begin{cases} 1, & \text{if } |t_i^{(0)}|>c, \\ 0, & \text{if } |t_i^{(0)}|\leq c. \end{cases}$$

(a) Show that

$$\mathrm{E}[(1/M)\textstyle\sum_{i\in T} X_i] = (M_T/M)\Pr(|t|>c) \leq \Pr(|t|>c),$$

where $t \sim St(100)$.

(b) Based on the observation that [when $(1/M)\sum_{i=1}^M X_i > 0$]

$$\mathrm{FDP} = \frac{(1/M)\sum_{i\in T} X_i}{(1/M)\sum_{i=1}^M X_i},$$

on the reasoning that for large M the quantity $(1/M)\sum_{i=1}^M X_i$ can be regarded as a (strictly positive) constant, and on the result of Part (a), the quantity $M_T \Pr(|t|>c)/M_R$ can be regarded as an "estimator" of the FDR [= E(FDP)] and $M \Pr(|t|>c)/M_R$ can be regarded as an estimator of $\max_T$ FDR (Efron 2010, chap. 2)—if $M_R=0$, take the estimate of the FDR or of $\max_T$ FDR to be 0. Consider the application to the prostate data of the multiple-comparison procedure in the case where $c = c_{\dot{\gamma}}(k)$ and also in the case where $c = \bar{t}_{k\dot{\gamma}/(2M)}(100)$. Use the information provided by the entries in Table 7.5 to obtain an estimate of $\max_T$ FDR for each of these two cases. Do so for $\dot{\gamma} = .05, .10$, and .20 and for $k = 1, 5, 10$, and 20.

Exercise 34. Take the setting to be that of Part 6 of Section 7.8a (pertaining to the testing of $H_0\colon \mathbf{w} \in S_0$ versus $H_1\colon \mathbf{w} \in S_1$), and adopt the notation and terminology employed therein.

(a) Write p_0 for the random variable $p_0(\mathbf{y})$, and denote by $G_0(\cdot)$ the cdf of the conditional distribution of p_0 given that $\mathbf{w} \in S_0$. Further, take k and c to be the constants that appear in the definition of the critical function $\phi^*(\cdot)$, take $k''=[1+(\pi_1/\pi_0)k]^{-1}$, and take $\phi^{**}(\cdot)$ to be a critical function defined as follows:

$$\phi^{**}(\underline{y}) = \begin{cases} 1, & \text{when } p_0(\underline{y})<k'', \\ c, & \text{when } p_0(\underline{y})=k'', \\ 0, & \text{when } p_0(\underline{y})>k''. \end{cases}$$

Show that (1) $\phi^{**}(\underline{y}) = \phi^*(\underline{y})$ when $f(\underline{y})>0$, (2) that k'' equals the smallest scalar $\underline{p}_0$ for which $G_0(\underline{p}_0) \geq \dot{\gamma}$, and (3) that

$$c = \frac{\dot{\gamma} - \Pr(p_0<k'' \mid \mathbf{w}\in S_0)}{\Pr(p_0=k'' \mid \mathbf{w}\in S_0)} \quad \text{when } \Pr(p_0=k'' \mid \mathbf{w}\in S_0) > 0$$

—when $\Pr(p_0=k'' \mid \mathbf{w}\in S_0) = 0$, c can be chosen arbitrarily.

(b) Show that if the joint distribution of $\mathbf{w}$ and $\mathbf{y}$ is MVN, then there exists a version of the critical function $\phi^*(\cdot)$ defined by equalities (8.25) and (8.26) for which $\phi^*(\mathbf{y})$ depends on the value of $\mathbf{y}$ only through the value of $\tilde{\mathbf{w}}(\mathbf{y}) = \boldsymbol{\tau} + \mathbf{V}_{yw}'\mathbf{V}_y^{-1}\mathbf{y}$ (where $\boldsymbol{\tau} = \boldsymbol{\mu}_w - \mathbf{V}_{yw}'\mathbf{V}_y^{-1}\boldsymbol{\mu}_y$).

(c) Suppose that $M = 1$ and that $S_0 = \{\underline{w} : \ell \leq \underline{w} \leq u\}$, where ℓ and u are (known) constants (with $\ell < u$). Suppose also that the joint distribution of w and $\mathbf{y}$ is MVN and that $\mathbf{v}_{yw} \neq \mathbf{0}$. And letting $\tilde{w} = \tilde{w}(\mathbf{y}) = \tau + \mathbf{v}_{yw}'\mathbf{V}_y^{-1}\mathbf{y}$ (with $\tau = \mu_w - \mathbf{v}_{yw}'\mathbf{V}_y^{-1}\boldsymbol{\mu}_y$) and $\tilde{v} = v_w - \mathbf{v}_{yw}'\mathbf{V}_y^{-1}\mathbf{v}_{yw}$, define

$$d = d(\mathbf{y}) = F\{[u - \tilde{w}(\mathbf{y})]/\tilde{v}^{1/2}\} - F\{[\ell - \tilde{w}(\mathbf{y})]/\tilde{v}^{1/2}\},$$

where $F(\cdot)$ is the cdf of the $N(0, 1)$ distribution. Further, let

$$C_0 = \{\underline{\mathbf{y}} \in \mathcal{R}^N : d(\underline{\mathbf{y}}) < \ddot{d}\},$$

where $\ddot{d}$ is the lower $100\dot{\gamma}\%$ point of the distribution of the random variable d. Show that among all $\dot{\gamma}$-level tests of the null hypothesis H_0, the nonrandomized $\dot{\gamma}$-level test with critical region C_0 achieves maximum power.

Bibliographic and Supplementary Notes

§1. The term standard error is used herein to refer to the standard deviation of an estimator. In some presentations, this term is used to refer to what is herein referred to as an estimated standard error.

§3b. The results of Lemmas 7.3.1, 7.3.2, and 7.3.3 are valid for noninteger degrees of freedom as well as integer degrees of freedom. The practice adopted herein of distinguishing between equivariance and invariance is consistent with that adopted by, e.g., Lehmann and Romano (2005b).

§3c. In his account of the S method, Scheffé (1959) credits Holbrook Working and Harold Hotelling with having devised a special case of the method applicable to obtaining confidence intervals for the points on a regression line.

§3c, §3e, and §3f. The approach to the construction of simultaneous confidence intervals described herein is very similar to the approach described (primarily in the context of confidence bands) by Liu (2011).

§3e. The discussion of the use of Monte Carlo methods to approximate the percentage point $c_{\dot{\gamma}}$ or $c_{\dot{\gamma}}^*$ is patterned after the discussion of Edwards and Berry (1987, sec. 2). In particular, Lemmas 7.3.4 and 7.3.5 are nearly identical to Edwards and Berry's Lemmas 1 and 2.

§3e and §3f. Hsu and Nelson (1990 and 1998) described ways in which a control variate could be used to advantage in obtaining a Monte Carlo approximation to the upper $100\dot{\gamma}\%$ point of the distribution of a random variable like $\max_{\{\tilde{\boldsymbol{\delta}} \in \tilde{\Delta} :\, \tilde{\boldsymbol{\delta}} \neq \mathbf{0}\}} |\tilde{\boldsymbol{\delta}}'\mathbf{t}|/(\tilde{\boldsymbol{\delta}}'\tilde{\boldsymbol{\delta}})^{1/2}$ or $\max_{\{\tilde{\boldsymbol{\delta}} \in \tilde{\Delta}\}} |\tilde{\boldsymbol{\delta}}'\mathbf{t}|$. A control variate is a random variable having a distribution whose upper $100\dot{\gamma}\%$ point can be determined "exactly" and that is "related to" the distribution whose upper $100\dot{\gamma}\%$ point is to be approximated. Result (3.154) suggests that for purposes of obtaining a Monte Carlo approximation to $c_{\dot{\gamma}}$ [which is the upper $100\dot{\gamma}\%$ point of the distribution of $\max_{\{\tilde{\boldsymbol{\delta}} \in \tilde{\Delta} :\, \tilde{\boldsymbol{\delta}} \neq \mathbf{0}\}} |\tilde{\boldsymbol{\delta}}'\mathbf{t}|/(\tilde{\boldsymbol{\delta}}'\tilde{\boldsymbol{\delta}})^{1/2}$], the random variable $\mathbf{t}'\mathbf{t}/M_*$ [which has an $SF(M_*, N - P_*)$ distribution] could serve as a control variate. The $K = 599999$ draws used (in Subsection f) in obtaining a Monte Carlo approximation to $c_{.10}$ were such that subsequent to rearranging the 599999 sample values of $\mathbf{t}'\mathbf{t}/M_*$ from smallest to largest, the 540000th of these values was 2.317968; in actuality, this value is the upper 10.052% point of the $SF(10, 10)$ distribution, not the upper 10.000% point.

§3f. In computing the Monte Carlo approximations to $c_{.10}$ and $c_{.10}^*$, the value of $\max_{\{\mathbf{u} \in \mathcal{U}\}} |[\mathbf{x}(\mathbf{u})]'\mathbf{W}'\mathbf{t}|/\{[\mathbf{x}(\mathbf{u})]'(\mathbf{X}'\mathbf{X})^{-1}\mathbf{x}(\mathbf{u})\}^{1/2}$ and the value of $\max_{\{\mathbf{u} \in \mathcal{U}\}} |[\mathbf{x}(\mathbf{u})]'\mathbf{W}'\mathbf{t}|$ had to be computed for each of the 599999 draws from the distribution of $\mathbf{t}$. The value of $\max_{\{\mathbf{u} \in \mathcal{U}\}} |[\mathbf{x}(\mathbf{u})]'\mathbf{W}'\mathbf{t}|/\{[\mathbf{x}(\mathbf{u})]'(\mathbf{X}'\mathbf{X})^{-1}\mathbf{x}(\mathbf{u})\}^{1/2}$ was determined by (1) using an implementation (algorithm nlxb in the R package nlmrt) of a nonlinear least squares algorithm proposed (as a variant on the Marquardt procedure) by Nash (1990) to find the minimum value of the sum of squares $[\mathbf{t} - \lambda\mathbf{W}\mathbf{x}(\mathbf{u})]'[\mathbf{t} - \lambda\mathbf{W}\mathbf{x}(\mathbf{u})]$ with respect to λ and $\mathbf{u}$ (subject to the constraint $\mathbf{u} \in \mathcal{U}$) and by then (2) exploiting relationship (3.173). The value of $\max_{\{\mathbf{u} \in \mathcal{U}\}} |[\mathbf{x}(\mathbf{u})]'\mathbf{W}'\mathbf{t}|$ was determined by finding the maximum values of $[\mathbf{x}(\mathbf{u})]'\mathbf{W}'\mathbf{t}$ and $[\mathbf{x}(\mathbf{u})]'\mathbf{W}'(-\mathbf{t})$ with respect to $\mathbf{u}$ (subject to the constraint $\mathbf{u} \in \mathcal{U}$) and by taking the value of $\max_{\{\mathbf{u} \in \mathcal{U}\}} |[\mathbf{x}(\mathbf{u})]'\mathbf{W}'\mathbf{t}|$ to be the larger of these two values. The determination of the maximum value of each of the quantities $[\mathbf{x}(\mathbf{u})]'\mathbf{W}'\mathbf{t}$ and $[\mathbf{x}(\mathbf{u})]'\mathbf{W}'(-\mathbf{t})$ (subject to the constraint $\mathbf{u} \in \mathcal{U}$) was based on the observation that either the maximum value is attained at one of the 8 values of $\mathbf{u}$ such that $u_i = \pm 1$ $(i = 1, 2, 3)$ or it is attained at a value of $\mathbf{u}$ such that (1) one, two, or three of the components of $\mathbf{u}$ are less than 1 in absolute value, (2) the first-order partial derivatives of $[\mathbf{x}(\mathbf{u})]'\mathbf{W}'\mathbf{t}$ or $[\mathbf{x}(\mathbf{u})]'\mathbf{W}'(-\mathbf{t})$ with respect to these components equal 0, and (3) the matrix of second-order partial derivatives with respect to these components is negative definite—a square matrix $\mathbf{A}$ is said to be negative definite if $-\mathbf{A}$ is positive definite.

§4e. Refer to Chen, Hung, and Chen (2007) for some general discussion of the use of maximum average-power as a criterion for evaluating hypothesis tests.

§4e and Exercise 20. The result that the size-$\dot{\gamma}$ 2-sided t test is a UMP level-$\dot{\gamma}$ unbiased test (of H_0 or $\tilde{H}_0$ versus H_1 or $\tilde{H}_1$) can be regarded as a special case of results on UMP tests for exponential families like those discussed by Lehmann and Romano (2005b) in their chapters 4 and 5.

§7a, §7b, and Exercise 27. The procedures proposed by Lehmann and Romano (2005a) for the control of k-FWER served as "inspiration" for the results of Subsection a, for Exercise 27, and for some aspects of what is presented in Subsection b. Approaches similar to the approach proposed by Lehmann and Romano for controlling the rate of multiple false rejections were considered by Victor (1982) and by Hommel and Hoffmann (1988).

§7b. Various of the results presented in this subsection are closely related to the results of Westfall and Tobias (2007).

§7c. Benjamini and Hochberg's (1995) paper has achieved landmark status. It has had a considerable impact on statistical practice and has inspired a great deal of further research into multiple-comparison methods of a kind better suited for applications to microarray data (and other large-scale applications) than the more traditional kind of methods. The newer methods have proved to be popular, and their use in large-scale applications has proved to be considerably more effective than that of the more traditional methods. Nevertheless, it could be argued that even better results could be achieved by regarding and addressing the problem of screening or discovery in a way that is more in tune with the true nature of the problem (than simply treating the problem as one of multiple comparisons).

§7c and §7f. The microarray data introduced in Section 7.7c and used for illustrative purposes in Section 7.7f are the data referred to by Efron (2010, app. B) as the prostate data and are among the data made available by him on his website. Those data were obtained by preprocessing the "raw" data from a study by Singh et al. (2002). Direct information about the nature of the preprocessing does not seem to be available. Presumably, the preprocessing was similar in nature to that described by Dettling (2004) and applied by him to the results of the same study—like Efron, Dettling used the preprocessed data for illustrative purposes (and made them available via the internet). In both cases, the preprocessed data are such that the data for each of the 50 normal subjects and each of the 52 cancer patients has been centered and rescaled so that the average value is 0 and the sample variance equals 1. Additionally, it could be argued that (in formulating the objectives underlying the analysis of the prostate data in terms of multiple comparisons) it would be more realistic to take the null hypotheses to be $H_s^{(0)} : |\tau_s| \le \tau_s^{(0)}$ $(s = 1, 2, \ldots, 6033)$ and the alternative hypotheses to be $H_s^{(1)} : |\tau_s| > \tau_s^{(0)}$ $(s = 1, 2, \ldots, 6033)$, where $\tau_s^{(0)}$ is a "threshold" such that (absolute) values of τ_s smaller than $\tau_s^{(0)}$ are regarded as "unimportant"—this presumes the existence of enough knowledge about the underlying processes that a suitable threshold is identifiable.

§7d. As discussed in this section, a step-down procedure for controlling $\Pr(\text{FDP} > \kappa)$ at level ϵ can be obtained by taking (for $j = 1, 2, \ldots, M$) α_j to be of the form (7.65) (in which case α_j can be regarded as a function of $\dot{\gamma}$) and by taking the value of $\dot{\gamma}$ to be the largest value that satisfies condition (7.70). Instead of taking α_j to be of the form (7.65), we could take it to be of the form $\alpha_j = \max_{S \in \tilde{\Omega}^+_{[j\kappa]+1;j}} c_{\dot{\gamma}}([j\kappa]+1; S)$ (where $\tilde{\Omega}^+_{k;j}$ is as defined in Section 7.7b). This change could result in additional rejections (discoveries), though any such gains would come at the expense of greatly increased computational demands.

§7d and Exercises 28 and 30. The content of this section and these exercises is based to a considerable extent on the results of Lehmann and Romano (2005a, sec. 3).

§7e and Exercises 29 and 31. The content of this section and these exercises is based to a considerable extent on the results of Benjamini and Yekutieli (2001).

§7f. The results (on the total number of rejections or discoveries) reported in Table 7.5 for the conservative counterpart of the k-FWER multiple-comparison procedure (in the case where $\dot{\gamma} = .05$) differ somewhat from the results reported by Efron (2010, fig. 3.3). It is worth noting that the latter results are those for the case where the tests of the M null hypotheses are one-sided tests rather than those for the case where the tests are two-sided tests.

§8a. The approach taken (in the last 2 parts of Section 7.8a) in devising (in the context of prediction) multiple-comparison procedures is based on the use of the closure principle (as generalized from control of the FWER to control of the k-FWER). It would seem that this use of the closure principle (as generalized thusly) in devising multiple-comparison procedures could be extended to other settings.

Exercise 17 (b). The Bonferroni t-intervals [i.e., the intervals $A_1(\mathbf{y}), A_2(\mathbf{y}), \ldots, A_L(\mathbf{y})$ in the special case where $\dot{\gamma}_1 = \dot{\gamma}_2 = \cdots = \dot{\gamma}_L$] can be regarded as having been obtained from interval (3.94) (where $\boldsymbol{\delta} \in \Delta$ with $\Delta = \{\boldsymbol{\delta}_1, \boldsymbol{\delta}_2, \ldots, \boldsymbol{\delta}_L\}$) by replacing $c_{\dot{\gamma}}$ with $\bar{t}_{\dot{\gamma}/(2L)}(N-P_*)$, which constitutes an upper bound for $c_{\dot{\gamma}}$. As discussed by Fuchs and Sampson (1987), intervals for $\boldsymbol{\delta}_1'\boldsymbol{\tau}, \boldsymbol{\delta}_2'\boldsymbol{\tau}, \ldots, \boldsymbol{\delta}_L'\boldsymbol{\tau}$ that are superior to the Bonferroni t-intervals are obtainable from interval (3.94) by replacing $c_{\dot{\gamma}}$ with $\bar{t}_{[1-(1-\dot{\gamma})^{1/L}]/2}(N-P_*)$, which is a tighter upper bound for $c_{\dot{\gamma}}$ than $\bar{t}_{\dot{\gamma}/(2L)}(N-P_*)$. And an even greater improvement can be effected by replacing $c_{\dot{\gamma}}$ with the upper $100\dot{\gamma}\%$ point, say $\bar{t}^*_{\dot{\gamma}}(L,\, N-P_*)$, of the distribution of the random variable $\max(|t_1|, |t_2|, \ldots, |t_L|)$, where $t_1, t_2, \ldots, t_L$ are the elements of an L-dimensional random vector that has an $MVt(N-P_*, \mathbf{I}_L)$ distribution; the distribution of $\max(|t_1|, |t_2|, \ldots, |t_L|)$ is that of the Studentized maximum modulus, and its upper $100\dot{\gamma}\%$ point is an even tighter upper bound (for $c_{\dot{\gamma}}$) than $\bar{t}_{[1-(1-\dot{\gamma})^{1/L}]/2}(N-P_*)$. These alternatives to the Bonferroni t-intervals are based on the results of Šidák (1967 and 1968); refer, for example, to Khuri (2010, sec. 7.5.4) and to Graybill (1976, sec. 6.6) for some relevant details. As a simple example, consider the confidence intervals (3.143), for which the probability of simultaneous coverage is .90. In this example, $\dot{\gamma} = .10$, $L = 3$, $N-P_* = 10$, $\bar{t}^*_{\dot{\gamma}}(L,\, N-P_*) = c_{\dot{\gamma}} = 2.410$, $\bar{t}_{[1-(1-\dot{\gamma})^{1/L}]/2}(N-P_*) = 2.446$, and $\bar{t}_{\dot{\gamma}/(2L)}(N-P_*) = 2.466$.

Exercises 21 and 22 (b). The size-$\dot{\gamma}$ F test is optimal in the senses described in Exercises 21 and 22 (b) not only among size-$\dot{\gamma}$ similar tests, but among *all* tests whose size does not exceed $\dot{\gamma}$—refer, e.g., to Lehmann and Romano (2005b, chap. 8). The restriction to size-$\dot{\gamma}$ similar tests serves to facilitate the solution of these exercises.

References

Agresti, A. (2013), *Categorical Data Analysis* (3rd ed.), New York: Wiley.

Albert, A. (1972), *Regression and the Moore–Penrose Pseudoinverse*, New York: Academic Press.

Albert, A. (1976), "When Is a Sum of Squares an Analysis of Variance?," *The Annals of Statistics*, 4, 775–778.

Anderson, T. W., and Fang, K.-T. (1987), "Cochran's Theorem for Elliptically Contoured Distributions," *Sankhyā*, Series A, 49, 305–315.

Arnold, S. F. (1981), *The Theory of Linear Models and Multivariate Analysis*, New York: Wiley.

Atiqullah, M. (1962), "The Estimation of Residual Variance in Quadratically Balanced Least-Squares Problems and the Robustness of the F-Test," *Biometrika*, 49, 83–91.

Baker, J. A. (1997), "Integration over Spheres and the Divergence Theorem for Balls," *The American Mathematical Monthly*, 104, 36–47.

Bartle, R. G. (1976), *The Elements of Real Analysis* (2nd ed.), New York: Wiley.

Bartle, R. G., and Sherbert, D. R. (2011), *Introduction to Real Analysis* (4th ed.), Hoboken, NJ: Wiley.

Bates, D. M., and Watts, D. G. (1988), *Nonlinear Regression Analysis and Its Applications*, New York: Wiley.

Beaumont, R. A., and Pierce, R. S. (1963), *The Algebraic Foundations of Mathematics*, Reading, MA: Addison-Wesley.

Benjamini, Y., and Hochberg, Y. (1995), "Controlling the False Discovery Rate: a Practical and Powerful Approach to Multiple Testing," *Journal of the Royal Statistical Society*, Series B, 57, 289–300.

Benjamini, Y., and Yekutieli, D. (2001), "The Control of the False Discovery Rate in Multiple Testing Under Dependency," *The Annals of Statistics*, 29, 1165–1188.

Bennett, J. H., ed. (1990), *Statistical Inference and Analysis: Selected Correspondence of R. A. Fisher*, Oxford, U.K.: Clarendon Press.

Berger, J. O. (1985), *Statistical Decision Theory and Bayesian Analysis* (2nd ed.), New York: Springer-Verlag.

Bickel, P. J., and Doksum, K. A. (2001), *Mathematical Statistics: Basic Ideas and Selected Topics* (Vol. I, 2nd ed.), Upper Saddle River, NJ: Prentice-Hall.

Billingsley, P. (1995), *Probability and Measure* (3rd ed.), New York: Wiley.

Box, G. E. P., and Draper, N. R. (1987), *Empirical Model-Building and Response Surfaces*, New York: Wiley.

Bretz, F., Hothorn, T., and Westfall, P. (2011), *Multiple Comparisons Using R*, Boca Raton, FL: Chapman & Hall/CRC.

Cacoullos, T., and Koutras, M. (1984), "Quadratic Forms in Spherical Random Variables: Generalized Noncentral χ^2 Distribution," *Naval Research Logistics Quarterly*, 31, 447–461.

Carroll, R. J., and Ruppert, D. (1988), *Transformation and Weighting in Regression*, New York: Chapman & Hall.

Casella, G., and Berger, R. L. (2002), *Statistical Inference* (2nd ed.), Pacific Grove, CA: Duxbury.

Chen, C.-P. (2010), "Inequalities for the Euler–Mascheroni constant," *Applied Mathematics Letters*, 23, 161–164.

Chen, L.-A., Hung, H.-N., and Chen, C.-R. (2007), "Maximum Average-Power (MAP) Tests," *Communications in Statistics—Theory and Methods*, 36, 2237–2249.

Cochran, W. G. (1934), "The Distribution of Quadratic Forms in a Normal System, with Applications to the Analysis of Covariance," *Proceedings of the Cambridge Philosophical Society*, 30, 178–191.

Cornell, J. A. (2002), *Experiments with Mixtures: Designs, Models, and the Analysis of Mixture Data* (3rd ed.), New York: Wiley.

Cressie, N. A. C. (1993), *Statistics for Spatial Data* (rev. ed.), New York: Wiley.

David, H. A. (2009), "A Historical Note on Zero Correlation and Independence," *The American Statistician*, 63, 185–186.

Davidian, M., and Giltinan, D. M. (1995), *Nonlinear Models for Repeated Measurement Data*, London: Chapman & Hall.

Dawid, A. P. (1982), "The Well-Calibrated Bayesian" (with discussion), *Journal of the American Statistical Association*, 77, 605–613.

Dettling, M. (2004), "BagBoosting for Tumor Classification with Gene Expression Data," *Bioinformatics*, 20, 3583–3593.

Diggle, P. J., Heagerty, P., Liang, K.-Y., and Zeger, S. L. (2002), *Analysis of Longitudinal Data* (2nd ed.), Oxford, U.K.: Oxford University Press.

Driscoll, M. F. (1999), "An Improved Result Relating Quadratic Forms and Chi-Square Distributions," *The American Statistician*, 53, 273–275.

Driscoll, M. F., and Gundberg, W. R., Jr. (1986), "A History of the Development of Craig's Theorem," *The American Statistician*, 40, 65–70.

Driscoll, M. F., and Krasnicka, B. (1995), "An Accessible Proof of Craig's Theorem in the General Case," *The American Statistician*, 49, 59–62.

Durbin, B. P., Hardin, J. S., Hawkins, D. M., and Rocke, D. M. (2002), "A Variance-Stabilizing Transformation for Gene-Expression Microarray Data," *Bioinformatics*, 18, S105–S110.

Edwards, D., and Berry, J. J. (1987), "The Efficiency of Simulation-Based Multiple Comparisons," *Biometrics*, 43, 913–928.

Efron, B. (2010), *Large-Scale Inference: Empirical Bayes Methods for Estimation, Testing, and Prediction*, New York: Cambridge University Press.

Fang, K.-T., Kotz, S., and Ng, K.-W. (1990), *Symmetric Multivariate and Related Distributions*, London: Chapman & Hall.

Feller, W. (1971), *An Introduction to Probability Theory and Its Applications* (Vol. II, 2nd ed.), New York: Wiley.

Fuchs, C., and Sampson, A. R. (1987), "Simultaneous Confidence Intervals for the General Linear Model," *Biometrics*, 43, 457–469.

Gallant, A. R. (1987), *Nonlinear Statistical Models*, New York: Wiley.

Gentle, J. E. (1998), *Numerical Linear Algebra for Applications in Statistics*, New York: Springer-Verlag.

Golub, G. H., and Van Loan, C. F. (2013), *Matrix Computations* (4th ed.), Baltimore: The Johns Hopkins University Press.

Graybill, F. A. (1961), *An Introduction to Linear Statistical Models* (Vol. I), New York: McGraw-Hill.

Graybill, F. A. (1976), *Theory and Application of the Linear Model*, North Scituate, MA: Duxbury.

Grimmett, G., and Welsh, D. (1986), *Probability: An Introduction*, Oxford, U.K.: Oxford University Press.

Gupta, A. K., and Song, D. (1997), "L_p-Norm Spherical Distribution," *Journal of Statistical Planning and Inference*, 60, 241–260.

Hader, R. J., Harward, M. E., Mason, D. D., and Moore, D. P. (1957), "An Investigation of Some of the Relationships Between Copper, Iron, and Molybdenum in the Growth and Nutrition of Lettuce: I. Experimental Design and Statistical Methods for Characterizing the Response Surface," *Soil Science Society of America Proceedings*, 21, 59–64.

Hald, A. (1952), *Statistical Theory with Engineering Applications*, New York: Wiley.

Halmos, P. R. (1958), *Finite-Dimensional Vector Spaces* (2nd ed.), Princeton, NJ: Van Nostrand.

Hartigan, J. A. (1969), "Linear Bayesian Methods," *Journal of the Royal Statistical Society*, Series B, 31, 446–454.

Hartley, H. O. (1950), "The Maximum F-Ratio as a Short-Cut Test for Heterogeneity of Variance," *Biometrika*, 37, 308–312.

Harville, D. A. (1980), "Predictions for National Football League Games Via Linear-Model Methodology," *Journal of the American Statistical Association*, 75, 516–524.

Harville, D. A. (1985), "Decomposition of Prediction Error," *Journal of the American Statistical Association*, 80, 132–138.

Harville, D. A. (1997), *Matrix Algebra from a Statistician's Perspective*, New York: Springer-Verlag.

Harville, D. A. (2003a), "The Expected Value of a Conditional Variance: an Upper Bound," *Journal of Statistical Computation and Simulation*, 73, 609–612.

Harville, D. A. (2003b), "The Selection or Seeding of College Basketball or Football Teams for Postseason Competition," *Journal of the American Statistical Association*, 98, 17–27.

Harville, D. A. (2014), "The Need for More Emphasis on Prediction: a 'Nondenominational' Model-Based Approach" (with discussion), *The American Statistician*, 68, 71–92.

Harville, D. A., and Kempthorne, O. (1997), "An Alternative Way to Establish the Necessity Part of the Classical Result on the Statistical Independence of Quadratic Forms," *Linear Algebra and Its Applications*, 264 (Sixth Special Issue on Linear Algebra and Statistics), 205–215.

Henderson, C. R. (1984), *Applications of Linear Models in Animal Breeding*, Guelph, ON: University of Guelph.

Hinkelmann, K., and Kempthorne, O. (2008), *Design and Analysis of Experiments, Volume I: Introduction to Experimental Design* (2nd ed.), Hoboken, NJ: Wiley.

Hodges, J. L., Jr., and Lehmann, E. L. (1951), "Some Applications of the Cramér–Rao Inequality," in *Proceedings of the Second Berkeley Symposium on Mathematical Statistics and Probability*, ed. J. Neyman, Berkeley and Los Angeles: University of California Press, pp. 13–22.

Holm, S. (1979), "A Simple Sequentially Rejective Multiple Test Procedure," *Scandinavian Journal of Statistics*, 6, 65–70.

Hommel, G., and Hoffmann, T. (1988), "Controlled Uncertainty," in *Multiple Hypotheses Testing*, eds. P. Bauer, G. Hommel, and E. Sonnemann, Heidelberg: Springer, pp. 154–161.

Hsu, J. C., and Nelson, B. L. (1990), "Control Variates for Quantile Estimation," *Management Science*, 36, 835–851.

Hsu, J. C., and Nelson, B. (1998), "Multiple Comparisons in the General Linear Model," *Journal of Computational and Graphical Statistics*, 7, 23–41.

Jensen, D. R. (1985), "Multivariate Distributions," in *Encyclopedia of Statistical Sciences* (Vol. 6), eds. S. Kotz, N. L. Johnson, and C. B. Read, New York: Wiley, pp. 43–55.

Johnson, N. L., Kotz, S., and Balakrishnan, N. (1995), *Continuous Univariate Distributions* (Vol. 2, 2nd ed.), New York: Wiley.

Karlin, S., and Rinott, Y. (1980), "Classes of Orderings of Measures and Related Correlation Inequalities. I. Multivariate Totally Positive Distributions," *Journal of Multivariate Analysis*, 10, 467–498.

Karlin, S., and Rinott, Y. (1981), "Total Positivity Properties of Absolute Value Multinormal Variables with Applications to Confidence Interval Estimates and Related Probabilistic Inequalities," *The Annals of Statistics*, 9, 1035–1049.

Kempthorne, O. (1980), "The Term *Design Matrix*" (letter to the editor), *The American Statistician*, 34, 249.

Khuri, A. I. (1992), "Response Surface Models with Random Block Effects," *Technometrics*, 34, 26–37.

Khuri, A. I. (1999), "A Necessary Condition for a Quadratic Form to Have a Chi-Squared Distribution: an Accessible Proof," *International Journal of Mathematical Education in Science and Technology*, 30, 335–339.

Khuri, A. I. (2010), *Linear Model Methodology*, Boca Raton, FL: Chapman & Hall/CRC.

Kollo, T., and von Rosen, D. (2005), *Advanced Multivariate Statistics with Matrices*, Dordrecht, The Netherlands: Springer.

Laha, R. G. (1956), "On the Stochastic Independence of Two Second-Degree Polynomial Statistics in Normally Distributed Variates," *The Annals of Mathematical Statistics*, 27, 790–796.

Laird, N. (2004), *Analysis of Longitudinal and Cluster-Correlated Data*—Volume 8 in the *NSF-CBMS Regional Conference Series in Probability and Statistics*, Beachwood, OH: Institute of Mathematical Statistics.

LaMotte, L. R. (2007), "A Direct Derivation of the REML Likelihood Function," *Statistical Papers*, 48, 321–327.

Lehmann, E. L. (1986), *Testing Statistical Hypotheses* (2nd ed.), New York: Wiley.

Lehmann, E. L., and Casella, G. (1998), *Theory of Point Estimation* (2nd ed.), New York: Springer-Verlag.

Lehmann, E. L., and Romano, J. P. (2005a), "Generalizations of the Familywise Error Rate," *The Annals of Statistics*, 33, 1138–1154.

Lehmann, E. L., and Romano, J. P. (2005b), *Testing Statistical Hypotheses* (3rd ed.), New York: Springer.

Littell, R. C., Milliken, G. A., Stroup, W. W., Wolfinger, R. D., and Schabenberger, O. (2006), *SAS® System for Mixed Models* (2nd ed.), Cary, NC: SAS Institute Inc.

Liu, W. (2011), *Simultaneous Inference in Regression*, Boca Raton, FL: Chapman & Hall/CRC.

Luenberger, D. G., and Ye, Y. (2016), *Linear and Nonlinear Programming* (4th ed.), New York: Springer.

McCullagh, P., and Nelder, J. A. (1989), *Generalized Linear Models* (2nd ed.), London: Chapman & Hall.

McCulloch, C. E., Searle, S. R., and Neuhaus, J. M. (2008), *Generalized, Linear, and Mixed Models* (2nd ed.), Hoboken, NJ: Wiley.

Milliken, G. A., and Johnson, D. E. (2009), *Analysis of Messy Data, Volume I: Designed Experiments* (2nd ed.), Boca Raton, FL: Chapman & Hall/CRC.

Moore, D. P., Harward, M. E., Mason, D. D., Hader, R. J., Lott, W. L., and Jackson, W. A. (1957), "An Investigation of Some of the Relationships Between Copper, Iron, and Molybdenum in the Growth and Nutrition of Lettuce: II. Response Surfaces of Growth and Accumulations of Cu and Fe," *Soil Science Society of America Proceedings*, 21, 65–74.

Müller, A. (2001), "Stochastic Ordering of Multivariate Normal Distributions," *Annals of the Institute of Statistical Mathematics*, 53, 567–575.

Myers, R. H., Montgomery, D. C., and Anderson-Cook, C. M. (2016), *Response Surface Methodology: Process and Product Optimization Using Designed Experiments* (4th ed.), Hoboken, NJ: Wiley.

Nash, J. C. (1990), *Compact Numerical Methods for Computers: Linear Algebra and Function Minimisation* (2nd ed.), Bristol, England: Adam Hilger/Institute of Physics Publications.

Nocedal, J., and Wright, S. J. (2006), *Numerical Optimization* (2nd ed.), New York: Springer.

Ogawa, J. (1950), "On the Independence of Quadratic Forms in a Non-Central Normal System," *Osaka Mathematical Journal*, 2, 151–159.

Ogawa, J., and Olkin, I. (2008), "A Tale of Two Countries: the Craig–Sakamoto–Matusita Theorem," *Journal of Statistical Planning and Inference*, 138, 3419–3428.

Parzen, E. (1960), *Modern Probability Theory and Its Applications*, New York: Wiley.

Patterson, H. D., and Thompson, R. (1971), "Recovery of Inter-Block Information When Block Sizes Are Unequal," *Biometrika*, 58, 545–554.

Pawitan, Y. (2001), *In All Likelihood: Statistical Modelling and Inference Using Likelihood*, New York: Oxford University Press.

Pinheiro, J. C., and Bates, D. M. (2000), *Mixed-Effects Models in S and S-PLUS*, New York: Springer-Verlag.

Plackett, R. L. (1972), "Studies in the History of Probability and Statistics. XXIX: The Discovery of the Method of Least Squares," *Biometrika*, 59, 239–251.

Potthoff, R. F., and Roy, S. N. (1964), "A Generalized Multivariate Analysis of Variance Model Useful Especially for Growth Curve Problems," *Biometrika*, 51, 313–326.

Rao, C. R. (1965), *Linear Statistical Inference and Its Applications*, New York: Wiley.

Rao, C. R. (1973), *Linear Statistical Inference and Its Applications* (2nd ed.), New York: Wiley.

Rao, C. R., and Mitra, S. K. (1971), *Generalized Inverse of Matrices and Its Applications*, New York: Wiley.

Ravishanker, N., and Dey, D. K. (2002), *A First Course in Linear Model Theory*, Boca Raton, FL: Chapman & Hall/CRC.

Reid, J. G., and Driscoll, M. F. (1988), "An Accessible Proof of Craig's Theorem in the Noncentral Case," *The American Statistician*, 42, 139–142.

Sanders, W. L., and Horn, S. P. (1994), "The Tennessee Value-Added Assessment System (TVAAS): Mixed-Model Methodology in Educational Assessment," *Journal of Personnel Evaluation in Education*, 8, 299–311.

Sarkar, S. K. (2008), "On the Simes Inequality and Its Generalization," in *Beyond Parametrics in Interdisciplinary Research: Festschrift in Honor of Professor Pranab K. Sen*, eds. N. Balakrishnan, E. A. Pea, and M. J. Silvapulle, Beachwood, OH: Institute of Mathematical Statistics, pp. 231-242.

Schabenberger, O., and Gotway, C. A. (2005), *Statistical Methods for Spatial Data Analysis*, Boca Raton, FL: Chapman & Hall/CRC.

Scheffé, H. (1953), "A Method for Judging All Contrasts in the Analysis of Variance," *Biometrika*, 40, 87–104.

Scheffé, H. (1959), *The Analysis of Variance*, New York: Wiley.

Schervish, M. J. (1995), *Theory of Statistics*, New York: Springer-Verlag.

Schmidt, R. H., Illingworth, B. L., Deng, J. C., and Cornell, J. A. (1979), "Multiple Regression and Response Surface Analysis of the Effects of Calcium Chloride and Cysteine on Heat-Induced Whey Protein Gelation," *Journal of Agricultural and Food Chemistry*, 27, 529–532.

Seal, H. L. (1967), "The Historical Development of the Gauss Linear Model," *Biometrika*, 54, 1–24.

Searle, S. R. (1971), *Linear Models*, New York: Wiley.

Sen, P. K. (1989), "The Mean-Median-Mode Inequality and Noncentral Chi Square Distributions," *Sankhyā*, Series A, 51, 106-114.

Severini, T. A. (2000), *Likelihood Methods in Statistics*, New York: Oxford University Press.

Shanbhag, D. N. (1968), "Some Remarks Concerning Khatri's Result on Quadratic Forms," *Biometrika*, 55, 593–595.

Shao, J. (2010), *Mathematical Statistics* (2nd ed.), New York: Springer-Verlag.

Šidák, Z. (1967), "Rectangular Confidence Regions for the Means of Multivariate Normal Distributions," *Journal of the American Statistical Association*, 62, 626–633.

Šidák, Z. (1968), "On Multivariate Normal Probabilities of Rectangles: Their Dependence on Correlations," *The Annals of Mathematical Statistics*, 39, 1425–1434.

Singh, D., Febbo, P. G., Ross, K., Jackson, D. G., Manola, J., Ladd, C., Tamayo, P., Renshaw, A. A., D'Amico, A. V., Richie, J. P., Lander, E. S., Loda, M., Kantoff, P. W., Golub, T. R., and Sellers, W. R. (2002), "Gene Expression Correlates of Clinical Prostate Cancer Behavior," *Cancer Cell*, 1, 203–209.

Snedecor, G. W., and Cochran, W. G. (1989), *Statistical Methods* (8th ed.), Ames, IA: Iowa State University Press.

Sprott, D. A. (1975), "Marginal and Conditional Sufficiency," *Biometrika*, 62, 599–605.

Stigler, S. M. (1986), *The History of Statistics: The Measurement of Uncertainty Before 1900*, Cambridge, MA: Belknap Press of Harvard University Press.

Stigler, S. M. (1999), *Statistics on the Table: The History of Statistical Concepts and Methods*, Cambridge, MA: Harvard University Press.

Student (1908), "The Probable Error of a Mean," *Biometrika*, 6, 1–25.

Thompson, W. A., Jr. (1962), "The Problem of Negative Estimates of Variance Components," *The Annals of Mathematical Statistics*, 33, 273–289.

Trefethen, L. N., and Bau, D., III (1997), *Numerical Linear Algebra*, Philadelphia: Society for Industrial and Applied Mathematics.

Verbyla, A. P. (1990), "A Conditional Derivation of Residual Maximum Likelihood," *Australian Journal of Statistics*, 32, 227–230.

Victor, N. (1982), "Exploratory Data Analysis and Clinical Research," *Methods of Information in Medicine*, 21, 53–54.

Westfall, P. H., and Tobias, R. D. (2007), "Multiple Testing of General Contrasts: Truncated Closure and the Extended Shaffer–Royen Method," *Journal of the American Statistical Association*, 102, 487–494.

Wolfowitz, J. (1949), "The Power of the Classical Tests Associated with the Normal Distribution," *The Annals of Mathematical Statistics*, 20, 540–551.

Woods, H., Steinour, H. H., and Starke, H. R. (1932), "Effect of Composition of Portland Cement on Heat Evolved During Hardening," *Industrial and Engineering Chemistry*, 24, 1207–1214.

Zabell, S. L. (2008), "On Student's 1908 Article 'The Probable Error of a Mean' " (with discussion), *Journal of the American Statistical Association*, 103, 1–20.

Zacks, S. (1971), *The Theory of Statistical Inference*, New York: Wiley.

Zhang, L., Bi, H., Cheng, P., and Davis, C. J. (2004), "Modeling Spatial Variation in Tree Diameter-Height Relationships," *Forest Ecology and Management*, 189, 317–329.

Index

D

M

N

Q

R

Milton Keynes UK
Ingram Content Group UK Ltd.
UKHW050307111024
449327UK00043B/2090

9 780367 572037